Natural Science in Archaeology

Series Editors

The last three decades have seen a steady growth of application of natural scientific methods to archaeology. The interdisciplinary approach of archaeometry has found increasing appreciation by the archaeologists and is now considered indispensable and an integral part of archaeological studies. Interdisciplinary collaboration requires a multidisciplinary background. It is becoming increasingly difficult for the individual to grasp the whole field of archaeometry with its rapid developments. The aim of the series *Natural Science in Archaeology* is to bridge this information gap at the interface between archaeology and science. The individual volumes cover a broad spectrum of physical, chemical, geological, and biological techniques applied to archaeology as well as to palaeoanthropology with the interested nonspecialist in mind.

The single monographs cover:

- large fields of research
- specific methods of general interest (archaeometric methods of dating, material analysis, environmental reconstruction, geophysical prospecting, remote sensing and data processing)
- materials of interest to the archaeologist, such as sediments, soils, metal and nonmetal artifacts, animal and plant remains and other organic residues
- practical aspects such as sampling and data interpretation
- case studies, to demonstrate the potential and limitations of the various techniques.

More information about this series at http://www.springer.com/series/3703

Andreas Hauptmann

Archaeometallurgy – Materials Science Aspects

Springer

Andreas Hauptmann
Haus der Archäologien, Archaeometallurgy
Deutsches Bergbau-Museum / Ruhr University
Bochum, Germany

ISSN 1613-9712
Natural Science in Archaeology
ISBN 978-3-030-50366-6 ISBN 978-3-030-50367-3 (eBook)
https://doi.org/10.1007/978-3-030-50367-3

This Springer imprint is published by the registered company Springer Nature Switzerland AG.
The registered company address is: Gewerbestrasse 11, 6330 Cham, Switzerland

Dedicated to the pioneering scholars in archaeometallurgy Hans-Gert Bachmann, Friedrich Begemann and Robert Maddin who passed away recently.

Acknowledgements

This volume is the result of years of research at the Deutsches Bergbau-Museum Bochum (DBM) and teaching activities in the Institutes of Archaeological Sciences and the Institute of Geology, Mineralogy and Geophysics at the Ruhr University Bochum.

As a former researcher at the DBM, I had access to excellent working opportunities. This concerned the very generous establishment of a chemical–mineralogical research laboratory and generous working conditions, which were made possible by the directors of the DBM, Bergassessor Hans Günther Conrad (†), Prof. Dr. Rainer Slotta and Prof. Dr. Stephan Brüggerhoff, and the vice director Prof. Dr. Thomas Stöllner. I owe them a great debt of gratitude for supporting any research activities. In terms of time, this was in the early stages when the DBM became a member of the Leibniz Association. It was also at a time when new discoveries in archaeometallurgy were being made in many countries, funded by numerous prominent research institutions (Volkswagen Foundation, German Research Foundation, Thyssen Foundation, Gerda Henkel Foundation, Institute of Aegean Prehistory, INSTAP, Philadelphia) and the German Archaeological Institute. I was able to participate actively and passively in many of these research projects.

Scientifically I was especially supported by the archaeologist Prof. Dr. Gerd Weisgerber (†) of the DBM. I learned a lot from the late scientists Prof. Dr. Hans-Gert Bachmann, Prof. Dr. Friedrich Begemann and Prof. Dr. Bob Maddin. This volume is dedicated to the last three gentlemen.

I was able to hold many meaningful discussions with guest scientists of the DBM. This concerned Prof. Dr. Robert Maddin, then Honorary Curator of Archaeological Science, Peabody Museum, Harvard and, as a Humboldt Foundation Senior Science Fellow, guest in Bochum for several months in 1991. The focal points of the metallurgist were investigations on the history of iron and steel as well as that of copper. Prof. Dr. Marc Pearce from the Institute of Archaeology, University of Nottingham, United Kingdom spent several months of 2016/2017 as a Leverhulme International Academic Fellowship holder at the DBM and the Institute of Archaeological Science of the Ruhr University in Bochum. Together with my longtime colleague Prof. Dr. Ünsal Yalçın I have carried out several research projects and developed new ideas. With Prof. Dr. Sabine Klein, previously Goethe University Frankfurt a.M., I also cooperated on several research projects, most recently on gold, silver, copper and bronzes from the Royal Cemetery of Ur, Mesopotamia.

These interactions continued with endless fruitful intellectual discussions during my teaching activities on archaeometallurgy and archaeometry. I had close contact with all the PhD students at Bochum through the years to work on an academic project in archaeometallurgy. My deep thanks to them: all their names should be addressed here collectively.

Furthermore, I received help from many people I met at the Ruhr University and at institutions outside the university. It is a pleasure for me to mention Dr. Norbert Hanel, Institute for Archaeological Science, Ruhr University Bochum, who in endless discussions showed an extraordinary willingness to familiarise himself with this new topic for him as an archaeologist. He corrected a large part of the book. I have had many discussions with Prof. Dr. Paul Craddock, formerly British Museum, London, and with my colleague Prof. Dr. Thomas Stöllner, also head of the Institute of Archaeological Science, Ruhr University Bochum. He was one of the few scholars in Germany who established Archaeometry and Archaeometallurgy as official courses of studies at the University.

Within the DBM I have been accompanied by a number of colleagues, whom it is my joyful duty to thank. These are:

Dr. Michael Bode, Dr. Ian Cierny (†), Dr. Jenny Garner, Mrs. Isika Heuchel-Pede, Dr. Moritz Jansen, Dipl.-Min. Dirk Kirchner, Mrs. Sandra Kruse (formerly Morszeck), Mrs. Regina Kutz, Andreas Ludwig (who over the years has produced the best thin section preparations for microscopic examinations), Dr. Stephen Merkel, Prof. Dr. Michael Prange, Prof. Dr. Thilo Rehren (previously DBM), Dr. Eveline Salzmann, Dr. Peter Thomas, Heinz-Werner Voß, Mr. Georg Wange.

The following students helped in endless daily problems with great patience: Oliver Stegemeier, Mitja Musberg and Julien Villatte.

I am very much obliged to Miriam Skowronek, who with great patience did the entire translation of this volume. Their help was generously financially supported by the Institute of Aegean Prehistory (INSTAP) in Philadelphia, USA. And, last but not least, I gratefully acknowledge the attendance of Christoph Hauptmann and Tobias Skowronek for making and qualifying drawings and graphs published in this book.

Andreas Hauptmann

Contents

1 Introduction: Archaeology and Archaeometallurgy 1
- 1.1 Archaeometallurgy as Its Own Discipline 2
- 1.2 The Concept of Archaeometallurgy 5
- 1.3 Goals of This Volume 6

2 Stages of Early Metallurgical Activities 9
- 2.1 Diffusionism Versus Autonomous Developments 9
- 2.2 Ore Geology and Environmental Change 11
- 2.3 Metallurgical Developments 12
  - 2.3.1 Preliminary Stage 13
  - 2.3.2 Initial Phase: The Neolithic Period 14
  - 2.3.3 Innovation and Consolidation Phases: Early to Late Chalcolithic and Early Bronze Age I 16
  - 2.3.4 Industrial Phase: Developed Bronze and Iron Age and Later 17
- 2.4 Chaîne Opératoire and Metallurgical Chain 19

3 Ancient Ore Deposits 21
- 3.1 Introduction 21
- 3.2 What Is an Ore Deposit? 22
  - 3.2.1 Magmatic Ore Deposits 23
  - 3.2.2 Sedimentary/Surficial Ore Deposits 24
  - 3.2.3 Diagenetic-Hydrothermal Ore Deposits 25
  - 3.2.4 Metamorphic Ore Deposits 25
  - 3.2.5 From Metal Provinces to Metallogenetic Belts . . 25
  - 3.2.6 Monomineralisation Versus Polymetallic Ore Deposits 26
  - 3.2.7 The Design and Morphology of Ore Deposits . . . 27
- 3.3 Ore Minerals Intergrowths 29
  - 3.3.1 Ore Textures 29
- 3.4 Spatial Geographic Distribution of Ore Deposits in the Old World 32
  - 3.4.1 Plate Tectonics and Ore Deposits 32
  - 3.4.2 The Tethyan Eurasian Metallogenic Belt (TEMB) 34
  - 3.4.3 Polymetallic Deposits and Lead–Zinc Deposits in the Paleozoic of Europe 37
  - 3.4.4 The Altaids and Uralids 38

3.4.5 Ore Deposits in Old Cratons 39
3.4.6 Copper Deposits 39
3.4.7 Gold 57
3.4.8 Platinum-Group Minerals 69
3.4.9 Silver 74
3.4.10 Lead and Zinc 85
3.4.11 Tin 97
3.4.12 Iron 111
3.4.13 Arsenic, Antimony, Bismuth 122
3.4.14 Mercury 129
3.5 Oxidation and Secondary Enrichment Zones 132
3.5.1 Old Workings in the Gossan 132
3.5.2 The Formation of the Oxidation Zone: The Gossan 133
3.5.3 The Secondary Enrichment Zone (Cementation Zone) 139
3.5.4 Protore: The Primary Ore Body 139
3.5.5 Alternative Models of Gossan Formations 140
3.5.6 Secondary Zoning Versus Metallurgical Models . . 140
3.5.7 Special Ores in the Oxidation and Cementation Zones 141
3.6 Placer Deposits 143
3.6.1 General Remarks 143
3.6.2 Formation of Placers 144
3.6.3 Gold Placers 147
3.6.4 Gold Placers and Platinum-Group Minerals (PGM) 150
3.6.5 Tin Placers 152
3.6.6 Iron Placers 153
3.7 Mining Ore Deposits: Geological Factors, Economy and Politics 153
3.7.1 Early Stages of Exploitation 153
3.7.2 Geological Factors and Innovations 155
3.8 The Beneficiation 161
3.8.1 General Remarks 161
3.8.2 Dry Treatment 162
3.8.3 Wet Dressing 165
3.9 Sampling Ancient Ore Deposits 169

4 Basic Physical–Chemical Principles of Ancient Metallurgy . . 177
4.1 Introduction 177
4.2 Silicate and Metal Phase Diagrams 178
4.2.1 Binary Systems 178
4.2.2 Ternary Systems 185
4.2.3 Three-Dimensional Presentation in Tetrahedron . . 186
4.3 Micro-equilibria and Partial (S)melting 188
4.4 Firing Conditions 190
4.4.1 Boudouard-Equilibrium 190

4.4.2 Colours by Burning 192
4.4.3 Redox Reactions 192
5 Archaeometallurgical Slags and Other Debris 199
5.1 Introduction 199
5.2 Slags in the Archaeological Context 200
5.2.1 Prospecting Slag Heaps 201
5.2.2 Slagging Versus Slagless Metallurgy 203
5.2.3 Slag Heaps of Metal Smelting 204
5.3 The Investigation of Archaeometallurgical Slags 215
5.3.1 Questions and Strategies 217
5.3.2 Sampling Slags and Analytical Approach 218
5.4 Types of Slags 222
5.4.1 The Oldest Slags 222
5.4.2 Crucible Slags 224
5.4.3 Slag Cakes, Matte Smelting 228
5.4.4 Slag Cakes and Slag Foam 231
5.4.5 Dross 232
5.4.6 Cinder 233
5.4.7 Tap Slags 233
5.4.8 Plate Slags 235
5.4.9 Furnace Slags (Hearth-Bottom Slags) 236
5.4.10 Crushed Slags 238
5.4.11 "Free Silica Slags" 240
5.4.12 Iron Working Slags 243
5.4.13 Fluxes 245
5.4.14 Ceramic Slags 245
5.5 The Chemical Composition of Ancient Slags 246
5.5.1 Iron Slags 246
5.5.2 Copper Slags 248
5.5.3 Lead–Silver Slags 250
5.5.4 Tin Slags 253
5.5.5 Antimony Slags 255
5.6 Important Slag Phases 256
5.6.1 Slag Phases and Phase Associations 256
5.7 Sulphide Mattes 271
5.7.1 Sulphide Phases 272
5.7.2 The System Cu–Fe–S 275
5.7.3 Sulphidic Mattes and Silicate Slag 277
5.8 Speiss 281
5.8.1 Composition of Speiss 281
5.8.2 Speiss and Matte 284
5.8.3 Archaeological Evidence 284
5.9 Cupellation Remains 286
5.9.1 Composition of Litharge 286
5.9.2 Archaeological Evidence 287
6 Making Metals: Ancient Metallurgical Processes 295
6.1 General Aspects 295

6.1.1 Reconstructing Ancient Smelting Processes 295
6.1.2 Metallurgical Installations 299
6.2 The Metallurgy of Copper 305
6.2.1 Copper Artefacts 305
6.2.2 Classification of Copper Ores: Metallurgical Aspects 309
6.2.3 The Extractive Metallurgy of Copper 310
6.3 The Metallurgy of Gold 318
6.3.1 Gold Artefacts 319
6.3.2 Ancient Metallurgical Processes of Gold 323
6.3.3 Gilding Through the Ages 330
6.4 The Metallurgy of Silver and Lead 332
6.4.1 Silver Artefacts 332
6.4.2 Lead Artefacts 334
6.4.3 Classification of Silver Ores: Metallurgical Aspects 336
6.4.4 The Metallurgy of Silver 337
6.4.5 The Metallurgy of Lead 344
6.5 The Metallurgy of Tin 348
6.5.1 Tin Artefacts 348
6.5.2 Ancient Metallurgical Processes 351
6.5.3 Tinning 356
6.6 The Metallurgy of Iron 357
6.6.1 Iron Artefacts 358
6.6.2 Metallurgical Processes 360
6.6.3 Researches on Ancient Iron Production 367
6.7 The Metallurgy of Antimony 370
6.7.1 Archaeological Evidence 370
6.7.2 Production and Processing 370
6.8 The Metallurgy of Zinc 373
6.8.1 Rare Finds of Ancient Zinc 374
6.8.2 Production and Processing 375
6.9 The Metallurgy of Mercury 377
6.9.1 Cinnabar and Mercury in Ancient Times 377
6.9.2 Production and Processing 378
7 Metals and Alloys 381
7.1 Ancient Alloys: General Remarks 381
7.1.1 The Earliest Alloys 382
7.1.2 Physical Properties 383
7.2 Copper-Based Alloys 383
7.2.1 Classifications and Properties 383
7.2.2 Arsenical Copper 385
7.2.3 Antimonial Copper and Bronze 393
7.2.4 Tin Bronze 395
7.2.5 Brass 401
7.2.6 Iron in Copper 406
7.3 Gold and Gold Alloys 410

7.3.1 "Pure" Gold 410
7.3.2 Gold Alloys 411
7.4 Silver Alloys 420
7.4.1 Aurian Silver 420
7.4.2 Copper-Containing Silver 421
7.4.3 Silver–Copper Alloys 422
7.5 Iron Alloys 424
7.5.1 The Late Coming of Iron and Steel 424
7.5.2 Iron, Steel and Cast Iron 425
7.5.3 The Iron–Carbon Phase Diagram 427
7.5.4 Phosphorous Iron 430

8 Macro- and Microstructure of Metals 433
8.1 Macrostructure 433
8.1.1 Conventional Radiography 433
8.1.2 Computed Tomography (CT) 434
8.2 Microstructure 435
8.2.1 Metallography 435
8.2.2 Neutron Diffractography 442

9 Ethnographic Evidence and Artisanal Metal Production 445
9.1 General Remarks 445
9.2 Gold Exploitation in Africa 446
9.3 Iron in Africa 448
9.4 Copper Smelting in Himalaya 451
9.5 Pictures of Metal Making in Japan and China 453
9.6 Bronze Casting in India 454

10 Experimental Archaeometallurgy 459
10.1 General Remarks 459
10.2 Early Copper Smelting 460
10.2.1 (Re-)Melting Native Copper 460
10.2.2 Microstructural Investigations 461
10.2.3 Crucible Smelting 462
10.3 Casting Copper Oxhide Ingots 464
10.4 Early Tin Assaying and Production 465
10.4.1 Kestel and Göltepe 466
10.4.2 Hampshire, Cornwall 467
10.5 Lead Smelting 468

11 Provenance Studies 471
11.1 Introduction 471
11.2 Trace Element Analysis 472
11.2.1 Grouping of Elements 472
11.2.2 Developments 473
11.2.3 Distributions of Trace Elements 475
11.2.4 Silver and Lead 476
11.2.5 Gold 477
11.2.6 Iron 478

11.3 Isotope Analysis 480
11.3.1 Lead Isotope Analysis 480
11.3.2 Tin Isotopy 495
11.3.3 Copper Isotopy 499
11.3.4 Osmium Isotopy 502

Glossary 507

Bibliography 517

Subject Index 581

Geographical Index 589

Abbreviations

Temperature

°C	temperature in degree Celsius
K	absolute temperature in Kelvin scale

Symbols of Scale Units

pm	pico = 10^{-12}
n	nano = 10^{-9}
μ	micro = 10^{-6}
m	milli = 10^{-3}
c	centi = 10^{-2}
k	kilo = 10^{3}
M	mega = 10^{6}

Length

μm	micrometre
mm	millimetre
m	metre
km	kilometre

Weights

wt.%	weight percent
ppm	parts per million
ppb	parts per billion
μg/g	milligram per gram (= ppm)
g	gram
kg	kilogram
t	ton

Times

s	second
min	minute
h	hour
d	day
a	year
Ma	million years
AD	Anno Domini
BC	Before Christ

Symbols

α, β, γ, δ, ε	Chemical components in a system having different physical and crystallographic states.
ΔG°	Standard free energy of formation
ΔH	Enthalpy of reaction
σ	Standard deviation

Chemical Elements

Ac	actinium
Ag	silver
Al	aluminium
Am	americium
Ar	argon
As	arsenic
At	astatine
Au	gold
Ba	barium
Be	beryllium
B	boron
Bh	bohrium
Bi	bismuth
Bk	berkelium
Br	bromine
C	carbon
Ca	calcium
Cd	cadmium
Ce	cerium
Cl	chlorine
Cf	californium
Cm	curium
Cn	copernicium
Co	cobalt
Cr	chromium
Cs	caesium
Cu	copper

Db	dubnium
Ds	darmstadtium
Dy	dysprosium
Er	erbium
Es	einsteinium
Eu	europium
F	fluorine
Fe	iron
Fl	flerovium
Fm	fermium
Fr	francium
Ga	gallium
Gd	gadolinium
Ge	germanium
H	hydrogen
He	helium
Hf	hafnium
Hg	mercury
Ho	holmium
Hs	hassium
I	iodine
In	indium
Ir	iridium
K	potassium
Kr	krypton
La	lanthanum
Li	lithium
Lr	lawrencium
Lu	lutetium
Lv	livermorium
Mc	moscovium
Md	mendelevium
Mg	magnesium
Mn	manganese
Mo	molybdenum
Mt	meitnerium
N	nitrogen
Na	sodium
Nb	niobium
Nd	neodymium
Ne	neon
Nh	nihonium
Ni	nickel
No	nobelium
Np	neptunium
O	oxygen
Og	oganesson
Os	osmium

P	phosphorus
Pa	protactinium
Pb	lead
Pd	palladium
Pm	promethium
Po	polonium
Pr	praseodymium
Pt	platinum
Pu	plutonium
Ra	radium
Rb	rubidium
Rg	roentgenium
Rh	rhodium
Re	rhenium
Rf	rutherfordium
Rn	radon
Ru	ruthenium
S	sulphur
Sb	antimony
Sc	scandium
Se	selenium
Sg	seaborgium
Si	silica
Sm	samarium
Sn	tin
Sr	strontium
Ta	tantalum
Tb	terbium
Tc	technetium
Te	tellurium
Th	thorium
Ti	titanium
Tl	thallium
Tm	thulium
Ts	tennessine
U	uranium
V	vanadium
W	tungsten
Xe	xenon
Y	yttrium
Yb	ytterbium
Zn	zinc
Zr	zirconium
REE	Rare earth elements. A set of 17 chemical elements in the periodic table (specifically the fifteen lanthanides, as well as scandium and yttrium). These are Ce, Dy, Er, Eu, Gd, Ho, La, Lu, Nd, Pr, Pm, Sm, Sc, Tb, Tm, Yb, Y.

1 Introduction: Archaeology and Archaeometallurgy

Excavations of archaeological sites usually yield a variety of finds. Varying amounts of inorganic and organic finds incur. These are man-made objects such as stone tools, ceramic vessels, jewellery, weapons, glass objects; often waste products of daily human activities as well. In the category of organic finds, food residue or remains of human burials are being excavated. Particularly important and usually occurring in large quantities are ceramics. These artefacts are usually salvaged in archaeological contexts, such as architectural remains, hearths, houses or graves.

Findings and features of metallurgical activities are hardly classified in this canon, apart from metal artefacts themselves.

Although Eggert (2001) clearly defines the sites of raw material extraction and processing in his opus *Prähistorische Archäologie*, he does not detail the findings of metallurgical activities any further. Old mines, ores, slags, intermediates such as ingots, smelting sites, relics of smelting furnaces, blacksmith workshops, technical ceramics, etc. are not mentioned. The topic of raw material extraction as the basis for economic developments is hardly mentioned, at least not in German textbooks on classical archaeology or Roman provinces (Hölscher 2006; Borbein et al. 2000; Fischer 2001). Only Renfrew and Bahn (2000) discuss the extraction of metals by metallurgical melting processes in their textbook *Archaeology: Theories Methods and Practice*, and slags are explicitly mentioned as well. But even scientifically oriented manuals of methods of archeometry, such as that of Ciliberto and Spoto (2000) or Brothwell and Pollard (2001) do not touch this topic.

Only metal artefacts have been analysed in large numbers in recent decades and are available, for example in the former, now greatly expanded "Stuttgart database" (Junghans et al. 1960, 1968, 1974; Hartmann 1970, 1982; Krause 2003). For some years now, databases containing thousands of lead isotope analyses have also been published, of which the Gales-Stos-Gales were published in the Oxalid database, others are available in the Curt Engelhorn Center for Archaeometry in Mannheim or—as heritage of the activities of Friedrich Begemann and Sigrid Strecker—in the laboratories of the Deutsches Bergbau-Museum in Bochum.

In order to grasp the details of the frequently quoted *chaîne opératoire* of metallurgy (Fig. 1.1) as fully as possible, not only exact knowledge of ore deposits (near the surface) is required, but also correct and appropriate methods of material analysis of archaeological finds and features of waste, intermediates and final products that are produced in this chain. After all, since the Chalcolithic, metallurgical activities to produce and process metals from ore have been amongst the most significant developments in human history and have always left tangible material relics, be they within settlements or far away, in the vicinity of ore deposits. Hundreds and thousands of hoards with tons of bars and ingots of various shapes, finished metal objects, weapons, utensils and

A. Hauptmann, *Archaeometallurgy – Materials Science Aspects*, Natural Science in Archaeology,
https://doi.org/10.1007/978-3-030-50367-3_1

jewellery were produced. The production requires the implementation of metallurgical processes, which in turn is associated with the production of waste (slag), the extent of which generally exceeds that of metal artefacts considerably. Thus, it is almost the norm that remains of metallurgical activities are found in excavations of settlements. Slag dumps that have accumulated over millennia close to a mining area can contain hundreds of thousands of tons (cf. Sect. 5.1).

1.1 Archaeometallurgy as Its Own Discipline

Nevertheless, the investigation of metallurgical finds emerged as its own discipline: archaeometallurgy.

Decisive impulses for this were made in the 1960s through the extensive expeditions of Theodor Wertime and his team to countries of Western and Central Asia (Iran, Afghanistan, Turkey). Their goal had been, among others, to research the early metallurgy. In the early 1970s, Beno Rothenberg and his team began fieldwork on the Sinai Peninsula, Timna and the Iberian Peninsula. Beno Rothenberg has gained the interest of a number of mining engineers, geoscientists, chemists and metallurgists (H.G. Bachmann, H. G. Conrad, P. T. Craddock, A. Lupu, R. F. Tylecote and others) to collaborate in the exploration of early copper mining in these regions. He was the first to arouse interest in the industrial remains, especially of the Iron Age metallurgists, who had left considerable slag heaps as by-products of the extraction of copper in Timna, a field that was then still largely unattractive in archaeology. He used the example of Timna to push the term industrial archaeology, which had previously only been focused on the modern era, back in time by millennia (Rothenberg 1973; Slotta 1982).

The collaboration of scientists in these projects, who were focused on modern aspects of their fields of specialisation, at first included the risk of transferring concepts of modern technology to archaeological finds and features. These were things that needed precise adjustment. Nevertheless, the experience of modern metallurgists has hardly been and still cannot be dispensed with, especially if they have worked in areas of small-scale mining activities in developing countries (Priester et al. 1993). Ronald F. Tylecote and Paul T. Craddock may well be the most well-known scientists today who have dedicated their careers to archaeometallurgy and published crucial book publications (Craddock 1995; Tylecote 1992a, b). Radomir Pleiner has devoted himself to the complex subject of early iron extraction through the extensive excavation of iron smelting works and blacksmith workshops. He has recorded his life's work in two groundbreaking books (Pleiner 2000, 2006).

Almost simultaneously, activities concerning the study of ancient metallurgy developed in the United States, which in addition to the study of metal artefacts themselves posed the question of the ore deposits from which these early metals were produced. They were based in many ways on the work and inspiration of material scientist Cyril Stanley Smith. In Philadelphia, Robert Maddin, James Muhly, Vincent Pigott and Tamara Stech teamed up to form an interdisciplinary group, dedicated to the topic of provenance studies in the *Mesopotamian Metal Project*. Muhly dedicated his standard work to the two main metals: copper and tin (Muhly 1973).

Given these few examples, which mention only a small number of scientists and activities, it is hard to understand why this new discipline should have manoeuvred itself into a ghetto, as Killick (2015) had feared. The organisation of conference series like *Archaeometallurgy in Europe* or *The Beginnings of Metallurgy in Asia* (*BUMA*) are indicative of a very rich field of expertise, which must not only be seen from a regional point of view but are also of great thematic interest. Archaeometallurgy is a field of expertise that has emerged, just like Egyptology or Near Eastern Archaeology, due to rich finds and features from Classical Archaeology.

Killick's (2015) accusations that archaeometallurgists were too focused on their methods and

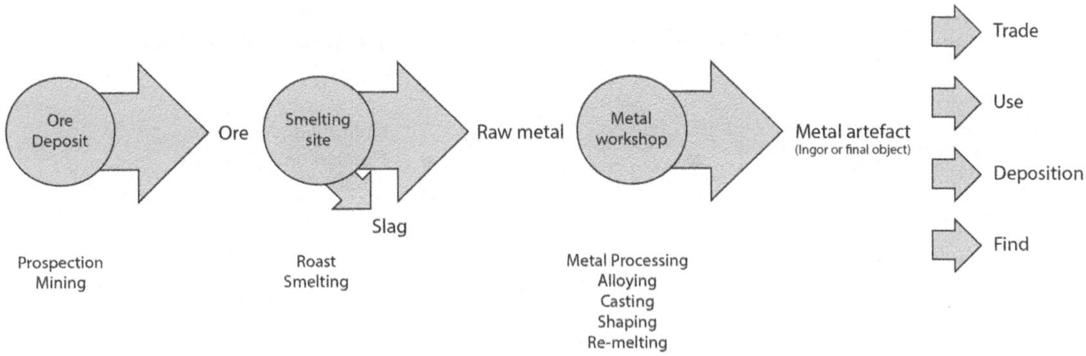

Fig. 1.1 The metallurgical chain from ore to metal is a logical sequence in a double sense: it shows the *chaîne opèratoire* of human activities within metal production, and, in addition, it provides information on the material transformations of ores to smelting sites, of semi-products to metal workshops, the making of artefacts and their distribution, and where they may finally be unearthed in an archaeological context. Slags are mainly formed by smelting ores or by melting metal (e.g. by casting, alloying or refining processes). Modified after Hauptmann (2007)

do not pay enough attention to greater archaeological questions is only conditionally acceptable. It is true that the field of archaeometallurgy has evolved from archaeological issues. However, the discipline is based heavily on techniques from the field of material analysis, and tests for the application of suitable methods are appropriate and correct, as they may differ significantly from those of modern research. The appropriate chemical and physical methods exist but must be adjusted to the specific problems and issues of archaeometallurgy (Fig. 1.2). This takes time and requires experience.

The scientist must be attuned to the specific problems of early metal extraction and processing. These may well be different from those of modern geosciences, especially in fields of ore mineralogy and economic geology. Modern economic geology focuses on genetic and economic issues in a supra-regional context. Of archaeometallurgical interest on the other hand is a reconstruction of the quality of mined ores in the near-surface area ("Gossan"). The isotope analyses of lead have a different application in modern geosciences than in archaeometallurgy.

In principle, there are two basic problems.

The first problem is the still prevalent lack of knowledge in archaeology today, about the enormous economic and social importance that the extraction and processing of ores and metals has historically had in human history on a global scale. Many archaeologists are reluctant to approach this topic. Most are interested in the final products, the unique and striking metal artefacts of gold, copper, bronze or iron found at archaeological sites. The composition and distribution of ores in their deposits, as they were available to man in ancient times, technological and social problems as well as questions about the old mining endeavours and metal extractions, have met with interest only in a few universities and non-university institutions.

This is not to argue that the generalist archaeologists are not also responsible for this lack of communication. Too many (especially in the United States—Muhly 1980, 102) lack specific training in archaeological science to use archaeometallurgical knowledge and data, and improve the training of archaeologists.

Archaeology, however, is doubtlessly one of those disciplines that transferred most from other fields (Parzinger 2016). In contrast, chemists and geoscientists are hardly trained in interdisciplinary approaches as archaeologists are. Artioli and Angelini (2011) in a rather negative sense even suggest the contact between mineralogy and archaeometry as a "fatal attraction".

The second problem concerns the handling of chemical–physical data and mineralogical observations. Due to modern processing techniques and analytical possibilities, modern devices such as the scanning electron microscope

Geoscience/Metallurgy		Archaeometallurgy
Geological developments up to – 4,5 Ma	GEOCHRONOLOGY DATING	Cultural developments Precise dating back to 10.000 BC
Genetic systematics Economics Global	ORE DEPOSITS	Oxidation zone Morphology ore deposits
Main elements Ores	GEOCHEMISTRY ORE PETROLOGY	Ore mineralogy Intergrowth Trace elements
Earth's evolution	ISOTOPE ANALYSIS	Provenance studies
Structure/development of earth's crust Exploration Paleomagnetism	GEOPHYSICS	Prospection of archaeological remains Archeomagnetism
Industrial processes Equilibria	PHYSICO-CHEMICAL PRINCIPLES	Artisanal craftsmanship Microequilibria

Fig. 1.2 This overview shows which questions modern geoscientific and metallurgical research is currently focusing on and which are of importance in archaeometallurgy

can produce large amounts of data quickly (and often in a manageable financial framework). Muhly (1995) noted "Most archaeologists welcome scientific evidence but abhor scientific controversy. They want answers: if those answers derive from some of the most sophisticated techniques known to science, so much the better, but it is the answer, not the technique that interests the archaeologist". The problem is that archaeological scientists often seem more interested in the technique than in the specific answers that it can give to archaeological problems. Analytical data from material science are only of use to archaeology if it answers specific questions ("what does it mean?").

Pearce (2016) notes that many analyses have been carried out using inappropriate techniques (perhaps because a specialist in the local university offered their services for free) and without necessarily a clear research question; indeed, the lack of a clear research question is perhaps often the reason why inappropriate techniques were chosen. In this sense, the words of Georg Kossack are to be reminded of, who is of the opinion "...that the archaeologist may only use the scientific findings obtained from a relevant find material for his interpretation, if he is familiar with the methods and theoretical prerequisites of the corresponding disciplines... ." (Kossack 1986).

Even though a considerable amount of interdisciplinary research has not necessarily deserved this label (Samida and Eggert 2013), the last three decades have seen a steady growth of the application of natural scientific methods to archaeology. The interdisciplinary approach of archaeometry has found increasing appreciation by the archaeologists and is now considered indispensable and an integral part of archaeological studies. With their opus, *The Coming of the Age of Iron*, Wertime and Muhly (1980) successfully created an extraordinary interdisciplinary synthesis of modern metallurgy, archaeology and ethnoarchaeology.

Interdisciplinary collaboration between archaeologists, natural scientists and engineers requires a multidisciplinary background. It is becoming increasingly difficult for the individual to grasp the whole field of archaeometry including archaeometallurgy with its rapid developments. The aim of the series of the numerous books *Natural Science in Archaeology* published during the last 25 years by Günther Wagner and coworkers is to bridge this information gap at the interface between archaeology and science. The individual volumes cover a broad spectrum of physical, chemical, geological, metallurgical and biological techniques applied to archaeology as well as to paleoanthropology with the interested non-specialist in mind. This strategy has been continued by Roberts and Thornton (2014), whose work sheds light on the global development of early metallurgy.

1.2 The Concept of Archaeometallurgy

It is one of the urgent desiderata of archaeometallurgy to investigate the areas of origin of ores and metals, to track down centres of metal production and, in further steps, to add research on trade zones, trade routes, hubs, actors and political-social mechanisms. These are provenance studies.

However, this is not possible until archaeological activities and observations have been identified and their significance recognised. Analytical methods should then be used and geoscientific, chemical-analytical and isotopic data should be produced to create fingerprints of metals and ore deposits. Only with such results facts can be established and hypotheses and theories can be formulated.

Analytical results in databases are to be put in a supra-regional context. Comparable to the principles of regional settlement analysis, a recording of geological and geographical factors for the distribution of raw material resources, which have always determined the course of old trade routes, has to be added as well. Here is just an exemplary mention of the enormously resourceful fold mountains of the Mesozoic Tethyan Eurasian Metallogenic belt of the Alpine Himalayan orogenic system ("TEMB"), which extends from the Alps across the Balkans and Turkey and Iran to the Himalayas and Southeast Asia (Jankovič 1997).

Geoarchaeological work in this sense has been conducted by, amongst others, Given and Knapp (2003) in Cyprus. In a summary of a milestone meeting of his graduate school *Raw Materials, Innovation, Technology of Ancient Cultures (RITaK)* in Bochum, Thomas Stöllner (2013) presented more aspects of mining regions as economy-scapes and generally continued them in research and teaching.

There are some works that shed light on the key issue of geography: Marshall (2015), with his prudent analysis of "Prisoners of Geography", has outlined in general how geographical "favour factors" have always guided political developments. He sums it up, inter alia, with the sentence "Governments come and go, the Hindukush remains". In his book *Tying the Threads of Eurasia*, Wilkinson (2014) factually analyses the natural landscape features and resource distributions along the ancient course of the Romantic Silk Road.

The production and use of metals requires a series of different human activities in a technological sense. They are connected with chemical and physical transformations of materials during the change from ore into metal. These interactions can be summarised—parallel to the *chaîne opèratoire* (see Stöllner 2003a, b)—in a metallurgical chain (Figs. 1.1 and 1.2). The activities start at the ore deposit with the mining of ore, followed by technological steps used in beneficiation, smelting processes, where slag, raw metal and other intermediate products are made. Different methods are used to reconstruct reaction vessels used in this process. Metal is subsequently treated in different steps—if necessary alloyed with other metals—until a final product is achieved (e.g. ingot, axe, chisel, etc.). This product might be traded and may later be found in an individual archaeological context. It is important to differentiate in this chain of interactions between mining, smelting and further processing of metallurgical products because each of these steps in the production line is controlled by different factors. Mining is closely bound to the ore deposit, while smelting and the sometimes very complex processing are more dependent on the context of settlements, and therefore were possibly carried out under different social conditions.

The chain of these activities shows that miners and metallurgists have always had an astonishing amount of know-how. This starts with ore mining, with the knowledge of which ores are distributed in the rock in what way, and how they can be enriched by suitable treatment. Metallurgists had the expertise to produce metals through high-temperature processes at over 1000 °C and to produce predefined composites from these precast objects. Parzinger (2016) emphasises that in the Bronze Age, in addition to training fixed institutions with the so-called elites, there must have been artisanal experts who had knowledge of the production and distribution of crafted products. Without specialists, a social structure of increasing complexity can neither be formed nor maintained. This applies to the extraction of raw materials as well as the architectural construction of large buildings, complex drainage systems such as in Mesopotamia or monumental cult sites or tombs.

To decode these craft processes from archaeological finds and features today, a diverse spectrum of knowledge is required. Archaeologists and scientists alike still often have to work hypothetically in archaeometallurgy and in mining archaeology, resulting in many potential misinterpretations of archaeological legacies due to a lack of technological expertise.

To sum up, the main questions and problems of archaeometallurgy which arise from the metallurgical chain are the following:

1. Reconstructions of technological processes and craftsmanship applied to produce metal, ingots and finally objects.
2. Distribution of metal from a source, i.e. the reconstruction of ancient transports and trade networks and, vice versa, provenance studies to find the source from where the raw materials of a metal object originated.

These questions are connected with:

3. The chronology of mining and metallurgy over the millennia.
   They are followed by archaeological topics such as:
4. The spatial organisation and social pattern of mining and metal production.
5. (Over-)regional cultural and economic impact.

1.3 Goals of This Volume

The purpose of the book at hand is to work on the experience gained in fieldwork, be that surveys or archaeological excavations, and discoveries in the early extraction of metals. Concepts are to be pointed out and the appropriate materials explained. This includes presenting the spectrum of archaeometallurgy in spite of and especially because of the specialisation of its facets, in a way that is intelligible to all readers.

It is not the primary goal of this volume to discuss applied (scientific) methods. Only if a discipline has emerged as a particularly important process, it followed up upon. For example, this is true, for metallography, for isotope analysis or experimental archaeometallurgy.

Geosciences and metallurgy, on the other hand, play a special role in the processing of archaeometallurgical finds. It has already been emphasised that a methodical fine-tuning of modern aspects and methods on the archaeological finds and findings is to be made (Fig. 1.2).

Archaeologists intending to engage in archaeometallurgy require basic knowledge of ores and ore deposits, slags, and the production and properties of metals. Confusion of ore and metal, of smelting, melting and casting, of furnaces and crucibles, of metals and alloys, are poorly suited to tackle further issues, like to reconstruct the organisational pattern of metal production and its socio-economic context, or the spatial distribution of mines, smelting sites and metal workshops in a mining district in certain periods.

This book focuses on ores and (smelting) slags, which are among the most common finds of archaeology. It will be shown why and in what way the composition of ores—the raw material base for the extraction of metals—has influenced the technological know-how of early humanity and that it acted as a stimulus for innovations in mining and metallurgy. Here, the transition from oxide to sulphidic ores is especially crucial. How can you prove when miners reached the rich-ore or cementation zone, when did they reach the primary ore body? Ores, in their variety and diversity, are usually intergrown. They rarely occur in nature in "pure" form. Rather, e.g. parageneses of copper + lead + silver + zinc + arsenic ores, or of tin + copper + arsenic can be found, which are intergrown in varying concentrations with gangue or host rock. Often a co-smelting is needed there.

Although the geochemistry and mineralogy of ore deposits near the surface, in which old mining has been particularly intensive, have been repeatedly described in terms of archaeometallurgical aspects, they often require intensive processing. So far, the influence of the morphology of mineralisation on the structure of old mines has not been sufficiently recognised. This is the reason why mineralogy and geology of ores and ore deposits are discussed in more detail in text and pictures. Only the shape and design of mineralisations may help to decipher the strategy of ancient mining activities and may contribute to find potential raw sources.

How did slags form? How can the supra-regionally similar composition of slags be understood? Is this a technology that has spread in the sense of diffusionist models worldwide, or is it based on chemical–physical reactions that occur automatically in thermal processes? Are the steps of metal production, classified according to modern criteria, fully transferable to the old practices? The reconstruction of old metallurgical techniques requires basic chemical–physical data, which are also explained in phase diagrams. But in many cases, they only allow limited interpretations. Here, experimental experiences are just as important as the observation of ethnological practices in developing countries.

This book is mainly written for students (and their professors) of archaeology, geosciences, chemistry, metallurgy and material science who are dealing with the complex questions of metal production. In the field and later in the laboratory, they are often faced with materials and situations whose approach and interpretation are not always easy to understand. Basics of neighbouring sciences are needed everywhere but are not always easy to obtain. This concerns often analytical facilities. Ores in their geological context as well as slags on their heaps and other metallurgical products may not only be seen from a modern scientific point of view. They should be related to specific (pre-)historic circumstances and the cultural context.

2 Stages of Early Metallurgical Activities

Even though, from a global perspective, the developmental stages of early metallurgy show significant chronological and geographical differences, surprisingly many common technological basics can be observed again and again, despite the considerable spatial distances. This is made obvious by many publications, of which only a few can be mentioned here. Robert Maddin followed the topic many years ago with the organisation of the already mentioned conference series *The Beginning of the Use of Metals and Alloys* (*BUMA*), of which, e.g. the work of Maddin (1988) has to be mentioned. This volume contains articles that show the extraction and distribution of ores and metals in Europe, the Middle East, South-west and South-east Asia, South America and Africa. Recently, Roberts and Thornton (2014), presented a thoroughly comprehensive global overview of the development of early metallurgy with their opus *Archaeometallurgy in Global Perspective*, highlighting the current state of research in this field.

However, these comparisons quickly make clear that the earliest traces and activities of the use, extraction and distribution of metals are located in the Eurasian and Near and Middle Eastern regions and are probably best-researched there. For this reason, a short presentation of the topic will be presented here by way of example from these regions.

2.1 Diffusionism Versus Autonomous Developments

For a very long time, the hypothesis was that the metallurgy of copper as the earliest metal, embedded in a cultural context of civilisations, were brought via the so-called diffusionists model to Europe from the culturally more advanced Near and Middle East, more specifically from the area of the Fertile Crescent (Childe 1928; Wertime 1967). Indeed, the copper artefacts of Aşıklı Höyük (Esin 1993) and Çayönü Tepesi in south-east Anatolia (Muhly 1991), which were likely produced from native copper (Yalçın and Pernicka 1999; Maddin et al. 1999), have long been considered to be securely dated archaeological indicators for the earliest metallurgy in Anatolia. The artefacts of both sites date back to the Pre-Pottery Neolithic (eighth millennium BC). However, these findings are not substantiated by corresponding evidence of a broader copper metallurgy.

After the new calibrated radiocarbon chronology showed that this has to be corrected, Renfrew (1969) suggested, also based upon the context of the New Archaeology, that metallurgy developed at multiple centres at various time periods, also independently in the Balkans and in southern Spain/Portugal (Fig. 2.1). In fact, at Belovode, a Vinča culture site in Eastern Serbia (5000–4650 BC), the discovery of copper slags, minerals, ores

A. Hauptmann, *Archaeometallurgy – Materials Science Aspects*, Natural Science in Archaeology,
https://doi.org/10.1007/978-3-030-50367-3_2

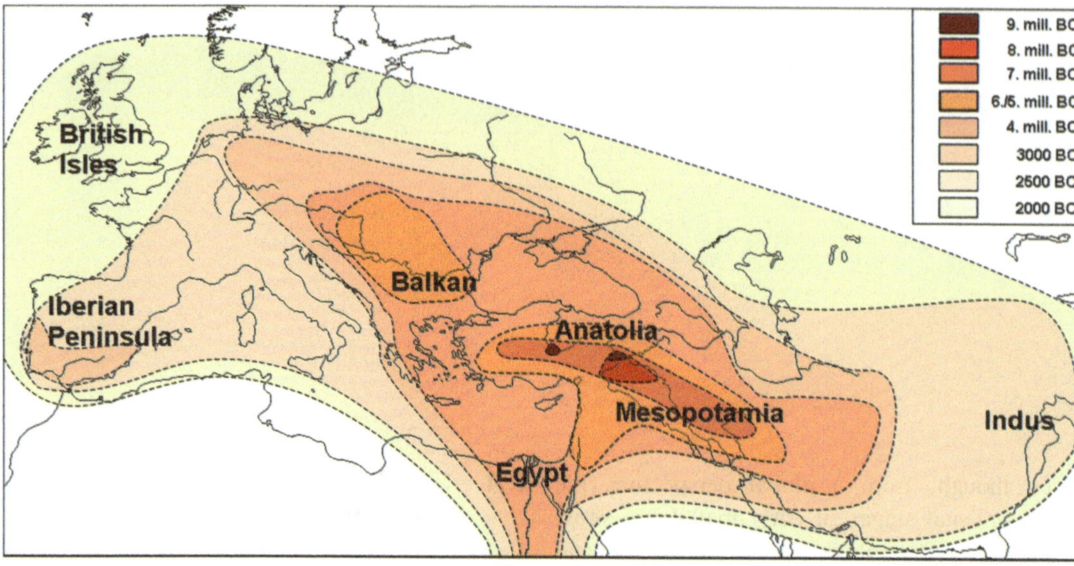

Fig. 2.1 Isochrone map showing the origins and diffusion of copper metallurgy in the Old World. Based on the calibrated radiocarbon chronology, and according to new finds possibly several independent centres of earliest metallurgy are visible. Modified after Yalçın (2017)

and artefacts provided the earliest direct evidence for the smelting of copper ores to date on the Balkan peninsula. Slags are usually the most convincing indicators for the smelting of ores. These smelting activities took place contemporarily with the emergence of the first cast copper objects (Radivojević et al. 2010). Excavations at Pločnik, also a Vinča cultural site, recovered a piece of tin bronze from an occupation layer dated to the mid fifth millennium BC. They were found to derive from the smelting of mixed copper–tin ores (Radivojević et al. 2013). At the Iberian Peninsula, evidence for copper smelting appears at Cerro Virtud, Almeria at 4910–4460 BC (Montero Ruiz and Ruíz Taboada 1996), and at Tal-i Iblis in south-east Iran at the first half of the fifth millennium BC (Dougherty and Caldwell 1966; Thornton 2009a, b; Frame 2012).

This model of mutually independent developments in metallurgy is also continuing in other areas of Europe, albeit temporally offset (Pearce 2015). Therefore, this author also criticises the recently presented diffusionism concept by Roberts et al. (2009), based on Childe and Wertime, which postulated again a development of metallurgy in a region in eastern Anatolia and northern Iraq, from where it spread to other areas. This concept would postulate that copper metallurgy requires highly specialised knowledge. This is certainly being too sceptical about the capacity of human societies for innovations (Pearce 2015) and neglects the easy way to smelt copper and to cast it into artefacts, as Laschimke and Burger (2017, 2018) have demonstrated experimentally. He argues this new paradigm of diffusionism to be an *a priori* model based on a minimalist's interpretation of the evidence. The single-origin model for copper metallurgy would be unproven.

Simple practical-crafted sequences especially of copper metallurgy have been compiled by Forbes (1972), Renfrew (1969) and Wertime (1973) (Fig. 2.2), as compiled by Pernicka (1990). These, however, are theoretical models, that only derive from metallurgical aspects. Ottaway (1994) expanded the model of the loop of Archaeometallurgy, by summarising and rounding off the occurrence of the metals, their production, use and societal purpose up to and including their decay. Pernicka (1990, 2013) focuses on the production of various metals and their alloys in chronologically ordered time periods (copper and arsenic–copper, gold, lead and silver, tin and tin–bronze, iron).

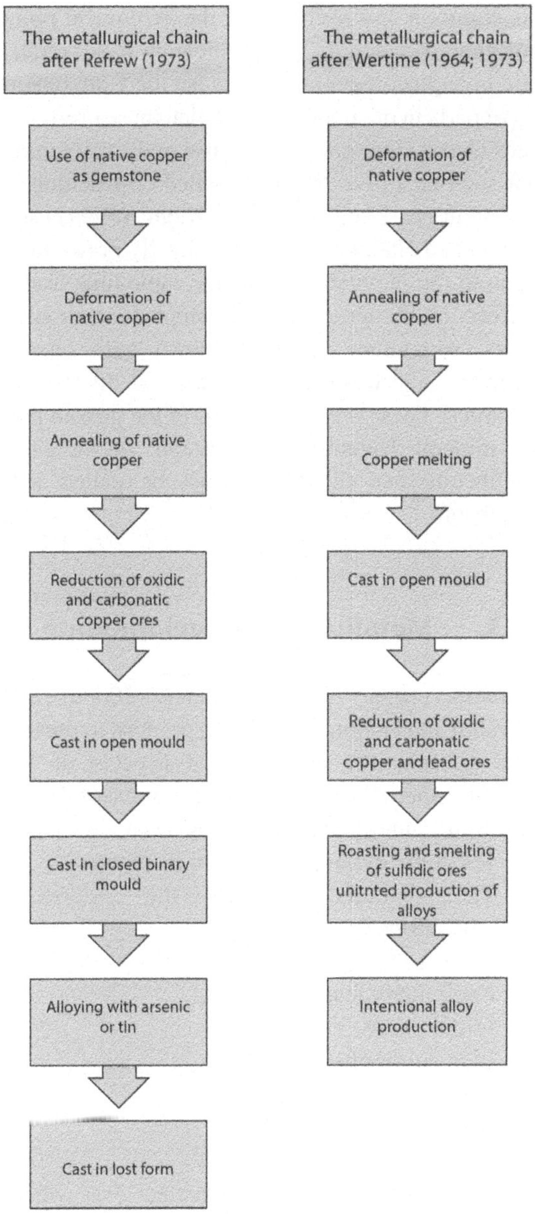

Fig. 2.2 The schematics designed for the order and development of craftsmanship and metallurgy based on the processing of copper (and lead) ores. Modified from Renfrew (1969) and Wertime 1967, 1973)

2.2 Ore Geology and Environmental Change

The development of metallurgy is dependent on the distribution and composition of ore resources and geographic-cultural favour-regions. On one hand, there are metallogenic provinces with numerous rich ore deposits, on the other hand, there are regions that have comparatively few or even no ore bodies.

It is known that in Europe, in western and Middle Asia, sequences of climatic and vegetational changes from glacial periods to an increase in temperature and rainfall occurred around the start of the Holocene period (Roberts 1998). These developments, e.g. led to the rise of agriculture during the Neolithic period in the Fertile Crescent and beyond. But they also had significant geological and geomorphological consequences, which were not only important for the first use of metallic raw materials but are still of economic importance today. In many parts of the world, quaternary sediments contain valuable deposits of gold, tin ores, diamonds and titanium ores, which could be mined as placer deposits (Catt 1992). Due to the change between arid and humid climates, the enrichment of placers was favoured. It can be assumed that placers of gold and tin were available to man in abundance before the beginning of an "industrial" use in the developed Bronze Age (see Borg 2014).

If one were to project the distribution of ore deposits of these regions into the isochrone map of Fig. 2.1, one would conclude that the earliest areas of metallurgy are those regions where mineral deposits are of particular density. In fact, throughout Europe, from the Iberian Peninsula, over the low mountain ranges of France, the fold mountains of the Alps, the Carpathian arch to west Asia (Tauriden and Pontiden in Anatolien, Zagros mountain range in Iran, Afghanistan, Pakistan) this region is the so-called *Tethyan Eurasian metallogenic belt* (abbreviated: TEMB; Jankovič 1997), which eventually continues into the Himalayas and further east, making it a unique geological phenomenon of global scale in the Old World (cf. Sect. 3.4). In this supraregional area, there exist thousands and thousands of ore deposits. The isochron map in Fig. 2.1 shows that the temporal developmental stages of metallurgy also run approximately in the direction of the TEMB. On the other hand, the map is limited to the North African coast in the south, with the exception of Egypt. This may be due to the fact that in the sedimentary complexes of the African-Arabian continental plates the number of ore deposits decreases dramatically compared to the TEMB. In the south of Mesopotamia, there are no ore deposits either. In addition, the focus of the research was more likely directed at the cultural realms of the Old World. Towards the north, the metal-using cultures spread slowly across Europe over the periods of time.

Surprisingly, in discussing the earliest development of (copper) metallurgy, the topic has almost exclusively been the processing of native copper. This is particularly the case for Anatolia (Yalçın and Pernicka 1999; Maddin et al. 1999), Iran (Maczek et al. 1952), the Balkans (Pernicka et al. 1993), and the Iberian Peninsula (Rovira and Montero 2003). From a geological point of view, occurrences of native copper are quite common in these metallogenetic provinces. In the discussions about the earliest copper, however, the incredible masses of malachite, cuprite or atacamite in the gossan (iron hat) of ore deposits has been disregarded, such as, e.g. in the deposits of Tsumeb in Namibia or those in Chile, from where it has become world famous. In order to obtain copper from these secondary minerals, they do not even have to be exposed to very high temperatures. Only for the melting of copper one must reach at least about 1100 °C. The fact that materials have already been treated with high temperatures in the Neolithic Age is shown by finds from the widely used artificial lime plaster in the Middle East (see below). Molten lead was found as well, such as Yarim Tepe's lead bracelet in northern Iraq, dated to the sixth millennium (Moorey 1994).

A geological fact is should be mentioned when archaeological considerations of the individual stages of development are discussed. From the use of native metals, if this use existed at all as an individual stage (cf. Sect. 3.7), over the use of so-called oxidic ores to sulphur-containing and sulphidic ores, there are transitions in nature. A schematic separation of individual ore compositions in the near-surface area of deposits is common and useful from the geological point of view. A categorical classification is not realistic in archaeometallurgy because there are extensive parts in ore deposits, in which these ore-types are closely intergrown. The use and exploitation of such ores of a deposit is called co-smelting in metallurgy. Lechtman and Klein (1999) and Klein and Hauptmann (2016) already draw attention to the inevitable and at the same time useful effects of this process. The composition of such ores, sometimes also intergrown with quartz, limonite and hostrock, are ideal self-fluxing ore charges. These are often found in the gossan part of mineral deposits, so discussions on the early deliberate use of fluxes must be called into question.

2.3 Metallurgical Developments

Strahm (1994, 2007), Stöllner (2008a) and Strahm and Hauptmann (2009) propose to present the geoscientific–technological aspects of the *chaîne opèratoire* in their real archaeological–cultural context and to name several successive stages in the evolution of metal extraction.

The authors suggest the following classification:

- Preliminary stage
- Initial phase
- Innovation phase
- Consolidation phase
- Industrial phase

This classification is not distinct everywhere nor possible in every way. The authors emphasise that there may well be deviations. Nevertheless, it is still possible that the Preliminary Stage can be coherently defined. Equally, the relevance and validity of the last phase, the Industrial Phase, can hardly be doubted. The division between the Innovation Phase and the Consolidation Phase may create some controversy. We will therefore also present these two development steps together.

2.3.1 Preliminary Stage

The decisive prerequisite for the use of metallic raw materials, be it to use metals that occur directly in nature, or to produce metal from ores, is fire. Clear evidence for the earliest controlled use of fire, i.e. fire sustained over days, exists in Europe and the Middle East since the Paleolithic (see overview in Schiegl 1997; Meignen et al. 2007). In this epoch begins an obvious phase of experimentation in which man exposed the most diverse substances to a fire treatment, e.g. obsidian, flint or chalcedony to change their physical properties. Likewise, colour changes could be achieved by heat treatment, e.g. in agate, carnelian or iron hydroxides, the latter for use as a pigment. In the Middle East, limestone, calcite or gypsum was burnt to lime (CaO) at numerous localities to produce lime plaster or cement by pozzolanic reactions, which has been used since Pre-Pottery Neolithic, Phase B (approximately 8700–7000 BC, abbreviated as PPNB) for the design of art objects such as the figures of Ain Ghazal in Jordan or the skulls from Jericho, but also in architecture, for the production of plastered floors in Asikli or Çayönü Tepesi. A compilation of this topic can be found in Hauptmann and Yalçın (2000).

The earliest use of native copper and of green copper minerals, often malachite, at the outcrops of ore deposits, seems not to have been oriented towards the production of the metal itself. Green minerals represented a new colour of the Neolithic period. However, these gemstones, commonly referred to as greenstone in archaeology, were not monotonously shaped out of malachite, but, depending on availability, they used a whole range of green-coloured minerals. As Hauptmann (2004) demonstrated with the example of the greenstones of the PPNB-settlement of Basta in Jordan, they were composed of turquoise from Sinai, malachite, atacamite, chrysocolla, plancheite, emerald green feldspar amazonite, chrom-bearing green fluorapatite and green calcite, which have been traded from various sources. Nothing is known about the frequently cited early use of azurite and blue beads made of this mineral. Azurite is a blue mineral, and this colour did not have any meaning, nor are any finds of azurite (pigments), with a few exceptions, firmly proven from this period. For many finds, it seems that they were merely additional raw materials which were worked using already known and widely accepted techniques of stone manipulation. This included simply hammering and heating (annealing) in order to improve the properties for shaping the material into small artefacts. However, if native copper had been reshaped into artefacts, it is difficult to understand why malachite should not be smelted to metal as well. The fact that the production of malachite copper beads could not have been a big undertaking is shown by the experimental work of Laschimke und Burger (2017), who have demonstrated that the production of small pellets, melted from native copper or from malachite, and the small awls and serial beads of Asikli Höyük forged from them, is a simple crafting method, provided that one looks into the starting material a little. This experience is contrary to the considerations of Childe (1928). He believed that metallurgical processes and techniques were so difficult to master, that they only could have been discovered once (see Pearce 2015).

It is striking that the earliest use of copper and greenstones falls geographically in those areas where the so-called Neolithic Revolution spread with the advent of agriculture and livestock. It is the area of the Fertile Crescent, which runs from Mesopotamia in the east across the desert margins south of the mountain ranges of eastern Anatolia to the southern Levant.

The first metals then appear in PPNB exactly in these regions, where ore deposits can be found, namely on the southern edge and in Central Anatolia and in the mountain ranges of today's Iraq and Iran (Esin 1996; Maddin et al. 1999; Pigott 1999; Schoop 1995; Yalçın and Pernicka 1999). The most famous locality is Çayönü Tepesi, where, together with hundreds of malachite beads, over a hundred small copper objects were found (Özdoĝan and Özdoĝan 1999). Metallographic analysis has shown that these small awls, and beads only a few mm in size, were made of solid copper, were cold worked and shaped at moderate temperatures. They were not melted or cast.

Green artefacts were found along the Fertile Crescent in northern and southern Iraq (in the cave of Shanidar, Ali Kosh), southwest Anatolia (Çayönü Tepesi, Hallan Çemi) and in numerous other sites in the Near East ascribed to the PPN period. In the Balkans, such objects have been found, for example, in Lepenski Vir near Majdanpek in Serbia (Srejoviċ 1975), Iernut and Cernica (Romania, Comşa 1991). They were also documented from the Chasséen in southern France (Servelle and Servelle 1991) and the copper beads, from Monument XII in Sion (Petit-Chasseur, Wallis, Switzerland) may also have the same significance. They were found not only close to ore deposits, but they were traded over large distances too. Green-coloured (copper) ores were also ground for cosmetic and colouration purposes. Examples for the use of green powders are the mask of Nahal Hemar at the southern end of the Dead Sea (Israel; Bar-Yosef and Alon 1988), and the green eyeliners of the figurines of Ain Ghazal in Jordan (Tubb 1985).

2.3.2 Initial Phase: The Neolithic Period

As late as the Middle to Late Neolithic period, people had access to almost untouched, presumably abundant outcrops of ore deposits that are rare today, largely existing in remote areas, far from civilisation, e.g. in areas of Africa, Australia, Siberia or South America. In the Old World, signs of mining of such ore outcrops from this early period are hardly detectable. As in the previous Preliminary Stage, this stage of development is almost exclusively dedicated to copper. Gold as well as lead, silver or tin does not play a role yet. The earliest evidence of copper mining activity currently known can be found on the Balkan Peninsula. The mine of Ai Bunar in today's Bulgaria dates to the second half of the fifth millennium, almost simultaneously to that of Rudna Glava in Serbia (Pernicka et al. 1993; Weisgerber and Pernicka 1995). However, Glumac and Todd (1991a, b) correctly point out that prehistoric mining does not necessarily have to be associated with metallurgy.

The beginning of pyrometallurgy, i.e. the treatment of ores under high temperatures or (naturally occurring) metals, and especially extractive metallurgy, started in the period between 6000 and 5000 BC.

The macehead of Can Hasan (Anatolien) is of particular importance. It is dated to around 6000 BC (French 1962). The macehead has sometimes been interpreted as the earliest metal object produced by casting. According to research by Yalçın (1998) however, the macehead was forged from solid copper at high temperatures, which means that in terms of technological development, this crafting step is situated prior to casting. The fact that inclusions of isolated silver grains could be identified in the copper matrix shows that the artefact never reached the liquid state.

Not quite comparable to copper is lead, which is very easy to extract from its oxidic or sulphidic ores. It melts at 327 °C. From the technological point of view, therefore, it is not surprising that a bead of lead from the sixth millennium was found in Yarim Tepe in northern Iraq (Moorey 1994). However, this is an isolated find. It remains a question if this could be an indication for the earliest lead smelting—even before copper smelting (Craddock 1995; Potts 1997).

No secure evidence for copper objects produced by casting exists prior to the fifth millennium BC. The first cast copper artefacts, dependent in their production on controlled temperatures exceeding 1000 °C, emerge throughout Eurasia in the early fifth millennium BC. This is evidenced by the copper chisels and needles of Mersin/Yumuktepe, layer XVI (Yalçın 2000a, b). He shows via metallographic and chemo-analytical investigations that the artefacts were cast from liquid copper.

According to the current state of knowledge, the earliest smelting of copper is known from Belovodo from the middle Danube basin in Serbia and dated to the Vinča culture (5400–4500 BC) (Radivojević et al. 2010), although there are also some hints in the same time horizon from Anatolia. In these regions and beyond, evidence of copper smelting of similar age or even older can be expected. The best evidence of targeted smelting of copper ores are generally slags or slagged crucibles, although there are exceptions. The authors examined some small slag pieces from Belovode (pieces with a weight of 4 g and a size of 2 cm) dating to the late sixth millennium. Among others, these slags contain delafossite and magnetite, phases typical of very early copper production. Although the amounts of copper produced here were only sufficient for small artefacts, these are the earliest securely dated remains of copper smelting, which are also simultaneously considered as a development of extractive metallurgy on the Balkan peninsula, geographically independent of Anatolia. In addition to the slags, some pieces of secondary copper ores were also found, so that the ensemble of the finds is a bit richer than that of Çatal Höyük in Anatolia. It should be noted that no remnants of crucibles were found.

Glumac and Todd (1991a) report archaeometallurgical finds of Vinča-age settlements of the fifth millennium BC (chronology see Glumac and Todd 1991b) in the middle Danube valley in Serbia. This is for once Selevac, not far from Belovodo, where small (2 cm) slag pieces with dendrites of cuprite in glassy matrix and with small droplets of copper were found as well. From the simultaneous, nearby Gornja Tuzla and in Pločnik small ceramic vessels were excavated. Since the small size of these ceramic vessels seems to fit the dimension of the small slag pieces, the authors assumed that these could be crucibles for ore smelting.

In Zengővárkony, also in the middle Danube valley, but located in today's southwest Hungary, regionally outside the area of Vinča culture, finds of small pieces of slags with cuprite and cassiterite and various mixed, unspecific copper-bearing phases and various mixed, unspecific CuO-SnO_2-FeO-CaO-SiO_2-phases were made. They likely date to the fifth century (for dating see Pernicka et al. 1997). Glumac and Todd (1991a) concluded from these analyses very quickly that cassiterite has been mixed into the copper. This would be the earliest indication of a deliberate mixture of two types of metals to make an alloy, by far. The use of any natural polymetallic Cu-Sn-ores such as stannite (Cu_2FeSnS_4), and the consequence of smelting such ores were excluded. For the problem of cassiterite in copper-rich slag, see Sect. 5.5.4.

Later, in a joint venture between archaeologists and geoscientists, Radivojević et al. (2013) report a tin-rich copper foil from the fifth millennium from Pločnik. It was analysed by a portable X-ray fluorescence spectrometer (pXRF). The authors suggested reasonably that the foil may have been smelted from a mixed ore mixture of chalcopyrite and stannite.

The archaeological evidence of the earliest extractive metallurgy in Eurasia and the Middle East is so rare that one should rather speak of an experimental phase of metallurgy. They are often indistinct and controversial, which could also be due to analytical–interpretive uncertainties of archaeological finds in laboratories.

One of the finds, which also represents these problems, comes from the Transcaucasus, where a green ring bead and a few bits of azurite from the neighbouring ore deposits David Garedji and Madneuli were found in the Neolithic settlement of Aruchlo in South Georgia. Aruchlo dates to the sixth millennium BC. The bead was first interpreted as an externally corroded metal-bead of a copper-tin alloy via a pXRF in the National Museum in Tbilisi (Georgia). Later it was identified at the DBM by scanning electron microscopy with an energy-dispersive X-ray analysis (SEM-EDX) as a bead that was fully composed of (par-)atacamite with containing some clay minerals. Evidence of secondary corrosion from soil deposition could not be excluded but was not convincing. Tin contents could not be confirmed. In Aruchlo, no remains of crucibles were found. The most convincing interpretation at the moment is that this is a green bead.

Garfinkel et al. (2014) report the finding of an awl in a grave Tel Tsaf in the Jordan valley, which is "probably associated with the burial" dated to the late sixth millennium BC. The assumption of the authors was that the awl had been cast. Measurements via pXRF indicated a tin bronze with 3.5–7 wt.% Sn. More research is needed here.

Tiny fragments of "slags" of Çatal Höyük (Anatolia), layer VI (c. 6500 BC) proved to be similarly problematic to Aruchlo. They were first described by Neuninger et al. (1964). They contain, amongst others, the typical slag phases delafossite and magnetite. Mellaart (1966) has not excavated any (s)melting crucibles either. The slags are highly controversial as products of extractive metallurgy, due in part to the absence of fayalite (Tylecote 1986). However, this argument has not been *conditio sine qua non* for the formation of smelting slags since the investigations of Hauptmann (2007) on Chalcolithic–Early Bronze Age copper slags from Faynan. Today, the finds of Çatal Höyük are interpreted as accidental products of a fire in a settlement (Radivojević et al. 2017). The jewellery, which was first mentioned as a lead bead in Çatal Höyük, i.e. made of lead produced by melting metallurgy, was later identified as galena (PbS), i.e. as a natural mineral (Sperl 1990).

In all cases, it has been shown that in this initial phase of metallurgy all metallurgical activities were carried out on a very small scale within settlements. This means that ore has been taken from the outcrops of natural mineralisations over sometimes considerable distances into villages where it has been further processed. This applies especially to those regions where sources of raw materials are rare. This organisational model, commonly referred to as domestic mode production (Hauptmann and Weisgerber 1996; Golden 2014), can be traced supra-regionally to the Chalcolithic and Early Bronze Age periods, and it is only changing fundamentally with the production of metal artefacts on a larger scale.

The Initial Phase is followed by a stage of development in which the extraction of ores and processing of metals experiences a new step.

2.3.3 Innovation and Consolidation Phases: Early to Late Chalcolithic and Early Bronze Age I

Metal objects were sometimes produced by simple processes, sometimes in rather complicated techniques. These certainly came back to the experience and knowledge of pottery making, a related form of transformation of materials by fire (Hansen 2016). Experimenting with a few, more common metals and making not only simple tools but also jewellery and prestige objects had already a history in the fifth and fourth millennium. The fourth millennium BC was a particularly dynamic period characterised by a large number of technical innovations. These included, e.g. the wheel and the cart, the domestication of the horse as well as the plow. The wool sheep triggered a textile revolution, which was surpassed only later by artificial fibres.

The abundance of minerals at outcrops of ore deposits, at gossans, as well as the experimentation with natural resources and the processing of metals at high temperatures, casting and alloying have led to special developments in some favoured regions and the first metallurgical centres, with locally noticeable, massive productions which arose there. According to the just described observations of early ore transportations from raw source to settlements and despite the so-called domestic mode production, such centres may also have formed in regions that are not located in the immediate vicinity of deposits through trade and exchange of the material.

Over the densely populated Eurasian continent, a network of trade relations was developing that grew ever closer, and a cultural and knowledge exchange across large areas increased. And so, the number of "creative" centres grew. As can be seen in the analysis of geographic locations, urban civilisation has always developed where, induced by climate and water, favourable conditions for nutritional foundations existed. Marshall (2015) formulated these connections with the title of his book *Prisoners of Geography*.

For the Near East, various authors (Roberts et al. 2009; Pernicka 2013) named regions where metallurgical activities have emerged since the fifth millennium. These are the Iranian highlands west and south of Tehran (Tepe Zageh, Cheshme Ali) and the south-eastern foothills of the Zagros mountains (Tal-i Iblis), then in south-east Anatolia along the middle Euphrates in the Taurus mountains (Norsuntepe) and in the southern Levant. With the exception of the Levant, these parts of the Tethyan Eurasian metallogenic belt have a large number of rich ore deposits. The southern Levant, however, can be considered a highly frequented transit region between Anatolia, the Caucasus and Iran on the one hand, and Egypt, on the other. The material evidence for an independent invention of metallurgy in Iberia is weak (Roberts et al. 2009).

However, the much higher inventory of metal artefacts in south-eastern Europe between the late sixth millennium and about 3000 BC suggests that this region had also a position as a major innovation centre. As mentioned above, the earliest evidence of smelting copper ores in Serbia and Bulgaria dates back to around 5000 BC, which is several hundred years before it starts in the southern Levant (Golden 2014). The finds of heavy and material-consuming tools are evidence that the raw material copper, as well as that of gold, must have been present in large amounts, while silver and lead have not yet played a significant role. On the Balkan peninsula, one of the early metallurgical centres is the necropolis of Varna I, which dates back to between 4550 and 4450 BC (Leusch et al. 2014). It is of supra-regional importance not only for its unique discoveries of gold objects but for the interpretation of the emergence of complex social structures that took place in the Copper Age not only in south-eastern Europe but throughout the Old World. In total, over 3000 gold artefacts were excavated in 320 graves of Varna, many pectorals, jewellery and ceremonial weapons. Gold artefacts were created in mass production, even deliberately manufactured gold–copper alloys were found. The enormously rich finds show that there must have been unusually large amounts of this precious metal in the wider area, although researchers have not been able to find its origin.

The heavy implement horizon of the fifth and fourth millennium (Strahm 1994), which extends from south-eastern Europe across the Danube countries to Central Europe, was obviously not limited to one cultural area, but represented an intercultural phenomenon, and can possibly also be associated with the dense, widespread occurrence of ore deposits in the Carpathian region, in the Slovak Ore Mountains (Schalk 1998) and in the Alpine area.

But also later, in the Middle East, there existed impressive metallurgical centres as well, that date back to the fourth millennium. Therefore, a chronological supremacy of metallurgy in one region or another is difficult. The largest, best-known hoard is that of Nahal Mishmar in the Judean Desert in Israel. It comprises more than 400 artefacts made on one hand from comparatively pure copper from Wadi Arabah deposits, and on the other from complex copper–arsenic–antimony alloys of unknown origin (Tadmor et al. 1995). They are among the earliest evidence of lost wax castings. They are excellent evidence of a highly stylised processing of non-ferrous metals at this time. Stylistically, they can be attributed to the Ghassul culture. Shalev (1991) shows a dissemination of similar objects in neighbouring Be'er Sheva culture group.

During this time period, the earliest and most widely used alloy was arsenical copper. It appears in south-eastern Europe as well as in Anatolia, the Transcaucasus and the Levant as early as the fourth millennium. Arsenical copper has also been found in the Far East since the transition of the fourth/third millennium BC (Linduff and Mei 2014). In this period, these alloys were most likely a random product based on the smelting of naturally occurring mixed ores. Section 3.5.8 explains that many copper deposits already contain arsenic-containing ores in near-surface areas, which inevitably lead to the production of natural alloys when they are used.

2.3.4 Industrial Phase: Developed Bronze and Iron Age and Later

In this stage of development, innovations can be recognised almost everywhere in mining and metallurgy. However, the Bronze Age with the finds of many thousands of artefacts of copper, gold, silver, tin and lead and their alloys have not developed because humans have discovered new sources of raw materials (contra O'Brien 2013). It could rather be suggested that most of the mineral deposits were known long before. Only because of increasing economic demands, new technologies have been developed and large quantities of metals even from lower-grade ores were produced.

Mining activities proceeded to an exploitation of ores in larger depths. Deeper shafts had to be constructed, the coursing of the mines had to be improved. In humid regions, drainage of the pits was necessary. The most powerful renewals and expansions in mining, which are currently archaeologically tangible, date back to the Roman Empire, which at the turn of the century and the centuries thereafter extended to the countries around the Mediterranean. One of the most impressive technological constructions for draining was found in the Roman mines of Rio Tínto on the south-western Iberian Peninsula. Underground, eight pairs of water wheels with diametres of more than 4.6 m each were installed here to raise the mine water by about 30 m to the surface (Weisgerber 1979).

These enormous developments in mining led to an increased use of low-grade ores of sulphidic composition, associated with a change in chemical compositions, which was to be mastered. Metallurgically, this was not only connected with the construction of roasting facilities but above all with smelting in larger furnaces, which also resulted in a significantly increased production of slag. Smelting of ores was therefore now different than in the epochs before, in the immediate vicinity of the raw sources, where mining activities went on. Craddock (1995) described this stage as a slag smelting process, which contrasts with the few slags found in the eras before. It is a step towards an extensive, in places almost industrial metal production. As shown by the millennia-long copper production at Faynan, the production of blister copper at the deposits, and their further processing within more remote settlements, was associated with new organisational patterns that led to social specialisation of labour (Hauptmann and Löffler 2013).

According to the state of research, after the long use of small crucibles, the first smelting furnaces appeared in the developed Early Bronze Age (third millennium BC), in the Eastern Mediterranean and the Middle East.

These were simply constructed wind-powered furnaces that were in use at geographically wind-exposed positions. They reached a volume of about 50 l. Gale et al. (1985) observed Early Bronze Age smelting sites with relics of such wind-powered installations on some Aegean islands. Hauptmann (2007) discovered numerous Early Bronze Age smelting sites in Faynan (Jordan) where several dozen wind-powered furnaces were excavated. Avner et al. (1994) reported exposed smelting sites in southern Wadi Arabah, and Abdel-Motelib et al. (2012) found archaeological evidence of wind-powered furnaces in southern Sinai. In Arisman, on the Iranian Plateau, a smelting furnace was excavated as part of a German–Iranian research project (Vatandoust et al. 2011), whose construction is reminiscent of the wind-powered furnaces of Faynan.

Smelting furnaces with artificial air supply via tuyères, connected to bellows, are a technological innovation that followed the wind-powered furnaces and that spread supra-regionally in the developed Bronze Age.

The further mental step of mixing different metals or further manipulating finished metal artefacts may perhaps derive well from the Neolithic knowledge of pottery making (Hansen 2016). For example, tempering of clay in pottery production, i.e. the mixing of two different materials (Hansen-Streily 2000), or the treatment of surfaces of ceramic vessels, may have served as models for metallurgical processes. Only by tempering, the firing of clay pots became possible without cracking. But this hypothesis also means that the tangible innovations in this late epoch of metallurgy are the result of developments whose actual "inventions" are temporally very distant.

If you mix copper with tin in a ratio of about 10:1, you get tin bronze. This alloy has become globally important and has given its name to a whole epoch of human history. Tin bronzes were produced in amounts of tons. They appeared in the Old World in the middle of the third millennium BC, apparently first in the Near East, in regions sometimes geographically far from tin ore deposits (Pernicka 1998). Later, the abundance of tin bronzes is increasing enormously, becoming the most common and widespread ancient alloy, and in the Late Bronze Age, as evidenced for example by the discovery of the shipwreck of Uluburun (Yalçın et al. 2005), finds of tin ingots are emerging increasingly.

Ideally composed tin bronzes are not regularly found in the early stages, intentionally made alloys are controversial. Intergrowths of stannite + cassiterite + chalcopyrite are well known in geoscience of ore deposits. Such ores can very well be processed into tin bronzes. The occurrence of mixed tin–copper ores in ore deposits is possibly underestimated in archaeology (see Sect. 3.5.6).

Intensive and extensive metallurgical activities affected the daily life of human beings. (S)melting, metalsmithing and alloying techniques, patinations, metal-organic issues, metal-mediclinics and pharmaceutics were used frequently. A wide variety of metals were blended, and (copper-based) alloys with tin, lead and zinc found use as jewellery as well.

Far-reaching social manipulation was achieved by iron, which led to fundamental changes with the beginning of the Iron Age. It partly replaced copper and bronzes and became a metal used throughout the day-to-day life. In the middle of the first millennium BC, ferritic iron and steel were predominantly produced throughout the world. It was understood to optimise the physical properties of weapons and tools through targeted carburisation of iron.

As far as gold is concerned, the refining process of salt cementation (parting), discovered on the basis of the excavations in Sardis (West Anatolia), had achieved great economic importance (Ramage and Craddock 2000). The Roman Empire began with an unprecedented territorial expansion and associated growth in gold exploitation in the Iberian Peninsula, the Balkans and Britannia. Tons of raw gold were subjected to the elaborate refining process. Considerable quantities of gold were, of course, used for the jewellery and luxury goods of rich people, as well as the imperial representation. Goldsmithing reached a high level. The bulk of the gold, however, went to the state mints.

2.4 Chaîne Opératoire and Metallurgical Chain

After the chronological models of the development of metallurgy over the millennia, a practical model from ore to metal is discussed, which covers the most important steps in a metallurgical chain. It was referred to as chaîne opératoire by Leroi-Gourhan (1943) and Pfaffenberg (1992). This chain shows the mining of ores from an ore deposit and the extraction of metal by smelting processes, their shaping into commodities or jewellery and their distribution by the trade. It is outlined in Fig. 1.1 with the individual steps and products.

It contains three basic aspects:

1. It reproduces the step-by-step production generally required for metal production. This includes the mining of ores with various tools and mining techniques (chisels, hammers, fire setting) and their enrichment by beneficiation. Ore concentrates are often smelted several times to produce mat from sulphidic ores and to co-melt prills and small lumps of metal to bars and ingots. This produces waste products, especially slags. The ingots of one metal are alloyed with another metal if necessary, and finished products are produced by casting and/or (repeated) heating (forging).
2. It contains information on the corresponding chemo-physical material transformations caused by the various enrichment and high-temperature processes on the way from ore to the production of metal. For a better understanding of archaeometallurgical finds, knowledge and transfer of parameters of modern metal technology is necessary, but not in every case transferable to archaeological finds. These steps are discussed in Chap. 6 in the various metallurgical processes of the individual metals.
3. It provides the basis for the underlying social issues and problems of mining and metal

extraction over time. This is particularly evident in the organisation of copper production in the southern Levant. Even in the Chalcolithic period and in the beginning stages of the Early Bronze Age, small quantities of ore were smelted down to metal within villages or settlement communities. Examples (Shiqmim, Abu Matar, Maadi, Asqelon, Zambujal and others) show that ore was traded for several hundred kilometres from the sources of raw materials during these periods (see Hauptmann 2007 for a review; Gauß 2015; Stöllner 2008b). Only with the beginning of mass production in the developed Bronze Age in the middle of the fourth millennium is the production of metal in smelting furnaces in the immediate vicinity of the mineralisation carried out.

Figure 1.1 shows the most important cornerstones of this idealistic chain. The (pre-) historical reality of the craft was probably much more complex. Due to the generally difficult archaeological features, only in exceptional cases all the individual steps can be shown in detail or even the entire *chaîne opératoire* recorded. The finds from the Early Bronze Age copper processing workshop of Khirbat al-Hamra in the ore district of Faynan by Levy et al. (2002) show the complexity of the processing steps of already recovered copper up to the production of ingots alone. To reconstruct all these steps is the concern of (archaeologically oriented) mining archaeology and (scientifically oriented) archaeometallurgy. A typological study of archaeometallurgical finds would not be sufficient, a scientific analysis of the material is necessary. But even analytical studies provide only limited information. They could decipher chemo-physical parameters, but rarely every single step of craftsmanship. More so when intermediate or final products (blooms, ingots, bars, final artefacts) are involved in the production process, or when products of metallurgical work are being recycled (recycling of slags, repeated re-melting of metal).

For an advanced understanding of ancient techniques, it is recommended to use textual and figurative sources. Gaius Pliny Secundus wrote in his opus Naturalis Historia, written around 77 AD, in 2 out of 37 books (books 33 and 34) on the nature of metals and their alloys. There are also several sources from the Medieval and especially from the Renaissance period in Europe. Very informative are the books written by Theophilus Presbyter (twelfth century AD), Georg Agricola (1556 AD), Vanoccio Biringuccio (1540 AD) and Lazarus Ercker (1574 AD), who described in detail the smelting and melting techniques used between the twelfth and sixteenth centuries AD. John Percy (1861/1880) published traditional techniques still used by metallurgists in the nineteenth century. They contain a large number of observations concerning mining and metallurgical processes. There are detailed commentaries and re-interpretations, e.g. by Bartels et al. (2007). Al-Hamdani, a Yemenite metallurgist of the tenth century AD, wrote an exceptionally detailed book on gold and silver metallurgy in the Islamic world.

The second source for better interpretations of archaeometallurgical finds are based on ethnographic studies of base metal and iron production (Gowland 1899; Celis 1991; Anfinset 1996, 2011; Bisson 2000). Further information is presented in Chap. 9. The third source is experimental studies as they were performed by Tylecote et al. (1971), Merkel (1990), Bamberger and Wincierz (1990), Herdits (1993), Hanning et al. (2010), Heeb (2014), Heeb and Ottaway (2014), Day and Doonan (2007), Hauptmann et al. (2015), Laschimke and Burger (2015) and Hauptmann et al. (2016). These will be discussed in Chap. 10.

These aspects allow the courses of metallurgical processes and technologies to be observed more in detail from the very beginning until the making of final objects. All objects and products, controlled by individual craftsmanship and by chemo-physical laws, from slag and refractories to casting moulds, etc., can be better assigned to specific steps in metal production. This will be shown in more detail in Chaps. 5 and 6.

3 Ancient Ore Deposits

3.1 Introduction

Mineral deposits have supplied useful and valuable material for human consumption long before they became objects of scientific curiosity or commercial exploitation (Misra 2000). In fact, the earliest human interest in rocks and beautifully coloured minerals was probably because of the easy accessibility at the surface, e.g. of the earthy red hematite, malachite, gold and gemstones in placers. As an example might be the almost industrial use of red pigments for thousands and thousands of rock paintings in the National Park of Serra da Capivara in eastern Brazil some 35,000 years ago (Guidon and Delibrias 1986). The red pigment hematite used in masses for the paintings must have been exploited from the huge Itabirite Banded Iron Formation.

Beside any polymetallic mineralisations, Misra (2000) emphasises the basic necessity of "capstone courses" in mineral deposits to better understand the meaning and importance of (metallic) raw sources—in modern times and in archaeology and archaeometallurgy as well.

Chapter 2 explained why the development of metallurgy in different regions of the Old World and beyond began at different times. But it developed everywhere, once begun, surprisingly uniform. These differences, and at the same time these uniformities, certainly have also socio-cultural, economic and political reasons. On the other hand, they are based on features and specific criteria of the geology of ore deposits to occur in similar ways everywhere. The early stages of metallurgy and its further technological development continue to depend on the regional availability and the quality of ores. We are dealing with regions in which abundant mineral deposits occur in large numbers, such as in the TEMB. Contrary to that, there are areas that have little or no raw material resources, as was and still is the case in Mesopotamia or the northern African coastal areas. This regional distribution of ore deposits will be described in more detail in Sect. 3.4.

In the "old" deposits copper, tin, iron, lead and silver ores and gold were mined. Furthermore, antimony ores were known as rarities, and only in Roman times mercury and zinc ores were added as well. Arsenic, cobalt and nickel were unknown as individual ores in prehistoric times. They were only produced in the Middle Ages as their own (semi-)metals. Mining archaeologist's problem is that the ores exploited in ancient mines in ancient times are gone; what is presented to the scholar today are ores left due to their low quality.

All of these metals and semi-metals are present in different concentrations in the earth's crust. This is shown in Table 3.1 with the specification of the so-called Clarke values.

This chapter deals with a selection of ore deposits that we know have played a role in (pre)-historic periods mainly of the Old World. First of

A. Hauptmann, *Archaeometallurgy – Materials Science Aspects*, Natural Science in Archaeology,
https://doi.org/10.1007/978-3-030-50367-3_3

Table 3.1 The metals and semi-metals that were usable in antiquity are enriched in different concentrations in the earth's crust

Element	Concentration (ppm)
Copper	55
Iron	56,300
Lead	12.5
Silver	0.07
Gold	0.004
Tin	2
Arsenic	1.8
Antimony	0.2
Mercury	0.08
Zinc	70

They are called Clarke values. Values given in parts per million (ppm). After Taylor (1964)

all, these are basically only those that were outcroppings at the surface. This chapter highlights aspects of importance especially to early mining, early metal technology, and early metallurgical developments.

Important factors are pure descriptions of ore bodies, their shape and their geological bedding in order to understand strategies of ancient mining activities. This aspect of ore deposits corresponds to the descriptive approach as published in geoscience in the 1960s (Schneiderhöhn 1962; Huttenlocher and Ramdohr 1965). Are they massive mineralisations or ore breccia, where ore intensively intergrown with host rocks? Are there any clearly defined hydrothermal veins, such as at the Mitterberg (Austria) or in the Harz Mountains (Germany)? Do the host rocks consist of soft sandstone, dolomite or shales, such as the Timna and Faynan copper deposits (Israel and Jordan), where the layered copper mineralisations can also form diffuse impregnations? Or are the mineralisations embedded in hard host rock as in the tin ore in the granite in the Ore Mountains (Erzgebirge) in Germany or in Cornwall in southwest England? This chapter deals primarily with qualitative and quantitative mineralisations, as well as the geology and geochemistry of near-surface parts of deposits. These are the oxidation and reduction zones of sulphide deposits, which have been critical for early mining, as well as secondary enrichments of ores and other minerals in placer deposits. From a metallurgical point of view, it is important to know whether predominantly oxidic or sulphidic or respectively sulphurous ores were present, if massive rich ores were available and how extensive native metals were present.

Of particular importance is the frequency of ore minerals in deposits. Did ore shoots exist which were exploited? It will be shown by a number of examples that in "old" deposits, which were already published from the eighteenth century and later (Klaproth 1795–1815; Von Naumann 1852; Wilke 1952; Ferguson 1950) ore minerals like native copper, auricupride (Cu_3Au), küstelite (AuAg), silver ores such as the nicknamed *Buttermilcherz* and *Gänsekötigerz* (buttermilk ore and goose shit ore; these are mostly chlorargyrite, AgCl) or stromeyerite (CuAgS) were available in part in considerable quantities, which are presented in today's textbooks only as mineralogical peculiarities. In this chapter, less attention is paid to the concepts of modern deposit research of the genesis of ore deposits, as well as modern economic considerations.

3.2 What Is an Ore Deposit?

In general, ore deposits are those natural enrichments of usable minerals and rocks in the earth's crust, which are considerable for economic extraction in terms of size and content. If they include metal-containing rocks and mineral mixtures, i.e. ores, they are called ore deposits. In mineralogy, ores are usually understood as pure ore minerals, such as galena (PbS), sphalerite (ZnS) or chalcopyrite ($CuFeS_2$). For ore geologists and miners, ores are a mixture of ore minerals, e.g. chalcopyrite + pyrite + galenite, which are intergrown in varying quantities with non-metal containing minerals such as quartz and calcite. These minerals that cannot be used for metal extraction are called gangue (Pohl 2005). With technical methods and economic benefits, metals or their compounds can be extracted from the ore deposits. If the ore enrichment is too small to degrade, it is called ore occurrence.

Economic aspects have fluctuated tremendously over millennia. The geo-resources of Iran will serve as an example. According to Momenzadeh (2004), more than 400 copper deposits and occurrences are known in the country. However, currently, only about seven of those are being mined at the moment. In Afghanistan, Soviet geologists counted about 135 tin deposits in the 1970s, of which only two seemed suitable for further investigation (Wolfart and Wittekindt 1980). We may assume that copper ore deposits with <1 wt.% Cu, as they are mined today in South America, were of no interest in prehistoric times. On the other hand, ore deposits, which are of no importance today, may well have been used in ancient times. One can even assume that most of the smaller ore deposits were previously used, provided that they outcropped at the surface. Sverchkov (2009) has illustrated this impressively in his presentation on the abundance of mineral deposits in Uzbekistan. The modern assessment of metal contents of ore deposits is therefore generally not necessarily useful and often misleading for the assessment and understanding of (pre-)historically used mineral deposits. Rich ores remain unconsidered and are not mentioned in the sampling of modern prospectors. And yet, the mineralogical literature of the past centuries (Leonhard 1860; Von Naumann 1852 and others) makes it obvious that the importance of rich ores is probably underestimated. Bartels (2014) discusses this with reference to silver ores in the Harz Mountains and in the Ore Mountains. This illustrates a fundamental problem of archaeometallurgy, which is the exceeding difficulty to grasp quantitatively and qualitatively on the ores mined in (pre-)historic time (Ixer 1999; Baron et al. 2014).

If mining activities are selectively focused on only certain ores of a deposit, the approach and designation of one and the same deposit may change from prehistoric to modern times. Bartels (2014) presents this via the example of the deposits in the Harz Mountains and in the Ore Mountains, where lead and silver ores were mined in ancient times. Zinc was probably only mined in the Middle Ages and in the modern era. Zinc, however, is paragenetically connected with the other two ores. Although such ore deposits are often covered by oxidation zones with conspicuous, abundant calamine zones, zinc was largely unknown before the Middle Ages. Sporadic finds of prehistoric brass are known from the Middle East but were probably coincidental. However, zinc was deliberately mass produced in Roman times and crafted into jewellery, weapons and equipment (Hanel and Bode 2016). It can be assumed that the striking secondary calamine mineralisations have hardly been overlooked in the (pre-)historical mining of lead and silver from the deposits of Laurion and Thasos (Greece) or on the island of Sardinia (Italy).

The traditional genetic classification of ore deposits is mainly based on the classification of rocks and distinguishes magmatic, sedimentary and metamorphic deposits. New discoveries, e.g. in plate tectonics, ocean exploration, deep-sea drilling and isotope geochemistry have all led to new genetic concepts in recent years and decades.

We distinguish geologically several types of ore deposits:

3.2.1 Magmatic Ore Deposits

(a) Liquidmagmatic ore deposits. They are formed by the immiscibility of oxidic and sulphidic ores from siliceous liquids. Belonging to this group are chromite-, platinum- and nickel-copper deposits in (ultra-) basic rocks. Smaller occurrences of this category can be found on the island of Cyprus and in Oman.

(b) Pegmatitic-pneumatolytic ore deposits. They are formed from remaining melts and gases at high temperatures (about 350–500 °C) and high pressures. They form hydrothermal veins and impregnations in coarse-grained host rocks. Pegmatites can include archaeometallurgically important zinc ores, which are intergrown with wolfram ores, tourmaline, but also with copper and arsenic ores. Gemstones like emerald, topaz and ruby have to be included here as well.

(c) Hydrothermal veins, stockwork or porphyric mineralisations. They form through the precipitation of hot (= hydrothermal), mineral-rich solutions in rifts and cracks of crashed blocks of host rocks. This way they form m-sized vein-like lodes with zonal structure of ores. Very common are epigenetic veins, which formed after the surrounding rock and generally have sharp contacts. Often, due to tectonic activities, they can be cracked into a number of parallel-running bunch of veins (Fig. 3.1). Well-known lodes are the Pb-Ag-Co-Ni-Bi veins in the Ore Mountains, the Pb-Zn-Ag lodes in the Harz Mountains, the copper-lead-siderite-lodes in the Siegerland area, and the copper ores at the Mitterberg (Austria), the numerous polymetallic veins of the Iberian Peninsula or of the Iranian plateau. If recent hydrothermal solutions reach the surface, they may blow out as geysers or hot springs.

In impregnation deposits, ores in their finest form occur in many cracks and clefts in intensely fissured rock complexes. Mineralisations occur in magmatic as well as in sedimentary rocks. These are the Porphyry Copper ore deposits, which mostly occur at the west coast of South America in Chile. The world's largest copper deposits can be found there at Chuquicamata and La Escondida.

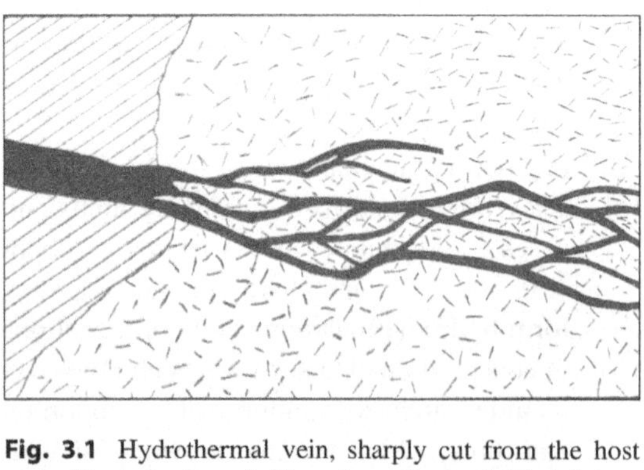

Fig. 3.1 Hydrothermal vein, sharply cut from the host rock. By tectonic activities, the vein is divided into a swarm of small veinlets. Mine *Gottlob Morgengang*, Freiberg mining province. From Beck and Berg (1922)

3.2.2 Sedimentary/Surficial Ore Deposits

(a) Supergene enrichment. Secondary (supergene) enrichments in sulphidic ore deposits with the formation of oxidation and cementation (secondary enrichment) zones are a supra-regional phenomenon. As this type of deposit formation is of enormous importance for early mining and metallurgy, it will be discussed separately and in detail in Sect. 3.5. The process of secondary (supergene) enrichment applies to most of ore bodies, if in varying dimensions. Through elevation and erosion, a deposit may enter circulating water, which is able to remove parts of metals above the groundwater level. Dissolved metals will precipitate at the groundwater level, which can often lead to considerable metal enrichments in the secondary enrichment zone.

(b) Residual ore deposits. To this category belong alluvial placer deposits. Most well-known in this category are gold and tin placers with cassiterite. Tin placers are of considerable importance for the early production of tin, although the mining of primary Sn-Cu-deposits (see Sect. 3.4.11) should not be underestimated. Tin placers may contain gold as well. Magnetite placers, who might have played a part in Colchis (Georgia), belong in this category also.

(c) Residual bauxite. By profound weathering of aluminium and siliceous rocks, bauxite deposits can form, which may lead to significant gold accumulations. Famous gold deposits of such type are those in West and South Africa, which provide excellent examples of artisanal mining and metallurgy, which are very important for the understanding of early gold mining (Werthmann 2009; Borg 2014) (cf. Sect. 9.4). Further examples are laterites high in iron and nickel, which may form ubiquitous ferricretes. These are (and have been in the geological past) widespread all over Africa in tropical areas and were often used for iron smelting until the recent past. Nickel-bearing laterites are especially products of the weathering of peridotites. Here, As-enrichment can occur as well (Evans 1992).

(d) Autochthonous iron and manganese mineralisation are chemical and partly biogenic marine and terrestrial formations. The bog iron ores of the Northern European Plain and in Scandinavia as well as the oolithic iron ore deposits on the Swabian Alb in southern Germany played a significant role in (pre-)historical times and are associated with this category. The manganese nodules and crusts of the deep sea are part of this category as well.

3.2.3 Diagenetic-Hydrothermal Ore Deposits

(a) The Kupferschiefer type in northern Europe, which covers a range of about 600,000 km^2, belongs to this type of deposit. In some areas, it contains very high concentrations of non-ferrous metals and silver. The Kupferschiefer of Zechstein in Germany and Poland was possibly of great importance in the Bronze Age with its sedimentary sulphidic ores (Borg et al. 2012).

(b) Lead–zinc deposits in (marine) carbonate rocks belong to the Mississippi Valley type (MTV). Known deposits of this type are very abundant in extensive lime and dolomite formations in the world. They occur throughout the western and middle Mediterranean, where they have been important sources for lead, zinc and silver in ancient times. Important localities are Laurion, Sardinia, Ibiza, Cartagena, Thasos, Algier, and Tunis.

(c) Sedimentary-exhalative or syngenetic-hydrothermal deposits. This type of deposit includes volcanic massive sulphide ore deposits (VMS), which have become known especially in Cyprus and Oman. Their genesis is comparable to the recent deposition of sulphidic ore sludges by the so-called black smokers on Mid-Ocean Ridges in the submarine region. These were discovered only a few decades ago as a result of oceanic exploration (Hannington et al. 1998).

3.2.4 Metamorphic Ore Deposits

After their genesis, all of these types of deposits may be transformed in mineral content and form by epigenetic and metamorphic processes, i.e. by the influence of pressures and high temperatures. Metamorphic processes lead to recrystallisation of ores, to precipitation of new ones as well as to accumulations and enrichments due to mass shifts. Often deposits of volcanic-sedimentary origin (VMS deposits) are affected by later deformations and metamorphic processes, e.g. those in the Proterozoic Arab Shield or within the gold aureols in the crystalline basement of Upper Egypt and Sudan. Hundreds of these gold deposits in these areas were extensively exploited in ancient times and led to the enormous richness in gold in ancient Egypt (Klemm and Klemm 2013).

3.2.5 From Metal Provinces to Metallogenetic Belts

Following Petrascheck (1965), some metallogenetic terms have to be further distinguished. Their knowledge is for the assignment of metal artefacts to ore deposits (provenance studies), by trace element and lead isotope analysis, very helpful.

A *metal province* includes all deposits of the same metal in (regional) geological units. This is especially true for the worldwide distributed tin provinces. In Central Europe, this is the tin province of Cornwall–Brittany–Iberian Peninsula, then in Southeast Asia from southern China, Thailand to Malaysia, the tin belt in Central Asia from Uzbekistan/Tadzhikistan to China–

Mongolia and the tin belts in South America and Africa.

A *metallogenetic province* or *ore province* comprises the totality of ore deposits formed in a geological section of the earth's crust during a particular geological era. Furthermore, they exhibit material relations. Examples of metallogenetic provinces are, e.g. the Ore Mountains on the border of Germany to the Czech Republic, the Harz Mountains or the deposits of the Slovak Ore Mountains. They are usually larger than 100,000 km^2. Belonging to one ore province as well are the sedimentary copper deposits in the Cambrian sandstones and dolomites in the southern Levant (Faynan, Timna) (Segev et al. 1992) and in southern Sinai (Umm Bogma), which played an important role in prehistoric times in this area (Hauptmann 2007; Abdel-Motelib et al. 2012), or the copper deposits in the Samail Ophiolite Complex in Oman (Goettler et al. 1976; Hauptmann et al. 1988; Peters 2000). The copper deposits of Murgul in northeastern Anatolia, with an area of about 400 km^2, belong to an ore province as well (Lutz 1990).

Important *metallogenetic epochs*, on the other hand, are geological periods in which certain types of ore deposits were formed on a supra-regional level. Thus, the genesis of volcanogenic massive sulphide deposits (VMS), such as Cyprus, Ergani Maden, Rio Tinto (Iberian Peninsula), Murgul (northeast Turkey) or Madneuli (southern Georgia) dates into the Jurassic/Cretaceous period. Many gold placer deposits worldwide date to the end of the Tertiary and the beginning of the Quaternary.

A *metallogenetic district* or *metallogenetic zone* is a spatially separable part of an ore province. Deposits of a metallogenetic district are usually formed almost simultaneously and are closely related in terms of substance and genetics. An example of this is the copper districts of Faynan and Timna.

Metallogenetic belts refer to crustal parts of comparable geological evolution that may contain several metallogenetic provinces. Their length can be many thousands of kilometres and their deposits do not have to be formed at the same time. Well-known examples are the global east-west running mountain belt of the Tethyan Eurasian metallogenic belt (TEMB) as well as the extraordinarily deposit-rich regions in the steppe belt between the Urals and China. Metallogenetic regions are discussed in Sect. 3.4.

3.2.6 Monomineralisation Versus Polymetallic Ore Deposits

It should be noted that ore deposits only exceptionally are mono-mineralised. This includes, e.g. the already mentioned bog iron ores, which consist almost exclusively of iron hydroxides and can be found in all parts of the world. In northern Europe (Germany, Poland, Scandinavia), there are many hundreds of smaller deposits that have been exploited for iron production since the Roman Empire (Ganzelewski 2000; Jöns 1997). Bowles et al. (2000) report in detail about the connections of the Viking expansion in the eleventh/twelfth century with the iron production in this region. Mention has already been made of the extensive ferricretes in many African countries. The banded iron formation in der Eastern Desert of Egypt and Sudan consists almost entirely of hematite (Khalil and El-Shazly 2012) and therefore they resemble the so-called Itabirite ores in the Iron Quadrangle of Brazil. While it is well known that the use of the many "gold only" deposits in this area was the basis of the enormous gold wealth of the Egyptian pharaohs, but it is not clear if and to what extent the iron ore deposits were mined. "Gold only" deposits are also known from the Tauern Mountains in the Austrian Alps, which, strictly speaking, consist of gold–silver alloys like those in Egypt, as well (Boyle 1979). Monomineralic are the mercury deposits of the Almaden in Spain, New Almaden in California, Tuscany or Turkey. They consist of cinnabar (HgS) with occasional contents of native mercury (Pohl 2005).

Much more common are *polymetallic deposits*, which consist of various chemically and mineralogically different metal ores. In ancient times, hardly all ores of a polymetallic deposit could be exploited, the technological means to produce all their metals were hardly available. Westner et al. (2016) show that it can be difficult to detect the desired "target metal" in the analytical processing of slags from the smelting of polymetallic ores.

Fahlores and the variety of other sulphosalts (sulphides of Cu, Pb, Ag, As, Sb, Zn, Fe) occur as minor components along with galena and sphalerite. In prehistoric times, fahlores were used for copper production. Considerable silver contents in these ores were later extracted as well. Ore deposits containing predominantly fahlores, as they occur in Cabrières (Languedoc/France) or in the Inn valley (Austria) which have been mined since the late Neolithic and the Chalcolithic, are rare. Laurion's polymetallic deposit in Greece, which has been described mainly as a lead–silver deposit since the Bronze Age and antiquity, also contains copper ores, which were apparently also mined in the Early Bronze Age (Gale et al. 2009). The ore deposit of Ross Island, which is considered to be the oldest copper mine in the British Isles, also has lead, zinc, arsenic and iron ores next to a range of copper ores (O'Brien 1999). The deposits in the copper district of Faynan–Timna–Sinai, important geo-resources for copper since the fourth millennium, are ultimately polymetallic, as they contain manganese ore in large quantities and also contain lead and phosphorus ores—even if the latter were not exploited in ancient times. Copper ore deposits such as, e.g. those of Madneuli in Georgia also contain lead, arsenic and zinc ores as well as gold.

The lead–silver–zinc deposits of Laurion as well as those of Iglesiente/Sulcis in Sardinia contain, in addition to the main constituents of lead ore, also significant amounts of zinc ores. These accumulate in the iron hat where they may form economically viable batches of calamine. Extensive, thick areas of oxidation zones are documented for both Laurion (Skarpelis and Argyraki 2009) and the lead–zinc deposits of Iglesiente (Aversa et al. 2002; Boni et al. 2003). Their use is very obvious due to the amount of brass objects since Roman times but is not yet documented.

3.2.7 The Design and Morphology of Ore Deposits

Misra (2000) pointed out that the style of mineralisation refers to the pattern of distribution of ore minerals in a host rock, and it varies from being very subtle (even invisible to the naked eye as in some gold deposits such as Sakdrissi or Treŝ Minas in Spain) to quite pronounced (as in the massive sulphide copper deposits in Cyprus, Oman or Rio Tinto). The shapes of mineral deposits are also highly variable, from concordant tabular and stratiform to discordant veins and breccia bodies or filling of karstic cavities as in Puig de s'Argentera at the island of Ibiza in the Mediterranean (Table 3.2). These examples are showing how the style of mineralisation and morphology of an ore deposit are directly affecting the shape, design and construction of ancient mines.

It can be argued quite well that the numerous, sometimes small, mining niches of Puig de s'Argentera in Ibiza (Hermanns 2015; Fig. 3.2) or from the silver mine on Siphnos did not result from fire-setting alone (Weisgerber 1985) (Fig. 3.70), but are primarily due to the irregularly shaped ore pockets, which are dictated by the geometry of karst cavities.

The ensemble of numerous medieval tunnels on the Altenberg in the Siegerland (Zeiler et al. 2018), on the other hand, is based on a tectonically fanned, steeply standing hydrothermal gangue swarm, as shown in Fig. 3.1. A similar situation is also shown on the well-known altarpiece of the Annaberger Bergaltar (lastly Slotta 2012). It displays an iconic representation of silver mining in the Ore Mountains in the sixteenth century AD. In addition to detailed descriptions of ore processing, metallurgical processes and coinage, more than 10 pits and mine entrances can be seen. Hundreds of hydrothermal veins, often situated within metres next to each other, occur in many areas of the Ore Mountains (Wagenbreth and Wächtler 1986). On the Mitterberg, however, the old mining is clearly limited to a few veins, in contrast to the wide distribution of many Bronze Age melting places.

Table 3.2 Some commonly used terms for the style of mineralisation and morphology of ore deposits

Stockwork
An interlacing network of small and narrow (commonly measured in centimetres) close-spaced ore-bearing veinlets traversing the rock. Typical examples: Footwall alteration zone of Volcanic-hosted Massive Sulfide Deposits. Cyprus, Oman; Rio Tinto, Sakdrissi (Georgia, Au)
Massive
Mineralisation comprising >50 vol.% of the host rock. Typical examples: Volcanic-hosted Massive Sulphide lenses. Cyprus, Oman, Rio Tinto, Aljustrel/Iberian Peninsula
Vein-type
Mineralisations in veins or bunches of veins, commonly discordant to the host rock layering (depositional). Typical examples: Base- and precious metal veins (polymetallic, Cu, Zn, Pb, Ag, Fe). Mitterberg (Cu), Siegerland (Fe, Cu, Pb, Zn), Freiberg (Pb, Zn, Ag, As, Bi, Co, Ni), Harz mountains (Pb, Zn). See Fig. 3.1
Stratiform
Mineralisation confined to a specific bed and, thus, broadly conformable to the host rock layering (depositional). Typical examples: Kupferschiefer-type stratiform deposits (cf. Fig. 3.4.1g), Faynan and Timna (Cu) transition to strata-bound type
Stratabound
Mineralisation discordant to host rock layering (depositional), but restricted to a particular stratigraphic interval. Typical examples: Mineralised breccia bodies in Mississippi-Valley-type deposits. Filling of karst cavities. Puig de s'Argentera, Ibiza (see Fig. 3.2), Laurion, Siphnos, Thasos, Sardinia (Pb, Zn, Ag), Silesia district (Poland), Inntal valley (fahlore, Austria)

Modified after Misra (2000)

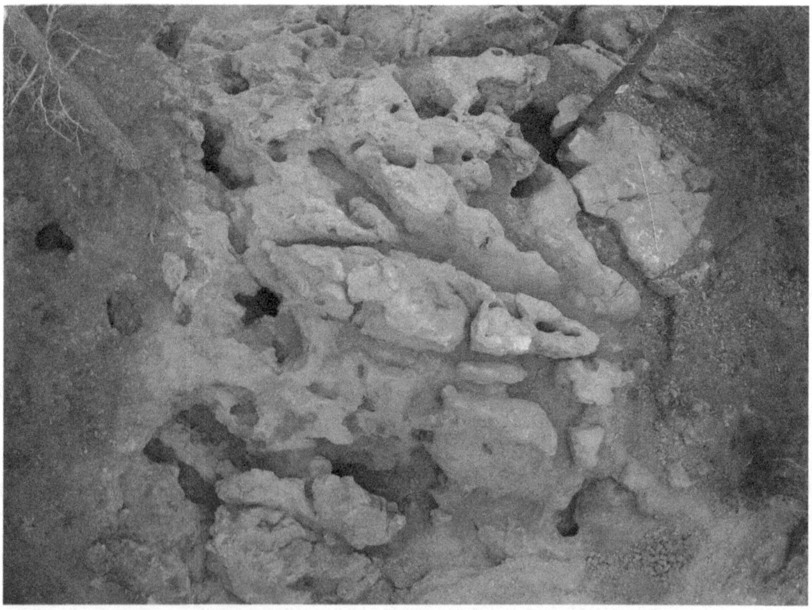

Fig. 3.2 Puig de s'Argentera, Ibiza, Balearic Islands (Spain). Aerial view of the ancient silver mine. The mineralisations were bound to karst cavities. It is questionable if the cavities are man-made, e.g. by firesetting, or are of natural origin. Photo courtesy M.H. Hermanns, German Archaeological Institut

Stöllner (2013) details the coherent, systematic mining of copper ores in the steep hydrothermal vein in the Arthurstollen.

3.3 Ore Minerals Intergrowths

3.3.1 Ore Textures

Ores occur in nature in intergrowths of ore minerals in varying ratios with gangue and host rock. They are usually not monomineralic. Of importance, apart from the chemical composition of the mineral components themselves, is the arrangement of individual components, the texture. In contrast, the structure is determined by the shape of the individual mineral components and by their mutual geometric relationships. Textures range in scale from megascopic scale in the field to (ultra-)microscopic scale. Furthermore, the fabric is important, i.e. the arrangement of its components in space (Ramdohr 1975; Craig 2001).

The knowledge of ore intergrowths is not only important for genetic interpretations of ore deposits but in our case especially for the beneficiation of ores: which and how many tools and methods were necessary to extract certain ores (Fig. 3.3).

Gangue and host rocks are generally unworkable material in archaeometallurgy. On the other hand, they can be of considerable importance as a (natural) fluxing agents. An example is the

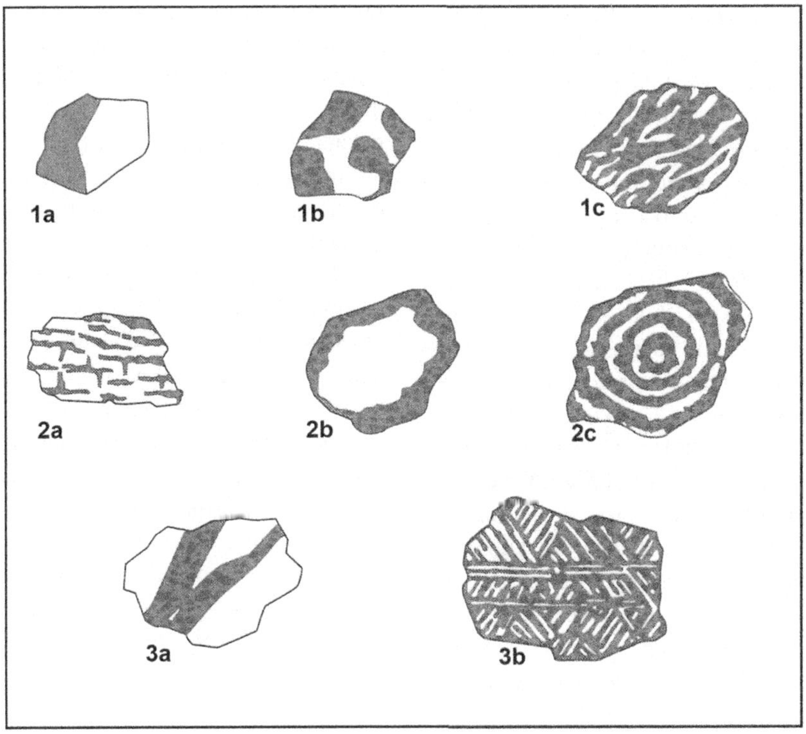

Fig. 3.3 A classification of basic intergrowth of ore textures (and, in our case of phases in sulphidic mattes). (**1a**) Simple intergrowth or locking type; rectilinear or gently curved grain boundaries. Most common, many examples. (**1b**) Mottled spotty or amoeba-type locking or intergrowth. Simple, common pattern, many examples. (**1c**) Graphic, myrmecitic or eutectic type. Wide-spread examples are chalcopyrite and bornite. (**2a**) Disseminated, emulsion-like or drop-like type. Fahlore in galena. (**2b**) Coated, mantled, enveloped, corona- or rim-like. Widespread, e.g. chalcocite or covellite around pyrite, galena. (**2c**) Concentric-spherulitic, multiple shell-type. Galena, chalcopyrite, bornite, cerussite-limonite and many other examples. (**3a**) Vein-, stringer-like or sandwich-type. E.g. gold veinlets in pyrite. (**3b**) Network-, boxwork- or Widmannstätten-type. Very characteristic for meteoritic iron. Less common, e.g. hematite-ilmenite, bornite-chalcopyrite. Here of greater importance: characteristic intergrowths of chalcopyrite–bornite almost as a rule in Cu–Fe-matte. Modified after Amstutz (1960)

intergrowth of malachite with limonite, quartz and clay minerals: the latter three are excellent slag formers which separate from the copper precipitated from malachite in smelting processes. The simplified chemical formulas for these reactions are:

$$2\,FeOOH + SiO_2 + 2\,AlO(OH) \rightarrow$$

$$Fe_2SiO_4 + Al_2O_3 + 2H_2O$$

$$Cu_2(OH)_2CO_3 + CO \rightarrow 2CuO + H_2O + CO_2 + CO$$

$$CuO + CO \rightarrow Cu + CO_2$$

The accumulation and degree of intergrowth of ores and/or metals also depends on their concentration in the earth's crust (cf. Sect. 3.1). The noble metal gold, which is only present in extremely low concentrations, will only form in exceptional cases larger dendritic crystals or accumulations of nuggets in placer deposits. Normally gold is only distributed in finest flakes in veins, often barely or non-recognisable to the naked eye (the so-called non-visible gold). Cassiterite, the most important starting ore for bronze production, hardly ever occurs in larger crystal aggregates than a few centimetres. Often cassiterite is distributed finely dispersed in the gangue and host rock, as is knows in the typical greisening. In the tin deposits of Kestel in Anatolia, the tin ores are present in extremely fine crystals as well in a mixture of hematite, limonite and clay minerals, which required extremely elaborate beneficiation (cf. Chap. 10).

For the mining of silver ores, it is crucial in which form these occur: Today we know that silver-rich fahlores can be included as the finest dust in galena. Silver-bearing rich ores like argentite, pyrargyrite, or stephanite may occur in more massive lumps only in the secondary enrichment zones of sulphidic ore deposits. However, today only intergrowths of galena with silver-containing fahlores in the mm to micrometre scale are known, it has long been assumed that even in ancient times silver was only extracted from galena. This theory is supported by lead inclusions of smelted galena analysed in slags from Pandjhir (Afghanistan). They contain up to 6000 ppm of silver (Merkel et al. 2013), which is very high. However, in the Middle Ages, definitely very rich silver ores were mined in the Harz Mountains and the Ore Mountains (Germany) as well as in the Slovak Ore Mountains in many mines besides galena (Bartels 2014), so that the model silver-containing galena should be reconsidered for its general validity. With great caution, such observations should be transferred as models to other regions where such detailed geological and mining–archaeological investigations have not been conducted, but in which they are generally possible.

Copper, lead and iron ores, whose distribution in the earth's crust can reach 10,000 times or more than that of gold, are present in many ore deposits in massive sections of concentrated ores, sometimes weighing tons, which have to undergo far fewer treatment processes.

This is illustrated in Fig. 3.4, in which ores, intensively intergrown with the host rock and of different masses, are shown. For example, the galena (Fig. 3.4a), as it was likely available to the miners in Cartagena or Mazarron in the Iberian Peninsula (Spain), or in Laurion (Greece) in ancient times, or the chalcopyrite from the deposit of Mitterberg in Austria we denominate as massive ores (Fig. 3.4b). Fahlores (Fig. 3.4c) occur in the Mitterberg in hand-sized pieces as well. Often more massive are the veins and bags of fahlore from the prehistoric mining area of Schwaz, as they are found today. In order to recover such ores quantitatively, a much more complex beneficiation must be carried out. The same applies to the mm-large inclusions of silver-containing galena in a banded lode of quartz, barite and fluorite. The piece is a cross-section of the hydrothermal vein of the *Gottesehre* mine in the Black Forest (Fig. 3.4d), where mining has been proven since the Middle Ages. Occasionally, finds of silver flakes (native silver) were also made here. There is no gold visible in the photo of the vein from the Sakdrisi gold mines in South Georgia (Fig. 3.4e). The crushed, finely fanned up veins of hematite and quartz—a so-called stockwerk-like mineralisation—contain merely non-visible gold today. About 120 kg of material had to be taken from the vein to extract a total of a few grams of gold (Fig. 3.4f). It cannot be ruled out that real

Fig. 3.4 Ores are intergrown to varying degrees with neighbouring rocks. The figure shows a sequence from massive to finely distributed ores and their parageneses in the neighbouring rock or gangue. Depending on the degree of intergrowth, simple or complicated processing of the ore is required. (**a**) Massive galena with some spots

bonanzas have been the decisive stimulus for the beginning of mining, but numerous processing tools prove that this probably was the exception and the ancient mineralisation did not look much different from today. In this case, also the old beneficiation was laborious. Gold containing lumps were taken from the mine and had to be finely ground. Afterwards, it was sieved and the gold was then washed out by panning.

3.4 Spatial Geographic Distribution of Ore Deposits in the Old World

Geodynamic setting affects the composition of ore deposits. This provides information about mineralogical assemblages and geological ages of formation of ore deposits. These two aspects are essential when it comes to the chemical and especially the lead isotope analysis of archaeological artefacts for provenance studies. For a general overview of paleo-locations of ancient terranes and super terranes as well as developments within plate tectonic models see, e.g. Evans (1992), Robb (2005) and Dill (2010).

The global distribution of mineral deposits shows emphases due to the geological evolution of the earth. In metallogenic provinces or zones in global tectonic units, large numbers of rich ore bodies occur; on the other hand, there are regions that have comparatively few or no ore bodies. These geological circumstances, at least in the early stages, led to different temporal and technological developmental processes in metallurgy.

The purpose of this chapter is not only to show some regions where mineral deposits occur in unbelievable density, but also to show how intensively they have been exploited throughout the ages. This may be the basis for estimating quantities of metal that might have been produced in ancient times. O'Brian (2007) has done this for the production of copper using the example of some known deposits.

3.4.1 Plate Tectonics and Ore Deposits

Fundamental for the understanding of the global distribution and genesis of ore deposits is the theory of plate tectonics. Today, it is known in geoscience that the earth's lithosphere is broken in about a dozen plates (Press and Siever 2003), which are sliding, colliding or separating from each other as they are moving across the interior of the earth. This movement of individual plates or tectonic bulk structures is closely linked to the genesis of ore deposits.

Evans (1992) or Pohl (2005) describe three types of plate boundaries:

1. Spreading centres with typical mid-ocean ridges and divergent continents. Mid-ocean ridges are located in the Atlantic Ocean or the eastern Pacific Rim.
2. Converging or colliding plate boundaries with typical mountain ranges, magmatic belts, and possibly deep-sea channels. A well-known example for the collision of two plates is the Himalaya mountain range. It represents the collision of the Indian subcontinental and the Eurasian plate. It continuous to the west as the Tethyan Eurasian metallogenic belt (TEMB).
3. Transform faults. Diverging plate boundaries on the seafloor are linked by active basaltic

Fig. 3.4 (continued) of gangue consisting of limestone, Linares (Mazarron), Spain. Sample size: 8 × 11 cm. (**b**) Massive chalcopyrite intergrown with siderite and a band of dolomite and siderite, Mitterberg, main lode, Hauptgang, Austria. Sample size: 13 × 17 cm. (**c**) Cm-sized vein of fahlore, embedded in dolomite, Eiblschrofen, Schwaz, Austria. Sample size 13 × 16 cm. (**d**) Section of a hydrothermal vein with tiny spots of (silver-containing) galena embedded in bands of fluorite, barite, quartz. Mine "Gottesehre", St. Blasien, Black Forest, Germany. Sample size: 9 × 12 cm. (**e**) Non-visible gold particles distributed in the stockwork-like mineralisation, Sakdrisi, Georgia. Width of vein ca. 1 m. (**f**) Gold particles extracted by beneficiation of ca. 120 kg of gold-containing mineralisation, Sakdrissi, Georgia. Size of gold grains <1 mm. (**g**) Fine-grained polymetallic ore mineralisation of chalcopyrite, sphalerite, galena and others, showing fine laminated layers of ores and sedimentary host rock. Rammelsberg, Harz area, Germany magmatism, which in geology is called seafloor spreading. Early stages of plate separation can also be observed on present continents. They are characterised by elongated trench valleys, so-called rift valleys, volcanic activity and earthquakes. Rock material is raised from the earth's mantle. One of the known examples is the East African Rift, which extends from East Africa over the Red Sea and north to the Anatolian fold mountains. In the northern part of this Rift Valley, the copper deposits of Faynan and Timna are located.

In 1977, scientists from the Scripps Institute of Oceanography, San Diego, discovered hot mineral springs, the so-called Black Smokers and White Smokers, at the seafloor of the East Pacific Ridge at a depth of over 2000 m. This provided new evidence for an improved understanding of the genesis of ore deposits and for the discovery of new deposits. The smokers were also found on other mid-ocean ridges, that is, on diverging plate boundaries, e.g. in the Atlantic. They produced enormous amounts of sulphidic ores with high in copper, lead, silver, zinc, iron and other metals, even gold is precipitated (Fig. 3.5).

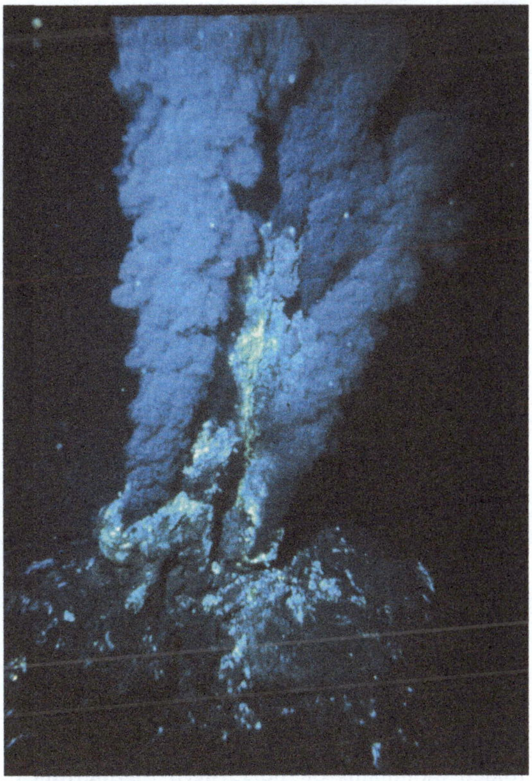

Fig. 3.5 A Black Smoker is also known as an underwater hot spring found on the ocean floor. Black Smokers are formed in fields of hundreds of metres wide. When they come in contact with cold ocean water, many minerals precipitate, forming a black chimney-like structure around each vent. Now Black Smokers are known to exist in the Atlantic and Pacific Oceans at an average depth of 2100 m. Photo: Public domain; U.S. National Oceanic and Atmospheric Administration

As it became clear that today's spreading centres are creating rich mineral deposits due to the impact of countless Black and White Smokers on the sea floor, geologists searched for remains of paleo-sea floor transported to continental plates, as they may also contain valuable minerals. Such paleo-sea floor remains are called ophiolites. These are parts of oceanic crust and the topmost mantle, which were transported as nappes or tectonic dislocations in continental direction. They consist of a sequence of basalts, andesites (sheeted dyke complex), gabbros, diorites (cumulative sequence) and ultrabasic rocks such as peridotites, serpentinites, dunites and harzburgites (Press and Sievers 2003). Such mineral-rich rock sequences are known from the island of Cyprus, where rich copper deposits played a major economic role in the Middle and Late Bronze Age, not only in the Mediterranean but possibly also far beyond. The copper deposits of Oman were the basis for raw materials supply for Mesopotamia in the third millennium BC (Begemann and Schmitt-Strecker 2009; Begemann et al. 2010; Salzmann 2019). Ophiolites are furthermore known from the Urals, Kazakhstan, or the Eastern Desert of Egypt and Nubia, where hundreds of gold deposits formed the basis of Egyptian wealth. The most important ophiolites are of Mesozoic age (Jurassic, Cretaceous), in Egypt and in the Altaiden often also of Paleozoic age (Devonian, Carboniferous).

Converging plate boundaries, where oceanic plates are colliding, are forming subduction zones. The overlapped plate is subducted and lowered, i.e. pulled into the underlying mantle, where it is finally melted again. This process is associated with significant tectonic activity and volcanism. Here new mountain ranges are formed. Subduction zones are located between the oceanic Nazca Plate and the South American continent. There, the Atacama deep-water channels and the chain of volcanoes in the Andes are located. This zone is characterised by a large amount of Cu-Pb-Zn-Au-Fe-Mn deposits, many of which are called "giants", since they belong to the world's largest deposits of the earth (Sillitoe 2012). Important for a striking global enrichment of ore deposits is the collision of the Eurasian plate in the north as well as the Arab-African and the Indian plate in the south. This is the so-called Tethyan Eurasian metallogenic belt (TEMB), which will be discussed below.

At active and paleo-boundaries of converging plates, at collision zones of oceanic and continental plates, many more sulphidic ore deposits of hydrothermal or magmatic origin occur. In the Old World, there are examples in the alpine folded mountains of the Eastern Mediterranean (Greece, Anatolia) to Iran, Afghanistan and Pakistan. On the west coast of South America, the world's largest copper ore deposits are located in the Cordilleras in Chile, Peru. These are so-called porphyry copper ores, a type of deposit that (Robb 2005) designates as "high tonnage–low grade". But they contain enormously rich oxidation zones. There are also vein deposits high in lead, silver and gold.

At transform faults, tectonic plates are sliding past each other. No lithospheric material is created or destroyed. An example of such plate tectonics is the San Andreas Fault in California. There, the Pacific Plate glides past the North American.

Considering these new models of reservoir formation, however, one must take into account that geologically very young sedimentation areas have completely covered any occurring ore deposits and thus rendered them unusable for exploitation in ancient times. Examples include Mesopotamia, the eastern Iran desert area.

Each of the explained metallogenic provinces show specific mineralogical and geochemical patterns. They were formed at different ages and herewith they provide a very useful trait for discriminating between various ores and metal objects as it allows the use of lead isotope analysis. Provenance studies based upon isotope analyses, geochemistry and mineralogy will be treated in detail in the discussion of trade connections (Chap. 11).

3.4.2 The Tethyan Eurasian Metallogenic Belt (TEMB)

The Tethyan Eurasian metallogenic belt of the Alpine-Himalayan orogenic system plays a major role in the distribution of ore deposits in the Old World. The TEMB has a global size of around 10,000 km and was first described by Janković (1997) as a specific metallogenetic zone. In Europe, it includes the mountain ranges of the western Mediterranean, the Alps and the Balkanides, then continues on over the Taurids and Pontids on the Asia Minor Peninsula, the Transcaucasus, the Zagros Mountains in the Iranian high plateau and finally to Pakistan, Afghanistan, the Hindukush, Pamir, the Tibetan Highlands, Burma and Sumatra, to connect with the Circumpacific Belt (Fig. 3.6). This fold-mountain belt, whose formation can mainly be

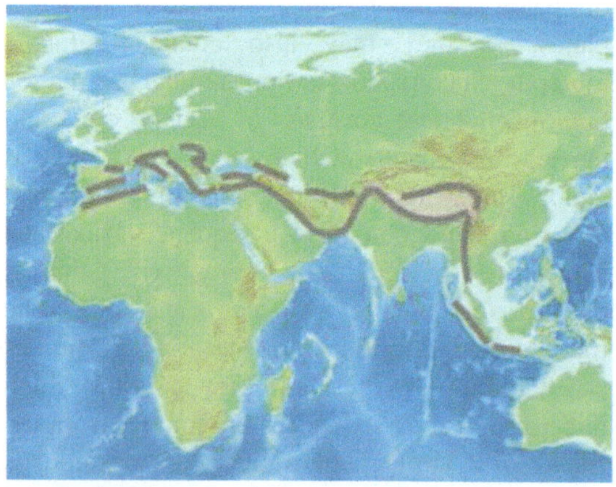

Fig. 3.6 Schematic global running of the Tethyan Eurasian metallogenic belt (TEMB) of the Alpine-Himalayan orogenic systems (after Storetvedt 1990). This geological unit is especially rich in base and noble metal resources. See also Fig. 3.7

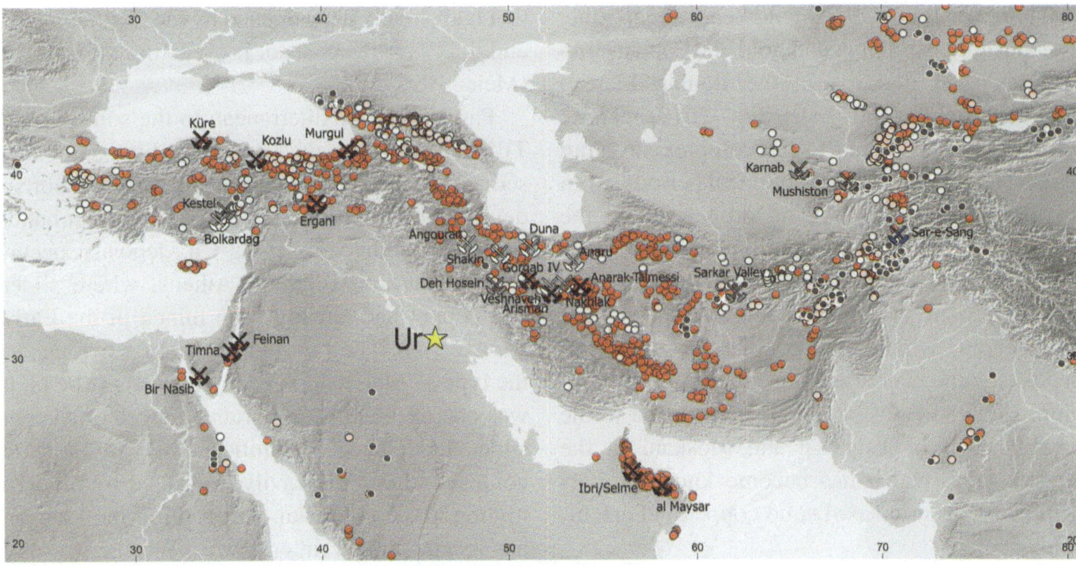

Fig. 3.7 Copper and silver ore deposits in the Middle East, geologically embedded in the Tethyan Eurasian metallogenic belt (see Fig. 3.6). Note the enormous density of metal enrichment in this geological unit. From Salzmann (2019)

attributed to the Jurassic and Cretaceous periods but which continues to this day, is the result of complex geological events that can be explained in terms of plate tectonics. The collision of the Eurasian with the Afro-Arabian and Indian continental plates during the closure of the Tethys Sea resulted in collision processes, subductions, rifting processes and accretions of many precambrian and Paleozoic microcontinent sites. The TEMB is therefore by no means homogeneous, and the ore deposits can display very different geological ages (Bertrand et al. 2014). During the formation of the TEMB oceanic subduction and obduction processes took place, which led to the formation of deposits in ophiolite complexes, such as, the volcanogenic massive sulphide deposits of the Troodos Ophiolite Complex in Cyprus (Delaloye et al. 1980).

On Cyprus, some 90 copper deposits and mineralisations are known which were exploited in ancient times with the focus on the Late Bronze Age and the Roman period (Stos-Gale et al. 1998a, b). Among the many base metal and noble metal deposits in Anatolia, 200 show remains of ancient mining and metallurgy (Wagner et al. 1989). The northern concentration of the ore deposits is located in the orogenic belt of the Inner and Outer Taurid mountain range in south-eastern Turkey (Öztunali 1989). Particularly in the region of the upper Euphrates, several non-ferrous metal deposits occur, e.g. those in the Keban area. A large number of copper deposits occur in the ophiolite rock sequence of the Outer Taurids, along the Zagros Line. The ore deposits of Ergani Maden, Siirt/Madenköy, Karadere, and Cüngüs-Midye are located here (Çağatay 1978). This series continues towards the south-west. In the ophiolitic rocks of Hatay and Baer Bassit, and of the Kizildağ (Dilek et al. 1991), at the Syrian-Turkish border, the entire copper mineralised ophiolitic sequence from the ultrabasic basement up to volcanic rocks is exposed. It forms an ensemble with the Troodos Ophiolite Complex in Cyprus.

Three world-class VMS-copper deposits ("giants") are located in the TEMB which were exploited in ancient times as well. These are Murgul (Lutz et al. 1994) and Ergani Maden (Seeliger et al. 1985) in Anatolia and Reko Diq in the triangle Iran–Pakistan–Afghanistan (Sillitoe 2012). The fourth giant is Rio Tinto at the western end of the Iberian Peninsula.

There are hundreds of gold–lead–silver–zinc, iron and copper, copper and gold and copper–tin–tungsten–gold deposits in the Iranian Plateau (Momenzadeh 2004; Salzmann 2019). These have formed according to the complex structure of the TEMB in different geological periods, from the Paleozoic to the Alpine orogenesis in the Cretaceous. Many of the approximately 600 (silver-bearing) lead–zinc occurrences were created here. There are around 400 copper deposits and occurrences, which in part, like the other ore deposits, have been used since the fourth millennium (Fig. 3.7). Of particular importance is the large deposit of Talmessi and Meskani in the Anarak area, which has become known for its significant occurrence of solid copper and arsenical copper.

Thousands of rich non-ferrous and noble metal ore outcrops as well as iron ore deposits occur in the European part. More than 1500 smaller ore deposits have been counted in the eastern Alpine region alone, in which the copper ore deposits of Mitterberg (Salzburg) and the unique fahlore deposits from Schwaz-Brixlegg are especially well known (Ager 1980). In the Carpathian Basin, the formation of the large golden and non-ferrous metal deposits ("Golden Quadrangle") was associated with tertiary andesitic and rhyolitic volcanism. In the Aegean and in the Tyrrhenian Sea (Mount Etna, Vesuvius) this magmatism lasted until the Holocene. The pyrite–magnetite–hematite deposits on the island of Elba, the mercury deposit of Monte Amiata in Tuscany and the Pb-Ag-Zn-Au deposits at Rodalquivir and Mazaron in southern Spain belong to this era. Surveys and excavations have revealed hundreds of mines and smelting workshops as well as remains associated with the (pre-)historic exploitation of copper, iron, gold, lead, silver, tin, and other ores. For numerous deposits, a clear age allocation is often difficult, due to its complex development.

The mineral riches of the Iberian Peninsula already led the Phoenicians and the Carthaginians to settle in Spain and Portugal. After assuming power, the Romans could occupy the peninsula that had already been opened by the Carthaginians and thus save the costs of prospecting themselves (Meier 1995).

Parts of the Mediterranean to the south of the TEMB are largely covered by carbonate sediments, which in Italy are extremely poor in deposits. In Serbia and Greece, on the other hand, some important Pb-Zn-Ag-Cu deposits occur, e.g. in Laurion, south of Athens, where silver and copper had already been mined in the Early Bronze Age. The Laurion ore district belongs to the type of Mississippi Valley deposits (MVT), it was one of the most important sources of ancient silver and played a significant role in Hellenic coinage. Although heavily destroyed by modern mining, the early metal extraction in the Laurion has been studied intensively (Weisgerber and Heinrich 1983; Rehren et al. 1999a, b; Lohmann 2005, Lohmann in press). The same is true for Sardinia. In the southwest of this Italian island, at Iglesiente, extensive MVT Pb–Zn deposits are located. Also here, modern mining activities destroyed many ancient remains of exploitation and made a mining archaeological research very difficult (Hanel and Morstadt 2014).

Other metasomatic MVT deposits occur in Trepča, a deposit in Serbia that is already in the Alpine fold region. The large-scale Pb–Zn–Ag deposits of Iglesias, Sardinia, as well as the Cu deposits occurring there, belong to a spatially defined metallogenetic province, which has been used since the Bronze Age. The Cambrian age of many of these deposits is based on the geological structure of the Adriatic plate, a northern spur of the continental crust of the Sahara slab.

Interestingly enough, the geographic distribution of ore deposits in the TEMB roughly corresponds to the course of cultural history in Eurasia: In contrast to Africa and America, whose north-south oriented axes extend over several temperate climatic and vegetation zones, here is an east-west axis, which is located in mostly temperate climates. Here, humans had access to a large-scale, inexpensive, natural transport network with river valleys and wide plains (Primas 2007).

3.4.3 Polymetallic Deposits and Lead–Zinc Deposits in the Paleozoic of Europe

The Variszic Mountain Range in central Europe and the Altaides in central Asia are also exceptionally rich in ore bodies, especially in gold and tin. Central Europe and large parts of southwestern Europe are underlain by the Variscan orogen of the Paleozoic, which adjoins the TEMB to the north. It consists of a complex mosaic of Precambrian microcontinents. This orogen is covered by large areas of younger sediments.

There are numerous deposits from the Paleozoic.

A metallogenic province, which was of importance for the (pre-)historical time is that of the Saxon-Bohemian Ore Mountain (Erzgebirge)/ Krušné hory on the border of Germany to the Czech Republic. More than 1000 polymetallic ore veins run through the crystalline basement rocks containing a large variety of Ag, Pb, Zn, Cu, Co, As, Sb, Bi and U ores. Characteristically there are many silver-rich sulpho-arsenides and native silver. Silver mining was described probably earlier than the twelfth century AD, and tin mining is going to be discussed for (pre-)historic periods. The mineralogy and geology of the mining district in the Ore Mountains have been researched with exemplary intensity since the sixteenth century AD and have contributed significantly to the foundations of today's mineralogy. The Ore Mountain district internationally is the cradle of modern mineralogy. In addition to these hydrothermal veins, the Ore Mountains have a rich tin–tungsten deposit, which along with the tin–copper deposits of Cornwall and Brittany are perhaps among an important raw material source of tin in ancient times. Unfortunately, clear archaeological evidence is open.

The Harz mountain complex in Germany possesses a large number of polymetallic ore deposits. The Harz itself is carboniferous, i.e. it originates during the formation of the orogeny 300–320 million years ago. The origin of the hydrothermal Zn–Pb–Cu–Ag ore veins of the upper Harz district is not fully understood (Stedingk 2004). Galena and sphalerite are the main components of the veins, chalcopyrite is subordinated. The veins of the Harz complex have been exploited at least since the medieval period. The Rammelsberg high-grade ore deposit, one of the world's major synsedimentary metal resources, has a comparable mineral content as the hydrothermal veins of the upper Harz district. It is believed that the medieval miners had access to much greater amounts of rich ores than is the case today (Asmus 2012).

In the Variscan basement of the Upper Rhine Graben and in the Cévennes, other metallogenic provinces exist. The density and composition of ore veins at Sainte-Marie-aux-Mines in the French Vosges is similar to the situation at Freiberg in the Ore Mountains (Mahe-Le Carler et al. 2010). The many hydrothermal veins in the Siegerland are to be mentioned here. The Siegerland with its more than a thousand veins and vein swarms (Bornhardt 1910) lies within one of the most important tectonic structures of the Rhenish Slate Mountains (Kirnbauer and Hucko 2011). Here, until recently, iron, lead, zinc, copper and silver were mined from complex ore parageneses.

The predominantly hydrothermal veins in the Black Forest are also to be mentioned here. Goldenberg (1996) has described the archaeometallurgy of these veins and documents the use of ore deposits since Roman times, then very intensively in the mediaeval period. Markl (2015a–2017) described in great detail geology, ore mining and mining history in four volumes.

The mining district of the Iberian Pyrite Belt (IPB) belongs to the largest ore deposits in the world. These deposits date to the Carboniferous period. It houses eight of the 20 largest volcanogenic massive sulphide (VMS) deposits, including Rio Tinto, Aljustrel, Neves Corvo. The name Rio Tinto derives from one of the most heavily polluted rivers, whose water is reddened by weathering and oxidation of Fe sulphides and contaminated by heavy metals (Fig. 3.16). Predominant ore is pyrite. Since the Chalcolithic/ Early Bronze Age, but especially in Roman times, copper, silver and gold ores have been mined, which have been enriched in the oxidation zone and in ore shoots (Marignac et al. 2003). The Neves Corvo deposit also produces high-grade cassiterite (and stannite) in large quantities (Neukirchen and Ries 2016).

Ross Island is a stratabound copper–lead–zinc–arsenic plus minor silver and cobalt mineralisation hosted in Carboniferous limestone across the eastern part of Ireland. As in the deposits of Schwaz-Brixlegg (Austria), the ore content mainly consists of fahlores. Ross Island has been mined since the Early Bronze Age (O'Brian 2007).

3.4.4 The Altaids and Uralids

For reasons of the high richness of raw sources, also the Altaids and Uralids should shortly be mentioned.

Politically, this territory is located between the Russian Federation, Kazakhstan, Kyrgyzstan, Uzbekistan, Tajikistan, western Mongolia, and China. The Altaids constitute an orogen collage lying between the cratons of East Europe (Baltikum), Siberia and North China. They are one of the largest and most economically important of the crustal blocks that comprise the Eurasian landmass (Yakubchuk et al. 2001). The Altaids consist of Paleozoic rocks. The Altaids are separated from the Tethyan Eurasian metallogenic Belt. The microcontinent of Kazakhstan is located in the south-central area of this huge fold belt, north of the Tarim River and the plain of Turan. It consists of a geologically complex agglomerate of Precambrian to Paleozoic microcontinents and island arches that have been welded together. Especially ores from copper and tin deposits located in the sequence of these geologically old rocks are of special interest for the ancient trade routes to the west because they should be printed by significant "old" lead isotope ratios.

The most prominent tectonic characteristic of Paleozoic Kazakhstan are two volcanic belts of Devonian and late Paleozoic age, which were bent together to form a gigantic horseshoe-shaped structure (Filippova et al. 2002).

The Altaids host vast numbers of ore deposits, many of world-class, including gold, copper, silver, lead–zinc, nickel, and tin. It is especially the gold province of the Tien Shan which is the richest gold province in central Asia and one of the largest gold provinces in the world. These deposits are part of the gold belt of the Altaids. Muruntau in Uzbekistan is the largest gold resource in Eurasia. Ore deposits in central Asia have been exploited since the third millennium BC. The name "Altai" can be translated with *The golden mountains*. Geologically, the rich metal endowment of the Altaids is a result of a prolonged and complex history of crustal growths and deformation (Yakubchuk et al. 2001, 2002). Sverchkov (2009) reports that relicts of "old" mining and metallurgy have been found at many of the countless ore deposits in Uzbekistan since the beginning of geological exploration in the nineteenth century, which at least indicates extraction in the Middle Ages, but would make previous mining of ores quite likely. Often, the polymetallic mineralisations did not allow any clarification on what had actually mined there.

The Ural Mountains, the geographical border between Eastern Europe and Asia, characterises an ancient suture or seam zone formed by the collision of the Siberian-Altaidic and Eastern European platforms in the Paleozoic era about 250–300 million years ago. The individual mountain ranges of the Urals correspond to completed episodes of mountain formation in which mighty layers of sedimentary rocks were pressed into very long, north-south-facing folds, later unfolded and partly permeated by magmatic rock. Like the fold mountains of the Altaides, the Urals are also notedly rich in metal deposits.

Among the numerous ore deposits, the sedimentary deposit of Kargaly, southwest of the Urals, played a substantial role in the prehistoric copper supply. It has been investigated by E. Chernykh since the 1980s. The ancient copper district of Kargaly encompasses about 500 km^2 (Chernykh 2003) and is part of the gigantic sedimentary copper district of Dzhezkazghan (Yakubchuk et al. 2001) in the large steppe and forest-steppe regions of central Kazakhstan.

Copper extraction in Kargaly was demonstrably practiced in the early and late Bronze Age.

3.4.5 Ore Deposits in Old Cratons

South of the TEMB the Afro-Arabian continental blocks are located. In wide areas of the Levant and North Africa, they are covered by kilometre-thick sediments consisting of carbonates and sandstone. In the eastern Mediterranean, these sandstone deposits are commonly referred to as Nubian sandstone. The southern part of the Mediterranean is largely free of mineralisation except for some Pb-Zn and Fe mineralisation.

In the southern Levant and on the southwestern Sinai Peninsula, two large copper districts have already been mentioned, namely Faynan and Timna in the Wadi Arabah, and Bir Nasib/Umm Bogma in the Sinai Peninsula. They belong to a supra-regional province of sedimentary deposits formed by the erosion of Precambrian veins and subsequent deposition in a lagoonal milieu at the borderline of the Cambrian sediments to the Precambrian basement. In the southern part of the Sinai, it consists of the Precambrian crystalline basement of the Arab-African table. Abdel-Motelib et al. (2012) describe vein-like hydrothermal copper deposits that were also exploited in prehistoric times. As shown by slag heaps distributed in this regions there must have been a considerable copper production as old as the Early Bronze Age, and a later interest in the time of the New Kingdom in the second half of the second millennium BC and the in the Islamic period, when old slags were consequently recycled. Otherwise, the northern part of the Sinai is an area extremely poor in resources.

The Eastern Desert in Egypt consists of Precambrian crystalline rocks of the Arabic-African table as well. In the northern part, only a few insignificant copper mineralisation occurs as hydrothermal veins, such as in Wadi Dara. Iron ores of the Precambrian banded iron formation (BIF) were not exploited in ancient times. The area of the southern Eastern Desert of Egypt to the Sudan, however, is characterised by a huge gold aureole. Klemm and Klemm (2013) mapped in this area many hundreds of ancient gold-quartz veins which were exploited since the fourth millennium BC until at least the Early Arabic period from the ninth to eleventh century AD. This gold aureole reaches an extension to the crystalline basement of the western Arabian Peninsula. Ancient metal production seems to have concentrated on gold in the area of the Gulf of Aqaba. At least, this is the impression gained from a survey of 29 Roman-Islamic gold mines at Al-Wajh, Daba and Al-Jadida, which has been carried out by Kisnawi et al. (1983). Traces of early copper production have been found at three different sites: Shim at-Tasa (Shanks 1936), Imsayea and Az-Zuwaydiyah at Al-Disa (Kisnawi et al. 1983). They indicate ancient mining and smelting of small sedimentary ore deposits in sandstone. In northwestern Hijaz in Saudi Arabia, ore deposits occur along with fault systems linked geologically with the Rift Valley (Sillitoe 1979). Numerous ore deposits are located in the Ad-Dawadimi district in the central Saudi-Arabian crystalline basement, but little is known about size and quantities of these occurrences, i.e. if they ever reached any importance in ancient times.

3.4.6 Copper Deposits

With 70 ppm in the solid earth crust, copper (Cu) belongs to the relatively rare elements. It is ranked 25th in the frequency list.

The largest and most important ore deposits today are located in Chile, and in the United States. Here, more than about 20% of the known world's reserves can be found. Furthermore, copper ore is mined in Africa, Australia, China, Canada, Indonesia, South America, Russia and Poland (Hedenquist et al. 2012). The occurrences in Central Europe, which were particularly important in the Bronze Age, are all exhausted under modern economic aspects. Among the world giants are the European Kupferschiefer Deposits (Borg et al. 2012).

Today, copper deposits are sometimes considered to be worthy of mining even at levels of

0.3 wt.%, while in the Bronze Age, copper ores were surely used selectively with 10–15 wt.% copper.

3.4.6.1 Copper Minerals

The dominant geochemical property of copper is its great affinity for sulphur: it is chalcophile and forms a large number of sulphidic ores. On the other hand, copper is easily dissolved and forms mostly clearly green to blue coloured, divalent mineral compounds (Cu^{2+}) such as malachite, brochantite, chrysocolla, azurite, and others. In nature, Cu^{+}-compounds, such as cuprite, are formed less frequently, but on the other hand, they are often found in metallurgical contexts. Copper is very mobile in the earth's crust but is precipitated quickly, especially in the presence of sulphur or carbonates.

Some of the most important copper minerals are listed in Table 3.3. Altogether, many more than 150 copper minerals or copper-containing minerals are known. Details for copper minerals (as well as for the other minerals of the following sections) can be found in mineralogical textbooks (Ramdohr 1975; Wenk and Bulakh 2004; Okrusch and Matthes 2005; Rapp 2009, and others).

Table 3.3 Some of the more important and common copper minerals and their copper contents

Mineral name	Chemical formula	Cu (wt.%)
Native copper	Cu	99
Cuprite	Cu_2O	88
Tenorite	CuO	80
Chalcocite	Cu_2S	79
Digenite	Cu_9S_5	78
Covellite	CuS	66
Bornite	Cu_5FeS_4	63
Atacamite	$Cu_2(OH)_3Cl$	59
Malachite	$Cu_2(OH)_2CO_3$	57
Brochantite	$Cu_4(OH)_6SO_4$	56
Azurite	$Cu_3(OH)_2[CO_3]_2$	55
Antlerite	$Cu_3(OH)_4SO_4$	53
Enargite	Cu_3AsS_4	48
Chalcopyrite	$CuFeS_2$	34
Chalkanthite	$CuSO_4 . 5H_2O$	25
Cubanite	$CuFe_2S_3$	23
Tennantite	$Cu_{12}As_4S_{13}$	42–52
Tetraedrite	$Cu_{12}Sb_4S_{13}$	30–45
Chrysocolla	$CuSiO_3$ + aq.	30–36

Values of copper concentrations in these minerals after Doelter (1912–1932). They may vary according to various solid solutions

Native Copper and Its Alloys

Of all these copper minerals, native copper rather rarely occurs as a predominant mineral in ore deposits. Regionally it is present in many ore deposits in lesser quantities, especially in the oxidation zone of sulphidic copper deposits, where it can occur together with native silver as well as with cuprite and with copper sulphides such as digenite and chalcocite. One of the few exceptions, and at the same time the most prominent modern example of a deposit of predominantly native copper, is that of Lake Superior in the United States, where hundreds of tons of metallic copper were mined in modern times.

In the Old World, the deposits of Talmessi and Meskani in the Anarak area of Iran are commonly mentioned and referred to as the stimulus for the earliest metallurgy in the region (Pigott 1999). Here, native copper occurs in large quantities together with different copper arsenides. These copper arsenides are a natural assemblage of alloys of algodonite (with a varying chemical formula of $Cu_{5-8}As$ with 80–87 wt.% Cu), domeykite (Cu_3As with 72 wt.% Cu), and of an As-saturated copper-solid solution named whitneyite (Ramdohr 1975). The vein-shaped deposits are embedded in Eocene volcanic rocks, and therefore are geologically very young (Momenzadeh 2004).

Rich deposits of native copper are also found in Anatolia, where Yalçın and Ipek (2016) and Yalçın (2020) have conducted very intensive (mining-) archaeological and archaeometallurgical research in Derekutuğun. As native copper, like other natural metals, holds a special position in early metallurgy, it will be discussed in more detail in Sect. 3.6 (Native Metals).

Copper and Copper–Iron Sulphide Minerals

Chalcopyrite is the most important and almost omnipresent sulphidic copper ore in the overwhelming part of copper deposits worldwide. It, therefore, has a key position, both theoretically and practically, in early metal extraction. Chalcopyrite is often associated with pyrite and other Cu-bearing sulphidic As, Sb, Bi, Ni, Pb ores, such as the fahlores tennantite and tetrahedrite, and arsenopyrite, enargite, bournonite and others. The consequence is that in mining extractions, what is subjected to metallurgical processes is more often a polymetallic mineral mixture (*ore–cocktail*, Klein and Hauptmann 2016), whereas a monomineralic chalcopyrite ore is rare. The smelting of arsenic and antimony rich ores, in particular, can have considerable metallurgical consequences (cf. Sect. 6.2).

Due to the easy weathering of its iron contents, chalcopyrite quickly converts to limonite (FeOOH), forming a brown crust blended with secondary copper minerals such as malachite. With increasing weathering, the removal of iron forces chalcopyrite to decay and the Cu-rich bornite is formed. Through precipitation of copper solutions in the region of supergene enrichments digenite, chalcocine and covellite are formed.

Cu–Fe mixed crystals of the composition chalcopyrite–bornite up to pure Cu-sulphides are also formed in the smelting of sulphide ores in copper matte. In the investigation of archaeometallurgical finds, a distinction between natural and artificial Chalcopyrite is not always trivial (see Sect. 5.7, Matte).

Sulphosalts

The tetrahedrite group minerals are a solid-solution series between the As-rich tetrahedrite and the Sb-rich tennantite. These are called fahlores. They are widely distributed and they are accessory minerals in almost all sulphidic copper deposits and even in their oxidation zones. It is impossible to count the ore deposits where they are occurring. Ixer and Pattrick (2003) wrote a useful overview of the mineralogy, genesis, and occurrence of fahlores along with the sulphosalts of the enargite group. Fahlores may be high in silver (1–10 wt.%) and can contain up to 48 wt.% silver in extreme cases.

The extended formula notation is therefore:

- Tennantite: $(Cu,Ag)_6Cu_4(Fe,Zn,Hg)_2As_4S_{13}$
- Tetraedrite: $(Cu,Ag)_6Cu_4(Fe,Ag,Zn)_2Sb_4S_{13}$

In addition, they can contain bismuth and nickel in low concentrations. Tetrahedrite and tennantite may occur together in the same ore deposit and can be associated with many base-metal sulphides such as chalcopyrite, galena, and sphalerite and ores of the pyrargyrite–proustite series (see Sect. 3.4.9, Silver deposits). It may occur as a minor constituent, such as inclusions exsolved from galena. This causes the "argentiferous galena" of so many ore deposits.

The arsenic sulphide called enargite Cu_3AsS_4 is a guide mineral of copper ore deposits low in iron. It occurs in paragenesis with fahlores, e.g. in the Inn Valley and in countries of the Balkan Peninsula. It often contains levels of Sb up to several percentage by weight and forms a solid solution of the enargite group with the end member famatinite (Cu_3SbS_4). Enargite is found in very large quantities in the porphyry copper deposits, in the deposits of the Circumpacific belt, such as in the Andes (Chuciquamata), in the Rocky Mountains, in Indonesia and the Philippines. Here lies the island of Luzon, which is the namesake of the Sb-carrying enargite luzonite. Neuninger et al. (1970) write about experimental fahlore smelting based on traditional smelting on Luzon in the nineteenth century AD.

Secondary Copper Ores

The numerous secondary copper minerals with their green, blue, and red colours are among the most prominent and widely used ore minerals. It can be assumed that these surface-open minerals played a crucial role in prospecting for ore deposits in ancient times (see Sect. 3.5, Oxidation zone).

Secondary copper ores crystallise as oxides, carbonates, sulphates, chlorides, silicates, phosphates, arsenates and sulphides. Which minerals will actually crystallise depends on the acidity of the soil (pH) in the copper deposit and its oxidation–reduction potential (Eh). There are basically four areas to divide: oxidising and acidic, oxidising and alkaline, reducing and acidic, reducing and alkaline (Fig. 3.8).

Malachite is the most widely used secondary copper ore in the oxidation zone of sulphidic

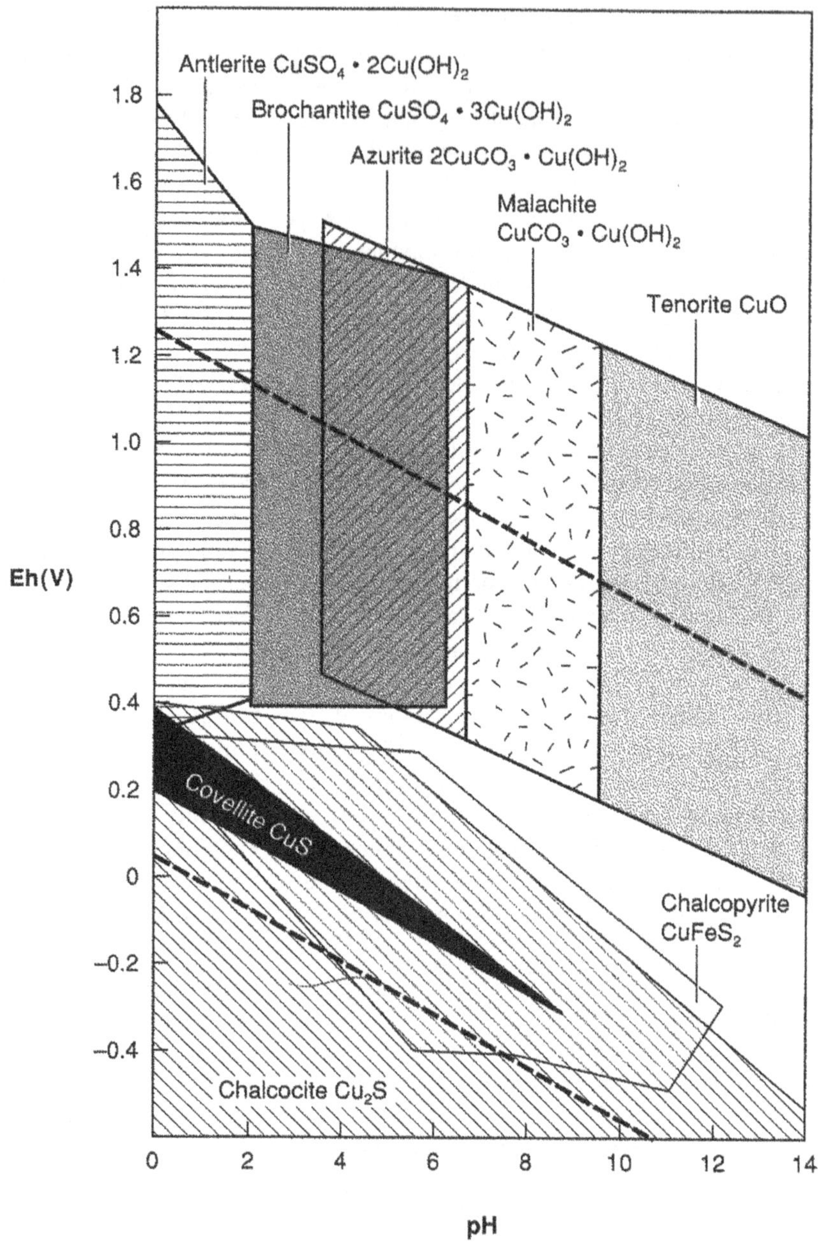

Fig. 3.8 Formation of primary and secondary copper minerals as a function of acidity and redox-conditions of the host rock. Hatched lines indicate the upper and lower stability range of water. The diagram shows that secondary copper minerals such as azurite and malachite etc. precipitate at high Eh (V)-values in a large range of alkaline to acid environment (pH-values). Sulphidic minerals (chalcopyrite, covellite, chalcocite) precipitate at low Eh-values (<0.4) in an alkalic environment dominated by low pH-values (c. 12–8). Diagram after Pourbaix (1977) and Garrels and Christ (1965). Modified after Schweizer (1994)

copper ore deposits. It is so common because copper mineralisation is very often embedded in carbonate rocks, which are virtually omnipresent in nature.

The copper arsenate olivenite ($Cu_2(OH/AsO_4)$) forms easily from the Cu-As-sulphide enargite. It occurs in particularly large quantities in the copper deposits of the Andes, where it had a significant influence on the early production of arsenical copper (Lechtman and Klein 1999).

Mixed Cu-Sn-Fe ores (e.g. stannite or *bell-metal* ore (Cu_2SnFeS_4)) are decaying in the oxidation zone to minerals like mushistonite

($(Cu,Zn,Fe)Sn(OH)_6$) and Cu and Fe sulphates and cassiterite (SnO_2). Under arid or moderately humid climatic conditions, copper minerals remain in place during weathering, so that an intense green mixture with tin oxide can be present (Barnes 1974). In addition, the parageneses of tin and copper mineralisation can present intense intergrowths of chalcopyrite with cassiterite and stannite, which can have the same effect.

3.4.6.2 Copper Deposits

Of the numerous deposit types in which copper-bearing ores dominate, only a few geologically and mineralogically different structured are discussed. These are playing a special role in archaeometallurgy.

These are the following examples:

- Hydrothermal copper veins (Mitterberg, Austria).
- Fahlore deposits (Schwaz/Brixlegg, Cabrières, Ross Island).
- Volcanogenic Massive Sulphide deposits (Cyprus, Oman).
- Sedimentary stratified copper deposits (Kargaly, southern Urals of Russia, Faynan and Timna, Wadi Arabah).

It should be *expressis verbis* noted that these classifications are simplified schemata. Many ore deposits have a much more complex genesis.

The total number of copper deposits in Eurasian countries alone is enormous (see Fig. 3.7). It is not possible to describe them in their entirety. With a few selected examples, the enormous intensity and consequence of the exploitation of ore mining areas over the ages should be demonstrated.

Hydrothermal Copper Veins

The Mitterberg District

Of the more than 100 copper deposits in the Alps (Cierny 1997), the ore district of Mitterberg is one of those with the greatest importance. The Mitterberg hydrothermal copper-bearing veins are located at the northern margin of the Northern Greywacke Zone in the Eastern Alps, south of the city of Salzburg in Austria. The copper district consists of several more or less vertically dipping veins. The Mitterberger Main Load and East Lode in the northern area are the most important of these. The thickness varies between 0.2 and 4 m. The length of the minable ore is some 3.5 km. The host rock of the veins consists of rather soft and friable shales, phyllite and quartzite of lower Paleozoic age. The mineralisation is connected with the Alpidic folding in the Cretaceous age. Mitterberg is a polymetallic ore deposit, but chalcopyrite is the predominating ore. It occurs in massive lots, but it was also mined in thin veins from the relatively soft host rock. The copper sulphides occur up to the outcrops of the veins. Due to the extensive glaciation of the Alps during the ice ages, as well as strong erosion activities, only a moderately developed oxidation zone has formed on the Mitterberg. It can, therefore, be assumed that sulphide ores were mined and processed even in the earliest phase of exploitation.

In addition to chalcopyrite, pyrite occurs. Out of the countless examples of copper deposits, where fahlores occur in limited quantities but are added with As, Sb, Ni and Ag concentrations to specific chemical fingerprints for the final metal composition, the copper deposit of Mitterberg in Austria should be mentioned. Alongside the main ore chalcopyrite, the Mitterberg also contains Ni-rich ores such as gersdorffite (NiAsS) and millerite (NiS), as well as Ni-rich tetrahedrites in the upper depths of the veins (Weber et al. 1972) (Fig. 3.9). It was intensively discussed in the past as a possible source of the widely distributed Bronze Age ingot torques (Ösenhalsringe) in the Alpine foreland, and even in a recent investigation, Lutz (2016) left a provenance of these artefacts from Mitterberg undecided. In the upper part of the ore deposit, also uraninite (UO_2) occurs, which shows a geologic age of 90 million years and indicates stages of younger mineralisations. This mineral is of major importance for the interpretation of lead isotopy of the Mitterberg copper ore. Minerals in the gangue are quartz, dolomite ($CaMg(CO_3)_2$), siderite ($FeCO_3$) and ankerite ($Ca,Fe(CO_3)_2$) (Paar 1993).

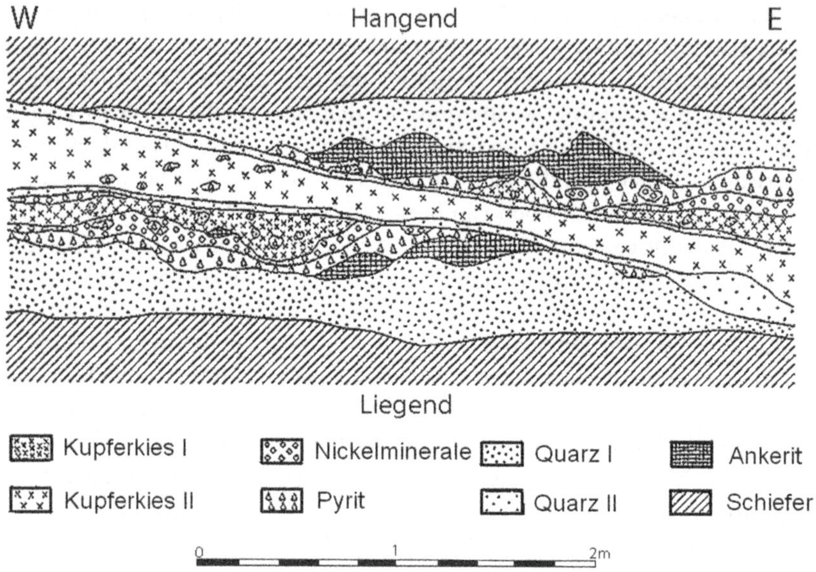

Fig. 3.9 Mitterberg (Austria), Main Load, mine 21, ½, seventh floor. Section through the hydrothermal vein showing various layers and associations of mineralisations. Note the predominating chalcopyrite (Kupferkies I, II), and note the mineralisation of nickel compounds. Nickel is an element which is closely following the copper during smelting operations. From Bernhard (1965)

In prehistoric times, ores from the veins were mined on a large scale and were already smelted to copper in the Bronze Age. This is evidenced by numerous chains of old surface depressions which are crossing the area, as well as by relics of old galleries, which repeatedly were cut by modern mining operations. The ancient mining and smelting of ores on the Mitterberg has been explored for about 100 years and is considered one of the few prime examples of mining–archaeological and archaeometallurgical research in general (Zschocke and Preuschen 1932; Stöllner 2009b). By now nearly 200 Bronze Age smelting sites, mining sites and ore dressing areas are known, which prove an enormous production of the metal (Fig. 3.10). From the late Early Bronze Age, and especially in the Middle and Late Bronze Age, Mitterberg was an important supplier of copper to the Central European region.

Fahlore Deposits

In special cases, fahlores may occur monomineralic and may dominate mineralisations of ore deposits. In many cases, fahlore deposits are embedded in carbonate bedrock. In Europe, these deposits are of interest because they are discussed as important raw material bases for the Early Bronze Age copper types with chemical As-Sb-Ag-Ni-Ag signatures, as defined by Junghans et al. (1960, 1968) (Krause 2003; Höppner et al. 2005).

Schwaz/Brixlegg

The mining district of Schwaz/Brixlegg in the Lower Inn Valley, east of the city of Innsbruck, with its numerous prehistoric mines at Moosschrofen and Eibelschrofen and many other localities (see Fig. 3.11) is one of the few deposits with the antimony-rich tetrahedrite as the predominant ore and has to be regarded as exceptional (Goldenberg 2014b).

The fahlore deposits of the Inn Valley are fahlores in the Schwazer dolomite of the Northern Greywacke zone, formed by syngenetic–hydrothermal enrichment in the Lower Devonian sedimentation area of the Schwazer dolomite. The deposit is affected by a low metamorphosis. The fahlore district in the Inn Valley is the largest of its kind in the Eastern Alps. The ores occur as massive mineralisations along veins and fissures (Fig. 3.12), as well as in brecciated ore bodies and in concordant layer enrichments (Schulz and Schroll 1997). The genesis of the deposits is still controversial. The storage of ores in stratabound bodies and layers suggests both a sedimentary

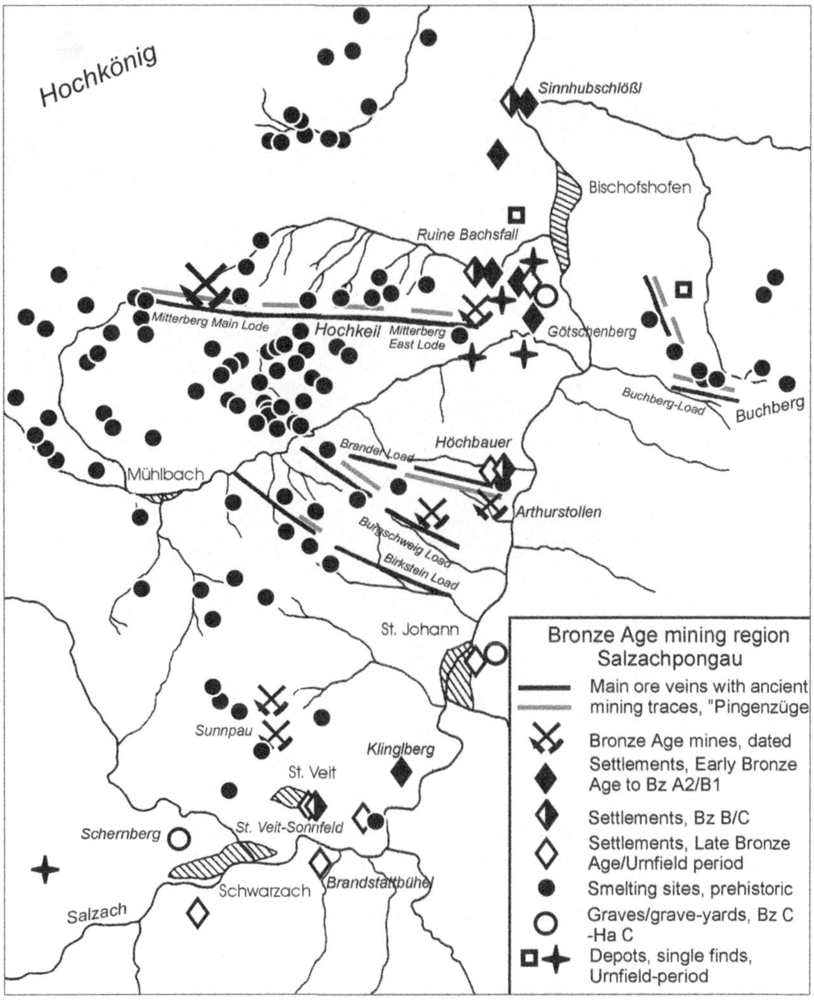

Fig. 3.10 Salzachpongau, Mitterberg area (Austria). Bronze Age mining and copper smelting sites mainly of the second millennium BC. Note that the nearly 200 smelting sites, prehistoric mining sites are located roughly in the same areas showing distinct production areas. In between ore dressing (beneficiation) sites were found. The area is about 200 km^2. From Stöllner (2003b)

deposit genesis, while discordant veins are interpreted as hydrothermal veins. Antimony-rich and silver-containing fahlores (tetrahedrite) are predominant (Krismer et al. 2011c). They were already mined in the Neolithic period, and later in larger scales in the Early Bronze Age at the localities of Mariahilfsbergl and Kiechlberg (Krismer et al. 2013).

Since the Bronze Age, the fahlore deposits of Schwaz and Brixlegg were part of the most important copper suppliers which were of significance even beyond the Alps. In one of the most prominent mining–archaeological interdisciplinary research programmes in Austria (History of Mining Activities in the Tyrol and Adjacent Areas, HIMAT) the early mining and metal extraction of this region is being researched since 2007 (Goldenberg et al. 2011). Many of the Early Bronze Age copper artefacts recovered in numerous hoards in the northern Alpine foothills were made from the so-called fahlore copper. In the Middle Ages and in the beginning of modern times (fourteenth/fifteenth century AD) there was an enormous boom in mining and metallurgy in this area. In addition to copper, the silver contained therein was extracted as well.

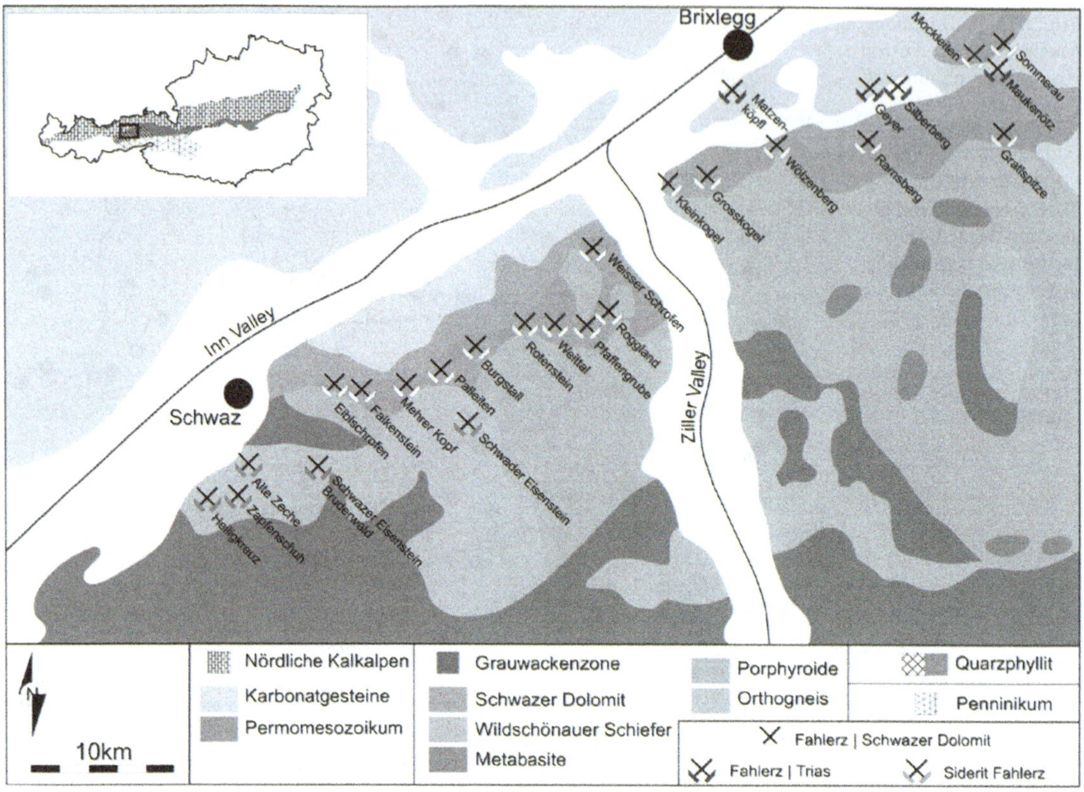

Fig. 3.11 The district of fahlore deposits in the Inn Valley (Austria). The ore deposits were exploited in recent years from 25 modern mines in the region of the cities Schwaz and Brixlegg. They are mainly embedded in the Schwazer dolomite complex. Goldenberg (2014b) explored more than the double of prehistoric mines in the Inn Valley

The most comprehensive overview is detailed in the so-called Schwazer Bergbuch, which was edited by Bartels and Bingener (2006). Several years ago, the legendary mining district attracted attention due to the rockslide at Eiblschrofen. Presumably due to the (pre-)historic mining activities, a massive hillside slumping occurred, which caused thousands of cubic metres of rock to slip on the mountainside.

Ross Island

Ross Island is unique among the 30 or so known Bronze Age mines in Britain. It is located in the Killarney National Park on the Island of Ireland. The first mining activity is associated with Beaker pottery and has been securely dated to 2500–2000 BC, but similar to the Inn Valley deposits it probably began in the Final Neolithic/Copper Age period (O'Brian 2007).

The Ross Island polymetallic ore deposit exploited in the Bronze Age consist of arsenic-rich assemblages of chalcopyrite - tennantite ± arsenopyrite ± bornite. The veinlets comprise a complex intergrowth of the main components chalcopyrite and silver-poor tennantite (with up to 0.3 wt.% Ag) plus a little bit of the today rare mineral strohmeyerite (AgCuS). Most of the mineralisations occurred as intermittent beddings with a thickness of 3 m. These are embedded in about 20 m thick, highly brecciated, dolomitised, partially silicified and intensely recrystallised Lower Carboniferous limestones, close to the present-day land surface (Ixer and Pattrick 2003). The mineralisation within this limestone

Fig. 3.12 Eiblschrofen, Schwaz (Austria). Vein of an antimony-bearing fahlore (tetraedrite), embedded in dolomite. The sample was found at a freshly bulldozzed path in the forest. Width of the piece 12 cm. Photo: A. Hauptmann, Deutsches Bergbau-Museum

is exposed on the modern land surface. Most of the mineralisations do not show any development of gossans.

All these ores show, that Ross Island had sufficient quantities to produce arsenical copper for the British Isles in the earliest periods. The richness of the ore deposit of Ross Island is further attested by the sale of 4153 tons of dressed copper ore to the Swansea smelters during mining operations between 1804 and 1829 (Ixer 2004).

Cabrières

Another mining district with importance for the early period, with argentiferous tetrahedrite often associated with chalcopyrite and secondary ores, is located near Cabrières/Herault in the Languedoc in southern France. Due to tectonic activities, the geology of Cabrières is particularly complex. The ores are located in a region of complex geological units of Paleozoic age, often at the border between metamorphised plutonic and volcanic rocks and sedimentary units with limestone, chalk and dolomite. The ores are bound to quartz-bearing veins. Evidence of mining from the beginning of the third millennium BC and some remains of the smelting of ores were discovered at the sites of Vallarade, La Capitelle du Broum, Pioch Farrus and La Roussignole by Ambert (1999). Archaeometallurgical investigations of Prange and Ambert (2005) by trace element and lead isotope analysis aimed to establish the relation between ores and the copper types described by Junghans et al. (1960, 1968) in southern France. Ambert et al. (2005) present a metallurgical workshop in a settlement dating back to 3100–2400 BC in Capitelle du Broum.

Volcanogenic Massive Sulphide Deposits (VMS)

These ore deposits are related to the precipitation of metals from hydrothermal solutions circulating in volcanically active submarine environments. The black smokers, which were discovered only recently at the sea floors at Mid-Ocean Ridges in the Pacific, Atlantic and Indian Oceans, represent a direct analogy for the type of VMS deposits, namely the ophiolite-hosted copper deposits such as those of the Troodos Massif, Cyprus, of the Samail ophiolite complex of Oman, and some deposits in Anatolia. These ore deposits are a transition between hydrothermal to sedimentary ore-forming processes. As explained above, they were formed at mid-ocean ridges and were moved, together with new oceanic crust, through plate tectonic activities.

Cyprus

On the island of Cyprus in the eastern Mediterranean, copper deposits have been intensively researched, both geoscientifically and archaeologically. More than 90 copper ore deposits and mineralisations make Cyprus the most important copper district in the region and beyond (Stos-Gale et al. 1998a, b) (Fig. 3.13). The estimates indicate that about four million tons of slag have been produced on the island (Zwicker 1985). They have been exploited since at least the beginning of the Middle Bronze Age (Belgiorno 2000). One peak of copper production at Cyprus is dated to between 1600 and 1200 BC when Cyprus had a monopoly for copper all over the Mediterranean. It is situated between the decline of activities at Faynan at the end of the third millennium BCE and a renaissance at the end of the second millennium BC. A second summit of copper production

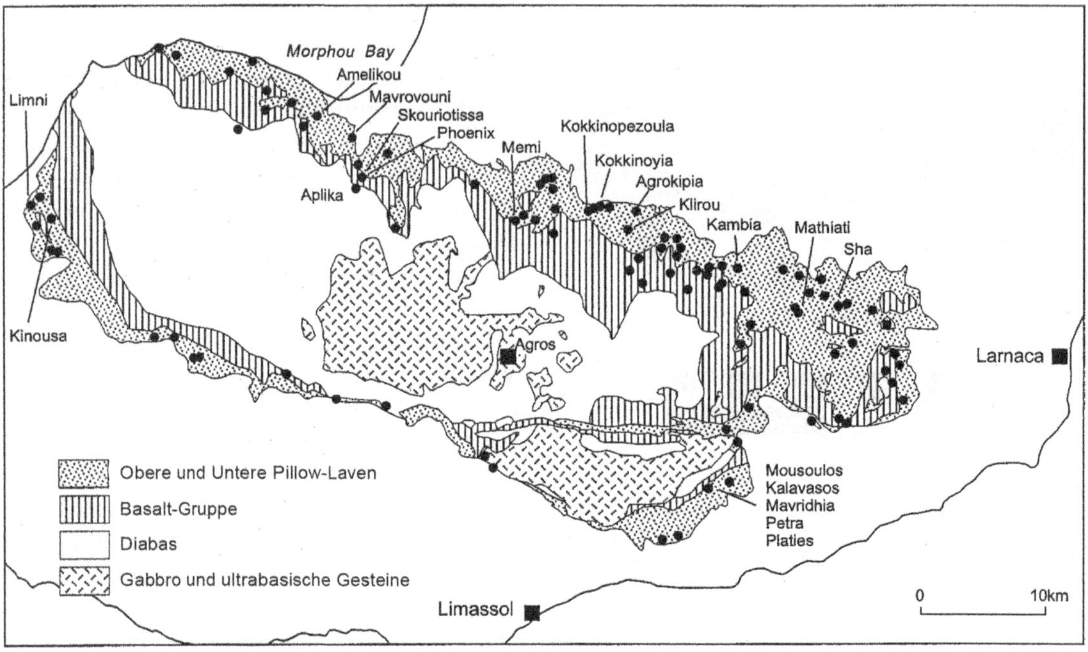

Fig. 3.13 Location of Volcanic Massive Sulphide copper deposits (VMS) in the Troodos ophiolite complex of Cyprus. More than 90 sulphide occurrences are known in the ultramafic rocks of the Basal Group and in the Upper and Lower Pillow Lavas surrounding the intrusive complex at the core of the ophiolite. Their exploitation during the Bronze Age lead to the monopoly of Cyprus in the Mediterranean and beyond. From Hannington et al. (1998)

is dated to the Roman–Byzantine period (Kassianidou 2008).

All copper deposits of the island of Cyprus contain massive sulphides of volcano-sedimentary origin, which formed at a Mid-Ocean Ridge and were moved, together with new oceanic crust, through plate tectonic activities (Sillitoe 1972). The deposits are associated with (ultra-)basic rock and occur in all lithologic units of the Troodos ophiolite complex. They date to the Cretaceous period. The prototype of the massive sulphide deposits (Constantinou 1982; Sillitoe 1972) belongs to the Pillow-Lava Unit. Different mineralisation types occur under aspects of exploitation. Typical are massive mineralisation lenses. Stratigraphically below are network-like stockwork mineralisations (Fig. 3.14). As outlined in Sect. 3.5, VMS at Cyprus used to show extensive gossans with extraordinary yellow-orange colouring by various iron minerals. These gossans are in part of submarine origin (Herzig et al. 1991). The ore deposits are mineralogically and geochemically rather monotonous. The ore consists predominantly of pyrite with erratic occurrences of Cu–Fe sulphides and occasionally sphalerite.

Cyprus provided the first sulphidic ores to ancient copper smelters and offers a model case to study the technology of smelting those ores in ancient times (Given and Knapp 2003). Ni, As, Co and Pb exist only in traces (Constantinou 1982). In contrast to Ergani Maden, bright red gossans exist in Cyprus, in which ancient mining was carried out intensively. Cementation zones developed locally with rich copper sulphides (covellite, chalcocite, bornite) and/or cuprite. Stos-Gale and Gale (1994) describe copper ores from the cementation zone of the Kambia mine as containing more than 30 wt.% Cu. Some archaeometallurgical importance must be given to the so-called stockwork mineralisation where chalcopyrite and Co–Ni–As–Fe ores appear in the rooting zones of the massive sulphide deposits, meaning in the gabbros and periodotites underneath the volcanic rocks (Constantinou 1982).

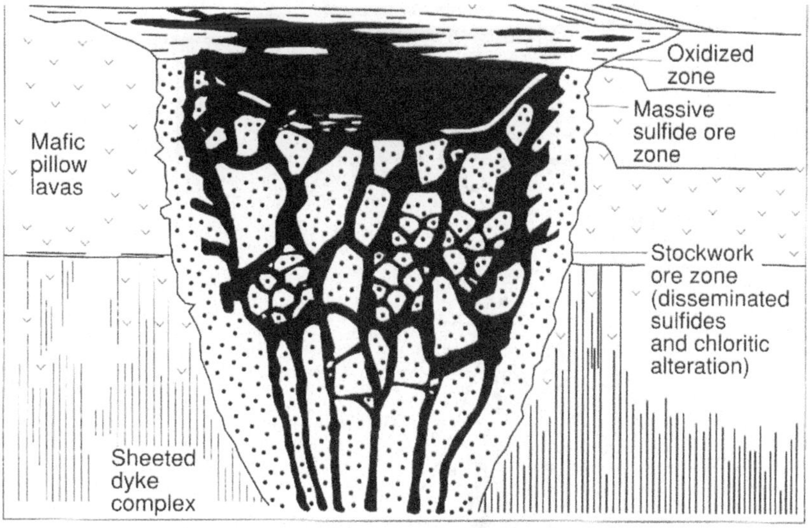

Fig. 3.14 Schematic stratigraphy of the Troodos ophiolite complex and overlying sediments at Cyprus showing the copper mineralisations in various formations. Note massive sulphide deposits and the network-like stockwork mineralisations below. Modified after Robb (2005)

Such ore mineralisations have been reported from Limassol (Panayiotou 1980). These have been mined at least since the Hellenistic–Roman period. The existence of manganese-containing and Cu-rich sediments played an important role in the smelting of ores. This so-called umber or devil's mud occurs in lenses together with radiolites and claystones formed stratigraphically above the pillow-lavas. The devil's mud contains considerable concentrations of noble metals (gold up to 14 ppm, silver up to 100 ppm; Bear 1963; Kindermann 2005). Gold occurs in the finest flakes of native gold, silver is bound to jarosite (see Sects. 3.4.7 and 3.4.9). Amazingly, and in contrast to Rio Tinto on the Iberian Peninsula, no ancient mining activities were discovered so far to exploit the noble metals from the devil's mud.

Oman

Almost identical to Cyprus in their origin and geological setting are the copper deposits in the Samail ophiolite complex which form large parts of the Oman Mountain Range of the south-east Arabian Peninsula. It is probably the largest such lithic complex worldwide. It represents an almost complete section of the Cretaceous oceanic crust and uppermost mantle. Its age is dated to 85–95 million years (Coleman 1981), i.e. they originate from the same geological time period as the ore deposits from Cyprus. In total, about 150 copper ore deposits and smaller prospects have been discovered in the Oman Mountain range (Weisgerber 1978, 1980, 1981; Hauptmann 1985; Begemann et al. 2010) (Fig. 3.15). Lens-shaped to vein-formed, irregular mineralisations with amounts of secondary copper minerals exist at the basis of the ophiolite complex. On the island of Masirah, 40 mining and smelting sites with mostly small copper mineralisations were discovered (Weisgerber and Al-Shanfari 2013).

The volcanic sequence of the Samail ophiolite complex in the north of Oman encloses a few huge VMS deposits. These giants genetically belong to the Cyprus type, i.e. they are co-volcanic or syngenetically deposited between volcanic rocks and the diabase dykes. Especially the brightly coloured gossans at Lasail, Aarja and Bayda in the Wadi Jizzi discontinuity (Fig. 3.60a, Sect. 3.5), consisting of masses of goethite, limonite, jarosite, jasper, malachite and chrysocolla, with a depth at Lasail of c. 40 m, and those of Rakah and Tawi Rakah in the Ghuzayn, discontinuity will have been conspicuous features for ancient miners, just as they are for modern prospection purposes. In these ore deposits, a disseminated and brecciated type of sulphide (*cockscomb texture*) is dominant. The gossan at least of Lasail is interpreted as of submarine origin (Herzig et al. 1991).

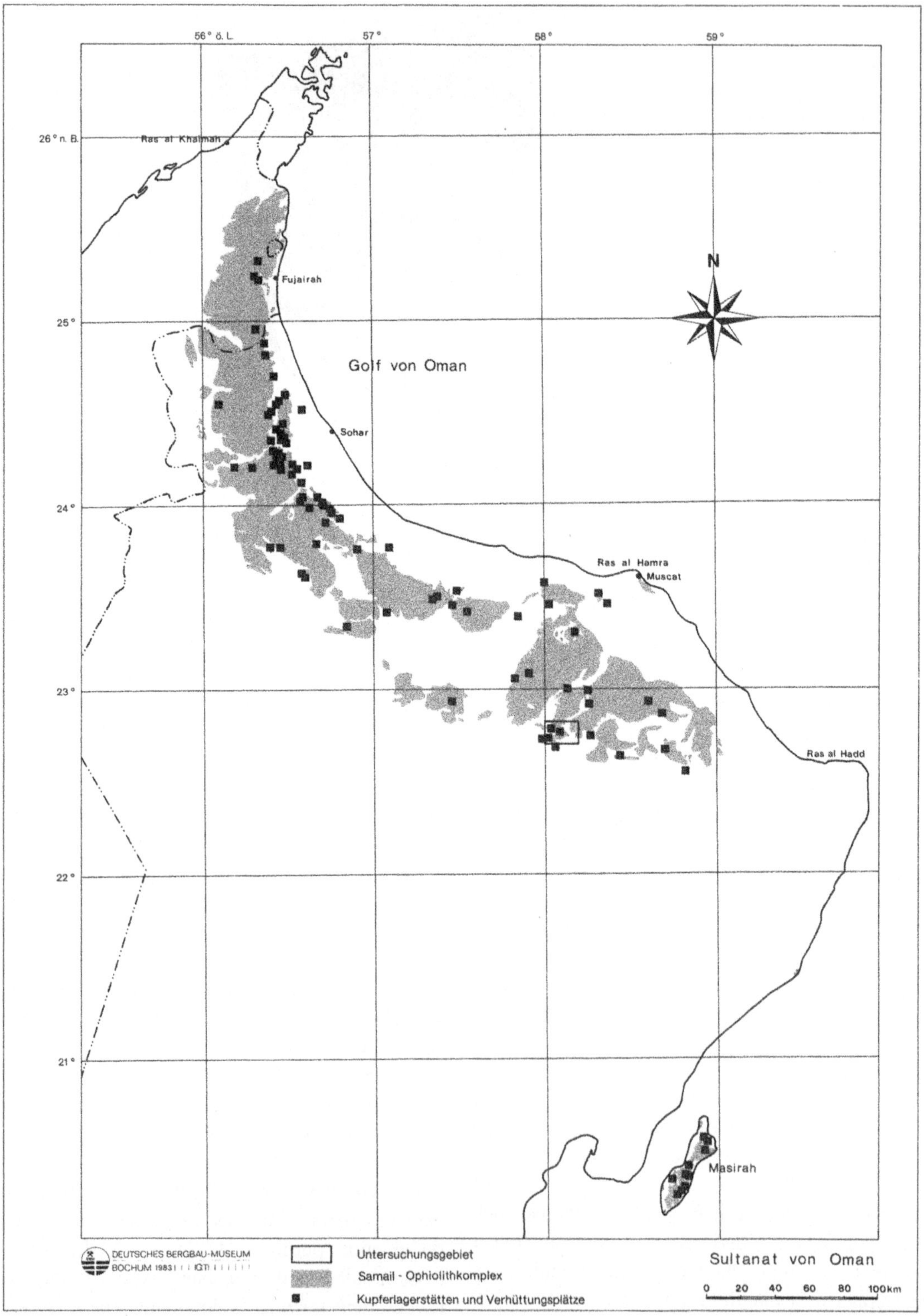

Fig. 3.15 Volcanic Massive Sulphide copper deposits (VMS) and other mineralisations in the Samail Ophiolite Complex of Oman (southeast Arabian Peninsula) (shaded area). Signs outside the shaded area are some different ore deposits and, in parts, ancient smelting sites. Altogether, about 150 ore deposits were found in Oman. Map after Hauptmann (1985)

At Lasail, ancient mining activities seem to have been concentrated towards the central part of the ore deposit. The mineralisation is dominated by pyrite and pyrrhotite, and by minor amounts of chalcopyrite. In this ore deposit, at a depth of c. 40–50 m, local enrichments of chalcocite (Cu_2S) and bornite (Cu_5FeS_4) were found in a gallery of possibly Islamic date (Weisgerber 1987a, b), cut by modern underground mining activities. Sphalerite (ZnS), galena (PbS), tennantite (Cu_3AsS_3) and Co–Fe–Ni minerals are rare and should be considered as accessory minerals. Minor gold concentrations occur in these deposits. They were exploited recently at the ore deposit of Rakah. Ancient mining of gold is not found. All prospects show intensive and extensive staining by green secondary copper minerals such as malachite, brochantite ($Cu_4(OH)_6/SO_4$), chalcanthite ($CuSO_4 \cdot 5H_2O$), atacamite ($Cu_2(OH)_3Cl$) and chrysocolla ($CuSiO_3 \cdot nH_2O$) the ophiolitic rocks.

Anatolia

Also in Anatolia, volcanogenic massive sulphide deposits occur, similar to those on the island of Cyprus and Oman. They are embedded in a series of ophiolitic rocks of upper Cretaceous age (diabase, ultramafic rocks, gabbros, shales, radiolites; Helke 1964; Evans 1992). The VMS ore deposit of Ergani Maden in southeast Anatolia was considered the largest copper mine in Turkey until 1994. Modern mining activities have obviously destroyed most of the traces of ancient metal extraction.

Two massive sulphide lenses overlie sulphidic impregnations in altered chloritic host rock. Ergani Maden has a geochemically similar pattern to the deposits in Cyprus. It is difficult to differentiate between them by means of their lead-isotope composition. The composition of the ore body is monotonous too, with pyrite, pyrrhotine, chalcopyrite, magnetite, hematite and subordinated sphalerite. As fahlores are frequent minerals in small quantities in many copper deposits, it is not surprising that Tylecote (1976) mentions As and Sb fahlores in the cementation zone of Ergani Maden.

It has been suggested repeatedly that Ergani Maden was the most important prehistoric copper source in Anatolia. Reasons for this suggestion were the convenient location on a trading route between central Anatolia and northern Mesopotamia through the Tigris valley, the rich occurrence of native copper (Wagner et al. 1989) as well as oxidic copper ores and observations made on ancient mines in the beginning of the twentieth century (Pilz 1917). Almost 50 years ago, Tylecote (1976) also reported old mines in the oxidation and in the cementation zones, and the Heidelberg team described ancient mines and slag heaps in the surroundings of the modern open-pit mines. Today there are only small exposures of the ore body preserved and statements about the composition of the ancient ore body would be mostly speculative. A cementation zone with Cu–Fe sulphides (bornite, covellite etc.) can be observed only in a few places. In summary, the destruction of ancient mining traces by modern mining might be the reason, why not yet one secure prehistoric date can be connected with Ergani Maden.

Hundreds of malachite beads were found in the Pre-Pottery Neolithic settlement of Çayönü Tepesi, which is located only some 50 km to Ergani. Özdoĝan and Özdoĝan 1999) suggest that these were made of the striking green copper ores collected from the gossan of Ergani Maden and, herewith this ore source was exploited as early as the eighth millennium BC. Schmitt-Strecker et al. (1992) state that slags and ores found at the Late Chalcolithic/Early Bronze Age site of Norsuntepe in eastern Anatolia and Early Bronze Age metal artefacts from Hassek Höyük at the upper Euphrates, are identical in the lead isotope pattern with Ergani, and concluded that this ore deposit was used in the fourth and third millennium BC as well.

A deposit of great significance is Murgul in northeastern Turkey. Already in prehistoric times, it could be considered as an economical "giant" near the border to Georgia. Murgul is also a VMS deposit, as found in Ergani Maden, Cyprus and Oman. The traces of prehistoric mining and early smelting are largely destroyed, but Lutz (1990) was able to prove that copper has been mined in

Murgul since the fourth millennium BC. In the Bronze Age, gold was very likely mined as well. Murgul was probably one of the key copper resources of the Georgian Bronze Age (Dschaparidze 2001) and also played an important role at Arslantepe around 3000 BC (Hauptmann et al. 2002a).

Besides the ore deposit of Ergani Maden and Murgul, the allegedly most comprehensive study of ancient ore deposits from the specific viewpoint of ancient metal production has been carried out by the Heidelberg–Mainz group. The study took place in the Aegean and in Anatolia during the 1980s and 1990s (Seeliger et al. 1985; Pernicka 1987; Wagner and Öztunali 2000; Pernicka et al. 2003). In Anatolia, 91 ancient copper deposits were discovered. Thirty-six out these deposits showed safely dated prehistoric mining activities. These sites were investigated under aspects of geology, (mining-) archaeology, and archaeometallurgy. Samples of ore, slag and metal were collected and analysed in the laboratory for chemical, mineralogical and isotope compositions. The Max-Planck-Institut für Chemie (Mainz) carried out lead isotope analyses, which was not only ground-breaking research but also became a standard tool for archaeometallurgical research (e.g. Begemann et al. 1992, 1994; Schmitt-Strecker et al. 1992). Yalçın (2000a, b, c, 2008) continued surveys and excavation works especially in northern Anatolia around the famous Early Bronze Age settlement of Alaca Höyük and at the native copper bearing mine of Derekutuğun.

A number of smaller Cu-deposits between the upper Euphrates and Lake Van (Seeliger et al. 1985; Wagner et al. 1989) also belong to the series of ophiolitic rocks. However, most of them show no obvious signs of ancient mining and/or slag heaps. Nevertheless, some of these deposits might have been exploited already during the Neolithic period (Schoop 1995), because they contain native copper (e.g. Kirmizitarla near Ergani Maden). All of these deposits are located in the southeast Anatolian Neolithic settlement area, where an early use of native copper has been suggested (Muhly 1991; Maddin et al. 1999).

Iberian Pyrite Belt (IBP)

The Iberian Pyrite Belt is located in the southwestern part of the Iberian Peninsula. This tectonically defined region extends from north-west to south-east and has a length of c. 250 km. This belt hosts more than 90 polymetallic deposits which form a metallographic province. All these deposits belong to the type of Volcanogenic Massive Sulphide deposits. The IPB is one of the most outstanding ore provinces of the world. It hosts the largest concentration of Volcanogenic Massive Sulphide deposits worldwide, including several world-class and giant deposits such as Río Tinto in Spain and Neves Corvo close to the ancient mine of Aljustrel in Portugal (Inverno et al. 2015).

Main components are sulphidic copper and zinc ores with some lead and silver. In contrast to Oman, Cyprus and Anatolia, which are dominated by copper ores, the deposits of the Iberian Pyrite Belt have a pronounced polymetallic character. There are lenticular masses of pyrite with chalcopyrite, galena and sphalerite as main components, but hundreds of minerals both of primary sulphidic and of secondary origin as accessory minerals. This is in accordance with the trace elements pattern which shows elevated concentrations of gold and silver, in addition of tin, cobalt, mercury, and bismuth. The ore deposits of the IPB are embedded in Late Devonian to Middle Carboniferous host rocks. Palaeofaults and synsedimentary structures acted as channels and discharge traps for the metallic liquids and to the gossans developed over the VMS. The IPB was even called a "monster" in terms of relative metal weight (Leistel et al. 1998). One of the most characteristic features of IBP deposits is their unusually powerful gossans (Velasco Roldan et al. 2013).

A number of famous ancient ore deposits are located in the IBP, e.g. Aljustrel, Tharsis, and especially the most famous ore deposits of Rio Tinto (Salkield 1987; Anguilano 2012). Descriptions of the IPB with its deposits, especially those in Rio Tinto, can also be found in Pryor et al. (1972), Amorós et al. (1981), Barriga (1990), Leistel et al. (1998).

Fig. 3.16 Rio Tinto, Huelva province, south-western Iberian Peninsula. The red river received its name from the extensive red acid mine drainage. Rivers like that of the Iberian Pyrite Belt may contain between 4 and >30 wt. % total ferric iron (colour!), minor amounts of Cu, As, Zn in the lower percentage levels) and trace elements of Pb, Sb, Ni, Cd (Martín-Crespo et al. 2019). The acidic waters contain free sulphuric acid (pH 2–2.6). Photo: courtesy G. Goldenberg, University of Innsbruck, Austria

The Rio Tinto sulphide deposits (*verbatim* the red river, see Fig. 3.16) are located c. 90 km northeast of the city of Huelva. Cerro Colorado (red hill) is the central feature of the mineralised zone at Rio Tinto. Covered by a huge gossan cap, at the base of which are the most extensive ancient workings for silver in the world, flanked by more recent workings consisting of three open-pits and three underground mines. The mineralisation is erratic and of three types (Pryor et al. 1972):

1. The gossan, formed by the oxidation of pyritic ores, has a thickness of 20–40 m. The material is high in silver (30–50 ppm) and gold (2–3 ppm).
2. In the cementation zone (secondary enrichment zone), copper is enriched and silver. These are both descending metals from the gossan, as well as oreshoots from the underlying stockwork mineralisation, which could have acted as feeder channels for syngenetic sulphides (Pryor et al.1972).
3. Copper is present in the primary ore body.

Mining activity in this area probably dates to the Chalcolithic/Early Bronze Age period at Chinflon, a mine and smelting site near Cerro Salomon, with a peak between the third and the first century BC (Rothenberg and Blanco-Freijeiro 1981). Later phases of metal production are illustrated by the large amount of slags (c. 6 Mio tons) at Corta Lago. The modern open pit in Cerro Colorado revealed the most intensive ancient mining traces at the base of the oxidation zone in the transition to the cementation zone (Leistel et al. (1998) (Fig. 3.35 in Sect. 3.4.9).

Sedimentary Copper Deposits

These are major sources not only for copper itself, but also for other metals such as iron, manganese and silver, and in modern times also for cobalt. The deposits are chemical precipitates in sedimentary host rocks, to which large quantities of metal have been added in solution.

The Kupferschiefer

Some of the largest and most important sediment-hosted stratabound copper deposits, spanning large areas of Germany and Poland, belong to the European Kupferschiefer (in English language copper shale) (Borg et al. 2012). This is a metre-thick layer of black $C_{organic}$-rich shale. All the deposits included in this geological formation are of Permian age. They were formed by the precipitation of metal- and sulphide-rich hydrothermal solutions that have reacted with sediments on the seabed. But also epigenetic vein mineralisations occur in parts, the so-called "Kobalt-Rücken" (cobalt-ridges). Geochemically, Cu, Pb, Zn, Ag ores are predominating, minor elements are Co, Ni and as trace elements Au, Pt, Pd.

According to many written sources, ores from the Kupferschiefer were exploited in the twelfth century AD, but have been of major importance for the European metal economic scenery in the sixteenth century AD (Bingener et al. 2012). An earlier use of this enormous rich deposit seems to be highly probable, but there is no hard evidence for that. Otto and Witter (1952) could only express the suspicion that the Early Bronze Age artefacts, with their As, Sb and Ni contents of the Aunjetitzer culture, could also be attributed to the Kupferschiefer ore deposits in the German low mountain range. New geochemical–isotopic analyses, triggered by the findings of the famous Nebra Sky Disc, could also only detect certain geochemical similarities with Mansfeld mineralisation (Frotzscher 2009). This is probably due to the fact that the geochemical bandwidth of the entire Kupferschiefer and the numerous mineral deposits of the German low mountain range are barely palpable in their full range.

Kargaly

The copper deposit of Kargaly is also a giant and, undoubtedly, a unique metallurgical complex. Its uniqueness is absolutely evident if we compare it not only with neighbouring mining centres but also with numerous mining complexes in northern Eurasia. Many features distinguish the Kargaly deposit from other copper producing centres in ancient Eurasia: the extreme antiquity, huge number of mining-works traces and the abundance of impressive archaeological and historical sites.

The prehistoric mining and metallurgy center of Kargaly is located on the south-western flank of the Urals in the northern area of the Great Eurasian steppe (Chernykh 2003). Spanning about 500 km^2, it is one of the largest copper deposits ever mined in ancient times. Copper production started in the fourth millennium and reached gigantic proportions in the Late Bronze Age. Today, around 30.000 remains of mining activity are visible. The amount of copper smelted in the Bronze Age is of the order of 50–100 thousand tons. The sedimentary ore deposit is of Permian age. The seam-like mineralisation is situated partly only a few metres below the surface and was mined in a very similar manner as in Timna, Israel, through closely juxtaposed shafts with a branched system of tunnels and chamber-pillar constructions. The mineralisation consists of oxidic ores, which are stratiform embedded in sandstone and marl. Sulphidic ores are almost completely absent. Larger slag heaps are missing; surprisingly, copper seems to have only been partially melted in Kargaly itself.

Wadi Arabah and Sinai

In the southern Levant and on the south-western Sinai Peninsula, two copper ore districts have to be mentioned, namely Faynan (Jordan)–Timna (Israel) in Wadi Arabah and Umm Bogma–Serabit el-Khadim (Sinai). They are part of a supra-regional ore province of sedimentary–epigenetic deposits formed by the erosion of Precambrian veins and subsequent deposition in a lagoonal milieu at the borderline of the Cambrian sediments to the Precambrian basement.

Timna and Faynan

Today, due to the lateral displacement of the Wadi Arabah Rift Valley, the two copper ore districts are separated by a distance of about 105 km. Timna is located on the western side of Wadi Arabah in the south, Faynan on the eastern side further north. The paleography shows that they once formed a joined coherent entity in the Cambrian period (Segev 1987; Segev and Sass 1989).

By far the richest mineralisation both in Timna and Faynan occurs in Cambrian dolostones. At Timna, this formation is mainly buried in great depths and, with but one small exception, was not used in ancient times. At Faynan, these dolostones belong to the Cambrian Burj-Limestone Formation and are also called Dolomite-Limestone-Shale Unit (DLS). This is a copper–manganese mineralisation. Here, this formation is exposed to the surface. Considerable epigenetic stratabound mineralisations also occur in the Cambrian sediments of the Massive Brown Sandstone (MBS) at Faynan and in the Amir Formation at Timna. In summary, the lithostratigraphy of Cambrian sediments at Timna and Faynan is similar as reported by Segev and Sass (1989) and Segev et al. (1992). For the development history of metallurgy in the region, the composition of ores likely played a crucial role, because in both deposits, there are almost exclusively secondary copper ores such as various chrysocolla, (par-)atacamite, malachite, dioptase and plancheite (Hauptmann 2007) Intense intergrowths of cuprite and Fe hydroxides (tile ore) are just as typical of mineralisation on the western side of the Arabah as cuprified plant remains (Hauptmann and Wagner 2007). Copper sulphides occur less prominently in Timna and Faynan. Concentrations of uranium in these ores and sediments are reasons for large ranges of lead isotope ratios.

The decades of research by Beno Rothenberg in Timna since the 1960s are among the pioneering achievements of modern archaeometallurgy. It was Rothenberg, who, via the slag heaps that occur in this area, first took up the topic of prehistoric industrial archaeology and—based on the works of Zschocke and Preuschen (1932)—along with several international research teams, began exploring early mining activities and early metallurgical practices (Rothenberg 1973, 1988, 1990; Conrad et al. 1980). Similar to Kargaly, in Timna the copper mineralisation embedded in carbonated sandstone is located only a few metres below the surface. They were reached via closely adjacent shafts and mined in irregular chamber-pillar constructions. In Timna, several thousands of shafts were counted in an area of about 20 km^2 (Conrad et al. 1980) (Fig. 3.17).

The richest and at the same time the one of these two sedimentary ore districts, which evidently acquired the greatest economic importance, especially in the Early Bronze Age and the Iron Age, is Faynan. Faynan has been known as a prehistoric ore deposit for a long time but has been intensively studied over recent decades under (mining) archaeological and archaeometallurgical aspects. After the first modern survey work by Kind (1965), the Deutsche Bergbau-Museum developed a systematic chronological investigation of mining and metallurgical technologies (Weisgerber 2006; Hauptmann 2007; Hauptmann and Löffler 2013). This was deepened by a major international research project of Thomas Levy and his team from the University of San Diego California. They focused on the Iron Age smelting sites in the Faynan district (Levy et al. 2014). Later, Löffler (in prep) investigated all finds and findings from the ancient mines and he counted so far a minimum >369 deep and ample constructed ancient mines in the district of Faynan.

Geochemical–isotopic studies on the export of Faynan copper in various periods were carried out. Faynan copper was not only one of the dominant metals in the Chalcolithic and Bronze Age in the southern Levant (Tadmor et al. 1995; Hauptmann et al. 2015), its tracks were most likely traced even to Mesopotamia in the third (Hauptmann et al. 2016), and to the Aegean in the first millennium BC. (Kiderlen et al. 2016). Graeme Barker and his colleagues studied the extensive desertification of the Wadi Faynan landscape in extensive geoarchaeological work and summarised these findings in a

Fig. 3.17 Timna, Israel. Alluvial fans in the Timna Valley (Lieder 1980) are covered by countless, densely packed and backfilled shafts (*Tellerpingen*). The horizontally oriented ore horizon, located a few metres below, was made accessible by these shafts. In the underground, the ore was exploited in chamberlike galleries. Overall, the number of shafts in the entire Timna Valley is estimated at several thousand. Photo: A. Hauptmann, Deutsches Bergbau-Museum

comprehensive compendium (Barker et al. 2007). On the basis of Hauptmann (2007), Thomas Levy and his team conducted an intensive mining archaeological–metallurgical survey in the Faynan area and carried out extensive excavations at Khirbat Hamra Ifdan, Khirbat en-Nahas and other localities (Levy et al. 2002, 2014).

The formation highest in copper in Faynan is the Dolomite-Limestone-Shale Unit (DLS), where secondary copper minerals occur intergrown with manganese oxides. The assignment of ores to the geologically old formation of the Cambrian is reflected in a clear cluster of lead isotopic compositions (Hauptmann 2007). This formation was the most important source of copper especially in the Early Bronze Age II/III and Iron Age (Fig. 3.18). Mineralisations in the Massive Brown Sandstone (MBS) are epigenetically deformed and occur in cracks, joints and fissures. These mineralisations are free of manganese. The lead isotopic composition of copper ores differs from that of the DLS and displays a significant spread due to uranium contents in the sediments. These ores were mined in different epochs. Mines dating to the Chalcolithic/Early Bronze Age I can be found here. But intensive Roman mining activity has been proven as well. These technological developments are displayed in Fig. 3.18. The temporally abrupt change of mining from the DLS to the MBS and vice versa is archaeologically fully clarified.

Sinai

The desert of the Sinai Peninsula is a triangular landmass between the Gulf of Suez and the Gulf of Aqaba. Abdel-Motelib et al. (2012) report an archaeometallurgical survey on the Sinai Peninsula and in the Eastern Desert of Egypt.

Fig. 3.18 Wadi Khalid, Faynan (Jordan). Secondary copper ores (malachite, atacamite, chrysocolla, dioptase) mixed with manganese oxides, embedded in shales of the Dolomite-Shale Unit have been the raw source for Early Bronze Age to Iron Age miners. It is a polymetallic self-fluxing copper ore easy to be smelted. Photo: G. Weisgerber (†), Deutsches Bergbau-Museum Bochum

The most important ore deposits are located at the border zone of the sedimentary formations, which cover the entire northern part of the Sinai and the Precambrian basement in the south. These are sedimentary copper–manganese mineralisations in the region of Um Bogma, Bir Nasib and Serabit El-Khadim. The geological setup of the copper deposits is comparable with Timna and Faynan. They are embedded in Cambrian to Carboniferous sandstones. The ores are composed exclusively of secondary copper ores intergrown with ferrous manganese hydroxides. Of special importance are the ancient mines of Serabit El-Khadim and Wadi Maghara, where the most eligible copper phosphate turquoise ($CuAl_6[(OH)_2/PO_4]_4$) was mined in ancient times. Turquoise was one of the most important gemstones in ancient Egypt. These mines were described by several explorers since the nineteenth century AD (Abdel Motelib et al. 2012).

Unfortunately, the mines are not known which have been the sources for the enormous copper production at Bir Nasib. At this smelting site, some 100,000 tons of copper slag have been produced. Bir Nasib is among the largest smelting sites in the Eastern Mediterranean and testifies to a supra-regional center for copper production during the Late Bronze Age/Iron Age I, probably much earlier. Copper produced from the sedimentary deposits on the Sinai Peninsula was exported as early as the late fourth millennium BC to the settlement of Maadi in the Nile Delta (Pernicka and Hauptmann 1989; Hauptmann 2017).

In the southern part, the Sinai consists of the Precambrian crystalline basement of the Arabic-African shield. Abdel-Motelib et al. (2012) describe some hydrothermal copper veins, which were exploited in prehistoric times as well. However, they are unimportant in extent and mineral content. Overall, with but the exception of Bir Nasib, the Sinai is a rather resource-poor area on the periphery of the rich ore deposits in the north and east of the Mediterranean, which has hardly played a role in the time of the New Kingdoms. Manganese, which is mined on the Sinai to this day, played no role in ancient times.

3.4.7 Gold

Gold (Au) is the noblest chemical element. It is siderophile, i.e. it has an affinity to iron. The mean concentration of gold in the earth's crust is about 0.004 ppm. This is a factor of around 10,000 less than copper. Nevertheless, gold deposits and even the smallest occurrences of gold in the earth's crust are extremely widespread worldwide. Today, a gold deposit is mined if it contains about 3 ppm, depending on the type of deposit and the recoverability of the deposits. The total production of gold worldwide has been estimated at around 164,000 tons since the beginnings until today. That would be a cube with an edge length of more than 20 m (Pohl 2005; Butt and Hough 2009).

3.4.7.1 Gold Minerals

Due to its noble character, gold is found in most ore deposits in native form. The most common natural alloy components of gold are silver and then copper, which, if deliberately added in larger proportions, are called "coinage metals".

Free gold or visible gold comprises particles of native gold visible with the naked eye such as feathery crystals, dendrites or nuggets. Normally it occurs in tiny, mm-sized grains. Much smaller is mustard gold, named after an old miner's term in Germany, and it consists of tiny tinsels embedded in earthy and soft yellow-brown limonite. This material is formed by the decomposition of gold-bearing pyrite in the oxidation zone of sulphidic ore deposits (see Sect. 3.5 oxidation zone). Large nuggets may reach weights of more than 100 kg.

Refractory gold or invisible gold is not recognisable with the naked eye. It is dispersed as a solid solution or submicroscopic inclusions in quartz, in sulphide minerals, e.g. in pyrite or arsenopyrite, and in various sulphosalts. Refractory gold has a grain size of ca. 5 μm and below. Therefore, more sophisticated processing technology is required for extraction. In ancient times this was mainly roasting, careful and laborious mechanical beneficiation, and extraction by panning operations or amalgamation.

A selection of more familiar gold minerals is listed in Table 3.4. Spiridonov and Yanakieva (2009) differentiate gold minerals, of which some are listed in this table into Au-(Ag-)Hg-, Au-Pd- and Au-Pt compounds. Of further importance are Au-(Ag) compounds with tellurium and selenium. Refer to the noted literature for further information.

Colours

Depending on the content of silver, copper and other alloying elements, gold changes its colour. Colours of gold and gold alloys were already used very accurately in the Bronze Age as distinguishing criteria of different gold grades (Waetzold 1985; Hauptmann et al. 2018, cf. Sect. 7.3). Mostly pure gold has a golden yellow metallic lustre with a slight reddish tone. With increasing silver content, gold becomes lighter. Electrum, with a composition of about $Au_{55}Ag_{45,}$ is silvery white. Auricupride ($Au_{40}Cu_{60}$) has a strong red hue and resembles the colour of copper. An exact discussion of these colour gradings can be found in Cretu and van der Lingen (1999).

Gold, Electrum and Aurian Silver

The most common occurrence in primary gold deposits is a natural Au–Ag alloy. Under natural conditions, gold and silver form a chemically solid solution series of miscibility (Samusikov 2002). According to the nomenclature, gold itself may contain up to 30 wt.% Ag (fineness 1000–700). The composition of the frequently occurring electrum contains 30–70 wt.% Ag (fineness 700–300). The mineral küstelite contains 70–90 wt.% Ag (fineness 300–100). This is aurian silver with 10–30 wt.% of gold. This is a very rare mineral, and it is reported only

Table 3.4 Some of the more important and common gold minerals and their contents of silver and copper

Mineral name	Chemical formula	Au (wt.%)	Ag (wt.%)	Cu (wt.%)
Native gold	AuAg	2–20	1–2	
Electrum	AuAg	50–80	≥ 20–50	
Auricupride	$AuCu_3$	ca. 40		ca. 60
Küstelite	AuAg	10–30	70–90	
Cuproauride	CuAu			
Krennerite	$(Au,Ag)Te_2$			
Calaverite	$AuTe_2$			
Petzite	Ag_3AuTe_2			
Sylvanite	$AuAgTe_4$			
Fischesserite	Ag_3AuSe_2			
Weischanite	$Au,Ag)_{1.2}Hg_{0.8}$			
Gold amalgam	γ-$(Au,Ag)_2Hg_3$			

They may vary according to various solid solutions. Data after Doelter (1912–1932), Ramdohr (1975), Hartmann (1982), and Tylecote (1987). Compilation of more than 20 gold minerals in Spiridonov and Yanakieva (2009) and in LAPIS special issue 2 (1992)

from a limited number of deposits in the Middle and the Far East. In this mineral, gold substitutes for silver, and a complete series extends from aurian silver, argentian gold (electrum) to gold.

According to modern textbooks, native aurian silver is extremely rare and almost vanished from the geological record. It is, however, reported by early mineralogists such as R. Jameson (1821) who mentioned compositions with 72 wt.% Ag and 28 wt.% Au in hydrothermal veins. He noted that such ores would widely be exhausted in Europe. Further information about aurian silver is cited in Fordyce and Alchorne (1779) who analysed aurian silver with 28 wt.% gold from the silver deposit at Kongsberg, Norway, and analyses are reported also from Rath vom (1879), which showed gold contents in silver from there varying from 27 to 50 wt.%.

Aurian silver occurs almost exclusively in gold deposits, and especially in gold-quartz veins and lodes. In Egypt, where most of the prehistoric artefacts consisting of aurian silver were found (Gale and Stos-Gale 1981b; Hauptmann and von Bohlen 2011), these veins are embedded in Precambrian crystalline rocks and have a high geological age.

A large number of artefacts consisting of aurian silver with Au-concentrations in silver of up to 50% were found in Egypt (Stos-Fertner and Gale 1978; Gale and Stos-Gale 1981b; Hauptmann and von Bohlen 2011). The Gales analysed a large number of Egyptian silver artefacts and found similar continuous gradations between silver, electrum and gold in objects from the Pre- and Protodynastic periods as well as from the Old Kingdom and Middle Kingdom. In addition, they state that copper was a ubiquitous contaminant in Egyptian gold and silver objects, being commonly in the range of a few percentage up to 10 wt.%. Iron and zinc are continuously associated elements in the percentage levels as well.

From the metallurgical point of view, artefacts consisting of aurian silver must not necessarily originate from ore deposits bearing aurian silver. They can also be produced from occasionally occurring mixed gold–silver–copper–iron–zinc mineral associations (Pohl 2005). Klemm and Klemm (2013) note that one typical indicator for the presence of gold in hydrothermal veins in Egypt is malachite, a material commonly used in the Predynastic period. Therefore, it remains uncertain whether or not copper was deliberately added to the alloy to debase the noble metal, to change the colour or to lower the melting point of the metal.

Copper Containing Gold

It is suggested that most samples of native gold contain only traces of copper, probably substituting in the lattice. Hartmann (1970) stated in his extensive study on the chemical analyses of archaeological gold artefacts from all over Europe, that natural gold would not exceed copper contents of 1.5–2 wt.%. In a similar matter, Tylecote (1987) and Mohen (1990) set the limit of copper in gold at 1 wt.%.

However, after Ferguson (1950), primary gold may contain up to 20% Cu. Natural alloys that high in copper are named cuproauride (CuAu) and auricupride ($AuCu_3$). Cuproauride seems to be rare in nature. But according to Ramdohr (1975), auricupride may occur rather frequently in primary gold deposits. Auricupride was found in gold deposits in Armenia, Russia, Cyprus, Switzerland, Chile, and additional localities in Argentina, South Africa, Australia, the Czech Republic and Greece. Auricupride might once have been enriched in ancient untouched gold deposits. Its occurrence next to gold could be a reasonable explanation for ubiquitous copper concentrations in the lower percentage level in prehistoric gold objects. Copper, like silver, is leached out of gold in sedimentary deposits.

Amalgam—Natural or Anthropogeneous?

Amalgam is an alloy of mercury (Hg) with another metal. Many natural metals can form amalgams with mercury. Gold also has a high affinity for mercury (Hg). The differentiation between natural gold amalgams (Moon 2001) and waste products of historic gold washers, who won the precious metal through the amalgamation process, is not easy. Gold with Hg levels of 1.2–6 wt.% were observed in the paleo-placer of Witwatersrand, demonstrating that gold of such composition can be natural (Oberthür and Saager 1986). Hauptmann et al. (2010) analysed mercurial gold grains in gold placers, in the vicinity of the (pre-)historical gold mine of Sakdrisi in the Transcaucasus, Georgia. According to reports of historical gold mining, it cannot be ruled out that natural gold has been contaminated by anthropogenic gold amalgam. In alluvial placer deposits in Scotland, gold was detected with mercury contents of 2.6–8 wt.% as well (Leake et al. 1998). Mercury is a more common minor component of gold than copper. Dilabio et al. (1988) report on worldwide anthropogenic contaminants in soap deposits. They observed gold spheres of the order of 0.x mm, which indicate local refining of gold in historical periods. The Hg-rich gold flakes from Georgia have a characteristic porous microstructure (Fig. 3.19), as is typical for amalgamated gold. It arises when, after a conscious enrichment

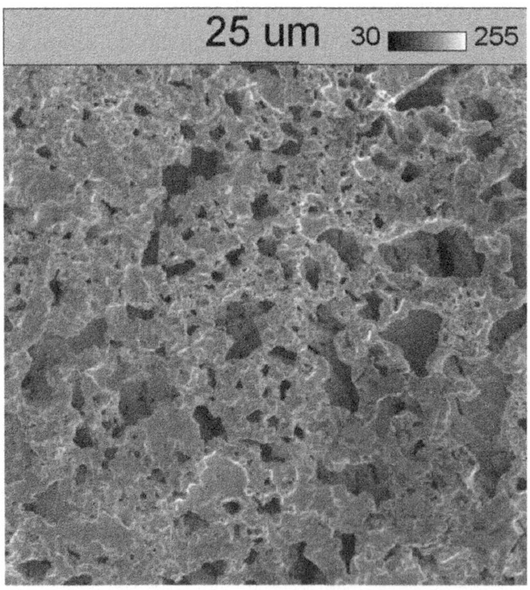

Fig. 3.19 Gold district of Sakdrisi-Bolnisi (Transcaucasus, Georgia). Grains of gold–mercury amalgam from the creek Kazruleti. The gold–silver–mercury alloy is 80–82 wt.% Au, 3.5–5.5 wt.% Ag and 14–15 wt.% Hg (semiquantitative analyses by SEM). The porous structure points to remains of technical gold processing by amalgamation. However, in this region, mercury-bearing mineralisation is known also. From Hauptmann et al. (2010)

of gold with mercury by humans, it is expelled again by a heating process. A clear distinction between anthropogenic and natural, Hg-rich gold flakes is not always possible. Due to interference between $^{204,202}Hg$ and ^{204}Pb in the measurement of lead isotope ratios, a representation of the ^{204}Pb cannot be given for Hg levels in gold.

Gold-(Silver-)Tellurides

In primary ore deposits, gold is often associated with tellurium. A typical mineral is calaverite. The tellurides krennerite, petzite, and sylvanite are typical minerals of the subvolcanic gold veins of Roşia Montană in the so-called Golden Quadrangle in modern Romania (Boyle 1979; Hauptmann et al. 1995).

3.4.7.2 Gold Deposits

According to the mode of occurrence, gold can be classified practically into two major categories:

1. Gold occurring in veins embedded in solid rock is called primary or mountain gold (German *Berggold).*
2. Gold from eluvial or alluvial placer deposits.

Before presenting these two groupings with some examples, a map of the Near and Middle East, showing the surprising density of gold occurrences in some areas of this part of Eurasia, is given (Fig. 3.20). It is reminiscent of the title of a publication by Borg (2014): "Gold is where you find it".

Archaic Orogene Gold-Quartz Veins

Many of the large number of gold deposits of this type occur in the context of Precambrian basements, where they are bound to Greenstone Belts or banded iron formations (BIF) (Groves et al. 2003). Here, the so-called "old gold-quartz veins" of the Nubian Desert occur (Klemm and Klemm 2013).

A most famous example of this type of ore deposit are the gold deposits in ancient Egypt and Nubia, which gave rise to the enormous richness of the Pharaos. The Precambrian basement of the Arabian-Nubian shield hosts around

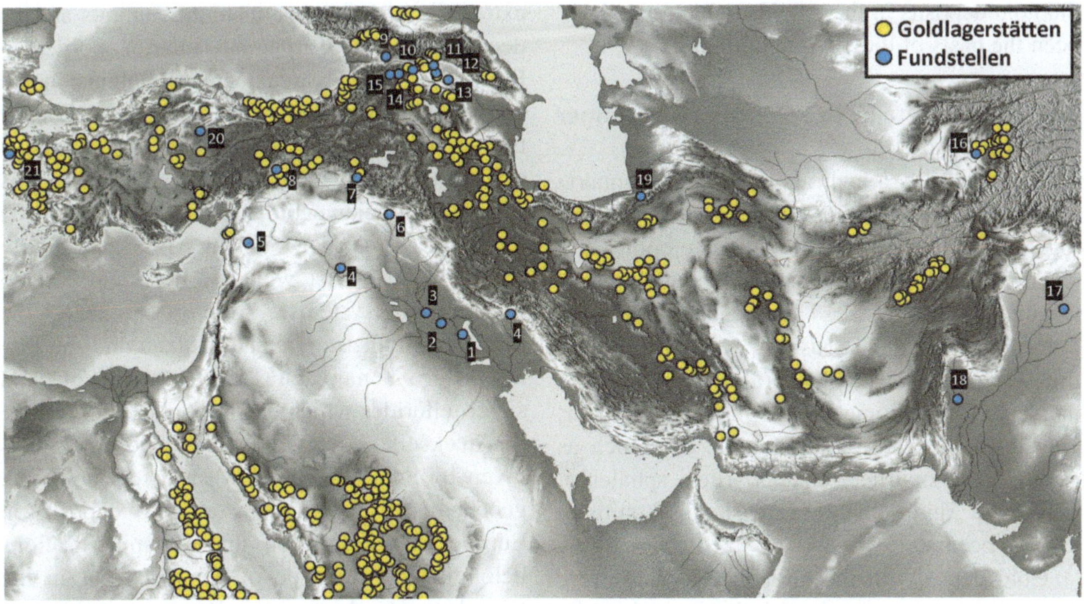

Fig. 3.20 Map of gold deposits and occurrences and localities of early gold artefacts in the Near and Middle East. Note the densities of gold occurrences in the Tethyan Eurasian metallogenic belt (TEMB), in the Precambrian rocks of Egypt/Nubia and in the Saudi Arabian Peninsula. Localities of early gold artefacts are: Mesopotamia: (1) Ur, (2) Nippur, (3) Kish, (4) Mari, (5) Ebla, (6) Tepe Gawra, (7) Basur Höyük, (8) Arslantepe; Transcaucasia: (9) Satschere, (10) Martkopi, (11) Ananauri, (12) Znori, (13) Hasansu, (14) Bedeni, (15) Tsalka; other regions: (16) Shortugai, (17) Harappa, (18) Mohenjo-Daro, (19) Tepe Hissar, (20) Alaca Höyük, (21) Troia. From Jansen (2019).

250 gold production sites. The gold deposits are confined to quartz-mineralised shear zones in ophiolite sequences (serpentinites, basaltic rocks), which were formed between 700 and 800 million years ago. Other gold deposits are associated with rhyolitic-andesitic volcanic rocks and with older gneisses. Due to the geological age, all the rocks were affected by metamorphosis to various extent. Finely dispersed gold particles were mobilised and agglomerated to larger particles. Gold occurs together with sulphide minerals such as pyrite, pyrrhotite, chalcopyrite, galena, sphalerite and their supergene alteration products. In the context of all these mineralisations, alluvial gold placer deposits are formed, which may contain platinum group minerals. It is estimated that during 6000 years of gold production in Egypt's history, a maximum of 6000 kg of gold were mined and produced (Klemm et al. 2001).

Archaic orogenetic gold-quartz veins frequently occur in the old shields of Canada and Australia. Also, the so-called *Tauerngold* of the Alps in Central Europe, which was mined at least since the Middle Ages, falls into this category, as well as the gold of the Kolar district in Mysore (India).

The richest gold province in central Eurasia occurs in the late Paleozoic fold and thrust belts of the Tien Shan, a component of the giant Altaid orogenic collage (Yakubchuk et al. 2002). Extending through Uzbekistan, Tajikistan, Kyrgyzstan, and continuing into western China, the Tien Shan hosts an array of world-class gold deposits. These include late paleozoic orogenic type gold deposits, such as Muruntau and Kumtor. Two of the world's ten biggest gold resources. Muruntau (Uzbekistan) was also a prehistoric mine for the semi-gemstone turquoise (Plotinskaya et al. 2006). In addition, there are the gold deposits in the Altaids, which may have been the basis of the Scythian gold.

Gold in Polymetallic Ore Deposits of Younger Geological Ages

Gangue-shaped hydrothermal gold-quartz mineralisations, which are closely related to geologically young subvolcanic intrusive bodies and dacitic and andesitic volcanics, are widespread throughout the world. The ores are bound to volcanic fissures and crushing zones and sometimes brecciated. Their mineral content includes native gold with sometimes high levels of silver, embedded in quartz and sulphidic ores such as pyrite, arsenopyrite, chalcopyrite, sphalerite, polybasite and others. Occasionally stibnite (Sb_2S_3) occurs.

Gold deposits in the Greater Caucasus were formed in the context of the collision of the Arabian with the Eurasian plate in Early to Middle Jurassic (Kekelia et al. 2008). More than 60 gold-quartz veins and placers were found in the Svanetia district. Gold-bearing veins are often associated with As- and Sb-rich ores. Especially placers were exploited at least during the last decades. Whether these placers were the source of the famous myth of the Golden Vlies, which led to the gold-richness of the Colchis, is not impossible but remains unproven for the time being.

Generally, in hydrothermal alteration processes in adjacent rocks of gold-bearing quartz veins, reactions lead to the formation of pyrite. Subsequent oxidation of pyrite produces an earthy, limestone-rich mass in which gold is embedded as mustard gold. This type of gold was mined in the old days (see Sect. 3.5).

As gold is losing much of its silver content during oxidation, the purity of gold is increasing. Furthermore, since chalcopyrite originally present in the deposit is converted to malachite during oxidation, which partially incorporates iron in the secondary minerals, the mined gold will be relatively pure but will take up low levels of copper and iron (Klemm et al. 2009).

Also typical are Au–Ag tellurides. This type of deposit includes many historic, often extraordinarily rich ore districts. Famous are the deposits in the Carpathian interior border in Romania (Transylvanian Ore Mountains in the Golden Quadrangle), which were mined since the early Bronze Age (Makkay 1995; Cauuet et al. 2003) and which could have been the basis for the Thracian gold of the Greeks. Rosia Montana (Apuseni Mountains) is a particularly rich deposit district with porphyry and gangue mineralised systems. Amongst others, the minerals krennerite, petzite, stutzite and the Pb–Sb–Au–Te–S nagyagite are important gold and silver tellurides which can be found here.

An excellent overview of this deposit district is given by Schmiderer (2008), who discusses the geochemistry and geology of gold deposits in Europe. He also mentions that gold minerlisations belonging to this type of deposit occur in the Slovakian Ore Mountains. However, gold from these deposits are free of tellurides.

Volcanic Massive Sulphide Deposits (VMS)

According to Herzig et al. (1991) in Cyprus, sulphidic VMS deposits with distinct oxidation zones and gossans contain gold regularly. This is also applicable for the submarine sulphide mineralisations ("black smokers") of the mid-ocean ridges in the Atlantic Ocean. The genesis of the deposits on Cyprus is ascribed to such "black smokers". On Cyprus, gold as well as silver can be found in the *devil's mud* formations (see Sect. 3.5).

Further examples of such gold mineralisations, which played a role in ancient times, are the sulphide deposits of Madneuli and David Garedji and the surroundings in southern Georgia (Transcaucasus). According to Gialli (2013), these deposits are genetically associated with the VMS as well. Studies by Stöllner (unpublished data Bochum) suggest old, likely prehistoric mining in the gossan of the Madneuli deposit. In the immediate vicinity of Madneuli in Sakdrisi, prehistoric gold mining in mineralised vein systems in volcanic rocks was investigated by T. Stöllner and I. Gambaschidze (Stöllner et al. 2010; Stöllner and Gambaschidze 2011) (cf. also Hauptmann et al. 2010). Here, iron-rich gold-

Fig. 3.21 Sakdrisi, Transcaucasus (Georgia). Gold bearing vein from the supergene altered zone of the close-by located VMS-deposit of Madneuli. Note the enrichment of reddish hematite in the middle of the vein. The host rock is heavily silicified and shows boxwork with ovoid cells. These probably implicate pyrite as the most likely original sulphide (cf. Taylor 2011). The yellowish-reddish colour of the sample is due to the formation of limonite and jarosite. Photo: A. Hauptmann, Deutsches Bergbau-Museum Bochum

quartz hematite veins appear (Fig. 3.21). Here the oldest mining of gold (fourth millennium) was found.

So far, however, there is no further clear evidence of early gold extraction near the surface of VMS deposits. Copper extraction from the fourth millennium BC has been proven in the deposit of Murgul in the north-east of Anatolia (Lutz et al. 1994). Gold, however, was not searched for and not found. At Rio Tinto's VMS deposit, gold is currently being mined at the bottom of the oxidation zone, but no evidence of ancient gold mining has been found. A similar picture emerges in the VMS deposit Tawi Raki in the Samail Ophiolite Complex in Oman. At least in early Islamic times, the deposit was used as a rich resource for copper, as indicated by the huge slag heaps of Tawi Raki (Hauptmann 1985; Weisgerber 1991). Modern prospecting also proves gold contents in mineralisation (unknown reference), whose use for the early period is not proven.

In a general sense, the question as to whether gold has been mined in old VMS deposits still remains. It is possible, that simply not enough is known about the field evidence and analytical data of early gold mining to be sure.

Placer Deposits

Gold deposits in sediments are widespread throughout the world, namely those that (1) accumulate by chemical weathering in residual placers, and (2) eluvial or alluvial-fluviatile placers. In most cases, (pre-)historically relevant gold deposits are associated with placer deposits.

Residual Deposits and Supergene Enrichment

The genesis of these deposits is based on chemical weathering and displacement of light rock components. Gold in surface-near areas of lateritic weathering zones are often bound to rocks of greenstone belts. Such deposits are formed in areas of good drainage, allowing the oxidation to penetrate deep into the ground and induce a wide-ranging solution of (earth-) alkalis. Silver concentrations in primary gold is dissolved (see below), but also gold is partially solved as Au^{+} and Au^{3+}, and reprecipitated as a much purer authigenic gold. There is an accumulation of relatively pure gold (Fig. 3.22), where extraordinarily large nuggets and idiomorphic gold crystals are formed. Gold nuggets often contain inclusions of surrounding silicate components. They can, therefore, differ chemically from primary gold. Kamenov et al. (2013) proved this based on the basis of lead isotope analyses. At the same time, they have made the important observation that, in some gold deposits, both primary and placer gold, lead may be included as a tracer element suitable for lead isotope analysis. It is therefore possible to measure lead isotope compositions for provenance studies (see also Sect. 11.2).

Solution and precipitation of gold is a function of redox-conditions (Eh) and of the acidity/alkalinity in soils and rocks (pH). Solution and precipitation is controlled, e.g. by organic matters, humic acids, Cl-, B-, J-concentrations, and by cyanides, but it is especially controlled by redox-conditions, as shown in Fig. 3.23. The precipitation of gold may occur in the colloid state with Fe-, and Mn-colloides (see also:

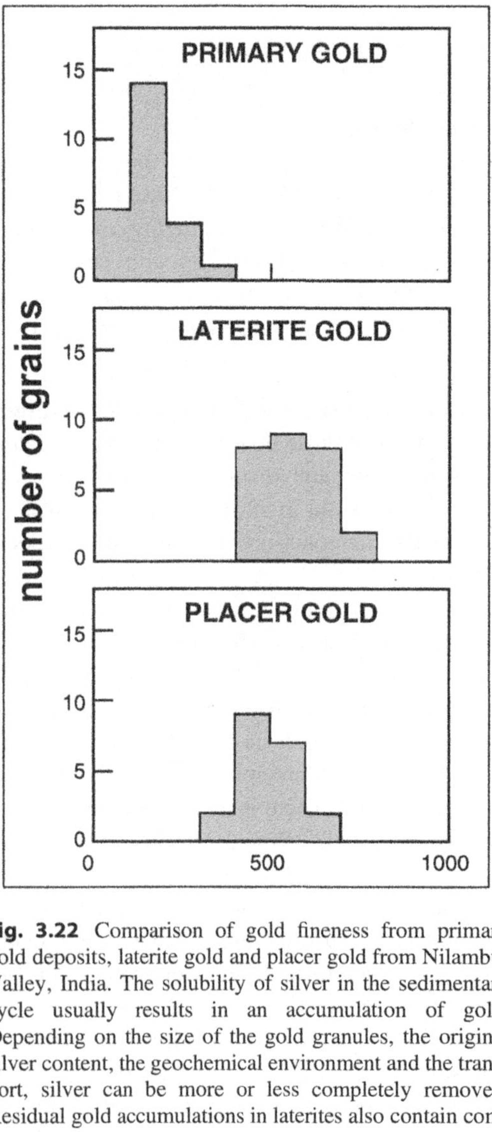

Fig. 3.22 Comparison of gold fineness from primary gold deposits, laterite gold and placer gold from Nilambur Valley, India. The solubility of silver in the sedimentary cycle usually results in an accumulation of gold. Depending on the size of the gold granules, the original silver content, the geochemical environment and the transport, silver can be more or less completely removed. Residual gold accumulations in laterites also contain comparatively pure gold. Modified after Santosh et al. (1992)

Cassius'scher Goldpurpur in ruby red glass making).

Gold-bearing laterites occur, e.g. at Ada Tepe in the Eastern Rhodopes in Thracia, where gold was mined since the Bronze Age (Popov and Jockenhövel 2010). Gold may also be enriched in the context of ophiolites in listwaenites, as they are common, e.g. in Turkey (Bayburtoğlu and Yıldırım 2008). But their potential in ancient times is not known.

Extensive gold-bearing laterites occur in many countries in western Africa. In Ghana, Mali, Burkina Faso and other countries of the African continent, the use of abundant gold resources has recently led to the legendary riches of local tribes and outstanding artisanal craftsmanship. Artisanal mining, processing and processing techniques, as well as diverse social and political aspects, are well documented by Werthmann (2000, 2003a, b, 2006) and Borg (2014) and represent one of the best ethnographic studies on contemporary gold mining (see Sect. 9.4).

Eluvial and Alluvial Placers

The weathering and erosion of primary gold deposits cause gold to pass into soils, rubble, or into streams and rivers, where it undergoes many physical and chemical changes. The grain size of native gold in placers and intensely weathered rocks is mostly <500 μm. Rarely, nuggets are formed which are >5 mm or larger.

Eluvial placer, also known as residual placer, is formed by a short sediment transport downhill from a primary deposit, with gold being transported from its near-surface area (Fig. 3.24). In the process, a slight enrichment of gold and its weathering-resistant components form in a matrix of loose detritus immediately above the genuine rocks. They are usually located not far from the primary sources of raw materials. The material is eventually taken up by streams and rivers and, due to their high density, further enriched to fluvial or alluvial placers. Both types can merge into one another. Very often, river courses change due to geological and geomorphological activities, creating fossil alluvial terraces that bear gold.

One of the best known and most impressive examples of predominantly eluvial placer deposits are the gold deposits in northwestern Spain, notably Las Medulas. In Roman times, rich fossil placers were mined here in gigantic extents.

Herodotus (VI, 46–47) reports on the gold mines exploited in antiquity on the island of Thasos, where "the hill was completely ransacked by the searching for gold". In the mines, gold was

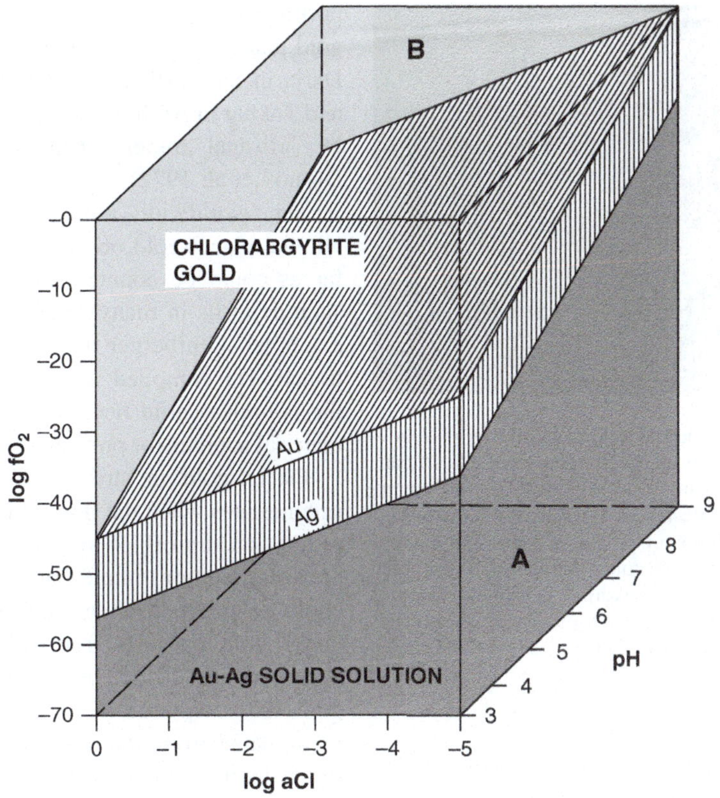

Fig. 3.23 Simplified Au-Ag-Cl-O phase diagram. It shows the field of stability of a gold–silver solid solution in field A. In this field the oxygen concentration of the water (log fO_2) is low. De-alloying gold–silver grains connected with the formation of silver-chloride (AgCl) takes place with increasing oxygen concentrations (B) (Krupp and Weiser 1992; Möller 1995a, b). The remaining gold will have a spongy texture as shown in Figs. 3.19 and 3.27. Changing concentrations of oxygen in the system is connected with electrochemical reactions

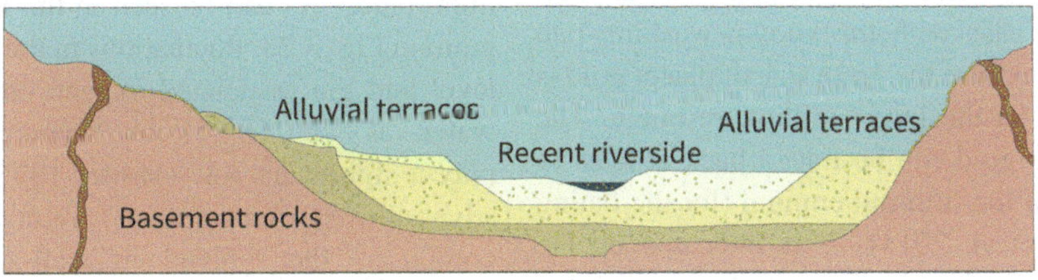

Fig. 3.24 Model of eluvial and alluvial enrichments of gold from primary gold-bearing veins. For example, eluvial accumulations occur below the outcrop of a gold-quartz vein. These in turn accumulate in the fluviatile region to gold-bearing alluvial sediments. If the course of a river changes, fossil (gold-bearing) river terraces arise. Such a situation would also be typical for, e.g. tin placers with cassiterite. Modified after Pohl (2005)

accumulated in karstic sediments in dolomites, with gold embedded in a calcareous limonite matrix. Following the genesis, the gold mineralisations are part of eluvial accumulations of the precious metal in karst cavities (Vavelidis et al. 1988). Vavelidis et al. (1985) suggest a comparable genesis for the gold mineralisations on the island of Siphnos.

Fluviatile or alluvial placers occur in active waters as well as in older river terraces. They are located at morphologically extraordinary places where flow velocity changes, i.e. behind

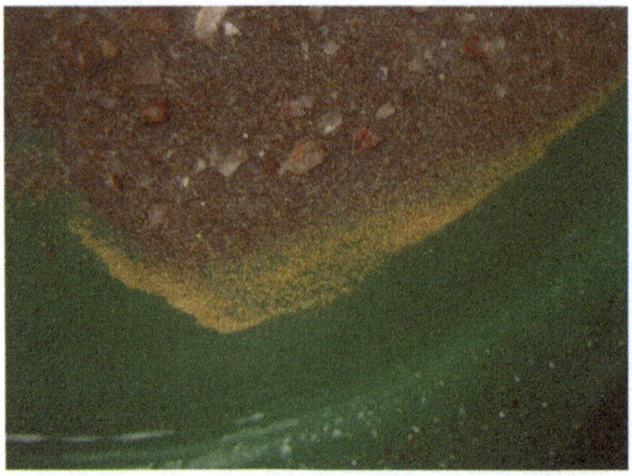

Fig. 3.25 The extraction of gold from alluvial placers is in the simplest case by the well-known gold wash pan, which was and still is in use worldwide. There are countless comparisons of artisanal gold washers, but there is no archaeological evidence of washing pans. The picture shows the accumulation of gold from the sand of a river terrace in the area of Sakdrisi in South Georgia (Transcaucasus) during a research project in 2009 of the Deutsches Bergbau-Museum Bochum. Photo: Alexander Omiadze, Tbilisi

crossbars or boulders, at the inner edge of ox bows, on gravel and sandbanks. They extend over several hundred metres, sometimes there are placers that extend over kilometres.

There are recent accumulations of gold on the surface which led to gold rushes in the historic past and do so even still today (Holliday 1999). Gold production from alluvial placers is done in the simplest case with the known washing pan (Fig. 3.25), because the grain size of placer gold is usually only in the lower millimetre range. The use of placer gold can be traced back at least to the middle of the fifth millennium (for the dating see Leusch et al. 2014), as is indicated by the decoration of the great dish of the chalcolithic necropolis of Varna (grave 4) with almost dust-sized gold granules (Éluère and Raub 1991). The same is likely true for another bowl from the same grave (Lichardus 1988).

Just as common are fossil placer deposits from various geological eras. The most famous fossil placer deposit is Witwatersrand in South Africa, which formed in the Precambrian. Tertiary fossil eluvial gold placers, which were of great importance in history, are those of Las Medulas in the northwest of the Iberian Peninsula. The fossil gold placers of Samti at the terraces of the Amur Darja in the northeast provinces of Badakhshan and Takhar in Afghanistan are exploited presently by artisanal miners (Chmyriov et al. 1973; Dronov et al. 1972).

The worldwide number of recent eluvial-alluvial placer-gold occurrences is enormous. As far as could be counted, there are hundreds of such deposits in many countries. To pick out an example, Lehrberger (1995) should be mentioned. He compiled a very good overview of the countless gold occurrences as potential raw material sources for prehistoric objects in Europe.

Placer gold in streams and rivers typically accumulates when the flow changes. An example of this is a slip-off slope of a river bend. Or there are accumulations in mechanical traps, such as in churn holes, where a mechanical separation of the dense gold particles from less dense minerals occurs (Borg 2014). In the sedimentary cycle, gold is transported both mechanically and is precipitated from solutions. This can sometimes result in the formation of huge nuggets (Sect. 3.7). They can reach weights of up to several dozens of kilograms, as evidenced by finds from Russia, North and South America, Africa and Australia.

The natural solubility of gold itself and the frequent gold–silver alloys are of considerable importance. As can be seen in the Pourbaix diagram in Fig. 3.23, fluctuations in the groundwater level and the associated oxygen contents of the water, as well as (electro-)chemical processes play a significant role (Möller 1995a, b). Gold–silver alloys (solid solutions) are not stable under conditions that prevail in well-aerated rivers waters (Krupp and Weiser 1992). The silver content in gold will react to form silver chloride and will simultaneously be leached out due to its aqueous solubility. The remaining detrital gold grains are heterogeneous, with a silver-rich core and a silver-depleted, spongey rim (Fig. 3.26). Such a honeycomb texture is also observed in gold–silver artefacts, indicating that such corrosion phenomena may not only be of natural origin but have also been artificially produced by a deliberately applied gold–silver parting. Meeks

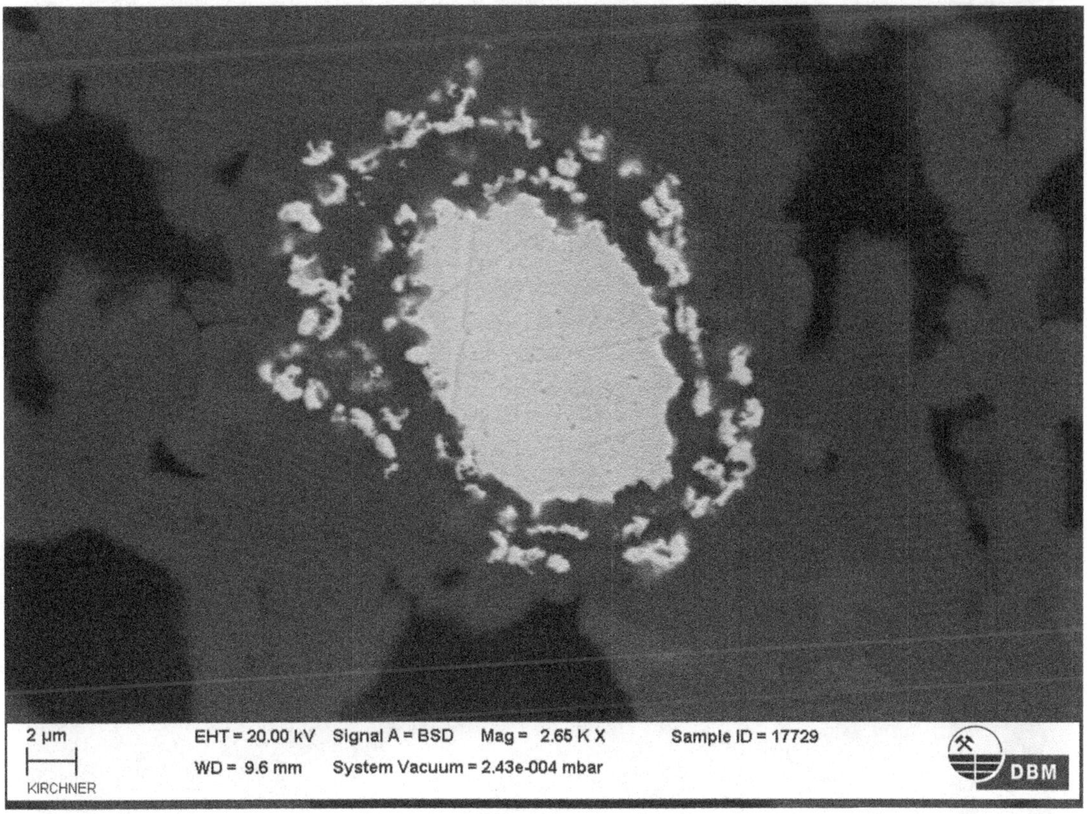

Fig. 3.26 GEO-28/75b, Gold mine Sakdrisi (Georgia). Microporosity of a gold nugget. The gold–silver grain, embedded in secondary limonitic ore, shows marginal dissolution of silver. In 3D-version it would provide a porous shaped surface. SEM picture, backscattered mode. From Jansen (2019)

(2000) and Geçkinli et al. (2000) observed this in gold finds from Sardis at the Pactolus (western Anatolia) from the first half of the first millennium BC (see also Sect. 6.3).

In places where gold solubility is slightly depressed by low oxygen fugacity, e.g. in bottom sediments, gold may precipitate from solutions to form authigenic pure gold. This mechanism may be more significant in the formation of alluvial gold than the classical mechanism of gravity enrichment of detrital, allogenic gold grains. Hence, gold enrichment and refinement is a by-product of the natural weathering and biocatalysation process (Lawrence and Griffin 1994). More thorough gold dissolution and reprecipitation is also associated with permanent water-saturated saline environment. The formation of authigenic gold might also be supported in golden artefacts buried for some millennia in a saline environment such as in the sabkhas of lower Mesopotamia (Fig. 3.27).

In general, gold from placer deposits, as well as laterite gold, has lower silver contents than primary deposits (Fig. 3.22). Hence, placer gold shows a higher fineness than gold extracted from a primary hydrothermal vein. Primary gold is typically only 70–90 wt.% pure, while placer gold may contain between 95 and 99 wt.% gold.

However, the gold content of silver is not necessarily an indicator of its origin from primary deposits. The gold artefacts of the royal tombs of Ur, Mesopotamia contain from 10 to 50 wt.% Ag (Hauptmann et al. 2015). Yet it is not clearly verifiable, that such artefacts can always be addressed as deliberately produced alloys. The vast majority of the artefacts, as evidenced by the inclusion of platinum group minerals, are derived from sedimentary gold deposits.

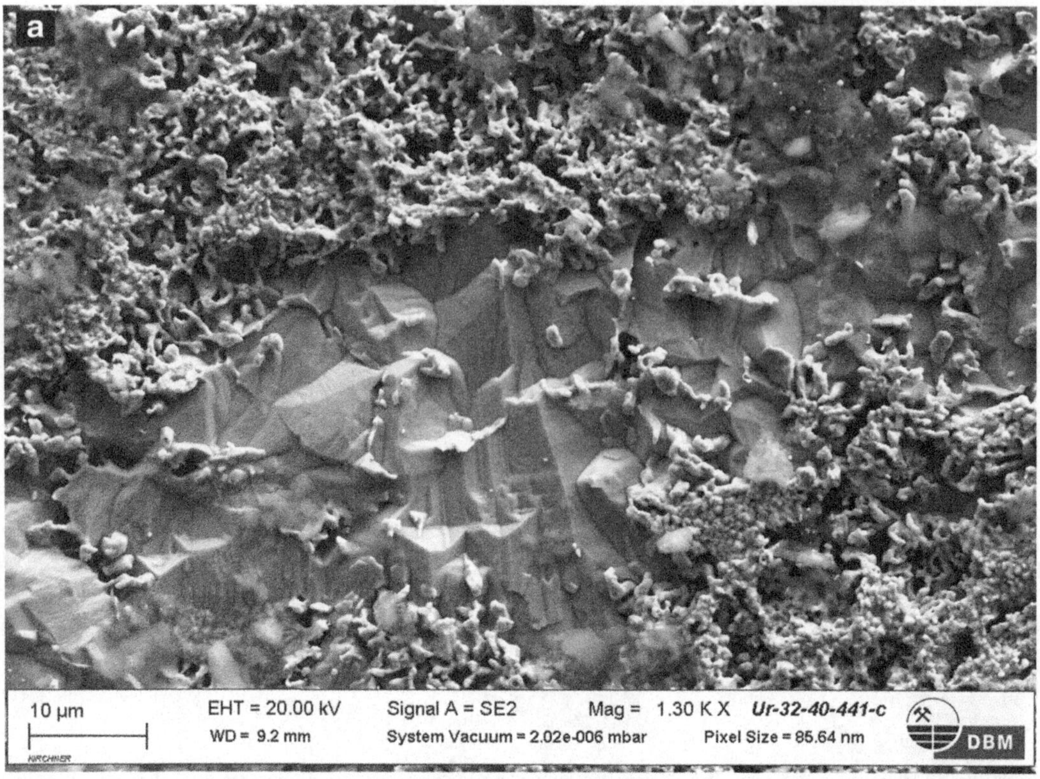

Fig. 3.27 Ur, Mesopotamia, Royal Tomb PG 1847/R, sample 32-40-441c. Cellular honeycomb texture in a hair ribbon. The spongy surface layers can result from silver dissolution from an Au–Ag alloy formation of budlike protrusions (**a**) and networks (**b**) of authigenic gold (see circles) on the surface of gold–silver crystals. Not in every case a distinction between authigenic gold and dissolution of silver textures are distinguishable. SEM picture, secondary electron mode. From: Hauptmann et al. (2018)

Fig. 3.28 Mamulo, Sakdrisi-Bolnisi-District, Georgia (Transcaucasus), sample (GEO-29/1). Placer gold from a close-by located primary gold source in granitic rocks. The gold grains are mixed with heavy minerals such as magnetite (black), secondary copper minerals (blue) and quartz (white). These are partly included in the gold grains. The concentrate indicates that also placer gold contaminated in such a way could have been produced and smelted in prehistoric times. Copper minerals could have led to increased concentrations in gold. Size of the picture c. 5 mm. Photo: Dirk Kirchner, Deutsches Bergbau-Museum Bochum

Frequent associated minerals of placer gold, e.g. magnetite, ilmenite, zircon, titanite, cassiterite, arsenides, tellurides or copper minerals (Fig. 3.28), are incorporated into the metal in placer deposits as inclusions mechanically, but especially by the chemical precipitation of gold (authigenic gold) (Hauptmann et al. 1995, 2010). The same applies to minerals of the platinum group elements (PGE: osmium, iridium, ruthenium, platinum) and their trace elements (Ga, Sb, Bi), but also to clay minerals or quartz. During subsequent processing of the gold, in (s)melting processes, corresponding elements are incorporated into the gold and can be used as local tracers. The close association of gold with iron minerals runs like a red thread through the geology of gold. The widespread iron, be it in the form of pyrite, limonite, hematite or magnetite, may therefore be a practical geological guide to gold (Schade 2001). Above all however, minerals of the platinum group elements (PGM), which often occur as inclusions in gold artefacts were melted from placer gold. In the Old World, prehistoric artefacts with PGM are common (Junk and Pernicka 2003; Jansen et al. 2016). Known such artefacts are, e.g. from the Pactolus (Ramage and Craddock 2000), from Egypt (Ogden 1976), they were found in Ur, Mesopotamia (Jansen et al. 2016) and in Takht-i Kuwad at the Oxus river (border between Afghanistan and Tajikistan; Mongiatti et al. 2010). However, famous are especially the many so-called gold–platinum artefacts from Colombia and Ecuador (Scott 2012a, b).

3.4.8 Platinum-Group Minerals

3.4.8.1 Platinum-Group Elements (PGE)

The chemically closely related noble metals of the VIII subgroup of the Periodic Table, summarised under the abbreviation PGE, comprise the elements iridium (Ir), osmium (Os), palladium (Pd), platinum (Pt), rhodium (Rh) and ruthenium (Ru). Their average content in the earth's crust is very low, well below 1 ppm (Pt: 0.01 ppm; Rh: 0.001 ppm; Ir: 0.0002 ppm; Ru: 0.001 ppm; Os: 0.0002 ppm. PGE can be subdivided into two subgroups: (1) the iridium–platinum group elements (IPGEs: Os, Ir, Ru) and (2) the palladium–platinum group elements (PPGEs: Rh, Pt, Pd) (see also Economou-Eliopoulos 2010).

The PGE plus gold are often collectively called noble elements. These metals have that name for two reasons. First, they are rare. Second, they are unreactive and are stable in their metallic state. Their rarity is in part a consequence of their siderophilic character. They have similar geochemical behaviour and tend to occur together in the geological environment in (ultra-) basic magmatic rocks in ophiolites and carbonatites. In addition, they show a slight chalcophile character. This is expressed by chemical compounds with As, Sb, Bi, S, Se and Te (Pohl 2005). PGE also occur in alluvial sediments. From the economical point of view, the precious PGE are sold today at higher prices even than gold.

3.4.8.2 Platinum-Group Minerals (PGM)

PGM exist in a native state or are alloyed together. More often they may contain Pd, Ir, Os, Rh, Ru, Cu, Au and Fe in the range of 10–30 wt.% PGE (Falbe and Regitz 1996). Platinum, for example, is usually alloyed with Fe (< 10–25 wt.% Fe). In addition, there are compounds of Pt, Pd and Ru with sulphur, arsenic and antimony.

Some important minerals of the platinum-group elements are listed in Table 3.5.

Native platinum is rare in nature, but it occurs. It is a silvery-white, very ductile metal, which is usually found in irregular or rolled granules in the fine grain range up to nuggets with weights of several kilograms. Iron-rich varieties are called iron platinum. Other PGEs can be added as alloys as well. Nuggets of native platinum were collected, e.g. from the most famous placer deposits at Nizhni Tagil, Urals. In 1843, a 12-pound platinum nugget was found at this locality. The second largest platinum nugget (6.2 kg) from this locality is being kept in the collections of the Natural History Museum in Vienna (Weiser 2004). This nugget ranks as one of the largest nuggets ever found. It also consists of a PtFe alloy.

Sulphidic PGMs, such as the steel-grey cooperite, which can contain up to 17 wt.% Pd (Cabri 2002) and PGM arsenides, with the most well-known of them being sperrylite, occur in basic and (ultra-)basic magmatic rocks such as the Bushveld-Complex and especially in placer deposits related to such rocks. Sperrylite may contain small amounts of Rh, Fe, Pd and Sb.

Main PGMs are composed of iridium–osmium–ruthenium alloys and, to a lesser amount, of Pt–Fe alloys. The composition of the PGMs are located generally quite well to the miscibility gap of rutheniridosmine according to the revised nomenclature of Harris and Cabri (1991) from many localities all over the world such as Burma, Papua New Guinea, the Urals, Russia, and Alberta (western Canada) (overview in Economou-Eliopoulos 2010). This trend is also confirmed by the investigation of PGM inclusions in prehistoric gold artefacts by Jansen et al. (2016) (see below).

PGE antimonides, e.g. geversite (40 wt.% Pt), and the rare stibiopalladinit (Pd_3Sb, rhombic, 70 wt.% Pt) were detected in nickel pyrrhotite (Cu-Ni-) deposits in ultramafic rocks, where PtFe alloys were observed, too.

3.4.8.3 Platinum Deposits

In their review of PGE deposits, Economou-Eliopoulos (2010) reports that a primary enrichment of platinum metals to mineable deposits worldwide is bound to chromite deposits in ultrabasic rocks (peridotites, harzburgites, dunites, but also basalts). This is in marked contrast to granitic rocks, where PGE levels are much lower.

One of the most prominent and economically important deposits today is the chromite-horizon in the Bushveld-Complex in South Africa. It is a magmatic intrusive body. It was formed by a so-called hot spot of the earth's mantle. Due to the layered bands of chromite, it belongs to the class of layered intrusions type deposits. The PGE

Table 3.5 Minerals of the platinum-group elements (PGE)

Mineral name	Chemical formula	Concentration of PGE (wt.%)
Platin	(Pt,Fe)	79–96 Pt
Sperrylite	$PtAs_2$	57 Pt
Cooperite	PtS	86 Pt
Stibiopalladinite	Pd_3Sb	70 Pt
Osmiridium	(Os,Ir)	100 Os,Ir
Braggite	(Pt,Pd,Ni)S	62.6 Pt, 17.1 Pd
Laurite	RuS_2	61.2 Ru
Geversite	$PtSb_2$	44.5 Pt
Stibiopalladinite	Pd_3Sb	68.6 Pd

Given are concentrations of platinum (Pt), osmium (Os), iridium (Ir), palladium (Pd), rhodium (Rh) and ruthenium (Ru) in weight-%. Data after Mineralienatlas Lexikon (Internet)

contents are in the range of 3.5–19 ppm. Primary PGM mineralisation is associated with chromite lenses. Mineralogically, the PGE are bound to chromium spinel ($Fe^{2+}Cr_2O_4$), pentlandite ($(Ni, Fe)_9S_8$), chalcopyrite ($CuFeS_2$), pyrrhotite (FeS), nickelite (NiAs), and gersdorffite (NiAsS). Similar deposits can be found in the Great Dyke in Zimbabwe, and in the Stillwater Complex (USA). For the regions of Old World, such layered intrusions are without meaning.

Chromite deposits are also formed in the context of ophiolites, that is, with rock sequences of oceanic lithosphere, which were obducted as covering slabs over existing rock complexes in the Paleozoic, but very often also in the Mesozoic. Again, chromites are bound to (serpendinised) ultrabasic rocks. This type of deposit includes occurrences of chromites of economic significance in the Middle Urals and Alaska (Alaskan-Ural type deposits). PGM in ophiolitic complexes frequently occur in countries of the Tethyan Eurasian metallogenic belt (TEMB).

The most famous and largest ophiolite complexes in the Old World are located in Oman and Cyprus. In Oman, some of the many copper mineralisations are connected with (Cr-) bearing peridotites, and in Cyprus, there is only one single occurrence embedded in ultrabasic rocks. However, numerous smaller ophiolites occur on the Balkan Peninsula, in Anatolia and in Iran, where their chromite occurrences or ultrabasic rocks are discussed as PGM sources. In the Philippines and Papua New Guinea, PGM-placer is found in the context of ophiolites as well (Weiser and Bachmann 1999).

Detrital platinum minerals are now recognised in geosciences as reliable tracers for exogenous PGE/PGM rearrangements (Borg 1996, 1997; Weiser and Bachmann 1999, u.a.). In some cases, they can gain local economic concentrations in (sub) recent or fossil river sediments and form so-called platinum placers. The discovery of the Bushveld's rich platinum deposits in South Africa, almost 80 years ago, was based on a placer prospecting that noted the silver-shining grains of "platinum".

Of importance for archaeometallurgy among the PGM/PGE deposits are primarily the accumulations of PGM in placer deposits bound to ophiolitic rock sequences, especially if they occur together with alluvial gold accumulations.

The most well-known placers of this type are located in Nizhny Tagil in the Middle Urals Mountains in Russia. There, even today PGMs are mostly mined from secondary placers that can be several tens, even more than a hundred kilometres long (Borg 1996; Weiser 2004). In such placers, however, gold accumulations also occur in geographical proximity, leading to the formation of mixed gold PGM placer deposits. Such PGM gold placers are common and were described by Cabri et al. (1996) in many countries. In economic geology, other PGM-placers are known in Russia Far East as well (Gornostayev et al. 2000; Shcheka et al. 2004).

PGM-gold placers close to five large ophiolitic complexes with chromite deposits were found in western Turkey, all of which contain PGMs with high Ir, Os and Ru contents. In these ophiolitic terranes, some mesothermal gold veins occur (Elevatorski 1996), so that placer deposits with both gold and PGMs can be expected in these regions. In the Hindukush region on the border of Afghanistan to Tajikistan and Pakistan, ophiolite complexes have become known, too (Benham et al. 2010; Jansen et al. 2016). Given the abundance of ophiolites in the TEMB mountain ranges from the Balkans to the Himalayas, it would not be surprising if further PGM gold placers were to be found. In such a case, one might cautiously question the unique selling point of the Pactalos.

Harris and Cabri (1991) and Weiser and Bachmann (1999) describe the compositions of PGM in the quaternary system Ru-Os-Ir-Pt. They note that many of the analysed samples are from a supraregional context in the osmium–iridium domain and plot only a limited number of Pt-rich compositions in the diagram (Fig. 3.29).

3.4.8.4 Inclusions of Platinum-Group Minerals in Gold Artefacts

The high specific weight and corrosion resistance of PGMs are prerequisites for their occurrence in placer deposits along with gold in the

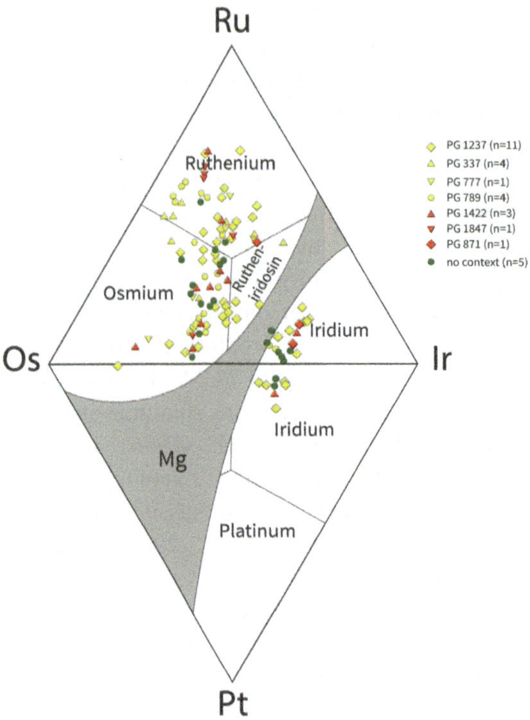

Fig. 3.29 Composition of platin-group-element alloys in the quaternary system Ru-Os-Ir-Pt as constructed by Harris and Cabri (1991). The miscibility gaps in both ternary systems are based on a combination of analyses of PGE alloys and experimental data available. Plotted are compositions of PGM inclusions in gold artefacts from the Early Dynastic Royal cemetery of Ur, Mesopotamia, as analysed by Jansen et al. (2016). These plots are in the same range of compositions as natural PGM from various localities in Russia, California, Oregon and Brazil (Weiser (2004). Note the overwhelming number of alloys in the field of osmium and iridium, while Pt-rich compositions are rare. Abbreviation: Mg = Miscibility gap

sedimentary cycle. Once such placers have been used to recover gold, PGMs can be mixed with the gold and incorporated into the gold during subsequent melting processes. They can then be found in artefacts. However, gold flakes or nuggets can also contain PGM micro-inclusions in the sedimentary range, which are incorporated by the formation of authigenic gold.

PGM micro-inclusions have been observed in many gold artefacts not only from the Old World but also from South America. Scott (2012a, b) describes gold objects covered by a layer of biphasic platinum-gold "alloys" or PGM micro-inclusions from pre-Columbian cultures (fifteenth to sixteenth century AD) of Colombia, Ecuador, and Peru. The goldsmiths at that time were able to weld the PGM granules to the gold flakes, producing a (macroscopically) relatively homogeneous, bright, malleable metal alloy. This could not be melted again and had a whitish-silver-like colour. A platinum content of approximately 15% already results in a light grey colour.

Technologically, PGM micro-inclusions survive reprocessing and metallurgical processes. They are not completely dissolved in gold due to their significantly higher melting points compared to gold and gold–silver alloys. Platinum has a very high melting point at 1772 °C. Inclusions of PGM in gold artefacts occur again and again, such as in Lydian gold coins from Pactolus in western Anatolia (Craddock 2000d), but also in Egypt and in numerous gold artefacts from Bronze Age localities in Mesopotamia (Meeks and Tite 1980; Hauptmann et al. 2018) or in Celtic gold coins (Junk and Pernicka 2003).

In archaeometallurgy of the Old World, gold placers with PGMs are especially known from the Pactolus river in western Turkey. This region is intensively archaeologically explored by Ramage and Craddock (2000). Here, at Sardis, the metallurgical separation of gold and silver (parting) was proved for the middle of the first millennium BC, under King Croesus. PGM of the compositions described were suggested previously to be characteristic for gold from the Pactolus River, Western Anatolia (Young 1972). Moesta (1986) even concluded that an Ir-rich Pt–Ir alloy would be indicative of gold of the Pactolus river. Because comparable PGM inclusions in many gold artefacts were analysed from the Royal tombs of Ur, Mesopotamia (Figs. 3.30 and 3.31), this led to the wrong conclusion that these would originate from the Pactolus. Jansen et al. (2016, 2018a, b) were able to correct these errors by osmium isotopy (see Sect. 11.3.4).

Provenance Studies

The statement of Meeks and Tite (1980), that PGM micro-inclusions in gold artefacts are indicators of the origin of gold from placer deposits, is probably just as correct as their

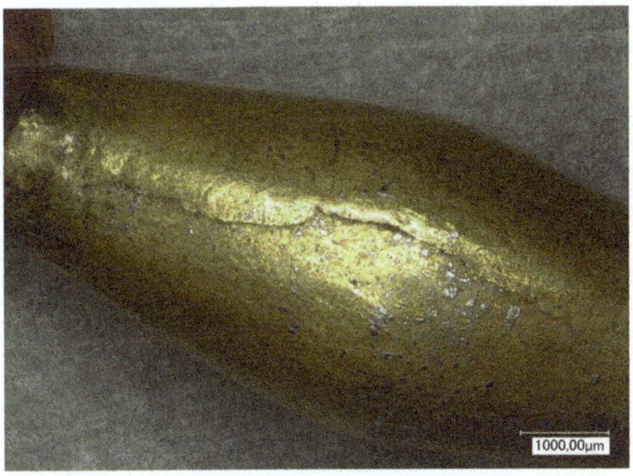

Fig. 3.30 Royal tombs of Ur, Mesopotamia, sample 30-12-561. Golden bead from a string consisting also of beads from lapis lazuli and carnelian from PG 1116. Note the dozens of grey platinum–group minerals in the gold bead. Scale 1000 mm. Digital micrograph. Photo: M. Jansen, Penn Museum, Philadelphia

suggestion that such inclusions are hardly usable as tracers for an exact origin of gold deposits. It was explained that PGM micro-inclusions are geologically restricted to ore deposits in ultrabasic rock complexes. If such inclusions are identified, then one can assume a provenance in a regional environment of ultrabasites. However, such a rough differentiation is often insufficient for archaeometallurgy, especially in regions where, e.g. ophiolites occur more frequently.

Therefore, studies have been performed to characterise PGM inclusions geochemically by their contents of the platinum group elements platinum, osmium, iridium, ruthenium (Meeks and Tite 1980). It was later found, that the compositions of PGM inclusions match the frequency distribution of Os-Ir-Pt-Ru as expected, as was described by Cabri et al. (1996). The authors

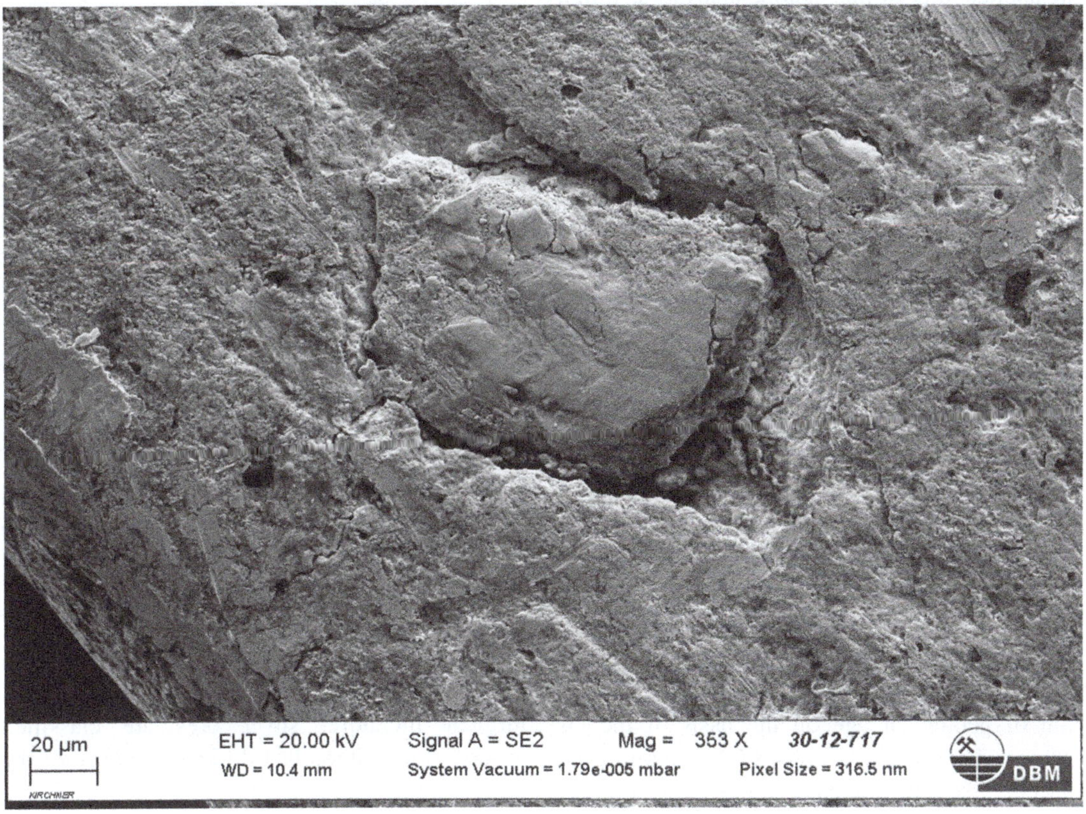

Fig. 3.31 Royal tombs of Ur, Mesopotamia, sample 30-12-717. Angular-shaped, idiomorphic platinum group mineral (PGM) inclusion in a golden hair ribbon from the Royal tombs of Ur. Due to its high melting point, the PGM was not liquified during the processing of the gold object. SEM picture, secondary electron detection mode, scale 20 mm. From Jansen et al. (2016)

analysed more than a hundred PGM inclusions in gold artefacts from the collections of the British Museum, London, by means of energy-dispersive X-ray fluorescence spectrometry on a scanning electron microscope, and found predominantly compositions with iridium, osmium and ruthenium, while platinum itself occurred very subordinately. They note that the compositional variations of PGM inclusions within a single object would be so large, that the inclusions would be too inaccurate, and therefore unusable as tracers for the identification of gold deposits in an archaeological context. Therefore, no further considerations on the origin of the gold are made in the publication.

The same difficulties are described by Junk and Pernicka (2003) in the investigation of PGM inclusions in Celtic gold coins. For the first time, they analysed $^{187}Os/^{188}Os$-isotopic ratios, to test the applicability of this isotope analysis as a tool for provenance studies of gold. Osmium isotope analyses have already been used in the geosciences with the aim of detecting the evolution of (ultra-)basic rocks. Bowles et al. (2000) point out that PGM inclusions differ from surrounding mantle rocks and can be altered by sedimentary influences. In summary, Junk and Pernicka (2003) came to the conclusion that even the measurement of osmium isotopes would not exceed the possibilities of geochemical analysis and would therefore not be suitable as a tracer either.

Jansen et al. (2016) are critical of this statement. They show possible origins of gold based on osmium isotope analyses of PGM inclusions in gold artefacts from the royal tombs of Ur in Mesopotamia, either from placers close to Paleozoic ophiolites like Samti of Samti (Takhar) in Northern Afghanistan near the border to Tadjikistan or Zarshouran (Western Azerbaijan) in Iran. The basis of these findings is the evaluation of geological age on the basis of the Os isotope. They exclude other PGM-leading gold placers, like the old Neoproterozoic ophiolites from the Eastern Desert type (750–800 Ma), Egypt, and the geological young Mesozoic ophiolites from the Samail complex (96 Ma) in Oman. Thus, in combination with other geochemical and of mineralogical tracers, and in combination with archaeological features (close-by occurrence of lapis lazuli in north-east Afganistan), the Os isotope ratio is a very valuable source for provenance studies.

3.4.9 Silver

Silver (Ag) is a chemical element typically associated with sulphur, i.e. it is chalcophile. The concentration of silver in the earth's crust is approximately 0.079 ppm. Thus, it is situated at the 69th place in the frequency table of the elements. Silver has the highest geochemical abundance among the noble metals (20% more than gold) in the earth crust. Today, silver is mainly produced in Bolivia, Mexico, the USA, Canada and China. In 2011, the world consumption of silver for industrial purposes and for the production of coins was about 170,000 t.

The most popular variety of silver today is sterling silver. This is an alloy with 92.5 wt.% silver, the rest is usually copper.

Hydrothermal cementation reduces silver from its compounds to the elemental state and, at the same time, oxidises less noble metals present in solutions to higher valence states (e.g. $Fe^{2+} \rightarrow Fe^{3+}$). This mostly happens in the cementation zone of deposits.

3.4.9.1 Silver Minerals

To date, 167 ores are known that contain silver in different concentrations. Among those are a number of argentiferous lead ores, which is of significance, as by that the existence and exploitation of silver this is mineralogically as well as metallurgically closely linked to that of lead (Bachmann 1993). For this reason, in archaeological–metallurgical literature, silver ores are classified according to criteria that differ from those of the mineralogical classification.

Bachmann (1993) classified three major categories of ores used for silver production since the Bronze Age:

(a) Lead-based ore. He refers to galena, cerussite and anglesite which are the most common ores used for silver production.
(b) Fahlore. Fahlore such as tetrahedrite and tennantite can contain substantial amounts of silver and were utilised for silver production at least since the medieval period. They are low in lead.
(c) Jarosite ore. Jarosites are complex hydroxylsulphates of iron and are typically found in the gossan of sulphidic ore deposit. They may contain Na, K, Pb, Ag, Cu. Generally, they are low in lead.

This classification shows that it is a simplification of modern "technical mineralogy" to speak only of the silver content in galena and to monotonously cite it as the basis for the extraction of the precious metal in ancient times. The same applies to cerussite and jarosite. Any approach must, therefore, be cautious.

In the following, some important silver ores are discussed, which approximate the classification of Bachmann (1993). Table 3.6 lists some fairly common silver minerals and silver-carrying minerals. It becomes clear that many of them are also geochemically bound to As, Sb and Bi in addition to the already mentioned lead.

Table 3.6 Some of the important silver ores and silver-containing ores and their silver contents

Mineral name	Chemical formula	Ag (wt.%)
Native silver	$Ag_{(Au,Cu,Hg,As,Sb,Bi)}$	up to 100% Ag
Argentite	Ag_2S	87
Cerargyrite	AgCl	75
Dyskrasite	Ag_3Sb	73
Proustite	Ag_3AsS_3	65
Pyrargyrite	Ag_3SbS_3	60
Stephanite	Ag_5SbS_4	68
Strohmeyerite	CuAgS	52–53
Polybasite	$(Ag,Cu)_{16}Sb_2S_{11}$	up to 74
Chlorargyrite	AgCl	75
Schapbachite	$AgBiS_2$	28
Miargyrite	$AgSbS_2$	37
Argentiferous ores		
Tetrahedrite	$(Cu,Ag)_3Sb_{3.25}$	2–4, up to 20
Tennantite	$Cu_3AsS_{3.25}$	ditto
Freibergite	$Cu_3(Ag,Sb)S_{3.25}$	33–47.8
Argentojarosite	$AgFe^{3}$ $^{+}{}_{3}(SO_4)_2(OH)_6$	19
Galena	PbS	0.01–1
Cerussite	$PbCO_3$	traces
Anglesite	$PbSO_4$	traces

Data after Doelter (1912–1932), Ramdohr (1975), Hartmann (1982)

Native Silver and Intermetallic Compounds

In many ore deposits native silver occurs in the transition zone between oxidation zone and secondary enrichment (cementation) zone in iron-rich context (limonite), and is formed by the following reaction (Pohl 2005):

$$Fe^{2+} + Ag^{+}{}_{\text{Lösung}} \leftrightarrow Fe^{3+} + Ag^{0}.$$

Overall, silver is rare. If at all, it crystallises in hydrothermal veins in moss-, wire- or plate-like structures. Idiomorphic silver *crystals* are rare. However, native silver can occur in some ore districts, such as in Herrerias and neighbouring deposits in south-eastern Spain, where intensive exploitation of native silver and silver chlorides is believed to have occurred in the El Argar culture (approximately 2200–1550 BC) (Murillo-Barroso et al. 2014). In Herrerias, native silver occurs with mercury (Hg) concentrations in the percentage range.

Native silver is often embedded in the gangue (calcite, fluorite, barite, quartz) or in the host rock. However, as known from the Kongsberg deposit (Norway), it can also form arm-thick wires and massive lots, up to several dozens of kilograms in weight. Kongsbergite is a silver variety with approximately 5 wt.% Hg. In the sedimentary cycle, silver is rather rare compared to gold. Overall, in (pre-)historic times, native silver does not have nearly the same importance as native gold does.

In nature, gold and silver form a continuous solid solution, from silver through aurian silver to argentian gold including electrum and gold (cf. Sect. 3.4.7). In contrast, silver almost does not alloy with copper at room temperature. Gold–silver alloys with about 30–45 wt.% silver are called electrum, which has a silvery-white colour. Aurian silver occurs almost exclusively in gold deposits, and especially in gold-quartz veins, lodes etc. (Wedepohl 1968–1978). In the Mansfeld *Kupferschiefer*, one of the largest ore deposits in Central Europe, native silver often occurs in thin sheets, in addition to copper. The *Kupferschiefer* was probably already mined in the Bronze Age (Otto and Witter 1952).

Silver Halogenides

Compounds of silver with chloride, bromide, iodide occur in the gossan of lead–zinc–copper deposits in arid and humid climates, or close to coastlines of the sea, affected by the halogenides of the air. These ores have the superscription silver horn ores. Silver does not form carbonates.

Cerargyrite (AgCl) (also called chlorargyrite or horn silver) belongs to this group. It was an overall important silver ore in the gossan of ore deposits but seems to be totally exploited today (Machatschki 1953; Ramdohr 1975). It forms greenish-grey crusts and is not easily recognisable in the field due to its inconspicuous blackish-grey colour and has certainly often been overlooked. Because the mineral is often mixed with powdery limonite and decomposed galena and forms earthy masses, it bears a resemblance to bird excrement, which earned it the nickname *Gänsekötigerz* (goose shit ore) in the mining linguistic. Contemporary authors also report *Buttermilcherz* (buttermilk ore), which could be drawn from passage cavities with trowels as a grey, liquid pulp. Mineralogically, these ores are a mixture of clay minerals and chlorargyrite (Wilke 1952). For the historical time, the extraction of silver-bearing rich ores has been rather underestimated.

Very similar to cerargyrite, but much rarer are bromargyrites (AgBr) and iodoargyrites (AgJ). Courcier (2014) analysed bromargyrites as corrosion crusts on silver artefacts from Azerbaijan, apparently formed by soil deposition.

Rich Sulphidic Silver Ores

The abundance of rich sulphidic silver ores among minerals is relatively high.

Argentite (Ag_2S) occurs mainly in the upper part of the cementation zone, in part still in the oxidation zone of sulphidic lead–silver deposits. Sometimes silver shoots of argentite associated with native silver, strohmeyerite, fahlores and sulphosalts occur in large masses (see below). Most famous is a block of several tons found in 1477 AD in the ancient mine of St. Georg at Schneeberg, Ore Mountains (Wagenbreth et al. 1990; see also below). Significant amounts of argentite along with native silver have been reported from several deposits in Mexico (Betechtin 1968; Ramdohr 1975). Otherwise, argentite as well as cerargyrite is difficult to differentiate as silver ore in the geological context, especially if it is intergrown with limonite and is present in tiny inclusions. This is often seen in Mississippi Valley-type deposits.

Strohmeyerite (CuAgS), the old German *Kupfersilberglanz*, is found in small quantities in numerous localities. This is true also for other Ag-rich ores with copper, whose silver content, according to the analyses of Domeyko, can be around 3–29 wt.%. In the classical old mining city of Freiberg (Ore Mountains, Germany), CuAg ores with almost 20 wt.% Ag have been found, similarly so at snake mountain in Siberia, as well as in Rudelstadt in Silesia. Presumably, the ore was formerly present in large quantities, because Von Naumann (1852) writes that it is used as a rich silver and copper ore.

Silver is also the main component of many sulphosalts (pyrargyrite and proustite, the German *Rotgültigerze*, stephanite, miargyrite etc., see Table 3.6). Thus, the ore schapbachite can lead to significant Bi-contents in silver ores (Ramdohr 1975).

Fahlores

Fahlores (tennantite, tetrahedrite with 2–4 wt.% Ag, in freibergite also up to 48 wt.% Ag; for an overview of fahlores see Ixer and Pattrick 2003) are mineralogically not only copper minerals (Fig. 3.4 1c), but also extremely important silver carriers in numerous deposits. Silver contents are higher in Cu-Sb fahlores (tetrahedrite) than in Cu-As fahlores (tennantite). Particularly silver-rich fahlores accumulate in the area of the cementation zone (Schneiderhöhn 1962). Fahlores are mostly, but not always, closely associated with galenite (PbS). They form microscopic inclusions in galenite. They are the reason that galenite is often referred to as a silver-rich ore.

Famous exceptions to the fahlore–galenite assemblage are almost monomineral fahlore deposits in the Inn Valley (Austria) (Krismer et al. 2011b), in Cabriéres (Ambert 1999) and in the Slovak Ore Mountains (Schalk 1998). In prehistoric times, however, they were always mined as copper ores. The produced fahlore–copper varieties have been classified as separate metal groups in the project "Studies on the Beginnings of Metallurgy" (*"Studien zu den Anfängen der Metallurgie"*) and subsequent projects (Krause 2003). It was not until the fifteenth century AD that these deposits were mined on a larger scale for silver (Westermann 1986).

Argentiferous Galenite

Galenite occurs in many deposits along with chalcopyrite, pyrite and sphalerite. It was and is today, one of the economically most important silver ores, since larger bulk samples can often contain 0.3 wt.%, sometimes >1 wt.% of silver. In fact, however, only up to 0.1 wt.% Ag can be solved in the crystal structure at room temperature. The majority of silver is embedded in galena as silver-rich fahlores, proustite and pyrargyrite, argentite, sterling silver and others. These minerals are generally obvious only as microscopic inclusions. All in all, argentiferous galenite may therefore be an economic important silver ore. However, if one resorts to older mineralogical literature, one notices that, in addition to galena, ore shoots may contain silver minerals that were used in ancient times (see below).

Argentiferous Cerussite

Cerussite ($PbCO_3$) is the most widespread oxidation product of galena. If galena had inclusions of (sulphidic) silver ores, these were taken over as the finest impregnations in the actually light-transparent cerussite. Cerussite itself is crystallographically pure (Keim et al. 2016), it may contain tiny inclusions of cerargyrite and native silver, of covelline, and more rarely of chalcocite. Close to the secondary enrichment zone, these ores may be enriched in a large scale in the ore deposit, but already visible in the hand specimen (Ramdohr 1975). Such inclusions are visible even by the naked eye. It therefore could have been utilised in antiquity as a silver ore in a similar way as galena. For a long time, surface-near cerussite was probably underestimated as a mineral of major economic importance in "ancient" ore deposits. Cerussite is a common mineral, e.g. in the ancient lead–silver–zinc–copper deposit of Laurion, Attica (Greece) (Skarpelis and Argyraki 2009), and Iglesias, Sardinia (Italy) (Boni et al. 2003). Numerous lead–zinc–silver deposits, showing mixed sulphidic-oxidic mineralisations, are also known from the Iranian Plateau. A most famous example is the mine of Nakhlak (Vatandoust 2004; see Fig. 3.41a, b). Most of these are embedded in carbonate rocks. Meyers (2003) assumes that silver was extracted first from surface-near silver ores, then from silver-bearing secondary lead ores and finally from silver containing primary galena (Fig. 3.32).

Jarosite

Other ore minerals that have perhaps been underestimated as important and widespread silver carriers so far are the water-free sulphates of the series jarosite–beudantite $(K,Na,Fe)_3(OH)_6(SO_4)_2)$, whose varieties argentojarosite, argentian plumbojarosite and argentian beudantite can contain significant amounts of silver and lead. The minerals of the jarosite–beudantite series are typical, widespread minerals from the lower part of gossans of pyritic ore deposits and, therefore, they are intensively intergrown with (hydr-)oxidic iron ores such as limonite/goethite and hematite (Fig. 3.33). In addition, they all strongly resemble the ochre to reddish coloured mixtures of these iron minerals from the oxidation zone and can therefore easily be confused with these. So they hardly look like silver ores, and the smelting of these ores could be connected with amounts of slags. Craddock (2013a) wonders why the Phoenicians, who ran intensive mining on silver-bearing jarosites in Rio Tinto, first recognised them as an economically interesting source of raw materials, where neither before nor later did the Spaniards and the British would have exploited them.

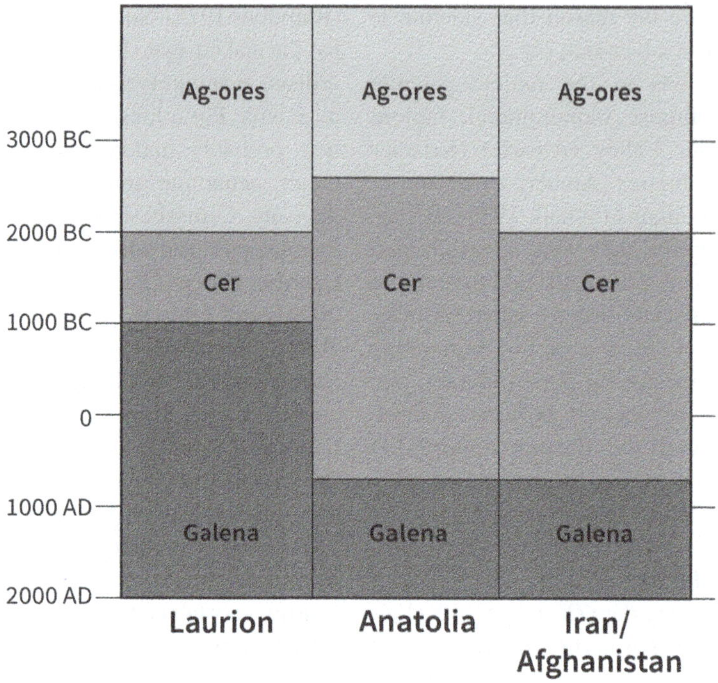

Fig. 3.32 Schematic representation of the development of silver producing technology in the ancient world, in Laurion and similar sites in the Aegean, Anatolia, Iran and Afghanistan. The model suggests possible changes from a technology that first uses silver-rich ores, followed by silver-bearing oxidised lead ores, and finally primary sulphidic lead ores. Note the problems of (non-)silver contents in cerussite as described in Sect. 3.4.10. Dates are approximate and based upon elemental compositions of dated silver artefacts. Modified after Meyers (2003)

Fig. 3.33 Baranco Jaroso, Cuevas Almanzora, Almeria (Spain). Silver-bearing jarosite intergrown with a reddish mixture of limonite/goethite and hematite and quartz. The tiny vitreous dull lustered crystals are hard to distinguish from the iron ores (size of the section c. 10 cm). The piece from the Museo Geominero (IGME) in Madrid might be basically comparable to jarosite from Rio Tinto. Photo by A. Hauptmann, Deutsches Bergbau-Museum Bochum

There are examples of ancient silver mining from Ag-rich jarosites, especially from Corta Lago, Rio Tinto (Spain), while astonishingly enough, there has not been any evidence for the exploitation of jarosites occurring in the *devil's mud* on Cyprus.

3.4.9.2 Silver Deposits

A modern assessment of metal contents in ore deposits is generally of limited use and misleading for the understanding of (pre-) historically mineral deposits. It has already been pointed out that this is based primarily on very different sampling (cf. Sect. 3.9). Although silver-rich ore minerals are nowadays mostly observed only on the micro-scale, there are significant written references to massive occurrences, which were deliberately used in ancient times, for example by Albertus Magnus (thirteenth century AD) and Georg Agricola (sixteenth century AD) for Germany, Al-Hamdani (tenth century AD) for Yemen.

Silver appears in a variety of different deposit types, of which epigenetic-hydrothermal ore deposits are far more prevalent. Characteristic gangue minerals of silver ores are rather carbonic minerals and baryte ($BaSO_4$) and less frequently quartz. This is in contrast to gold deposits. A simplified summary of the genetic organisation (Pohl 2005) in relation to "old" deposits is as follows:

1. Hydrothermal veins associated with volcanic rocks and subvolcanic intrusions.
2. Massive sulphide deposits containing silver as a minor component. Metasomatic deposits with Ag-rich lead–zinc ores.

Hydrothermal Veins

The first includes many polymetallic deposits (Pb-Zn-Ag and Ag-Co-Ni-As-Bi-U veins) in Central Europe, e.g. in the Saxon-Bohemian Ore Mountains, in the Harz Mountains, in the Siegerland area, Black Forest, in the Slovak Ore Mountains, and those rather monomineralic deposits located in the Schwazer district in the Inn Valley (Tyrol). The country rocks of these mineralisations are mostly quartz and/or calcite. The main ores are galenite, sphalerite, fahlores and other Ag-rich sulphosalts. Pohl (2005) assigns them to the largest known silver producers.

It must be emphasised at this point that hydrothermal veins are only conditionally present as monotonous-homogeneous mineralisations. The example Freiberg in Saxony shows how hydrothermal veins can be fanned out by tectonic activities (Fig. 3.1). In a similar fashion, this can be determined in the Siegerland (Germany). Reconstructions of mining activities on the locality of Altenberg, where, in the late Middle Ages, numerous shafts were sunk within the settlement alone, suggest a similar fragmentation and fanning of the veins as in Freiberg (Dahm et al. 1998; Kirnbauer and Hucko 2011; Zeiler et al. 2018).

Bartels (2014) maintains that mining in the early medieval Harz mountains (Germany) proves that miners were looking for the so-called bonanzas or high-grade silver ore shoots that could have a silver content in the range of 10 wt.%: only in the seventeenth century AD did silver production based on galena become economically viable in the Harz mountains. He takes up the problem of silver-rich ores for medieval silver mining in Central Europe, where he refers to, e.g. Leonhard (1860), who shows in his *"Grundzüge der Mineralogie"* (Fundamentals of Mineralogy), that the precious metal silver comes from rich silver ores, as listed in Table 3.6. On native silver, he writes about pound-to-centner occurrences in veins in the Ore Mountains, Bohemia, Hungary and France, about the discovery of inch-wide druses, which were filled with hair and wire silver. He emphasises how rich in silver the mines of Mexico, Peru, and Chile are. The most impressive silver ore finds ever dates to the year 1477 AD. It was retrieved from the "Sankt Georg" mine in Schneeberg in the Ore Mountains. Samples of ore proved the famous ore deposit as a massive deposit of mainly argentite and sterling silver. However, the exact calculations of the size and weight of this find vary between 2 t and 20 t (Thalheim 1990; Wagenbreth et al. 1990; Bartels 2014).

Ore shoots are also known from the hydrothermal veins in the Siegerland area. In 1784 AD, in the pit "Plätze", or as it was later called,

"Heinrichsegen" near Littfeld in Siegerland (Germany), pyrargyrite with silver contents of 50 wt.% were found. This was discussed in more detail by Kirnbauer and Hucko (2011).

In view of the close association of silver and lead ore in depository science and metallurgy, Bartels (2014) correctly points out that the fact that galena was mined, and the conclusion that it contained some silver, should by no means automatically and for any given era indicate that silver was the main product, with lead as a by-product.

The largest silver ore body of this type in the world is the Cerro Rico de Potosi (Bolivia), which carries as its most important silver ores tetrahedrite, pyrargyrite and miargyrite and has so far supplied between 30,000 t and 60,000 t of silver. It was one of the richest silver sources for the Spaniards since the sixteenth century AD. In addition, besides cassiterite, stannite, pyrite and chalcopyrite, arsenopyrites occur as well (Bartos 2000). Mining for silver probably started around 1000 AD. During the colonial era silver was extracted by amalgamation. Large amounts of mercury came from Huancavelica in Peru.

Massive Sulphide Deposits

As far as the second point of Pohl (2005) is concerned: the type of massive sulphide deposits with silver as by-product (Pb-Zn-Ag ores) in carbonates includes the prototype of the Mississippi Valley-type (MVT) deposits. Such deposits are located in Anatolia and on the Iranian Plateau, as well as the ore district of Laurion in Attica (Greece) and on the island of Sardinia (Italy). In China, such occurrences are known as well. Apart from Keban and Laurion, these deposits are discussed in Sect. 3.4.10 as part of the lead–zinc–silver deposits.

Two genetically distinct polymetallic deposits, which have played prominent roles as silver sources in the history of early metallurgy in the Old World, are presented here in more detail. These are Laurion and Rio Tinto. In addition, a problem with silver-containing jarosites, affecting the deposits of Cyprus, will be mentioned.

Laurion

In the eastern Mediterranean some carbonate-hosted lead–zinc–silver deposits are known, namely those on the islands of Thasos (Gialoglu et al. 1988) and Sifnos (Vavelidis et al. 1985; see Sect. 3.4.10). At least on Sifnos mining has been demonstrated since the Early Bronze Age (Weisgerber 1985), and silver from Sifnos was also traded—albeit to a limited extent—as well as Laurion's in the third millennium (Gale and Stos-Gale 1981a). On Thasos, the production of lead and silver seems to start in the Iron Age (Hauptmann et al. 1988). So far at least, there is no evidence for previous activities. In addition, there are numerous, in part significant lead–zinc–silver deposits in southern Anatolia (Yener et al. 1991).

One of the most important silver deposits in the Old World was that of Laurion in Attica (Greece). The deposit of Laurion became known as a silver source of antiquity. It is said to have been mined for galena as a main source for silver for Athens in the first millennium BC. According to various authors (Ardaillon 1867; Marinos and Petrascheck 1956; Conophagos 1980; Lohmann 2005), it is a prime example of ancient experience and knowledge of mining prospection and exploration, processing and metallurgy of silver, although many details of these activities have not been clarified yet. Repeatedly and monotonously, the silver-bearer of Laurion is mentioned as argentiferous galenite. This raises the question as to what extent this is indeed compatible with the observations set out by Bartels (2014), and if there were not at least occasional silver-rich ore shoots, that are hardly detectable today.

Laurion is an ore district with an extension of approximately 80 km^2 (Fig. 3.34a). It is part of the Hellenides Domain formed along the Africa-Eurasia convergent plate boundary active for the last 80 million years (Scheffer 2017). Geologically speaking, this ore district is not a pure silver deposit, but a polymetallic one (Pb-Zn-Cu-Ag-Fe) (Marinos 1982). The carbonate-hosted ore deposit (karstic cavities in limestone, dolomite with amounts of baryte) is of hydrothermal-metasomatic origin and is located at three interfaces between marbles, schists and shale sequences (Fig. 3.34b). Sulphides were much more abundant along the third contact mineralisations than in higher stratigraphic levels.

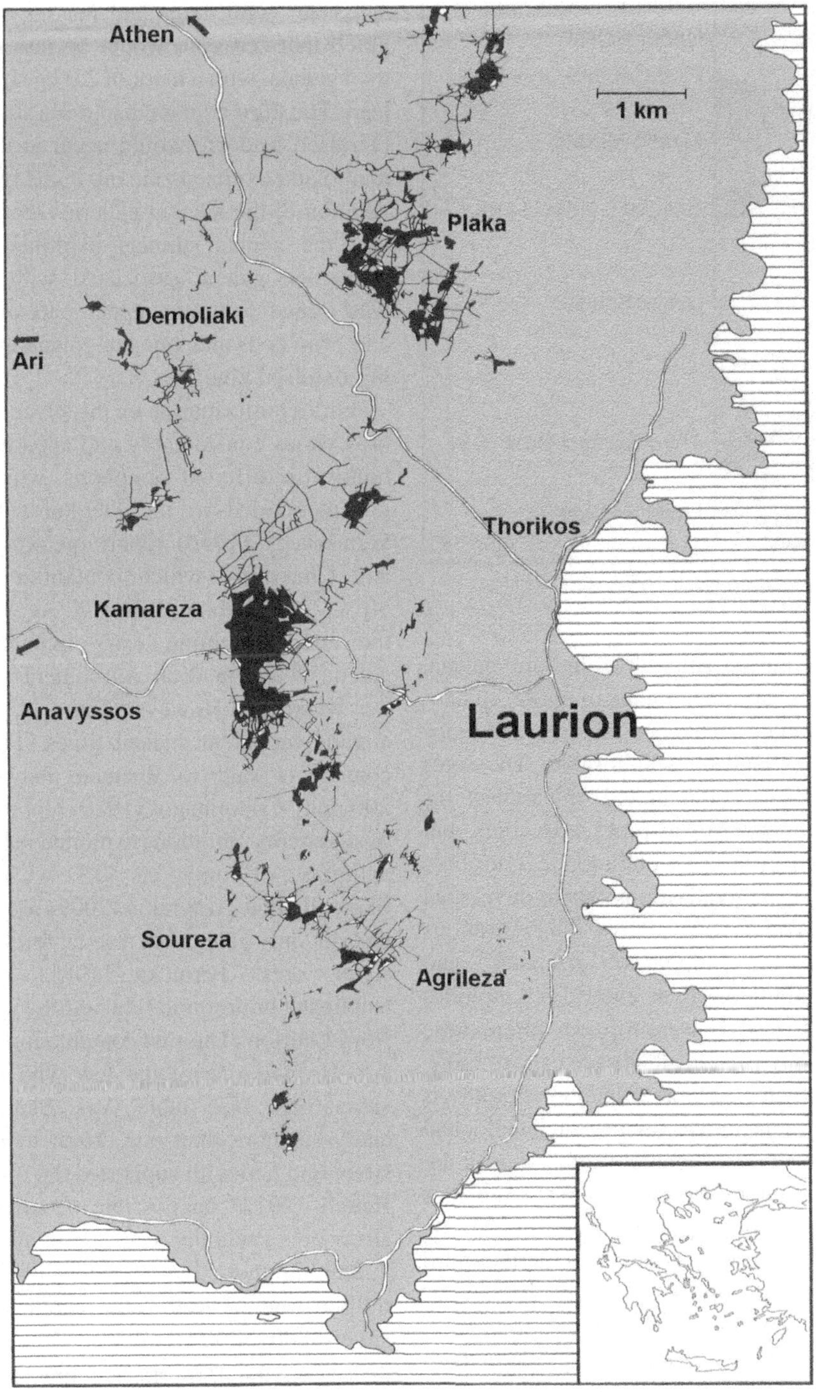

Fig. 3.34 (**a**) Map of the ancient mining district of Laurion (after Conophagos 1980). Note the large extent of ca. 80 km^2 of this mining/ore district. Three hundred and fifty ancient shafts and galleries are estimated by Pernicka (1987). (**b**) Laurion, Greece. Schematic stratigraphic section illustrating the Mesozoic sequences of marbles and schists and the main ore bodies. C3a represents the lenses and veins of ore in the lower few metres of the Lower Schist; C3b is the main manto, and C3c is the so-called stockwork mineralisation. From Kepper (2005)

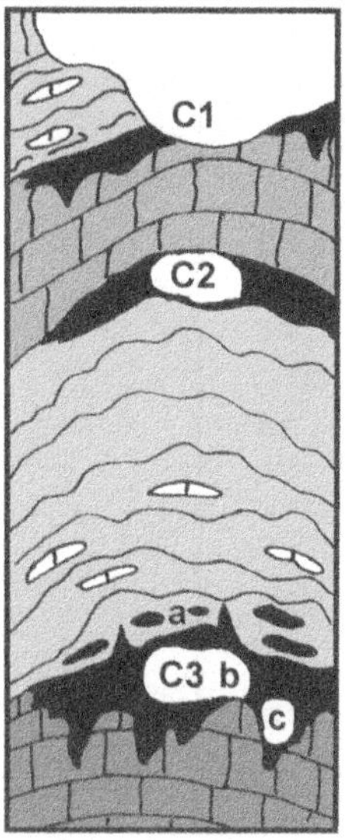

Fig. 3.34 (continued)

The predominant sulphidic ores are galena, sphalerite with varying amounts of pyrite, chalcopyrite, tennantite-tetrahedrite-fahlores and various (NiAs-containing) sulphosalts. The ores occur in massive stratabound mineralisations (so-called mantos), in ore pipes and veins, but also in karst caves mixed with limonite. Laurion's near-sea karst mineralisations have developed extensive and diverse oxidative mineralisations with carbonates, sulphates, arsenates and silicates. In addition, silver chlorides, sulphides and rarely native silver were formed. Often, these are very soft ore mixtures that were easy to mine.

Exactly which ores were mined in antiquity is hardly comprehendible with archaeological means. The only reasonably useful source are reports published by French mining companies about the old mining industry in Laurion in the late nineteenth century, especially in the third contact. The geologist Kepper (2005) is evaluating these reports and combines them with modern geoscientific research. According to these reports, the Compagnie Français des Mines du Laurium (CFML) recovered zinc from the enormous reserves of calamine and sphalerite left intact by the ancient Greek miners. According to Huet (1878), the third contact mineralisation (karst fillings, mantos) is carrying "considerable cerussite" with veins and masses of galena. He noted that cerussite would be less rich in silver than galena, with a tenor of 2.0 kg silver per ton of lead. The third contact manto is a zoned ore body. Oxidised lead ore would occur at the top of the manto and oxidised zinc ore in the basal part. The location of the ancient galleries seems to suggest that the ancient miners exploited mainly the silver-rich galena associated with cerussite. It was mined from the upper part of the mantos and from beds and veins of galena in the underlying oxidised zinc ores.

Further information on the silver content of the ores varies considerably and apparently depends highly on different samplings, which is usually not mentioned in the literature. Marinos and Petrascheck (1956) report on ores around Ari and Charvalos, which contain a Pb level of 30 wt.% and about 1.6 wt.% Ag. Important for the antique mining was obviously only the galena, which had an Ag content of 40 ppm to 2.5 wt.% Ag. However, these rich loads were already mined in ancient times. Today, the Ag content is said to fluctuate between 40 and 200 ppm. Conophagos (1980) reports estimations of silver concentrations exploited in "lead ores" in antiquity of only ca. 0.1 wt.%. However, Skarpelis and Argyrakis (2009) are pointing out that not only galena but also fahlores are the main silver ores. Pernicka (1981) detected the sulphosalt bournonite ($PbCuSbS_5$) next to fahlore from Laurion. The two contain almost 5000 ppm Ag. He was one of the few who detected rich silver ores (see table Ag). Skarpelis (2007) analysed more than one dozen minerals rich in silver and herewith supported the observations of Bartels (2014) on possible occurrences of rich silver ores generally available in ancient times.

In total, more than 3500 tons of silver and 2–3 million tons of lead with changing amounts of silver (800–3000 ppm) were mined in the ore district of Laurion. In the nineteenth century, about 10.5 million tons of "old" slags on 35 heaps were still known, which likely have been recycled recently and melted down. They contained approximately 7 wt.% Pb and 140 ppm Ag (Conophagos 1980). While Meixner and Paar (1982) report that in the middle of the twentieth century AD just 160 g silver per ton lead was extracted from the ore, there are reports from Cordella (1869–1871) who wrote that 30 kg silver per ton lead was extracted!

In Laurion, silver production likely started already in the Early Bronze Age, around 2900 BC (Mussche Herman 2006), but definitively in the Late Bronze Age. The local silver mining was the basis for the economic well-being of Athens (Kalcyk 1982; Lohmann 2005; Gale and Stos-Gale 1981b).

Based on field surveys in ancient mines, and by lead isotope analyses, Gale et al. (2009) were able to demonstrate that not only silver or lead were mined in the Bronze Age, but copper as well for making copper-based artefacts, thus breaking the picture of a monomineralically used deposit. Stimulation for the use of copper ores was the strikingly coloured minerals malachite and azurite in the oxidation zone of Laurion. The abundance of secondary minerals near the surface of this ore province alone indicates that this is a polymetallic deposit (Marinos 1982), as well as bismuth, arsenic, nickel, cobalt and gold.

Not to be underestimated is the amount of oxide minerals in the gossan of the deposit, which in any case have been inevitably available in ancient times. Skarpelis and Argyraki (2009) expressly refer to the zinc-rich oxidation zone and describe the abundant occurrence of non-sulphidic coating of the walls of karstic cavities in the marble. Predominant zinc ores are the mixture of calamine, i.e. smithsonite ($ZnCO_3$), hydrozincite ($Zn_5(OH)_6(CO_3)_2$) and hemimorphite ($Zn_4[(OH)_2|Si_2O_7]\cdot 2H_2O$) which are intergrown with cerussite, earthy limonite and hematite. The authors estimate an extension of the oxidation zone of approximately 270 m and emphasise that sulphidic ore bodies unaffected by oxidation are uncommon. It would be surprising if the ancients did not realise the geological setting and mineralogical features of the Fe- and Zn-rich ore in the upper part not only of this ore deposit. Pliny merely reports that at Laurion zinc oxide was carefully recovered from the walls of the flue for medical purposes—the basis of the zinc ointment. In modern times until the twentieth century AD not only lead and silver have been mined, but also zinc and copper.

About 250 sophisticated so-called ore washeries testify that ore and intermediate products, such as litharge as well, were carefully processed to extract silver in antiquity (Rehren et al. 2002; Papadimitriou 2012, 2016). Further problems and literature concerning the history of mining are summarised in Weisgerber and Heinrich (1983) as well as in Lohmann (2005). Rehren et al. (2002), in an analytical study of fine-grained tailings from the ore washeries used in antiquity, suggested that argentiferous galena was the main ore mined in ancient times. They demonstrated by microscopical texture analysis that cerussite-encrustations would have been performed by 2500 years of soil weathering around grains of galena.

Rio Tinto

Silver-bearing jarosite was mined in Roman, and probably already in Phoenician times in the ore deposits of the Iberian Pyrite Belt, especially in Rio Tinto. Jarosite in Rio Tinto contains up to 1000 ppm Ag (Salkield 1987). As silver was rather more soluble than other elements, it tended to be precipitated in a metre-thick layer along the lower contact of the oxidised zone of the gossan in the form of massive (argento-) jarosite and other complex minerals. Jarosite at the base of the gossans above the sulphide ore body, e.g. at Corta Lago and Salomon is commonly marked by an earthy layer of up to 1.5-m thickness which is rich in gold, silver, lead, antimony, bismuth and selenium. This layer is locally honeycombed with ancient Roman workings for silver and lead, and perhaps gold (Fig. 3.35).

Also Amstutz (1960), a geologist specialised in ore deposits, suspects that in analogy to Peruvian deposits, Rio Tinto would also have had larger pockets of silver-bearing jarosites that were already exploited in antiquity. Analysis performed by Amorós et al. (1981) on a scanning electron microscope (SEM) verified tiny globular grains of cerargyrite in plumbojarosite. The old workings were observed precisely along this interface and it is suggested that they extracted

Fig. 3.35 Corta Lago, Rio Tinto, Spain. The modern open cast mining activities show honeycombed layers of galleries at the base of the red oxidised gossan. This earthy layer of jarosite is approximately 1.5-m thick and it is extraordinarily rich in noble metals. The date of the galleries is varying. Weisgerber (Archive Deutsches Bergbau-Museum) and Leistel et al. (1998) suggested partly Roman age, Hunt-Oriz (personal comm., 10/2017) and Salkield (1987) proposes mining constructions from the nineteenth century AD. Both considerations might be true. Photo from Jean-Marc Leistel, Paris

silver from the jarosite. It is possible that they also exploited the gold concentrations in this layer.

Jarosite is in part intergrown with quartz and a limonite–hematite matrix. It has already been pointed out that by consistency and colour, they can easily be confused with pure iron minerals. Williams (1950) and Salkield (1987) suggest that, instead of the term jarosite, they should be referred to by the rather mining-linguistically term *jarositic earths*. This would include samples with very high SiO_2 contents. The existence of Pb–Ag-rich jarosites with Fe-rich minerals and quartz, represents a metallurgically ideal starting point from the jarosite-rich layer of Rio Tinto, with an excess of quartz or quartz-rich materials easily distinguishable again as unmelted components in slags (see Sect. 6.4). In addition, the exploitation of silver ores associated with Pb-bearing minerals in ancient times is of major archaeometallurgical importance. It shows that it was obviously not necessary to add foreign lead for lead coating silver (s)melting as suggested by Bachmann (1993).

Anguilano et al. (2010b) focus on the exploitation of the northern lode of Cerro Salomon and in particular on the stratigraphic section of Corta Lago, spanning between the Phoenician period up to the second century AD, mainly highlighting the Roman period (200 BC–200 AD).

Silver enrichments in jarosite and other minerals at the basis of the gossan, such as the one in Rio Tinto, were also observed in the gossan of the gold–silver mine of Rouez in the North West of France (Guiollard 1993). Again, gold and silver (up to 33 ppm Au, up to 0.3 wt.% Ag) form a metre-sized layer below the oxidised zone of the gossan, above the cementation zone. Radiocarbon dating suggests that it was exactly this layer that was exploited in the Roman period (Fig. 3.36). The pin-point accuracy with which the silver and gold-bearing layer was mined both in Rouez, as well as in Rio Tinto, can be regarded as a masterpiece of mining ventures!

In addition, another two smaller silver sources have to be mentioned which are of importance in

Fig. 3.36 Ancient galleries of Roman age at the silver-gold mine of Rouez, Champagne, France. Note the well-calculated exploitation of silver, perhaps also of gold from a metre-sized layer at the bottom of the oxidised part of the gossan. From Guiollard (1993)

archaeometallurgy of the Eastern Mediterranean. This is Cyprus and Keban.

Cyprus

Also at the island of Cyprus, which was the major centre of copper production in the Mediterranean over millennia, silver does occur along with gold in a layer of jarosite, which exists between the gossan and the sulphidic ore body. In the 1930s, 5 tons of gold and 21 tons of silver were extracted from jarosite (Constantinou 1982), but for the time being, there is no evidence for prehistoric silver production at Cyprus.

Jarosite was the very common mineral in the gossans which were capping most of the (copper) sulphide deposits of Cyprus. In some deposits, the gossans were up to 25-meters thick. This layer was called *devil's mud*.

It is somewhat surprising that, in contrast to the situation in Rio Tinto, the silver-bearing jarosite layers of the deposits were obviously not used in ancient times. Numerous ancient shafts and galleries penetrate the gossans for exploration and exploitation of the deposits and went through these layers rich in jarosite.

Keban

Keban is an important karstic Pb–Ag deposit in south-eastern Turkey, located at the upper Euphrates. It is associated with alkaline intrusions of Eocene magmatites (Öztunali 1989). As proven by finds of litharge and lead–silver slags from nearby Fatmalı Kalecik (Hess et al. 1998), and of prehistoric hammer stones, lead and silver, possibly also (As-containing) copper and gold, have been mined as early as the fourth millennium BC (Wagner et al. 1989; Palmieri et al. 1996a, b).

3.4.10 Lead and Zinc

Geochemically, lead (Pb) and zinc (Zn)—and also silver (Ag)—are linked together in polymetallic mineralisations. Because of this, they are discussed together here. The proportions of the individual metals vary in individual deposits as a function of geochemical environment, geological setting and climatic factors. It is possible that lead and zinc deposits, which were at first known only as silver deposits in the

Table 3.7 Some important lead ores

Mineral name	Chemical formula	Pb (wt.%)
Native lead	Pb	100
Lithargite, massicotite	PbO	92.8
Galenite or galena	PbS	87
Cerussite	$PbCO_3$	77
Anglesite	$PbSO_4$	68
Boulangerite	$Pb_5Sb_4S_{11}$	58
Jamesonite	$Pb_4(Sb,Fe)_6S_{14}$	40.2
Pyromorphite	$Pb_5(PO_4)_3Cl$	76

Values of lead concentrations after Doelter (1912–1932) and modern mineral data. They may vary according to various solid solutions

pre-(historical) context, may have been mined later for lead and afterwards also for zinc. But it is also conceivable that, due to a lack of knowledge of such ore deposits, misinterpretations came up in archaeology or archaeometallurgy.

Lead occurs in the earth's crust at a concentration of about 12.5 ppm. Lead is a soft greyish metal and can easily be deformed. Its melting point is very low, it is at 327 °C. Lead is a decidedly chalcophile metal, that is, it has a high affinity to enter into chemical compounds with sulphur in nature. With a density of 11.34 g/cm^3, it is one of the heaviest metals that played a role in ancient times. Thus, the isotopes ^{206}Pb, ^{207}Pb and ^{208}Pb are the heaviest stable atoms. These isotopes are the end products of three natural radioactive series of elements. Lead isotopes are of eminent importance in archaeometallurgy, as they play a crucial role in provenance studies of metal artefacts. Because of the abundance of lead-bearing ores, almost all non-ferrous artefacts have lead levels sufficient to measure these isotopes.

Zinc is present in the earth's crust at about 70 ppm. It is as common as nickel (75 ppm) and almost six times more abundant than lead. This puts it in 25th place in the frequency table of the elements and is one of the more common metals. Zinc, like lead and copper, is a chalcophile element.

3.4.10.1 Lead Minerals

There are many lead compounds. After all, there are about 350 minerals with lead in the chemical formula. Therefore, the number of sulphidic, oxidic and secondary lead minerals is unusually high. Only a handful of minerals that are more widely distributed are listed in Table 3.7.

Native Lead

Native lead is very rare (Ramdohr 1975), and it does not seem to play any role in archaeometallurgy. Only in few ore deposits, lumps of native lead may occur, e.g. up to 50 kg were found in the lead deposit in Långban, Sweden. However, conditions required to form lead are rather exceptional in oxidation zones of ore deposits.

Lithargite

Apart from native lead, the two lead minerals lithargite and massicotite, which differ only in their crystallographic modification (lithargite: tetragonal, massicot: orthorhombic), have the highest lead content (up to 92.8 wt.%). They are very rare in nature. However, lead oxides form regularly in the metallurgy of lead.

Galena

Galena or galenite is likely the most important lead ore of the lead deposits, which are common worldwide. An example of a galena mineralisation is shown in Fig. 3.37a. Many of these deposits were of importance for antiquity. Galena often, but by no means always, contains some amount of silver. Silver contents can reach about 0.1–1 wt.%, occasionally up to almost 10 wt.% (Foord et al. 1988; Renock and Becker

Fig. 3.37 (**a**) Rohdenhaus, North-Rhine Westphalia (Germany). Irregularly shaped massive galena in lime stone. It is characteristic of a karstic mineralisation. Width of the sample 20 cm. Photo: H.W. Voß, Deutsches Bergbau-Museum. (**b**) Barranco Jaroso, Cuevas de Almanzora, Almeria (Spain). Massive formation of cm-thick layers of well crystallised white cerussite by weathering of galena in the gossan of the lead–zinc deposit. The lead carbonate was formed on a layer of limonite. Width of the sample c. 0.5 m. Museo Geominero, Madrid. Photo: A. Hauptmann, Deutsches Bergbau-Museum Bochum

2011). Silver is present in galena as (sub-) microscopic inclusions of silver-bearing minerals such as fahlores or argentite (George et al. 2015). Machatschki (1953) notes that, for this reason, galena is the most important ore for silver mining today after a possible extensive exploitation of actual rich silver ores. The content of other metals, such as Zn, Fe, Sb, Bi, Ag and Cu can often be explained with finest intergrowths with other ore minerals (Frenzel et al. 1973; Clark and Sillitoe 1971).

Hence, galena is often associated with other base metal sulphides such as sphalerite, pyrite, chalcopyrite and others. Silver-containing galena with approximately 0.15 wt.% silver was suggested to have been exploited in the medieval mine of Melle, France (Téreygol 2002). For the medieval silver production, this is extremely low and is in discord with written evidences (Bartels 2014; Al-Hamdānī, c. tenth century AD). (Medieval) mining of galena itself must not always be connected with silver production. In Roman Britain, galena was extensively mined and smelted (Craddock 1995), but it is thought that any associated silver production played little to no role. Galena is not always a silver-bearing sulphide.

Lead Minerals of the Oxidation Zone

Further, there are a number of secondary lead minerals near the surface of sulphide ore deposits, of which the bright, whitish cerussite is the most widespread, then follow anglesite and pyromorphite. Cerussite is a quite frequent mineral (Fig. 3.37b) in the carbonate-hosted Pb–Zn–Ag deposits of Sardinia (Exel 1986), but also in the most famous ancient lead–silver–zinc–copper deposit of Laurion, Attica (Greece) and many others (Skarpelis and Argyraki 2009; see also Sect. 3.4.9). In few ore deposits, cerussite reaches large quantities of the ore. It was mined until modern times in Karatau in southern Kasachstan, and it is an important ore in Leadville (Colorado), Broken Hill (Rhodesia) (Betechtin 1968). However, exposed to corrosion, galena is more stable than sphalerite. It is quickly covered by dense skins of secondary lead minerals. Therefore, galena is often embedded in carbonaceous ore of zinc (Pohl 2005).

Keim et al. (2016) note that only slightly more than 10% of the original silver content in galena is contained in the secondary minerals. Hence, cerussite and anglesite are virtually very low in silver. This leads to the conclusion that silver released during weathering of argentiferous galena or other Pb-sulphides is likely to be removed as soluble ions/complexes or reprecipitated as distinct silver phases, e.g. native silver or acanthite. These in turn are often included as isolated inclusions in cerussite, so that it can be dyed black. In his work, Kepper (2005) gives an overview of the compound and colour, and in which intergrowth cerussite and galena were observed by the miners of the nineteenth century in the ancient pits in Laurion.

The formation of chlorargyrite is only possible in environments with elevated silver and chloride concentrations.

3.4.10.2 Zinc Minerals

Zinc is not found in nature in elemental form.

There are only a few zinc minerals which are of importance in archaeometallurgy. These are frequently associated with galena (PbS) and also with copper ores.

Sphalerite

The most common zinc ore is sphalerite or wurtzite (ZnS) (Table 3.8). It often contains, just like secondary zinc minerals, iron (Fe),

Table 3.8 Important zinc ores in archaeometallurgy and their metal content after Doelter (1912–1932)

Mineral name	Chemical composition	Zn content (wt.%)
Sphalerite, Wurtzite	ZnS	67
Franklinite	$ZnFe_2O_4$	16.6
Zincite	ZnO	80
Willemite	Zn_2SiO_4	58
Smithsonite	$ZnCO_3$	52
Hemimorphite	$Zn_4[(OH)_2|Si_2O_7]\cdot 2H_2O$	54
Hydrozincite	$Zn_5(OH)_6(CO_3)_2$	59
Aurichalcite	$(Zn,Cu)_5(OH)_6(CO_3)_2$	44.9
Rosasite	$(Cu,Zn)_2CO_3(OH)_2$	14.7
Adamite	Zn_2AsO_4	45.6
Sauconite	$Na_{0.3}(Mg,Zn)_3(Si,Al)_4O_{10}(OH)_2\cdot 4H_2O$	$\leq$34

Most of the zinc minerals used in ancient times were non-sulphidic ores. Zinc sulphide (sphalerite) was generally not used before the Medieval period manganese (Mn), magnesium (Mg) and traces of cadmium (Cd). The geochemistry of zinc in the hypogenic cycle of mineral formation is very close to galena, which means that sphalerite is often very intensely intergrown with galena (and chalcopyrite), as shown by the ore from Rammelsberg in the Harz (Germany) (cf. Sect. 3.3, Fig. 3.4g). It is rare to encounter sulphidic ores that are completely free of sphalerite. For example, the widespread paragenesis of zinc and lead minerals is one of the foundations of so-called polymetallic ores (Smirnov 1954).

This makes it difficult to separate the two minerals and has already required elaborate beneficiation in antiquity. Examples include the ores of Laurion in Greece. Here, in the second half of the first millennium BC, many complex ore processing plants such as circular mills (*Kollergang*) or so-called washtables were designed to separate silver-containing lead ores or silver ores from the zinc blende (cf. Sect. 3.9) (Nomicos 2020). These facilities may also have been designed to process silver-bearing litharge, which incurred in significant quantities in Laurion during the smelting of galena (Papadimitriou 2012, 2016, 2017).

Secondary Zinc Minerals

In the exogenous cycle lead and zinc behave very differently. While lead is one of the least mobile elements in the oxidation zone (see above), zinc is one of the most mobile. Sphalerite is one of the most easily oxidised sulphides and there are several secondary zinc minerals in the oxidation zone of zinc deposits. Here, in contact with calcareous rocks, carbonates or silicates crystallise with (OH)-groups.

Most commonly, the minerals smithsonite, hemimorphite, hydrozincite and various Zn-rich clays such, e.g. sauconites are formed. Sometimes they are predominant in ore deposits and are even visible—if at all—in the scale of hand specimens. They are summarised as oxidic zinc ores under the collective name calamine. This is the technical-mineralogical collective term for various Zn-carbonates and Zn-silicates (with varying amounts of (secondary) lead ores). These seemingly colourless whitish and inconspicuous masses of calamine are difficult to recognise in the gossan of zinc deposits, quite unlike the strikingly green secondary copper minerals (Fig. 3.38a).

Taylor (2011) has compiled a very informative collection of photographs that reveal the difficult

Fig. 3.38 (**a**) Dossena Paglio-Pignolino, Bergamo (northern Italy). Whitish crusts of hydrozincite and green aurichalcite embedded in a porous earthy brown limonite. This zinc-oxidic ore deposit was probably exploited as early as in Roman times. Photo: H.W. Voss, Deutsches Bergbau-Museum Bochum. (**b**) Masua Montecani, Sardinia (Italy). Cadmium-rich, yellow stalagmitic smithsonite from the upper part of the Pb–Zn deposit in the south-west of the island. Even if this striking model of mineral is restricted to few spots, it may have been a tracer for zinc minerals in the deposit. Courtesy of M. Boni, Cagliari (Sardinia)

identification of secondary oxidic zinc minerals in the gossan in general, let alone the identification of individual zinc minerals. The earthy zinc minerals are hardly identifiable in the intergrowth with those of large quantities of limonite. The Zn-silicate hemimorphite is therefore not easily recognisable, because it only forms tiny, mm-sized crystals of glassy lustrous, turbid, colourless, sometimes whitish or brown colour. The same applies to the mineral hydrozincite, which occurs only in the rough, crypto-crystalline, earthy form, in inconspicuous pure, grey and yellow-white colours. Any colouration of this rather colourless calamine is based in many other zinc deposits on a substitution of zinc by Cd, Fe, Mn, Mg and Cu in the chemical formula of the mineral. Like smithsonite, it may form banded crusts or be stalactite-like, and like many other Zn minerals, it may also contain cadmium. Smithsonite is found very often in most non-sulphide zinc deposits. In the lead–zinc deposit of Iglesiente (Sardinia, Mediterranean), greenish-bluish smithsonites occurs near the surface, above the galena mineralisation (Valera and Valera 2005). As an exception, strongly yellow coloured crusts and stalactites may be found (Boni et al. 2003) (Fig. 3.38b).

It is astonishing that there is still no mining–archaeological evidence for an early extraction of zinc ores, because in Roman times the source of the brass artefacts, which are mass produced in the Imperium Romanum and whose production relies on the production of zinc ores, is still largely unknown.

The reddish-brown-yellow zincian clay mineral with the name sauconite is easily overlooked, but it seems to be most common and is closely associated with smithsonite, hemimorphite and hydrozincite (Taylor 2011).

Aurichalcite occasionally appears fused in small clusters of needle-shaped crystals with limonite and calcite. Aurichalcite is a zinc–copper hydroxycarbonate, usually found as a secondary mineral in copper and in zinc deposits. It is the only Zn mineral that has a distinct greenish-bluish colour, due to the substitution of zinc with copper. Aurichalcite can easily be mistaken for malachite. Theoretically, brass could be directly melted from the Cu-containing zinc carbonate aurichalcite. The same applies to the mineral rosasite, which also occurs with the trio of minerals smithsonite, hemimorphite and hydrozincite, but is much rarer.

Metamorphic oxidic zinc ores such as franklinite, zincite or willemite are limited to a few occurrences. They hardly play a role in archaeometallurgy.

3.4.10.3 Lead–Zinc Deposits (Silver Bearing)

Limestone and dolomite are the most widespread host rocks of lead–zinc deposits. In many, but not in all occurrences, silver and copper are enriched.

Genetically, the lead–zinc–silver deposits are classified into several groups (Pohl 2005), of which only a few are mentioned here:

1. Mississippi Valley Type deposits (MVT). This type of deposit results from the precipitation of hydrothermal solutions in carbonate rocks.
2. Hydrothermal vein deposits in different host rocks with a predominantly quartz-rich gangue.
3. Sedimentary exhalative deposits (SEDEX deposits). They are formed by the deposition of ore-containing solutions in a marine environment.

Mississippi Valley Type deposits (MVT)

Lead–zinc deposits are largely among the MVT. They are also called Carbonate Replacement deposits (CRD).

As described, e.g. in an overview by Paradis et al. (2007), MVT deposits are epigenetic, stratabound, carbonate-hosted ore bodies composed predominantly of galena and sphalerite with changing concentrations of silver minerals. The deposits are so-named because several classic MVT deposits are located in carbonate rocks within the drainage basin of the Mississippi River in the central United States. The deposits occur mainly in lime- and dolostone as open-space fillings, collapse breccias, and/or as replacement of the carbonate host rock (Sangster 1988).

The composition of ores from lead–zinc deposits is rather simple, consisting predominantly of galena, sphalerite, pyrite and/or marcasite and their secondary oxidation ores. The ores, together with Fe-hydroxides, hematite and jarosite, are filling open karstic caves or breccias They have different shapes and designs, they are layered, form veins or form ferruginous, "earthy" and hemimorphite-rich clay. The weathering of these ores very often leads to the formation of economically interesting oxidic zinc ores with the main component calamine. Silver minerals are incorporated into galena in sulphidic form or are present as tiny sulphide or chloride precipitates in the mineral mixture of the oxidation zones.

Due to widespread limestone formations in the Mediterranean (Cerny 1989), karstic ore deposits have played an important role as raw material sources for the mining of lead, silver, zinc, iron, copper and gold. They are however bound to the alpinotype folded carbonate rock and can be tracked in a broad belt from the Iberian Peninsula through southern France and Italy, the Alps, over the Balkan Peninsula and up to Anatolia. Well-known deposits of this type are Bleiberg (Austria), Sulcis-Iglesias (Sardinia), Cabrières, Laurion, Thasos, Sifnos (Greece) and a number of deposits in southeast Anatolia such as the Zeytindağ near Keban. The Pb–Zn–Ag deposits of Laurion, Sulcis-Iglesias and Sierra Morena are structurally similar because they are genetically comparable. The incredible and legendary wealth of Laurion was perhaps not only based on a particularly rich mineralisation but on an intensive exploitation of the deposit by a "large army of workers" (Lauffer 1979) in the second half of the first millennium BC. For the Romans Laurion was therefore no longer economically viable. They focused their activities on corresponding deposits in Spain (see below).

From the modern economical point of view, MVT deposits tend to be relatively low grade, averaging perhaps 4–8 wt.% of lead + zinc, although examples of higher-grade ore bodies or portions of ore bodies are known. Lead–zinc ratios of MVT deposits range from lead-dominant to zinc-dominant with a majority of deposits displaying (Zn / Zn + Pb) ratios ~0.7. Today, the silver content of the ore is negligible and the metal is usually not recovered. However, there are impressive examples of silver extraction from MVT in ancient times.

Hydrothermal Veins

Hydrothermal ore veins with sulphidic Pb–Zn–Ag mineralisations are hardly economically interesting anymore. But they were important in ancient times or in the Middle Ages. In Germany, the many polymetallic Pb–Zn–Ag–(Cu-) ore veins in the Harz (Asmus 2012), the Ore Mountains (Wagenbreth et al. 1990), the Siegerland (Kirnbauer and Hucko 2011) and in the Black Forest (Goldenberg 1996) were of importance.

Non-ferrous metal and iron ore deposits of the *Rheinisches Schiefergebirges*, e.g. in the Siegerland area (Germany), have been mined since the La Tène period and in the Middle Ages (Kirnbauer 1998; Kirnbauer and Hucko 2011; Kronz and Keesmann 2005). Interesting is especially the geological structure of the hydrothermal veins. They sometimes reach a maximum length of 12 km. In total, several thousand veins are known (Kirnbauer and Hucko 2011). Bornhardt (1910) called them vein swarms (*Gangschwärme*). Genetically, these related parts are larger vein tracts that were divided, shifted and distorted by tectonic activities. There was enormous rifting and uplifting. The Siegerland is located within one of the most important supra-regional tectonic structures of the *Rheinisches Schiefergebirges*.

The structures of the Siegerländer ore veins are by no means a stand-alone feature. In the Harz, too, the hydrothermal veins are sometimes severely tectonically disturbed (Buschendorf et al. 1971). Gangues and ores are often brecciated and form tension gashes, where the mechanical quality of the host rock plays a role in the ore formation that is not to be underestimated: shale tends to display significant plasticity, greywacke, however, is hard and brittle.

For reconstructing and understanding early mining technologies and strategies, such complex hydrothermal mineral deposits with their numerous mineralogical ore parageneses are confusing and difficult to solve without the knowledge of depository scientists. Globally speaking, these examples are probably not unique, but can also be found in other hydrothermal deposits. They are just not described in enough detail or are overlooked.

It has already been pointed out that a separation between lead and zinc can occur in the oxidation zone. Large (2001) defines non-sulphide zinc deposits in which smithsonite and hydrozincite together with hemimorphite are the principal zinc-bearing minerals in the upper part of ore deposits, which are very low, or sometimes even void of lead. Remaining lead in the oxidation zone is often intergrown with cerussite. The investigation of the deposit of al-Jabali, Yemen (Merkel et al. 2016) shows that silver can be mined then as well. Overall, there is a variety of ore deposits ranging from quasi-monomineral Zn mineralisations to polymetallic Zn–Pb–Ag–Cu parageneses.

Non-sulphide zinc dominated deposits are those of calamine-dominant deposits in Europe and the Middle East. Famous deposits of major importance in archaeometallurgy are located in Sardinia, Laurion, the European Alpine Zn-district, Spain, Greece, Turkey, Iran, and in the south-western corner of the Arabian Peninsula (Yemen). For Europe, a geological overview is written by Boni et al. (2003), and an archaeometallurgical discussion, mainly of lead isotope compositions, is published by Bode (2008) and Bode et al. (2009). For a long time, surface-near cerussite-galenite mineralisations were probably underestimated as an economically important part of "ancient" ore deposits.

Sedimentary Exhalative deposits (SEDEX)
SEDEX deposits are formed by the discharge of hot, mineral-bearing hydrothermal solutions on the ocean floor. Here, fine-grained ore minerals (mainly sulphides) are precipitated and deposited in layered, banded ore bodies, which are alternatively stored with marine sedimentary rocks. The transition to purely volcanogenic massive sulphide deposits (VMS) is fluid. The latter, however, are almost exclusively connected with volcanic rocks. Zinc-rich deposits of lead, copper and also gold in ophiolites, such as in Cyprus, Oman or in the Iberian Pyrite Belt (Rio Tinto) belong to this type.

3.4.10.4 Local Specifics

Great Britain
Many Carboniferous MVT-type ore deposits occur in Wales, Shropshire, the Lake District, and the Northern and South Pennine ore fields (Bevins 2010). Primarily those in Wales were mined for lead at the start of the common era. Lead isotopic and epigraphic studies on lead ingots found off the coast of Corsica have shown that they originated in the Mendip Hills or Flintshire and were transported to Rome around 200 AD (Hanel et al. 2013a, b).

The lead was generally low in silver contents. Mineralisations of lead with copper and silver are only dominant in the mine of Cwmystwyth in Central Wales, where they were mined in the Bronze Age. Frequently, the near-surface sulphide ores have been converted to anglesite.

Germany
In Germany, there are some known zinc deposits with calamine ores. Despite the overprinting by modern mining activities the ore deposit of Stolberg/Aachen belongs possibly to the largest source for zinc. Next comes Wiesloch near Heidelberg. For many years, Stolberg's non-sulphidic zinc deposit was the focus of consideration for the possible supply of ores for the early brass production (Hanel and Bode 2016).

The MVT deposit of Wiesloch was examined thoroughly. The oxidation zone of the carbonate-hosted lead–zinc deposit has seen mining activities for at least 2000 years (Hildebrandt 1998; Pfaff et al. 2010). It was exploited for lead and silver ores since Roman times.

Other dominant ores in these two deposits are sphalerite, galenite, cerussite and limonite. Kirchheimer (1977) and Hildebrandt (1998) state the silver contents of the galenite to be around 300–400 ppm, sometimes even up to 940 ppm. The gangue is made up of barite, dolomite, calcite and gypsum. The ores occur in the form of nodules and nests together with fibrous blende. Significant arsenic and antimony levels of up to 3 wt.% were measured. In the near-surface oxidation zone, the carbonate ore minerals dominate. Calamine can sometimes be quite ferrous and forms rough crusts on galena (Ostwald and Lieber 1957).

Hildebrandt (1998) suspects the calamine discovery in Germania, mentioned by Pliny, to be in Wiesloch. He argues that the Aachen-Stolberg area, to which this text passage usually is assumed to refer to (e.g. v. Petrikovits 1958; Weisgerber 1993), was part of the province of *Gallia Belgica* back then. Weisgerber (1993) emphasises that the interpretation of the text passage by Pliny is still subject to discussion.

Of interest is a statement by the experienced geologist Kirchheimer (1977). He reports that the oldest mining in Wiesloch was that of silver-bearing galena. In old mines, the galena had been exploited selectively and the calamine ores were untouched or had been discarded as a dead rock. It is not easy to prove this, but the aspect of such selective mining has not been considered in any other research so far. However, Kirchheimer's findings are confirmed by lead isotopic analyses of slags and ores from Wiesloch by Ströbele et al. (2015). He writes: "Since nearly all smelting relics plot well within the Wiesloch ore field, we conclude that during the High Middle Ages as well as Roman times, only local ore was smelted at Wiesloch. Also, almost all Roman objects overlap with the Wiesloch ore field, which indicates that silver production from argentiferous ores was indeed practiced during Roman times". All these smelting activities resulted in large slag dumps of ~ 400,000 t of which 360,000 t result from the smelting of argentiferous calamine and galena, from the late tenth to the late twelfth century AD (Hildebrandt 1998).

Iberian Peninsula

The Iberian Peninsula, the old *Hispania,* was the main producing country in antiquity, not just for lead, in both the Roman Republic and the Roman Empire. There, sophisticated mining and metal extraction was practiced, as seen in Rio Tínto, Las Médulas, Linares, and Cartagena-Mazzarón (overview in Meier 1995; Domergue 2008). In Roman times, no area produced as much lead (Rothenhöfer and Hanel 2013), and according to Pliny (Nat. Hist. XXXIV.30) and Strabo (Geographica III.2.8) also silver, gold, copper and iron, as the south of the Iberian Peninsula. Hunt Ortiz (2003) shows in his book on prehistoric mining and metal production in the south-western Iberian Peninsula that in the Chalcolithic and Bronze Age, metallurgy had already begun there.

The southern part of the Iberian Peninsula is amongst the most intensely mineralised regions of Europe, with a great variety of ore districts containing primarily Pb–Zn–Ag ores (Boni et al. 2003). The most important economic districts in ancient times are the Sierra de Cartagena-Mazarrón, Sierra Morena and Sierra Almagrera at the south-eastern coastline.

The deposits of the Sierra Morena and the Sierra de Cartagena belong to the world's largest sulphide deposits and carry mainly Cu, Pb, Zn, Ag ores, as well as gold (Sáez et al. 1996; Leistel et al. 1998). The total historic production of the Cartagena-Mazarrón zinc–lead–silver deposit is estimated at 2.5 million tons of lead ore, grading 15 wt.% lead and 150 ppm silver. Remarkably, the zinc-rich ore in the uppermost part of the ore deposit at a depth of 25–100 m was left largely untouched. Partial mining of this ore type was not undertaken before the 1950s/1960s, when modern flotation techniques became available to separate the lead and zinc sulphides. This ore district primarily consists of Variscan sheeted vein swarms and stockwork mineralisations hosted within silicified dacites. The principal minerals are sphalerite and galena associated with lesser amounts of marcasite, pyrite, chalcopyrite and silver-rich copper sulphosalts. As it is usually the case, silver is intimately associated with galena. In the middle of the nineteenth century AD some small copper prospects developed.

Two important silver mines are those of Linares and Peñarroya. In Linares, Roman lead/silver mining is proven by the so-called stone of Linares, on which a relief of miners from the third century AD can be seen (e.g. Domergue 2008).

This relief is on exhibition in the Deutsches Bergbau-Museum Bochum.

Sardinia (Italy)

On the island of Sardinia, which played an important role not only as a trading centre in the middle of the Mediterranean but also as a raw material base in the Bronze Age and especially in Phoenician times in the first half of the first millennium BC, many extraordinarily rich lead–zinc–silver deposits can be found next to some smaller copper and iron deposits (Valera et al. 2002; Boni and Large 2003; Boni et al. 2003; Valera and Valera 2005; Hanel and Morstadt 2014). Since the Sardinian ores are mentioned neither by Greek nor by early Roman authors, Nriagu (1983) and Kylander et al. (2005) assume, based on their research of impurities of lead in peat cores, that the mining stopped, or was severely limited between the reign of the Carthaginians and the Roman Empire (between the sixth and the third century BC). In their opinion, the lead deposits of Sardinia only played a role in lead trading up to classical antiquity.

The most productive area, with almost 50 ore deposits that have been exploited by mining, is located in Iglesiente in the south-west of the island (Exel 1986). All mines are shut down. The lead–zinc sulphide district (SEDEX and MVT), predominantly located in karst cavities of Cambrian limestone is covered by a thick oxidation zone. Valera et al. (2005) estimate the gossan, e.g. of the Monteponi deposit, to have extended tens of thousand cubic metres, consisting mainly of masses of cerussite, anglesite, calamine (smithsonite, hemimorphite), mixed with limonite and associated with galena (Fig. 3.39). Tools used in mining (metal buckets) and other finds from the Roman period were found in galleries in depths of 94–157 m (pers. commun. N. Hanel, March 2019).

Two types of mineralisations have to be distinguished. The first type, which occurs in large, still economically interesting quantities, is called "Calamine Terre". It consists predominantly of a fine-grained mixture of smithsonite, hydrozincite and hemimorphite in ferruginous, "earthy" clay

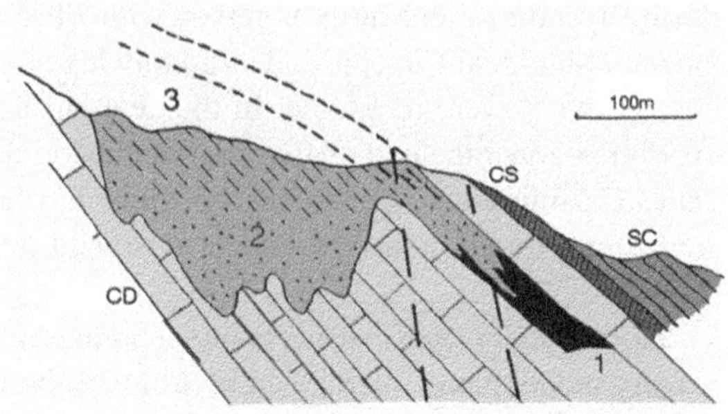

Fig. 3.39 Simplified cross section from a calamine-type deposit at Monteponi in southwest Sardinia (Italy). Note the extensive mineralisation of secondary ores in the upper part of the ore deposit which belongs to the Mississippi Valley-Type deposit! Abbreviations: CD = Cambrian limestone and dolomite; CS = Calcschists; SC = Cabitza schists; 1 = primary lead and zinc sulphides; 2 = smithsonite and minor cerussite, 3 = hemimorphite and cerussite. From Boni and Large (2003), modified after Zuffardi (1982)

(Boni et al. 2003; Aversa et al. 2002), which concentrates on the upper parts of each deposit on this island. The mineralisation is bound in irregularly shaped karst caves and mantos which have extensions in the range of meter to ten metres. In addition, stalagmites and palm-sized concretions of almost pure smithsonite are more common, which is an excellent starting material for the production of zinc or brass (Fig. 3.38a, b). In addition, (secondary) lead ores, and very occasionally native copper and native silver occur. This massive supergene mineralisation of the Monteponi deposit is considered to be one of the most important secondary zinc–lead mineralisations.

The primary mineralisations of Monteponi (Type 2, primary lead and zinc sulphides in Fig. 3.39) consist mainly of argentiferous galena ("Ricchi in Argento") with fahlore inclusions as silver ores, sphalerite and pyrite in changing amounts. They contain up to 0.4–0.6 wt.% Ag in the galena (Assereto et al. 1976).

There is no evidence in the deposit that the calamine mineralisation of the oxidation zone was used in antiquity. Only Domergue (2008) is of the opinion that silver-bearing galena was likely mined in pre-Roman times. Overall, however, the old mining industry in Sardinia remains a research desideratum.

As far as the zinc is concerned, the systematic extraction of zinc and the production of brass must have been known at least since Roman times (Weisgerber 2007). There are thousands of brass artefacts from the Roman period for which zinc was essential. At Aléria, off the coast of Corsica, 21 Roman brass ingots weighing a total of 92.5 kg were recovered (Hanel and Bode 2016). Therefore, it is easy to imagine that this near-surface non-sulphidic zinc mineralisation was used at that time. The fact that zinc was an expensive metal (but cheaper than lead) has already been pointed out by Craddock (1990).

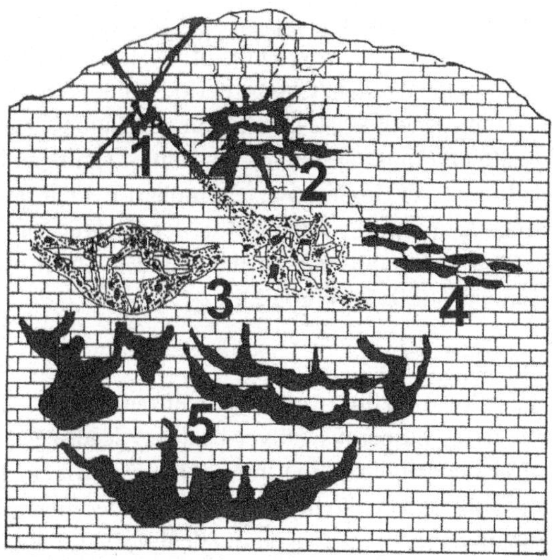

Fig. 3.40 Schematic representation of the various irregular forms of lead–zinc–silver mineralisations in the karstic deposit of Agios Sostis on Sifnos. (**1**) Along disturbances and gaps. (**2**) At joints, cracks and crevices. (**3**) Brecciated mineralisation. (**4**) Network of parallel cavity fillings (so-called mantos). (**5**) Pocket-like, stockwork, tubular. These mineralisation types are comparable to iron karst mineralisations as shown in Fig. 3.51. The mining exploitation of these ores in the Bronze Age resulted in very irregularly shaped cavities. Modified after Vavelidis et al. (1985). See also phenomena of karst-mineralisation at Laurion as described by Skarpelis and Argyraki (2009)

Sifnos, Agios Sostis (Greece)

Geologically speaking, the pit of Agios Sostis on the Aegean island of Sifnos is a lead–zinc–silver deposit as well. Using physical dating methods, it was determined that the mine had been active since the first half of the third millennium BC (Weisgerber 1985). Vavelidis et al. (1985) give a very good geological overview of the mineralisation in the marble, which is comparable in geological and mineralogical composition with many other deposits of this carbonate-hosted Mississippi Valley-Type deposit (Leach and Taylor 2009).

A characteristic feature of the deposit discussed here again are the irregularly shaped ore bodies, which are always discordantly enclosed in the marble. As can be seen from Fig. 3.40, in the old deposit of Agios Sostis, these run parallel and diagonal to the stratification at a short distance, in pocket- and tube-like shapes (mantos) and form column and crack fillings. Stockwork mineralisations occur with enclosed marble blocks and breccias. The mineralisations are present in different solidified forms, ranging from ochre, earthy to sintered, solid lots. The mineralogical composition shows a large variety of ores. Limonite, hematite, cerussite, smithsonite, hemimorphite, manganese oxides occur as main components, jarosite, tetrahedrite (Sb-fahlore) are subordinated, galena and sphalerite are only rare, and argentite and proustite/pyrargyrite sulphides are only accessories. The lead–zinc–silver deposits contain up to 0.7 wt.% silver, with high silver contents being detected especially in jarosite enrichments. Striking are Sb contents, which can reach over 20 wt.%. The study of slags confirmed the extraction of silver from the polymetallic ores (Pernicka et al. 1985).

The shapes of the karst mineralisations shown in Fig. 3.40 allow an understanding of the irregular shapes of the mining excavation cavities. Weisgerber (1985) showed by mining–archaeological workings how the mineralisations were mined in irregular *Pingen* and pits. In the older mining period (middle of the third millennium BC), depending on the location of the mineralisation, formless shafts, irregular adits, galleries and chambers were constructed, which, according to the definitions of mining language, flow smoothly into one another. It was worked in bulky, mostly rounded chambers in a close-fitting space. System and strategy of mining were merely to follow the earthy, soft mineralisations in every possible way. Only in the later mining phase in the Archaic/Classic period (middle of the first millennium BC), the mineralisation was mined in a more systematic system of shafts and galleries. An impressive example of such an irregular mining system of a lead–silver–zinc deposit was published for the island of Ibiza by Hermanns (2015) (Fig. 3.2).

Iran

Vatandoust (2004) and Momenzadeh (2004) report on numerous (silver-bearing) lead–zinc–silver deposits from the Iranian Plateau. Most of these are of MVT-type and are embedded in carbonate rocks of various geological ages. Of special importance are the two ancient localities of Arisman and the ore deposit of Nakhlak. At the large smelting and habitation site of Arisman, the raw source(s) used for smelting are difficult to identify. Litharge found at Arisman (fourth/third millennium BC) is suggested to have been produced from the ore of the mine of Nakhlak (Pernicka et al. 2011). From the archaeological point of view, the ore deposit of Nakhlak was at least mined in antiquity (Stöllner et al. 2004a, b).

The silver-bearing lead deposit of Nakhlak in central Iran is among the richest in the countries from the Eastern Mediterranean to the Middle East. This deposit is part of a mineralisation zone that stretches along the desert Lut and Kavir to the Alborz Mountains in the north. It is predominantly embedded in Cretaceous carbonate rocks. The ore bodies form vein-like mineralisations (Fig. 3.41a), but also the typical pocket-shaped karst cavities or breccias. There are about 28 such ore bodies known, which reach on average 1–5 m thickness (Holzer and Ghasemipour 1973; Stöllner et al. 2004a, b). Dominant ores are galena and cerussite, which occur in equal quantities. Added to this are barite, pyrite and very little sphalerite. Of importance for early metal extraction is, once again, the near-surface area of the deposit, where cerussite, anglesite, and occasionally hemimorphite are described (Fig. 3.41b). Here, they are intensely intergrown with galena and represent a prime example of raw materials that are suitable for a

Fig. 3.41 (**a**) Nakhlak (Iran). Subvertical hydrothermal vein-like mineralisation of silver-bearing galena, cerussite, and gangue of whitish calcite. The mineralisation has a thickness of 1.5–2 m. It is brecciated, and the carbonaceous host rock is mixed with the ores. Photo: Gerd Weisgerber (†), Deutsches Bergbau-Museum Bochum. (**b**) Nakhlak (Iran). Set of small crystals of galena and cerussite with some brown limonite on calcitic host rock. The intergrowth between these "oxidic" and sulphidic minerals represents an ideal raw material for metallurgical co-smelting processes. The width of the sample is 15 cm. Photo: Heinz-Werner Voß, Deutsches Bergbau-Museum Bochum

metallurgical co-smelting. Silver occurs in inclusions of fahlores in the lead minerals. Modern measurements of ore concentrates result in a lead content of approximately 65% with 70 ppm Silber.

India

A large ore district with several sulphide Zn–Pb–Cu–Ag deposits (Zawar, Dariba, Agucha) is located in north-western India in the Aravalli Hills. The early extraction of zinc from these extensive sources was studied by Paul Craddock (Craddock et al. 1990; Craddock and Eckstein 2003). The mineralisations are embedded in metamorphic dolomites and other sedimentary rocks of the Precambrian. This mountain range stretches over 1000 km approximately northeast from Gujarat to Delhi principally through Rajasthan. The type of ore deposit is a geologically old MVT.

Old mining was discovered. Ores consisting of primarily of sphalerite, then Ag-containing galena, pyrite, and chalcopyrite were already mined in the late first millennium BC and then from 1300 to 1800 AD. The authors claim that the production of zinc was mainly performed using sulphidic zinc ore and not secondary zinc ores as it was practised in the Old World. Galena was separated by handpicking.

Of great interest is that in other parts of this mountain range, copper deposits that are of comparable geological age can be found, which were apparently already mined in prehistoric times (Begemann et al. 2010)

Al-Jabali

Another source of silver known for its early exploitation is located in the south-west of the Arabian Peninsula. The ore deposit of al-Jabali (Yemen) is a carbonate-hosted zinc–lead–silver mineralisation. The primary sulphide deposit shows features of both the Mississippi Valley and Carbonate Replacement type deposit. Nearly all the ore is oxidised, but relics of galena can be found in the upper part of the ore deposit (Al-Ganad et al. 1994). According to al-Hamdani, the ore deposit was of major importance in the medieval period and silver was exploited around 942 AD (Allan 1979). The mine is under exploitation today. An overview of this deposit and its exploitation of silver is described in Merkel et al. (2016). Parageneses of such ores as in al-Jabali are suitable raw material bases from which zinc is obtained as a random by-product in the silver extraction (Craddock 1990).

Otherwise, many of the publications in archaeometallurgy on zinc deposits are confined to discussions of the sulphide mineralisations and rarely make mention on non-sulphide mineral zones, which were so economically important in the early stages of metallurgy. In general, astonishingly few sources of raw material for zinc are known in view of the supraregional, enormous finds of brass ornaments and tools as well as the brass coins of the Roman Empire. In the context of the provenance studies of the 21 Roman brass bars of Aléria (Corsica), Hanel and Bode (2016) at least discuss about 20 possible zinc–lead deposits as possible sources of raw materials from the area of the north-western Roman provinces.

3.4.11 Tin

The mean concentration of tin (Sn) in the earth's crust is about 35 ppm. This shows that tin is very rare compared to other metals such as copper, iron and zinc. Geochemically, tin is comparatively incompatible and lithophilic with weakly chalcophilic tendencies.

Currently, more than 60% of the world-tin production comes from the "Tin Belt"of south-eastern Asia (Pohl 2005) (Fig. 3.44a). The most important producing nation for tin is China, followed by Peru and Bolivia. Today, tin concentrations in ores of primary deposits of 0.3–1.2 wt.% Sn are considered to be economically profitable, while placers are used even below 0.01 wt.% Sn. Given a sufficient grain size, cassiterite is enriched by simple gravitational treatment. However, very fine *Staubzinn* (German for dust-tin) and complex sulphosalts with stannite require greater efforts in the beneficiation (Pohl 2005).

Tin is a resource for a variety of industrial applications, e.g. for the production of solders, sheets, chemicals and pigments. Tin alloys are part of everyday life. The annual world consumption of tin is about 300,000 t.

3.4.11.1 Some Problems of Ancient Tin

In the prehistoric and early history, the greatest importance of tin ores was first as a natural and then as a deliberate alloying constituent of copper for the production of tin bronze. Tin bronze was, so to speak, a strategic metal. But finding Bronze Age tin production and recognising it as a technical process is very difficult. In archaeology and archaeometallurgy, this, as well as the provenance of tin, is formulated as the "tin problem". This is partly because of the geologically-geographically limited occurrences of tin provinces (see below). On the other hand, it is due to the tin mineralisation itself. In contrast to the usually clearly coloured copper ores or gold, tin ores such as cassiterite or stannite are harder to recognise at fieldwork due to their inconspicuous colouring. Especially cassiterite, which itself is dark brown in colour, is often masked by dark minerals in solid rock, such as tourmaline, wolframite, which are closely linked to it paragenetically. On the other hand, tin-wolframite veins in solid rocks, interspersed with bright yellow secondary tin minerals such as varlamoffite, could have been recognised as a striking geological phenomenon. In the Bronze Age, there had already been thousands of years of mining experience in regards to extracting resources from hard rock. Therefore, this cannot be considered as a valid reason *not* to mine tin ores from their primary deposits, but to use only placer tin with cassiterite due to "simple" mining reasons. Also in placer deposits, cassiterite hardly stands out in colour when mixed with dark minerals such as magnetite- or pyroxene-bearing sands or with garnet, monazite, hematite, rutile, titanite and others. The already mentioned *Staubzinn* is virtually invisible.

Prospecting them is easier if tin ores are paragenetically connected with copper minerals, because, under weathering conditions of oxidation zones, all copper–tin mixed ores and the omnipresent chalcopyrite tend to form greenish coloured secondary Cu^{2+}-containing minerals. In such cases, at least for the initial stage of bronze metallurgy, one may consider increased use of such blends to produce the coveted alloy.

But there is hardly any information about the grade of the richness of tin ores used in ancient times. Although cassiterite or stannite can be found in more massive lots or in clearly crystallised aggregates, this seems more likely to be limited to exceptional cases. According to present knowledge, the massive occurrence of mushistonite ($CuSn(OH)_6$) in its homonymous type locality is more likely to be an exception as well (Alimov et al. 1998). The prehistoric extraction of tin ores from other deposits in Central Asia seems to have been more difficult. Parzinger and Boroffka (2003) report, e.g. from Karnab (Uzbekistan) that cassiterite is barely visible with the naked eye in the veins. The size of crystals analysed in the laboratories have reached only about 0.2 mm.

Elaborate enrichment processes, such as perhaps those in Karnab (Uzbekistan; Garner 2014), seem to have been necessary for the prehistoric tin extraction from the Kestel deposit in southeastern Turkey as well (Yener et al. 2003; Yener 2009). The resulting extremely fine-grained waste dumps, as hardly tangible, were likely weathered during a period of approximately 5000 years. There are perhaps parallels to the difficult archaeological field evidence of methods of tin mining and processing in historical periods at Rooiberg in South Africa (Chirikure et al. 2007). In any case, the idea of Kestel as a gold mine seems to be eliminated, because Hanilçi et al. (2017) were able to detect a significant accumulation of tin (up to 244 ppm) in soil samples, especially from the neighbouring settlement of Göltepe, while gold only reached a maximum of only 50 ppb.

The historic-modern extraction of tin ores in Cornwall, both in mines and placer deposits, is a remarkable illustration of the enormous effort of beneficiation required to enrich tin ores (Penhallurick 1986; Greeves 2016). There are clear parallels to the (early) gold extraction.

Table 3.9 Some important and some rather unknown tin ores

Mineral name	Chemical formula	Sn (wt.%)	Cu (wt.%)
Cassiterite	SnO_2	79	
Herzenbergite	SnS	79	
Varlamoffite	$(Sn,Fe)(O,OH)_2$	65	
Malayaite	$CaSnSiO_5$	58	
Mushistonite	$CuSn(OH)_6$	42	13
Teallite	$PbSnS_2$		37.5
Stannite	Cu_2FeSnS_4	26.6	30
Hocartite	Ag_2FeSnS_4	26.3	
Cylindrite	$Pb_3Sn_4Sb_2S_{14}$	25.7	
Franckeite	$Pb_5Sn_3Sb_2S_{14}$	17	
Mawsonite	$Cu_3(Fe,Sn)S_4$	14	44

Values of tin concentrations after Doelter (1912–1932) and modern mineral data. They may vary according to various solid solutions. Note especially mixed Cu–Sn ores such as stannite, mushistonite and mawsonite! Further 29 mixed Cu–Sn ores are listed in Rovira and Montero (2003)

3.4.11.2 Tin Minerals

In total, only 34 minerals are known that contain tin in various concentrations. That is less than a tenth of the ores that contain silver or copper. Some of the most important tin minerals are listed in Table 3.9.

Native Tin

Native Tin is extremely rare, it occurs only as an accessory mineral in some deposits (Ramdohr 1975). It did not play any role in early metallurgy. Similarly rare are tin sulphides such as cylindrite and teallite, with the exception of the huge deposits in Bolivia (Potosi), where they can occur in fairly large quantities.

Cassiterite

Cassiterite is the most common and important tin ore. It is one of the most weathering resistant minerals. Since cassiterite is very hard, chemically difficult to dissolve and has a high specific weight (6.98–7.1 g/cm^3), it accumulates frequently in placer deposits. Primary cassiterite occurs in close context with granitic rocks.

Cassiterite can sometimes appear in its deposits in more massive lots. It has a number of knick-names. The often short prismatic and knee-like crooked crystal twins were referred to in traditional mining terms as "elbow twin" (German *Visiergraupen* or *Zinngraupen*), while cassiterite in a short, glass-head-like formation was called "wood tin" (German *Holzzinn*). Very often, however, and very typically, the tin oxide is intergrown with granitic rocks in the finest grain area (mm-sized). These impregnations are called *Zwitter*, which occur in the context of altered granites (so-called greisen), typically in the Saxon Ore Mountains or in Cornwall (Fig. 3.42a). The mining of such mineralisations requires a huge amount of treatment.

Garner (2014) was able to recover a limited number of stone tools during her field work from the Iron Age mines of Karnab in Uzbekistan (not to compare at all with the finds from Göltepe!), she suspects the existence of more massive lots of tin ores, which had been exhausted in ancient times.

Perhaps the estimates of metallurgist Bryan Earl are correct, who states that for the deposits of Cornwall "...occasionally a lump of cassiterite is found in a lode, but this is not at all common. Rich ore holds about 5.0% metal, good ore about 2.0%, and skilled miners in western England work down to about 0.2% from stream detrital deposits..." (Earl and Yener 1995).

Cassiterite itself is especially geochemically pure. While it might contain some Fe, Ti, Ta, Nb, Mn, As and V (Rapp et al. 1999; Dill

Fig. 3.42 (**a**) Cligga Head, Perranporth, Cornwall. Mineral paragenesis of stannite + cassiterite + arsenopyrite + wolframite + tourmaline (fine black mixture) in quartz-rich pegmatitic matrix (qz). This paragenesis is also characteristic of other tin deposits. Width of the sample c, 25 cm. (**b**) Poldice Mine, Redruth, Cornwall. Massive sample of cassiterite, chalcopyrite and arsenopyrite. This sample shows the intensive intergrowth of tin–copper–arsenic ores in the tin belt of Cornwall. Width of the sample c. 6 cm. Abbreviations: *apy* arsenopyrite, *ccp* chalcopyrite, *cst* cassiterite, *qz* quartz, *sta* stannite, *tur* tourmaline, *wlf* wolframite. Both samples from mineral collection Dale A. Foster, Falmouth, Cornwall. Photos: Dale A. Foster

2010), it only contains a few parts per million of lead (Begemann et al. 1999), which, compared to the generally higher levels of lead in copper ores, may be of importance for archaeometallurgical provenance studies on the origin of Bronze Age tin (see Sect. 11.3.1).

Although cassiterite is the most important tin ore today and presumably has been so since the Bronze Age, its association with As–Cu–Ag–Pb–Zn–Bi ores and their secondary counterparts in oxidation zones of ore deposits, not limited to the Initial phase of tin metallurgy, must not be underestimated. Especially common is arsenopyrite as well as chalcopyrite, bismuthinite, emplectite, pyrite, sphalerite, fahlore, stannite, galenite and covellite. These paragenesis occur in numerous hydrothermal veins. The ores are regularly referred to in the geoscientific literature, e.g. at the tin deposits in Potosi (Bolivia), those of Cornwall or the Ore Mountains. However, with a few exceptions (Iberian Peninsula; Rovira and Montero 2003) they are often overlooked in the archaeological literature.

In 3.42b, an example of a massive piece of cassiterite + chalcopyrite + arsenopyrite is shown, which was found as a representative sample in the Poldice Mine, Redruth (Cornwall) in May 2015.

Stannite

Stannite is the primary sulphide with the contents of Cu and Fe. The intergrowths of stannite with chalcopyrite and other sulphides of non-ferrous metals is the reason why its chemical composition may differ significantly from its ideal composition. Ramdohr (1975) reports that stannite can often contain Zn, Ag and Pb percentages. In Alimov et al. (1998) microprobe analyses are presented which show that high zinc contents can replace the Fe in the stannite.

It occurs locally enriched in hydrothermal veins in addition to cassiterite and sphalerite, (Ramdohr 1975; Pohl 2005). In Fig. 3.42a such a mineral mixture is demonstrated from Cornwall. Six hundred and forty-seven ore deposits with stannite are listed on the internet min.dat. Worldwide.

Information about the distribution of stannite is contradictory. This is regrettable because this ore could have played an especially important role in early metallurgy. It is possible, however, that the ore has been largely exploited these days and is therefore described as a rare mineral. Von Naumann (1852) still described stannite as the most abundant mineral locally, e.g. in the ore deposits of Cornwall and in the Ore Mountains.

Stannite, in the geological context of many tin-rich provinces all over the world (Bolivia, Peru, the Saxo-Bohemian Ore Mountains, Cornwall, Japan, the Balkans, and Middle Asia) is regularly intergrown with chalcopyrite, arsenopyrite and fahlore.

Ramdohr (1975) emphasises the widespread distribution of stannite, but at the same time points out that, in tin deposits, it generally occurs only in small amounts. Exceptions supposedly exist, such as in the ore provinces of North-East Siberia, then on the Yukon and in the Alatau Mountains in Tienschan and in the Ferghana Basin. Furthermore, mawsonite must be mentioned, which is described together with copper ores from Tasmania, and stannoidite ($Cu_5(Fe,Zn)_2SnS_8$), which occurs with copper ores, stannite and native silver in Japan.

In the Mušiston ore district in the Zeravšan mountains in Tadzhikistan stannite forms massive mineralisations, which is associated with quartz (Alimov et al. 1998). Already Barnes (1974) and Wertime (1968) suggested that the role of stannite and/or its secondary minerals could have been underestimated, and pointed out that the smelting of stannite would have yielded a "natural" tin bronze. A production of such a "natural" tin bronze was also suggested by Radivojević et al. (2013) who described tin bronzes from localities in Serbia from a time period of approximately 6500 years ago. However, it remains unclear on which ore base these tin bronzes were smelted: from primary sulphides or from weathering products of stannite. The finds from Serbia date about 2000 years earlier than the earliest tin bronzes that appear in Mesopotamia in the middle of the third millennium (Pernicka 1998). In any case, the statements of Alimov et al. (1998) and Killick (2014), that stannite could hardly have been an ore base that played a role in early metallurgy, must remain unresolved.

The former authors argue, after only a short field stay in Mušiston, that the old mining had clearly been concentrated on the oxidation zone of this deposit, and that predominantly secondary minerals of tin and copper had been exploited (mushistonite, malachite, azurite, cassiterite and varlamoffite). But they doubt that this site would have played an important role in prehistoric tin mining, as the deposit would be located in an almost inaccessible high altitude location at a height of 3000 m. However, later detailed mining–archaeological investigations by Garner (2014) prove Bronze Age mining even in larger depths, and large waste dumps of mushistonite were left. Prehistoric mining in extremely remote high-altitude areas is not unique. For example, lapis lazuli was exploited in quantities in the mountains of Sar-i Sang in northern Afghanistan in mines located in heights of >2600 m (Weisgerber 2004a). In the third millennium, lapis lazuli was traded over thousands of kilometres to Mesopotamia. Shemakhanskaya (1991) describes stannite from Mušiston as well and suggests, that this CuSn mixed ore could have found use in the production of "old" bronzes.

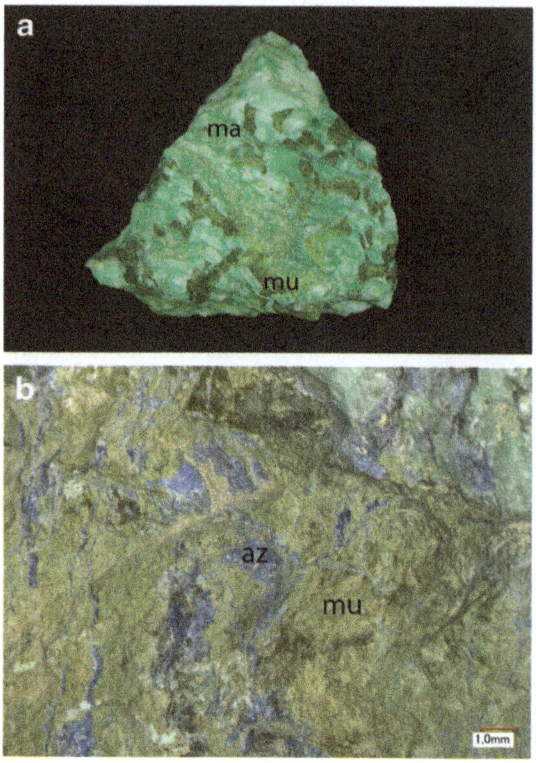

Fig. 3.43 Mušhiston, Tadzhikistan. (**a**) Mushistonite together with malachite (Ma) on quartz. This Cu–Sn mixed ore (13 wt.% Cu, 13 wt.% Sn, 6 wt.% Fe, 1 wt.% As 1 wt.%) is a "natural bronze ore" (**b**) Mushistonite, interspersed with fine veins of azurite, with some quartz. Abbreviations: *az* azurite, *ma* malachite, *mu* mushistonite, Photos: A. Hauptmann, Deutsches Bergbau Museum

Mushistonite and Other Secondary (Copper-) Tin Ores

Stannite decomposes easily in the oxidation zone, whereby mushistonite can be formed. Without naming the mineral, already Patterson (1971) suspected that weathering processes of tin ores may have provided a key material that led early metallurgists to the discovery of bronze. Mushistonite, with its striking olive-green colouration, might then have played a special role in old tin–bronze metallurgy as a hydroxide Cu–Sn compound.

The type locality of this mineral lies in the ore province of Mušiston in Tadzhikistan. Otherwise, however, the mineral is rare. So far it is known it has been found in less than 10 localities worldwide. Mushistonite occurs in nests and relatively finely distributed in carbonate rocks or in quartz. Thus, massive samples have the same colour attractiveness in the oxidation zone as, e.g. malachite. This is clearly visible in Fig. 3.43a, as well as the fine blue veins of azurite (Fig. 3.43b). Tin ores from mushistonite were subjected to chemical analyses (Alimov et al. 1998; Garner 2014). The tin contents range from just 3 to 30 wt.%, and the Cu contents of mixed Sn–Cu ores of just 3 to 17 wt.%. In addition, in a similar manner as in stannite, Zn and Fe contents were measured in the lower percentage range and As and Ag contents in the 0.x wt.% range. Therefore, it is not devious to suppose that the beginnings of tin metallurgy revolved around the use of "green stones". Here, as with stannite, the chemical composition is a "natural bronze ore". The lead contents are very low (<0.009–1 wt.%), but can sometimes still be used as a geochemical fingerprint.

A number of other Cu–Sn mixed ores are known (see Table 3.9). Rovira and Montero (2003) show that in numerous ore deposits of the Iberian Peninsula mixed Cu–Sn ores can be found also, from which low-alloyed Sn bronzes could have been smelted in prehistoric times. This probably also applies to the early metallurgy of many other areas where mixed primary sulphides of Sn–Cu rich ores were found as described above.

3.4.11.3 Tin Deposits

Tin deposits provide an excellent example of how an element with its ores can be almost globally confined to a few metallogenetic provinces (Fig. 3.44a, b). However, the location of the large tin belts includes most of the large, economically interesting deposits. It does not include the many small occurrences of tin ores that might have been of importance in ancient times too. This concerns, e.g. occurrences in Turkey, Iran, Afghanistan and in India.

Most tin provinces are of post-Cambrian age; they only formed in certain epochs. Remarkable is the similarity of the deposits all over the world: More than 99% of the tin comes from deposits that are directly or indirectly linked to granitic rocks or their (sub-)volcanic equivalents (Hosking 1979; Lehmann 1990). Both are due to magmatic activities in the area of subduction zones or are situated in the area of colliding continents.

Earl and Özbal (1996) state that tin ore is classified today as very rich if it holds 5 wt.% or more of metal, rich if about 3 wt.%, and good if 1.5 wt.% is present. Ore that contains 0.2 wt.% or even less has been mined from placer deposits.

The geological classification of tin deposits is complex (Hosking 1979, 1988; Lehmann 1990). For our purposes, we propose a simplified classification which, for practical reasons, differentiates between three types of deposits that are eligible for early mining (Pohl 2005). The first two types are often called primary or mountain tin, while the third is called placer tin:

(a) Tin veins connected with granitic rocks. Classic examples of this deposit type are the tin occurrences in Cornwall, Brittany, the Ore Mountains and the Malaysian Peninsula. In pure (pegmatitic) tin veins, cassiterite occurs together with zinnwaldite, a tin-bearing mica, quartz, tourmaline (schorl), wolframite, scheelite, tantalite and fluorite (CaF_2). Among these primary deposits, tin-greisen and associated veins of cassiterite play the most important role by far. Formations of greisen are transformations of rocks by acid fluids at high temperatures. Feldspar becomes unstable and is replaced by topaz, mica, quartz, tourmaline and cassiterite. A distinction is made between a coarse-grained variety, the actual greisen, and a fine-grained variety, which was called a *Zwitter* by the Saxon miners. *Stockscheider*-like textures describe cm-sized, rough-pegmatitic parts in coarser formation than the surrounding rock. The composition of ore veins with cassiterite and wolframite with greisen zones in St. Agnes granite, Cornwall, is exemplified by the example of Cligga Head (Fig. 3.45a). Cassiterite impregnations in granite, the so-called *Zwitter*, are from the same locality in Cornwall and are shown in Fig. 3.45b.

(b) Hydrothermal veins often show a vertical zoning ("telescoping"), where pure tin ores dominate in the deepest parts of the veins, i.e. close to the granite intrusions, and upwards, so with increasing distance from these, copper and lead ores may replace the tin (Fig. 3.46). Here overlapping occur, so that large parts of the mineralisations consist of polymetallic ores (Sn–Cu–Pb–Zn–Fe–

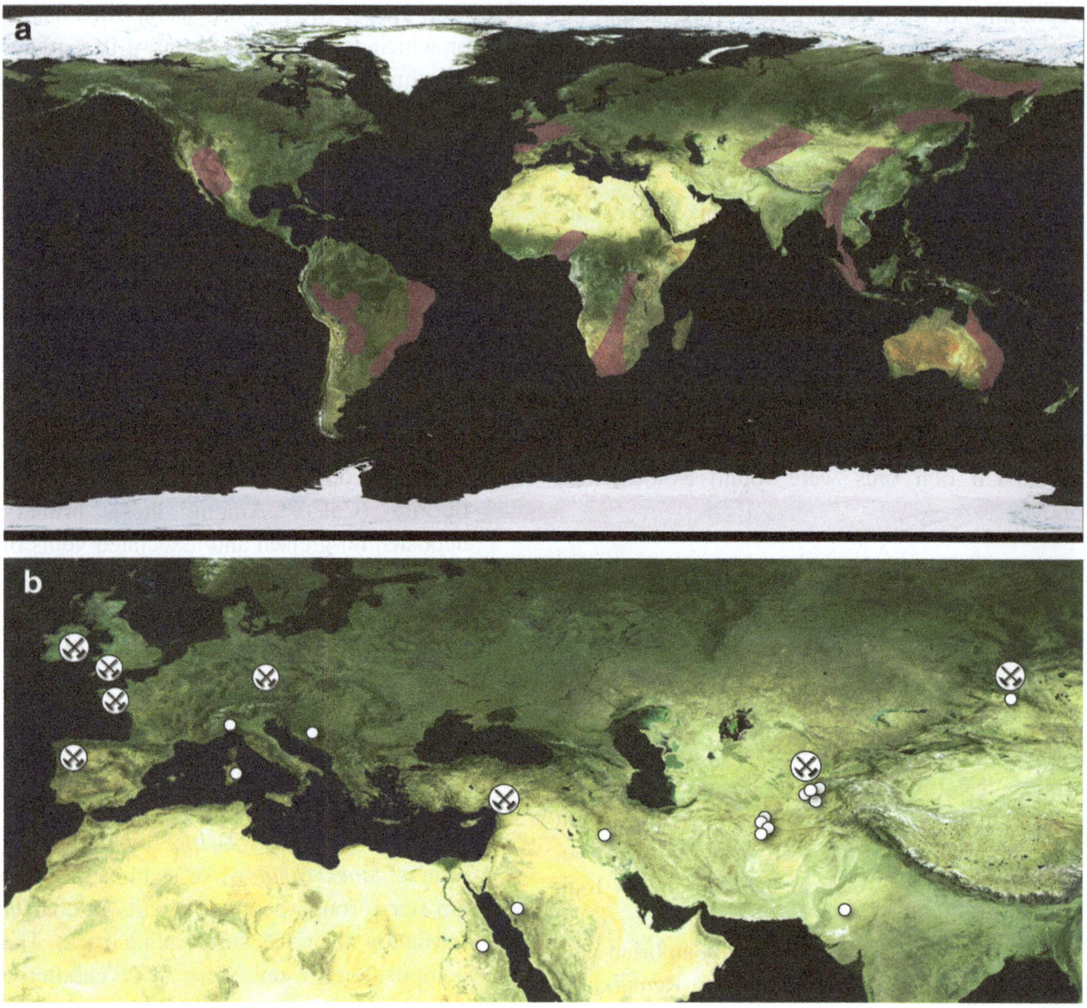

Fig. 3.44 (**a**) The large tin deposits are limited to only a few metallogenetic provinces worldwide, whose distribution is largely concentrated on (former) continental margins. In the Old World, they extend from Europe over the countries of Central Asia to China and Mongolia. In the New World, tin occurs on the western and eastern edges of the South American continent. Accordingly, a tin belt can be observed in western Africa. The Southeast Asian belt from Thailand to the south down to Malaysia is one of the largest and still economically important ones in the world. Similarly, the East Australian tin belt is of great importance. However, the location of the tin belts does not include the many smaller deposits of tin that exist in many countries around the world. Modified after Evans (1992). (**b**) As a section, tin deposits in Europe and the Middle East. A number of these deposits have been investigated (see text). These are indicated by crossed hammer and pick

As–Bi). Such blends are of the utmost importance in prehistoric tin archaeometallurgy, because they can be used to produce tin bronzes and, moreover, they may contain arsenic and other impurities. Copper–tin sulphides are easily recognisable the moment they weather to secondary minerals and the copper causes a green colour (see above). As already described for the paragenesis of the individual minerals, there is by no means a complete geochemical separation of the tin from copper. Cassiterite parageneses with Cu–Ag–Pb–Zn–Bi ores are known from tin deposits supra-regionally, e.g. in Cornwall (El Shazley et al. 1957), China (Tanelli and a

b

Fig. 3.45 Cligga Head, Cornwall. (**a**) Greisen-bordered sheeted veins of tourmaline, wolframite and cassiterite (black) in the St. Agnes-Cligga Head granite. The *Greisen* consists of quartz and muscovite replacing feldspar. (**b**) Impregnation of cassiterite, the so-called *Zwitter*, in the dense Greisen in the St. Agnes-Cligga Head granite. Photos: A. Hauptmann, Deutsches Bergbau Museum

Lattanzi 1985), in Bolivia and Peru (Potosi; Bartos 2000), Japan (Taylor 1979), in Tuscany (Venerandi-Pirri and Zuffardi 1982) and in Afghanistan (Cox 1986; Taylor 1979). In particular, the paragenesis of tin ores with arsenopyrite is of interest in that it may provide a conclusive explanation for the abundant As bronzes of the Bronze Age. Cu–Sn polymetallic ores are also known from numerous deposits in the Iberian Peninsula.

Polymetallic tin veins bound to intermediate volcanic rocks like andesite, dacite and porphyrites are found in the South American Cordilleras in Peru and Bolivia. Especially the deposit Cerro Rico de Potosi should be mentioned here (Bartos 2000). The ore deposit of Potosi is very rich in silver. The tin minerals here are stannite and other polymetallic Cu–Sn–Sb–Ag–Bi–Pb ores. As far as the paragenesis of tin with silver ores is concerned, the tin provinces in Central Asia and in the north-eastern part of Asian Russia (Yakutia) need to be mentioned, where quartz–carbonate–sulphide veins and stockwork zones embedded in volcanic rocks are described (Indolev and Nevoisa 1974). Polymetallic ores, stannite and/or cassiterite as subordinated components are also found in volcanogenic

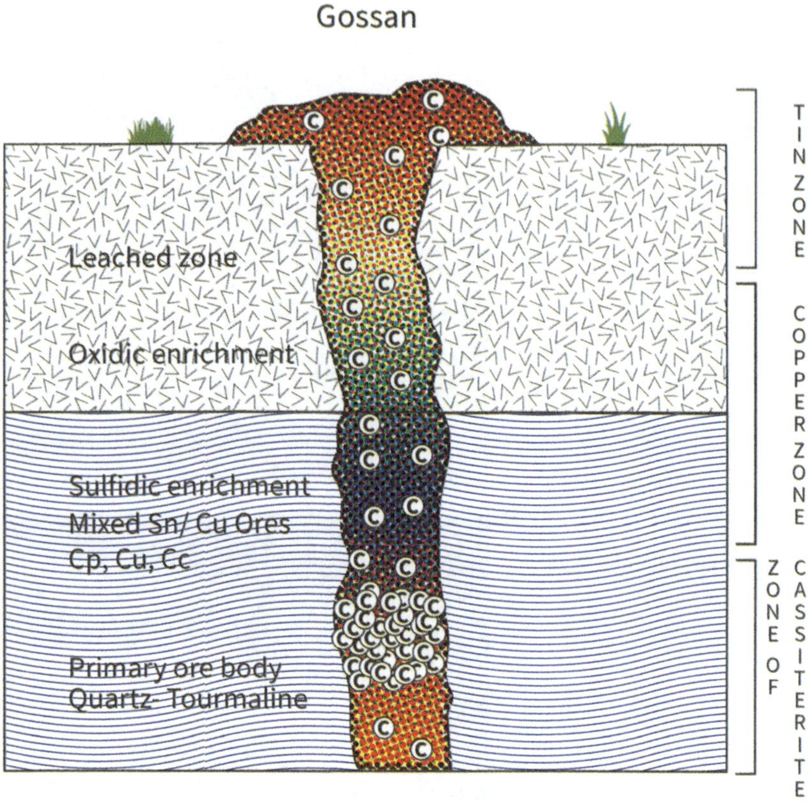

Fig. 3.46 Vertical zoning of tin-bearing hydrothermal veins (telescoping) exemplified by the tin deposits of southwest England. Typically, in the lower part of the veins (= close to granite intrusions), the pure tin ore cassiterite is predominating in association with quartz, tourmaline and fluorite. The composition of the veins is changing to their upper parts (= larger distance to granite intrusion). Due to geological overlapping, many tin deposits show associations of polymetallic, mixed Sn–Cu–Pb–Zn–As ores. After Hosking (1951), Evans (1992)

massive sulphide deposits (VMS), e.g. in the deposit of Neves Corvo in Portugal (Relvas et al. 2006).

Pre-colonial tin mines with a mineral assemblage of tourmaline, cassiterite, Fe-carbonates, pyrrhotite, chalcopyrite, sphalerite and Ni-sulphides were investigated by Recknagel (1908) and recently by Chirikure et al. (2007) from Rooiberg in South Africa. The latter author suggests that it is possible that tin bronze was produced by smelting copper ore contaminated with cassiterite.

(c) Tin placers are formed by the erosion of primary tin deposits and accumulation in fluviatile and coastal areas, sometimes together with gold (cf. Sect. 3.6). Recent tin placers are existing, as well as fossil tin placers, which are often covered by sediments. Fossil placers are usually embedded in solidified sediments, the extraction of which is just as costly as mining them from solid rock. The main mineral here is cassiterite, which also occurs very fine-grained, as "tin dust". This is comparable to the fineness of gold. Extensive tin placers have been mined since the Bronze Age, e.g. in Cornwall (Penhallurick 1986), but are still found to a considerable extent in southeast Asia today. According to modern analyses, the tin content in placers can be as low as 0.001 wt.% (Pohl 2005), without questioning their profitability. In the Ore Mountains (Germany) and its foothills, tin placers were common and eponymous. An example is the town name Seiffen, which is equal to placer. They were mined at least since the Middle Ages, as described by Georg Agricola (1556). Only in the placer, a substantial geochemical separation between copper and tin can be observed, since corresponding mixed ores are far less stable in the sedimentary cycle than cassiterite. Accompanying minerals such as tourmaline, wolframite or nonferrous metals are decomposed faster than cassiterite as well.

3.4.11.4 Regional Specifics

For the purposes of this book and in the cultural context of the Old World, only a few examples of tin deposits and occurrences in Europe and Eurasia will be mentioned to highlight some specific archaeometallurgical aspects. The most important points are:

- When and where were tin ores mined? And from monomineral cassiterite or polymetallic deposits?
- Which deposits have been of greater economic importance in prehistoric times?
- What are the reasons for the difficult verifiability of (pre-)historical tin extraction?

Central and Western Europe

In Central Europe, the largest tin provinces are Cornwall, the Ore Mountains, Brittany and the Iberian Peninsula. For these regions, there is written and mining–archaeological evidence of the mining of primary tin and of placers, but only for medieval periods and modern mining. Early in archaeological research, these tin placers have indeed been discussed as early sources for tin, due to the immediate geographical vicinity of the prehistoric cultural development in these regions. Borg and Pernicka (2017) emphasise that in Cornwall tin has in all probability already been mined in the early Bronze Age. The high proportion of tin bronzes in the metal inventory of the Wessex culture seems to be an indication of this.

Although Cornwall is world-renowned for the production of cassiterite, it has produced a substantial tonnage of copper, as well as eleven other metals/metalloids and other minerals, e.g. from polymetallic mineralisations of tin with arsenic, lead, zinc, wolfram and silver. Only in southern England, there seems to be evidence of prehistoric tin mining and metal production since the Early Bronze Age. Comprehensive work has been presented by Penhallurick (1986). Roden (1985) discusses possible origins of tin mining in the first century BC in Belerion (the Greek name for Cornwall) following the written evidence of Diodor (V, 22). It is not clear, however, whether cassiterite-bearing placers were mined here, or mountain tin, which may contain tin-containing polymetallic ores in addition to "pure" tin ores. Mining–historical evidence is missing.

In the Ore Mountains, apparently not only tin placers were mined. Their earliest mining is, however, hardly detectable before the twelfth century AD. Finds of Bronze Age pottery that might suggest the possibility of an earlier mining are exceptional only (Bartelheim et al. 1998). Currently, the prehistoric use of tin deposits in the Ore Mountains cannot yet be proven scientifically (via lead isotope analyses) (Niederschlag et al. 2003).

The tin province in Brittany is geologically comparable to Cornwall, and extensive placer deposits can be found here. Gallo-Roman underground mining is evidenced by various deposits (Roden 1985). However, this statement offers no indication as to whether tin placers with cassiterite were mined, or primary mineralisations characterised by polymetallic tin ores. Otherwise, old mining is difficult or impossible to date here, just like at any other deposit.

In the north and northwest and in the central region of the Iberian Peninsula, numerous tin deposits are located, which mainly bear cassiterite. Of particular interest is the occurrence of cassiterite of Cerro de San Cristobal in western Spain (Merideth 1998a). In the quartz veins, mm- to cm-sized crystals of cassiterite appear in paragenesis with the black tourmaline schorl and wolframite. Slags are not known from this locality, but stone hammers and a number of anvil stones and rubbing stones have been found. Pliny (34th book) mentions that the Iberian tin deposits played an important role in the supply of tin to the ancient world. But a distribution map of the deposits used at that time has not yet been created (Domergue 1970). In the north of Portugal, settlement patterns of the Late Bronze Age may be in line with the geography of the tin deposits. Tin slags were found here, too, as well as various pits and tin placers, stone and bronze axes of the Middle and Late Bronze Age (summary in Roden 1985) and crucible fragments (Merideth 1998a, b). Reference is also made to the use of polymetallic copper-tin ores in Spain during the Bronze Age (Rovira an Montero 2003).

Perhaps especially at smaller tin deposits, prehistoric activities can be proved. They are of no meaning for economic considerations today, and therefore modern activities may not have covered or destroyed prehistoric traces. But clear archaeological evidence has been found only at very few tin deposits and after extensive research, such as in Central Asia or in Kestel/Göltepe (Anatolia; see below).

Huska Mason et al. (2014) and Mason et al. (2016) report slight tin enrichment at Mount Cer in western Serbia. There, Bronze Age settlements are said to be located in geographical context with enrichments of cassiterite in the rivers Milinska and Cernica. Unfortunately, tin sources in the Balkan region would be poorly documented, but several small fluvial placer deposits have been described in western Serbia. They concluded that in prehistoric times, similar to Kestel or Deh Hosein in Iran, tin could have been recovered from these placer deposits. Problematic in their discussion is the fact that they cannot report on any tools that were used for ore processing.

Somewhat elusive as well is the mining of tin ores in the ore province of Campigliese, near Populonia (Tuscany, Italy). Tin ores (cassiterite and Sn–W–As–Cu–Pb ores) in low concentration mixed with hematite and limonite are known from there (Venerandi and Zuffardi 1982). On the basis of recent written records, Pampaloni (2017) suggests a prehistoric tin extraction: There are indeed prehistoric settlements in the region, but any mining–archaeological evidence is said to have been destroyed by modern mining. Tizzoni (1999) has doubted that before. His reasoning, however, is implausible. He doubts that the low content of cassiterite would have been mined before Roman times. He posits that it would have been unlikely that the tin ore had been recognised as such at any earlier time than that, and that there would hardly have been any knowledge of smelting the ores to metal.

In their publication, Radivojević et al. (2013) presented two tin bronzes from the Vinča cultural sites of Pločnik and Gomolovain in southern Serbia. These are the earliest tin bronzes presently known. This publication also compiles a series of tin bronzes from the Balkans (Bulgaria, Serbia) up to the Early Bronze Age. The authors analysed significant levels of Pb, As, Ni, Co, Fe and Au and interpret these data as a production from stannite-rich ores with varying levels of fahlores. They state that such tin occurrences are widespread throughout the Balkans. Unfortunately, there is no mining–archaeological evidence, no processing tools, and no remains of any crucibles from this early tin extraction.

The Ancient Near East and Eurasia

The Early Bronze Age use and processing of tin from the deposit of Kestel and the nearby settlement of Göltepe in the southeast of Turkey has long been controversially discussed in archaeometallurgy. Age measurements provide dates between 2874 and 2133 BC (Yener et al. 1989a, b, 2003; Yener and Vandiver 1993).

Geologically, tin mineralisations are proven beyond doubt. The tin occurrence is tied to hematite mineralisations, similar to Tuscany. It is in close proximity to granite intrusions and pegmatitic veins that penetrate gneisses, limestone and quartzitic rocks of the Camard Formations (Cagatay and Pehlivan 1988). The extraordinarily small grain size of the tin minerals (<150 µm) and the low tin concentrations in the deposit itself (<1 wt.%; Earl and Özbal 1996) were the main reason for the doubts about any importance of the Kestel deposit in ancient times. It may be worth considering that possibly existing rich ore zones have been exploited already. It has already been pointed out above that small grain sizes and low tin contents are generally characteristic for many tin deposits and require special, extensive processing and beneficiation (Earl 1985). After all, since time immemorial, also gold has been extracted from various deposits in minute concentrations with great effort.

In contrast, higher tin contents (0.2–0.4 wt.%) were analysed in the processed hematite-rich sands at Göltepe, and the thin slag layers inside of crucibles found in the settlement almost reached 4 wt.%. The unbelievable amount of processing tools (about 50.000) and crucibles found in Göltepe is of the utmost interest to archaeometallurgy: Yener et al. (2003) report huge amounts of crucible fragments, demonstrating large-scale processing of the metal in small vessels. This quantity of crucible fragments is unparalleled so far. However, an estimated production volume of several tons of tin in Kestel or Göltepe may be too high, and the measurements of many kilometres of prehistoric galleries in the mines at Kestel is only roughly estimated from the publications (Willies 1992), but not quite verifiable to this extent.

Besides Kestel/Göltepe, tin ores were also found in Kayseri-Hisarcık in central Anatolia. Iron–tin–arsenic ores with cassiterite were deposited by fumaroles within soft, easily carved material, allowing the tin mineral to be readily extracted, even with antler and horn (Yalçın and Özbal 2009; Yener et al. 2015). Early Bronze Age II-III and even earlier pottery was found in context with some mines.

For the early bronze metallurgy, the Central Asian tin belt is of major importance, ranging from Tadzhikistan, Uzbekistan (Karnab, Changalli, Mušiston) and Kazakhstan to China and Mongolia (Fig. 3.44a, b). Especially the results of Alimov et al. (1998), Parzinger and Boroffka (2003), Cierny and Weisgerber (2003); Cierny et al. (2005) and Garner (2014) should be referenced here, who, in a long-term research project, have dealt mainly with the western part of it. The prehistoric exploitation of these tin deposits, described from Central Asia, is of considerable importance to the archaeometallurgy of the ancient world. For a long time, the deposits of the Southeast Asian tin belt in Malaysia and Thailand were repeatedly considered as potential tin suppliers of the European Bronze Age. Reasons for this were on the one hand their current economic importance, on the other hand, due to political reasons, the tin deposits in Central Asia were not accessible to research projects of the Western world. Karnab in Uzbekistan is a typical example of a tin deposit that genetically bound to granite. The veins are in the context with granite intrusions, so they consist of relatively pure tin ores (cassiterite). They are in contrast to Mušiston, which features rich Cu–Sn mixed mineralisations.

Another important mining district is located in eastern Kazakhstan, which was studied between 2003 and 2008 in a Kazakh-German research project and published in an exhibition catalogue (Stöllner and Samašev 2013). The tin deposits, again in the margins of rising granites, are geologically one of the oldest known. Malchenko and Ermolov (2006) describe tin deposits in the Paleozoic fold mountain belt of the Altai Mountains (Devonian/Carboniferous). The main tin bearing deposits are reported from the east and north-east of the country. Quartz–wolframite–cassiterite deposits are located in the central part of Kazakhstan. Evidence of early mining activities there (Early Bronze Age) is reported by Stöllner et al. (2013a, b). In this work, chemical and lead isotopic analyses of non-ferrous metal and tin deposits and tin bronzes were published, which are worked out in detail by Gontscharov (2019). Corresponding to the geological location of the deposits, the lead isotope data in part revealed very high geological ages ($^{208}Pb/^{206}Pb > 2.11$ und 2.12). They thus fit the isotope pattern of the oldest (Early Bronze Age) tin-bronzes from Mesopotamia (Begemann and Schmitt-Strecker 2009). In Central Asia and the Altai region, tin, gold, lapis lazuli and nephrite, which together mark the beginning of the Bronze Age in Mesopotamia and Anatolia, occur together (Alimov et al. 1998).

A not to be underestimated role in the earliest supply of tin for the Old World may have been played by Afghanistan. The northern part of the country belonged to the Bronze Age culture of the Bactria Margiana Archaeological Complex (BMAC), which, between about 2300–1700 BC, included the area of eastern Turkmenistan, southern Uzbekistan and western Tadzhikistan (Lyonnet 2001). The centre was the area of upper Amur Darya (Oxus River). In Afghanistan, the archaeological state of research on early tin is still a long way off, but geological and economic raw material investigations by the Federal Institute for Geosciences and Natural Resources as well as the Soviet Geological Survey and the US Geological Survey have at least described the geological layouts of ore deposits. According to Wolfart and Wittekindt (1980), about 135 localities with tin ores were found. The tin contents are often <1 wt.%, only four of the deposits are of modern economic interest. These tin deposits are bound to geologically young granitic intrusions of the early Alpidic context. Following Cox (1986) and Taylor (1979), in Afghanistan, the entire genetic variety of tin deposits occurs, as explained above for Cornwall, the Ore Mountains, Malaysia and Potosi. Veins with cassiterite + wolframite + arsenopyrite were found, but cassiterite with hematite and typical telescoping of cassiterite to parageneses with galena + chalcopyrite + stannite + bismuthinite as well. The authors generally mention "ancient workings". Many mineralisations are located in the Kandahar area (Arghandab) in the desert areas in the south of the country (Fig. 3.47), then there

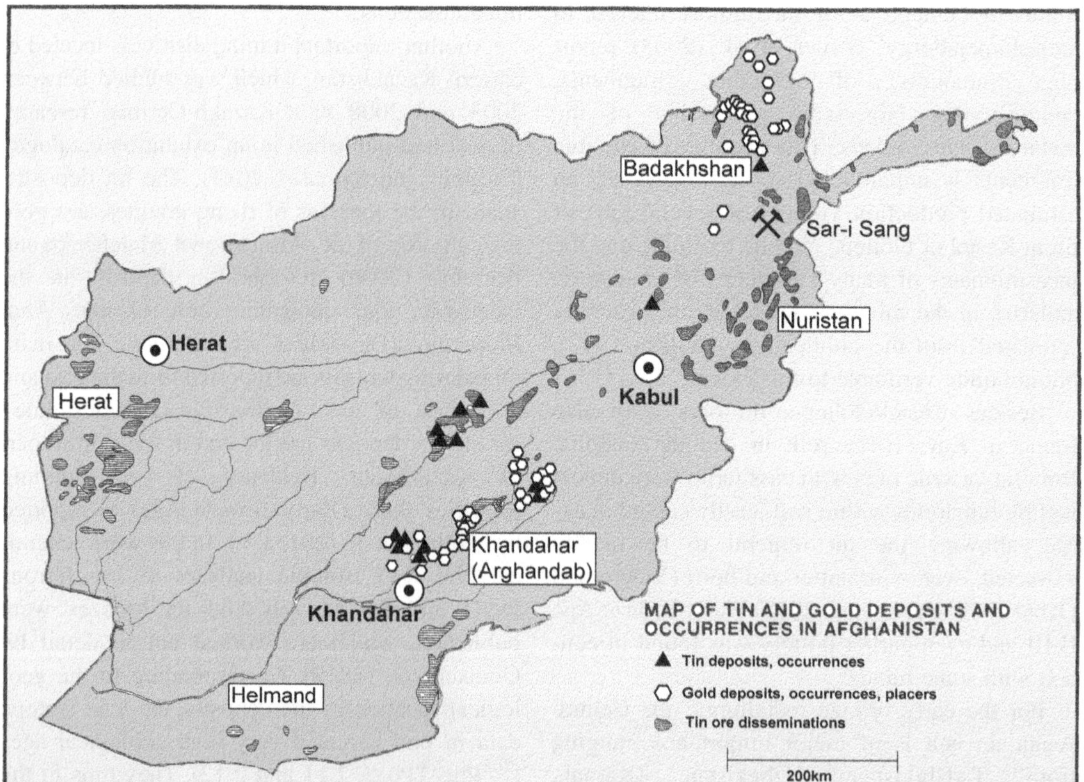

Fig. 3.47 Tin and gold deposits in Afghanistan. Note the "Tin ore disseminations" in the Herat area. It is likely that this ore province extends westward to Iran. The old mining area of lapis lazuli near Sar-i Sang, which is often quoted as the "anchor" in the reconstruction of trade relations of the (Early) Bronze Age, is marked for the sake of orientation. Modified after data from Shareq et al. (1977) and Pigott (1999)

are deposits in the Herat zone in the west, and the deposits in Nuristan and Badakhshan in the north. Here cassiterite occurs in several deposits along with chalcopyrite and other non-ferrous sulphides. In the Band-e Medirah area (Herat) in Afghanistan, tin mineralisations are described with copper, lead and zinc contents of 0.2–1 wt. %, respectively. A number of larger tin occurrences are shown along with gold occurrences.

In the discussion about the prehistoric trade routes of tin, which, according to written sources, was traded from the East to the Middle East, Iran, though rich in resources, attracted surprisingly little attention. This may be due to the fact that deposits of tin ores there have become known only recently. Nezafati et al. (2009) report about the prehistoric deposit of Deh Hosein (preliminarily dating to the first half of the second century BC), which revitalised the discussion about possible tin mining in the Iranian Plateau. Deh Hosein is located 45 km south-west of the city of Arak at the eastern edge of the Zagros Mountains. Interestingly, this deposit has a polymetallic mineralisation of cassiterite + pyrite + chalcopyrite + arsenopyrite + stannite, and not a pure tin–wolfram mineralisation, as in the greisen of the Ore Mountains or in Cornwall. But in Deh Hosein, cassiterite only occurs in tiny crystals of 250 μm, similarly fine-grained as the tin ores observed in Kestel and the Kazakh deposits. The discussion about this is controversial. On one hand, it is discussed that larger aggregates have been completely exploited, on the other hand, e.g. Kestel, as a potential supplier of tin, is questioned as much as the deposits in Tuscany due to the fineness of its tin ores. Further research is needed.

Further, Momenzadeh (2004) also reports on polymetallic Cu–Sn–W–Au mineralisations in pre-Cretaceous magmatic-metamorphic rocks. Such mineralisations occur, e.g. in eastern Iran, near the Afghan border, and apparently follow and join the mineralisations in the Herat (Fig. 3.47). They are barely known to modern mining. Such mineralisations were apparently more widespread in the past.

3.4.12 Iron

Iron (Fe) is the fourth most abundant element in the earth's crust (5.76 wt.%) after oxygen, silicon and aluminium and is enriched by a factor of 10,000 compared to copper (55 ppm). Iron occurs in many siliceous, sulphidic and oxidic minerals as a major or minor element.

Iron is a predominantly siderophile element; it often forms iron oxides (and hydroxides). But iron is chalcophile as well and crystallises as sulphides. Of particular importance for the formation of ore deposits is the varying degree of oxidation of iron in nature: It occurs in three valence levels, these are: Fe^0 (native Fe), Fe^{2+} (ferrous iron), and Fe^{3+} (ferric iron). The latter valence level, the oxidised iron, is immobile. This leads to a fixation of large amounts of iron in the exogenous cycle of rocks. A large amount of iron ore is formed by weathering. Therefore, the geochemical behaviour of iron in this environment is particularly important (Pohl 2005).

3.4.12.1 Iron Minerals

Table 3.10 lists a number of iron ores that are of greater importance in archaeometallurgy.

Table 3.10 Important iron minerals, their chemical composition and iron content

Mineral	Chemical formula	Fe (wt.%)
Native iron	Fe	99
Kamacite	(Fe,Ni)	92–97
Taenite	(Fe,Ni)	35–80
Magnetite	Fe_3O_4	72
Hematite	α-Fe_2O_3	70
Maghemite	γ-Fe_2O_3	70
Goethite	α-FeOOH	62
Lepidocrocite	γ-FeOOH	62
Siderite	$FeCO_3$	48.2
Pyrite	FeS_2	46
Marcasite	FeS_2	46
Pyrrhotite	FeS	61–63
Ankerite	$(Ca,Mg,Fe)CO_3$	$\leq$15
Ilmenite	$FeTiO_3$	37.3
Titanomagnetite	Fe_2TiO_4	Changing
Vivianite	$Fe_3^{2+}(PO_4)_2\ 8H_2O$	Changing

Values of iron concentrations after Doelter (1912–1932) and modern mineral data

Terrestrial (Native) Iron

In contrast to meteoritic iron, terrestrial iron occurs extremely rarely and only in (ultra-)basic rocks, which have formed only under specific reducing conditions, e.g. by resorption of coal in a magma by a penetration of basaltic veins through coal deposits. These are "natural smelting products", as formulated by Ramdohr (1975).

Terrestrial iron was found as a curiosity in Buehl near Kassel (Germany), where small particles occur in alkaline volcanics. In basalts from the island of Disko (Greenland), on the other hand, terrestrial iron blocks of 10 kg to 25 tons occur.

Buchwald (2005) distinguishes two types of terrestrial iron in Greenland:

1. Massive blocks of nickel cast iron with more than 1.5 wt.% C.
2. Minute grains of malleable low-carbon nickel–iron (< 0.2 wt.% C, 2–4 wt.% Ni), distributed in the basaltic rock.

Meteoritic Iron

There are several thousand finds of meteorites worldwide, most of which are siliceous rock meteorites or rock-iron meteorites. More than 800 iron meteorites are known (Okrusch and Matthes 2005). Meteorites are of cosmic origin. There are numerous iron meteorites with weights up to tens of tons, e.g. the Hoba meteorite in Namibia (Fig. 3.48) or the 20-tons iron meteorite

Fig. 3.48 Grootfontein, Otavi-Mountains (Namibia). The Hoba meteorite weighs about 60 tons and is the largest meteorite of the world. It consists of 82.4 wt.% Fe, 16.4 wt.% Ni and 076 wt.% Co. Other elements present in traces are carbon, sulphur, chromium, copper, zinc, gallium, germanium and iridium (Schneider 2008). The Hoba meteorite is classified as an ataxite. Photo: A. Hauptmann, Deutsches Bergbau-Museum Bochum

from Agpalilik, Greenland (Buchwald 1992). Characteristic feature of iron meteorites are their high nickel contents (flat contents of approximately 5–11 wt.%), they are therefore natural iron–nickel alloys. The principal iron–nickel alloy minerals in iron meteorites are kamacite and taenite.

Fe–Ni meteorites are classified according to their nickel content. Low-nickel meteorites contain 5–6.5 wt.% Ni and are made up by kamacite, it is a single-phased Fe–Ni alloy. The lower limit of 5 wt.% is hardly undercut. Meteorites with 6.5–13 wt.% Ni or more consist of two phases, namely kamacite und taenite (30–70 wt.% Ni).

Kamacite and taenite form the distinctive beam-like Widmannstätten structure (Fig. 3.49). These rough octahedral lamellae gave these most common iron meteorites the name octahedrites. Hexahedrites are very rare, contain less nickel and consist entirely of kamacite. Kamacite consists of α-Fe with only 5 wt.% Ni, and taenite consisting of γ-Fe high of nickel. The intergranular mass between these two minerals is called plessite. It is a fine-grained mixture of kamacite and taenite. A third group of iron meteorites consists of a very fine-grained kamacite taenite structure and contains between 16 and 18 wt.% Ni. In iron meteorites constantly varying amounts of cobalt and chromium are present. Rare but typical minerals in iron meteorites are the silicate coesite (high-pressure modification of SiO_2), schreibersite $(Fe,Ni,Co)_3P$, and cementite (Fe_3C). Overall, most meteorites have a different chemical composition than the rocks in the earth's crust. Apart from the metals mentioned, they also have elevated contents of noble metals such as iridium (Ir), osmium (Os), rhenium (Re), gallium (Ga) and germanium (Ge) (Lovering et al. 1957; Heide 1988). It may be possible to use these elements as criteria for distinguishing between meteoritic and terrestrial iron, provided that it is possible to apply precise trace element analysis to archaeological artefacts.

After entering the earth's atmosphere, iron meteorites easily form a weathering crust of magnetite, maghemite and goethite. The weathering can continue to the inside, with the Fe–Ni alloy merging into a mixture of goethite and the rare minerals akaganeite (β-FeOOH) and hibbingite ($Fe_2(OH)_3Cl$) with relic patterns of the original metals. In that case, they are no longer clearly distinguishable from corroded terrestrial iron,

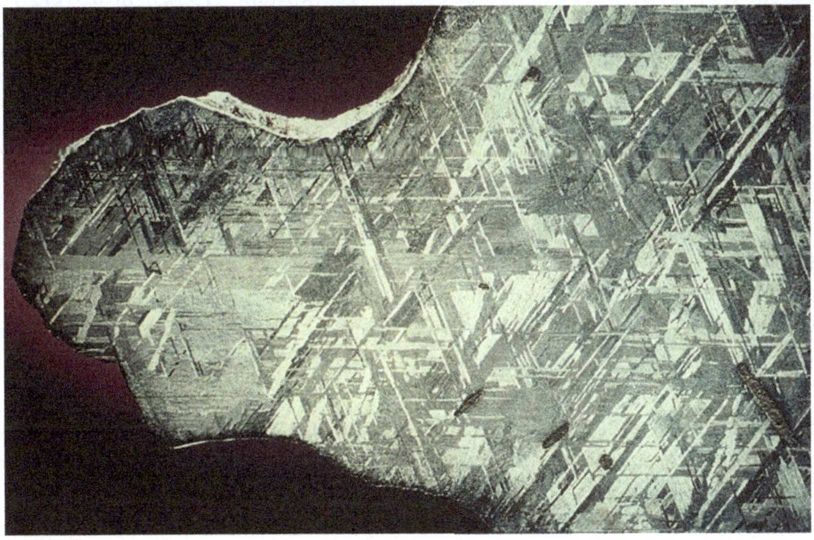

Fig. 3.49 Grootfontein, Otavi-Mountains (Namibia). Fragment of the world's largest Hoba iron meteorite. Note the Widmannstätten structure, which is a bulky texture consisting of kamacite with rims of taenite. The filling in between is named plessite. Collection of the German Mining Museum Bochum. Photo: A. Hauptmann, Deutsches Bergbau-Museum Bochum

especially when meteoritic iron has been forged, resulting in a disorder of the Widmannstätten structure.

Meteoritic Versus Terrestrial Iron

The identification of meteoritic and its differentiation from terrestric iron is not only of geological interest to locate earlier impacts of iron meteorites. It is also of essential importance in archaeology to reconstruct whether (pre-) historical iron objects were made of meteoritic iron or melted from nickel-rich iron ores.

From the technological point of view, it seems possible to shape meteoritic iron hot to artefacts within simple installations, but also to hammer it in cold state. This is also true for terrestrial iron. Such iron was still being used by the Inuit in Greenland in the nineteenth century (Buchwald 2005). Characteristic of meteoritic iron is that even by repeated forging the typical Widmannstätten structure is damaged, but not completely destroyed. However, the author explained how iron meteorites almost lose their structure by weathering. In the best case, they might be recognised by the presence of minute surviving patches of kamacite and taenite and/or particles of schreibersite, cementite or chromite ($FeCr_2O_4$), but this is often doubtful (Johnson et al. 2013). Nickel and cobalt may still be identified in meteoritic shale that is more than a hundred thousand years old.

According to the present state of analytical investigations (see Buchwald 2005; Pernicka 1990), due to the currently limited conservational reasons, meteoritic, terrestrial, and iron produced by melting processes, cannot always be clearly distinguished. Destructive sampling for microscopic investigations or for analytical investigations by mass spectrometry in most cases is not possible. Therefore, publications on the early use of meteoritic iron for the production of artefacts in prehistoric times are controversial. This is discussed in Pernicka (1990), who describes a series of iron artefacts from the fourth and third millennium from the Near East, southwest Asia (Iran, Mesopotamia, Anatolia, Egypt, Crete, Syria, see Coghlan 1977; Waldbaum 1980), which are said to have been made of meteoritic iron. Here, he announces doubt. The conclusions have been made on the basis of their nickel concentrations, analysed either quantitatively or qualitatively and sometimes also on the basis of their Widmannstätten structures. The discussion has recently rekindled in the study of small tubular beads from approximately 3200 BC Gerzeh in northern Egypt. Based on non-invasive studies by Johnson et al. (2013) and Rehren et al. (2013), the beads of meteoritic iron are said to have been formed by hammering and rolling. Due to the high age of the finds, a production of iron by melt metallurgical processes was ruled out.

The nickel content is often used as a distinguishing feature of meteoritic iron. Using examples from Greenland, Buchwald (2005) shows that the local terrestrial iron reaches a maximum of 4 wt.% Ni, which is below the meteoritic iron. Cobalt, with 0.1–0.4%, lies under the meteoritic iron as well. As a further distinguishing criterion, osmium isotopy has also been used. After a long discussion, an occurrence of massive iron hydroxides in Tettenwang, Bavaria, has been identified as weathered meteoritic iron, due to the isotope ratios of $^{187}Os/^{186}Os$ (Appel and Garges 1991).

Piaskowski (1992), however, who studied a series of Ni-bearing iron artefacts metallographically and chemically, proposed a completely different alternative for their formation. In his publication, he concludes that the spread of "meteoritic" iron artefacts may have been exaggerated in many cases and that they could very well have been smelted from correspondingly composed iron ores. This proposal makes sense because such mineralisations indeed exist. They occur in nickeliferous laterites and will be discussed below.

Limonite

The widespread Fe-hydroxide limonite consists mainly of the two varieties goethite and lepidocrocite. It is the iron ore that had by far the greatest spread as a source material for the extraction of iron or steel in early metallurgy. It is the final product of all weathering processes of iron ores and it is the most widely distributed iron ore. Pyrite alters readily to hydrated iron oxide. Limonite is the primary ore of pisolites, oolites or of bean ore (Fig. 3.50) and pea ores. Mixed with clay, it forms the pigment of yellow ochre. It should not be confused with laterite, though both are composed of goethite. A common variation of limonite is the globular, botryoidal goethite that results from the crystallisation of gels. If such crystal aggregates are formed from hematite, they are called kidney ore.

Fig. 3.50 Heidenheim, Swabian Alb, SW-Germany. (**a**) Bean ore (name!) or pisolitic iron ore is a characteristic ore type in south-western Germany and in Switzerland. The bean- or pea-sized globules are often baked together to larger units. (**b**) Under the microscope the beans show clearly the concentric structure of the ore. They are made up of limonite with some clay and quartz grains. Width of the picture ca. 3 mm. In case the globules are only mm-sized the ore is named oolitic iron ore or Minette-iron ore. This type of ore has a large distribution in Alsace-Lorraine. It was exploited up until the middle of the twentieth century AD. Photos: (**a**) M. Jansen, Penn Museum, Philadelphia; (**b**) Ü. Yalçın; Deutsches Bergbau-Museum Bochum

Hematite

Hematite is also very common. According to grain size and appearance, different trivial names are used, especially in Germany, such as *Eisenglanz*, *Roteisenerz* or *Blutstein*. Specularite or itabirite, which in ancient times was a highly desired pigment for cosmetics, consists of fine-grained, silvery and metallic shining micaceous or tiny crystals. Fine-grained, massive hematite could also be of black-deep purple-red colour, as it occurs in the famous iron ore deposit at Mugharat el-Wardeh (Jordan) (Alamri and Hauptmann 2013). Such ore was exploited as early as the fourth millennium for making maceheads and pigments. One of the hundreds of maceheads of the Nahal Mishmar hoard is made of hematite which in its mineralogical design and colour is identical with Mugharat el-Wardeh. Hematite is mostly associated with hydrothermal ore deposits.

An earthy variety of hematite is red ochre. It contains varying amounts of clay (20–70 vol.%). Hematite controls the intensity of the colour. Red ochre is a mixture of anhydrous iron oxide (usually hematite) and clay (or weathered feldspar with or without lime content, or Al-containing silicates, oxides and hydroxides). The iron oxide content should be at least around 30 wt.%. The mixture is solid in nature, feels clayey and greasy, has good adhesion and is water repellent. Red ochre is in principle a red variety of yellow to brown-yellow ochre. Instead of iron hydroxides, it contains anhydrous iron oxide.

Mining for red ochre belongs to the earliest mining activities to use mineral raw sources. In South Africa, mining for red ochre (specularite) at Ngwenya, Swaziland (Dart and Beaumont 1969) dates back some 41,000 years, as well as at some other localities. In Europe, red ochre was mined about 20,000 years ago (Koukouli-Chrysanthaki and Weisgerber 1999) at Tzines on the island of Thasos (Greece), and in the Black Forest, Germany, some 5000 BC, during the Bandkeramischen Kultur (Zimmermann and Goldenberg 1991). In the mines of Tzines fine-grained, loose and colour-intensive hematite was mined, which had accumulated in irregular karst cavities in metre-thick pockets (Weisgerber and Cierny 2008). Tzines was exploited as a raw material source for iron ore until the middle of the twentieth century AD.

Also the caves in the mountainous region of Serra da Capivara (state of Piauí, Brazil) with their thousands of spectacular red rock paintings are among the oldest traces of human existence in South America. The drawings were estimated to be c. 35,000 years old (Guidon and Delibrias 1986). In 1991, UNESCO placed the location on the list of World Heritage Sites. The creation of the drawings must have required enormous amounts of dye. Here it is striking that the Serra da Capivara is located geographically close to the vast deposits of Itabira, where immense deposits of hematitic ores exist.

Other palaeolithic mines for exploiting red ochre are known from Rydno (Poland; Królik and Schild 1999) and from Lovas (Hungary; Sajó et al. 2015).

Magnetite

Next to meteoritic iron, magnetite is the highest grade of iron ore, it contains 70 wt.% Fe and contains ferrous and ferric iron ($Fe^{2+}Fe_2{}^{3+}O_4$). It has an interesting property which might have been known in ancient times: it is a natural magnet. As an accessory mineral magnetite occurs in many igneous rocks but it could be enriched in hydrothermal veins. Due to its stability against weathering magnetite and its solid solutions can be occasionally enriched in placers and laterites and bauxites.

While hematite is an iron ore with a simple chemical composition that reveals little information about its origin, magnetite can uptake significant amounts of Mg and Al, and transition elements (e.g. mainly Ti, Mn, Cr, Co, Ni, etc.) in the crystal lattice. These elements are possible indications for the provenance of this Fe oxide (Benvenuti et al. 2013). Of major importance is the formation of titanomagnetite (Fe_2TiO_4). It also may form solid solutions with hematite by taking up Fe^{3+} ions to form maghemite (Deer et al. 1967).

Pyrite/Marcasite

Pyrite and marcasite are minerals of the same chemical composition with different crystallographic features. They both are "jacks of all trades" and occur in almost all geological environments and are associated with other sulphides or oxides in quartz veins, sedimentary rock, and metamorphic rock, as well as in coal beds, and as a replacement mineral in fossils. Pyrite may contain small amounts of arsenic and gold and is often associated with chalcopyrite. In such cases, pyrite was and is exploited because of these metals. Marcasite may occur along with pentlandite ($(Fe,Ni)_9S_8$) and other copper ores.

In ancient times, pyrite and marcasite were used for making fire. Using a hard piece of chert, small fragments can be chipped from a crystal which immediately become inflamed and are sparking due to the oxidation to iron oxide. Once exposed to the surface of ore deposits, pyrite and marcasite, and pyrrhotite as well are weathered quickly to Fe-hydroxides (limonite, FeOOH). Sulphur will form sulphuric acid (H_2SO_4) which can be smelled while working in pyrite dominated mines. The largest mines made up of copper-containing or gold-bearing fine-grained pyrite are Rio Tinto and Cyprus. Estimated reserves at Rio Tinto are a billion tons of ore.

3.4.12.2 Iron Deposits

Due to its wide spread, there are countless iron ore deposits worldwide, in sometimes extremely dense spatial distribution, and can occur in various sizes. In most cases, these are "only iron" deposits. The density of iron ore deposits is made obvious, e.g. in Germany, or more precisely in the former Prussian state, where at the beginning of the twentieth century, 18.000 mining fields were lent for iron mining (Einecke and Köhler 1910). It can be assumed that there are similarly many deposits in neighbouring and other countries. This clarifies that iron ore deposits are much more common than all other non-ferrous metal and precious metal deposits. The requirements of the ancient iron metallurgists, therefore, were not dependent on a single large iron ore deposits, possibly in a larger distance, but could easily be met from the countless very small outcrops of bog ore or limonite in the close vicinity of his settlement (Pleiner 2000). This is probably the main reason which decisively pushed the distribution of iron metallurgy over the Old World.

Economically important world reserves today are located primarily in Brazil, where the largest iron accumulations in the world are located. The type of deposit occurring there belongs to the "Banded Iron Formation"(abbreviated BIF) sedimentary ores, which have formed predominantly in the Proterozoic period. The BIFs are layered, they consist of iron oxides and quartz layers, can consist of almost pure hematite and take a lateral extent of tens of thousands of square kilometres. Famous are the vast deposits of Itabira, Brasilia, which named the variety of hematite itabirite. Possibly the raw sources of Itabira have been the base for the red colour pigments of Capirava, where many thousands of Palaeolithic petroglyphs were found (see above).

BIFs are common worldwide and mined in many countries in North and South America, Greenland, Australia, Africa, India and Russia (Pohl 2005).

Laterites

Laterites are products of intense weathering of iron-bearing rocks in warm, tropical climates. Aside from quartz, which originates from the source rock and is hardly soluble, they consist of clay minerals, goethite, hematite and gibbsite (γ-$Al(OH)_3$). The iron oxides cause the usually reddish-brown colour of the laterites. The uppermost part of the laterite is covered by an iron-rich cap, the so-called ferricretes. Such iron accumulations can lead to the formation of widespread terrestrial-sedimentary iron ore deposits, which are formed close to surfaces and have been easily accessible to (pre-)historic metallurgists. They are exceptionally geochemically uniform and only in parts they contain P_2O_5 and MnO contents as vivianite and various Mn hydroxides. The archaeological evidence of mining such mineralisations is difficult. Shallow opencast workings would quickly be obliterated by subsequent cultivation or simply by soil erosion, leaving no traces in the archaeological record (Pleiner 2000).

The weathering of ultrabasic rocks (peridotite, serpentinite) allows nickel and cobalt to be enriched in laterites. The type locality for such deposits is New Caledonia, where 2–5 wt.% Ni and more were analysed in ferricretes and the underlying saprolite zone (discussion in Pohl 2005). Ni-rich laterites may also contain traces of Cu, As, Zn, Co and Cr as relics of original (sulphidic) mineralisation in peridotites. Lateritic iron ores occur in almost all continents of the world in different geological periods (Schneiderhöhn 1962). Here we refer in particular to deposits in the Balkans, in Turkey, and in large parts of Africa (Eliopoulos and Economou-Eliopolous 2000; Eliopoulos et al. 2012). Their formation is a result of warm, humid climates in the geological past (discussion in Butt and Cluzel 2013). Eliopoulos et al. (2012) show microscopically small inclusions of Ni-magnetites ($(Fe, Ni)_3O_4$) and chromites ($FeCr_2O_4$) in the laterites of the Gordes deposit (Turkey) which, given appropriate reducing conditions during smelting, are optically and chemically identical to those described by Johnson et al. (2013). Neither this author nor Rehren et al. (2013) consider nickel-bearing laterites as reference material in their discussion on the genesis of the assumed meteoritic iron beads from Gherzeh.

The enrichment in laterites especially of nickel and cobalt are of interest in archaeometallurgy because they may represent a widespread alternative source of raw material to the controversial interpretations of artefacts of meteoritic iron. Photos (1989) experimentally produced nickel-rich iron by smelting lateritic ores and she pointed to the frequent occurrence of those ore deposits in Greece.

Siliceous, Al_2O_3-bearing and carbonic gangues and host rocks of lateritic iron ores are usually slagged during smelting, as, e.g. the widespread manganese, which does not cause any difficulties as it is a very good slag former due to its high stability as an oxide. Other elements or impurities are often incorporated as alloying

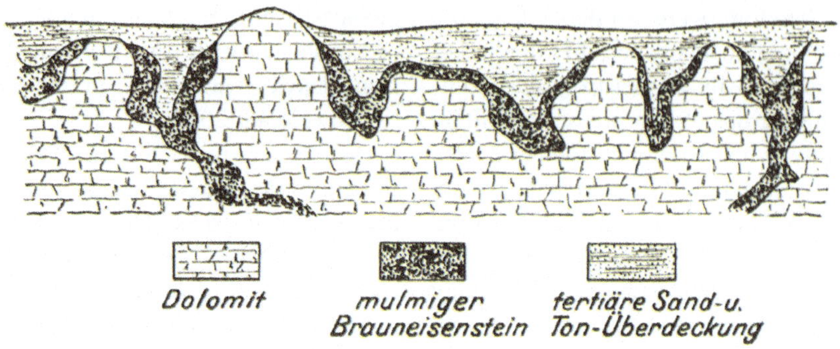

Fig. 3.51 Section through an iron mineralisation of Karst-ores von Amberg-Auerbach, southern Germany, showing characteristic enrichment of limonitic ore in pockets and caves. Surface-near mineralisations like that were easy to exploit in ancient times but were only of limited importance for modern economics. From Klockmann (1908)

constituents or in mechanical form in the resulting metal and can change the quality of the iron produced. Especially phosphorus and sulphur are to be mentioned here. Phosphorus goes into the pig iron if this is formed next to ferritic iron (see Sects. 6.6 and 7.1). It forms Fe–P alloys and makes the iron brittle. Furthermore, sulphur contents in the ore should be rather low, as it also makes the iron brittle.

In Africa, lateritic iron ores are evident in a broad zone stretching from western Africa to the east to Kenya, Tanzania and Uganda, and also in northern Africa, and has also been documented in Nigeria, Chad and the Sudan (summarised in Widdowson 2007). Further south, ferricrusts occur in the Congo, Angola, Zambia, Zimbabwe, Mozambique, South Africa and Madagascar. It seems that *lateritic* ferricrusts were the most likely source of iron ores in many regions in Africa.

Pisolitic Iron Ores

Bean ores and Minette ores are a widespread sedimentary type of such deposits that occur throughout Europe and in other countries. They were mined from the Iron Age until Medieval times, e.g. in Lorraine, on the Swabian and Franconian Alb in southern Germany, Switzerland, in Belgium or in the south-eastern Alps. The iron ore deposits and the iron production in southwest Germany, especially in the area of the Swabian Alb, were researched in detail in a long-term research project of the Baden-Württemberg state department for the preservation of historic monuments (1995, 2003, 2005). Here, ores with up to 60 wt.% FeO were available. Often they contained manganese levels up to 2 wt.% as well as some phosphorus (0.1–0.2 wt.% P_2O_5). On the Swabian and Franconian Alb in southern Germany, the ore bodies copy the sometimes peculiarly shaped surfaces of the underlying limestone or dolomite. They belong to the type of mineralisations in Karsts (Fig. 3.51).

Bog Iron Ores

Of sedimentary origin is also a group of iron ores, which, depending on their environment, are referred to as bog iron ores or lake ores. They have a supra-regional distribution and occur in swamp and lake areas in the entire northern European lowlands, from France to Denmark, Sweden, the northern part of Germany and Poland to Russia (Fig. 3.52a). These are irregular precipitates of mostly poorly crystallised or amorphous iron hydroxides (in addition to the Fe-phosphate vivianite) in river head areas, swamps, wet meadows, lakes (lake ores in Scandinavian countries). Concretions or bank-like mineralisations are formed, which lie close to the surface with a thickness of about one-half to one metre (Fig. 3.52b) and are embedded in sand, clay or even in gravel. They often extend over many hundreds of square kilometres. The iron contents reach approximately 45–50 wt.% Fe_2O_3.

a

b

Fig. 3.52 Bog iron ores. (**a**) Nudow, near Potsdam-Mittelmark (Germany). The ores are located directly on the surface, under the turf. From Sitschick et al. (2005). (**b**) Buttjebüll, Bordelum, Schleswig-Holstein (Germany). Layers, lines and incrustations of bog iron are widespread in the German lowlands from the border of Denmark to wide areas in Poland. The incrustations have a thickness of c. 1 m. They are formed directly under the present surface. These bog iron ores were exploited in the Roman period. In the picture, the world-famous mining archaeologist, the late Gerd Weisgerber is studying the layers of bog iron ore. Photo: A. Hauptmann, Deutsches Bergbau-Museum Bochum

In northern Germany, on the border to Denmark, near Joldelund, the extraction of iron from bog iron deposits during the Roman Empire was investigated in an interdisciplinary archaeological–archaeometallurgical project (Jöns 1997). This exemplary project included studies of depository science as well as the extraction methods of iron from these types of ore. In addition, geoarchaeological investigations in the sense of the human–environment spiral, i.e. investigations into the history of vegetation and landscape were made in order to study the intrusion of metal-producing humans on the landscape and its environment. The spatial density in

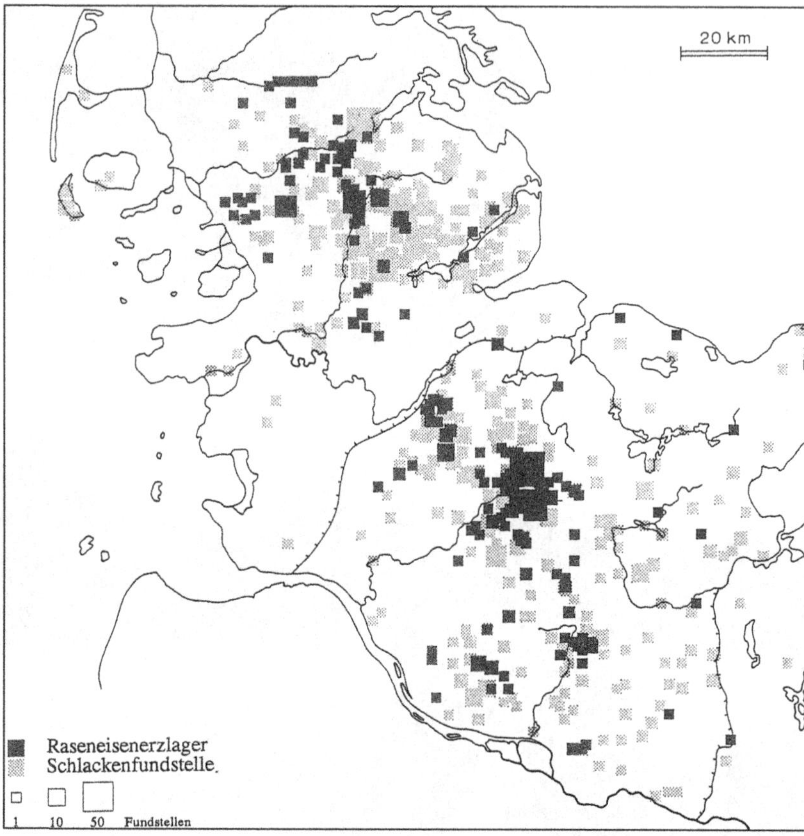

Fig. 3.53 In the northern European lowlands, there are countless smaller deposits of bog iron ores. The high density of deposits is shown by the example of Schleswig-Holstein. The map shows the occurrence of bog iron ores and those localities where this iron ore was mined from the Roman Empire until the twentieth century. This could be classified temporally based on the slag finds. From Jöns (1992/1993)

which bog iron ores can occur is illustrated by the example of northern Germany in Fig. 3.53, which may be representative of many other regions.

The formation of many bog iron ore deposits took place in recent geological (and historical) past, a process that can still be observed today at various localities in rivers and lakes by continuous precipitation of brownish coloured limonite ("Lake-ores" in Sweden). Bog iron ores are made of porous limonite, which can easily be dehydrated by roasting and afterwards reduced. Similar to all terrestrial sedimentary iron ore deposits, these ores have quite high Mn and P contents, the latter in the form of vivianite, but they are otherwise extremely poor in siderophilic trace elements such as Ni, Co, Ag, As and the like (cf. Sect. 11.2) who would follow the metal in the smelting of iron ores and thus would be useful for provenance studies.

Kidney Ores

The world-famous kidney ores of the Lahn-Dill area (Germany) are also sedimentary iron ore deposits. As Kronz (2005a) summarises, the volcanic-exhalative-synsedimary layers formed by the following geological processes. At the time of the older Devonian, the ocean floor existing in this area was dotted with warm springs from which hematites and SiO_2 precipitated in colloidal state. Strata bound layers of siliceous iron ores were formed. In addition, limestone precipitated. There was a close relationship of the ore bodies to volcanic rocks. The ores themselves consist mainly of dense, fine-grained hematite, which, compared to limonite, is much less suitable for smelting. There were two types of ores for iron production: The first type is predominantly quartz-rich hematite ores (siliceous kidney ore with c. 50 wt.% Fe and 17 wt.% SiO_2).

The second type is hematite rich in calcite (calcareous kidney ore with 30–35 wt.% Fe, 10–15 wt.% SiO_2 and c. 20 wt.% CaO). Following the old bloomery process for smelting, siliceous kidney ore was used purposefully, while the calcareous kidney ore was only used in later periods in blast furnace iron production. The geological later Carboniferous siliceous kidney ores are also of a volcanogenic-exhalative-synsedimentary genesis.

The early extraction of iron ores from the Lahn-Dill area was explored by Jockenhövel and Willms (2005) in an interdisciplinary project that carried out archaeological, archaeometallurgical and botanical investigations in the model area of the Dietzhölze valley. The kidney ores in the Lahn Dill area were already mined in the middle of the first millennium BC. However, a large part of the 350 slag sites mapped in this area date to an iron production in the thirteenth/fourteenth century AD.

Hydrothermal Iron Ore Deposits

Here we have to mention the hydrothermal iron ore (siderite) veins in the Siegerland area, which were mined since the La Tène period, in the Middle Ages and until the middle of the last century. They too are meaningless today. They show a significant vertical zoning, caused by primary hydrothermal telescoping, as well as by secondary zoning. The latter was the reason that the near-surface areas of the veins are deeply weathered to iron (hydr-)oxides and secondary base metal minerals. With increasing depth, the iron ores quickly turn into siderite. The brown iron was probably the starting point of prehistoric iron smelting in Siegerland (Gilles 1957a, b; Schneiderhöhn 1962).

The well over a hundred hydrothermal veins of this ore district (Fig. 3.54) have a polymetallic character: They are associated with Cu, Pb, Ag, Zn, Co and nickel ores. Even native copper was found in relatively small amounts in Siegerland.

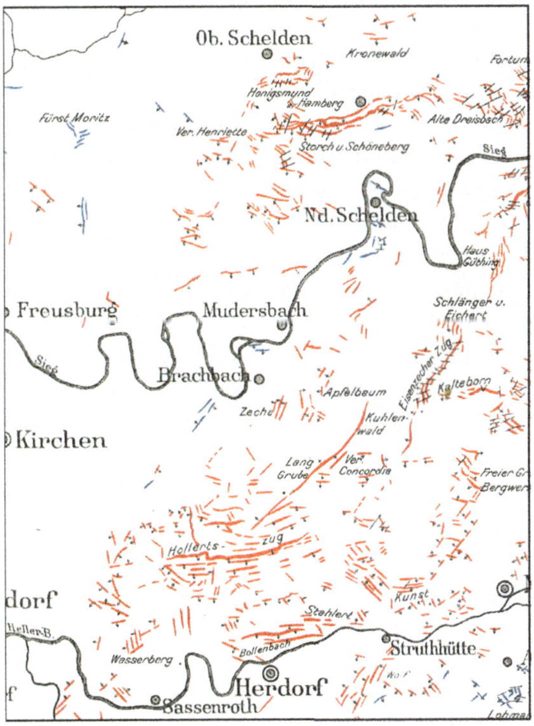

Fig. 3.54 Hydrothermal polymetallic veins in the area of Siegerland near the city of Siegen. In this ore district, much more than 100 ore veins were counted. Predominant iron ore is siderite, at the outcrops transformed to hematite and limonite. In addition, the veins contain Cu–Pb–Ag–Zn–Co–Ni ores. From Einecke and Köhler (1910)

Chalcopyrite can be found in most of the siderite veins. All of these metals can be valuable tracers for provenance studies of Siegerland iron. In particular, the comparatively high levels of Pb are excellent for characterising the Siegerländer veins and the iron produced from them. Salzmann (2013) was able to assign a number of prehistoric iron objects to this region by lead isotope analyses. Provenancing iron objects by lead isotope analyses, undoubtedly the most powerful tool for provenance studies in archaeometallurgy (cf. Sect. 11.3.1) may not be of much help given the generally low amounts of lead in iron objects.

Of hydrothermal origin is also the iron mineralisation on the island of Elba (Tuscany, Italy), one of the most important iron sources in the Mediterranean area since the first millennium BC (Corretti and Benvenuti 2001; Benvenuti et al. 2013). Stratiform to pod-like to vein-type iron ore mineralisations are made up of lamellar or micaceous hematite, magnetite and pyrite commonly associated with quartz, feldspar and chlorite. They are preferentially embedded in phyllite-quartz-formations of the Carboniferous-Triassic age Tuscany Nappe, close to the contact with hanging carbonates. Pyrite may occur locally in masses, decomposed to limonite near the surface. The hematitic ore usually contains around 10 wt.% SiO_2 but is very low in phosphorous. However, the iron ore from Elba was found to contain up to 0.49 wt.% W and 0.8 wt.% Sn which provides a distinctive geochemical signature in comparison to other European iron ore deposits (Benvenuti et al. 2013). The authors suggest the two elements to be valuable markers for provenance studies to trace the distribution of Elba iron ore, or iron smelted from this ore, respectively, in the Mediterranean. Iron from Elba was certainly smelted at the nearby coast of Populonia by the Etruscan inhabitants (Benvenuti et al. 2000). Iron smelted from this ore deposit was extensively traded throughout the Mediterranean and elsewhere from antiquity to modern times. The iron production was appreciated by ancient writers such as Varro, Vergil, Strabo and Pliny the Elder.

A number of iron ore deposits of various tonnage, with varying contents of manganese, can be found in the eastern Alps, Austria (Pohl and Belocky 1994). Today, only the Erzberg mine, Styria is operating (Fig. 3.55). The ore deposit is of metasomatic-hydrothermal origin, embedded in limestone of Devonic age. It was formed by a replacement of limestone by hydrothermal reaction of Fe-rich solutions. Main ores are siderite and ankerite which occur in huge quantities, and, similar to the siderite veins of Siegerland, subordinated pyrite, chalcopyrite, fahlore and cinnabar (Pohl 2005: 125). Erzberg is the world largest ore deposit made up of carbonatic Fe ores. In the oxidation zone, siderite was converted to limonite. These ores can have Fe contents of more than 50 wt.%. However, except for small remnants, the oxidation zone is fully exploited today.

Another important siderite deposit occurs at Hüttenberg, Carinthia, which is located close to the Erzberg. Smelting and working at this ore deposit yielded the famous *Ferrum Noricum*, produced in Roman times. Both the Erzberg and the Hüttenberg ore bodies are made of fine- to coarse-grained, light brown sparry siderite and are hosted by Paleozoic marble and paragneiss formations of the Austroalpine Units.

Placer Deposits

Placer deposits with iron ores, which are common in gold, tin-stone or platinum and platinum group elements, are rare. One of the few coastal placers known for prehistoric Fe mining is the one in the Colchis (Georgia) (Twaltschrelidze 2001). This is where placers with magnetite and titanomagnetite occur in a coastal strip of 50 km on the eastern Black Sea. The richest lie between the estuaries of the rivers Supsa and Natanebi. It has long been believed that the extraction of iron began here at the end of the second millennium (Pleiner 2000) and that the Colchis iron can be considered the oldest iron in the Old World. That this is not the case is explained in Sect. 6.6.3.

3.4.13 Arsenic, Antimony, Bismuth

Arsenic (As), antimony (Sb) and bismuth (Bi) are semimetals. Semimetals are such chemical elements, which have an intermediate position

Fig. 3.55 Erzberg, near the locality of Hüttenberg, Upper Styria (Austria). View of the opencast mine. Main ores of the Erzberg are siderite and ankerite and some base metals (see text). The oxidation zone, which is mainly exploited and vanished, was made up of limonite. Erzberg is supposed to be the raw source of the famous *Ferrum Noricum*. Photo: oepg2008.unileoben.ac.at

between metals and non-metals in their properties. None of the three elements is particularly abundant in the earth's crust (As = 1.8 ppm; Sb = 0.5 ppm; Bi = 48 ppb; Emsley 1991) although numerous minerals contain them as major or minor constituents. Thus, in the earth's crust, they occur much more often than gold and silver but are a lot rarer than copper and lead.

Arsenic, antimony and bismuth are chalcophile elements with slightly siderophile tendencies. This means that they like to occur in sulphidic minerals, but to a lesser degree can also undergo intermetallic compounds.

3.4.13.1 Minerals and Ore Deposits of Arsenic

In nature, arsenic is present in more than 90 mineral arsenates. It is widely distributed in ores of various ore deposits. Arsenic, as a by-product, occurs very frequently in copper and iron ore deposits, where it is bound to a series of ores or also forms its own arsenic minerals, i.e. minerals which consist predominantly of arsenic. Particularly interesting for archaeometallurgy is the geochemical-mineralogical association of arsenic with copper and tin mineralisations, since it provides a common explanation for both the frequently observed arsenic contents in early copper artefacts as well as in tin bronzes.

As-rich minerals are of no importance for the formation of ore deposits. Arsenic shows a particularly high affinity for nickel and cobalt and is therefore associated with corresponding ores, especially in hydrothermal veins. Arsenopyrite is particularly interesting as a carrier of gold: it is predominantly present as tiny inclusions.

Table 3.11 Some important arsenic minerals, their chemical composition and As-contents

Mineral name	Chemical composition	As (wt.%)
Native arsenic (*Scherbenkobalt*)	As	99
Skutterudite	$CoAs_3$	79.5
Chloantite	$NiAs_3$	74.6
Rammelsbergite	$NiAs_2$	71.9
Smaltite	$(Co,Fe,Ni)As_2$	71.9
Löllingite	$FeAs_2$	73
Realgar ("ruby sulphur")	AsS	70
Orpiment	As_2S_3	61.9
Niccolite	NiAs	56
Arsenopyrite	FeAsS	46
Cobaltite	(Co,Fe)AsS	45.3
Gersdorffite	NiAsS	45.3
Scorodite	$FeAsO_4{\cdot}2H_2O$	32.4
Arseniosiderite	$Ca_3Fe_4(OH)_6(AsO_4)_4\ {\cdot}3H_2O$	30
Domeykite	Cu_3As	28.3
Adamite	$Zn_2(OH)(AsO_4)$	26.5
Olivenite	$Cu_2(OH)(AsO_4)$	26
Pharmacosiderite	$KFe_4(OH)_4(AsO_4)_3\ {\cdot}7H_2O$	25
Annabergite	$Ni_3(AsO_4)_2\ {\cdot}8H_2O$	25
Erythrite	$Co_3(AsO4)_2\ 8H_2O$	25
Tennantite	$Cu_{12}As_4S_{13}$	21
Enargite	Cu_3AsS_4	19
Algodonite	Cu_6As	16.4
Proustite	Ag_3AsS_3	15.2
Whitneyite	Cu_9As	11.6
Beudantite	$PbFe(OH)_6(SO_4)(AsO_4)$	10.5 Changing
Pharmacolite	$CaHAsO_4{\cdot}2H_2O$	Changing
Clinoclas	$Cu_3(AsO_4)(OH)_3$	Changing

Values of arsenic concentrations after Doelter (1912–1932) and modern mineral data. They may vary according to various solid solutions. Note the high number of minerals with main and minor arsenic contents

The most important of arsenic ores are compiled in Table 3.11.

Of the minerals listed in Table 3.11, very few, if any, have been mined as potential arsenic ores in pre- and early history and antiquity: Ores that would be suitable would be native arsenic, then arsenopyrite as well as the strikingly yellow-orange coloured arsenic sulphides realgar and orpiment. Due to their relative rarity, all other ores are hardly considered as arsenic sources and have led at best to the known arsenic contents in various early copper varieties, which, however, are always only in the lower percentage range or less. Occasionally, the smelting of As-rich ores such as fahlores, enargite, or other sulphides intergrown with copper ores also produced various speisses (cf. Sect. 5.7), which are then found as waste products (Craddock et al. 1987; Kassianidou 1998), or next to "more pure" copper or sulphide droplets as microscopically detectable inclusions in slags. This is due to their deposits, so it can occur supra-regionally. To postulate this a technology of its own is difficult (Thornton 2009a) and has not yet been substantiated.

Since arsenic could not be produced as an independently used material prior to the Medieval period, this chapter therefore refrains from writing a section on arsenic deposits.

Native Arsenic, Cu- and Ni-Arsenides

Native arsenic used to be described as an accessory mineral from some hydrothermal veins

(Quiring 1946), where it formed concentrically banded or "shelly" arsenic. Its occurrence was mainly in the near-surface area. In Germany native arsenic was given the mining term *Scherbenkobalt*. Native arsenic may contain antimony and may form solid solutions up to SbAs (Ramdohr and Strunz 1978).

The copper arsenides algodonite, domeykite and whitneyite mostly occur as a mixture and may have been underestimated in quantity for a long time. They are known from the "old" copper ore deposits of Talmessi-Anarak in Iran, where they may have been mined in the initial stages of metallurgy and are therefore often seen as the "starting material" of copper metallurgy (Heskel and Lamberg-Karlovski 1980). In the nineteenth century, the arsenides were mined in large quantities in Chile (Alpers and Brimhall 1989). Gold and silver are often linked to arsenides as well as to arsenic-bearing sulphides. The Ni-arsenide niccolite is rather rare.

Arsenic-Sulphides

Arsenopyrite and other Cu–As ores, along with cassiterite, stannite and chalcopyrite, are among common ores in Cornwall and at Altenberg in the Ore Mountains (Germany). In the nineteenth century, they contributed significantly to the As production in the United Kingdom (Quiring 1946). The fahlore tennantite is one of the most common arsenic ores next to arsenopyrite and enargite. Like the Sb-Fahlerz tetrahedrite, it often has silver contents (up to approximately 2 wt.%, Quiring 1946).

Among the arsenic ores, realgar and orpiment occupy a special position. In the Old World, their rare deposits are concentrated in the young fold mountains of the Tethyan Eurasian Metallogenic Belt (TEMB). In ancient Iran and Azerbaijan, these As-sulphides were well known. However, they were not used in metallurgy, but as disinfectants and remedies. In Anatolia, in the Caucasus and in China there are deposits with realgar and orpiment as well.

Arsenates and Enrichment in Fe-Hydroxides

Supergene oxidation of arsenopyrite or alternatively As-rich minerals associated with copper (fahlores, enargite) can lead to the formation of greenish coloured Cu-arsenates. They occur within gossans and leached cappings and can easily be mistaken for "pure" secondary copper minerals. A very useful overview of these minerals including instructive pictures are compiled in Taylor (2011). There are a large number of arsenates in combination with a wide range of other metals (Fe, Cu, Pb, Zn, Ni, Co). The most prolific arsenic secondary mineral is scorodite, which is characterised by pale green colour and granular arborescent texture.

A peculiarity of arsenopyrite is its frequent limonisation, whereby it loses its colour and thus becomes unrecognisable. Thus, inconspicuous rusty brown incrustations and colloform aggregates of hydrous ferric oxides and arsenates of Cu–Fe can be found in gossans, which may contain up to 10–15 wt.% As. The sorption of arsenic in iron hydroxides can lead to an accumulation on surfaces or in the groundwater (Tretner 2002). Iron hydroxides such as goethite and lepidocrocite, or, more simply, limonitic iron ores are in turn widely distributed in soils globally. A particular example of this is the bog iron ores in the northern German lowlands and southern Scandinavia, which consist of these minerals. As-strong limonites are particularly strongly represented in the oxidation zone of sulphidic ore deposits. Therefore, the source of arsenic is considered to be the weathering of As-containing iron sulphides under oxidic conditions.

A characteristic feature for enrichments of near-surface arsenic is a transregional lateritisation, which developed not only during the Pliocene and Quaternary under subtropical conditions. It can be observed in Central and southern Europe, in the Near and Middle East, and in large parts of Africa as well. It also affected outcrops of ore deposits, where minerals such as pyrite, As-rich fahlores and arsenopyrite of the primary ore body were transformed into "soft iron ores" in which As increased, forming rusty brown incrustations rich in minerals also such as arseniosiderite and beudantite. Hence, it was the ferric and arsenical copper ore in the topmost parts of ore deposits that were exploited by the ancient miners.

Interestingly, As-rich iron hydroxides also occur in soils, mine tailings, waste dump profiles and other types of anthropogenic deposits (Ixer 1999; Taylor 2011). They were identified, e.g. from the extensive mineralised outcrops and from dumps at the well-known copper deposit of Špania Dolina in the Slovakia Ore Mountains which was exploited in ancient times (Števko and Sejkora 2012). In the ore of the gossans of the Montagne Noir, France, more than 5 wt.% As were detected (Krawczyk 1991).

In Volcanogenic Massive Sulphide deposits (VMS), as, e.g. those of Cyprus, near Ergani Maden (Anatolia) or Rio Tinto in the Iberian Pyrite Belt, arsenic contents in surface-near areas stay relatively low, usually ≤ 1 wt.% (Luyken and Heller 1938), which may be due to small admixtures of fahlores. In contrast, many copper mineralisations in the basal rock suites of ophiolites can contain significantly high levels of As and Ni. When their copper is smelted, this leads to the characteristic CuAsNi copper types, as, e.g. on the Arslantepe of the third millennium (Hauptmann et al. 2002a) or at sites in the Samail Ophiolite in Oman (Begemann et al. 2010).

Dill et al. (2010) report on considerable As-enrichments in untouched relics of gossans left by the ancient miners at Mega Livadi, Serifos, Greece. The goethite-bearing "limonite" was the first product of supergene alteration of the primary ore during the late Miocene. This soft iron ore, which formed close to the present surface and was worked in prehistoric times, contained significant concentrations of Zn, Pb, Cu and S. In addition, a conspicuous concentration of arsenic was present that has formed in connection with the precipitation of ferric hydroxide. Afterwards, in the abandoned shafts and galleries of the ancient mines, a mining-related "anthropogenic" mineralisation was formed which was different from the original mineral association: As-rich ores were exploited previously, and Cu-humates and -carbonates were formed instead much lower in As than the original rock.

This example may have a key position in explaining arsenic copper, which was widely used during the Bronze Age. It cannot be ruled out that As-enriched Fe-hydroxides and the numerous green coloured secondary As minerals in the once-pristine gossans of ore deposits have been available as raw material in larger quantities. Thus, in the fourth and fifth millennium BC and during the Bronze Age, they could have been an important source of material for the production of arsenic copper in the early stages of metallurgical development, which can be found all over the world in metal artefacts, especially in the Old World. After their almost quantitative exploitation, these near-surface ore mineralisations are difficult to find today. A comparable observation was made by Borg (2010) on another metal, namely gold. Potential accumulations of gold in placer deposits after the end of the Ice Ages, which had originally been available to the Bronze Age miners and metallurgists (cf. Sect. 3.6.3).

3.4.13.2 Mineralogy of Antimony

There are about 100 minerals in nature containing antimony. Some of the most common are listed in Table 3.12. Many of them are quite rare and have hardly ever been mined as actual antimony ores.

Native Antimony

According to chemical analyses by Doelter (1918), native antimony is quite pure and only occasionally contains several percentage of arsenic. It rarely forms a solid solution with arsenic. It has a high affinity for sulphur and likes to form compounds with, e.g. copper, lead and silver. Yet, embedded in ores as tiny flakes, it may be more prevalent than commonly thought (Ramdohr 1975).

The modification important in mineralogy or in archaeometallurgy is the so-called grey antimony. It is silver-white to tin-coloured and has a Mohs hardness of only 3–3.5, so it is comparatively very soft, yet still brittle. Because of these unfavourable properties, one cannot deform native antimony (as opposed to copper, lead, silver or gold) when cold. Today, antimony is almost never used alone but is always an alloy component to increase the hardness of soft metals such as lead, copper or zinc. Pure antimony is needed in modern semiconductor technology.

The earliest finds of native antimony, only partly corroded to senarmontite, come from Tell

Table 3.12 Some of the more important ores of antimony with higher concentrations of Sb

Mineral name	Chemical formula	Sb-content (wt.%)
Native antimony	Sb	100
Dyskrasite	Ag_3Sb	15.7
Breithauptite	$NiSb$	67.1
Allemontite	$SbAs$	max. 73.9
Kermesite	Sb_2S_2O	75
Antimonite (Stibnite)	Sb_2S_3	71.6
Tetrahedrite	Cu_3SbS_3	24.8
Bournonite	$PbCuSbS_3$	24.7
Polybasite	$(Ag,Cu)_{16}Sb_2S_{11}$	9.4
Valentinite	Sb_2O_3	83.5
Senarmontite	Sb_4O_6	83.5
Stephanite	$5Ag_2S \cdot Sb_2S_3$	15.2
Pyrargyrite	Ag_3SbS_3	22.3

In nature, altogether, more than 100 minerals contain antimony down to the lower level of percentages and below. After Doelter (1912–1932)

Abu Hgeira in Syria/Mesopotamia from the third millennium BC. These are bead fragments from children's graves (personal comm. R. Wartke 1999). It is unclear whether it is a natural or artificially produced mineral: It is unlikely that the preparation of metallic antimony from stibnite was a process that was practiced in antiquity (Moorey 1994), or even in prehistory. A discussion of prehistoric artefacts made of so-called antimony can be read in Sect. 6.7.

Although we know little about the amounts of native antimony available in ancient times in near-surface areas of deposits, the presence of this mineral is supported by some older mining reports. Martin Heinrich Klaproth (1802) reports on native antimony from Andreasberg in the Ore Mountains, and further finds are known from Příbram (Czech Republic), from Peru and from Canada.

Stibnite

Stibnite or antimonite was economically more important and was used more often. Stibnite forms shiny metallic, coarse radial or dense masses concentrated on nodules and layers and irregularly confined to single ore shoots. Beautiful druses with long acicular crystals appear, such as in Baia Mare in Romania as well as in the deposits in China or in Japan. When stibnite weathers, it forms a greyish-white to yellow-brown crust of valentinite and senarmontite. The antimony artefacts found in the Caucasus are always covered by a whitish crust of these minerals (Hauptmann and Gambaschidze 2001).

Due to its distinctive metallic lustre and its smoothness, powdered stibnite in fat has a history of usage as a black cosmetic pigment since c. 3000 BC, e.g. as an eyeliner throughout the ancient Near and Middle East (Moorey 1994; Priesner and Figala 1998). Today it is called kohl. A number of such decorative items have been found, e.g. at Hasanlu period IVB, i.e. in the Iron Age period (ninth century BC; Fleming et al. 2011).

Antimony Bearing Fahlores

Due to their silver content, the main foundation in ancient times were fahlores, and they still are today. The main silver bearer is a variety of Sb-tetrahedrite called freibergite, which can contain up to 18 wt.% silver (see also Sect. 3.4.6).

3.4.13.3 Antimony Deposits

The geochemical cycle of antimony, similar to that of arsenic and mercury, is characterised by more near-surface hydrothermal systems. According to Pohl (2005), the following types of antimony deposits can be distinguished:

- Sedimentary-exhalative ore deposits of antimonite, sometimes intergrown with the Ca-W-mineral scheelite ($CaWO_4$) and with gold. Such deposits are often found in China.
- Metasomatic ore bodies in limestone, often in contact with shales and in parts strongly silicified. These gold-free antimony-veins are usually simply composited stibnite-bearing veins of quartz with changing amounts of pyrite. These types of deposits include mineralisations in Serbia, Romania (Maramures or the mining district around Baia Mare, the former Hungarian Nagybánya), and those in Tuscany, Kadamzhai, Kyrgyzstan, Hunan, China).
- Volcanogenetic-epithermal veins with antimony ores in context with granites close to gold deposits. Here, for example, the deposit of La Lucette in eastern Brittany should be mentioned (Geffroy 1980), from which at least gold was obtained since prehistoric times. It is not clear whether antimony was mined in ancient times as well. This subheading also includes antimony deposits together with non-ferrous metals, e.g. in the Western Carpathians. Subvolcanic stibnite veins with tungsten and tin are found in Potosi, Bolivia.

In the Near and Middle East, antimony deposits are rather rare. There are a few in Egypt, Turkey and Central Iran near the famous copper deposit of Anarak. There are some large (modern) antimony mines and in Turkey in particular is a significant producer of antimony in the world today but there is no clear evidence that these deposits have been exploited in antiquity. Ancient workings were only mentioned.

There are some ore deposits with antimonite in the Great Caucasus, exposed to the surface (Godrabelidze 1933), which obviously were exploited in ancient times (Gambaschidze and Gambaschidze 1995). More ancient objects of antimony and antimony bronzes as well have been found in the Caucasus than in the entire Near and Middle East (Gambaschidze et al. 2001).

Antimony deposits in Georgia consist mainly of stibnite partly associated with pyrite and chalcopyrite. They are insignificant for today's economic conditions but were all prospected at the beginning of the nineteenth century. In the mountains of the Kasbegi region, some deposits of pure antimonite have been found in the Eli mountains and near to Sioni massifs, in the area of Shirvani near Dschutta and in the Branat-Zkaro near Gorisziche, at an altitude of 2000–3000 m. The second district with antimony mineralisation is located in Racha. The largest antimony mines are Zopkhito and Sagebi at the headwaters of the river Rioni in the area of Racha. Radiocarbon data from these mines prove activities of exploitation in the Late Bronze Age at about 1500 BC, but even more ancient exploitation is easily possible and a question of research. Dyson (1964) proves evidence of contact between Caucasia and northern Mesopotamia in the ninth century BC. At the Tschweschri, a tributary of the Rioni, as well as Ghebi, Kwemo-Sakao and Licheta, antimonite was discovered, which occasionally occurs together with pyrite, arsenopyrite and small amounts of copper and lead sulphides and has corresponding levels of minor constituents. The author's own analyses of antimonite from Sopchito have resulted in arsenic contents in the range of 1–2 wt.%, which already catch the eye by the yellow and orange-red hues of realgar and orpiment. Otherwise, the trace element contents were far below 1 wt.%. The third antimony district is located in Svaneti, about 150 km northwest of Kutaisi. Near the village of Lalchara, antimonite was found, also frequently associated with pyrite, galena and chalcopyrite.

Some antimony deposits also gained importance because of their gold content. Dill (1993) gives as examples deposits from Bavaria (Germany), from the Alps and from the Slovak Ore Mountains (Banská Štiavnica, Slovakia). Especially here the extraction of antimony was of enormous importance because of its gold content.

Table 3.13 Some important minerals of bismuth ores and their metal contents

Mineral name	Chemical formula	Bi content (wt.%)
Native bismuth	Bi	100
Bismuthinite, bismutine	Bi_2S_3	81.3
Bismite	Bi_2O_3	89.7
Bismutite	$Bi_2(CO_3)O_2$	81.96
Maldonite	Au_2Bi	34.7
Tetradymite	Bi_2Te_2S	59.3
Galenobismutite	$PbBi_2S_4$	55.5
Emplektite	$CuBiS_2$	62.1
Matildite	$AgBiS_2$	54.9

There are >200 Bi-containing minerals. Values of bismuth concentrations after Doelter (1912–1932) and modern mineral data

3.4.13.4 Geochemistry and Mineralogy of Bismuth

In archaeometallurgy, bismuth only has a role as a trace element, which can be analysed in metal artefacts.

In an overview, Clissold (2007) wrote that for such a rare element, the mineralogy of bismuth is surprisingly diverse. Some 205 minerals that contain essential amounts of bismuth are currently known. Only a few of these minerals are listed in Table 3.13. Of the entity of Bi-minerals 61 are secondary phases. Primary phases vary from comparatively "simple" to exceedingly complex compositions. Ramdohr (1975) states that in its occurrence it is a typical "throughrunner".

Native bismuth is the most widespread (but not most abundant!) naturally occurring metal. Bismuthinite is also a common ore, but only in very small amounts. It occurs in tin, silver and gold veins. An interesting observation is that the two minerals were observed to be placer minerals, e.g. in Saxony and in Bolivia. Generally, the majority of the Bi minerals, whether of primary or secondary origin, are extremely rare and only a small number are found repeatedly in nature.

Bismuth sulphide minerals are found only a little more frequently in hydrothermal deposits. Important sources of bismuth are lead, zinc and copper ores that have bismuth as an "impurity". Similar as to how native bismuth is, from the mineralogical point of view, a "throughrunner", geochemically speaking it is extensively used as a pathfinder element in geochemical exploration for a variety of such deposits and others.

Bismuth is present in a wide variety of ore types, principally associated with high-temperature acid magmatic rocks, in wolfram-, copper- and lead-related skarn deposits and in hydrothermal vein systems (Pohl 2005). Bi-bearing hydrothermal veins with Bi–Co–Ni–As–Ag–U ores are known from Schneeberg, Ore Mountains. It is, together with molybdenum, wolfram, tin and other accessory elements, often associated with gold mineralisation (Baker et al. 2005).

Important Bi-bearing ore deposits of economic importance today are located in China, Vietnam, Canada, Bolivia, Mexico, Japan and other countries.

3.4.14 Mercury

The average mercury (Hg) content in the earth's crust is 0.02 ppm. Hence, it belongs to the rather rare elements. Due to one specific physical property, it has an outstanding importance among the metals: it is liquid at room temperature. Its melting point is at −38.9 °C, its boiling point is at only 357 °C.

3.4.14.1 Minerals of Mercury

The number of high-grade mercury minerals is rather low. But mercury as a trace element is present in many sulphide ores, as it can be substituted by Cu, Ag, Au, Zn, Cd, Bi, Sb, Pb, and other metals as "vaporising" metal (Pohl 2005).

Important mercury minerals are listed in Table 3.14.

Table 3.14 Important mercury ores and their metal contents

Mineral name	Chemical formula	Hg content (wt.%)
Native mercury	Hg	100
Montroydite	HgO	93
Cinnabarite	HgS	86
Metacinnabarite	HgS	86
Landsbergite	AgHg	74
Kolymite	Cu_7Hg_6	73
Kalomel	Hg_2Cl_2	38
Moschelite	Hg_2J_2	38
Polhemusite	(Zn,Hg)S	38
Schwazite	$(Cu,Hg)SbS_{1-4}$	29
Livingstonite	$HgSb_4S_8$	22

Values of mercury concentrations in these minerals after Doelter (1912–1932) and modern mineral data. They may vary according to various solid solutions

Native Mercury

Native mercury, a liquid mineral, is an ore mineral! It occurs as a minor, but an omnipresent component in all mercury ore deposits together with cinnabarite (Fig. 3.56).

Its alloys with other metals, some of which are also found in nature, are called amalgam, e.g. (moschel-)landsbergite, and kolymite. The Hg–Ag amalgam landsbergite comes from the locality of Landsberg, Palatinate (Germany). Mercury also forms halogenides, e.g. moschelite and the Hg-oxide montroydite.

More widespread than any other mercury minerals, (meta-)cinnabarite are the most important ones. Cinnabarite is found in nature in a light to cochineal red colour. It is often darkened due to impurities. The actually red pigment darkens under natural light and gradually transforms into a black, chemically identical modification of metacinnabarite. This modification of the cinnabarite occurs as a seemingly amorphous black powder or coating. The mineral, in the German language also known as *Lebererz*, consists of cryptocrystalline masses which contain abundant foreign earthy and bituminous admixtures (idrialite, an asphalt-like hydrocarbon, ca. C_3H_2). Cinnabarite is resistant to weathering and accumulates locally in placers, which has certainly favoured the discovery of the mineral as a dye. Sometimes mercury is found together with antimony and/or arsenic

Fig. 3.56 Almaden, Pozo San Teodoro, Spain. Silvery shining droplets of mercury (mm-scale) on the host rock of rather grey meta-cinnabarite from the largest mercury only deposit of the world. It is the only metal that is liquid at room temperatures. The association of the two minerals means that mercury must have been known when cinnabar was used as a red pigment in ancient times. A distillation process was not necessary. Photo Andreas Hauptmann, Deutsches Bergbau-Museum Bochum

minerals. For the use of mercury and Hg-minerals, see Sect. 6.9.

3.4.14.2 Ore Deposits of Mercury

Mercury deposits are much rarer compared to copper, lead or silver deposits.

They usually only consist of (meta-)cinnabarite. Bitumens are characteristic companions. Characteristic gangues of cinnabarite deposits are furthermore silicified rocks close to (sub-)volcanic rocks, carbonates and clays. For a detailed genetic classification of mercury ore deposits see Dill (2010).

The origin of mercury deposits is extraordinarily complex, but fundamentally related to its physical property, that it is a very mobile, "vaporising" metal that occurs in many sulphide ores as a trace element (Pohl 2005). Gold may also contain Hg, which sometimes makes it difficult to distinguish between placer gold and gold obtained by amalgamating processes carried out by gold panners (Fig. 3.19; Hauptmann et al. 2010). The detection of Hg in copper is suggested by Pernicka (1990) as a possible distinguishing criterion for artefacts made from native copper and those produced by a smelting process.

Cinnabarite is one of the few sulphides whose genesis can still be observed today, from moderately tempered sulphide-containing hot springs, such as in Sulfur Bank, (California). Submarine hydrothermal vents with high Hg levels were also observed off the coast of New Zealand (Stoffers et al. 1999). This also explains the genesis of many mercury deposits, which occur in quite different shapes and designs like vein swarms, impregnations and replacements of various host rocks. Clays and above all organically rich black shales (e.g. Mansfeld copper shale and other oil-bearing shales) have significantly higher Hg contents than magmatic rocks. This indicates that Hg is likely mobilised in the gas phase with hydrocarbons, rather than in any aqueous phases (Pohl 2005).

The richest mercury deposit in the world, with about 30% of world reserves, is the area of Almaden in Central Spain (Pohl 2005). At Almaden, syndiagenetic-stratiform HgS mineralisations occur in the Silurian Criadero quartzite, a metamorphic sandstone. Almaden is a mercury only deposit. The main ore mineral is cinnabarine, but native mercury occurs as well (Fig. 3.56). Within the mine native, mercury used to drop down from the galleries and form puddles. Hence, we are not certain when it is reported in archaeological texts whether mercury itself or cinnabar as a red pigment were exploited at least since the Bronze Age.

There are two more major mercury deposits in Europe. The first one is located at Monte Amiata in Tuscany (Italy), the second one at Idrija (Slovenia). At Monte Amiata, hot springs in the context of recent volcanism deposited cinnabarite in sandstone, limestone and volcanic rocks. Other mercury deposits, e.g. Idrija, are still genetically unclear. However, bituminous sediments seem to play a role. Idrija contains stratiform, sometimes syngenetic orebodies in pyroclastic and sedimentary rocks. In addition to cinnabarite, there are also individual large bags of native (liquid!) mercury (Lavric and Spangenberg 2003). Overall the mine has contributed around 13 wt.% of global mercury production, and its mercury was one of the purest (99.99 wt.%) of any mercury mines (Giumlia-Mair 2009). Further mercury mineralisations are located in Landsberg in the Palatinate and in Stahlberg (Germany), where cinnabarite occurs together with silver and minerals, such as moschel-landsbergite, in acidic volcanic rocks.

As detailed compiled in Giumlia-Mair (2009), contrary to Central Europe, mercury and/or cinnabarite have been used in the Mediterranean since prehistoric times. Mining is known in Suplja Stena in Serbia during the Neolithic and Bronze Age. In Tuscany, traces of mining and stone hammers indicate prehistoric extraction. Cinnabarite was used as a dye in a Neolithic burial. In Russia, thousands of stone hammers were found at a mercury deposit near Nikitovska (Ukraine).

There are also a number of mercury deposits that have played a role in the cultural history of the Middle East and Central Asia. At Haliköy Ödemis near Ephesus (western Anatolia), cinnabarite occurs associated with gold and antimony mineralisations together with barite (Segerstrom and Stotelmeyer 1909).

Mercury ores were known in Iran at least since the second half of the first millennium BC and in the Islamic period. Deposits of cinnabar occur, e.g. in eastern Iran (Ghorbani 2013), located on the ancient trade route from south-eastern Afghanistan to the south via Shahr-i Sokhta/Sistan, the western foothills of the Chagai Hills (Baluchistan), to the south at Tepe Yahya in the Kerman (Kohl and Lyonnet 2008). There is a cinnabar deposit of supra-regional importance in Khaidarkan (Kyrgyzstan, Middle Asia). Other rich cinnabar deposits that are still of significance today are located in the Tien Shan and Fergana districts. Nothing is known about their significance in earlier epochs.

3.5 Oxidation and Secondary Enrichment Zones

3.5.1 Old Workings in the Gossan

One of the most important characteristics of ore deposits on the earth's surface was in prehistoric times as it is today, the so-called iron hat of ore deposits, which is alternatively referred to as gossan or leached capping. Geologically, this is the outcrop of the oxidation zone of sulphidic ore deposits. The supergene enrichment zone or cementation zone occurs at the base of the oxidised part of an ore deposit.

Due to its striking colour, mostly massive reddish red-brown to orange, more or less obscured by an often-lean vegetation cover, it was and is an essential basis for any prospecting and exploration activity. The mining term iron hat very aptly describes the accumulation of reddish iron oxides and hydroxides on the outcrop, especially of copper deposits, but also of lead–zinc–silver, tin or iron ore deposits.

In the Old World, there are almost no outcrops of ore deposits untouched from ancient workings. Usually, they have been more or less completely exploited in earlier periods. These ancient mining activities near the surface are called *Alter Mann* (= goaf, old workings) in German. The possibly oldest description of "ancient" mining activities comes from the Greek historian and geographer Agatharchides in the second century BC. In his description of the inhumane conditions of the Ptolemaic gold mining in the Eastern Desert of Egypt, he mentioned that long before his time people used copper chisels (and not iron tools!) for mining. We know also from medieval sources that old workings were already an unwelcome legacy of earlier mining, and even then, they made the exploitation of any ore deposit directly from the surface impossible (Weisgerber 1996).

It is a general problem of mining–archaeology and archaeometallurgy to capture the originally existing mineral inventory of a mineralisation in one such *Alter Mann* qualitatively and quantitatively. In relation to our present problem of assessing "ancient" ore deposits, this means that all we have remaining for further study are the residues of already exploited mineralisations. It is therefore difficult to assess the likelihood of native metals such as copper, silver and gold being available for exploitation at the very beginning of the metal ages, or the richness of ore shoots in a lode.

However, valuable estimations are possible using the literature on mining from the last two centuries, and from ore deposits discovered and exploited in recent times: Structure and geochemistry of pristine iron hats are described in detail from locations in the USA, Chile, Australia, Namibia, (South) Africa, South America, Middle Asia or Russia, where new ore deposits are opened up right up to the present day (Smirnov 1954; Robb 2005; Taylor 2011). They are very good guidelines for the reconstruction of (pre-)historical oxidation zones in the Old World because very little is known about the oxidation zones of many deposits in regions of, for example, the early sophisticated civilisations of the Middle East to Central Asia.

An example is the copper deposit of Tsumeb in Namibia. Kuntz (1904) reports on the almost fantastically rich outcrops on the surface of this copper deposit (Fig. 3.57). In the nineteenth century, copper was still being smelted with primitive methods from these rich outcroppings in the Otavi highlands.

The rich oxidation zones (supergene zones) and secondary enrichment zones (cementation or

Fig. 3.57 Tsumeb (Namibia). One of the richest ore outbreaks that existed lately: The 12-m high and 40 m in diameter measuring malachite hill from the copper ore deposit of Tsumeb. The photo was made around 1905 and shows the almost massive, hardly mined malachite mineralisation shortly before the exploitation of the deposit by modern industrial mining. Photo: Courtesy Tsumeb-Museum, Tsumeb (Namibia)

hypogene zones) of sulphidic ore deposits encouraged numerous researchers to investigate the shape, design and mineral content of the deposits as early as the nineteenth century AD. Of considerable interest in archaeometallurgy are the reports of ores that are barely even known today, made for example by the mineralogists Glocker (1829), Von Naumann (1852) or Leonhard (1860). One of many other minerals examples of this is the ore stromeyerite (CuAgS), which is now almost a rarity, but its abundance in "ancient" deposits could have played a significant role in assessing the early extraction of CuAg alloys (Hauptmann et al. 2002a). Other silver-rich sulphides, whose silver contents reach partly more than 50 wt.%, are being mentioned. Naumann reports that the stromeyerite was used as a rich silver and copper ore. In the old textbooks, numerous sites with stromeyerite, as well as its past economic importance, are mentioned.

3.5.2 The Formation of the Oxidation Zone: The Gossan

On landside, the oxidation of sulphidic ores takes places above and close to the groundwater level. The general profile of the near-surface weathering zone of such ore deposits as the end product of a long and intense exposure to atmospheric agents leads to a series of chemical-physical decompositions of sulphidic ores. Land-formed gossans and those formed in an oxygen-rich seawater have a similar bulk mineralogy. The formation of gossans in a submarine environment, however, will be explained later in Sect. 3.5.5.

Most of the time, the mineral inventory of ore deposits, corresponding to their often polymetallic character, is versatile. Therefore, leached cappings and oxidation zones of base metal deposits may exhibit an enormous variety of Cu-, As-, Ni-, Zn-, Pb-, Ag-, Sb-, U-bearing and rich ores (Barnes 1974; Williams 1990; Dill 2010; Taylor 2011), so that the range of ongoing reactions is versatile as well.

An oxidation zone can be divided into several parts (Fig. 3.58). Taylor (2011) gives an overview of the structure and composition of the areas from gossan to supergene enrichment zones to the primary ore body:

1. Leached outcrop with quartz and Fe-(hydr-)oxides and other secondary ore minerals (leached capping, part of the gossan).

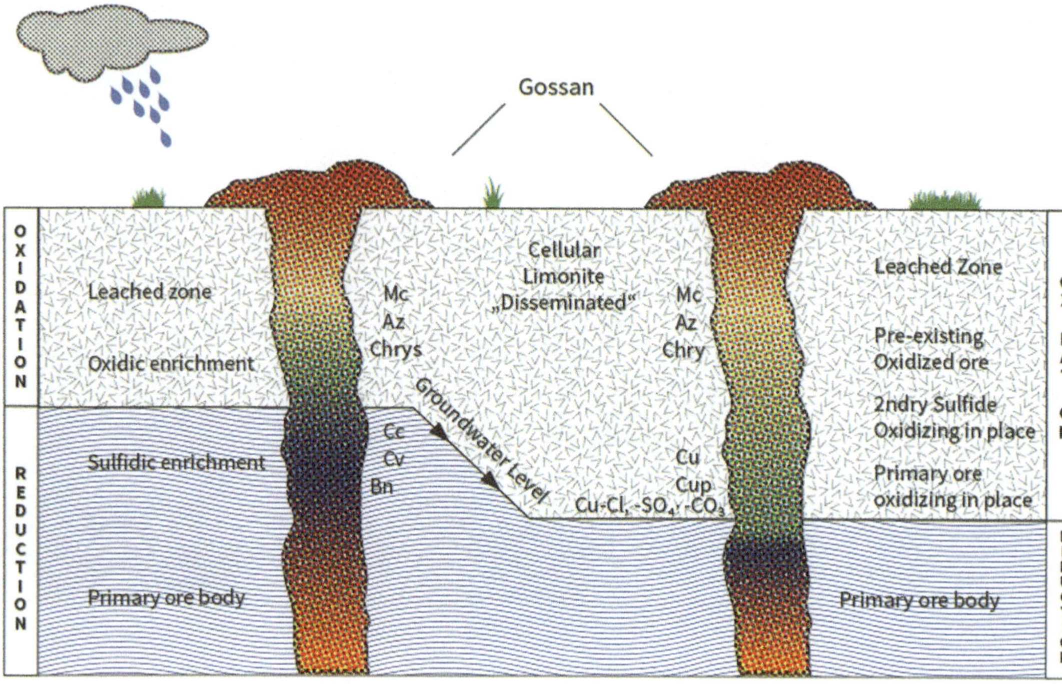

Fig. 3.58 Schematic sketches of oxidation and cementation zones of vein-like shaped sulphide ore bodies near the surface. Note the fluent transitions between the gossan, oxidation and cementation zones to the primary ore bodies under different climatic conditions. Important are relics of sulphidic ores close to the surface. The dashed pattern shows hydrothermally altered host rock. Here, brecciated mineralisations and impregnations may be formed. Modified after Barnes (1974) and Hauptmann (2007)

2. Below are one enrichment of Fe-(hydr-)oxides with increasing secondary ore minerals (gossan *sensu stricto*).
3. Metal-rich sulphides in the supergene enrichment or cementation zone.
4. Primary sulphides in the original ore body or protore.

These zones are completely different from each other, both by their mineralogical and chemical compositions. Overall, the mineralisations of the secondary zone are mostly oriented vertically to the surface. Their sequence is used in the archaeometallurgy as an example for the individual stages of development of early metallurgy. This is a viable model, although many ore deposits follow different geological-genetic factors, such as sedimentary copper deposits. It should be noted, however, that this model by no means has to be limited to vein-like mineralisations, as shown in Fig. 3.58. More extensive, larger sulphidic ore bodies, as on Cyprus, in Oman, in the Iberian Pyrite Belt (Rio Tinto) or in South America, show this sequence of mineralisation as well. Depending on geographic and climatic conditions, these have developed to different extents at various depths and show different ore contents.

Taylor (2011) describes genetic factors of near-surface zones in words and images:

- Solubility of individual minerals with respect to their stability field.
- Acidity/alkalinity (pH-values) of the local groundwaters from the top to the depth.
- Factors relating to oxygen availability and oxidation potentials (Eh-values).
- Factors relating to water availability (seasonal variations, rain).
- Climatic conditions (humid atmosphere, arid conditions, high mountains).
- Tectonic activities (uplifting, burial).

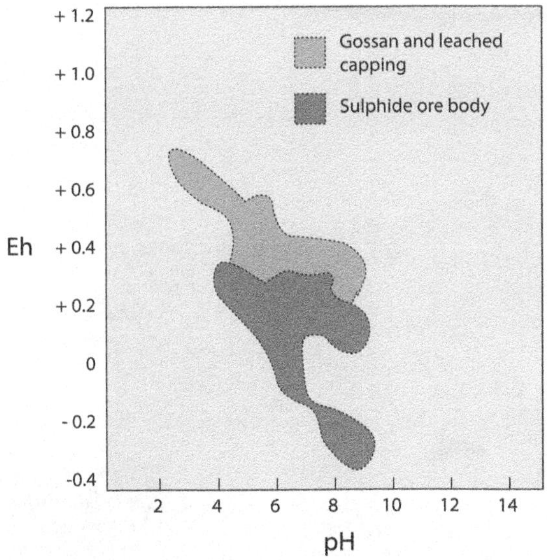

Fig. 3.59 Simplified representation of the oxygen availability, and oxidation potentials (Eh-values) versus the acidity/alkalinity (pH-values) of the local groundwaters in a sulphidic ore body. Modified after Taylor (2011)

These factors are dominating at low temperatures. The diagram of the oxidation profile (Eh) versus the acidity/alkalinity (pH) (Fig. 3.59) shows the general stability ranges of the gossan and leached capping and the primary ore deposit. The oxidation potential Eh is large at and near the surface and decreases with increasing depth. It becomes neutral and negative near and below the groundwater table in the sulfidic ore body.

In addition, sulphide orebodies could react by electrochemical reactions between the sulphides and the groundwater as effective electronic conductors. This would support a corrosion of sulphides. Recent studies concerning supergene oxidation of copper- and of iron ores have established that sulphide oxidation would occur as an electrochemical process mediated also biochemically by specific bacteria, such as the *Thiobacillus ferrioxidans* (Pohl 2005).

Important equations for the decomposition of iron ores, such as the most widespread pyrite in the oxidation zone of a sulphidic deposit, are the following:

$$2\,FeS_2 + H_2O + 7\,O_2 \rightarrow 2\,Fe^{2+}SO_4 \cdot aq + 2\,H_2SO_4$$

$$2\,Fe^{2+}SO_4 \cdot aq + H_2SO_4 + 0.5\,O_2 \rightarrow Fe_2{}^{3+}(SO_4)_3 \cdot aq + H_2O$$

These equations demonstrate that the pyritic system generates a major oxidising component in the form of short-living ferric (Fe^{3+})-sulphate and stable iron hydroxide (limonite, FeOOH), which in combination with the diluted acid leads to a dissolution of other metal sulphides, e.g. of chalcopyrite:

$$CuFeS_2 + 2\,H_2SO_4 + O_2 \rightarrow CuSO_4 \cdot aq + Fe^{2+}SO_4 \cdot aq + 2\,H_2O$$

Copper and, similarly, zinc, silver and other metals (see below) dissolve and are precipitated again at greater depths:

$$\underset{\text{chacopyrite}}{CuFeS_2} + Cu^{2+}SO_4 \cdot aq \rightleftharpoons \underset{\text{covellite}}{2\,CuS} + Fe^{2+}SO_4 \cdot aq$$

A gossan or a leached cap is popularly viewed as the product of repeated and sustained immersion of the ores and host rocks within a diluted acid bath. This effect is envisaged as taking large volumes of iron into solution which re-precipitates near the surface as masses of red-brown-orange iron (hydr-)oxides and converts host rock into various combinations of pale clays, quartz and iron (hydr-)oxides (Fig. 3.60).

Leached Cappings

These factors create a solution and oxidation of minerals at the surface and form the leached cappings. Sulphides are altered into water-soluble sulphates and carbonates, which seep down with the water. This produces a leaching and depletion of several elements near the surface. Associated with this, in the uppermost part of the gossan, through the interaction of acid waters with the host rock of the ores, there is a considerable accumulation of quartz, either in the form of chert or vuggy silica. Such quartz-rich constituents remain in this part of the oxidation zone in the form of cellular to skeletal relics (boxwork texture). They are usually intergrown

Fig. 3.60 (**a**) Lasail (Sultanate of Oman). View over the limonite dominated leached capping and gossan of the copper deposit of Lasail in the north of Oman after it was left at mining in Islamic times. The dimensions of

with relics of original sulphidic minerals and often contain bright green secondary copper minerals. This means that mixed oxidic and sulphidic ores could have already been used in the early stages of metallurgy, before more or less "pure" sulphidic ores were used in larger depths.

Gossan Sensu Stricto

In minor depths below the leached capping, in ore bodies with pyrite and chalcopyrite often accumulate significant amounts of limonite (FeOOH) and form the so-called "iron hat" or gossan of the deposit. It is usually conspicuously rusty-red coloured and was and still is an unmistakable sign for any prospectors in geology, mining archaeology and archaeometallurgy. As pointed out by Taylor (2011) this iron capping is named in the simplistic way as a "limonite" bearing zone. He points out that this ferruginous material is variable in detailed composition. It is essentially a fine-grained mass of amorphous hydrated iron oxides, intermixed with other fine-grained components such as quartz and manganese oxides. The main iron (hydr-)oxides are hematite (Fe_2O_3), goethite (α-$Fe^{3+}O(OH)$) and very rare lepidocrocite ($FeO(OH)$). This iron-rich, often siliceous material may contain other fine-grained minerals such as clay particles, jarosite ($KFe_3(SO)_4(OH)_6$), gypsum ($CaSO_4$), alunite ($KAl_3(SO_4)_2(OH)_6$) or scorodite (($FeAsO_4 \cdot 2H_2O$). Especially jarosite-rich layers (cf. Sect. 3.4.9) are mixed with limonite and hard to recognise.

Extensive iron hats occur, e.g. frequently in the south-east Arabian Peninsula and in the Mediterranean island of Cyprus (Fig. 3.60). It has to be noted that texture and composition of materials from a natural gossan is very similar to that of intentionally roasted ore, and a distinction may be quite difficult with single samples in the laboratory. Pryor et al. (1972) argued that the gossan could be regarded as the natural equivalent of the cinders produced industrially by the roasting of pyrite ores.

The pure ore (German: *Derberz*) itself shows a transition to increasing intergrowth with the "softened" host rock which is easily mineable. Due to hydrothermal alteration, the host rock is often decomposed and mineralised in a corona around the real ore body for several tens or hundreds of metres or more. In addition, faulting and fracturing are leading to transitions of mineralisations to host rock. A significant example of interspersed, low-grade mineralisations in heavily cracked and crumbled serpentinitic host rock are the copper ore veins exploited in prehistoric times in Oman. These are tectonically created "stockwork-mineralisations" of massive sulphide deposits.

Ancient mining activities following mineralisations due to tectonic events are the outcrop of the gold mine at Sakdrisi, Georgia (Fig. 3.61).

Below the leached cappings and depending upon the composition of host rocks (volcanic and magmatic rocks, carbonaceous and other sedimentary rocks), the original assemblages and local groundwater, as explained by Taylor (2011) above, a wide range of oxides, hydroxides, silicates, sulphates, phosphates, chlorides and other compounds are formed. The oxidation zone of copper deposits contains predominantly oxidised ores such as malachite, which may precipitate in large amounts, provided that carbonaceous host rocks are dominating. An example is

Fig. 3.60 (continued) mining become clear through the leached siliceous boulder left-over gate, formerly an adit or part of a gallery, which was driven into the original leached capping and gossan. On the margins, some 100,000–150,000 tons of slag were smelted (Hauptmann 1985). Note the striking bright orange-red colour of the sediments of the mine tailings which filled the ancient mines in the middle. Lasail is a volcanogenic (pyritic) massive sulphide deposit (VMS, see Sect. 3.6). (**b**) The mine Mathias, the island of Cyprus (Greece) shows the limonite dominated gossan of the volcanogenic massive sulphide copper deposit (VMS). The part shown in this picture is dominated by "soft iron ores" in the gossan. Both photos: A. Hauptmann, Deutsches Bergbau-Museum Bochum

Fig. 3.61 Sakdrisi (Georgia). The silicified, hematitic reddish gossan of the Bronze Age gold mine at Sakdrisi is heavily cracked by fault zones which were mineralised. Ancient mining activities were following these mineralised "veins", pockets and cavities and left a "blocky" outcrop. This aerial photograph was taken in October 2005, after the surface near excavation by T. Stöllner. Today, the ancient mine is destroyed by modern activities to exploit the gold of this deposit. Photo: T. Stöllner, Deutsches Bergbau-Museum Bochum

the massive occurrence of this mineral in Tsumeb. But malachite occurs also in the sedimentary copper ore deposits of Katanga in massive amounts, so that even today malachite figures can often be found in the tourism business.

If siliceous host rocks are dominating, chrysocolla ($CuSiO_3 \cdot aq$) will be formed. Furthermore, azurite and cuprite occur, as well as, more often than reported in the geological literature, native copper. Barnes (1974) reports native copper from regions of south-eastern Anatolia. He attributes such abundant resources to multiple fluctuations in groundwater levels in the course of alternating glacial periods, as well as interglacial ascent and descent cementation (cf. Fig. 3.58).

It should be emphasised that remains of sulphides such as chalcopyrite, covellite (CuS), and chalcocite (Cu_2S) are present almost regularly in the gossan. In the lower part of the oxidation zone, there is a zone of alteration in which complex mixtures of weathering products of duffy, earthy consistency are present, as for example the black, earthy *Kupferpecherz* or the red coloured tile ore, which consist of cuprite (Cu_2O), limonite and changing contents of chalcopyrite.

As described, supergene enrichment is generally designated by assemblages of quartz, iron (hydr-)oxides such as hematite, goethite/limonite ("soft iron ores"), and clay minerals. These three minerals are the basic materials for the formation of Fe-rich silicate slags of metallurgical operations worldwide. Mixed in ideal proportions, they provide a self-fluxing ore for the ancient metallurgist. H_2O-bearing iron sulphates often occur in vast amounts. Gossans on Cyprus and in Oman may contain ferromanganese umber.

There is an expressive example for such ores in the great Rio Tinto pyritic copper deposits in Huelva, in the Iberian Pyrite Belt, or on the island of Cyprus. The supergene features and evolution of gossan cappings of the massive sulphide deposits of Rio Tinto were investigated by Velasco Roldan et al. (2013). They consist of minerals, e.g. melanterite ($FeSO_4 \cdot 7H_2O$) and jarosite. Jarosites are Fe hydroxysulphates of complex composition with Na, K, NH_4, Ag and Pb as additional cations (e.g. plumbojarosite,

$PbFe_6((OH)_6(SO_4)_2)_2$). Beudantite ($PbFe_3(OH)_6/SO_4/AsO_4$) forms often. These secondary iron minerals permeate the iron hat and the oxidation zone in earthy, loose masses, which were very easy to mine.

According to reports, e.g. by Salkield (1987), the Rio Tinto copper deposits were capped by extensive hematitic and limonitic gossans up to 300 m thick. These contained practically no copper but only traces of silver and gold. These noble metals were slightly enriched immediately below the gossans in a clearly demarcated earthy zone consisting of plumbojarosite and beudantite (Pryor et al. 1972).

Below this zone, the massive pyrite was leached for c. 10 m or more, and this was followed by an enriched chalcocite zone, also enriched in gold and silver. This was exactly the zone where Roman mining activities were observed (Fig. 3.35). Below follow immense pyritic lenses containing little chalcopyrite, galena and sphalerite in which only traces of gold were present. Most of the mining has been in the pyritic lenses. Recently, however, mining of the 15 million tons of slightly auriferous gossan and an underlying enriched copper zone has been in progress at Rio Tinto (Pryor et al. 1972).

3.5.3 The Secondary Enrichment Zone (Cementation Zone)

In deeper parts of oxidation zones down into the level of the groundwater table, sulphides accumulate. This area is characterised by oxygen deficiency, i.e. by a reducing environment and an excess of fresh, undecomposed sulphides and is called cementation zone or secondary enrichment zone. Here, sulphates in solution are precipitated, depending on their position in the galvanic series of metals. In this case, noble sulphates are deposited at the expense of base metals.

The following reactions are representative of other metals as well:

$$5\,FeS_2 + 14\,CuSO_4 \cdot aq + 12\,H_2O \rightarrow$$
$$7\,Cu_2S + 5\,FeSO_4 \cdot aq + 12\,H_2SO_4$$
$$CuFeS_2 + CuSO_4 \cdot aq \rightarrow 2CuS + FeSO_4 \cdot aq$$
$$PbS + CuSO_4 \cdot aq \rightarrow CuS + PbSO_4 \cdot aq$$

This means that pyrite and solutions of Cu-sulphate are reacting in water to the precipitation of high-grade chalcocite, chalcopyrite, and covellite, and galena and Cu-sulphate are transformed to the minerals covellite and anglesite. Enargite (Cu_3AsS_4) is formed by an intensive dissolution and removal of iron. Somewhat above the reducing cementation zone native gold precipitates, and elements such as silver, nickel, and the like, which are more noble than copper in the galvanic series, accumulate here as sulphides.

Since cementation zones are bound to the groundwater table, they are usually spatially limited. The mining of all the rich ores from the cementation zone can, therefore, lead to a short heyday, which can be over quite quickly.

3.5.4 Protore: The Primary Ore Body

Below the cementation zone is the unchanged, primary sulfidic orebody, which always contains ores of significantly lower grade. From a mining point of view, the transition from the rich cementation zone to the primary orebody has often been associated with significant economic problems (see Sect. 3.7). The zone of primary sulphides in the protore is the base material of the iron hat. Supergene reactions, i.e. the uptake of the base metals from the solution may begin even below the groundwater table, sometimes much lower, when its redox potential (Eh) increases to a level where primary sulphides are no longer stable (Fig. 3.59).

Within sulphide ore bodies containing exclusively less noble metals such as lead, zinc, iron, cobalt, nickel and others, a comparable zone of cementation is not developed. There is a blurry transition between the oxidation zone and the unweathered sulphides of the primary orebody.

3.5.5 Alternative Models of Gossan Formations

In the traditional view, supergene and gangue minerals are the products of reactions between hypogene sulphide minerals and descending, acidic meteoric waters; these processes take place at or near the groundwater surface in subaerial environments. Herzig et al. (1991) argue, after investigations of deep-sea Transatlantic Geo-Traverse (TAG) fields, that hydrothermal sulphide deposits formed on ancient mid-ocean spreading axes reveal a second type of gossan formation, namely the oxidation of sulphide minerals by interaction with oxygen-rich ambient seawater. This process is referred to as "seafloor weathering". Examples of such gossans occur mainly on Cyprus and in the Oman Mountain range. At Skouriotissa or Kokkinoyia near Mitsero, Cyprus, bedded ochre overlying massive sulphides has been reinterpreted as a submarine gossan.

In VMS deposits copper and other metals are mobilised from primary massive sulphide ore and re-precipitated at depth. This precipitation of Cu sulphides with high Cu/S ratios can produce an economically significant copper enrichment at the reactive redox boundary between massive sulphide protore and downward penetrating fluids. The supergene enrichment "blanket" is overlain by an intensely altered and leached Fe-rich gossan. It is composed mainly of hematite, goethite, jarosite and clay and may be considerably enriched in Au, Ag, Pb, As, Hg and barite.

3.5.6 Secondary Zoning Versus Metallurgical Models

The exploitation of "secondary zoning" of sulphide ore deposits, i.e. the formation of an oxidation zone with an iron hat, containing, e.g. mainly oxidic ores and some native copper and a secondary enrichment zone (cementation zone) below with sulphidic ores, has been the principal basis for models and hypotheses about phases of metallurgical development. It has sometimes been presented in the best Darwinistic evolutionary manner (Forbes 1971). Strahm (1994) has diversified this model, as has been explained in Chap. 1.

In many mining districts, the mining of metal-enriched layers of the cementation zones have caused considerable (but short) economic flowering. Muhly (1980) argued that in the case of Late Bronze Age Cyprus it would have been probably the exploitation of the secondary enrichment zones that inspired real technological jumps and an enormous increase of copper production. This spatially limited zoning, as well as the so-called ore shoots at tectonic fault zones, led to rushes in mining (in the German language the *Berggeschrei*), which, however, were temporary in most cases. Reaching the primary ore body at a deeper level was often followed by an economic decline.

This "genealogical tree of metallurgical evolution" is not valid in all cases. Doubt and criticism of it was formulated in a paper by Stöllner (2019a, b).

Indeed, there are a number of factors to consider that may influence the described idealised sequence of leached capping, gossan to the enrichment of sulphidic ores in the cementation zone. These are for example:

- Geochemical–mineralogical features.
- Geographical–climatic features (see above).
- Different types of ore deposits.

Geochemical–Mineralogical Features

There is a feature which affects the extent of secondary enrichment zones. If the redox-potential increases near groundwater level and sulphides become unstable and start to decompose, metal and sulphate ions are released into solution. This results in porous volumes that are partially or even completely filled with other secondary minerals.

This association of minerals beyond stability boundaries are supported by considerable discrepancies found between the natural mineral assemblages and what the (theoretically calculated) Eh-pH-diagram (Pourbaix diagram) predicts. It arises from the fact that at low temperatures prevailing at the earth's surface, reaction rates are often so low as to allow phases to coexist metastably (Sato 1992). It complicates the ideal sequence of oxidation/cementation zone.

Geographical and Climatic Features

The extent of oxidation zones is also subject to regional variations, as they are determined by the composition of host rocks, climatic factors and the geomorphological environment. Outcrops are generally well developed in all (sub-)tropical and (semi-)arid regions (Schneiderhöhn 1962). This is true for countries and regions in the Old World such as the Balkans, Anatolia, Iran and Afghanistan as well as in South America (Chile, Peru) or South Africa. Oxidation zones, especially of copper deposits, are well known in Central Europe.

In contrast, they are largely absent in cold areas or in the high mountains of northern Europe, where rainfall often predominantly occurs in solid form. Thus, in the copper deposits of the Alps almost no secondary zoning is developed because already in prehistoric times this area was exposed to glacial climatics with high erosion rates, caused by the rugged relief of the mountain ranges and valleys. The same is true for the mountain ranges in Western Central Asia and Himalaya, where countless ore deposits are located as well.

Different Types of Ore Deposits

The model cannot be transferred to the geological situation in the southern Levant, or to Kargaly, because these sedimentary ore deposits do not comply with the previously described type, just as, e.g. the karstic ore deposits do not conform to that type either. In the sedimentary copper ore deposits in the Wadi Arabah and on Sinai a "lateral zoning", exemplified by different geochemical and mineralogical zones, can be observed. However, as the ores are mainly oxidic, zoning is not of such great importance as the "secondary zoning" in sulphide deposits. Considering all this evidence, it seems to be clear that the development of metallurgy might have depended on the individual composition of the particular ore deposit and was thus so complex, that single evolutionistic steps of the model can hardly be associated with specific historical periods. The concept of metallurgical development built on "secondary zoning", which was characterised by the change of compositions with increasing depth, has hardly any validity in this region.

3.5.7 Special Ores in the Oxidation and Cementation Zones

Many major and trace elements from a "classical" vertically positioned original sulphide ore body are released in the oxidation zone, and they are moving to varying concentrations mostly downwards to the cementation zone. This is associated with significant material shifts. This different behaviour of elements or minerals must be taken into account when sampling deposits or ancient mines, in case that a sampling of geochemically representative ore samples should be achieved for provenance studies.

In the following section, just a few metals which are significant for archaeometallurgy in this context are addressed again (details in the chapters of the individual metals in Chap. 3).

Gold

By the decomposition of gold-containing pyrite or arsenopyrite to sulphates gold will precipitate as metal. It forms the so-called mustard gold embedded in limonite.

Residual concentrations of precious metals in gossans as native gold or electrum and a variety of silver minerals can be economically important in Volcanogenic Massive Sulphide deposits (VMS) and stockwork mineralisations even to this day. A prehistoric exploitation of gold from the oxidation zone of such deposits is proved in the Transcaucasus in southern Georgia for the ore deposits of Sakdrisi and Madneuli (Hauptmann et al. 2010; Stöllner et al. 2010; Jansen 2019), and in Armenia for the ore deposits of Sotk and Fioletovo (Wolf et al. 2011).

Noble metals occur in gossans of other deposits as well. In those of the Iberian Pyrite Belt in Spain in Rio Tinto (Velasco Roldan et al. 2013), and also in the copper deposits of Cyprus, residual concentrations of native gold, electrum and silver minerals are present which are bound to secondary jarosite minerals (Na,K, $Fe_3(OH)_6(SO_4)_2$) and related minerals such as argentojarosite ($(Ag,Fe_3(OH)_6(SO_4)_2$ and plumbojarosite ($PbFe_6(OH)_6(SO_4)_2$) (more details also in the Sect. 3.4.9).

Lead

Lead behaves in contradictory ways. It is one of the least mobile elements of sulphide ores (Smirnov 1954), which can accumulate even in the oxidation zone due to the dissolution of other metals, especially zinc ores, which are primarily in paragenesis with lead ore. The old mining rule applies: if the oxidation zone contains ores that are uneconomic in their lead content, then the sulphide ores of the actual ore body will not be profitable for lead either. Lead does not accumulate in the cementation zone.

On the other hand, Smirnov (1954) states that galena is slightly unstable under surface conditions and secondary lead minerals were formed by its decomposition. However, this decomposition decreases on further oxidation, because the resulting oxidation products have a vanishingly low solubility. They are forming encrustations around galena which inhibits further oxidation. As a result, even in highly developed oxidation zones where hardly any other sulphides are found, galena is still relatively common. This is of metallurgical significance insofar as oxides and sulphides occur intergrown.

Cerussite ($PbCO_3$), then anglesite ($PbSO_4$) and mimetite ($Pb_5(AsO_4)_3Cl$) are among the secondary lead minerals in the oxidation zone. Above all, cerussite can contain the finest residual inclusions of galena or acanthite (Ag_2S). For this reason, cerussite was mined in the past as *Schwarzbleierz* (German for black lead ore). Cerussite and anglesite are obviously free of silver (Keim et al. 2016).

Silver

Basically, silver releases from oxidised galena. Under arid conditions, silver is easily fixed within the oxidation zone and forms halogenides and sulphides, and it is bound in jarositic ores (Sillitoe 2009; Taylor 2011). Cerargyite (AgCl) is the most common secondary silver mineral. Jarosites are often mixed with limonite and they are hard to recognise with the naked eye. For the time being, it is not clear to which extent these precious minerals were of economic importance everywhere in ancient times. However, it is known that layers with silver-containing jarosite were obviously exploited in large quantities at Corta Lago/Rio Tinto (Spain) (Salkield 1987; Craddock 2013a).

As it is reported by Constantinou (1982) and Kindermann (2005), on Cyprus, silver does occur along with gold in jarosite and limonite layers called *devil's mud* which is found between the gossan and at the top of sulfidic ore bodies. The *devil's mud* is a very fine-grained yellowish layer consisting mainly of varieties of jarosite, gypsum, minute particles of opal ($SiO_2 \cdot aq$), nano-scale gold particles, and free sulphuric acid (H_2SO_4). Mineralisations similar to the *devil's mud* do occur frequently in VMS deposits all over the world, but they are named differently.

It was mentioned above that cerussite may contain silver. It, therefore, could have been that it was used as a silver ore in antiquity in a similar way as galena. For a long time, surface-near cerussite was probably underestimated as a mineral of major economic importance in "ancient" ore deposits. Cerussite is a common mineral in the ancient lead–silver deposit of Laurion, Attika (Greece) or Cartagena (Spain). Numerous lead–zinc–silver deposits are known from the Iranian Plateau (Vatandoust 2004). Most of these are embedded in carbonate rock and contain (silver-bearing) cerussite. Meyers (2003) convincingly argues that silver was extracted first from surface-near silver ores, then from silver-bearing secondary lead ores and finally from silver-containing primary galena (Fig. 3.32).

On the other hand, there is often an enrichment of silver in the lower part of the oxidation zone and also in the secondary enrichment zone. In the secondary enrichment zone, it appears as native silver. Supergene silver enrichment with the precipitation of acanthite and argentian chalcocite-group minerals or native silver is well documented, e.g. in hydrothermal lead–zinc–silver deposits of the rainy Harz Mountain range, the Ore Mountains, or the sedimentary deposit of the European Kupferschiefer in the Upper Permian formation (all located in Germany) (Ramdohr 1975).

Zinc

Zinc sulphates, which are formed in the oxidation zone of the widespread sphalerite, are very readily soluble. However, in dolostones, zinc sulphate quickly precipitates again in the form of carbonates. Frequently, zinc silicates are also formed. In the oxidation zone, zinc largely separates from lead, although ores of both metals are intergrown together in the primary orebody. Zinc is enriched above the lead-rich parts. The separation does not always have to be complete (Bourgarit and Thomas 2015; Werner 1981; Doelter 1912–1932) and can be reflected for example in lead contents of brass artefacts (vgl. Kap. 7.2.5). This is of major importance for provenance studies of brass artefacts.

Enrichment of oxidic zinc minerals in the supergene parts of carbonate-hosted Pb–Zn ore deposits (Mississippi Valley Type) are known in the Mediterranean in Sardinia (Italy) (Boni et al. 2003; Valera and Valera 2005). Historically they were a possible source of zinc from this district, but no archaeological remains of ancient mining were discovered until now.

The same is true for Laurion in Attica (Greece). The ancient miners at Laurion were obviously interested in mining of silver-rich lead ores and in silver and lead production in ancient times (Conophagos 1980). Globular aggregates of calamine ore coating the walls of karstic cavities in the marble were exploited in the nineteenth and twentieth centuries. Why no in (pre-)historic periods? The oxidised zone of this ore deposit observable today may exceed 270 m in thickness (Skarpelis and Argyraki 2009).

Arsenic

Under oxidising conditions, iron (hydr-)oxides are controlling the solubility of arsenic. Arsenic is concentrated near the surface. It can be assumed that the near-surface enrichment of arsenic was crucial for the early supra-regional production of arsenic-containing copper. It was pointed out in Sect. 3.4.13, that amorphous hydrous ferric oxides containing "invisible" arsenates of iron and rusty brown incrustations of Ca-Fe-arsenates are frequently present in the iron hat.

Tin

Since cassiterite is one of the most weather-resistant minerals and due to the dissolution of other metals, a slight accumulation of tin occurs in the oxidation zone of tin bearing (sulphidic) deposits (Hosking 1988; see Fig. 3.46).

Less resistant is stannite (Cu_2FeSnS_4), which, as a sulphur-bearing ore, decomposes under oxidising conditions. However, stannite occurs in polymetallic sulphidic deposits, as they occur in Potosi (Bolivia) and Peru, but it was also mined in the Ore Mountains in Germany. The occurrence of the CuSn mixed ore mushistonite ((Cu, Zn, Fe)$Sn(OH)_6$), which is of essential importance in archaeometallurgy, is apparently quite rare. Initially, it was only known from its type locality in Tajikistan. Mushistonite is formed by weathering from stannite in the oxidation zone of tin deposits. This means that the mineral can also occur in the oxidation zones of Sn–Cu deposits that have developed with a comparable primary mineral inventory under similar climatic conditions. A detailed study of deposits of Eurasian tin belts would be of interest here, as well as those in Africa and South America.

3.6 Placer Deposits

3.6.1 General Remarks

Placer deposits are mechanically formed concentrations of heavy, weather-resistant minerals that result from the erosion of primary deposits and are formed in fluvial and coastal areas. Marine placers are formed by the accumulation of heavy mineral grains in the surf or beach area (see Sect. 3.4.7).

Geologically young placer deposits are dating to the Quaternary or Tertiary periods and should be distinguished from fossil placers embedded in geologically older rock formations. Geologically young placer deposits were the source for mankind in prehistoric and historic time periods. Gold, tin minerals (cassiterite) and sometimes magnetite and platinum-group minerals are enriched in placers. Important placer minerals are also diamonds and other gemstones such as garnet or ruby.

Condition for the enrichment of minerals in placers is not only their mechanical and/or chemical resistance to weathering and transport abrasion but also higher density than associated minerals. Gold and cassiterite have a significantly higher density than quartz. This means that these two can be easily separated from quartz by panning.

Some ores and minerals from gangue and host rock found in placer deposits are listed in Table 3.15.

3.6.2 Formation of Placers

The formation of a placer deposit is a process of sorting heavy minerals from light minerals mainly by fluvial water systems, rarely in beach or eolian context. A fluvial placers are fed via several "tributers" located in the catchment area to a "trunk river". The catchment area may consist of various geological typical for specific heavy minerals. Acidic magmatic rocks may contribute cassiterite for the alluvial environment, (ultra-) basic rock may contribute platinum-group minerals (Fig. 3.62).

The emergence and spread of placer deposits, both today and in the geological past, is mainly the result of different geomorphological processes on the earth's surface, in the vicinity of corresponding raw sources. In addition to a variety of other variables, tectonics, geodynamic settings and climatic conditions play a major role (e.g. weathering, rate of erosion, nature of sediment supply, opportunities for sediment reworking). Above all, the global climatic conditions and geomorphical processes during the Tertiary and especially in the Quaternary

Table 3.15 Some of few heavy minerals found in placer deposits which are of major importance in archaeometallurgy

Mineral	Chemical formula	Density (g/cm^3)
Placer ore minerals		
Native gold	Au(Ag)	15.5–19.3
Native platinum	Pt	14–19
Magnetite	Fe_3O_4	5.2
Hematite	Fe_2O_3	5.26
Cassiterite	SnO_2	6.8–7.1
Chalcopyrite	$CuFeS_2$	4.1–4.3
Other placer minerals		
Wolframite	$(Fe,Mn)WO_4$	7.2–7.7
Scheelite	$CaWO_4$	6.1
Rutile	TiO_2	4.2
Ilmenite	$FeTiO_3$	4–5
Quartz	SiO_2	2.65
Diamond	C	3.5

These are weather-resistant minerals. Chalcopyrite may be included here as a special case. It is only stable under special climatic conditions. In comparison, a few other minerals are cited which are often found as placer minerals. The variety of density is due to the variety in the chemical compositions. Native platinum is representative for all platinum-group minerals

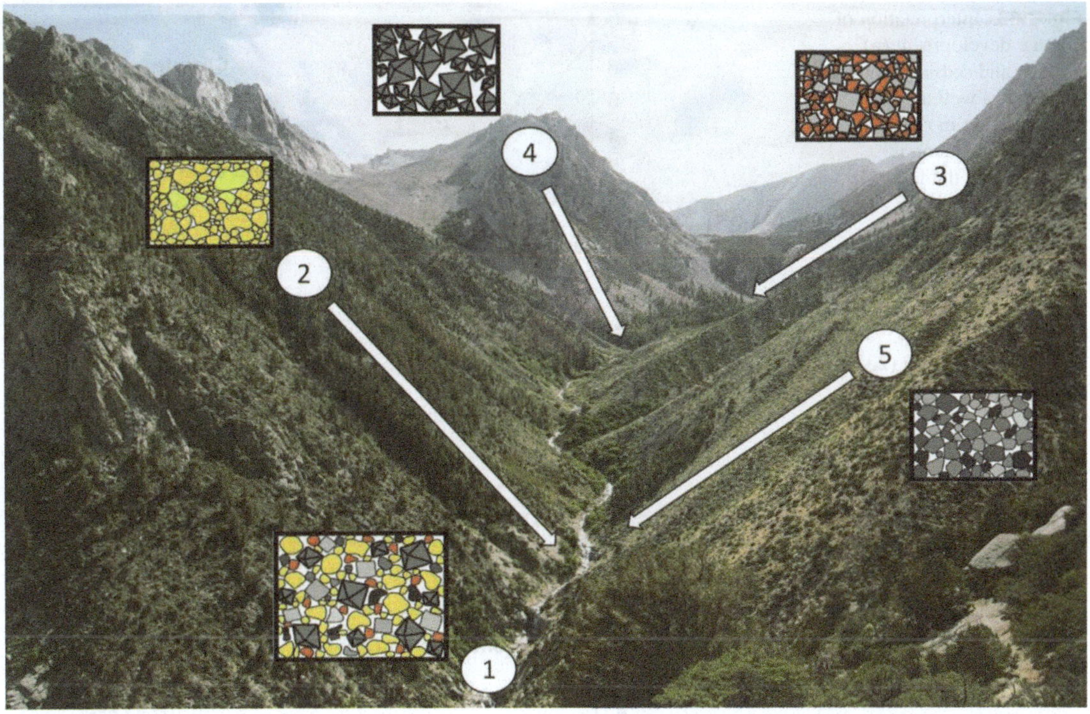

Fig. 3.62 The catchment area for an alluvial placer deposit may be composed of various types of rocks and ore deposits which contribute to the composition of the placer in the foreground (**1**). (**2**) Primary gold deposit, the main supplier of the placer; (**3**) Cassiterite from mineralisations in acid magmatic rocks; (**4**) Cu, Pb, and other metals from various deposits; (**5**) Platinum-Group (PGM) metals from (ultra-)basic rocks. All these metals may affect the composition of the gold (s)melted from such a placer. From Jansen (2019)

periods were what led to the formation of gold and tin placers (Sutherland 1985; Evans 1992; Camm and Croot 1994).

They were important and coveted raw sources, not only in (pre-)historical times. Alluvial placers were also the reason for all the gold rushes in the late nineteenth century AD in North America (California and on the Klondike and Yukon river in Alaska), and in Australia (Coolgardie). Gold rushes in Brazil (Serra Pelada) began at the end of the twentieth century AD. This also applies to gold rushes in western and South Africa (Burkina Faso, Tanzania). This will be detailed in Chap. 6.

3.6.2.1 Climate and Geomorphology

Rich gold placers do occur in formerly glaciated areas, where rocks and ore outcrops were decomposed by frost weathering. In these regions, the retreat of glaciers after the last glacial epoch led to the relocation of ore-bearing debris, with gold subsequently accumulating in alluvial placers. In such a sedimentary cycle, gold also undergoes authigenic precipitation from solutions, allowing larger nuggets to form (Falconer and Craw 2009; Lawrance 2009).

This young (geological) period of gold enrichment offered the Bronze Age prospectors a rich potential of raw sources, after a long period of anthropogenically undisturbed gold accumulation. Borg (2010) notes in the graphic of Fig. 3.63 that not only the Bronze Age landscape of Central Europe but also wide areas of the mountains of the Tethyan Eurasian Metallogenic Belt and the Eurasian steppe land had developed largely under the influence of postglacial climatic conditions since the maximum of the last glaciations before about 18,000–20,000 years. Here, large areas were under the influence of permafrost and thereby experienced a profound

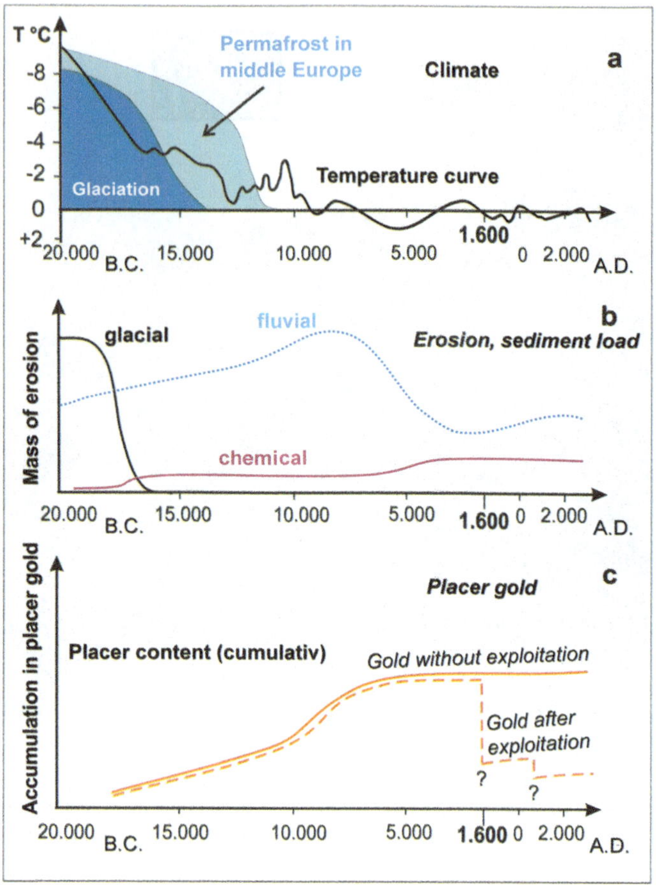

Fig. 3.63 Interpretation of climate development (**a**), erosion and sedimentation rates (**b**) as well as gold accumulation in river placers (**c**) in Central Europe since the last glaciation. Clearly recognisable are the increased accumulation rates of placer gold in prehistoric times after the fading of permafrost. Modified after Borg (2010)

weathering and mineral decomposition of the upcoming rock and the (local) upcoming occurrences of gold.

The profound weathering rate initially contrasted only with relatively low rates of erosion, as the permafrost conditions fixed the debris of weathering so that it was not available for erosion. Subsequently, especially in the sub-boreal region (in central Europe around 2500–1500 BC), increased precipitation rates led to significantly increased fluviatile-mechanical erosion rates of the disintegrated debris in the high and low mountain ranges. This, in turn, caused large amounts of sediment in streams and rivers. During this time, the accumulation rate of placer gold from the eroded and fluviatile transported glacial and postglacial weathered debris may have been significantly higher than today's accumulation rate.

In Cornwall, the quaternary tin placers (Camm and Croot 1994) are only partially formed by alluvial displacement of abrasive debris. In a study of cassiterite crystals from tin placers, Camm and Hosking (1984) observed that many cassiterite grains show no signs of abrasion. This conflicts with the general models of fluviatile placer development by a rapid downslope of material from primary ore sources. They suggested that an "in-situ" erosion system of tin bearing granites would have existed in the periglacial land system of the Pleistocene which lead to residue tin enrichments in situ, above the primary sources of tin in the granite batholiths.

Fig. 3.64 Tambopata National Park, Peru. Illegal artisanal exploitation of alluvial gold placers along the upper Amazonas river within the protected forests of the National Park. Large parts of the forest were untimbered for gold washing. Planet Labs., Inc. (Public domain) via Wikimedia Commons

3.6.3 Gold Placers

Gold is the most ubiquitous of all alluvial placer deposits. A very impressive example of exploitation of gold from a placer is shown from the upper Amazonas area (Fig. 3.64). Boyle (1979) suggests that the majority of the c. 0.1 million tons of gold ever exploited has come from young placers and from fossil placers. The recovery of gold by panning never leads to pure gold concentrates. Various heavy minerals cannot be completely separated from the gold due to a similar density and herewith they are affecting the trace element pattern by (s)melting of the concentrate, which results in the entry of various trace elements (Hauptmann et al. 2010). In most cases, gold can be washed out in small grains in the millimetre range (Fig. 3.65). Due to their soft, malleable and ductile properties they are generally deformed and flattened.

It has been explained in Sect. 3.4.7 that gold is also slightly soluble so that it can lead to the formation of idiomorphic gold crystals in the sedimentary context. Exceptionally large nuggets (Fig. 3.66) can be found, which can be extremely useful tracers for prospecting the metal. In addition, gold nuggets have always been fascinating triggers for gold rushes everywhere.

Las Médulas

In Europe, the Iberian Peninsula is particularly rich in gold. It was the most important gold-producing region of the Roman Empire. The most important gold district is concentrated in the north-western part of the peninsula, where more than 800 historical gold mining sites

Fig. 3.65 Gravel quarry Kimm Wabern-Zennern (Hesse, Germany), river Eder. Placer gold, washed in the 1990s by Dr. W. Homann, Dortmund. Note the slightly rounded gold grains and flakes shaped by erosion and re-deposition of gold. The placer gold concentrate contains a low number of quartz grains and magnetite crystals washed along with the gold. Size of the grains c. $\leq$ 1 mm. Keyence Digital Microscopy. Collection Deutsches Bergbau-Museum Bochum. Photo: T. Skowronek, Deutsches Bergbau-Museum Bochum

(personal comm. Sánchez-Palencia (2017) were operated in Roman times. A geographic section of this rich auriferous region is presented in Fig. 3.67. The enormous imprints and changes left in the landscape are still visible today.

One of the largest gold placers exploited in antiquity was Las Médulas in the province of Galicia. For information about Roman gold mining and production activities Davies (1935), Quiring (1948) and Domergue (2008) offer a wealth of information. Catalogues of Roman precious metal mines in Spain and Portugal were published by Sánchez-Palencia (2000), and by Pérez-Garcia et al. (2000).

Las Médulas was of greatest economic importance for the Roman period. Gold mining was so gigantic and impressive that it was already recorded by the Roman writer Pliny in an excellent description in the Naturalis Historia (33, 74–76). It is one of the largest and most impressive examples of early extraction of raw materials in the Old World. In 1997 the opencast mining site of Las Médulas was included in the UNESCO World Heritage List as an outstanding example of innovative mining technology.

Gold occurs mainly in the Tertiary (Miocene) Santalla Formation, which is embedded in a 100-m thick sequence of conglomerates, gravel, sand, and gravel with a clayey matrix. The gold content is here between 20 and 100 mg/m^3 (Pérez-Garcia et al. 2000). The enrichment of gold is caused by a deep-reaching lateritic

Fig. 3.66 Gold nuggets, Birch Creek, Atlin (British Columbia) (right nugget), and Ophir, New South Wales, (Australia) (left nugget). The nugget from Ophir (44.5 g) is irregularly shaped and at least partly recrystallised. The one from Birch Creek (54.4 g) is much more smoothed at the surface due to mechanical transportation. Collection Deutsches Bergbau-Museum Bochum. Photo: M. Ganzelewski, Deutsches Bergbau-Museum Bochum

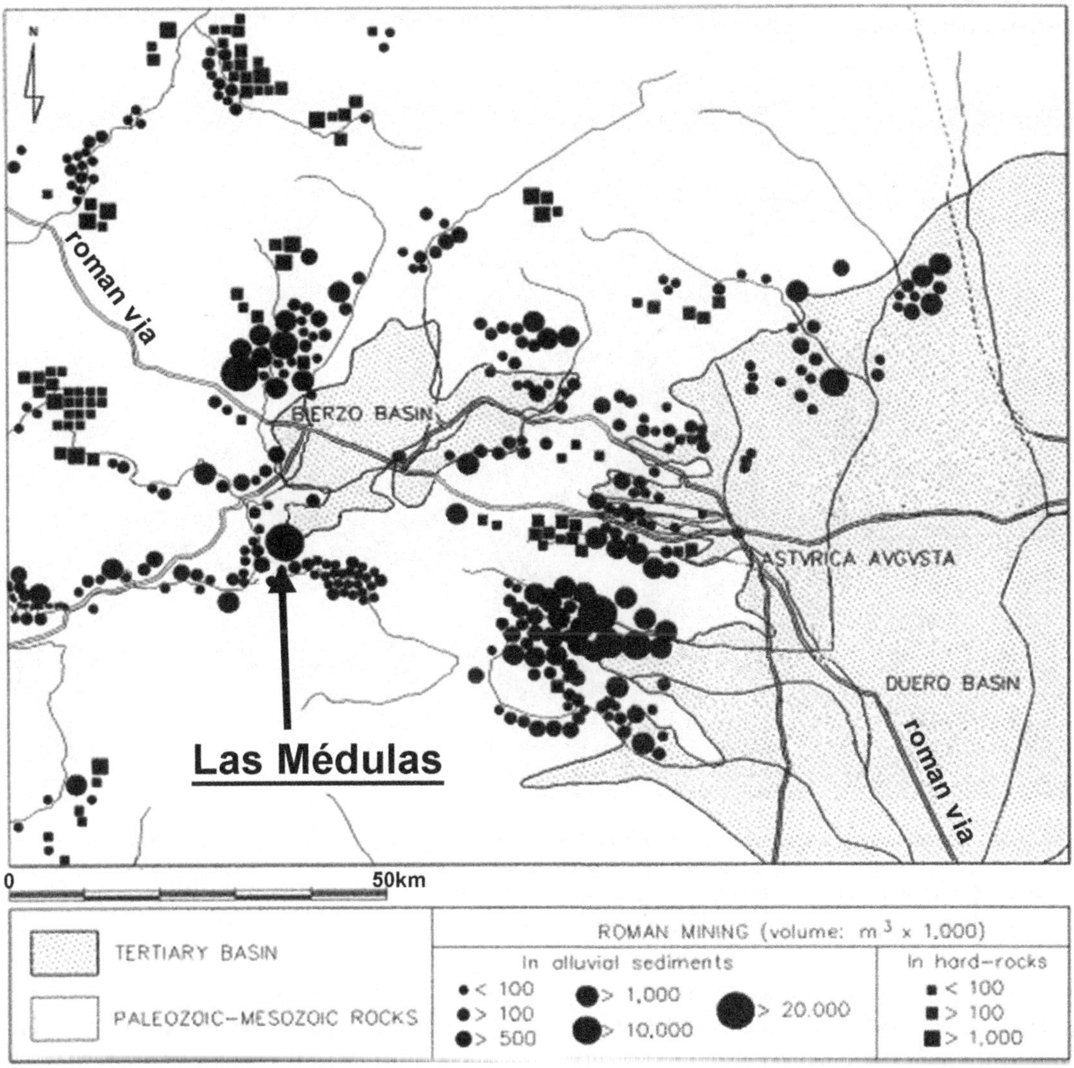

Fig. 3.67 Roman gold mining sites in the north-western part of the Iberian Peninsula. Note the enormous density of mines (> 800, see text) which decidedly contributed to the Roman monetary system, e.g. of Augustus, who established the gold standards known as the "Aureus". From Pérez-Garcia et al. (2000)

weathering of the pre-Miocene crystalline basement. Gold was eroded from gold-bearing quartz veins and enriched by a solution-precipitation process. These pre-Miocene lateritic soils were deeply eroded during the Alpine uplift. Free gold passed to alluvial fans as detrital particles and/or in the colloidal state (Pérez-Garcia et al. 2000). Accordingly, in Las Médulas there are gold grains with a fineness of above 995 (see above).

The gold was washed out of the sediment with gigantic expenditure by hydraulic actions. Water required was brought about by channel systems from distances up to 40 km. It is estimated that within approximately 200 years, around 100 million tons of sediment were moved (Fig. 3.68) and relocated in Las Médulas: Mountains were taken down and valleys were filled up, creating artificial lakes. This led to serious changes in the landscape. Pliny reports an annual production of about 6500 kg of gold from the gold mines of Asturias, Galicia and Lusitania. This would result in a total yield of around 1635 tons over a period of around 250 years (Sánchez-Palencia Ramos 1979). On

Fig. 3.68 Las Médulas, Galicia, north-west Iberian Peninsula. Aerial view of the opencast Roman gold mine. Almost 100 million tons of gold-bearing sediments were washed out by hydraulic mining and changed the landscape over kilometres. Due to these unique destructions, Las Médulas was nominated to a UNESCO World Heritage. Photo: A. Hauptmann, Deutsches Bergbau-Museum Bochum

the basis of geological estimates with an average content of 50 mg/m^3, Hérail and Pérez-Garcia (1989) and Sánchez-Palencia et al. (2000) calculated a total production of 4.217 kg of gold from the pits of Las Médulas per year. In large parts, the gold was washed out in sluices. However, there are also considerations, according to which mercury was used to extract particularly the very fine-grained gold dust, after the amalgamation process (oral comm. Sánchez-Palencia et al. 2017). This is acceptable for several reasons. Mercury was known in Roman times, and Almaden's mercury deposit is only 450 km from Las Médulas. Archaeological evidence for an amalgamation process is currently difficult, as a geochemical comparison with gold artefacts from Las Médulas is difficult too.

3.6.4 Gold Placers and Platinum-Group Minerals (PGM)

Like gold, minerals of the platinum-group metals (Ru, Os, Ir, Pt, see Fig. 3.29) are released by weathering from (ultra-)basic rocks such as peridotites, dunites and serpentinites from ophiolitic complexes and transported in the sedimentary cycle. This is why PGMs are often concentrated along with gold in alluvial placers and at the base of laterites (Pohl 2005).

Subsequently, gold was panned together with PGM, and such a heavy metal fraction were later co-(s)melted. Due to their high melting points, PGMs will remain as concrete micro-nuggets in gold. Platinum itself will be solved in the liquid gold during smelting placer gold (Schmiderer 2008; Sanadze and Labartkava 1961).

A spectacular example where both gold and PGM have been concentrated into a deeply weathered lateritic profile is the Serra Pelada deposit in the Carajás mineral province of Brazil (Moroni et al. 2001). This deposit was discovered and exploited until the recent past by mining companies and thousands of private artisanal miners (*garimpeiros*) who took advantage of the rich and easily extractable laterite-hosted ore. The exploitation of gold from this mine is documented in a fascinating and thrilling way by photos from Sebastião Salgado (1993, 2019).

The gold placers of Pactolus in western Anatolia are well known. This gold also became known within research for its inclusions of platinum group element minerals (see above) (Young 1972; Craddock 2000c; Junk and Pernicka 2003). The gold of the Pactolus founded the legendary wealth of the Lydian kings in Sardis. The central figure is the legendary rich King Croesus, who ruled there in the sixth century BC and who achieved “world renown” through the introduction of pure gold coins via the introduction of gold–silver separation (parting, see Sect. 6.3) into the currency system of the time.

For a long time, archaeological finds of gold with PGM inclusions from sites in the Middle East were all erroneously attributed to the Pactolus (Young 1972). However, the close context of PGMs with ophiolitic rocks means that they can occur quite frequently, geographically speaking, as ophiolites are quite common in the TEMB from the Alps to the Himalayas across the Balkans, Anatolia, Iran and Afghanistan (Economou-Eliopoulos 2010). Hence, the geographic assignment of gold artefacts seems to be difficult after metallographic or geochemical measurements alone (Junk and Pernicka 2003).

However, Jansen et al. (2017, 2018b), in their investigations of PGM-bearing gold artefacts from the Early Dynastic III period Royal Tombs of Ur in Mesopotamia, were able to make a northern Afghanistan origin probable, thus refuting Young’s (1972) considerations. Because of their combination of PGMs and of tin concentrations, it was evident that the resources came from a placer deposit. The fact that the gold of Ur was always associated with the blue semi-precious lapis lazuli in the context of the finds supported clear the provenance from Afghanistan. Osmium isotope measurements of PGMs in the gold artefacts showed identical compositions with the PGM gold from the border area to Tajikistan (cf. Sect. 3.4.8).

These famous gold placers were mined at the border river between Afghanistan and Tajikistan, at Amur Darya in Bactria. The trade routes of northern Afghanistan traversed this mineral-rich region. This is probably also the source of the Achaemenid Oxus treasure (Curtis 2012) and the Hellenistic gold treasure of Tillya Tepe in northern Afghanistan (Schiltz 2010). West of Badakshan, near the lapis lazuli deposit, in north-eastern Afghanistan (Takhar Province), near Samti (Chah-i Ab area) lies a vast area of gold placers. These placers are 8-km long, 900–1700-m wide and have a width of about 28 m. Inventories are estimated at 20–25 t gold. Additional primary and secondary gold deposits lie between Kandahar and Zabul. In total, nearly 100 gold occurrences are listed by the Afghanistan Geological Survey (see also Wolfart and Wittekindt 1980), sometimes in the context of tin deposits, such as Nuristan in East Afghanistan. Further, there are numerous gold deposits at the upper reaches of the Indus in the border area Afghanistan–China–Pakistan–Tajikistan and in the Fergana Basin in Tajikistan, at the Amur Darya, where ultrabasites and granitic rocks occur in close spatial proximity (Seltmann and Porter 2005).

These gold deposits are good examples of a typical long-lasting exploitation of placers. After all, they were exploited from the third millennium BC to the beginning of the Common Era, and then probably in Islamic times. It is believed that the gold of the Early Bronze Age in Mesopotamia was also negotiated from further east, namely from Dardistan in north-western Pakistan to Ur.

Herodotus (3rd book, 102–105) may also have referred to these regions when referring to the giant gold ants, which, in reality, were probably groundhogs (Peissel 1984). From this gold province, considerable tribute was paid to the Persian kings.

3.6.5 Tin Placers

Primary occurrences of tin minerals with major cassiterite are formed in felsic rocks such as granites. After the erosion of such rocks, cassiterite might be deposited and enriched in placers. Under special geological environment, it might even be mixed with platinum-group minerals and especially with gold.

This combination of tin and gold in placer is a very important source of information for the assessment of Early Bronze Age gold artefacts from Central Europe. Hartmann (1970, 1982), in parallel to the research project of Junghans et al. (1960, 1968, 1974), analysed about 4000 objects, most of which also had low concentrations of tin. He called this gold type A3. He suspected that this type of gold was alluvial panned gold and suggested the golden quadrangle in the Carpathian arc as the region of origin.

Borg and Pernicka (2017) however come to a different conclusion. After evaluating numerous literature sources (Penhallurick 1986; Camm 1995) they describe the importance of the gold wealth of the Carnon River in Cornwall's tin district, where gold was obtained in significant quantities as a side product during the extraction of cassiterite (tin) from alluvial river placers from prehistoric to modern times. Extensive geochemical and lead-isotopic investigations by Borg (2010), Pernicka (2010) and Borg and Pernicka (2017) finally identified the geochemical similarity of Hartmann's A3 gold with the gold from the Cornwall tin placers and concluded that this gold was the source of the gold intarsias of the early Bronze Age Nebra disc.

In smelting processes, cassiterite will dissolve in gold, unlike PGMs. In such cases, bulk analyses in chemical composition will lead to elevated levels of tin compared to natural gold.

Depending on the type of weathering and transport, the external form of cassiterite crystals is formed in the sedimentary context: In alluvial placer deposits, they are successively rounded off to small spherulites. Camm and Croot (1994) show in their investigations that, on the other hand, idiomorphic crystal forms of cassiterite in Cornwall placer deposits at least partially rule out a formation in alluvial placer deposits, but rather rely on in situ formation, as described below.

Due to its solubility and the tendency of rapid re-precipitation, the formation of authigenic gold occurs (Falconer and Craw 2009), so that nuggets can form in the sedimentary environment. This effect is not observed in cassiterite which is much more chemically stable than gold.

Attention should be paid to two statements that show how differently tin placers are valued as raw material sources in modern times and for archaeometallurgy. The geologist Hosking (1988) writes in the current book "Geology of Tin Deposits" with global claim: "Modern placers are by far the most important sources of tin".

Barnes (1974), however, points out some important practical problems of a possible early discovery and use of tin placer deposits. He writes analogously: "Even in a tin vein, cassiterite is not normally an abundant mineral and the grades worked today probably average well <2 wt.%. In placers grades are very much lower (Pohl 2005: $\leq$ 0.01 wt.%) and the cassiterite occurs as a sand or grit-sized material intimately mixed with other granular minerals of very similar appearance such as … wolframite, tourmaline, magnetite, hematite, garnet, topaz, monazite, and rutile plus silica sand. These granular minerals fill the spaces between larger gravel pebbles, cobbles and boulders which lie on the bedrock at the base of a river sediment. To collect such material, and inadvertently, or even purposely, add it to a furnace charge is quite beyond the realms of credibility. Unlike gold, where veins were found probably by tracing placers back to their source, tin placers were almost certainly discovered after cassiterite had been identified before in epigenetic ores and then, perhaps, by analogy with gold, traced to river beds".

However, Penhallurick (1986) emphasises that so-called tin grounds can form in the lower area of a tin placer, resting on the solid rock at the base of a tin stream.

3.6.6 Iron Placers

Magnetite placers are formed by accumulations of this mineral, which is washed out of acid and basic (volcanic) rocks. Sands enriched in magnetite may contain TiO_2 values in the lower percentage range because of the variety of titanomagnetite (Fe_2TiO_4). Magnetic sands occur, e.g. in many places on the Italian west coast (Schneiderhöhn 1962). Famous are the (Ti-) magnetite placers on the east coast of the Black Sea in the Colchis (Twaltschrelidze 2001). They extend over a strip of about 50 km. The richest lie between the estuaries of the rivers Supsa and Natanebi. This mineralisation is believed for prehistoric mining since the eleventh to sixth centuries BC (Chachutajšvili 2009).

3.7 Mining Ore Deposits: Geological Factors, Economy and Politics

In the Old World, most of the ore deposits rich in gold, silver, lead, copper, tin, antimony and iron exposed at the surfaces have been mined by man to exploit metal resources periodically throughout the last 4000 or 5000 years, or even longer. These activities, together with the extraction of metals, are a prominent example of how in humanity, the relationship between technology and nature is intimately interlinked, at the latest since the Neolithic period. For these reasons, mining is named a techno-natural system by Asrih et al. (2019).

Many instructive examples are known today after surveys, field works and excavations of many scholars in the last decades in ore districts, mines and smelting sites in Europe and Eurasia. These are, to select only a few examples, e.g. on the Iberian Peninsula (Rothenberg and Blanco-Freijeiro 1981; Rovira 2004; Hunt Ortiz 2003; Domergue 2008 and others), in the Balkans (Pernicka et al. 1993; Gale et al. 1991; Cauuet et al. 2003), in the Alpine regions (Weisgerber and Goldenberg 2004; Oeggl and Prast 2009; Turck et al. 2019), in Anatolia (Wagner and Öztunali 2000; Seeliger et al. 1985; Palmieri et al. 1993; Klaunzer 2017), in Iran (Stöllner et al. 2004a, b; Vatandoust et al. 2011), in Oman (Hauptmann 1985) and in the Wadi Arabah (Rothenberg 1990; Hauptmann 2007; Levy et al. 2014; Löffler in prep). Kargaly in the south of Russia was investigated by Černych (2003).

Many of these ore deposits were exploited over centuries and millennia often in changing booms and peaks and periods of decreased activities depending on political, financial and geological reasons. Stöllner (2003b) named periods of intensive exploitations "Imprinting Phases". In every case, mining activities to exploit ores from an ore deposit is a long-lasting process (see below), almost as a rule over many generations. Very often, the statement "exhausted" is used prematurely for a certain period of decreasing activities. A compendium of the systematics of mining ancient (ore) deposits is published by Stöllner (2008b).

3.7.1 Early Stages of Exploitation

3.7.1.1 Prospection

Prior to the commencement of actual mining activities, prospection activities are required. They involve two sorts of work, namely the prospecting of new mineralisation in a given study area and the exploration of the structural composition of an ore deposit in its geological context (Weisgerber 1989).

In the search for ore deposits on the basis of clearly recognisable terrain features, it has always been easier for humans to work in low-vegetation areas such as the Mediterranean countries, the Near and Middle East and the steppe areas in Central Asia.

An outcrop of an ore deposit, the iron hat (gossan, see Sect. 3.5), with its distinctive reddish-brown accumulations of iron hydroxides and oxides is one of the most striking features of mineral outcrops. The solubility of near-surface minerals from the iron hat in meteoric water and their precipitation in soils and streams can lead to very striking staining. The reddish streams coloured by the huge hat zones of the deposits

Fig. 3.69 Sungun, Northern Iran. Copper mineralisation at the mouth of an ancient gallery. The greenish-bluish colour, caused by solution of copper minerals and re-deposition just outside the mine, is a striking feature for prospecting this metal deposit. Photo: G. Weisgerber (†), Deutsches Bergbau-Museum Bochum

in the Rio Tinto area in the Iberian Pyrite Belt, which gave the locality its name, are a very eloquent example of this (Fig. 3.16). If minerals precipitate out of copper-containing solutions, it leads to striking green colourations of surrounding soils and rocks (Fig. 3.69). Noticeable changes in streams and rivers are also found around salt deposits when their waters turn into natural, tasty brines.

Prospecting new ore deposits in forest regions, such as in Central Europe, was more difficult. Here, however, there are useful indications of possible concentrations of copper, lead, zinc and other heavy metal accumulations in the soil via the flora. A useful marker of orientation was the appearance and distribution of metallophytes (chalcophytes). These are plants that endure a high natural or anthropogenically caused supply of local heavy metals. They usually have only a low vegetation density. Examples of metallophytes are, e.g. the zinc violet (*Violetea calaminariae* and *Viola guestphalica*), a subspecies of the sea thrift (*Armeria maritima ssp. Halleri*) or the so-called copper viola (*Silene vulgaris*). An overview of metallophytes can be found at Punz (2001) and Knolle and Wilfried (2016).

Metallophytes do not only grow on natural near-surface mineralisation and ore outcrops. Studies on the appearance of metallophytes in the mining regions of the Harz Mountains, in the Ore Mountains or in the Black Forest (Germany) and in the mining regions of Portugal (Goldenberg 2014a, b; Weiler and Nover 2008) showed that they may also be useful for locating older mining waste dumps, sand debris from stamp mills in floodplains, slag heaps and other metallurgical debris. These facts are of fundamental importance for modern fieldwork of mining archaeology and archaeometallurgy.

The enrichment of heavy metals from old mining, ore beneficiation and metallurgical activities, such as lead, zinc, nickel, copper, chromium, cobalt and cadmium in mine tailings, in colluvia and alluvial sediments and the formation of acid mine drainage, can be detected almost everywhere.

They represent valuable geoarchives to reconstruct the sorts of ores mined, as Dobler (2000) was able to show for the Harz district and Beckmann (2007) for the Upper Palatinate in Germany. But they also may create enormous contaminations of soil, water and air environments in the Iberian Pyrite Belt (Rio Tinto) as investigate Martín-Crespo et al. (2019).

3.7.1.2 Mining Laws

Early prospecting activities also included exploration and assessment of mineral deposits and their economic value. Therefore, a detachment of the mineral wealth from personal property and a transfer to a general legal–political claim by ruling systems is likely to have occurred rather early (Stöllner 2008a; Asrih et al. 2019).

It can be assumed, e.g. that the extraction of silver from the ore district of Laurion in the second half of the first millennium BC, was controlled by the Athenian government, as it would otherwise hardly be possible to organise the numbers of tens of thousands of miners (slaves) and specialised metallurgists (Lauffer 1979). Prisoners were used as miners in Roman times as well (Geerlings 1983, 1985; Hirt 2010). Written testimonies about orderly mining laws do not exist. Among the earliest mining laws are the ancient Vipasca tablets, which report, e.g. on the exploitation of the gold mine of Três Minas, Portugal, by the Roman governmental administration (Wahl 1988).

As discussed by Schattner (2017), in antiquity both government-organised mining and metal extraction was carried out on an industrial scale, e.g. in Rio Tinto, Las Medulas or Laurion, as well as small-scale mining, e.g. lead–silver mining in Ibiza. At best, it is conceivable that this small deposit was controlled politically due to its geostrategic location in the Mediterranean. Today, small-scale or artisanal mining means the extraction of mineral resources with simple, manual, non-industrial methods (Grießl 2014). Probably small-scale mining of ores, as well as that of gemstones, may have had a much larger extent in ancient times than is the case in Europe today. Today, small-scale mining and metal extraction using simple technical equipment is widespread in almost all countries of Latin America and Africa (Priester et al. (1993).

The first consolidation of written lore for mediaeval mining began in the late twelfth century AD. It is related to the proclamation and implementation of the royal claim to silver and salt mines. Details of the mining law (so-called *Bergregal*) are discussed by Stöllner (2008a, b) and Asrih et al. (2019).

Interesting are ethnographic observations noted by Werthmann (2009) to recent gold mining (gold rushes) in Burkina Faso, West Africa, when thousands of people flocked to sites of gold nugget finds and, within a short time, mined and processed gold ore and disappeared again within a few months. Werthmann aptly described this as "hit and run" behaviour (see Sects. 9.1 and 9.2). Rising mining activities after the discovery of the first nuggets were initially completely disorganised, which quickly led to criminal excesses. However, local systems developed rapidly, resulting in orderly clan stakeouts.

As far as other recent gold rushes are concerned, there have always been some famous examples in the recent past. In the nineteenth century, the age of the Industrial Revolution, and also in the twentieth century, thousands of people left for Alaska, California, Australia and New Zealand to seek their fortune by gold panning. Comparable developments were observed there.

3.7.2 Geological Factors and Innovations

In the Old World, from the Iberian Peninsula in the West to Southwest Asia in the East, there are no outcrops of ore deposits which have remained unworked in ancient times. Usually, they have been more or less completely exploited in earlier periods, at least in their upper parts. We know also from mediaeval sources, such as Georg Agricola (1556), that old workings were already an unwelcome legacy of earlier mining and even then made the exploitation of any ore deposit directly from the surface impossible.

The strategy to open a mine and to exploit an ore deposits shows large varieties and is primarily controlled by the shape and design of a mineralisation. Basic structures of a mine, i.e. shafts, galleries and chamber-pillar constructions are following the spatial extent of the ore bodies. This, in turn, means a mutual dependence on geological conditions and applied technologies (Asrih et al. 2019). It is thus the primary task, on an interdisciplinary basis between geoscientists and archaeologists, to correctly record the course of mineralisation in length and depth and to define on this basis the strategy of the development of this ore deposit. It can be assumed that in ancient times, due to the very simple mining tools, each deposit was mined with the least possible expenditure of energy.

(Horizontal) synsedimentary mineralisations have a different spatial constellation than (vertical) hydrothermal veins (see Sect. 3.2) with

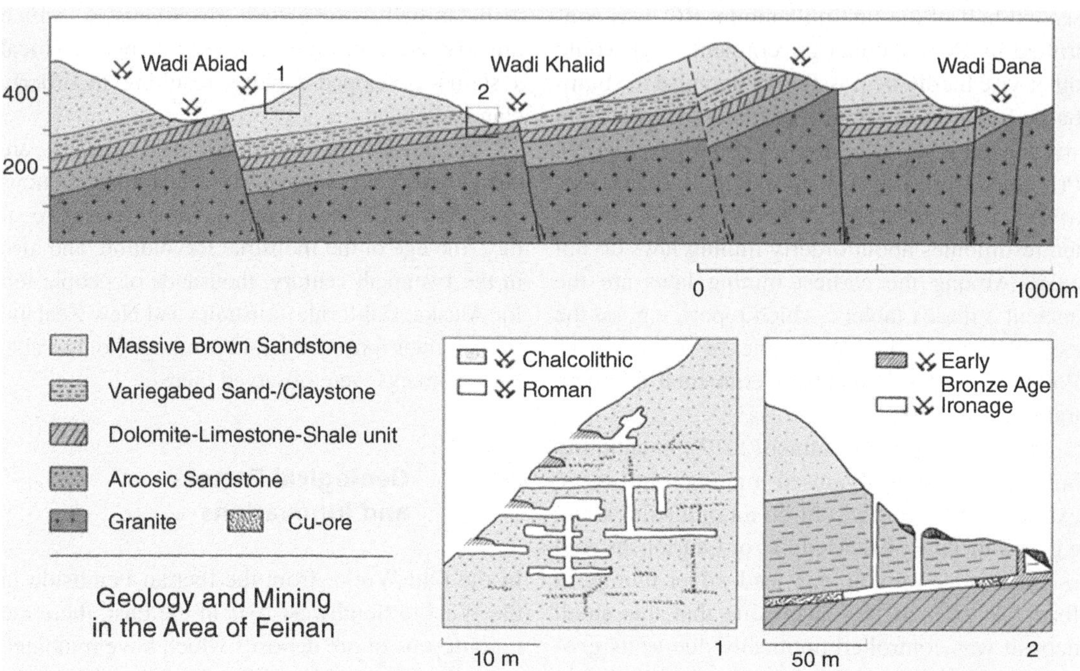

Fig. 3.70 Schematic illustration of the mining developments in the Faynan copper district. The geological section Wadi Abiad–Wadi Dana shows two ore-bearing formations. They have been exploited over many time periods from the fourth millennium to the Islamic period at the twelfth century AD. Different technologies of exploitation systems have been applied in each respective period. From Hauptmann (2007)

banded mineralisations and gangues. These are in turn different from mineralised breccias in limestone or dolostones or impregnations of host rocks. Each of these configurations can also be differently deformed by tectonic faults and slips (see Sect. 3.4). The mineralogical structure of sulphidic ore deposits can be decisively affected by secondary zoning and the formation of oxidation and secondary enrichment zones (Sect. 3.5).

Until about the Roman period, miners had always tried to follow a mineralisation in the solid host rock as closely as possible by the naked eye. Mineralisations, e.g. of copper and lead–zinc ores in paleokarsts or hydrothermal veins, are clearly visible from the host rocks. Examples of this are the lead–zinc–silver mineralisation such as in Nakhlak (Iran) (Stöllner et al. 2004a, b) or in the Mitterberg (Austria) (Stöllner et al. 2009b).

The example of the stratiform sedimentary copper deposits of Faynan (Jordan) (Hauptmann 2007) or from Timna (Israel) (Rothenberg 1990) clearly showed how plausibly old mining developments of these deposits were carried out in various epochs.

In Fig. 3.70 it is shown that in Faynan (Jordan) geologically different mineralised zones have been exploited in different epochs (Hauptmann 2007). The most interesting observation is the development of Early Bronze Age to Iron Age mining from the third to the first millennium BC in Wadi Khalid. Mining activities in this wadi can be described as follows: Following the seam-like ore mineralisations of the Dolomite-Limestone-Shale Unit ("DLS") from their outcrops, which are dipping slightly towards the northwest, room and pillar constructions were dug. They had a height of 1–1.5 m and a lateral extent of ca. 30 × 55 m. The mines followed the ore deposit down to a depth of only 15–20 m. The Early Bronze Age waste tailings can be quite clearly distinguished from those left by the Iron Age mining activities. While the former, almost without exception, are positioned close to the outcrops of the DLS, the Iron Age heaps are normally located at the foot or even at the slope of steep mountain ranges, thus illustrating an exploitation of copper ores by much deeper shafts.

These mines demonstrate a long duration and continued re-initiation of the mining activities, which slowly penetrated deeper and deeper into the mountains. The extension of sometimes tectonically attacked DLS-seam was well known to the miners. In the Iron Age, they were thus forced to exploit those parts of the DLS, where the ore had not been completely removed previously, by much deeper shafts in the steeply rising valley sides. Double shafts had been constructed, one being sunk for climbing and the other for raising the ore. On the upper course of the Wadi Khalid, a double shaft with a depth of c. 70 m (reconstructed from survey data), perhaps for hauling and ventilation, has been found (Weisgerber 1989).

Other ore deposits pose significant problems here, namely the multiple lead–zinc(−silver) mineralisation found in carbonaceous rocks and dolostones in the Mediterranean Basin. They are referred to as paleokarstic mineralisations. Genetically, they are part of the Mississippi-Valley-Type (MVT) deposits. Such mineralisations are irregularly shaped, forming pockets, nests and mineralised cavities (cf. Fig. 3.40). The shape of vein-like mineralisations as well as brecciated ores is determined by tectonic activities. These karst cavities are filled with earthy limonitic material that may be solidified to varying degrees by calcification. Accordingly, ancient mining activities left irregularly shaped chambers and galleries. In much field work, it is unclear which cavities were created by mining-strategic activities, and which are just remnants of the exploitation of natural karstic mineralisations. Examples of such mineralisations can be observed, e.g. at Agios Sostis on Siphnos (Aegean island in Greece, Fig. 3.71) (Vavelidis et al. 1985), the lead–zinc and gold mines of Thasos (Vavelidis et al. 1988), on Argentera (the island of Ibiza in Spain, Fig. 3.2) (Hermanns 2015), in the tin mine of Kestel (Yener et al. 2003), or in Laurion (Greece) (Skarpelis and Argyraki 2009).

The different types of ore deposits and zones of mineralisation, i.e. the mining of mineralisations in the iron hat, over the oxidation and cementation zone in hydrothermal vein reservoirs, can be used conditionally to reconstruct developments of technologically diverse new facilities and innovations (Stöllner 2019a).

To simplify, it can be stated that highly enriched and less complex ores could be obtained with the least possible human and technical effort. The more complex the compositions, and the lower the metal concentrations in the ores, the greater must be the quantities of material recovered and processed, the human resources and the range of productions needed in order to still be economical (Bartels 1997). Without doubt, a difficult technical cliff was probably the transition from the rich cementation zone to the primary ore body in hydrothermal veins, which typically contains lower percentage ore (cf. Sect. 3.5).

Many hydrothermal veins of the European low mountain ranges, such as those found in the Harz mountains, the Ore Mountains or the Black Forest (Germany), in the Czech Republic and the Vosges (France), may probably have been mined before the Mediaeval period. Unfortunately, there are no older documents from many of these mineralisations (Bachmann et al. 2002/2003; Téreygol 2002; Gauthier et al. 2015). Ores have possibly been exploited directly by drifting galleries within the veins because that were the most interesting areas economically, and also the geologically most viable for the heading (Asrih et al. 2019). For draining the mines, special draining galleries have been driven (in German language *Erbstollen*), often a work through several generations of miners. The technology of draining a mine by galleries or water lifting by wheels or vessels led to an exploitation of ores by successive increasing depths. As shown by the water wheels found in Rio Tinto, about 4 m in diametre, the drainage of mines was already carried out in Roman times, with sophisticated technology (Weisgerber 1979).

As shown by the development of mediaeval mining industries in Central Europe (Bartels

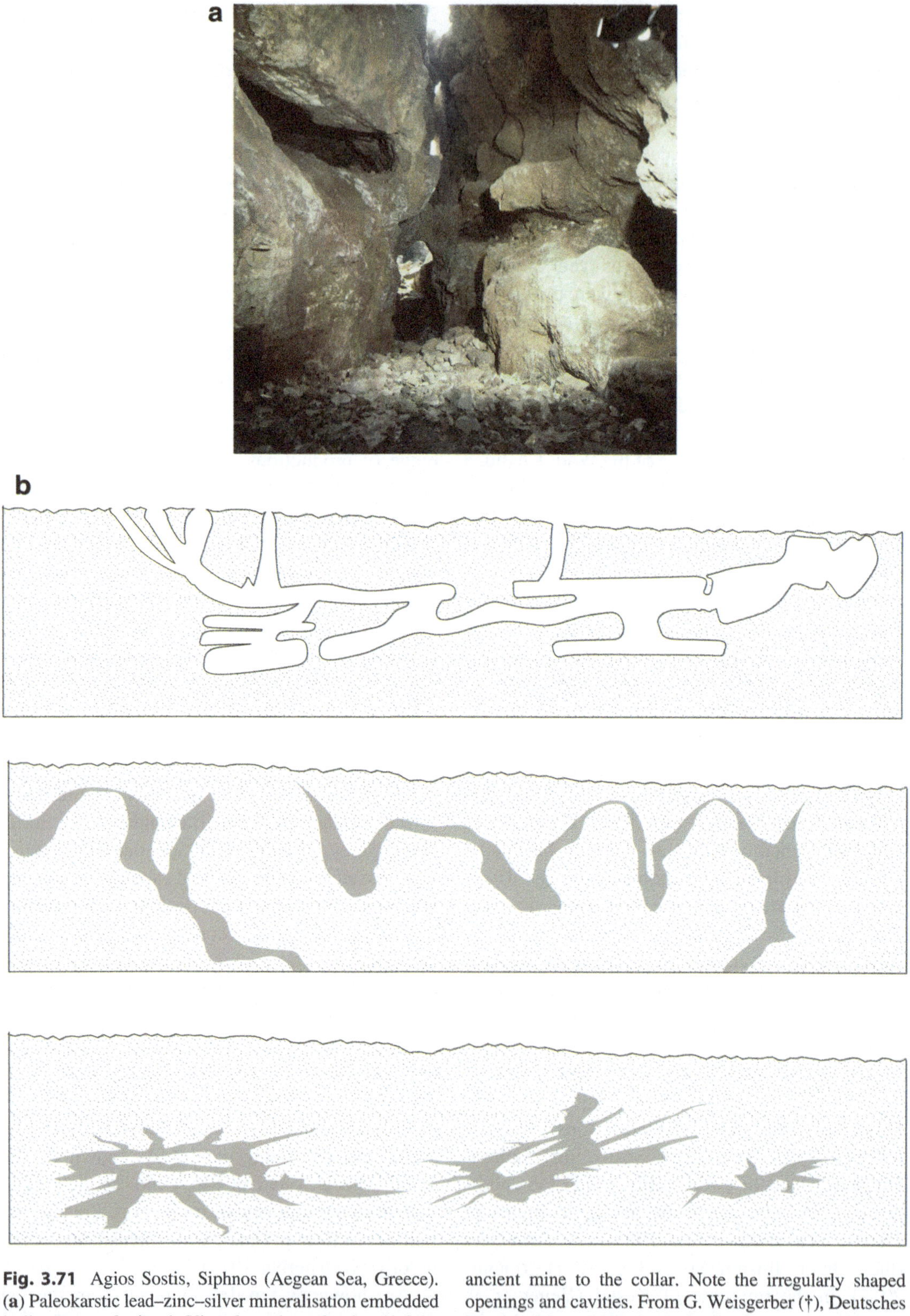

Fig. 3.71 Agios Sostis, Siphnos (Aegean Sea, Greece). (a) Paleokarstic lead–zinc–silver mineralisation embedded in limestone of mine 1. View from the today level of the ancient mine to the collar. Note the irregularly shaped openings and cavities. From G. Weisgerber (†), Deutsches Bergbau-Museum Bochum. (b) Surveyed plan of the same

1997), a decision on innovations in mining as well as in metallurgy can be politically initiated. On the other hand, as production and trading of raw materials may have an enormous strategic level, metal production is also highly exposed to political disruptions. The consequences are regional production relocations, competitions, structural crises. The introduction of new metals and alloys, such as the transition from arsenic copper to tin bronze, or the spread of iron or steel belong to this problem area as well. Examples of a relocation of metal production can be seen in the Early Bronze Age of the north-eastern Aegean, but also in Mesopotamia. Lead isotope analyses of Early Bronze Age artefacts were able to prove (Muhly et al. 1991) that with the coming of tin bronze in the middle of the third millennium BC, new, up until now unknown ore deposits were exploited, and foreign metal was found in the economic cycle in south-west Asia.

For fine-grained mineralisations that are visually elusive ("invisible gold" in Sakdrisi (Georgia), or tin ores in hematite in Kestel (Turkey)), mining activities have to be described as a continuous assay process (pers. comm. Timberlake 12/2017), in which the richness of the ores had to be checked with great effort, in whatever way possible, before further advancement of mineralisation could be considered.

3.7.2.1 Ore Shoots and Bonanzas

Geologically unusual ore enrichments, the so-called "bonanzas", can form in the upper parts of mineral deposits. They are caused by the migration and precipitation of various ores, e.g. from pyrite or copper ores in deeper parts of a deposit. In this way, an enrichment below the gossan may happen. Alternatively, they are formed by precipitation and concentrations of metal-rich solutions at the groundwater level. This can lead to massive accumulation of copper, but also of cobalt, nickel and silver. These metal enrichments include so-called "ore shoots". These are usually elongated, chimney-, pipe- or ribbon-like masses of rich ores which are preferably built at the crossing of two or more veins or any kind of rock openings. Similar to ore shoots are pay streaks, a term which also refers to enrichment of ores in placers.

Such unusual mineralisations occurred and still do occur from time and time again. From the last centuries, there are numerous reports about silver ores. Thus, in 1477 AD, in the St. Georg mine in Schneeberg in the Ore Mountains (Germany), a silver ore mass measuring 2 × 4 m and weighing 8 tons was found. According to the tradition, a table was made from it on the spot, which was then used to cater to Duke Albrecht (Wagenbreth et al. 1990). In the silver deposit of Kongsberg/Norway, lumps of silver with several 100 kg weight were found repeatedly. In 1820, an ore mass of native silver weighing more than 1 ton was found in Sonora, Mexico.

In the history of mining and metal production, the exploitation of the cementation zone itself has time and again led to economic heydays that lasted until the underlying poorer ore body was reached. For example, the finds of rich, massive lots of silver ores with native silver in the mediaeval Ore Mountains and in the Harz Mountains in the sixteenth century led to the so-called *Berggeschrei* throughout Europe, which attracted miners from all over the world and caused an economic boom lasting only a few decades (Bartels 1997). It has been proven that the transition to poorer ores of the underlying primary ore body has resulted in significant economic structural crises. It is possible that the abundant appearance of oxhide ingots in the Mediterranean area after 1250 BC, which, after the measurement

Fig. 3.71 (continued) mine confirms the observation made by photo a: The shape of the mining cavities are always affected by the shape and design of an ore deposit. Such erratic mineralisations are transferred from Figs. 3.40 and 3.51

of lead isotope concentrations, originated in the reservoir area of Apliki in Cyprus (Stos-Gale et al. 1997), is due to a similar phenomenon. Shortly after, in the late eleventh century BC, the production of these oxhide ingots ceased.

The idea of silver contents in galena not only in the Upper Harz deposits but also in comparable deposits worldwide, is described by Bartels (2014). He describes this as a procedural repertoire of ore mining and mineral processing.

Under modern industrial circumstances, it is understandable that only a few cubic metres of mined ore containing 25 wt.% or even 50 wt.% of silver, perceived as accessory minerals, are of interest for genetic or museological purposes. These parts, however, are hardly mined deliberately today. These ore shoots are spectacular exceptions. But as long as hammerheads, mallets, hoisting irons, crowbars, wedges, shovels, and other hand tools were used, silver riches hardly disappeared in a large amount of low silver-bearing galena. Rather, they were deliberately mined separately (Rehren et al. 1999a, b) and, as a result, miners held extremely valuable material in their hands: A so-called Berg-Zentner argentite of 110 pounds, in a favourable case, contained nearly 48 kg of silver, which is equivalent to about 4.0–4.35 kg of gold, since gold:silver ratio ranged, with some fluctuations, from 1:11 to 1:12 (Pfister 2010; Soetbeer 1879).

In fact, in industrial-scale ore mining operations predominantly low silver galena is present, so that overall a relatively high silver yield can be achieved in view of high production rates. But it leads to erroneous assessments when such production figures are transferred to (pre-) historical or mediaeval practices. Using the example of the Sankt Andreasberg Mine in the Harz Mountains, Bartels (2014) further points out that sampling of lead ores low in silver must lead to the result that the deposit is extremely poor in silver. However, only when a quantitatively less representative silver-rich minerals are included in an estimation, are potentials regarding a silver extraction more clearly recognisable. This applies not only to many comparable lead–silver deposits, but also, e.g. for the estimation of tin deposits (see Sect. 3.4.11).

3.7.2.2 Depletion of Reserves

A recurring question is when a deposit actually was exhausted, leading to a ceasing of any mining activities. When Strabo writes (9, 1, 23) that the silver mines of Laurion are now exhausted (in Augustian times) and the workers have begun to reclaim mountain and slag heaps, this statement was, as we know today, incorrect. The deposit was mined in the following centuries and millennia until the recent past with new technological methods. Only the assay limit, defined by the economic conditions at that time, was reached. It is not necessarily dependent just on the geological conditions and can change (several times) over time.

Why mining and metal production have come to a standstill in certain ore mines can hardly be answered in a monocausal way. The history of mining has shown that the rarest reason for this has been a complete depletion of mineralisation. Investigations in the mediaeval mining district of the Harz Mountains (Bartels et al. 2007) indicate operational, technological and economic reasons associated with increasing depth of exploitation, which possibly led to the end of mining. Here again, the transitions from oxidic to (rich) sulphidic ores of the secondary enrichment zone and subsequently the transition to the lower grade primary ore body have to be mentioned. Also, water and ventilation problems could have been major problems which finally limited exploitation of an orebody. An impressive archaeological example for the limitation of mining technology was found at Faynan (Jordan). As mentioned above, the 70-m deep Iron Age shaft in the Wadi Khalid, Faynan, was discovered as a single super-achievement of ancient mining. It was certainly the upper limit for entering and hauling ore from the very depth (Hauptmann 2007).

It was probably political unrest in the eastern Mediterranean, but not the depletion of deposits, which caused the breakdown of the Cyprus copper monopoly with its numerous ore deposits in the Mediterranean around 1150 BC—and at the same time led to a huge boom in metal production in Faynan (Hauptmann 2007). This industrial copper production in the much smaller ore district lasted for a few centuries. The reason why it finally came to a standstill could also have been mining–technological reasons. The deposit of Faynan was certainly not exhausted. This is proven by the mining activities in Roman times around the beginning of the Common Era.

3.8 The Beneficiation

3.8.1 General Remarks

The raw products, i.e. the ores exploited from a miner could often not be used directly for subsequent smelting. Ores are generally concentrated by crushing, grinding, milling or flotation to reduce grain size and liberate ore minerals to separate gangue and host rock from the real ore. As pointed out by Stöllner (2019b), this important step of processing or beneficiation was multiply underestimated in research of mining archaeology and archaeometallurgy.

Products of beneficiation may be transferred to smelting sites to optimise slagging as required by the metallurgists. In addition, in old and mediaeval to modern times, ancient slags and other metallurgical intermediates, such as chunks from roasting processes or fragments of litharge, have been processed to extract metal or other valuable inclusions. Beneficiation resulted in an automatic concentration and homogenisation of the ore to be extracted. This is an important step in the *chaîne opératoire*, moving from mostly inhomogeneous composed mineral deposits to more or less precisely assembled metals and alloys.

On the other hand, heaps of waste tailings were produced containing fine-grained gangue and host rock and residues of ores. In case sulphidic ores have been beneficiated highly contaminated acid mine drainage were left in case of rainfalls and rivers (Martín-Crespo et al. 2019) leading to considerable environmental problems for human beings, flora and fauna.

Although the original qualities of surface-exposed parts of ore deposits can hardly be estimated anymore, there are a number of possibilities to reconstruct handicrafts and methods of beneficiations in ancient times.

Dispersed and low-grade gold ore has to be treated differently than massively occurring galena. Extracting of non-ferrous metals and iron ores, old miners had to work centimetre-sized or a bit larger ores. This does not apply to finely dispersed tin ores also, which have been detected in hematite ore and in the host rock in Kestel and Göltepe in south-eastern Anatolia (Yener et al. 2003; Yener 2009). The existence of (processed) high-grade iron ores as probably used to produce the overall existing fayalitic tap slag (Rennfeuer-schlacken) with c. 70 wt.% FeO are often enigmatic. How were low-grade lateritic iron ores processed to reach Fe levels suitable to produce iron (Keesmann et al. 1983). Or have there been ferricretes available?

Generally, tools and equipment show considerable variation over time, but their mode of operation often has technologically identical features. Beginning with prehistoric times until medieval times, the beneficiation of ore was carried out with comparatively simple mechanical means, both by dry treatment and wet processing. It is a valuable help for any reconstruction that even today in many countries where artisanal mining is still practiced, simple conventional hand sorting, grinding and crushing is carried out to enrich ores. The processing of ore using chemical agents was unknown until the Middle Ages. Effects of flotation were presumably exceptions (see below). Today flotation with milk or (organic) oils is still used in artisanal small-scale mining today (Wotruba and Riedel 2006). This cannot be ruled out for ancient times either. Magnetic graders were unknown.

Ores of various metals may be intergrown intensively, whose simultaneous extraction by metallurgical processes in ancient times was hardly possible and mostly undesirable. Examples are lead and zinc ores, intergrown with copper or silver ores. They occur in Laurion (Greece), Cartagena (Spain), Rammelsberg and Harz (Germany) as well as in the ores of Mitterberg (Austria). It is not clear when zinc ores could be separated effectively. Lead and copper ores were smelted to metal in the Chalcolithic and earlier, zinc ores in large quantities were probably used since Roman times. There are a number of different Bronze Age metal artefacts, mostly from Iran, which contain several percentage of zinc. Thornton and Ehlers (2003) suspects that zinc had been deliberately extracted as a metal this early in history. Hanel and Bode (2016) on the other hand believe that this metal could only be successfully extracted from its ores and used as an alloying metal at the beginning of the Common Era.

Interesting publications on modern and old methods of beneficiation are Taggart (1945) and Lucas (2006). Useful overviews and sequences of processing in antiquity can be found in Pliny (Nat. Hist. XXXIII) from the first century AD. They reflect basic steps of beneficiation of ores, starting with the mechanical comminution to the subsequent separation of ore, gangue and host rocks due to different densities with water. This is referred to repeatedly in the (pre-)historical context (Rosumek 1982; Stöllner 2003a, b, 2019b), whether they are simple versions of processing or technologically complex installations that have evolved over time.

Stages of dry beneficiation are:

1. Hand sorting of ore.
2. Crushing of hand-sorted ore to prepare pea-sized material.
3. Sorting by sieving.
4. Fine grinding to get powdered ground stock. Stages 3 and 4 blend into each other. The step of the following wet mechanical beneficiation is:
5. Panning of the powdered ground stock to separate components of gangue and host rock
6. Sluice boxes and sorts of jigging installations

After these preparations, concentrated ore is ready for metallurgical processes.

3.8.2 Dry Treatment

Hand Sorting of Exploited Ore

This processing step was usually carried out by hand on larger pieces, either inside of a mine or just next to the adit, or in the immediate vicinity. Particularly rich ores were separated, extracted and processed on their own. The rest remained in the mine as stowing.

Crushing of Hand Sorted Ore

This further step of dry processing was crushing and grinding, i.e. comminution with a hammer or a hammer stone, and again sorting by hand. In prehistoric times, the most common tools for this task were made of stone. At almost all sites in the area of early mines countless fist-sized, occasionally also much larger such hammer stones have been found (Fig. 3.72a). They are usually made of hard rocks, are rounded by their use or have approximately the shape of edge-rounded cubes with dents on the side surfaces. In addition, in the archaeological context, there are anvil stones and plates or corresponding fixed equipment from larger rock slabs or lumps, at which, depending on the size (up to the metre range) one to several dozens of small, bowl-like depressions have formed, which in their appearance are reminiscent of egg stands. These depressions were created by the crushing of ore (Fig. 3.72b).

Almost more frequently, such hammer stones and anvil stones are found on ancient slag heaps, where they were used not only for crushing ores, but also for crushing slag or, in the case of Laurion, for crushing silver-bearing litharge (Papadimitriou 2012). A processing of slags as a separate process step is often overlooked in archaeometallurgy. Since in old metallurgical processes, the separation of metal or metal sulphides (matte) from slag or corresponding intermediate products was hardly complete, varying amounts of valuables remained in the slags as inclusions. These were mechanically crushed and possibly treated using water, to exploit those remaining inclusions. In Sect. 5.4.10 (crushed slags), the field results of this preparation are discussed in detail. There are two regions in the Old World that should be mentioned as type localities for slag processing on a larger scale.

On the one hand, these are the many processed slag heaps of Bronze Age copper mining in Trentino (Italy) (Cierny 2008; Silvestri et al. 2019). On the other hand, there are the distinctive heaps of processed slags of the order of several thousand tons at Khirbat en-Nahas in the ore

Fig. 3.72 (a) Khirbat en-Nahas, Faynan (Jordan). Hammerstones found at the large Iron Age slag heap. Hammer stones like these were found at many (pre-) historic sites all over the world, near ancient mines, and on slag heaps. Slag as well as ore could be crushed to mm-sized butches. (b) Wadi Salh, Oman. Anvil stones with a number of small, irregularly ordered depressions which are reminiscent of an egg stand. They are results of beneficiation by hands. Next to this anvil stone, a number of hammerstones were found. These finds were associated with a slag heap the age of which is proposed by Wagner and Yule to be in the Early Bronze Age as well as in the Iron Age using thermoluminescence dating. Photos: A. Hauptmann, Deutsches Bergbau-Museum Bochum

Fig. 3.73 Riberinha, Três Minas, Portugal. One of the many hundreds of granitic anvil stones of quadrostyle stamp mills from the locality of Covas. This anvil stone was used for beneficiation of gold out of solid rock. The weight is 360 kg, the sizes are ca. 96 × 44 × 44 cm with 4–5 partially overlapping recesses. These trough-like recesses have been created by the intact impacts of probably four stamps. It cannot be ruled out that several of these hammering-troughs were arranged side by side in use. Photo: G. Weisgerber (†), Deutsches Bergbau-Museum

district of Faynan (Jordan) (Bachmann and Hauptmann 1984).

As a technological development of the pounding technique with hand-held hammerstones and anvil stones, complexly designed, have been mechanically operated stamp mills, which were probably used in the second half of the first millennium BC or at the beginning of the Common Era in the Mediterranean.

A prominent example of stamp mills can be found in the gold-bearing district of Campo de Jales, near Três Minas, in northern Portugal (Wahl 1993). Pliny (Nat. Hist. XXXIII) reports about this Roman achievement in the first century AD. Here, mechanically operated stamp mills, likely with four-column stamp mills, were used for the processing of gold from the host rock. Numerous standardised blocks of granitic rocks were found with regularly arranged small trough-shaped depressions (Fig. 3.73), which served as anvils and apparently could be replaced. It cannot be ruled out that several of these stamp mills were used side by side.

Interestingly, these granitic anvil stones have also been found at smelting sites, so they were also used for crushing slags. After crushing, the ore was possibly ground in circular mills. Circular mills were also found next to the worn granite anvils of the stamp mills in the area of Três Minas or Campo de Jales. Finally, the crushed ore was ground to a particle size of about 1 mm.

Where the mechanised stamp mills were invented and used on a larger scale must ultimately remain open. Wahl-Clerici et al. (2012) suggest that the finds of approximately 750 Roman anvil stones in Três Minas and other localities could be sufficient evidence of an invention of mechanised stamp mills at this time. Such

Fig. 3.74 Demoliaki, Attica (Laurion, Greece). (**a**) Relics of a circular mill, a so-called edge runner mill or Chilean mill for the processing of lead–silver–zinc ores from the classic time of silver mining in the Laurion ore district (mid first millennium BC). (**b**) The edge runner mill, composed of several dolomite blocks, is characterised by a series of closely and regularly spaced depressions, possibly formed by a type of rotating stamp mill. Photos: G. Weisgerber (†), Deutsches Bergbau-Museum Bochum

a standardised stamp mill with four indentations on the four sides has also been found near the Roman gold mines of Dolaucothi in Wales from the period between the first and second century AD (Meier 2013).

For the processing of raw ore by crushing, grinding and milling, the principle of the kollergang (also called edge runner mill or Chilean mill) has to be added here. The edge runner mill is known today as a grinder for crushing and mixing of stones and ores. One or two heavy, upright stone discs, the so-called runners, rotate in a groove around a vertical axis and thus grind the contents. Edge runner mills of simple designs are still found in various countries where small-scale mining is going on (Priester et al. 1993).

Possible prototypes of edge runner mills can be found in Hellenistic Laurion (Greece) (Papadimitriou 2016, 2017; Nomicos 2020). If the runners (round disks) of the edge runner mills are equipped with heavy stampers of stone like the stamp mill, a similar comminution effect as in Três Minas might be possible. In this way, the typical, closely adjacent depressions may have formed on the surface of the circular dolomite blocks (Fig. 3.74a, b). These systems were formerly called "helicoidal washeries" by Conophagos (1980). However, whether this was actually a wet or a dry processing must remain unknown.

Craddock (1995) stresses that it is altogether surprising that stamp mills did not seem to have been built again in Europe until the late Medieval period. Only Georg Agricola, in his book *De re metallica* from 1556 AD, mentions a stamp mill with 3–4 stampers for the treatment of raw ore. However, hydraulic-mechanical stamp mills have been in use until very recently. This is illustrated in impressive photographic documentation by Slotta et al. (2001) from the historical gold mining in Transylvania (Romania).

Sorting by Sieving

Sieving the crushed material is a logical step in the further processing of ores since ore processing techniques never produce an even-grained fraction of materials. Depending on the degree of intergrowth and hardness of individual ore minerals and minerals of the gangue and host rock, different grain-size fractions are produced, which must be separated. The tendency is for minerals with a higher specific gravity to accumulate in the fine-grain fraction. The sieving of ores is best done with water, but it is also possible in the dry state. Apart from the reports of Pliny, nothing is known about the use of sieves in the beneficiation of ores in the Old World.

Grinding to A "Pulverised" Mill Base

For many ores, after richer Parts of ore have taken out, the comminution of poorer ores and subsequent fine grinding is required. This applies to fine-grained polymetallic ores, above all to gold.

In ancient times, grinding mills, also called querns with flat slabs of hard rock measuring about 30 × 60 cm, and rotary querns were used. Naturally, beneficiation was more intense and complex the finer the ore had grown together with the host rock. Examples of grinding mills for the beneficiation of gold ores were published by Klemm and Klemm (2013) from New Kingdom Egypt (sixteenth to eleventh century BC) but were also in use in artisanal beneficiation of gold today in Tanzania (Borg 2014).

Rotary or round mills are frequently found in ancient Egypt. These consisted of two round, interlocking discs, centrally fixed, made of as hard as possible rock, with diametres of c. 4–6 decimetres. The space between the two discs was filled with pre-crushed ore and milled into powdery mill base. Many such finds have been made at numerous sites near gold mines in Egypt (Klemm and Klemm 2013). They seem to have proven to be the best tools for beneficiation across the ages, from the New Kingdom into the Early Arab period (seventh to eleventh centuries AD).

To ancient Egypt comparable tools are still in use today. In Japan, gold production is known from the mines of the island Sado Kinzan, where activities from the eighteenth century were published in numerous illustrative woodcuts. Winkelmann (1964) published and commented on these woodcuts. One of these illustrations shows the fine comminution of the gold ore and its washing out (Fig. 3.75). Fine grinding of the ore is carried out in man-powered rotary mills, which were also used supra-regionally and across time in cereal grinding (Lucas 2006) and were found in very similar form in ancient Laurion (Lohmann in press). The accumulation of gold then takes place in gold wash troughs and by manual washing in water-filled barrels.

3.8.3 Wet Dressing

Panning

After dry comminution work, stamping or pounding, a very fine-grained ore concentrate was produced by grinding, which would then be washed. The ore was panned. The principle of separating (pre-crushed) minerals according to their different density with water (Table 3.15) is exploited. In the simplest, most basic step, the glitters of gold are enriched by panning and circling in washbowls with water, flushing out the specifically lighter admixtures (Fig. 3.76, see also Fig. 3.25).

In addition to the separation of different grain sizes, sieving devices are used as well. The material is slurried, an enrichment of heavier ore particles in the finer sieve fractions takes place. In Roman times, as one knows from many installations on the Iberian Peninsula, the enormous demand for water was covered by dams and aqueducts.

If water was not available in sufficient quantity, it was possible to carry out a separation of different ore particles by means of wind processing. Very impressive images and explanations of this process were made in traditional gold mining in Burkina Faso by Werthmann (2006).

An optimal separation of the valuable minerals was likely never reached because sediments (as well as ground ore) usually contain heavy minerals or other ore components, which are difficult to separate in the concentrate (see Fig. 3.62). Hauptmann et al. (2010) have shown this with secondary copper minerals in concentrated placer gold. Gold washing in pans is also still a common, simple method of prospecting for gold today. It is also used almost worldwide in unconventional small-scale mining

Fig. 3.75 Fine grain crushing and washing of gold ore by as illustrated from the gold production at Sado in Japan around 1720 AD. Note the crushing in a man-powered rotary mill as used in ancient times for cereal grinding. Illustration from the book *Altjapanischer Goldbergbau* (Winkelmann 1964). This picture was previously published by Conophagos (1980) due to its similarity with installations for beneficiation found at the ancient Laurion (Greece)

in remote areas (Myanmar: Bennett 2017; Africa: Werthmann 2000; Borg 2014; Pakistan: Internet; Peru: Stöllner 2011a, b). Until the twentieth century, placer gold was washed out with simple methods even in the Golden Square in the Carpathian arc (Slotta et al. (2001).

The same is true for alluvial tin. A very easy method was used by Earl (1985) to wash cassiterite from ore and to prepare it for assaying in small portions by means of a slightly dented shovel. The principle is similar to gold panning and has been used until recently for an assaying of tin ores in Cornwall. In this way, the author has shown an easy method according to which tin ore could have been extracted from the prehistoric mine of Kestel in Anatolia (Yener et al. 2003).

Such a simple method was already emphasised by Krusch (1907). He is of the opinion that for a small-scale processing of tin ores, for prospection work, only a mortar, a pan and a balance are needed. If larger quantities are to be analysed, next to edge runner mills or stone breakers, the most primitive settling tanks are used.

Sluice Boxes and Sorts of Jigging Installations
Since prehistoric times, these simple methods of exploitation, beneficiation and of prospection activities, presumably changed rapidly to slightly larger scales depending on regional

Fig. 3.76 River Inguri, Svanetia, Great Caucasus (Georgia). Gold panning by a Georgian inhabitant at the river Inguri, near the village of Ieli in 2007. Photo: A. Hauptmann, Deutsches Bergbau-Museum Bochum

circumstances. This next step was the use of sluice boxes, which were known in ancient times. They consisted of wooden troughs, the floors of which were provided with transverse bridges or covered with coarse materials, fine twigs, sheep or cattle skins. The heavy gold particles were trapped by the obstacles placed in the troughs and remained there. In the ancient world, these fur-lined troughs gave rise to the legend of the Golden Fleece in the Caucasus.

The construction of wet beneficiations has been and continues to be extraordinarily varied and resourceful and has changed again and again over time and localities.

In a gigantic extent, gold from the gravel layers of Las Médulas in north-western Spain was washed out in Roman times using vast amounts of water (*ruina montium,* cf. Fig. 3.68). Another common wet mechanical process was and is table washing. A slightly inclined plate with or without corrugations was fed from the top corner with the ore-bearing slurry. The plate is shocked while a fine curtain of water is added in the direction of plate tilt. Due to the impacts of the shocks (or by individual water thrusts) the heavy ore particles are carried further than the lighter components of the gangue or the host rock so that a separation between ore and rock is achieved. An ore-rich slick is created. The remains of the slurries from these processes were possibly further separated from ore particles, which were carried along, by further sedimentation in wet-tyes.

Buddle boxes or chests were obviously common supra-regionally and across the ages. A particularly impressive example of this was excavated by Stöllner (2019b) at the Troiboden (Austria). It is a series of 15 wooden chests, dating to the Bronze Age of the fifteenth/fourteenth century BC. The Troiboden is a high moor area on the main course of copper mineralisation of the Mitterberg copper deposit in Austria (Fig. 3.77). Here, close to the place of prehistoric mining, lie the heaps of dry and wet mechanical treatment processes. In these wooden boxes (or jigging installations, Georg Agricola *Planherd*) several washing and concentration processes of finely crushed, chatty ores were performed. This was performed by slight stirring. This moving enabled a flotation just over the plank constructed in the middle of the tye and led to an enrichment of very fine-grained slurry of heavy minerals. Perhaps finely grinded ores were concentrated as well, as they would in a slurry tank. These wooden boxes may be comparable to the mud basins used in pottery.

Almost identical constructions of wet-tyes have become known from medieval silver

Fig. 3.77 Troiboden, Mitterberg, Austria. Example of one of the many wooden boxes from the Bronze Age processing of sulphidic copper ore in the Mitterberg area. The plank constructed in the middle of the tye had possibly the function for flotation fine-grained slurry of ore. From Stöllner (2019b)

production in the Ore Mountains of Saxony and Bohemia in the Czech Republic (Derner et al. 2016). The excavations have shown the processing steps to separate ores low in silver and lead from the valuable silver minerals or silver-bearing galena. The authors emphasise that given a bulk content of approximately 1 wt.% of silver this was certainly a painstaking procedure in which large masses of ore had to be processed first in order to even achieve a countable profit. Rich ores with a silver content of 20–60 wt.% were only rarely found. The main function of wet processing was the production of an ore concentrate by multi-stage washing. For polymetallic, fine-grained ores, such a concentration was obviously not possible by manual separation. The ore first had to be processed by a pounding and grinding process before it was delivered to the washing.

In the Hellenistic Laurion, complex washing systems were constructed to produce clean washed ore, the so-called *ergasteria*, which consisted of levelled areas, water channels and wet-tyes for depositing the ground ore. Overall, about 250 remnants of washing sites are known from Laurion (Rehren et al. 2002). Here, not only finely ground ore but also (silver-containing) cupellation residues (litharge, PbO; Papadimitriou 2012) were separated from the host rock using running water. A special construction of these *ergasteria* enabled that water used for wet processing in this dry region to be re-cycled to "fresh" water again. A convincing description of these plants can be found in Papadimitriou (2017).

The fact that the construction of such processing plants has been adopted centuries later has been shown by gold washing from the sixteenth century AD as documented by Lazarus

Fig. 3.78 Woodcut showing a typical gold panning scenery and installation according to Lazarus Ercker (1580). (**A**) The miners throwing the washed-out ore on a jigging sieve. (**B**) The jigging sieve. (**C**) The washer who adjusts the jigging sieve. (**D**) The installation of the buddling ditches. (**E**) The sluices. (**F**) The worker stirring up the slurry in both installations. (**G**) Barrel for further washing the turbidity. From Lazarus Ercker 1580, Freiberger Forschungshefte D34, Berlin 1960

Ercker (1580/1960; Fig. 3.78). Strikingly similar installations to Laurion are documented from Japanese gold processing as part of the artisanal gold mining and smelting from the beginning of the eighteenth century AD (Winkelmann 1964; Chap. 9, Fig. 9.9). An almost modern illustration comes from Siegerland, Germany, from the eighteenth century (Bartels 1994).

3.9 Sampling Ancient Ore Deposits

Sampling ancient ore deposits differs in many points from usual methods of sampling modern ore deposits in today's geoscience. These are defined according to other distribution criteria defined again by mathematical-statistical considerations in large-scale geological structures and carried out in entirely different dimensions (Fig. 3.79).

Gatzweiler and Akin (1993) divide several strategies of sampling modern ore deposits.

Grab samples are taken during the very early stages of exploration when conspicuous minerals or mineralisations have been discovered and the question arises: what is it? As indicated by its name, this sampling method is rather unsophisticated and results are not intended to be representative. Also, in archaeometallurgy, during field surveys in ancient mining districts, grab samples taken from ancient mining tailings, ancient mines, from slag heaps or from excavations, are often the only possibility to get an idea of what kind of raw material was exploited and processed at all. On the other hand, this kind of sampling approaches at best the selective way of ancient exploitation

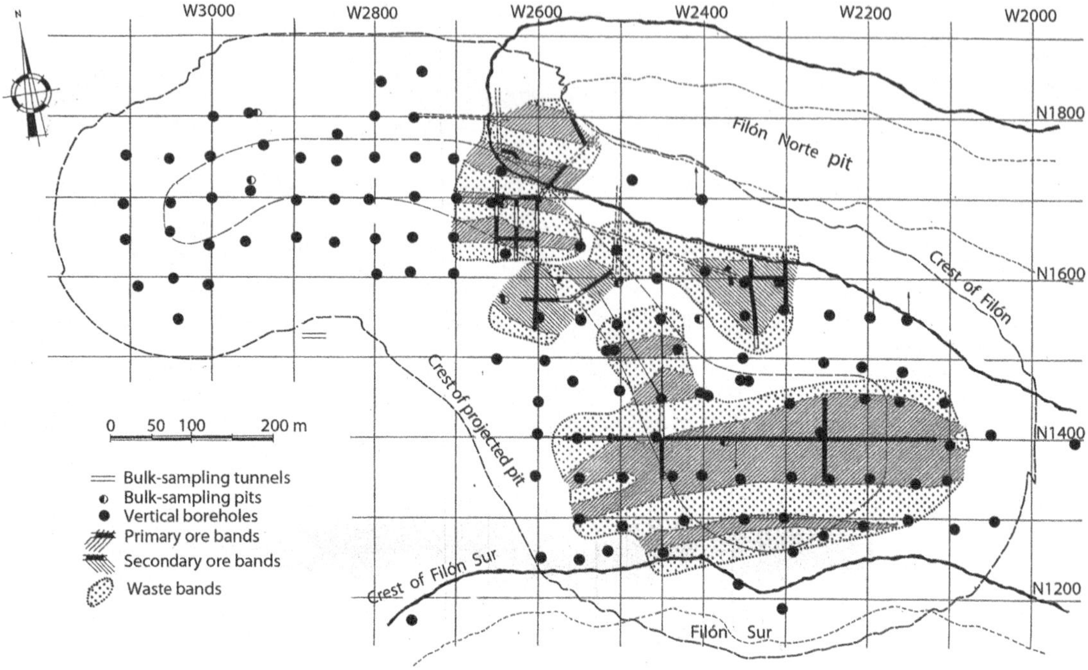

Fig. 3.79 The sampling of a "modern"ore body follows mathematic-statistical grids. The ore body of Cerro Colorado, Rio Tinto, was investigated for its copper concentrations by a regular network of boreholes and bulk sampling tunnels. Simplified after Pryor et al. (1972)

when preferable rich ores were taken out of an ore deposit.

The most common method is channel sampling applied to outcropped mineralisations and underground exposures, at least concerning base metals. The linear slots are typically cm-deep and up to several metres long. Sample weight is roughly between 2 and 10 kg. In most cases, large parts of gangue material and host rocks are included in the samples but the analyses of such materials provide useful overviews on the metal content of an ore deposit.

Other modern methods of large-scale sampling are directed towards the shape and extent of the whole ore body. They are used to comprise dimensions of about 10 to 10^3 m.

Modern feasibility studies are mostly calculated under aspects of today's economy, based on the geological origin of deposits. Such analyses are of no great value for the assessment of prehistoric metal extraction, as average percentages calculated in that way do not cover the possibility of any existence of rich ores. It would be difficult, therefore, to form an opinion as to whether an ore deposit with an average of 3–4 wt.% Cu content, considered quite rich nowadays, could have been exploited as a copper deposit in prehistoric times.

Not every modern metallogenetic map, therefore, may give reliable information about the possible intensity of ancient mining. In modern textbooks, ore deposits are often associated with metals which have no connection to the (pre-) historic raw material sources. What was mined for copper in the Bronze Age could possibly be addressed as a lead–silver–zinc deposit today. Aspects relevant for early periods such as, e.g. the existence of rich ores available for the prehistoric miners, the abundance and composition of ores in the gossan, or analyses of those trace elements, which end up in the metal after smelting, are of limited interest for modern geologists. The deposit of Rudna Glava in Romania, where secondary copper ores were mined in the Neolithic (Jovanović 1988), is treated as a skarn deposit with magnetite, pyrrhotite and chalcopyrite in the comprehensive overview of ore deposits of Europe (Dunning et al. 1989). These ores did not play a role in prehistoric times.

Hence, any mining archaeological and archaeometallurgical investigation should start with a mineralogical and geological study of an ore deposit, respectively, of the mining site. The aim of sampling ores in ancient mining districts or even within ancient mines is to get information on quality and quantity of ores mined and processed into metal in certain archaeological or historical periods. The analysis of ore samples (or also of slags, see below) allows us to reconstruct applied technologies. It is also the most important basis for any provenance study which is not possible without adequate knowledge of ore deposits. Ore samples taken from ancient mines or mining sites may provide valuable information on texture, mineralogical, chemical and isotopic compositions.

There are several possibilities for samplings:

1. Ore samples should be taken, wherever possible, directly from the ancient mines.
2. Sampling of mine tailings or from waste dumps of processing ores in the context of ancient mines. However, these usually consist of economically low-grade material. A (re-) processing of old waste dumps in later times can never be excluded.
3. Sampling of ores on slag heaps. Often ores have been transported from the mines to smelting sites or furnaces. As with the mine tailings, it is also difficult to find more than economically uninteresting ores there.
4. Sampling of inclusions of metal or sulphides in slags. This is probably the most suitable material for the characterisation of an ore deposit by trace element and isotope analyses.

How Many Ores?

Sample type, size and number are subject to considerable fluctuations and are dependent on randomness. Unfortunately, many mining archaeological and archaeometallurgical investigations are based on insufficient number, size and quality of samples. Stos-Gale (2001a, b) notes that a characterisation of an (ancient) ore deposit by lead isotope analyses would not be possible with <100 samples. The author relates to deposits of Cyprus. Gale and Stos-Gale (2000) already emphasised before that it is not an easy matter to establish the number and limits of the range of lead isotope compositions for particular ore deposits.

Filled Ancient Mines

A fundamental problem of sampling is that almost always and everywhere, whether inside ancient mines, mine tailings or on waste dumps, one finds only ore that has *not* been mined in ancient times or has been left behind as worthless material. Material available is material left by the miners after activities came to an end. Ixer (2004) points out that it is a well-known irony that one of the difficult issues to find and to recognise at an abandoned mining site is the ore that was being exploited. Mining–archaeological research in Europe and the Near East during the last decades has shown that well-known, still economically viable ore deposits had already been exploited in ancient times at least in their upper part. In Germany, the term for these old workings is *Alter Mann*. It is therefore important to extract the greatest amount of information from such material through analytical investigations. Archaeometallurgy is to procure samples of appropriate ores from ancient deposits in sufficient numbers for statistical calculations and mass estimations, or in order to compare them with the analyses of artefacts. For archaeometallurgy, ore deposits should not only be analysed for main metal contents, but also undergo a more thorough geological, mineralogical, geochemical and isotope examination.

Sampling inside ancient mines is often difficult and requires archaeological excavations. Almost as a rule, they are more or less filled with sand and clay and are not easily accessible. This limits sampling due to the time and resources needed for excavation. In some cases, when ore deposits are re-evaluated, ancient mines were cut by modern prospection galleries or underground workings as is the case at the ancient copper mines of Mitterberg near Salzburg, at the gold mine of Sakrisi in Georgia (Stöllner et al. 2010;

Gambashidze and Stöllner 2016) or at Faynan (Jordan) (Hauptmann 2007). These are exceptions and lucky cases that allow the extraction of ore samples in a sufficient and targeted manner. However, within a mine, one has to distinguish between ancient prospection galleries or prospection shafts, which are leading to the mineralisation, and which are usually void of ore, and the real working faces where the desired ore was extracted. There are strong indications that rich ores were preferentially exploited in ancient time, and that they were even sorted out within the mines themselves (Stöllner 2003b). Such a concept is especially easy to understand when ancient miners reached enrichment zones of ore deposits below the oxidation zone (Hauptmann 2007). Nevertheless, it must be decided on a case-by-case basis whether a targeted sampling of riches is made or whether one chooses to undertake a run-of-the-mill ore sampling in order to possibly make calculations of mined metal quantities. In any case, this requires that the extent of the mined volumes is known.

Waste Dumps

Samples taken from waste dumps or tailings comprise discarded wall rock, gangue, sub-economic mineralised samples or those containing unwanted ores. They are anthropogenically selected and do not necessarily represent ore exploited in ancient times. They are corroded by meteoric water, by the access of air, by changing redox conditions and electronegativity of soils (Dill 2010). They were built at least partly *after* ancient mining activities and are possibly different from ore that was exploited in ancient times. Rich ores will be missing, and the original metal concentrations will be distorted. Minerals formed on the surface of mining heaps or on the surface of old mines can, according to the classification of primary and secondary ores, be called "tertiary" ores (Ixer 1999). They often have striking colours: white, black, or green and blue copper-rich coatings are formed. Although such mineralisation can certainly be used to prospect old pits and heaps (Fig. 3.69) and to assist in the first identification of the sort of mined ores, as has been the case in Laurion (see above), or in the field work of the German Archaeological Institute and the Deutsches Bergbau-Museum in Veshnaveh in Iran (Stöllner 2004). However, such ores are not necessarily identical to those mined in prehistoric times.

Ore Geology

During the sampling of an orebody, primary and secondary vertical or lateral zoning has to be considered. This phenomenon is observed in hydrothermal veins and massive sulphide ore deposits of volcanic origin, which occur in the fold belts of the Eastern Mediterranean mountain ranges. They often outcrop at the surface and extend vertically down to differing depths (see Sect. 3.5). It is especially important to identify oxidic (secondary) and sulphidic ores and native metals and to check the availability of rich ores, even if these might have been exceptions and locally restricted. Ore samples may provide insight on whether ancient mining was restricted to exploit surface-near mineralisations in the oxidation zone of a gossan, or penetrated the secondary enrichment zone in a greater depth.

Of special interest is the uppermost pristine surface outcrop of an orebody an ancient miner was faced with. Dill (2010) carried out an analytical study, which comments on the problem of ore formation in anthropogenically altered surfaces. He discusses the changes of the primary ore body of Mega Livadi, Serifos, the subsequent formation of ores in the oxidation zone, and the formation of ores formed on surfaces after the Early Bronze Age mining activities (3325–2890 BC; "tertiary ores"). The primary ore body was a skarn-type Cu–As–Sb–Fe–S deposit. Predominant minerals were galenite, sphalerite, fahlores, pyrite and arsenopyrite. During the Miocene, a supergene mineralisation in the gossan was formed by strong alteration. Its main constituent was a "soft iron ore". This was exposed to the present surface. It was the first material that was exploited in ancient times. It consisted of a soft goethite and hematite bearing limonite with significant amounts of Cu, Zn, Pb and Sb. Typical minerals found at the *pre*-Early Bronze Age surface revealed a considerable enrichment of arsenic at the topsoil of the gossan. Arsenate mineral assemblages such as beudantite, arseniosiderite, scorodite and pharmacosiderite were common to topsoil horizons. An enrichment of As in limonite, and at surfaces was found to be a common feature. It was also observed, e.g. by Bowell (1994). In his study, he determined an enrichment of arsenic in natural clay-rich soils with iron oxy-hydroxides and oxide minerals, taken from the Ashanti mine, Ghana, from 2 to 35,600 ppm. This enrichment was also observed in the oxidised surface portion of the mining tailings. At Mega Livadi, the anthropogenic mineralisation formed at the surfaces of the *post*-EBA mining activities, inside the ancient galleries were a number of K-Cu-oxalate minerals, malachite and azurite. The study on the mineralisation in the uppermost part of the Late Miocene bridges the gap between mineralogical studies of anthropogenic mining residues (post-mining mineral assemblages) and minerals resulting from descending meteoritic water. We deal with a natural supergene mineralisation (gossan) grading into a post-mining anthropogenic mineralisation ("tertiary", see above). A sampling of this mineralisation—both within old pits, where they are often found as wallpaper-like coatings, and outside the pits, in streams and gullies of tailings as shown in Fig. 3.69, are not very suitable for analytical investigations.

Quantities

In the ideal case, sampling old ore deposits is also aimed at making a calculation of quantities of metal produced and traded from there. They are thus a very important basis for an estimate of their economic significance and emanation. This can concern an entire ore district, sometimes even just a single mine.

In the developed stages of metal extraction in the Old World, at least since the developed Early Bronze Age, a large-scale production of copper and bronze had begun. Smelting of ores took place in the immediate vicinity of the ore deposits and was associated with the production of masses of slags that were left at their place of origin as waste products. An excellent material for our purposes is available, which gives information about the sort of metal extracted. Since a complete separation of metal or sulphides from the slag could never be achieved in ancient times, these materials are almost always to be found as inclusions in siliceous slag (Fig. 3.80). As high-temperature products, they are altered from the parent ore, yet they are a direct link between the source of raw material and the final product. The big advantage is that these slag inclusions, unlike a single ore sample, represent a larger average of an ore charge. The prerequisites of such a sampling principle is that slag is (almost) almost completely homogenised compared with ore.

We will show that ore in ancient times was not self-evidently smelted together with purposely added flux that may have had a different chemical or isotopic composition. Perhaps it is not necessary to discuss this problem at all.

At least siliceous fluxes tend to have low lead contents only to decisively impair a lead isotope pattern of ore. Generally, it is interesting to ask whether fluxes were used in archaeometallurgy at all, from outside of an orebody, at least until the Late Bronze Age.

Noel Gale and Zofia A. Stos-Gale (e.g. Stos-Gale et al. 1998b), in their enormous project of Late Bronze Age copper oxhide ingots in the Mediterranean and beyond, discussed this problem in detail. After their research had shown that the largest part of copper ingots by far came from Cyprus, they not only analysed hundreds of ore samples for their lead isotopic composition. In addition, they collected inclusions of copper prills and matte embedded in slags to characterise the compositions of the Cypriot ores from different sites and ore bodies, respectively.

Contradictory statements exist about the mineral contents in the near-surface areas of the massive sulphide deposits of Cyprus. Constantinou (1982) asserts that there is almost no malachite or azurite available, so his statement that Cyprus would have been one of the earliest copper

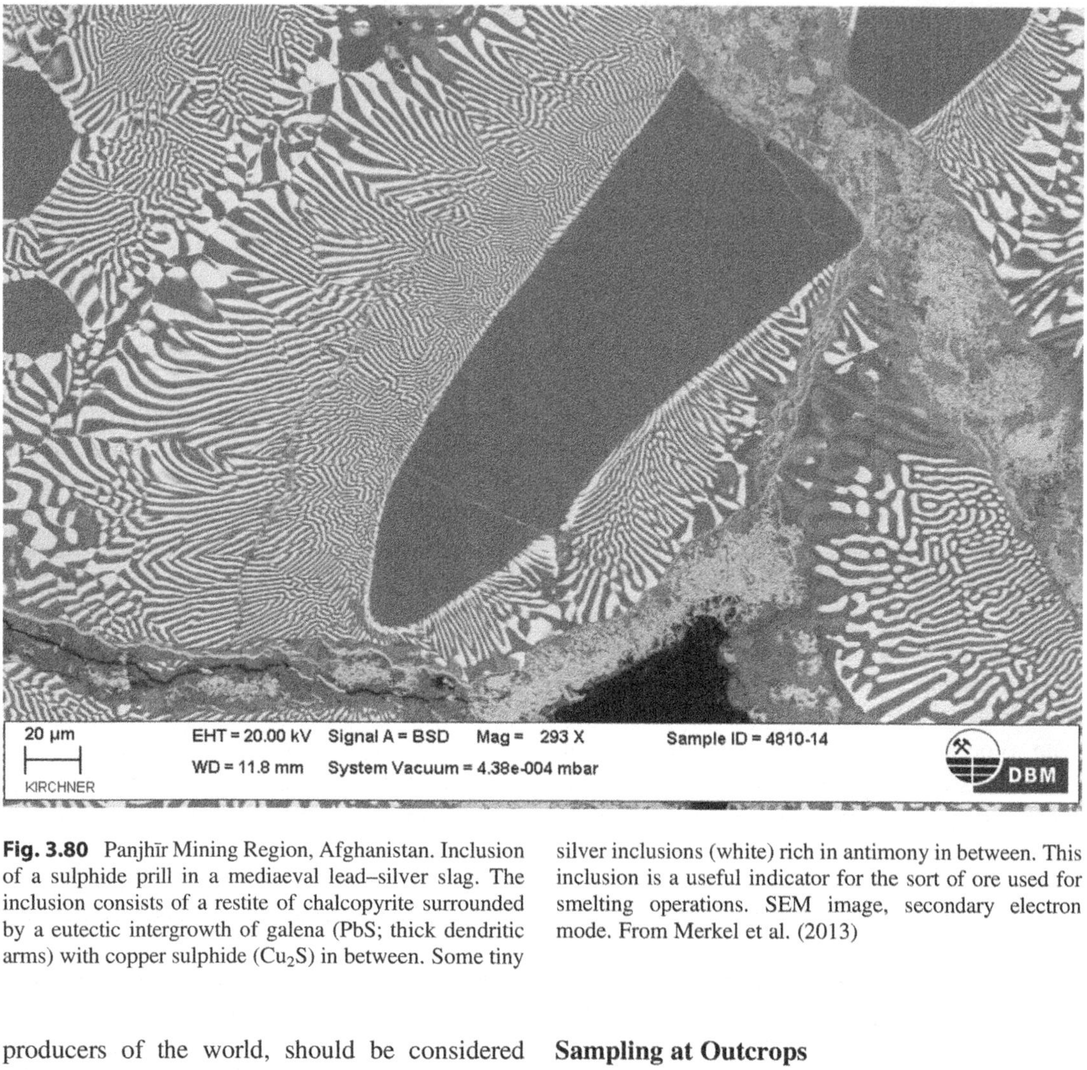

Fig. 3.80 Panjhīr Mining Region, Afghanistan. Inclusion of a sulphide prill in a mediaeval lead–silver slag. The inclusion consists of a restite of chalcopyrite surrounded by a eutectic intergrowth of galena (PbS; thick dendritic arms) with copper sulphide (Cu_2S) in between. Some tiny silver inclusions (white) rich in antimony in between. This inclusion is a useful indicator for the sort of ore used for smelting operations. SEM image, secondary electron mode. From Merkel et al. (2013)

producers of the world, should be considered critically. But Stos-Gale and Gale (1994) carried out surveys and found these oxidic ores in large enough quantities to have been of importance for the Early Bronze Age. It is more difficult to assess the role native copper played in Cyprus. It can still be found today in small, filigree dendrites, but its geochemical and isotopic composition is not compatible with that of the few Chalcolithic artefacts, which have been found on the island. Gale (1991) assumes therefore that the metal has been imported from other sources. At times the economic importance and use of the Cu-sulphate in Cyprus might have been overestimated, although metallurgically there is no problem in reducing these to copper.

Sampling at Outcrops

At Faynan (Jordan), it was possible to deal with these problems rather satisfactorily, as the modern prospecting work had not only opened up a number of old mines but also created a number of fresh outcrops. No modern mining destroyed the remains of ancient mining and metallurgy. At Faynan, some 50 ore samples, 85 slag samples, and 60 metal droplets and lumps of metals were collected in an area of c. 20 × 25 km (Fig. 3.81) (Hauptmann 2007). All this material was collected from a series of natural outcrops, partly strikingly exposed by natural erosion processes, from mines, tailings and from a number of smelting sites of various ages. In this way, it was possible to characterise the copper that was

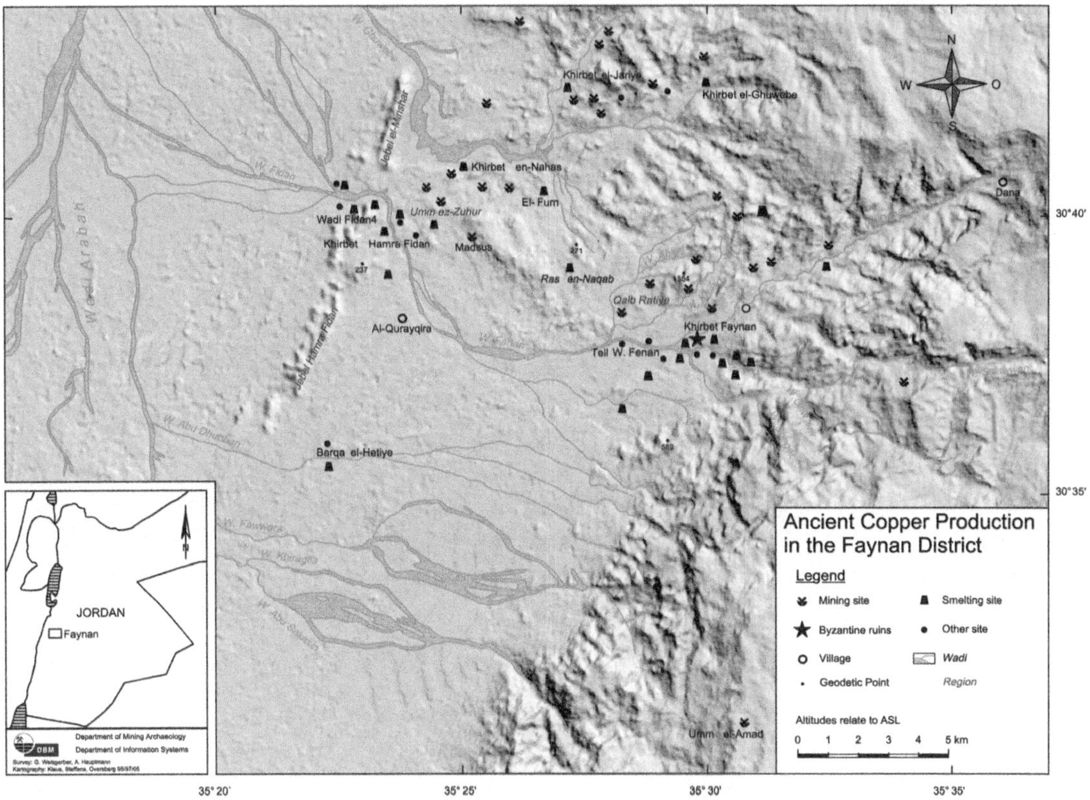

Fig. 3.81 Copper ore district of Faynan, Jordan. The sampling of an "ancient" ore body is generally restricted to archaeological sites such as settlements, mines and smelting sites. In this case, ore from natural outcrops and "archaeological" ores were collected together with slags to characterise the copper produced from the ore district. From Hauptmann (2007)

extracted from the ancient Faynan ore district under all necessary aspects. In a very similar way, in the ancient copper mining district of Timna (Israel), fresh ores for analytical purposes exposed by heavy erosion since the Late Bronze Age/Iron Age and slags from well-dated heaps could be collected and investigated (Bachmann and Rothenberg 1980; Hauptmann and Horowitz 1980; Conrad et al. 1980; Asael 2010, Hauptmann 2007; Segal et al. 2015). Faynan and Timna belong to well-investigated examples in mining archaeology and archaeometallurgy.

Basic Physical–Chemical Principles of Ancient Metallurgy

4

4.1 Introduction

In materials science, metallurgy, in geoscience, and in ceramic science it is an established procedure to express the composition of an assembly of different phases through a phase diagram. Phase diagrams are simplified chemical systems with known formation temperatures (high temperatures in metallurgy) of individual phases. They may be useful for understanding more complex phase relations in chemically complex melts.

The term phase has several meanings. First, we distinguish the term used in archaeology, which indicates a chronological period. In materials science, however, a phase is understood to describe a substance of defined chemical composition, which can be distinguished by physical–analytical methods (X-ray diffractometry, (electron-) microscopy, etc.). In the solid state, phases may be different types of crystals. But a melt is also a phase, just like a fluid phase or a gas phase.

By contrast, components are merely chemical constituents that make up phases. In geoscience, mostly oxidic components are used. In metallurgy, sulphidic components are also used. But complete mineral compositions can also be components if this is helpful in the discussion of chemical systems. The phase diagram of forsterite—fayalite consists of MgO, FeO and SiO_2, which are combined into the phases or components Mg_2SiO_4 and Fe_2SiO_4. This familiar diagram is discussed in Sect. 4.2.

In archaeometallurgy, different diagrams are used according to the complexity of the main chemical–mineralogical components:

- Material that consists of two main components is represented in binary diagrams.
- Three variables can be represented as vertices of an equilateral triangle.
- For systems with four variables, projections in a tetrahedron can be used.

In this chapter, the discussion of phase diagrams will be limited to some selected examples more frequently used in archaeometallurgy. It will contain phase diagrams, on the one hand, with oxidic-siliceous phases such as Fe-oxides, SiO_2, Al_2O_3 and CaO, on the other hand with some metallic phases such as Au, Ag, Cu, Fe, non-metallic phases such as S, O and C and others. More phase diagrams will be presented in various chapters, e.g. when alloys and slag systems of more complex composition will be discussed.

It should be stressed at this point that the diagrams discussed in the following chapters refer to equilibrium conditions in the laboratory. A state of equilibrium indicates that the concentrations of the substances involved in given external conditions (temperature, pressure, redox conditions) do not change over any length of time (Salmang et al. 2007). In ancient times, under the available materials and conditions,

A. Hauptmann, *Archaeometallurgy – Materials Science Aspects*, Natural Science in Archaeology,
https://doi.org/10.1007/978-3-030-50367-3_4

these have hardly been achieved. The gas atmosphere of ancient metallurgical processes, controlled by the combustion of the charcoal, was subject to strong fluctuations during a (s-)melting process.

Reaction vessels were heated up quickly, ore, gangue and host rock were smelted (completely or in parts), liquefied (siliceous-oxidic) slag rapidly cooled again. Metals were liquefied, poured into moulds, mixed with other metals, cooled, reheated for formation and hardening or to alloy them with other elements such as iron with carbon. In the liquid state, many metals are soluble in each other and form homogeneous melts. However, they may often differentiate into several phases in the solid state.

Bachmann (1982b) points out that the plot of complex slag analyses, reduced to three or four main components in ternary or quaternary systems would only lead to meaningful interpretations if there is a reasonable correspondence between theoretical and actual phase contents. Research so far has shown that slag, if solidified from homogenous liquids, forms parageneses during cooling, which also appear in comparable natural rocks. He, therefore, emphasises the necessity of mineralogical phase analysis next to chemical analysis for archaeometallurgy. In his studies, he included X-ray diffraction analyses of all major components of ancient slags (fayalite, spinel, e.g. Fe_3O_4 or $MgAl_2O_4$, and pyroxene, e.g. $CaFeSi_2O_8$).

The phase rule is the basis for the systems that are to be discussed below, with several components that are required for the understanding of siliceous and sulphidic slag melts as well as molten metals. Introductory works on state, equilibrium or phase diagrams can be found in numerous materials textbooks (e.g. Hansen and Anderko 1958; Slag Atlas 1995; Askeland 1996; Hummel 1998; Salmang et al. 2007; Markl 2015a).

The knowledge of phase diagrams provides information to better understand and to reconstruct conditions for the formation of liquids, i.e. of smelting operations:

1. Compositions of the starting load.
2. (S-)melting temperatures. It is known that in old smelting furnaces with human muscle power and charcoal as fuel, temperatures above c. 1200 °C were reached only punctually so that this average is a reasonable basis for old pyrometallurgical reactions.
3. Duration of heating. In contrast to modern blast furnace processes, which can last for days or weeks, in archaeometallurgy, we calculate smelting processes to have been in the order of hours only or to a maximum of 1–1.5 days (including cooling).
4. Degree of liquefication. How much of a load has been liquefied? This means: how much of the load has been melted and how much remained unmelted in the solid state?
5. Redox conditions. The composition of Fe-rich silicate melts and metals is affected significantly by the gas atmosphere (CO/CO_2). Gold and silver, e.g. as noble metals, are stable over wide ranges and are not oxidised. Other metals, metal oxides and silicates are very sensitive to the composition of the gas atmosphere. Iron-containing materials may have varying oxidation states: Fe^0, Fe^{2+} und Fe^{3+}.

4.2 Silicate and Metal Phase Diagrams

4.2.1 Binary Systems

Physical properties of materials consisting of two main components can be discussed in binary phase diagrams. We distinguish three different possibilities:

1. Complete miscibility of the components. In this case, we call it an isomorphic binary phase diagram.
2. Limited miscibility with eutectic crystallisation. Here, it is a eutectic binary phase diagram.

3. Immiscibility of two components in the liquid and solid state. These are monotectic phase diagrams.

Systems with two components are usually presented by plotting temperature on the ordinate (*x*-axis) in a rectangular coordinate system and the composition of the two phases on the abscissa (*y*-axis). The latter is given in weight %, mol % or atom % (each from 0 to 100%) of a component.

In archaeometallurgy, there are a number of particular binary phase diagrams of metals and alloys which are of importance. Important examples for binary metal alloys are copper-alloys with arsenic, tin, lead, zinc, iron-alloys with carbon, gold-alloys with silver and copper, and lead–silver. In silicate-oxidic slags, there are the most predominant systems forsterite–fayalite, and the systems iron-oxide-SiO_2.

4.2.1.1 Complete Miscibility: Isomorphic Phase Diagrams

The simplest case of a binary system is the formation of a continuous solid solution between two components. This is an isomorphic binary system. There are two important binary phase diagrams for archaeometallurgy. This is the system gold–silver and forsterite–fayalite.

System Gold–Silver

Gold and silver are completely miscible (Hansen and Anderko 1958). These metals are a classical example of a binary system with a continuous series of solid solutions (Fig. 4.1a). In comparison with other binary systems of solid solutions, the distance between the liquidus and solidus lines is very narrow. The melting point of gold at 1063 °C is lowering to that of silver at 961 °C.

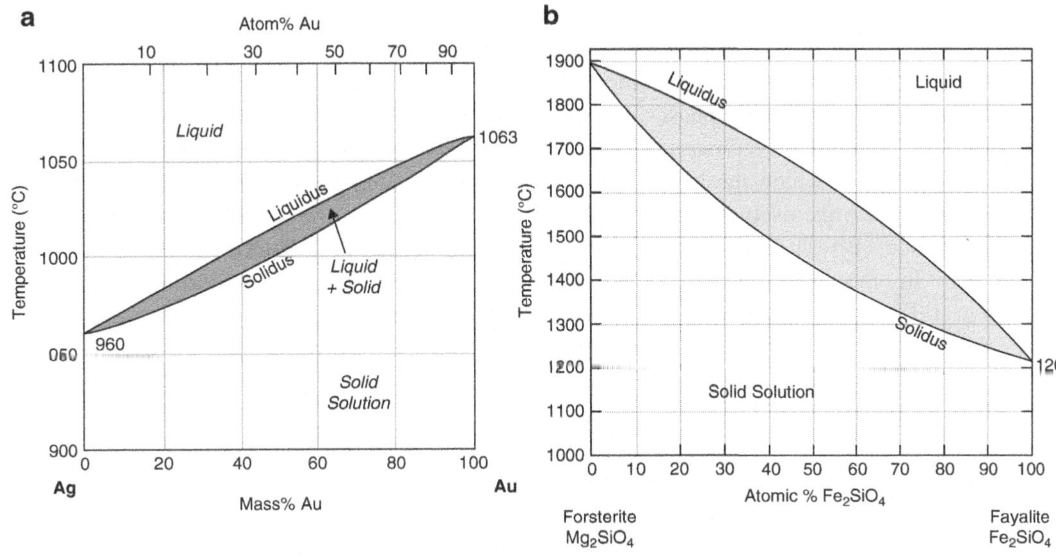

Fig. 4.1 (**a**) Gold and silver at high temperatures form a continuous series of solid solutions. Note that the convex liquidus line and the concave solidus line are closely connected. They remember the shape of a very thin "cigar". Given are the weight-% in the upper abscissa and the atom-% in the lower abscissa. From Hansen and Anderko (1958). (**b**) The binary phase diagram forsterite–fayalite (Mg_2SiO_4–Fe_2SiO_4) is an example of the physical behaviour of three components (MgO, FeO, SiO_2). It is also a solid solution diagram which, however, consists of a wider "cigar-shaped" area made up by two curves. The top curve is called the liquidus and the bottom is called the solidus. These define the compositions of solid and liquid phases in equilibrium with each other, i.e. the compositions of a variety of Mg–Fe-rich olivine crystals. The Fe-rich variety, a fayalite rich olivine, often occurs as the predominant phase in ancient smelting slags. According to Bowen and Schairer (1935)

To melt a gold–silver alloy of certain composition in order to cast an artefact, it must be heated above the respective liquidus temperature. The solidification process begins when the temperature falls below this line again. In Fig. 4.1a it becomes obvious that a gold–silver alloy with 65 wt.% gold and 35 wt.% silver (wt.% = Mass %), converted that would be 50 atom% of silver, i.e. an alloy in which both metals are present in the same ratio, has a liquidus temperature of 1020 °C. The solidus temperature here is 1000 °C. The solidus temperature corresponds to the bottom line of the state diagram. Between liquidus and solidus temperature lies the solidification range in which a liquid and a solid phase coexist, in this case a mixed crystal. During cooling, the solidification process is completed when the solidus line is crossed. All materials that are used for any processing at high temperatures must not be heated above the solidus temperature.

System Forsterite–Fayalite

This system is another example which comprises two widespread siliceous slag phases of the olivine-solid solution forsterite–fayalite (Mg_2SiO_4–Fe_2SiO_4) in which the Mg^{2+} and the Fe^{2+}-ions can replace themselves completely in the crystal lattice (Bowen and Schairer 1935) (Fig. 4.1b). In the pure system, forsterite crystallises at 1890 °C, fayalite at 1205 °C. In reality, slag melts always contain various other components in different concentrations (CaO, MnO, BaO, Al_2O_3, Na_2O, K_2O), which contribute to a lowering of the melting points. Any thermometry based on pure phases is only possible with approximations in ancient slags.

In the system forsterite–fayalite predominantly Fe-rich compounds occur in ancient slags. An olivine of fayalitic composition is the main component of most bloomery (and smithing) slags and of slags from base metal smelting as well.

In this diagram, the upper line of the "cigar-shaped" region is the liquidus line. This line represents the beginning of the solidification of a melt or the end of a melting process. Above this line, the system is fully liquid. The bottom line is the solidus line. It indicates the completion of the solidification process of a melt or, alternatively, in a "view from below" the beginning of melting. Below this line, the system is fully solidified. It is crystallised. In the region between the liquidus and solidus lines, a liquid and a solid phase coexist in a semi-solid state. In phase diagrams, the solid phases are generally identified with a Greek letter, e.g. with α or β, etc.

The liquefaction in this system is simple. As temperatures are rising from low temperatures, a homogeneously composed solid Fe-rich olivine of composition 80 wt.% fayalite initially melts partially when it reaches the solidus line at a temperature of 1250 °C. A FeO-richer melt with 95 wt.% fayalite is formed, which is in equilibrium with the remaining solid phase. As temperatures continue to rise, finally, a complete melt forms above the liquidus line at 1425 °C. This is indicated by a connecting line called a conode. Throughout the system, all compositions are miscible with each other, both in the liquid and the solid range. This is an isomorphic system.

During cooling a melt of the same composition, a crystal of olivine crystallises at 1425 °C first with 50 wt.% forsterite and 50 wt.% fayalite, which becomes more fayalite-rich with decreasing temperatures until it solidifies at a temperature of 1260 °C. If the cooling takes place very quickly, as can be assumed for many old metallurgical operations, no equilibrium state will be established by the diffusion of the corresponding components. Nevertheless, even in such circumstances, a core of MgO-rich forsterite and a FeO-rich edge, fayalite is formed. Olivine shows such a zonal construction in many slags (Fig. 4.2a, b).

Another binary system with a complete solid solution series of great importance in archaeometallurgy is the system gold–silver, which is discussed in Sect. 7.3.

4.2.1.2 Limited Miscibility: Eutectic Phase Diagrams

If two liquidus lines are meeting in a valley whose bottom forms a horizontal phase boundary, then it is a eutectic composition. The eutectic composition is the composition of the lowest melting temperatures throughout the system. Below the

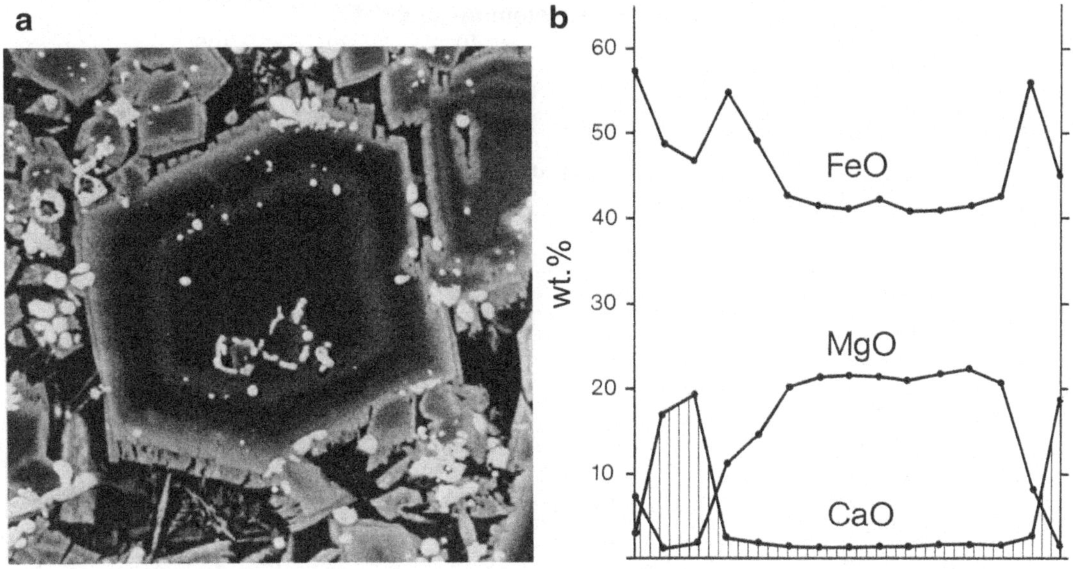

Fig. 4.2 Al-Maysar 1 (Oman), sample OM 350/1. (**a**) Fayalite from a Bronze Age copper tap slag. The crystal shows strong (multiple) zonation and a distinct fibrous rim of CaO- and FeO-rich olivines. Scanning electron microscopy (WDX), secondary electron mode. Width of the picture 210 μm. (**b**) Microprobe-profile through the crystal of fayalite from Fig. 4.2a. Note the MgO-rich core which tends to almost zero at the rim. Instead, FeO- and CaO-rich compositions are prevailing at the rim. From Hauptmann (1985)

solidus line, crystals are formed in any solid-state composition, which are present almost everywhere as a solid solution in a chatty eutectic intergrowth.

Two aspects are important in archaeometallurgy. For metallic systems, e.g. silver–copper (Ag–Cu, see below) or iron–carbon (Fe–C; Sect. 7.5.3), crystallisation in the course of cooling is of importance. Eutectic crystallisation affects hardness and strength. Furthermore, varying compositions between silver and copper affect the colours of alloys. In the case of siliceous slag melts on the other hand, it is important to get the ore charge that is to be heated up as close to a eutectic composition as possible, in order to optimise the melting of the entire charge at a low-temperature area of about 1200 °C.

System Silver–Copper

A limited miscibility between two components with eutectic crystallisation occurs, e.g. in the system of silver–copper (Ag–Cu), which was of importance in the old jewellery industry, and it still is, today (Fig. 4.3a). This means, that silver and copper are only soluble in each other in every ratio if they are in the fully liquid state. In the solid state, on the other hand, there is only a very limited solubility of the two components. This system is discussed in the compendium of Hansen and Anderko (1958).

When adding small amounts of copper to silver, a solid solution, called α-phase, will precipitate. The highest amount of copper that can be dissolved in silver is only 8.8 wt.%, the rest is 91.2 wt.% silver. This occurs at 780 °C. At any other temperature, lower or higher, the solubility is less. For example, the solubility of Cu in Ag at 400 °C is only 1.2 wt.%. At lower temperatures than c. 400 °C the solubility is essentially nil. A similar behaviour is observed when silver is added to copper. At 780 °C, the solubility of Ag in Cu is at a maximum of 8 wt.%. This solid solution is called β-phase. At 28.1 wt.% Cu the liquid can pass directly from the liquid to the

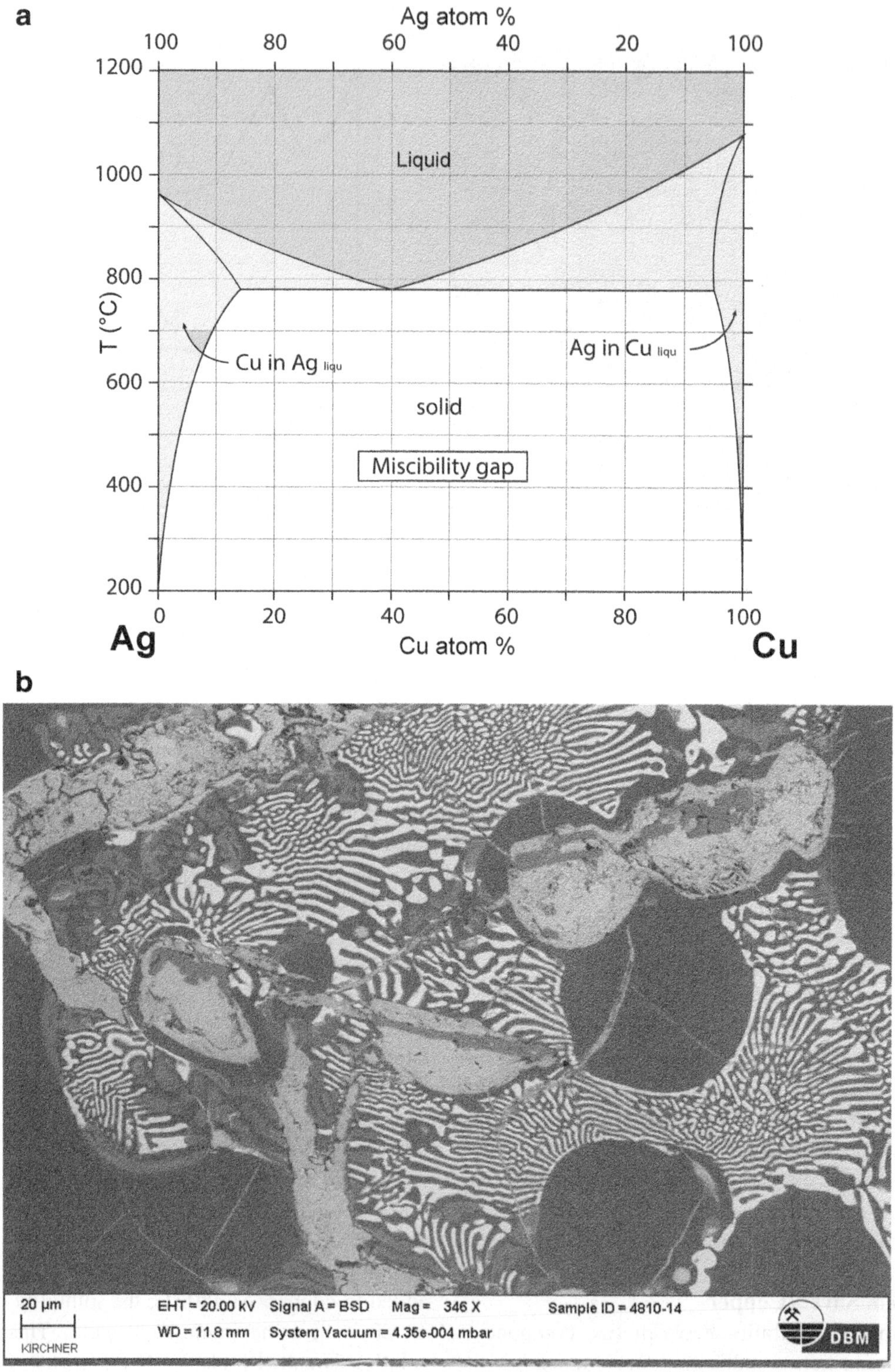

Fig. 4.3 (**a**) The binary copper–silver system shows that due to the eutectic decomposition at 780 °C the homogeneous liquid phase decomposes into copper-rich and silver-rich phases. There is an extended miscibility gap between the two components. There is a very small solubility of copper in silver and vice versa. After Hansen and Anderko (1958). (**b**) Eutectic intergrowth of copper and silver. Detail of the sample shown in Fig. 3.80

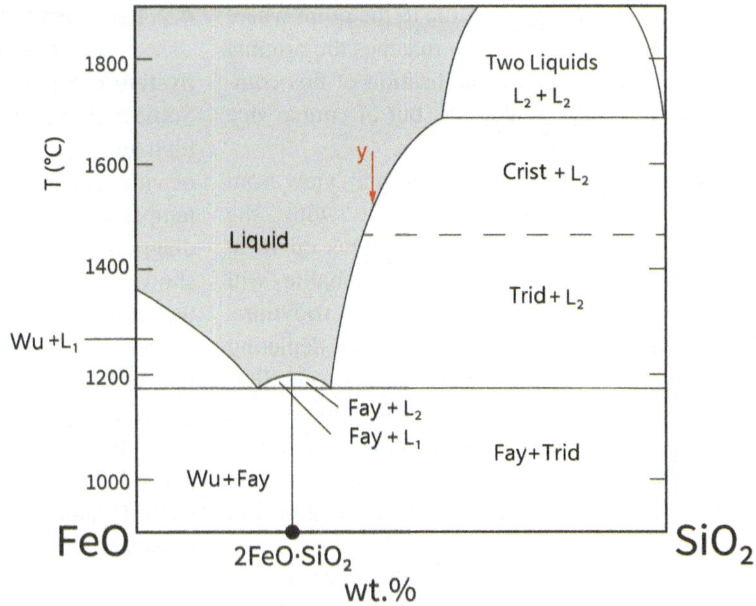

Fig. 4.4 The binary system FeO–SiO_2 after Bowen and Schairer (1932). Note the low-melting area at c. 30 wt.% SiO_2 and 70 wt.% FeO where the wide-spread fayalite is crystallising at 1205 °C. In case of a slight excess of SiO_2 in the charged ore (marked by y), the liquidus temperature is rising to a height that was not reachable in ancient processes. *crist* cristobalite, *trid* tridymite (both high-temperature modifications of SiO_2), *fay* Fayalite, *wu* wuestite

solid. This composition solidifies at 780 °C. This is a lower temperature than either of its constituents. It is called the eutectic composition. The mixture precipitated from the liquid consists of a heterogeneous mixture of fine plates of α- and β-phase interspersed in each other (Fig. 4.3b).

Upon reaching the solidus temperature, all other compositions still contain different amounts of liquid melt, which then solidify to a fine-grained mixture of the smallest mixed crystals. An intense intergrowth occurs, which is called eutectic microstructure.

Alloys that contain less solute than the eutectic composition, i.e. <28.1 wt.% Cu, are called hypoeutectic. Compositions between 28.1 and 92 wt.% Cu are called hypereutectic.

The practical implications of this mixture of two phases on the strength of materials and of colour will be discussed in Sect. 7.4.

System Wuestite–Silica

The liquidus temperatures in the system wuestite–silica (FeO–SiO_2) is particularly important for slag formation in iron metallurgy, but also for nonferrous metallurgy (Fig. 4.4). This system was investigated by Bowen and Schairer (1932).

We will first consider what happens when a low tempered solid ore of the following composition is going to be melted ("view from below"): 35 wt.% SiO_2 and 65 wt.% FeO. This is slightly left from the ideal composition of fayalite. If this is expressed in minerals, then this is quartz and (partially) reduced Fe^{2+} oxide, the wuestite. The melting process takes place under reducing conditions. We assume that, at several hundred degrees celsius, in this mixture some metallic iron is precipitated, but the metal content is ignored here. When the temperature rises to 1178 °C, the eutectic temperature is reached, and the ore begins to melt. The first melt composition is about 22 wt.% fayalite and 78 wt.% wuestite in the solid state (w + L_1). The ore remains at this temperature for so long until all the quartz (indicated here as the high-temperature SiO_2-modification tridymite) is melted. Then, with further energy input, the melt composition moves along the liquidus line to the upper left. The melt constantly changes its composition and becomes richer in FeO due to the melting of wuestite, but it is constantly in equilibrium with crystalline FeO. FeO continues to melt as the temperature increases, and the melt continues to develop along the liquidus line to the point where the melt composition exactly matches the original ore composition. The crystallisation of this composition is exactly analogous, but of course vice versa.

Let us consider the system from a "view from above". When cooling a melt with the composition Y, it reaches the liquidus curve at about 1400 °C and initially, cristobalite will crystallyse, which quickly converts to tridymite. As the coexisting melt is continuously depleting in SiO_2, it moves diagonally away from SiO_2 on the liquidus line to the bottom left. At the eutectic at 1178 °C, the first fayalite crystallises. There, the entire melt crystallises as a eutectic mixture with a typical texture of fayalite and wuestite.

It can be seen in this system that right next to the composition of the low melting fayalite a slight increase of only a few wt.% of SiO_2, the liquidus–solidus temperatures are increasing dramatically. The high-temperature modifications of SiO_2, tridymite and cristobalite, only liquefy at c. 1700 °C and separate into two liquids.

Metallurgically, this means that if an ore charge already contains a slightly too high quartz content, pieces of quartz will be found un-decomposed as "resisters" or "relicts" in the ancient slag (Muan 1955) (see Sect. 4.2.3). They were not melted, because the temperatures to liquefy such compositions were not reached in ancient times. This is particularly true for working under slightly oxidising conditions.

However, this is also true in a similar way to other systems with divalent oxides such as CaO, MgO and MnO with SiO_2, which also only melt in a narrow range of variation at comparatively low temperatures. However, this does not apply to PbO, which has low melting compositions with SiO_2 over a wide range. By contrast, an excess of Fe-oxides in the ore charge has no dramatic consequences, as the liquidus/solidus temperature only rises slowly to 1.369 °C. This means that it only takes a moderate increase in temperature to reach a melt in a liquid state. Compared to the larger crystals of fayalite which crystallised previously, a fine-grained mixture of fayalite and wuestite is now formed.

4.2.1.3 Immiscibility: Monotectic Phase Diagrams

System Copper–Lead

Some metals and oxides or silicates do not mix at high temperatures. This is similar to oil and water or vinegar and oil, which are immiscible at room temperature and form an emulsion. In a phase diagram, such a liquid-state miscibility gap is shown by a dome-shaped liquidus line. Within this dome, two liquids coexist.

One such type of a binary melting system is that of copper–lead (Fig. 4.5) (Chakrabarti and Laughlin 1984). Here, already in the liquid state at c. 1100 °C, a separation into two liquid phases ($L_1 + L_2$) occurs, a miscibility gap is formed. At 955 °C and 47 wt.% lead the system has a monotectic reaction. From Cu-richer melts, copper crystallises at 1087 °C (the melting point of copper), which coexists with the melt L_1.

Furthermore, it becomes obvious that at 955 °C, a Pb-rich melt and Cu crystals are separated. Complete solidification of this melt occurs at 327 °C. Below this temperature, the

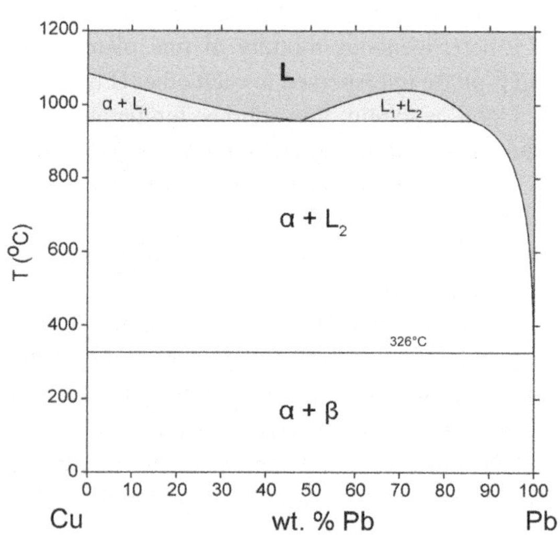

Fig. 4.5 The binary system copper–lead (Cu–Pb) after Chakrabarti and Laughlin (1984). Note the large miscibility gap between the two metals. The most important feature of this diagram is the monotectic invariant reaction at 47 wt.% Pb at 955 °C. Left of this (monotectic) point a copper solid solution low in lead coexists with a liquid L_1 rich in lead. Right of this point two liquids (L_1 and L_2) are coexisting, the one rich in copper, the other one rich in lead

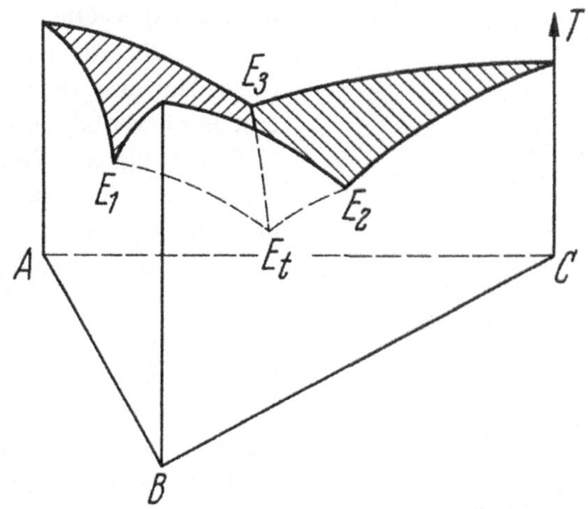

Fig. 4.6 Spatial representation of a ternary system with a eutectic (E_t). This results from the combination of the three eutectics E_1, E_2 and E_3 of the binary systems, which form cotectic lines in the spatial representation. The height of the diagram corresponds to the rising temperatures (T). Modified after Salmang et al. (2007)

two solids copper and lead exist separate from each other.

4.2.2 Ternary Systems

A material system made up of just two components is not very common in archaeometallurgy if one is excepting the "pure" siliceous fayalite slags or intentionally produced binary alloys such as, e.g. copper–tin or the widespread natural gold–silver alloys.

Almost all materials, whether slag or metals, are usually composed more complex. For discussing melting and solidification behaviour at various temperature ranges, we, therefore, need a three-dimensional representation. Markl (2015a, b) compares such a spatial model aptly with a geographic map in which the temperature lines are shown as the contour lines. What was a field in the binary system now becomes a melting space, liquidus lines become surfaces and eutectic points are defined by cotectic lines, in which several phases are in equilibrium with each other and with the melt. Finally, there are thermal minima in which two or three cotectic lines are meeting. A ternary phase diagram with components A, B and C is thus constructed from the three binary boundary systems A–B, A–C and B–C. It is self-evident to represent these systems in an equilateral triangle and to apply the temperatures perpendicularly thereto. This leads to a spatial representation, as can be seen for a simple case in Fig. 4.6.

However, in a multi-phase system, this type of presentation can become confusing. Therefore, the system is generally projected at the two-dimensional level (Fig. 4.7). Points of equal liquidus temperatures are combined to form isotherms. In Fig. 4.7, precipitation or primary fields (e.g. AE_1, E_tE_3 for component A), which are confined by field boundaries (e.g. E_1, E_t and E_3E_t). Each congruent melting compound is within its precipitation field. Often the fields boundaries, the cotectic lines, are provided with arrows in the direction of falling temperatures.

Ternary systems in isosceles triangles are used to represent chemical compositions of single metals and mineral phases, but further also for the projection of melt—solid state equilibria, which refer to these mineral phases. As a result, the temperature range in which the solidification of this composition has taken place can at least approximately be determined.

Such plots can be calculated in concentration triangles in the two-dimensional range, as shown

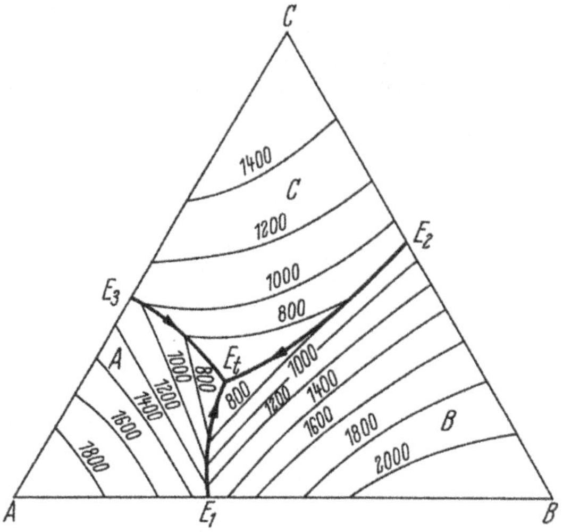

Fig. 4.7 Two-dimensional representation of a ternary system. Note the course of the isotherms within the precipitation fields *A*, *B* and *C*. The three eutectics of the edge systems E_1, E_2 and E_3 run as melt channels to the eutectic E_3, which represents the lowest solidification point. Modified after Salmang et al. (2007)

in Fig. 4.8. On the sides the percentages of the constituent parts, reduced to the main components, can be read, whose sum ($A + B + C$) must be 100%. This is a reduced analysis. The choice of concentrations may vary (mol %, weight %, atom %).

System FeO–SiO$_2$–Al$_2$O$_3$

In the system to be discussed now, FeO–SiO$_2$–Al$_2$O$_3$ (Fig. 4.9), the mineral phases fayalite (Fe_2SiO_4 = 2FeO·SiO$_2$), ferrosilite ($FeSiO_3$ = FeO·SiO$_2$), hercynite ($FeAl_2O_4$ = FeO·Al$_2$O$_3$), mullite ($Al_6Si_2O_{13}$ = 3Al$_2$O$_3$·2SiO$_2$) and iron cordierite ($Fe_2Al_4Si_5O_{18}$ = 2FeO·2Al$_2$O$_3$·5SiO$_2$) occur. However, the crystallisation of iron cordierite has never been observed in ancient slags.

Due to the composition of ore charges to be smelted (e.g., limonite + quartz + clay minerals + ore ± fluxing agents), and due to the only slightly reducing firing conditions in ancient smelting furnaces, we use to find high concentrations of iron oxides ($Fe^{2+,3+}$) and silicates in archaeometallurgical slags with the predominant mineralogical constituent fayalite. In most cases, Al_2O_3 is present in addition as a further component so that the bulk chemistry will make up to 80 wt.% or more. Hence, a plot into the ternary system FeO + SiO$_2$ + Al$_2$O$_3$ can be done. The plot in the system FeO–SiO$_2$–Al$_2$O$_3$ is shown here (Fig. 4.9) in a greatly simplified way.

4.2.3 Three-Dimensional Presentation in Tetrahedron

In addition, ancient (s)melting slags produced from polymetallic ores and complex gangue and host rocks may contain changing amounts of oxides such as CaO, MgO, BaO, and Na$_2$O, K$_2$O and also P$_2$O$_5$, CuO, PbO, SnO or ZnO. Complex compositions like that can be plot in four isolated and joined triangles. It is also possible to use a three-dimensional representation in a tetrahedron with four major components. For this purpose, Keesmann and Schmitz designed an

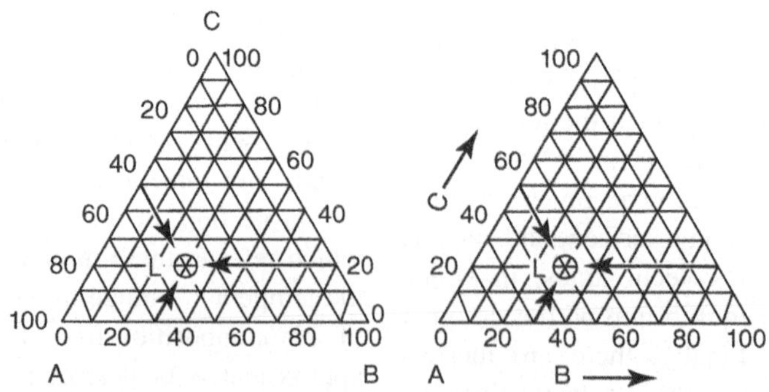

Fig. 4.8 Representation of a point *L* with the composition of the concentration of the substance $A = 50$, $B = 30$ and $C = 20$ ($A_{50}B_{30}C_{20}$) in a ternary system. Along each axis, the concentrations of 0–100 are plotted

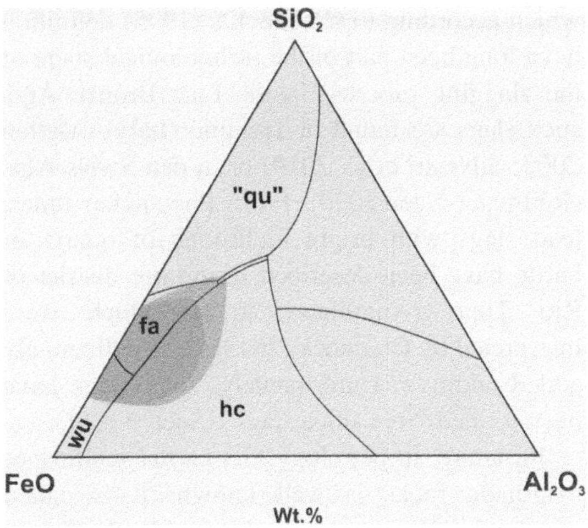

Fig. 4.9 Average bulk chemical compositions of ancient slag from various supra-regional iron smelting sites (small dark area) and copper smelting sites (larger medium shaded area) shown in the ternary system $FeO–Al_2O_3–SiO_2$ (weight percent) (Osborn and Muan 1964). The compositions match mainly the eutectic area of the binary system $FeO–SiO_2$ at fayalite (= fa) and wuestite (= wu) (see Fig. 4.4). It is surrounded by steeply arising areas of quartz ("qu") and spinel (hc = hercynite). These are components that crystallise from very high temperatures, or, vice versa, they are stable up to high temperatures. Compositions of slags after data from Sinai (Egypt), Timna (Israel) (Bachmann 1978, 1982b), Oman (Hauptmann 1985), Congo (Keesmann 1989), Lahn-Dill area, Germany (Kronz 1997), Joldelund, Germany (Ganzelewski 2000), Rio Tinto, Spain (Anguilano 2012) and unpublished data from the DBM. From Hauptmann (2007)

analysis programme (www.monkeybreadsoftware.de/Analysentetraeder).

In order to make simplifications, the oxides mentioned can be added to the main constituents of the systems mentioned ($FeO–SiO_2–Al_2O_3–CaO$) according to their geochemical compatibility. The following abbreviations are used in the ceramic industry:

- $S = SiO_2 + P_2O_5$
- $C = CaO + (Na_2O, K_2O)$
- $A = Al_2O_3$
- $f = FeO + MgO + MnO$

These were also used in the Slag Atlas (1995). Overall, however, a presentation of analytical data in a tetrahedron has not prevailed in materials science.

An estimation of solidification temperatures by plots of analytical data into a tetrahedron is not a simple approach. One can only try to see the spatial representation with a projection onto the isotherms of the tetrahedral faces, i.e. to combine the individual ternary systems. This was done by Keesmann and Schmitz with plots of the chemical compositions of copper slags from different countries into the system S–C–A–f (Fig. 4.10).

It shows a tendency from characteristic iron-rich slags to iron-poor variations, which lie approximately in the left half of the triangle C–S–A. As a guide, the plots at the bottom of the tetrahedron can be consulted. These compositions have in common that they are estimated to be in low-melt ranges. The varying compositions are the result of changing host rocks of the mineralisation and different treatment technologies.

Compositions of individual (normalised) mineral phases, calculated from the chemical bulk composition of slag samples, can also be graphically plotted in a tetrahedron. Hezarkhani

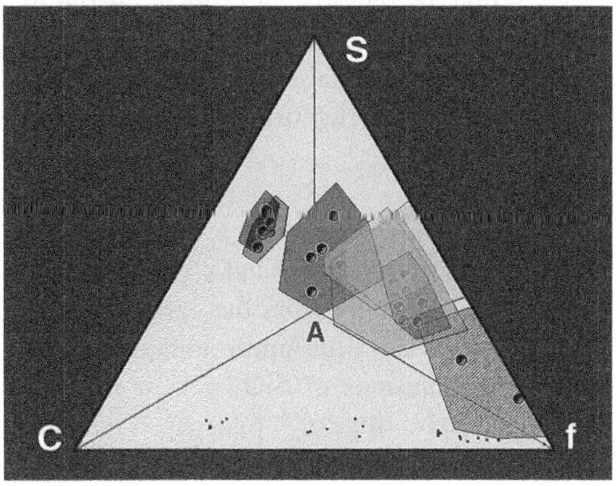

Fig. 4.10 Composition of copper slags from different regions (Iran, Niger, Wadi Arabah, Faynan) and different temporal positions (Bronze Age to about Middle Ages), represented in the quaternary system C–S–A–f. All slags are roughly following low-melt areas as they would be represented in the various ternary systems. For spatial orientation, the data can also be seen in plots in the plane C–A–f. From Keesmann and Schmitz (www.monkeybreadsoftware.de/Analysentetraeder)

and Keesmann (1996) have shown such calculations of copper slag samples from Iran. They have demonstrated how manageable and comprehensible representations can be made from a confusing amount of chemical analysis data. In addition, the equilibrium parageneses and the phase combinations observed in a micrograph can be plotted.

4.3 Micro-equilibria and Partial (S)melting

The concept of viewing archaeometallurgical slag a priori as material solidified from a *completely* liquefied melt has to be re-considered and examined—particularly for the early stages of metallurgy. Slag from this period, and also from later times often contain undecomposed inclusions, and their proper interpretation is of fundamental importance.

There are two possibilities:

1. Such inclusions may be remnants of deliberately added fluxes for optimising a smelting process.
2. They may be relics of an original charge that did not completely dissipate during heating. This has been discussed in Sect. 4.2 with the melting of an Fe-rich ore using the phase system $FeO–SiO_2$.

It will be shown in Chap. 5 that meanwhile, after many excavations, hand-sized slag cakes were found especially from the early periods of metallurgy in the fifth/fourth millennium BC, containing inclusions of SiO_2-rich or Fe-oxide-rich material (quartz, feldspar, magnetite agglomerations). Examples are mainly from known from the Near East (Faynan (Jordan)), Tell Hujayrat near Aqaba (Jordan), Nevali Çori, Arslantepe and Murgul (eastern Anatolia), Shahr-i Sokhta (east of the Iranian plateau) (see summary in Hauptmann 2007). An exceptional example is shown in Fig. 4.11.

However, slags with unmelted mineral and rock inclusions can also be found in later epochs, which according to Craddock's (1995) definition have long been part of the technological stage of the slagging process. In the Late Bronze Age, such slags are found in Trentino (Italy) (Metten 2003; Silvestri et al. 2019) or in den Swiss Alps (Reitmaier-Naef 2020). From Phoenician times, lead slags with bright inclusions of quartz or barite have been described in the ore district of Rio Tinto (Anguilano 2012), which were interpreted by Craddock (2013a) as a deliberately added additive. Unfortunately, such slags have been termed "free silica slags" (Sect. 5.4.11).

In nature, in petrology, the partial melting of magmatic rocks is well known. It is called metatexis (Wimmenauer 1985). Such partial melting leads to the formation of the so-called metatects in which, in varying proportions, unmelted light or dark restites are embedded. Very difficult to liquify are SiO_2-rich components such as quartz or feldspar of the original rock or minerals such as biotite, hornblende or chromite.

In a small scale, in archaeology, this concept is similar to ancient technologies of slags and ceramics. Heimann and Magetti (2014) are writing: "Since the firing of a ceramic body is a short-term process, the time the object was subjected to the maximum firing temperatures is rarely sufficient to attain an equilibrium situation with respect to the phase transitions initiated by the thermal treatment. In particular, calcareous ceramics fired for only a few minutes to a maximum temperature below 800–850 °C in a typical bonfire retains its calcite content even though the thermodynamic decomposition temperature has been exceeded".

It is therefore far more informative to study the formation, especially of very early slag during gradual heating of the ore charge from the first liquefying of the components at lower temperatures to a complete liquefaction of the charge. In smelting vessels, especially from the early stages of metallurgy, the temperature as well as the time of firing may often not be sufficient to melt the charged material completely.

A charge composed of SiO_2, CaO, Fe-oxides, Al_2O_3, Na_2O, K_2O, P_2O_5, Cu_2O, chlorides and sulphates in various mineralogical components

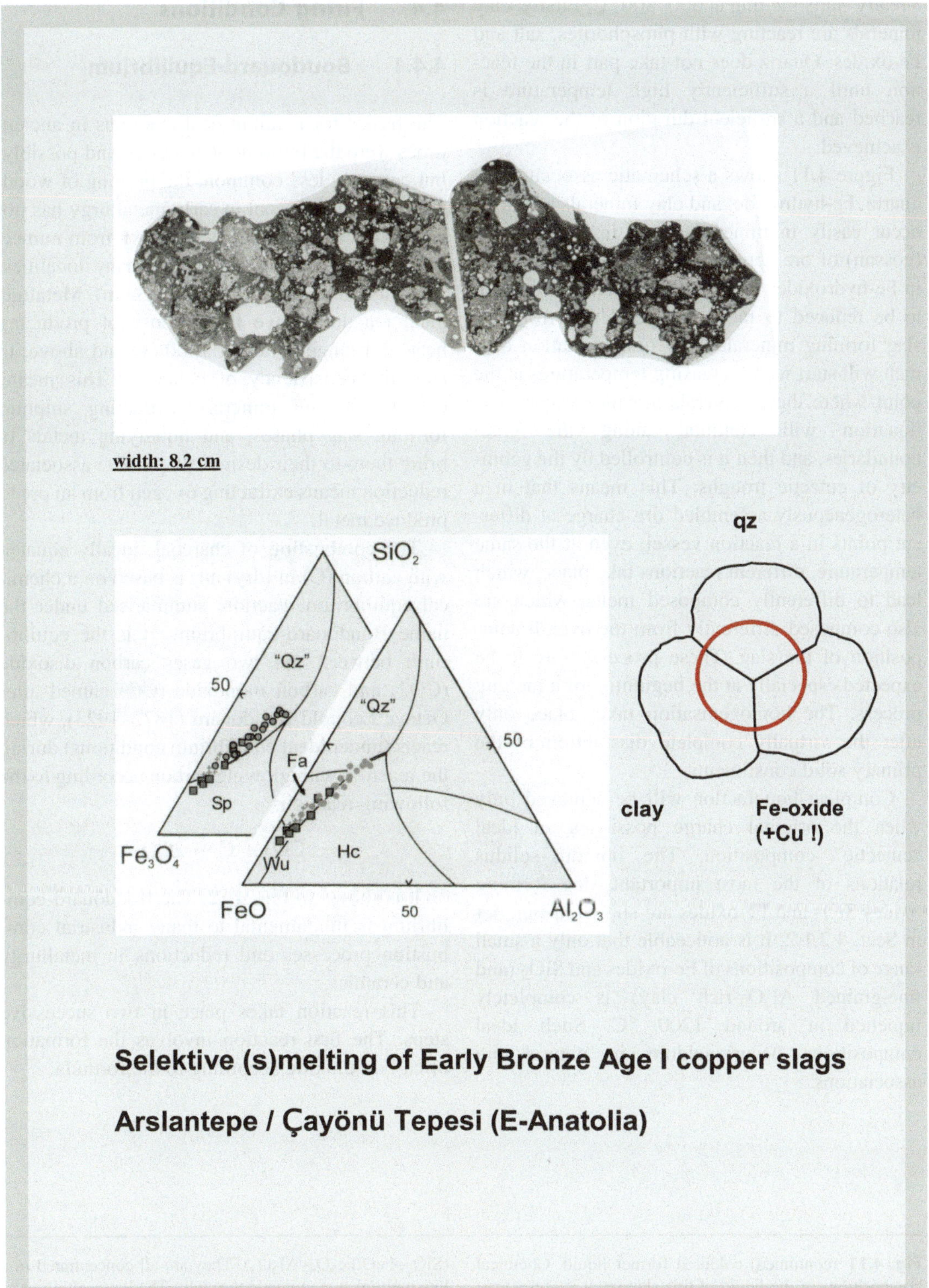

Fig. 4.11 Partial (s-)melting of Early Bronze Age copper slag. The picture shows a section through an Early Bronze Age copper slag cake from Arslantepe/Malatya (eastern Anatolia, level VIB2, first half of third millennium BC). The porous slag cake is full of unmelted fragments of quartz, Fe-oxides and host rock, embedded in a dark-

already starts melting at 600–800 °C. Firstly, clay minerals are reacting with phosphorites, salt and Fe-oxides. Quartz does not take part in the reaction until a sufficiently high temperature is reached and a sufficient duration of the reaction is achieved.

Figure 4.11 shows a schematic association of quartz, Fe-hydroxide, and clay minerals as it may occur easily in mineralisations in the iron hat (gossan) of ore deposits. Copper ore is included in Fe-hydroxides and in this model, it is thought to be reduced to metal independently from the slag forming minerals. The first formation of a melt will start with increasing temperatures at the point where these minerals are intergrown. The liquation will continue along the grain boundaries, and then it is controlled by the geometry of eutectic troughs. This means that in a heterogeneously assembled ore charge at different points in a reaction vessel, even at the same temperature, different reactions take place, which lead to differently composed melts, which are also composed differently from the overall composition of the slag. These processes are to be expected especially at the beginning of a melting process. The homogenisation takes place only after the virtually complete dissolution of the primary solid constituents.

Complete liquefaction will be achieved only when the original charge possesses an ideal "eutectic" composition. The liquidus–solidus relations of the most important slag forming oxides SiO_2 and Fe-oxides are shown in Fig. 4.4 in Sect. 4.2.1.2. It is noticeable that only a small range of compositions of Fe-oxides and SiO_2 (and fine-grained Al_2O_3-rich clay) is completely liquefied at around 1200 °C. Such ideal compositions occur seldom in natural ore associations.

4.4 Firing Conditions

4.4.1 Boudouard-Equilibrium

The basics for metallurgical reactions in ancient times were the burning of charcoal, and possibly, but certainly less common, the burning of wood. The use of black coal in early metallurgy has not been proven. Charcoal, as is known from numerous finds of slag inclusions in many localities, was used in the lumpy form (2–3 cm). Metallurgical reactions have the purpose of producing heat in a range of about 1000 °C and above, to alter the consistency of materials. This means dehydrating ore minerals, extracting sulphur, forming slag phases, and liquefying metals to bring them to their desired shape. The associated reduction means extracting oxygen from an ore to produce metal.

The combustion of charcoal, ideally equated with carbon (C) in (dry) air, is based on a chemical equilibrium reaction, summarised under the name Boudouard-equilibrium. It is the equilibrium between the two gases carbon disoxide (CO2) and carbon monoxide (CO) named after Octave Leopold Boudouard (1872–1923), which reacts (under ideal equilibrium conditions) during the reaction with glowing carbon according to the following reaction:

$$CO_2 + C \rightarrow 2CO$$

It is shown in Fig. 4.12. The Boudouard-equilibrium is fundamental to many industrial combustion processes and reductions in metallurgy and ceramics.

This reaction takes place in two successive steps. The first reaction involves the formation of carbon dioxide according to the formula:

Fig. 4.11 (continued) coloured former liquid. Chemical compositions of the liquid of this slag (spot analyses) are plotted along with those of a comparable slag sample from nearby located Çayöny Tepesi in two ternary systems (SiO_2–FeO/Fe_3O_4–Al_2O_3). They are all concentrated in a low melting area (eutectic through). The low melting parts of the system are mineralogically represented by the grain boundaries of the ore quartz–Fe-oxide–clay (Al_2O_3)

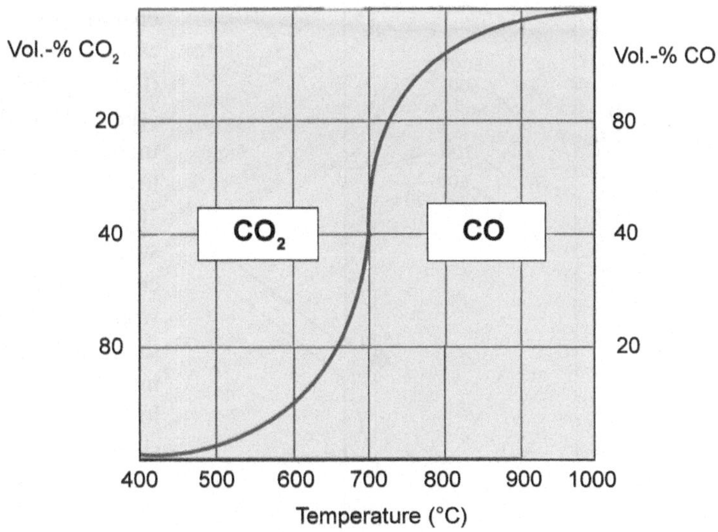

Fig. 4.12 Distribution of the two gases carbon dioxide and carbon monoxide in Boudouard-Equilibrium at medium atmospheric pressure of the atmosphere at sea level (pressure 1013 bar). Note the increase in carbon monoxide at high temperatures. After O.L. Boudouard (1872–1923)

$$C + O_2 \rightarrow CO_2 \rightarrow \Delta H\left(298.15^{\circ}K\right) = -94.05\ \text{kcal mol}^{-1} \qquad (I)$$

This reaction is associated with very high thermal effects, it is exothermic. This reaction is extremely fast, it creates great heat in the reaction vessel. It can be seen from the diagram that initially, at even lower temperatures (around 400 °C), only the carbon dioxide is stable. That changes with increasing temperatures. Increasingly hot gases are flowing through the (remaining) charcoal bed, the carbon dioxide will be reduced to carbon monoxide according to the following formula:

$$CO_2 + C \rightarrow 2CO \rightarrow \Delta H\left(298.15^{\circ}K\right) = +41.21\ \text{kcal mol}^{-1} \qquad (II)$$

This reaction is significantly slower than the oxidation of charcoal (C) to carbon dioxide and consumes energy, it is endothermic. In addition, it depends on the size of the charcoal pieces and their volume in reaction vessels (Rehder 2000). At about 680 °C, the two gases are present in equal parts. With higher temperatures, the equilibrium shifts to the right side of the reaction equation. At 1000 °C and above, the equilibrium is on the side of CO. This means that at high temperatures almost only pure carbon monoxide is stable. Straube (1996) explains that the ratio of CO and CO_2 depends on the temperature generated, which in turn depends on the supply of oxygen. The higher the supply of oxygen, the more complete the combustion of charcoal and the greater the CO_2 content in the combustion gas.

In a bloomery furnace of about 1.5 m height, with artificial air supply via tuyères and bellows, which was widespread in the early copper as well as in iron metallurgy, and which was fired with lumpy charcoal, both the gas composition and temperature are far away from any equilibrium (Rehder 1989, 2000). They vary and thus, differently reducing conditions can prevail at different points in the reaction vessels. Thus, the charcoal directly at the nozzle will burn mainly to CO_2. Here is the largest heat range with 1300–1500 ° C (Tylecote and Boydell 1978; Pleiner 2000; Rehder 2000) (Fig. 4.13). With increasing distance from the nozzle, the CO content increases due to the decreasing oxygen supply and the temperatures decrease.

On the other hand, this means that in small crucibles with a diametre of 15–20 cm, the redox conditions are very weak, unless they are used in a larger charcoal bed. It is thus very unlikely that a

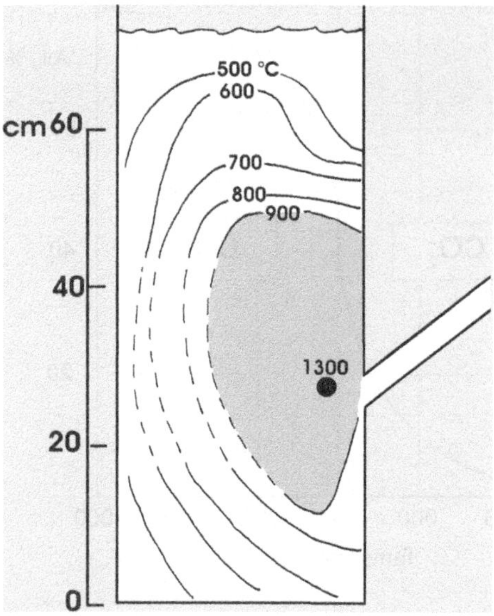

Fig. 4.13 Distribution of temperature in a shaft furnace constructed for experimental copper smelting. The isotherms, given in °C, show a maximum close to the tuyère set into the furnace for artificial air supply. After Tylecote and Boydell (1978). Straube (1996), Pleiner (2000) and Rehder (2000) calculate an even higher focus of temperature

Boudouard-equilibrium, which is based on a C-surplus, will ever be reached. Much more probable is that CO will be oxidised to CO_2 through a surplus of oxygen by blowpipes.

4.4.2 Colours by Burning

The gas atmosphere prevailing during the combustion of charcoal in a metallurgical process influences the colouring of the walls of the reaction vessel or furnaces, which were usually made of iron oxide-containing ceramics. As described by Noll and Heimann (2016), they may contain in addition organic impurities derived either from the clay starting material or added as a tempering agent to the clay during moulding. These are burned out in the first stages of the firing process under still oxidising conditions, wherein the ceramic material, if it has iron contents, is coloured red by the oxidation to hematite (Fe_2O_3). The transition to reducing conditions of the gas atmosphere, which is associated with a change in the valence of the iron, is also associated with a colour change. The red hematite ($Fe^{3+}{}_2O_3$) will be reduced to black magnetite ($Fe^{2+}(Fe^{3+})_2O_4$) (Fig. 4.14). As the temperature increases, there is a tendency to transition to low-grade oxides of iron, e.g. wuestite ("$Fe^{2+}O$"). However, a black or grey colouration can still be caused by incompletely burnt organic components, or by very fine particles of carbon, which are formed during combustion according to the formula $2CO \rightarrow C + CO_2$ and can penetrate into the ceramic (Rehder 2000). In ceramics, the low-grade iron is partially present in Fe^{2+} silicates.

4.4.3 Redox Reactions

The reduction is a chemical reaction in which the oxygen is removed from a compound, e.g. from a metal oxide, and the metal oxide is reduced to metal. To remove the oxygen, e.g. from the iron (II) oxide, so to reduce the iron (II) oxide to iron, a reducing agent is needed. This agent combines with oxygen during the reduction, it is oxidised. This also applies to copper and lead. The simplified formulas are:

$$Fe^{2+}O + CO \rightarrow Fe^0 + CO_2$$

$$Cu^{2+}O + CO \rightarrow Cu^0 + CO_2$$

$$Pb^{2+}O + CO \rightarrow Pb^0 + CO_2$$

At the same time, the reducing agent, the carbon monoxide CO, is oxidised to carbon dioxide CO_2. The term reduction is therefore closely linked to the concept of oxidation. This is why these chemical reactions are called redox reactions.

Ancient metals, their alloys and also archaeometallurgical slags have generally been produced under high-temperature and varying reducing conditions, which were determined by the combustion of charcoal or wood. They were affected by the oxygen pressure in the gas atmosphere, controlled by the CO/CO_2 ratio.

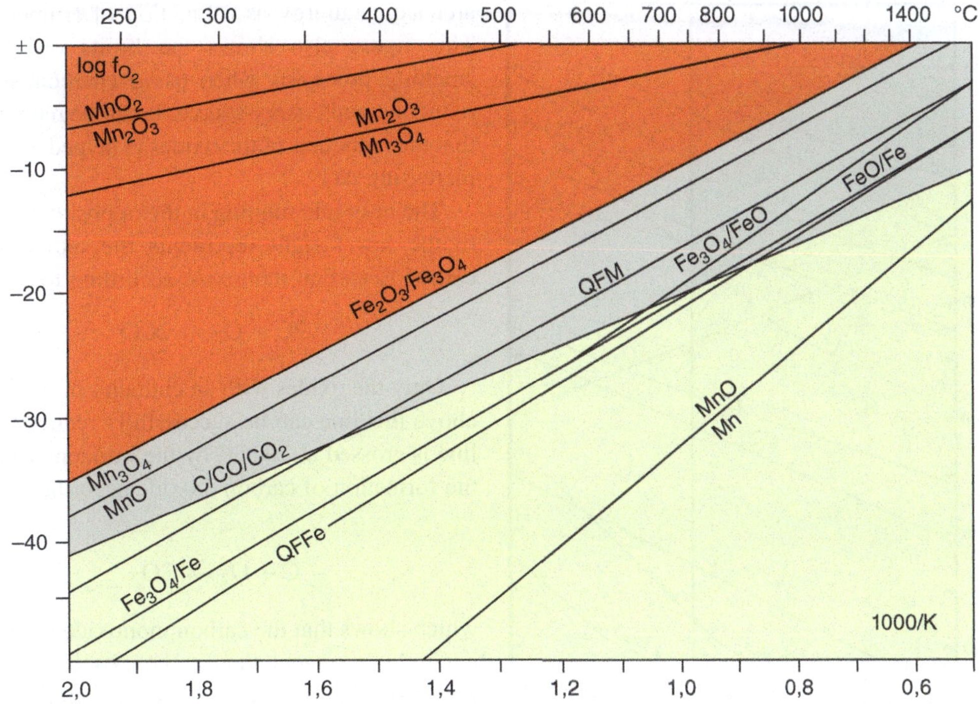

Fig. 4.14 Colour change in ferrous ceramics under oxidising and reducing conditions of the gas atmosphere in a reaction vessel, shown as a function of oxygen fugacity (= partial oxygen pressure, pO_2) and temperature. The colour change from red to grey takes place along the buffer equilibrium Fe_2O_3–Fe_3O_4. Shown is also the limit of the furnace atmosphere in wood/charcoal firing, or, in other words, the buffer equilibrium C/CO/CO_2. Modified after Noll and Heimann (2016)

Depending on the geometry and design of the crucibles or furnaces used, the shape, size and number of tuyères used, as well as depending on applied crafting techniques, these conditions within the reaction vessels themselves were subject to fluctuations over time. In the early days of metallurgy, from the fifth to the third millennium BC, only a few metals (copper, lead, tin) were smelted from their ores. This happened in smaller, open crucibles, with little coverage of fuel and under only slightly reducing conditions. Later, around the first millennium BC, iron, and occasionally zinc, were produced under significantly more reducing conditions. At that time long larger shaft furnaces of 1–2 m height were known.

4.4.3.1 The Ellingham Diagram

The different reducibility of metal oxides or semi-metal oxides (arsenic, antimony, bismuth) to metals can be read in the so-called Ellingham diagram (Fig. 4.15). The Ellingham diagram, named after its first editor (Ellingham 1944), can be read in the following way.

It is a graph showing the stability of oxidic compounds depending on their thermodynamic conditions. It is usually used to evaluate the ease or difficulty to reduce pure metal oxides to metals depending on temperature and the composition of the gas atmosphere in metallurgical processes. In this diagram, the curves of the nobler metals are higher than those of the less noble ones, the latter correspondingly having a higher oxygen affinity and being difficult to reduce.

All lines that represent reaction equations between metal oxides and metals run to the top right. This means that in principle, in the upper left part, the stability range of oxides, in the lower right part, the one of metals is shown. This in turn

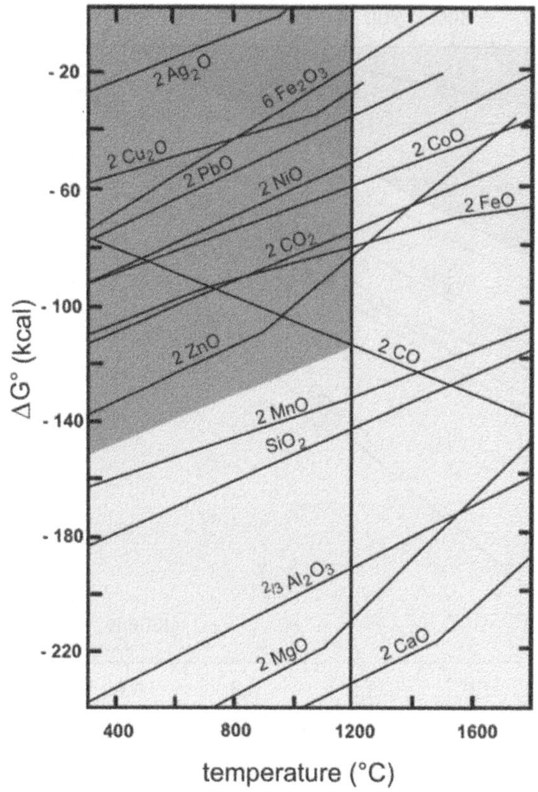

Fig. 4.15 Free enthalpy (energy)—temperature chart ($\Delta G°$) for metal and oxide formation. The limits of ancient smelting processes are set by the temperature of 1200 °C, which can generally be reached as an average maximum and the line of the reaction $2C + O_2 \rightarrow 2CO$. Only those oxides can be reduced to metal (darker field), whose enthalpies of fusion are above the crossing-point of these two lines; otherwise, they will become parts of slag compounds, e.g. MnO. Modified after data of Kubaschewski and Alcock (1979)

means that the Ellingham diagram indicates an increasing stability of oxides to lower positions in the diagram. That is, oxides with higher negative values of free enthalpy (energy) of formation $\Delta G°$ are more difficult to reduce. The reduction of oxides to metal is easier to perform at higher temperatures, while the less negative $\Delta G°$ reflecting a decreased oxide stability.

The different reducibility, or the ease with which metal compounds can be reduced to metal, is based upon the free enthalpy (energy) of formation ($\Delta G°$) as it is shown in Fig. 4.15 for some selected oxides important in archaeometallurgy as a function of temperature. This figure also shows the limits of ancient smelting processes when using charcoal as the reduction and energy source. It can be recognised that the reduction of the oxides is helped along by increasing $\Delta G°$.

The only line running in the opposite direction to the lower right represents the oxidation of carbon to carbon monoxide according to

$$2C + O_2 \rightarrow 2CO$$

Only the oxides with an enthalpy of formation above this line can be successfully reduced. This line is crossed at 700 °C by the horizontal line of the formation of carbon dioxide according to:

$$C + O_2 \rightarrow CO_2$$

which shows that the carbon monoxide is a stronger reducing agent than carbon itself. It is this reduction that forms, e.g. the so-called sponge iron in a cooler zone of the furnace (Charles 1980).

If you, for example, have a mixture of Cu-oxide, Fe-oxide and quartz, and want to extract the copper as a metal while slagging the iron, the slag must melt at a temperature at which the Cu-oxide is already reduced to metal, yet the Fe-oxide must still be present as such, allowing it to react with the quartz, forming one or more slag phases.

At low reducing conditions and average temperatures, however, Cu-oxides react with Fe-oxide and thus escapes a metal formation. Phases like cuprite ($Cu^{+}{}_2O$) and/or delafossite ($Cu^{+}Fe^{3+}O_2$) are formed. This effect has been observed in slags of early metallurgy, as in Wadi Fidan/Faynan (Jordan), Hujayrat al-Ghuzlan (Jordan), Tell Magass (Jordan) and other sites in the Old World (Hauptmann 2007; Bourgarit 2007).

By contrast, at higher smelting temperatures and at the same time highly reducing conditions, the Fe-oxides may also be reduced to metal, so that iron and copper would deposit simultaneously and combine to form an almost unusable alloy (Craddock and Meeks 1987).

Many metals (Au, Ag, Cu, As, Sb, Ni, Co, Sn, Zn, Fe) are lying in the upper left corner in the Ellingham diagram and are therefore reducible into metal and can be separated from the slag. However, the noble metals gold and silver are not oxidised and remain in the metallic state, even at high temperatures. The enrichment of these elements during the smelting of oxide ores increases in the sequence from below to above, as shown in Fig. 4.15.

In contrast, it can be seen that a number of oxides below the crossing-point of the two lines cannot be reduced to metal. These are, e.g. SiO_2, Al_2O_3, MnO, MgO and CaO. They could not be reduced to metal because they need extremely strong reducing conditions which the ancient metallurgists were not able to produce. Instead, these metals are captured as oxides or silicates in the slags. We can reconstruct the redox conditions because most of the ancient slags are iron-rich silicate slags that reacted very sensitively to the oxygen content during their formation. This leads to the crystallisation of mineral phases with differing $Fe^{2+,3+}$ or $Cu^{1+,2+}$-ratios at a constant chemical composition, identifiable directly from the mineralogical composition in the slag. Minerals that can be used as oxygen barometers are listed in Table 4.1.

The use of the Ellingham diagram in archaeometallurgy is limited by several parameters. This is the reducibility of metals, which is limited downwards with the reaction line $2Zn + O_2 \rightarrow 2ZnO$. For metals below this line (Mn, Si, Al, Mg, Ca) are not known in ancient times. Another limiting parameter is the temperature to reach in ancient reaction vessels. It was explained in Sect. 4.1 that a reasonable average temperature in ancient smelting furnaces could be approximately 1200 °C. The oxides reducible under the conditions of burning fuels like wood or charcoal are therefore limited to the upper left-hand corner of the diagram.

It should also be emphasised that the Ellingham diagram only applies to pure oxides or metals. A formation of mixed phases leads to a deviation from the shown relationships and can be represented in other diagrams. Another disadvantage is that only statements about the theoretically achievable equilibrium states are possible, while the path and time for adjusting the equilibrium, i.e. the crucial area of reaction kinetics, remains ignored.

Table 4.1 Important mineral phases, metal and metal-oxide-inclusions in ancient slags that can be used to reconstruct the redox condition of metallurgical operations (oxibarometers)

Fe-rich slag	
Fayalite	$Fe_2{}^{2+}SiO_4$
Laihunite	$Fe^{2+}Fe^{3+}{}_2(SiO_4)_2$
Iscorite	$Fe^{2+}{}_5\ Fe^{3+}{}_2SiO_{10}$
Ferrosilite	$(Fe^{2+},Mg)_2Si_2O_6$
(Pyroxferroite)	"$Fe^{2+}SiO_3$"
Wuestite	"$Fe^{2+}O$"
Magnetite	$Fe^{2+}(Fe^{3+})_2O_4$
Ilmenite	$Fe^{2+}TiO_3$
Hematite	$Fe^{3+}{}_2O_3$
Mn-rich slag	
Braunite	$Mn^{2+}Mn^{3+}{}_6SiO_{12}$
Tephroite	$Mn^{2+}{}_2SiO_4$
Mn-Pyroxenoide	« $MnSiO_3$ »
Hausmannite	$Mn^{2+}Mn^{3+}{}_2O_4$
Partridgite	$Mn^{3+}{}_2O_3$
Other metal oxides	
Delafossite	$Cu^{+}Fe^{3+}O_2$
Cuprite	$Cu^{+}{}_2O$
Tenorite	$Cu^{2+}O$
Cuprospinell	$Cu^{2+}Fe^{3+}{}_2O_4$
Lithargite/massicote	$Pb^{2+}O$
Mennige	$Pb^{2+,4+}{}_3O_4$
Cassiterite	$Sn^{4+}O_2$

They are rich in metal which shows several valences. In a similar way, several metals and metal oxides can also be used. Compare Ellington diagram in Fig. 4.12. Note that no noble metals (gold, silver) are included in this list

4.4.3.2 Siliceous and Oxide Compounds of Iron and Other Metals

The enthalpy of formation (ΔG°) is only valid for pure oxides, so the smelting of natural mixed ores with gangue and host rock might be divergent from the reactions mentioned in the Ellingham diagram because it leads to a fractionation of different elements into slag and metal.

Nevertheless, oxidic and silicate metal compounds with multiple valence states are also good indicators for reconstructing redox conditions. They are also good oxygen barometers. The valences of these compounds as well vary depending on the redox conditions and lead with constant chemical composition to different mineralogical phase inventory not only in metallic inclusions.

In comparatively "pure" iron-rich silicate slags, it is the iron content ($Fe^{0,2+,3+}$) in different phases which reacts directly to changes in the oxygen content in the gas atmosphere of the furnace. In contrast, calcium, magnesium- or manganese-rich silicate slags are less susceptible to variations in oxygen content in the gas atmosphere of ovens. This is because the Mn valences ($Mn^{2+,3+,4+}$) have much larger variation widths than those of $Fe^{2+,3+}$, and therefore also manganese-rich silicate slags. Many archaeometallurgical slags also contain copper phases with different valences ($Cu^{0,+,2+}$).

Metal or oxide inclusions of metals and silicate compounds can be used as oxibarometers as well. Such mineral phases and metals are listed in Table 4.1. The intergrowth of these phases allows further fine gradations of the redox ratios and thus also further interpretations of craft processes (Keesmann 1989; Hauptmann 2000; Kronz and Keesmann 2005). Further comments on this will be made in Chap. 5.

In iron-rich silicate slags, Fe-oxides are formed at higher redox conditions at the expense of Fe-silicates. This interaction is exemplified by this reaction:

$$\underset{\text{fayalite}}{2Fe^{2+}{}_2SiO_4} + 2O_2 \leftrightarrow \underset{\text{magnetite}}{2Fe^{2+,3+}{}_3O_4} + 2SiO_2$$

The equation shows that at higher O_2 contents in the gas atmosphere of a smelting furnace, instead of fayalite, SiO_2-richer mineral phases with varying $Fe^{2+/3+}$ -ratios are formed (e.g. clinopyroxenes such as ferrosilite or hedenbergite), which are then associated with magnetite.

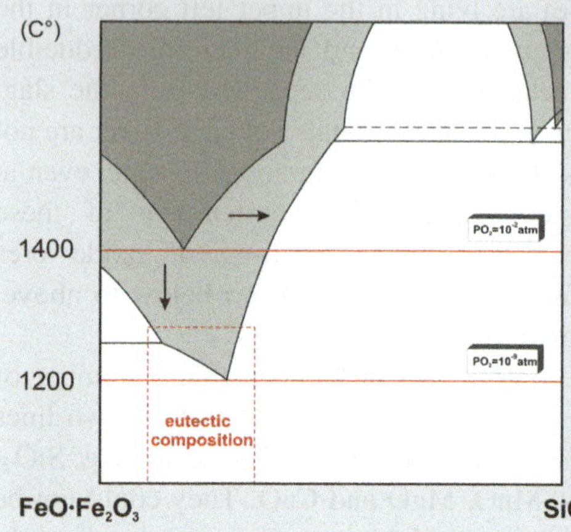

Fig. 4.16 The binary system FeO/Fe_3O_4–SiO_2 shows that Fe^{2+} -rich silicate melts in the temperature range of 1200 °C form only in a limited range of compositions and under strongly reducing conditions ($pO_2 = 10^{-9}$ atm). When the oxygen content in the firing atmosphere increases ($pO_2 = 10^{-4}$ atm), magnetite increasingly forms, and the eutectic increases. It also shows that the geometry of the system remains essentially the same. From Hauptmann (2007)

Fayalite is therefore an important and effective oxibarometer. This phase is rare in slags of the earliest metallurgy. In contrast, Bronze Age slags, especially from copper metallurgy, are typically characterised by the phase association fayalite + magnetite (Fig. 4.16). Slags of later iron production following the bloomery process contain fayalite + wuestite ("$Fe^{2+}O$") as main components and are in equilibrium with metallic iron ($Fe^{\pm 0}$). Reduced to these main components, this is classified in the system Fe–Si–O. These equilibria via the formation of Fe-rich solid solutions are called buffer equilibria. The above reaction fayalite–magnetite–quartz is referred to as FMQ or QFM buffer. This means that the mineral phases fayalite, quartz and magnetite define ("buffer") an oxygen partial pressure that is in equilibrium with them at a certain temperature. If oxygen from outside is added to a system of fayalite, magnetite and quartz, fayalite will be

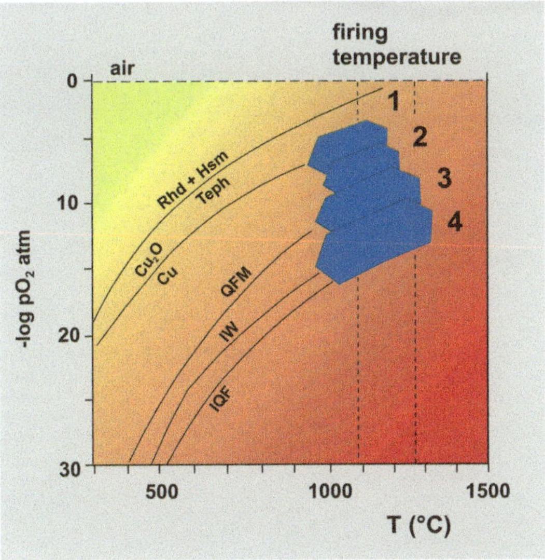

Fig. 4.17 Estimated ranges of slag formation from different chronological periods in Faynan (Jordan), depicted in a pO_2/T-diagram using the Fe–Si–O, Mn–Si–O and Cu–O buffer equilibria. The periods are Early Bronze Age I (1), Early Bronze Age II/III (2), Iron Age (3), Mameluk period (4). The variety of slag formations were based upon the mineralogical identification of fayalite (F), hausmannite (Hsm), iron (I), magnetite (M), quartz (Q), rhodonite (Rhd), tephroite (Teph), wuestite (W). After data of Eugster and Wones (1962). From Hauptmann (2007)

consumed until the original equilibrium value of pO_2 is restored. That is the buffering effect of this reaction.

In petrology, the buffer equilibrium of the system Fe–Si–O, which is further supplemented by the reaction

$$\underset{\text{hematite}}{3Fe^{3+}{}_2O_3} + O_2 \leftrightarrow \underset{\text{magnetite}}{2Fe^{2+,3+}{}_3O_4} + 0.5O_2$$

is a very important tool, because many geological systems at the same temperature have a similar oxygen partial pressure (p_{O2}). The equations are known as the Bowen–Fenner Trend.

The crystallisation of phases with differing $Fe^{2+/3+}$ and $Mn^{2+/3+}$ ratios are identifiable directly from the mineralogical phase content in the slag. For the identification of the redox conditions Lutz (1990) and Metten (2003) in addition used Mößbauer-spectroscopy. Based upon micro-mineralogical observations Hauptmann (2007) found varying redox conditions in copper slags from the Early Bronze Age (fourth/third millennium BC) to the Mamluk period (eighth–twelfth century AD) at Faynan (Jordan) (Fig. 4.17).

5 Archaeometallurgical Slags and Other Debris

5.1 Introduction

The most important information carriers for the reconstruction of old metallurgical processes are slags. They are usually found as worthless waste at the place of their creation. They are extremely resistant to weathering. In contrast to ceramics, slags, as they are not guiding fossils in the narrower sense, have been given only minor attention in archaeology, although they are among the most common finds in excavations. Similar to ancient ceramics, slags are a memory of the composition of the precursor's materials, i.e. ore, intergrown with gangue and host rock, perhaps as a result of the use of fluxes intentionally added to a furnace charge. In addition, all physical parameters of their formation are frozen in them, i.e. firing temperature and duration, redox conditions and the formation of the liquid itself.

According to modern understanding, slags are deliberately produced waste products of the smelting of ores (Biswas and Davenport 1980; Rosenqvist 1974). They are meant to separate the gangue and the host rock of the ore in the liquid state from the resulting metal or matte, possibly with the addition of fluxes, as well as to protect the molten metal from oxidation. Ideal slags are low in metal and have a low viscosity. Modern slags are generally tapped out of the furnace in a completely liquid state. Slag formation is fundamental for the efficiency of pyrometallurgical operations.

This definition, with some limitations, is also useful for (pre-) historical slags. However, the term "slag" in the *Goslarer Bergrecht* (Goslar mountain law) around 1.360 AD (Frölich 1953) where it is used in different ways, also appears helpful here. Once it appears in the enumeration of the geography of smelting factories as "*der Schlacken*", undoubtedly indicative of slag heaps. Further, the term slag is used for all materials that have been (s)melted before, both waste in the sense of a slag heap, and still usable scrap metal that are in line with intermediate products and (usable) waste, as they arise in the usual way in a smelting factory (copper matte, lead matte, furnace slag, etc.). Even in the eighteenth century AD, a distinction is made between "rich slags" and "poor slags", which in no way represent slags in the modern sense, but refer to intermediates with different metal contents.

Fundamentally, in ancient times we have to distinguish between smelting slags of the metals which were known at that time, i.e. slags of iron and base metals such as copper, lead, silver, tin, antimony and slags of further processing of these base metals (crucible slags, melting slags) (Tylecote 1976) as well as the characteristic smithing slags from iron processing. In the archaeological context, since the Bronze Age, many smelting slags occur in quantities of hundreds of kilograms to thousands of tons in the immediate vicinity of the ore deposits and document mass production, others are found only very rarely.

A. Hauptmann, *Archaeometallurgy – Materials Science Aspects*, Natural Science in Archaeology,
https://doi.org/10.1007/978-3-030-50367-3_5

Table 5.1 Possible objects and materials from ancient slag heaps and the resulting information concerning dating and reconstructing of (s)melting processes

Material	Information	Technique	Comments
Ore, fluxing agents, host rocks	Composition of the charged ore, metal in progress, provenance of ore	Chemical and mineralogical analysis (ICP-OES, XRD, microscopy, etc.), LIA	Ores in ancient smelting sites are often waste. Sampling in ancient mines more promising.
Fuel	Dating, fuel supply, palaeoclimate, material input for slag formation.	^{14}C-dating; Botanical studies (SEM) Chemical analysis of ashes.	Charcoal usually available at all smelting sites
Refractory material Blowpipes, tuyères, fragments of crucibles, casting moulds, furnaces	Temperatures and duration of process Composition of refractory materials Metal composition	SEM studies on vitrification; Petrographic studies; ICP-OES, (p)XRF, SEM/EDS	As precise as possible spatial localisation of samples in reaction vessels. Mixing up with slagged charged ores possible.
Slag and intermediate materials Remains of roasting, matte, speiss, litharge	Temperatures, redox-conditions, viscosity, different steps in metal production, sort of metal, efficiency of smelting, provenance of components of charge	Chemical and mineralogical analysis (AAS, ICP-OES, XRD, (p)XRF, microscopy, etc.) LIA	Ever present and important material at old smelting sites and workshops
Metal Ingots, final objects	Efficiency of smelting process, parameters of (s)melting process, differentiation of raw material-alloy, provenance of metal	Analyses of trace elements (ICP-OES, NAA) Metallography LIA	Exist as inclusions in slag. Larger pieces of metal could be waste.
Tools for crushing Slags Hammer stones, anvil stones	Efficiency of smelting process	Petrographic studies	To find in many (pre-) historic slag heaps

Abbreviations: *^{14}C* radiocarbon dating, *HR-ICP-MS* high-resolution inductively coupled plasma mass spectrometer, *MC-ICP-MS* multicollector-ICP-mass spectrometer, *EDS* energy dispersive spectrometry (spot analysis), *SEM* scanning electron microscopy, *NAA* neutron activation analysis, *XRD* X-ray diffractometry, *(p)XRF* (portable) X-ray fluorescence spectrometry, *LIA* lead isotope analysis. Microscopy includes Light microscopy and Keyence Digital microscopy. From Craddock (1995) and Hauptmann (2007)

As pointed out by Craddock (1995) and Hauptmann (2007), next to slag, other waste materials from smelting processes can be found at ancient smelting sites. These include ore, which can give information about the raw material, fragments of smelting furnaces and crucibles, tuyères and other refractory materials made of ceramics, and remains of firing material such as charcoal which is of eminent importance for dating smelting sites. It is noticeable that furnace fragments and tuyères have often been found in large numbers. All of those contain information on ancient technologies and on the organisation of smelting sites. Table 5.1 shows an overview of the possible finds in slag heaps, their potential information content and methods of analyses.

5.2 Slags in the Archaeological Context

In the 1960s, there was a growing interest in the technical aspects of archaeology, the so-called industrial archaeology (Slotta 1982) and thus an interest in relics that are beyond the generally aesthetically and purely historically oriented finds of archaeology. This category mainly includes slags. They are usually the best indicators in the search for evidence of early metal production because they are the major and most weathering resistant waste products of smelting operations. As mentioned, slags have been discarded over the course of millennia from single finds and smallest heaps to tons developing into larger and larger mounds. Thus, they are the best proof for long-term enterprises of smelting ores and the production of metal from geo-resources.

It was Theodor Wertime, who in 1966 and 1968 undertook two research trips to Iran together with the scientists Ronald F. Tylecote, Radomir Pleiner and Beno Rothenberg who were interested in archaeometallurgy (Wertime 1967, 1968, 1973). The goal of these trips was to explore the history of (pre-)historical pyrotechnics. They visited numerous sites on the Iranian Plateau with interesting historical technical documentation, which, at the same time, were interesting testimonies of traditional metallurgy. They found numerous slag heaps, but in many cases hardly any prehistoric evidence. The interest in documentation and investigation of slags or slag heaps, which had developed in the meantime, was then further significantly deepened by the work of Beno Rothenberg (1972a, b), who, in the 1960s, mapped hundreds of settlements and old smelting places in southern Israel, i.e. in Timna and its surroundings and on the Sinai Peninsula and in this way drew attention to this archaeological find genus.

From the global overview of archaeometallurgy presented by Roberts and Thornton (2014) it becomes clear that in the last 50 years in many countries and continents of the world not only metal objects but also countless slag heaps of different ages have been prospected, found and further investigated in the laboratory, from prehistoric to mediaeval metal extraction to dumps of (sub-) recent extraction of iron in Scandinavia or Africa. They testify to the surprisingly intensive worldwide use of geo-resources of mankind in different periods of time to an extent that could hardly have been imagined before. Many new ore or mining districts have been discovered. All of them decisively changed the image of early metal extraction from the late Middle Ages to (pre-)historical time. Not least of all, this was a decisive impetus for further intensive research activities in archaeometallurgy that continue to this day.

5.2.1 Prospecting Slag Heaps

In archaeological prospecting work and surveys, slag heaps were and are comparatively easy to find archaeological evidence. If the heaps have not been reprocessed by later recycling to recover residual metal contents or to use them for architectural purposes, they may provide an easier indication of the extent of early metal production than the far more difficult and elaborate excavations of old mines. Sometimes they are already visible at the surface of the earth without excavation, especially when voluminous heaps occur in low-vegetation areas or in the immediate vicinity of supra-regionally large ore deposits ("giants"). In many cases, however, slag heaps have to be salvaged through archaeological excavations.

Like in the prospection of ore deposits, a valuable basic and cheap method for surveys for ancient smelting sites is the prospection for special vegetation. Due to the accumulation of heavy metal contents (Cu, Pb, Cd, Zn, Co, Ni, etc.) in slag, vegetation can be prevented, or at least only a special sparse vegetation can be allowed. Also, similar to the prospection of ores, the so-called pointer plants can locally characterise the vegetation, e.g. the yellow calamine violet (*Viola calaminaria*) or the creeping rice paddy mustard (*Arabidopsis halleri*). Heavy metal contents in old slags can also lead to considerable environmental damage, as demonstrated in Wiesloch (Germany) (Hildebrandt 1998) or in Příbram (Czech Republic) (Ettler et al. 2001).

In addition, there are new geophysical prospecting methods that can be used not only for the search but also for non-destructive exploration of old smelting places.

As sophisticated prospecting techniques, proton magnetometers have been used for some years (Goldenberg 1990; Kruse et al. 1997; Humphris and Carey 2016) (Fig. 5.1). For this kind of analysis, ferromagnetic materials (metallic iron, wuestite, "FeO", magnetite, Fe_3O_4, maghemite, α-Fe_2O_3), which are present in almost all ancient slags (see Sect. 5.6), are

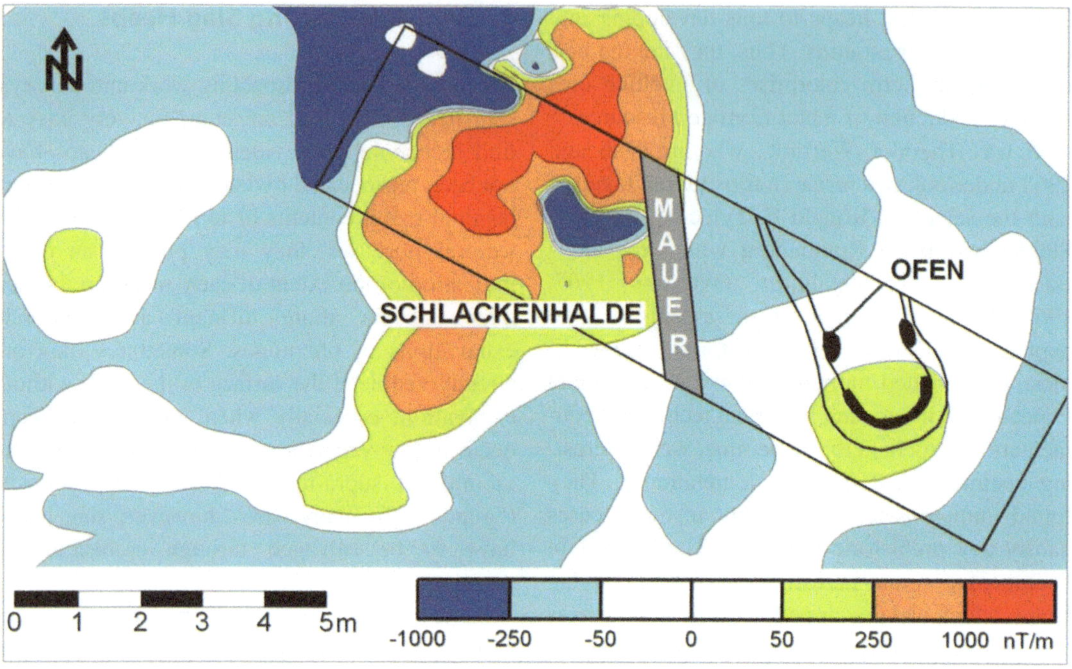

Fig. 5.1 Isomalic map (geomagnetic vertical gradient) showing the results of at the archaeological excavation at Hüttenberg (Styria) where the famous Ferrum Noricum was produced. Measurements at trench D1 (grid 0.5 × 0.5 m). Courtesy G. Walach/B. Cech)

particularly useful. Thus, it seems to be a very promising method to gain information about the existence, structure and organisation of smelting sites before any archaeological excavations are carried out. Practical tests have however shown that the method can only be appropriately put to use for the clear-cut prospection of slag heaps. Basic experience was collected especially with slag heaps from iron smelting, e.g. by geophysical prospection work at Roman iron smelting sites at Joldelund, Germany (Kruse et al. 2000).

The metallurgist Preßlinger (2002) and the geophysicist Walach (2004, 2017), who worked closely together in Austria on the early metallurgy of Styria, emphasise, however, that geophysical prospecting is particularly promising when combined with other physical prospecting methods and techniques and coupled with archaeological excavations. In their work, they have particularly intensively studied the structure of individual Bronze Age metallurgical smelting places in this region, i.e. location and arrangement of slag heaps, smelting furnaces, roasting beds and other fireplaces. From the large number of findings, a very interesting overall picture has emerged about the surprising extent, distribution and supraregional importance of this Bronze Age copper region in Styria.

Humphris and Carey (2016) applied this interdisciplinary approach with modern methods in a study of iron smelting slags in Meroe, Sudan. Meroe was the capital of the Kingdom of Kush in the time of c. 300 BC to 350 CE. Numerous slag heaps had been found in the vicinity of the royal city. The aim of their investigations was to determine the total volume of the slag heaps, to determine the density of the slag beds and finally to estimate the application of the iron on the basis of analytical data. They rightly point out that this cannot be done with surface-level topographical work alone, but methods must be used to reliably determine the volume of slag produced.

The methods used by the authors in Meroe include:

- A GIS-supported topographic survey.
- Classical geoprospection methods to survey even the heaps that are invisible from the surface and that were covered in the arid area of the sand desert of Meroe.
- Gradiometry and electrical resistivity transects. The gradiometer data allowed the spatial definition of the slag heap based on magnetic signature, whilst the electrical resistivity modelling allowed the depths of the deposits to be estimated. The electrical resistivity model clearly identified the variety of different objects and materials to be found usually within slag heaps next to slags themselves (Table 5.1). The electrical resistivity model clearly identified different components within the slag heap, which the excavated data demonstrated to be an earlier building. This was not evident from the surface survey.
- Excavation and quantitative sampling of slags. From the excavation data, the total volume of metallurgical debris excavated from the trenches was calculated. With the data from the quantified archaeometallurgical material, it was possible to produce a density model for each trench.

5.2.2 Slagging Versus Slagless Metallurgy

The major slag heaps, partly of gigantic dimensions, derived mostly from copper, lead–silver and from iron smelting activities. More rare are slag heaps from the extraction of metals from complex polymetallic ores. Exceedingly rare are slags from the extraction of tin and antimony, none have ever been found from zinc, arsenic or nickel.

Without exception, the large, extensive slag heaps date back to periods of time, which according to Craddock (1995) are characterised by a metallurgically developed slagging process. This process is characterised by the smelting of comparatively low-grade ores with simultaneously high accumulation of slags. Hostrock and gangue fused with the ore formed liquid slags that separated more or less cleanly from the resulting metal or metal sulphide and were left as a waste product at the smelting sites. In the Old World, the slagging process generally begins in the Middle to Late Bronze Age. Obviously only in exceptional cases does it begin a little earlier. The development of this smelting process represents a revolutionary technological development, without which the large quantities of metals of these epochs could not have been produced. The earliest slag heaps as evidence of an intended slag smelting process to date come from Murgul (north-eastern Anatolia, middle fourth millennium, Lutz et al. 1994) and from Faynan, Jordan (middle of the third millennium BC, Hauptmann 2007).

The *slagging process* is in contrast to localities and regions where metal was produced in the initial stages, i.e. in the late Neolithic to Chalcolithic/Early Bronze Age I periods (depending on the cultural context fifth to third millennium BC).

Here, the situation in Wales and Ireland must be mentioned. These are areas that do in fact have plenty of potential in regard to ore deposits. The intensity of early mining and the simultaneous absence of any smelting slags do not seem easily explainable. Regarding this, Craddock (1993) writes: "To anyone with experience of early mines and production sites the most noticeable feature of the mines in the British Isles is the total absence of slag". Craddock (1990) suspects that a smelting of very rich (oxidic) ores could have played a role in this, during which almost no slags would have been produced, which of course are difficult to find in the archaeological context. Perhaps a bit too blatantly, he describes the lack of finds of slags with the term *slagless metallurgy*.

Another way to explain this terminology is in the organisation of ore smelting in the early stages of metallurgy. In these early epochs, ore was usually not smelted to metal directly at the deposit itself or in the immediate context of mining activities but negotiated in parts over large geographical distances to settlements. Here, they were smelted in a domestic-scale mode of metal production. Examples of this model exist in the Old World of the Iranian Plateau (Arisman: Pernicka et al. 2011), from Anatolia (Çatal Höyük: Radivojević et al. 2017), from the southern Levant and from Egypt (Hauptmann 2007; Abdel-Motelib et al. 2012; Shiqmim: Shalev and Northover 1987; Tell Abu Matar: Golden 2014), the Balkan Peninsula (Belovode: Radivojević 2013; Radivojević et al. 2010), from the alpine region (Bartelheim et al. 2002) to the Iberian Peninsula (Zambujal: Gauß 2015). An overview of the localities in Central and Southern Europe and the Middle East with a discussion of the slag finds was compiled by Bourgarit (2007).

5.2.3 Slag Heaps of Metal Smelting

It was primarily slag heaps of non-ferrous metallurgy from the developed periods of the Bronze Age and later periods, which are among the most striking evidence of early metal production. They are mainly situated in the wide, low-vegetation regions of the Orient and the Iberian Peninsula. The largest and most impressive heaps are located in Corta Lago/Rio Tinto in the south-western Iberian Peninsula and on the island of Cyprus in the eastern Mediterranean. Both are slag heaps that have formed over millennia at outcrops of ore deposits of non-ferrous metal deposits that are among the largest in the world.

If such heaps are disclosed in profiles, they offer very good opportunities to date the different periods of metal extraction. They prove that the use of geo-resources or the smelting of ores are usually not period-limited, but are generally long-term enterprises. The profile of the lead–silver slags of Corta Lago (Anguilano et al. 2010b) or of Skouriotissa in Cyprus (Kassianidou 2003a) was archaeologically and scientifically dated to a period from the Late Bronze Age to Roman times. These are all slag heaps that, in their extent, take on an "industrial scale".

Corta Lago

Among the most impressive and largest heaps of lead–silver and copper slags are those of Corta Lago. According to information from Rothenberg and Blanco-Freijeiro (1981), the gigantic heaps have piled up to an approximately 8-m-high heap and extend over a length of 500 m (Fig. 5.2). The early ore production and metal production was discussed many times, e.g. by Davies (1935), Williams (1934, 1950), Salkield (1970), Rothenberg and Blanco-Freijeiro (1981), as well as recently by Anguilano et al. (2010b) and Anguilano (2012). Estimates of Corta Lago's slag production show considerable variation, which is not surprising in view of these enormous amounts. The heaps of Corta Lago, as well as those of Skouriotissa, are significantly disturbed by current industrial activities. Moreover, all data are based on surface-level estimates only and could not be examined more precisely, as in Meroe. Previous authors (Rothenberg and Blanco-Freijeiro 1981), following a systematic sampling of the slag, estimated a total of more than 15 million tons. Salkield (1987), on the other hand, calculated a total of 9 million tons of slag produced by silver metallurgy (only) in Rio Tinto. Anguilano (2012) is more cautious and thinks a slag amount of 6 million tons is more realistic.

Altogether, the site of Corta Lago is very important from the archaeometallurgical point of view, presenting metallurgical activities of approximately 1300 years, dated by a unique collection of ores, slags, furnace linings, tuyères, metal fragments and pottery (Anguilano 2012). It stretches from the Late Bronze Age (twelfth to ninth century BC) to the Roman period in the second century AD.

Cyprus

Amongst the largest (pre-)historic slag heaps of early copper mining in the Old World are those on the island of Cyprus. It has been estimated that there are approximately 4 million tons of copper slag (Constantinou 1982). Stos-Gale et al. (1998b) have compiled a list of all these slag

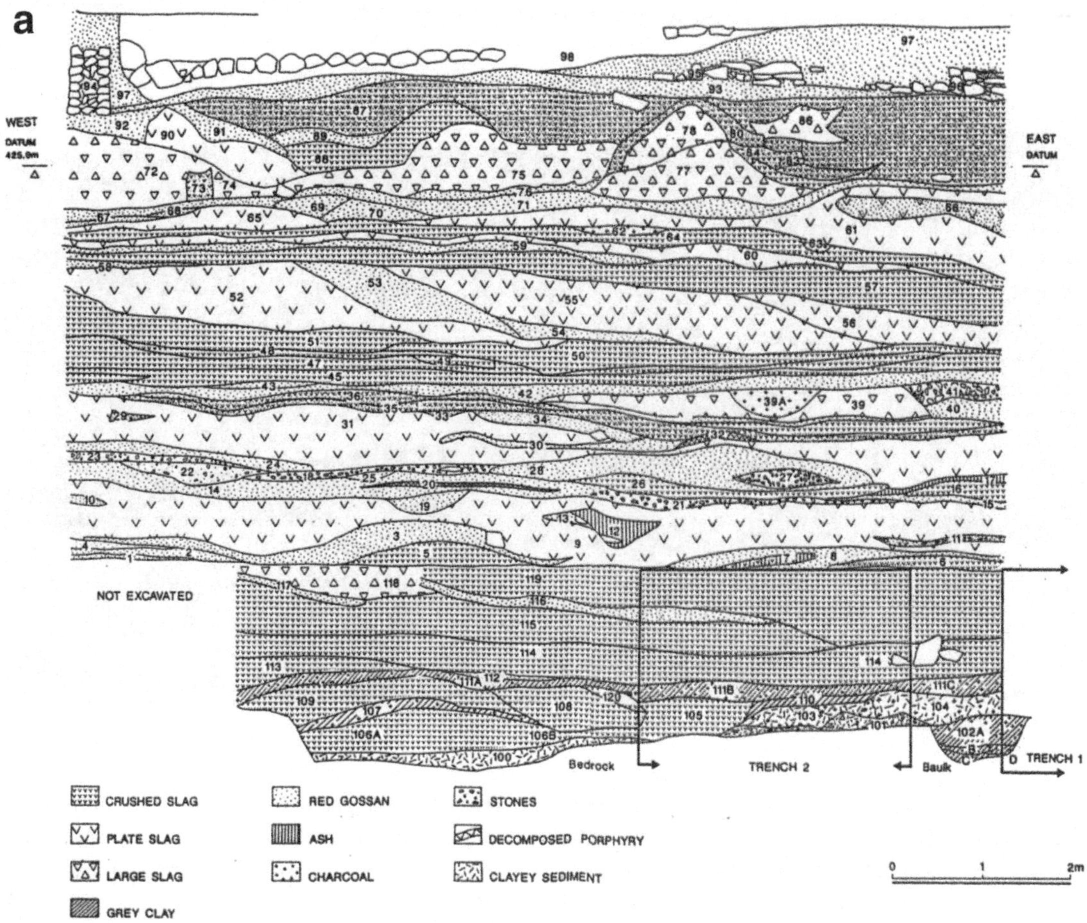

Fig. 5.2 The slag heap of Corta Lago, resulting from the lead and silver and perhaps also from copper production in the south-western edge of the Iberian Peninsula. The profile was cut by modern mining of the open cast mine nearby. As described by Craddock (2013a), the profile has 121 different layers. (**a**) Several sorts of slag types are distinguished macroscopically. Note that a part of the slag heap consists of crushed slag! From Rothenberg and Blanco-Freijeiro (1981), redrawn from Anguilano et al. (2010b). (**b**) Photograph of the same profile from G. Weisgerber (†), Deutsches Bergbau-Museum Bochum

heaps, most of which have also been analysed for their lead isotopic composition. Eighty-five smelting sites are distributed over the periphery of the Troodos Mountains, where most of the larger copper ore deposits are concentrated. In addition, there are another 19 slag heaps, which are located far from known mineralisation. The ore deposits of Cyprus are of volcanogenic massive sulphide deposits (Sect. 3.4.6). The largest slag heap is located at Skouriotissa. Next to the ore deposit, which is exploited today in a huge combined opencast and underground mine, more than an estimated 1 million tons of slag were found (Lavender 1962). A part of this gigantic amount is exposed as an impressive profile (Fig. 5.3). At Petra, in the Kalavassos mining area, about 0.3 million tons are accumulated. Surprisingly, there are almost no other archaeological finds in these slag heaps. Although some pottery can be found, which of course allows dating in the field, but remnants of smelting furnaces or tuyères are rare. Constantinou (1982) mentions finds of stone tools that he assigns to the processing of ores from the huge,

b

Fig. 5.2 (continued)

relatively poor slag heaps near Skourotissa. Kassianidou (2003a) sees it differently. She reports that at many of the smelting sites in Cyprus there are varying amounts of crushed slags in addition to the larger slag pieces. To smash larger slags to particle sizes of <0.5 cm requires numerous stone tools (see Sect. 3.9). Crushed slags are nothing special on old slag heaps. They are described in the following section.

The limited amount of metallurgical finds in slag heaps can likely be attributed to difficult circumstances in excavations. The furnaces were probably located outside the heaps. Perhaps, as was the case, e.g. at the mediaeval smelting places in Oman, the furnaces were carved into the surrounding rocks. On the Iron Age slag heaps in the Faynan area, countless fragments of furnace ceramics were found, as well as hundreds of tuyère nozzles (Hauptmann 2007; Levy et al. 2014).

Two characteristic types of slags from Cyprus are described by Kassianidou (2003):

- The late Roman heaps in Cyprus consist mainly of slag cakes of about 10-cm-thick tap slag with weights up to 60 kg. They have a size of c. half a square metre (Fig. 5.3) and show very clear flow structures (cf. Sect. 5.4). Generally, they are composed of several layers, which indicates several tappings of the liquid slag from the furnaces. Such slags can be observed from many other localities where, in the past, metal slags were produced in the smelting of metal.
- Slags from Late Bronze Age Politiko Phorades are plano-convex cakes with a diametre of approximately 25 cm. The rim at the slightly concave bottom shows the size of a matte cake (CuFe-sulphides) that was produced. This type of slag is identical to matte smelting slags as represented by Early Bronze Age Shahr-i Sokhta (Iran) (Hauptmann et al. 2003), Islamic slags from Mullaq (Oman) and from Abu Zainah (Oman) (see also Sect. 5.4.7). At the slag heap of Politiko Phorades, tuyères were found as well.

Fig. 5.3 Skouriotissa, Northwest of Cyprus (Greece) (**a**) Profile of the slag heap of Skouriotissa, cut by road constructions. The profile provides a height of the slag heap of c. 25 m. Excursion of the University of Nicosia 1981. (**b**) The largest amount of slags consists of large, up to 60 kg heavy, iron-rich fluid slags. It was produced in Roman–Byzantine times. The slag heaps of Skouriotissa are by far the largest in the eastern Mediterranean region. Photos: (**a**) G. Weisgerber (†), Deutsches Bergbau-Museum, Bochum. (**b**) A. Hauptmann, Deutsches Bergbau-Museum Bochum

The Alps

One of the most important of the European mining regions, which evidences the intensive and consistent use of copper deposits in prehistoric times to the recent past, is the Alpine region from the area south and west of Vienna to the French Alps in the west. It is one of the best-studied mining regions worldwide. Building on the prospecting work in the first decades of the twentieth century by Zschocke and Preuschen (1932) and later by Pittioni (1957), and especially in the context of the special research area "History of Mining Activities in the Tyrol and Adjacent Areas—Impact on Environment and Human Societies" (abbreviated HiMAT) a large number of smelting sites or slag heaps and mines were mapped out, which indicate an enormous copper production since the Early Bronze Age over many centuries and show a close social connection of the population groups within the alpine regions (Stöllner et al. 2016). These activities were then further pursued in the research institutions supported by Deutschland, Austria and Switzerland (DACH). As shown by chemical and lead isotope analyses (Lutz 2016), copper production in the Eastern and Central Alps during

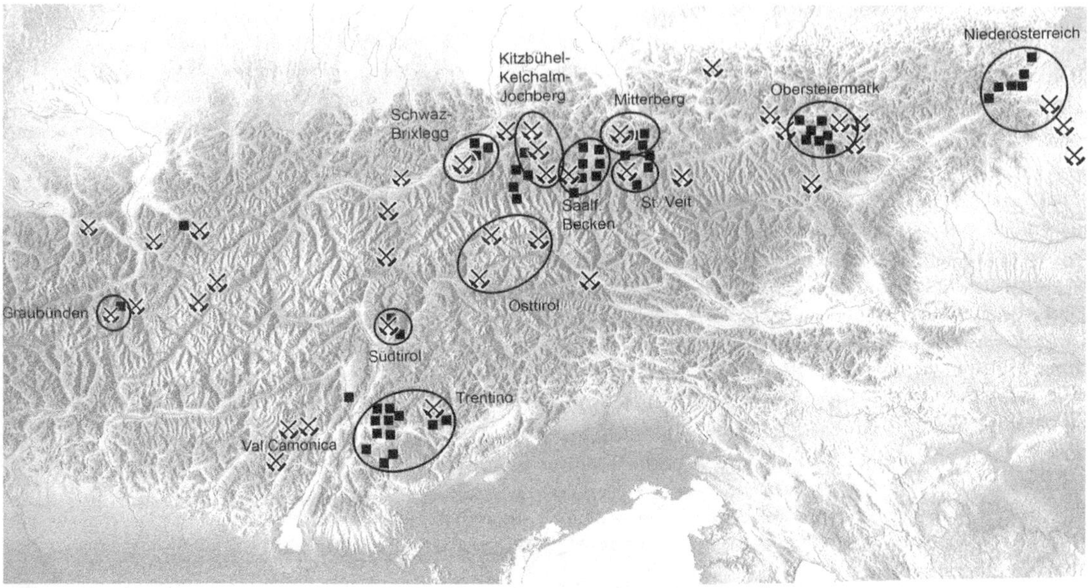

Fig. 5.4 Distribution of copper districts in the eastern and in the southern Alps. Hammer and wedge-symbols are representing groups of mines. Black squares are representing dozens or even hundreds of smelting sites. At Mitterber c. 200 smelting sites were discovered (see Fig. 3.10), the same dimension also in Trentino. From Stöllner (2003b)

the Bronze Age and early Iron Age apparently played a major role in the supply of metals in Central Europe. In this epoch, the Alpine economy changed: mining and metallurgy transformed parts of the originally agricultural landscape to mining regions with "early industrial" structures.

At many of these smelting sites, smelting furnaces and roasting plants were excavated. In Lower Austria, 202 smelting sites were found (Klemm 2015), in the district of Mitterberg about 200 (Stöllner 2003a, b; Fig. 5.4), at the Jochberg near Kitzbühl approximately 40 (Goldenberg 2014a, b), in the Swiss Grisons 80 smelting sites (Reitmaier-Naef 2019), and in Trentino (northern Italy) more than 200 smelting sites (Cierny 2008; Silvestri et al. 2019). In Trentino, region-specific metallurgical techniques seem to have developed, possibly due to deposits, namely the massive occurrence of plate slags and the almost consistent processing of larger slags to crushed slags using water (see Sect. 5.4). From the late Early Bronze Age, and especially during the Middle Bronze Age, the Mitterberg district established a pre-eminence as a copper producer (Stöllner 2013). An estimated 20,000 tons of copper were produced here, mainly from chalcopyrite (the "East-Alpine copper" of the SAM project).

In the Inn Valley only a few smelting sites were found (Goldenberg 2014a, b). This is surprising at first, because in the Inn valley alone, in the area between Schwaz and Brixlegg, at least 25 prehistoric mines were discovered (Krismer et al. 2011b, c). This comparatively low number of smelting sites may be due to the fact that old smelting sites are covered and destroyed by modern mining. The mines are located in a steep terrain on the southern edge of the shallow Inn Valley. In the case that the smelting of the ore was carried out in the vicinity of the mines, a burial of slag heaps due to mudflows, debris pockets and ponding is obvious (oral comm. G. Goldenberg 9/2016).

Anatolia

Between 1983 and 1986, the Heidelberg research team of Günther A. Wagner, Ernst Pernicka, Friedrich Begemann and co-workers discovered almost 200 smelting sites in Anatolia and the Aegean, which were very intensively and consistently analysed in the field as well as in the laboratory (Pernicka et al. 1984; Seeliger et al. 1985; Wagner et al. 1986, 1989; Pernicka 1987; Wagner and Öztunali 2000). Among other things, the largest database of lead isotope ratios of slags and ores of Anatolia was produced here. At many of these sites old mining was found as well. In Anatolia, 91 of the deposits were mined for copper, around 50 of which show prehistoric activity. The deposits themselves are very often polymetallic and contain copper–lead–zinc–silver ores. Obviously, it is not always clear whether copper or lead–silver has been mined. Of the lead–silver deposits, Gümüşhane (northern Anatolia) is the most important lead–silver deposit of the eastern Pontids. In Gümüş in the Bolkardağ area in southeast Anatolia, slag heaps with a tonnage of c. 100,000 tons and hundreds of mining adits were observed. The galena exploited here has a silver content of up to 0.4 wt.%. The mining activities were dated back to the third millennium BC, while the slag heaps, in contrast, provides a dating approximately to c. 1000 AD. The Bolkardağ area has been repeatedly discussed in archaeological literature as an ancient source of silver and gold (Moorey 1994).

Faynan (Jordan)

The case study of Faynan (Jordan) can show how the organisation pattern of copper extraction has changed from its beginnings in the fourth millennium to mass production in the Iron Age and later in Roman times (Hauptmann and Löffler 2013). In the mining district, regionally limited to 20 × 25 km, with the neighbouring smelting sites of Khirbat en-Nahas and Wadi el-Jariye, 24 slag heaps have been found which date from the Early Bronze Age (late fourth millennium BC) to Early Islamic Period (Hauptmann 2007). Based on the packing factor of 0.8, and a mean density of 3.5 g/cm^3 as proposed by Bachmann (1982b), we calculated a tonnage of slags of 150,000—200,000 tons. This calculation ranges from a few kilograms of slag, produced on a small scale in the Early Bronze Age settlement of Wadi Fidan 4, to the four Iron Age heaps with c. 100,000–150,000 tons (Khirbat en-Nahas, Khirbat el-Jariye, Khirbat al-Ghuwayba, Faynan 5, c. 1100–800 BC). Khirbat en-Nahas with c. 50,000–60,000 tons is the largest smelting site in the Faynan district. In contrast to the Roman–Byzantine slag heaps of Skouriotissa, abundant tuyère nozzles and coarse-grained fragments of smelting furnaces were found on the Iron Age heaps in the Faynan area.

In Faynan there are about 14 (scattered) slag heaps from the Early Bronze Age II/III (middle of the third millennium BC), which total nearly 5000 tons (Fig. 5.5). They are located in the immediate vicinity of the mineralisation and the Early Bronze Age mines. This is currently the largest amount of slag that has ever been found for this period. Similarly, large slag heaps of similar age are found only in Wadi Ba′Ba on the west coast of the Sinai Peninsula. Dating to the Early or Middle Bronze Age is likely, but not certain (Abdel-Motelib et al. 2012). In Faynan, the first melting furnaces which were operated by natural wind were found in the Early Bronze Age II/III (Hauptmann 2007). Khirbat en-Nahas, Faynan (Fig. 5.6). Aerial view of the huge copper smelting site. Most of the copper slag was produced in the Iron Age (tenth/ninth century BC) (Levy et al. 2014).

5.2.3.1 Oman

In Oman, many surface-level slag heaps supra-regionally date to the centuries of the Mediaeval period and the Early Modern period. These often cover older activities, which only reached smaller proportions. In the surveys conducted by Weisgerber and his team in the late 1980s in the Oman Mountain Ranges, a large proportion of the 117 copper smelting sites with a total of around 300,000 tons of slag date to the Early Islamic period and later (Hauptmann 1985). The slag heaps at the copper ore deposits of Lasail and Semdah in the north of the country alone comprised an estimated 100,000–150,000 or 80,000 tons, respectively (Hauptmann 1985; Weisgerber 1978; see Fig. 3.60a in Sect. 3.5.1).

Fig. 5.5 Faynan 9, Faynan district (Jordan). Slag heap from Early Bronze Age II/III copper smelting. This slag heap is one of 14 smelting sites from this period which are the earliest known in the Near East. Photo: G. Weisgerber (†), Deutsches Bergbau-Museum Bochum

The heaps were still visible some years ago before they were destroyed by modern mining activities.

These voluminous slag heaps often covered previous activities and they were also evidence that slag heaps have been recycled already in ancient times, e.g. Iron Age slag heaps were recycled not much later. Remnants of Bronze Age slag heaps were detected in comparatively few cases. Slags of the Bronze Age copper mining, which are proved so impressively by the finds of large quantities of copper ingots on the one hand (Weisgerber 1980), on the other hand by textual evidence from Mesopotamia (Reiter 1997), could only be found by archaeological excavations.

There are other examples of impressive slag heaps showing how intensively and consistently geo-resources have been used in ancient times to produce metal. This includes ore deposits and smaller occurrences that never had the dimensions of "Giants". This is proof that in ancient times ores from many smaller deposits were exploited, which are hardly mentioned anymore in modern geology.

Lead–Silver Slags

As for slag heaps from early lead–silver metallurgy, there are almost exclusively examples from the High and Late Middle Ages in Europe, with only a few exceptions such as Corta Lago in the province of Rio Tinto, Laurion, Kosovo, Gümüshan (Anatolia) or some smaller Aegean sites (e.g., Hellenistic to Roman–Byzantine activities on the island of Thasos and Sifnos (Greece); Hauptmann et al. 1988). It will be explained later in Sect. 6.4 that the special conditions of smelting lead ores at low temperatures are a decisive reason for that.

One of the major mediaeval ore districts is the Harz mountains (Germany). Klappauf (2000), in 25 years of extensive surveying, mapped some 800 smelting sites of copper–lead–silver production

Fig. 5.6 Khirbat en-Nahas, Faynan district (Jordan). Aerial photograph of the settlement and smelting site of the "ruins of copper", as the name can be translated. The site is dated mainly to the Iron Age (tenth to ninth century BC). The quantity of slag produced at this site has been estimated at c. 50,000–60,000 tons. The mines from where the ore was transported to Khirbat en-Nahas are located in the surrounding wadis. Photo: courtesy T.E. Levy, University of California San Diego

in the western Harz district (Fig. 5.7). The number of smelting sites for the entire Harz area was estimated to be 2500 sites. One hundred and fifty of the sites in the western Harz were securely dated by historical methods, pottery and radiocarbon dates mainly to the twelfth century AD.

Serbia and Kosovo

Close to the present territory of Serbia and Kosovo lies the former Roman province of Moesia. It extended over several hundred kilometres in a west-easterly direction south of the Lower Danube.

In Moesia Superior, the British Museum carried out a survey in the area of Mount Kosmaj south of Belgrade in Moesia Superior (Serbia) in the 1980s (Merkel 2007) to investigate the early lead–silver mining in this resource-rich country. The focus was particularly on the Roman lead–silver mining, which was already reported by Davies (1935). The author accepted an earlier estimate from 1892 which "reckoned" some 1,000,000 tons of total slag in the region around Stojnik, Guberevac and Babe and some 5000 mining shafts as a basis for Imperial Roman silver production. The economic development of the Moesia Superior for the exploitation of lead and silver is related to military occupation south of the Roman frontier line of legionary forts (*Limes*) along the Danube (Dušanić 1977).

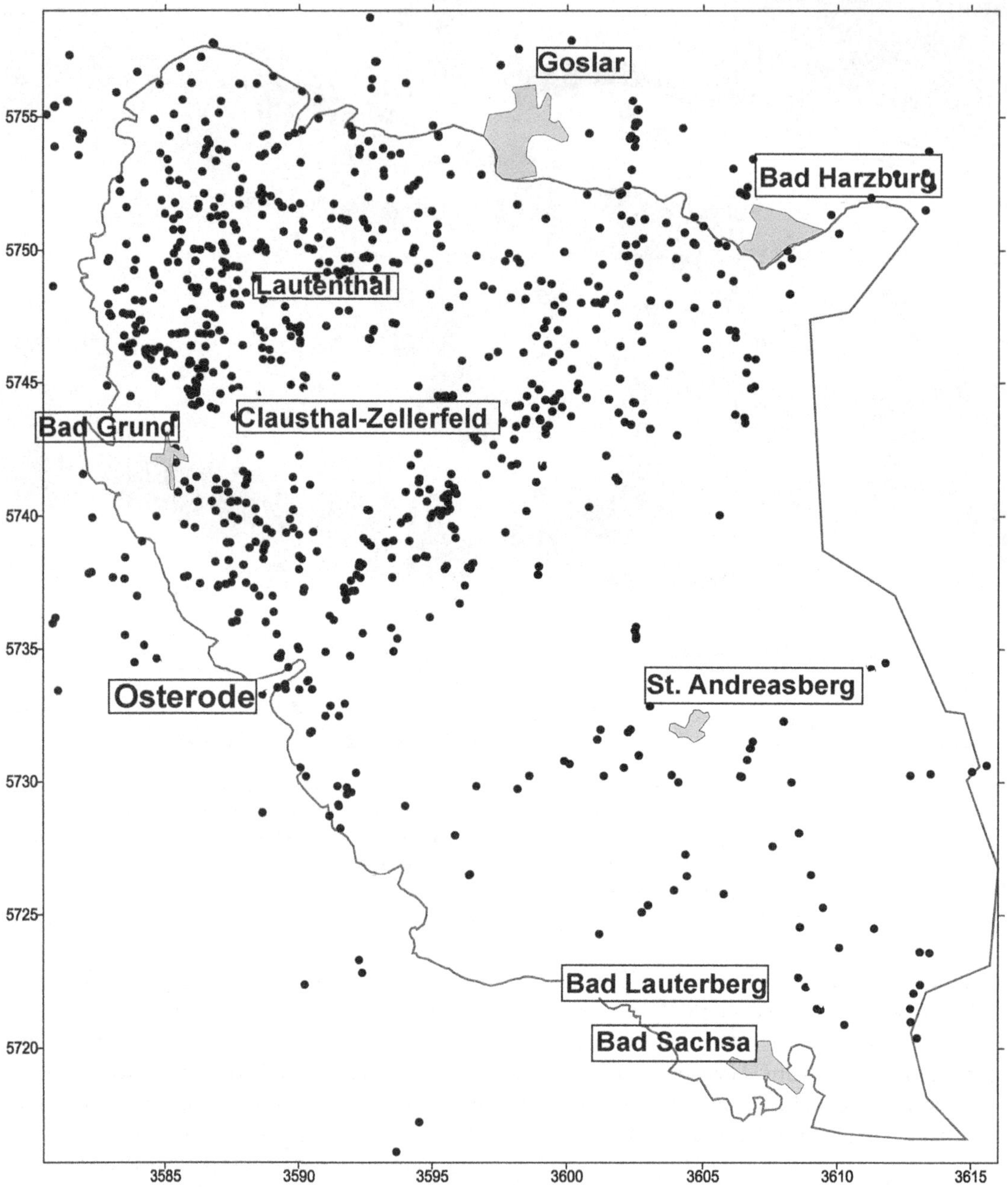

Fig. 5.7 Harz district (Germany). Location of some 800 mediaeval smelting sites in the western Harz district with the locations of some of the major modern cities. Note the city of Goslar, where the large copper–lead–zinc–silver deposit of Rammelsberg is located. The twelfth century AD lead–silver production site of Huneberg near Bad Harzburg was excavated by Klappauf (2000). Archaeometallurgical finds from the Huneberg were analytically studied by Asmus (2012). The eastern part of the Harz district was not mapped due to the separation of Germany until 1990. Modified map from Klappauf (2000)

Further metallurgical aspects of silver production in Moesia Superior concerning ore grade and smelting efficiency have been based upon modern geological reports and reworking of the ancient slag. However, the Roman (and mediaeval?) slags from near Mt. Kosmaj have been reprocessed at various times in the twentieth century AD to recover lead and silver. Therefore, possibly large parts of the estimated total have been removed. The full extent of modern reworking is uncertain, and therefore, the original amount of Roman slag will be uncertain as well. As a consequence, the mining districts centred at Rio Tinto (Spain) must still represent the highest Roman production levels for silver in Middle Europe.

The archaeometallurgical finds of the excavations at Stojnik (slags, litharge, metal remains) have not yet been investigated. Merkel (2007) quite rightly emphasises that analytical details are of great interest in making comparisons with other early lead–silver manufacturing facilities, e.g. Laurion or Corta Lago/Rio Tinto.

In an interdisciplinary research project between the Deutsches Bergbau-Museum and the Archaeological Service in Kosovo, archaeological and archaeometallurgical research is being carried out intensively (Gassmann et al. 2015; Körlin and Gassmann 2016; Westner et al. 2016). In particular, the early silver production in the Zhegovc Mountains in the vicinity of the Roman city of Ulpiana in Kosovo, near the town of Pristina is being explored. The area is also part of the Roman province of Moesia Superior. Ulpiana is considered as an administrative centre, if not as a kind of capital. In fact, it is believed that Ulpiana may have been the central location of an ore district (last in Petrović 2007).

In this area, in the mountains in the hinterland of the Roman city of Ulpiana, are numerous polymetallic ore deposits with arsenical lead–silver–copper ores. The remains of early mining in this region are overwhelming. In an area of 30×30 km^2 hundreds of waste dumps have been mapped, which often are located close to each other and form real "lunar landscapes".

In contrast, the slag heaps that can still be found today, serving as evidence for the smelting of ores and a silver production, do not reach such dramatic proportions. So far, about 30 slag heaps have been found in the study area, most of which are mediaeval-modern, and only a few could be dated to Roman times. Not all heaps reach a thickness of one or several metres, as is the case with Bostan. There are clearly visible profiles in the modern villages of Gadish Kishnapole, Mirash Novo, Cernice and Ponesh. The tonnages are very difficult to estimate, as the slags were partially worn down to re-smelt them again. However, the amounts in the district tend to be in the range of hundreds of thousands of tons, the thickness of the heaps can reach several metres (oral comm. G. Gassmann 1/2018).

The reason for the rather restrained evidence of Roman metallurgy in Kosovo is therefore due to the extensive recycling of the slags. This is obviously a common problem that needs to be considered in any archaeometallurgical research, in excavations and in surveys. Reasons for recycling are recent economic interests in possibly high levels of silver primarily in early lead–silver slags.

Recycling of Lead–Silver Slags: Laurion

Other, originally much more voluminous slag heaps of lead–silver mining from ancient times, apparently no longer exist today. One of the most impressive examples of radical, modern recycling of old slags is Laurion/Attica (Greece) (Conophagos 1980). In the nineteenth century, vast amounts of antique lead–silver slags were removed and re-melted down here, leading to the nearly total destruction of the smelting installations and the most important evidence of early silver extraction (Weisgerber and Heinrich 1983). The same is true for the slag heaps of the early lead–silver mining in Iglesias/Sardinia, which have been almost completely reprocessed (Hanel and Morstadt 2014), making it extremely difficult to reconstruct early metal extraction.

The mining area of Laurion with its vast number of ancient mines, with its many processing plants, with its settlements and graves, with its road network and sanctuaries belongs to the most important mining-historical monuments of the Old World (cf. Sect. 3.4.9.2 and Lohmann 2005). At Laurion, an early estimate of the total quantity of slag was 1,550,000 tons, but estimates for production by Conophagos (1980) and Bachmann (1982b) were 1,400,000 tons of lead with 3,500 tons silver produced in the Classical Period in Greece. Davies (1935) reported a high estimate of some 2,000,000 tons of ancient slag. The Greek historian and geographer Strabo (9,1,123) already reports recycling of silver-rich slags in the Hellenistic and then Augustinian periods, that is, in the first century BC.

5.2.3.2 Bloomery Slags

Due to the globally extremely widespread deposits and occurrences of bog iron ores (see Sect. 3.4.12), there are also numerous piles of iron slags at countless locations throughout the world, but they hardly reach a greater thickness. There are no known heaps that reach dimensions equal to those in Corta Lago or Cyprus. A detailed summary can be found at Buchwald (2005) and Pleiner (2000).

Of all the countless slag heaps, only a few examples are mentioned here. Extensive and radical exploitation of surface-near bog iron ores were found all over the lowlands of Denmark, Scandinavia, northern Germany and Poland during the Roman period (c. first century BC to fourth century AD) where hundreds and hundreds of smelting sites are described by Bielenin and Woyda (1978), Jöns (1997), Voss and Smekalova (2003) (Figs. 5.8 and 5.9). Overviews are discussed by Pleiner (2000) and Buchwald (2005).

This is followed by slag heaps in the mining areas in the Siegerland (Stöllner 2003b), in the Lahn-Dill area (Jockenhövel and Willms 2005; in both ore districts also base-metal slags can be found), Noricum (Austria, Straube 1996), Morvan (Bourgogne, Mangin et al. 1992), Sweden, Poland (compiled by Pleiner 2000) and then in many localities and regions in the Near East, Central and East Asia, Africa and the Indian subcontinent (Humphries and Rehren 2013).

Many of these smelting sites have been investigated by prospecting methods and subsequent excavations. Systematic annual excavations at the Danish Iron Age production site at Snorup, Jutland, revealed more than 5000 slag blocks at one production site (Voss and Smekalova 2003). At other smelting sites, dozens or even hundreds of slag blocks were found. One slag block represented one production unit. It resulted from a bloomery process performed in a shaft furnace above the pit in which the bulk slag was collected. The large slag blocks weigh between c. 30 and 200 kg, even up to 500 kg, and have often been compared to the shape and size of an elephant's foot. Each slag block obviously represents a furnace run of several days. When a pit was filled up a new pit was dug and—perhaps—if possible the clay shaft of the furnace was moved and reused over a new pit.

Jöns (1997) and Haffner et al. (2000) published two volumes on the early iron production at Joldelund in the northern part of Germany. They led an interdisciplinary team of researchers and worked on the natural environment of this area rich in iron ore. They used physical prospecting methods to detect the slag heaps or slags produced in this Roman period (Fig. 5.9) and carried out extensive archaeological excavations. From these works, a very interesting distribution map was created, illustrating the extensive deposits of bog iron ore which were mined and smelted in the plains extending from Holland, Germany, Denmark to Poland and southern Scandinavia (Fig. 3.53).

The geographical extent of early iron production in the Holy Cross Mountains in Poland is overwhelming. Organised slag-pit furnace fields are characteristic of this region and reached their summit in the second to third century AD. Two hundred and thirty-one slag sites were found there, averaging about 90–100 slag blocks each (Figs. 5.8 and 5.9). Such iron workings would have produced 1.5–1.8 tons of iron from 18 tons of ore and 20 tons of charcoal. Altogether more than 4000 such slag blocks were found throughout the area of the Vistula curve (Bielenin and Woyda 1978). Orzechowski (1994) assumed that very skilled groups of smelters would have been working during this time period, presumably with trading links to the region of the Roman frontier to the south.

There is another example of intensive exploitation of iron ores in the mediaeval period. The

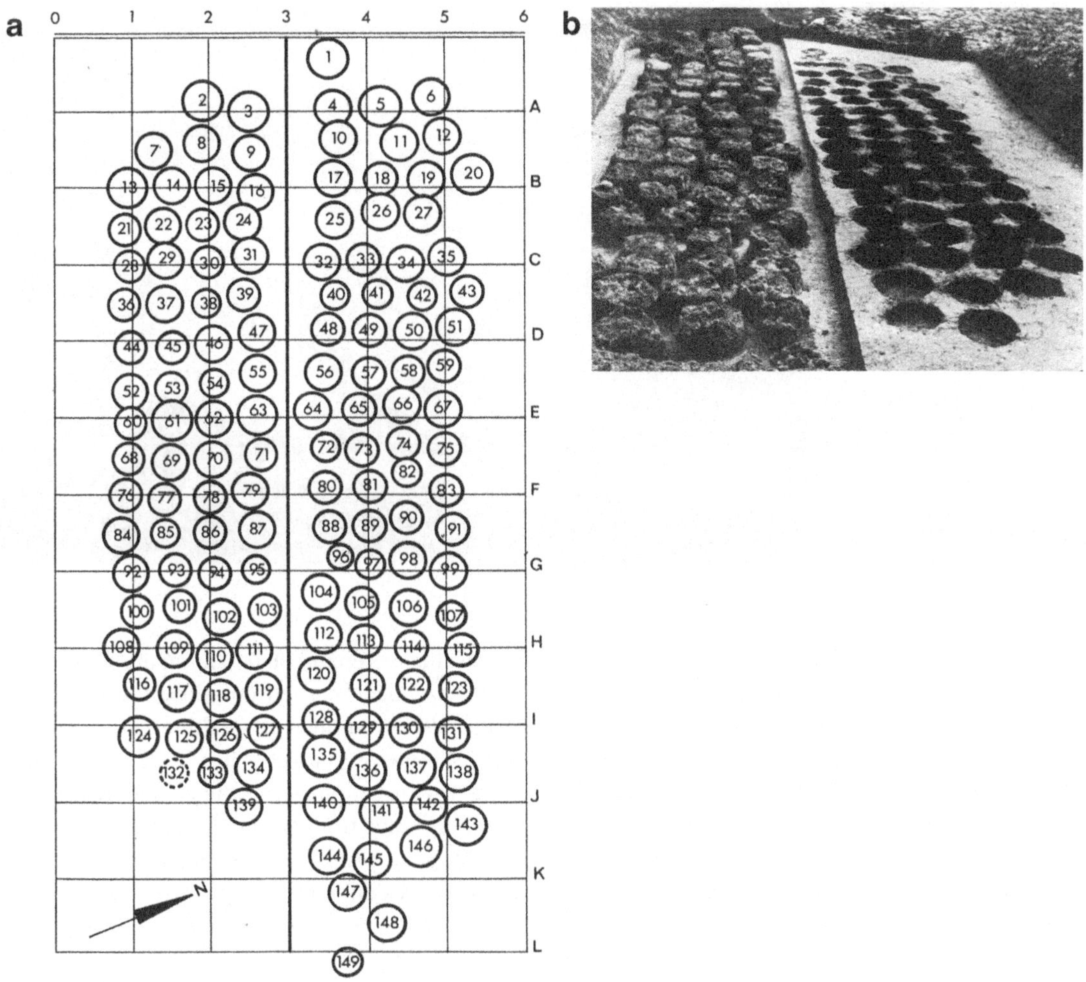

Fig. 5.8 Słupia Stara, Opatów, locus 3 (Holy Cross Mountains, Poland). Slag-pit furnace fields of late Roman iron production. One hundred and forty-nine positions of slag-pit furnaces were excavated. (**a**) excavated slag blocks, (**b**) slag pits where slag blocks were extracted. Photograph courtesy of K. Bielenin (†)

submarine volcanic red ironstones (hematite, Fe_2O_3) of the "Lahn-Dill"-type as well as the siderite veins ($FeCO_3$) of the Siegerland in middle Germany were intensively used in the Middle Ages. Three hundred slag heaps with a calculated tonnage of 7500 tons and sites related to early modern iron and base metal production were discovered (Kronz 1997). The author investigated the formation of slags from this area under exemplary mineralogical–petrological aspects and was thus able to extend the understanding of fayalitic melts from previous investigations by Schürmann (1958), Osann (1971) and Knau (1952). Later, together with Keesmann (Kronz and Keesmann 2005), he published an informative overview of the production of ferrous and non-ferrous metals in the work published by Jockenhövel and Willms (2005) on the Dietzhölze-Valley in the Lahn Dill area in Germany.

5.3 The Investigation of Archaeometallurgical Slags

The modern scientific investigation of slag is a relatively young field of research in

Fig. 5.9 Joldelund (northern Germany). Slag block of late Roman iron production using bog iron as characteristic for the Lowlands of Denmark, northern Germany and Poland. The slag block was formed in a hole below the shaft furnace construction to collect the slag below the bloom. The picture shows the flows of the slag into the pit and the plane above where the bloom was accumulated. The weight of this slag block is c. 26 kg. Photo: G. Weisgerber (†), Deutsches Bergbau-Museum Bochum

archaeometallurgy. It developed alongside the "New Archaeology" in the late 1960s, when archaeologists recognised the potential of scientific methods as an integral part of their research strategy. It is bound to the perception that the production of metal and the distribution of metals was the basis of economic wealth in many regions.

It was also in the time period that Beno Rothenberg (1972a, 1990a) started pioneering work on ancient copper mines and slag heaps at the famous site of Timna at the western margin of the Wadi Arabah (Israel). He recognised the potential of the industrial remains and organised interdisciplinary teams of scientists to investigate the archaeological findings. Lupu and Rothenberg (1970), Tylecote (1976), and Bachmann (1978, 1982a, b, c) modelled the first hypotheses on early copper smelting and melting for this locality. Bachmann analysed mineralogical phase content and chemical compositions of slags, calculated their viscosity and identified metal inclusions to reconstruct the technological parameters of ancient smelting processes.

"Slag—indicators of archaeometallurgical processes" was the title of a synopsis concerning the study of ancient slag by Bachmann (1978). It was the summary and continuation of work which had been carried out since the beginning of the 1970s on ancient smelting slags: Morton and Wingrove (1969a, b, 1972) published some investigations on Roman and Mediaeval bloomery slags from England based upon modern metallurgical principles. They analysed mineralogical phases and chemical compositions of such slags and discussed their composition in phase diagrams. Milton et al. (1976) and his team analysed a few slag samples from Timna (Israel) using the same principles and analytical methods.

The emphasis of Bachmann's work followed this concept. In accordance with modern metallurgical techniques, he suggested to determine the parameters for slag basicity and viscosity, and to estimate melting ranges via the use of phase diagrams and thus assess the efficiency of the initial charging and the amount of control over the smelting process. Bachmann presented a calculation programme for the determination of this data and, in a second publication (Bachmann 1978), added another programme to calculate the viscosity. His programmes were based on copper slags produced during an advanced state of metallurgy. The investigated samples came from the western margin of the Wadi Arabah and from various sites on the Sinai Peninsula. They dated for the most part from the Late Bronze Age (fourteenth to eleventh centuries BC) and even later periods. He interpreted older material, which had been dated preliminarily into the Bronze Age or Chalcolithic period (Bachmann 1978; Rothenberg 1973), very cautiously and assigned it to a "trial-and-error" phase because the composition of the material deviated considerably from the bulk composition of younger slag.

Intensive mineralogical–petrographic studies by Kronz and Keesmann (2005) have broadened our knowledge of the mineralogy and petrology of archaeometallurgical slag, best exemplified by an investigation of mediaeval iron production in the Lahn-Dill-area in Germany.

In the early stage of slag research, scholars mainly collected conspicuous examples of tap slag (i.e. slag that had been "tapped" from the furnace during or at the end of a smelting process) from larger slag heaps of later periods, i.e. from the Late Bronze Age and later periods of more advanced technology. These tap slags originated mainly from the production of copper and iron, rarely from lead–silver. Often it happened that the archaeological context was not fully included in these investigations. The occurrence of the almost omnipresent phenomenon of crushed slags to extract entrapped prills of metal or sulphide for re-melting was often neglected in interpretations. Metallurgical remains that were produced in the early periods of metallurgy, e.g. slags from early tin and bronze production, are extremely rare in the archaeological record, as are remains from gold and silver processing.

5.3.1 Questions and Strategies

The scientific investigation of archaeometallurgical slag involves asking questions about technological aspects, about the ore charged and its provenance. These are the basis for reconstructing economic and craft-related systems in the past.

It can be summarised that scientific analyses of slag enable us to understand the following aspects:

- Which metal or alloy was produced by a metallurgical operation that led to the formation of a specific slag?
- What was the nature of the ore that was originally smelted? Was it oxidic ore or sulphur-containing (sulphidic) ore? What was the percentage of host rock and gangue intergrown with the ore? Can we find evidence for a deliberate addition of fluxes, or have self-fluxing ores been smelted? Mineralogical and chemical analyses may provide information on the nature of ores. Despite the fractionation of elements during the change of ore to metal by slag formation, perhaps through the addition of fluxing agents and the actual removal of the metal itself, it is feasible to obtain information about the inclusions of metal or sulphides left in the slag (Hauptmann 2007). We possibly can identify slags from smelting processes versus those from melting or refining metals.
- Principles of early extractive metallurgy, alloying and treatment of metals, meaning the physico-chemical or thermodynamic processes (firing temperatures, duration of the process, gas atmosphere controlled by the shape and design of a furnace and air supply). But the present stage of research is not sufficient to reconstruct skilled manual work of the artisans' craft in detail. The wide variation of technical processes known to be used in traditional iron smelting in Africa gives a very good example of the diversity of possible techniques (Schmidt 1996; Childs and Herbert 2005). What were the firing conditions that led to the formation of slags?
- But a meticulous analysis of all the facts definitely allows us to outline ancient technical processes and that should certainly be the aim of all archaeometallurgical research. Espelund's (1991) statement: "we know the principles but not the practice, the ancients knew the practice

but not the principle" is perhaps a little bit too pessimistic about today's analytical abilities.

- Main, minor and trace elements of the silicate slag matrix can all give information about the composition and nature of the ore and its gangues. This is of particular interest when smelting was not carried out in the immediate vicinity of the ore deposit, but when the ore had been transported over larger distances away from the mines. This problem is especially important for the initial stages of metallurgy in the fourth and third millennia BC, not only in the Near East but possibly throughout the world—when smelting of ores was normally carried out inside settlements.
- Slag not only enables us to reconstruct ancient smelting processes, over the course of provenance studies they very advantageously offer the possibility of studying raw material sources. When no archaeological ores are available to investigate the chemical and the Pb-isotopic fingerprint, slags could also be good candidates for metal provenance studies (Baron et al. 2014).

All of these investigations and aspects of scientific slag investigation provide valuable information to decipher a suite of technological practices of ancient metalworkers and the amount of metal produced in certain time periods. However, in order to reconstruct manual details, and to understand organisational patterns at least in part, social complexities of metal making, slag investigations in addition need to definitely include archaeological and geological sources, ethnographic observations and experimental studies.

We can learn a lot from written sources such as Theophilus Presbyter (twelfth century AD), Georg Agricola (1556) and Vannoccio Biringuccio (1540), who described in detail the complexity of smelting and melting techniques used between the twelfth and sixteenth centuries AD. And Georg Agricola (1556), in his work on mining and metallurgy, describes 22 different recipes for performing a single metallurgical process, namely the liquation process, i.e. the addition of lead to silver-bearing copper, to extract the silver.

John Percy (1861/1880) published traditional techniques still used by metallurgists in the nineteenth century. There are many other sources like that. Of special importance are ethnographic studies, such as in Africa where many such studies on iron and copper smelting and processing, and on bronze making were undertaken (Bisson 2000; Förster 1987). Further information about this topic is discussed in Chap. 9.

Last but not least, it is a great help to understand and to identify archaeometallurgical finds by comparing them with results from experimental metal production. One of the most interesting studies was undertaken by Hanning and Pils (2011), who replicated the Chalcolithic copper smelting exemplified by the finds from Zambujal on the Iberian Peninsula. She also practised smelting of chalcopyrite ore in a furnace according to the archaeological evidence of the Middle and Late Bronze Age Mitterberg (Austria) (Hanning in prep). This will be discussed in further detail in Chap. 10.

5.3.2 Sampling Slags and Analytical Approach

In the case of mining-archaeological or archaeometallurgically oriented fieldwork, old slags are often sampled with special consideration given to slag pieces with an exclusive appearance, which are then (isolated from the overall context) transported to the laboratory table. These are often fragments of tap slags with clear flow structures, which in their composition are in most cases and almost independent of their age ideal Fe silicate slags with the main component fayalite (Fe_2SiO_4). Due to their comparatively low solidification temperature and low viscosity, they are a very suitable type of slag that separates well from metal or sulphide contents during a melting process.

Not all types of slags and metallurgical remains mentioned in Table 5.1 are properly identified during archaeological fieldwork. Hence, representative sampling of archaeometallurgical remains often fragmentary. Scientific data from the analytical laboratory need to be integrated better into the archaeological researches.

Crushed slags, often processed already in ancient times to extract small metal inclusions, are often overlooked. However, they may form masses of heaps and dumps. Numerous striking examples have been excavated in Trentino (northern Italy) where they were dated to the Late Bronze Age (Cierny 2008; Silvestri et al. 2019). Bachmann and Hauptmann (1984) described Iron Age heaps of crushed slag at Faynan (Jordan). In case crushed slags are only subordinated parts of slag heaps, they are usually mixed with sand and clay and interspersed between larger slag cakes. They are overlooked. The function of crushed slags will be explained in Sect. 5.4.10.

If possible, slags have to be collected from a secure stratigraphic–archaeological context, either from slag heaps located near ore deposits or near ancient mines, or, alternatively, from excavations of workshops in settlements containing slags and other metallurgical debris. If the archaeological context is not fully included in the study, slags are isolated finds, and reconstructions of technological developments or organisational patterns in metal production will be made more difficult.

5.3.2.1 Macroscopic Evaluation

The macroscopic analysis of the texture of slags, meaning the orientation and distribution of phases or other identical structural constituents (gas bubbles) in their spatial association, has rarely been carried out before and was underestimated. Already by naked eyes or light magnification, in a cm-scale, a simplified classification of slags is possible. The reason has been simply that the available finds had too large dimensions for thin sectioning or mounted sample preparation. Slag produced in small-scale operations, which often lead to inhomogeneous compositions, can however be easily studied for fabric analysis. In iron metallurgy, this is true for fist-sized, dome-shaped slag, produced throughout the Old World, as a by-product during iron processing (so-called plano-convex bottoms, PCB, or smithing slag, see Pleiner 2000).

Often slags were not completely liquified and may contain fragments of heat-resistant particles of gangue (quartz, feldspar, baryte) or coarse-grained pieces of host rock which did not, or only partly, participate in reactions during the smelting process. Also, inclusions of partly liquefied retour slags may occur, and inclusions of charcoal are important sources for radiocarbon-dating slag. Botanic investigations may contribute to estimations of environmental damages by uprooting. In Chalcolithic and Early Bronze Age copper metallurgy, hand-sized slag cakes are frequently to find. These are in size comparable to later smithing slags in iron metallurgy (see 5.4.12). Such slags usually result from smelting in small reaction vessels (e.g. in crucibles). They differ from the smelting slags of later periods in their rather high content of trapped metal, and they also contain many inclusions of un-decomposed fragments of high refractive gangue (Fig. 5.10a, b).

Inclusions (millimetre- to centimetre-sized) in siliceous slags of metal or sulphides are valuable indications for the sort of metal to be produced. In the Old World, there are occurring sometimes slags that show impressions of sulphidic cakes at the bottom side (Fig. 5.10b).

Following this approach, the slag should be subjected afterwards to a series of scientific analyses to identify bulk chemistry, specific mineralogical and metallurgical phases, and the composition of lead isotope analyses. These can be accomplished through a number of different analytical techniques. Possibilities of modern materials analytical methods are compiled, e.g. at Artioli and Angelini (2011).

5.3.2.2 Microscopic Scale

Due to rapid cooling archaeometallurgical slags are frozen in a glassy state or consist more general of a fine-grained mixture of different phases. They have a complex composition, depending on the bulk chemistry, and depending on the redox conditions, because they are usually rich

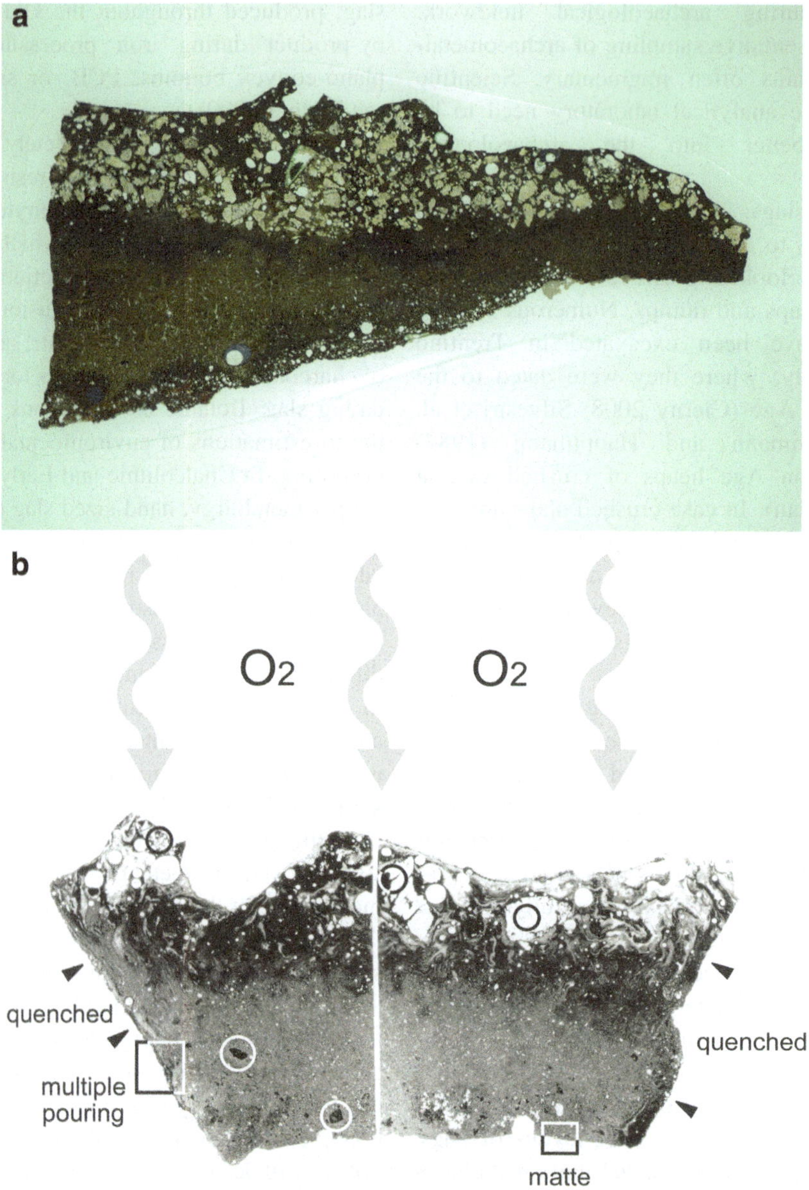

Fig. 5.10 (**a**) Tepe Hissar, Iran, Early Bronze Age. Section through a copper smelting slag (half of a slag cake with a diametre of ca. 17 cm). The main component of the lower dark part that was fully liquefied is fayalite, forming a typical bulky spinifex texture. The slag was probably tapped from a furnace into a fore-hearth, where reactions between slag and the clay bottom of the hearth formed a rim of gas bubbles. The upper part of the slag is covered by a layer of light inclusions of quartz and gangue of the ore. The texture looks like a breccia. The ore obviously was crushed to a grain size of a few mm. The dark and the light layers of the slag were welded together. The contact zone is rich in magnetite and suggests a sudden presence of air caused by a hiatus during tapping. From Hauptmann (2007). (**b**) Shahr-i Sokhta (Iran), Early Bronze Age (mid third millennium BC). Section of a round slag cake from composed of two thin sections. It shows a horizontal division into two parts. The upper part is rich in white inclusions of gangue/ore (O) and gas bubbles embedded in a dark matrix. Here, the slag is enriched in magnetite and glass that formed in the presence of air during cooling. Note the streaks of glass. The lower parts are almost fully crystallised; it contains predominantly hedenbergite and reveals inclusions of ore (O). The side surface is quenched and indicates that the slag was poured into an irregularly formed cold ground. Multiple layering of magnetite may result from multiple pouring. The rim at the bottom shows the size of the original regulus of matte and copper that could not be fully separated from the slag, as remains of matte at the bottom show. The diametre of the upper size of the slag: 8 cm. From Hauptmann et al. (2003)

in iron. This will be discussed in Sect. 5.6. Therefore, optical microscopy under transmitted and reflected (polarising) light of surface polished or petrographic thin sections of the slags is an indispensable method, particularly for the determination of slag phases and texture analysis. Optical microscopy can be performed using different magnifications (40× to 500×).

While metal inclusions (or those that bear a concentration of metal, such as sulphide inclusions) in slags can be identified without major problems, it is often difficult to determine siliceous or oxidic mineralogical phases by optical microscopy alone, with the exception of main components such as fayalites, magnetite or wuestite and pyroxene (see Sects. 5.6 and 5.7). Sometimes, as was the case at Faynan (Jordan), or with lead–silver smelting slags from various localities in the Old World, the slags have even solidified in a glassy state, although such "glasses" show transitions from the cryptocrystalline up to a microcrystalline state formed in residual melts.

In these cases, X-ray-diffraction of a powdered slag sample can help to identify all phases. Scanning electron microscopy (SEM) and electron microprobe analysis (EMPA) have gained increasing importance in slag analysis, as they provide fast and reliable (quantitative) spot-analyses for silicate, oxide and metal phase identification for main, minor and trace elements. Such analyses, however, require a thorough mineralogical knowledge. SEM and EPMA were applied successfully, e.g. by Kronz and Keesmann (2005) to precisely characterise mineralogical phases in iron slags from the Dietzhölze area (Lahn district, Germany), and by Westner et al. (2016) investigating Roman and mediaeval slags from silver production in Kosovo (Balkan Peninsula) or by Ettler et al. (2009a, b) studying mediaeval silvers slags from the Czech Republic. The purpose of these investigations was to understand the formation of silicate melts to obtain information about the physico-chemical parameters of the smelting operation and the cooling history of the slags.

5.3.2.3 Chemical Analysis

It has proven advisable to carry out bulk chemical analyses, including trace elements, by conventional analytical methods such as X-ray fluorescence spectroscopy (XRF), optical emission spectroscopy coupled with a plasma source (ICP-OES). Alternatively, analyses can be performed using laser-ablation ICP-mass spectrometry (LA-ICP-MS) which also provides an analysis of isotope compositions.

A precondition for such analyses is a sufficient, representative quantity of material. This varies between a few grams and a small part of a kilogram depending on the original size of the sample. If the archaeological context allows for it, it proves prudent to analyse a whole series of samples to obtain good statistical results.

The interpretation of bulk chemical analysis of slags can be impaired by the heterogeneous composition of a slag. This is particularly the case when the slag still contains un-decomposed constituents of the original ore charge. These were often observed in copper smelting slags from the early stages of metallurgy, in Late Bronze Age copper smelting slags from the Alps, in those from Phoenician lead–silver slag from the south-west of the Iberian Peninsula (Fig. 5.28), and in slags from iron smelting.

In such cases, it does not make sense to chemically analyse pulverised bulk samples. It is more useful to analyse small areas of slags in the microscale. This can be done by energy dispersive X-ray analysis in a scanning electron microscope or with an electron microprobe with thin sections or mounted samples.

An alternative method is the use of portable X-ray fluorescence spectrometers (pXRF). The pXRF is based on an energy-dispersive system and is thus applicable for qualitative and semi-quantitative inorganic identification of major and minor elements. Spot analyses of 3 or 8 mm can be performed with this device. The unquestionable advantage of this method is that it can be used outside the laboratory in museums and in situ at archaeological excavations. It can be used also in the identification of eye-size inclusions or heterogeneities of materials. pXRF was of decisive importance for measuring dozens of gold artefacts from the Early Dynastic III Royal tombs of Ur Mesopotamia (Hauptmann et al. 2018).

However, it has to be noted that having the low energy to penetrate materials the pXRF just enables to provide analyses of surfaces, and not from the interior parts. This reveals problems for proper interpretation if, e.g. metals are segregating or enriched during metallurgical processes.

As pointed out by Pearce (2018) many archaeologists and museum curators do not seem to understand that pXRF data do not reflect detailed bulk compositions. They cannot substitute for proper high-resolution analytical methods for statistically firm bulk or trace elements or even isotope analyses. Instead, they often are postulating non-destructive analytical work.

But if the information required do not only involve petrological aspects, the value of the information produced by the analysis is not lost. Hence, the criticism to use chemical bulk analysis for slag investigation as expressed by Keesmann and Hilgart (1992) is understandable in connection with their mineralogical study of non-homogeneous (smithing) slag. In these cases, it was very helpful to analyse single sections in the micro-scale. This method has the added advantage that the composition of inclusions and the surrounding liquified slag can be studied separately.

5.4 Types of Slags

This section discusses a classification of slag types by external characteristics. The outer typology of slags can already allow a first approach to understand ancient smelting processes and further processing of metals and give indications on furnace, hearth or crucible constructions. The variety of these constructions is high, but they are made up of recurring basic elements (Tylecote 1976, 1980; Pleiner 2000). This outer typology of slags can occur in (s)melting processes of different metals and may overlap: Tap slags occur, e.g. in iron production after the bloomery process, so they are referred to as "bloomery slags", but they are also found in copper, lead–silver or younger tin production. In iron metallurgy, a distinction is made between reheating slags of a bloom and smithing slags, which are produced during the further processing and shaping of iron and steel (Pleiner 2000).

The outer typology of slag finds may also help for a rough dating of slag heaps or certain layers of heaps, at least within a limited mining district.

5.4.1 The Oldest Slags

Copper is a metal that, very early on, was smelted from high-grade green oxidic copper ores. As a rule, these slags are rarely found close to ore deposits, but rather in workshop areas inside of settlements. Obviously following a tradition of trading patterns of Neolithic periods, ores could have been transported from a raw source to villages, often over distances of more than 100 km (Hauptmann et al. 1993; Hauptmann 2000). In the Near East, the transport of valuable ores from a source to a settlement parallels the tradition of moving exotic materials such as obsidian, silex, red, black and green pigments, tiny beads and tools made of native copper over supra-regional distances. Already in the eighth millennium, in the Pre-Pottery Neolithic, high-temperature processes were known. In long-lasting firings, limestone was burned into lime (CaO), lime plaster (i.e. a mixture of lime with sand, volcanic ashes or pozzolana, and water, with changing hydraulic properties), which was used to shape artistic figurines or to create plane surfaces in architecture. Kingery et al. (1988) investigated this with the example of Çayönü Tepesi in Southeast Anatolia and Hauptmann and Yalcin (2000) using the example of Asikli Höyük. Native copper items recovered in settlements in southern Anatolia were already formed by heat treatment at that time as well (Maddin et al. 1999; Yalçın and Pernicka 1999).

Considering that materials could already be consciously manipulated by high-temperature processes at this early stage, and given the very limited inventory, it is difficult to define clear criteria for intentional smelting processes over randomly heated materials, especially for those epochs in which the extractive metallurgy had not yet been established.

This is illustrated by the example of small slag fragments from the Late Neolithic-Chalcolithic settlement of Çatal Höyük in south-eastern Anatolia. During his excavations, Mellaart (1966, 1967) found green and blue pigments in graves. He also found a few millimetre-sized pieces of slag and about a dozen corroded green beads. The finds date back to around 6500 BC, about 1500 years earlier than other finds of copper slags. The earliest evidence of extractive metallurgy dates to the turn of the sixth to fifth millennium BC in the Balkans (Radivojević et al. 2010), and probably around this time in the Near East (Thornton 2014).

Neuninger et al. (1964) analysed these probable slags. They were inhomogeneous. The team identified cuprite (Cu_2O), tenorite (CuO), pseudomorphs of delafossite ($CuFeO_2$), magnetite, finely divided spheres of metallic copper, and porous copper, apparently precipitated by solid-state reduction. These phases were embedded in an Fe-containing siliceous, but consistently porous matrix. According to this phase inventory, which is very typical for slags, the authors described it as metallurgical copper slags. However, these investigations and their interpretations were long disputed. In a thorough study, Radivojević et al. (2017) confirmed a comparable phase inventory for further slags from the same locality. However, they are convinced that the high-temperature debris must be oxidic copper minerals of pigments that have formed by chance due to the action of fire. This interpretation is expedient and convincing, given the extremely low amount of slags and the fact that Mellaart apparently has found no evidence of metallurgical installations.

The collection of slags up to the time of the Chalcolithic remains sparse and is limited to single specimens. Pearce (2015), in a critical discussion of the diffusionism paradigm versus a multiple centres metallurgical development model in the Old World, reports that the earliest tiny slags which prove (perhaps only by chance) pyrometallurgical extraction of copper from ores, were found in the Vinča period settlement of Belovode (Serbia in the Balkans). They date from the beginning of the fifth millennium BC (Radivojević et al. 2010; Radivojević 2013). These slags reach only a size of 2 cm with a weight of only a few grams. Characteristically, like those of Çatal Höyük, they contain considerable amounts of cuprite, both in massive lots, shaped as dendrites and in eutectic intergrowth with metallic copper. These phases are crystallised from a melt, as are the frequent delafossite (Fig. 5.11), magnetite, and the siliceous phases leucite, clinopyroxene, fayalite and a glassy matrix (for chemical composition of these phases see Sect. 5.6). However, the inhomogeneous texture of the slag shreds is clearly reminiscent of "dross". Radivojević (2013) at least reports a slagged shard, which could have been a crucible fragment.

In south-eastern Iran, on the other hand, the first evidence for the smelting of copper ore, resulting in very small amounts of cm-sized slags, comes from the first half of the fifth millennium BC Tal-i Iblis (Dougherty and Caldwell 1966; Thornton 2009a). In addition, a numerous very old slags were intensively studied from eastern Anatolia and from the southern Levant (Çukur and Kunç 1989; Hauptmann et al. 1993 for Nevali Çori; Lutz et al. 1994 for Murgul; Shugar 2000 for Abu Matar). At Brixlegg (Eastern Alps, Austria), slags from fahlore smelting were discovered in Neolithic layers dating to the middle of the fifth millennium BC (Bartelheim et al. 2002; Höppner et al. 2005). A summary can be found in Bourgarit (2007).

The archaeological findings of the earliest smelting of copper ore are also extremely rare elsewhere. Perhaps the expectation of finding crucibles associated with metallurgical processes in the sense of smelting may not be right. Experimental archaeometallurgy (see Chap. 10) has made a decisive contribution to understanding the earliest, and probably the simplest, smelting processes. Experiments have shown that even without a crucible, metallic copper can be produced from ore in a charcoal bed in simple depressions in the soil (Fasnacht 1999; Timberlake 2018). Timberlake produced a small slag cake about 10 cm in diametre, weighing 250 g, interspersed

Fig. 5.11 Belovode, eastern Serbia (Vinča settlement), sample 21. Copper slag from the earliest smelting activities, according to cal. ^{14}C-data 7000 years ago. Lathes of delafossite and fine dendrites of cuprite in a dark, glassy matrix in the left part. Massive parts of cuprite and magnetite (mid grey) on the right side. Scanning electron microscope, backscattered mode. Photo courtesy by Radivojević et al. (2010).

with millimetre-sized inclusions of metal. It consisted of CaMg-clinopyroxene and CaMgFe-olivine, delafossite and glass. The copper droplets were recovered after mechanical treatment of the slag and re-melted in a crucible to produce larger pieces of metal.

If any finds of slags are made, then it is not surprising, that in addition to the already described phase structure and the texture of the generally only partially liquefied slags (see below at dross), a large fraction of the volume of these is almost always interspersed with relics of the hostrock, and especially with refractory inclusions (quartz, feldspar, chromite, barite, periclase (MgO)) (Hauptmann et al. 1993; Hauptmann 2007) (Fig. 4.11). Often, iron oxides (mostly magnetites) are found in differently shaped agglomerations as relics of original Fe-rich ore components in these slags. They stand out clearly from phases that are crystallised from a melt (clinopyroxene, melilithe, olivine, i.e., fayalite) and may also be reminiscent of hammer scales in iron working slags.

5.4.2 Crucible Slags

Crucible slags are those which have been produced in small ceramic reaction vessels, in crucibles, whether in open or closed crucibles (Tylecote 1976). Accordingly, the amount of charge is thus very limited in comparison to smelting slags from larger smelting furnaces. Only a few grams or kilograms of slag can be found, if any at all. Archaeological finds and findings on crucible slag can therefore be difficult. Above all, when addressing such crucible slags, care should be taken that completely preserved pieces are present instead of fragmented parts.

Informative publications on crucible slag can be found at Tylecote (1982a, b), Modarressi-Tehrani (2004), Rehren (2003), Bayley and Rehren (2007) and Furger (2018). Crucible processes, according to their descriptions, comprise a wide range of possible metallurgical operations:

1. Smelting from ore to metal, such as copper and tin ores
2. Simple melting of smaller metal pieces (metal prills, scrap metal) to get slightly larger amounts of metal for casting of artefacts
3. Refinement of raw metals by alloying (tin bronzes, brass, steel)
4. Refining and (crucible) assaying in the metallurgy of (noble) metals

The size of crucibles can range from a few cubic centimetres to several litres. Considering the range of operations to which crucibles are used, correspondingly occurring slags can be composed completely differently. Crucibles that were open at the top, heated from above, or even from their lower part by blowpipes or bellows are slagged on the inside. Many crucibles are also slagged on the outside as a reaction of ceramic material and wood or charcoal ash.

Tylecote (1982a, b) describes major components that contribute to the chemical composition of crucible slags. They are formed by a comparatively intensive reaction of ceramic walls with wood or charcoal ash, whereby CaO and alkalis (Na_2O, K_2O) are introduced. The ratio of SiO_2 + Al_2O_3/FeO + MnO—is high. In addition, all of the metals or alloys that are melted in the crucibles are involved in the formation of slag. These are copper, tin, zinc, lead, arsenic, silver and gold or their alloy components. Ferrous and siliceous components are derived from adherent slag residues in and on metals from the previous melting process. This is demonstrated on finds of two approximately cupcake-sized agglomerations of scrap metal and slagged crucible sherds from regionally independent localities, namely the Early Bronze Age settlement of Khirbat Hamra Ifdan in the Faynan district (Jordan) (Levy et al. 2002), and from the simultaneous settlement in Shahdad (East Iran) (Fig. 5.12).

Smelting Ore

The earliest crucibles were apparently initially used for the smelting of ores. They date back to

Fig. 5.12 Collection of copper prills and small pieces of scrap metal as found in Early Bronze Age metal workshops. These are evidence for re-melting small pieces of metal to obtain larger batches. By re-melting, adhering relics of slag could be removed and the metal was refined in this way. (**a**) The "cupcake" excavated at the Early Bronze Age metal workshop III–IV of Khirbet Hamra Ifdan (Faynan, Jordan) (Levy et al. 2002). (**b**) A similar "cupcake" found at the contemporaneous settlement of Shahdad (Eastern Iran), building D, room 10, late third millennium BC. Photo: A. Opel, Deutsches Bergbau-Museum Bochum

the early Chalcolithic. They were open and were used in larger fireplaces and heated from above by means of blowpipes. Craddock (1995) described in a general way that the so-called crucible slags are about nut-sized, tend to be lighter and more heterogeneous in comparison with later well-crystallised smelting slags. They have a porous texture with many more entrapped gas bubbles, and their analysis shows that they are usually formed under much more oxidising conditions. Craddock is convinced that crucible slags can be found everywhere in all possible settlements since the beginning of the metal ages, but due to difficulties of findings, he cannot prove this exactly. Sometimes, in archaeological literature, cinders and crucibles from the earliest epochs are readily cited (see Kassianidou and Knapp 2005; Giardino 2000), but if slags are missing then it is vaguely argued that pure ores were melted here.

As they were heated from above, slags formed in crucibles crystallised under slightly reducing conditions only. At copper smelting, in addition to the formation of metal, the formation of higher-order oxides such as magnetite, delafossite, cuprite and glass occurs regularly. Cu-oxides, due to their amounts, have almost become "guiding phases" for the earliest smelting processes (Bourgarit 2007).

At the Chalcolithic settlement of Zambujal (Portugal), a few nut-sized copper smelting slags, ceramic crucible fragments and metal objects indicate copper smelting and casting as early as the late fourth/beginning of the third millennium BC (Gauß 2015). The very heterogeneously composed slags were very similar to those from Belovode and contained magnetite, cuprospinel ($Cu^{+}Fe^{3+}{}_2O_4$), delafossite, occasional small crystals of fayalite, a glass matrix and inclusions of copper, CuFe sulphides and iron arsenides. In addition, numerous unmelted quartz inclusions have been observed. This means that ideal firing conditions have only partially been achieved here, i.e. that the ore was melted only under slightly reducing conditions and only for a comparatively short time. Craddock et al. (2007b) called this "On the Edge of Success".

According to their experimental work, Hanning et al. (2010) suggested that these would be crucible slags. On the basis of the findings, they carried out a series of smelting experiments in crucibles with a diametre of slightly more than 10 cm, which were filled with copper ores and heated in a hearth, in a charcoal bed, from above by means of blowpipes. As they used very pure ores for their experiments, hardly any slag was produced indeed.

Melting Prills, Scrap

The small amounts of crucible slags from sites of the earliest metal ages in the Alpine foreland (Pfyn and Bisamberg in Switzerland, Maggetti et al. 1991; Rehren 2004) also indicate a re-melting of copper, i.e. a re-melting of prills and small lumps of copper into larger units of metal-rich slags or of metal with adhering residues of slag. Ethnographic descriptions are known from recent Africa. From Katanga, Bisson (2000) describes how pieces of scrap metal of copper were co-melted in crucibles set into the lower part of a larger furnace.

In the Late Chalcolithic to Early Bronze Age I settlements of Wadi Fidan 4 (Faynan area; Hauptmann 2007), at Tall Magass (Khalil 2009) and Hujayrat al-Ghuzlan in the southern Wadi Arabah in Jordan (Eichmann et al. 2009) only a handful nut- to hand-sized slags were produced at each site (Hauptmann et al. 2009).

In contrast to slags from later periods, these early slags from the Levant and Anatolia usually have extremely high metal contents (up to above 50 wt.%, Hauptmann et al. 2003; Fig. 5.13). To recover such inclusions, the slags may have been systematically crushed. Fe-oxide contents are around 20–40 wt.%. They contradict the modern definition of slags. Fayalite typically occurs in smelting slags of later periods and is rare in crucible slags. These are rather rich in high melting, refractory inclusions. The rather difficult circumstances of the finds gave rise to the assumption of a "slagless metallurgy" (Craddock 1990).

Refining

Due to the meagre inventory, there are no archaeological finds and findings from the early stages of metallurgy that could prove a deliberate production of tin bronzes by mixing copper metal and tin.

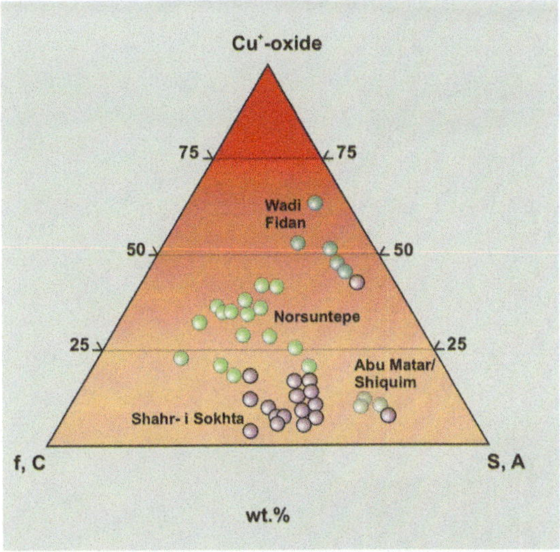

Fig. 5.13 Copper contents in slags from several localities in the Near East and from Shar-I Sokhta (Iran) dated to the Chalcolithic/Early Bronze Age periods, plotted in the ternary system Cu + Cu-oxide—f,C—S,A (reduces analyses). Note the enormous high copper losses in these slags, and the decrease from Chalcolithic samples to those of the developed Early Bronze Age. *f* Fe-oxide, *C* CaO, *S* SiO_2, *A* Al_2O_3. From Hauptmann et al. (2003)

Crucible slags from processing metals were first described in the Middle to Late Bronze Age (tin) bronze workshops of Nichoria (Cooke and Nielsen 1978), Isthmia (Rostoker et al. 1983), Olympia and Athens of the classical period (Zwicker 1985). In their work, Cooke and Nielsen (1978) equate the term crucible slags to melting slags. According to their observations, adherent remains of ceramic crucible sherds are the most convincing feature of crucible slags. The authors state that metallic copper with up to several wt.% of tin forms homogeneous solid solutions. On the other hand, when these are formed to oxides in the slags under weakly reducing conditions, these metals become largely separated. In a crucible, Cu-oxides (cuprite, delafossite) low in tin (about 2 wt.%) are crystallising, and Sn-oxides low in copper. Often, these slags are formed on the edge of a crucible and show transitions to ceramics. Such crucible slags are chemically composed of alkali-aluminium silicates with metal contents resulting from the reaction between metal/oxide melt with the crucible wall, charcoal ash and siliceous fluxes.

Refining, Assaying

In this section, we include the extraction of tin based on the finds and findings of the old mine of Kestel and the nearby settlement of Göltepe, because here we can find the most convincing comparisons with modern assaying. By (crucible) assaying we mean the analysis of metal contents in an ore in a very small scale. Bryan Earl experimentally showed tin melting in crucibles (Earl 1985, 1986; Earl and Özbal 1996; Earl in Yener et al. 2003), as it was practised as an assaying process in Cornwall until recently. The principle of this melting of small ore portions corresponds convincingly to what was found in Kestel and Göltepe. There are very few ancient findings for early tin metallurgy. All the more interesting are the observations described here, which are drastically different from the much better-researched processes of copper and iron metallurgy.

The Early Bronze Age Kestel mine and Göltepe has been researched by Aslihan Yener (Yener 2000, 2009; Yener and Vandiver 1993; Yener et al. 2003). Both date to the middle of the third millennium. The localities are located near Bolkardağ in the Taurus Mountains in Anatolia. The early production of tin is still a fundamental research desideratum, and so, after the discovery of tin ores and their extraction and processing into metal in Anatolia, there was fierce scientific discussion.

An estimated 5000 ground stone tools testify to an unprecedentedly intensive treatment of tin-containing hematite from the Kestel pit. Several hundred fragments of bowl-shaped crucibles (diametre 20–40 cm) show a smelting of the tin ore in many small portions. Almost no slag was found except for a few cubic millimetre-sized pieces. However, the inside of the crucibles was covered with thin layers of silicate slag interspersed with needles of cassiterite (SnO_2).

Such an extraction of tin in crucibles is underpinned by a very rare closed find from Askaraly II in Kazakhstan (Stöllner et al.

Fig. 5.14 Askaraly II (Kazachstan), sample KZ-424. Fragment of a tin crucible with whitish coloured slag on the inside. The fragment comes from the cultural stratum of the Mastan-Baj I settlement (Andronovo-Fedorovka culture), which dates back to the second half of the second millennium BC. Note the white crust of the crucible, which is reminiscent of maiolica. In the background cm-division. Photo: G. Wange, Deutsches Bergbau-Museum Bochum

2013a). Three ceramic fragments of crucibles were recovered here. They come from the cultural stratum of the settlement Mastau-Baj I (Andronovo-Fedorovka culture). Thus, they date to the second half of the second millennium BC. The inside of one of the fragments was covered by a rough whitish crust (Fig. 5.14). In this thin, white silicate slag crust of the crucible, the same cassiteritic needles were identified (Fig. 5.15) as were observed by Yener and Vandiver (1993) in the crucible fragments of Göltepe. Slags were not found.

As an example of crucible slags from silver refining, finds from the Roman fort of Xanten (Germany) should be mentioned. These indicate that a silversmith was working in this military settlement. The slags were formed in crucibles closed by a top. The predominant phases of these slags are lead-bearing silica glasses with (calcium-) lead silicates (~$(Ca,Pb)_2SiO_4$), melilite and kalsilite (see Sect. 5.6). In addition, cuprite was observed in these slags, which is the reason for the intense red colouration of these slags (Rehren and Hauptmann 1995; Fig. 5.16). Inclusions of silver-rich Cu-Ag-Pb alloys are present everywhere. Chemistry and phase content of the slags indicate the extraction of silver from polymetallic ores. Adhering remnants of ceramic linings allow an addressing as crucible slags. Unusually, however, is the amount of slag compared to the rather small volume of the crucible used for this process. In casting crucibles, usually only very small amounts of slag are produced. The authors suggest that used cupel material was re-melted, with the addition of sand, to recover the silver contents. The resulting Ag-Cu-regulus, with the addition of lead, was processed in a further step to pure silver (Fig. 5.16).

5.4.3 Slag Cakes, Matte Smelting

From the Early Bronze Age to the late Mediaeval period, there occurs a type of slag in the Old World that is closely associated with the production of mattes, i.e. a recovery of copper, lead and/or silver from (polymetallic) sulphidic ores through a previous concentration to sulphidic melts.

Slag cakes are formed, which show as a characteristic feature on its underside negative impressions of round reguli. Mm-thick crusts and drop-sized inclusions of various sulphides can sometimes be found. Depending on the size and weight of the slag cakes, and depending on their metal contents, these segregated matte

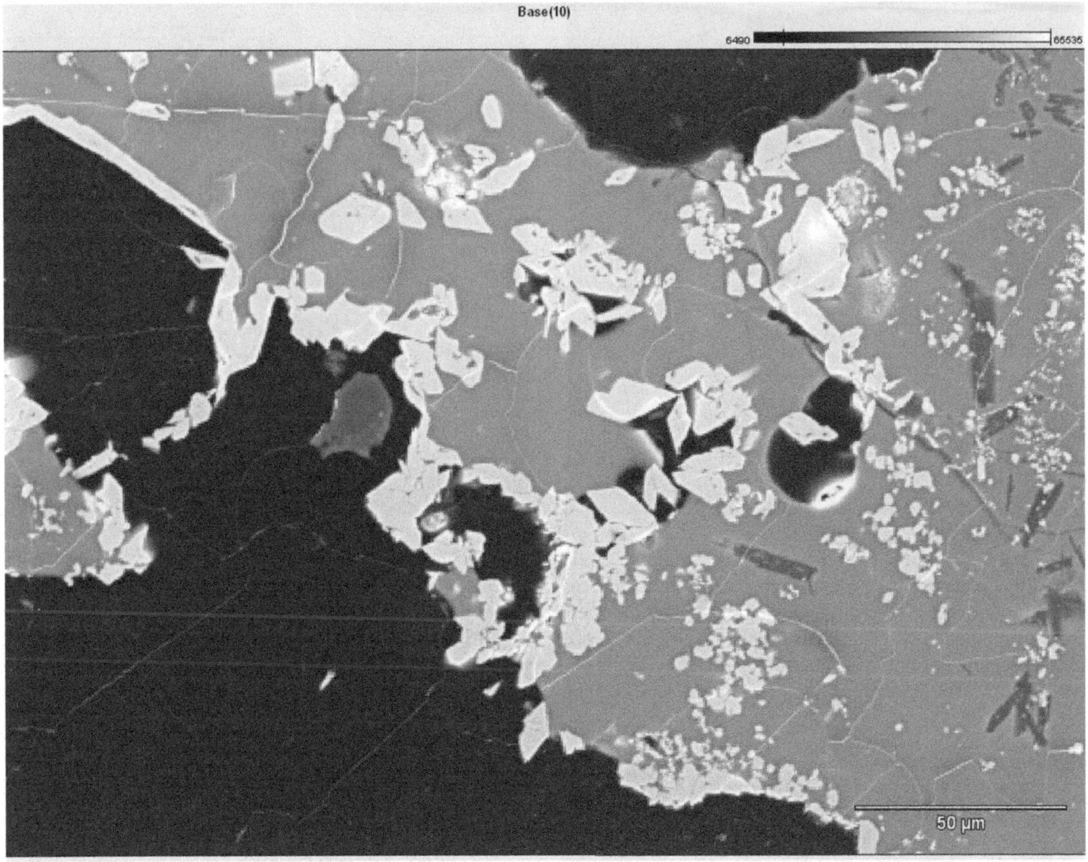

Fig. 5.15 Askaraly II (Kazachstan), Probe KZ-424. Slag incrustation in the crucible fragment shown in Fig. 5.14. Finely dispersed cassiterite crystals (white) in glassy silicate matrix (dark). Scanning electron microscope, backscattered mode. Photo: D. Kirchner, Deutsches Bergabu-Museum Bochum

impressions are also different in size and can also contain a segregation of metals. Due to the fairly clean separation of the sulphide cakes overall, these slags must have reached a very low viscous state.

Selected examples of this type of slag show that matte smelting has a deep (pre-)historical metallurgical background. These are as follows:

- Shahr-i Sokhta (eastern Iran, Sistan Basin): Within the settlements of the Sistan Basin, especially in Shahr-i Sokhta, Tepe Mochi and Shahdad, copper slags have been found. The metallurgical activities in Shahr-i Sokhta date from 2700 to 2500 BC (Tosi 1983; Hauptmann and Weisgerber 1980; Hauptmann et al. 2003; Meier 2015). These slag cakes are all about 2 cm thick and have a diametre of 8–12 cm (Fig. 5.10). Their weight is at c. 350–400 g. They consist mainly of clinopyroxene and in the upper part of increasing amounts of magnetite and unmelted quartz-rich constituents. They have apparently been tapped in a good liquid state, together with high-grade sulphide melts with copper fractions (Fig. 5.17), into a small pit in front of the smelting furnace. The segregation of the sulphides and the metal from the siliceous slag most likely already happened within the furnace but was completed in the forehearth during cooling. The evidence of a

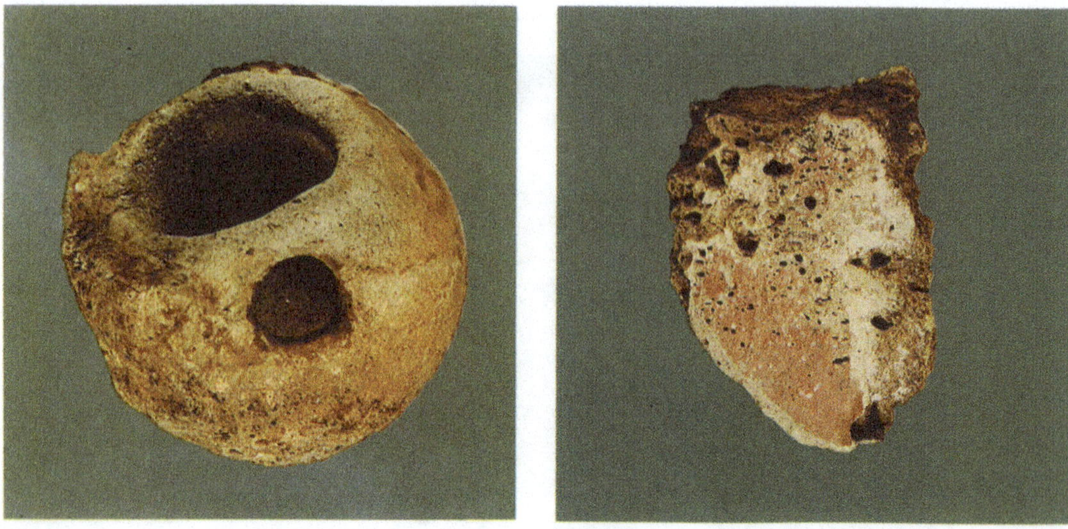

Fig. 5.16 Xanten (Colonia Ulpia Traiana, CUT; Germany), Roman settlement and military camp. (**a**) Small crucible for recycling silver-rich waste products from silver metallurgy. In contrast to the spherical exterior of the crucible (width 5 cm), the interior is rather cylindrical in shape. The function of the opening (13 mm) on the side is unclear. (**b**) Fragment of a glassy crucible slag. Lefthand the sherd of the ceramic crucible. The red colour of the slag is caused by cuprite. Note the impression of the regulus at the lower part of the slag. From Rehren and Hauptmann (1995)

Fig. 5.17 Shahr-i Sokhta (Iran, 2700–2500 BC), Sample SSDV4. Sulphidic matte cake mixed with an exsolved copper metal regulus was found in the archaeological context of the slag cake SS 141. Due to corrosion, the metal regulus is partly transformed to cuprite. Photo: G. Weisgerber (†), Deutsches Bergbau-Museum

slag cake tapped in a forehearth means that such constructions must have been around already in the middle of the third millennium BC.

- Politiko Phorades (Cyprus). The copper smelting site of Politiko Phorades is dated to the earliest phase of the Late Bronze Age (LCI, i.e. 1650–1550 BC). Plano-concave slag cakes of a pronounced fayalitic composition with a diametre of c. 40 cm at the top and a well-defined concavity at the bottom were recovered (Kassianidou 2003a, b, 2012). It is evident that the concavity was left behind when a solid layer of matte had been removed (Fig. 5.18). Unfortunately, the purity of the matte cake was not observed. The weight of the slag is probably ten times higher than that at Shahr-i Sokhta.
- Kition and Enkomi (Cyprus). In the metal workshops of these two Late Bronze Age cities, copper slags were found which show a comparable design as from Politiko Phorades (Hauptmann et al. 2003). These slags resulted from re-melting a "copper bearing conglomerate" rich in Cu-(Fe-) sulphides which would have been sent to the cities for further metallurgical treatment, i.e. for smelting or melting. Matte with changing amounts of metallic copper was produced in the workshops of Kition

Fig. 5.18 Politiko Phorades, Cyprus. Lower part of a copper slag with a negative impression of a matte cake. The slag is from the earliest phase of the Late Bronze Age (1650–1550 BC). At the upper part is has a diametre of c. 40 cm. From Kassianidou (2012)

and Enkomi. It could be proven that slag high in Cu-(Fe-) sulphides was crushed and re-melted again. Due to the composition of the ores exploited from the massive sulphide deposits in the Troodos Ophiolite Complex at Cyprus, the sulphide inclusions are very low in trace element concentrations.

- Mullaq (Oman). This copper smelting site is mainly dated to the Early Islamic period (c. ninth/tenth century AD). Many fragments of plano-concave slag cakes similar to those described from Cyprus have been observed. They have weights of >1 kg. Hauptmann (1985) investigated the mineral phase content of these slags (olivine, hercynite, clinopyroxene) and the composition of sulphidic relics adhering at the bottom of the cakes and droplets inside the slag. These consisted mainly of CuFe-sulphides with a composition of chalcopyrite–bornite solid solutions. Again, such a composition is in good accordance with the sulphidic ore body of the primary ores of the massive sulphide deposits in the Samail Ophiolite Complex of Oman.
- Tawi Aarja (northern Oman). Heaps with slag cakes with diametres of c. 25–30 cm and a weight of 13–16 kg can be found in several localities in the north of Oman, near the modern copper mines of Lasail, Bayda and Semdah. They most likely date around 1200 AD or later, so they are slightly younger than the slags from Mullaq. They, like the other mentioned slag cakes, show negative impressions of matte cakes on their underside. Hauptmann (1985) identified mixed-composited CuFe-sulphides with a composition of chalcopyrite bornite solid solutions, as in the slags of Mullaq. However, in contrast to the above examples, it should not be assumed here that these slags solidified within a shaft furnace, as they were constructed in Oman during the Islamic period (Weisgerber 1987b), but in a so-called bowl furnace. In this, no ore was smelted, but recycled sulphide-rich slag boulders, as they occurred in Mullaq.

5.4.4 Slag Cakes and Slag Foam

Flat slag cakes with a size of about 10–20 cm and a weight of 250–300 g were found at numerous Early Bronze Age I–III sites in the Near and Middle East (Arslantepe, Norsuntepe, Nevali Çori, Çayönü Tepesi, Murgul, Hujayrat al-Ghuzlan; Hauptmann et al. 1993, 2003). The characteristic feature of these slags are their transitions from semi-solid, bumpy small cakes to foam-like pieces with vesicular cavities down to the lower centimetre range. They have a blister-like surface. Many of these slags, most of which are fragmentary (Fig. 5.19), do not show any clear criteria by which to define a top or bottom, i.e. they are irregularly shaped on both sides. They are interspersed with refractory inclusions of ore-bearing quartz, feldspar and hostrock, magnetite agglomerations, charcoal and gas

Fig. 5.19 Arslantepe (Anatolia), Area D7(9–13), Loc. 139, 149, Early Bronze Age I (VIB2), sample TR-8/11. Section through a slag cake. The texture of the slag shows that it was only partially liquefied. It is interspersed with refractory inclusions of ore-bearing quartz, feldspar and host rock, magnetite agglomerations, charcoal and gas bubbles. Macro-photography of several thin sections of the entire section of a slag cake. Width of the whole slag 8.2 cm. Photo: A. Hauptmann, Deutsches Bergbau-Museum

bubbles. Remains or impressions of charcoal are distributed throughout the cakes. They show similarities with the texture of blister copper.

The scarred surfaces as well as the entire texture of these slags are probably caused by a partial melting of the batch and by degassing of the charge and the charcoal bed.

5.4.5 Dross

This type of slaggy material is described in historical metal textbooks (e.g. Schnabel 1890) as a mixture of slag, semi-products, metal inclusions, fragments of fuels and fragments of furnace linings. This material, in the German language called *Gekrätz*, can be collected when a furnace is cleaned after a smelting operation. Dross is collected to be re-(s)melted again due to its high metal content. Dross is also formed as a mass of solid impurities floating on a molten metal or dispersed in the metal, such as in wrought iron. It forms on the surface of low-melting-point metals such as tin, lead, zinc, copper or their alloys in crucibles or casting moulds by oxidation of the metal(s). In the ideal case, dross is mechanically skimmed off from the metal.

Small quantities, only a few kilograms of dross incrustations, prills and drop like lumps were found at Timna, site 200, in a melting-casting crucible site, and were described in detail by (Craddock and Freestone 1988). It consisted of agglomerations of not fully liquefied slag prills and drops that appeared to have been skimmed off some molten mass of metal. A part of the slags from Timna 200, which were described as casting or crucible slags are in fact dross which remains as an incrustation on the inside of a crucible, especially around the spout, but some of the slag is skimmed off the surface of the molten metal and discarded. It is found as small lumps, and in quite small quantities at the casting sites.

A distinction between dross and slag foam is difficult. We cannot exclude that finds from other ancient workshops, which were addressed as crucible slags, would have been better addressed as dross—and vice versa.

The term dross is furthermore applied to describe the texture and composition of the Vinča culture site (sixth millennium BC) slags from Belovode in eastern Serbia (Radivojević et al. 2010). These tiny, only cm-sized droplets of slags were produced in the earliest stages of extractive metallurgy. The authors observed an inhomogeneous, fine-grained mixture of dehydrated and only partially reduced copper minerals (cuprite, Cu_2O, delafossite $CuFeO_2$, tenorite, CuO) and gangue materials (limonite, maghemite, γ-Fe_2O_3), liquefied parts of silicate slags (pyroxene, leucite, fayalite) with magnetite and some metallic copper.

Today, dross is removed from wrought iron by hammering and later rolling (David 1969). From tin and lead the dross can be removed by adding sodium hydroxide pellets, which dissolve the oxides and form a slag.

5.4.6 Cinder

In many metallurgical operations, drossy solid material collects on the top of the molten slag or metal, and when removed resembles a mass of only partially liquified material, often intermixed with slag (Morton and Wingrove 1969a, b). Since this material has never reached a molten or free-flowing state in the furnaces, it should be cinder.

The definition of cinder suggests that it is a drossy material insoluble in the molten slag. It might be solved with progressive temperatures are reached. It therefore follows that bloomery cinder will differ in constitution from that found on the top of charcoal blast furnace slags, and this again will differ from that found on slag from later blast furnaces where even higher temperatures were obtained. In the bloomery process, the slag is either tapped off into a slag pit from the hearth tap hole or left in the hearth and removed when cold as "furnace bottoms". The upper surface of these furnace bottoms might well include drossy material which, in effect, is cinder, and the two terms cinder and slag should not be confused. They are structurally different and also differ in chemical and mineralogical composition. Cinder found on the top of slags tapped from the blast furnace, particularly from those of the charcoal era, includes many readily identifiable constituents such as charcoal, prills of iron, glassy slag, oxides and sulphides of iron, and many other insoluble phases (Morton and Wingrove 1969a, b).

5.4.7 Tap Slags

Bloomery or tap slag from the smelting of iron and of non-ferrous ore minerals represent the mass of many smelting sites from the Late Bronze Age onwards to the Mediaeval period. They have been most intensively studied, both from the archaeological point of view and by scientific methods (Bachmann and Rothenberg 1980; Bachmann 1982a, b; Morton and Wingrove 1969a, b, 1972; Pleiner 2000; Kronz 1997; Kronz and Keesmann 2005). They are formed during the tapping of low-viscosity melts from smelting furnaces and solidify depending on regional characteristics of furnace construction and process techniques in the form of pancakes, tapered slag packages, annular or cone-shaped slags.

Most of these slags were tapped from a smelting furnace as a waste product and referred to as tapped slag or tap slag. The running of these low-viscosity and easy-flowing liquid slags into a forehearth or into a slag pit just below the smelting furnace itself was called *Rennen* or *Rinnen* in German language. This was a widely used term connected with the *Rennfeuer-verfahren*, *Rennprozess*, *Rennofen* of the iron metallurgy (Beck 1884). The English term for this process is bloomery process. Bloomery, bloomer iron and bloomery slag all denote the direct reduction of iron ore. According to the original meaning of the German word, however, *Rennfeuerschlacken* may originate from the smelting of base metals also.

They have been typologically characterised (Bachmann 1982a; Sperl 1980). Typical are bead-like or rope-shaped flow structures, which are often compared to the structure of the volcanic Pahoehoe lava (Fig. 5.20). The underside is rough and has adherent particles of soil on which the slag was drained. Often there are inclusions of charcoal suitable for precise radiocarbon-dating. The slags reach weights up to 60 kg, their surfaces are smooth. Impressive flow slags from late

Fig. 5.20 Copper tap slag, OM 149. Zahir, Oman, probably c. 2200 BC. Typical flow structure of a very low-viscosity tap slag. The individual surges from the melting furnace result in a smooth, beaded surface or rope-like deformed by pushing together the still viscous skin. These forms are called pahoehoe lavas in geological lava flows. A thin-walled bubble cavity is broken up. The slag shown comes from the Bronze Age copper metallurgy and displays large, thin-walled bubbles. Width of the slag 12 cm. Photo: A. Hauptmann, Deutsches Bergbau-Museum Bochum

Roman times can be found in Skouriotissa in Cyprus, where dumps with about 2 million tons of slag occupy the probably largest copper production of this epoch (Fig. 5.21). They consist of layers of superimposed melts and drops, sometimes in pillow form in cross section.

Chemically speaking, most of the tap slags consist of iron-rich silicate slags with the main components fayalite (Fe_2SiO_4) and wuestite ("FeO") for iron slags and magnetite (Fe_3O_4) for non-ferrous metal slags. With this composition, they are in the low-melting range of iron-rich silicate melts, which is set at c. 1200 ° C (see Fig. 4.16). Typical is a so-called spinifex structure of skeletal, thin-plate to leaf-shaped or needle-like crystals of the fayalite, which suggests rapid cooling. Since tap slags exhibit cooling borders of magnetite, iscorite and hematite visible already in the macro scale, they do not necessarily have to come from a tapping of a single furnace run, but may have accumulated from individual tappings (Fig. 5.22). Tap slags

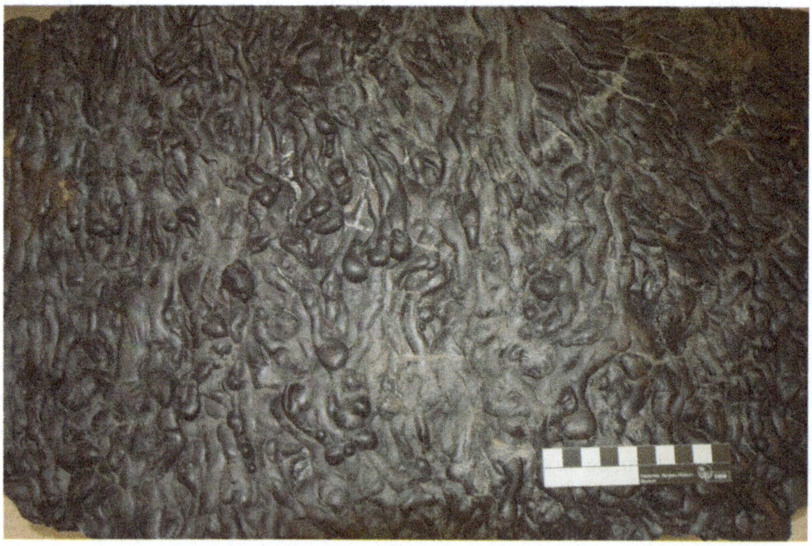

Fig. 5.21 Copper tap slag, Skouriotissa, Cyprus. Tap slags from the copper smelting site of Skouriotissa may reach a weight of up to 60 kg. Usually, they consist of several layers that indicate a multiple tapping. The slags shown in this picture are not safely dated, but most of the slags from this site date to the Late Roman–Byzantine periods. Photo: A. Hauptmann, Deutsches Bergbau-Museum Bochum

Fig. 5.22 Lengerich, Teutoburg Forest (Germany), sample no. D-104/9. Thin section of an iron smelting tap slag. Several layers in an iron silicate slag consisting of fayalite. In between dark cooling rims are visible which are enriched in iron oxides (magnetite, hematite, iscorite). Size of the sample: 3.2 cm. Photo: A. Hauptmann, Deutsches Bergbau-Museum Bochum

are usually low in metals (Cu-contents <1–2 wt. %). This means that the value carriers of the non-ferrous metals have separated quite well from the slag. The high concentrations of iron oxides (c. 60 wt.%) in bloomery slags was a major disadvantage in iron metallurgy, as the ores necessarily had to have correspondingly high iron contents in order to enable the extraction of iron at all.

Vertically flowing melts are to be designated as tap slags as well, which are formed by smelting of iron ores in deepened pits below shaft furnaces. Such slag-pit blocks, dating from the late Latène period to the Roman Empire, are described in detail from Joldelund in northern Germany (Ganzelewski 2000), but are spread over the entire northern German lowlands, Denmark and Poland (Bielenin 1976; Bielenin and Woyda 1978), where slag fields are described with thousands of slag-pit blocks (Holy Cross Mountains, Mazovia) (Figs. 5.8 and 5.9). Probably over 5000 tons of iron were produced here.

5.4.8 Plate Slags

Plate slags are a special variety of extremely fluid melts that typically appear in early copper metallurgy. They are plane parallel slags with a layer thicknesses of only c. 0.2–1 cm. Due to their small thickness, they are very fragile and, in the archaeological context, can therefore only be found as fragments in the size of a few cm^2. The surfaces are smooth and sometimes wrinkled, the undersides bulge in places, but are also smooth. Rounded rims are bulged also. On average (in a cross-section), it can be seen that they are extraordinarily homogeneous and ideally do not contain any inclusions of refractory batch constituents (Fig. 5.23).

The term plate slag (*Plattenschlacke*) comes from the Alpine region. In context with hearth-bottom slags ("slag cake") and with crushed slags ("slag sand"), plate slags were described first from the Late Bronze Age Austrian Alpine region (Klose 1916; Preßlinger et al. 1980). There, they can be found in large quantities, primarily at the smelting sites in South Tyrol and Trentino (Cierny et al. 1995; Silvestri et al. 2019; Reitmaier-Naef 2020). A particular example is Luserna Platz von Motze, where nearly 9000 fragments of plate slag were recovered (Silvestri et al. 2015). Plate slags can also be found at other Alpine smelting sites of the Late Bronze Age, but nowhere in the amounts as in Trentino.

Otherwise, archaeometallurgical plate slags can not only be found in the archaeological context of the (southern) Alps, but seem to be a global phenomenon, not only in copper metallurgy but in every ancient base metal metallurgy. They were found, e.g. at Rio Tinto (Anguilano 2012), in the Harz mining district (Germany; Klappauf 2000; Fig. 5.23b), Nepal (Anfinset 1996), Laurion (Greece; Conophagos 1980), Tepe Hissar (Iran; Thornton et al. 2009), and other localities. Heimbruch (1990) analysed high

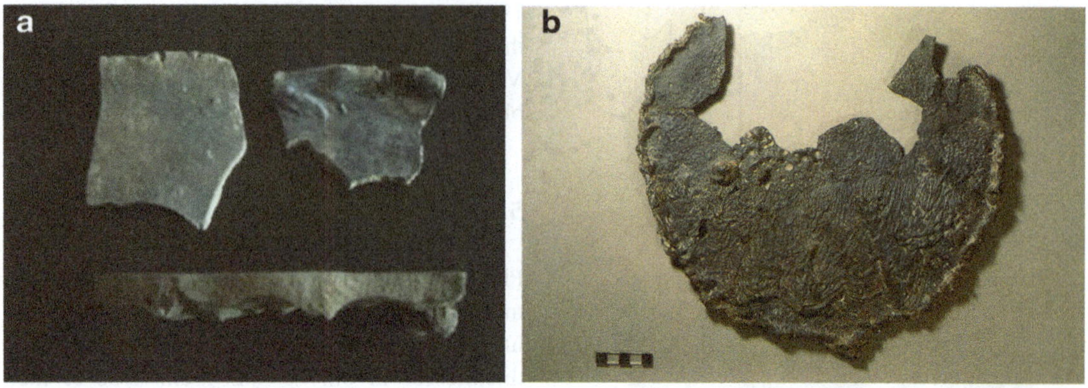

Fig. 5.23 (**a**) Luserna Platz von Mozze, Trentino (northern Italy). Fragments of some Late Bronze Age copper plate slags. Note the smooth surface of the thin plate slags (below 1 cm), and the small protrusions on the bottom. The slag has a pronounced fayalitic composition and shows inclusions of Cu-Fe-sulphides. Photo: E. Silvestri, Beni Culturali, Trento. (**b**) Kunzenloch near Lerbach, Harz-District (Germany). Plate slag from copper smelting, ninth to eleventh century AD. The slag solidified in the shape of a small circle, with a wrinkled surface and thickened rims. It is a eutectic composite, low-viscosity tap slag that solidifies like (**a**) outside the furnace. Photo: C.S. Fuchs, formerly Niedersächsische Bodendenkmalpflege, Montanarchäologie, Goslar

levels of zinc, barium and lead in the plate slag of the Harz Mountains. She proposes a smelting of copper ores from the ore deposit of Rammelsberg. We also noticed the formation of plate slag as a result of experimental archaeology (Bamberger and Wincierz 1990).

The plate slags of the Alps are Fe-silicate slags with more than 90 Vol.-% of fayalitic olivine. They are therefore products of nearly ideal melts of eutectic composition with extremely low viscosity. Next to fayalite, the most predominant phases to observe in plate slags are Fe-rich spinel, a glassy matrix and changing amounts of inclusions of CuFe-sulphides (chalcopyrite, bornite, digenite, Silvestri et al. 2019; Reitmaier-Naef 2020). Inclusions of native copper were also found. The morphology of olivine crystals is needle-like with a pronounced lattice construction. They are solidified not only perpendicular to the surface, but also throughout the section in a spinifex structure (Fig. 5.24a). Thin crust of magnetite and hematite on the surface, solidified with milk-skin wrinkles, are proof of rapid cooling in air, i.e. outside of the smelting furnace. Occasionally, plate slags may contain partially liquefied inclusions of slags added as flux, also ore-bearing magnetite agglomerates and quartz inclusions in the gangue (Fig. 5.24b). They can thus reach greater thickness and their composition can approach a little bit the composition of furnace slags.

The formation of plate slags is continuously debated. Piel et al. (1992) and Metten (2003) argue that, e.g. the two different types of slags, namely plate slag and furnace slag occurring at many sites in the Alps could be the results of one single smelting process. They interpret both kinds of slag as (a) eutectic molten material, which was separated by tapping from (b) the remaining charge or furnace slag that remained as a restite or resister along with metal or matte in the furnace. This model is in good accordance with the results of experimental archaeometallurgy of Bamberger and Wincierz (1990). Hence, it is reasonable to suggest that the wide geographic distribution of this type of slag results not only from special local craftsmanship but is rather the product of given chemo-physical conditions during smelting and slag formation which are similar everywhere.

5.4.9 Furnace Slags (Hearth-Bottom Slags)

This type of slag is found on slag heaps in the form of lumpy, fist-sized or often more than head-

Fig. 5.24 The macroscopic view of thin sections shows plate slags of different "purity". (**a**) Luserna, Trentino (northern Italy). Example of a slag cooled from a complete liquid. The fayalitic olivine is orientated in needle-like crystals vertically to the surfaces and forms a spinifex-texture. (**b**) Acqua Fredda, Trentino. The slag is inhomogeneously composed. Remains and inclusions of partially melted slags with unmelted inclusions of quartz shows a transition from plate slag to furnace slags. The inclusions of quartz contain droplets of sulphidic CuFe-ores. Photos: (**a**) E. Silvestri, Soprintendenza per i Beni archaeologici Trento; (**b**) Metten, formerly Deutsches Bergbau-Museum Bochum

sized chunks up to patty-shaped forms. Often, they are broken into smaller chunks, either intentionally or due to weathering. They can reach original weights of well over 10 kg. They are porous, sometimes also tuffaceous and contain inclusions of charcoal, unmelted mineral and rock fragments in the cm range and relics of slags that have been re-melted. Hearth-bottom slags (German nickname *Ofensau*, Pleiner 2000) are formed when the tap hole is made above the furnace bottom. Hence, this type of slag provides useful information about the geometry of the lower part of a furnace. Hearth-bottom slags are commonly found on heaps not only of early iron, but also of copper production.

Approximately 50–60 cm-sized flat hearth-bottom slags were found on heaps of the Late Bronze Age copper production (c. 1300–1100 BC) in Trentino and in Grisons (Metten 2003; Cierny 2008; Silvestri et al. 2015;

Fig. 5.25 Acqua Fredda, Trentino (Italy). A Late Bronze Age hearth bottom from copper smelting. The slag is only partially liquefied. Note the large number of inclusions of unmelted quartz-rich copper ores (open circle) which are often weathered to sulphates. Quartz is a typical resister while fragments of plate slags (rightwards arrow) are added for fluxing. The diametre of the entire slag is 35 cm, the weight is approximately 6 kg. Photo: H.W. Voss, Deutsches Bergbau-Museum Bochum

Reitmaier-Naef 2020; Fig. 5.25). Their surfaces also show unmelted ore fragments as well as inclusions of sulphide copper ores, which are largely weathered to sulphates, quartz-rich, ore-bearing constituents and slag pieces added most probably as fluxing agents.

The slags from the Phoenician slag heaps of Monte Romero site 56, described as "ball-shaped" slags by Rothenberg and Blanco-Freijeiro (1981), could possibly be understood as a type of hearth-bottom slag as well. This plano-convex slag cake measures approx. 20 cm in diametre. The upper side shows a flat surface. The slag is extremely porous and has a low density. Some contain many inclusions of quartz and baryte. For these reasons, they were called with the nickname "free silica slags" (cf. Sect. 5.4.11).

Smelting experiments have shown that conditions for the formation of furnace slags depend on the process techniques. They indicate too short a reaction time of the charge in the hottest part of a smelting furnace or too rapid sagging of the charge components, insufficient firing temperatures or too high a content of refractory components in the charge (Bamberger and Wincierz 1990; Merkel 1990). The microtexture shows considerable inhomogeneities. Hearth-bottom slags contain extensive agglomerates of Fe-oxides (wuestite, magnetite), mixed with mostly idiomorphic fayalite, in which relics of undecomposed ores, especially quartz, can be embedded (Ganzelewski 2000). They are to be understood as "restites" from unreduced and sintered charge residues (Metten 2003). Larger amounts of hearth-bottom slags were found in the context of Celtic iron production in the Siegerland (Gilles 1957a), the Imperial Iron production in northern Germany and around (Ganzelewski 2000) and the Late Bronze Age copper mining in the Alps (Metten 2003; Silvestri et al. 2019; Reitmaier-Naef 2020).

5.4.10 Crushed Slags

This type of slag describes fine-grained (mm-sized) crushed, also pea- to nut-sized slag with a corresponding fine grain content, especially on early slag sites of Bronze and Iron Age copper metallurgy, sometimes in dimensions of hundreds and thousands of tons. In small quantities, crushed slag—provided that it has been recovered and documented as finds at all—has also been found in many distinctly older localities, which were involved in metal extraction. In Austria, when research on early slags started some decades ago, they had the nickname slag sand (*Schlackensand*). The role of mechanically reworked/processed slag, however, has not been sufficiently considered in the reconstruction of metallurgical processes. This is also because the sampling of ancient slags from the archaeological context, that is, from heaps or excavations, was focussed a long time just on larger slag cakes and neglected such sand-like material, if not present in a dominant amount.

Crushing of slags has been carried out as a routine process step in ancient times almost everywhere to recover metal or matte inclusions as value components included in siliceous, rubbishy slags. The reason for such reworking was the fact that a complete, quantitative separation of slag and metal (or matte) was not possible in the early periods of metal smelting. A perfect technical process with a complete liquefaction of the charge in question has never been attested; it is an ideal expectation of modern technology (Hauptmann et al. 2003). Logically, crushing slags to extract metal or matte prills must have been followed by a subsequent (repeated) remelting of prills and lumps produced in a first smelting stage. As described in Sect. 11.2, these activities finally would lead to a homogenisation of metal produced from one and the same or from various ore deposits.

Crushed slags can be found almost regularly at (pre-)industrial smelting sites worldwide. Even until recently, they have been an integral part of metallurgical craftsmanship in artisanal metal production. Washing of crushed slags to extract tiny prills of metal is shown in the ethnographical context of historical copper production in Japan (1801, Lewin and Hauptmann 1984).

They can already be found at smelting sites of early copper extraction (fifth/fourth millennium BC), albeit in small, often overlooked quantities. Masses of crushed slag were observed at so many smelting sites from the Bronze and Iron Age but were mentioned in the literature often with just one word.

Heaps of crushed slags (*Schlackensand*) can be found at many smelting sites in the Alps. Particularly striking examples of systematic slag processing are well known of many of the approximately 250 Late Bronze Age smelters in the Trentino Copper Ore district (northern Italy, approximately 1200 BC) (Fig. 5.26). Grinding stones and hammer stones were frequently found in the context of the slag heaps. Very fine-grained slag sand has also been observed in the Middle East for tempering Bronze Age pottery (Herdits 1993). Manasse et al. (2001) report about a crushing of slags to mechanically optimise the extraction of metal from possibly Roman age slags at Campiglia Marittima in northern Italy. In South America, it seems that crushed slag belongs to a well-established stage in early metal production as well. Shimada and Craig (2013) report a mechanical processing of slags from the Sícan smelting sites in the Batán Grande Mining and Metallurgical Complex in Peru (900–1375 AD). Anfinset (2011) mentioned the use of crushed slags for re-melting in the artisanal copper smelting in Nepal.

The most impressive example of slag heaps consisting of crushed slags can be found in the Faynan mining district (Jordan). Enormous heaps of pounded and crushed slag are characteristics of the Iron Age smelting sites at Khirbat en-Nahas, and at some other smelting sites of this copper production period in the Faynan area (Bachmann and Hauptmann 1984; Hauptmann 2007; Ben-Yosef et al. 2014). All these slag heaps are a mix of small and large blocks of tap slags and crushed slags, fragments of furnace walls and hundreds of tuyères. Several thousand tons consisting solely of crushed slags occur at Khirbat en-Nahas (Fig. 5.27). The heaps contain amounts of fine-grained charcoal. Once again, tools used for crushing were hammer stones, anvil stones and mortars cut into the sandstone. At Faynan, the grit was not only used as a grog for

Fig. 5.26 Segonzano, Trentino (northern Italy). The copper smelting site of Segonzano seems to be characteristic of the many Late Bronze Age smelting sites in the southern Alps (c. 250). Most of the slags are crushed to slag sand as it is called in popular saying. Crushed slag was produced to extract CuFe-sulphide and copper inclusions get stuck in large slag cakes, and to use it for re-melting. The smelting site of Segonzano contains some hundred tons of slag. Photo: A. Hauptmann, Deutsches Bergbau-Mueum Bochum

clay-made smelting furnaces and/or tuyères, but most probably to enhance smelting processes.

The smelting experiments of Merkel (1990), following the reconstruction of the Iron Age furnaces of Timna 30, and using local ores, clearly showed that it is exactly those slag quantities that solidify within the furnace, which have the highest residual contents of metal or of metal sulphides. Therefore, it is likely that they were targeted as objects of value and thus were crushed. A contemporaneous crushing of Timna slags is also documented there by the tempering of furnace walls and tuyères with crushed slag, as well as by the tempering of Iron Age ceramics, which was found also in the Negev.

5.4.11 "Free Silica Slags"

The term "free silica slags" "was first used in archaeometallurgy by Rothenberg and Blanco-Freijeiro (1981). The authors used the term to refer to such slags that contain bright, unmelted "quartz" as inclusions in a dark liquefied matrix. The type locality of the "free silica slags" was Monte Romero, site 56, in the Rio Tinto area in south-western Spain. The slags had a diametre of about 20 cm and some were apparently very porous. They were called "ball shaped" silver slags. In addition, they showed inclusions of metallic lead. They dated to the eighth to seventh century BC. "Free silica slags" were suggested

Fig. 5.27 Khirbat en-Nahas, Faynan (Jordan). Several thousand tons of crushed slag have been found along the slopes around the Iron Age smelting site. Still copper-containing slag was mechanically reprocessed here and the metal prills were washed out by water. Photo: A. Hauptmann, Deutsches Bergbau-Museum Bochum

to be a highly distinctive feature of the treatment of jarosite and other complex silver-rich polymetallic ores.

Since that time, slags with quartz-rich inclusions have been found in numerous other localities in the southwestern Iberian Peninsula (Overviews in Kassianidou 1992; Hunt Ortiz 2003; Craddock 2013a). They are associated with the further extraction of silver in Phoenician times and are referred to as an innovation.

The term "free silica slags" is now stereotypically used for such slags that have bright inclusions, referred to as "quartz". However, the term "free silica slags" and the interpretations of their origin, as discussed so far, are complex.

Keesmann (1993) at first correctly points out that the term, petrographically speaking, is incorrect and misleading. "Free silica slag" is a term one would use for a SiO_2-saturated slag melt, from which cristobalite and/or tridymite crystallise as their own phase. However, these would crystallise in conjunction with the other slag phases, and not in the form of massive, isolated crystal aggregates in the size of several millimetres to centimetres as inclusions in solidified slag melts. In the relevant slags, the inclusions are embedded in fayalite- or lead silicate-rich melts, which do not show any logical sequence of crystals.

The denomination "ball-shaped" silver slag has been misleadingly simplified over time. According to the original data, the "free-silica slags" from Monte Romero had a flat surface and convex shaped undersides and had a striking porosity (Rothenberg and Blanco-Freijeiro 1981). So they are bun-shaped shaped slags, as they have been identified at many old smelting sites as pit or furnace slags of copper and iron metallurgy (e.g. Metten (2003); Piel et al. (1992); Silvestri et al. (2015); Reitmaier-Naef (2020) for Bronze Age copper slags from Trentino, Italy, Switzerland; Garner (2010) and Menic (2016) for the iron district of Siegerland, Germany; Hauptmann (1985) for early copper production from Oman).

Kassianidou (1992) and Craddock (2013a, b, c) are convinced that unmelted inclusions of quartz or baryte in the “free silica slags” of Monte Romero were deliberately added as flux agents during the smelting of the ores. They argue that neither the ores found in the archaeological context nor in the depository context contained quartz or barite. Craddock (2013a) suggests for the “free silica slags” from Monte Romero that crushed quartz is added to a molten slag at a relatively late stage in the process. The fresh conditions of the quartz fragments in the slag shows that the molten slag must have been rapidly cooled and set, terminating the process. The reason for this addition is not clear and several explanations have been advanced (Tylecote 1987). As the quartz is unreacted, the silica can clearly have played no part in the chemistry of the process, and so must be physical. The effect of adding the cold quartz in such quantities must have been to quickly cool the slag, making it much stiffer.

However, geological and geochemical surveys in the area of Monte Romero have shown that massive quartz-rich parts (“hornstone”, silicified porphyries, jaspers) as well as barytes are present in the gossan of the mineralisation (Williams 1950). Fine-grained jarosite rich layers in Rio Tinto occur at the lower part of the gossan. They are mixed with goethite and hematite and quartz, barite and chlorite. This material was locally termed as “silica chert” and was used as a flux for copper smelting in the industrial age of the nineteenth century AD. Martín-Crespo et al. (2019) emphasise that there are still high proportions of primary (original) ore present in the modern mine of Monte Romero, and that the waste material of modern mining has high contents of barytes and quartz, which are among the typical mineral assemblages of the deposit. In the many streams of the area, quartz-rich sands can be found.

Later investigations on “free silica slags” by Huelva showed that the inclusions are not made of pure quartz fragments, but of fine-grained quartz mosaics similar to quartzite, sometimes banded and interfused by thin streaks of limonite, forming phenocrysts with sharp borders (unpublished data Bochum) (Fig. 5.28). The grains are angular to well rounded as they occur in nature in the sedimentary context. If broken, the fragments may show a flinty or conchoidal fracture. The liquid part of these slags is powdered by magnetite agglomeration and here and there tiny bright white inclusions of silver in the microscopic scale are observable.

The macro-texture of these slags is reminiscent of natural conglomerates. The grains do not show any phenomena of absorption by the liquid slag like as it can be seen in slags from the Alpine region or, alternatively, in natural quartz-rhyolites, and they do not contain inclusions of any (sulphidic) ores. Obviously, the grains were in contact with the liquid slag for only a very short time before solidification. This makes no sense for a flux agent.

Keesmann (1993) analysed samples of “jarosite” from Rio Tinto and found that these, in addition to jarosite itself, contained strongly varying proportions of goethite/hematite and sometimes large amounts of quartz. The existence of Pb-Ag-rich jarosite with Fe-rich minerals and quartz represents an ideal metallurgical starting ore. An excess of quartz or quartz-rich materials can easily be found as unmelted components in slags.

The “pure” quartz inclusions in the slags of Monte Romero are in contrast to the unfused quartz fragments in Bronze Age copper slags from the ore district of Trentino (Metten 2003; Silvestri et al. 2019; Reitmaier-Naef 2020). These almost regularly contain relics of sulphidic ores (Piel et al. 1992) and are to be interpreted as unmelted relics of sulphidic copper ores in quartz-rich gangue. In ancient slags from other regions not only unmelted or partially melted inclusions of refractive mineral phases, such as quartz, were found. A variety of host rocks or minerals from the gangue, chromite, barite or periclase (MgO), were identified as undecomposed resisters—also in tapped slags—in supra-

Fig. 5.28 Huelva, Calle del Puerto 9 (Spain). Slag fragment of a nicknamed "free silica slag". The macro-texture of the slag shows that the slags have a considerable porous volume. Inclusions of quartz-like material (qz) are mostly rounded, partly broken and have sharp borders. They probably originate from the sedimentary context (sand). They do not show any interaction or reaction with the surrounding slag (dark material). Slag matrix is powdered with agglomerations of magnetite. Inclusions of silver were observed. Photo: A. Hauptmann, Deutsches Bergbau-Museum Bochum. Courtesy of C. M. Hernández

regional context of extractive metallurgy (Hauptmann 1985 for Oman; Hauptmann et al. 1993 for Arslantepe and Nevali Cori in Anatolia; Hauptmann et al. 2003 for Shahr-i Sokhta, Iran; Hauptmann 2007 for Tepe Hissar, Iran). In slags from the alpine region, inclusions of quartz-rich materials sometimes reacted intensively with the liquid slag (Metten 2003). The borders of quartz grains show phenomena of tempering, absorption and formation of high-temperature-modifications of quartz such as cristobalite and tridymite (Fig. qz.crist.absorp.). Impurities are dissolved and form eutectic-like textures (Fig. 5.29).

At this point, unanswered questions remain. However, for the interpretation of early metallurgical techniques, it is essential to find out whether they are actually inclusions of purposely added fluxes or unfused relics of the originally used charge (see also section fluxes in this chapter). These two models are as contrary as a "low technology" against a "high technology".

5.4.12 Iron Working Slags

Iron working slags are a widespread, regularly type of finds from settlements of different time periods, from the distribution of iron during the La Tène period to the Industrial Revolution. They are usually referred to as smithing slags, although they may originate from various metallurgical operations of iron processing. In Haithabu several hundred smithing slags were found (Westphalen 1989), equally large quantities in the oppidum of Manching (Schäfer 2013). Since Roman times, they have also been typical finds in the vicinity of mines, since until the use of gunpowder in the seventeenth century and long after that, miner's iron tools were forged, shaped and repaired in the immediate surroundings. So-called mountain forges were found, e.g. in Faynan (Jordan) (Hauptmann et al. 1991) but are also known from the mediaeval mining in the Erzgebirge (Eckstein et al. 1994) and from the Black Forest (Germany) (Goldenberg 1990, 1996b).

Fig. 5.29 Vela di Valbusa, Trentino (northern Italy). Inclusion of quartz grains in an Early Bronze Age copper slag. The shining grains of quartz show reactions of absorption by the liquid Fe-silicate slag and formation of needles of cristobalite/tridymite at the border and formations of cristobalite (*Ballengefüge*) within a few quartz grains. Photo by A. Hauptmann, Deutsches Bergbau-Museum Bochum. Courtesy M. Pearce, University of Nottingham, UK

Fig. 5.30 Tawi Aarja (Oman). Plano-convex shaped smithing slag ("PCB") probably from the Early Islamic period (c. twelfth century AD). The cake is slightly curved up in direction of the tuyère. Photo: G. Weisgerber (†), Deutsches Bergbau-Museum Bochum

Smithing slags have the design of palm-sized, plano-convex cakes with a rough surface. They are "rusty" and rich in bubbles (Fig. 5.30). Smithing slags can easily be mistaken for plano-convex hearth bottoms (Pleiner 2000) or reheating slags (Gilles 1957a, b). They form in the ash bed of a forge below the nozzle opening when an iron object is shaped. Chemically and mineralogically, they resemble bloomery slags. They are iron-rich silicate slags with main components fayalite, wuestite, magnetite, leucite and other characteristic silicates. Typical inclusions within smithing slags are fragments of hammer-scale (Goldenberg 1990), which form as an oxidation layer on the surface of iron objects during heating, and then fall into the slag, as well as relics of quartz sand. A characteristic feature of smithing slags is their layered structure. The individual layers form during the individual steps of the blacksmithing operations.

Fundamental research on the distinction between smelting slags and slags of further processing of iron was done by Keesmann and Hilgart (1992), Kronz (1997) as well as Kronz and Keesmann (2005). A detailed compendium of all sorts of bloomery slags and smithing slags in iron metallurgy was written by Pleiner (2000).

Of the type of further processing, slags are also slags from the Tumulus culture field at Hillesheim (Germany). The only cm-sized fragments (^{14}C-dated to the fourth century BC) could originate from rewelding small iron granules or smaller pieces of iron, or from the shaping of blooms (Kronz 2003).

In copper metallurgy, slag with a size comparable to smithing slag also occurs. These are addressed in the chapter on slag cakes (Sects. 5.4.4 and 5.4.12). Without exception, the slag originates from the Chalcolithic period and the beginning of the Early Bronze Age and has been smelted in small reaction vessels, i.e. in crucibles. They differ from the smelting slag of later periods in their still very high content of metal.

5.4.13 Fluxes

To put this in a wider context in archaeometallurgy, the concept of fluxes versus partially smelting is comparable with a concept of "low-tech" versus a "high-tech"—technology.

In modern metallurgical processing techniques, the ideal compositions of slags are controlled by the (calculated) optimum ratio of ore: flux. This concept has long been paradigmatically transferred to ancient metal production. Optically detected inclusions in old slags or seemingly chemical irregularities were quickly identified as intentionally added fluxes. That may be true in many cases. However, like the use of temper in ancient ceramics, the decision about the origin of fluxes for slag formation in early metallurgy is not easy and clear-cut (Noll and Heimann 2016). The identification of fluxes is particularly difficult if the coarse-grained particles included in liquid slags consist of a mixture of quartz, feldspar, baryte and rock fragments that are inherent in many natural mineralisations of ore deposits anyway. In case such particles provide inclusions of ore in the microscopic scale, as it is, e.g. in the case of Early and Late Bronze Age copper slags from Trentino (northern Italy, Metten 2003), there is no reason to claim the use of fluxes.

Evidence of the use of fluxes in ancient smelting processes must be based on sound knowledge of the mineralogical and chemical composition of the source material and it must be made safe by archaeological evidence. Zwicker (unpubl manscript) was convinced that the chlorine content in Timna slags, as well as their CaO content, was due to the use of flux agents. This was not tenable. The chlorine levels were due to the use of the copper chloride atacamite ($Cu_2(OH)_3Cl$) as well as the influence of thousands of years of soil storage in the saline environment in the Timna Valley. The CaO contents were due to the intergrowth of copper ores with calcite ($CaCO_3$) (Bachmann 1978). Comparatively uncertain is the discussion about inclusions of quartz and barite in Phoenician slags of Monte Romero in the southwestern Iberian Pyrite Belt. This has been discussed in the section "free silica slags" (Sect. 5.4.13).

A clear example of the deliberate use of fluxes in Iron Age copper production was excavated in Timna (Israel, second half of the first millennium BC). There, mainly oxidic copper ores were mined from the sandstone-dominated Amir Formation. The archaeological evidence showed that right next to the smelting furnaces of site 30 still filled storage containers of black pieces of manganese ores were found (Rothenberg 1990b). These ores were transported from the Timna Formation outcrops (Segev and Sass 1989) from a distance of a few kilometres to the smelting place; they did not come from the Iron Age mines. The intentionally crushed manganese ores were mixed with the copper ores to enhance during smelting the formation of low viscosity tap slags (Bachmann and Rothenberg 1980).

5.4.14 Ceramic Slags

Frequently, in the metallurgical context, ceramic slags can also be found. They form during the (s) melting processes by overheating of metallurgical installations made of clay, such as furnace walls and linings, crucibles, tuyères or other technical ceramics. A high degree of fusion of the clearly light ceramic slags regularly leads to macroscopically visible glazing and blistering (Keesmann and Hellermann 1989), so that they have a frothy look. They are often coloured white-grey to greenish and contain, in addition to inclusions of rock fragments, feldspar or quartz used for tempering, especially mullite ($Al_6Si_2O_{11}$). Quartz is often converted to its high-temperature modifications cristobalite and tridymite. Ceramic slags are significantly poorer in Fe-oxides compared to metallurgical slags, but contain higher levels of SiO_2, Al_2O_3 and K_2O. Depending on the effect of atmospheric oxygen and iron oxide contents, they also show a colour zoning from reddish brown (outside) to grey to colourless (inside, strongly reducing) (Keesmann and Heege 1990). If these slags react with the aggressive metallurgical melt, transitions to these can form. This often leads to the resorption of furnace walls. It has the effect of forcing the formation of a eutectic melt by adding CaO, Al_2O_3 and SiO_2

from the clay lining of a metallurgical installation. These mixing processes have been used repeatedly in metallurgy. This can be tracked over long periods of time (Kronz 2000).

Ceramic slags are also regularly found waste products of pottery making and are therefore unmistakable indicators for the localisation of pottery workshops. They are caused by too fast heating up the pottery kiln or too high furnace temperatures during pottery making. In the worst case, a vitrification and the beginning of slagging occurs at around 700 °C. Typically, mullite also occurs here. In metallurgy, ceramic slags from furnace or crucible walls occasionally show transitions to metallurgical slags.

5.5 The Chemical Composition of Ancient Slags

Processes of metal extraction from the earliest beginnings to the Mediaeval period, and sometimes even to the present day, have generally led to the formation of slags with relatively narrow chemical compositions. The basic trends are iron-rich silicate slags with high FeO and SiO_2 contents, as explained in Sect. 4.2. FeO can be replaced by varying levels of CaO, MgO, MnO or Al_2O_3. The alkali contents (K_2O, Na_2O) are generally low. It is also shown that in old slags PbO, ZnO or SnO can reach very high concentrations.

In most cases, the reference system for ancient silicate slags are systems with the components FeO/Fe_2O3—SiO_2—CaO—Al_2O_3 at atmospheric pressure. Bachmann (1978) suggested the combined system SiO_2—$CaAl_2O_3$—FeO.

5.5.1 Iron Slags

5.5.1.1 Bloomery Tap Slags

Chemical composition and mineralogy of many archaeometallurgical slags are amazingly consistent in their concentrations of $Fe^{2+}O$ and other divalent metal oxides such as MnO, MgO and CaO. It is a characteristic feature of bloomery slags of iron production since its first summit in the Iron Age until the mediaeval period, and in artisanal iron making until the recent past. The average of $Fe^{2+}O$ content is approximately 60–65 wt.% and of SiO_2 (+ Al_2O_3) about 35–40 wt.%. Examples for chemical compositions of bloomery slags are listed in Table 5.2.

This extremely high iron content is no evidence for a primitive technology, i.e. for a high loss of iron in slag. On the contrary, it testifies to

Table 5.2 Chemical composition of bloomery slags from the Swabian Alb (southwest Germany)

	Herma-ringen	Langenau	Essingen	Metzingen	Frickenhausen
SiO_2	12.48	8.42	17.27	42.3	39.2
TiO_2	1.17	0.46	0.36	0.6	0.69
Al_2O_3	13.6	6.12	5.22	12.2	15.6
FeO	57.15	74.05	60.98	4.41	8.61
MnO	0.41	0.63	4.19	1.08	0.85
MgO	0.42	0.4	0.3	7.45	4.7
CaO	3.23	6.25	0.62	27.5	26.4
Na_2O	0.19	0.05	0.08	n.n.	0.15
K_2O	0.48	0.22	0.38	1.78	2.16
P_2O_5	0.44	1.26	0.69	0.41	1.23
Total	89.57	97.86	90.09	97.73	99.59

The smelting sites where the slags were collected from date to the first/second century AD (Hermaringen), first/third century AD (Essingen), and second/third century AD (Langenau). For comparison, Fe-low slags from blast furnace slags from the sixth/ninth century AD (Frickenhausen), and from eleventh/thirteenth century AD (Metzingen) are shown. Note the remarkable decrease of FeO in the Middle Ages! Values given in weight-%. From Yalçın and Hauptmann (1995, 2003) the high skill of an old metallurgist. It is primarily caused by the formation of a low melting liquid during the heating of Fe-rich charges in a furnace. As shown in Fig. 4.4 such a liquid, with the predominant phase of Fe-rich olivine (fayalite) is limited to a comparatively narrow composition. Next to fayalite, bloomery slags may contain Fe-oxide such as wuestite as the second characteristic mineralogical phase, and partially magnetite, when the liquid slag was exposed to higher oxidising conditions. Slags with little or no wuestite indicate an even more efficient process (Tylecote et al. 1971). In addition, the fayalitic melts have a significantly lower viscosity than those with higher SiO_2 contents and are thus easier to separate from the resulting metal (Keesmann and Hilgart 1992). The viscosity is connected with low melting/crystallisation of the slag and can be estimated by plotting the bulk chemical composition of a slag into the binary phase system $FeO–SiO_2$ or better into the ternary phase system $FeO–CaO–SiO_2$ which is part of the quaternary system $FeO–CaO–SiO_2–Al_2O_3$. The lowest melting temperatures in these systems are around 1050 °C, which may be lowered even further by some low-melting alkali components such as K_2O and Na_2O.

From the phase diagram $FeO–SiO_2$ alone, it can be concluded that, if the ratio $FeO:SiO_2$ of the charge is not in the ideal range, changing proportions of Fe-oxide incrustations are formed, either as accretions on top of tuyères, adhering to the furnace walls, or they are visible in the slag itself as unmelted concretions. Quartz inclusions may also remain as unmelted constituents in the slag. In this way, furnace slags are formed, which were described in Sect. 5.4.9.

In iron metallurgy, mainly, but not exclusively in connection with fayalitic slags of bloomery smelting, carbon poor ferritic iron will mainly be produced (cf. Sect. 6.6). In case of decreasing FeO-concentrations in slags, as can be observed in mediaeval slags, the carbon concentrations in iron are increasing and steel or cast iron will be produced. The continuous transition from bloomery processes to blast furnace iron production (seventh to twelfth century AD) and the change of the composition of slags were investigated at the Swabian Jura in southern Germany by Yalçın and Hauptmann (1995) (Fig. 5.31).

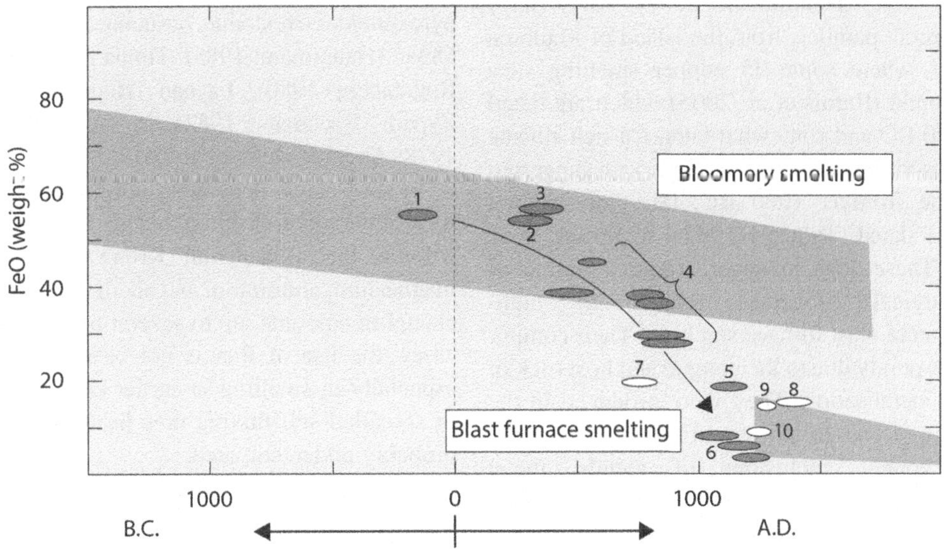

Fig. 5.31 Iron oxide concentrations of slags from bloomery smelting and blast furnace smelting exemplified by slags from the Swabian Alb, from Sauerland (Germany), and from Sweden (after Gilles 1957a, modified by Yalçın and Hauptmann (1995)). Note that in these regions, there was a continuous transition between the two metallurgical techniques and not a technological breakthrough as is suggested often

The later methods of blast furnace smelting worked with much lower-iron slags. FeO levels in the slags of modern blast furnaces are even lower than 1 wt.%.

5.5.1.2 Blast Furnace Slags

Silicate slags extremely high in CaO and MgO (20–40 wt.% CaO, 4–8 wt.% MgO, 30–40 wt.% SiO_2; see Table 5.2) and corresponding dominant phases of melilite and/or merwinite (Ca, Mg-silicates), typical for modern slags of blast furnace and base-metal smelting (Obenauer 1954), occurred at mediaeval-early modern smelting sites in the foreland of the Swabian Alb (Yalçın and Hauptmann 1995) and in the Upper Rhine area (Gassmann et al. 2005) and originate from the beginnings of pig iron production. They are referred to as blast furnace slags. Other examples of blast furnace slags of the early Middle Ages occur, e.g. in Lapphyttan (Sweden) (Björkenstam and Fornander 1985).

However, there are also chemically comparable slags from quite different periods. These are finds of copper ore smelting. They come from Wadi Fidan 4 (Faynan) and date to the Early Bronze Age (Hauptmann 2007). They have more recent parallels from the island of Mallorca (Spain), where some 35 copper smelting sites were found (Ramis et al. 2005) which are dated to 2.000 BC and somewhat later. Ca-rich silicate slags discovered at Afunfun (Niger) are suggested to come from c. 1000 BC, but they are not securely dated (Tylecote 1982a, b; Killick et al. 1988). These slags, however, are not a product of metallurgical processes related to high-temperature blast furnace smelting. Their composition is purely due to the gangue and host rock of their mineralisation. They were produced in the smelting of ores from dolomitic-sandy host rock. They contain inclusions of metals, metal sulphides (matte) or of speiss, which are indications for the composition of the smelted ores. Moreover, not all of these slags were completely liquefied.

5.5.2 Copper Slags

Copper slags can be very similar to bloomery slags, both in terms of their main chemical composition and their mineralogical phase content. The main elements in the composition of copper slag are shown in Fig. 4.9 and for comparison, the bloomery slag from iron extraction is also shown. They can hardly be distinguished from each other by chemical analysis alone. The main components in the two sorts are fayalite together with wuestite and magnetite, pyroxene and glass are quite common as well. Copper slags, compared to bloomery slags of iron metallurgy, are usually formed under less reducing conditions. Therefore, magnetite (Fe_3O_4) or Fe-rich spinel such as hercynite ($FeAl_2O_4$) outweighs wuestite in copper slags. Frequently occurring metallic copper prills or inclusions of copper-rich matte can facilitate the distinction.

In accordance with the geological environment of numerous sedimentary copper mineralisations with manganese ores (Sect. 3.5.1), manganese-rich silicate slags in which FeO is replaced by MnO are also found. MnO contents in such slags are between 20 and 40 wt.% (Fig. 5.32). Main phases are tephroite (Mn_2SiO_4) or manganese-rich pyroxenoides (rhodonite, bustamite, etc.), as, e.g. in Oman (Hauptmann 1985), Timna (Bachmann and Rothenberg 1980), Faynan (Hauptmann 2007), Zypern (Bachmann 1982a, b), Sinai (Bachmann 1978; Abdel-Motelib et al. 2012; Pfeiffer 2013).

In such slags, depending not only on the host rock of the mineralisation, but also on the composition of the furnace wall (Kronz 1997) calcium, magnesium, aluminium and alkali oxides occur in changing amounts, up to several percent. In most cases, the use of flux is not clearly detectable, especially in smelting in earlier epochs. The use of so-called self-fluxing ores in ancient times is probably underestimated.

5.5.2.1 Slags from Fahlore-Smelting

The term "fahlore metals", coined by Otto and Witter (1952), already suggests—indirectly—that there must have been early use of fahlores in Europe as a raw material for copper extraction.

However, slags that are undoubtedly attributable to production of copper from fahlores (e.g., tennantites, tetraedrites, and also enargite) which are high in arsenic or antimony but very low in iron (Table 3.3) are extremely rare. In the Inn Valley in Austria, with its numerous fahlore deposits which were examined very intensively within the framework of the HIMAT-Project (Sect. 3.5.1), some stray finds of slags from layers of the outgoing Neolithic have been found (Bartelheim et al. 2002). They were found during excavations at the Mariahilfbergl near Brixlegg and date to the Münchshöfen culture (fifth/fourth millennium BC). Scarce finds have also been made at La Capital du Broum in the well-known Chalcolithic Copper mining district of Cabrières (France) (Bourgarit et al. 2003).

Also from the early phases of the Bronze Age (fourth/second millennium BC) and the Late Bronze Age (late second/early first millennium BC), slags suitable for the reconstruction of smelting processes of fahlores are extremely rare. Here, the excavations at Kiechlberg (Krismer et al. 2013; 3800–2000 BC) as well as those at Buchberg near Schwaz (Martinek 1996; Schubert and Pernicka 2013; 2020–1770 BC) must be mentioned. It is probably also an exception that Goldenberg (2014a, b, 2015), during the excavations in the Schwaz/Brixlegg area (excavations near Mauk), found only very few slag heaps from the Late Bronze Age, which clearly prove a smelting of fahlores. The slags, at least in Europe, reflect the geology of the fahlore deposits as they occur in the Inn Valley, but also at Cabrières in southern France (Ambert 1999), where fahlore deposits were also used in very early periods.

The Inn Valley is part of the Northern Calcareous Alps. A number of fahlore deposits are located here, synoptically called "Pb-Zn-(fahlore-) mining district" by Schulz and Schroll (1997) (Fig. 3.11). Typical ores are tennantite and tedraedrite and their solid solutions, partly very rich in silver. Silver-rich minerals such as pyrargyrite (Ag_3SbS_3) have also been detected. Pyrite is partly intergrown with fahlores in quantities. Host rocks are dolomites of the Greywacke Zone. The gangue of the mineralisation is quartz, partly with baryte. Comparable mineralisations are described by Ambert (1999) from the ore district of Cabrières (France).

Bourgarit et al. (2003), Krismer et al. (2013) and Schubert and Pernicka (2013) examined some prehistoric slags from the sites mentioned. These early slags are consistently only partially liquefied. Due to the chemical composition of the dolomitic host rock, they are distinctly rich in CaO (up to >10 wt.%) and MgO (up to 9 wt.%). The authors analysed samples with significantly lower FeO contents compared to the fayalite slags from numerous other sites worldwide. Corresponding to the rather highly oxidised phase inventory, the authors give the iron content of these slags as Fe_3O_4 values of c. 20 wt.%. The gangue was responsible for SiO_2 values of c. 30–40 wt.%. Cu, Sb, As, Ag, S are minor constituents in all samples. Antimony and arsenic are in the lower percentage levels due to sulphidic and metallic inclusions. However, along with silver, they are clearly concentrated in inclusions of copper. There are (eutectic) intergrowths between copper and arsenic or antimony. In addition, inclusions of speise (Fe–As) could be detected. Sulphidic droplets, in contrast, are very low in As and Sb and correspond mainly to compositions of the ternary system Cu–Fe–S. The composition of the metal points towards that which Otto and Witter (1952) described as fahlore metal.

The silicate-oxide phase assemblage of the slags from the Inn Valley and from Cabrères as well consists mainly of silicates high in Ca, Mg, Zn, but low in Fe such as clinopyroxene, olivine, melilite-group phases (åkermanite), leucite, (Mg, Fe, Zn)-spinel and silicate glass.

The rarity of these slag finds can have multiple reasons. In the Inn valley, it is not clear whether possible heaps are buried because of the extremely steep Alpine terrain. An alternative is that the smelting of fahlores in the early phase of the use of these deposits did not take place directly at the outcrops of the ore deposits, but that the ores were transported to settlements in distance and smelted there. It is possible that

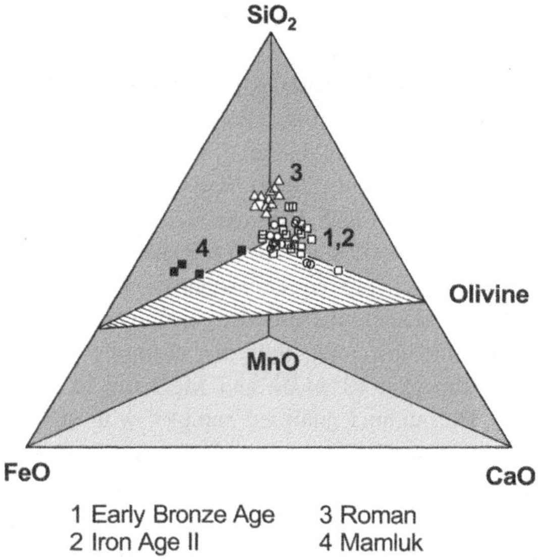

Fig. 5.32 Chemical bulk analyses of manganese-rich slags from Faynan in the system SiO_2–FeO–MnO–CaO. One can recognise clusters, which correlate with certain time periods during the history of metal production in this area. Similar compositions of manganese-rich silicate slags were also found in the Timna area (Bachmann and Rothenberg 1980). From Hauptmann (2007)

not all settlements from these periods are recorded yet.

5.5.3 Lead–Silver Slags

Ancient slags from lead–silver smelting differ—as far as they are archaeologically tangible—in their chemical and mineralogical composition from early slags of iron and copper technology. These differences are mainly due to the fact that lead ores can be smelted at unequal lower temperatures. This will be discussed in more detail in Sect. 6.4. Like all slags, lead–silver slags are very much dependent on the composition of the ores, their gangue and host rock, and these are different from many of copper and iron ores. First, it should be noted that the metals lead, silver and zinc are geochemically and paragenetically linked in deposits and therefore also occur together in slags.

Lead–silver slags may be moderately high in FeO and SiO_2 contents. However, high CaO contents, which can reach >13 wt.% and more are typical (Table 5.3). In Laurion, CaO-rich slags with such high values are very typical (Bachmann 1982a, b; Tsaimou et al. 2015). CaO and MgO levels are typical of lime and dolomite host rocks as they occur in ore deposits of the Mississippi Valley type (MVT; Pohl 2005) occurring in karstic cavities such as on Sardinia or at Laurion.

Slags from lead smelting, as for example from the mediaeval Harz (Klappauf 2000) and the Erzgebirge (Eckstein et al. 1994) are rich in CaO, BaO, PbO and ZnO.

Baryte, quartz and pyrite or Fe-oxides appear as typical gangues of such lead–silver–zinc ores. Indeed, high BaO levels in smelting slags are therefore generally an important indication of Pb-Ag technology. BaO levels occasionally exceed the CaO levels. BaO-rich slags are known from the Roman lead–silver mining from Rio Tinto (Craddock et al. 1985; Keesmann 1993) and from the Hellenistic period on the island of Thasos (Hauptmann et al. 1988). Copper mineralisations can in exceptional cases contain barite as well and thus be the cause for Ba-bearing slags.

ZnO levels in smelting slags are an important indication of Pb-Ag technology as well, as sphalerite (ZnS) is also a typical gangue mineral of corresponding ores. ZnO contents can reach >30 wt.%.

Lead levels are often very high in lead–silver slags. Ettler et al. (2009a, b) reports of mediaeval lead–silver slags with up to >34 wt.% lead. Usually, lead in slags form Pb-silicates (next to inclusions of lead itself). The proportion of silicate-bound lead can reach high levels and lead silicates cause enormous reductions in melting temperatures. Microprobe analysis of different phases in slags of the early lead–silver recovery show that in the slags lead is often present in metallic form and segregates like galenes from the siliceous-oxide melt. Characteristics of early slags of lead–silver mining on Thasos are finger-sized segregations of metallic lead (Hauptmann et al. 1988). They also contain droplets of zinc sulphide (sphalerite) and other polymetallic sulphides and metallic phases and alloys. Such losses of lead have

Table 5.3 Bulk chemistry of some ancient lead–silver slags from the Laurion area (Greece), modified after Tsaimou et al. (2015)

wt.%	S1	S2	S3	S4	S5	S6
SiO_2	26.1	26.1	24.3	24.3	24.7	25.4
CaO	15.6	15.5	16.2	14.2	13.6	13.3
Al_2O_3	4.70	4.65	4.59	3.02	4.63	4.40
FeO	12.7	13.9	15.3	9.17	10.4	16.9
MgO	2.45	3.10	2.27	2.04	2.85	2.87
K_2O	1.17	1.14	1.04	0.80	1.22	1.30
Na_2O	0.61	0.54	0.47	0.40	0.61	0.54
ZnO	9.26	10.08	9.29	21.99	18.48	9.68
PbO	22.5	20.3	21.2	19.5	18.5	20.0
MnO	0.66	1.05	1.06	1.16	1.19	1.16
Ag	0.001523	0.00033	0.00088	0.000692	0.00038	0.00057
As	0.19	0.07	0.08	0.97	0.05	0.05

Main components of these slags are SiO_2, CaO, FeO, ZnO, and especially of PbO which forms a surprisingly, extremely high constituent

probably been tolerated in smelting to avoid over-reduction of Fe-oxides to iron and their formation of so-called hard-heads. These are found occasionally (Hauptmann et al. 1988).

In the context of bulk analyses of lead–silver slags, it is advisable to check if the lead content should be calculated as $Pb^{\pm 0}$, or in the form of oxides or silicates (Pb^{2+}, Pb^{4+}). Zinc is generally not reduced to metal, but it contributes to the formation of silicate or sulphidic mineral phases.

Due to their Pb contents, a fast-melting, even at relatively low temperatures, is characteristic for lead–silver slags, so that hard-to-melt components remain unresolved for a long time. Therefore, they often show a glassy, obsidian-like solidification with densely seeded, mm-sized inclusions of quartz, barite or feldspar (Fig. 5.28; Hauptmann et al. 1988; Ettler et al. 2009a, b). This glassy texture is not only due to a rapid cooling. Glass likes to form in abundance between the crystalline phases. In addition to SiO_2, FeO, and Al_2O_3 it can also absorb very high amounts of BaO (<10 up to 25 wt.%) and PbO (c. 20 wt.%). Due to their large cation size, lead and barium are retained in the glass as network-modifying cations. Lead behaves as an "incompatible element" and is concentrated in the residual glass matrix (Ettler et al. 2009a, b).

On the other hand, the chemistry of lead–silver slags leads to a characteristic phase and slag structure, which is often typically needle-like. Typical phases of lead–silver slags are CaO-rich phases such as akermanite ($Ca_2MgSi_2O_7$) and clinopyroxene (mainly hedenbergite, $CaFe^{2+}Si_2O_6$), (Fe-rich) olivine, spinel, glass. In Ba-rich slags hyalophane ($(Ba,K)Al_2Si_2O_8$), celsian ($BaAl_2Si_2O_8$) and in special cases, andremeyerite ($BaFe_2Si_2O_7$) occur. ZnO contents in lead–silver slags facilitate the formation of the Zn-rich spinel gahnite ($ZnAl_2O_4$), hardystonite (member of the melilite group, $Ca_2ZnSi_2O_7$) and willemite (olivine group, Zn_2SiO_4) (Ettler et al. 2009a, b).

It has already been mentioned that, unlike slags of early copper and iron smelting, those from early lead–silver metallurgy are much less likely to be found at field survey. This may be because these metals were generally produced in smaller amounts in ancient times. Another reason for the limited inventory of finds is also the fact that lead–silver slags were often reprocessed at the end of the nineteenth or at the beginning of the twentieth century AD due to their silver content. Examples for such activities are known from the lead mining areas in south-western Sardinia (Italy) and from Laurion (Greece), where tens of thousands of tons of slag existed originally (Hanel and Morstadt 2014). Conophagos (1980) synoptically reports of more than 1.5 million tons of slag from Laurion, which were located there prior to the nineteenth century AD. In both areas, early lead–silver slags were eliminated almost completely. The same is true for the Bronze to

Iron Age lead–silver slags of Sifnos, albeit on a much smaller scale (Pernicka et al. 1985). An assessment of early (pre-)Roman lead and silver extraction from the corresponding ore deposits based on the amounts of slags is therefore difficult. A reasonable idea to solving this problem is the approach of isotope analysis and archaeological finds such as, e.g. ingots.

Slags from Sardinia that were addressed as "early" (undated) contained 9–14 wt.% lead and up to and more than 0.1 wt.% silver (von Maltzan 1869; Sella 1999), so more silver than what was extractable from the lead–silver deposits. Back then, between 10 and 25 wt.% lead was analysed from the ancient slags of the Laurion area (summary in Kalcyk 1982). Little is known about their silver content. Silver contents are typical in old slags, albeit they are mostly very low. From the analysis of nineteen slag samples from Laurion, Bachmann (1982b) reported 17 samples with <0.005 wt.% Ag and only two other samples with 0.05 wt.% and 0.59 wt.% Ag. In comparison to Rio Tinto, Rothenberg and Blanco-Freijeiro (1981) report analytical data for 120 slag samples. 49 samples contained <0.005 wt.% Ag. From the remaining analyses, only seven samples represent exceptions with higher concentrations from 0.06 to 0.27 wt.% Ag.

In the low mountain ranges of Germany and its neighbouring countries, there are a number of lead–silver deposits, the use of which is likely or sometimes proven already in pre-Roman times (Rothenhöfer 2003), such as the lead–silver production of Lüderich (Körlin and Gechter 2003). However, numerous of these mining activities are proven for the Middle Ages and early modern times by archaeological excavations and recorded in numerous written traditions (Theophilus, Georg Agricola, Sebastian Münster). In the region of the Erzgebirge, lead–silver slags from the thirteenth and fourteenth centuries AD were found at Freiberg and Annaberg, but these contained only a few wt.% of PbO, and silver concentrations in the range of several tens to a maximum of 300 ppm (Eckstein et al. 1994). Rehren et al. (1999a) report on mediaeval lead–silver smelting in the Siegerland in West Germany. The few analyses seem to prove that the silver contents are very low at only 30 ppm. Even PbO does not even reach 10 wt.%.

In the Czech mining area of Příbram, individual layers of slags from the fourteenth/fifteenth century AD were found, containing 25 wt.% lead and almost 0.3 wt.% silver (Ettler et al. 2009a, b). Apparently, most of these slags were later recycled, between 1850 and 1950, because of their silver content.

In the mining area of Panjihir in the Hindukush Mountains (Afghanistan), the silver content of 440–460 ppm and lead inclusions in slags with 0.6 wt.% Ag were measured in slags. Merkel et al. (2013) report "slag heaps", but there is nothing known about the real extent of these heaps. The dating of these slags is uncertain. A prehistoric age is possible, but the absolute heyday of silver extraction was in the tenth century AD, according to historical sources. The Ag contents were correlated with high levels of Cu, Sb, Bi and As, due to a significant content of silver-containing fahlores in the galena. Fahlores often contain 2–4 wt. % Ag, sometimes up to 36 wt.%.

Probably the earliest lead–silver slags include the few slag finds of the small chalcolithic hamlet Fatmalı-Kalecik in the Keban salvage area in the Upper Euphrates (Eastern Anatolia) (Hess et al. 1998). The main components of the slags are SiO_2 (c. 30 wt.%), FeO and PbO are at the same level at c. 30 wt.%. The BaO-concentration is below 1 wt.%. The slags contain up to 170 ppm silver which can sporadically be seen as tiny inclusions. The composition of the slags from this site results from the smelting of ores extremely rich in lead with minor concentrations of arsenic. The ore minerals must have been more or less of oxidic nature as the sulphur concentration is rather low. Obviously, this ore came from the surface near the gossan of an ore deposit. The lead content in the slag was not totally reduced to metal. A high amount was bound mainly in green-coloured lead-silicate glass. Measurements by SEM-EDX-analyses revealed, next to other components, (CaO, FeO, Al_2O_3, ZnO) 44–48 wt.% PbO and 31–32 wt. % SiO_2 in the glass. Slags of such compositions are liquefied as low as 800 °C and below (Kuxmann and Fischer 1974).

5.5.4 Tin Slags

Although tin has played a supra-regional role as the most important alloy since the Early Bronze Age in the Old World, and although there are larger deposits of tin in Europe and Central Asia (Sect. 3.4.11), there are surprisingly few tin slags from ancient times. Hence, too detailed reconstructions for the extraction and smelting of this metal in antiquity and before that period are difficult. They are based largely on studies of slags from modern times (Earl 1986; Tylecote et al. 1989; Chirikure et al. 2010; Heimann et al. 2010). Tylecote (1987) and Merideth (1998a) already pointed to the rarity of ancient tin slags. They attributed this to the fact that one could achieve a very high concentration of the starting material in the extraction of this metal from tin placers with the main mineral cassiterite, depending on the skill in beneficiation. In the best cases, at any following metallurgical operations, slag would hardly occur at all.

Tylecote et al. (1989) emphasised that tin smelting in the "blowing house", i.e. in a free-standing shaft furnace in the south-western part of England—yields, ultimately, very little slag. In that tin field, only a few slag dumps are known from the blowing process and the largest of these has been estimated to have held less than 1000 kg of material. It must be kept in mind that that thousands of tons of tin have been smelted with this technique in the West of England.

The extremely high rarity of tin slags is also observed in old areas of hard rock underground mining of tin. While the extensive work of Alimov et al. (1998) or Garner (2014) provided valuable new information about the ancient tin mining in Central Asia, which must have been instrumental in supplying the Old World with tin or tin bronzes, the works contain no evidence of slags. Extensive tin placers, such as those in Cornwall or in the Erzgebirge, are obviously not known there. Similar observations were also made by Stöllner et al. (2013a, b), which report only about *one* slightly slagged crucible fragment which proved tin metallurgy at Askaraly II in East Kazakhstan from around 1600 BC—from a region belonging to the tin-rich regions of Central Asia!

In their studies on tin mining and metallurgy in the Early Bronze Age Göltepe (Anatolia) Yener and Vandiver (1993) point out that only slag layers in the order of a few millimetres were found on the inside of—admittedly countless—crucibles. This, in turn, fits in well with the low levels of tin minerals in the nearby Kestel mine and would support the concept of intensive reprocessing, as proposed by Tylecote and Merideth.

In material science, tin slags have been investigated in various ways (Rapp 1977; Yener and Vandiver 1993; Tylecote et al. 1989). The most intensive and detailed investigations of tin slags were performed by Miller and Hall (2008) and Chirikure et al. (2010). They report on scarcely distributed glassy looking tin slags collected from the surface in the precolonial Rooiberg Valley region (South Africa, middle of the second millennium *AD*). These slags originate from an ore deposit with a mineral assemblage of tourmaline, cassiterite, Fe-carbonates, sulphides such as pyrrhotite, chalcopyrite, sphalerite in context with the geologically famous Bushveld complex. Note the paragenesis of SnCu ores!

5.5.4.1 Tin Concentrations

Old tin slags often have very high levels of tin, which are reported in the literature as Sn^0, $Sn^{2+}O$ or $Sn^{4+}O_2$, depending on how they are measured and calculated. Tin can be contained in slags in all these valences. They are tolerated in the smelting so as to avoid, just as is the case in lead and silver production, an over-reduction of Fe-oxides to iron and its reaction with tin to so-called hard-heads ($FeSn_2$). In metallurgical reactions, SnO is tightly chemically bound to FeO, as both have a very similar oxygen affinity. In addition to the SnO reduction, a reduction of Fe-oxide from the slag takes place simultaneously. This leads to the formation of unwanted FeSn alloys (SnFe, $SnFe_2$), the so-called hard-heads. The higher the FeO concentration in the slags, the higher the amounts of hard-heads produced in the fabrication of a low-SN slag. Only for slags with low FeO contents, i.e. for CaO-rich silicate slags, the production of raw tin without the formation of hard-heads is possible (Möller 1971).

It must be noted that pure cassiterite from tin placers contains 78 wt.% Sn. Tylecote et al. (1989) are convinced, that the ancient smelters used ores dressed up to give at least 65 wt. % tin. Tin slags in the United Kingdom are characterised by very high levels of SnO (10–40 wt.%) (Tylecote 1987). In modern tin extraction practices in England slags with high tin oxide concentrations have been re-smelted (Charles 1980). This may also be the reason why tin slags are so rarely found there from ancient times. Another example is the tin slags from Rooiberg. They show between 10 and sometimes up to 50 wt.% SnO_2. The higher values indicate that tin smelting could easily be inefficient. It is also probable that such slags were recycled.

Otherwise, tin slags of various origins consist of the main components CaO, SiO_2, Al_2O3 and FeO. The projection of these values into the quaternary system shows a high range of variation, especially of FeO (c. 6–40 wt.%) and of SiO_2 (c. 20–65 wt.%; Tylecote et al. 1989; Chirikure et al. 2010) (Fig. 5.33).

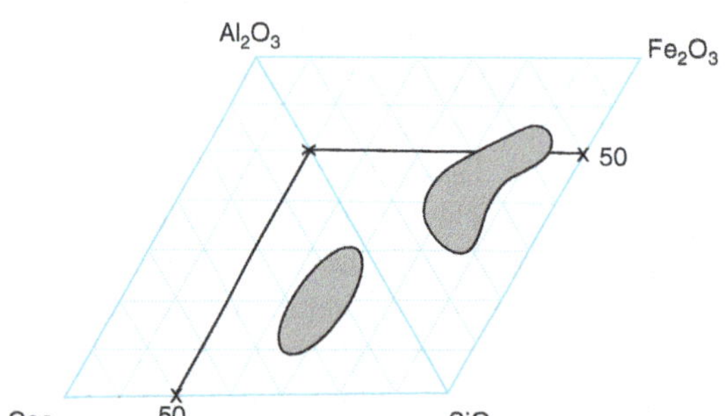

Fig. 5.33 Composition of tin slags from various localities (south-west England; Rooiberg, South Africa) and of various time periods plotted in the system $CaO–SiO_2–Al_2O_3–Fe_2O_3$. Note the large variety of Fe_2O_3- and of SiO_2-values. Fe_2O_3-calculation even though the presence of spinel and fayalite suggests that large amounts existent as Fe^{2+}. Slags contain mostly between 11 and 40 wt.% SnO. SnO overwhelmingly is in the divalent state and is a glass-forming element. After data from Chirikure et al. (2010)

Tin oxide accumulates in the glassy phase of tin slags so that a frequent and abundant formation of tin-containing silicate glasses (with up to 50 wt.% SnO) is observed, sometimes in part with CaO and FeO contents. The glasses are also rich in SiO_2 and Al_2O_3. Network-formers of the glass are mainly SiO_2, SnO and SnO_2. They are predominant components and may reach 80 wt.% (Chirikure et al. 2010). Miller and Hall (2008) analysed glass in tin slags with 50 wt.% SnO_2 but, admittedly, they did not exclude that this value would also include droplets of tin oxide embedded in glass. High tin contents, in the form of SnO_2, can produce a characteristic colouration similar to that of the white-coloured tin glazes, as they are known, for example, from the Renaissance-era Maiolika in Europe. In Bronze Age crucibles from Askaraly, Kazakhstan, whitish slag crusts are present in crucibles, with the main constituent being Ca-feldspar anorthite ($CaAl_2Si_2O_8$) (Fig. 5.14).

Glassy slags devoid of crystalline inclusions are, in fact, rare. In many slags, copious amounts of devitrification products were found, such as olivine of predominantly hortonolitic composition (approximately 50:50 forsterite: fayalite), plagioclase, primary and secondary skeletal cassiterite, and in particular Sn-containing spinels (Heimann et al. 2010). Rovira et al. (2009) note, however, that the CaSn silicate malayite ($CaSnSiO_5$) can form in such slags.

If tin is smelted from alluvial heavy mineral placers, as in Rooiberg or in Cornwall and Devon (Great Britain), then, due to frequent associations of cassiterite with Ti-magnetite and zirconium ($ZrSiO_4$), corresponding phases can be expected in the placers. Following this, tin slags from such placers contain Ti-augite, zircon, rutile (TiO_2), ilmenite ($FeTiO_3$) and zirkelite ($(Fe,Ca)O{\cdot}2(Zr,Ti)O_2$). The TiO_2-content reaches up to >10 wt. %, the ZrO_2-contents up to 9 wt.% (Tylecote et al. 1989; Chirikure et al. 2010). This is reflected accordingly in the phases of the slags. However, the most common phase is cassiterite which is crystallising from the liquid and is not a relic of the charged ore. Because of the generally rapid cooling of slags, it forms colourless, needle-shaped to rhombohedral crystals, quite different from the compact prisms found in granitic rocks in nature, which crystallise much more slowly. Commonly, µm- to mm-sized, spheroidal metallic tin inclusions can be observed as well, which in turn often include inclusions of intermetallic SnFe phases.

It is also noteworthy that boron never exceeded the rather high detection limit of about 2 mass % B_2O_3, pointing to an effective beneficiation of the ore by separating cassiterite from the surrounding tourmaline, presumably by crushing and subsequent panning in water (Chirikure et al. 2010). But it is very typical that tin slags from Cornwall and Devon contain more than 1 wt.% antimony.

The glass matrix almost invariably contains thermally fractured remains of dissolving quartz grains as many slags from base-metal and iron smelting do. They are in part transformed to the SiO_2-high temperature-modification cristobalite and/or tridymite (see Sect. 5.6.1). This fact may establish a minimum processing temperature as the equilibrium temperature of the transformation of (pure) high-quartz to high-cristobalite/tridymite is in the range of 1050 °C (Flörke 1959; see also Heimann 1977). The transformation rate is catalytically promoted by a variety of impurities (e.g. alkalies, vanadium ions) and the formation of the high-temperature-modifications already may take place at temperatures as low as 800 °C (Bruhns and Fischer 2000).

5.5.5 Antimony Slags

Smelting residues accumulated in the production of antimony in ancient times are as rare as those of the recent past.

The earliest relics were found in a workshop of Sagebis Dsiris in Racha (Georgia) near an antimonite deposit (Hauptmann and Gambaschidze 2001). They date to the middle of the second millennium BC. This epoch corresponds to the first frequent occurrence of antimony artefacts in Georgia.

The slags of Sagebis Dsiris are surprisingly whitish coloured and are composed heterogeneously. Micrographs and X-ray diffraction show that they consist of various Sb-oxides such as cervantite (Sb_2O_4), valentinite (Sb_2O_3) and stibiconite ($Sb^{3+}Sb^{5+}{}_2O_6(OH)$). These antimony oxides are in part needle shaped. In-between, grains of quartz are embedded.

These may be perhaps relics of a roasting process, because to convert antimony sulphide from the ore or from the antimony crudum, a roasting process is required. In this process, various antimony oxides are already produced at 275 ° C, at about 400 ° C Sb_2O_4 is formed ("antimony ash") (Fig. 5.34).

In addition, Degryse et al. (2015) describe slags from Sagebis Dsiris consisting of quartz and metallic antimony, according to SEM/EDX measurements. Presumably, these are also Sb oxides.

These finds are only partially comparable with those from Sulzburg (Germany). From this locality Goldenberg (1994) and Siebenschock et al. (1996) describe smelting residues of antimony ores from around the fifteenth/sixteenth century AD. This is a porous, sintered mass of the gangue of quartz mineralisation and the antimony crudum (Sb_2S_3). They aptly characterise these residues as the remains of a liquation process, from which the Sb sulphide has precipitated from the quartz at very low temperatures (600–700 °C). Microscopic examination of this Sb sulphide revealed a markedly different picture compared to the starting ore, although it was chemically identical. The antimony crudum, which in the archaeological finding is often adhered to the remnants of the ceramic crucible wall, loses the twin lamination typical in the original ore by recrystallisation. The newly formed Sb_2S_3 crystals of the antimony crudum are mostly needle shaped.

Tylecote (1983) describes the slags of antimony smelting from the eighteenth century AD in Scotland. The gangue of the antimony mineralisation was separated by liquation of the latter and slag "cakes" were made and discarded. These have been Fe-silicate slags (c. 43 wt.% FeO, c. 35 wt.% SiO_2, c. 2–3 wt.% Sb) which contained inclusions of iron and antimony sulphides. The author furthermore mentions

Fig. 5.34 Sample Geo-101/1, Sagebis Dsiri, Racha, Georgia. Antimony slag consisting of crystals of quartz and, in-between, various antimony oxides. According to X-ray diffractometry (no. 4252/11) these oxides are cervantite, Sb_2O_4; valentinite, $Sb_6O_{13} \approx Sb_2O_3$; and stibiconite, $Sb^{3+}Sb^{5+}{}_2O_6(OH)$. Photo: A. Hauptmann, Deutsches Bergbau-Museum Bochum

Sb-rich slags from Greece, but at the same time, he emphasises his doubts about these being slags from an antimony production.

5.6 Important Slag Phases

In contrast to naturally occurring minerals, the crystals occurring in siliceous-oxidic slags of archaeometallurgy are to be referred to as phases, because they are artificial technical products. This is also true for other materials related to archaeometallurgical operations, such as all types of (technical) ceramics. In most cases, the slag phases are of comparable chemical composition to naturally occurring minerals, so that the names of the latter can be used.

Slags contain inclusions of metals, metal sulphides (matte, Sect. 5.7) or of speiss (noble metal containing intermediate of arsenides, antimonides, Sect. 5.8), which give indications about the composition of smelted ores. Slag minerals are only those that have been formed from slag phases through natural corrosion. Bachmann (1982a, b, c) studied this on slags from Laurion, which were stored in a submarine environment. Occasionally, however, sulphides may also be formed in soil bedding in anaerobic and sulphur-bearing environments, as shown by Rehren (2004) on Pfyn's Neolithic copper crucibles in Switzerland. He was able to prove that the newly formed rims of copper-iron sulphides on the inner side of the crucibles had nothing to do with the smelting of ores.

An extensive overview of secondary slag minerals was compiled by Kolitsch et al. (2013).

5.6.1 Slag Phases and Phase Associations

In the following section, the most important slag phases formed by metallurgical high-temperatures processes, which were observed in archaeometallurgical slags, are discussed. Their chemical composition is shown in Table 5.4.

The vast majority of smelting slags that have been produced worldwide since about the Middle to Late Bronze Age and the Iron Age consist of iron silicates, in particular Fe-rich olivine (fayalite) and Fe-rich clinopyroxene (hedenbergite). This applies to slags from Europe, Asia, Africa and

Table 5.4 Selected mineralogical phases (silicates, oxides) observed in archaeometallurgical slags

Silicates	
Olivine	
Fayalite	Fe_2SiO_4
Forsterite	Mg_2SiO_4
Kirschsteinite	$(CaFe)_2SiO_4$
Knebelite	$(MnFe)_2SiO_4$
Laihunite (Ferrifayalite)	$Fe^{2+}Fe^{3+}{}_2(SiO_4)_2$
Merwinite	$Ca_3Mg(SiO_4)_2$
Monticellite	$CaMgSiO_4$
Tephroite	Mn_2SiO_4
Willemite	Zn2SiO4
Iscorite	$Fe^{2+}{}_5Fe^{3+}{}_2SiO_{10}$
Clinopyroxene	
Augite	$(Ca,Mg,Fe)_2Si_2O_6^{[}$
Diopside	$CaMgSi_2O_6$
Hedenbergite	$CaFe^{2+}Si_2O_6$
Melilite	
Gehlenite	$Ca_2(Al,Mg)(Al,Si)SiO_7$
Åkermanite	$Ca_2MgSi_2O_7$
Melilite	$(Ca,Na)_2(Mg,Al)Si_2O_7$
Andremeyerite	$BaFe_2Si_2O_7$
Feldspar	
Anorthite	$CaAl_2Si_2O_8$
Celsian	$BaAl_2Si_2O_8$
Feldsparoide	
Leucite	$KAlSi_2O_6$
Kalsilite	$KAlSiO_4$
Wollastonite	$CaSiO_3$
Mullite	"$Al_6SiO_2O_{13}$"
Alite	Ca_3SiO_5
Oxides	
Cuprite	$Cu^+{}_2O$
Delafossite	$CuFeO_2$
Hematite	Fe_2O_3
Periclase	MgO
Quartz and its high-temperature modifications	
Cristobalite, Tridymite	SiO_2
Magnetite (Spinel in the larger sense)	Fe_3O_4
Tenorite	$Cu^{2+}O$
Wuestite	"FeO"

Given are simplified chemical formulas of the phases. These, in most cases, occur as solid solutions in various concentrations. Some of the chemical formulas are given in a simplified way after Trojer (1963)

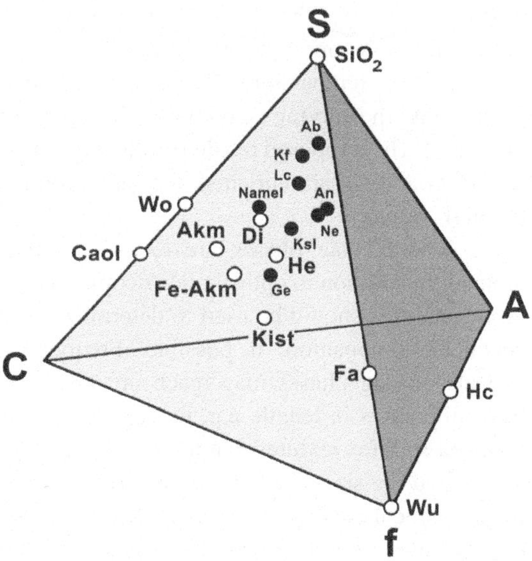

Fig. 5.35 Important phases in iron rich silicate slag, shown in the system $CaO–FeO–Al_2O_3–SiO_2$ (C–f–A–S). *Ab, Kf, An* feldspars (albite, kalifeldspar, anorthite); *Lc, Ksl, Ne* feldspathoids (leucite, kalsilite, nepheline); *Namel, Ge* melilites (Na-melilite, gehlenite); *Akm, Fe-Akm* åkermanites, ferroåkermanite; *Di, He* clinopyroxenes (diopside, hedenbergite); *Wo, Caol* calcium silicates (wollastonite); *Kist, Fa* olivines (kirschsteinite, fayalite); *Wu* wuestite; *Hc* hercynite; *SiO_2* cristobalite, tridymite. Modified after Keesmann (1989), from Hauptmann (2007)

New World countries. In base metal slags they contain varying proportions of Fe-rich spinels (magnetite), in iron slags wuestite.

Such slags or slag phases are no longer produced in modern metallurgical processes.

The composition of (silicate and oxidic) slag phases and their associations have been discussed in their mineralogical-petrographic aspects by Keesmann (1989, 1993). The choice of a suitable system focussed on the quarternary system $CaO–SiO_2–FeO–Al_2O_3$ (Fig. 5.35). He differentiates, for example, between the paragenesis fayalite—wuestite—hercynite in iron smelting slag with the projection into the system $SiO_2–FeO–Al_2O_3$ and clinopyroxene-bearing parageneses in the system $CaO–FeO–SiO_2$. A mathematically based correction of the chemical analysis is necessary for a more accurate estimate of the crystallisation temperature. Keesmann accounts for this with the varying amount of "glass" within the complexly composed residual melt in the slag, where Al and Ca are enriched and thus produce too high values.

In general, slag phases are extremely fine-grained. In addition to optical microscopy, microprobe analyses should be used to determine their chemical composition, if possible. Frequently, crystals of slag phases may reach up only to a few millimetres in length, e.g. in slags with pronounced spinifex textures, but more often they do not exceed the size of a few micrometres. Therefore, chemical analyses of slag phases measured by electron microprobe may possibly be contaminated by surrounding crystals or by glass.

The mineralogical phase inventory, as well as the total chemical composition of archaeometallurgical slags, is determined primarily by the composition of the furnace charge. In most cases, material rich in iron ores or quartz from the gossan of sulphidic ore deposits dominates, or from gangue and host rock. The use of deliberately added fluxes should always be critically examined, especially for slags from earlier eras. Material influences of ceramic crucible or furnace walls or tuyères are possible.

5.6.1.1 Olivines

Olivine has an extended range of solid solutions according to (Fe, Mg, Ca, Mn, $Zn)_2SiO_4$. Fe-rich olivine (fayalite, Fe_2SiO_4) is the dominating phase in very many Fe-rich smelting slags. Fayalite is an easily crystallising slag phase and, depending on cooling rate and oxygen fugacity of the liquid slag, provides an almost limitless variety of shapes and textures. In case of a slow cooling of the liquid it crystallises in idiomorphic, granular individuals, often cotectic intergrown with wuestite (Fig. 5.36a). At a fast cooling rate fayalite crystallises in (elongated) skeletons and feathery dendrites and often displays a so-called spinifex texture (Fig. 5.36b). This name has its origin in petrology, where it refers to the crystallisation of volcanics, which were exposed to very rapid cooling rates. In between there are, according to the descriptions of Donaldson (1976), all transitions from (hyp-)idiomorphic crystals, various skeletal forms to dendritic shaped crystals. The texture can change very fast within one and the same slag cake. Fayalite crystals are colourless, have a high refraction of light and are easily recognisable by their distinct interference colours (green–yellow–orange–violet).

Pure fayalite contains 70.5 wt.% FeO and 29.5 wt.% SiO_2. It is an end member of the

Fig. 5.36 (**a**) Kition, Cyprus (Greece), sample Cy-1/1b. Late Bronze Age copper smelting slag with skeletal crystals of fayalite. Width of the picture c. 2 mm. (**b**) Kition, Cyprus (Greece), sample Cy-1/1d. Late Bronze Age copper smelting slag with very thin and long needles of fayalite forming a spinifex texture. Width of the picture c. 2 mm. (**a, b**) Surface-polished thin sections, transmitting light, crossed polarisers. From Hauptmann (2011)

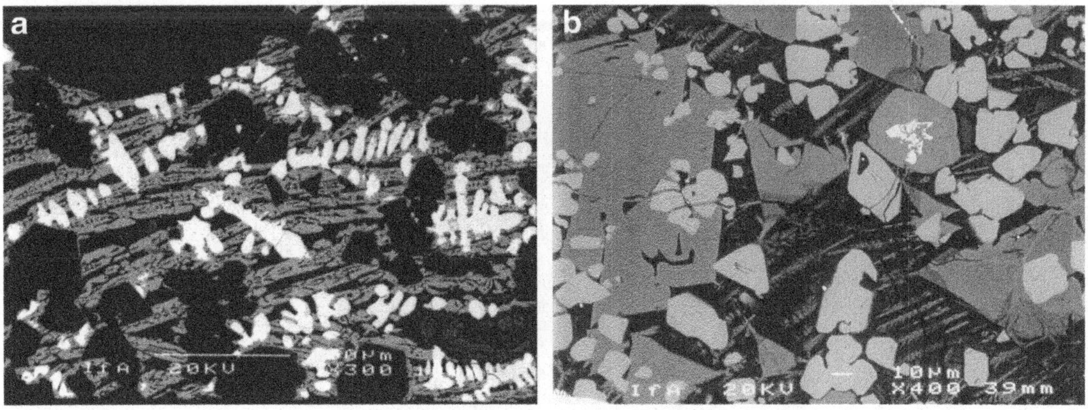

Fig. 5.37 (**a**) Großkuchen, Swabian Alb (Germany), sample D-4/2. Bloomery slag from the third/fourth century AD. The main component is fayalite (grey), crystallised as elongated skeletons or as chains. Fayalite is associated with dendrites of wuestite (white). Its stained surface shows that it is in parts oxidised to magnetite. (**b**) Çayönü Tepesi (eastern Anatolia), sample 84/27 O 4–15. Early Bronze Age copper smelting slag. Primary crystallisation of idiomorphic crystals of magnetite (grey–white) associated with fayalite (grey) which surrounds the magnetite. Fayalite is crystallised also in a second stage as thin needle-skeletons. Both pictures SEM-pictures, backscattered mode. Note scale inside the pictures. Fotos: (**a**) From Yalçın and Hauptmann (1995); (**b**) T. Rehren/A. Hauptmann, Deutsches Bergbau-Museum Bochum

consistent solid solutions fayalite (Fe_2SiO_4)—forsterite (Mg_2SiO_4)—tephroite (Mn_2SiO_4) (Fig. 5.32). The fourth end member, larnite (Ca_2SiO_4), has not been observed in archaeometallurgical slags. In iron slags, Fe-rich olivine is associated with wuestite and thus shows more reducing conditions than in copper slags, in which the phase association fayalite + magnetite is the rule (Fig. 5.37a, b).

Crystals of fayalite may show pronounced zoning with MgO rich cores and FeO and CaO enrichments towards the rims (Fig. 5.38).

Ca-rich olivine has been described from many slags of various metallurgical processes. At the rims of these olivine crystals sharply separated, beard-like intergrowths are formed (Fig. 5.38). They occur in prehistoric copper slags from Oman (Hauptmann 1985) or in Islamic (high mediaeval) slags in Iran (Hezarkhani and Keesmann 1996). In slags of iron metallurgy (smelting and processing) Ca-rich olivine has been observed as well (Kronz 1997). Reasons for the formation of Ca-rich olivine are essentially the overall composition of the slag melts and the oxygen partial pressure of the gas atmosphere in the reaction vessel. Low oxygen partial pressure promotes increased Ca incorporation in olivine (Snyder and Carmichael 1992). CaMg-monticellite only forms exceptionally. The most Ca-rich variants of olivine correspond in their compositions to mixed crystals between monticellite ($CaMgSiO_4$) and kirschsteinite ($CaFeSiO_4$). Between the forsterite–fayalite solid solution and kirschsteinite, there exists a miscibility gap (Mukhopadhyay and Lindsley 1983), as shown in Fig. 5.39.

Olivine in slags contains no or very little aluminium. Olivine analyses with Al_2O_3—contents of 2–6 wt.% (Yalçın and Hauptmann 1995) are likely due to other phases mixed in the residual liquid.

In copper metallurgy, manganese-rich slag melts containing tephroite (Mn_2SiO_4) and knebelite ($MnFeSiO_4$) as predominant phases are common. They are based on correspondingly composed Mn/Fe-rich copper ores. Examples are slags that have been melted from sedimentary-epigenetic deposits in the southern Levant (Faynan, Timna) and on the Sinai Peninsula (Abdel-Motelib et al. 2012; Hauptmann 2007;

Fig. 5.38 Maysar 1, Oman, sample 350/1. Early Bronze Age copper slag showing pronounced zoning of two crystals of fayalite-rich olivine. The core is rich in MgO, the sharply formed fibrous rim consists of two phases, namely CaO- and FeO-rich olivines. This exsolution is shown in Fig. 5.39. Between the idiomorphic crystals of olivine crystals of magnetite (grey) and droplets of copper (white). SEM-picture, secondary electron mode. Width of the line c. 100 µm. Photo: A. Hauptmann, Deutsches Bergbau-Museum Bochum

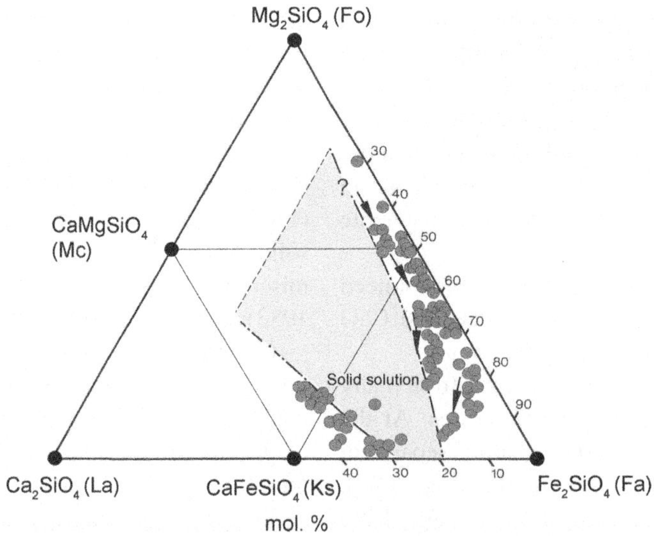

Fig. 5.39 Composition of Fe-rich olivines from various copper slags and smithing slags plotted in the ternary system forsterite (Fo)—fayalite (Fa)—larnite (La). Arrows show the tendency of crystallisation following the cooling of the slags. Note the miscibility gap between the two solid solutions forsterite—fayalite and kirschsteinite (Ks)—monticellite (Mc). From Hauptmann (1985)

Krawczyk 1986), as well as from Cyprus (Bachmann 1982a, b) and Oman (Hauptmann 1985). The MnO contents of these slags are between 15 and 50 wt.%, depending on the composition of the parent ore (Fig. 5.32). The corresponding olivine in these slags can reach up to 60 wt.% MnO. Microscopically, they are hardly distinguishable from fayalite.

Zn-olivine of the composition of willemite (Zn_2SiO_4) is found in lead and copper slags from the smelting of Zn-rich ores. This phase is easy to identify by its remarkable pleochroic colours (red violet to blue violet). A solubility of Zn in olivine and immiscibility between olivine and willemite was assumed by Faber (1936) and later shown in experiments (Sarver and Hummel 1962; Ericsson and Filippidis 1986). According to these authors, olivine can contain up to 18 mol. % Zn_2SiO_4 at ~1000 °C. Willemite occurs in mediaeval slags from the Pb-Zn-deposit of Wiesloch (Germany) (Ströbele et al. 2010).

5.6.1.2 Ferrifayalite (Laihunite)

In the literature, olivine containing Fe^{3+} is referred to as either laihunite or ferrifayalite ($Fe^{2+}Fe^{3+}{}_2(SiO_4)_2$) (Fig. 5.40). However, the definition of both phases is not explicit (Ginzburg et al. 1962; Fleischer et al. 1977; Laihunite Research Group 1976). Schäfer (1983) refers to ferrifayalite/laihunite as an intergrowth of fayalite and laihunite with the chemical formula $Fe^{2+}Fe^{3+}{}_2(SiO_4)_2$ and thus defines the mineral as a variety of fayalite with 10–40 wt.% Fe_2O_3. Ferrifayalite was found in copper slags from Trentino, namely slag cakes which obviously solidified slowly within a furnace (Fig. 5.41; Metten 2003). Under the microscope, she observed dark brown to black, opaque thick prismatic crystals that show a lighter grey in reflected light than the fayalite. In many cases, they are intergrown with fayalite in a domain-like way. She suggests the ferrifayalite to rather be a product of oxidation at high temperatures after the primary formation of fayalite than a primary precipitation of ferrifayalite itself. She could exclude a subsequent corrosion of fayalite via soil storage after the melting process.

5.6.1.3 Iscorite

Iscorite ($Fe^{2+}{}_5Fe^{3+}{}_2SiO_{10}$) was first described in modern iron smelting slags from South Africa (Smuts et al. 1969). It was observed there in strikingly Fe-rich silicate slags, which were formed at the bottom of a reheating furnace in steel production (FeO ca. 70–75 wt.%; SiO_2 ca. 20 wt.%). Mineralogically, iscorite-bearing

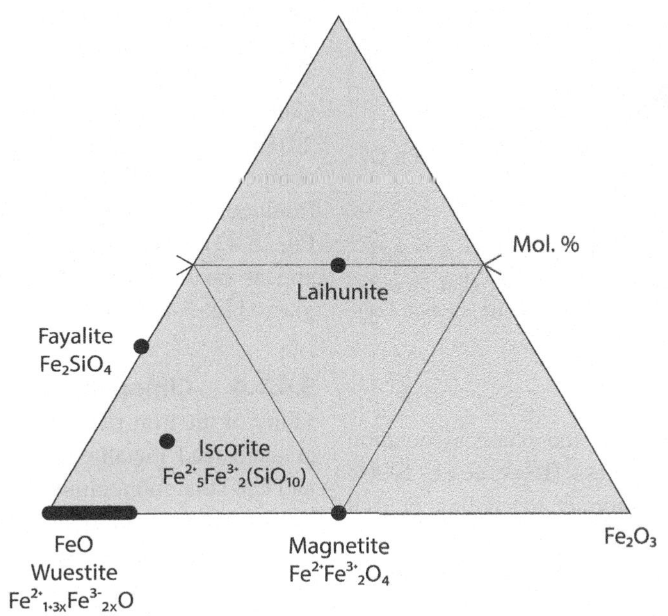

Fig. 5.40 Fe-Oxides and Fe-silicates in archaeometallurgical slags shown in the ternary system $FeO–Fe_2O_3–SiO_2$. Modified after Rose et al. (1990)

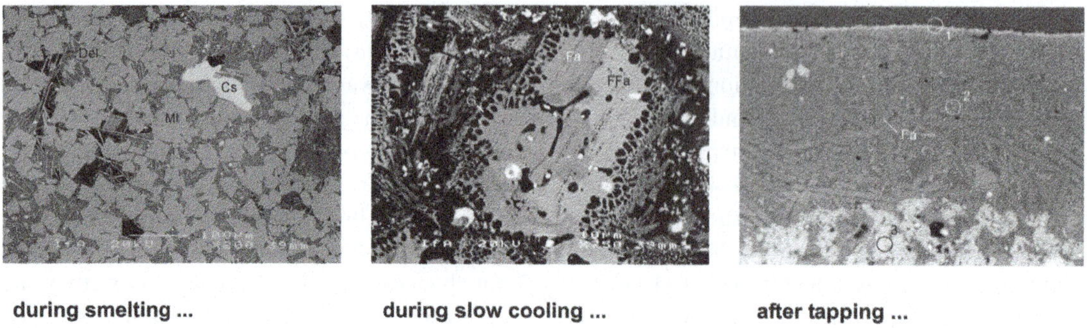

Fig. 5.41 Acqua Fredda, Trento (Italy), Late Bronze Age slags. Phenomena of oxidation during smelting copper ores. (**a**) Smelting under rather oxidising conditions leads to primary precipitation of magnetite (Mt) surrounded by needles of delafossite. Inclusions of CuFe-sulphides (Cs). (**b**) Fayalite—ferrifayalite intergrowth (Fa, FFa) in an idiomorphic crystal. The rim shows eutectic intergrowth with glass (G). (**c**) Texture of the upper part of a heterogeneous plate slag. The micrograph shows their different types of magnetite: (1) magnetite forms at the surface during cooling under oxygen influence, (2) fine-grained magnetite crystallises between elongated fayalite (Fa), (3) dissolving agglomeration of magnetite. SEM-pictures, secondary electron mode. From Metten (2003)

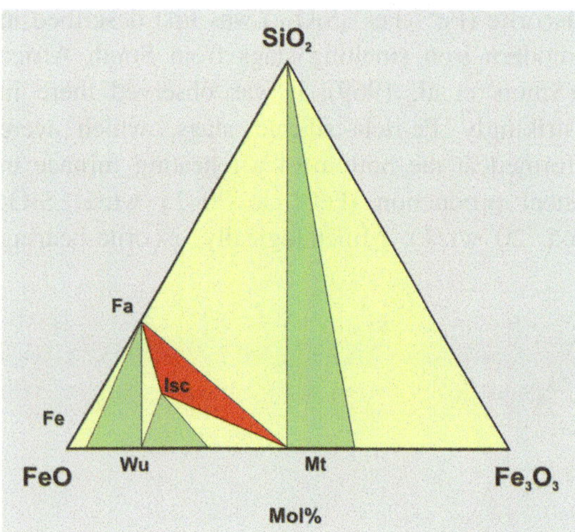

Fig. 5.42 Range of stability of iscorite ($Fe^{2+}{}_5Fe^{3+}{}_2SiO_{10}$) and its phase association with fayalite (Fa), wuestite (Wu) and magnetite (Mt), plotted in the system FeO–Fe_3O_4–SiO_2. From Keesmann (1989)

slags are characterised by the phase association wuestite + fayalite + glass (Rose et al. 1990). Iscorite is a reaction product in cooling Fe-rich silicate melts and contains both Fe^{2+} and Fe^{3+} (Figs. 5.40 and 5.42). It forms according to the reaction:

$$Fe^{2+}Fe^{3+}{}_2O_4 + 2Fe^{2+}O + Fe^{2+}{}_2SiO_4 \rightarrow Fe^{2+}{}_5Fe^{3+}{}_2SiO_{10}$$

The detection of iscorite in thin sections poses no particular difficulties because its strip- to tabular-shaped crystals are highly anisotropic. Iscorite has been diagnosed as a minor component or accessory in numerous archaeometallurgical slags of iron, copper and Pb–Zn technology (Keesmann 1989; Rose et al. 1990; Yalçın and Hauptmann 1995; Kronz 1997; Ganzelewski 2000; Metten 2003; Ströbele et al. 2010). Iscorite occasionally even occurs as a main component, as in the slags of the Imperial Iron Production in Joldelund (Schleswig-Holstein, Fig. 5.43). There, strip-shaped iscorite crystals appear on magnetite in contact with K-Al-rich glass (Fig. 5.43).

5.6.1.4 Clinopyroxene

Many of the iron-rich slags, especially from copper and lead metallurgy, contain so much SiO_2 and CaO that monoclinic clinopyroxene (Ca-rich) is formed. It is usually less abundant than olivine, but it can also be the main constituent in slags, as was observed by Krawczyk (1986) in samples from Timna (Israel, Late Bronze Age), Beer Ora

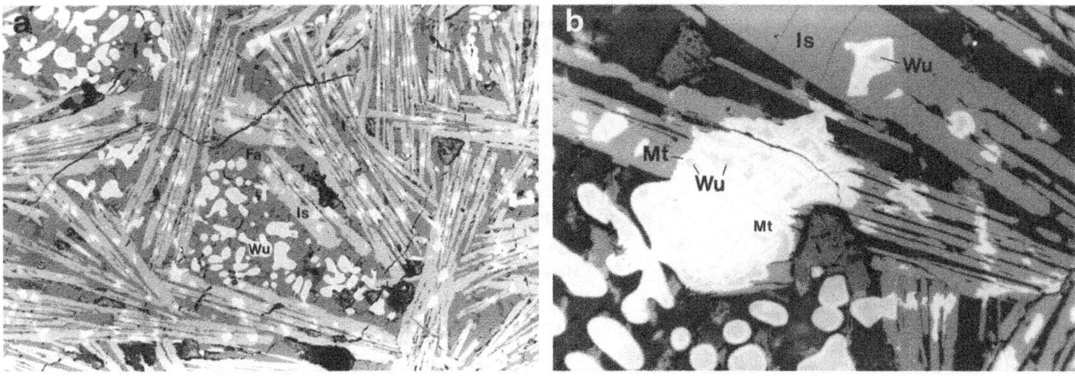

Fig. 5.43 (**a**) Kammberg, Joldelund (northern Germany), sample D-27/11b. Tapslag from Roman iron smelting site. The main component is lath-like iscorite (Is). Wuestite (Wu) reacted with fayalite (Fa) to form iscorite. Wuestite is partly oxidised to magnetite (Mt) Surface-polished thin section, reflected light. Width of the picture 0.8 mm. From Ganzelewski (2000), Deutsches Bergbau-Museum Bochum. (**b**) Cayönü Tepesi, SE-Anatolien, sample CT-84/270 4-15. Hand sized slag cake from copper smelting, Early Bronze Age I (c. 3000 BC). Lath-shaped crystals of iscorite sitting on idiomorphic round magnetites. Iscorite results from the reaction between magnetite and siliceous melt. It precipitates before fayalite (dark grey laths). This characteristic phase assemblage is formed under low reducing conditions as they might be significant for copper slags of early stages of metallurgy. Surface-polished thin section reflected light. Width of picture ca. 1.5 mm. Oilimmersion. Photo: A. Hauptmann, Deutsches Bergbau-Museum

(dto., Byzantine to Early Islamic period) and Eilat (dto., undated). Kronz (1997) notes that clinopyroxene does not occur in iron metallurgy.

Clinopyroxene is colourless in thin section. However, in most cases, it has a pronounced pleochroism with colours ranging from orange-yellow dark brown to bottle-green already in polarised light with feathery-tufted skeletal crystals or needle-shaped crystal forms to idiomorphic individuals (Hauptmann 1985; Hezarkhani and Keesmann 1996). Especially iron-rich pyroxenes are partially intertwined with magnetite. The examination of these intergrowths in detail shows that the pyroxene sometimes finely lamellar and includes in this skeleton both spinel and glass.

The chemical composition of clinopyroxenes can be shown in the concentration system $CaSiO_3$–$FeSiO_3$–$MgSiO_3$ (wollastonite–enstatite–ferrosilite) (Fig. 5.44). Depending on the Mg or Fe content of the slag melt, their chemical composition is mostly between diopside and hedenbergite with a focus on hedenbergite and ferrohedenbergite.

In addition, clinopyroxene in ancient slags may contain high levels of Al_2O_3 (Tschermak's molecule, cf. Hauptmann 1985). Microprobe measurements of clinopyroxene also showed various proportions of Fe^{3+} and Cu^{2+}. Ettler et al. (2001) found up to 0.20 wt.% PbO in clinopyroxene in historical slags from Příbram (Czech Republic). The amount of Zn was significantly higher, up to 3.12 wt.% ZnO, corresponding to zincian hedenbergite (Morimoto et al. 1988). The average amount of Zn in clinopyroxenes is 2.26 wt.% ZnO. Hezarkhani and Keesmann (1996) point out that pyroxenes are often intensely intergrown with magnetite in the micro range. Here are problems with the determination of microprobe analyses.

5.6.1.5 Pyroxenoide

A certain special position is occupied by pyroxene-like phases, namely Mn-rich pyroxenoids, which crystallise in slags in which FeO is largely replaced by MnO (c. MnO 15–30 wt.%). The phases to be classified here are bustamite ($(Mn,Ca)_3(Si_3O_9)$ and rhodonite $(Ca,Mn)_4Si_5O_{15}$) (Hauptmann 2007).

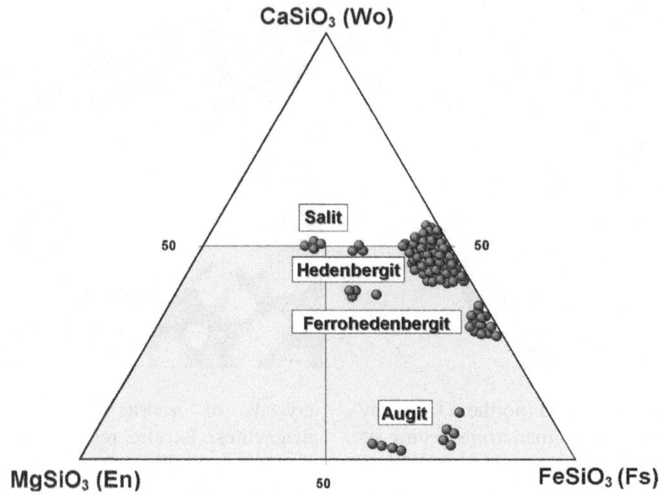

Fig. 5.44 Clinopyroxene in ancient slags from various archaeological sites and regions (Oman, Timna, Anatolien, Iran), plotted in the ternary system wollastonite (Wo)—enstatite (En)—ferrosilite (Fs) after Deer et al. (1978). Most of the clinopyroxene has a composition close to hedenbergite ($CaFeSi_2O_6$) and Ferrohedenbergite. Only a few measurements show a tendency to CaMg-rich compositions (salite) and to augite. According to data from Hauptmann (1985), Krawczyk (1986), Hess (1998), Hezarkhani and Keesmann (1996). *Wo* wollastonite, *En* enstatite, *Fs* ferrosilite. Values given in Mol.%

5.6.1.6 Melilites

Melilites are silicates with the general formula $X_2YZ_2O_7$ or $(Ca,Na,K)_2(Mg,Fe^{2+},Fe^{3+},Al)(Si,Al)_2O_7$. They are very complex solid solutions of the end members åkermanite—Fe-åkermanite—gehlenite—Fe-gehlenite—K-Na-melilite. Zn-rich melilite is called hardystonite.

In modern times, melilites are important components of modern blast furnace slags and concretes. But they also occur in prehistoric non-ferrous metal slags. Here, the following parageneses can be distinguished (Hezarkhani and Keesmann 1996; Hauptmann 2007):

- Melilite slags
- Melilite–olivine slags
- Melilite–olivine–pyroxene slags

Melilite crystals show predominantly zoning. The core is higher in MgO and becomes Fe-richer towards the outside. It forms in long prisms or in planar slabs. Often, it crystallises in a typical peg structure with a "middle juncture" consisting of inclusions. Frequently, in ancient slags Ca–Mg melilites are crystallising with only moderate iron (åkermanite) or zinc contents. In fayalite-rich slags, melilite was not observed.

A prerequisite for the formation of melilites in slags is a high Ca enrichment at a moderately high SiO_2 content in the charged material (Hezarkhani and Keesmann 1996). This is the case in a smelting of ores embedded in calciferous or dolomitic gangue or host rock. This includes many iron ore deposits like, e.g. the Erzberg (Styria, Austria) or the bean ores of the Swabian Alb (southern Germany). Copper ore in such a gangue was also observed in Afunfun (Nigeria). Smelting these ores produced melilite bearing base metal slags (Keesmann 1985). At Faynan (Jordan), copper mineralisations in sandy-dolomitic rocks were exploited in the fourth millennium BC (Fig. 5.45). In these early slags, melilite associated with clinopyroxene was observed (Hauptmann 2007). The same applies to Mallorcan copper slags of the Bronze Age (Ramis et al. 2005). Ca–Mg-bearing melilites, sometimes with Zn contents, were discovered in slags of prehistoric fahlore mining in the Inn Valley, Tyrol (Austria) (Bartelheim et al. 2002; Krismer et al. 2013). The gangue of the silver-

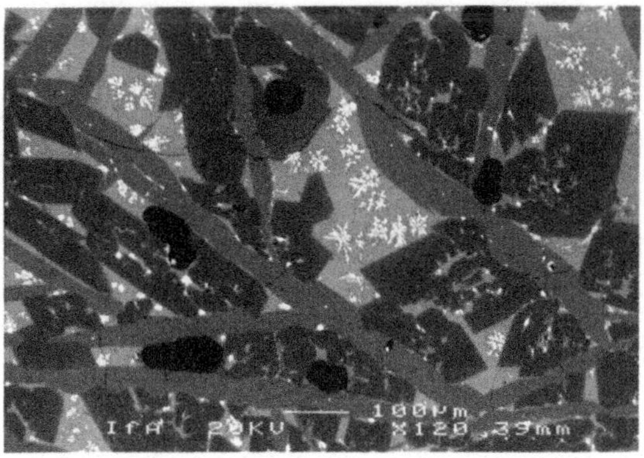

Fig. 5.45 Wadi Fidan 4, fourth millennium, Faynan copper district (Jordan), sample JD-8/29. Melilite-olivine-slag. Tabular crystals of merwinite (olivine, $Ca_3Mg(SiO4)_2$, Mw) and rhombohedric åkermanite (Å). Within the matrix dissolution of dendritic cuprite in glass. SEM-picture, secondary electron mode. From Hauptmann (2007)

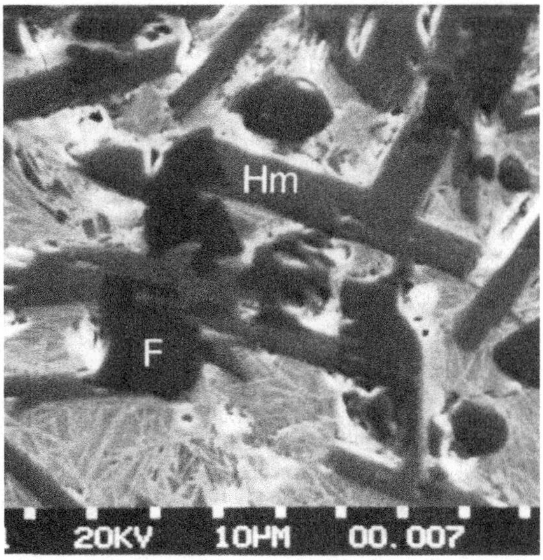

Fig. 5.46 Skres, island of Thasos (Greece), sample TH-14/1. Slag from a smelting site with 3000–5000 tons dated to the Hellenistic-Roman period. As indicated by crystals of hematite (Hm) the slag was highly oxidised and possibly corroded. Note the whitish needles of lead silicates filling the voids. A Zn-containing spinel (franklinite) is marked by F. SEM, secondary electron mode, scale = 10 µm. From Hauptmann et al. (1988)

bearing tennantite–tetraedrite deposits consists of the Devonian Schwaz dolomite of the northern Greywacke zone. At the Swabian Alb, in mediaeval iron slags, microlites of Fe-gehlenite—Fe-åkermanite, formed by devitrification have been observed by Yalçın and Hauptmann (1995).

Hardystonite is a rare calcium zinc silicate mineral. It is frequently found in lead smelting slags and is associated with willemite $Zn_2(SiO)_4$.

5.6.1.7 Andremeyerite

The phase occurs in old slags of lead–silver metallurgy in varying amounts. Andremeyerite was described in Roman slags of Rio Tinto (Craddock et al. 1985) and by Hauptmann (1988) in Hellenistic slags from the Greek island of Thasos, where it partly was the main constituent.

Mineralogically, andremeyerite belongs to the family of melilites. It crystallises in needle-like to tabular crystals. It is colourless to emerald green/brownish in the microsection and shows very strong pleochroism. In addition to hyalophane and a glassy residual liquid, it may contain up to 0.7 wt.% PbO. It occurs associated with (Ca-rich) olivine, hyalophane, clinopyroxene and Zn-containing spinel (Fig. 5.46).

5.6.1.8 Feldspars

Feldspars of the solid solution series anorthite ($CaAl_2Si_2O_8$)—albite ($NaAlSi_3O_8$)—orthoklas ($KAlSi_3O_8$) are rather rare constituents in archaeometallurgical slags. Plagioclas is a synonym for the anorthite—albite mineral series.

In iron metallurgy, anorthite-rich feldspar was occasionally found in slags in the Lahn-Dill area (Germany) (Kronz 1997). The author explains the formation of this phase with increased SiO_2 and Al_2O_3 contents in the residual liquid by crystallisation of SiO_2 + magnetite under oxidising conditions instead of fayalite or reduction of fayalite to $Fe^0 + SiO_2$. Feldspar in fayalitic slags may contain up to c. 2 wt.% FeO (Kronz and Keesmann 2005).

In slags with high BaO contents, which are primarily associated with the lead–silver metallurgy and have emerged from deposits with high levels of baryte in the gangue, the BaO-feldspar hyalophane can occur as the main constituent.

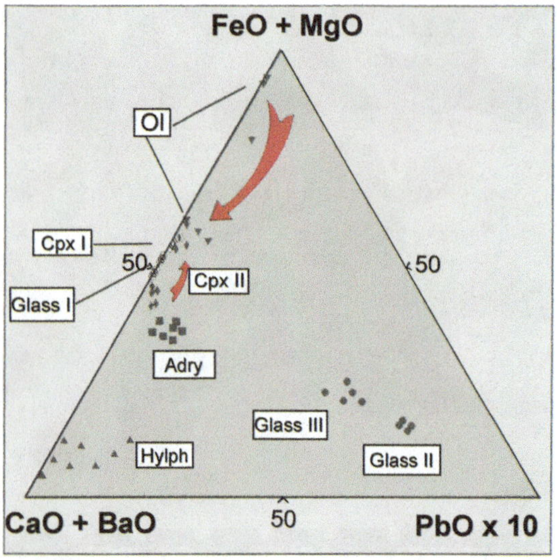

Fig. 5.47 Chemical composition of slag phases in lead–silver slags from Thasos (Greece) plotted in the ternary system CaO + BaO – FeO + MgO – PbOx10. Note the PbO-concentrations analysed in andremeyerite (Adry) and hyalophane (Hylph)! Glas I and clinopyroxene (Cepx I) were observed in a mostly glassy solidified slag. Glass II, III are phases of the residual liquid in more complete crystallised slags. Plotted in mol-%. From Hauptmann et al. (1988)

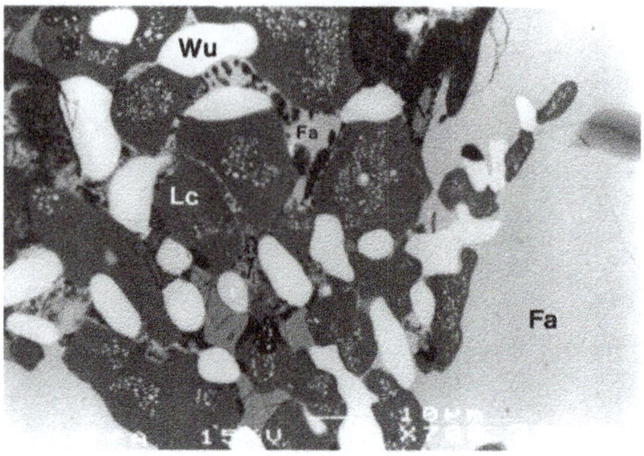

Fig. 5.48 Smithing slag, Kammberg/Joldelund (Germany), sample D-27/86 g. Between fayalite (Fa) and wuestite (Wu) crystals of leucite (Lc) precipitated. From these dots of Fe oxides are exsolved which form cotectic intergrowths. Between the leucite fayalite in a second generation is growing. SEM-picture, backscattered mode. From Ganzelewski (2000)

Hyalophane ((K,Ba)$Al_2Si_2O_8$) is a solid solution between orthoclase ($Ca_2Al_2Si_2O_8$) and the final phase celsian ($BaAl_2Si_2O_8$).

Hyalophane was observed in BaO-rich (about 2–19 wt.%) Hellenistic slags of the Aegean island of Thasos as a regularly occurring phase (Hauptmann et al. 1988). It has also been observed in mediaeval slags of lead–silver metallurgy in the Ore Mountains (Germany) (Eckstein et al. 1994) as well as in slags of Phoenician age from Monte Romero/Rio Tinto (Kassianidou et al. 1995). Hyalophane contains 30–40 wt.% BaO. Faber (1936) and Trojer (1963) also describe hyalophane in modern slags of lead metallurgy. Remarkable are, in contrast to natural feldspars, Fe-, Pb- and Zn-contents in the level of percentages. The phase forms thin slab-like crystals and needle-like crystals (Fig. 5.47). It crystallises before or cotectically with Fe-rich olivine, the crystals of which it penetrates or on which it sits "fur-like" (Hauptmann et al. 1988). Hyalophane shows twinning according to the albite law, typical are inclusions of glass in a zoned arrangement. Except for the glassy residual liquid hyalophane in slags from the island of Thasos, it is the phase that has the highest lead concentrations (Pb^{2+} up to 0.5 wt.%).

5.6.1.9 Leucite

In iron-rich slags intergranular, i.e. in the residual melt leucite ($KAlSi_2O_6$) often occurs in idiomorphic to anhedral crystals with often drop- to rod-shaped inclusions of iron oxides. It is relatively easy to identify microscopically due to its low reflection in incident light and also due to these inclusions in the form of so-called scoriaceous garlands. Often its idiomorphic crystals are surrounded by a leucite hemisphere in cotectic intergrowths with various oxide (wuestite, magnetite) and silicate phases (Fig. 5.48). If this intergrowth covers the entire crystal, then the phase is difficult to identify.

Leucite is the main potassium carrier of fayalite slags, whereby the potassium content was probably introduced into the slag melt essentially by the fuel ash and by the melting furnace wall (Kronz 1997). This is the reason why leucite is found mainly in processing slags from iron metallurgy. The crystallisation of leucite in the residual melt corresponds to the effect of K_2O as the melting point-lowering component. Leucite is also common in mediaeval Pb–Zn slags from Wiesloch (Germany) (Ströbele et al. 2010).

5.6.1.10 Mullite

The chemical formula of mullite may be given in a simplified way as $Al_6Si_2O_{13}$ (Trojer 1963). In reality, its composition shows considerable variations. In modern materials, mullite is a predominant phase in chamotte (fire clay) or in porcelains.

In archaeometallurgy, mullite is a rarely occurring silicate, which was previously observed only in furnace walls of mediaeval iron smelting in the Lahn-Dill area (Kronz 1997) and in the adjacent Siegerland (Gassmann and Yalçın 2009). The phase occurs as columnar crystals with a length of about 10 µm. To what extent the occurrence of mullite could point to a systematic selection of Al_2O_3-rich clay to improve the thermal stability of furnace walls, or whether corresponding clays occur in these regions anyway, must remain open.

Fig. 5.49 Wadi Ghwair, Faynan copper district (Jordan), Early Bronze Age II/III, sample JF-21/1a. Glassy copper slag with streaks of flow structure. Beginning crystallisation of cuprite and metallic copper in the colloidal state gives the slag in reality a reddish tint. Transmitting light, width of the picture ca. 1 mm. Photo: A. Hauptmann, Deutsches Bergbau-Museum

5.6.1.11 Alite

This phase, alite (Ca_3SiO_5) is known from the modern industry of Portland cement clinker, is described by Eckstein et al. (1994) in mediaeval lead–silver slags from the Ore Mountains close to the border of Germany—Czech Republic. Given the melting point of alite, which is above 1700 °C (Trojer 1963), they suspect that this phase, finely intergrown with fluorite (CaF_2), has been formed by the decomposition of a complex calcium fluorosilicate. Fluorite is apparently secondary.

5.6.1.12 Glassy Slag and Siliceous Residual Liquid

If amounts of glass or microcrystalline phases in the residual liquid in slags are analysed, then one realises that these are much lower in FeO than the corresponding bulk compositions. Aluminium, sodium, potassium and phosphorus are enriched, while the percentage of magnesium is greatly reduced.

In addition to crystalline slags, very glass-rich types can be found on numerous slag sites. These are observed in copper slags, but also occur in mediaeval lead–silver metallurgy. Some are glasses with incipient crystallisation along flow marks (Fig. 5.49). In crystalline slags, residual melt is partly also glassy with transitions to microcrystalline texture.

5.6.1.13 Spinel

In most ancient slags, Fe-rich spinel (magnetite) is the first phase to crystallise and can act as a nucleus for wuestite dendrites or even iscorite. It can also be embedded in olivine. Spinel forms euhedral crystals. The composition of the crystalline individuals is in many cases inhomogeneous, whether by zoning or by oriented inclusions (lamellae).

Using the example of copper slags from Iran, Hezarkhani and Keesmann (1996) show that spinel is characterised by high contents of Fe and Al. These are essentially mixed crystals between magnetite ($FeFe_2O_4$) and hercynite ($FeAl_2O_4$). The zonings may be different in various samples: the core might be high in magnetite composition and the rim hercynite-rich and vice versa. Furthermore, lamellar inclusions with magnetitic compositions in a matrix and a cover of spinel rich in hercynite have been observed.

There were also found minor amounts of magnesioferrite ($MgFe_2O_4$), jacobsite ($MnFe_2O_4$), ulvite (Fe_2TiO_4) and cuprospinel ($CuFe_2O_4$). In Mn-rich slag, e.g. from the Faynan ore district (Jordan), microprobe analyses of hausmannite ($Mn^{2+}Mn^{3+}{}_2O_4$) showed that this spinel contains up to several wt.% Al_2O_3 and Mn oxides may be substituted by up to 27 wt.% Fe_2O_3 (Hauptmann 2007). These oxides do not always show exsolutions but remain in the homogeneous state. In nature, such compositions are not stable but usually decompose and form lamellae.

In Mediaeval slags from the lead–zinc ore deposit of Wiesloch (Germany) spinel may contain changing amounts of Al_2O_3 and ZnO. Main end members are gahnite ($ZnAl_2O_4$, cores) and pure magnetite (rims) (Ströbele et al. 2010).

Particular attention is paid to cuprospinels, which typically occur in early copper slags. The solubility of Cu2+ in magnetite, under slightly reducing conditions at 900–1000 °C, leads to the formation of a $CuFe_2O_4$–Fe_3O_4 solid solution (Gadalla et al. 1962/1963). During cooling, eutectic decay forms a crystallisation of delafossite + magnetite (+ hematite) (Yamaguchi and Shirashi 1971). This is shown in Fig. 5.51a. Cu-rich phases, from the metallurgical point of view, can be considered as evidence of a need for improvement in a smelting process. These phases also show strong zoning. MgO, MnO, FeO and Al_2O_3 decrease with increasing CuO content. This means that an accumulation of the component of cuprospinel is observed from the inside out. Accordingly, the sums of the Al and MgFe spinels decrease towards the outside.

Magnetitic spinel in copper slags contains inclusions and is intergrown with metallic copper, digenite/chalcocite ($\pm$ Cu_2S), delafossite ($CuFeO_2$) and cuprite (Cu_2O).

5.6.1.14 Wuestite

Wuestite does not form euhedral crystals but occurs as dendritic aggregates or as small spheres. Spheres are rare and are usually crystallising early, dendrites are common and are often late or the last phase to crystallise. The most common impurities are Al (<6.88 wt.%) and Zn (<13.19 wt.%) (Keesmann 1989; Kronz 1997).

Wuestite is a dominating phase in Fe-rich slags and can be formed in dependence of the oxygen partial pressure (pO_2) between metallic iron and magnetite, respectively spinels high in magnetite components. Depending on the chemical composition wuestite can crystallise from the melt before fayalite and form a cotectic with this phase (Fig. 5.50a, b).

The pictures show that many of the slags did not solidify homogeneously but can consist of different areas with multiple micro-equilibria. In contact with charcoal, newly formed iron and dendritically formed wuestite may be in equilibrium (Fig. 5.50a), forming idiomorphic and dendritic crystals of magnetite near the surface. In a transition zone, wuestite becomes heterogeneous: oriented exsolution lamellae or somewhat lighter reflective transition spots in wuestite form (Fig. 5.43b). Sometimes, complexes of exsolved wuestite—magnetite can be observed, growing idiomorphic and homogeneously as magnetite in the melt. This indicates a fast-acting oxidation.

5.6.1.15 Periclase

While CaO does not form individual phases in ancient slags and crystallises, e.g. as a Ca-rich monticellite–kirschsteinite–forsterite series in olivine, periclase (MgO) may form an independent phase in ancient slags. It is precipitated as a product of the decomposition of dolomite when copper ores are smelted embedded in such a host rock according to

$$MgCO_3 \rightarrow MgO + CO_2$$

Periclase was observed in Chalcolithic/Early Bronze Age I slags found in the copper mining district of Faynan (Jordan) (Hauptmann 2007), and Krismer et al. (2011c) found it as distinct inclusion in magnetite agglomerations in Late

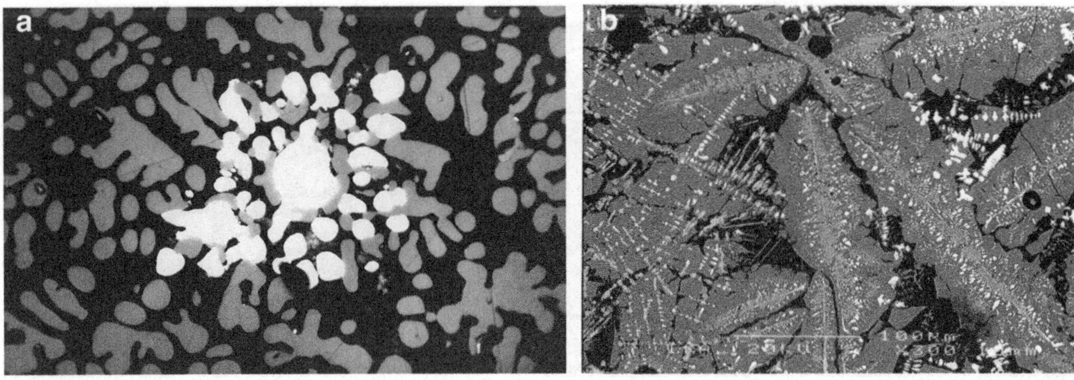

Fig. 5.50 (**a**) Essingen-Weiherwiesen, Swabian Alb, sample D-3/22c. Primary crystallised dendrites of grey wuestite (Wü) in a dark matrix of fayalite and glass. Wuestite is partially reduced to white ferritic iron (Fe) which, in the semi-liquid state, agglomerates to larger droplets. Reflected light, oil immersion. Width of picture c. 0.6 mm. From: Yalçın and Hauptmann (1995). (**b**) Lauchheim, Swabian Alb, sample D-14/2a. Fe-rich iron slag with wuestite (white) in two generations: Rounded dendrites as primary crystals, and cotectic intergrowth with fayalite. Some of the fayalite crystals provide a white weld of wuestite. Minor constituents are fine slablike to lath-shaped iscorite and overwhelmigly interstitial glassy ichor. SEM-picture, backscattered mode. Scale: 0.1 mm. From: Aus Yalçın and Hauptmann (1995, 285)

Neolithic slags from fahlore smelting. Dolomite may occur as a widespread host rock of copper, lead–silver–zinc or iron ores.

In nature, periclase occurs in contact metamorphic rocks and is a major component of most basic refractory bricks. It is a cubic form of magnesium oxide (MgO). In nature, it usually forms a solid solution with wuestite (FeO) and is then referred to as ferropericlase or magnesiowuestite (Trojer 1963).

5.6.1.16 Delafossite

Many slags from the early stages of metallurgy contain delafossite ($CuFeO_2$) which occurs as rims and needles surrounding spinel aggregates (Fig. 5.51a). Delafossite is formed by the following oxidation reaction:

$$Cu_{Metal} + 1/3\ Fe_3O_4 + 1/3\ O_2 \rightarrow Cu^+Fe^{3+}O_2$$

In slags with high Cu contents (>10 wt.%), as, e.g. in the Chalcolithic and Early Bronze Age periods of southeast Anatolia and the southern Levant to the Iberian Peninsula (Bourgarit 2007), the crystallisation of fayalite is often suppressed. The reason for this is usually a too weakly reducing gas atmosphere during the melting process.

Instead, Cu and Cu/Fe-oxides such as cuprite, delafossite and cuprospinel crystallise in oriented intergrowth with magnetite. Delafossite occurs in tabular, sometimes also curved crystals with striking bi-reflection or anisotropy and thus can be easily identified. Delafossite, crystallised from the melt, is also commonly present in eutectic intergrowth with cuprite (Fig. 5.51b). As silicate phases, in association with these oxides, clinopyroxene or glass is formed instead of fayalite. The two Cu/Fe-oxides are guide minerals, just as the fayalite, for slags from later eras.

The occurrence of delafossite + cuprite + magnetite is almost exclusively limited to copper slags formed under only slightly reducing conditions. Such conditions predominate in the smelting of copper ores in small reaction vessels or in the further processing (remelting and/or alloying) of copper in a crucible under a thin charcoal cover. This coincides with the

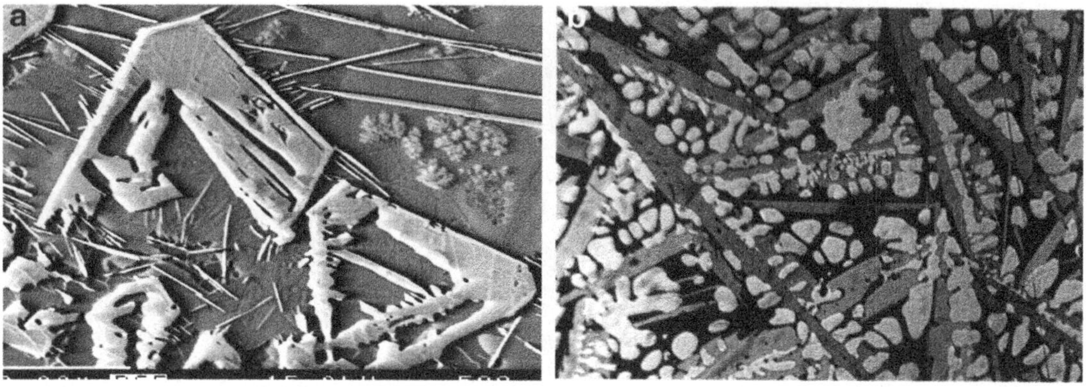

Fig. 5.51 Wadi Fidan 4, Faynan ore district (Jordan), sample JD-8/13a. Fourth millennium BC copper slag. (**a**) Magnetite with rims and exsolution lamellae of delafossite which also forms long, thin needles. Precipitation of cuprite in the residual liquid. (**b**) Delafossite in eutectic intergrowth with cuprite. Both pictures SEM, backscattered mode. From Hauptmann (2007)

widespread occurrence of this phase in modern converter slag (Trojer 1951), where it is also stable under relatively strong oxidising conditions.

5.6.1.17 Cuprite

In copper slags inclusions of cuprite ($Cu^{+}{}_2O$) are observed. The medium-grey cuprite is easily identifiable microscopically by its strikingly red internal reflexes. Cuprite can crystallise in different ways.

Cuprite, as a corrosion product of metallic copper inclusions, formed by soil storage, is widely distributed in ancient slags of copper metallurgy. In these slags, spherical inclusions of cuprite occur with finely dispersed flakes of metallic copper, and ring-shaped crusts, which in turn encase small copper droplets, appear. The original metal, which has segregated from the silicate melt with copper sulphide, has decomposed to cuprite in changing extent. Such intergrowths have been observed in Late Bronze Age slags from Kition's workshop in Cyprus, where former metal droplets are completely decomposed to cuprite, while the much more stable Cu sulphide is preserved as such (Hauptmann 2011).

Cuprite which crystallises directly from the liquid state next to delafossite and magnetite needs to be distinguished (Fig. 5.51b). Here, coarse-grained dendrites appear, which probably formed by liquefaction of reduced and dehydrated malachite. Crystallisation of cuprite from the melt requires high firing temperatures and only weakly reducing conditions. Cuprite is the Cu oxide which is stable at high temperatures and melts only at 1228 °C. The Cu^{2+}-valent tenorite ($Cu^{2+}O$) decomposes when heated from 1000 °C in Cu + oxide (cuprite) and oxygen and is therefore generally not found as a slag phase.

Another form in which cuprite occurs was found in glassy solidified slags and glassy fractions of crystalline slags. Copper and primary cuprite sometimes appear in the finest distribution, up to the colloidal range (Fig. 5.49). This was observed in early Bronze Age slags from Faynan (Hauptmann 2007), which thereby obtained a typical reddish colour. In slags from Iran finest inclusions of copper or cuprit-rich slags were described as well (Hezarkhani and Keesmann 1996).

Again another form in which cuprite is found in old slags are inclusions of copper droplets, which are surrounded by a hem-like copper sulphide (± Cu_2S). In slags of Shahr-i Sokhta, Iran, dendrites of sulphide are embedded in a eutectic matrix of cuprite + copper sulphides. Hauptmann et al. (2003) justify this intergrowth with a slight desulphurisation of the sulphides by the only

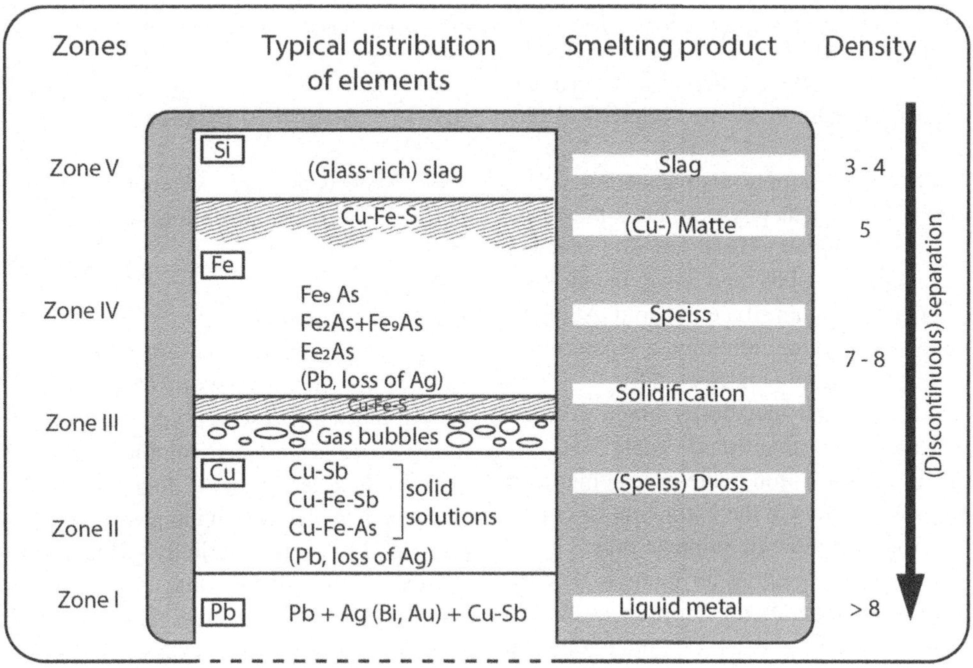

Fig. 5.52 Structure and composition of siliceous slag, matte and speiss according to their increasing density and mutual immiscibility. With higher arsenic and/or antimony contents a separation of speiss occurs. Modified after Keesmann (1993)

slightly reducing firing during the smelting process. This corresponds to the principle of a roast reaction (cf. Sect. 6.2).

5.7 Sulphide Mattes

With the word matte, we describe mixed, molten sulphides of heavy metals, preferably those of copper and iron, but also of lead, which separate from the siliceous slag during smelting processes (Tafel 1929). They occasionally contain some oxides. Sulphides and siliceous slags are miscible only to a very limited extent.

The sulphides have a significantly higher density than the surrounding siliceous slag with which they were produced with. The density of slags is around 4–5 g/cm^3. The density of (copper) matte is about 4.8–5.8 g/cm^3 and is even higher in the case of corresponding lead contents. Lead matte has a density of 7.58 g/cm^3. Accordingly, ideally there is a gravitational separation between the two in the liquid state. Keesmann (1993) has presented this—although it is in practice incomplete—separation of matte and slag on the basis of finds from Monte Romero (Iberian Peninsula) in a clear graphic (Fig. 5.52).

This model is hardly true for crucible metallurgy, as described from the Chalcolithic and Early Bronze Age periods, since in smelting processes in these early epochs of metallurgy charged materials not always have been completely liquified. A separation of copper, speiss and matte from each other and from only partially liquified slag is conceivable due to their different melting points. Refractory phases with high melting points will long remain in the solid state (cf. Figs. 5.19 and 5.29 in Sect. 5.3) and thus will not settle down.

In contrast, phases with low melting points will flow through a partially solidified, porous framework of slag components and accumulate on the bottom of a crucible or form larger aggregates within the slag. This could explain at least a qualitative separation of individual phases.

This process seems to be comparable with a deliberate melting of sulphides from siliceous gangue, as suggested by Metten (2003) for the Bronze Age copper production in the southern Alpine region.

Copper sulphides may also contain varying proportions of other metals as long as they have an affinity for sulphur. These are iron, arsenic, antimony, nickel, cobalt and zinc. In addition, mattes can even contain silver and gold. At higher levels of arsenic and/or antimony, a separation of speiss is more common (cf. Sect. 5.8).

In modern copper metallurgy, three different copper mattes are distinguished (Table 5.5).

Copper or copper–iron matte or lead matte are intermediate products in the extraction of copper or silver from sulphidic or sulphate ores. One of the key benefits of smelting for matte is its rapid liquefaction and separation even prior to the formation of a siliceous slag. The relatively low melting points of the sulphides also contribute to the separation of mattes. They too will be discussed below.

5.7.1 Sulphide Phases

Table 5.6 lists the most common sulphide phases observed in archaeometallurgical and in modern slags.

5.7.1.1 Colours

In slags of non-ferrous metallurgy, mm- to cm-sized drops or lumps of Cu- and CuFe- and Fe-sulphides are observed very frequently. In the microscopic scale, such sulphide inclusions can be identified easily due to their high reflectivity compared to the surrounding silicate and oxide phases. They are strikingly white, yellowish, brown, purple or grey-blue in colour. In light-optical microscopy, a direct dependence between copper and iron contents of sulphides and their colours is already visible, which allows a rough

Table 5.5 Typical analysis of modern mattes after Habashi (1978)

	Low-grade matte	Intermediate-grade matte	High-grade matte
Cu	10–30	30–45	45–79
Fe	37–50	25–37	45–79
S	22–24	24–27	20–22
Phases	Ccp, Bn Po	Bn, Ccp	Cc, Bn

Values given in weight-%. *Bn* bornite (Cu_5FeS_4), *Cc* chalcocite (Cu_2S), *Ccp* chalcopyrite ($CuFeS_2$), *Po* pyrrhotite ($Fe_{1-x}S$)

Table 5.6 Sulphide phases in mattes in archaeometallurgical and modern slags (Ettler et al. 2003; Hauptmann 1985; Westner et al. 2016)

Phase name	Chemical formula	Cu (wt.%)	Fe (wt.%)	Others (wt.%)
Bornite	Cu_5FeS_4	63.3	11.1	–
Idaite	Cu_3FeS_4	50.9	14.9	–
Chalcocite, digenite	$Cu_{9+x}S_5$	80	–	–
Chalcopyrite	$CuFeS_4$	34.6	30.4	–
Chalcopyrrhotine (Isocubanite)	$CuFe_2S_3$	23.4	41.2	–
Covellite	CuS	66.5	–	–
Cubanite	$CuFe_2S_3$	23.4	41.2	–
Cuprostibite	Cu_2Sb	51.1	–	48.9 (Sb)
Galenite	PbS	–	–	86.6 (Pb)
Koutekite	Cu_5As_2	68	–	–
Pyrrhotite	$Fe_{1-x}S$	$\leq$4.5	60.4	–
Sphalerite, wurzite	ZnS	–	14.7	51.6 (Zn)
Talnakhite	$Cu_9Fe_8S_{16}$	37.2	21.8	7.6 (Ni)

Many of these sulphides form solid solution crystal series with varying chemical composition. Here, the ideal chemical compositions are listed

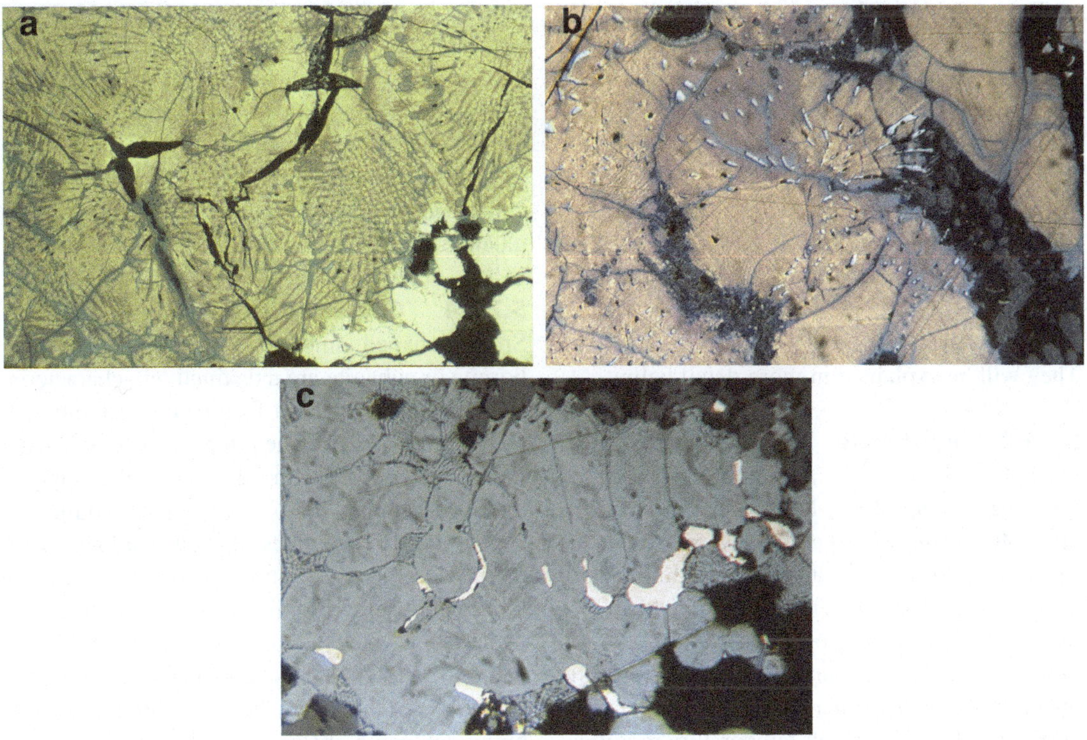

Fig. 5.53 Inclusions of matte with varying iron contents in archaeological copper slags. The chemical composition, i.e. the iron content of the matte phase, can be estimated microscopically by their colours and textures: (**a**) Aarja (Oman), sample OM 453, around twelfth century AD. The main component is chalcopyrite (rich yellow) with a dotted "firework exsolution" of magnetite and lamellar exsolution of bornite. In the lower right-hand corner pale yellow pyrrhotite can be seen. Image width: c. 0.3 mm. Estimated Fe content: c. 50 wt.%. (**b**) Mullaq (Oman), sample OM 338, ca. around twelfth century AD. Brown-red bornite predominates here as well, with a comparable exsolution of magnetite and lamellar exsolution texture of chalcopyrite. Along the grain boundaries formation of blue covellite by weathering. Image width: c. 0.6 mm. Estimated Fe content: c. 65 wt.%. (**c**) Maysar 1 (Oman), sample 214, Early Bronze Age. The intense grey-blue colour reflects almost pure chalcocite. Slight separation of brownish lamellae from bornite. Symplectic intergrowth of Cu-sulphide with cuprite. In the intermediate grain area metallic copper. Image width: c. 0.3 mm. Estimated Fe content: c. 85–90 wt.%. Reflected light, oil immersion. Photos A. Hauptmann, Deutsches Bergbau-Museum Bochum

estimation of the Cu:Fe ratio in polished sections. They are shown in Fig. 5.53.

Fe-sulphide inclusions, consisting of pyrrhotite, show a bright, light yellow to pale brownish pink colour. Pyrrhotite has an Fe content of 60–64 wt.%. The copper content is low, it does not exceed 4.5 wt.% (Yund and Kullerud 1966).

If the copper content of the matte is increasing, high-temperature chalcopyrite solid solution crystals are formed, which have a deep, more yellowish hue. This mixed crystal has a Cu content of 22.7–37.6 wt.%, the Fe content is around 30–42 wt.% (Sugaki et al. 1975).

The originally bright yellow, Cu-rich bornite solid solution crystal has 56–75 wt.% Cu and 7–17 wt.% Fe. It quickly changes its colour to a reddish-purple hue, in which it is usually identifiable.

If pure matte is present, blue or grey appearing chalcocite or digenite is formed. These phases may contain a maximum of up to 12 wt.% Fe. They are also found in many mattes along grain boundaries and cracks, where they are formed by corrosion. This is where the deeply blue-coloured covellite with its striking interference colours appears.

The sulphide phases of mattes are also characterised by high-temperature textural features, which in most cases make them easily distinguishable from similarly composed natural ores. These are (lamellar) separations caused by the decay of originally homogeneous phases or symplectic intergrowths by eutectic deposition of different sulphide phases. They were described, e.g. by Hauptmann (1985), Hezarkhani and Keesmann (1996), Schreiner (2002), Metten (2003), Addis (2013) and Westner et al. (2016). They will be explained in more detail below.

5.7.1.2 Pyrrhotite

Pyrrhotite ($Fe_{1-x}S$) is a commonly observed high-temperature phase. It occurs as round prills up to 40 µm in size. Any surplus of Cu in the formula may be exsolved and forms thin rims of chalcopyrite. Pyrrhotite may be embedded in a matrix texture of chalcopyrite—bornite exsolution lamellae (Hauptmann 1985) or as symplectic intergrowths with sphalerite/wurtzite (Ettler et al. 2003). Pyrrhotite is a high-temperature modification in the system Fe–S, stable in the temperature range from 308 to 1190 °C (Craig and Scott 1974). Pyrite itself breaks down at 743 °C.

Due to the high affinity of iron to oxygen compared to copper (which has a high affinity to sulphur), iron forms preferably Fe oxides (magnetite, Fe_3O_4) and, herewith, causes an enrichment of matte in base metals. The pyrrhotite–magnetite association exhibits either an equilibrium relationship with large crystals, as observed by Ettler et al. (2009a, b) in modern lead slags, or a symplectic intergrowth (with chalcopyrite), as observed by Hauptmann (1985) in copper slags from Oman. The melting temperature of the association pyrrhotite-magnetite is slightly above 1050 °C only (Naldrett 1969). Symplectic intergrowth of pyrrhotite and magnetite is equilibrated between 1030 and 1050 °C.

5.7.1.3 Copper–Iron Sulphides

Copper-containing sulphides exhibit large ranges of chemical compositions, beginning with the composition of chalcopyrite to bornite (Cu_5FeS_4) to digenite/chalcocite (Cu_9S_5). All these phases form solid solutions at high temperatures (Schlegel and Schüller 1952; Yund and Kullerud 1966; Sugaki et al. 1975). During cooling, these homogeneous solid solution mixed crystals are disintegrated and form characteristic lamellar exsolution textures.

Symplectic intergrowths occur as well, especially in adhesions of chalcopyrite ($CuFeS_2$) with galenite (PbS) (see below).

5.7.1.4 Chalcopyrite

Spindle-shaped lamellar textures of chalcopyrite, bornite or cubanite are a distinctively characteristic, reliable recognition feature for metallurgical matte pieces, also when they would occur in inclusions of speiss. In addition, CuFe-sulphide melts may contain up to 30 wt.% Fe_3O_4 (Gilchrist 1989), which separates during cooling. Very often, therefore, separations of magnetite can be seen, whose grey droplets penetrate the lamellae from centres in radial beams (Fig. 5.52). In nature, such lamellar systems occur in chalcopyrite as well. However, these are limited only to highly tempered chalcopyrites (Ramdohr 1975).

The formation of lamellae textures is due to separations or eutectic intergrowths of Cu–Fe–S phases from a sulphide melt, which is homogeneously composed at high temperatures (about >1200 °C). During cooling, the stability fields of the individual mixed crystal fields of pyrrhotite, chalcopyrite and bornite are reduced.

5.7.1.5 Magnetite

This oxidic phase is a widespread accessory mineral in sulphidic mattes. The formation within sulphide inclusions is due to the higher affinity of Fe to oxygen compared to Cu (which has a high affinity to sulphur). Fe forms preferably Fe oxide when oxygen is available and, herewith, causes an enrichment of matte. In addition, as can be seen in the microscopic view, Fe-sulphide is often associated with Fe-oxide.

5.7.1.6 Bornite

Bornite (Cu_5FeS_4) is the most common phase in base metal mattes. At temperatures of 350 °C, bornite solid solutions show an extent that goes far beyond a stoichiometric composition. Upon cooling, this mixed crystal decomposes lamellar depending on its composition with separation of chalcopyrite and/or chalcocite/digenite). Sugaki et al. (1975) note that bornite solid solutions under equilibrium are paragenetically linked to covellite.

5.7.1.7 Digenite and Chalcocite

High-grade matte consists of the crystallographic modifications of $Cu_{9+x}S$ (digenite, chalcocite). This "pure" composition rather rarely occurs in ancient slags. Often, these copper sulphides still contain low levels of iron, which separate into brown-grey lamellae of bornite in the blue-grey matrix (Fig. 5.53c). Such lamellar systems are also observed in natural sulphides (Ramdohr 1975); yet they are so typical for most of mattes in slags that they can be established as a characterising feature. Due to their small size, an exact chemical analysis of the lamellae is hardly possible.

5.7.1.8 Covellite

Covellite (CuS), under equilibrium conditions, occurs in paragenesis with bornite. However, covellite occurs more commonly as a product of corrosion. All Cu-containing sulphides in rocks show chalcocite/digenite and covellite along cracks and as shells around metal inclusions.

5.7.1.9 Galenite

Galenite (PbS) can be found in old slags with relatively high lead and sulphur content, but it generally occurs only with subordinate importance. Lead often is slagged as silicate. Nevertheless, old lead–silver slags contain droplets of galenite which, depending on the ratio of original polymetallic ores, may be intergrown with pyrrhotite, chalcocite and various FeAs- and FeSb-sulphides and are associated with metallic lead. Interstitial (Zn,Fe)S inclusions associated with PbS were also detected in slags from the smelting of polymetallic ores (Westner et al. 2016). In Hellenistic lead–silver slags of the island of Thasos, inclusions of lead matte contain levels of As, Sb and Ag in the lower tenth of a percent range (Hauptmann et al. 1988).

Galenite (PbS), in the liquid state, forms solid solution crystals with copper and with CuFe-sulphides, which separate with decreasing temperatures. If these are present in excess, cores are formed which are surrounded by eutectic-symplectic intergrowths with PbS.

Examples of intergrowths between galenite and CuFe-sulphides are not uncommon. Rehren et al. (1999a, b) described them from mediaeval slags from the Altenberg in the Siegerland (Germany). Merkel et al. (2013) have also identified them from mediaeval Pandjhīr (Afghanistan) slags observed in the extraction of silver from polymetallic PbCuAsSb ores (Fig. 5.54). With excess PbS, dendrites of PbS are formed in a matrix of CuFe sulphides and PbS, which are also fused together in a eutectic microstructure. Westner et al. (2016) were able to observe this in Roman lead–silver slags from Kosovo.

5.7.2 The System Cu–Fe–S

The main phenomenon of this ternary system is its extended miscibility gap, which extends almost continuously between the sulphide axis Cu_2S–bornite–chalcopyrite–pyrrhotite and the Cu–Fe axis in the low sulphur part of the system (Fig. 5.55). It has been extensively explored by Schlegel and Schüller (1952) and Krivsky and Schuhmann (1957) because of its global importance in extracting copper from its sulphide ores by matte smelting.

The Cu–Fe–S system shows two crucial metallurgical aspects.

The first is that sulphidic ores during smelting are moved onto the Cu_2S–FeS axis by light desulphurisation. A matte is formed which can have contents spanning the whole width between these two phases. The second most important aspect is the separation of metals and metal alloys (Cu, CuFe, γ- and δ-Fe) from the sulphides.

The phase associations during the cooling of the sulphide melt are shown in the binary system Cu_2S–FeS (Fig. 5.56). On the copper-rich side,

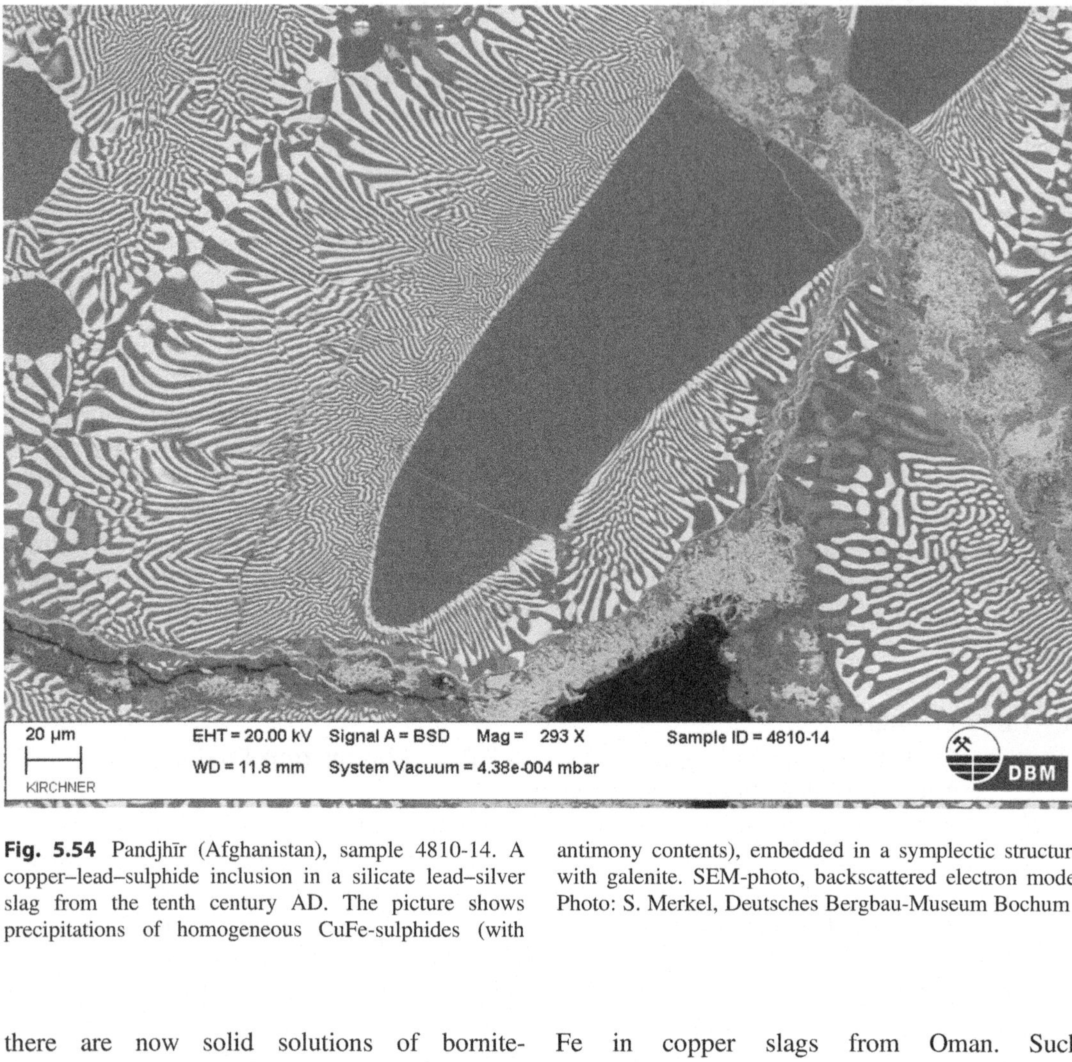

Fig. 5.54 Pandjhīr (Afghanistan), sample 4810-14. A copper–lead–sulphide inclusion in a silicate lead–silver slag from the tenth century AD. The picture shows precipitations of homogeneous CuFe-sulphides (with antimony contents), embedded in a symplectic structure with galenite. SEM-photo, backscattered electron mode. Photo: S. Merkel, Deutsches Bergbau-Museum Bochum

there are now solid solutions of bornite-chalcopyrite with exsolutions of chalcocite/digenite and on the other side of bornite-chalcopyrite with pyrrhotite. Some aspects have to be emphasised here. The eutectic here is at 60 wt.% FeS and 925–940 °C (Chang et al. 1976). If the liquid matte solidifies under equilibrium conditions, very Cu-rich stones can coexist with Fe-sulphides.

From the course of the tie lines (conodes) in the Cu–Fe–S system, on the other hand, it can be seen that from a copper matte with as little as 12 wt.% Fe, metallic copper will precipitate without iron exsolving. In contrast, desulphurisation of the bornite already leads to the precipitation of biphasic CuFe alloys. Hauptmann (1985) detected corresponding alloys with 15–55 wt.% Fe in copper slags from Oman. Such Fe-containing alloys are referred to as furnace sow or salamander.

Precipitation of copper from high-grade matte by desulphurisation can be observed in archaeometallurgical slags as inclusions. In fact, it takes the shape of (sub-)microscopic, hair or wire whiskers from the sulphidic matrix, which increasingly are getting compact. This was first demonstrated by Habashi (1978) in the context of investigations of modern matte smelting. Experimentally, the precipitation of copper whiskers from matte was shown in the simulation of primitive smelting processes of sulphidic copper ores by Craddock et al. (2007b) and later by Hanning (in prep.) on smelting experiments for Bronze Age smelting according to the archaeological

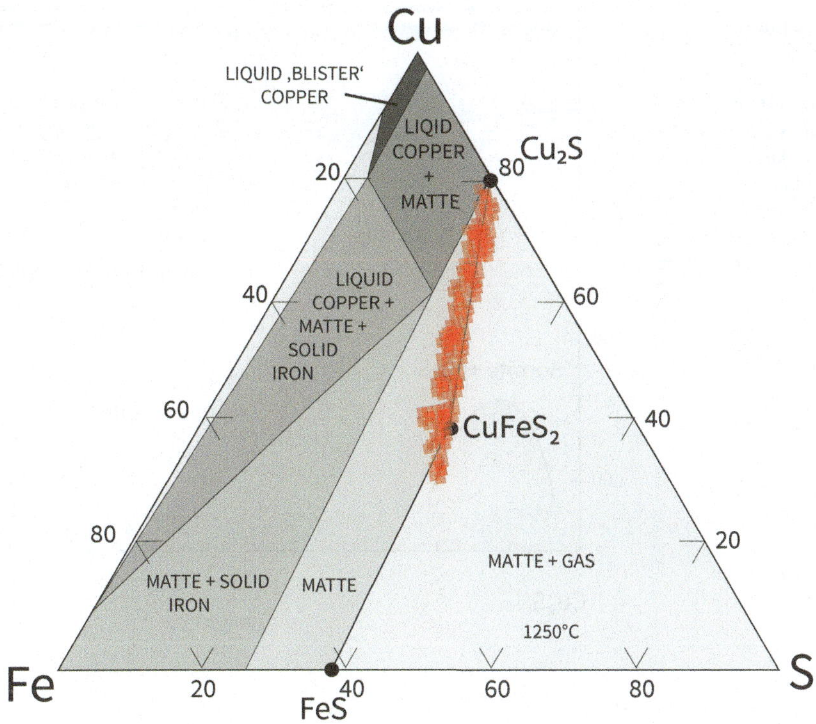

Fig. 5.55 Chemical composition of sulphide inclusions in archaeometallurgical copper slags from various localities, represented in the ternary system Cu–Fe–S at 1250 °C (Schlegel and Schüller 1952; Krivsky and Schuhmann 1957). The design of the single fields showing the associations of phases stable at this temperature is only slightly changing at lower temperatures. Note the archaeometallurgical relevant matte- and metal-compositions. The majority of the sulphide inclusions have high iron values. The connection lines (conodes) between the sulphide side and the metal side of the immiscibility gap clearly show that these compositions would exclude a precipitation of copper, i.e. these slags are not suitable to produce copper. Exceptions are a few sulphide inclusions close to the composition of chalcocite ($Cu_{9+x}S_5$) which would allow the precipitation of metallic copper without iron. The plot of Cu-rich matte within the miscibility gap might be due to the exsolution of metal in the submicroscopic scale. Values are given in weight percent. Data from Hauptmann (1985; Oman), Hess (1998, Arslantepe, Anatolia), Hezarkhani and Keesmann (1996, Iran), Metten (2003, Addis 2013; Westner et al. 2016; Reitmaier-Naef 2019; Silvestri et al. 2015, 2019; Trentino, Grisons)

findings from the copper district of Mitterberg. At the Deutsches Bergbau-Museum Bochum, investigations of matte-copper intergrowths of these experiments were carried out by scanning electron microscopy (Fig. 5.57).

5.7.3 Sulphidic Mattes and Silicate Slag

Due to the predominant and supra-regional importance of copper metallurgy in the pre- and protohistoric eras of the Old and the New World, slags are particularly often associated with copper-bearing sulphide inclusions. However, there are also lead- and zinc-bearing sulphide inclusions from lead smelting and silver mining (Merkel et al. 2013; Westner et al. 2016). Matte of lead is scarcer in slags because lead has a lower affinity for sulphur, it is more likely to be slagged as a silicate. Lead matte produced in old lead–silver metallurgy is formed to a greater extent, the lower the Cu_2S content and the higher the FeS content of the stone (Tafel 1929).

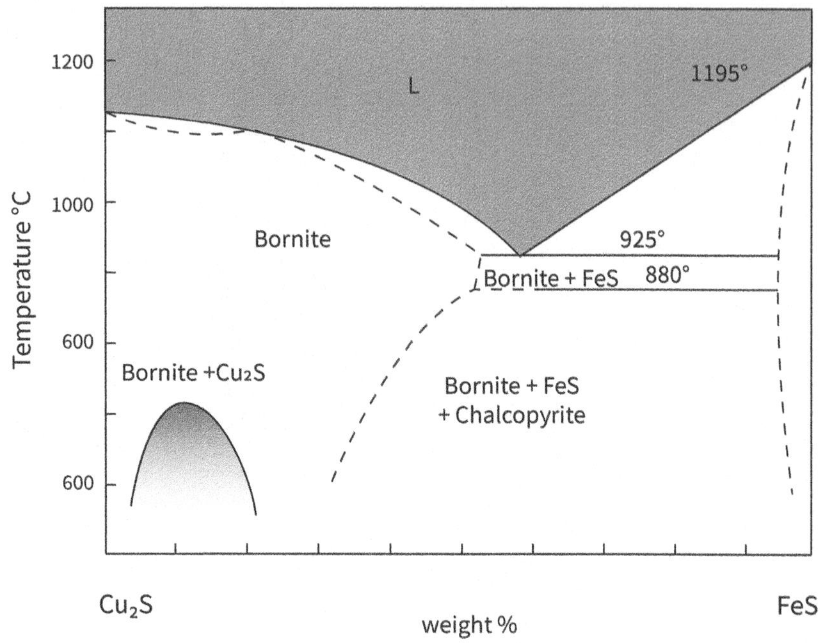

Fig. 5.56 In the phase system Cu_2S–FeS compositions of mattes from the narrow system Cu–Fe–S are shown. It is striking that over large areas especially bornite coexists with varying proportions of chalcopyrite, FeS and $\pm Cu_2S$, which corresponds to the (microscopic) composition of mattes in many slags. System according to Schlegel and Schüller (1952) and Chang et al. (1976)

Sulphide inclusions in siliceous slags, even if identified next to inclusions of metal droplets, are often interpreted as an indication of intended sulphide melting. Such inference should be drawn with caution, since it is precisely in the early stages of metallurgy in prehistoric times that oxidic ores with naturally lower levels of sulphides from the oxidation zone of ore deposits have been smelted. They even occur in considerable quantities in sedimentary copper ore deposits with predominantly sulphur-free minerals, e.g. in Faynan (Jordan) or Timna (Israel) (Hauptmann 2007; Asael 2010). The goal in these epochs was certainly the production of metal, not that of matte, which would have to be subjected to a roasting process. The fact that the value carrier matte probably was not recognised here, and was thus discarded, is indicated by garbage pieces from smelting sites in Oman (Hauptmann 1985).

Matte is also formed by smelting secondary sulphate-containing ores. In the near-surface area of sulfidic copper ore deposits, antlerite ($Cu_3(OH)_4SO_4$) and brochantite ($Cu_4(OH)_6SO_4$) occur in (semi-)arid areas in larger amounts, even in large masses. Examples include the vast, economically significant copper deposits of the Andes with their extensive oxidation zones (Alpers and Brimhall 1989). The sulphates also occur in the copper mining area of Oman, which was so important for the pre- and early Middle East history (Hauptmann 1985; Begemann et al. 2010). Near-surface areas of lead–silver–zinc deposits, on the other hand, are not so rich in sulphates. Anglesite ($PbSO_4$, cf. Sect. 3.5.5, Table 3.7) is rather rare. Cerussite ($PbCO_3$) is more likely to form here as a secondary ore, as it abundantly occurs in the deposit of Nakhlak in Central Iran.

Beside of such an association, copper sulphides can occasionally be formed on archaeological finds by soil storage over millennia in anaerobic and sulphurous environment. Rehren (2004) showed that copper sulphides inside of crucibles of the Pfyner culture (late Neolithic, fourth millennium BC, Switzerland) would not be evidence for sulphide smelting but would result from the reaction of sulphur with copper incrustations in the crucibles.

In silicate slags two types of sulphide inclusions have to be distinguished:

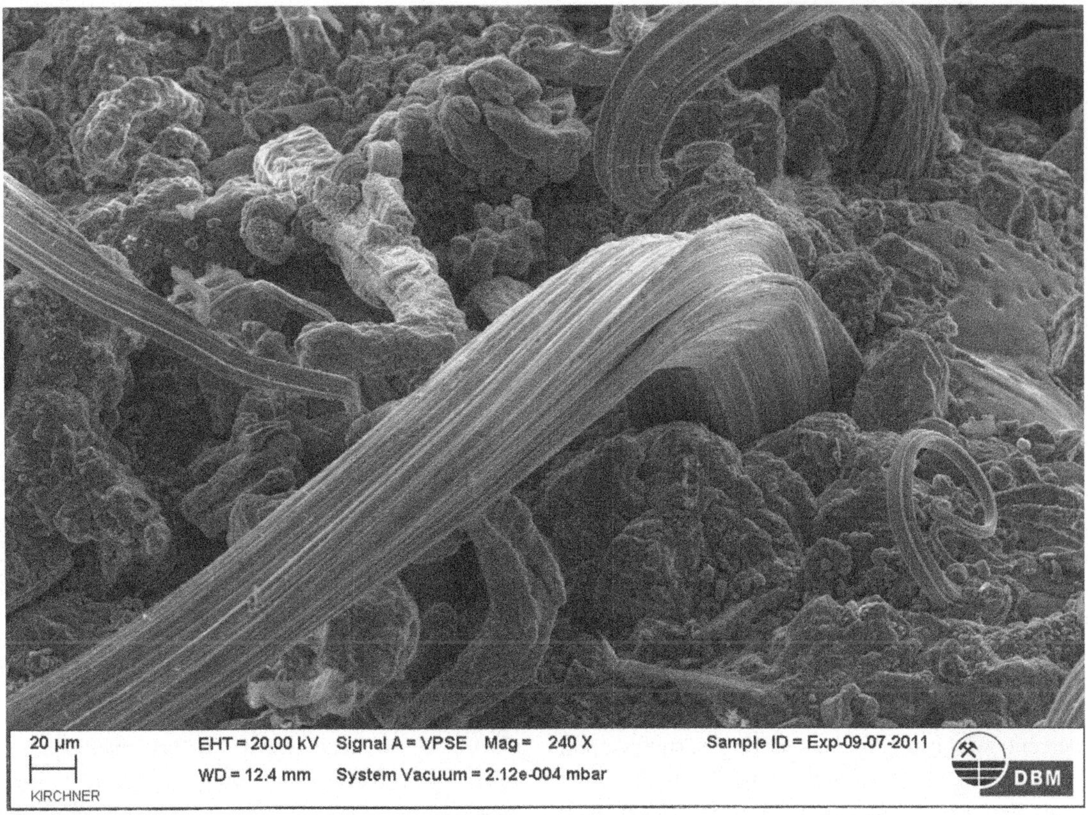

Fig. 5.57 Precipitation of curved, wire-shaped copper-whiskers from a dense matte of chalcocite/digenite. This matte was produced experimentally by E. Hanning in 2011 in a smelting of sulphide copper ores according to the Bronze Age model Mitterberg (Austria). Metallic copper precipitated from a homogeneous melt during cooling connected with (slight) desulphurisation of the matte according to the miscibility gap in the system Cu–Fe–S. SEM-picture, backscattered electron mode. Photo: D. Kirchner, Deutsches Bergbau-Museum Bochum

- Larger, round droplets (≥mm-size and larger) or plate-like precipitates of CuFe- with or without Pb-sulphides, which often surround crescent-shaped smaller metal droplets. They are particles of the former sulphidic ores attached in the siliceous slag.
- Sulphidic flakes (µm-size) embedded interstitially between fayalite or clinopoyroxene crystals. These are precipitates of Fe-sulphides from the liquid silicate melt.

Large, round droplet-shaped sulphide inclusions were quickly liquefied and agglomerated already at the beginning smelting processes and thus lead to a separation from the emerging slag. Rosenqvist (1974) and Frost et al. (2002) noticed that liquefaction of mattes can already start in a temperature range of 500–600 °C. Silicate slag formation usually needs a longer time period of heat access so that liquefied sulphides and metal may be separated soon from the successively liquefied slag. This is further supported by the low viscosity of the matte, which is far below that of a silicate melt. Frost et al. (2002) experimentally have found that in this temperature range, e.g. arsenopyrite (FeAsS) + pyrite (FeS_2) react with each other to pyrrhotite ($Fe_{1-x}S$) and to an AsS-melt. A comparatively slight metallurgical enrichment of sulphides from an ore mixture is that of antimony sulphide. Antimony sulphide already melts at 546 °C, other sulphidic antimony ores melt in a comparably low temperature range. In the Middle Ages and later, the enrichment of antimony sulphide was carried out in clay pots, a melting furnace was not required (cf. Sect. 6.8).

A pure matte of equal parts of copper and iron only melts at ±900 °C (Yund and Kullerud 1966), but Cu- and Pb-containing sulphides can begin to melt at around 500 °C (Craig and Kullerud 1967). Generally, melt formation of polymetallic ores with high Cu and Pb contents occurs at temperatures between approximately 600 and 700 °C. In addition, contents of arsenic, antimony, nickel, cobalt, zinc, lead and silver introduced from the ores are further facilitating the formation of liquid matte and are reducing the melting points of mattes.

What concerns sulphidic flakes of matte in the µm-size the following considerations are helpful. During the melting process, at temperatures of about 1200 °C, a limited interaction occurs between the sulphidic and silicate melts, which is different for the individual sulphides (Maclean 1969). In basic slags with high FeO contents, e.g. pyrrhotite ($Fe_{1-x}S$) can dissolve up to a few percent, which precipitates as tiny particles upon cooling of the slag melt. Due to their small size, they are not separated from the silicate melt. Such µm-sized sulphide inclusions are a widespread phenomenon, which was observed in slags from several ancient smelting sites, e.g. in Iran (Hezarkhani and Keesmann 1996), in Bronze Age/Iron Age slags from the Alpine region (Reitmaier-Naef 2019; Silvestri et al. 2019), in Roman slags of silver mining in Kosovo (Westner et al. 2016) or in slags from Cyprus (Hauptmann 2011).

In the archaeological context massive sulphide pieces as well as those of smaller metal lumps are extremely rare but are occasionally salvaged (Fig. 5.58). Also, at copper melting sites, slags are found whose underside has negative impressions from which copper sulphide cake has detached. Examples of this have already been described from Shahr-i Sokhta (Hauptmann and Weisgerber 1980), Cyprus (Kassianidou 2012; Hauptmann et al. 2003) or Oman (Hauptmann 1985).

Copper matte may also contain low levels of arsenic and antimony, provided that correspondingly composed ores have been smelted, that is, e.g. fahlores. If the solving properties of these (semi-)metals in the stone are exceeded, speiss (CuAs, CuSb) can form. These arsenides or antimonides can also have significant iron contents, provided that ores with chalcopyrite have been processed. The association of stones and speiss is observed frequently and will be discussed in Sect. 5.8.

Fig. 5.58 (**a**) Shahr-i Sokhta (Iran), third millennium BC, sample 0500999, further circumstances of the find are unknown. Round oval regulus made of very high-quality copper sulphide (white metal, ~ Cu_2S), which encases a core of cuprite. This was probably originally made of metal. The regulus has settled quite completely from slag, as shown in Fig. 5.10. Total diametre 6.1 cm, cuprite core 4.5 × 3.5 cm. According to Hauptmann and Weisgerber (1980). (**b**) Maysar 1 (Oman), sample OM 360. Demixing of high-quality copper sulphide (white metal) with a copper core, partly converted to cuprite, from a Bronze Age slag. According to Hauptmann (1985)

5.8 Speiss

The word speiss describes the compounds of the heavy metals iron, copper, nickel and cobalt with major amounts of arsenic and/or antimony. They are called arsenides or antimonides. Speiss and sulphidic matte are only slightly soluble in each other. There is a large miscibility gap (Fig. 5.60). Matte contains only a few percent of these semimetals while speiss contains only a few percent sulphur. In speiss, the precious metals gold, silver and platinum can be enriched depending on the starting ore (Tafel and Wagenmann 1951). From a metallurgical point of view, speiss is a rather unwanted product that forms in copper and lead–silver production. It prevents the tapping of metal and it is difficult to separate speiss from lead because speiss is very hard and brittle and can only be processed by casting (Tafel and Wagenmann 1951; Rosenqvist 1974). It was obviously difficult in ancient times to extract any contents of precious metals from speiss.

Fresh speiss is a material similar in appearance to silvery shining metal, similar to cast iron. But it starts oxidising and decomposing in the air within hours. Then it is friable, dark brown and rusty in colour. It often looks like limonite. Hence, speiss is not a common find in archaeometallurgy as is slag, copper metal or iron itself.

5.8.1 Composition of Speiss

Based on the bulk composition, two types of speisses can be distinguished:

- Ferrous speiss, which is composed of Fe–As phases and antimonides. Thornton et al. (2009) define a ferrous speiss as one with a base-metal content of <2–3 wt.%. Fe_2As is the most common ferrous speiss.
- Base-metal speiss, which is a mixture of Cu-, Ni-, Fe-, Ag-, Pb-arsenides and -antimonides.

If the primary ore smelted contained in addition sulphidic zinc minerals, small inclusions of sphalerite (ZnS) in the sulphide rims occasionally are evident (Krismer et al. 2013). In their publication on modern lead production, Ettler et al. (2009a, b) present a whole range of Fe–As–Sb–Pb–Ni–Sn–Cu–Zn intermetallic speiss composites which were produced by smelting of polymetallic ores from Příbram (Czech Republic).

Tin containing stannides with heavy metals are also speisses, but there is no evidence in the archaeometallurgical context.

The most common ferrous speiss (c. 35 wt.% As, 65 wt.% Fe) forms long laths, which do not show straight grain boundaries (Fig. 5.59) but may show eutectic intergrowths with (As-containing) α-iron. FeAs, the Fe-rich modification of the natural mineral westerveldite ((Fe, Ni,Co)As) is always heavily oxidised. Other phases are löllingite ($FeAs_2$), troilite ($Fe_{1-x}S$) and native arsenic. There are also eutectic intergrowths of Fe_2As and other phases of the system Fe–As–S which are difficult to identify by SEM or even by microprobe measurements. Hence, Photos et al. (1989) analysed the chemical compositions of main phases in speiss from Palaia Kavala, and the bulk compositions of eutectic intergrowths (Table 5.7). Craddock et al. (1987) analysed speiss from several sites in the Rio Tinto area (Iberian Peninsula).

The melting point of Fe_2As is as low as 919 °C, FeAs is ca. 1027 °C. The eutectic Fe_2As–Fe at 70 wt.% Fe is at 835 °C (Hansen and Anderko 1958).

Typically, noble metals such as platinum, gold and silver are enriched in speiss in the liquid state, and it may cause considerable problems in silver-smelting operations. This was described in the sixteenth century AD by Georg Agricola (1556). According to Ercker (1580), speiss can be desilvered only in a separate and very complex smelting operation. Based on these reports, Craddock et al. (1985, 1987) proposed a speiss technology in which argentiferous speiss from the Roman Rio Tinto (Spain) could have been desilvered. Finds of speiss described by Kassianidou et al. (1995) from Monte Romero in south-western Spain (first half of the sixth century BC) were perhaps material from which silver was extracted too (Craddock et al. 1987)

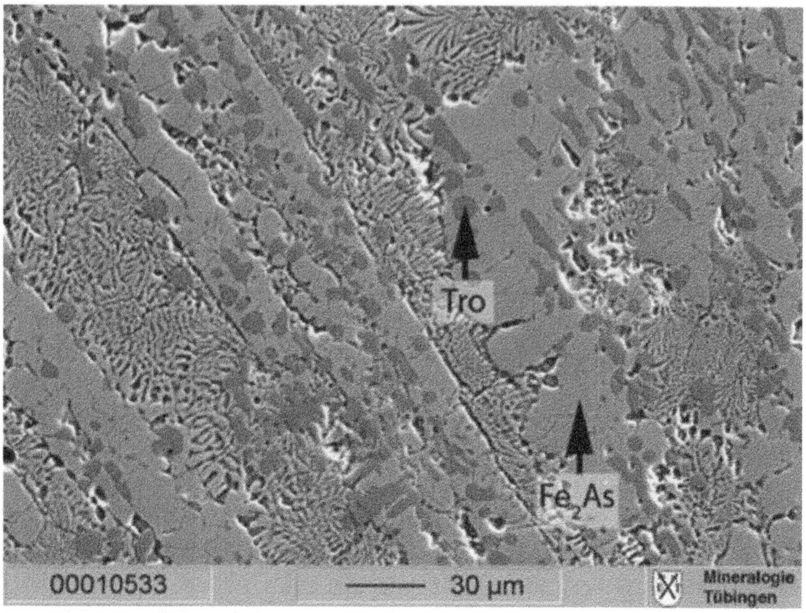

Fig. 5.59 Wiesloch (Germany), late tenth to late twelfth century AD. Typical speiss-texture with long laths of Fe_2As, and a eutectic intergrowth of Fe_2As and probably α-iron. Dark grey droplets embedded in Fe_2As are troilite ($Fe_{1-x}S$; Tro). These may otherwise heavily corrode and would give the speiss a reddish rusty tint. SEM picture, secondary electron mode. From Ströbele et al. (2010)

Table 5.7 Chemical composition of finds of speiss from slag heaps at Corta Lago, Rio Tinto (seventeenth/eighteenth century AD), and from the Roman site of Las Arenillas, close to Rio Tinto

Speiss chemistry	Fe	As	Sb	Cu	S	Pb	Ag	Sn
Corta Lago								
CL-79 T2 layer 4 17	33.5	39	19.5	n.a.	n.a.	2.3	1	4.3
CL-79 T2 Box 17	50.5	40.5	6	n.a.	n.a.	0.83	0.27	1.7
Las Arenillas								
RT 103 HP 507 site E	72.7	18.6	7.2	n.a.	n.a.	0.6	<0.02	0.6
Palaia Kavala								
Elephteroupolis-1	53.7	33.1	3.92	2.01	3.99	1.23	0.007	n.a.
Kechrokampos-4sp	55.7	35.1-	n.d.	n.d.	0.83	0.41	0.006	n.a.
Kechrokampos 1sp.main	51.9	33.7	2.92	3.24	n.d.	n.d.	n.d.	n.a.
Kechrokampos-1sp.eut	79.2	9.4	0.4	0.83	n.d.	n.d.	n.d.	n.a.
Makrychori-7sp.main	86.4	10.2	1.36	0.27	0.17	n.d.	n.d.	n.a.
Makrychori-7sp.eut	57.6	35.2	3.78	0.19	0.18	n.d.	n.d.	n.a.

Speiss from both sites is made up of Fe-arsenides with changing amounts of antimony, and some lead. Speiss from Corta Lago contains some silver. Source: Craddock et al. (1987). The Ottoman period speiss from Palaia Kavala (Northeastern Greece) was published by Photos et al. (1989). Values are given in weight %

by complicated metallurgical processes. This hypothesis, however, was refused by Kassianidou (1998). It is a point of discussion whether speiss was utilised in these ancient epochs for desilvering.

Speiss generally indicates the use of arsenopyrite (FeAsS) or at least portions of this mineral in the ore charge. Unfortunately, it is unknown whether the precipitation of speiss simply occurs if the charge is rich in arsenic or antimony or if there was a special "speiss technology" which enabled the smelter to produce speiss selectively to bind arsenic and antimony. Thornton et al. (2009) and Thornton and Rehren

(2009) postulate a "speiss technology" dating back to the third millennium BC in Iran. In case this suggestion would turn out to be true the question arises if Craddock would not be on the right side.

Speiss shows all transitions to a furnace sow or salamander in the iron metallurgy on the one hand, and to various compositions of mother alloys on the other hand (Tafel and Wagenmann 1951). With increasing copper content the composition of speiss turns to the so-called "mother alloys" or parent solid solution and finally to impurities in metal, which in archaeometallurgy are the "fahlore-metals" of the SAM-project (Junghans et al. 1968).

5.8.1.1 Speiss as Parent Solid Solution?

Just as controversial as the potential desilveration of speiss in ancient times is the question of whether speiss could have functioned as a mother alloy in prehistoric times. A mother alloy is an intermediate product that contains a high concentration of the alloying partner in the parent metal. It is melted with the base metal to produce the actual alloy.

Such a hypothesis was formulated by Zwicker (1980). Zwicker comes to this result due to his smelting experiments with ores from the settlement of the Chalcolithic Norsuntepe (fourth millennium BC) on the Upper Euphrates. He was able to produce a copper speiss from oxidised ores with high concentrations of copper, antimony and arsenic. This had the function of a mother alloy because he could then melt it with pure copper to a corresponding copper alloy with low levels of the two elements. Therefore, the author does not consider such a process unlikely even for prehistoric times.

Due to the elemental ratios, Tadmor et al. (1995) consider the melting of a parent solid solution alloy with pure copper to be conceivable for a part of the artefacts of the Nahal Mishmar treasure, but they only see limited analytical evidence for such a thesis. From a metallurgical point of view, this use seems feasible, since nickel and arsenic-speiss are soluble in copper (Rostoker and Dvorak 1991). The relatively easy processing is also an argument for Tylecote (1987) that a production of arsenic-containing copper could have taken place from a mother alloy.

Thornton et al. (2009) suggested that speiss, due to its high arsenic concentrations, was produced deliberately at Early Bronze Age Tepe Hissar in the northern center of Iran in order to add it to copper to improve its casting properties. They investigated inclusions of Fe-speiss in slags. On the other hand, this cannot be proved for the contemporaneous site of Shahr-i Sokhta in southeastern Iran. For if the FeAs speiss found at Shahr-i Sokhta had been added as a parent alloy to the copper produced there, there would have been dangerous that the iron content in the copper produced would have increased as well. That was not observable (Hauptmann et al. 2003). The speiss there is instead interpreted as a random by-product of a piece of copper ore with high levels of arsenopyrite.

From Arslantepe, smelting slags from Early Bronze Age IB (ca. 3000–2800 BC) are described which contain nickel and copper speiss with high levels of arsenic (24–49 wt.%) and nickel (34–65 wt.%) or copper (61–62 wt.%) (Hess 1998). They are interpreted as possible parent alloys for the production of NiAs-containing copper artefacts.

This concept is based on the comparatively low melting points of the speiss. In their eutectic composition, the melting temperature of arsenic or antimony is about 700 °C. The presence of significant amounts of lead will lower the melting temperature further and will increase the freezing range for alloys near the eutectic. These properties likely made speiss-rich alloys very suitable for complex castings such as those from the Nahal Mishmar hoard (Tadmor et al. 1995).

From the Early Bronze Age site of Arisman (northwest Iran) a mixture of FeAs-speisses and copper ores or metal was observed by Rehren et al. (2012). They report analyses of various slag types. Approximately 20 tons of slag provide evidence of sustained and large-scale production of metal, using both furnaces and crucibles. Based on their analytical results they assumed that both iron-arsenic speiss, and arsenical copper were produced, apparently side-by-side at the same site but using different ores and processes.

The authors offered a tentative outline of the metallurgical processes involved in the production of these arsenic alloys and an explanation of why speiss needed to be produced from arsenopyrite in a separate step. They suggest that the speiss would then be added either to secondary copper ore or to separately smelted copper metal, in order to finally produce arsenical copper in a regular and well-controlled process.

5.8.2 Speiss and Matte

Speiss and matte are not soluble and are separated due to their various specific weight (cf. Fig. 5.51). This is caused by the extensive immiscibility between the two materials as shown in the ternary diagram. Speiss is enriched below the matte. It is also immiscible with silicate slag, and also with liquid lead. As it has a specific gravity lower than lead and higher than matte, during the smelting of a complex polymetallic lead ore, up to four separate layers may be formed in the furnace.

In the ternary system (Cu + Fe) – (As + Sb) – S (Fig. 5.60, Lesniak 1991) the constellations of mattes and speiss compositions are shown. Arsenic and antimony are combined in the phase diagram in Fig. 5.60 because they show very similar geochemical behaviour and can both be replaced without any change in the configuration of the miscibility gap (Lesniak 1991). For similar reasons, copper and iron are also combined. The association of speiss and matte and their very limited intergrowths show that there is an extensive miscibility gap between them. Speiss is formed when the ores to be melted contain sufficient amounts of arsenic and/or antimony (Ravitz and Fisher 1936). In contrast, matte forms when ore with high sulphur contents is smelted.

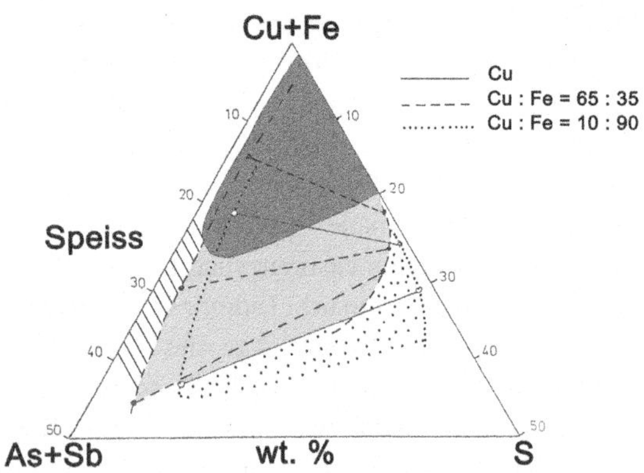

Fig. 5.60 The ternary system (Cu + Fe) – (As + Sb) – S shows the miscibility gaps between sulphur-rich Cu/Fe-matte and the location of speisses. With increasing Fe-concentrations the miscibility gap is considerably enlarged towards the (As + Sb) – S line. Values given in wt.%, temperature 1200 °C. Marked is the miscibility gap. Modified after Kleinheisterkamp (1944), Lesniak (1991) and Moesta (2004)

In Sect. 5.7 it was shown in the discussion of the system Cu–Fe–S that the configuration of the miscibility gap between sulphide and metal-rich side hardly changes with changing sulphide or metal contents. Kleinheisterkamp (1944, 1948) points out that low lead contents have no particular influence on the equilibria of the matte–speiss equilibrium, nor does the silver. Both accumulate in the metal phase.

The composition of speisses lies on the left side of the triangle approximately in the middle of the binary subsystem Cu + Fe and As + Sb. If the iron content in the system predominates, then the speiss has transitions to furnace sows or salamanders of iron metallurgy, if copper or lead predominate, copper or lead lumps are formed (Tafel 1929). With increasing copper content, the composition of speiss turns to deficit alloys, and finally to impurities in the metal.

5.8.3 Archaeological Evidence

In early metallurgy, speiss was produced throughout the ages, and there is a continuous discussion whether or not it was intentionally made as a sort of parent alloy, or it happened as a waste product during smelting ores. In an overview, Thornton et al. (2009) observed that speiss is known from many prehistoric and historic smelting sites of copper production and lead–silver production as well and may occur as lumps of grams to several kilograms.

The earliest evidence dates in numerous countries to the Chalcolithic periods.

In central and western Europe, copper- and lead-rich speiss was produced as early as the

Chalcolithic period (Keesmann and Moreno Onorato 1999; Müller et al. 2004). Similar to speiss in composition are numerous prestige metal objects with up to 25 wt.% arsenic and/or antimony found in the Chalcolithic southern Levant (Tadmor et al. 1995), and Early Bronze Age copper objects with up to 7 wt.% arsenic in the Transcaucasus (Meliksetian et al. 2011).

Heavily corroded Fe-speiss was found at third millennium BC Shahr-i Sokhta (Iran) with 41% Fe, and 18 wt.% As (Hauptmann et al. 2003), and at Early Bronze Age Poros Katsambas (Crete, Greece) with 53 wt.% Fe, and 39 wt.% As (Doonan et al. 2007). Both finds exhibit a eutectic texture near the typical composition of Fe_2AS–Fe.

Speiss was found in the copper workshops of Early Bronze Age Maysar-1 (Oman, unpublished data Hauptmann) and Jericho (Khalil and Bachmann 1981). A similar find is available from Trieben (Austria) (Preßlinger and Eibner 1993). It is a flat lump of copper and of matte, and in between a thin layer rich in FeAsSb (Fig. 5.61). It is an inclusion in a Bronze Age copper slag. It was obviously formed by the smelting of fahlores or their corrosion productions, respectively. The different layers of copper metal, matte and speiss represent an ideal sequence of separation between the three phases as shown in Figs. 5.52 and 5.60.

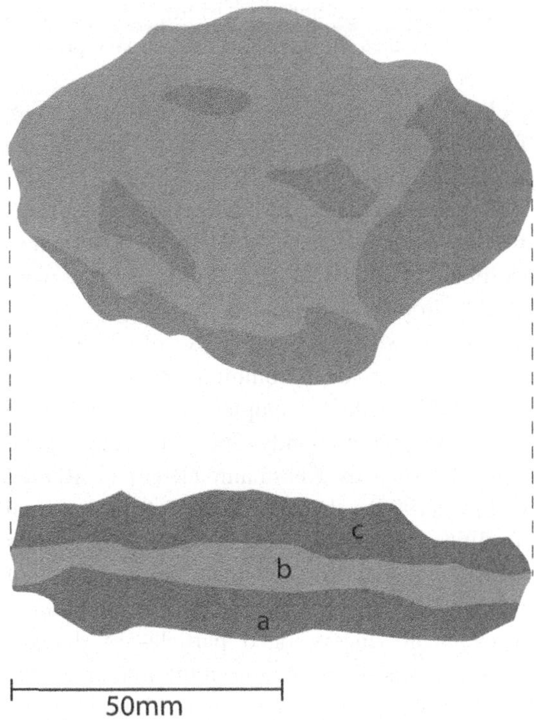

Fig. 5.61 Versunkene Kirche, Trieben (Austria). Late Bronze Age copper regulus probably produced from fahlores. It consists of three different layers: (**a**) Raw copper with 80 wt.% Cu, 5 wt.% Fe, 9 wt.% As, 6 wt. % Sb. (**b**) Iron-arsenic-rich layer with 38 wt.% Fe, 12 wt.% As. (**c**) Matte with 47 wt. % Cu, 27 wt. % Fe, 4 wt.% As, 2 wt. % Sb, 19 wt.% S. From Preßlinger and Eibner (1993)

Speiss related to the production of silver was found at the seventh century BC Monte Romero (south-western Spain; Kassianidou 1992) along with matte and slag intergrowths. Superimposed were siliceous slag, below was matte, then speiss and as the heaviest phase metal (Copper, lead, silver, varying antimony, bismuth and gold contents) (Fig. 5.52). Such a layer sequence is idealised. In the reality of archaeometallurgical finds, however, there is hardly an ideal, complete gravitational separation of the individual phases. Speiss also contains matte produced in low concentrations during smelting. In this way, finds from, e.g. Monte Romera show numerous inclusions of silver-bearing lead as well (Keesmann 1993, see below). And the incomplete separation of slags and iron arsenide is evident in Early Bronze Age slag in Iran, where Thornton et al. (2009) observed such inclusions.

The most pertinent speiss, in terms of their iron dominance, is a Hittite find from Boğazköy (Muhly et al. 1985). Here, a very similar complex mixture of metallic arsenical iron, intergrown with iron arsenide and iron sulphide, was interpreted as evidence for the smelting of an arsenical iron ore such as arsenopyrite. However, the resulting iron metal must have been useless, due to the extreme brittleness of the alloy. At present, no positive interpretation is available as to the reason for the existence of this find. It has been described in the context of a number of iron ingots found together in a room, but no direct evidence relates the speiss to the primary production of this metal.

Later, base-metal speiss was traded in ingot form during the Late Bronze Age, and in the Roman period (Tylecote 1987).

Speiss was found in mediaeval Pb–Ag-slags from Wiesloch near Heidelberg (Germany) (Ströbele et al. 2010). These finds occur as very small droplets suspended in slags, or as irregularly shaped aggregates of 5–30 mm diametre. The most common phase is Fe_2As (Fig. 5.59). Other examples of archaeological speiss are predominantly finds from base metal contexts, such as Keesmann (1999) or Rehren et al. (1999a, b) mentioned from several localities. In these cases, however, the speiss invariably contains considerable amounts of base metals, either copper and/or lead, both as mechanically incorporated particles of discreet phases and as solid solution in the iron arsenide.

The production of speiss is not limited to the Old World. The production of a Cu–As–Pb alloy with around 15–20 wt.% As during the first millennium BC has been reported by Mei (2000) from Nulasi, western China, together with the analysis of a Cu–As–Pb buckle found in the Hami region, containing 25 wt.% As.

5.9 Cupellation Remains

Litharge (PbO) is the most representative waste material that is evidence for the cupellation production step. It is formed by the deliberate oxidation of (argentiferous) metallic lead.

Cupellation is a process to extract and separate the precious metal from silver-bearing lead. The principle is based on a prior enrichment of silver in the lead by a smelting process and a subsequent separation of the two metals in the cupellation process in cupels. This is due to the easier oxidation of lead compared to silver. In Sect. 6.4, the extraction of silver from lead is discussed in detail.

5.9.1 Composition of Litharge

Litharge consists of crystalline lead oxide (PbO), which occurs in the modifications massicot (α-PbO) and lithargite (β-PbO) (Trojer 1963). Lithargite consists of large, platy crystals of bright red colour (Fig. 5.62), massicot appears more yellowish. The red colour comes to light when fragments are cut. Archaeological samples of litharge are usually at least partly weathered to whitish coloured lead (hydro-) carbonates such as (hydro-) cerussite, massicot, and lead glass and contain more than 75 wt.% lead. Due to the enormously high specific weight of lead (11.34 g/cm^3), any finds of litharge are extraordinarily heavy.

The chemical composition of finds of litharge is affected by two materials:

- The composition of the ores smelted in the previous process
- The material of the cupellation hearth in which litharge accumulates during the cupellation process

In literature often it is referred to as finds of "pure" litharge, which can be confused with the composition of used cupellation hearths (Tylecote 1987). In Table 5.8 a few examples of compositions of litharge and cupellation hearths are compiled.

Table 5.8. Chemical composition of pieces of litharges found at the Roman vicus of Lüderich (Fig. 5.62)

In the beginning, cupellation hearths were usually made of a SiO_2-poor or -free mixture of lime, clay, marl and plant ash and they sucked significant amounts of litharge, perhaps also some lead and thus silver, into their pore spaces. Litharge often not only has a certain copper content but also contains CaO, Al_2O_3, SiO_2 and MgO. It was not until the Middle Ages that bone ash was used to construct the cupellation hearth (Craddock 1995). Such a material has led to increased accumulation of phosphorus (P_2O_5). Instead, Middleton et al. (2000) found in their studies of cupels from Sardis that they were made using special talc (steatite).

Fig. 5.62 Lüderich (near Cologne, Germany), sample D-102/19b. Fragment of a flat, porous fragment of a cupel from extraction of argentiferous lead in a Roman vicus (above). The surface of the sample is covered by a whitish crust of secondary lead minerals such as (hydro-) cerussite, while the inner part shows an intensive reddish colour of lithargite (PbO). Below a thin section showing red lath-shaped crystals of lithargite. The texture traces the slackly material of the cupel where the liquid lithargite was infiltrated. The black point is metallic lead. From Bode (2008), Deutsches Bergbau-Museum

It was not undesirable for litharge to be sucked into the crucible in large quantities, as this could have a positive effect on the effectiveness of the process or on the purity of the extracted silver. The silver- and lead-containing kiln, when it was no longer functional and needed to be replaced, may have been smashed to pieces and returned to the shaft furnace process (Keesmann 1991, 1993) to recycle the silver. Silver contents of almost 0.3 wt.% were analysed by Pernicka et al. (1998) in the Uruk-age Habuba Kabira (upper Euphrates) in cupels. Cupels from Sardis contained silver contents up to 1 wt.% (Middleton et al. 2000). Furthermore, Cu, As and Sb can occur in the area of first/tenth of a percent (Pernicka et al. 1985).

5.9.2 Archaeological Evidence

In cupellation, lead oxide (PbO) is a characteristic, unmistakable waste product. It is easy to identify in archaeological evidence, albeit in different shapes and colours. Fragments of litharge, rarely completely preserved can be found in whitish weathered pieces, which are reddish coloured inside and strikingly heavy.

Within the scope of ancient lead–silver production, remains of litharge, which arise during the extraction of silver from lead, are found mainly in workshop areas within settlements. Other remains found associated with workings with (argentiferous) lead should include fragments of lead ore, refractory-ceramic furnaces, slags and crucibles as well as possible washing/beneficiation installations to concentrate the ore.

Goldenberg (1996a) distinguishes between three different sorts of finds, which may be formed during the separation of silver from lead through cupellation:

1. Made of pure litharge, finds of tool-remains form during the removal of litharge from the surface of a liquid silver-bearing lead-bath. These are about 8–10 cm large, husk-shaped

Table 5.8 Chemical composition of pieces of litharges found at the Roman vicus of Lüderich (Fig. 5.62) and from Silchester

Inv.-nos. DBM	D-102/19a	D-102/19b	D-102/34	D-102/38	Silchester
SiO_2	0.73	0.38	4.19	0.38	0.42
TiO_2	0.02	n.d.	0.56	0.02	n.a.
Al_2O_3	0.35	0.07	3.05	0.21	0.23
Fe_2O_3	0.03	n.d.	0.38	n.d.	0.23
MnO	0.02	n.d.	0.12	n. d.	n.a.
Na_2O	n. d.	n.d.	n. d.	n. d.	n.a.
MgO	0.09	n.d.	0.17	0.01	n.a.
CaO	1.00	0.05	2.86	0.02	11.8
P_2O_5	0.04	0.08	0.29	0.02	7.86
S	0.11	0.02	0.62	0.04	n.a.
Pb	89.8	87.7	73.4	87.4	50.4
Cu	0.32	0.15	0.33	0.15	18
Cr	230	180	290	170	n.a.
Zn	140	10	2700	20	n.a.
As	50	40	150	40	n.a.
Ag	<0.001	<0.001	32	<0.001	0.13 wt.%
Sb	n.d.	n.d.	1200	n.d.	n.a.
Ba	n.d.	n.d.	n.d.	n.d.	n.a.
Ni	n.d.	n.d.	n.d.	n.d.	n.a.
Σwt.%	92.5	88.4	86.2	88.3	89.07

Due to their low CaO-concentration sample D-102/19b consists of a fragment of comparably pure litharge, while D-102/34 are probably fragments of cupellation hearths which were made of CaO-, Al_2O_3- and SiO_2-rich material (Bode 2008). Another example from Silchester has extremely high concentrations of CaO and P_2O_5. This sample is a fragment of a cupellation hearth made of lime and bone ash (Tylecote 1987). This sample is not safely dated to the Roman period. Values in wt.% until Pb. Otherwise (from Cr below) in parts per million (ppm). *n.d.* not detected

cones of lithargite, which have a hole in the longitudinal axis. They were formed during the removal of the liquid litharge with an (iron-) rod during the cupping process. Lithargite coated the rod in multiple separate layers and in this way a cone was formed slowly. The rod was then pulled out of the cone. Such cones were found at many sites of Roman-Greek antiquity. Examples have become known from the Hellenistic Laurion (Greece; Conophagos (1980), from the approximately simultaneous lead–silver production of Skres on the island of Thasos (Greece, Fig. 5.63; Hauptmann et al. 1988), from Rio Tinto (southwest Iberian Peninsula; Craddock 1995), but also from localities in the north, for example from a Roman vicus at Dornheim/Wiesloch (Hensen 2001).

Fig. 5.63 Skres, Island of Thasos (Greece), Lead–silver smelting site. Cone made of litharge for the removal of the liquid PbO-bath during cupellation. The cone formed in several layers around a thin iron (?)—rod. The shape and size of the cone correspond exactly to the specimens that Conophagos (1980) presented from Laurion. Photo from Hauptmann et al. (1988)

2. Hearth lining impregnated with litharge. In the earliest stages, the metallurgical facility where the cupellation process was performed was probably made of simple holes dug in the ground, lined with a mixture of special clays, lime or marl that were then litharge soaked in. The term litharge cake best describes the transition between the tool finds of relatively clean litharge and the actual cupels.

Scarce relics of cupels as evidence for the extraction of silver can already be found in the fifth millennium in Asia Minor, later in roughly palm-sized fragments. Helwing (2014) gives an overview of finds in Anatolia, Mesopotamia and the Iranian high plateau, although the finds are not always specified according to the classification presented here. Among the finds are finds of litharge from Tappeh Sialk (3660–3520 BC; Nezafati and Pernicka 2006, 2012) and from Tappeh Hissar in the northern central Iran (Hissar II–III period, i.e. fourth/third millennium BC; and from Tappeh Shoghali in the southeastern part of Iran (Sialk III–IV period; 3350–3000 BC; Nezafati and Hessari 2017). In Turkmenistan, relics of the cupellation process were found at Ilgynly Depe (Salvatori et al. 2009).

Evidence for silver production by cupellation also comes from eastern Anatolia. These are finds from Habuba Kabira (northern Syria, upper Euphrates river). They date from the second half of the fourth millennium BC, i.e. the late Uruk period (Kohlmeyer 1994; Pernicka et al. 1998). Fragments of litharge and lead–silver rich slags from Fatmali-Kalecik (southeastern Anatolia) were dated to the late fifth/early fourth millennium BC; Hess et al. 1998). This belongs to the Late Chalcolithic period or late Ubaid/early Uruk period. The authors suggest the finds as a decisive feature for the understanding of the incipient stages of lead–silver metallurgy. Silvery lead ores were first smelted and afterwards processed in a second stage inside of settlements.

In the case of Arisman (Sialk III–IV period, late fourth/early third millennium BC) it is especially obvious that here—isolated from all lead–silver mineralisations—the precious-metal-bearing lead was desilvered on a large scale (Pernicka et al. 2011). What is astonishing is the observation that silver-bearing lead ores were transported over long distances to settlements not only in the metropolitan area of Mesopotamia but also in the central Iranian highlands. Of interest are the oval-shaped cupel bowls found in Arisman, often found side by side in rows of two to three (Helwing 2011). They can measure up to c. 8 × 30 cm (Fig. 5.64a). At the Early Helladic I site Lambrika (Greece) many hundreds of round litharge cakes with some 10 impressions, cm-sized, were excavated (Fig. 5.64b). The function of these impressions is not absolutely clear, possibly there was an enrichment of silver (oral lecture E. Andrikou, Bochum 11/2019).

As an interval to the mediaeval cupellation herds, Tylecote (1987) suggests hearth-like, covered facilities (Fig. 5.65). In the mediaeval to early-modern period, indicated, e.g. by the descriptions of Georg Agricola, real kilns with diametres of >1 m were built.

Thereafter, with the proliferation of silver and lead, the finds at least of the lower parts of cupels become more frequent and larger.

One of the most impressive findings is the numerous cupels that have been excavated in Sardis in western Anatolia (Ramage 2000). They date to the middle of the first millennium BC. Very characteristic are cupellation hearths, which have recesses on the surface caused by the exsolution of silver (gleam of silver). Pliny (XXX, 95) describes exactly this effect from other cupels in his books Historia Naturalis and calls it "… ut oleum aquis", from the floating of silver on litharge. Rehren and Klappauf (1995) explain the physico-chemical background of this effect using the mediaeval cupel fragment from the Harz as an example (Fig. 5.66).

Fig. 5.64 (**a**) Arisman (central Iran), late fourth/early third millennium BC. Three whitish coloured cupellation hearth linings soaked with litharge from the settlement of Arisman I. They were found in the workshop middens of the Sialk III–IV periods. The cupels were arranged next to each other in order to carry out cupellation of several batches of lead–silver on a large scale and with considerable experiences. Photo: T. Stöllner, Deutsches Bergbau-Museum Bochum. (**b**) Lambrika (Greece), Early Helladic I (3200–2650 BC) workshop. Arch. Museum Bravnon no. 4759. Round cake of litharge with 10 impressions where possibly the silver was concentrated and separated from the lead oxide. Courtesy of Dr. Eleni Andrikon, Head of the Ephorate for east Attica. Copyright Hellenic Ministry of Culture and Sports

Another outstanding find is the cupel of Murcia. This round, slightly vaulted cupel has a diametre of 43 cm (Fig. 5.66a). The cupel of Murcia was created during the desilvering of the lead ore of the nearby lead–silver deposit of Carthagena.

An equally large oval-shaped 35 × 42 cm cup was found in the Laurion deposit district

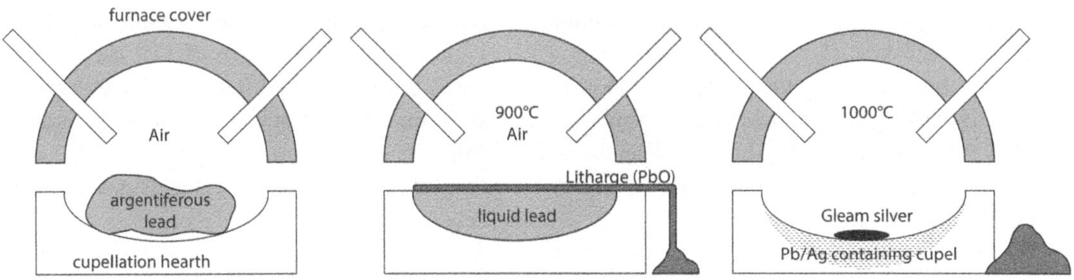

Fig. 5.65 Sketch of a cupellation hearth to extract the noble metal from argentiferous lead. It was probably used since prehistoric epochs. At around 900 °C liquid surface formed litharge was pulled from the liquid lead. When all the (argentiferous) lead was oxidised and partly sucked in the bottom of the cupellation hearth, finally the gleam of silver appeared. Modified after Goldenberg and Steuer (2004)

a

b

Fig. 5.66 Litharge cakes found from the archaeological context. (**a**) Pedro del Pinatar, Murcia (Spain) (Collection Paya). Typical roundish shape of an ancient cupel, made of litharge (litharge), which has been absorbed by the underlying shell of lime, marl, clay and ash. Both materials have probably reacted with each other. Outer diametre of the cupel 43 cm. Photo: C. Domergue (2008). (**b**) Litharge cake from Sardis from the middle of the first millennium BC (Ramage 2000). Note the impression on the surface of the cake, where once the molten silver (gleam silver) sat like “oil on water”

(Conophagos 1980), and there are also reports of litharge cakes, in particular of cupellation hearths fragments, of the Aegean islands of Thasos and Sifnos, which is not surprising because the islands often have smaller lead–silver deposits (Hauptmann et al. 1988; Pernicka et al. 1985).

In the context of litharge cakes, slags with CaMg-silicates (clinopyroxenes), glassy lead silicates, magnetites and inclusions of silver are occasionally found (Hess et al. 1998; Pernicka et al. 1998). Such slags, as well as "lead-free solidified masses" which accumulate during the cleaning of the hearth, were in the Middle Ages and later referred to as *Geschur* or *Gekrätz* (Goldenberg 1996a).

3. During the last stage of the cupellation process, the silver swims on the surface of the litharge. However, a complete separation of silver has generally not been possible in the past, so that silver inclusions can repeatedly be detected in the litharge cakes (Fig. 5.67). Conversely, residues of lead in silver artefacts can be used as evidence for a cupellation process.
4. Sediments with high proportions of very finely processed litharge were found in large quantities in and around the so-called ore-washing facility in Thorikos; Demoliaki and other localities in the Laurion area (Greece) (Rehren et al. 1999a, b, 2002, Papadimitriou 2012, 2016, 2017). They mainly date to the period after the fourth century BC. Further investigations of soil samples from the ergasterion of Ari prove a large content of crushed litharge inclusions as well (Fig. 5.68).

This mixture of pulverised litharge was firstly used as a temper for the production of watertight plaster of cisterns. Later, in Roman times, finely processed litharge entered the medical trade as a cure, as described by Pliny (XXXIII, 108–109).

Far more reasonable, however, is the assumption of recycling the litharge to recover silver residuals. As already mentioned, it can be assumed that even in the Hellenistic era, the cupellation process failed to completely separate silver from the litharge. Rehren et al. (2002) estimate silver contents in litharge in the area of 0.0*x* wt.%. That is still the first/tenth of the mined lead–silver ore. Since the silver was contained in mechanically incorporated particles in the litharge of the cupel, it could be extracted by a fine treatment and subsequent washing. On the other hand, the crushed litharge could have been melted down again. The possibilities of reusing the crushed litharge from the Laurion area are examined in detail by the authors.

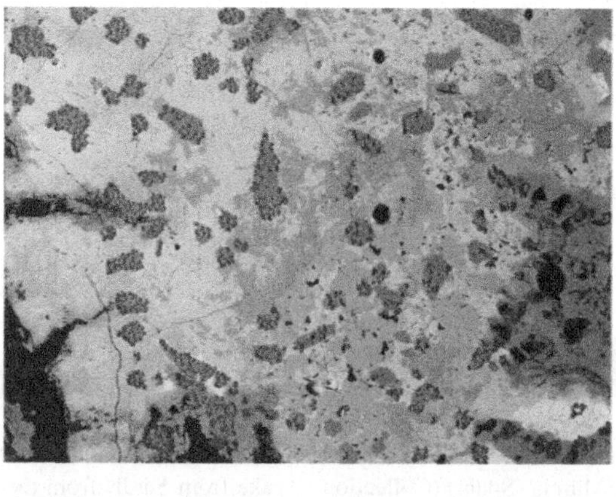

Fig. 5.67 Fatmali-Kalecik, eastern Anatolia, late fourth millennium BC. Section of litharge no. TR 24/7. Inclusions of silver in lead oxide. Small black spots of silver chloride are scattered in the silver metal (grey). Scanning Electron Microscope, backscattered mode. Width of image about 0.41 mm. Photo from Hess et al. (1998)

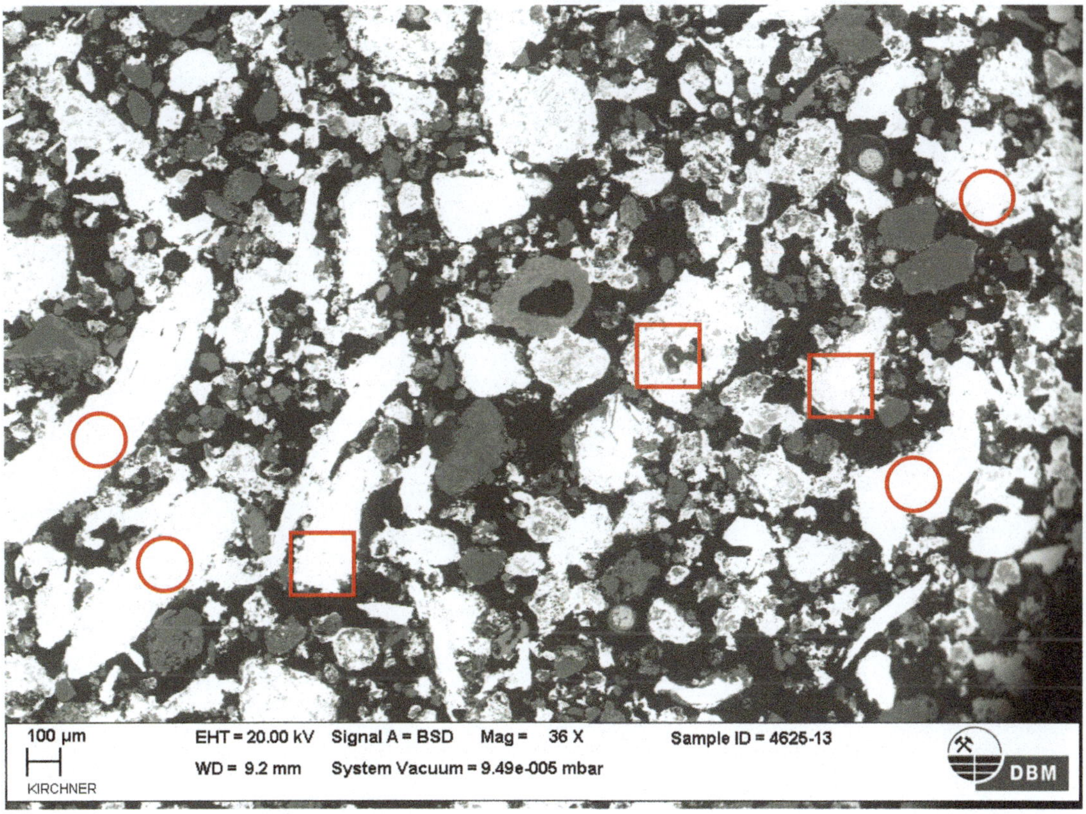

Fig. 5.68 Ari, Laurion mining district (Greece), sample ARI03 (DBM no. 4625/13). Fine-grained crushed litharge in a soil sample ("washed sand") found close to the kollergang of Ari. The litharge is shown in masses of white angular pieces (red circles). These are in parts decomposed to cerussite ($PbCO_3$) which used to form a serrated ring around the litharge (red quadrants). Scanning electron micrograph, secondary image mode. Photo: D. Kirchner, Deutsches Bergbau-Museum Bochum

Rehren et al. (1999a, b) state that the apparent absence of significant quantities of litharge in the Laurion can herewith be discussed in view of fresh evidence. It is known now with certainty that significant quantities of litharge were processed on an industrial scale similar to the washing of ore, using the very same installations and, hence, techniques.

6 Making Metals: Ancient Metallurgical Processes

6.1 General Aspects

Of the 60 metals and semimetals that are known today, only 9 were known from antiquity to the Roman period, namely copper, gold, silver, lead, tin, iron antimony, mercury and zinc. The semimetal arsenic, which occurs regularly in ancient artefacts from the beginning of extractive metallurgy, was unknown, as was nickel. It was only in the nineteenth and twentieth centuries AD that the number of metals produced by smelting increased considerably.

The metals known in (pre-) historical time, their earliest use and some important physical properties are summarised in Table 6.1.

To recover metals, most ores must be smelted. The recovered raw metal is further processed in subsequent steps. It is therefore important to distinguish between smelting and metalworking as two fundamentally independent steps in the craft industry. This and the following chapters focus on the different methods of ore smelting up to the formation of alloys. The metals zinc and mercury were not produced by smelting processes. Due to their low melting and boiling points, they were obtained by distillation and subsequent condensation.

On the other hand, the further processing of metal into everyday objects or weapons or the often very delicate, sophisticated methods of producing decorative objects are largely excluded (punching, engraving, granulation and inlay work a. o.).

As discussed in Sect. 4.3, three essential parameters are of basic importance for the successful smelting of ores:

1. The production of heat up to at least about 1100–1300 °C.
2. The production of reducing conditions in smelting furnaces, vessels or crucibles (surplus of carbon and/or carbon monoxide (CO) within the gas atmosphere).
3. Sufficient time of firing to guarantee the transformation of ore to metal, and a formation of slag.

As discussed in Sect. 4.3 up until the time of the Industrial Revolution in the second half of the eighteenth century AD, the smelting of ore was achieved by burning charcoal and perhaps to a lesser content by hard coal, wood and other biomasses.

6.1.1 Reconstructing Ancient Smelting Processes

Due to the great regional variation of the mineral content and the extent of near-surface parts of ore deposits, any archaeometallurgical research should be conducted as detailed is possible

A. Hauptmann, *Archaeometallurgy – Materials Science Aspects*, Natural Science in Archaeology,
https://doi.org/10.1007/978-3-030-50367-3_6

Table 6.1 Metal production from ores in ancient times (enlarged and modified after Keesmann 1989)

Copper (since eighth/fifth millennium BC)
• Native copper ore is rare, copper ores rather widely distributed, easy to recognise. Co-smelting oxidic–sulphidic ores.
• Liquid metal easy to produce by reduction from ores.
• Production of artefacts with reasonable material properties for alloys with arsenic, tin, zinc, lead.
• Easy to work (casting and smithing).
Gold (since fifth millennium BC)
• Native gold occurs only as gold/silver alloys. Rare.
• Native gold resp. gold/silver alloys directly ready for use.
• Liquid metal easy to produce by simple melting.
• Easy to work (casting and shaping).
Silver (since fourth/third millennium BC)
• Native silver ore very rare, rich silver ores rare and difficult to find. Silver-containing lead ores.
• Native silver directly ready for use.
• Extraction of silver from ores in the liquid state needs multistage processes.
• Metal easy to work (casting and shaping).
Lead (since fifth/third millennium BC)
• Rich lead ores widely distributed and easy to find.
• Liquid metal production by reduction from ores easy.
• Alloying with copper (bronze), tin.
• Soft metal easy to work.
Tin (since third millennium BC)
• Suitable ores for tin production rare and difficult to find.
• Liquid metal production by reduction from ores difficult.
• Alloying with copper (bronze), lead.
• Easy to work.
Iron (since third/second millennium BC)
• Massive iron ores widely distributed and easy to find.
• Metal production by reduction from ores easy, but in the solid state.
• Difficult to work.
• Main purpose: Alloying with carbon (steel, cast iron). High variety of alloys.
Antimony (since third/second millennium BC)
• Antimony ores rare, a local specialty. Rich antimony ores are rare.
• Liquid metal easy to produce by simple smelting.
• Alloying with copper.
Zinc (since late Hellenistic period)
• (secondary) zinc ores frequent, easy to find.
• Liquid metal difficult to produce from ores by simple smelting. Distillation.
• Alloying with copper (brass).
Mercury (since second half of first millennium)
• Native mercury and mercury ores are very rare. Local specialty.
• Liquid metal difficult to produce by distillation.
• Working in the liquid state at room temperature (alloying and parting).

concerning the investigation of minerals in order to reconstruct ancient pyrometallurgical processes. Obtaining metal from the secondary zoning could obviously have epoch-making effects. Instructive examples have been done by Barnes (1974), Charles (1980), Wagner et al. (1989) or Ixer and Pattrick (2003).

The development of early mining and metallurgical technologies has not been happening in clearly definite phases, they developed mostly in several transition phases. Such "bridging technologies" include, e.g. the co-smelting of mixed oxidic and sulphidic ores (Rostoker et al. 1989a). Such mixtures may not be underestimated in their importance for the smelting of ores. As pointed out by Lechtman and Klein (1999), the intermediate metallurgic co-smelting of mixed oxidic and sulphidic ores was even a simple technology.

In ancient times, if metallurgical processes proved to be useful, they quickly enlarged "industrial" dimensions and evolved into so-called imprinting phases (Stöllner 2008a, b). These are reflected in the archaeological findings. Especially from such imprinting phases many archaeometallurgical finds have been investigated, which has led to a certain simplification of the interpretation of early technologies. This is also true, e.g. for the role of sulphide inclusions in slags, which often are quickly considered as useful evidence for an intentional smelting of sulphide ores.

Generally, a reconstruction of early metallurgical processes is based on principles of modern technologies to be found in modern textbooks (e.g. Tafel 1929; Tafel and Wagenmann 1951; Rosenqvist 1974; Biswas and Davenport 1980), and a reconstruction of ancient smelting processes is usually possible after thermochemical processes.

However, the transfer of such modern concepts to archaeological findings also harbours dangers. In ancient times, compared to what can be found in modern textbooks, more diverse and different processing steps were taken to extract metals. The enormous variety of metallurgical stages comes from medieval and early modern writings. Bartels et al. (2007) discussed this topic using the reports of Theophilus Presbyter (twelfth century AD) and of Georg Agricola (sixteenth century). The reconstruction of practical, handcrafting details is rather elusive, because smelting furnaces or crucibles (see below) are mostly destroyed or only preserved fragmentarily. Shape and design, firing and thermodynamics of ancient smelting furnaces did not allow an ideal processing until the Middle Ages. Consequences were often only partial melting of the batches, which produced "breccia-like" slags with abundant undecomposed inclusions. Addressing such inclusions as fluxes should only be done with great care. A quantitative separation of metal from slag or intermediate products was difficult and required additional (mechanical) processing, found in all early metal techniques. Technological innovations then led to increasingly extensive charging of smelting furnaces and thus to larger amounts of slag. The supraregional and diachronic similarity of iron-rich silicate slags has its origin primarily in physicochemical mechanisms of liquefaction of comparable ore materials at high temperatures. It is not primarily an expression of technological-craft skills.

The most important source of information are slags, which are almost regularly present in the archaeological context as waste products of smelting processes and further processing metals in a hearth or crucible. "Frozen" in them is much information about the ores used and their formation in reaction vessels. This information can be decoded with scientific concepts and methods. This was discussed in Chap. 5.

Important information on the technology as well as the provenance of metal can be obtained from the juxtaposition of minor and trace elements as well as the ratios of the lead isotopes of ore and metal (Pernicka 1990, 1999, 2014c). However, as mentioned above, an approach to technical details based on archaeological findings or scientific procedures is often limited. Here, ethnographic comparisons (cf. Chap. 9) or experimental work (cf. Chap. 10) can be very useful. For artisanal iron smelting, there are numerous examples known from Africa (Schmidt 1996; Tylecote et al. 1971; Kense 1983; Celis 1991). East and western Africa (Tanzania, Ghana, and Burkina Faso) provide detailed evidence of gold production (Winkelmann 1964; Werthmann 2009; Borg 2014). Descriptions and instructive pictorial representations about traditional gold metallurgy and non-ferrous metal smelting are known from nineteenth century AD Japan (Gowland 1899, 1901, 1906, 1914; Hauptmann und Lewin 1984), and Anfinset (2011) describes traditional copper smelting from Nepal. Experimental research is mainly available for copper and iron (Hanning et al. 2015; Straube 1996; s. u.).

Generally, the following stages can be listed for early metal production.

6.1.1.1 Roasting

The actual smelting process can be preceded by one or several roasting processes, which are carried out in the presence of oxygen at temperatures of about 400–700 °C for several hours, even for several days. Most probably roasting processes have been integral parts of early smelting not only of copper ores, but also of other ores.

The purpose of roasting ores is to remove the intergranular water as well as crystallographic fixed OH- and CO_2 concentrations from the ore. In addition, sulphide and sulphate contents should be removed or at least reduced by oxidising them to SO_2- gases. Hereby, the porous volume of the ore is enlarged and improves all reactions of reduction in a furnace, and breaks up the cohesion, so that the ore can be crushed more easily. Furthermore, volatile elements such as arsenic, antimony or bismuth can be removed. Once started, a roasting process usually continues on its own, it is an exothermic process.

Roasting processes are also used to desulphurise sulphidic intermediate products such as matte. The archaeological proof of roasting facilities seems to be especially difficult for the earliest periods of metal extraction, and there are only a few cases that prove such a process definitively (see Sect. 6.2).

6.1.1.2 Smelting

The effects of smelting are based on high-temperature reactions in which ores are decomposed and unusable components such as sulphur, oxygen, iron and other oxidic and siliceous components are separated from metal. This causes a chemical change of the original ore in major, minor and trace element contents. Since pure ores only in exceptional cases were used, but usually were intergrown with gangues and host rocks, slag is produced as a waste product.

6.1.1.3 Single-Step Smelting

Seen from chemical aspects solely, sulphur-free ores can be smelted to metal in a one-step high-temperature process under reducing conditions. Gold, copper and iron, which may occur in native form in nature, can also be melted in single-stage process. Gold and copper melt at somewhat >1000 °C, copper was already processed in the late ninth millennium in south-east Anatolia by cold and hot deformation (Maddin et al. 1999; Yalçın and Pernicka 1999) as was gold in the ending fifth millennium BC in Varna (Ivanov 1991). Artefacts believed to be made from meteoritic iron were possibly produced in the late third millennium BC in the Middle East (Buchwald 2005). The smelting of ores generally begins at the turn of the Neolithic/Chalcolithic period. Typical and widespread oxidic ores are malachite ($Cu_2(OH)_2CO_3$), azurite ($Cu_3(OH)_2[CO_3]_2$), limonite or goethite (FeOOH), magnetite (Fe_3O_4), but also cassiterite (SnO_2) may be treated in single-stage operations.

In Timna (Israel) and Faynan (Jordan) prehistoric smelting of almost pure oxidic copper ores has been exemplarily treated (Bachmann and Rothenberg 1980; Hauptmann 2007) and a theoretical single-stage smelting process would have been very likely to extract the metal. However, this does not mean that primarily only metal prills and lumps were produced which have to be remelted into larger units in several subsequent operations, be it copper, iron or lead. It may have been possible to add flux to improve the formation of slag, but this is often postulated too lightly for early metallurgy.

6.1.1.4 Co-Smelting

Next to a theoretic single-step smelting of oxidic ores, also co-smelting of oxidic ores with sulphides may result in a precipitation of metal by roast-reactions. This is especially true for copper ores (see above). Generally, in the upper part of ore deposits oxidic copper ores (i.e. oxides, carbonates, sulphates and chlorides of copper) may have a clear surplus, but very often they are intergrown with remains of primary sulphidic ores such as chalcopyrite, chalcocite and others (see Sect. 3.6). The same is true for lead ores, as exemplified by Nakhlak (Iran), where mixed lead carbonate and sulphidic lead ores (galenite) are mixed intensively (Fig. 3.41b; Klein and Hauptmann 2020). Lechtman and Klein (1999) performed experimental work on a wide range co-smelting of oxidic: sulpharsenide ore mixtures. The sulphidic ores have not been roasted before smelting. They yielded coherent copper-arsenic alloy ingots.

6.1.1.5 Multistage Smelting

When smelting sulphidic respective sulphur-containing ores, smelting should be preceded by a roasting process. The subsequent smelting process can be carried out in closed shaft furnaces under reducing conditions. If the sulphur content is completely removed by roasting, the ore can be reduced directly to metal. If the ore still contains sulphur, matte (in most cases CuFeS matte) with different amounts of copper will be produced. This effect can be used selectively in a deliberate matte smelting. The matte is roasted and smelted again under reducing conditions. This is repeated until the finally produced white metal (Cu_2S) is converted to copper in a final smelt. Multistage smelting of sulphidic copper ores was intensively investigated in the Alpine region in Austria (Eibner 1982; Goldenberg 2004), northern Italy (Cierny 2008; Silvestri et al. 2015) and Switzerland (Reitmaier-Naef 2020). The constant presence of slag heaps with similar types of slag, roasting beds and smelting furnaces throughout the Alpine area indicates widespread standardised metallurgical processes in the Middle to Late Bronze Age.

Part of the multistage smelting is also the extraction of silver from argentiferous ores such as galena or jarosites. Here, after a (reducing) smelting process, another (oxidising) process is required, the cupellation.

6.1.1.6 Distillation and Condensation

There are three metals that are not obtained by the traditional melting processes in a crucible or bloom furnace because of their low melting and boiling points. These are arsenic, mercury and zinc, which were known in ancient times, albeit some of them only in the Middle Ages. These metals are transition directly from the solid state of their ores into the gaseous state (vapour), without liquefication. This process is called distillation. By subsequent cooling, a condensate is generated, which is the transition of a substance from the gaseous to the liquid or solid state. This process is called condensation.

6.1.2 Metallurgical Installations

6.1.2.1 Crucibles

It has long been suggested that in the early stages of metallurgy, ore was (s)melted in ceramic crucibles with a volume of perhaps 200–500 ccm, since in such small reaction vessels firing conditions were more easily controllable (Hauptmann and Weisgerber 1996). Crucibles would have been set in a fire pit-type furnace filled with glowing charcoal, fired by blowpipes. Crucible processes have opened the way for the development of new processes on a larger scale or for new alloys (Craddock 2000b). Tellingly, across Europe and beyond, the earliest smelting of ores, as well as subsequent processing, was being carried out within settlement communities (Hauptmann and Weisgerber 1996).

From the Middle East, crucibles for smelting copper ores are known since the end of the sixth millennium from the excavation at Tal-i Iblis (Pigott 1999; Frame 2012). From the Chalcolithic (fourth millennium BC) such crucibles are described from Faynan (Jordan) (Hauptmann 2007; Hauptmann et al. 1996; Weisgerber 2004c; Craddock et al. 2007b). Neolithic-Chalcolithic crucibles are known from the cultures of Pfyn, Altheim and Mondsee in the Alps in Switzerland (Maggetti et al. 1991; Strahm 1994), and also from Chalcolithic settlements of the Iberian Peninsula (Los Millares; Keesmann and Moreno Onorato 1999; Zambujal; Gauß 2015).

However, it is not always clear whether all these ceramic crucibles were actually used for smelting ores, although this is quite possible from a technical point of view (Tylecote et al. 1974; Pigott 1999). According to Bayley and Rehren (2007), however, over time there are three additional technical processes in the foreground, for which crucibles have been used:

1. In cementation processes, solid metal reacts with a gaseous phase (vapour phase produces chemical transformations). In one case cementation is effectively alloying, in other refining. Three examples: the parting of silver from gold (Sect. 6.3), brass making (Sect. 7.2.5) and crucible steel making (Sect. 7.5). When parting was performed for the first time is not definitely certain. Bayley (1990) published ceramic crucibles from the tenth/eleventh century AD, which were used in England for a parting of gold. Ramage and Craddock (2000) provided evidence for parting in crucibles in the middle of the first millennium BC.
2. Fire assaying is the testing of an unknown substance by subjecting samples of them to a series of chemical or metallurgical operations carried out in small crucibles, probably developed during the Middle Ages in the context of mining and coin production. According to the archaeological evidence, tin smelting was performed in crucibles at Early Bronze Age Kestel/Göltepe (see Sect. 6.5).
3. A melting operation is the third and most common functionally defined process, beginning from the Neolithic period up to the present day. Melting of solid copper is generally conceivable. Laschimke and Burger (2018) suggest that the more angular trough-shaped crucibles of the Pfyn culture (Fig. 6.1) were used to melt smaller metal portions, lumps and prills, for example to produce ingots.

The authors do not include a smelting of ores into this classification. This may be explained by the fact that the insides of many earliest crucibles are covered, if at all, only by very thin, mm-thick slag incrustations (Craddock et al. 2007b). Melting experiments have shown that, e.g. copper in metallic form can be produced from small-sized ores when they are loosely placed in a pit in the ground filled with glowing wood or charcoal (Fasnacht 1999; Timberlake 2007). From these activities, just blowpipe nozzles are found very rarely (Fig. 6.2), while depressions in the soil are hard to be found in the archaeological context.

6.1.2.2 Roasting Installations

In the region of Kitzbühel/Jochberg (Austria) today, above all due to the many years of research by Richard Pittioni, some 40 prehistoric copper smelting sites are known. These evidences for copper production fall into the early Urnfield era (fourteenth/thirteenth century BC). Results of mining-archaeological investigations on one of the numerous smelting sites at Jochberg yielded the evidence of two superimposed stone settings with an inner width of c. 1 m (Goldenberg 2004). These stone settings were filled with an accumulation of iron oxides (hematite and limonite) and charcoal, which stood out clearly from the surrounding soil by their reddish colour. The close proximity to four smelting furnaces and slags suggested the identification of these as roasting beds (Fig. 6.3). These findings are among the earliest evidence of deliberate pretreatment of ores by roasting prior to smelting.

Later, more sophisticated constructions of larger roasting stalls come from Oman, where copper has been mined since early prehistoric times on many exposures of ore deposits. Weisgerber (1980) finds distinct remains of roasting stalls, some set into a masonry wall and others set into the local bedrock, from numerous early Islamic smelting sites (Fig. 6.4), as they are pictured some 400 years later in Georg Agricolas work from the sixteenth century. They have a floor plan of about 1.5 × 2.5 m and are about 1 m high. In front of the roasting stalls are heaps of fine-grained powdery hematite, limonite, rock fragments as well as copper and iron sulphates. The grouping of several roasting stalls next to

Fig. 6.1 Kanton Thurgau (Switzerland). Three pottery vessels used for metallurgical operations from Niederwil, Gachnang-Egelsee and Hüttwilen from the Pfyner Culture (fourth millennium BC). These vessels are termed by Leuzinger (1997) as casting moulds ("Gusstiegel"). According to its shape and design, just the left one with the spout might have been a casting mould. The other two moulds on the right side of the picture may possibly be used for remelting copper prills to larger units. Foto: Daniel Steiner, AATG, http://www.archaeologie.tg.ch. Courtesy Urs Leuzinger, Thurgau

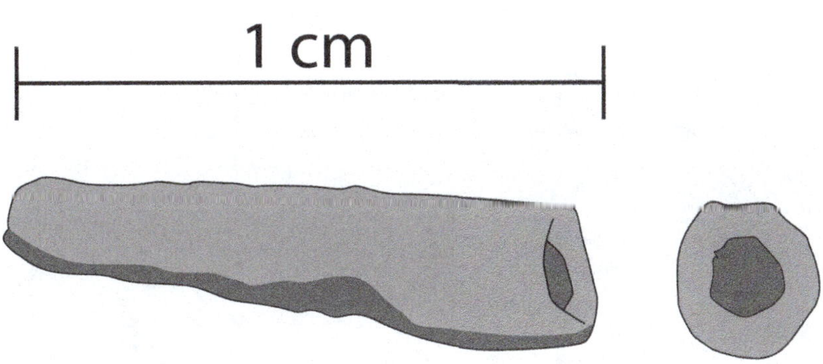

Fig. 6.2 Peştera Caprelor, Cheile Turzii (Rumania, Bodrogkeressztúr-culture, fifth/fourth millennia BC). Nozzle of a blowpipe from a workshop of the Chalcolithic layers for shaping gold artefacts in the gold-rich Transylvania. It is an example of the smallest tuyère nozzles known. It was made of bone and was used for very small-scaled metallurgical operations comparable to blowpipe proofs. Note the scale: the solid material of the nozzle was the top perhaps of a wooden blowpipe. From Lazarovici et al. (2012)

each other can be explained by the fact that roasting processes usually had to be repeated several times in order to achieve the desired low sulphur content of the ore batch as precisely as possible.

6.1.2.3 Smelting Furnaces

Only in the developed Early Bronze Age, with a burgeoning metal boom, smelting of ores is shifting close to the exposures of deposits themselves. The Faynan (Jordan) copper ore district is

Fig. 6.3 Jochberg/Kitzbühel (Austria), Urnfield period. View of a roasting bed constructed by a setting of stones. It has a width of c. 1 m. Note the reddish colour of the filling in the roasting bed caused by various high valence iron oxides formed by heating under oxidising conditions. From Goldenberg (2004)

Fig. 6.4 Tawi Aarja (Oman). Three early Islamic roasting stalls (c. ninth/tenth century AD). The roasting stalls have an open front. They were constructed close to the exposure of the gossan of the same-titled copper deposit. In Oman, roasting stalls are always grouped in units of three or four examples. Photo: G. Weisgerber (†), Deutsches Bergbau-Museum Bochum

Fig. 6.5 Faynan 9, Faynan (Jordan), middle of the third millennium BC. Preserved backside of two wind-powered copper smelting furnaces. These furnaces show up to 27 layers in their backside which proves repeated smelting operations at one and the same place. The furnaces are part of a larger battery of smelting furnaces at the top of an Early Bronze Age slag heap. Photo: A. Hauptmann, Deutsches Bergbau-Museum Bochum

home to the earliest smelting furnaces operated by natural wind (Hauptmann 2007; Fig. 6.5). They date to the middle of the third millennium BC. It is not clear here how high a possible furnace shaft was constructed over the only fragmentarily preserved furnace bottoms.

The reconstruction of ancient smelting furnaces can be problematic due to their generally poor state of preservation. Shafts of smelting furnaces were apparently destroyed regularly after each furnace run to recover leftover pieces of metal (copper, iron globules) in the furnace and perhaps also to remove residues of slags. Often, only their basis is preserved and thus they are hardly distinguishable from hearths. However, Fasnacht (2006) was able to meaningfully reconstruct a freestanding smelting furnace based on many fragments from his excavations in Agia Varvara Almyras (Cyprus) (Fig. 6.6). It had the shape of an upwardly tapering cone with a height of about 1 m. The oven fragments date mainly in the cypro-classical time (approximately 400 BC). At the base was an opening where tuyères were inserted. Several hundred fragments of tuyères were found at the site.

Tuyères, commonly used equipment for driving (s)melting furnaces by artificial air supply connected with bellows were used since the Late Bronze Age. Later, they were also used in iron metallurgy. It seems to be the rule, that in archaeological excavations, especially the highly vitrified nozzles are to be found, because these are the parts of tuyères which are most resistant to corrosion, while the shape and design of bellows must often subject of speculations. The nozzles were the parts of tuyères which extended in the inner part of furnaces (see Fig. 4.13). Impressive examples of Late Bronze Age and Iron Age tuyères were excavated at Politiko Phorades (Cyprus), at Faynan, Timna and Bir Nasib/Wadi Nasib (Sinai) (Fig. 6.7). At these sites, tuyères have been produced in masses. At Politiko Phorades 500 fragments of tuyères were excavated (Kassianidou 2003a), at the two copper districts in the Wadi Arabah the number of nozzles of tuyères found have not been counted (Rothenberg 1990; Hauptmann 2007; Levy et al. 2014), and masses of tuyères have been found at smelting sites on the Sinai Peninsula (Abdel-Motelib et al. 2012).

Fig. 6.6 Reconstruction of a copper smelting furnace using fragments from an excavation at Agia Varvara-Almyra (Cyprus). The furnace had a height of c. 1 m, the opening at the base of the conus was used for the tuyères. It was filled with clay for the reconstruction. The fragments of the furnace were dated approximately 400 BC. From: Fasnacht (2006)

Smelting furnaces from the Middle or Late Bronze Age copper metallurgy of the Alps come from Fennhals (Alto Adige, northern Italy) (Hauser and Nothdurfter 1988), from Mitterberg (Eibner 1982), from Jochberg in North Tyrol (Goldenberg 2004) and from Trentino (Cierny 2008). According to ethnographic comparisons and melting experiments, it is probable that after finishing a smelting process at least the furnace front was destroyed in order to remove metal (residues), slags etc. In addition, high-temperature damage during smelting had to be expected. Like smelting furnaces, installations for roasting of ores and/or matte (roasting pits and roasting stalls) are archaeologically verifiable from the Middle and Late Bronze Age Alpine region. A "typical" Alpine smelting site of the Bronze Age usually consisted of three basic components: roasting pit, smelting furnace and slag heap (Hanning et al. 2015).

Fig. 6.7 Wadi Nasib-Wadi Sih (Sinai Peninsula, Egypt), copper smelting site. Heavily vitrified tuyères nozzles dated to the Late Bronze and Iron Age. Note the multiple reused tuyères. The diameter of the air hole was 1.5–2 cm. Tuyères like these are characteristic for copper production of this time period on Sinai. Photo: A. Hauptmann, Deutsches Bergbau-Museum Bochum

The best-preserved smelting furnaces come from the smelting of iron and date between c. 150 BC and 500 AD. These are the so-called slag-pit furnaces with no construction for the tapping of slag but instead, the slag formed a block underneath the bloom in the shaft. Many thousands of excavated slag-pit furnaces are known to represent the iron metallurgy of *Germania Libera* and other non-Roman territories. The reconstruction of an iron slag-pit furnace before and after the smelting is shown in Fig. 6.8. Pleiner (2000) compiled a very comprehensive overview and classification of archaeologically excavated smelting furnaces from Europe.

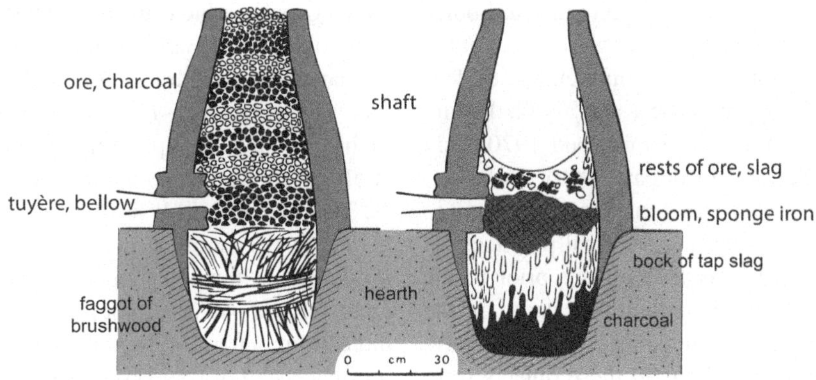

Fig. 6.8 Type of slag-pit furnace as found in large parts of Europe. The reconstruction shows this furnace before and after smelting. The slag block that forms under the furnace and replaced the organic material after smelting is a tap slag which was covered by a mixture of bloom and slag. See also Fig. 5.9. Modified after Pleiner (2000)

6.2 The Metallurgy of Copper

Copper (Cu) is the most widespread and most widely used metal of the ancient world. It is a reddish shiny, soft, ductile metal. It melts in a temperature range comparable to gold and silver. These temperatures and other physical properties of these three metals are summarised in Table 6.2. Because copper forms easily natural and artificial alloys with numerous other metals, it has only been used as an independent metal at the beginning of its use. When exposed to air, copper oxidises very easily to green coloured weathering products. When considering metallurgical operations, it is important that the metal is chalcophile, i.e. it has a high affinity for sulphur and forms various compounds with it. This is in contrast to iron, which apart from pyrite (FeS) and pyrrhotite ($Fe_{1-x}S$), mostly forms oxidic compounds.

Its name was given to copper by the Romans. The metal was given the Latin term "aes cyprium", that means "ore from Cyprus", since in antiquity copper ores were mainly mined in Cyprus. "Cyprium" later became "cuprum" from which the elemental symbol "Cu" emerged. In German-speaking countries, cuprum finally became "Kupfer".

6.2.1 Copper Artefacts

6.2.1.1 Distinguishing Artefacts of Native Copper

Under this striking title, Maddin et al. (1980) collected criteria that made it possible to distinguish between artefacts made from native copper without a smelting process and those that were smelted from ores. They addressed one of the most important problems of the earliest copper metallurgy, namely, when did the use of native copper begin, first as an exotic material to create artefacts without smelting processes, and when did extractive metallurgy emerge?

Table 6.2 Main physical properties of copper in comparison to silver and gold. The hardness is given as Diamond Pyramid Number (DPN) as used in the Vickers hardness test

	Cu	Ag	Au
Density (g/cm^3)	8.92–8.96	10.5	15.5–19.3
Hardness (DPN)	2.5–3	2.5–3	2.5–3
Melting point (°C)	1083	962	1063

A large number of small objects such as hooks, awls and beads made of copper were found at the Pre-Pottery Neolithic sites (ninth/eighth millennium BC) of Çayönü Tepesi and Aşıklı Höyük in Anatolia (Çambel and Braidwood 1970; Esin 1993; Özdoĝan and Özdoĝan 1999). However, the Neolithic period was considered too early to speculate on smelting-metallurgical practices. In fact, the presence of copper objects prior to the coming of pottery, in a cultural stage of the "incipiency of food producing" was inconceivable. It was not clear how all these objects were manufactured.

It was assumed that they would have been made from lumps of native copper available on many ore deposits in Anatolia and shaped by hammering in the cold and hot stage. However, they initiated a vivid discussion of whether it would be possible to distinguish artefacts made from native copper from those produced by smelting ores.

The first considerations go back to Otto and Witter (1952), who tried to differentiate the discussed types of copper based on their trace element pattern. According to their proposals, chemically very pure copper was generally considered to be native copper. Junghans et al. (1974) followed this proposal and in the project "*Studien zu den Anfängen der Metallurgie*" ("Studies on the beginnings of metallurgy") and they classified copper types N and E00 from southeastern Europe as native copper. This proposal was later confirmed by Pernicka et al. (1997).

However, the degree of purity is not a reliable criterion for identification of native copper. The old copper districts of Faynan (Jordan) and Timna (Israel) can be used as counterexamples. The copper smelted from the mineralisation of the sandstone in Faynan is unusually very pure (Hauptmann et al. 1992; Hauptmann 2007); the copper produced in Timna also shows a very high degree of purity (Leese et al. 1985/1986). A comparison of native copper with smelted copper from these localities shows that the low concentrations of Ag, As, Sb and Au are not reliable distinguishing criteria. Only artefacts that have cobalt and nickel contents of about 100–1000 ppm are most likely smelted. Artefacts from Oman, which were smelted during the Chalcolithic and Early Bronze Age periods, are characterised by high As and Ni contents (As $\leq$1 wt.%; Ni ~1 wt.%), corresponding to their ore base in ophiolitic rocks (Begemann et al. 2010). The Co content is around 0.1–1 wt. %. The chemical analysis of silver concentrations is not always useful because the redox behaviour is similar to that of copper. However, isolated inclusions of silver in copper are a convincing indication for a non-smelting process (cf. macehead of Can Hasan, see below).

On the other hand, native copper itself may contain varying impurities, so that a distinction by chemical means is not always explicit. This is also shown in a study by Rapp (1988), who analysed the trace element content of 1065 samples of native copper from various deposits in the world. The result showed varying contaminations with As, Sb, Ag, Fe, Si and Ni.

Arsenical copper is problematic. Early copper artefacts produced by smelting of ores are predominantly arsenic. Those of Ilipinar (western Anatolia) from the Chalcolithic (Begemann et al. 1994) show As contents >1 wt.%. In Ai Bunar (Bulgaria), smelted copper contains arsenic in the percentage range (Pernicka et al. 1997). Apart from the copper types N and E00 of Junghans et al. (1974), arsenical copper was widely used as early as in the fifth/fourth and later in the third millennium BC. The fact that native copper can already contain high As contents (about 5 wt.%) is proven by the deposit of Talmessi (Iran), which is famous in archaeometallurgy, where intermetallic minerals such as domeykite (Cu_3As), whitneyite (Cu_9As) and algodonite (Cu_6As) occur often. There are similar deposits in the area of the Zagros mountains, in the Taurus mountains and on the Balkan peninsula (Schneiderhöhn 1962). Although arsenic and antimony are similar in geochemical terms, the latter occurs in native copper only in very low amounts. Cu antimonides, e.g. horsdorfite (Cu_6Sb), rarely occur in nature.

An apparently good discriminator for distinguishing between native and smelted copper is probably the volatile mercury (Pernicka 1990).

The median of 16 Cu objects from Çayönü Tepesi is 10 ppm. Important results have resulted from its laboratory tests: Heating copper with 200 ppm Hg for 15 min at 800 °C caused no change. Prolonged heating at 900 °C showed the first reductions to 150 ppm. Only when the copper was completely melted, the Hg content dropped down to 0.7 ppm.

Metallographically, there are more distinct criteria for distinguishing the two types of copper. The macehead of Can Hasan in southern Anatolia may represent an example for a transition. Originally, it was assumed that it had been cast from native copper. However, this is surprising in terms of the chronology (c. 6000 BC), because the first cast artefacts in Anatolia did not appear until the second half of the sixth millennium (Yalçın 1998), and in southeastern Europe and Iran they are unequivocally traceable in the fifth millennium. However, Yalçın (1998) was able to prove metallographically that the macehead was not cast, but merely heated and then shaped. Evidence for this interpretation are traces of the recrystallisation of copper (see Chap. 8) and isolated inclusions of silver, which would have dissolved in copper in a liquid state.

In case round droplets of sulphide inclusions could be determined in a polished section under the light microscope, it is a secure indicator that this piece solidified from the liquid state. Copper and copper sulphide differentiate from each other because of an extensive miscibility gap.

Another characteristic of molten copper is droplets of cuprite inclusions. Native copper that did not reach the liquid state is free of those inclusions. However, it is necessary to delineate weathering-related or genetically formed natural cuprite in paragenesis with copper, which likes to penetrate into the metal along cracks and from surfaces, hinting at idioblastic forms (Ramdohr 1975).

Cuprite inclusions were described by Yalçın (2000b) in his investigation of copper needles and chisels from layer XVI of the settlement of Mersin (southern Anatolia) (c. 5000 BC), thus proving their origin from smelted copper, i.e. proving the use of extractive metallurgy. This

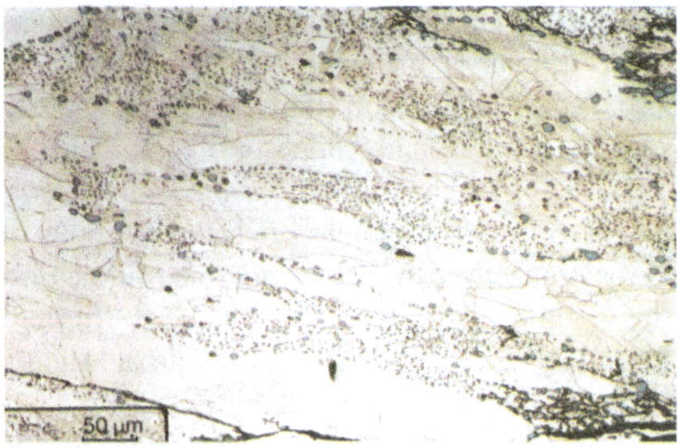

Fig. 6.9 Mersin–Yumuktepe (Anatolia). Chisel from c. 5000 BC. Sample MY 1323 The picture shows a hypoeutectic copper-Cu_2O-structure with 0.18% oxygen. Visible are primary copper crystals with the (Cu + Cu_2O)—eutectic intergrowths. This copper was produced by smelting copper ore. The crystals form a long, stretched net. The beginning of recrystallisation (formation of twinings) indicates a heating of the artefact for further shaping. Mounted sample, etched with $NH_4OH \cdot 2\,H_2O_2$. Reflected light. From Yalçın (2000a, b)

refers to a typical metal structure known in modern refined copper that has not been deoxidised. It still contains oxygen, which forms Cu-oxide (cuprite). There is an extensive miscibility gap between Cu and Cu_2O in the liquid state. The eutectic is at 0.39 wt.% oxygen or 3.5 wt.% Cu_2O. Such a composition crystallises in a eutectic microstructure. In the artefacts of Mersin, a structure can be seen that lies below this eutectic, on the copper-rich side. Copper crystallises in dendritic forms, in between in the eutectic intergrowth of Cu–Cu_2O (Fig. 6.9).

6.2.1.2 Ingots

Some of the most important copper ingots from ancient times should be mentioned here, which are significant in that they take an important key position between ore deposit and finished product in any provenance studies. Not only do they provide interesting insights into metallurgical craftsmanship, but they can also provide an excellent average composition of metal, provided that the ingots were produced within an ore district. However, if scrap metal, produced at other locations, was circulated, any provenance studies are almost hopeless. For casting of ingots, not only products of the "primary" smelting process were used, but also scrap metal, workshop waste, previously spoiled castings and copper prills, which are extracted by mechanical processing from slags.

There are two examples of such collections of Early Bronze Age metal pieces in the Middle East (third millennium BC), from Shahdad (Eastern Iran) (Meier 2015) and from Khirbat Hamra Ifdan (Faynan, Jordan) (Levy et al. 2002) (Fig. 5.12). In these cases, according to the archaeological findings, we can suggest that these scrap metal collections are of local origin.

Since the Early Bronze Age in the Middle East (mid-third millennium BC), copper has been sold as ingots into the supra-regional trade. By the example of Al Maysar (Oman), we know that copper ingots in the Umm an-Nar time (2200–2000 BC) had round, plano-convex shapes with a diameter of c. 12–14 cm and a weight of about 0.6—almost 2 kg (Weisgerber 1981) (Fig. 6.10). The same type of copper ingots was also found at Al-Aqir near Bahlā'(Weisgerber and Yule 2003), some hundred kilometres away from Al Maysar. This shows that plano-convex ingots were apparently widely distributed in Oman. They were obviously traded above all to Mesopotamia (Salzmann 2019).

In Central Europe, copper has been used and processed since the Neolithic (fourth millennium BC). Since the Early Bronze Age (beginning of

Fig. 6.10 Al Maysar (Oman), Maysar 1, house 43 (DA 4840). Hoard of 22 plano-convex copper ingots and fragments with a total weight of >6 kg. Photo: G. Weisgerber (†), Deutsches Bergbau-Museum Bochum

the second millennium BC) copper ingots have been negotiated in the shape of ingot torques (*Ösenringbarren*). These are a common shape of metal in the Alpine Uplands. However, there are more *Ösenringbarren* than finished products of the same composition, so that their interpretation as ingots seems to be at least questionable (Lutz 2016). The largest Early Bronze Age deposit comes from the Bavarian Mauthausen. A total of 700 ingot torques were found, they weigh between 180 and 220 g. The total weight is 150 kg.

Plano-convex ingots were also circulated somewhat later in the Middle Bronze Age (1600–1300 BC) in western Europe (Bachmann et al. 2002/2003). In the Voralpengebiet some 60 completely conserved plano-convex ("Gusskuchen") were found and c. 2000 fragments with a weight of 300 kg (Lutz 2016). In many cases, such plano-convex cast cakes (Ø 15–20 cm) have been previously interpreted as a final product accumulated at the bottom in a smelting furnace. However, they are products of an independent casting process. Their weight was usually about 1500–3000 grams, sometimes up to 8 kg. Like the Middle East, they also came portioned, i.e. they were smashed or sawn into halves, quarters or even smaller pieces.

In the Late Bronze Age, the so-called oxhide ingots of the island of Cyprus gained widespread importance in the Mediterranean and far beyond. The most important find of oxhide ingots (354 ingots with a total weight of about 10 tons) comes from the shipwreck of Uluburun, which sank around 1330 BC off the southern coast of Anatolia. The most extensive descriptions of this shipwreck have been compiled by Yalçın et al. (2005). The plate-shaped oxhide ingots with their typical four bulges (Fig. 6.11) had a weight of approx. 25–30 kg. According to the research of Laschimke and Burger (2011) and Hauptmann et al. (2016), the shape of these ingots is a result of a casting technique: copper was cast from four crucibles into sand moulds, resulting in the four bulges of the ingot. There was no (ceramic) crucible known in the Late Bronze Age that would have been able to resist heating of such an amount of metal to the liquid state, to transport it, and

Fig. 6.11 In the shipwreck of Uluburun (south coast of Anatolia, 1.360 BC), among many other objects, 354 oxhide ingots of copper were recovered. After their recovery, they were taken and stockpiled in the Museum of Bodrum. Photo: A. Hauptmann, Deutsches Bergbau-Museum Bochum

perform a casting. Metallographic investigations on drill-cores (Hauptmann et al. 2002a) have shown that the oxhide ingots of Uluburun consisted of qualitatively very poor copper. They contained amounts of mm-sized slag shreds, which proved that separating the slag from the metal could not have been perfect.

In Roman times, hundreds of flat, round copper cakes (Ø approx. 25–55 cm, 20–30 kg, rarely up to 90 kg) were sold in the western part of the empire. Many were recovered from shipwrecks on the coast of Languedoc-Roussillon in France or Corsica (Fig. 6.12). Klein et al. (2007) analysed many of these ingots, which apparently originate from mineral deposits in south-western Spain (Rio Tinto or Ossa Morena Zone) and were meant to be brought to Rome (Domergue 1990).

6.2.2 Classification of Copper Ores: Metallurgical Aspects

According to modern metallurgical processes, as presented by Tafel and Wagenmann (1951) or Czedik-Eysenberg (1958), the extraction of copper from its ores can be classified simply as follows:

1. Simple reducing smelting of
   - Native copper.
   - Oxidic or chloridic copper ores (malachite, azurite, atacamite). In deposits like Tsumeb (Namibia), Katanga, Chucicamata and El Salvador (Chile) massive amounts of rich secondary copper ores occur, sometimes also in the Old World, e.g., in Faynan (Jordan), Timna (Israel).

   From these ores, copper can be produced in a 1-stage chemical–metallurgical process.
2. Smelting of sulphidic ores.
   - Chalcopyrite, bornite etc. are the most widespread sulphidic copper ores.

   Multistage processes are required here (roast-reaction and roast-reduction process). Many investigations pertaining to this were performed in the eastern and southern Alps (see Sect. 6.1).
3. Smelting of fahlores.
   - Tennantite and tetrahedrite.
   - Enargite.

Fig. 6.12 Chiclana de la Frontera, Cádiz (Spain), first century AD. Round cakes of Roman copper ingots ("flan tronconique") and lead ingots from the shipwreck Peció del Cobre Altogether, 28 copper ingots and 18 lead ingots were excavated from the shipwreck (Vallespìn Gómez 1986). The copper ingots had weights between 20 and 30 kg, the lead ingots between 33 and 61 kg. The ingots are archived in the Museo de Cádiz. Photo: F. Teichner, University of Marburg

The smelting of fahlores and other copper ores with high arsenic and/or antimony contents requires a multistage processing. The removal of arsenic, antimony or even silver is difficult.

For this classification, it should again be emphasised that transitions of ore compositions did indeed exist, depending on the deposit (oxidation zone), which makes various processing techniques likely.

6.2.3 The Extractive Metallurgy of Copper

6.2.3.1 Smelting of Oxidic Ores

The smelting of high-grade secondary copper ores perhaps followed a traditional workshop recipe that goes back to the Neolithic/Chalcolithic period when smelting activities were carried out inside of settlements, and not necessarily elsewhere outside, closer to the mines. The most frequent secondary "oxidic" copper ore by far is malachite, which will be used here to explain basic reactions.

Chemically and physically, the reduction of malachite to cuprite and finally to metallic copper takes place according to the following equations:

$$Cu_2\left[(OH)_2CO_3\right] + CO \rightarrow Cu_2O + 2\,CO_2 + H_2O$$

$$Cu_2O + CO \rightarrow 2\,Cu + CO_2$$

The first reaction shows that malachite is dehydrated and decarbonated to Cu-oxide at temperatures of about 250 °C. The reduction to metallic copper takes place already at 800–900 °C, below the melting point of copper. Therefore, the metal can also separate in a solid state. It then forms a porous, sponge-like mass and is not liquefied (Fig. 6.13).

During this early period, copper was smelted either directly in simple depressions in the ground in a charcoal bed, or alternatively in small

Fig. 6.13 Metallic copper that formed in experimental copper smelting attempts with oxidic ores in the solid state below the melting point (1.083 °C). The grain shape of the ore has hardly changed. The metal was not liquefied. Typical is the spongy, porous structure of copper. Courtesy Simon Timberlake, Cambridge/Bochum 2/2018. Photo: A. Hauptmann, Deutsches Bergbau-Museum Bochum

crucibles filled with glowing charcoal in simple depressions or set in simple stonework. Hanning et al. (2010) have experimentally reconstructed this in a similar way as numerous experiments before them (overview in Hauptmann 2007). They worked strictly on the basis of archaeological finds and findings from the fortified chalcolithic settlement of Zambujal (Portugal, c. 3000 BC). They used up to six bellows nozzles to reach the temperature required to produce liquid copper. Similar results are reported by Timberlake (2007) and Craddock et al. (2007b), who heated oxidic copper ores in crucibles filled with glowing charcoal in simple pits in the ground.

Under these conditions, there is generally the danger of inhibited reduction and/or rapid (re)-oxidation of copper to cuprite. Reasons for this may be O_2 excess in the gas composition due to over-activities of blowpipes or delayed reactions of CO_2 with the fuel to carbon monoxide ($CO_2 + C \rightarrow 2\ CO$) (Hauptmann 2007). There are, as explained in Sect. 4.2.3, a number of micro-equilibria next to each other.

When slags are formed due to the intergrowth of malachite and other secondary ores with gangue and hostrock, high amounts of cuprite (Cu_2O), delafossite ($Cu^{+}Fe^{3+}O_2$) and magnetite ($Fe^{2+,3+}{}_3O_4$) will precipitate. These phases can be seen almost like index phases for the earliest smelting of copper ores, in a geographically wide area (see Sect. 5.3) (Bourgarit 2007).

6.2.3.2 Co-smelting of Oxidic-Sulphidic Ores

Oxidation zones of sulphidic ore deposits regularly contain mixed ores of malachite, azurite, chrysocolla, atacamite or cerussite, intergrown with iron (hydr)-oxides (limonite and hematite) and relics of sulphidic ores such as chalcopyrite, bornite, galenite or pyrite etc. Since mining activities in ancient times generally focused on near-surface parts of ore deposits, such a varied raw material base for the production of metal needs to be considered.

Rostoker et al. (1989a, b) have compiled a series of reactions for Cu-rich ores on the basis of thermodynamic data, according to which copper can be produced from inter-reactions of such mixtures. Two of these reactions are:

$$3\ Cu_2O + FeS \rightarrow FeO + 6\ Cu + SO_2 \uparrow$$

$$5\ CuO + CuFeS_2 \rightarrow FeO + 6\ Cu + 2\ SO_2 \uparrow$$

Such co-smelting involves the inter-reaction of thermally decomposed malachite, and to sulphate or CuFe oxides oxidised chalcopyrite (Rostoker und Sadowski 1980). The same is true for lead ores, and also remains of CuFePb-mattes belong to such a material. FeO is slagged with the gangue or the host rock. The authors have carried out corresponding experimental work in crucibles. By doing so, they were able to demonstrate a one-step co-smelting of mixed ores for the production of metallic copper, which is a very sensible archaeometallurgical transition phase towards multistage extraction of metal from pure sulphide ores.

6.2.3.3 Smelting Sulphidic Ores

At the present time, the early technology of sulphide ore smelting, which is chronologically approximately situated in the Middle Bronze Ages, cannot be completely decoded by archaeological findings alone. All of the models developed so far are essentially based on modern or historical processes, although the evaluation of metallurgical literature from the late nineteenth and twentieth centuries (e.g. Percy 1861; Schnabel 1901; Tafel und Wagenmann 1951; Czedik-Eysenberg 1958) provided important information, as in the books of these authors comparatively simple process steps are described. This is particularly true for metallurgical operations for the extraction of copper, lead, silver and gold in Japan (Gowland 1899, 1901; Lewin und Hauptmann 1984).

In principle, the smelting of sulphide ores must be based on multistage processes based on alternating roasting and melting of ores and mattes. Tafel und Wagenmann (1951) distinguish between the following processing techniques:

1. Matte smelting (German "Steinarbeit").
   (a) Partial or dead roasting of the charge.
   (b) Smelting in shaft furnaces for matte and slag.
2. Black copper smelting.
   (a) Partial roasting followed by roast-reaction smelting under slightly reducing conditions ("English or Welsh process").
   (b) Dead roasting with a subsequent reduction in a shaft furnace ("German process").

As emphasised by Percy (1861), these processes often include a number of individual steps. For copper recovery according to the Welsh process, for example six or more individual operations are required. It may, therefore, be difficult to classify prehistoric smelting processes into any of these processing techniques. However, there have probably been a number of different techniques that have led to simplified versions of these techniques, which are difficult to reconstruct using analytical methods alone. Here, experimental archaeology and ethnographic observations are of considerable importance.

6.2.3.4 Matte Smelting: Partial and Dead Roasting (German: Steinarbeit)

By roasting sulphidic ores or matte, these are desulphurised, forming sulphur-containing exhaust gases (SO_2, CO and CO_2), which may also contain arsenic- and antimony-oxides, lead- or zinc-oxides depending on the ore composition. The resulting metal oxides are further treated in a reducing process step. Using the example of iron sulphides present in ore or matte, the desulphurisation and oxidation during a roasting process become clear:

$$3\,FeS_2 + 8\,O_2 \rightarrow Fe_3O_4 + 6\,SO_2\uparrow$$

$$4\,FeS_2 + 11\,O_2 \rightarrow 2\,Fe_2O_3 + 8\,SO_2\uparrow$$

Due to the higher affinity of copper to sulphur, this initially remains stable as a sulphide. During the roasting of chalcopyrite, therefore, only a partial desulphurisation occurs:

$$5\,CuFeS_2 + 2\,O_2 \rightarrow Cu_5FeS_4 + 4\,FeO + 3\,SO_2\uparrow$$

$$2\,Cu_5FeS_4 + O_2 \rightarrow 5\,Cu_2S + 2\,FeS + SO_2\uparrow$$

By partial roasting of chalcopyrite, bornite (Cu_5FeS_4), digenite/chalcocite, pyrrhotite and iron oxides (magnetite and hematite) are formed. FeS is desulphurised to Fe-oxides according to the reactions above. The formation of Fe-oxides thus leads to Cu-sulphide enrichment. This is especially important for the roasting of matte produced in previous smelting operations (see below). In addition, a number of different sulphates are formed.

If the copper ore is dead roasted, all sulphides are desulphurised to Fe-oxides and Cu-oxides.

$$CuFeS_2 + 3\,O_2 \rightarrow FeO + CuO + 2\,SO_2\uparrow$$

Emerging Fe-oxides are slagged during the subsequent smelting process.

As described in Sect. 6.1 archaeologically uniquely identifiable installations for roasting sulphide copper ores have been found rarely in the archaeological context. Examples are in the

Alpine region (Goldenberg 2004) and in Oman (Hauptmann 1985; Weisgerber 1987b).

Of importance are the investigations of Kunstmann (2004), who analysed archaeological material from a roasting bed of the Late Bronze Age site Flitzen II in the Paltental (Styria, Austria). Based on the fired phase content of the clay mineralogy (chlorite, illite and muscovite), he was able to establish a roasting temperature range of approximately 600–700 °C. Enrichments of arsenic and antimony indicated processing of fahlores. The presence of fine-grained slag particles (magnetite, pyroxene, olivine and quartz) in a sieve fraction could even be indicative of a roasting of matte (interspersed with slag). The red colouration could be traced back to fine grains of hematite (α-Fe_2O_3).

6.2.3.5 Matte Smelting: Smelting in Shaft Furnaces for Matte and Slag

Matte is copper–iron matte, which is an intermediate product in the extraction of copper from sulphide ores. One advantage of matte smelting is the relatively low melting points of the mattes which makes it possible to smelt the mixed sulphide ore at a lower temperature than would be required for the corresponding metals. Thus, a matte with equal amounts of iron and copper melts below 1000 °C whereas an alloy of the same metals would require more than 1400 °C. Matte smelting thus lowers the fuel requirements and also makes the separation from the liquified slag easier. Copper sulphide may also be processed to produce metallic copper directly, without the use of any reducing agent.

Important reactions in matte melting carried out in a neutral or slightly reducing atmosphere (Biswas and Davenport 1980) are

$$4\,CuFeS_2 + 7\,O_2 + 2\,FeS_2 \rightarrow 2\,Cu_2S + 4\,FeS + 2\,FeO + 6\,SO_2\uparrow$$

Cu_2S and FeS are to be regarded as components of the matte. FeO, in turn, is slagged with SiO_2. In further steps, matte formed during the smelting process is roasted again until the sulphur is completely removed:

$$Cu_2S + 2\,FeS + 5\,O_2 \rightarrow 2\,CuO + 2\,FeO + 3\,SO_2\uparrow$$

Hereby formed Cu-oxides are smelted under reducing conditions as described in the process above. The resulting Fe oxide is gradually slagged.

Cu-(Fe-)-sulphide phases found in the archaeological context are compiled in Fig. 5.55.

6.2.3.6 Matte Smelting in Bronze Age Alps

Archaeometallurgical investigations on the process of sulphide copper smelting were carried out by Zschocke and Preuschen (1932) who dedicated to this topic for the first time on the Bronze Age Mitterberg in the Salzburgerland (Austria). Important investigations were then carried out in close accordance with the archaeological context by Eibner (1972, 1974, 1982). His reconstruction of early copper production was strongly inspired by the "German process" described by Georg Agricola and later by Czedik-Eysenberg (1958). A description is given by Hanning et al. (2015).

In his exemplary research activities on the Mitterberg, Thomas Stöllner focused on a very broad mining and processing and the supraregional importance of this deposit in Central Europe. He conducted a long-term research project, which he summarised with his colleagues in numerous publications (e.g. Stöllner 2013, Stöllner et al. 2016). As a very new aspect for reconstructing ancient copper production, he investigated for the first time extensively the complex steps of beneficiation of ores from Mitterberg (cf. Sect. 3.9).

As far as the reconstruction of the old metallurgical process techniques is concerned, the investigations focused on the best available archaeological finds, the slags of the many smelting sites and, less frequently, the smelting furnaces and roasting facilities. This resulted in lively, differently oriented controversies, which showed that purely scientific statements are indeed essential, but limited without additional ethnological and experimental information and difficult in many aspects for reconstruction of manual processing techniques (Reitmaier-Naef 2020).

Three different types of slag can be found on the Late Bronze Age to early Iron Age smelting sites of the Alps (coarse, massive, flat slag and in parts in Trentino also slag sand, see Sect. 5.3 and Fig. 5.23 and others). In the analytical study of matte inclusions in these various slags from the smelting site Acqua Fredda in Trentino (northern Italy), Metten (2003) demonstrates that none of these types of slag shows a trend of copper enrichment. She concludes that all of these slag types must come from one process flow. An assignment to different smelting processes based on their Cu/Fe sulphides is not possible. Silvestri et al. (2019) reach a similar conclusion. She has carried out about 150 analyses of sulphide inclusions in plate slags and other slag types from smelting sites in the Trentino area (Italy). She has found that in plate slags—no matter their origin—there are almost only sulphide inclusions with compositions of chalcopyrite with a tendency to bornite. Covellite inclusions are extremely rare. Thus, she could not detect any accumulation of copper depending on various types of slag either. The actual enrichment of the matte and a deposition of black copper must, therefore, have taken place somewhere else. So far, there is no archaeological evidence available.

Addis et al. (2016, 2017) and Reitmaier-Naef (2020) come to conflicting notions. They observe an accumulation of copper in the mattes from the slag cakes to the plate slags and conclude that the slags are the result of different process steps. These considerations, however, are not consistent with the analyses performed by Addis (2013), so the problem must remain unsolved. Ultimately, the conversion of matte to metal is not clear.

The interpretation of quartz inclusions in slag cakes as a deliberate addition of flux to a FeO saturated melt, as presented in Addis et al. (2016, 2017) is questionable. There is no convincing evidence for that. The problem is reminiscent of the "free silica slags" discussed in Sect. 5.4.11. Relics of quartz inclusions with remains of sulphidic ores have already been interpreted as such in many other slags. Quartz inclusions in slags, both in Trentino and in Grisons (Reitmaier-Naef 2019), have large amounts of sulphide inclusions and can, therefore, be easily interpreted as relics of a gangue of mineralisation, as they occur in the hydrothermal veins with chalcopyrite in the locality Tingherla. This deposit is located in the middle of the ore district of Trentino (Metten 2003). On the other hand, crushed plate slags in Late Bronze Age slag cakes can be interpreted as intentionally added flux, as in the slag cakes in Trentino (cf. Fig. 5.25).

6.2.3.7 Black Copper Smelting: Partial Roasting/Roast-Reaction ("English or Welsh Process")

This smelting process was performed under slightly reducing conditions only. It was and still is the most common method of copper extraction from sulphide ores.

In case of a limited partial roasting, the smelting process would be called roast-reaction smelting. In the early phase of extractive metallurgy, the production of copper, as well as that of lead (consciously or unconsciously), proceeded probably according to the roast-reaction process.

There are the following observations:

- In the near-surface part of sulphidic ore deposits, mixtures of oxidic and sulphidic copper and/or lead ores are present in most cases. Examples have been shown in the chapter oxidation zones of mineral deposits (3.6).
- Slag phases like delafossite ($CuFeO_2$) or cuprite (Cu_2O) in many slags of the fifth to third millennium BC in the Near and Middle East (Hauptmann 2007) indicate smelting processes that have probably out been carried in small and open-shaped reaction vessels under only slightly reducing conditions.

Bourgarit (2007) reviewed the phase inventory of slags from the Chalcolithic period from the Iberian Peninsula to the Middle East as well as corresponding ore deposits, as far as information on the mineral content in the oxidation zone were available. He agreed that these two criteria are valid at almost all localities, indicating a comparable solid basis for a roast-reaction as a principal metallurgical process.

The basic reactions of the roast-reaction process are:

$$Cu_2S + 3/2\ O_2 \rightarrow Cu_2O + SO_2 \uparrow$$

$$Cu_2S + 2\ Cu_2O \rightarrow 6\ Cu + SO_2 \uparrow$$

This means that liquefied Cu-sulphides (ore and matte) are oxidised with subsequent precipitation of metal.

Metallic copper readily collects in the sulphide phase, it is associated with the sulphide in the siliceous slag. After tapping, the two phases often form a flat cake at the bottom of the slag. The negative impression of it can be seen clearly (Figs. 5.10 and 5.18). The sulphide phase more or less completely envelops the metal (Fig. 5.17). In the co-smelting of oxidic and sulphidic ores, the quantitative proportion of copper is increased if the process is carried out under slightly oxidising conditions.

Intergrowths of copper with Cu- or Cu(Fe)-sulphides do not necessarily have to be interpreted as evidence of an intentional matte smelting, which would require mining in the cementation zone of the primary ore body. Slags with sulphide and metal inclusions, especially from the Chalcolithic and the beginning of the Bronze Age, are rather useful evidence for a smelting of oxidic copper ores with shares of sulphides from the oxidation zone (Fig. 6.14). It has already been pointed out that this is an interesting proof of a transitional phase in technology from pure oxide to sulphide smelting (Rostoker et al. 1983).

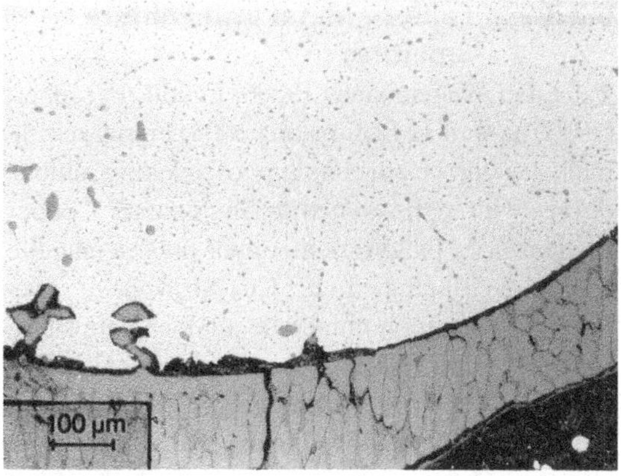

Fig. 6.14 Shahr-i Sokhta (Iran), Early Bronze Age (mid third millennium BC), sample SS 141 Roast-reaction in the microscopic view. Section of a copper droplet embedded in a slag cake surrounded by a rim of copper sulphide close to Cu_2S. The rim, in turn, is slightly desulphurised and consists of dendrites of Cu_2S embedded in a matrix of Cu_2S + Cu_2O. Mounted sample, reflected light. From Hauptmann et al. (2003)

6.2.3.8 Black Copper Smelting: Dead Roasting with Subsequent Reduction in a Shaft Furnace ("German Process")

In principle, the smelting of copper under stronger reducing conditions has been known since about the Late Iron Age to the Roman period, in which the metal has been produced almost to an industrial extent. Reductive smelting is comparable to the so-called bloomery smelting (German *Rennfeuerverfahren*) of iron metallurgy in Europe (cf. Sect. 6.6). Accordingly, during and after the smelting process, iron-rich silicate slags with the main constituent fayalite should be tapped from the furnace as completely as possible in the liquid state. But Czedik-Eysenberg (1958) rightly emphasises that in reductive smelting as well as in matte smelting, the furnaces should not be constructed too high, because otherwise due to an excess of carbon monoxide (CO) in the gas atmosphere iron-containing copper or even iron salamander would be deposited.

This model is confirmed to a limited extent by archaeological evidence, such as the size and design of tap slags on slag heaps. For example, since the advent of smelting furnaces in earlier periods, e.g. the developed Early Bronze Age, there has been regular evidence that varying proportions of slags have solidified together with metal or mattes within the smelting furnaces and, just like later the bloom in iron metallurgy, were recovered by breaking the furnace (Metten 2003; Hauptmann 2007; Cierny 2008). According to observations made by Theophilus Presbyter (c. 1.125 AD), even in the twelfth century AD, the molten copper is allowed to cool in the furnace and then recovered in the solid state.

6.2.3.9 Fahlore Metal and Fahlore Smelting

Tafel and Wagenmann (1951), Czedik-Eysenberg (1958) as well as Bachmann (2003) emphasise the multistage process technology of smelting fahlores or ores with high As- and/or Sb-contents. Copper, smelted from fahlore-rich copper ores (tennantite, $Cu_{12}As_4S_{13}$, tetrahedrite, $Cu_{12}Sb_4S_{13}$) or even from the rare monomineralic fahlore deposits at Brixlegg (Inn Valley, Austria) or at Cabrières (France), is rich in the elements As+Sb + Ni + Ag. In the classification of Junghans et al. (1968, 1974) many artefacts of the Neolithic and the Early Bronze Age of Central Europe, which show striking contents of these elements (0.5 – >10 wt.%) are characterised as "fahlore metals". So in ancient times, in principle, there were two types of copper. One was characterised by small amounts of trace elements, the other variety was made of copper with high levels of those elements.

There are other examples of "fahlore metals" that have gained importance in archaeometallurgy and have been intensively studied. Many of the Chalcolithic artefacts from the hoard of Nahal Mishmar west of the Dead Sea (Israel) (Bar-Adon 1980) have extremely high levels of arsenic and antimony. This hoard is an absolute rarity not only in the entire Middle East. The analytical studies by Tadmor et al. (1995) have shown that in this hoard, consisting of 416 metal artefacts (251 maceheads, 92 standards, hammers and axes), many copper artefacts show concentrations of 1–25 wt.% antimony and 0.4–15 wt.% arsenic. In these predominantly ceremonial devices, the silver contents reach up to 1.2 wt.% and correlate with Sb and also with As and Bi. A production of these artefacts from fahlores or their weathered ores is likely. The source of such ores is presumed to be eastern Anatolia or (Trans-) Caucasus, but it is still unknown.

In ancient times, fahlore metals do not necessarily have been obtained by sophisticated smelting techniques as was assumed by Strahm (1994) and Krause (2003) and described for the developed Bronze Age and the late Middle Ages by Bachmann (2003). They can also be produced under "primitive" conditions, in simple depressions in the soil, which are filled with charcoal. This is impressively proven by archaeological findings from the Neolithic Mariahilfsbergl/Brixlegg (Bartelheim et al. 2002) and the Chalcolithic Cabrières (Bourgarit et al. 2003). On Ross Island, fahlores were also smelted in the late Neolithic (O'Brien 1999) while the smelting of chalcopyrite was obviously not successful during this time period.

According to experimental work (Neuninger et al. 1970; Happ et al. 1996; Bourgarit et al. 2003; Lohrscheider et al. 2003) and examination of metallurgical waste products (Bayley and Eckstein 1997) the process of fahlore smelting differs from that of other ores due to insensibility to lightly stronger oxidising firing. This allows a smelting of fahlores in simple hearth pits. The authors produced "fahlore metals" with low As and Sb contents in simple pits in the ground, as well as in simple depressions set in stone. They relied on ethnographic observations made by Zschocke and Preuschen (1932). On the Philippine island of Luzon, fahlores were still smelted with charcoal in simple bowl furnaces in the nineteenth century AD, producing copper, slag and white metal (matte), which was melted down again.

It must be kept in mind that prehistoric smelting of fahlores was not conditioned to their silver content, as was the case in the late Middle Ages, but only to copper. In the necessarily multistage smelting of the iron-containing chalcopyrite ($CuFeS_2$) there are technical cliffs. It is important to ensure that during the process, a not too high oxygen supply (such as near the nozzles of tuyères) leads to the precipitation of magnetite and the liquid slag is solidified. This is represented in Fig. 4.17 (stage no. 3) by the equilibrium quartz–fayalite–magnetite (QFM). On the other hand, smelting should not be carried out under too reducing conditions, since then iron oxide would not be absorbed by the siliceous slag, but would be reduced to metal, which would mix with the copper and render it useless. This is represented in Fig. 4.17 (stage no. 3 and 4) by the equilibrium iron–wuestite (IW). Therefore, iron once again is an important indicator of the smelting process, a key parameter. Overall, it does not seem convincing when a "fahlore technology" is postulated as a complex procedure of its own.

Physical chemistry shows that there is a technically self-running process for the recovery of useful metallic phases, especially for mixtures of sulphidic fahlores and their weathering products. This is shown by the phase diagram Cu + Fe–As +Sb–S (Fig. 5.60). It shows that between a sulphidic phase with up to c. 20 wt. % As and/or antimony (which can be equated due to their similar geochemical behaviour) and a Cu-rich metal phase, there is a miscibility gap quite like, if not as extensive, as the one in the system Cu-Fe-S, which is important for the metallurgy of corresponding Cu-Fe ores (see Sect. 5.7). If an ore mixture is smelted under oxidising conditions, a sulphidic phase forms, from which a Cu-rich, As, Sb-containing alloy can be deposited. These Cu–As–Sb alloys are referred to as deficit alloys (*Mangellegierungen*) in metallurgy. In our case, these are the "fahlore metals". To the left of the closing miscibility gap is the field of Cu-rich speisses, which cannot be used as a metallic material because of their brittleness (see Sect. 5.7).

As far as the fahlores are concerned, smelting under rather low-reducing conditions, which was obviously characteristic of early processes, must have had only an advantageous effect on such ores. The desulphurisation of the sulphides associated with the process leads quickly and easily to the precipitation of copper. From this copper arsenic and antimony were difficult to remove completely. Under these conditions, only vicious and often only semi-molten slags with high levels of oxidic components (magnetite, delafossite etc.) were formed, as they actually appear in archaeology.

The fact that "fahlore metals" tended to have other physical properties than copper with low levels of trace elements, was apparently clearly distinguished by ancient metallurgists. One could distinguish between these varieties. However, the assessment of As- and Sb-containing copper species under physical properties is ambivalent (cf. Sect. 7.2).

6.2.3.10 Refining

Although it is repeatedly postulated, above all in archaeological literature, so far there is no convincing indication that copper was deliberately refined in the Bronze or Iron Age. A production of crude copper must, therefore, be assumed for these periods. This means that a purity level of approx. 95–97 wt.% copper was hardly exceeded. For metallurgically speaking, liquid crude copper still contains various reaction gases, especially SO_2, which formed during the beginning of smelting processes due to the action of water vapour and other gases. Such gas formation during cooling leads to the formation of bubbles both in the metal, as well as in the resulting slag. In copper, such bubbles are frozen during cooling and cause the blistered surface of the metal (blister copper). In the microscale, the oxygen content can be seen in inclusions of cuprite (Cu_2O). Inclusions of copper sulphides (Cu_2S) are also common. It must also be assumed that complete separation of the copper from the slag was by no means guaranteed in ancient times, so that the copper almost always contained incorporated or attached particles of slags, just as slags contained amounts of copper inclusions, which probably often led to the systematic treatment of slags in early periods (Hauptmann et al. 2002a).

For a long time, refining of copper has been achieved by "poling". Greenwood is immersed in the liquid copper bath, whereby gas inclusions are expelled. Today, this is done by electrolytic refining through the converter process. However, the poled copper also contains dissolved oxygen, which makes the copper brittle in the solid state. It must, therefore, be removed by a further reduction step with charcoal or possibly by addition of a substance with high oxygen affinity (deoxidant). This process is known in the German language as *Zähpolen*. According to Tafel and Wagenmann (1951) contents of Fe, Pb and S are reduced quickly, that of other elements only slowly.

In ancient times, however, a cleansing of copper was perhaps achieved by repeated pouring from one vessel into another. It may remain open whether this can be seen as a separate step of refining.

As pointed out by Levy et al. (2002), a repeated re-melting and recycling of copper lumps and prills into larger units, as practised at Early Bronze Age Khirbat al-Hamri, Faynan (Jordan) is probably identical with the production of washed or purified copper as described in third millennium BC cuneiform tablets from Ebla in Syria (Reiter 1997). This was called "urudu-luhha", in contrast to *urudzi,* which means unpurified copper. It was suggested earlier that such a purification would mean deliberate refining of metal in the modern metallurgical sense (Waetzold and Bachmann 1984). As shown by the phase contents of slags and copper, this was not the case at Khirbat al-Hamri, and there is no evidence as yet for definite refining at any other of the ancient Near Eastern EBA workshops.

6.2.3.11 Smelting Furnaces

Herdits and Löcker (2004) emphasise in their article on Bronze Age copper smelters in the Mitterberg "chalcopyrite district" (cf. Fig. 3.10) that, at least in the Late Bronze Age, the design and architecture of smelting plants and the external typology of slags in the eastern Alps are very uniform. This indicates an expansion area from the Swiss Grisons via South Tyrol, Trentino, North Tyrol, the Mitterberg district in the Salzburger Land via Styria to Lower Austria. The knowledge of the smelting of sulphide copper ores, whether they were chalcopyrite-dominated sulphides or fahlores, was apparently so perfectly standardised that they were applied consistently throughout the eastern Alpine region.

Impressive excavations were those in Acqua Fredda in the Southern Alps (Italy), which have been carried out since the late 1970s (summary in Cierny 2008). The battery of 6 and then 3 other Late Bronze Age smelting furnaces (fourteenth to ninth centuries BC) seem to fulfil the recommendations of Czedik-Eysenberg (1958). The stone set furnaces have rectangular floor plans with dimensions of approx. 0.45 × 0.6 m with a height of only 0.5–0.7 m. Smelting conditions that were too reducing were hardly achieved. The slags produced were mainly divided into slag cakes and plate slags with transitions (cf. Sect. 5.3). In particular, chalcopyritic ores were smelted in these furnaces.

Similar to Acqua Fredda, furnaces for the smelting of fahlores were discovered and excavated by Goldenberg (2014a, b) on the Jochberg near Innsbruck in the Inn Valley (Austria). Three furnaces, also constructed in context, correspond largely to the archaeological findings from the Middle and Late Bronze Age known furnaces in the Alpine region (Fig. 6.15). The furnaces constructed on a rectangular plan had an inner diameter of about 30 × 40 cm and were about 80 cm high. Inside they were lined with clay. Not clear is whether these furnaces had a closed front. There is no archaeological evidence that the furnaces had once been walled up. This may correspond to the general lack of furnace fronts in the archaeological context in the Alpine region observed by Eibner (1982). In contrast, a few nozzle tubes were found. After the experimental reconstruction of Bronze Age copper metallurgy in the Mitterberg area, Hanning (2012) assumes that the furnaces were also walled up at the front and that the nozzle pipes were probably set in the front.

6.3 The Metallurgy of Gold

Gold (chemical symbol Au), next to copper and lead, is one of the first metals used by humans. From the craftsman's point of view this metal is a near-perfect material which is excelling all other metal known in ancient times: when pure, it never tarnishes, it has a high resistance against corrosion, it is not oxidised and, hence, it never loses its bright colour. Gold possesses a low hardness and a high density (Table 6.2, 71% heavier than lead). Gold starts to evaporate during heating already near its melting point at 1063 °C. Its low hardness (DPN = 2.5–3), the extremely high ductility and the fact that it can be found in nature in its native state (but always alloyed with silver) allowed the production of gold artefacts by simple hammering in the cold and hot state. It can be hammered to extremely thin foils, it can be drawn to long wires, it can easily be cut by a flint

Fig. 6.15 Jochberg (Nordtirol, Austria), fourteenth/thirteenth century BC (final Middle Bronze Age/Early Iron Age). Smelting furnaces 2 and 3 ("twin furnaces") for copper production. Picture from an advanced stage of excavation (Goldenberg 2014a). It must remain open whether these furnaces had an open front or a closed furnace front

blade, and when only slightly heated, it can be worked into beautiful designs.

6.3.1 Gold Artefacts

Among the large number of gold artefacts which were made in the (pre-) historical time from this precious metal, only a few examples are picked out here. It is less intended to address the enormous artistic diversity of gold processing, i.e. we will not go into the techniques of casting, forging, chiselling and granulation as it is the speciality of Armbruster (2000, 2012a, 2016, et cetera). Rather, it should be shown in what form gold and its natural alloys occur and how they can be manipulated metallurgically-chemically to change their compositions in certain targeted directions.

The earliest gold artefacts in the Eurasian world are younger than the oldest copper objects found in Anatolia, which date to the 9th/8th millennium BC (see Sect. 6.2).

The most ancient gold objects were found in a burial site of the Chalcolithic cemetery I at Varna (Bulgaria), near the west coast of the Black Sea. According to the new ^{14}C-calibration the gold dates to the middle of the fifth millennium BC (4550–4450 BC; Higham et al. 2007; Leusch et al. 2014). Altogether, ca. 3,000 gold objects with a weight of 6 kg were discovered as funeral givings. The metal did not have any monetary value. A few gold objects only from Varna were first chemically investigated by Hartmann (1982), Echt et al. (1991) and Éluère and Raub (1991). The last two authors were convinced in those days that the gold artefacts hardly reached the fully liquid state but were produced by hammering at temperatures just below the melting point of gold using the effect of sintering. Reasons for such a hypothesis were the observations of porous (surface) textures and inclusions of quartz grains and clay minerals in the artefacts. They suggested that the physical property of such a sintering was comparable with techniques also be used in modern gold metallurgy. Besides that, they found that tiny gold grains washed from a placer were used for the decoration of a plate at Varna.

A large number of gold artefacts from Varna was recently investigated in a serious study by Leusch et al. (2014). The chemical composition of the gold artefacts was mostly non-destructively measured by a portable X-ray spectrometer (pXRF), a few artefacts were measured almost non-destructively by laser ablation ICP-MS. Accordingly, the gold artefacts of Varna contained up to 45 wt.% silver. This is no indication of an intended alloy, because natural gold can contain silver concentrations up to 40–50 wt.%.

In numerous objects copper concentrations of 1.5–3.0 wt.% were found, in one case 36 wt.%. The last case is considered as a possible indication for intentional alloying, because AuCu minerals such as auricupride are rather rare in nature (Ramdohr 1975). According to Schmiderer (2008), copper concentrations in natural gold are usually <0.5 wt.% while Hauptmann et al. (2010, 2018) estimated the limit at <2 wt.% (see also Sect. 7.3.2). The armlets and globular beads from Varna are clear evidence that the gold or gold alloys, respectively, were cast using lost-wax casting.

In Israel, in the Nahal Qana Cave contained eight gold–silver rings with a diameter of about 2–3 cm and weights of 90–160 g, a total of almost 1 kg (Gopher et al. 1990; Shalev 1993). They date to the second half of the fourth millennium. Their purpose is not very clear, presumably they were gold bars. They were cast at or slightly below the melting point of pure gold at 1064 °C into ring-shaped open moulds. These rings contain up to about 30 wt.% silver, with a significant accumulation of gold on the surfaces of the rings. The surfaces of these rings show porosity and are hammered. The authors interpret these phenomena as intentionally performed depletion gilding. This is the only case that could be an indication of this extraordinary technology as early as the fourth millennium.

The fact that depletion gilding was also practiced in isolated cases a thousand years later in southern Mesopotamia in the second half of the third millennium BC has been proven by La Niece (1995) and Hauptmann et al. (2018) by investigations of gold artefacts from the royal tombs of Ur, namely of some gold–silver–copper alloys (one dagger, several chisels). It must be emphasised that the few gold–silver–copper alloys from Ur are very extraordinary finds of this type of alloy, otherwise widely distributed in South America, but only millennia later (see below).

In contrast, Hauptmann et al. (2018) interpret the surface porosity on gold–silver artefacts from the royal tombs of Ur (Ag = 20–40 wt.%) as naturally occurring corrosion phenomena that have formed over the course of millennia in the saline environment of southern Mesopotamia (Fig. 6.16). In the royal tombs of Ur, one of the world's three greatest treasures of gold objects was excavated by Sir Leonard Woolley between 1922 and 1934. This sensational find is today divided among the museums in Philadelphia (USA), the British Museum (London) and the National Museum in Baghdad (Iraq) (Zettler und Horne 1998). The treasure contains hundreds of gold and silver artefacts, hundreds of copper and bronze objects, carnelian and lapis lazuli beads, steatite stone vessels and many other materials. One of the distinguishing features of Ur's gold artefacts are their very common inclusions of platinum group minerals (PGM; Hauptmann and Klein 2016), which together with the lapis lazuli pearls represent outstanding provenance indicators and made it possible to source the gold from the deposits at the upper reaches of the Amur Darya in northern Afghanistan or Tajikistan (Jansen et al. 2018b). In addition, the PGM inclusions are very good indicators of the origin of gold from alluvial gold deposits, it is not gold mined from a primary deposit (see Sect. 7.3).

With the Greek legend of the "Golden Fleece", the precious metal in the region of the Caucasus has become world famous, at least in the field of mythology. It states that the king's son Iason sailed with a group of heroes—the Argonauts—on the ship "Argo" of Iolkos in Greece, to the Colchis (to present-day western Georgia) to demand from King Aietes the "Golden Fleece". He finally succeeded with the help of the king's daughter Medea, who lulled the dragon, who guarded this fleece.

The "Golden Fleece" stands primarily as a symbol for the production of gold from river placers, a very old process, which is also shown in the work of Agricola Georgius (1556). Here, a (sheep- or ox-) fur is placed in gold-bearing rivers and removed after some time. Due to the high specific gravity, gold flakes accumulate in the fur, while all lighter components of fine-grained sediments such as quartz and clay particles are washed off. The gold is washed out of the fur and treated further. The "Golden Fleece" is a symbol for the wealth of mineral resources, of metals that occur in Georgia. The myth has a true core: recent

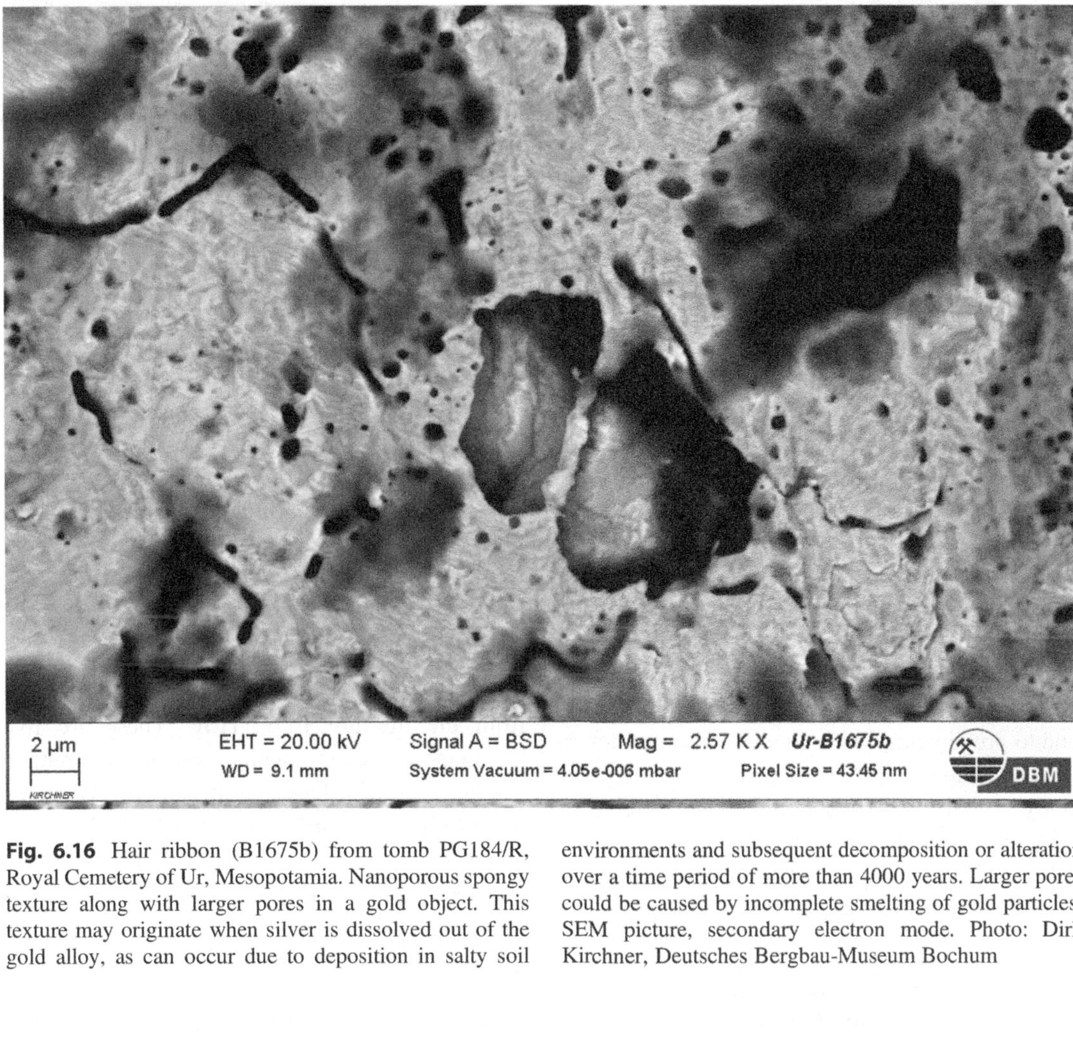

Fig. 6.16 Hair ribbon (B1675b) from tomb PG184/R, Royal Cemetery of Ur, Mesopotamia. Nanoporous spongy texture along with larger pores in a gold object. This texture may originate when silver is dissolved out of the gold alloy, as can occur due to deposition in salty soil environments and subsequent decomposition or alteration over a time period of more than 4000 years. Larger pores could be caused by incomplete smelting of gold particles. SEM picture, secondary electron mode. Photo: Dirk Kirchner, Deutsches Bergbau-Museum Bochum

excavations in the princely tombs near the city of Vani, the capital of the Kingdom of the Colchis, have indeed produced incredibly rich gold discoveries that support this legend (Kacharava and Kvirkvelia 2008). The graves date to about the first half of the first millennium BC, to the time of the Greece colonisation of the eastern Black Sea coast.

The gold of the Colchis is an outstanding highlight. However, there are much older gold finds in Georgia than those from the Hellenistic period. They date to the third, possibly even the fourth millennium BC (Lordkipanidze 1991) and come from Bronze Age tombs of the Kura-Araxes and Trialeti cultures. At this time, however, it still plays no economic role compared to a large number of bronze objects. Georgia's earliest gold is roughly comparable in time to eastern Anatolia, where, e.g. in the "royal" tomb of Arslantepe, among about 70 copper objects, numerous silver and silver–copper objects and a small spiral made of gold was found (Frangipane et al. 2001). The tomb dates to around 3000–2900 BC.

Like the gold artefacts of Ur or Varna, those from the (Trans-) Caucasus or Georgia contain between 1 and 40 wt.% of silver. This wide range of variation is quite similar to that of measurements of natural gold in this area, whether it be alluvial gold from placer deposits or gold mined from solid rock deposits, such as from the mine of Sakdrissi. The gold from the

(Trans-) Caucasus, however, contains no inclusions of platinum group elements (Jansen et al. 2018b; Jansen 2019).

The frequency of Bronze Age gold artefacts in south-west Europe is high in silver. Already Hartmann (1982) analysed more than 500 gold artefacts from the Iberian Peninsula. However, no analytical investigations were performed on natural gold sources in this area, and the analyses of objects were limited to the analysis of Au, Ag, Cu and occasionally Sn. For Hartmann (1982) the presence or absence of tin would identify gold from a placer deposit or from a gold deposit in solid rock, respectively. In addition, he argues for a local source in the eastern Mediterranean for the former (presence of tin) and a source for the latter (absence tin).

As published later by Nocete et al. (2014), altogether more than 4000 objects have been analysed up until today. However, there still remains a desiderate to characterise the numerous gold objects in an analytically more accurate way and to provenance raw sources in this area rich in gold deposits. He started to perform an intensive study from the analytical point of view to characterise the exact chronological context, geochemical and lead isotope composition of gold objects from the south-west Iberian Peninsula.

6.3.1.1 How Gold Was Traded in Ancient Times

The state in which gold was traded in ancient times is a field that has so far received little research. Based on the Sumerian and Old Babylonian administrative sources from the third and second millennium BC, Hauptmann et al. (2018) suspect that, at least in some parts of the Middle East, gold was traded in the form of "dust". This is a convincing assumption because in ancient times gold was mostly washed out of alluvial placers in the form of the finest of flakes, sometimes perhaps as small nuggets. This has also been proven analytically for the gold of the royal tombs of Ur (Jansen et al. 2016, 2018a, b). The fact is that, due to its extremely low concentrations of deposits, in all metallurgical operations and in further processing gold has always been handled in the smallest amounts compared to copper or iron. This is also meticulously recorded in various ancient texts (Waetzold 1985; Reiter 1997; Paoletti 2016). There are also ethnographic examples of trading "gold dust". Gold from abundant deposits in West Africa entered the long-distance trade in the form of tiny flakes, such as until recently in the Trans-Saharan trade in exchange for salt (Bovill 1968). This was perhaps based on traditions back to the Carthaginian period (Hahn 2014).

To deal with such small amounts, to weigh them, to buy or to sell, the trade form of fine gold flakes is best. From an archaeological point of view this is not tangible, it is just as difficult to find or to identify appropriate scales or gold weights.

6.3.1.2 Gold Ingots

The earliest artefacts identified as gold bars come from the treasure F of Troy. These are five bar-shaped ingots with regularly arranged notches, which looks almost like specified sections or portions (Treister 1996). They are about 10-cm long and have weights of almost exactly 10 g. They date to the Early Bronze Age around 2150 BC.

Gold ingots and moulds for such ingots are known later only from Roman times. In the Roman city on the Magdalensberg (Carinthia, Austria) a casting mould for a gold ingot with dimensions of >10 cm was found (Piccottini 1994). Gold ingots are presented by Wiegels (2014) in his review of Roman silver and gold ingots from Transylvania (Rumania) and from the Egyptian Aboukir/El Kubra. The bar-shaped ingots are around 15-cm long and have weights of around 300–500 g. These bars are stamped, they thus at least allow a temporal allocation. On the other hand, it is very questionable whether these stamps would also be a kind of purity seals. Only from one ingot the fineness of the gold was analytically proved. It has a fineness of 883/1000, i.e. it contained only 88.3 wt.% of gold. These ingots may possibly be referred to as transport ingots. They represent barely refined gold that was used for coinage.

In the trans-Saharan salt–gold trade in west Africa (Mali, Ghana and Burkina Faso), however, references to the production of unstamped coins were also found in the Islamic world (eleventh century AD and later) (Nixon et al. 2011). These are comparable to the pellet-like coins of the European Iron Age.

6.3.1.3 Gold Coinage

Bachmann (1999a) wrote a compendium on the distribution of gold and gold alloys, their metallurgical operations, and the history of gold coinage in the Old World: In Europe and Asia Minor, the advent of coinage results in a widespread of gold. The earliest coins were still made of natural gold–silver alloys. They date from about one hundred years after the invention of the parting process in Sardis (see below) in the middle of the first millennium BC. The countless Celtic gold coins, especially the hoard of 450 gold coins from the Celtic oppidum of Manching (Bavaria, third to first century BC) is a striking example for the wide distribution of gold coins (Gebhard et al. 1995). The content of gold varied, ranging from 30 to nearly 100 wt. %.

Roman gold coins were minted in Rome not before 216 BC during the time of the Second Punic War. The output was limited at first, but under Julius Caesar (100–44 BC) gold coins came into circulation on a larger scale.

6.3.2 Ancient Metallurgical Processes of Gold

For the production of gold, the following ancient techniques could possibly be used:

1. Roasting: Production process prior to any smelting.
2. Leaded smelting: Extraction of gold from low-grade gold ores.
3. Cupellation to separate gold from lead.
4. Amalgamation: Extraction of Au from ore concentrates.
5. Parting/cementation: The separation of gold and silver. A refinery process.
6. Gilding through the ages.

All these paragraphs are in a comprehensive overview described and discussed, e.g. by Bachmann (1999a, b).

6.3.2.1 Roasting

Roasting of gold-bearing ores, especially those obtained from the solid rock, has the sole purpose of loosening the grain structure of the ore and/or to reduce the moisture content before processing. Furthermore, sulphur contents of associated ores (pyrite) are decreased. Archeologically, however, a roasting process from gold metallurgy is proven nowhere, just as little as from regions of traditional or small-scale gold extraction, be it in Mesoamerica (Ruvalcaba Sil et al. 2009), Africa (Borg 2014) or Europe (Slotta et al. 2001).

A metallurgical recovery of gold–following complex steps of manual, mechanical or hydraulic beneficiation of gold–was probably carried out by direct melting processes of gold concentrates. Depending on the extent to which impurities still had to be removed, this could have been done by reducing or oxidising firing conditions. Gold is never oxidised, but impurities such as Fe-oxides, SiO_2, SnO_2, As–sulphides etc. are forming a "dross". To what extent additives of slag-forming oxides were used in early gold metallurgy is completely unknown.

According to the presumably small amounts, in which gold was always smelted and melted by metallurgical processes, ceramic vessels with small capacities only have presumably been used as reaction vessels.

6.3.2.2 Leaded Smelting: Extraction of Gold from Low-Grade Gold Ores

This process is used to extract gold from poor gold-bearing ores ("refractory ores"), for example, from gold-bearing pyrite, arsenopyrite and ores from the gossan of pyrite-rich deposits such as in Rio Tinto (Salkield 1987), on Cyprus "devil's mud" (up to 14 ppm Au, silver up to

100 ppm; see Sect. 3.5.1), or possibly also from the island of Sifnos (Greece; Pernicka and Wagner 1985). Gold contents are associated with fine-grained limonite batches, clay minerals and jarosites (KFe-hydrogen-sulphate with Au, Ag, Pb contents). Native gold and silver minerals occur in such geological units. Depending on the quantities involved, leaded smelting is carried out in smelting furnaces or in small crucibles.

The principle of leaded smelting is based on the following principles (see also Sect. 6.4): After a possible preliminary roasting lead, lead sulphide or lead oxide (possibly crushed waste of litharge or residues of cupels after the cupellation process) will be mixed with the gold-containing ores. The subsequent smelting process is conditioned to lead. Depending on the composition of the ores, the following end products are produced:

- Crude lead with noble metal contents.
- Speiss: As–Fe–Cu–Sb compounds.
- Matte.
- Slag: The resulting slag consists of lead silicates and base metal oxides.

This sequence of smelting products could be evidenced at old smelting sites by finds, e.g. at Monte Romero in Spain (Keesmann 1993). The role of speiss is controversial. It is not clear whether noble metals could be extracted from this intermediate product with old metallurgical processes (see Sect. 5.8). The most important noble metal collector is the crude lead. It is an old rule that the noble metal content increases with increasing density of the slag to the crude lead (cf. Fig. 5.52). The noble metal contents of gold and silver are separated from lead or lead oxide by cupellation. First of all, gold and silver, but also copper occur together. They have to be separated later, as shown in Sect. 6.4.

Archaeological evidence for leaded smelting in early gold metallurgy is more than ambiguous, and thus dating the beginnings of this process is difficult (Celauro et al. 2017). However, the fact that successful leaded smelting requires knowledge of the cupellation process is not a problem, since it has been known in the Euphrates region of eastern Anatolia since the fourth millennium BC (Hess et al. 1998; Pernicka et al. 1998). Photos et al. (1989) suspected an application of the method in Skapte Hyle (Pangaion area, Greece) in Roman–Hellenistic times, where she emphasised the numerous finds of gold and silver-containing speiss in addition to lead silicate slags. Harrison (1931) proposes gold production by means of leaded smelting for the Roman Três Minas (northern Portugal), since slags in the old premises of the mine were lead slags and he had apparently found litharge. The method seems to make sense according to the typology of the mineralisation, but unfortunately there is no reliable dating and no reliable archaeological evidence can be presented. In any case, the technological flow sheet of gold mining in Três Minas of Bachmann (1999b) on the basis of the descriptions of Pliny is very instructive. It shows the complexity of obtaining gold from comparatively poor ores starting from the extraction to the various metallurgical processes (Fig. 6.17).

6.3.2.3 Cupellation to Separate Gold from Lead

Next to finds of litharge (see above), indirect evidence for cupellation is low concentrations of lead in silver objects from the fourth/third millennium BC Byblos (Pernicka 1993). It would be logical to apply cupellation also for the extraction of gold, but the archaeological evidence is difficult. Of importance is that in the cupellation process, especially those elements are separated from the gold, which has a significantly non-noble character and is oxidised easily according to the following formula:

$$2\,\mathrm{Me} + \mathrm{O_2} \rightarrow 2\,\mathrm{MeO}.$$

These are metals such as copper, arsenic, antimony, nickel, zinc, bismuth or iron. It does not include silver, which therefore remains in the gold during the cupellation process!

It should be noted that the earliest textual reference to "purified gold" from the Sinkasid Palace in Uruk (Mesopotamia) dates from the early second millennium (Waetzold 1985). However, it is not clear what is actually meant by that, i.e. whether it is actually gold that has been

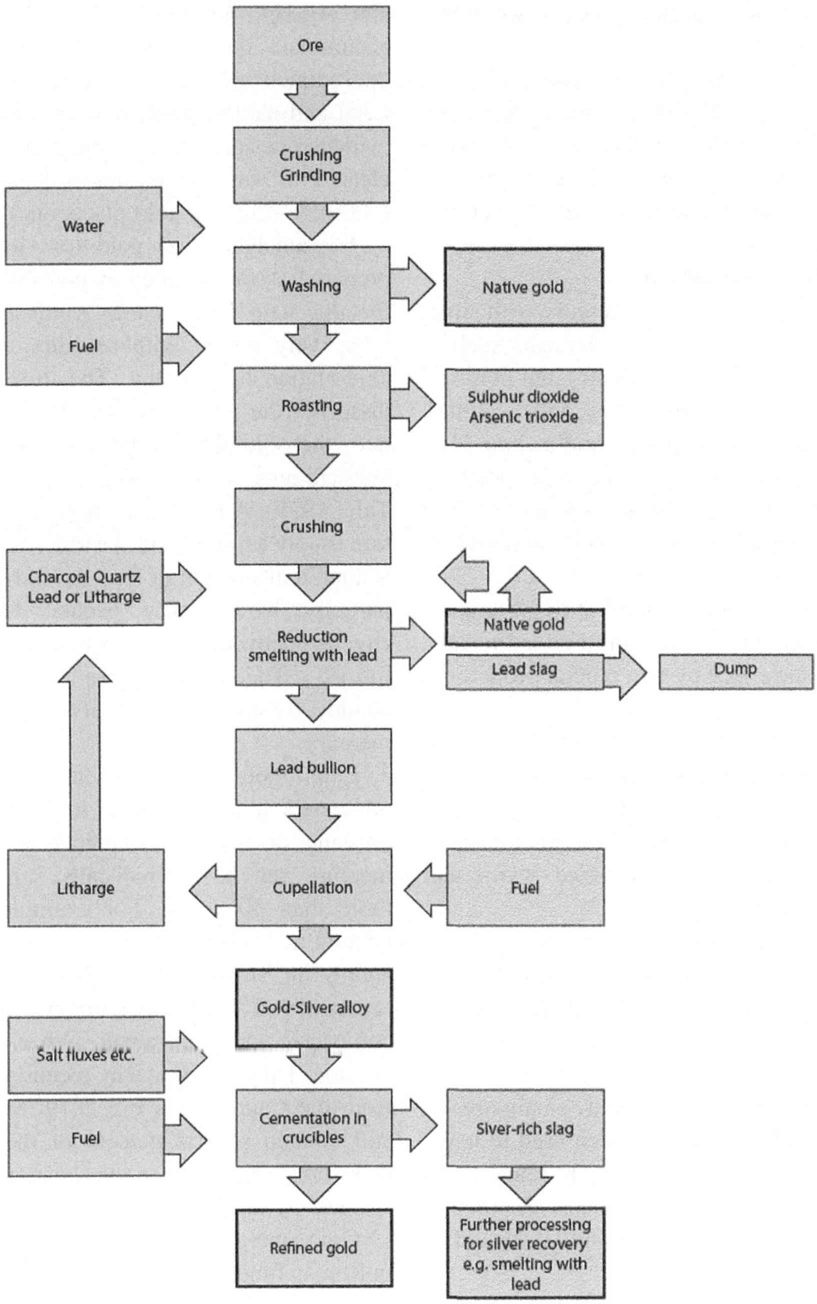

Fig. 6.17 Hypothetical flow sheet to illustrate the possible extractive metallurgy of refractory gold ores, exemplified for the situation of the ancient mine of Três Minas in northern Portugal in Roman times. It shows the complexity of different stages necessary to enrich the precious metal from a low-grade ore deposit. From Bachmann (1999b)

enriched by any metallurgical process (see below at parting/cementation).

The Greek geographer Agatharchides of Knidos (c. 200–120 BC) also describes a process reminiscent of a cupellation (Moesta 1986), but it is generally interpreted as a refining process. Again, what is meant here must remain unclear.

6.3.2.4 Amalgamation

Amalgams are compounds of mercury with silver and gold and with other heavy elements such as copper, lead, zinc, tin or palladium—but not iron! Mercury also forms amalgams with sodium, potassium, cadmium, antimony and arsenic. The low solubility of iron in mercury had the practical advantage that it could be transported in iron vessels, especially in Islamic times (Steuer et al. 2002).

Amalgams are liquid or solid at room temperature, depending on the amount of added metals. They are characterised by two crucial properties:

- They become softer and doughy with increasing temperature and solidify with decreasing temperature. This effect is used today in cadmium- or tin-containing silver amalgams in dentistry: Fillings are processed warm and solidify at body temperature.
- At several hundred degrees Celsius, the mercury can be almost quantitatively distilled off from the amalgams, so that metal remains. This is the basis of the amalgamation.

Due to their special behaviour, amalgams do not belong to alloys that have been used to make artefacts of any kind. Amalgams are rather to be understood as intermediates, which were mainly used for the extraction of gold and silver and as gilding alloys.

Amalgamation is used in the fire gilding of metal objects with gold dissolved in mercury. Due to the high affinity of gold as well as of silver to mercury, gold and silver amalgams such as weishanite $(Au,Ag)_{1.2}Hg_{0.8}$, gold amalgam γ-$(Au,Ag)_2Hg_3$ as well as α-, ß-, γ- amalgam = kongsbergite, para-schachnerite and moschellandsbergite (5–56 wt. % Hg) are also occur in nature. Compounds like Au_3Hg, Au_2Hg and Ag_5Hg_8 are likely artificial products that accumulate during amalgamation and entered the sedimentary cycle. A distinction between artificial and natural gold amalgams and associated evidence of applied amalgamation is not always clear. This was pointed out by Hauptmann et al. (2009) working on gold placers in the Caucasus.

For amalgamation gold (or silver) must be prepared as fine-grained as possible and be in a metallic state. Gold-bearing sands and sludge are intensively mixed with mercury after the first exploitation by panning. The thin, liquid metal dissolves the gold particles physically. In practice, the solubility of gold in mercury at low temperatures is very low (0.1–0.2 wt.%, see Tafel 1929). An amalgamation at room temperature is only a minor part of a true solution process. Rather, a dispersion of fine-grained gold or silver particles by mercury occurs, i.e. they are enveloped by mercury, so that a pasty mass is formed. The adhesiveness is increasing by already existing gold or silver contents in the mercury.

Liquid gold amalgam is easily separated from other rock particles due to its high density. By pressing in linen or leather bags, superfluous mercury can be mechanically separated up to more than 30 vol.%. For example, by simply heating and mixing vigorously in open pans, or simply in small vessels above a spirit stove (at about 360 °C), the mercury evaporates leaving compact crude gold, albeit with a varying pore volume. This was shown by recently panned gold from the Caucasus in Fig. 3.19. Such gold can still contain several percent of mercury and is therefore subjected to a pyrolysis, whereby gold is liquified and residual mercury evaporates.

Gold amalgams probably played a crucial role in Roman times, but certainly in the Middle Ages. Even today, mercury is used in many countries in artisanal gold production, such as in Burkina Faso or in Tanzania (Werthmann 2009; Borg 2014). Roman historians (Theophrast, 300 BC, Pliny, Naturalis Historia, Vitruv, first century BC) have stated that mercury was used in their time to extract the gold from the threads of gold-woven fabrics. The collaborators of the Projektgruppe Pliny (1993) interpret texts of Pliny the Elder who obviously describes the amalgamation process for the extraction of gold in detail. Lacerda and Salomons (1998) suggest that there are already serious environmental problems with mercury in Roman times. Around 77 AD, Romans imported about 5000 kg of mercury per year for amalgamation from Spain (Almaden) to Italy. This was later banned for health reasons, for the treatment (stamping of the amalgam mixture) was carried out by human or animal muscle power, i.e. slaves or donkeys, who quickly suffered from mercury poisoning (Minamata disease with damage to vision, hearing, coordination, growth retardation and muscle deformities) with regularly fatal consequences.

For medieval South America, mercury consumption has been recorded indirectly by Spaniards who send mercury from Almaden (Lacerda and Salomons 1998). Silver and gold production itself recorded data directly from the Patio process. From these information, the consumption of mercury can at least be estimated. In 1557 AD Spaniards introduced the Patio process in 1557 in Mexico, Peru, Chile also for silver recovery (Bersch 1898): Silver ore (chlorides, sulphides etc.) are ground with hostrocks, mercury is added at the same time and mixed by mules for 15–45 days. Amalgam is washed out with water and heated. It is estimated that during the colonial period of the Spaniards in central and northern South America, the consumption of mercury between 1570 and 1900 was approximately 196,000 tons, with an annual consumption of 290–1085 tons, which must have led to an enormous soil and air pollution at the same time. Just recently mercury was used in large quantities by thousands of garimpeiros for gold from lateritic environment production at Serra Pelada in eastern Carajás in Brazil (Moroni et al. 2001; Salgado 2019). It is also still used on a comparatively large scale in countries that still use traditional methods to extract gold, such as in Ghana, Tanzania or Peru. Otherwise, amalgamation is hardly used industrially anymore today, at most in conjunction with cyanidation. Among other best documentaries of the amalgamation process in all its details are the photographs of Bazil Roman in Transylvania around 1930 (Slotta et al. 2001) or those of Borg (2014) from Tanzania.

6.3.2.5 Parting/Cementation

Whether gold comes from an alluvial placer, or from a primary hydrothermal deposit, it contains silver, from a few to about 40–50 wt.%. In such high concentrations, it is called electrum. In contrast, the copper contents in natural gold (and in aurian silver) are always <1–1.5 wt.% only, usually 0.2 wt.% (see above). To separate the silver from the gold and enrich it, cementation is used.

The term cementation is somewhat confusing because it also describes the precipitation of a noble metal from a solution of its ion by another less noble metal. For example, copper precipitates out of a Cu-rich solution at iron objects, they are coated with copper. However, cementation, or, in other words parting, in solution was only possible after the invention of acids, i.e. in Medieval times.

When speaking of parting in antiquity, then dry parting is meant (Bachmann 1999b). This refining process to separate gold and silver is based on the ability of silver to form chlorides, sulphides and sulphates, as well as nitrates, which gold does not. Among the best-known cementation processes is the gold–silver separation by sodium chloride (NaCl). (Experimental) studies on the old cementation process, i.e. on parting, have been carried out often, and with different approaches. The most comprehensive reviews are the publications by Craddock (2000c, d) who, as part of the famous Sardis excavations, has compiled a global historical chronological overview of parting from its potential beginnings to the recent past. Then follows the publication of Celauro et al. (2017). The authors pursue a reconstruction of the various gold–silver separation methods on diverse experimental basis. Independently, both authors reach the conclusion that this metallurgical process may have been known in earlier eras than in the middle of the first millennium. Judging by the findings of Sardis, parting with Na-chloride had already reached a high technological standard in the middle of the first millennium, and Craddock (2000c) rightly wonders whether the method had not been known previously.

Parting methods, as described by Agatharchides, were first experimentally retraced by Notton (1974). Agatharchides, who visited Egypt in the late second century BC, wrote about gold mining and gold refining there. His descriptions were later adopted by Diodorus Siculus (60 BC).

Accordingly, AuAg alloys, chipped or as flakes (due to the larger surfaces), together with common table salt (NaCl) and brick dust (Ca-, Fe-oxides, SiO_2 and Al_2O_3) are heated up in the first experiment to c. 800 °C for multiple hours in small crucibles. In this case, silver and possibly also copper contents turn into chlorides, which evaporate. In simple terms, a parting of a gold–silver alloy with some copper using common table salt proceeds according to the simplified chemical reaction:

$$Au_{75}Ag_{20}Cu_5 \rightarrow Au + AgCl + CuCl_2$$

The added brick dust should serve as a slag former. However, only an enrichment of gold from 37.5 to 93 wt.% was achieved. In a second experiment, NaCl was added to a gold–silver alloy and heated in a sealed crucible for 5 days. Addition of lead or tin, as described in Agatharchides, was incomprehensible. If anything, this would reduce the gold yield. Addition of tin would form SnO_2, which halts the reaction and leads to the formation of Sn-containing gold. When lead is added, lead silicate slag forms, which may also stop the process. It is not clear here whether, with the addition of lead, a reaction with chlorides would take place at all, or whether Pb-chloride still would form.

Izawa and Nakanish (2014) suggest another function of a brick dust-like material. Their investigations are based on archaeological findings of the seventeenth century AD on the island of Sado in Japan and at the same time on pictorial representations of the gold–silver parting process of that time. Gold–silver parting was practised using large amounts of salt which must contain bittern. The mixture was heated to a maximum of c. 800 °C (i.e. about the melting point of rock salt at 801 °C) in a large number of clay pots (this as a source of Al, Si, Fe of the brick dust). A number of chemical reactions take place. NaCl reacts with the water of the bitter, forming Na ions and hydrogen chloride (gas). In the case of Sado, the original ceramic material contained quartz and meta-kaolinite as silica sources:

$$2\,NaCl + H_2O + 4\,SiO_2 + Al_2Si_2O_7 \rightarrow 2\,NaAlSi_3O_8 + 2\,HCl$$

Under the oxidising and temperature condition of the process Fe^{3+} concentrations of the ceramic material reacts with hydrogen chloride and forms Fe^{3+} chloride (gas):

$$3\,HCl + ½\,Fe_2O_3 \rightarrow FeCl_3 + 3/2\,H_2O$$

Fe^{3+}-chloride attacks silver and forms silver chloride (melt) and Fe^{2+}:

$$FeCl_3 + Ag \rightarrow FeCl_2 + AgCl$$

The cementation process of gold–silver parting is thus based on a series of gaseous reactions. It is practically and technically a quite complex metallurgical process.

The experiments by Wunderlich et al. (2014), conducted on the basis of Theophilus Presbyter, have shown that in order to reduce silver content in gold to <0.1 wt.%, a multistep process is required, which is laborious and involves significant gold losses. The authors have carried out three-stage experiments. Moreover, according to their measurements, it is difficult to bring the concentration range of copper in gold to the order of 0.01 wt.%. These analytical investigations were carried out by the named authors to verify the authenticity of the suspected Early Bronze Age gold finds of Bernstorf (southern Germany).

In gold metallurgy, numerous old writers (e.g. Dioscorides, Pliny, Galenos from Pergamon (129–199 AD), in the Leyden Papyros, in the Mappae Clavicula and in De Re Metallica of Georg Agricola) mention three substances when talking about the parting or cementation process. The first and most common one is *misy*, followed by *sory* and *calchitis*. As described by Celauro et al. (2017), these three names were mentioned by Galenos (Kuhn 1826) in his journey to the largest open-pit copper mine of the island Cyprus, Skouriotissa (Cyprus).

This is believed to be three several meter-thick layers of fine-grained clay and sulphate-rich sedimentary loose layers in the extended oxidation zone of the ore deposit. It cannot be ruled out that *misy* is the gold- and silver-containing *devil's mud*, which is characterised by jarosite, various Fe sulphates, limonite and clays (Kindermann 2005; see Sect. 3.6, Oxidation Zone). Other layers in the oxidation zone may correspond to the other terms (ochre, umber). These are the substances into which gold should be embedded in the cementation process to separate the silver from the gold.

An alternative to the gold–silver separating process is the use of Fe-sulphides, as described by Theophilus Presbyter in the twelfth century. In this process, to enrich gold, an AgAu alloy is heated with pyrite and the following reaction takes place:

$$3\,FeS_2 + 12\,AgAu + 2\,O_2 \rightarrow 6\,Ag_2S + 12\,Au + Fe_3O_4$$

The earliest archaeologically certain documentation of the refining of gold after the cementation process dates back to Sardis, the capital of the Kingdom of Lydia in western Anatolia, from the time of King Croesus, who ruled from 561 to 547 BC. Here the gold-bearing sediments of the river Pactolus were famous. The excavation results of Sardis are summarised in Ramage (2000). The archaeological evidence for this gold–silver parting included fragments of refractory sherds of crucibles and furnaces, tuyères, gold foils and gold flakes (gold dust), litharge cakes and other finds recovered from a workshop or temple district (Ramage 2000). They have been intensively studied with various analytical and metallographic methods. The original gold washed out of the Pactolus River had about 20 wt.% silver before cementation. In the excavation itself, there were also gold plates with 99 wt. % Au. This was clear evidence of a refining of gold using the cementation process. The platelets with the high gold contents showed microporous surfaces, as they are formed by the extraction of silver.

However, there is no evidence that this well-established innovation would not have been known before (Gebhard and Krause 2016). In contrast, it would be reasonable to consider a slow and continuous development. Neither cultures nor innovations developed suddenly, they always were looking back on a development and prehistory.

In all cases, these were chemical reactions that took place on thin gold foils and led to a surficial gold enrichment. This remembers the gilding process of the depletion-enrichment technique, as already applied in the third millennium BC at Ur, Mesopotamia (La Niece 1995; Hauptmann et al. 2018) (see Sect. 6.3.2.2).

As for a metallographic interpretation of the surfaces of cemented gold, uncertainties may exist because of limited possibilities of sampling. (silver-containing) gold foils with porous surfaces, as shown in Fig. 6.16, are not necessarily proof of a deliberate refining of gold. They can also be formed when embedded in a saline environment of soils for thousands of years, as suggested by Hauptmann et al. (2018). On the other hand, porous surfaces connected with an enrichment of gold on the surface were detected on the gold rings of Nahal Qana (Israel; Shalev 1993).

Quiring (1948) suspected that parting could have already been invented in the 11th dynasty in Egypt, i.e. at the beginning of the Middle Kingdom (after c. 2000 BC). The reason for this assumption is that gold artefacts of the 1st–10th dynasty (c. 3197–2000 BC) contain much more silver than those of the 11th dynasty (first Dynasty, 3197–2949 BC: 80–84 wt.% Au, 13–13.5 wt.% Ag). Gold artefacts of the 11th dynasty, on the other hand, are very pure. Copper contents are supposed by him to be evidence of intended alloys. Klemm and Klemm (2013) suggest another model. They wrote that in the early dynasties (possibly in the pre-dynastic period) the highly gold-containing placer gold was used predominantly, while in later periods (with the exception of the 11th dynasty) until the Roman period, it is likely that primary gold, extracted from ore by beneficiation, was used. This led to increased Ag contents. They suggest gold refining technology in Egypt as early as the New Kingdom (c. 1570–1069 BC) as a pioneering feat. Refining has been proven in Roman times anyway. Ultimately, the argument is difficult as long as there is no statistically secured base of the gold itself. And unfortunately, there is also very pure gold from the predynastic period, which, with the arguments above, could also be refined.

Altogether, Craddock (2000c, d) notes "that there is no (firm) evidence prior to the introduction of coinage for the use of a true cementation process".

6.3.3 Gilding Through the Ages

To produce gilding effectively, i.e. producing a relatively large gold-coloured area or achieving polychrome effects with the most sparing use of the valuable precious metal has long been a well-practiced and widely used method since the time when gold was first extracted and processed (Oddy 1981). The earliest attempt for gilding surfaces was observed in Varna. The gilding attempt was an application of gold dust (placer gold) presumably with plant glue on a clay dish. It is not clear to this day whether the gold dust was applied before or after firing the dish (Éluère and Raub 1991).

Historically detectable gold plating techniques are as follows (Hammer 1998/1999):

1. Gilding by mechanical attachment: This technique is the simplest and most original form and has been in use since the middle of the third millennium BC. Due to its high ductility, gold adapts well to the surface, and mechanical anchoring is achieved by folding, drawing in, beading and punching. In this purely mechanical anchoring, the adhesion is low and the risk of corrosion below the gold is large. For example, various rings from the royal tombs of Ur have been found to have corroded copper or bronze cores (Zettler and Horne 1998; Hauptmann et al. 2018).
2. Soldering of gold sheets or foils. Soldering a gold sheet (or a foil) with the base metal creates an adhesive, corrosion-resistant connection. Hard solders based on gold–silver, gold–copper and gold–silver–copper compounds are processed at temperatures of 700–800 °C. However, since gold diffuses into both the solder and the base metal at the temperature necessary for the soldering, a "gold loss" occurs in this type of plating technique. When using soft solders (based on gold–tin–solders) less diffusion takes place because of the lower soldering temperatures below 450 °C, but the resistance, stability properties are also lower.
3. Gold leafing. Prerequisite for the gold leafing is its availability and thus the knowledge to produce it. Gold leaf can be produced in a thickness between 1 and 6 μm. The gold consumption of gold leaf is low. Due to its extremely high ductility, thin foils with a thickness of 0.125 μm can be made in modern times, using pure gold. In the case of a 1-μm thick ancient gilding, 1 g of gold is enough for a surface of 500 cm^2. Since gold leaf can have defects and holes, metallic substrate can corrode. This is true, e.g. for gilded bronzes. When a gold leafing is heat treated, the bond to the substrate is enhanced by the diffusion of gold, but at the same time gold is lost. Leaf gold was not known in the fifth millennium BC. Extremely thin layers of gold (precipitated on glass) are getting transparent green at a thickness of c. 40 Å or blue to red violet at a thickness 15 Å. Reasons for making leaf gold (and not foils from other metals) are based upon the crystal structure. As the basic requirement for making leaf gold is the purity of the metal, early gilding techniques were plating rather than leafing. This is because in nature, in most cases gold contains some 5–40 wt.% silver. The making of leaf gold presupposes the application of refining (= parting) of gold.
4. Gilding by diffusion bonding. This is based on the property of the metal atoms to diffuse in the solid state. In close contact with the surface of two bare metal sheets, an experiment achieved such a large adhesion between a gold and a silver plate by diffusion of the

gold into the silver and vice versa, that further processing by forging and driving as well as stamping and punching were possible. This was done over one hour at 500 °C in a reducing atmosphere. The thickness ratio of the two metals is retained.

5. Fire gilding (or mercury gilding). The fire gilding uses a creamy gold amalgam, which is a mixture of the grey γ-phase Au_2Hg with mercury that can be used to coat an object. This is heated to evaporate the mercury content, leaving a thin layer of gold on the surface. For the production of the amalgam, mercury is used in considerable excess (Au : Hg = 1 : 7; corresponds to about 88 wt.% Hg), which is then removed again, e.g. by pressing in a leather or cloth bag. Anheuser (1999) states that the ideal temperature range for fire gilding is 250–350 °C. Below 250 °C the treatment time is too long, above 350 °C the Hg evaporates too fast and the layer is destroyed. A homogeneous formation of solid solution is in the range of the α-phase: the solubility limit of mercury in gold is about 13% at room temperature and increases to 20% at 419 °C. His analytical investigations of fire-gilded objects showed mercury contents of 5–25 wt. %, which showed no significant differences in colour.

   There are two possible techniques for the gilding of silver or copper objects using mercury (Lins and Oddy 1975). In the first, the surface of the base metal object is amalgamated by rubbing mercury onto it and then gold leaf is applied. In the second an amalgam of gold is made by either grinding gold dust in a mortar with mercury or by dissolving fragments of gold in boiling mercury and this amalgam is spread over the base metal surface. In either case, it is helpful to pretreat the surface with a solution of mercury salt. With both techniques, the final stage is to heat the object to evaporate the mercury and leave a continuous and strongly adherent film of gold on the surface. Although mercury is driven off by gentle heating during the final stage of the gilding process, detectable traces of mercury will always remain in the gilding layer. Based on these considerations which are supported by experimental work, Lins and Oddy (1975) investigated numerous archaeological and historical objects. They found that mercury gilding for the first time has been widely practised during the later Roman and Sassanide Empires of the third and fourth centuries AD in the west, and during the late Chou period in China of the third century BC in the far east.

6. Depletion gilding. This is a process of covering an object made of an alloy low in gold with a golden surface. Smith (1974) makes the point that this is not a metallurgical process, but it is an artificially accelerated corrosion process. Depletion gilding was a widely used technique in the pre-Columbian countries of Central and South America where this alloy was often mixed. The principle is based on the following effect: First, an alloy with varying amounts of gold is produced. Such an alloy is called tumbaga. As can be seen in the system Au–Ag–Cu (see Sect. 7.3.2) such silver-containing gold–copper alloys, when melted together, form a complete solid solution series throughout the entire range of compositions. Depending on the proportions of the components their colours can vary from orange-red to red-gold. Tumbaga alloys may have a melting or solidification point of only about 900 °C, an advantage that provides significant benefits to the metallurgist in their processing, especially during casting. In Fig. 7.14, Ruvalcaba Sil et al. (2009) compiled an overview of tumbaga alloys from Central and South America. After shaping the desired object, it is heated over an open flame so that the copper is selectively converted to Cu-oxide as a less noble component. It forms a black crust that is removed by hammering. Gold as precious metal, however, remains in the metallic state. This means that in the production of thin sheets from an Au–Cu ingot alternating hammering and heating automatically creates a gold-rich surface. Remains of Cu-oxide crust are selectively dissolved out in a broth of plant juices high in tannins or acids (oxalis, apple

and citric acid), salt (NaCl) and/or urine (ammonium carbonate). The result is a porous (spongy) surface layer, which is then compacted and smoothed by hammering and polishing. Ruvalcaba Sil et al. (2004) investigated to what extent these porous surface layers may be differentiated from natural corrosion layers.

The depletion process from tumbaga alloys has long been described as a typical pre-Columbian native American technique (Lechtman 1979; Scott 1983). Tumbaga alloys with their inherent enrichment properties swept through the Americas from Peru to Mexico and were in common use in that entire region when the Spaniards invaded Central and South America in the sixteenth century AD. This ultimately led to the conviction that it had been invented there. Gilding following the depletion gilding process, however, was known in the Old World long before its spread to South America. The application of this process has been observed on some gold–silver–copper artefacts from the royal tombs of Ur. It should not be excluded that it has found a much wider distribution than previously thought.

Aside from the technique of depletion gilding, other metallurgical experiments have been done in ancient times as well, such as in a depletion silvering made from mixed silver–copper alloys. In addition to a find from the Riemchen building in Uruk (Pernicka 1993), the procedure was proven on a dagger from the "royal" tomb from Arslantepe around 3000 BC (Hauptmann et al. 2003).

6.4 The Metallurgy of Silver and Lead

While the ore deposits lead and zinc have been dealt with together in Sect. 3.5.5, due to their frequent occurrence together in the formation of ore deposits, they must be separated from each other in this chapter on metallurgy. They are treated differently in order to extract them from their ores. Instead, silver's technological processes are very closely linked to those of lead, because this metal has a crucial property: it acts as a collector of precious metals. This is the main reason why silver and lead should be discussed here together.

Silver (chemical symbol Ag) is a whitish shiny, very malleable, ductile heavy metal with a density close to lead (Ag: 10.49 g/cm^3, Pb: 11.34 g/cm^3). Although it is one of the noble metals, its less noble properties are already evident in the variety of natural mineralogical connections. Silver has a pronounced tendency to compound with S, Sb, As, Se, Te and halides (Cl, Br). Silver has a melting point of 962 °C. With these properties, silver is a very suitable metal for making many differently shaped artefacts (wires, thin sheets, finger and earrings, bangles and cups). However, from the archaeological point of view, silver has the unpleasant property of rapidly and thoroughly corroding to form dark grey Ag-sulphides and -chlorides.

Lead (chemical symbol Pb) is a soft, blue-grey heavy metal, pure white on polished surfaces. It was used very early because of its ductility, its very low melting point (327 °C) and its ease of production by metallurgical operations. Lead, however, has not played such a major role in prehistoric metal ages as gold, silver, copper and tin did. No weapons could be made from it and only a few jewellery items were made. Lead is also easily confused with galena (PbS) because of its similar appearance. Both oxidise rapidly to the whitish carbonate cerussite ($PbCO_3$) when buried in the ground. Lead and its oxidic and sulphurous vapours are poisonous like those of many other heavy metals.

6.4.1 Silver Artefacts

The oldest finds of silver artefacts date back to the fourth millennium and come from the Middle East, namely Beycesultan, Korucutepe, Alishar Hüyük (Anatolia) and Tepe Sialk (Iran) (Ghirshman 1938; Wertime 1973).

Silver was used as an artist's material in jewellery and luxury items, but already in the middle of the third millennium BC silver for trade became a commodity. Overviews are compiled by Pernicka (1990); Kohlmeyer (1994) and Primas (1995). Prag (1978) reports 233 silver objects from the burial ground of Byblos (3800–3200 BC) and Moorey (1994) of 30 silver objects from the fourth and beginning third millennium in Egypt and Levant, of more than 20 from Uruk, Tell Brak, Susa and other localities in Mesopotamia and Anatolia. A silver fitting of boxes from the Naqada culture (predynastic Egypt, c. 3600 BC) contains 0.4 wt.% lead, i.e. this silver has most likely been extracted from lead by cupellation. Cuneiform texts describe the trade of silver from Anatolia to Mesopotamia. Among the many metal objects found in the royal tombs of Ur, Mesopotamia are hundreds of silver objects (Zettler und Horne 1998; Salzmann 2019).

While the extractive metallurgy of gold is technologically manageable, because it usually occurs as a native metal, that of silver is much more complex. Silver rarely appears in native form. It must be produced by combined smelting and cupellation processes from its ores (cf. Fig. 6.17). According to the current state of knowledge, where the origins of this technology are located is not very clear. In Central Europe, a larger use of silver first took place in der Bell Beaker culture at the end of the third millennium BC.

As in the Aegean, silver also appears almost simultaneously with the increased presence of lead in the Middle East (Gowland 1917/1918; Gale and Stos-Gale 1981a; Pernicka 1990). This is probably also an indication that the earliest silver objects were already made of cupellated silver (Krysko 1979). Hence, the metallurgical extraction of this noble metal is thus closely linked with that of lead, and thus with the proliferation of extractive metallurgy. With the earliest finds of cupels from the fourth millennium from Fatmali Kalecik and Habuba Kabira from the Near East, this is proven (see below).

On the south-eastern Iberian peninsula, a remarkable number of silver and silver-alloy artefacts were produced during the El Argar civilisation (2200–1550 BC) (almost 300 silver objects, 34 of a silver–copper alloy). However, no indications of silver extraction from lead ores via the cupellation process are known here before the first millennium BC (Hunt Ortiz 2003; Bartelheim 2007). It is suggested that in the Early and Middle Bronze Ages many silver artefacts have been made of native silver or silver chlorides which occur frequently in many deposits in these regions (Bartelheim et al. 2012; Murillo-Barroso et al. 2014). Later, in Roman times, silver was extracted in large quantities from many of the Empire's lead–silver deposits (Meier 1995), playing an important role as a coinage metal. As explained in Sect. 6.4.2, the accumulating lead was used for water pipes, in architecture and in the military field.

In antiquity, in contrast to copper and iron, silver had no practical significance, for example for the manufacture of tools or basic commodities. Like gold, however, it was a highly sought-after metal for the production of luxury articles, jewellery, the decoration of clothing, furniture, buildings and weapons. Therefore, already in classical times silver production had taken on industrial dimensions. Thus, the total amount of lead and silver produced from Laurion ores during the classical period was assumed very high. Conophagos (1980) estimates orders of 1.4 million tons of lead and 3500 tons of silver.

Similar proportions were then observed in the Roman Empire. In addition to the well-known silver producers from the ores of Cartagena (Spain), the noble metal was mainly obtained from Rio Tinto and Laurion. In Kosovo and in the area south of Belgrade (Upper Moesia, Serbia), around 1,000,000 tonnes of slag and about 5000 shafts are evidence of extensive lead and silver production. Davies (1935), Merkel (2007) and Domergue (2008) are authors, who published detailed reports of this.

6.4.1.1 Silver Ingots

Two stamped silver bars and some late antique gold bars were recovered in Transylvania (Romania) (Wiegels 2014). With the collapse of the Roman Empire, the production of silver came to an end. It is resumed only in the Carolingian period (eighth century AD), where many Roman mines were probably reopened. Such topics have been discussed by Téreygol (2002) for Melle (France) or Weisgerber (1995) for Trento (Italy).

At any rate, the demand for silver at this time leads to a long-distance trade in which the noble metal is also imported from countries in Arabia and Middle Asia (e.g. Afghanistan, Tajikistan and Uzbekistan) (Steuer 1987; Merkel et al. 2013, 2016). Steuer (1990, 1998) explains that a significant part of this was paid off as a tribute to the Vikings in the north. In Scandinavia and Baltic countries around 1500 treasure troves are known from this time, in which 150,000 Kufic coins (dirhams) were found. With the rich ore deposits in Middle Asia, but also in Spain, Africa, Greece and Persia, the Arabian empire was the largest silver producer in the Old World. In Scandinavia, Arab coins were apparently often remelted. In the tenth and eleventh centuries AD, a primitive currency developed in the Baltic and Eastern Europe, consisting of a mixture of hacked, broken and sometimes intact silver pieces of any shape, whose only value was in their weight. However, this metal, known as hacksilver, is possibly already encountered in prehistoric West Asia and in antiquity (Graham-Campbell 1982). In Khafajeh (Diyala region, Iraq) a silver treasure from the Uruk period (Early Dynastic IIIa, b) was found, which consists partly of silver objects (15 pieces and a string of pearls) and 21 remelted, shapeless silver objects, quite in the concept hacksilver (Delouga et al. 1967; own investigations). Balmuth and Thompson (2000) studied the geographic distribution of hoards of this uncoined silver.

It was only after the influx of silver from Arab countries came to an end in the twelfth century AD that mining for silver in Central Europe began to rise again (Bingener et al. 2012).

6.4.2 Lead Artefacts

The earliest lead finds, if they can be safely dated at all to the sixth millennium, are extremely rare and come from the Middle East. A compilation can be found at Moorey (1994). Since native lead, unlike native copper, is extremely rare in nature, it can be surmised that all of these earliest artefacts were shaped from smelted lead. Scientific material analyses are lacking in almost all of these examples. And so, it is not surprising that the jewellery in Çatal Höyük, initially mentioned as a lead bead, i.e. made of lead produced by smelting, was later identified as galena, the natural mineral itself (Sperl 1990). Later, in their research of archaeometallurgical finds of Tell Halula, Euphrates river (Middle to the Late Pre-Pottery Neolithic B, i.e. 8th/7th millennium), Molist et al. (2010) identified a bead as galena from the start. A bracelet found along with this bead was made of beads of native copper.

Early Bronze Age lead metal is very rare and the earliest lead objects were found quite sparingly. Reported are pearls, seals, decorated lead strips or "Syrian" bottles (Efe and Fidan 2006). Such lead objects occur in Anatolia, northern Iran and the Levante, although they were identified as such only by macroscopic examination. A perforated lead object attached to a wooden shaft from the Ashalim cave near Beersheba (Israel) dates to around 4000 BC and was the very first of the earliest objects being analytically investigated and proven to be lead. The analysis was performed with electron microprobe analysis on a clipped sample (Yahalom-Mack et al. 2015).

Probably the earliest collection of 38 cup-shaped lead artefacts dates from the Jamdat Nasr period (c. 3100–2900 BC) found during the Ur excavations. These are nine lead seals of amphorae. From the present stage of information, the lead isotope signature of the lead lids points most plausible to complex Anatolian or Persian metalliferous deposits (Klein and Hauptmann 2016, 2020).

6.4.2.1 Roman Lead Production

In the Roman Empire the usage behaviour of lead changed fundamentally. From then onwards, the production, distribution and treatment of lead must be considered as an important economic activity.

Even if the silver content of lead ores was most likely an important incentive for their mining: lead played a major role in the extraction of silver. Lead and silver have a high miscibility in the liquid state and most silver ores in antiquity were in fact argentiferous lead ores.

Fig. 6.18 Mal di Ventre, west coast of Sardinia (Italy). From a Roman shipwreck some 1000 lead ingots were discovered. The weight of such an ingot is between 30 and 40 kg, sometimes more than 70 kg. Altogether, a load of lead on this ship was c. 33 tons. All these ingots were epigraphically documented and in parts analysed for their trace element and lead isotope composition. The ingots are archived in the Civico Museo Archeologico "Giovanni Marongiu" in Cabras, Sardinia. Photo: courtesy Norbert Hanel, Cologne/Bochum

Masses of lead were mined in the Roman Empire. As a metal in everyday life, lead has found widespread use in Roman times. Three physical properties were crucial: high density, low melting point and easy deformability. Numerous shipwrecks in the western Mediterranean showed that tons of lead were shipped on sea to Rome as 30–40 kg, sometimes up to 75 kg heavy ingots (Fig. 6.18).

6.4.2.2 Ingots

Roman lead ingots represent a special sort of finds, which was investigated in detail by the group P. Rothenhöfer, N. Hanel, and M. Bode in an interdisciplinary research project (summaries in Rothenhöfer und Hanel 2013; Rothenhöfer et al. 2013). In total, in this long-standing research project *Corpus Massarum Plumbearum Romanarum* (CMPR), more than 2500 Roman lead ingots were investigated epigraphically and by chemical and lead isotope analyses for their provenance, revealing many details on social and economic organisations of imperial and private lead mining and trade. The ingots also provided valuable chronological information on the duration and scope of lead extraction at various locations—which is often difficult to grasp in the archaeological findings. The authors emphasise that no other metal, be it iron, copper or tin, offers similar comprehensive possibilities for epigraphic and at the same time natural scientific evaluations. No other metal has an even approximate number of massive ingots.

Rothenhöfer et al. (2013) give an overview of what lead was used for in these gigantic amounts in Roman times, aside from its importance as a silver-bearer and precious metal collector (see below). Lead was particularly important in construction engineering and architecture. It was used in large quantities as a filling of iron pluggings in stone blocks. For the construction of the Roman city gate of Porta Nigra in Trier (Germany) about 7 tons of lead were used (Schwinden 2001). Similar quantities were used for the production of pipes essential for the water supply of baths (Fig. 6.19), cities and other settlements. An estimated 10–15 thousand tons of lead were needed in Lyon for pipes, siphons and aqueduct systems (Trevor Hodge 1983). Also lead covers of roofs were needed. In addition to a wide variety of containers and vessels, numerous other everyday objects were produced, from the net sinker to lead labels, or in the military ranks, masses of lead sling bullets (Baatz 1990; Müller et al. 2014). The first appearance of a specialised lead-processing metalworking profession, the *plumbarii*, testifies to the everyday nature and diversity of the use of lead in the Roman Empire.

6.4.2.3 Medicals

An entirely different application of lead (as an inevitable "by product" of the production of silver from argentiferous lead) involves the use of lead compounds in medicine. It has been mentioned that lead is toxic, as are corresponding chemical compounds of the metal in solid form and as vapours. The manufacture of synthetic lead phases such as psimythion or synthetic lead carbonate is recorded in detail in the works of the Greek philosophus and natural scientist Theophrastus (fourth century BC), the Greek medical doctor Dioscorides (first century BC) or the natural scientist Pliny is certainly not specific for the Greek lead–silver deposit Laurion (Photos-Jones 2019, oral presentation Bochum). Unfortunately, up until now, there is no archaeological evidence of such a synthetic phase.

Fig. 6.19 Bath, Cornwall. Roman bathroom with lead pipe as a water pipe. Lead pipes were used throughout the Roman Empire as water pipes. A lead pipe of this quality needed about a ton of lead for a pipe length of 37.5 m. Photo: J. Hauptmann, Bochum

In his book on the natural sciences of the year 77 AD (Pliny Nat. Hist. 34, 166–178) Pliny has written extensively about the disinfecting and astringent effects of lead compounds, which, from a pharmacological and medical point of view, has been commented on by scientists of the Projektgruppe Pliny (1998).

6.4.3 Classification of Silver Ores: Metallurgical Aspects

Among the 60 known silver minerals and silver-bearing lead ores, galena and cerussite have always played a crucial role. That is still the case today. However, the importance of silver-rich ores in ancient times should not be underestimated. Bartels (2014) has commented on this. He also addresses the problems of assessing exploited near-surface parts of ore deposits. He postulates an originally much richer occurrence of native silver and silver halides, e.g. in the oxidation zone of hydrothermal veins in the Harz district and in the Ore Mountains (Germany), which were, however, only very intensively mined in the Middle Ages. As described, such rich mineralisations are also suggested for Spanish deposits.

Investigations to reconstruct the sort of (lead) ores used for the production of silver objects are discussed by Meyer (2003), who tries to correlates their trace element contents above all with native silver, simple "dry" silver ores such as argentite, cerargyrite, cerussite and galena (Fig. 3.32).

From a metallurgical point of view, the extraction of silver from corresponding ores can be classified as follows (Schnabel 1890; Tafel 1929):

1. Direct melting of silver from
   - Native silver (Ag).
   - Chlorargyrite (cerargyrite, AgCl) and bromargyrite (AgBr) in a one-stage (s)melting operation.

     No cupellation necessary. Chlorides and bromides evaporate during (s)melting.
2. Extracting silver from lead-rich ores.
   - Galena (PbS).
   - Cerussite ($PbCO_3$), anglesite ($PbSO_4$)

     Necessary: Two-stage metallurgical operation: smelting and cupellation.
3. Extracting silver from ores low in lead by adding lead as a collector (leaded smelting).
   - Ag-jarosite ($KFe_3(OH)_6(SO_4)_2$ with silver concentrations.
   - Ag-rich copper ores.
   - Ag-sulphides and sulphosalts

     Necessary: Two-stage metallurgical operation, smelting, liquation and cupellation.
4. Extraction of silver from natural gold–silver alloys.

   Necessary: Parting.

6.4.4 The Metallurgy of Silver

In summary, metallurgical processes to produce silver are closely connected with the metallurgy of lead. Silver is in addition closely connected with gold. It is well possible that silver was extracted in ancient times from gold-silver deposits as they occur, e.g. in the "Golden Quadrangle" in Carpathian Basin and elsewhere, but there is any archaeological evidence missing for the application of parting before the sixth century BC at Sardis that is necessary precondition for the extraction of silver from gold (Bachmann 1999a, b; Craddock 2000c,d; see Sect. 6.3 Gold).

6.4.4.1 Smelting

Metallurgically, it is comparatively easy to smelt rich silver ores. This can be done theoretically in a primitive one-step process, where an oxidation of the metal during firing, as in the case of copper, does not even have to be considered. However, it is very difficult to judge to what extent this has been done at all in ancient times. So far this has only been thought through on a theoretical basis (Meyers 2003), but there is no archaeological evidence for this. For the occurrence of ores high in silver see Sect. 3.5.4.

However, the extractive production of silver usually includes two steps. The first is the transformation of silver ore into a PbAg alloy (raw lead or argentiferous lead), either by the lead content of the ore itself or by the addition of lead when using lead-free or low lead ores. Lead has the property of absorbing silver from ores and metallurgical products. Depending on the nature of the ore, leaded smelting may be preceded by a roasting process. The required amounts of lead for lead smelting to extract silver in old times is likely underestimated. For example, in the late Middle Ages, around 1500 AD, lead was transported in unimagined quantities from lead deposits in the Harz district or in Cracow-Silesia to Thuringia, Saxony (Germany) or to Neusohl (Slovakia) for the liquation process (Blanchard 1992; v. Stromer 1995; Bartels 2000).

It is conspicuous that the earliest silver objects (4th/3rd millennium) are characterised by strikingly low gold contents (usually <0.1 wt.%). This definitely excludes the use of natural AuAg alloys for extraction of silver. In addition, they have relatively low levels of Pb, Bi and Ir. The extent to which chloride and bromide fractions of ores affect the composition of artefacts, or whether they completely evaporate during the smelting process, is difficult to assess. They may easily be attracted by silver from impurities in soil environments over archaeological time periods. This problem becomes clear from the discussion of correspondingly composed silver artefacts from Azerbaijan (Courcier 2014), and it may also apply to mercury levels (Hg) in silver artefacts. It is still unclear whether mercury could actually be a good indication of "limited", single-stage metallurgical operations without a following cupellation process. In ore deposits, mercury is a common element in native silver. Murillo-Barroso et al. (2014) report Hg levels in the percent range in native silver of deposits in south-eastern Spain. Contreras-Cortés et al. (2014) still find mercury levels in the 1/10-percent range in silver artefacts of the El Argar culture, which were probably produced from ores of these deposits.

According to Craddock (1995), a transition to the extraction of silver from less-rich ores could have developed between the third and second millennium BC, at least in the eastern Mediterranean and western Asia.

6.4.4.2 Extracting Silver from Lead Rich Ores

Near-surface lead mineralisation may contain changing amounts of cerussite ($PbCO_3$). For this ore alone, no pretreatment prior to the smelting process is required. Recently there seems to be a consensus that in the earliest stages, cerussite may have been the primary source for silver (Craddock 1995). This is a rather archaeometallurgical statement. That this ore mineral sensu stricto itself should have been the earliest major source of lead–silver mining, must be considered with caution. Although cerussite is a typical secondary mineral in the oxidation zones of lead deposits in carbonate terrain, it is regularly intergrown with galena. Pure cerussite is silver free (see Sect. 3.5.5), the silver contained in the galena in discrete particles is dissolved during weathering and occurs as mechanical admixtures in cerussite. The solubility of galena is ultimately very low (see Smirnov 1954). The weathering resistance of galena and its paragenesis with cerussite is evident, e.g. in the mineralisation of Nakhlak (Iran) (Fig. 3.41a,b).

Silver extraction from argentiferous galena is possible only through a multistep process, namely a previous roasting and smelting of galena and the extraction of silver out of lead by a second stage, the cupellation, as described below.

Since no roasting installations for lead ores were discovered in archaeology with certainty, Gowland (1917/1918) proposed just such as a heap roasting, as was still practiced in Japan in the nineteenth century AD. The extraction of argentiferous lead from a mixture of galena and cerussite can be performed without problems by the roast-reaction process:

$$2\,PbO + PbS \rightarrow 3\,Pb + SO_2 \uparrow$$

Which ores have been used and at what times cannot be established beyond doubt? This also applies to the famous lead ore deposit of Laurion, Attica, where silver was produced as early as the Early Bronze Age (Conophagos 1980).

6.4.4.3 Leaded Smelting

In the smelting of argentiferous galena or cerussite-rich ore, the addition of lead is not necessary. When smelting ores that contain silver but are lead-free or low in lead (e.g. jarosite or fahlores), lead must be added. The effect is based on the fact that in the liquid state silver is readily soluble in lead, which acts as a collector of precious metals: Silver, gold and other precious metals accumulate in lead, which have been incorporated into the material balance during charging. Other elements are slagged. Leaded smelting is still practiced in metallurgy today to extract gold, silver and platinum metals (Rosenqvist 1974). The product is called argentiferous lead. A separation of the precious metals takes place afterwards by cupellation, in which lead is oxidised, and the noble metals not.

Obviously, since the first millennium BC, lead was added for smelting as metallic lead or lithargite (PbO). This was shown by the Phoenician shipwreck of Mazarrón (southern coast of the Iberian Peninsula), with 2800 kg of lead ingots and wastes from cupellation activities aboard (Negueruela Martinez 2014), which obviously brought material to Rio Tinto for silver extraction from low-Pb ores.

If silver is a component of non-plumbic ores such as they occur in jarosite-bearing layers

($KFe_3(SO_4)_2(OH)_6$ with only low proportions of natrium, lead and silver) or other complex Ag-containing mixtures of ores, lead has to be added to extract the silver from host rock or gangue. Therefore, ancient silver separated from non-plumbic ores will have subordinated the geochemical signature of the original ores smelted, but ultimately they will have the (isotopic) signature of the lead added for cupellation. It is suggested that large amounts of silver were extracted from jarosite layers in the first millennium BC at Rio Tinto (Hunt Ortiz 2003). Due to the low lead content of the jarosite (which is associated with goethite and hematite ores in the gossan) lead had to be intentionally added for leaded smelting. At least during the Roman period lead ingots (40–60 kg) were transported to Rio Tinto. Three of such ingots were found in the Roman slag heaps (Salkield 1987; Rothenhöfer et al. 2016). Two of them showed the inscriptions Nova Carthago. Obviously, the ingots indicate a close trade relationship between the Rio Tinto mining district and the lead–silver mines near Cartagena, also located in the southwestern part of the Iberian Peninsula. It was previously pointed out that foreign lead had a crucial role already in the first step of the silver production (Craddock et al. 1985; Anguilano et al. 2010b).

The leaded smelting of silver-containing polymetallic sulphides such as the fahlores results not only in argentiferous lead and slag, but one of the final products of the reduction is also matte (Tafel 1929). Fahlores, however, were not used for silver before the Middle Ages. If ores contain larger amounts of arsenic or antimony, the so-called speiss is formed (cf. Sect. 5.8.2). It comprises metallic Fe- or Cu-arsenides and -antimonides. Speiss only has limited miscibility with matte and generally forms discrete particles or layers between matte and silicate slag (Figs. 5. 5.2 and 5.6.1). Argentiferous speiss from the Roman period was found at Las Arenillas, Rio Tinto, Spain (Craddock et al. 1987; Kassianidou 1998). However, there is no clear indication that speiss has been further processed in antiquity to extract silver contents (Kassianidou 1998).

In the smelting of very rich silver ores, such as they occur mainly in the cementation zones of mineral deposits, leaded smelting process in a shaft furnace is not necessary. Instead, rich ores such as argentite, cerargyrite, strohmeyerite or dyscrasite are added directly to the molten lead in the refining hearth (Schnabel 1890).

6.4.4.4 Liquation: The Extraction of Silver from Copper

Matte is a product of smelting sulphidic copper ores. It was shown that it was already known in antiquity and was processed further. However, silver concentrations in the matte have likely remained unrecognised for a long time. This would have developed differently, if it had been subjected to another leaded smelting.

Such a metallurgical operation is almost equivalent to the liquation process developed in the Middle Ages (Suhling 1976). In this process, which first allowed for the extraction of silver contents from copper ores in numerous ore deposits, and resulted in an unprecedented economic boom in Europe in the fifteenth/sixteenth centuries AD (Bartels and Klappauf 2012), silver-bearing crude copper is melted with lead. In the German language, this process is called *Seigerhüttenprozess*. The boom of mining in Europe between 1460 and 1560 AD is referred to as the "age of the liquation process" (Bingener et al. 2012). The triggering factor of this new metallurgical technology was the increase in silver, copper and lead production in Europe. The liquation process and other similar metallurgical processes (e.g. the "drying up process" or *Abdarrprozess*) were applied preferably in the mining districts of Saxony and Thuringia (Germany), in the Slovakian Ore Mountains and in Tyrol (Brixlegg/Rattenberg, Schwaz). This process was introduced to Japan in 1591 by a foreign merchant, probably a Portuguese. A comprehensive study of the European liquation process is contained in the work by Suhling (1976).

For the liquation process, it was necessary to add huge amounts of lead during (s)melting, some of which had to be transported over long distances from raw material sources to the liquation processing facilities (*Seigerhütten*). For example, these in the Mansfeld area (Saxony-Anhalt) had to import lead from the Harz, the ones in Fuggerau from Bleiberg (Carinthia, Austria) and the ones in the area of Banska Bystrica or Miskolc (Hungary) from the ore district of Cracow (Silesia, Poland).

The liquation process was possibly invented by companies from the German city of Nuremberg around 1450 AD and it was further technologically improved upon by Johann Thurzo from Cracow. For the first time, the liquation of copper is described in the well-known books by Vanoccio Biringuccio (1540), Agricola Georgius (1556) and Lazarus Ercker (1580).

In the liquation process, in a first step, argentiferous raw copper is melted with lead at temperatures around 1000 °C. This first step was called *Frischen*.

The principles of the liquation process are the following:

$$(\mathrm{Cu},\mathrm{Ag})_{\mathrm{liqu}} + \mathrm{Pb}_{\mathrm{sol}} \rightarrow (\mathrm{Cu},\mathrm{Pb},\mathrm{Ag})_{\mathrm{liqu}} \rightarrow \mathrm{Cu}_{\mathrm{sol}} + (\mathrm{Pb},\mathrm{Ag})_{\mathrm{liqu}}$$

A homogeneous melt is formed, because in the liquid state copper and lead are completely miscible (Fig. 4.5) If the melt cools down, both metals segregate. This effect is used, as well as the fact that the two metals have different temperatures of solidification (Cu = 1084 °C; Pb = 327 °C).

During cooling, copper solidifies first and the residual liquid becomes enriched in lead, which is dispersed between the dendrites of the copper crystals. The affinity of silver for lead greatly exceeds that of silver for copper, virtually all of the silver will be collected by the lead. Any composition between 36 and 87 wt.% lead will separate into two liquids. When the liquid contains less than 36 wt.% copper, solid copper crystallises directly from it. In both cases, the liquid discontinuously deposits most of its copper at 950 °C and loses the remainder progressively until the lead itself solidifies at 327 °C.

Depending on the silver concentration in copper, variable amounts of lead or lead oxide have to be added to the liquid copper. When the liquid CuPbAg alloy was tapped from the melting furnace it formed flat copper lead (−silver) cakes. These cakes (*Frischstücke*) solidified. Due to its low melting point, the now silver-enriched lead is dripping out of the copper into special liquation hearths and can be further processed. This is the liquation sensu stricto.

The copper–lead cakes were melted in a special liquation hearth at moderate temperatures. The low melting point of lead led to a quick partial liquefaction, which now contains the silver originally solved in the copper. The liquated argentiferous lead (*Saigerblei*) ran off. It was separated from the copper which remained now as "liquation residues" in the solid or pasty state. This residual spongy masses usually retained around 25–30 wt.% of lead and are called *Kienstöcke*. Due to the extraction of lead, the *Kienstöcke* of copper slumped down and thus they indicated the end of the liquation process.

However, as the *Kienstöcke* still contained a considerable level of (argentiferous) lead, they also contained other impurities. Also, the liquated lead contained some percent of copper. Therefore, the *Kienstöcke* had to undergo a further cleaning. They were heated up again but were not liquefied in a "drying" operation. Again, lead runs off the copper. This process was called *darren*. The copper was termed *Darrling*, and the rest of the lead was named *Darrkrätze*. The purified copper finally was refined in the liquid state in a *Garherd*. The rest of the impurities were oxidised and were slagged. *Garkupfer* was produced.

Argentiferous lead was collected for further processing in the cupellation hearth, where it was oxidised to PbO in the liquid state, while silver was separated in the metallic state (see below).

Altogether, a complete separation of impurities was not possible, and remains of argentiferous lead, bismuth, antimony and other elements remained in the copper. It was especially lead that hampered a processing of copper for making wires and sheet metal. In the fifteenth and sixteenth century, copper produced by liquation was lower priced. Lazarus Ercker (1580) reports that *Garkupfer* contains about 1 lot of silver per hundredweight of copper. According to Beierlein and Lange (1960), this corresponds to about 290–295 g/t (= parts per million) of silver. Similar data is mentioned by Georg Agricola. He said that the *Darrlinge* may contain 200–300 g/t of silver.

These data attest to an analytical precision that is utterly surprising and sensational for the sixteenth century AD. The ranges of silver concentrations of c. 300–600 ppm correspond to those we analysed in copper ingots from the famous shipwreck of *Bom Jesus* (Hauptmann et al. 2016). The ship sailed from Lisboa to western India and ended up off the coast of Namibia in 1533 AD. Almost 2000 copper ingots produced in the mining district of Slovakia were loaded on the Portuguese trade vessel (Knabe and Noli 2012). Also, the lead concentrations are in accordance with the difficulties to completely separate the lead from copper (Pb = 1–1.6 wt.%), as mentioned by Lazarus Ercker.

As described in the "Illustrated book on the smelting of copper" from Masuda Tsuna (1801) (edited by Shirakawa et al. (1983) and Lewin and Hauptmann (1984)), in the Japanese process (namban-buki), only charges of 100–150 lbs. (ca. 45–68 kg) were operated, and only 15% of lead were added to the copper to extract the silver. During the melting operation, the lead was squeezed out rapidly using an iron rod with an attached block of wood.

Altogether a complete separation of impurities was not possible, and remains of argentiferous lead, bismuth, antimony and other elements remained in the copper.

6.4.4.5 Cupellation

Argentiferous lead from leaded smelting (or directly obtained from smelting silver-containing lead ores) has collected the precious metals from the primary ore and separated it more or less from other ores and gangue components. The subsequent separation of silver from lead is then completed in a process that metallurgists call cupellation. Chemico-physically, this process relies on the selective oxidation of the lead from the silver-containing alloy, which in the liquid state are perfectly miscible.

The precious metal is separated in metallic form from lead oxide according to the formula:

$$(Pb, Ag) + CO_2 \rightarrow Ag + PbO + CO$$

The simplest method of cupellation is to place the silver-bearing lead in a small, tightly packed depression in the ground, in the cupel. Such cupels can, for example be set into a depression in a furnace floor with a slightly domed roof (Fig. 5.65). Later, walled flat masonry hearths lined with clay, limestone and ashes were used. Wood or charcoal was used as fuel. In the top heat, the lead melts very easily. However, the temperatures must exceed at least 888 °C, because this is the melting point of the litharge (PbO), which forms a honey-yellow-reddish melt. At the end of the cupellation process, the temperatures must even be increased to separate the silver in the liquid state.

Successively formed litharges must be stripped from the (liquid) surface. This could have been done with iron sticks, and cones of litharge are often found in workshop areas of old settlements (see Sect. 5.9). The silver remains in the liquid state and is enriched in the cupel. Gowland (1917/1918) and Tylecote (1986) are convinced that bone ash was already used in Roman times for the production of refining hearths, as is the case for medieval times. This early pre-medieval use of bone ash, which would have left remains of phosphorous in the litharge could not be supported by Craddock (1995) by analytical investigations. On the other hand, it has been proven that since prehistoric times, cupels consisted of silicate-poor, refractory material, such as marl or lime, as was customary up to the present century (Tafel and Wagenmann 1951). This material was used to avoid unwanted reactions of low melting lead silicates. In addition, porous material should facilitate the absorption of the liquid litharge.

Theoretically, the cupellation can be operated that intensively that pure silver is already produced in the first step when the lead is completely oxidised. In practice, however, silver enrichments have to be carried out in several steps, which are becoming smaller and smaller step by step. Obviously, cupellation is a very lengthy and tedious, many hours lasting process.

Evidently, from the Greek time to modern times, the custom has evolved to extract silver from lead in an (at least) two-step process (Conophagos 1980; Pernicka und Bachmann 1983). That seems to apply to silver extraction in Laurion. According to this, lead is first converted into lithargite in a large cupel until the silver retains a residual content of about 5 wt.% lead. This is followed by a second cupellation in smaller facilities, the so-called fine burning.

This means that the cupellation is by no means a complete separation of silver from lead or litharge in a single step. Papadimitriou (2012, 2016) believes that for this reason, extraordinary masses of litharge were produced in Laurion, which, due to their high residual silver content, have been prepared and recycled in ancient times as an integrated step in silver extraction. The mechanical separation of silver-rich particles happened, as the finds of prepared lithargite show, both in the circular mills and in the plane washeries (cf. Sect. 5.9).

In this context, it is interesting to note that Pliny writes in his book (Hist. Nat. 33, 31; 33, 47), that the cleanest, finest silver would come from the ore deposits (or workshops) of Spain, long before silver production from other provinces became economically significant. The Projektgruppe Pliny (1998) notes a calculation that from an argentiferous lead containing 1 wt. % silver, 100 kg of lead have to be oxidised and removed as litharge to extract 1 kg of the precious metal. According to Pliny, in Roman times even lead ore with only 0.5 wt.% Ag was smelted, so that the effort was even greater. From Laurion, it was reported that lead ore with only Ø = 0.1–0.4 wt.% Ag was treated, which is much less than in the (gold) silver deposits of Sifnos or Thasos (summarised by Gowland 1917/1918).

During the cupellation, only after the oxidation of the last lead residue, mirror-like silver becomes visible on the bottom of the cupel. Smelters call this moment gleam of silver. Descriptions of the cupellation, the gleam of silver as well as the construction of cupels can be found in Pliny (Hist. Nat. 33, 95–110). He describes a seemingly paradoxical effect. He observed that at the end of a cupellation process the heavy silver would float on top of the lighter litharge ("... ut oleum aquis..."). This effect is easy to explain and could be exemplified by finds from Sardis (Ramage 2000) or from the medieval Johanneser Kurhaus (Harz, Germany) (Rehren and Klappauf 1995) (Fig. 5.66b). It is based on the sorption of the litharge in the porous material of the cupel, from which the liquid silver is separated.

Cupellation as an efficient, almost quantitative method of separating noble and base metals is still a common technique today and is called fire assaying. The only difference to the technical-industrial cupellation process, instead of stripping or pouring the litharge, is the absorption of the liquid litharge by a porous crucible material. Today, in metal technology, cupolas are made of highly refractory sintered magnesia (magnesium oxide, MgO), which, because of their porosity, absorb litharge very well.

If geochemically largely pure lead–silver ores are processed, then the cupellation causes comparatively little difficulties according to the described metallurgical steps. Silver cupelled with lead would contain a residual Pb-concentration of between 0.05 and 2.5% and small amounts of bismuth (Meyers 2003). Silver artefacts with less than c. 0.05 wt.% Pb could, therefore, be an indication of non-cupelled silver.

Pb contents in silver >0.05 wt.% could be the result of:

1. Cupellation of lead smelted from an argentiferous lead ore.
2. Cupellation of a non-argentiferous lead added to an impure silver from a silver ore.
3. Silver smelted directly from a silver ore containing small amounts of lead.

Hence, higher lead levels in silver are consistent with cupellation, but are neither definite proof of the use of cupellation, nor of the use of argentiferous lead ores.

However, if more complex composed lead–silver ores were used, such as those described in Sects. 3.4.9 and 3.4.10, problems can arise. Copper or nickel contents, for example, form refractory, solid oxides during cupellation, which do not dissolve in the litharge (Chemikerausschuss 1964). It forms firm skins on the surface of the liquid lead, which hinders the further access of oxygen and the oxidation of the lead falters.

According to the smelting experiments of Pernicka (1981) and Pernicka and Bachmann (1983), in a two-stage process of smelting silver-containing ores and subsequent cupellation, the elements As, Sb, Zn, Sn and, in contrast to the above-mentioned findings, also copper are removed almost completely. Due to the geochemical behaviour of bismuth (Bi) during cupellation, nearly all the lead must be removed from the silver before bismuth can be oxidised and separated from the silver effectively. An important element that does not change its concentration from ore to metal is gold. Pernicka (1981) states that ores from Laurion contain between 0.003 and 0.26 wt.% Au, and Gale et al. (1990) published Au contents of Greek silver coins of the fifth to first century BC (as products from Laurion silver) in the order of 0.0016–0.21 wt.%.

Other precious metals such as Ru, Rh, Pd, Pt, Os and Ir also transfer completely to silver, with Os being very sensitive to metallurgical processes under oxidising conditions due to its high volatility. Their concentrations are, depending on the analytical method used, often below the detection limit.

Since the physical separation of silver from litharge and slag almost never went perfectly, and these contained varying amounts of silver droplets, recycling over time of small but silver-rich waste from these materials can be expected. It is not clear whether such a process could be compared with modern refining. Rehren and Hauptmann (1995) have investigated finds of small silver crucibles (Fig. 5.16), some small, silver-rich slags and bottoms of cupels from a workshop of the Roman Colonia Ulpia Traiana near Xanten (Germany) and they postulated such a recycling process.

It is likely that such small crucibles were also used for other purposes. Craddock (1989, 1995) describes small crucibles with similar capacities to those from Xanten from the third century AD Agucha in Rajasthan (India). Large side openings at these crucibles were let for air supply by blowpipes, with which the material inside the crucible was liquified. He suggests that such crucibles were used for fire assaying.

In the late Middle Ages Lazarus Ercker (1556 AD) described in his *Kleines Probierbuch* crafting details of cupellation for the determination of silver concentrations in lead ores and lead in a small scale. These may be comparable with many cupellation processes in ancient workshops. According to Ercker, a well-balanced aliquot is used for the quantitative silver determination of argentiferous lead. For treatment of small amounts rich in noble metals, a regulus of lead (*Bleikönig*) will be produced as an intermediate product that will be treated by cupellation in a crucible. In both cases, there may be an adding of high-purity lead, if the content of impurities such as copper is too high.

Both processes are based on the same principles of reactions, namely the selective oxidation of less noble metals. What is different in the way in which these less noble base oxides are removed. This is unlikely to be apparent in the archaeological context. However, it is possible to differentiate between the individual varieties of lithargite, as listed in Sect. 5.8.

6.4.4.6 The Pattinson Process

Cupelling was further developed up until modern times. The Pattinson process takes advantage of the property of the Pb–Ag system which, on slow cooling of argentiferous lead in the cupel, initially pure lead is crystallising, while a relatively noble metal-containing melt remains in the liquid state. During cooling down to 304 °C a eutectic is reached, which solidifies with 2.5 wt.% Ag (Tafel und Wagenmann 1951). By skimming off the crystallising lead, the melt can consequently be enriched in silver. Forbes (1971) understands Pliny in a way that implies that the basics of the Pattinson process could have already been known by the Romans. However, any proof of this thesis is missing.

6.4.4.7 The Patio Process

The principle of silver recovery after the patio process is the formation of silver amalgam by the addition of mercury to finely ground ore. The method is detailed in Schnabel (1901). This process became known by *Bartolomé de Medina* in *Pachuca, Mexico* in the middle of the sixteenth century AD, and it was used for silver extraction from ore in Central and Latin America. The centre of the process was San Dimas, Durango (Mexico). The ores there consisted predominantly of silver sulphides with small amounts of silver chlorides. The patio process has been particularly well developed for ores with less or no sphalerite, iron and copper ores, and low concentrations of arsenic and antimony.

In simple stamp mills and/or Chilean mills, mined silver ore was finely ground with the addition of water. This resulted in a kind of sludge. This then was brought to the "patio", the yard of the silver workshop. The sludge was mixed with 2–10 wt.% sodium chloride. Several tons of salt were needed annually for this process. The workers turned the mass over several times and spread it evenly. Then they let their mules trample it for several days. The workers then added about 2% of roasted or burned copper-containing pyrite (*magistral*). Now the sludge was doused with mercury from canvas bags and again trampled by the mules. The mass was then washed out in wooden barrels or masonry pits, so that ultimately silver amalgam remained due to its high density. To separate the silver amalgam, it was heated. The volatile mercury evaporated and silver remained.

This process resulted in too low a yield and was also very energy intensive. Since there was not enough suitable fuel, the production was finally stopped, while the silver production in Europe gained importance in the eighteenth and nineteenth centuries AD. Also in Europe, a similar process gained importance.

Here, silver ore was mixed with a solution with 10% sodium chloride. However, this mass was roasted in furnaces and then finely ground. Subsequently, the ore powder was mixed with water, fine-grained iron ore and mercury. While at the Patio process in the Americas mules processed the mixture, due to the mechanisation in Europe rotating barrels fulfilled the work. Again, heating was necessary to vaporise the mercury, leaving behind pure silver. In this method, the yield was higher, but still too low. This procedure has also been finished soon too.

It is not unreasonable to think that metallurgical processing similar to the basics of the Patio process was already known in Europe since Roman times and that not only silver, but also gold could have been obtained by the amalgamation process (cf. Sect. 6.3.2). Pliny describes in his book (Nat. Hist. 33, 99) a method that could be interpreted as such a process (Projektgruppe Pliny 1998). In the Iberian Peninsula, the lead–silver deposits in the Sierra Morena are geographically not far from the mercury deposit. Mercury was known in Roman times.

6.4.5 The Metallurgy of Lead

Physicochemically, several process techniques have to be distinguished in lead production. It is commonly believed that the most widely distributed galena (PbS) has always been the most important raw material for the extraction of lead from sulphidic ore. The standard method from its main mineral galena is roast-reduction, i.e. the sulphide is converted to oxide which again is reduced to the metal. Considering Georg Agricola's book IX open-hearth constructions in numerous European smelters, where sulphidic lead ores were smelted, it is hard to imagine a strong reducing smelting process. Lead production is so simple that its principles can easily be transferred to early eras. An alternative method therefore the roast-reaction, because in the oxidation zone, galena is frequently weathered to cerussite ($PbCO_3$), so that a mixture of these ores can also be assumed as a possible raw material.

However, another oxidic source of lead, namely lithargite (PbO), which was produced locally in masses as a silver-containing waste product of the cupellation process (see Sect. 6.4 Silver) and was recycled as finely ground material, should not be overlooked (Papadimitriou 2012, 2016, 2017).

The substances involved in the chemical reactions (PbS, PbO and Pb) are rather volatile, which sets a limit to extremely high temperatures.

6.4.5.1 Reduction of Oxidic Ores

The metallurgically easiest method of producing lead from its ores is based on the ore cerussite. The carbonate mineral decomposes at temperatures of 300–400 °C to PbO and carbon dioxide and can then be reduced to metal by carbon monoxide:

$$PbCO_3 \rightarrow PbO + CO_2 \uparrow$$

$$PbO + CO \rightarrow Pb + CO_2 \uparrow$$

However, there are almost no lead deposits known that consist of pure cerussite, so that these reactions are more theoretical. In almost all ore deposits, cerussite is intergrown with galena in varying quantities. In such cases, the best metallurgical approach is a roast-reaction process.

6.4.5.2 Roast-Reaction

The roast-reaction process can be applied to rich and pure mixed lead ores with only minor amounts of gangue. It has the advantage that the production can be completed theoretically in one operation. Tylecote (1962) has shown that even under primitive conditions galena can easily be reduced to lead without previous roasting. Hetherington (1980) has reconstructed this in experimental work. Lead is more readily extractable than copper from the abundant chalcopyrite ($CuFeS_2$), so it is generally believed that galena was the first mineral smelted by mankind. The roast-reaction process can be carried out even with wood as fuel, it does not need charcoal.

In the roast-reaction process, the ore is first partially roasted in open furnaces and then immediately reduced in the same reaction vessel (Tafel 1929). At moderately high temperatures (450–700 °C) galena is partially converted to lead oxide and then partially converted to sulphate by the roasting gas SO_2 (Tafel und Wagenmann 1951; Kellogg and Basu 1960; Willies 1982). The most important chemical reactions that occur in this process are:

$$2\,PbS + 3\,O_2 \rightarrow 2\,PbO + 2\,SO_2 \uparrow$$
$$(T \leq 400 - 450^{\circ}C)$$

$$2\,PbO + 2\,SO_2 + O_2 \rightarrow 2\,PbSO_4$$

If more air is supplied to the fuel and the temperature is increased, lead oxide and lead sulphate can react with the remaining lead sulphide. Hereby metallic lead is produced.

$$2\,PbO + PbS \rightarrow 3\,Pb + SO_2 \uparrow$$
$$(T = 600 - 700^{\circ}C)$$

$$PbSO_4 + PbS \rightarrow 2\,Pb + 2\,SO_2 \uparrow$$

In addition, the roast-reaction has a share of direct reaction according to

$$PbS + O_2 \rightarrow Pb + SO_2 \uparrow$$

The decisive factor for the roast-reaction process is therefore the reaction between PbO and PbS. In this temperature range, the lead oxide initially forms as yellow massicot, which is stable at >488 °C. It can be used as a yellow pigment. Red coloured lithargite is stable below this temperature. As in the case of the roast-reaction of the copper ores, this process cannot be carried out in shaft furnaces, since in such a case a premature reduction of PbO and $PbSO_4$ would take place.

The formation of metallic lead at the generally low temperatures of this process, however, is not associated with the formation of much slag, as they usually are formed in the liquid state and are usually found at many smelting sites of lead–silver production. This requires much higher temperatures. Required are metallurgical operations, such as those carried out in shaft furnaces.

6.4.5.3 Smelting Sulphidic Ores/Roast-Reduction

Generally, the two-stage process of roast-reduction is the second option for lead production. The first step of this process, the roasting of sulphidic lead ore (galena), proceeds according to the same chemical equations as mentioned above. The sulphidic lead ore is converted to litharge according to the equation:

$$2\,PbS + 3\,O_2 \rightarrow 2\,PbO + 2\,SO_2$$

The litharge is then reduced with carbon to metallic lead:

$$PbO + CO \rightarrow Pb + CO_2 \uparrow$$

Desulphurisation of lead ore is so quick and easy (Tafel and Wagenmann 1951) that it would not be surprising if it would not have known in ancient times.

The aim of this smelting work is to win as much raw lead (with the appropriate levels of precious metals, i.e. gold and silver) and to separate the gangue and the host rock from the ore as slag that is as liquid as possible. In order to do so, a careful beneficiation of the ore is necessary: it has to be finely ground, even it would occur in massive mineralisations to enhance the reaction between ore and carbon gases in the smelting vessel.

PbO is reduced to metal by the gaseous carbon monoxide (CO) even at temperatures <200 °C in relatively high parts of a shaft furnace, much more than iron oxides. To reduce these to metal, it would require significantly stronger reducing conditions, as can be seen in Fig. 4.15. Ideally, FeO should also act as a slag-forming agent in lead slags, thus contributing to the crystallisation of fayalite.

Nevertheless, even in ancient lead metallurgy, over-reductions happened. This means that the setting of the gas composition during smelting processes could not always be controlled exactly and everywhere and the iron oxide present in the gangue of the lead ore was reduced to metallic iron. This also happened in silver-lead smelting on the island of Thasos in the Aegean Sea (Greece), where in addition to lead inclusions Hellenistic lead slags also contain massive, fist-sized pieces of metallic iron (Hauptmann et al. 1988). These are the so-called hardheads or salamander.

6.4.5.4 Distribution and Behaviour of Elements

Pernicka and Bachmann (1983) investigated the distribution of main, minor and trace elements, exemplified for the lead–silver production at Laurion. When smelting silver-containing galena, chemical elements of the ore are changed due to two processes, namely the production of the raw lead and the subsequent silver extraction from the lead by cupellation.

In the first process (if slag formation occurs) slag-forming oxides such as SiO_2, Al_2O_3, Na_2O, K_2O, CaO, MgO, FeO, MnO, but also ZnO and possibly PbO are separated from metallic lead. The metals Cu, As, Sb, Bi, Sn and the precious metals Ag, Au and Ir will remain in the raw lead. If matte or speiss are produced, then these will also contain considerable concentrations of these elements. According to Tafel and Wagenmann (1951), bismuth is likely to accumulate in the raw lead during the smelting process. As and Sb accumulate in the speiss.

Silicate slags accumulating during the smelting of lead ore may easily reach lead concentrations >10 wt.%, sometimes even up to 30 wt.%. Some examples are those from Greece (for Siphnos and Thasos, see Hauptmann et al. (1988), for Laurion see Conophagos (1980)) from Spain (for Linares see Maréchal (1985)). They are discussed in Sect. 5.5.

For optimal processing, neither gangue material nor furnace wall should contain more than a few percent SiO_2, since PbO produced in a furnace tends to form easy melting lead silicates at higher temperatures. As shown in the system $PbO–SiO_2$ these lead silicates have low melting points in the temperature range between 700–800 °C (Geller et al. 1934). They “glue together” the batch glowing in the furnace, so that furnace gases cannot penetrate it and the lead ore cannot react any further. Although galena itself melts at 1114 °C, it is plastic and cohesive at furnace temperatures. In addition, there is a rapid resorption of the furnace wall.

Today lime or limestone is added for smelting of lead ores. Limestone decomposes to CaO at 850–950 °C. The solubility of lead oxide in the slag decreases as the content of CaO in the charge increases, that is, the formation of lead silicates in the slag is prevented, the amount of metallic lead is increasing. At the same time, the addition of CaO prevents losses of silver in the slag. The metal is enriched in lead. This was reproduced experimentally by Pérez et al. (2012).

This technological trick possibly resulted from experiences of the past. However, it cannot be proven whether the old metallurgists deliberately used lime in lead production. CaO contents in lead slags are not reliable indications of this. Rather, it should be taken into account that lead ore deposits (or Pb–Zn deposits) are often bound to carbonate rocks, at least in central Europe (e.g. Laurion, Sierra Morena, Siphnos, Ibiza, Kosovo, Upper Moesia, Alpine deposits; see Sects. 3.5.4 and 3.5.5). This means that together with the lead ore lime, dolomite ($CaMg(CO_3)_2$) and in karstic cavities also limonite (FeOOH) or their metamorphic transformation products can get into a charge. Sphalerite (ZnS) might also be added.

6.4.5.5 Smelting Furnaces

Archaeological finds and findings on metallurgical installations of lead extraction are extremely rare, in contrast to those of iron or copper metallurgy. Only slags and remains of cupels can be found (Sects. 5.4 and 5.8). Slags, however, an unmistakable sign for lead–silver production exist from pre-Christian times and are prove for proper smelting processes in shaft furnaces at temperatures of $\geq 1000\ ^\circ C$.

There are no roasting facilities known, as dug up, e.g. by Goldenberg (2004) from the Middle Bronze Age copper production at Jochberg (Austria). Remnants of smelting furnaces or shaft furnaces are also hardly appropriate to make realistic and reliable reconstructions of smelting furnaces, as Conophagos (1980) dared to do for the Hellenistic lead extraction in Laurion.

Worth mentioning is perhaps remnants of a lead smelting furnace from a late Roman villa at Scarcliffe Park, Duffield, Derbyshire-Nottinghamshire. They consist of simple, low, rectangular stones (Lane 1973, 1986; Willies 1982) (Fig. 6.20). This simple construction may have been maintained for a long time, at least in Great Britain. Tylecote (1986) published a similarly simple construction from the medieval Gunnerside, York (after Raistick 1927), which was retained in principle in many later experiments as a basic architectural unit (Hetherington 1980; Anguilano et al. 2010a,b). Also, Gowland (1900) suggests comparatively low, rounded formations of stones for the smelting of lead ores for the Greco–Hellenistic period.

Pictorial representations and descriptions of simple hearth constructions in which wood was used to smelt lead ores to metal at low temperatures can be seen in Georg Agricolas (1556, ninth book), but solely for Europe.

Interesting in this context is a particular type of furnace constructed from South America. At Potosi in Bolivia up until the sixteenth century for the smelting of silver-rich lead ores, the so-called Huayrachina, which was constructed differently (Van Buren und Cohen 2010, more literature there). This approximately 2 m high smelting furnace was characterised by a strongly perforated furnace wall, it was a wind furnace (Fig. 6.21). This smelting furnace shows very clearly how the smelting of lead ores can proceed according to the roasting reaction process under an excess of oxygen. Huayrachina furnaces have been in operation since pre-Columbian times. But they are still in use today in Bolivia when lead and silver are produced in small-scale artisanal productions (Rehren 2011).

Fig. 6.20 Simple lead smelting hearth associated with a late Roman villa at Scarcliffe Park, Duffield. Now in the Peak District Mining Museum, Mathlock Bath. From Craddock (1995)

6.5 The Metallurgy of Tin

With the onset of the Bronze Age in the Old World in the third and second millennium, after the initial attempts with arsenical copper, tin (chemical symbol Sn) was the most important and most sought-after utility metal next to copper. However, Bronze Age tin production is one of the most important research desiderata of early metal extraction, one of the hardest-to-grasp problems of archaeometallurgy. Where does the tin come from, where are the raw material sources of tin? How was tin smelted from its ores? And how was tin bronze made?

Like copper and other metals, tin got its name in Roman times. The name originates from the Latin *stannum*, also named as *plumbum album, plumbum candidum.* The Greek word is *κασσίτερος*. Tin is a silvery shining metal. Due to its extraordinary softness, it is usually associated with the planet Jupiter, more rarely Venus. It is extremely elastic and can be rolled out well into thin films. Even under prolonged storage and exposure of air, it does not dull. The melting point is very low. It is even lower than that of lead (Table 6.3).

6.5.1 Tin Artefacts

6.5.1.1 Rarity of Artefacts

Given the technological ease with which even Gowland (1899) observed how tin was smelted in Japan in the nineteenth century AD, the pronounced rarity of tin objects from prehistoric eras is surprising. The earliest tin object is a finger ring from Thermi/Lesbos dated to around 2600 BC (Thermi IV). It consists of twisted tin wire (Lamb 1936). It contains 76.3 wt.% tin and 23 wt.% iron (Begemann et al. 1992). It is not clear from their investigations whether the iron content is based on a corrosion crust or whether it is a tin-hardhead. In contrast, Yalçın (2016) published 2-cm large corroded metal fragments from the tombs of Alacahöyük (first half of the third millennium BC), which according to analytical measurements consisted of comparatively pure tin.

In view of the mass production of tin bronzes with the onset of the developed Bronze Age, this is surprising (Muhly 1973; Roden 1985; Penhallurick 1986). Overall, there is a considerable overweight of copper and bronze objects over the sparse finds of tin metal at the early

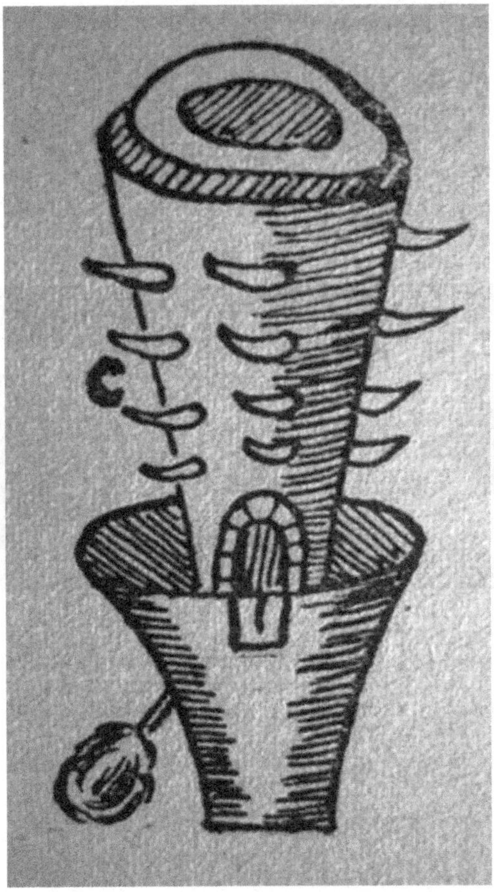

Fig. 6.21 Huayrachina furnace for the smelting of silver-bearing lead ores in Bolivia after Alonso Barba (1640). The drawing shows flames coming out of the numerous wind holes and the lead flowing out of the furnace. From Rehren (2011)

stages of the Bronze Age. The massive occurrence of marketable tin metal in the form of ingots in Central Europe is mainly limited to Iron Age and Roman contexts.

The tiny quantities of tin metal found led Charles (1975) to suggest that perhaps tin metal was not normally produced at all, but instead the beneficiated tin ore could have been co-smelted with copper ore or added straight to the molten copper and an alloy was produced directly.

On the other hand, a possible explanation for the situation has long been the so-called tin disease. This means that the usual, silver-white β-Sn solidified from the liquid decomposes at low temperatures (<13.2 °C) with increasing volume to a grey tin (α-Sn), which in turn decays to a whitish-grey powder.

$$\beta\text{–Sn} \rightarrow \alpha\text{–Sn} + 25\ \text{vol.}\%$$
$$(13.2 - 161^\circ\text{C}) \qquad (< 13.2^\circ\text{C})$$

This reaction is based on a crystallographic reorientation of tetragonal β-Sn to cubic α-Sn (diamond lattice). Another modification is rhombic γ-tin, which is stable above 161 °C (brittle tin). The reaction is slow, but can be accelerated by short-term cooling to −50 °C and then continues by itself. The tin disease begins by crystalline nuclei and forms pustules on the surface of the object. Theoretically, this may well have led to the destruction of tin artefacts over archaeological periods. On the other hand, decomposition of the tin is slowed down or prevented by Pb and Sb contents in tin artefacts, the so-called inhibitors. When bending a β-Sn rod, the mutually rubbing crystals cause a crunching sound, the "tin crackling".

In the middle and second half of the second millennium (Late Bronze Age, Early Iron Age), finds of tin artefacts become somewhat more common. Two tin objects date from the 18th Dynasty in Egypt (Lucas 1928). In western and northern Europe's Bronze Age small tin beads were found as well. For other finds from England, the Netherlands, northern Germany and Denmark it is not always sure if they are made of tin or possibly tin–lead or (tin plated) bronzes. Roden (1985) compiled a summary of this. Pernicka (1998) has written about the spread of tin bronzes in the Old World (see Sect. 7.2.3).

6.5.1.2 Bars and Ingots

As a possible explanation for the general missing of tin bars or ingots in the European early Bronze Age, Craddock et al. (2007b), like Charles (1975), suggests that already in the immediate vicinity of tin deposits and mines refined tin ore could have been added directly to copper smelts to produce bronzes, so that almost no tin itself ended up in supra-regional trades.

Table 6.3 Main physical properties of tin in comparison to lead. The hardness is given according to the Mohs scale

	Sn	Pb
Density (g/cm^3)	7.31 (ß-Sn)	11.34
	5.57 (α-Sn)	
Hardness	2.2–2.5	1.2
Melting point (°C)	213.9	327.5
Boiling point (°C)	233.7	174.9

Especially large amounts of tin ingots have been found in Late Bronze Age wrecks of merchant ships. The discovery of 99 tin ingots (weight c. 1 ton) or their fragments from the shipwreck of Uluburun (at Kas, southern coast of Turkey; Pulak 2005) changed the picture abruptly and prove that the metal was actually negotiated on a large scale as such in the Mediterranean (Pulak 2005). Very similar to copper ingots also the tin ingots were predominantly shaped in the form of oxhide ingots (Fig. 6.22). The findings of Uluburun triggered a wave of research activities. Nevertheless, even today it is not always possible to reliably determine the origin of these tin ingots by means of complex analytical methods such as trace element analysis and lead and tin isotope analysis (Sects. 11.2 and 11.3). However, it is clear that in ancient times tin was used predominantly for the production of tin bronze and only later for the tinning of copper artefacts.

Fig. 6.22 Uluburun, Tin ingot, Inv. no. KW 2922, Bodrum Museum, Turkey. Typical design of an oxhide shaped tin ingot. At the surface, the ingot is heavily corroded by the formation of calcium carbonate. Inside, it consists of very pure tin, as analysed by Hauptmann et al. (2002). Photo: G. Weisgerber (†), Deutsches Bergbau-Museum Bochum 1998

A larger cargo of copper and tin ingot was found in the Late Bronze Age shipwreck found off the coast of Cape Gelidonya near Kas (southern Anatolia). It dates back to around 1200 BC (Bass 1967). The rectangular tin ingots loaded on the ship can sometimes be extremely corroded, with the tin often being significantly substituted by calcium. This is due to oxidation and displacement of the tin by Ca-hydroxide, which is subsequently converted to carbonate. For example, tin ingots from Cape Gelidonya contain 71 wt.% CaO and 13.9 wt.% SnO_2 (Dykstra 1967).

Also, a few ingots were found in a shipwreck from the thirteenth century BC off the coast of Haifa (Kfar Samir) in Israel (Galili et al. 2013, 2015), which had loaded fragments of copper and tin ingots. Amongst the metal bars were 5 plano-convex and differently shaped tin ingots with weights between 2 and 4 kg. From here comes a rectangular tin ingot from the Iron Age, which bears a representation of the goddess Arethusa (Artzy 1983).

At Foraxi Nioi/Nuragus (Sardinia, Italy) in 1882, a Late Bronze Age complex of copper ingots, bronze scraps and about 10 kg of 1–3 cm pieces containing tin contents of about 76 wt.% was found. These pieces were addressed as cassiterite (SnO_2; Cambi 1959). There is no clear information on any finds of tin. In case that actually cassiterite was found, then it could be concluded that Furaxi Nioi's blacksmiths may have created bronze by co-smelting cassiterite in liquid copper. However, the published tin content may also be the value of a tin corrosion product, such as the finger ring of Thermi/Lesbos or the tin ingots of Cape Gelidonya.

In the western Mediterranean 18 tin ingots have been found near Port Vendres (south coast of France). They have typical handles and weigh between 3 and 10 kg. They date back to around

40–50 AD and are probably from the Iberian Peninsula (Colls et al. 1975). Two tin ingots of Roman age were found at Cap des Mèdes, east of Poquerolles (each about 30 kg) (Tchernia 1969) and several bars were discovered near Cabrera off the Balearic Islands (approx. 80 kg in total) (Aramburu-Zabala 2001).

6.5.2 Ancient Metallurgical Processes

Archaeological finds for the reconstruction of tin extraction or smelting are extremely rare. Thorndycraft et al. (2004) pointed out, that evidence of prehistoric tin mining even in the world-famous tin district of Dartmoor (southwest England) would be equivocal. Alluvial gravels would not have left any traces for mining and smelting sites also due to reworking in later periods. Furthermore, very little slag material was produced during smelting (Tylecote et al. 1989).

Therefore, any approach to decipher prehistoric metallurgical processes is possible only from ethnographic and traditional findings and experimental studies. Only a few tin mines can be dated securely, such as Askaraly in Kazakhstan, Karnab (Alimov et al. 1998; Cierny et al. 2005) and Čangali (Uzbekistan), Mušiston (Tajikistan) in Central Asia (Garner 2014). They date to at least the first half of the second millennium BC (Andronovo-Tazabag'jab period), but are probably even older. Numerous archaeological finds are proof for the prehistoric use of placer tin in Cornwall (Penhallurick 1986). Of the 135 tin deposits in Afghanistan which are geologically recorded (see Sect. 3.5.6) nothing is known.

It is generally believed that the exploitation of tin placers preceded mining in solid rock. Given the frequency of tin placers, this hypothesis makes sense, but probably represents a simplified view (see Garner 2014). Even since the Neolithic, mining technologies to exploit raw materials in solid rock have developed, so there is no reason not to postulate an early exploitation of tin deposits in primary deposits. Although tin ores in solid rocks are usually finely distributed, they are forming impregnations, i.e. they form so-called *Greisen* and *Zwitter*, so that a significant beneficiation of tin ores to reach at least the lower percent range for metallurgical tin production would have been necessary. However, this was perhaps also necessary for the extraction of tin ores from placers. Although their concentrations were higher than in gold in placers, real nuggets of cassiterite nuggets were probably only found in exceptional cases.

However, extraction of fine-grained tin ores is or was possible with simple means. In their experimental work on tin mineralisation of Kestel (Anatolia), Earl and Özbal (1996) have demonstrated a simple washing out of a heavy mineral concentrate on a paddle ("vanning") in a manner similar to that used in England up until recently for the determination of tin contents in the ore.

Fundamentals of the smelting of tin ores can be found in Pryce (1778), Louis (1911), Tafel (1929), Earl (1985, 1986) and Tylecote (1987).

6.5.2.1 Tin Smelting Furnaces?

Neither prehistoric eras nor antiquity has evidence of smelting furnaces in which tin ores of any kind could have been smelted. This could be explained by the following observations from more recent times. From Japan and south-east Asia Schnabel (1890) and Gowland (1899) describe examples of very simply constructed tin smelting furnaces, which were still in operation in the nineteenth century AD. These are simple, flat depressions set into the ground with dimensions of only about 30 × 30 cm. A tuyère was placed over it and a superstructure (vent) was constructed (Fig. 6.23). The smelting of tin must have been of the simplest kind under not strong reducing conditions. This primitive method was not a very efficient or economic one. Pryce (1778), describes similar constructions and smelting processes of tin smelting. It was just a ditch in the ground, and tin ore was thrown on a charcoal fire which probably was excited by a bellow. Metallic tin was precipitated in small lumps and prills, a resmelting was necessary. The consequence was a huge consumption of charcoal, and a great loss of metal prills and ore.

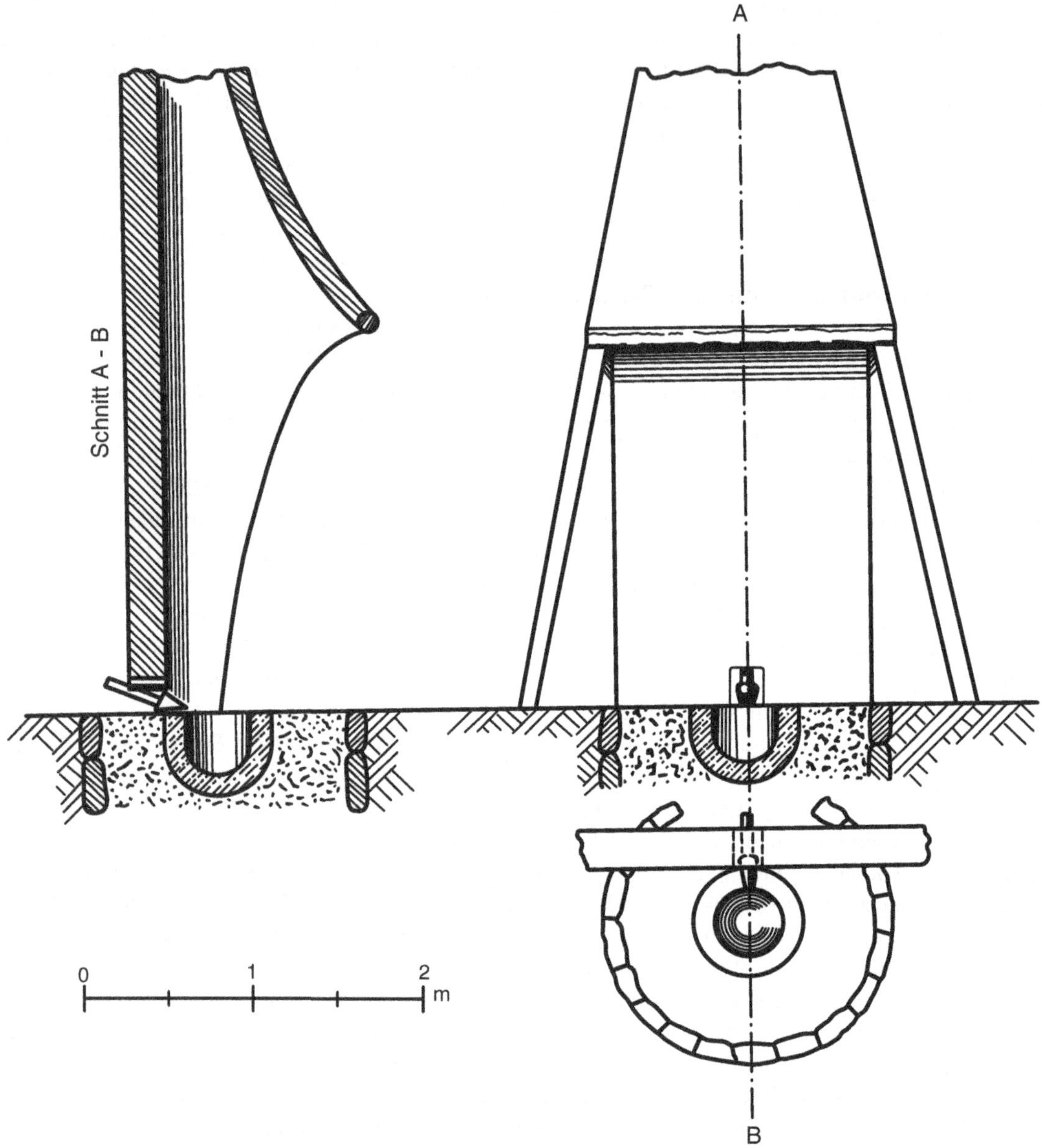

Fig. 6.23 Smelting furnace for tin from Japan as described from Gowland (1899). This type of furnace was also used for copper smelting

In the event that this type of tin smelting was operated even in prehistoric times, it is difficult to find exactly verifiable evidence for it: in contrast to copper prills, tin prills do not even lead to greenish weathering products and are difficult to find.

By contrast, images of shaft-shaped furnaces (blast or bloomery furnaces) are from the early modern period (e.g. in Georg Agricola 1556) (Fig. 6.24). The furnaces are very tight, but they were fired only by low burning flames and low air supply. This tin smelting also leads to the

Fig. 6.24 Blast or bloomery type furnace for smelting tin in Germany from Georg Agricola (1556, IX book)

formation of (tin-rich) slags, which are repeatedly smelted to extract the contained tin.

It can perhaps be assumed that in ancient times tin production was carried out in rather limited scales. The remelting of prills, lumps and slags was carried out in furnaces, but probably also in crucibles, which can certainly be compared to a fire assaying, as described in detail by Georg Agricola (1556, book IX). Earl (1985/1986) carried out experiments in small, about 10 cm high crucibles, whereby very little slag was obtained. In the attempts he used the vanning on a paddle for the enrichment of the tin ore, the melting of the ore itself in the crucibles is nothing else than a fire assaying.

6.5.2.2 Tin Crucibles

Archaeological finds of crucible fragments in context with tin metallurgy only exist in two localities, namely Göltepe (south-eastern Anatolia), and Askaraly in Kazakhstan.

The crucibles from the Early Bronze Age context of Göltepe have been described by Yener and Vandiver (1993). Yener (2009) emphasises in her work on the prehistoric tin mine of Kestel and the nearby settlement Göltepe that several thousand of crucible fragments were found there, so that a very extensive extraction of tin in small-scale reaction vessels was carried out. Yener and Vandiver (1993) emphasise that the crucibles were only covered by mm-thick slag layers on the inner edge.

The composition of the crucible of Göltepe is otherwise comparable to those of Askaraly II, which were analysed by Hauptmann (in: Stöllner et al. (2013a). They come from the cultural layer of the settlement Mastau-Baj (Andronovo-Fedorovka culture) and date back to the second half of the second millennium BC (Garner 2014), so they are younger than those of Göltepe. They were made of slightly leaned, ferrous ceramics and showed a characteristic, only mm-thick, whitish-coloured, pitted-glassy slag layer on the inside. Semi-quantitative analyses under the scanning electron microscope in elemental distribution images show a significant accumulation of tin in the slag crust (Figs. 5.14 and 5.15). The tin contents are based on finely dispersed crystals of cassiterite (SnO_2) and droplets of tin in a glassy matrix, which also contains submicroscopic particles on SnO_2 and tin. The cassiterite is not interpreted as a relic of the primary ore that was used, but is instead crystallised from a silicate melt. In the slag crusts, a further main constituent is Ca–feldspar anorthite ($CaAl_2Si_2O_8$). SnO_2 accumulates in the glass phase, so that in tin slags, a frequent and abundant formation of Sn-containing silicate glasses (with up to 50 wt. % SnO_2) with CaO and FeO contents is observed. High SnO_2 contents cause a similar coloration as the white coloured tin glazes, as they have become known from the Renaissance Maiolika. Surprisingly, the ostensible tin crucibles from Göltepe near Kestel, Turkey, show a very similar phase inventory (Yener and Vandiver 1993).

6.5.2.3 Smelting of Tin Ores

On the basis of these observations, the reduction of the tin ore cassiterite offers two possibilities for the formation of metallic tin and, at the same time some other Sn compounds:

1. By reduction of the cassiterite according to

$$SnO_{2\ solid} + CO \rightarrow SnO_{solid} + CO_2$$

$$SnO_{solid} + CO \rightarrow Sn_{liquid} + CO_2$$

2. By disproportionation of Sn^{2+} dissolved in a slag according to

$$SnO_{solid} \rightarrow SnO_{dissolved}$$

$$2\ SnO_{dissolved} \rightarrow Sn_{liquid} + SnO_{2\ solid}$$

In case tin oxide is mixed with siliceous gangue or host rock components either from solid rock or from a placer deposit, or from crucible ceramics, under reducing conditions required for tin smelting the SnO formed enters the glassy structure of a vitreous slag as a network forming component at concentrations >34 mol% (Carbó-Nóver and Williamson 1967; Karim and Holland 1995). During freezing of the liquid slag reoxidation or disproportionation of the Sn^{2+} to Sn^{4+} and Sn^0 occurs, and close to insoluble secondary SnO_2 metallic tin precipitates and a variety of crystalline compounds such as olivine (fayalite), feldspar and in particular spinels of a rich compositional variability (Ehrt et al. 2001).

$$2\ Sn^{2+} \rightarrow Sn^{4+} + Sn^0$$

The reaction has been confirmed experimentally to occur between 350 and 700 °C in SnO–SiO_2 glasses (Carbó Nóver and Williams 1967). According to Platteeuw and Meyer (1956) in the temperature range 300–1127 °C liquid tin + solid SnO_2 + gaseous SnO are coexisting as stable phases. The metallic tin formed by the disproportionation reaction shown is spread throughout the glassy slag in tiny spherical prills only a few micrometres across. All Rooiberg tin slags in South Africa contain these tiny tin prills and needles of SnO_2 (Grant 1994; Heimann et al. 2010), and the slag incrustation of the crucible from Askaraly contain them also (Fig. 5.15).

6.5.2.4 Hardheads

Because cassiterite is a very stable compound due to the high oxygen affinity of the tin, the reduction to metal requires a relatively very high reduction temperature and comparably strongly reducing conditions; they are thermodynamically similar to those of iron oxide—iron (Fig. 6.25) and thus somewhat more difficult to accomplish than those of copper. The reaction proceeds vigorously at about 900 °C, but the SnO_2 is reduced more slowly than FeO. In the case that tin ore is associated with Fe-bearing ores, there is always a risk of iron precipitation. This is the real difficulty in producing tin from its naturally occurring ores.

The diagram in Fig. 6.25 indicates clearly, why Wright (1982) wrote that tin smelting would be "an art whose object was to produce tin without producing iron". Iron ores such as magnetite are amongst the frequent contaminations of tin, at least in placer deposits. If iron oxide is present in a tin melt, either from the ore or added perhaps as a flux, the reaction rates of SnO and FeO with reducing agents are similar enough to make it very difficult to reduce all of the SnO without reducing some FeO. If iron is "pushed" into the furnace metal by strong reducing conditions, in order to recover as much tin as possible, then undesirable hardheads (SnFe alloys such as Fe_2Sn, Fe_3Sn_2, $FeSn_2$ and FeSn) will be formed (Smith 1996). Higher levels of impurities in tin can also be caused by co-reduction of foreign metals associated with the ore, such as Ni, Co, Pb, As, Sb and Bi. When smelting tin ores containing arsenic, the arsenic gets into the tin. Therefore, arsenic contents in prehistoric tin–bronzes can perhaps be traced back to the ore base. Basically, arsenic

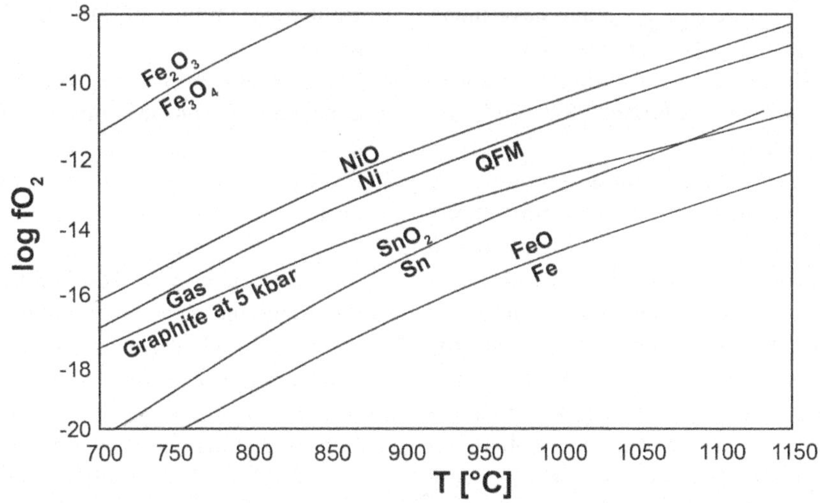

Fig. 6.25 Temperature—log fO_2 plot showing the Sn^0—SnO_2 pair compared with other geologically relevant oxygen buffer equilibria. The carbon—CO gas equilibria after French and Eugster (1965), and QFM after O'Neill and Wall (1987). Note the close vicinity of the Fe^0—FeO buffer equilibrium which clearly shows the ease of the precipitation of iron at the smelting of tin. Tin is stable in a slag containing fayalite. Abbreviations: Q = quartz; F = fayalite; M = magnetite. From Paparoni et al. (2010)

can be removed systematically before smelting by roasting. Arsenic levels in tin ores played a significant, problematic role in the historic extraction of tin in Cornwall (Sect. 3.5.6). However, this may not apply if pure, carefully prepared cassiterite is smelted. Begemann et al. (1999) have found that tin ores, especially pure cassiterite, contain extremely low levels of trace elements.

That may also apply to the tin ingots of Uluburun, which are made of very pure tin (Hauptmann et al. 2002a). Most early tin ingots notably from the eastern Mediterranean are remarkably pure. The iron concentrations in these ingots were mostly <0.012 wt.%. Significantly, hardhead has apparently not been encountered in the early ingots. Another reason for this in fact could have been that the very simplest non-slagging processes (low temperature, short smelting) were applied and these were not reducing enough for hardhead to form.

6.5.2.5 Tin Slags

A general difficulty in the smelting of tin ores embedded in siliceous gangue or host rock is the formation of SnO_2-rich slags, because the reduction of SnO_2 proceeds via the formation of SnO, which forms Sn-silicates in the presence of silicates:

$$SnO + SiO_2 \rightarrow SnSiO_3$$

Stannous silicate slag is one of the principal sources of tin loss (Smith 1996). The tin content in analyses of tin slags are often noted to be rather high. At Rooiberg, South Africa, Iron Age tin slags show between 10 and sometimes up to 50 wt.% SnO_2 (Miller and Hall 2008). The higher values indicate that tin smelting was inefficient. It is likely that such slags were recycled.

However, by adding CaO (and Al_2O_3), the SnO can be released from silicates according to the following equation:

$$SnSiO_3 + CaO \rightarrow CaO \cdot SiO_2 + SnO$$

Due to this reaction, slagging is better targeted with CaO to produce slags low in tin. In modern metallurgy, careful beneficiation and concentration of the ores are important to minimise the amount of slag produced. From early modern metallurgy and later it is known that tin slags of the first reducing smelting process often contain

10–30 wt.% Sn in metallic form and have therefore been systematically reprocessed (Tylecote et al. 1989). However, this also means that very pure concentrates of cassiterite (78.8 wt.% Sn) must have been used.

Given the fact that at equilibrium, some tin oxides always enter the slag, it is not entirely obvious why a slag-producing furnace would have produced a better yield of tin than a lower temperature smelt where no slag is produced (Grant 1994). In the latter case the gangue minerals, together with unreacted ore and charcoal, do not form a slag as such but rather a sintered matrix from which the molten tin separates by gravity. It would seem that recovery of tin at low temperature would be at least as efficient as in a slagging furnace. Perhaps this is precisely the reason for the astonishing rarity of prehistoric slags and the absence of tin smelting furnaces.

It has already been pointed out above that, as an alternative explanation for the very sparse finding of tin slag from the Chalcolithic or Early Bronze Age periods in Europe, a direct addition of tin ores into a copper bath for the production of bronze is considered. Following experimental work (Rostoker et al. 1983), this is just as possible as a melting together of metals. However, finds that have been examined so far, for example those of crucible fragments from Raquel Vilaca, Portugal (Merideth 1998a) are not enough to substantiate this thesis sufficiently. However, the production of tin in crucibles with subsequent mechanical separation by smashing the slag is ethnographically (Gowland 1899) and experimentally proven and was also known in England in the sixteenth century AD (Earl and Özbal 1996).

6.5.3 Tinning

In an overview, Meeks (1986) concludes that high concentrations of tin on the surface of copper or bronze artefacts can arise from three principal ways:

1. Deliberate tinning of bronze by wiping or dipping.
2. "Tin sweat" or inverse segregation, i.e. tin concentration on the metal surface during casting.
3. Selective corrosion of high tin bronze during burial.

In these three ways, different intermetallic CuSn compounds are formed, depending on the temperatures used. Most common are the ε-phase (Cu_3Sn) and the η-phase (Cu_6Sn). At any post-tinning heat treatment at high temperatures, the δ-phase ($Cu_{31}Sn_8$) and an $\alpha + \delta$ eutectic appear. Meeks (1986) studied experimentally these individual phases of the tinning effects and compared the results with archaeological samples by crystallographic-metallographic analysis and scanning electron microscopy.

The earliest tin layers on CuAsNi artefacts are described by Yalçın and Yalçın (2008). They report 8–10 μm thick tin layers on presumably Late Chalcolithic/Early Bronze Age spearheads of Tülintepe, Anatolia. They date to the time between 3400 and 2900 BC. Details of the tinning are not possible because information on tin phases are missing.

Tinned Etruscan and Greek mirrors date back to the fifth century BC, in Roman times tinned bronzes are widely used in the military field (helmets) and household, also as SnPb alloys for cooking pots. The earliest historical reference for tinning, however, is found only in Pliny (Nat. hist. 34, 162).

The practical techniques of tinning were published by Hammer (1998).

Because of its low melting point (232 °C), because of its resistance to (inorganic) acids and atmospheric influences to protect bronze vessels from corrosion, or for decorative reasons, tin has been used to surface-coating of copper and bronze artefacts, and in modern times on steel sheets and for producing containers in the food industry (cans). Tinning leads to a greyish to silvery metallic shining surface.

It is a simple grading up process of copper and bronze. Prerequisites are metallically bright, oxide-free surfaces of the objects to be tinned. Since lead was more readily available than tin, both metals have often been mixed. As a result, the melting point was even lowered (eutectic at 183 °C at 61.9 wt.% Sn). In Roman times up until the Middle Ages SnPb alloys of the ratio 80:20 wt.%, that is pewter, were used. By adding lead, however, the wetting is worsened because copper and lead are not soluble in each other. In contrast, copper is soluble in tin, so that tinning by immersion in a liquid tin bath leads to a complete tinning of the surface, but copper can go into solution by the formation of intermetallic compounds.

Liquid tin can also be applied to previously heated objects by means of a rag or a brush. Tinning leads to a grey coating with a slight shade of blue. Tinning can be followed by a heat treatment in a reducing atmosphere to produce a hard, shiny and silver-like layer as described by Pliny. He calls these objects "incoctilia". In this case, the hard intermetallic compounds such as Cu_6Sn_5 (η-Phase) and Cu_3Sn (ε-Phase) are formed by diffusion processes of the copper into the tin layer and vice versa. Such layers of white metal or pewter on the surface of "specula" are already described from Roman mirrors (Meeks 1995) and from the Merovingian period (coating 30–40 wt.% Sn, base metal c. 6 wt.% Sn; Raub 1986).

6.6 The Metallurgy of Iron

The transition from the Bronze Age to the Iron Age, and thus the dissemination of iron (chemical symbol Fe; for simplicity's sake, we use the name iron for the complex chemical–metallurgical Fe–C compounds) took place gradually in different regions of the earth's and in various epochs. It never was an abrupt change. Nonetheless, there was a growing interest in this new and very different kind of material, which had a strong influence on people's lives at the time. Technology, development and dissemination of iron metallurgy are presented in impressive overviews in the works of Nicodemi (2004), titled "The Civilisation of Iron. From Prehistory to the Third Millennium", as well as in Buchwald (2005) "Iron and steel in ancient times" and in Pleiner (2000, 2006) "Iron in Archaeology. The European Bloomery Smelters" and "Iron in Archaeology. Early European Blacksmiths".

Iron ores (mostly in oxidic form) were and are abundant worldwide and are available in massive lots. Iron was more difficult to produce than the previously known metals copper, gold, silver and their alloys due to its extremely high melting point and the required more reducing gas atmosphere. Those were produced as melts and poured in the liquid state. Iron was produced in the solid state during smelting, hot forged and cooled down at different speeds and tempered according to specific desired properties. In contrast to the metals mentioned, it could not be remelted. So it was probably the frequency of occurrence and the easy accessibility of the iron ores, which were finally responsible for an increasing distribution of this new material in the first millennium BC.

There is also a compelling model example of how iron metallurgy may have inadvertently evolved from early copper metallurgy. Roman (1990) and Gale et al. (1990) report that in Timna and other localities in the southern Arabah (Israel), roughly decimetre sized prehistoric copper lumps have been found, which, after further investigation, proved to be biphasic copper–iron alloys. Hauptmann (2007) also reports copper–iron alloys from the Iron Age Faynan (Jordan).

Such alloys were certainly undesirable wastes of copper smelting, which had two causes.

The first one is that the oxidic copper ores at the copper deposits of Timna and Faynan are not only found as the greenish secondary minerals such as, e.g. malachite, but that they frequently occur as reddish "tile ores". These "tile ores" are sandstone impregnated and cemented mixtures of cuprite (Cu_2O), limonite (FeOOH) and malachite. They were used as early as the fourth millennium BC in the Wadi Arabah area (Hauptmann and Wagner 2007; Hauptmann 2007). In general, copper ores are often associated with iron contents.

Also, chalcopyrite ($CuFeS_2$) plays a dominant role in the Bronze Age copper metallurgy worldwide and is thus a suitable starting material for the metallurgy of iron.

The second is, that smelting processes conditioned on copper were occasionally done under too reducing levels, so that any iron content in copper ore was reduced to metal too. This has also happened in early smelting furnaces which were higher constructed than those shown in Fig. 6.5. The course of the equilibrium curves Cu/Cu_2O and Fe/FeO in the Ellingham diagram (Figs. 4.15 and 4.17) shows that reducing conditions for making copper are not in a great distance to those necessary for making iron. Hence, the thermodynamic conditions have been already available.

6.6.1 Iron Artefacts

Apparently, at least in Central Europe, iron was first imported before the use of the ubiquitous iron ore deposits themselves began. This import happened in the form of ingots and bars. Only later did the knowledge of smelting of iron ore from the south reach this region, and it was not known before the Hallstatt or La Tène period, although rich iron ore deposits were available everywhere.

The earliest iron artefacts date long before the beginning of the Iron Age and come largely from Anatolia and the Middle East. It is believed that the Hittites played a pioneering role in early iron metallurgy (Yalçın 1999). He is correctly interpreting that both finding circumstances and written records indicate that the reduction to malleable iron from its ores and even the production of steel was already known before the beginning of the Iron Age. An overview of pre-Iron Age iron finds in Anatolia is compiled in Table 6.4. It is largely unknown whether the earliest iron artefacts were formed from meteoritic iron or produced by a smelting process, because in most cases no analytical studies are available (see Waldbaum 1999).

Most of these very early artefacts are precious symbols of power and jewellery. It was not until Tell Açana that iron was used to make weapons. All in all, this means a development from luxury material to the material of everyday life. As for the dagger of Alacahöyük with the so-called iron blade, it should be noted that the low Ni content

Table 6.4 Pre-Iron Age iron artefacts found in Anatolia

Locality	Description	Dating	Remarks
Tilmen Höyük	Bracelet	Beginning third mill.	n.a.
Alacahöyük	Dagger with golden handle		Ni low
	Needles with golden head		2.7 wt.% Ni
	Hanger	EBA II (2800–2500)	n.a.
	Crescent-shaped disk		2.4 wt.% Ni
	Fragment of knife		n.a.
Tarsus	Nodule of iron	2400–2100	n.a.
Alişar	Needle; fragment	1900–1700	n.a.
Acemhöyük	Dose of ivory with lining of iron	1900–1800	n.a.
Kusura	Fragment	1800–1600	n.a.
Alacahöyük	Some jewellery, objects, weapons	1800–1200	n.a. (2 met)
Boğazköy	Several objects, weapons	1450–1200	Ongoing an.
Tell Açana	400 iron weapons	1450–1200	n.a.
Korucutepe	Iron objects, bars	1400–1200	n.a.
?	Sword	1400–1200	Met
Beşiktepe	Blade	13rd/12th cent.	n.a.
Kaman-Kalehöyük	Objects, ores, slags	13rd/12th cent.	Ongoing an

Due to the lack of analytical investigations, the material used to shape these artefacts is often not known. In most cases, they are designated as "iron objects" due to their rusty colour. All data are BC. Abbreviations: EBA = Early Bronze Age; n.a. = no chemical analyses; met = metallography; ongoing an. = ongoing analytical studies already excludes meteoritic iron and rather makes the use of "smelted" iron likely (Pernicka 1990). So far, it has surprisingly been overlooked that the blade of the dagger, as can be seen in all photos available on the Internet, has a rusty-greenish weathering crust, as they are typical for the corrosion of biphasic CuFe alloys. This can be taken as an indication of the above-mentioned origin of iron metallurgy from copper metallurgy.

6.6.1.1 Meteoritic Iron?

The question of whether artefacts formed of meteoritic iron are among the earliest artefacts is as difficult to answer as to whether the earliest copper artefacts are made of native or smelted copper. Discussions can be found in Buchwald (1975, 1992, 2005), Waldbaum (1999) or Yalçın (1998, 1999). The decisive criterion for the identification of meteoritic iron artefacts is often seen in their nickel content, which in meteorites is in most cases between 5.5 and 11 wt.% (Buchwald 2005). The second criterion mentioned is the Widmannstätten structure which can be determined by metallographic methods.

Among the earliest iron artefacts are some tube-shaped beads excavated from grave pits at the prehistoric Gerzeh cemetery (c. 3200–3300 BC). These are suggested to be made of nickel-containing meteoritic iron (Rehren et al. 2013; Johnson et al. 2013). These, approximately 2 cm, fully corroded beads were apparently made by rolling flat platelets in a cold state. According to Johnson and his co-workers, the platelets contain relics of taenite (γ-(Fe,Ni) with c. 30–50 wt.% Ni), a component of the Widmanstätten structures common for meteorites. With the help of highly sensitive analytical instrumentations (Prompt γ-activation, PIXE, Neutron radiography), Rehren and his team found traces of Mn, Cu, As, Zn, Pb and Ge in the chemical composition, next to Fe and Ni.

A very critical opinion on the arguments for the use of meteoritic iron in those publications can be found on the website http://www.cheops-insider.homepage.t-online.de. The identification of the taenite component is called into question, another argument is the doubt about the workability of meteoritic iron in cold conditions to produce such thin platelets as from the Gerzeh cemetery. Buchwald (1975, 2005), Coghlan (1977) and Wayman (1987) do describe thin, worked platelets of meteoritic iron from the Arctic north-east America, but those are not folded and are larger.

Arguments that nickel contents in the lower percentage range would be a criterion for meteoritic iron, is doubtful. Nickel-containing iron should also be produced by smelting of appropriately composited iron-rich laterites (Piaskowski 1982; Photos 1989). Laterites occur (not only) in the geological context of ophiolites in the mountain ranges reaching from the Mediterranean to the Himalayas (cf. TEMB in Sect. 3.4), but also in wide regions in Africa. In addition, Eliopoulos et al. (2012) report on geochemical features of nickel–laterite deposits from the Balkan Peninsula and from Turkey. Theses laterites contain 35–70 wt. Fe_2O_3 and up to 4 wt.% Ni, in addition, several percent of chromium and arsenic, manganese, phosphorous, vanadium, cobalt, copper and zinc. The same is true for ferricretes that occur between the southern edge of the Sahara and the Kalahari. Laterites cover about 1/3 of the continent's surface (Germann et al. 1999), they also occur frequently in north-eastern Africa and beyond. As far as Egypt is concerned, from where the tube-shaped beads excavated from grave pits at the prehistoric Gerzeh cemetery, metallurgy of iron begins in the sixth century BC, embedded in the general archaeological-geographical framework.

6.6.1.2 Iron Bars and Ingots

Ingots are the raw material, which, formed through the art of the blacksmiths, finally were the shape of this material which was traded. Of the numerous local-specific forms (iron stocks, sword bars, socketed bars, block and rod ingots, axe-shaped bars and other shapes, see Pleiner (2006), two are especially noteworthy. These are bipyramidal iron bars and split blooms.

Bipyramidal iron bars or ingots are a typical early form of trade good of wrought iron in the Middle East. Probably the most famous hoard of such bars comes from Khorsabad in Iraq and dates to the eleventh century BC. In 1860 during an excavation at this locality c. 160 tons of bipyramidal iron bars, rings, chains and other implements were found (discussed in Pleiner and Bjorkman 1974). Obviously, this hoard was part of spoils and taxes in iron collected by Neo-Assyrian rulers from Anatolia. Bipyramidal iron bars were also discovered at Susa and Nimrud in Mesopotamia (Iraq). These iron bars are shown in Fig. 6.26.

Bipyramidal bars also appear in large quantities in Europe since the Late Bronze Age and the Hallstatt period in the eighth to sixth centuries BC. They then increasingly reached the north in the Celtic world and became widely used (Pleiner 2000).

A second widespread form of ingot is malleable iron blooms with weights of 1 kg up to the order of 8–11 kg (Pleiner 2003). They have been partially split once or several times in the hot state (Fig. 6.27) and are referred to as split blooms. They date from the late Late Bronze Age to the sixteenth century AD. In the Medieval period, such iron blooms reach 100 kg or more. They consisted mainly of ferritic iron (wrought iron) with varying proportions of steel and slag inclusions, as they are usually to observe in the blooms during smelting. These ingots often had a spongy texture due to the precipitation of iron during smelting in the solid state. They were not always compacted into solid metal. The purpose of partially splitting the blooms, therefore, was primarily as a means of quality inspection. In addition, one could better divide them into individual parts. It is not always clear whether these split blooms were produced from a single bloom or welded together from several smaller ones.

6.6.2 Metallurgical Processes

Globally and across time, the bloomery process has been the most crucial and widespread metallurgical process for extracting wrought iron. In Africa, the widespread application of bloomery iron smelting up until the recent past may be due to the common lateritic weathering soils with very Fe-rich ferricretes at the top. The most extensive treatise of the bloomery process and its worldwide dissemination in all archaeological aspects can be found in the compendia of Pleiner (2000).

In the German language, the early method of producing metallic iron from ore is referred to as the *Rennprozess* or *Rennfeuerverfahren* (Beck 1884). The name derives from *rinnen* or *zerrennen*. This refers to the tapping of a liquid, i.e. low-viscous iron-rich slag with the main component fayalite.

In contrast to the later developed, today modern methods, this was a so-called "Direct process" for the production of wrought iron, which consisted mainly of ferritic iron, but could also contain proportions of steel and cast iron (Straube 1986, 1996; Yalçın and Hauptmann 1995, 2003). Its leading basics have been described by many authors (e.g. Oelsen and Schürmann 1954; Schürmann 1958; Tylecote et al. 1971; Thomsen 1975; Avery and Schmidt 1979; Espelund 1992; Serneels 1993; Pleiner 2000).

The reduction of ores was performed in completely different constructions, beginning with ground-set pits or hearths with a single bellow, or shaft furnaces with a height of 1–2 m, operated by an induced draft. The process was discontinuous. To remove the iron bloom from the furnace after smelting, depending on its specific design, it had to either be demolished completely or opened at its front.

In some production districts, in addition to the shaft furnaces, open smelting hearths (bloomery hearths) survived that only had a diameter of 35–50 cm and were 20–40 cm deep. Here, air was blown in from above with bellows and curved nozzles. So called Catalan hearths of this type existed in the Pyrenees up until the nineteenth century (Percy 1864), which produced excellent iron or steel in blooms of up to 200 kg (Fig. 6.28). They can be regarded as a further development of the medieval bloomery hearths, which in turn were dated up to the eighth and ninth centuries. They were especially widespread in Sweden and Norway, where a high-quality iron was produced and negotiated over long distances.

In industrialised countries of the world, the bloomery process was replaced by the "Indirect process" around the seventeenth century AD,

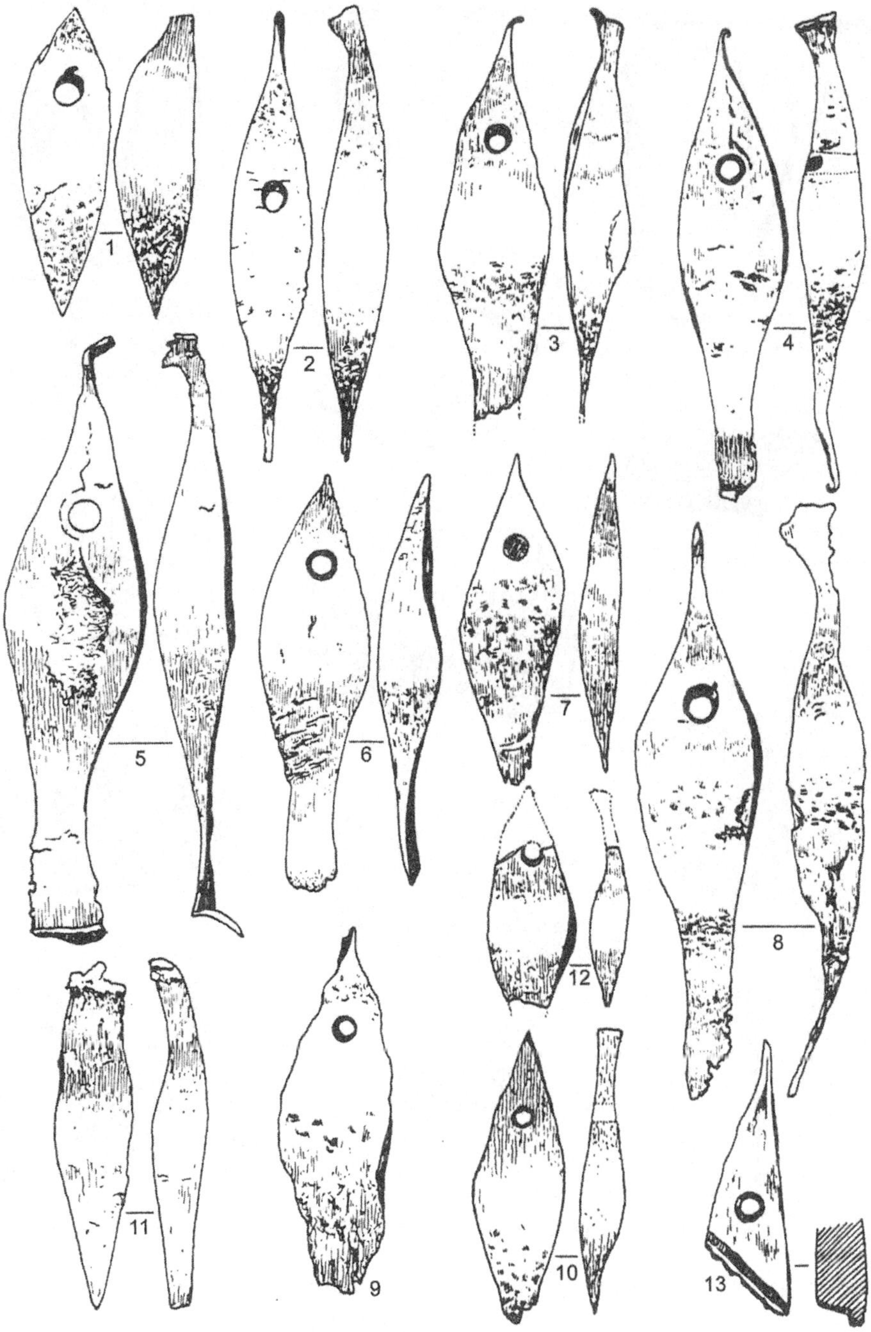

Fig. 6.26 Neo-Assyrian bipyramidal iron bars from Khorsabad, Nimrud, and Susa. Their weight was between 4 and 12 kg. From Pleiner (2006)

maybe even earlier. This process involves the modern production of carbon-rich pig iron, which is decarburised in a second subsequent process to wrought iron or steel. The bloomery process, however, has been widely used until just a few years ago as the most common technique in many parts of the world, wherever small-scale mining of iron ores and iron production has

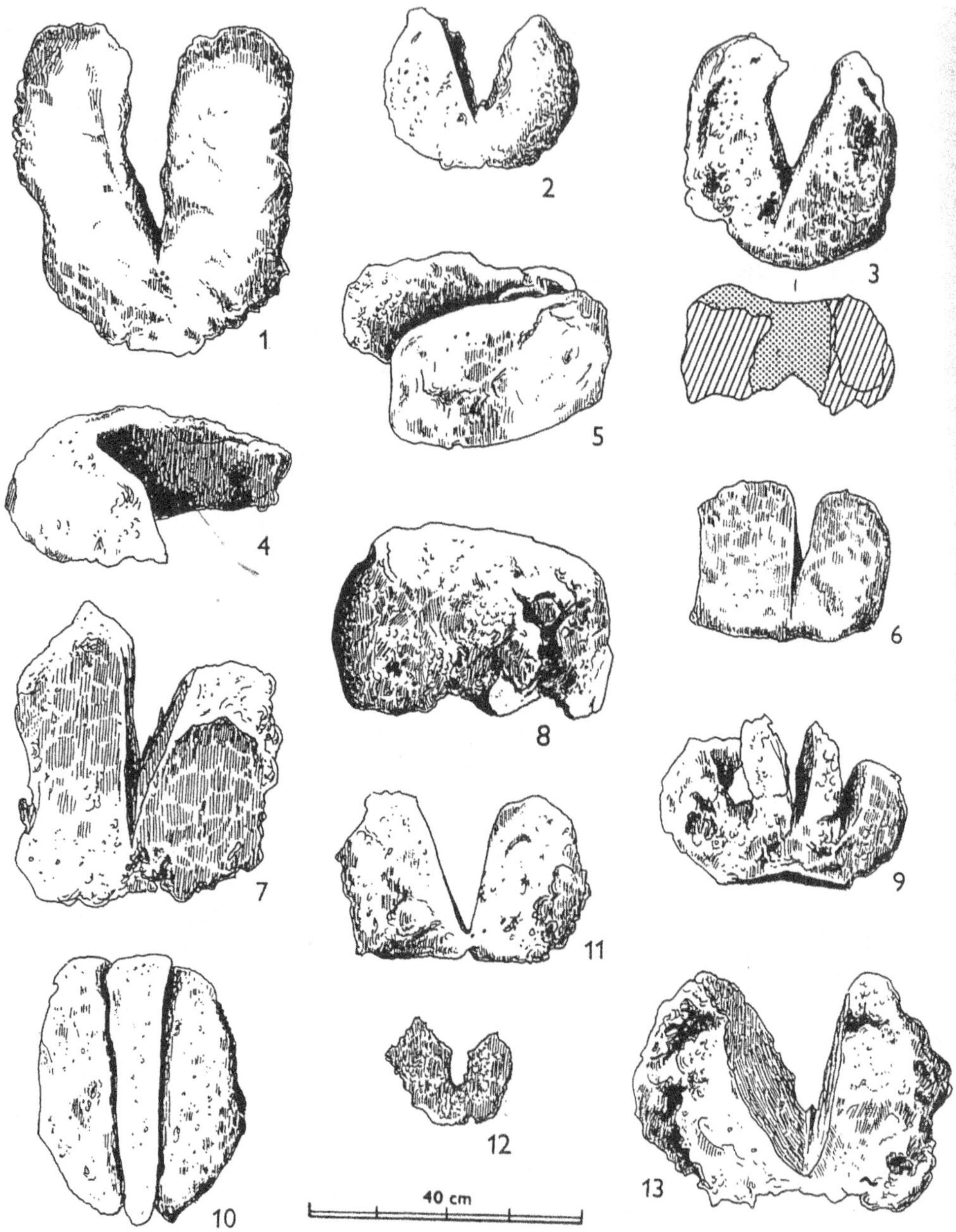

Fig. 6.27 Various split blooms from Europe and Russia from the first century AD (no. 1) with a weight of 9 kg to the thirteenth century with a weight of c. 4 kg (no. 5) to an early medieval split bloom with a weight of >12 kg (no. 7). Note the multiple split bloom from early medieval Sweden with a weight of >5 kg (no. 9). The size and weight of the split blooms give an idea of the sizes of blooms produced in various time periods. From Pleiner (2003)

been practiced. In Africa until the 1970s, iron was still produced with the bloomery process. This enabled important ethnographic studies (Avery 1982; Celis 1991; Eckert 1986).

However, we will also explain the important results of Straube (1986, 1996) who demonstrated that, in addition to wrought iron, steel could be produced in a bloomery furnace too. This is of

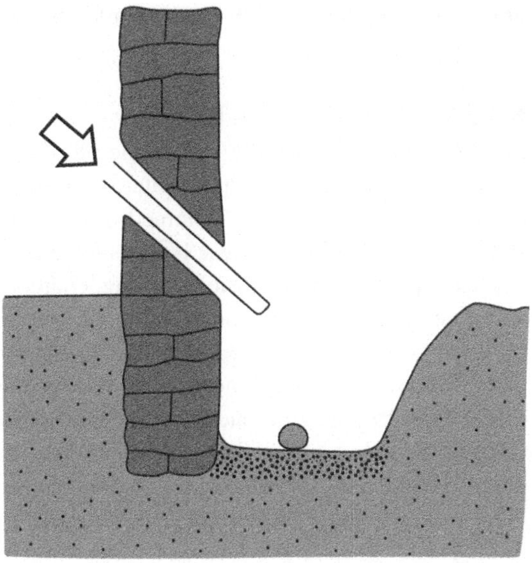

Fig. 6.28 The principal construction of a Catalan hearth. It is an effective example of the late bloomery process from the western part of the Mediterranean, especially in the Pyrenees. It consists of an open hearth with a depth of c. 1 m. Air was supplied by one vertical tube. The bloom was removed from above. The slag was tapped through a hole at the base of the construction. From Pleiner (2000)

fundamental importance because it reformulates the question of carburising ferritic iron to steel after its production in the smithing hearth. In the sole study of metal and slag, the importance of carbon in the bloomery process might be neglected. Carbon as it is contained in charcoal not only provides heat and CO-rich gas during combustion. Carbon is also soluble in iron. This can lead to different carbon levels in the iron, i.e. to different types of steel or cast iron (Espelund 1999).

Rehder (1989) had already emphasised "... that the idea that iron can be carburised simply by inserting it into a charcoal fire (in a forge), or that it becomes carburised by repeated heating for forging, cannot be supported...". A one-step steel production process used still in the nineteenth century AD is the "Tatara process" in Japan (Rostoker et al. 1989b). The authors claim that numerous other direct processes steels were known in the archaeological and ethnographic record in parts of Africa and India, e.g. in northern Cameroon (David et al. 1989).

Apart from the carbon content in iron, the phosphorus content also affects the quality of the iron, which in turn depends on the genesis of the ores. Bog iron ores, as they are widespread in the northern German-Polish lowlands (Ganzelewski 2000) and other sedimentary iron ores, such as the widespread Jurassic Dogger ores (Yalçın and Hauptmann 1995) always contain several percent of P_2O_5, which makes the iron "brittle" (Piaskowski 1965). Arsenic iron ores can also cause problems. 94% of the arsenic passes into the pig iron and forms an alloy with iron (Luyken and Heller 1938) which makes it brittle, even at low arsenic levels. Arsenic and iron form a speiss-like component.

The quality of iron is therefore also location-dependent, which is already mentioned in the oldest written sources.

6.6.2.1 Roasting

The first step in iron production is roasting. Taking goethite (FeOOH) as an example for an ore used to make iron (e.g. from a lateritic ore in Africa, or from bog iron ore in Middle Europe) we can describe the roasting process according to the following chemical equations:

$$2\ \mathrm{FeOOH} \rightarrow \mathrm{Fe_2O_3} + \mathrm{H_2O}$$

$$2\ \mathrm{Fe_3O_4} + \tfrac{1}{2}\ \mathrm{O_2} \rightarrow 3\ \mathrm{Fe_2O_3}$$

During this process, water evaporates at temperatures of ca. 200 °C, and α-Fe_2O_3 (hematite) is formed. The roasted material is magnetic due to the presence of minor quantities of maghemite (γ-Fe_2O_3) which is formed by the oxidation of magnetite. It gets a considerable porosity that promotes rapid reduction by carbon monoxide (CO) in the later stage of smelting. Iron ores (and also Fe-containing copper ores) treated by roasting have a typical reddish colour which might be a useful indicator to detect roasting in the archaeological record. Carbonate-bearing iron ores, such as the siderite ores ($FeCO_3$) from the

Erzberg in Styria (Austria), were also converted into hematite at about 400 °C:

$$2\ FeCO_3 + ½\ O_2 \rightarrow Fe_2O_3 + 2\ CO_2$$

If (hydr-) oxidic or hematitic iron ores are associated with sulphidic ores such as pyrite (FeS_2), as it is the case at the ore deposits at the island of Elba (Italy), the roasting has the additional effect to drive off the sulphur content—as it is usually the case with copper ores:

$$4\ FeS_2 + 11\ O_2 \rightarrow 2\ Fe_2O_3 + 8\ SO_2$$

Roasting is a solid-state reaction. At roasting, no silicate slag is produced. The roasting process is also suitable to remove arsenic from the ore.

6.6.2.2 The Bloomery Process

The next step in the production of iron is the smelting under reducing conditions. In a preheated bowl or shaft furnace iron ore and charcoal (or, alternatively, since Roman times, hard coal) are charged with roasted ore. The reduction of ore is performed in a temperature range between 800 and 1200 °C and can be described by the following principal reactions:

$$3\ Fe^{3+}{}_2O_3 + CO \rightarrow 2\ Fe^{2+,3+}{}_3O_4 + CO_2$$

$$Fe^{2+,3+}{}_3O_4 + CO \rightarrow 3\ Fe^{2+}O + CO_2$$

$$2\ Fe^{2+}O + 2\ CO \rightarrow 2\ Fe^0 + 2\ CO_2$$

The reduction of $Fe^{2+,3+}$-oxides to $Fe^{2+}O$ by CO again leads to an increasing porous volume by the loss of oxygen, and, herewith, accelerates reactions between the Fe-oxides and gaseous CO, and finally also between Fe^0 and CO to form Fe_3C. The main problems in the production of iron are therefore the reduction, the control of carbon and the discharge of low-viscosity slag. A reduction occurs when the ore is exposed long enough at high temperatures in a CO-rich atmosphere. Due to the high melting point at 1538 °C that, in ancient times could only be reached in exceptional cases, ferritic iron or low carburised steel—to the contrary of other metals such as gold, silver, copper and lead—could hardly reach the liquid state.

6.6.2.3 Products: The Bloom

In contrast to slag, metallic iron in the bloomery process only reaches the liquid state to a limited extent but is predominantly obtained in the solid state as a porous bloom. These blooms were usually interspersed with inclusions of slag, charcoal and remains of ore, which proved that the principle separation of bloom and slag was not complete. Blooms are already known from the decline of the Late Bronze Age and from the European Hallstatt period around 800 BC (Pleiner 2003). In order to make a useful piece of iron from this, the billet was first subjected to repeated hammering, folding and kneading while hot, successively compacting the porous iron and removing the inclusions. This extensive process required tremendous craftsmanship and a good deal of experience more than fusing together, e.g. (liquid) copper pieces did.

It is difficult to judge whether the solid α-iron was precipitated predominantly from the liquid FeO-rich melt or already at a lower temperature in a pure solid reaction. The structure of many examined samples allows both variants. In Fig. 5.50a, an agglomeration of drop-shaped α-iron (?) in slag is shown. The lowest temperature at which iron can be reduced from the ore is 633 °C. However, the working temperature of a bloomery furnace is >1000 °C, so it is well above that. The liquefaction of α-iron, however, presupposes temperatures of the order of 1600 °C, which, although theoretically possible, have probably only been achieved selectively under the firing conditions at that time (Tylecote et al. 1971; Avery and Schmidt 1979; Straube 1986). In the present case, it has not been examined whether this could rather be C-containing iron, i.e. steel.

The iron from the bloomery process is characterised by mostly low, but occasionally slightly higher carbon contents, which allow a classification as wrought iron or steel. Examples have been shown by Straube (1986, 1996) with his research on some of the 875,000 nails of the

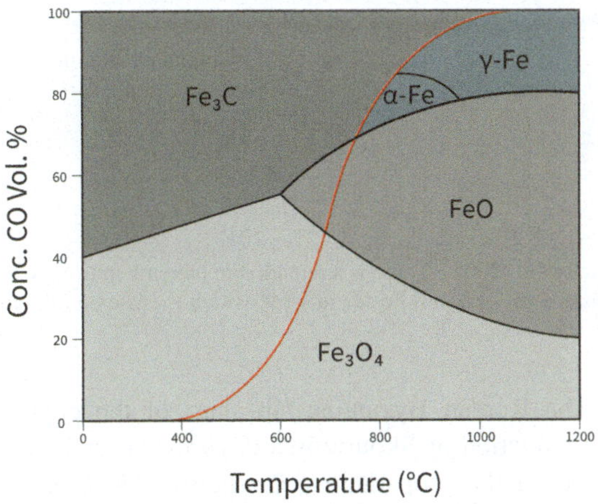

Fig. 6.29 The equilibrium diagram of Baur and Glaessner (1903) shows the areas of iron oxides to iron and the stages of carburisation up to cementite (Fe_3C). The red line represents the equation $CO_2 + C \rightarrow 2\,CO$ and the increasing concentrations in carbon monoxide. In the field of γ-Fe the stages of carburisation of iron of 0.01% C up to 0.4% C are running more or less parallel to the red line. Based on his smelting experiments Straube (1996) could demonstrate that in a bloomery process not only ferritic α-iron is produced but also C-containing steel and in parts cast iron were formed

famous Roman *Ferrum Noricum* in Magdalensberg (Austria). Yalçın and Hauptmann (1995) have shown similar results from their slag analyses of the Swabian Alb (Germany).

The Baur-Glaessner diagram (Baur and Glaessner 1903) (Fig. 6.29) shows the equilibria of iron oxides in the smelting of iron ores with carbon (charcoal) to α-Fe and γ-Fe and the precipitation of cementite (Fe_3C). On the one hand, the impact of carbon and the gas carbon monoxide, which is produced in the combustion, reduce the oxygen content in the furnace. On the other hand, they lead to a carburisation of iron.

Straube (1986, 1996) explains the individual areas of this diagram in a very detailed and understandable way. In the diagram, the red line is important because it marks the already discussed Boudouard equilibrium $CO_2 + C \rightarrow 2\,CO$. With different shades of grey, the fields of stability of the various iron oxides are indicated as a function of temperature (T) and the concentration of carbon monoxide (CO). Magnetite (Fe_3O_4) is stable in the lower part, wuestite ("FeO") is stable above. In order to reduce a magnetite (which is stable up to 570 °C) to wuestite, a gas mixture with 50% CO is required. Above wuestite, only α- and γ-iron with different C-contents are stable.

The equilibrium diagram also shows that between about 700 and 900 °C to the right of the red line, α-iron is reduced from wuestite, which only has a low solubility with carbon while at temperatures >900 °C, γ-iron is formed exclusively with a much higher solubility with carbon. This area is particularly important because different concentrations of C-containing iron (steel with 0.01–0.5 wt.% C) form here. Iron with more than about 1 wt.% C is undesirable because it is not forgeable. In case the gas atmosphere is saturated with carbon, i.e. CO, cementite (Fe_3C) is formed.

Straube (1986, 1996) thus showed with his smelting experiments at Magdalensberg (Austria) for the *Ferrum Noricum*, which was famous in Roman times, that in contrast to the then prevailing opinion, carbon-free soft iron is not the only product of bloomery processes, but that under appropriate working conditions a very extensive carburisation of the iron is possible. This carburisation would then decrease again, so that in this process even a temporary liquefaction of the metal occurs. According to the results of these experiments, it cannot be ruled out that the Norian smelters even managed to work towards a soft, medium-hard or hard steel in the furnace. Crew et al. (2011) report their smelting experiments, which they have carried out with bloomery furnaces: "... Cast iron can be produced in a bloomery furnace, in some quantity and without particular difficulties... Cast iron should be regarded as an inevitable by-product of bloomery smelting for high carbon steels rather than as an isolated and exceptional product...". Yalçın and Hauptmann (2003) show the gradual transition from wrought iron to cast iron, based on the analysis of slag finds from Celtic times to the High Middle Ages and modern times, combined with a decrease of Fe contents in slags from 50 to 75 wt.% to <20 wt.% (Fig. 5.31).

Table 6.5 Theoretical yield of iron from various compositions of iron ores

100 kg Iron ore	FeO	Fe_2O_3	Potential Fe-bloom (kg)
	70	78	None
	75	83	12
	80	89	25
	85	95	38
	90	100	51

Calculation modified after Pryce and Natapintu (2010) and Buchwald (2005). There is a remarkable increase in iron yield at smelting ores with 80% FeO and more. The calculation is based on FeO and Fe_2O_3 in wt.% which is close parallel to any roasted iron ores

6.6.2.4 Products: Slag

In practice, the equations discussed above are simplified. In most of the iron ores, no matter if these are bog ores, ferricretes or ores from the gossan of a hydrothermal vein, SiO_2 is present as sand or quartz. Usually during iron smelting almost all over the world, this is bound at least in larger parts to the formation of iron-rich silicate slag with the main component fayalite (Fe_2SiO_4). The formation of these slags, therefore, required a high iron oxide content in the ore. Fayalite has a composition of 70 wt.% FeO and 30 wt.% SiO_2. The significance of the ore quality is characterised by the so-called break point, as indicated in Table 6.5. Theoretically, with an ore of 70 wt.% FeO (= 77.9 wt.% Fe_2O_3), no production of iron would be expected, but at 75 wt.% FeO (= 83.5 wt.% Fe_2O_3) the maximum production of 12 kg iron could be possible. Bei 80 wt.% FeO one might extract c. 25 kg iron. That means increasing the ore quality by only a few percent can more than double the potential yield of useful metal. The best yield is achieved by using pure hematite for iron production which is nothing special in iron raw sources.

The bloomery process, due to the high iron contents of the slag, therefore included high losses. The reduction of iron oxide from the ore could not be allowed to be complete because of the formation of the slag. The production of iron-rich slag with always very similar composition is based on the worldwide similar starting material not only of iron but also of copper and other ores. Therefore, the name *Rennfeuerschlacke* is also applicable to the smelting of ores other than iron, e.g. for copper slags of the developed epochs. The Roman–Byzantine tap slags of the copper production of Skouriotissa (Cyprus) are not only externally comparable to *Rennfeuerschlacken*, but are also comparable in chemistry and phase content (if one disregards the valency of the total Fe-content: Fe-slags are dominated by wuestite (FeO), base metal slag by magnetite (Fe_3O_4)). Slags of the bloomery process are iron rich almost without exception (>45 wt.% FeO, cf. Table 6.6). They are often much more rich in iron than their possible starting ores, as they can be found in the field today! It is likely that ores were often enriched by beneficiation.

Bloomery slags, however, differ significantly from those from the production of liquid pig iron. The FeO contents in the slag of modern blast furnaces are below 1 wt.%.

The separation of the liquid slag from the bloom was done either by flowing of the slag into a pit below the furnace or by tapping it from the front side of the furnace (Fig. 6.6a, b). This process did not only spread throughout Europe during the time of the Roman emperors. It has been observed worldwide until recently (Humphries and Rehren 2013). Therefore, the slags are often found as differently sized heaps directly next to furnaces. Often, however, they were later reused, on the one hand as a fluxes for later smelting processes up until the industrial age, on the other hand as material for road construction.

Weights of blooms ranged from 2.5 to well over 10 kg, as evidenced by archaeological finds. Prior to any further processing, the bloom from shaft and bloomery furnaces had to be freed from and inclusions of slag and charcoal. This was done by

Table 6.6 Chemical bulk composition of iron smelting slags from various smelting sites dated to the La Tène period in the Siegerland area (Germany)

Sample no.	Locality	SiO_2	TiO_2	Al_2O_3	FeO	MnO	MgO	CaO	Na_2O	K_2O	P_2O_5	Total
Breccia undated												
D-118/02 7	Trüllesseifen	17.8	0.06	1.36	56.1	4.96	0.16	0.56	0.07	Nd	0.22	81.3
D-118/02 8	Trüllesseifen	25.6	0.08	1.66	52.8	5.40	0.27	0.74	0.08	Nd	0.22	86.8
Latène												
D-118/07 4a	Niederndorf	27.2	0.16	3.07	33.5	29.1	1.22	6.25	0.11	1.61	1.01	103.2
D-118/07 8a	Oberschelden	24.0	0.16	3.24	58.0	11.9	0.33	0.79	0.06	0.73	0.14	99.4
D-118/07 17a	Obersdorf	24.8	0.20	5.13	61.6	0.38	0.33	0.81	0.03	0.48	0.39	94.2
D-118/07 19c	Obersdorf	22.5	0.17	3.76	66.5	1.67	0.20	0.94	0.03	0.65	0.58	97.0
D-118/07 24a	Obersdorf	24.9	0.16	4.96	55.0	9.41	0.33	1.89	0.07	1.08	0.59	98.4
D-118/07 24b	Wilnsdorf	15.2	0.06	3.95	62.5	6.37	0.25	0.83	0.04	0.58	0.32	90.1
D-118/07 25a	Obersdorf	24.8	0.16	5.22	55.6	7.62	0.98	1.55	0.07	1.21	0.53	97.7
D-118/07 26a	Niederndorf	21.4	0.08	1.58	52.5	12.0	0.75	0.65	0.03	0.14	0.33	89.5
D-118/07 27a	Wilnsdorf	26.8	0.13	3.44	49.8	9.84	0.25	2.62	0.07	1.52	2.36	96.8
D-118/07 30b	Marienborn	24.2	0.20	6.82	52.3	7.59	0.37	1.72	0.08	1.27	0.55	95.1
D-118/07 41a	Wilnsdorf	23.4	0.14	5.59	62.0	0.76	0.38	1.09	0.06	1.10	0.31	94.8
D-118/07 43a	Wilnsdorf	27.3	0.14	4.69	49.6	9.73	0.32	1.41	0.13	1.61	Nd	94.9
D-118/07 44a	Mudersbach	20.8	0.14	3.84	55.7	6.14	0.61	3.79	0.08	1.37	1.48	94.0
D-118/07 52	Rinsdorf	19.6	0.12	3.53	59.6	7.02	0.39	1.96	0.10	0.84	0.68	93.8
D-118/02 3	Gerhardsseifen	27.9	0.19	3.78	52.2	8.22	0.78	2.09	0.14	Nd	0.24	95.5
D-118/02 6	Trüllesseifen	28.6	0.12	2.04	59.3	6.76	0.68	1.39	0.08	Nd	0.32	99.3

Note that the main components are FeO and SiO_2. A characteristic feature of the Siegerland slag is the high concentrations of MnO, which reaches up to 12 wt.%. The total of FeO + MnO nearly goes up to nearly 70 wt.%. From Gassmann and Yalçın (2009)

reheating and mechanical forging, allowing the re-liquefied slag to be squeezed out of the bloom. This process of reheating and forging again resulted in large losses of up to 70% of the iron bloom (Ganzelewski 2000). Large amounts of leaf-shaped hammer scales were produced. Hammer scale consists mainly of the Fe-oxides wuestite, magnetite and hematite. In addition, not to be underestimated amounts of smaller bloom fragments, intergrown with slag, were discarded or melted down again. When hammering out and reheating the bloom, the palm-sized plano-convex forged slags are produced ("PCB", cf. Sect. 5.3). All of this can be found in archaeological contexts (Serneels and Perret 2003).

6.6.3 Researches on Ancient Iron Production

6.6.3.1 Georgia and the Colchis

A complicated and finally surprising history of research concerns early iron metallurgy in the Colchis south of the Caucasus mountain range at the eastern coast of the Black Sea, in the current state of Georgia, which has long been considered the world's oldest production of this metal (Pleiner 2000). The territory of the historical Colchis is a subtropical environment with dense vegetation.

David Khakhutaisvili (1987) discovered more than 400 ancient smelting sites assigned to early iron production. These are distributed in six clusters between the vicinity of the city of Batumi to the reaches of the Great Caucasus and are located 15–30 km away from the coast of the Black Sea. Despite these distances, these iron-smelting sites have been associated with magnetite placers on the east coast of the Black Sea. Inanishvili (2007) commented somewhat more critically on this concept. In his work, he pointed out that the numerous hydrothermal oxidation zones of non-ferrous metal deposits in the (Trans-) Caucasus could possibly be viable iron resources for this case as well. Radiocarbon and archaeo-magnetic dates indicate that these sites were operating between the eleventh and the ninth century BC. Recently, the Supsa-Gubazeuli river system provided a series of dates from 1800 BC to 600 BC (Gilmour et al. 2014). If correct, these would be the earliest dates anywhere in the world for the production of iron and they would support the ancient suggestions of the earliest Greek historical sources (Herodotus, Xenophon and Strabo) that the Colchis industrial iron production would be the area where iron production originated.

Despite the pioneering work of the Georgians, a number of key questions remained, especially those relating to the technology, chronology and spatial distribution of this so-called ancient iron industry. In a Georgian–British research project of 2010–2012, Gilmour and his staff carried out fieldwork in the Colchis and visited numerous of these smelting sites. Gilmour (Gilmour et al. 2014) writes "...It was clear from XRF and SEM analytical work that most if not all the sites encountered were used for copper smelting...". Apparently, as has happened at some other localities, iron slags have long been confused with copper slags because of their high Fe contents, which may also be as high as in bloomery slags of iron smelting (see Sect. 6.2). Very similar conclusions were reached by Erb-Satullo et al. (2014) and lately also by Sergo Nadareishvili (unpublished data DBM), who investigated slags of the locations designated as iron smelting sites on the Choroki river on the border with Turkey and identified them as copper smelting slags instead. This would imply that the superlative of the oldest iron metallurgy in the Colchis should be revised again.

6.6.3.2 Iron Production in Central Europe

In Sect. 3.5.7, it has already been pointed out that for example in Germany, there are many hundreds of iron ore deposits, ranging from the south on the Swabian and Franconian Alb, then in the European low mountain ranges to the bog iron ore deposits in the northern German–Polish lowlands (Hingst 1983; Jöns 1997; Ganzelewski 2000). From the La Tène period to the late Middle Ages, they were systematically mined and smelted in hundreds of smelting sites, which, as with copper metallurgy, provide a vivid picture of the intensity of the old iron production. This continues geographically to the north, where countless deposits of bog iron ores have also been exploited in Denmark, Sweden and Norway (Bielenin 1983; Serning 1979; Serning et al. 1982; Modin et al. 1985; see also Fig. 3.53). For the early extraction of iron in Europe, a number of diverse interdisciplinary research projects have developed, which, in addition to archaeometallurgical and mining-archaeological studies, have applied a number of economy-archaeological and environmental aspects to the various mining landscapes. Using modern survey methods (geomagnetic prospecting, GIS-aided aerial photography) fieldwork was carried out and modern dating techniques were used to perform precise dating. Anthracological analyses of charcoal and pollen analysis were carried out as well as studies on the history of vegetation.

Of these many activities, only a few are mentioned here. Larger research projects for the smelting of iron ores in the La Tène period, the Roman Empire and also in the Middle Ages were carried out over many years in the Siegerland and in the Lahn-Dill area as well as in Joldelund in northern Germany. In the Siegerland, as in other

European low mountain regions, the evidence of iron production from the sixth and fifth century BC has condensed (Pleiner 2000; Garner 2010). In the early 1930s, only 18 prehistoric and 34 medieval smelting sites were known from fieldwork (Krasa 1955), in the post-war period, the number grew to about 180 prehistoric and 230 medieval places. Already in the early stages, in the 1950s and 160 s, German iron and steel engineers worked out the thermodynamic principles to the metallurgy of the bloomery process on the example of finds from the Siegerland (Schürmann 1958; Osann 1959; Sönnecken 1977) and created a solid basement for any further studies in these fields.

The most recent work has been published by Stöllner et al. (2009), Garner (2010) and Menic (2016). In these projects, the Siegerland area, with its hundreds of hydrothermal iron ore and base metal veins, was mapped as a politico-economic space using modern interdisciplinary methods.

On the basis of the thermodynamic data and calculations of bulk densities of slag heaps presented by Schürmann (1958) and later by Bachmann (1982b), repeated attempts were made to carry out quantity calculations (Ganzelewski 2000; Menic 2016). After all, an estimate of the quantities of metal produced is one of the fundamentally important issues of economic archaeology and econometrics, not only for the old iron production. This proved to be extraordinarily difficult because there are too many unknown factors in the archaeological findings. Almost as a rule, old slag heaps are disturbed, dismantled and recycled for architectural purposes (road construction, housing construction), there are uncertainties in the reconstruction of smelting furnaces and early process control, the estimation of mined and processed ores is difficult to grasp and the processing of the obtained pieces of metal (blooms and ingots) up to the trade contains considerable uncertain work steps. Data presented so far (Ganzelewski 2000 for Joldelund; Menic 2016 for the Siegerland; Hauptmann 2007 for Faynan) should, therefore, be treated as very rough estimates only.

The estimation of metal quantities directly affects the issue of fuel supply. This concerns the use of possibly special types of wood or charcoal, their size when incinerated in crucibles or smelting furnaces, or in the further processing of raw metal or the carburising of blooms. In his book "Mastery and Uses of Fire in Antiquity", Rehder (2000) has summarised practical and thermodynamic considerations on the combustion of carbon in ancient metallurgy, from working with blowpipes on small crucibles to larger smelting furnaces and carburising of iron. Smettan (1995) considers interventions in the La Tène to medieval vegetation in the Swabian Alb as possibly due to metallurgical activities.

In (semi-) arid areas the question about the sources of the amounts of wood used in metallurgy arose again and again. It has, therefore, become almost routine for botanists to carry out investigations on charcoals that are regularly found at smelting sites, also as inclusions in slags. Micro-recordings can determine types of wood, which were preferably used for smelting ores. This provides valuable information on the vegetation in mining-landscapes or alternatively for the procurement of wood.

In addition, ores, slags and metal finds from Siegerland were used for lead isotope analyses. It was investigated to what extent this method could also be used in iron metallurgy as a tool for provenance studies. In copper metallurgy, provenance studies using lead isotope measurements have long since become routine (see Sect. 11. 3.1). The first provenance studies in iron metallurgy were carried out by Schwab et al. (2006) on iron tools from the Celtic oppidum of Manching (Germany). The results were difficult to understand, especially as these tools were most probably made from very Pb-poor bog-iron ores. In the case of mineralisation in the Siegerland, however, this seemed to be more promising, as the veins contain almost entirely polymetallic, including lead-containing mineralisation. It could be proven that lead can be found as a siderophile element in the iron, just like Co, Ni and As. In addition, it was proved that the blooms of iron extracted from the veins would contain lead (Salzmann 2013).

6.7 The Metallurgy of Antimony

The metalloid antimony (chemical symbol Sb) is an extremely brittle material with a flaky, crystalline habit. There are three modifications:

1. Metallic or grey antimony. This is the most stable modification. It is bluish-white and has a metallic lustre, which tarnishes to black over time in air.
2. Black antimony. This is a black powder, which is formed by rapid cooling of antimony vapour.
3. Explosive antimony is produced by electrolysis of antimony (3+) salts as finely divided, amorphous powder. Scratching induces an explosive reaction, transforming it into the grey, metallic antimony.

In archaeometallurgy, only the modifications nos. 1 and 2 are of any importance.

6.7.1 Archaeological Evidence

In archaeometallurgy, antimony is a rather rare material compared to copper, gold or iron. While it is often present in low concentrations as a minor and trace element in copper artefacts, artefacts of pure antimony are a rarity in prehistoric times (Nicholson and Shaw 2000; Moorey 1994). There are also only individual, sometimes controversial findings from antimony from Mesopotamia (e.g. from Tello, the Sumerian Girsu, Early Bronze Age, and from Assur, c. 2000 BC), whose origin may possibly be found in the (Trans-) Caucasus (Hauptmann and Gambaschidze 2001).

Because in this area there is not only a striking accumulation of copper alloys with antimony (Colchis axes and maceheads with 10–20 wt.% Sb; Gambaschidze et al. 2001) and artefacts from antimony itself (pear necklaces, buttons, wangler) from the middle of the second millennium BC), there are especially numerous ore deposits of antimony (Sect. 3.4.13). This means that even at this time antimony was smelted from its ores, even if so far no evidence for it exists save for very few antimony slags (cf. Sect. 5.5.5; Gambaschidze et al. 2001; Hauptmann and Gambaschidze 2001).

Selimkhanov (1975) mentioned artefacts made of antimony from the Late Bronze Age tombs of Redkin Lager, (Trans-) Caucasia. Other beads, buttons and pendants made of antimony were reported from a Koban tomb near Tbilisi, a third millennium BC tomb in Velikent, Daguestan, and a second millennium one in Mamai-Koutan, also Daguestan and Jerablus Tahtani, Syria (Shortland 2002). Objects made of antimony are therefore reasonably common, at least on some sites in the second and third millennium of the Caucasus.

In prehistory and early history, antimony aroused people's interest, especially in the Orient, because it was also used in powdered form as a black, strongly colouring pigment in cosmetics (Potts 1997).

From the European Middle Ages, there are detailed written reports on the antimony production of Teophilus Presbyter (twelfth century AD), as well as pictorial representations in Georg Agricola (1556 AD) and Lazarus Ercker (1556 AD).

A more recent archaeological finding with corresponding crucible fragments for the production of *antimonium crudum* (see below) has been reported by Siebenschock et al. (1996) from a region close to the small antimonite deposit near Sulzburg in the southern Black Forest (Germany). These activities date back to the time between fourteenth and sixteenth century AD.

6.7.2 Production and Processing

Antimony is easily extracted from its ores, with antimonite (Sb_2S_3) being the most important and abundant ore. The peculiarity of the metallurgy of the antimony lies in the low melting point of the metal and its compounds as well as in the volatility of its oxides and sulphides. Antimony itself melts at 630.5 °C, the natural sulphide ore antimonite

(Sb_2S_3) melts at 546 °C, its oxidation products valentinite (Sb_2O_3) and senarmontite (Sb_4O_6) melt in a vacuum at 656 °C. When Sb_2S_3 is heated while exposed to the air, oxidation starts at 190 °C to produce $Sb^{3+}{}_2O_3$. This oxidation process is most active at 340 °C. Even with sufficient air supply, further oxidation to $Sb^{4+}{}_2O_4$ occurs only when Sb_2S_3 is no longer present.

This means that antimony can be obtained at relatively low temperatures.

Out of poor ores, antimony is enriched by volatilisation or antimony-bearing ore is concentrated by smelting it without contact to air. This process, referred to as liquation, on which a good part of other early metallurgical processes were based, such as copper, lead or copper production by matte smelting (cf. Sect. 5.7), which is technologically situated before the actual slag smelting *senso strictu*.

According to these traditions and findings, the production of antimony since the Middle Ages was carried out in three stages, the principle and methodology of which could probably also apply to earlier epochs. After Siebenschock et al. (1996) and Anderson (2012) these stages can be summarised as follows:

1. Liquation to produce crude antimony (*antimonium crudum*).
2. Roasting of *antimonium crudum* for desulphurisation and production of antimony oxide.
3. Reduction of antimony oxide by carbon to antimony.

Liquation

Antimonite bearing ores, which are intergrown with host rock or gangue, are heated to about 550–600 °C, i.e. above the melting point of antimonite. This melts out of the refractory gangue components and drips off like "sealing wax" (Quiring 1945). The temperature may not reach the boiling point of antimonite (857 °C). This *Seiger* method, as shown in the pictures of Georg Agricola, was performed in small elongated pots (in the German language called *Aludel*) as shown in Fig. 6.30. Although this figure refers to the extraction of mercury from cinnabar (HgS), the structure of these facilities is also applicable to antimony extraction. In each case two such vessels, one of which served as a crucible, the other as a collecting vessel, formed a smelting unit. The upper crucible was filled with pellets of ore with a piece size of 1–4 cm, covered with a lid and placed on the second, usually slightly smaller collecting vessel. Several such smelting units were usually arranged in rows and heated simultaneously. Siebenschock et al. (1996) reconstructed the principle of the method as shown in Fig. 6.31.

The liquation process is based upon the transformation of solid Sb-sulphide to liquid Sb-sulphide according to the equation:

$$Sb_2S_{3(solid)} \rightarrow Sb_2S_{3(liquid)}$$

The molten sulphide is collected in containers. The liquated product is called crude antimony or liquated or needle antimony.

Roasting

The roasting of the ores and *antimonium crudum* carried out under air supply leads to a desulphurisation. Depending on air supply and temperature either volatile antimony trioxide (Sb_2O_3) and/or non-volatile antimony tetroxide (Sb_2O_4) are formed according to the following formulas:

$$2\,Sb_2S_3 + 9\,O_2 \rightarrow 2\,Sb_2O_3 + 6\,SO_2$$

$$2\,Sb_2O_3 + O_2 \rightarrow 2\,Sb_2O_4$$

The temperatures required for the reaction are low. The thermal reaction starts at 275 °C and runs completely and quickly at about 350–400 °C (Quiring 1945). As described the volatile antimony trioxide is recovered in condensing pipes (*Aludels*). Roasting and volatilisation are affected almost simultaneously by heating the ore, mixed with coal or charcoal. If the volatilisation conditions are too oxidising, the non-volatile antimony tetroxide (Sb_2O_4) is formed and the recovery of antimony as antimony trioxide is diminished.

The removal of antimony as volatilised trioxide is the only pyrometallurgical method suitable for low-grade ores.

Fig. 6.30 Production of mercury from cinnabarite (HgS) after Georg Agricola (1556 AD, book IX). The method can also be applied to the production of *antimonium crudum*. A battery of stacked, bulbous ceramic pots filled with ore is covered by long, heated (not burning!) wooden poles, which are being prepared in the background. The ore in the pots must not be overheated due to its low melting point. The sulphide ore drips into the bottom pot and collects there as ingot. It is being liquated out

Antimony trioxide, easily formed during roasting (mineralogically senarmontite, see Sect. 3.5.7), is also often and easily formed as a natural weathering product of solid antimony and antimonite. It can therefore not be ruled out that the antimony buttons and beads from the second millennium BC Great Caucasus in Georgia, which today are almost always covered by a whitish-grey to yellowish crust (Gambaschidze et al. 2001; Hauptmann and Gambaschidze 2001) already contained a coating of this mineral during its manufacture, although it is more likely that the silver colour of the semi-metal was the prime target in the production of the artefacts.

Reduction

Theoretically, if pure antimony ores or pure *antimonium crudum* is available, antimony may

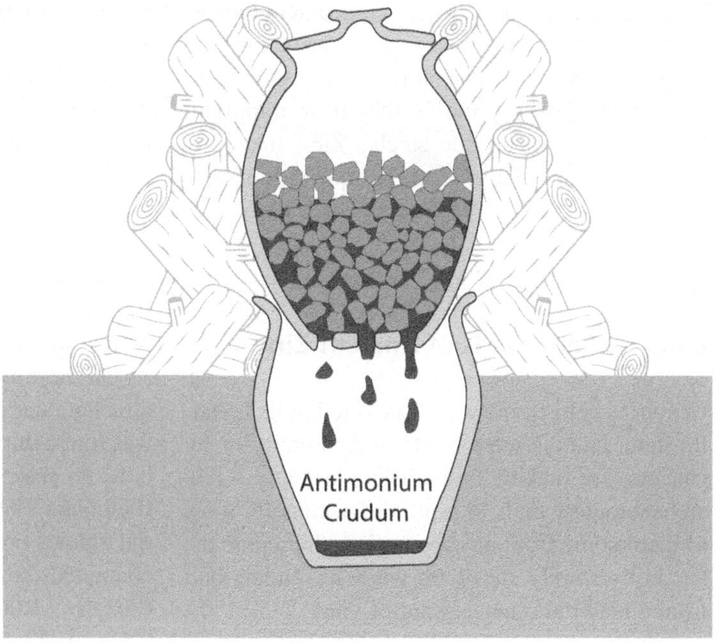

Fig. 6.31 Reconstruction of a melting unit for the extraction of *antimonium crudum* after the ceramic finds of Sulzburg (Black Forest, Southwest Germany). Modified after Siebenschock et al. (1996)

already happen around c. 700 °C according to the following formulas:

$$Sb_2S_3 + 9\,O_2 \rightarrow 2\,Sb_2O_3 + 6\,SO_2$$

$$2\,Sb_2O_3 + Sb_2S_3 \rightarrow 6\,Sb + 3\,SO_2$$

From the oxides of antimony metallic antimony can be generated by reduction with charcoal or wood in crucibles or in smelting furnaces.

$$Sb_2O_3 + 3\,CO \rightarrow 2\,Sb + 3\,CO_2$$

If no pure ores are available and therefore a slag formation is required, halite (NaCl), sodium carbonate ($Na_2CO_3 \cdot 10\,H_2O$), potassium carbonate ($(KAlSO_4)_2 \cdot 12\,H_2O$) or sodium sulphate ($Na_2SO_4 \cdot 10\,H_2O$) could be added to facilitate slag formation, as described in Georg Agricola (1556 AD; book IX).

Another possibility of antimony extraction from rich ores or from crude antimony, which does not require a roasting process, is the process of precipitation, in which the higher affinity of sulphur to other metals, in this case of iron, is used. It is based on the following reaction:

$$Sb_2S_3 + 3\,Fe \rightarrow 2\,Sb + 3\,FeS$$

The process was already known in the sixteenth century AD and was also described by Agricola. This process consists of heating molten antimony sulphide in crucibles with slightly more iron, in the form of scrap, than antimony.

6.8 The Metallurgy of Zinc

This metal (chemical symbol Zn) is a bluish-white, highly lustrous metal with a density of 7.13 g/cm^3. It is quickly covered with a thin layer of oxide and carbonate when exposed to air, which makes it dull. The smelting point is at only 419.5 °C, the vapourisation point at 907 °C. This does not mean, however, that zinc vapour is not formed before this temperature is reached. Even at 745 °C, the vapour pressure of zinc is 0.16 bar (Herbenar et al. 1950). Nielen (2006) proved this experimentally.

Zinc is easy to deform even at moderate heat. The name derives from the German name *Zincken* = *Zacken* (which means jag or spike), as zinc ores often have a “jagged” appearance.

Zinc readily forms alloys with other metals. Best known is the alloy with copper, forming brass. But zinc can also form alloys with gold and silver. Today, zinc is the most important anti-corrosion metal for steels. Zinc has only really been detectable since antiquity, but on a larger scale it has been produced since the Middle Ages.

6.8.1 Rare Finds of Ancient Zinc

Centuries before zinc was discovered in its metallic state, its ores were used—intentionally or by chance—for making brass. There are only a few archaeological finds of zinc, unfortunately, most of them come from unsafe stratigraphic contexts, are not securely dated or were misunderstood from a material science point of view.

Small amounts of metallic zinc can occur as by-products in the flues of lead–zinc–silver smelting furnaces. They could easily survive there, even though most of it is reoxidised quickly (see next section). Such an occurrence is exemplified by the discovery of a small lump of metallic zinc at the Roman lead mines at Charterhouse on Mendip, Somerset (England) in archaeological levels dated to the first century AD (Todd 1996).

One of the rare old zinc objects is a small rolled sheet from the Agora of Athens from the fourth or second century BC (Farnsworth et al. 1949). Rough chemical analyses showed that the piece contained Pb = 1.34 wt.%, Cd = 0.06 wt.%, Cu = 0.0055 wt.%, Fe = 0.0016 wt.%. Craddock (1990) sees this small sheet metal as an inadvertently produced piece of the lead–silver metallurgical operations from the nearby Laurion ore deposit, where also zinc ores occur. According to Pliny (Nat. Hist., 34th book) zinc oxide was carefully recovered, similar to Charterhouse, from the walls of the flues of furnaces and was used for medical purposes—the basis for the familiar zinc ointment. Pliny refers to it as *Lauriotis*. However, Craddock categorises a targeted use of metallic zinc as unclear: "... Thus we have a rather unexpected scenario for the early manufacture of brass and zinc. Sometime in the first millennium BC smiths at the lead-silver mines recognised that the droplets of white metal (mocksilver) which collected in the furnace flues could be added to copper to give *orichalkos*, a golden copper alloy (brass) ... which had hither to be obtained from natural mixed copper-zinc ores...".

Also a controversial is a zinc statuette from the Roman Dacia, it is classified only as a likely-to-be-real find (Fellmann 1991).

On the Engehalbinsel near Bern, Switzerland, a rectangular votive tablet made of metallic zinc was found during surveys with a metal detector. It bore an inscription identified as Roman–Celtic (Fellmann 1991). Rehren (1996) carried out material studies on this tablet and found that the zinc contained small amounts of minor elements (Pb = 1.06 wt.%; Fe = 0.287 wt.%; Cd = 1040 ppm; Cu = 1.090 ppm). The iron content concentrated on the marginal areas of the tablet. It was believed that the zinc tablet was cast in an iron mould, which, however, was not known in Roman times. It is judged very critically whether the zinc tablet is actually Roman, or if it is not a recently produced modern piece of zinc (Craddock 2009).

Zinc as the sole metal was apparently one of the last of the more common metals that have been smelted in ancient times. Evidence that zinc production was known in the ancient world is rare and confirms the exceptional extraction processes. The reports in Strabo's Geographica (XIII.1.56), in which he speaks of "false silver" (*pseudargyros*) in connection with the manufacture of brass (Caley 1964; Grothe 1971), are questionable.

In the Zawar district in North West India, zinc ores may have been mined from a polymetallic Pb–Zn deposit with apparently low levels of lead ores since the second half of the first millennium BC (Craddock et al. 1989). However, the authors cannot provide evidence of zinc production from this early period. This is only proven securely for later periods. It was not until the fourteenth century AD that zinc was produced on an industrial scale in Zawar (see below).

6.8.2 Production and Processing

In principle, it is not difficult to reduce oxidic zinc ores to metallic zinc. The problem is just to collect it as metal. Thus, if one were to try to smelt zinc ores in a traditionally constructed smelting furnace, zinc would rise along the shaft in the vapour phase and quickly oxidise back to a whitish zinc oxide, rather than pooling in the lower part of the furnace as a liquid metal.

In general, zinc ores must first be roasted to remove sulphur contents of primary ores such as sphalerite (ZnS) or carbonate contents of secondary ores such as smithsonite ($ZnCO_3$) or calamine (all oxidic zinc ores). The metal is obtained by transfer of the ores into the oxide and subsequent very strong reduction with (char-) coal at about 1000–1300 °C.

In contrast to most of the other metals that played a role in antiquity, however, this does not lead to a reduction to metal in the liquid state, but via the gas phase, similar to arsenic and mercury. Zinc evaporates, but is quickly reoxidised and settles in the upper part of a smelting furnace as ZnO. The corresponding chemical reaction is:

$$ZnO + CO \leftrightarrow Zn + CO_2 \ \Delta H\ (298.15\ K) = -115.6\ \text{kcal mol}^{-1}$$

This is a strictly endothermic reaction. This means that the reaction between the zinc oxide and the carbon monoxide at high temperatures is clearly on the left side. This in turn means that the carbon dioxide oxidises the resulting zinc quickly, unless there is an excess of pure carbon. In this case, CO_2 would react with C:

$$C + CO_2 \leftrightarrow 2\ CO$$

and carbon monoxide would form again. The Ellingham diagram (Fig. 4.15) indicates that the reduction to zinc occurs at high temperatures under conditions similar to those of the reduction of FeO to Fe^0.

Therefore, the reduction to zinc must take place in hermetically sealed vessels or retorts and the zinc vapour must be condensed to the liquid metal in special devices without exposure to air.

For producing the metal zinc a special kind of furnace very different from those used for making copper or iron is necessary. This furnace was equipped with retorts. Retorts were shaped as crucible-like hollow containers using refractory clay. The retorts were filled with ground zinc ore and charcoal and were closed with clay. At one end a retort was equipped with a thin tube-like opening for flowing off of carbon oxide gases and zinc vapour. They were heated from outside all around up to c. 1100 °C. Vapour of zinc was collected in a vessel below.

An example of such constructions is the distillation retorts reconstructed by Craddock et al. (1989, 2017), as they were able to show after their excavations in Zawar (Rajasthan, North West India). The retort shown in Fig. 6.32 dates from the fourteenth century AD and represents a copy of many dozens that, for the firing process, were placed close to each other and upside down in a perforated clay floor. The retorts were heated from the outside. The metallic zinc dripped out of the distillation retorts and was collected in a vessel, which was fixed below it. In Zawar several furnaces were excavated with dozens of fourteenth-century distillation retorts. Detailed descriptions of excavated industrialised zinc production from the fourteenth century AD in North West India and scientific examination of those materials are published by Craddock (2017a, b) and Middleton et al. (2017). Due to their research, they are convinced that the technological roots of zinc production lie in India.

Craddock et al. (1989) comment: " … To successfully smelt zinc a very different apparatus was necessary, the koshthi. The form of this seems to have been inspired by either the humble pottery kiln, or more likely the furnace used for producing brass both of which are sometimes divided by a perforated floor. Of course the principal of operation is totally different in the pottery kiln or cementation furnace. There is fire in the lower chamber and the flames go through the perforation in the floor… .".

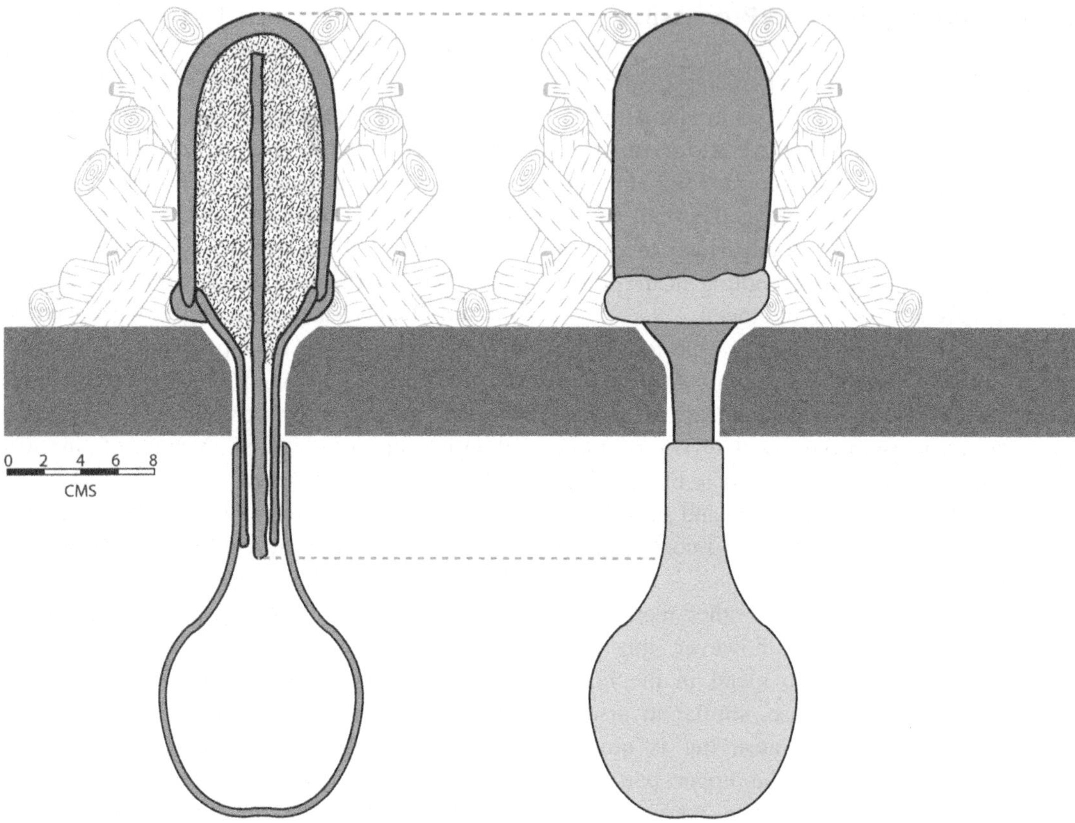

Fig. 6.32 Reconstruction (section and from outside) of a loaded distillation retort filled with calcined zinc ore and ingredients from fourteenth-century Zawar, North West India. The retort was shaped like an open-ended clay vessel. After filling the retort with calcined zinc ore, a funnel-shaped clay condenser was fitted over the open end and sealed with clay. Note the small stick inserted to form the central channel and to hold the ore pieces in place when the retort was inverted into the furnace. In the drawing, a hypothetic vessel is fixed to the distillation retort where the metallic zinc dripped down. A dozen or more of such retorts were placed in a cupola superstructure furnace with a perforated floor. Above the floor, the retorts were heated. Modified from Craddock (1989, 2017) and Marqués Sierra (2018)

Evidence of distillation processes under strictly reducing conditions and at high temperatures is unknown from the ancient world; only Pliny and Dioskuridos (both first century AD) seem to know distillation methods, and only in part, e.g. in the production of mercury (Grothe 1971). It can be assumed that, in ancient times, metallic zinc may have been produced in small quantities in the smelting of zinc-bearing lead ores, copper and iron ores as well as in the production of brass.

In Europe, Albertus Magnus (c. 1269 AD) reports on the production of zinc in the thirteenth century. In the sixteenth century AD, zinc was also mined in small quantities on the Rammelsberg in the Harz district (Germany). The smelting of the intensively intergrown sulphidic silver–copper–lead–zinc ore produced metallic zinc, which, according to Watson (1786), precipitated in the flue. The metal, which for a long time did not have a name as such, is identical to the substance that Georg

Agricola (1556) called *Contrafeth* or *Conterfei*. The name zinc first appears with Paracelsus (1922–1933). He called it a bastard of metals. The metallic nature of zinc was first recognised by the chemist Löhneyß GE von (1690), who lived in Goslar, Harz district (Germany). The industrial production of zinc as a separate metal began in Central Europe in the eighteenth century.

Craddock et al. (1989) note that the only metal (other than zinc) condensed from a vapour during the smelting process was mercury. The conditions for this were much less rigorous since just heating the mercury ore cinnabar (HgS) at about 700–800 °C would cause it to decompose releasing mercury vapour that could be easily condensed on a cold surface in the atmosphere with no danger of reoxidation. The construction of the retorts was very similar.

6.9 The Metallurgy of Mercury

Mercury (chemical symbol Hg) is a silvery-white metal. Due to one specific physical property, it has an outstanding importance among the metals: it is liquid at room temperature. Its melting point is at −38.9 °C. Due to the low boiling point (357 °C), the vapour pressure above liquid mercury at room temperatures is high enough to cause symptoms of poisoning when inhaled by humans. These temperature ranges of liquid mercury allow it theoretically to be found in archaeological excavations—if at all!—no matter whether in cooler or hot climate regions. Its density is extremely high (13.55 g/cm^3; for comparison gold 19.3 g/cm^3). Mercury has high surface tension and likes to roll up into small globules.

Names

The Latin name of native mercury is *argentum vivum*. In contrast, mercury produced from cinnabar (HgS) was called *hydrargyrum* (Pliny, Nat. Hist. XXXIII). It is not clear if these two sorts of mercury were recognised as materials identical to each other. The name cinnabar originates from the Greek word *kinnabari* ("dragon blood"). This is related to the red coloured resin of dragons blood tree (*Dracaena draco)*.

6.9.1 Cinnabar and Mercury in Ancient Times

6.9.1.1 Cinnabar

In prehistory, mercury must have been known to at least those miners who mined the intensely red pigment vermilion (i.e. cinnabar and HgS), at least since the Neolithic (Weisgerber 2003). In many ore deposits, the metal already separates from this sulphide, droplet for droplet (Beck and Berg 1922). However, the use of mercury could not be reliably proven for prehistoric periods. In contrast to cinnabar, mercury would be not easy to find and recover it in archaeological excavations.

The use of cinnabar as a red colour pigment has been known in history for a long time. It was first used in Mesopotamia and ancient Egypt, albeit rarely (Moorey 1994; Lucas and Harris 1967). In the Royal tombs of Ur cinnabarite was found in traces on a skull of a young woman in the grave of PG 1237, the Great Death Pit. As suggested by Baadsgaard et al. (2011) it was perhaps used as a preservative for embalming bodies, but Hauptmann et al. (2015) favoured the strong colouring and decorative effect of this rare mineral at the burial.

Cinnabar was used in mural paintings in Anatolia at least since the Neolithic (Kosay and Gültekin 1949). This probably points to a prehistoric use of the numerous local deposits in Asia Minor.

The Romans mined the red ore in large quantities in Almadén (Sisapo), Spain (Pliny, Nat. Hist. XXXIII, 118). Up until the recent time, Almadén was the largest mercury mine in the world as well. But the mercury deposits of Monte Amiata in Tuscany (Italy) or Idria (Slovenia) may have been mined in Roman times too, although the archaeological evidence is difficult to find (Giumlia-Mair 2009). The pigment extracted from cinnabar, the vermilion, was very popular with the Romans, but also very expensive.

6.9.1.2 Mercury

Craddock (1995) suggests that the use of mercury itself possibly seems to have begun in the last centuries of the first millennium BC in both the Old World, in countries of Middle Asia (cf. Sect. 3.4.14) and in China. In the Old World, mercury is mentioned in the works of the Greek and Roman people Aristotle, Theophrastus, Dioscorides and Vitruvius. They report that the production of mercury from cinnabar would have been done by pressure and vinegar. Woolley (1938) found over a kilogram of mercury at the Phoenician settlement of Al-Mina at the Mediterranean coast in southern Turkey. He believed that mercury had been used for the extraction of gold. By the fourth century BC the technique was established.

Pliny (Nat. Hist. XXXIII, 20, 42) describes in his work that mercury was needed for mercurian gilding. This means that distillation of mercury from cinnabar was done. This is also described (XXXIII, 118–125). For mercurian gilding, objects were coated with amalgam, an Au–Hg alloy, after which the mercury was evaporated (Lins and Oddy 1975). The use of gold and mercury, and the knowledge of the amalgamation effect also very likely involved the extraction of gold (and also of silver) from corresponding ores by this method (cf. Sect. 3.8). In post-antiquity, in the tenth century AD, Al-Hamdani is the first one to describe gold extraction by amalgamation with mercury (Dunlop 1957).

The only archaeological finds of mercury in Germany are from the Viking settlement of Haithabu (north Germany, c. 770–1066 AD). In several places, altogether 287 g of the liquid metal were found (Steuer et al. 2002). Like many of the silver coins found there, the mercury likely comes from the numerous deposits in Central Asia. In any case, shards of spheroidal, thick-walled ceramic transport bombs were also found in Haithabu, as well as in Tajikistan. Presumably, the mercury was imported for the gilding and silvering of objects. A little later, in the twelfth century AD, amalgamation is also known to Theophilus Presbyter.

In China, mercury seems to have first been used in the latter part of the first millennium BC (Craddock 1995). However, there it is not clear if mercury was used to extract gold before the medieval period. But plating gold onto silver or copper by amalgamation is known by the fourth century BC.

The gigantic use and loss of mercury in the modern gold extraction by the amalgamation process (cf. Sect. 6.3) by the Spanish Conquista in south-western America, as well as the historical current gold production in Brazil or California, for instance, resulted in dramatic environmental damages (Alpers et al. 2005; de Lacerda and Salomons 1998).

6.9.1.3 Poisonous Mercury

The ancient authors are sceptical about the use of mercury and cinnabar in medicine, as the toxicity of the two was well known (Pliny, Nat. Hist. X XXIII; Vitruv, Ten books on architecture, VII, 8–9). Nevertheless, both were used in ancient times as a cure-all (due to their toxicity, however, with corresponding negative consequences). In the fifth century AD, the mercury compound sublimate (Hg^{II}-chloride) was known. Paracelsus was the first physician to produce precipitates and basic mercury salts in the sixteenth century AD and used them as a cure.

6.9.2 Production and Processing

Mercury, like zinc (and later arsenic), is extracted via the vapour phase. These (semi-) metals are characterised by their low melting and boiling points.

The extraction of mercury from its ores is divided into two sections (Tafel 1929):

1. The decomposition of cinnabar (HgS), respectively, the production of mercury vapour.
2. The condensation of mercury vapour.

Production of Mercury Vapour

Mercury is extracted almost exclusively from its sulphidic ore cinnabar. When HgS is heated while exposed to air, which constitutes a roasting process, the following reaction takes place:

Fig. 6.33 Almadén (Spain), mining area of San Teodoro. Planes with dozens of chains of aludels fixed to each other for the distillation of mercury. This arrangement is between the furnaces and the condensation arrangements on the other side. Mercury precipitated from vapour and was collected in the middle and lowest part of the plane. Photo: R. Slotta, Deutsches Bergbau-Museum Bochum

$$HgS + O_2 \rightarrow Hg + SO_2 \uparrow$$

This reaction proceeds vigorously at temperatures between 350 and 400 °C. In order to achieve a complete distilling of the Hg, work has to be conducted at least at its boiling point of 357 °C. If the oxide HgO forms during the roasting process, this, in contrast to zinc oxide, decomposes rapidly to metallic mercury at higher temperatures:

$$2\,HgO \leftrightarrow 2\,Hg + O_2$$

Since cinnabar, however, also already sublimates at 580 °C, it is possible, by heating above this temperature, to expel the mercury as a sulphide, which is then decomposed via exposure to air. This is therefore not a reaction between a fine or coarse-grained solid (HgS) and a gaseous component (air, O_2) depending on the preparation, but a reaction between two gases. This is much more complete than the former.

Condensation of Mercury Vapour

The condensation of the mercury vapours takes place in aludels. An aludel is a pear-shaped vessel open at both ends so that several such vessels may be fitted into one another to form a series. The arrangement of aludels was already shown in Figs. 6.30 and 6.33 while discussing the production of antimony and zinc. Since the mercury, once passed from the vapour phase to the liquid state, readily separates because of its high specific gravity, good cooling is important. From the Middle Ages onwards, these condensing units were made of aludels until the recent past. The aludel arrangements from Almadén (Spain) are world-famous (Hauptmann and Slotta 1979), which have been included as technical monuments in the UNESCO World Heritage List (Fig. 6.33).

7 Metals and Alloys

7.1 Ancient Alloys: General Remarks

Pure metals are rarely used in modern technology. In most cases, alloys are used. These are very intimate, deliberately produced mixtures of a base metal and other metals, or today also non-metals, e.g. with silicon. The base metal for a tin bronze or for brass is copper, alloying elements are tin and zinc. Steels are alloys of iron with carbon.

By the number of elements contained in an alloy (as opposed to the low concentrated impurities) one distinguishes binary alloys, e.g. the very early arsenical copper (CuAs), the Bronze Age tin bronzes (CuSn), as well as the Iron Age FeC alloys. Ternary alloys, for example are AuAgCu alloys (tumbaga). There are also quaternary alloys, such as the Roman lead–tin brass or gunmetal (PbSnCuZn). They are listed in more detail in Table 7.1. Chemical analyses may identify the presence of particular alloying metals in sufficient concentrations to exclude that they are not just a background impurities.

For technical reasons, alloys have therefore always been of far greater importance than pure metals because they have the following physical properties:

- Higher hardness
- Lower melting temperatures, which improved the casting process
- Colour changes

Table 7.1 The following categories of ancient alloys were found in ancient times

Gold–silver (electrum)	Au + Ag
Tumbaga	Au + Cu ± Ag
Silver–gold (aurian silver)	Ag + Au
Arsenical copper	Cu + As
Antimonal copper (bronze)	Cu + Sb
Tin bronze	Cu + Sn
Leaded bronze	Cu + Sn + Pb
Brass	Cu + Zn ± Pb
Gunmetal	Cu + Zn + Sn ± Pb
Steel	Fe + C
Cast iron	Fe + C

One exception to this is gold. Since about the middle of the first millennium BC, it was produced by parting from the naturally occurring AuAg alloys as pure as possible (Ramage and Craddock 2000). Whether or not earlier parting is possible is discussed by Pernicka (2014a, b, 2018).

Deliberate Alloying or by Chance?

In classical times and later, alloys have been mostly produced by purposefully co-melting individual "pure" alloy components. It must be emphasised, however, 7that in the early stages of metallurgical developments, prior to the globalisation of the Old World in the Early Bronze Age, the composition of alloys was probably decisively impacted by the composition of ores. Metals such as iron, arsenic, cobalt, nickel, lead, silver, zinc and/or antimony were inadvertently introduced into an alloy because of

A. Hauptmann, *Archaeometallurgy – Materials Science Aspects*, Natural Science in Archaeology,
https://doi.org/10.1007/978-3-030-50367-3_7

smelting polymetallic ores. This means that, for example arsenic- and nickel-containing copper as well as low tin bronzes or brass could result from smelting such ores.

In archaeometallurgy no guidelines of modern metallurgy should be adopted, where a composition is already considered an alloy with just 1 wt.% of foreign metals. Here it is a precondition that all alloying parts are available in pure form. For archaeometallurgy, one should follow the suggestion of Tylecote (1991), who refers to (deliberately mixed) bronzes only as such at a foreign metal content of 4 wt.% and above.

Furthermore, the geographic distribution of raw sources is a relevant factor for identifying ancient alloys. This applies above all to the development and distribution of tin bronzes. In Fig. 3.44 (Sect. 3.5.6), it is shown that tin deposits are relatively rare. Of supra-regional importance were tin deposits of Cornwall and those in Central Asia. Unfortunately, little is known about the prehistory of tin deposits in the Ore Mountain Range (Germany). Those in Turkey are well investigated (Kestel and Göltepe, Yener 2009), but they were not of supra-regional importance. The currently earliest known bronzes of the 3rd millennium BC originate from a region that does not have any ore resources, namely Mesopotamia. The nearest tin deposits are located about 1000 km away in Afghanistan or Central Asia, copper deposits maybe some closer at the Iranian Plateau.

The following chapter discusses a series of alloys in more or less chronological order, as suggested by Charles (1980), who sees the coming of alloys as a logical metallurgical sequence. The earliest use of native copper in the Neolithic in the Old World (9th/8th millennium BC) is superseded by gold alloys, followed by arsenical copper and copper–antimony alloys. After that, tin bronzes are spreading, followed by iron or steel.

7.1.1 The Earliest Alloys

Among the earliest alloys are the numerous gold–silver artefacts excavated from the graves of Varna dating back to the end of the 5th millennium BC (Leusch et al. 2014). Then there are the eight gold rings of Nahal Qana (Israel) dating to the second half of the 4th millennium (Gopher et al. 1990). They also consist of Au–Ag alloys with up to approximately 30 wt.% silver. It is almost certain that in both cases the artefacts consist of natural gold–silver alloys (see Sect. 7.3). At Varna one ring-idol was found with 50 wt.% gold, 14 wt.% silver and 36 wt.% copper. This artefact was most probably deliberately made because of its reddish colour.

The hoard of Nahal Mishmar (Israel, c. 3500 BC; Bar-Adon 1980; Tadmor et al. 1995) is one of the most prominent examples of the earliest deliberately produced copper-based alloys. It consists of 416 metal artefacts. 251 of these artefacts are roughly globular and elongated (piriform) maceheads, eleven are disk-shaped maceheads. Furthermore, scepters, standards, maceheads and crowns belong to the hoard, as well as 19 tools and tool-shaped artefacts (chisels, axes and adzes). The last group of finds consists of unalloyed copper. It was found to be of local origin and was smelted from ores in the area of Faynan. By contrast, most artefacts consist of Cu-based AsSb alloys with up to 15 wt.% As and up to 25 wt.% Sb. One AsNi-rich alloy containing 8.6 wt.% Ni and 4.1 wt.% has been found as well. Chemical and lead isotope analyses have shown that the AsSbCu alloys were not mixed together from a compound of pure copper and an As–Sb–rich alloy. On the contrary, apparently complex fahlores (tennantite, Cu_3SbS_3, or tetrahedrite, Cu_3AsS_3) or their weathering products were the starting materials for the artefacts. Such ores do not occur in the Levant and allow the conclusion that these alloys originate from regions like the Iranian Plateau or the Caucasus.

The early production of brass from mixed ores such as zinc-containing copper ores or zinc-containing ores + copper can be cited as an example as well. Corresponding resources are rarely scattered geographically, far less than copper deposits. Production of brass by purposeful mixing of the metals zinc and copper was produced later. Comments regarding this are made in Sect. 7.2.5.

7.1.2 Physical Properties

Pure metals should not be used for any materials in which mechanical properties play a role, they have to be hardened. Hardening can be achieved by thermomechanical treatment, namely by cold work hardening. Subsequent tempering at temperatures of 600 °C–800 °C to induce the precipitation of microstructures such as the formation of solid solutions, eutectic crystallisations, segregation behaviour, and cooling rates (Hornbogen and Warlimont 1991). The strengthening effect is thereby completely or at least partially cancelled. Through the combination of multiple deformations and heat treatments, the shaping of metals can be combined with the targeted adjustment of the desired material properties. Some metals or metalloids, for example arsenic, also act as deoxidiser.

On the other hand, the higher hardness of alloys is a result of lattice defects in the alloyed material, because foreign atoms with different atomic radii than the parent substance are incorporated into the crystal lattice. Lattice tensions emerge, the hardness of the alloy increases. In arsenical copper solid solution strengthening remains very limited up to a concentration of c. 4 wt.% As, the hardness increases only insignificantly (Lechtman 1996; Budd and Ottaway 1991). Most of the early arsenical copper artefacts are in this range. Therefore, their concept of a deliberate alloying of early copper artefacts with arsenic is hardly convincing.

In total, at alloying the following phenomena occur (Askeland 1996; Hummel 1998):

1. The components of the alloys do (almost) not start chemical reactions and provide only limited solubility. They are present as "pure" components intergrown in various textures. Examples of such alloys are CuPb, CuFe, PbFe, and SnFe.
2. The components are reacting with each other, leading to the classical intermetallic compounds.
3. The components form mixed phases with each other. This also applies to the solid state. These are mixed crystals or solid solutions.
4. Mixtures of several phases of the three mentioned types can arise. Thus, in alloys of copper with about 40–45 wt.% zinc and about 1 wt.% lead, a CuZn solid solution phase with about 37 wt.% Zn (α-solid solution) is formed, as well as an intermetallic compound CuZn (β-solid solution) and "pure" lead in the form of inclusions.

Types 1 and 4 are referred to as heterogeneous or multi-phase alloys, types 2 and 3 as homogeneous or single-phase alloys.

7.2 Copper-Based Alloys

7.2.1 Classifications and Properties

Thanks to the work of many scientists over the past 70 years, there are now many thousands of analyses of copper alloys available for further research. Important databanks of (pre-) historic copper-based alloys are published by Junghans et al. (1960, 1968, 1974) and Krause (2003). Craddock (1976, 1977, 1978) published on copper alloys of the Archaic, Classical and Hellenistic Greeks. In the Berlin databank, Riederer (2001, 2002) published metal analyses on Roman copper alloys and those of the Central European Bronze Age as well as Etruscan, Sardinian, Greek, Egyptian and Near Eastern objects.

With the proliferation of copper as the earliest metal used worldwide, alloys have been rapidly formed, many of which may have been unintentional at first, but can then be quickly considered as intended mixtures of metals. In relatively late cultures, complex alloy systems occur often. Based on these developments it was often attempted to set up a nomenclature for antique alloys. Despite the tremendous variety of chemical compositions, it has been found that there are four basic principal alloying metals of copper. These are arsenic, tin, lead, zinc, and only subordinated antimony.

These influence the physical properties of copper in different ways. Arsenical copper is very easy to work with. Up to a few weight-percent contents, it can be forged and annealed as many times as required for shaping without rendering the alloy unusable. Tin bronzes increase the strength of this very malleable alloy, but also decreases its plastic behaviour with increasing contents of tin and finally leads to brittleness and high hardness. At the same time, the melting point decreases, which improves castability. Additions of lead facilitate casting and therefore provide suitable casting alloys. Copper with zinc content is called brass (the antique *aurichalcum*) and is well castable and malleable.

Bayley (1990) applies modern criteria for a classification of antique alloys: In Classical Antiquity, copper was alloyed with three metals only: tin, lead, and zinc. Any other metals present in minute amounts were usually unintentional impurities. Of these metals, tin was the most important alloying component prior to the first century BC, when zinc began to be alloyed with copper on a large scale to make a golden-coloured alloy, today called brass. Copper alloyed with significant amounts both of tin and brass is a gunmetal.

The relationship between alloy name and chemical composition is demonstrated in the ternary diagram in Fig. 7.1. In this diagram, the contents of these three metals are added together and normalised to 100%. The closer you get to a corner, the higher is the proportion of that element. Brass is an alloy with >5 wt.% Zn, and a gunmetal is an alloy containing several percent of both tin and zinc. Indeed, there is an increasing body of evidence to suggest that, from the 1st century BC onwards, in many cultural environments, brass became strongly associated with the Roman Imperialism (Ponting 2002, 2010).

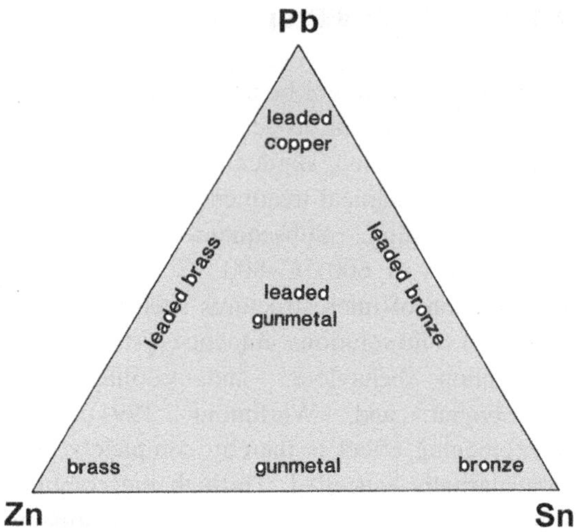

Fig. 7.1 Copper-based Pb–Zn–Sn ternary diagram showing the relationship between alloy names and composition for copper alloys (Bayley 1990)

Practical Applications

In their extensive work of the Roman–Germanic metalwork, Voß et al. (1998) and Hammer and Voß (2000) describe how ancient craftsmen who worked in the processing of metals knew how to distinguish and judge different alloys according to their practical usage. These are colour, castability, forgeability, machinability, but also the sound of objects. They show that the recurring correlations of preferred compositions for certain types of objects indicate that an understanding of different alloys already well existed in antiquity. However, this is not regionally limited. In her work on metal artefacts in Mexico, Hosler (1994) described how in the time from the sixth to seventeenth centuries metal objects were differentiated based on the criteria of sound and colour.

Especially in the works on the Roman–Germanic artefacts, it is shown how important simple, nature-connected test methods are and how they can answer these questions better and faster than modern devices can today. This certainly applies to traditional metalworking activities in many countries where metal is produced on a small scale (cf. Chap. 9), even today.

The practical approach of metal processing at high temperatures becomes particularly clear in the discolouration of iron. The colours of iron in a forge fire range from yellow to blue to grey between c. 180 °C and 400 °C. The annealing colour (600 °C–1200 °C) ranges from dark red to white.

According to the fundamental physical and optical properties of these individual alloys and their different colours, there were already different groups of craftsmen in ancient times (Petrikovits 1981), which have often survived to modern times (bell founder, brass founder, bronze founder, coppersmith, bronzesmith). The Roman craftsman obviously knew exactly which alloy was suitable for forging, and which was suitable for casting. He most probably had a deep knowledge of the properties of metals and alloys and applied suitable testing methods (Furger and Riederer 1995). If this was clearly understandable, in the written evidence it is sometimes doubtful. Pliny (Nat. Hist. 33 XXXIII and XXXIV), in his discussion on copper and its alloys, already exclusively uses the term *Aes*. But that can mean either copper, bronze or brass. He distinguished between *Aes Regulare* (bar copper, forging copper) and *Aes Caldare* (cast copper).

Therefore, Hammer and Voß (2000) have proposed a grouping based on the requirements for processing and use of copper and its alloys. Six groups were suggested including the three main alloying elements copper, tin, lead and zinc.

These individual groups are:

1. Copper: It is characterised by very good forgeability, but poor castability. Its forgeability is suitable for embossing sheet metal, the poor castability does not allow for castings with contours. The products are copper coloured. It can contain a maximum of 5% admixtures mainly of tin, zinc or lead.
2. Tin bronze: Characterised by easy castability, but difficult forgeability. It contains 5–14 wt.% Sn, with increasing tin content the casting properties are improved and the forgeability is decreased. Forging is only possible in the cold state and with repeated annealing. Items made of such a bronze have a high strength and the colour is bronze look. While in the Bronze Age and in antiquity tin bronzes with c. 10 wt.% Sn are widely distributed, so-called forging alloys with tin contents of often only 6 wt.% are used in modern times.
3. High-alloy tin bronze with >14 wt.% Sn: It represents a hard casting alloy used for bells and mirrors. It is so hard that only grinding is possible. When polished it will be shiny as a mirror. The tin concentration in the two-phase diagram Cu–Sn represents a limiting content above which the brittle δ-phase occurs.
4. Cast bronze, lead bronze (Pb > 5 wt.%): It is characterised by very good castability and is therefore used for the casting of objects with high acutance (statuettes). This alloy is not malleable, but its surface can be worked well (engraving, turning).
5. *Aurichalcum* with at least 5 to a max. of c. 28 wt.% zinc: This alloy (brass) is easy to cast and forgeable cold and hot. According to Pliny, it is one of the noble varieties of *Aes* (Pliny Nat. Hist. XXXIV: 4). The excellent processing properties make the antique brass one of the most sought-after alloys. In addition, the golden yellow colour is appealing. Brass in the true sense with its characteristic golden yellow colour contains at least 10 wt.% zinc. Such alloys are widely used in Roman times. Roman sesterces and Hemmoor buckets belong to this group. Zinc can be replaced by tin to a limited extent, but the limit of the α-area in the system Cu–Sn–Zn must not be exceeded.
6. Mixed bronzes are zinc, lead and tin alloys, containing at least about 3 wt.% of each element. It is deformable at low levels, at higher levels it is easily castable.

In accordance with these practical subtleties in the processing of alloys, their complexity becomes clear. There is still a very detailed distinction between copper-based alloys, so that one can discuss whether the common term "bronze" is actually sufficient for the characterisation of the material.

7.2.2 Arsenical Copper

7.2.2.1 The Spread of Arsenical Copper

Copper artefacts with concentrations of arsenic of about 1–2 wt.% and slightly above are widespread all over the Near and Middle East, in the early cultures of Central Europe (Junghans et al.

1960, 1968, 1974; Schubert 1981; Budd and Ottaway 1991), beyond in the North Pontic areas and Eurasian Steppe Belt (Chernykh 1992, 2011), in the Far East (Mei 2009), and later in South America (Lechtman 1980) as well.

This alloy was produced over so many centuries, it is named "arsenical copper", less frequently "arsenical bronze" which would not fit the original meaning of "bronze" as a deliberately produced alloy. Arsenical copper is the dominant alloy in the earliest development of metallurgy, typically occurring in a period between the use of unalloyed copper in the Neolithic period and tin bronzes in the well-developed Late Bronze Age. It is, in fact, the alloy of the Chalcolithic period and the Early Bronze Age. In the Old World, arsenical copper first appears at the transition of the 5th/4th millennium with the beginning of copper ore smelting. It marks the beginning of extractive metallurgy. This alloy was the reason for Selimkhanov (1977), who observed a distribution of enormous amounts of arsenical copper artefacts in the (Trans-) Caucasus area, to coin the term "Kupfer-Arsenzeit" (Copper–Arsenic Age) as a preliminary stage of the real Bronze Age. In fact, especially in the Near Eastern Early Bronze Age, this term is much more reliable because tin bronzes are very rare at this time.

That arsenical copper marks the beginning of smelting copper ores in the Near East is strongly supported by the studies of Begemann et al. (1994) on metal artefacts dating to the 4th millennium BC from Ilipinar in western Anatolia. They demonstrated that these artefacts, which have concentrations of arsenic ranging from 1.4 to 8 wt.% differ from metal objects of Neolithic origin, which have exceptionally low concentrations of trace elements. Neolithic objects had been made from native copper, while the ones from Ilipinar were made of copper smelted from ores with varying arsenic contents.

The actual origin of this arsenical alloy, however, still remains a matter of great controversy. There has been a tendency to see arsenical copper as representing a technologically distinct, more advanced phase of metallurgical development, with the implication that arsenical copper was deliberately produced and that the alloyed metal was known to be superior in comparison to pure (native) copper (Charles 1967). However, "... there is little evidence that copper containing about 1 or 1.5 wt.% arsenic would have offered any real advantages to prehistoric metalworkers (i.e., worldwide) ...". This observation also of Budd and Ottaway (1991) would be supported by Muhly (2006) who stated that there is no conclusive evidence that arsenic was recognised as a separate metal in ancient times, and no ancient language has a word for "arsenic".

Copper–arsenic alloys usually range from concentrations of much less than one percent to about 8–9 wt.%. Alloys with higher arsenic contents are rare, but they do exist. Examples of such compositions were excavated, e.g. in the context in the Great Caucasus area. At Late Bronze Age Lori Berd in Armenia, south of the Great Caucasus, buttons and rings of a bracelet with copper-based alloys used for bimetallic ornamental objects consisting of black copper alloys with up to 27.6 wt.% As were analysed embedded in matrices of tin bronzes (Meliksetian et al. 2011). Not too far away, in the northern Cis-Caucasian Digora culture, similar artefacts were found. At Faskau, in the necropolis of Koban (15th–11th centuries BC), black coloured inlays of domeykite-like compositions (Cu_3As) in tin bronze objects were reported from earlier excavations (Born 1984). Originally, these inlays—like the ones of Lori Berd–had a silvery shining tint, similar to the colour of the below mentioned *white copper*.

It should be shortly mentioned here that dozens of homogeneously composed ceremonial artefacts such as maces of a combined copper–antimony–arsenic alloy were found in the 5th millennium BC hoard of the Nahal Mishmar cave in Israel (Tadmor et al. 1995). The copper-based alloys contained up to 25 wt.% antimony, and 15 wt.% arsenic (Fig. 7.2). This is the most prominent locality for these alloys from the Chalcolithic period, but similar artefacts of comparable age were also found in a number of localities in the southern Levant (Shalev 1991).

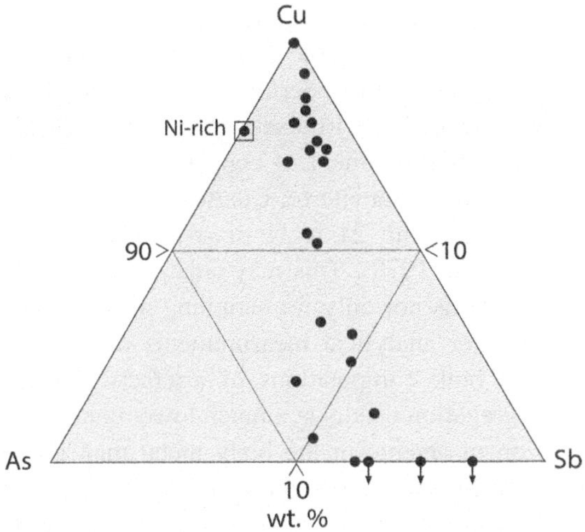

Fig. 7.2 Chemical composition of ceremonial artefacts from the Chalcolithic hoard of Nahal Mishmar, Israel, in the ternary system Cu–As–Sb. The plots show a good correlation between arsenic and antimony and probably indicate the smelting of a variation of fahlore-solid solution ($Cu_3(As,Sb)S_{3.25}$) or their secondary decomposed products, embedded in changing amounts in "pure" copper ores. Note the composition of the nickel-rich (Ni) elongated macehead. It contains 8.6 wt.% nickel. Data from Tadmor et al. (1995)

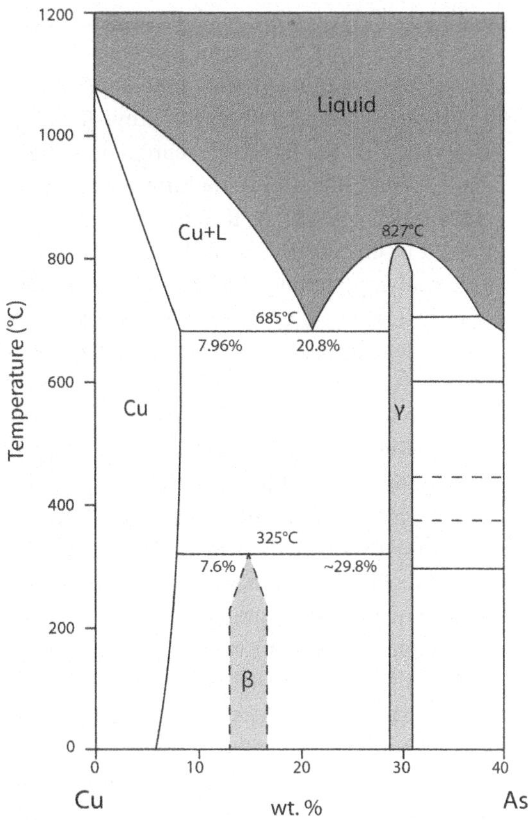

Fig. 7.3 Section of the binary phase diagram system of copper–arsenic (Cu–As). Modified after Subramanian and Laughlin (1998). The β-phase composition is c. $Cu_{6-9}As$, mineralogically identical with whitneyite, and the γ-phase is Cu_3As which corresponds with domeykite

7.2.2.2 The Copper–Arsenic Phase Diagram and Physical Properties

The equilibrium diagram for copper–arsenic (Fig. 7.3), especially the part close to copper, is the basic point for looking at the behaviour and properties of this alloy system. It shows the form in which the alloy exists for any combination of concentration and temperature. It must, however, be pointed out that it represents an ideal equilibrium situation in which both liquid and solid phases are homogeneous. In reality, non-equilibrium conditions are prevailing, cooling rates are high and the atomic diffusion rates, especially in the low-temperature area are slow, so that phase transformations are sluggish (Northover 1989; Pereira 2015). The solid–liquid interface becomes unstable and the result is often the characteristic tree-like, dendritic microstructure of cast metals.

As with many other solid solutions, arsenical copper also shows the following properties:

1. Arsenic, like antimony or tin contents in copper is slightly lowering the melting points of the alloy in comparison to pure copper. For instance, pure copper starts to solidify at 1084 °C; copper with 5 wt.% As starts to solidify at c. 1050 °C, and copper with 10 wt. % at c. 950 °C. The lowering of the temperatures of solidification leads to a slight improvement in casting properties. Dies (1967) suggests that under modern metallurgical conditions even less than 1 wt.% As would already result in a denser cast compared to arsenic-free copper.
2. By segregation, arsenical copper may discolour copper a little bit, giving to its alloys

a slightly lighter (white) colour at percentages as low as 2–3 wt.%. Arsenic contents of 4–12 wt.% create a golden tint, and copper with 12–18 wt.% As has a silvery to greyish colour. However, if an inverse segregation takes place, where the lower melting point of the arsenic-rich phase at the eutectic point is moved to the surface, then the alloy will provide a silvery-shining colour even in the range of only 3–5 wt.%, while the core is copper. At higher arsenic contents, the so-called *white copper* (c. 37–63 wt.% As), which is brittle, but very well polishable. It tarnishes very easily on contact with air and, like all As-rich compounds, turns black. When fresh, white copper is a bright alloy that was formerly used (mostly silvered) as a substitute for silver (von Prechtl and Karmarsch 1838). Since the middle of the 19th century, white copper has been replaced due to its toxicity by the copper–nickel–zinc alloy nickel silver. In any case, silvery shining surfaces of arsenical copper was intentionally created to raise the status symbolism of these objects (Mödlinger and Sabatini 2016).

3. Copper can dissolve only a few weight percent of arsenic, islands of Cu_3As will appear at lower temperatures and under equilibrium conditions (Budd and Ottaway 1991). The incorporation of As atoms in a crystal lattice of copper causes increased metal hardening because the two metals have different atomic radii (arsenic 115 picometer; copper 135 picometer). This results in a contraction of the crystal lattice of copper, which leads to tensions.
4. Because the slopes of the solidus and liquidus are very steep in the phase diagram, inverse segregation in real situations is so severe that the first signs of this phase appear in cast alloys with only a little over 1 wt.% arsenic (Budd and Ottaway 1991). It is a typical feature of arsenical copper. The arsenic-rich melt exudes this phase at the surface of the casting to create an almost uniform silver surface to the alloy. Inverse segregation or coring is a practical consequence of fast cooling of melts in a mould under non-equilibrium conditions. It leads to considerable differences in composition between the exterior and the centre of the casting. It leads to arsenic-rich layers outside with an enrichment of copper–arsenic phases such as whitneyite (c. Cu_9As) and domeykite (Cu_3As) with 21 wt.% or so (McKerell and Tylecote 1972). This may create considerable problems not only for sampling artefacts, but also for analytical measurements of surfaces and bulk compositions of artefacts. Inverse segregation occurs at a much lower concentration of arsenic in the body metal than is the case for tin bronze, probably due to the high vapour pressure of arsenic in the molten alloy (Meeks 1993).

Mödlinger and Sabatini (2016) state, that any thermal treatment (annealing) would have resulted in the loss of the crystallographic characteristics of inverse segregation. Therefore, any arsenical copper showing this phenomenon must either have been simply cast and not further altered, or was only cold deformed. Based on the dataset of 35,000 chemical analyses of Chalcolithic and Early Bronze Age As-bearing copper artefacts (Junghans et al. 1960, 1968, 1974; Krause 2003) the authors observed that the majority of arsenical copper objects with more than 2 wt.% As do not show a silvery surface at all. It is reasonable therefore to assume that objects with silvery surfaces are status items.

Which techniques were definitely applied in the Bronze Age, for example to create decorative details on metal figures is still being discussed controversially. A prominent example of this discussion is probably the Horoztepe (Anatolia) bull figurine dating to around 2100 BC (Fig. 7.4). Its origins are not entirely clear, because Horoztepe was looted before the excavation, and many metal artefacts were brought uncontrolled into the United States (pers comm. J.D. Muhly, November 2011). The frontal part of the bull has a silver-shining surface, which in the opinion of Stanley Smith (1973, 1981) consists of arsenic evaporated by a cementation process.

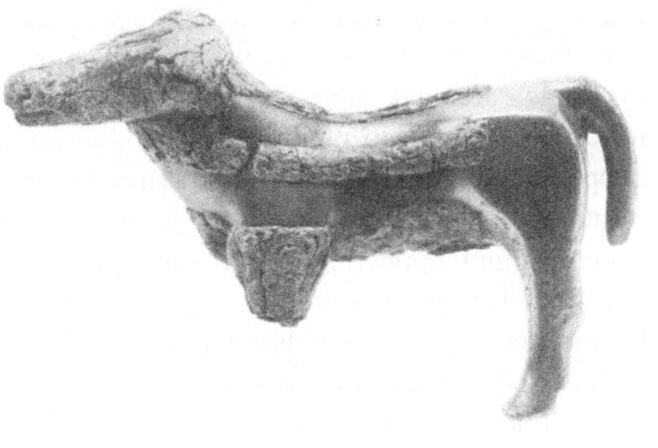

Fig. 7.4 Horoztepe (Anatolia), Hattian Culture, c. 2100 BC. Bronze bull figurine consisting of two various sorts of metal. The dark parts are made of a (leaded) tin bronze, while the whitish metal plating was created by arsenic-rich copper alloys. Length 12.2 cm, height 9.1 cm. After CS Smith (1981)

It was found that otherwise the hindquarters of this bull were cast in arsenical copper while the front legs are made of leaded tin bronze. In contrast, Scott (2002) concluded that "... the presence of the copper-arsenic phases whitneyite (mixture of algodonite ($Cu_{1-x}As_x$) and As-rich copper) and domeykite (Cu_3As) (analysed by X-ray diffraction method), suggest that surface segregation of arsenic is an explanation for the presence of these phases. Inverse segregation on casting is probably responsible for the coating in this case, since during the casting process some of the lower melting point constituents, such as the arsenic-rich phase, could be carried to the outer surface of the mold".

Figures such as the Horoztepe bull, but made of different metals, have parallels in the late 3rd millennium of Alaca Höyük (Anatolia) and Mesopotamia (Aruz 2003). A number of Early Cycladic II (ca. 2600–2400 BC) daggers show the same phenomenon, especially those said to be from Amorgos and now in the Ashmolean Museum (Oxford) (Sherratt 2000). See, however, these figures usually feature inlays made of silver or possibly electrum.

5. Arsenic added to copper functions as a deoxidant, taking up oxygen enriched in liquid copper, and is visible in the solid state as a mixture of cuprite copper. Arsenic would form arsenious oxide (As_2O_3). It is not soluble in copper and would separate as a sublimate (Charles 1967). Hence, arsenic, according to chemo-physical rules, improves the casting quality in comparison to pure copper. The experiences on this subject, however, are different for the early metallurgy and seem to depend strongly on the casting technique. While Zwicker (1991) observed that even 0.5 wt.% As reduces the porosity of copper under oxidizing conditions, (Budd and Ottaway 1991) had quite different experiences. They write that copper with 11 wt.% As poured into the sand is still porous. They could not observe a relationship between the amount of arsenic in copper and the overall soundness of the castings. Also, Kienlin et al. (2003) after studying Early Bronze Age flat axes from the northern alpine region doubt that arsenic in copper—at least in contents of a few percent—would have been of any importance. However, their encounter with this formation of strong porosity may have been caused by the contact of the melt with the sand. Hauptmann et al. (2015), with their experiments on the casting of oxhide ingots prove that the formation of H_2 and CO from the sand material can lead to the formation of considerable macropores in the solidifying melt.
6. Copper arsenic alloys have a range of reduced deformability in the temperature range between 500 °C and 700 °C (Dies 1967). These temperature ranges should, therefore, be avoided during hot deformation. On the other hand, cold processing in the area of the α-mixed crystal causes no difficulties.

7.2.2.3 Experimental Work

In order to record the physico-chemical processes of the formation of arsenical copper, i.e. to record the uptake of arsenic into the copper, experimental studies were carried out repeatedly. The primary goal of these experiments was to understand how arsenic had entered copper and what mechanical advantages arsenical copper had over pure copper in its application. Many smelting experiments showed that arsenical impurities present as natural constituents in ores contributed arsenic to the smelted copper (Tylecote et al. 1977; Tylecote 1980; Rostoker and Dvorak 1991). They yielded valuable information. However, arsenical copper with low arsenic levels sometimes led to contradictory results. These experiments and their varied results were discussed by Budd and Ottaway (1991); they concluded that copper low in arsenic, i.e. only a few percent, did not contribute to any remarkable improvement of the metal. Overall, however, the goal of these experiments was less to remove arsenic from the copper in order to work out a (physical) advantage of the tin bronzes.

McKerell and Tylecote (1972) show experimentally that the arsenic content in copper depends on the gas atmosphere during the melting process. Under reducing conditions, arsenic is almost completely absorbed by the copper. Once alloyed with copper, arsenic is difficult to remove. The importance of this observation for the development of early metallurgy is not to be underestimated, since this is probably the reason why arsenical copper significantly overlaps with the spread of tin bronzes over time. In contrast, under oxidising conditions, such as during the roasting of sulphidic ores, considerable amounts of arsenic volatilise as arsenious oxide, which smells pungently of garlic. Since arsenic and its chemical compounds are extremely toxic, but are, in varying quantities, always formed during the smelting of As-containing ores, this harmful property has also been considered as an argument for the decrease of arsenic copper and the increasing use of tin bronze in the European Bronze Age.

Pollard et al. (1990) show in their experiments that about 80% of the arsenic from the ores is already absorbed by diffusion into the copper in the solid state <1000 °C. They, therefore, argue that ultimately, the arsenic content in copper is determined by the rate of diffusion as well as by the amount of As evaporated. Rostoker and Dvorak (1991) found in their smelting experiments with malachite and realgar (AsS) that there are difficulties in smelting these materials. Realgar decomposes only slowly, making it difficult to rapidly dissolve As in the copper within one melt. In their experiments, they observed, in addition to the formation of As-containing copper, the formation of speiss and of Cu-matte, which was evidently formed by the reaction between the sulphur content of realgar and copper. All these phases were also retained in the resulting slag.

Co-smelting experiments to produce copper–arsenic alloys were carried out by Lechtman (1996) and Lechtman and Klein (1999) utilising ores collected in the Peruvian Andes. Smelting charges consisted of mixtures of oxidic copper ores with either Cu-sulpharsenides (simplified formula CuAsS) or Fe-sulpharsenides as they occur in the oxidation zones of the local deposits and how they can generally occur in the gossan of sulphidic deposits (cf. Sect. 3.6). Experiments included crucible and furnace techniques. The co-smelting experiments yielded coherent copper–arsenic alloy ingots. Two simplified reactions are:

1. Smelting of malachite + enargite:

$$8\ CuCO_3 + Cu_3AsS_4 \rightarrow Cu, As + 4SO_2 \uparrow + 8CO_2 \uparrow$$

2. Smelting of malachite + arsenopyrite:

$$3\ CuCO_3 + FeAsS \rightarrow Cu, As + FeO + SO_2 \uparrow + 3CO_2 \uparrow$$

The authors concluded that the Cu–arsenic alloys used in ancient Peruvian artefacts could have been made easily, deliberately or accidentally, by co-smelting procedures which were in all probability, sought by the ancient metallurgists for copper arsenic alloy production.

In order to explore the mechanisms of "silvery surfaces" as they are well known on the bull figurine of Horoztepe (Fig. 7.4) Mödlinger and Sabatini (2016) performed experimental work. Generally, their formation is suggested to result from inverse segregation. The authors suggest alternative processes such as a surface enrichment of arsenic by cementation and the formation of α-solid solution.

The second possibility of arsenic enrichment could be a post-depositional precipitation by the formation of a 50–500 μ thick As-rich layer caused by intentional corrosion or depletion of copper from the surface.

For further reading on these issues see also Hosler (1994), Charles (1975), Tylecote (1976), Muhly (1988), Ravich and Ryndina (1995).

7.2.2.4 Origin of Arsenical Copper: Impure Copper Versus Deliberate Alloy

It is still a point of discussion as to how arsenical copper has been formed. There are primarily two possibilities, which have not changed for over 50 years of research (Otto and Witter 1952; Charles 1967; Eaton and McKerrell 1976; Moorey and Schweizer 1972; Strahm 1994):

1. Arsenical copper smelted from As-containing ores
2. Deliberately produced arsenical copper, i.e. arsenical bronze.

As shown, the interpretation and differentiation between various arsenical coppers have focussed so far almost exclusively on the advantageous physical properties compared to metallic copper. Studies in materials science by Buchwald and Leisner (1990) and Budd and Ottaway (1991) suggest a reasonable demarcation between the two at concentrations of c. 2 wt.% As based on a measurable increase of ductility and hardness at this level. Also, Pernicka et al. (1990) suggest 2 wt.% as an approximate limit between naturally formed As–copper and intentionally produced As–bronze, even though they are fully aware of the problematic nature of this margin.

Arsenical Copper from Ores

The basis for this model is the fact that arsenic accumulates in ores and surface-near ore deposits. This also means an accumulation of other metals such as Cu, Sb, Zn, Pb in the "soft iron ores" in the gossan. There are a number of arsenates and Cu–arsenates, which show a greenish colour similar to pure secondary copper ores, e.g. olivenite ($Cu_2(OH,AsO_4)$), chenevixit ($Cu_2Fe_2(OH)_2AsO_4$), conichalcite $CaCu(AsO_4)(OH)$, scorodite ($FeAsO_4 \cdot 2H_2O$), annabergite ($Ni_3(AsO_4)_2 \cdot 8H_2O$) etc. (see Table 3.11). All these ores could have been exploited alongside malachite, brochantite, atacamite etc. Using such ores, arsenic could become inadvertently part of the smelting charge. Dill et al. (2010) have detected such minerals in the old mining galleries in the gossan of Mega Livadi on the island of Serifos (Greece), a few meters below ground only, and Taylor (2011) has shown the occurrence of these minerals in "modern" gossans in Australia, Africa and South America.

Mineral extraction of ores begins at the surface of an ore deposit, at the gossan. Therefore, in the early stages of extractive metallurgy, particularly these arsenic-containing parts must have been mined perhaps everywhere. It can be assumed that the arsenates are often associated with residual sulphides of primary As-rich ores in oxidation zones, so that local accumulations of oxide–sulphidic mixed salts can occur. These can reach massive proportions. In view of the in part enormous dimensions of oxidation zones, e.g. in the copper deposits of Rio Tinto in the southwestern Iberian Peninsula (Velasco et al. 2013), in the Iranian high plateau (Nezafati et al. 2008), or in the Andes (Hartley and Rice 2005), arsenical copper can be easily considered as a natural "dirty copper" (Lechtman 1996). A roasting of such ores prior to smelting is not necessary, and in such a case the arsenic as well as antimony is transferred almost completely from ore to metal (Tylecote et al. 1977).

Despite the supra-regional spread of arsenic copper, fluctuations occur. There are localities that are associated with artefacts with comparatively low As contents, and others, in which arsenic contents of multiple percent occur. From the geological view, arsenical copper high in arsenic could be produced by smelting of enargite or fahlores and related sulpharsenides or related secondary ores (see above). Sulphides may occur in relics in the gossan, but mainly in the transition of oxidation to secondary enrichment zone. Regrettably, only a few examples exist where metal analyses can be directly allocated with the appropriate original ores.

Copper districts very low in arsenic contents are the sedimentary deposits of Faynan (Jordan), Timna (Israel) and Sinai (Egypt). Ores from these deposits usually contain arsenic in the order of only 100 ppm (overview in Abdel-Motelib et al. 2012; Hauptmann 2007). Copper metal produced during the Chalcolithic in the southern Levant did not exceed the 40 ppm-level (Tadmor et al. 1995), Early Bronze Age bar ingots from Faynan yielded not more than 0.05–0.07 wt.% As (Hauptmann et al. 2015). Additionally, the more than one hundred copper oxide and bun ingots, most of which were produced from the massive, for the Late Bronze Age extremely important, volcanic-sedimentary sulphide ore deposits (MVS) on the island of Cyprus, usually contain only arsenic in the tenths of percent range or below (Hauptmann et al. 2002a), which can be ascribed to the already very low arsenic contents of this deposit type. In Serbia, Pernicka et al. (1993) and Schmitt-Strecker and Begemann (2005) made the observation that most of the Eneolithic objects were found to consist of remarkably pure copper. All later objects have higher concentrations of arsenic and antimony (c. 0.4–100 ppm, in few cases up to 0.5 wt.%). They could perhaps derive from the polymetallic ore deposit of Majdanpek.

Significantly higher arsenic levels, up to the lower percentage range, can be found in artefacts recovered from sites in Anatolia. Geographically-geologically, this is the region of polymetallic deposits in the context of the Tethyan Eurasian Metallogenic Belt (TEMB). The arsenic content of artefacts from Chalcolithic/Early Bronze Age strata at the urban settlement of Arslantepe, is in the lower percentage range (1–4 wt.% As). Swords and spearheads of the palace hoard (level VIA) and artefacts from the "royal" tomb (level VIB1) (Fig. 7.5) are mostly situated between arsenic contents of 2.5–5 wt.%, reaching up to 7 wt.% As (Caneva and Palmieri 1983; Hauptmann et al. 2002a). In the settlement itself,

Fig. 7.5 Arslantepe (Anatolia), "Royal" Tomb, Early Bronze Age I (level VIB, 3000–2900 BC). Adzes made of arsenical copper (2–3 wt.% As), probably imported from Transcaucasia. Lengths of the adzes 13.9 cm. Photo: Roberto Ceccacci. From Di Nocera et al. (2004)

arsenopyrite with up to 40 wt.% As was found as well, though lead isotope analyses lend support to archaeological suggestions that they were imported probably from Transcaucasia.

Chalcolithic artefacts from Norsuntepe at the upper Euphrat (Anatolia) (Pernicka et al. 2002), however, only contain As contents of $\leq$1 wt.%. In the Early to Middle Bronze Age, they only increase to 2 wt.%. Again, no agreement with the ore finds is possible. The arsenic content of all Norsuntepe artefacts poses a problem for the assignment to specific ores found in the excavations, since the As/Cu ratio of the ores is significantly lower.

Arsenical Copper Deliberately Produced

Numerous scholars suggest a production of arsenical copper by conscious smelting of As-rich polymetallic secondary copper ores or primary sulphides, or by adding pure As-sulphides to copper such as, e.g. realgar (As_4S_4), orpiment (As_2S_3), native arsenic, or, alternatively, an arsenic-rich mother alloys (Tylecote and Merkel 1985), or arsenic-rich speiss to copper (Thornton 2010).

Ore deposits with massive mineralisations of the striking realgar and orpiment are outcropping at numerous ore deposits worldwide, e.g. at the surface in the Great Caucasus (Twaltschrelidze 2001) or at Zarshuran in north-western Iran

(Mehrabi et al. 1999). They are located close to several copper prospects with evidence of prehistoric mining, but like native arsenic, no evidence of early mining has ever been discovered at these deposits.

A blend of speiss with copper in early metallurgy cannot be ruled out. In Sect. 5.8, numerous finds of speiss from countries in the Old World, which were made in the context of the (pre-) historical non-ferrous metallurgy, are described. They were found as inclusions in slag, but also as isolated pieces. Speiss is usually made of iron arsenides but can become so-called deficiency alloys (German *Mangellegierungen*) with increasing copper contents.

Intentional arsenic alloying is suggested for the Early Bronze Age I Poros Katsambas (Crete, Greece) (Doonan et al. forthcoming), where heat altered fragments of loellingite ($FeAs_2$) along with heavily slagged crucible fragments were found. The slag frequently contains CuAs prills with up to 40 wt.% As. In contrast, ingot fragments from associated contexts contain no detectable arsenic. It remains unclear, however, whether the ingots were made at this site or were imported from elsewhere. A comparison like that is doubtful. At the Early Bronze Age II site of Chrysokamino, also located on the island of Crete, copper prills embedded in some metallurgical slags show As- and Ni-concentrations in the lower percentage level and well above. As five tiny pieces of ores analysed from the sites are considerably lower in these two elements, a deliberate adding of As-, Ni-minerals is suggested (Bassiakos and Catapotis 2006). However, the source(s) which supplied Chrysokamino with ore are not known (Betancourt 2006a; Gale and Stos-Gale 2006), clear evidence is difficult. No copper ore deposits are known on the island of Crete.

At the Early Bronze Age site of Arisman (Iran) an estimated 20 tons of slag provide evidence of sustained and large-scale production of metal. Inclusions in the slag suggest that both speiss and arsenical copper could have been produced side-by-side at the same site but using different ores (Rehren et al. 2012). The authors are assuming a tentative separate production of speiss, which would have been added later to the copper (ore). Unfortunately, they could not provide any information about the exploited resources. No larger pieces of speiss were found as at, e.g. Monte Romero (Craddock et al. 1987).

The most convincing example of a possibly intentional production of arsenical copper is that of the island Kythnos (Greece), as detailed by Gale et al. (1985). They have analysed copper inclusions in Early Bronze Age slag As-concentrations, varying widely between 0.2 and 6.6 wt.%, which they can assign to compatible ores from the island. They thus provide confirmation that arsenical copper was smelted unintentionally, but after the crushing of the slag followed a deliberate choice of the copper prills rich in arsenic, in order to produce a suitable alloy. The arsenic-rich silver-coloured prills could easily be distinguished from the reddish pure copper prills by their colour. Results of lead isotope analyses made it plausible that As-containing copper artefacts of the hoard (from Kythnos) were also made from the same metal.

7.2.3 Antimonial Copper and Bronze

As Forbes (1950) remarks, many antimony-containing copper alloys have been found among the oldest copper alloys, but—with some exceptions—the percentage of antimony is so low that it can be ascribed to natural impurities of copper ores used, rather than to any addition of specific antimonial materials during the smelting of the copper.

This chapter, however, focuses on a previously little-studied area of archaeometallurgical alloys that of antimonial copper or bronze containing between c. 2 and 25 wt.% antimony.

7.2.3.1 Copper–Antimony Phase Diagram and Physical Properties

Perhaps one of the most obvious and desirable properties of copper–antimony alloys, especially for antimony contents above 10 wt.%, is their colour. Experimental preparation has shown that with antimony contents above 10 wt.% a golden colour is obtained with a silver touch at about 20 wt.% antimony (Pike et al. 1996).

Like arsenic, antimony is lowering the melting temperature of copper, but it is lowering much more: The copper-rich part of the phase diagram shows a eutectic point at 19 wt.% Sb at temperatures of only 645 °C (Dies 1967). Therefore, antimonial copper should have enormous casting advantages. In the solid-state antimony is partially soluble in copper up to 5.8 wt.%.

Copper–antimony alloys have a large solidification interval. Similar to copper arsenic alloys they have a pronounced shrinkage tendency and show crystal segregation. Discontinuous precipitation was reported for antimony concentrations of 9–10 wt.% and temperatures below 410 °C. The segregation temperatures are quite low, so that even at temperatures below 300 °C coarse lamellar segregation was observed. Concerning mechanical properties, Dies (1967) reports about the significantly bad influence of antimony on modern copper. He states that it is a harmful alloying element. The hardening effect of antimony on copper would be comparably low.

Alloying of copper with 3–7 wt.% Sb increases its hardness but with Sb contents above this it becomes brittle to the extent of forming cracks when worked (Charles 1980), because of the precipitation of intergranular phases. Prolonged heating of the alloy at about 650 °C in homogenisation experiments, produced more uniform microstructures and reduced hardness somewhat.

Due to the low increase of hardness and the brittleness, antimonial copper or bronze has poor mechanical qualities. The alloys are not suitable to come under stress during their use. In fact, in most cases, objects made of antimonial copper have decorative characters and were not used as weapons, chisels, or knives.

7.2.3.2 Origin of Antimonial Copper: Impure Copper Versus Deliberate Alloy

As already mentioned above, antimonial copper is known from the late 5th millennium BC Nahal Mishmar hoard found in the cave west of the Dead Sea (Israel). Tadmor et al. (1995) analysed several maces and maceheads and found varying concentrations of antimony (1–25 wt.%, see Fig. 7.2). In addition, the artefacts contain <1–15 wt.% arsenic and as minor and trace elements Ag, Bi, Co, Ni and Zn. These alloys were probably smelted from polymetallic ores such as fahlores or their weathering products. They are not products of deliberate alloying. As there are strikingly many complex antimony–arsenic copper alloys in the Caucasus, an origin in the fahlore deposits in the Gornaya Racha region in the Caucasus seems quite possible (Chernykh 1992; Gambaschidze et al. 2001; Hauptmann and Gambaschidze 2001).

Otherwise, the occurrence of high antimony copper artefacts low in the elements mentioned is rare. Other notable occurrences come from Velem St. Vid, western Hungary (Davies 1935), and from Hasanlu, Northwest Iran (Maclean et al. 1992). The distribution of antimonial copper, hence, is limited in space and time.

There are two explanations to produce high antimonial copper. The first is to co-smelt copper minerals with antimony contents (e.g. tetrahedrite) or (pure) antimony minerals (e.g. stibnite and Sb_2S_3). The second possibility is the addition of metallic antimony to copper. Either of these two possibilities are intentional alloying practices. These are not adventitious discoveries. In this case the alloys could be named antimonial bronzes.

At Velem St. Vid, these alloys were first discovered in the archaeological record during the mid-1890s, primarily from Late Bronze Age/Early Iron Age Hungarian contexts (Urnfield period). Velem St. Vid was one of the major metallurgical sites with ample metal workshops in the middle of the Danube Basin during the Late Bronze Age (Miske 1908). Initial identification by (wet) chemical analysis, identified the presence of high antimony copper alloys (18.6 wt.% Sb). Prior to this, antimony had only been discovered in the archaeological record as an unalloyed element. It most probably comes from the local source of antimonite near Schlaining.

Metallic antimony was used for making many small decorative items such as buttons and beads mostly found during excavations conducted in the

Great Caucasus, e.g. in the Late Bronze Age Koban cultural context (15th–8th century BC) in North Ossetia (Pike et al. 1996). As native antimony is rare in nature (Ramdohr 1975), and, on the other hand, the production of metallic antimony from stibnite (Sb_2S_3) is not too complicated and is proven for the Bronze Age in Georgia (Sect. 6.7), objects made of antimonial copper or bronze are deliberately produced alloys. These Late Bronze Age antimonial copper or bronze objects from the Koban-Culture area contain a significant amount of antimony. Quantitative analyses show 1–10 wt.% Sb in the majority of antimonial alloys (Pike 2002). There are also artefacts containing between 5 and 10 wt.% Sb, often in conjunction with 1–2 wt.% tin. Antimony was obviously used to mix it with copper for making these alloys.

Whether the use of antimonial bronze, and other alloys containing antimony, is more prevalent in the Koban artefacts than the southern part of the Great Caucasus (Pike 2002), must remain open, as in the western Caucasus numerous antimonial bronze objects were found with 12–20 wt. % Sb. They originate, among others, from Brili, Tbilisi, and from Middle Bronze Age graves in Bornighele in Meshketien (Gambaschidze and Gambaschidze 1995). It can also be assumed that antimonites from the Sophito deposit (Racha, Georgia/Caucasus) or antimony melted therefrom were used for the casting of antimony pearls or as an alloying additive. There are several deposits in the Caucasus with stibnite as the dominant ore (Sect. 3.4.2).

Many of the antimony pearls from Georgia consist of extremely pure antimony, in which the content of all measured trace elements is well below 1 wt.%. Only the S-contents reach 1 wt.%. Although there is very little archaeological evidence for the extraction of "pure" antimony, these artefacts have most likely been smelted from the original ore Stibnit (contra Hauptmann and Gambaschidze 2001). With the findings of antimony pearls from the burial ground of Brili, the evidence is tangibly close that in the mining area of Racha in the Middle Bronze Age, this semimetal may have been deliberately smelted even in larger quantities.

Apart from the fact that antimony from the Caucasus was presumably traded to Mesopotamia (in Ur, according to reports by Woolley in 1934, antimony was found as well), its use is also probable as an alloying agent in copper. This is hardly surprising, because in the Middle Bronze Age alloying techniques were already well known in the Middle East. This already becomes evident by the widespread tin bronzes.

The Georgian antimony bronzes also contain predominantly As contents in the percentage range, and only for a few it is below 1 wt.%. While As contents in copper artefacts are a pronounced characteristic of Bronze Age metal artefacts, and are almost consistent between 1% and 10%, arsenic can also be introduced via antimony or antimony ores such as stibnite.

However, some weapons and maceheads and a few of the Colchis axes consist of Cu–As–Sb bronzes with As and Sb in the percentage level (each up to c. 10 wt.%) (Hauptmann and Gambaschidze 2001). Similarly to the artefacts of Nahal Mishmar, these were likely smelted from fahlores.

Finally, it should be mentioned that Dayton (1978) described four beads from an Iron Age grave at Tell Farah (south) in Israel that were made of 66 wt.% tin and 33 wt.% antimony. An alloy of antimony and tin makes a whitish metal of bluish tint that is malleable and ductile. It may be confused with silver. It is evidence that antimony combines readily with copper and other metals to form alloys.

7.2.4 Tin Bronze

Tin bronze is an alloy of such importance, that it was eponymous for a whole era of human history. In the (pre-) historical context, the term "bronze" is used when the alloy contains more than 2 wt.% tin. In modern metallurgy, however, copper-based alloys with at least 60 wt.% copper are called bronze, as long as they are not attributable to the main alloying additive zinc. Metallurgically, the term is used today together with the preceding main alloying additives. Those are correctly addressed as aluminium, lead, manganese, or phosphorus bronzes. In modern metallurgy, pure metals are mixed to produce bronzes. Alloys can, therefore, be precisely designated as such, provided that they contain 1 wt.% of these additional components.

In the early metal ages, secondary impurity constituents of ores, gangues, and host rocks or deliberate fluxing agents could already enter the metal during the smelting processes, making such a clear demarcation much more difficult. Tylecote (1991), for example, suggests to speak of a bronze only at a content of at least 4 wt.% of tin (see also Pare 2000). But this has to be checked individually. The classification of bronzes was commented on in Sect. 7.2.

Tin bronzes show varying colours depending on their tin content. If the tin contents are below 10 wt.%, then the bronzes are reddish to tombac-coloured. If they rise to >10 wt.%, bronzes are coloured orange to yellow-gold-like. This is the usual colour of archaeological bronzes. At more than 20 wt.% tin, the bronze becomes whitish and loses the characteristic colour of a bronze. High-tin copper coins (22–27 wt.% Sn, the so-called mirror metal) have been used in the 1st century AD in England (Hedges 1964). Since they are silvery in colour, they imitate this metal. With these colour gradations, however, tin bronzes deviate significantly from those of copper or arsenical copper and thus represent a new metal. Bronze reaches a higher grade of hardness and is easier to cast than copper.

7.2.4.1 The Geographical Distribution of Tin Bronze

The question arises of how people came up with the idea of alloying copper with tin, as was the case in the developed Bronze Age, especially since tin is rare in comparison to copper (see Fig. 3.44). That is a tremendous technological achievement. And yet, people have managed to exploit tin deposits, to negotiate the metal over long distances to produce many tons of bronze. Tin bronze, therefore, is of greatest cultural and historical significance. Given its limited occurrences and the enormous demand for this metal, tin can be considered to be a strategic metal (Parzinger 2016).

The spread of bronze evidently did not just happen in one, namely the Eurasian cultural space, as shown in Fig. 7.6. This is a region where strategically important tin resources are sometimes very close, but in some cases very far from the centres of bronze metallurgy in central Europe. Pernicka (1998) and Radivojević et al. (2018) have shown that the locally nearby geological tin deposits in the Ore Mountain in eastern Germany, in Cornwall, and in Brittany in northwestern France could only be defined as the source for the pronouncedly rich bronze objects with noticeable difficulty, since prehistoric (mining) archaeological evidence is hard to find. On the basis of chemical and lead-isotopic analyses of many artefacts and ores, however, it could be shown that in the Bronze Age, the new metal was imported to the (central) European countries through extensive trade from Central Asia (Tajikistan, Kazakhstan, Uzbekistan, probably Afghanistan). In these countries, prehistoric tin mining has been proven (Parzinger and Boroffka 2003; Stöllner et al. (2013a, b); Garner 2014; Gontscharov 2019).

In addition, there are other countries and regions where geographical centres of bronze production have developed independently. These are on the one hand in northern Thailand and further north in the Yunnan Province in the southwest of China. Tin production is also proven from Japan and Africa. A valuable compilation can be found in Penhallurick (1986). The geological information about the Chinese tin deposits is incomplete. Hosking (1988) stresses, however, that many of them, as well as those of Cornwall, Afghanistan, etc., are copper bearing. Stech and Maddin (1988) and Pigott and Natapintu (1988) also describe geographically nearby copper deposits for Thailand. From Fig. 3.44 it can be seen that these areas of early bronze production, unlike those in the Old World, are located in extensive tin belts, where there were and are abundant tin deposits, both in primary deposits and in placers. The "strategic" metal transported over long distances in Eurasia was available here in large quantities.

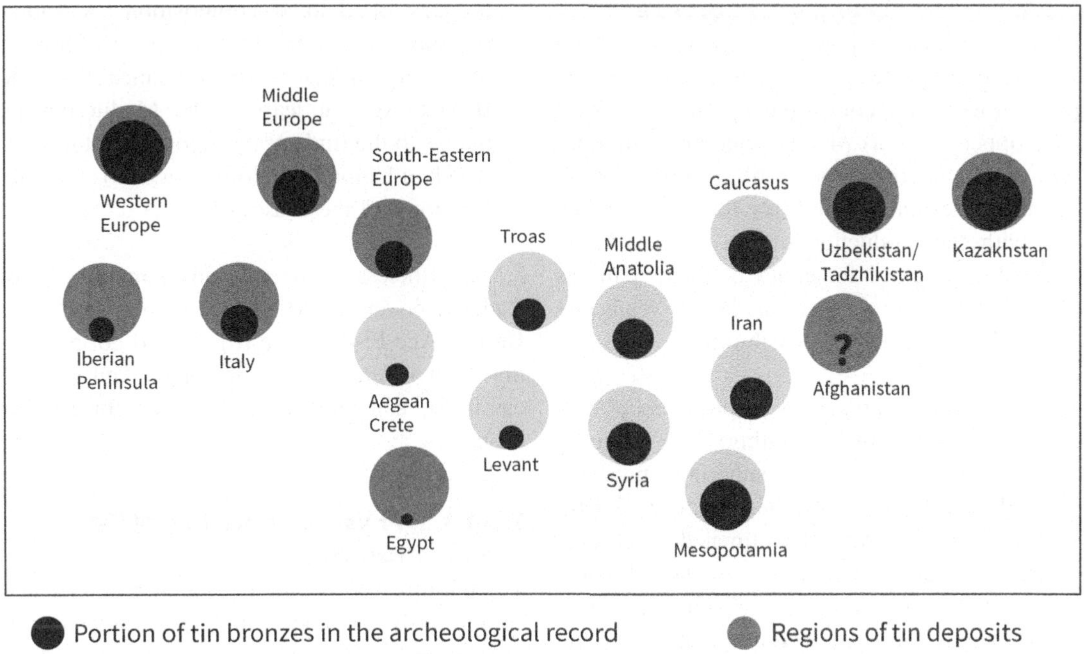

Fig. 7.6 Distribution of tin bronzes in the western part of the Old World and Middle Asia mainly between the 3rd and 2nd millennia BC. These data are compiled based on Spindler (1971), Wolfart and Wittekindt (1980), Chernykh (1992), Kaniuth (2006, 2007) and Stöllner et al. (2013a, b). The sizes of the dark circles are not scaled drawings. They should show trends. For the time being, the possible importance of more than 100 tin deposits in western Afghanistan is not investigated. Modified and enlarged after Pernicka (1998)

According to new radiocarbon data, in north-east Thailand the first centres of bronze technology developed approximately 2000 BC at the sites of Ban Chiang and Non Nok Tha (Stech and Maddin 1988; Higham et al. 2015).

High-tin bronzes containing >20 wt.% Sn present a striking phenomenon in the archaeological record. Pryce et al. (2014, 2017) have published a whole series of high-tin bronzes containing values of 30–40 wt.% tin. However, they had been measured by energy dispersive X-ray fluorescence spectroscopy, apparently on the surfaces of (corroded) fragments. The extent to which strong segregation effects of tin played a role here is to be apprehended.

The rapid proliferation of tin bronze at the expense of pure copper, or rather arsenical copper, represents a (technological) revolution associated with a significant improvement in tooling, equipment, jewellery, and weaponry manufacturing. However, large quantities of prestige objects and jewellery also show that the spread of tin bronze cannot be based solely on material science (see below), but was also based on aesthetic aspects, since tin bronze has a pleasant yellow to golden brown colour and was used at least in the early stages of their use only for pomp and prestige objects and not for everyday objects.

This discussion also deals with the question of what exactly was traded: tin and/or copper ore, tin ingots or pre-alloyed metal as ingots or final artefacts. This is connected with the still unanswered question, where ultimately the raw materials of tin bronzes originated.

Overall, there is a considerable preponderance of copper and bronze artefacts over sparse finds of tin metal throughout the Bronze Age. It has already been shown (Sect. 6.5) that the massive presence of tradable tin metal in the shape of ingots in Europe is limited to the Late Bronze Age and later contexts.

7.2.4.2 The Making of Tin Bronze

For a long time, the production of tin bronze, in what form it reached the end consumer, from the place of its manufacture—the locality of tin ores and copper ores—as well as when and where tin was added to copper, have been controversial. There are several possibilities for this, yet just like with the production of brass, there is no irrefutable evidence for either of them (Charles 1975, 1980; Rostoker et al. 1983; Waniczek 1986). There are several possibilities:

1. The earliest tin bronzes were produced as random alloys by smelting of stannite (Cu_2FeSnS_4) or its weathering products (varlamoffite, mushistonite, cf. Sect. 3.5.6). This should not be underestimated. Such ores are widespread not only in the classical localities of Europe (Cornwall, Ore Mountains, Brittany, Iberian Peninsula) in the surface near parts of tin deposits, but also in Central Asia. Mushistonite (proven for Tadjikistan, Garner 2014), with its green colour, resembles the green malachite, arguably the most common oxide copper ore from which copper was smelted.
2. Smelting of malachite + cassiterite. Parageneses of (secondary) tin and copper ores are common (see Sect. 3.5.6). The production of tin bronze by smelting these two ores has been proven experimentally. For this purpose, however, high-grade ore concentrates are required. Impure material containing constituents of gangue and host rock would have led to significant losses of tin and the formation of slag crusts. Prehistoric tin slags, however, are exceedingly rare.
3. Addition of cassiterite to liquid copper under reducing conditions. This is a cementation process in which the cassiterite is reduced to metal, which forms a continuous series of mixed crystals with copper.
4. Co-melting of the metals tin and copper. A decrease of melting temperatures (e.g. a copper alloy with 10 wt.% of tin will melt at 950 °C) allows a working at relatively low temperatures. It is the most likely and probably the most widely used method of producing tin bronzes of defined composition. A supra-regional production of tin bronzes by alloying of tin to copper is proven since the Late Bronze Age, at least in the Mediterranean, thanks to the finds of tin ingots from the shipwreck of Uluburun (1 ton of tin bars) and other locations off the coast of Israel.

Nevertheless, deep investigations of Rademakers et al. (2018) on crucibles from Late Bronze Age Pi-Ramesse (Egypt) do not exclude that next to the melting of copper and tin also cassiterite and copper were melted for making bronze.

7.2.4.3 Physical Properties of Tin Bronzes

The reasons why the era of arsenical copper was continuously replaced by tin bronzes at the beginning of copper smelting are generally stated both with aesthetic criteria and with the manipulation of material properties. It should first be noted that there are a number of factors in the production of arsenical copper, which was popular before tin bronze, which has made it difficult to achieve a targeted and safe composition of this alloy with constant physical components.

Charles (1967) supposed that tin as an alloying metal for making bronzes replaced arsenic because of the volatility and harm of arsenic vapours. These uncertainties also include the loss of arsenic in the roasting of ores, the changing composition of ores in the oxidation zone, changing redox conditions when smelting in crucibles and smaller furnaces, and the loss of arsenic in copper during high-temperature processing.

These criteria seem to be too simple, because arsenical copper was produced over many centuries, in some regions for two thousand years. Ravich and Ryndina (1995) suggest more convincingly that only the comparison of the properties of arsenical copper and tin bronzes and peculiarities of their metallurgy in different regions of the Ancient World will make it possible to understand why this transition took place. It will be shown by lead isotope analyses in Sect. 11.3.1 that tin bronzes indeed came to Europe from completely different ore deposits than those located in the Tethyan Eurasian Metallogenic Belt.

Casting Properties

Liquid copper absorbs oxygen from the air, which is exsolved during solidification. This leads during the crystallisation of Cu^+ oxides to the formation of bubbles and blow holes. As a result, the metal quickly becomes porous and brittle and its ductility is significantly impaired. A foamy surface is formed. Impressive examples of such copper are the late Bronze Age copper oxhide ingots, especially from the shipwreck of Uluburun (Hauptmann et al. 2002a; Yalçın 2008). Likewise, these effects could be observed experimentally by casting of such oxhide ingots (Laschimke and Burger 2011; Hauptmann et al. 2016). However, the role of arsenic as a de-oxidant in old copper objects was thoroughly questioned due to metallographic investigations by Kienlin et al. (2006).

This phenomenon does not appear with tin bronze, because tin as an antioxidant in the concentrations of about 10 wt.% is binding oxygen more effectively than arsenic in copper. In addition, tin bronzes (as well as arsenical copper does) have slightly lower melting points than copper (1084 °C). Whether these temperature differences have been an effective argument for better casting in ancient times is an open question. A tin bronze with 10 wt.% Sn solidifies just above 1000 °C, arsenical copper with 10 wt.% As solidifies at 950 °C.

Toughness and Hardness

As with almost any alloy formation, tin–bronze is also hardened by distortion of CuSn solid solution crystallisation structure during cooling of a melt. In the solid state, several different phases are formed depending on the temperature. Rapid cooling, as can be assumed for the handling of metals and alloys in ancient times, however, lacks the time for a rearrangement of individual solid solutions. No equilibrium is established and the alloy is still existing in equilibrium conditions of higher temperatures, even when it cooled down.

Moesta (1983) and Scott (1991) are describing the following complicated effects:

If a typical tin bronze with c. 10 wt.% Sn cools quickly from the melt of c. 1200 °C so that no equilibrium conditions can be set, then the alloy remains relatively soft and can be hammered while cold. However, if several steps of shaping an object require reheating, it becomes brittle. To avoid that, you have to do an extended annealing and quenching again in order to forge it in the cold state. On the other hand, annealing and slow cooling of a finally shaped artefact will result in a higher hardness, if brittleness is an acceptable side effect. Tin bronze and copper behave in exactly the opposite way than iron. Annealing and quenching make them malleable and workable in the cold stage. With iron, hardness and toughness like steel are produced by rapid cooling due to internal stresses.

Thus, in case a bronze object should be shaped in the cold state, annealing treatment is required. At suitable high temperatures and with corresponding long-term annealing this finally leads to a homogenisation of heterogeneous grain structures. It will be recrystallised more or less completely.

The knowledge and application of cold working and annealing of bronzes can be demonstrated exemplarily on the flanged axes of the Langquaid type. These axes have a wide distribution in southern and southwestern Germany, Switzerland and Austria (Abels 1972). These axes come from Lower Bavaria and date to the Early Bronze Age after about 1900 BC (Krause 1996). They regularly contain around 10 wt.% tin. Kienlin et al. (2004) examined such axes und observed a fully recrystallised microstructure with annealing twins pointing to a deformation by cold working before annealing. They point out that at least a two-phase forging process seems to have been the norm after the axes have been cast in a closed mould. This means that the axes were not forged primarily to give them their shape, but to increase their hardness.

However, the knowledge of annealing metal artefacts can be traced back to the Neolithic period where the earliest native copper was treated in Asikli Höyük and Çayönü Tepesi in Anatolia (Yalçın and Pernicka 1999). This included the knowledge of work hardening, but also the opposite effect, namely that at higher temperatures the ductility of metal objects could be improved again.

Another effect of the tin bronzes is the strong tendency for segregation. This means that solid solution crystals precipitated during cooling at first from a melt are lower in tin than the last ones. This leads to coring that can be homogenised by annealing. This is necessary if a cold deformation of an object is planned. Bronze objects also show the appearance of a reverse macrosegregation, i.e. higher tin contents at the surface of an artefact than at its inside. This phenomenon cannot be removed.

This may not be confused with deliberate tinning. Very early examples for tinning are some CuAsNi artefacts (sword and spearhead) from Tülintepe (Anatolia) which were manufactured at the late Chalcolithic/Early Bronze Age I period at the transition of the 4th/3rd millennium BC (Yalçın and Yalçın 2008). If correctly dated, this metallurgical technique could be predated in the Near and Middle East for a time period 5000 years ago.

Due to its wide distribution particularly interesting is the range of tin bronzes with tin contents between 10 and 15 wt.%. Up to this point, copper can dissolve tin as a homogeneous mixed crystal. It crystallises the α-phase from the melt. Theoretically, the solubility of tin in copper would decrease with decreasing temperature, and the more tin-rich ε-phase would precipitate. However, under normal working conditions and annealing times, and because of the retarded adjustment of phase equilibria in the copper–tin system (Chase et al. 2007), this enlarged area of solubility of tin in copper can be maintained during cooling.

7.2.4.4 Origin of Tin Bronze: Impurity Versus Deliberate Alloying

In spite of the above-mentioned debate, the question as to when an intentional tin alloy can be expected can, from an archaeological point of view, be answered more clearly than in the case of arsenical copper, at least as long as the standard tin bronzes are present in larger quantities with approximately 10 wt.% Sn. The widespread "classic" tin bronzes have this composition.

However, there are transitions and deviations, so a distinction between unintentionally imported tin contents and deliberately produced alloys is not always safe. As discussed above, the border between copper and bronze is assumed to be 1–2 wt.% tin, sometimes even >4 wt.%. Cleuziou and Berthou (1982) chose 0.5%, arguing that copper from the Iran–Afghanistan region would rarely be associated with a significant amount of tin. However, they are not aware about the compositions of copper or tin deposits in these regions.

Based on cuneiform tablets from Ebla, Müller-Karpe (1990a) assumed that tin contents of less than 1 wt.% in copper could also be intended additives. The tablets date to about the middle of the 3rd millennium BC, and refer to recipes and standard mixtures of copper and tin for the production of spearheads and other weapons (Van Lerberghe 1988; Waetzoldt and Bachmann 1984). They seem to indicate that by this time period the use of tin was perhaps an established practice in the production of metal artefacts. It is seen as a standard usage of a material that would only be possible if a reliable and accessible source for the procurement of this material would have been available.

In such debates, the knowledge of ore formation of tin and copper deposits is indispensible, which was discussed in Chap. 3. According to a close geochemical association of tin with tungsten, molybdenum, tantalum, beryllium and boron, it can be assumed that "pure" tin was produced from a majority of these resources. It should be noted, however, that tin ores are paragenetically associated with copper ores both near the surface of their deposits and at greater depths. This is true for many tin regions of the globe (Hutchison 1988), so it should be assumed that mixed copper–tin ores were melted into corresponding "natural" alloys, at least over certain periods of time. Also, copper deposits with tin minerals occur.

This does not only apply to the Balkan Peninsula, where Radivojević et al. (2013) described

Vinča Era copper-based artefacts (mid 5th millennium BC) from excavations in Pločnik (Serbia) with 11.7 wt.% tin. She strongly emphasises that this alloy is the result of a smelting of corresponding mixed ores.

Mixed CuSn ores were also found in Central Asia, Afghanistan, Cornwall, the Ore Mountains, Brittany, the Iberian Peninsula and South Africa. According to Wolfart and Wittekindt (1980) tin ores in Afghanistan are associated with chalcopyrite ($CuFeS_2$) in varying amounts (cf. Sect. 3.4.11).

The occurrence of the secondary CuSn ore mushistonite has already been reported. According to Recknagel (1908), cassiterite was found in close mineralogical paragenesis with copper ore at Rooiberg (South Africa) and thus it is possible that tin bronze was produced by smelting copper ore "contaminated" with cassiterite (Chirikure et al. 2007).

This means that in the earliest stages of metallurgy, tin bronze may not have been necessarily produced by intentional smelting of two ores or metals. This is supported by the fact that in the early phases of tin bronzes (in the Early Bronze Age in the middle of the 3rd millennium BC), copper artefacts are still relatively common. Finds of tin, however, are limited to very few exceptions.

It should also not be overlooked that hydrothermal tin deposits can also contain arsenopyrite in addition to Cu-bearing ores (Cox 1986), so that arsenic would possibly occur in the metal after smelting of such ores. Earl (1985, 1986) describes how during assaying of tin ores in Cornwall, unwanted amounts of arsenic can be removed by roasting the ore concentrates.

Tin bronzes occur in the Middle East in the middle of the 3rd millennium BC and spread over long distances from Mesopotamia via Anatolia to Europe in a surprisingly short time (Fig. 7.6). In Central Europe, the spread of tin bronzes starts two to three centuries later (Pernicka 1998). In northern Europe and the British Isles, bronze occurs between 2000 and 1800 BC (Pare 2000). Lead isotope analyses on the example of bronzes from Troia, Poliochni and Kastri in northwestern Anatolia showed that these tin bronzes could not have been produced locally but were imported from unknown sources (Muhly and Pernicka 1992). For they show a lead isotope pattern which indicates an origin of geologically much older resources than those found in north-western Anatolia! There is no evidence of a local production of copper and tin bronze artefacts.

The earliest bronze artefacts date from the Royal Tombs of Ur, Mesopotamia (c. 2400 BC). This region is void of any metal resources. Recent researches have been proven that, on the one hand, copper artefacts from Ur have been imported from mineral deposits in Oman (Begemann and Schmitt-Strecker 2009; Begemann et al. 2010; Salzmann 2019). On the other hand, lead isotope analyses of the bronze artefacts from Ur have shown that they originate from geologically very old ore deposits, not from Oman. This is a comparable situation as in the Troas.

7.2.5 Brass

Brass is the name for alloys of copper and zinc, provided that copper makes up at least 50 wt.% and zinc >5 wt.%. The brass of antiquity, the *aurichalcum* (greek *oreichalkos*), contains about 5–30 wt.% zinc. Modern brass contains 5–50 wt. % zinc, often with small amounts of aluminium, nickel, silicon or tin. The fact that *aurichalcum* actually meant our present-day brass is shown by Pliny's remark that it has the same quality (bonitas) as the sesterces and dupondies, which in the 1st century AD were known to have been made from brass with zinc contents of 17–23 wt. % (Pliny, Nat. Hist. 34,2).

"Pure" brass, in which the copper is only alloyed with zinc and contains all other elements in concentrations <1%, is very rare in ancient times (Riederer 2001). It more often contains varying amounts of tin and lead. Based on these minor components, various classifications have been worked out. Voß et al. (1998) discuss classifications for six groups of alloys that they plotted in ternary diagrams (Cu–Sn–Pb; Cu–Sn–Zn). They are based on a number of different physical properties (casting and forging properties). Bayley (1990), at the same time, compiles a group of seven copper-based alloys with the same metals zinc, lead and tin (Fig. 7.1). Among others, the lead- and tin-containing alloy of (leaded) gunmetal appears in this diagram. Riederer (2001, 2002) orders the diversity of Roman–Germanic brass species according to their increasing zinc content. Accordingly, brasses with less zinc containing tin and lead are more likely to be random products, which are also produced by melted scrap.

7.2.5.1 Ancient Brass

According to Thornton (2007) there are over 30 examples of prehistoric and protohistoric artefacts just from south-west Asia (eastern Mediterranean, Anatolia, Mesopotamia, Iranian Plateau, Central Asia and Caucasus) containing 5–8% zinc that has been reported in the literature. The author points out that a part of the artefacts may or may not be intentional alloys. Early artefacts of leaded brass were also found on the Iberian Peninsula. However, these are of a much younger age, the oldest dating back to the 6th century BC, others are younger. In any case, they are pre-Roman (Montero-Ruiz and Perea 2007).

In every case, it seems unlikely that the sites mentioned by the authors were centres of great metallurgical innovation. The formation of ancient brass or, if lead is present in the percentage level, of gunmetal, is accidental. This does not seem to be surprising, since an accidental production of brass is not unlikely, considering that the frequent association of copper, zinc and lead ores, as well as secondary minerals in ore deposits, can easily result they will find their way into the melt during smelting. There is no evidence that these early discoveries were made with a deliberate addition of zinc to copper. Zinc first became more known and was widely distributed in Roman times.

Brass Ingots and Bars

The Romans obviously produced brass on a large scale. Hanel and Bode (2016) have compiled an overview of antique loaf-shaped brass ingots, of which some are mentioned here.

Off the east coast of Corsica, near Aléria, 21 ingots of brass were recovered from a Roman shipwreck weighing each between 3 and 6 kg (Fig. 7.7). Based on inscriptions on accompanying lead ingots the ship could be dated to the end of the 2nd century AD (Hanel et al. 2013a, b). In the Thames Valley near Claydon Pike, a brass ingot was found during an excavation, which is supposed to resemble the Aléria bars in shape and design and is therefore presumably Roman (Bayley 1990; Weisgerber 2007).

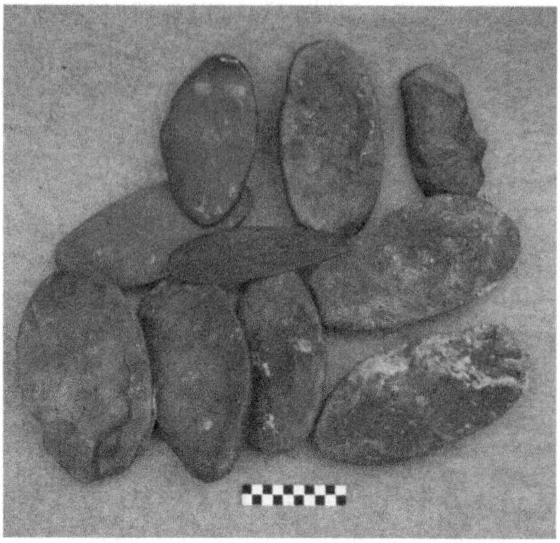

Fig. 7.7 Some of the 21 brass ingots found in the shipwreck off the eastern coast of Aléria, Corsica. They were dated to the end of the 2nd century AD. The weights of an ingot are varying mainly between 3 and 5 kg. Photo: Gerd Weisgerber (†), Deutsches Bergbau-Museum Bochum

Ancient ingots and bars of brass are otherwise rare. 38 rod and oval-shaped ingots of brass were salvaged from a shipwreck on the south coast of Sicily near Gela. They are significantly older. By pottery found there, they could be dated to the 6th century BC (Caponetti et al. 2017). The major components of the ingots are copper; zinc is between 16 and 26 wt.% and lead varies between 2 and 7 wt.%. Their weights are much below the one of the Roman Aléria ingots: most of them weigh between 300 and 600 g, only three are around 1 kg. The zinc content is compatible with the composition range of *aurichalcum*, the most precious alloy of this time.

In the shipwreck of Les Magnons B at the island of Embiez (Département Var, France) an unknown number of unspecified bars of brass was found next to lead ingots. The dating is vague (likely 4th–2nd century BC) (Dumas 1972; Parker 1992). The chronological classification of 57 bars of brass, which originate from the Rhine near Mainz, is still doubtful (Bachmann and Jockenhövel 1974).

25 Viking bars of brass were found at Hedeby (Germany), the northernmost point of the Carolingian empire. Analytical investigations revealed that they were standardised, high-quality brass alloys (mostly 21–23 wt.% Zn) (Merkel 2018). They are not consistent with ore deposits of the Rhine area, but are probably products of originating from ore deposits on the Balkans or Andalusia. These bars are considered to be a type of currency from the 9th century AD.

The Roman Brass Industry

The history of copper technology in Roman times undergoes a decisive, sudden expansion and new development, namely the targeted production of brass with zinc contents in the range of about 28–32 wt.% within a very short time. This happened during the reign of Augustus in the 1st century BC. However, the ancient writings hardly provide useful information on the production of brass. Only Theophilus Presbyter (around 1100 AD) has more precise rules on how to make brass. The provenance of the masses of zinc necessary for making brass in Roman times is an open question (see Sect. 3.4.10).

In contrast, numerous publications deal with the production of brass in Roman times and in the Middle Ages, from the analytical and experimental point of view (e.g. Bayley 1984, 1990; Rehren 1999b, 2003; Craddock and Eckstein 2003; Martinón-Torres & Rehren 2002; Bourgarit and Thomas 2015). Very useful overviews on the problems of zinc metallurgy and the making of brass are also published by Nielen (2006) and Hanel and Bode (2016).

The Roman *aurichalcum* is a term which implicates its improved casting quality and malleability and the increasing yellowish-golden colour. The brass may have also contained a few percent of lead and tin (see above). This alloy was used, especially in central Europe for the production of brooches, scabbards, belt plaques etc. Large statues, on the other hand, were rather cast from lead-rich tin bronzes.

In Asia Minor, King Mithridates VI from Pontos (c. 134–63 BC), presumably at the beginning of the eighties BC, minted coins of brass. Brass has been used more extensively since about 60 BC, according to recent research on weapons and equipment of Roman troops. During Caesar's siege of Alesia, Vercingetorix probably minted brass coins in the *oppidum*. The already mentioned brooches might have served as a material source. Cicero briefly mentions the term in his 44 BC "de officiis" when he complained about merchants who sold brass instead of gold. The historian Strabo reports on a brass production near Andeira, a not precisely located place somewhere in the Troas.

Another indication of the increased use of this metal at the end of the 1st century BC is provided by the brass coins introduced under Augustus in the course of his coin reform (*sestertii, dupondii, semisses*). During the imperial period, the zinc content of these imprints continuously declined, although it is uncertain whether this was due to a shortage, or a price increase of zinc or the zinc ore calamine.

A vividly discussed group of archaeological finds are the 150 so-called Hemmoor buckets from the 2nd/3rd century AD. These were found in a Germanic burial field near Cuxhaven (Germany) (Willers 1901). These are thin-walled, mostly cylindrical metal vessels, many of which are made of brass and some of tin bronze (Erdrich 1995). It was initially assumed that the buckets had been cast. However, the author was able to prove that they were each made from one piece of sheet metal by chasing and beating out and then reworked by turning. Voß et al. (1998) describe the texture of a wall of a Hemmoor bucket. It is a fine-grained mixture of solid solutions in a forged state, which was created by repeated hammering and annealing. The chemical composition is 10 wt.% zinc, 5 wt.% tin and 0.3 wt.% lead.

7.2.5.2 Physical Properties of Brass

Brass shows an interesting spectrum of colours: at zinc contents of 10 wt.% the reddish colour of copper still dominates, at around 20 wt.% it becomes bright red and at around 30 wt.% it changes to rich, golden shades (Hammer 1998). Traditionally, the reddish sorts of brass are referred to as tombac or red brass, without any precise delimitation being given.

Since Roman times, brass has been highly prised for its excellent casting and forging properties in comparison to tin bronze, as well as for its colour. Up to and slightly above c. 30 wt.% zinc, brass has a tensibility up to 50% and is easily deformable. It is therefore easy to forge and to deform otherwise. These are properties that tin bronzes do not have. The tensibility of those already falls to below 20% at tin contents of 10 wt.%. At tin contents of 14 wt.%, tin bronzes display a tensibility of almost zero, greatly hindering their malleability (Hammer 1998).

Another advantageous property of brass is, that in contrast to tin bronzes, the difference between liquidus and solidus temperatures is relatively small, so that significantly lower segregation effects may occur during casting.

The reason why, in Roman times, brass was produced with zinc contents of no more than 28–32 wt.% (Craddock 1978), could be the following:

- Up to a zinc content of about 30 wt.%, copper forms a homogeneous (α-phase) solid solution mixed crystal with zinc. Theoretically, this range can be extended up to 39 wt.% Zn. In practice, however, the first β-mixed crystals are already precipitated at just above 33 wt.% zinc (Brepohl 2016). Such alloys have a higher strength–which can certainly increase the utility value–without being brittle. However, during polishing, the β-mixed crystals are harder to remove so that they form a relief. This affects the beauty aspects of the metal, because one cannot achieve a brilliant lustre. The more β-mixed crystals are formed, the more difficult the deformation of the alloy is in the cold state.
- (Gold) smiths are predominantly interested in those sorts of brass, which only form α—mixed crystals, even upon rapid cooling. This would be brass with a maximum of 33 wt.% zinc. Even today, for vessels and bowls, for carving and for ornamental devices of all kinds, brass is used with usually no more than 33 wt.% Zn, because formability in manufacturing, strength in use, colour and polishability are in good proportion (Brepohl 2016).
- When Newbury et al. (2005) find that one can indeed produce brass with higher Zn contents, then the 28–32 wt.% Zn measured by Craddock (1978) were not due to technological limits, but instead they were intentional.

7.2.5.3 Making Brass

Statements in the literature about how brass was made in antiquity are inconsistent. Also, ancient writers hardly provide any useful information on this topic. Similar to tin bronze, there are basically three different ways to make brass. Furger and Riederer (1995) and Nielen (2006) describe them in the following way, but emphasise that there are no irrefutable arguments for any of the proposed options:

1. Co-melting of the metals copper and zinc in their metallic state, presumably by adding zinc to molten copper. This is thought to be unlikely in Europe before the 18th century AD.
2. Joint reduction of oxidic or carboniferous copper and zinc ores (when using sulphidic ores, these would first have to be roasted, i.e. to be converted into oxides).
3. Addition of oxidised zinc ores to metallic copper. This process is the so-called calamine process. It is a cementation.
4. Zinc was barely available in pure form in ancient times, as it is difficult to recover and isolate from ores. Brass (*aurichalcum*) was therefore not alloyed by mixing two metallic components, but by mixing liquid copper (*aes*) with additives, i.e. with zinc ore.

The second method, the co-reduction of copper and zinc ores, includes the greatest uncertainties, i.e. the largest variations in the composition of the final product. Because there are not only variabilities in locally different ore deposits, but also within a deposit itself. In addition, copper was smelted to metal as early as the Late Neolithic period and was probably traded as ingots and bars, hence, copper metal was available!

In the West, brass was produced according to the third method, the so-called calamine process, for at least 2000 years. This sophisticated cementation process took place in the solid state where oxidic zinc ores were interlayered with copper sheets and heated in a crucible with charcoal. The process was probably vaguely described by Pliny (Nat. Hist. 34, 2), then much more precisely by Theophilus Presbyter in the eleventh century and by Lazarus Ercker (1580). The oxidised zinc ore was added directly to molten copper. In this method, the so-called white calamine (see Fig. 3.38a, b) was used, which was mixed with copper shavings or chips and charcoal in sealed crucibles and heated to a maximum of about 1000 °C. Calamine is a technical collective term for carboniferous and siliceous, but always sulphur-free zinc ores of all kinds in changing conditions. These are hemimorphite ($Zn_4(OH)_2Si_2O_7{\cdot}H_2O$), smithsonite ($ZnCO_3$), hydrozincite ($Zn_5(OH)_3CO_3)_2$) and perhaps also willemite (Zn_2SiO_4).

The heating in crucibles at the boiling point of zinc of 907 °C (zinc vapour is already formed at 745 °C, cf. Sect. 6.8) finally leads to a release of the zinc, which, due to its high vapour pressure combines with copper by diffusion in the solid state (Fig. 7.8). This effect of the formation of solid solution mixed crystals with copper is the cementation process. According to this traditional method, an alloy with a maximum of approx. 28–32 wt.% zinc could be produced during a process of approximately 15 h (Craddock 1978; Newbury et al. 2005). It is assumed that this was about the maximum content that could be achieved with ancient methods. The zinc content in the copper formed during the cementation process is based on a thermodynamic equilibrium between the diffusion of the zinc into the copper and its evaporation from the brass, which sets in rapidly.

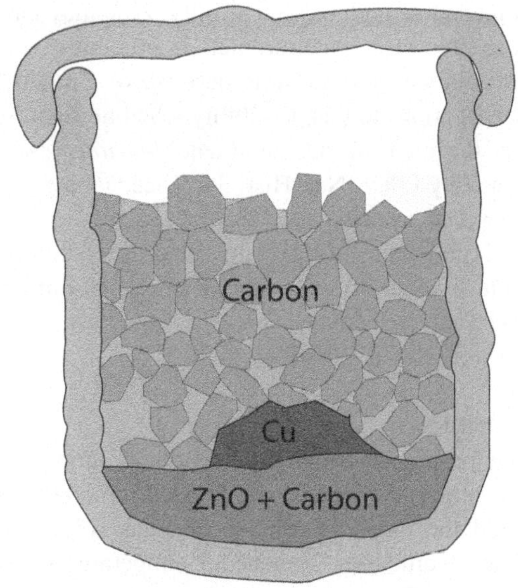

Fig. 7.8 Arrangement of crucible charge for cementation experiments to produce brass after Newbury et al. (2005). To duplicate earlier experimental work 75 g of treated powdered calamine (ZnO) was chosen for the cementation to 50 g of granulated copper in a closed crucible as shown in the sketch. These experiments are following the descriptions of Lazarus Ercker (1556)

Nielen (2006) emphasises in his reflections on zinc and brass that, among the early brass artefacts of the 2nd and 1st millennium BC, it is difficult to distinguish whether they were deliberately made using zinc ores, or whether the alloy was formed by processing copper ores high in zinc.

The crucial chemical processes in the production of zinc are described by the following equations:

$$ZnCO_3 \leftrightarrow ZnO + CO_2 \quad \text{(roasting)}$$
$$ZnO + CO \leftrightarrow Zn_{gaseous} + CO_2 \quad \text{(reduction)}$$

Even under slightly oxidising conditions, the zinc gas is reoxidised to zinc oxide, which forms a white smoke and a white precipitate on colder parts in a furnace (in German called *Ofengalmei*).

The metal zinc itself was probably largely unknown in antiquity due to its low boiling point and its high oxygen affinity. Zinc may have been present in small droplets in crucibles when the calamine process was performed. Craddock (1990) points to this possibility when he suspects that the medical zinc ointment (*lauriotis*) mentioned by Pliny (Nat. Hist. 34, 9, 225) may have been zinc droplets.

Experimental studies show that the traditional limit of about 30 wt.% zinc in brass can only be exceeded selectively on the smallest scale (Newbury et al. 2005).

Crucibles

The production of brass in Roman times was apparently carried out in closed crucibles to avoid the oxidation of the zinc gas (Fig. 7.8). When crucibles reached temperatures of c. 1000 °C, an excess of zinc vapour diffuses into the copper forming a brass.

At the Roman town of Colonia Ulpia Traiana (next to the modern city of Xanten, Germany), hundreds of only 3–4 cm-sized brass crucibles with lids were found from a 1st century AD context (Rehren 1998). They allowed the production of ingots with a maximum diameter of c. 1.5 cm. It has to remain unclear, however, whether or not high-zinc brass was produced in these small vessels. In a Roman camp near Neuss (Germany), Telle and Thönnissen (2006) found evidence that the brass already exceeded its melting point in the cementation crucibles. Noteworthy is the finding of five crucibles made of graphite in the Roman Aventicum, which are also associated with the production of brass (Hochuli-Gysel and Picon 1999). Craddock (1978) and Bayley (1984) point out that most of the crucibles were eventually smashed to enable a recycling of the trapped metal droplets. The cementation crucibles had diameters of only a few centimetres and were thin-walled (Bayley 1984; Rehren 1998; Telle and Thönnissen 2006). Nielen (2006) describes double-walled crucibles with clear width of 7–7.5 cm, which were salvaged from the Roman legion camp of Neuss near Cologne.

But amphora-shaped crucibles of more than half a meter in height are also known from the Roman Lyon. According to their size, the walls were about 2–3 cm thick (Picon et al. 1995). In all likelihood, the cementation crucibles used for the production of the brass of the Aléria bars had similar dimensions. If the two finds would have any relationship, then the casting of the Aléria bars of Corsica would have been performed in one step (Weisgerber 2007).

Roman cementation crucibles were brittle and not very heat-resistant, making the exact reconstruction of the vessel shapes difficult (Bayley 1984). The "loose" nature of thin crucible walls probably provided a necessary pressure balance during cementation (Rehren 1998).

High-zinc brass sheets produced by experimental works were in the order of 3–4 mm in diameter and ≤ 1 g in weight only. In contrast, historical accounts suggest that during the Medieval period, brass foundries were able to produce ingots of approximately 5 kg (Lazarus Ercker 1580).

7.2.6 Iron in Copper

The iron content of ancient copper and bronze artefacts is usually low, nearly always substantially <1 wt.%. Among the over 8000 analyses of prehistoric of European copper and bronze artefacts, Junghans et al. (1960, 1968, 1974) and Krause (2003) report no more than a few with more than 1 wt.% iron. Brown and Blint-Stoyle (1959) present more than 400 chemical analyses of British bronzes of the Middle and Late Bronze Ages, and they found the amount of iron would be in the same range. In Chinese bronzes from the same period, the average value for iron is 0.2 wt. %. Moorey (1994) gives chemical analytical results for 132 prehistoric objects from western Iran and finds only three with as much as 2.5 wt. % Fe.

The iron concentration in copper increases in many cultures with the onset of successive time periods when technological innovations affected metallurgy. In Europe, such metallurgical stages forward can be pinned down, e.g. by the spread of the slagging process in the middle of the 3rd millennium BC, in the Iron Age, in the Roman or in the Medieval periods (Craddock and Meeks 1987).

7.2.6.1 Visible Iron in Copper

The presence of isolated iron inclusions in varying quantities in ancient copper and bronze artefacts and slags has been described by several authors (Cooke and Aschenbrenner 1975; Sperl 1979; Charles 1980; Craddock and Meeks 1987; Hauptmann 1989). The reason for the formation of metallic iron is that of all the minor and trace elements regularly found in early copper, iron is the most dependent on the smelting process, and it precipitates as an own phase when copper contains just a few percent (see below). It is caused by the reduction from Fe^{2+} and Fe^{3+} in the ore or gangue to metallic iron (Fe^0). This is a function of the CO/CO_2-ratio in a furnace, i.e. the amount of charcoal, and stronger reducing conditions, and of higher temperatures. The iron ore itself is usually included in a charge through the intergrowth of ore with (limonitic) gangue or, alternatively, through the use of fluxing agents. Tylecote and Boydell (1978), Bamberger et al. (1986), Merkel (1990) and Bamberger and Wincierz (1990) observed that during the smelting of copper, over-reduction can lead to the formation of Cu–Fe alloys along with pure copper. These binary alloys are included in small particles in slag or may even agglomerate to larger units at the bottom of a furnace.

It should be noted, however, that such iron precipitations in the context of copper metallurgy or iron inclusions in copper have not become known from early periods. No α-Fe had been identified in Early Bronze Age coppers. In the Wadi Arabah adventitious production of iron in copper comes from Late Bronze Age Timna (14th–12th centuries BC) copper smelting (Roman 1990). "Pure" iron objects found from Timna have been analysed for the composition of their lead isotope ratios by Gale et al. (1990). They eliminated that these had anything in common with any of the Cu–Fe alloys. At Faynan, the earliest inclusions of α-Fe have been observed as inclusions in Iron Age II slag (1000 BC and younger), where Fe-concentrations are around 1–2.6 wt.% (Hauptmann 2007). Metallographic investigations showed that these iron contents are due to inclusions of (Cu-bearing) α-Fe as well as of Fe-phosphides.

A striking example of a dense bloom of α-iron with a weight of 15 kg was found at the Roman copper smelting site of Faynan 1 (Hauptmann 2007). The formation of metallic iron is not limited to copper production. As mentioned in Sects. 6.4 and 6.5, the precipitation of iron hardheads are known as a harmful by-products in ancient tin and lead production (Tylecote et al. 1989; Gale et al. 1990). Hauptmann et al. (1988) report on the find of a hardhead with a weight of several kilograms from the (Hellenistic) lead smelting site of Skres on the island of Thasos (Greece).

A special situation to understand Early Islamic copper metallurgy is the site of El-Furn (Faynan, Jordan). Heaps of nut-sized prills of two-phased CuFe alloys next to slag heaps were obviously waste material, which was produced next to pure copper and was no longer needed. During this medieval time period, Fe-rich copper was widely distributed in the Near East, e.g. at Beer Ora in the southern Arabah close to Timna (Craddock and Freestone 1988), in the southeast of the Arabian Peninsula (Hauptmann 1985), and at smelting sites in central Iran (Hezarkhani and Keesmann 1996). Two-phased copper–iron alloys seem even to have worldwide parallels in early metallurgy (a compilation can be found in Craddock and Meeks 1987; see also Hauptmann 1989).

In order to produce iron-free copper, ancient craftsmen either had to modify the technical parameters of the smelting process, or to lower the amount of iron in the smelted impure copper by means of a selective remelting. It does not seem too difficult to separate both metals in alloys such as the ones in El-Furn. Tylecote and Boydell (1978) managed to isolate iron with only 0.5 wt. % copper from a Cu–Fe alloy via simple remelting in open crucibles, and to rework it.

7.2.6.2 Aes Rude and Ramo Secco Bars

In numerous archaeological sites in Italy dated between the 6th and 3rd centuries BC, unworked, irregularly shaped lumps of copper-based materials were found, the so-called a*es rude*, which, according to archaeological considerations, were appreciated as a medium of exchange or as a pre-currency (Ingo et al. 2004a, b). *Aes rude*, i.e. crude copper in the shape of unformed and irregular lumps of copper-based materials, in some cases appear to be broken from cast as thick larger pieces. These metal artefacts were traded as items of value for centuries.

Aes rude consists of highly ferruginous copper, making them useless for producing other metal objects by casting or hot and cold working. Notwithstanding this intrinsic negative feature, the production of this intractable copper-based alloy was possibly deliberately carried out. It may have been a widely spread by-product of copper smelting perhaps in a new smelting furnace type, which was developed to optimise production volumes.

Figure 7.9 shows a scanning electron picture of a mounted sample of an *aes rude* from Sardinia (Ingo et al. 2004a). The micrograph shows a texture of solidification constituted by a large amount of α-iron grains randomly scattered or partially coalesced throughout the copper matrix where some black blowholes are present. A number of these α-iron grains have largely coalesced to form dendrites. They do not give any hint of a mechanical treatment after solidification.

Ingo et al. (2004a) studied c. 50 samples of a*es rude* from a hoard or a votive deposit found at Terrasebis near Cagliari (Sardinia, Italy). These finds were highly ferruginous copper. The iron content ranges from 13 to 27 wt.%, in most cases they are clustering between c. 20 and 27 wt.% (Fig. 7.9). Lead is at the lower percentage level, it is an impurity.

This data is in good agreement with the results obtained for nine a*es signatum* bars of the *ramo secco* type which are of more recent origin than a*es rude*. These *ramo secco* bars have iron contents ranging from 4 to 31 wt.% (Burnett et al. 1986; Craddock et al. 1987). Similarly to the compositions of the a*es rude* samples, they also show a cluster between c. 20 and 30 wt.%.

Also, based on this data we would not exclude that the clustered high iron levels in the a*es rude* as well as in the *ramo secco* bars resulted from a controlled smelting processes under strong reducing conditions and higher temperatures as compared to that commonly used for copper smelting.

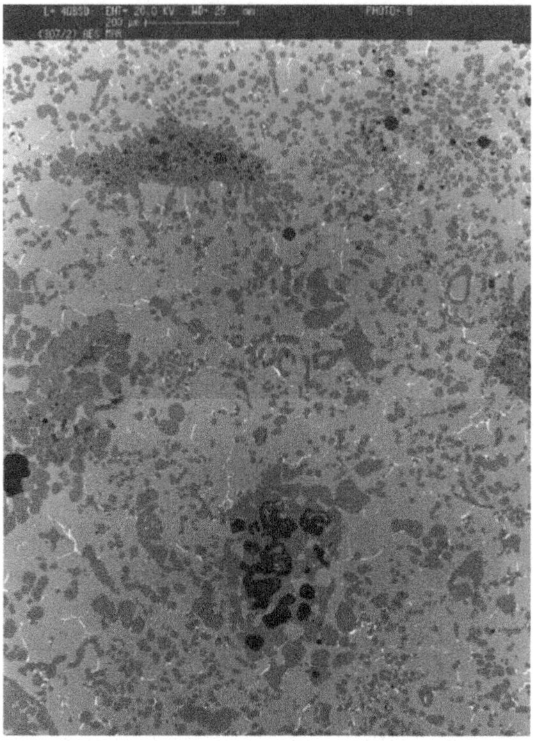

Fig. 7.9 *Aes rude*, Terrasebis, Sardinia (Italy), between sixth and third century BC. The micrograph proves a solidification texture constituted by a large amount of α-iron grains randomly scattered and partially coalesced to form dendrites throughout the copper matrix where some black blowholes are present. This material was not mechanically treated after solidification. Back-scattered electron image. Note the scale showing 200 μm. From Ingo et al. (2004a)

7.2.6.3 The Copper–Iron Equilibrium Phase Diagram

The copper–iron system is characterised by an extended miscibility gap (Fig. 7.10). In the copper-rich part of the system, homogeneous CuFe mixed crystals only form below the solidification temperature of copper at 1083 °C if the iron content does not exceed c. 3 wt.%. If the iron contents of the liquid are higher, iron separates in the form of γ-Fe, which transforms into its modification of α-iron below the temperature of 850 °C.

Microanalyses showed that both metals contain parts of the other when dissolved: iron regularly contains 5–7 wt.% copper, copper contains

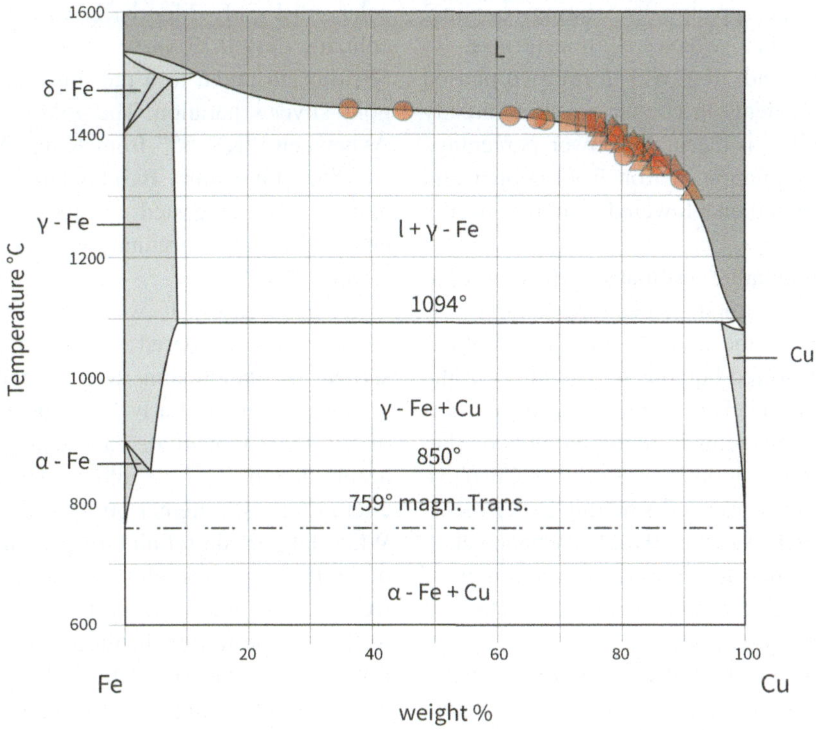

Fig. 7.10 The binary system copper–iron (Hansen and Anderko 1958) provides a large miscibility gap between the two components. Plotted are two-phase CuFe alloys from Early Islamic slags collected from various smelting sites in Oman and in Faynan (○) (Hauptmann 1985; Hauptmann 2007). In addition, 12 samples of *aes rude* (Δ) found in Sardinia from the beginning of the 6th century BC (Ingo et al. 2004a) and the composition of Roman *ramo secco* bars (□) from the 6th to 3rd century BC in Italy and two acquired from the art market (Hauptmann and Klein, unpubl. results) are indicated.

1.5–3 wt.% iron, in the case of El-Furn, Faynan, additionally $P < 1$ wt.%. Under the scanning electron microscope Fe-dots in size of 0.2 to about 2 mm are visible (Fig. 7.11), and, *vice versa*, there is some iron in the copper matrix. This observation is in good agreement with the copper–iron phase diagram that reveals the homogeneous CuFe-melt at temperatures around 1450 °C and above.

Two-phase copper–iron inclusions in old slags were variously used to estimate their liquidus temperatures (Milton et al. 1976; Cooke and Aschenbrenner 1975; Craddock and Meeks 1987). These estimates are based on the fact that such inclusions are crystallised from a liquid which, (in the pure) system is at temperatures of around 1400–1450 °C (Fig. 7.10). At these temperatures, iron is, for a large part, dissolved

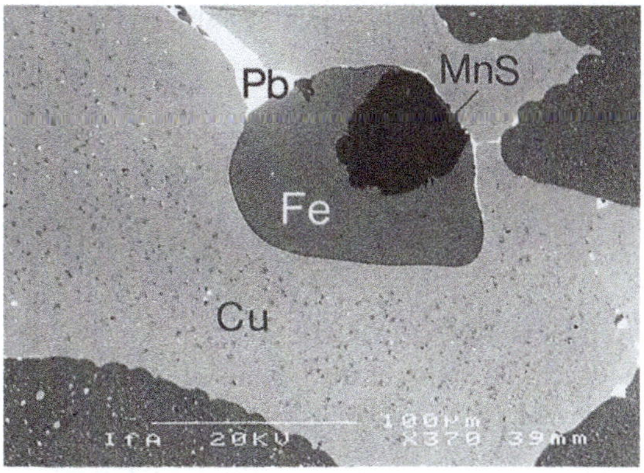

Fig. 7.11 Sample JD-6.4, El-Furn, Faynan (Jordan). Two-phase Cu–Fe alloy with some lead (pb) accumulated along grain boundaries and a droplet of manganese sulphide (MnS). Copper contains miniscule inclusions of α-iron (Fe). This, in turn, has inclusions of copper. Scanning electron picture, back-scattered electron mode. From Hauptmann (2007)

in copper. Iron precipitates during cooling, initially as γ-Fe, which is transformed to α-Fe. Copper and iron will form two-phased alloys. Iron contents in copper are thus already microscopically visible in the lower percentage area. The precipitation of iron from copper and vice versa is based on a lowered solubility during the cooling process.

These temperature estimates are somewhat problematic, as carbon, phosphorus and/or sulphur contents in the alloy as low as <1 wt.% may lead to lowered liquidus temperatures in the order of c. 100 °C. This would result in liquidus areas, which were actually found by plots of slag analyses in various ternary systems (see Chap. 5). On the other hand, several smelting experiments (Tholander 1986; Straube 1996; Kölschbach et al. 2000) have shown that furnace temperatures of around 1500 °C can sometimes be reached in "old" furnace constructions.

The presence of iron at concentrations higher than c. 3 wt.% is a very harmful impurity that impedes the cold and hot working of copper-based artefacts (Craddock and Meeks 1987). This negative effect is due to the above mentioned low iron solubility in copper at room temperature that induces the formation of iron islands dispersed in a copper matrix. The presence of these grains lowers the ductility of the copper material, making it impossible to cold or hot-hammer it for shaping an object.

7.3 Gold and Gold Alloys

7.3.1 "Pure" Gold

Artefacts of "Pure" Gold

Modern, electrolytically obtained gold has a purity of 99.99 wt.%, sometimes even 99.999 wt.% (Pieth 2019). Pernicka (2014a) points out that prehistoric gold artefacts of comparatively pure gold, but only with 99.9 wt.% are extremely rare. Of the 2800 analyses done by Hartmann (1970, 1982), only nine were analysed with this purity, and Perea et al. (2010) found only a single one in their analyses of 170 gold objects using a Scanning Electron Microscope. The Bronze Age solar disk of Moorbach/Aurich (Germany) contains only 0.23 wt.% silver, which Pernicka (2014a) attributed to a presumptive cementative gold–silver separation. The gold of the coffin of Akhenaten ("KV 55" from Kings Valley Tomb No. 55, 14th century BC) contains c. 2 wt.% Ag and is also suggested as gold that has been enriched by such a refining process (Klemm and Klemm 2013).

In both cases, it was not considered that edaphic, ground-based corrosion of gold, as shown in Hauptmann et al. (2018) for gold artefacts from the Early Dynastic Royal Tombs of Ur (Mesopotamia), can lead to a surface-accumulation of gold. Under certain circumstances, a higher purity can be pretended. While large-scale refining of gold has only been possible since the middle of the 1st millennium BC, as is known from Sardis, their origins may well go back further in time (Craddock 2000a; Ramage and Craddock 2000). The authenticity of the 99.99 wt.% Au pure gold ornaments from an ensemble of 20 plates from the Middle Bronze Age settlement of Bernstorf (Bavaria) is controversial (Pernicka 2014a, b, c; Gebhard et al. 2014). Bachmann (1999a, b) generally stated that most of the gold used for coinage in the Greek and Roman world was of "high purity".

Trace Elements in Gold

The overwhelming part of prehistoric gold artefacts can be discussed in the ternary system Au–Ag–Cu. In addition to high levels of silver, which characterizes many gold artefacts in the ten percent range, copper levels can range from 150 ppm to 1–1.5 wt.%, sometimes even up to a few percent. In deliberately produced alloys, copper contents can reach several tens of percent.

More than 5100 analyses of c. 5000 gold artefacts from all over Europe, performed by Hartmann (1970, 1982) using spark spectrometry, provided characteristic minor and trace element concentrations. Many archaeological gold artefacts are characterised by a trace element pattern comprising elements such as Fe, Ti, Sn, Mg, Ni, As, Sb, Bi, Pb, Pd, Pt and Bi (Ehser et al. 2011; Nocete et al. 2014; Pernicka 2014a, b, c). Later analyses added osmium and iridium as trace elements (Jansen et al. 2016; Jansen 2019). They are all in the range of a few (hundred) parts per million. Their concentration depends on the geological–geochemical characteristics of the gold deposits from which the metal was smelted (cf. Sect. 3.7). However, Jansen (2019) has stated that the trace element pattern of gold will also be considerably modified by any metallurgical processing such as remelting, soft-annealing or hammering.

The notable presence of titanium (Ti) in gold samples, e.g. from Cabezo Juré in Spain may be considered as an example of alluvial origin (Nocete et al. 2014). Although the reduction of Ti-oxide such as rutile (TiO_2) and ilmenite ($FeTiO_3$) to metallic titanium is not possible at temperatures that can be reached with charcoal, there are two possibilities for explanation:

1. That they originate from small inclusions of Ti-oxides.
2. That they originate from alluvial incorporations. These do not occur as oxides of high crystallinity, but rather as altered products on the surface (leucoxenes), presenting lower melting points. Leucoxene is a fine-grained corrosion product of TiFe minerals such as titanite, ilmenite, or perovskite. It is mainly composed of rutile or anatase (TiO_2) and Fe-oxides.

Even low values of platinum group elements such as platinum (Pt), palladium (Pd), osmium (Os), iridium (Ir) and ruthenium (Ru) may be used as indicators of the provenance of gold (Pernicka 2014a, b, c; Jansen et al. 2016; Nocete et al. 2014; see below) proposes that Pd values even of <1.2 ppm analysed in gold artefacts from Cabezo Juré (mid 3rd millennium BC) may be the origin of massive sulphide deposits of the Iberian Pyrite Belt in the southwest Iberian Peninsula.

Tin in Gold

It is generally accepted that tin concentrations in gold are a characteristic feature of alluvial placer gold (Hartmann 1970, 1982; Schmiderer 2008), because native gold and cassiterite (SnO_2) may be paragenetically associated. Both may be enriched in the heavy mineral fraction. Cassiterite may be co-smelted with gold and may cause a slight enrichment of tin in the noble metal. At Zambujal (Portugal) Hartmann (1982) characterised gold products with tin contents from 0.18 wt.% to 0.38 wt.% which obviously originated from gold placers containing cassiterite.

The behaviour of gold and cassiterite during sintering and melting conditions under changing conditions was studied experimentally by Éluère and Raub (1991) and Raub (1995), using alluvial as well as modern industrial gold and cassiterite. The experiments showed that impurities in gold are not removed by sintering. It showed that cassiterite would be slagged under oxidizing smelting conditions. Under reducing conditions, it would be efficiently converted to metallic tin.

It was initially assumed that tin could also have been introduced to gold by the inadvertent addition of bronze scrap. This assumption was apparently supported by the relatively high copper concentrations of many prehistoric gold objects (often more than 0.1 wt.%) compared with natural gold (less than 0.1 wt.% Cu in general, Schmiderer (2008).

7.3.2 Gold Alloys

Of fundamental importance for the early metallurgy of gold up to the modern era are alloyed with silver and with copper. Alloys with aluminium, palladium, platinum, nickel (white gold alloy) and other metals and non-metals were not known in earlier times. Occasionally, gold–silver–copper alloys may be of natural origin, but they may also have been deliberately mixed. This will be discussed below.

Generally, the concentrations of the alloying ingredients are given in both weight percent and atomic percent. Since the relative atomic masses ("atomic weights") of the metals are different, there are differences to the weight percentages. This information is not in thousandths, as the goldsmith and silversmith would use to do. They use to write that an alloy containing

Table 7.2 The purity of gold and gold alloys will be measured in various units

1. Carat. The gold content of an alloy in carats corresponds to the weight fraction of 24: >One-carat gold contains 1/24th weight components gold (= 4.167 wt.%). >Gold with 24 carat contains 24/24th weight components gold. However, for technical reasons, it is practically impossible to eliminate possible impurities. Thus, a purity of 99.99 wt.% is indicated for 24-carat gold.
2. Fineness. Alternatively, the fineness of gold is also written in parts of 1/1000. Three examples are: >Gold with a fineness of 585 consists of 585/1000 = 58.5 wt.% Au = 14 carat. >Gold with a fineness of 750 consists of 750/1000 = 75 wt.% Au = 18 carat. >Gold with a fineness of 999 consists of /1000 = 99.99 wt.% Au = 24 carat.
3. Ounces. Often the weight of gold, as well as that of other precious metals, is given not only in wt.% or in µg/g (= ppm), but also in troy ounces (1 oz.tr. = 31.1034768 g).

These units are (a) carat, and (b) fineness. The third unit, a troy ounce, is a unit for the weight of the noble metal. For details see McDonald and Sistare (1978)

58.5 wt.% of gold corresponds to an Au585 gold alloy. In terms of atomic percent, this is 43%. The calculation of gold alloys is explained in detail in Brepohl (2016).

Gold contents in alloys are in addition written in another well-known unit. All these different scales are discussed in McDonald and Sistare (1978). They are of importance, for example in the discussion of gold and gold alloys in the cuneiform texts of the 3rd and 2nd millennium BC (Waetzold 1985; Reiter 1997; Hauptmann et al. 2018). A brief overview is summarised in Table 7.2.

There are three physical conditions that have played a role in the use of gold alloys in ancient times:

1. Colours: There are clear colour gradations depending on the composition of the typical end members gold, silver and copper.
2. Working properties such as the lowering of temperatures compared with the pure metals.
3. Working properties such as an increase of hardness and malleability.

Colours of Gold Alloys

Colours are the most obvious difference as far as the jewellery owner would be concerned (Ogden 1993). It can, therefore, be assumed that the colours of gold alloys in the Au–Ag–Cu system have been thoroughly researched even since the pre- and protohistory of metallurgy and have had already different financial–economic values in ancient times (Reiter 1997; Hauptmann et al. 2018). These colours are shown Fig. 7.12. The colours shown in this diagram refer to fresh metal. It should be emphasised at this point that the different types of gold can change significantly under atmospheric influences and especially through soil corrosion over archaeological periods. For details see Brepohl (2016) and Sect. 6.3.1.

Pure gold is bright yellow. Low copper concentrations (c. 3–6 wt.%) do not change this colour. For orientation, reference is made to Kruger-Rand coins from South Africa. They consist of 91.7 wt.% gold, the rest is copper. Despite this copper content, the gold colour deviates only slightly from that of fine gold. Gold with concentrations of silver + copper in the dimension of c. 10 wt.% is yellow with a slight reddish tint. As the silver content increases, the alloy turns yellow, then green-yellow, then, between 40 and 50 wt.% of silver, it turns to a pale greenish–greyish tint. When silver is dominating, the colour changes to white. Pure silver is white. Gold with c. $\geq$ 10 wt.% copper gets an increasingly reddish colour, which quickly turns into intense coppery red. Copper itself still shows its typical red colour with a content of 30 wt.% silver and down to 40 wt.% copper. The most vivid colour change can be observed in alloys with c. 60 wt.% gold and changing silver and copper levels.

Depending on the composition, the colours of the Au–Cu alloys tend to be between pale red and bright red (Brepohl 2016). Alloys with high proportions of gold have orange shades while, surprisingly, alloys with approximately equal proportions by weight of gold and copper show pale reds. This is probably due to the

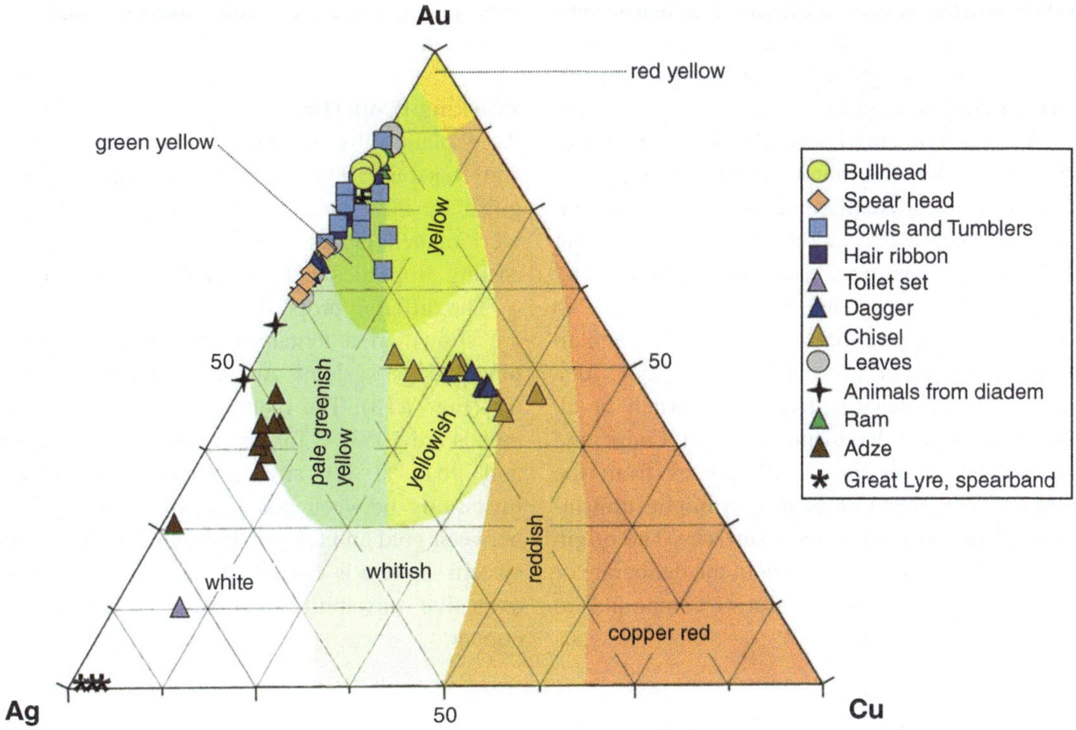

Fig. 7.12 Variations of gold colours resulting from specific alloy compositions of gold with silver and copper, visualised in the ternary system Au–Ag–Cu. The plots in the system are sorts of gold from the Royal Cemetery of Ur, Mesopotamia (middle of the 3rd millennium). See also Table 7.3. From Hauptmann et al. (2018)

crystallisation of the intermetallic compounds AuCu and $AuCu_3$ (which corresponds to a fineness of Au508).

In the Bronze Age, these colour variations were already a valuable assaying tool for the estimation of gold alloys using touchstones (Éluère 1986).

Tempering Colours and Tarnishes

Many ancient Egyptian objects made of gold–silver alloys bear a distinctive colouration that ranges from a pale reddish hue to a dark purple. These colourations are irregularly distributed in different colours, and therefore were explained as products of corrosion rather than of a deliberate patination (Frantz and Schorsch 1990). The authors state that "… The most notable examples of this kind are the gold-leaf decorations on the wood sarcophagus enclosures from the tomb of Tutankhamun, where areas of bright gold leaf are seen juxtaposed against areas of a dark purple colouration along irregular borders that would seem to have no relationship to an intended design …".

These products were composed of various gold–silver sulphides such as the pale-reddish AgAuS, identical with the natural mineral petrovskaite, and the black tarnished Ag_3AuS_2, which occurs in nature as the mineral uytenbogaardtite.

Reddish tarnishes, sometimes also greyish or black tarnishes are widely known as a general phenomenon on gold objects. They have been identified on gold–silver objects from the Royal Cemetery at Ur, from Uruk, from Mari, from Celtic artefacts, and from the Skyths (see Hauptmann et al. 2018).

SEM/EDS analyses on tarnished surficial areas on the surface of a hair ribbon from Ur revealed some concentrations of Fe (1.9 wt.%), K (1 wt.%), Na (0.6 wt.%), Si (4 wt.%) and S (2.4 wt.%). The formation of these tarnished areas on the surfaces of gold objects from Ur is not yet clear.

Long-term burial in soil, which is rich in iron salts can result in extensive surface effects. It could result in the leaching of silver and base metals and enrichment of gold.

We are not convinced yet, but we cannot exclude on the other hand that the so-called surface "pickling", usually in an acidic Fe-rich solution, was applied to colour the surface of gold. This is a common process employed by goldsmiths all over the world up to date to enhance the appearance of their wares and to change the colour of "cold" gold objects to a more warmer metallic hue (Wunderlich et al. 2014). It should be mentioned that Lucas and Harris (1967) observed gold pieces from the tomb of Tutankhamun bearing a bright, translucent red colouration on their surfaces. The origin of this colour may well reside in the deliberate or accidental addition of iron-bearing compounds to the gold (hydrated iron oxides, such as lepidocrocite).

Working Properties

As explained by McDonald and Sistare (1978) and Rapson (1990), ancient gold alloys distinguish by their basic mechanical properties and working behaviours, such as melting temperature, casting properties and malleability.

The different properties of the AuAgCu alloys are shown in a ternary system with the three binary edge systems Au–Ag, Au–Cu and Ag–Cu (Fig. 7.13). The melting points of the three metals are 1063 °C for gold, 1083 °C for copper and 961 °C for silver. There is a perfect miscibility between gold and silver, as well as between gold and copper. By contrast, the binary system Ag–Cu is essentially characterised by an extensive miscibility gap between silver and copper.

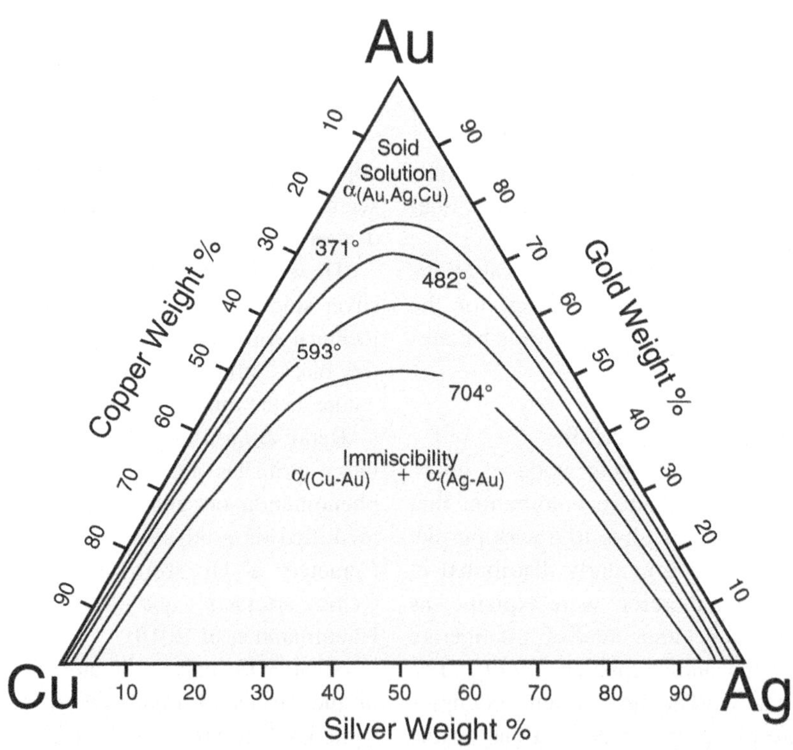

Fig. 7.13 Ternary phase diagram of gold–silver–copper (Au–Ag–Cu) showing some isothermal solid state boundaries of the immiscibility field. Note the large extent of the immiscibility field between Ag and Cu and its large extent into the system even to high concentrations of gold. After "Metals Handbook" 1973

7.3.2.1 Gold–Silver Alloys

Jewelleries and prestige items, made of this binary alloy with the main component gold and, subordinated silver (2–35 wt.%, rarely up to c. 50 wt.%), and additional low concentrations of copper are common all over the Old World since the very beginning of gold metallurgy in the late 5th millennium BC at Varna (Bulgaria). These are accepted to be natural alloys of gold (Hartmann 1970, 1982; Palmieri and Hauptmann 2000; Borg 2010; Hauptmann et al. 2010; Leusch et al. 2014 and others). Gold with silver concentrations in the range of c. 20–40 wt.% is called electrum (see Sect. 6.3).

Eight rings made of gold–silver alloys were found in the Nahal Qanah Cave (Israel) (Shalev 1993). They date from the second half of the 4th millennium BC (Gopher et al. 1990). These rings contain up to >30 wt.% silver, the copper content is mostly <1 wt.%. Gold–silver alloys with silver concentrations between 12 and 30 wt.% or more were published from the famous Royal Tombs of Early Dynastic Ur in Mesopotamia (c. 2500 BC; Plenderleith 1934; Hauptmann et al. 2018). Some examples of these alloys are compiled in Table 7.3.

Gold–silver alloys frequently occur as natural alloys both in primary gold deposits and in placers as well. Due to the general solubility of silver in the surface near environment, placer gold is frequently lower in silver than gold from primary sources (German *Berggold*, see Sect. 3.4.7.2).

To clearly distinguish native alloys from man-made alloys is not possible in every case. For aurian silver with 45 wt.% Ag in gold, as observed among the gold artefacts from the necropolis of Varna (Leusch et al. 2014) it is not clear if it is of natural origin or an anthropogeneous alloy. The same is true for an adze from the 3rd millennium BC Royal Cemetery of Ur, Mesopotamia, which has c. 50 wt.% of silver in gold, and, in addition, some 5 wt.% of copper (Hauptmann et al. 2018). In modern goldsmithing silver can be intentionally added to gold for economic or aesthetic reasons, and the same cannot be excluded for ancient metallurgy. As an example, definite sorts of gold and silver used for special purposes are known from cuneiform tablets (Waetzold 1985; Reiter 1997; Hauptmann et al. 2018).

Today, AuAg alloys of the composition Au750 are widely used in goldsmithing because they are almost as resistant as pure gold and have an excellent polishability, hardness and tensile strength. Therefore, they are suited for any kind of cold deformation. All alloys of this type are also easy to cast and solder (Wolters 1986). AuAg alloys can be used for granulation to solder granules. Here, one takes advantage of the melting interval between the solidus and liquidus lines, i.e. the temperature difference between the onset of melting and the end of melting, by heating the corresponding alloy until the first solid solution crystals begin to melt (see Fig. 4.1a, Chap. 4). Interestingly enough, a large number of Ur's gold artefacts have comparable compositions, be they bowls and tumblers, weapons or jewelleries (Hauptmann et al. 2018).

7.3.2.2 Gold–Silver–Copper Alloys

Gold Low in Copper

It is estimated that the copper content in natural gold would not exceed c. 1–1.5 wt.% (Hartmann

Table 7.3 Composition of a few selected gold–silver alloys from the Royal Tombs of Ur, Mesopotamia

Object	Inv. no. Penn	Au	Ag	Cu
Spearhead and bindings	B16690	69	30	1.0
Pin or hair ornament	B16908	77	20	0.9
Leaves from wreath	30-12-755	86	12	0.4
Chocker	30-12-443	83	15	0.9
Chocker	30-12-706	87	12	0.3

Many of the other artefacts made of these alloys match the natural composition of electrum, but it has to be open, whether or not they were deliberately mixed from different sorts of gold. Analyses were made using a handheld XRF-device (Hauptmann et al. 2018). Values are given in weight percent. See also Fig. 7.12. Penn = University of Pennsylvania Museum of Archaeology and Anthropology, Philadelphia (USA)

1970; Mohen 1990; Tylecote 1987). Recent analyses of gold nuggets from Central Europe even set the limit to 0.1 wt.% copper (Bachmann 1999b; Schmiderer 2008). Frequently, almost as a rule worldwide, and through the ages, copper concentrations in gold artefacts are exceeding this level for two or three percent or so (see overviews, e.g. in Bachmann 2006; Guerra and Rehren 2009; Barnes et al. 2015). Copper concentrations above this level are artificial alloys.

Gold artefacts with only a few percent copper are most probably due to the following reasons:

1. Gold and copper are geochemically associated, and green secondary copper minerals were used as pathfinders for gold prospection since the beginning of gold metallurgy (Klemm and Klemm 2013). In addition, there might have been a use of gold associated with Cu-containing gold minerals (e.g. auricupride, $AuCu_3$, see Sect. 3.4.7). The occurrence of Cu-containing gold minerals is apparently easily overlooked (Ramdohr 1975; Spiridonov and Yanakieva 2009).
2. The use of placer gold contaminated by copper minerals, as it was observed by Hauptmann et al. (2010) in the neighbourhood of the prehistoric gold mine of Sakdrissi in the Transcaucasus (Georgia) (Fig. 3.28).
3. Contamination of gold with copper during melting processes in crucibles previously used in copper or bronze metallurgy. Paoletti (in progress) emphasises, based on cuneiform texts from the 3rd millennium, that in the workshops of the early dynastic city of Ur (Mesopotamia, Iraq) gold and copper or bronze foundries had worked together in joint workshops. This textual interpretation is supported by investigations of Jansen (2019). Measurements of (Cu-containing) gold artefacts from the Royal Tombs of Ur showed a clear agreement with the lead isotope pattern of prehistoric copper artefacts from the same locality, whose origin from Oman could be proven (Begemann and Schmitt-Strecker 2009; Begemann et al. 2010).
4. In the recycling of gold artefacts that have been designed using Cu-containing solders, e.g. for attaching handle attachments to trays. It can be assumed that recycling of gold has often been carried out. That is an open problem. Often, accurate analytical procedures or localisation of measurements on gold objects are also not specified in the literature.

Tumbaga

One of the major differences in gold metallurgy between pre-Hispanic America and the Old World is the extent of intentional alloying of gold. In Central and South America, especially in old Colombia, exist large amounts of gold alloys with copper contents up to 40 wt.% (Fig. 7.14). These are called tumbaga or guanín. Often the tumbaga alloys contain silver. In most cases, this is due to the omnipresent silver content of natural gold, whether this comes from placers, or from solid rock mineralisations. The name tumbaga applies, however, no matter whether the alloy consists solely of gold and copper, or whether it has additional silver contents. Tumbaga alloys are manufactured artificially. The variable compositions of such alloys result in characteristic colours that vary from red-orange to red gold (see Fig. 7.12).

The earliest anthropogenic gold–silver–copper alloy currently known is a ring idol found in the necropolis of Varna I (Bulgaria) (Leusch et al. 2014). The necropolis dates to the Late Copper Age (Kodžadermen-Gumelnița-Karanovo (KGK) VI Cultural Complex, 4550–4450 BC). The idol contains 50 wt.% Au, 14 wt.% Ag and 36 wt.% Cu. The authors suggest that a natural gold–silver alloy originally containing about 22 wt.% Ag of a light greenish-yellow colour was mixed with copper to achieve a more yellowish colour, close to that of pure gold. However, if the gold has actually been manipulated in this way, this process has not prevailed because among the hundreds of gold artefacts from Varna, the idol is just a one-off.

Tumbaga alloys have often been further treated by depletion gilding to achieve a gilding of the surface (see Sect. 6.3.2.2). An example for

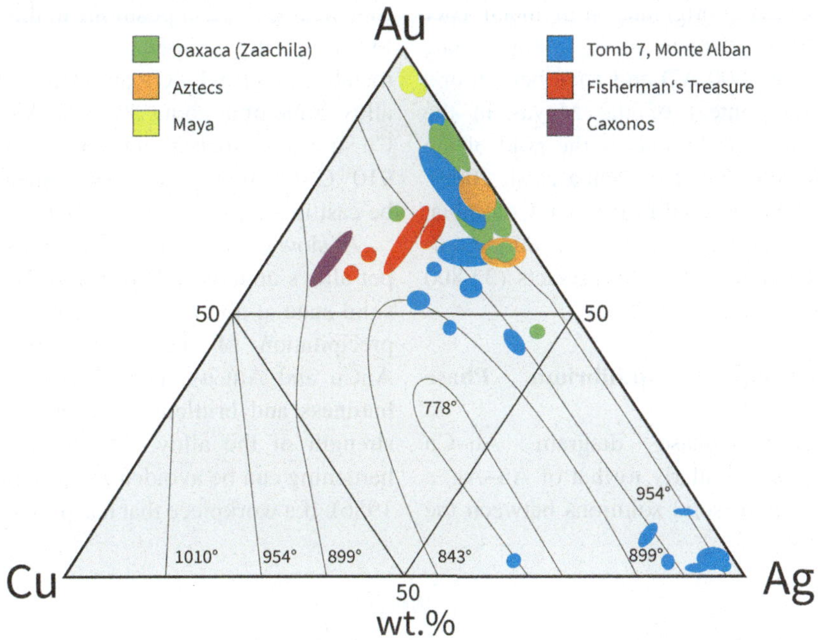

Fig. 7.14 Comparison of the elemental composition of gold artefacts from the main Mesoamerican areas. Note that most of the artefacts contain enormous amounts of copper (up to 40 wt.%) and of silver (up to 50 wt.%). These data are from a number of different localities, but they may be used as a reference for Mesoamerican style artefacts. Liquidus temperatures are shown in the ternary system Au–Ag–Cu. Note the low melting field of alloys beginning from the eutectic Ag–Cu at 767 °C towards the system Au–Cu at 889 °C. Modified after Ruvalcaba et al. (2009)

such a processing comes from the Royal Tombs of Ur in Mesopotamia (middle of the 3rd millennium BC), where a few daggers (Fig. 7.15) and chisels made of tumbaga alloys were found during the excavation (La Niece 1995; Hauptmann et al. 2016, 2018).

In the Greek and Roman world, the widely used gold–silver alloys generally had less than 5 wt.% copper. Such alloys are more suitable for mechanical deformation because they are more ductile. By contrast, tumbaga alloys have significantly lower melting points (Brepohl 2016) and are therefore more suitable for lost-wax casting. These differences are well reflected in the design of many artefacts. In South and Central America, for example artefacts with wire-like ornaments were cast and then formed through hammering (Scott and Seeley 1983).

The world famous, rich tumbaga gold artefacts from Mesoamerica (Mexico) and the northern South America (Colombia, Ecuador, Peru), the

Fig. 7.15 Royal Tombs of Ur, Mesopotamia, PG 1054. This dagger (Inv.-no. 30-12-550) is composed of several sorts of gold. The blade consists of a gold–silver–copper alloy gilded by a depletion gilding process. The golden surface is only in parts preserved (right). For the decoration of the hilt, whitish gold was used to manufacture some 400 nails, the collar and the sheets (left). Length of the dagger: 33 cm. A similar feature is to observe at the blade of a dagger from the Baghdad Museum. Photo: A. Hauptmann, Deutsches Bergbau-Museum Bochum

legendary Eldorado, originate from much later periods. In Mexico, many gold artefacts date back just to about 800 AD and have been found in the cultural context of the Mayas in the Yucatan region. Also famous is the gold of the Aztecs in Mexico City (Ruvalcaba et al. 2009). The Museo el Oro, located in Bogotá, Colombia, houses the world's largest and most unique collection of pre-Columbian gold artefacts (33,800 pieces, Lleras Perez et al. 2007).

The Gold–Copper Equilibrium Phase Diagram

The equilibrium phase diagram Au–Cu (Fig. 7.16) shows, similarly to that of Au–Ag, a continuous series of solid solutions between the two metals. All compositions in this system have lower melting temperatures than the two pure metals Au (1063 °C) and Cu (1083 °C). An alloy containing about 80 wt.% Au and 20 wt.% Cu shows the melting minimum in this system at 910 °C (Okamoto et al. 1987). Such an alloy can be cast much easier than the pure end members.

At slow cooling, suitable composed gold–copper alloys undergo a lattice transformation in the solid state at about <400 °C, which leads to the precipitation of the intermetallic compounds AuCu and $AuCu_3$. These cause an increase in hardness and brittleness and a reduction in the strength of the alloy. The usually undesirable hardening can be avoided by quenching (Wolters 1986). If a workpiece that has been quenched is to

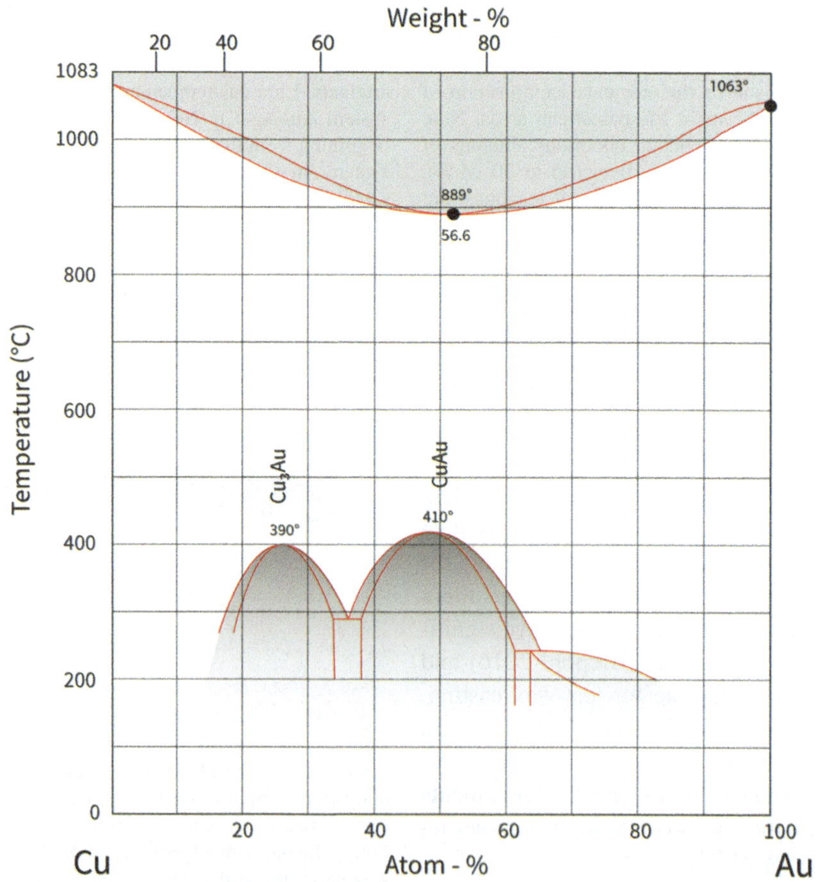

Fig. 7.16 Metallurgical phase diagram for the binary system Au–Cu, the tumbaga alloy *sensu stricto*. The solid solutions are deepening to the lowest melting point at 56.5 at.% Au (equal to 71 wt.% Au and 29 wt.% Cu) at 910 °C, some 150 °C lower than the melting points of gold or copper. Note the two intermetallic compounds of $AuCu_3$ and AuCu. From Okamoto et al. (1987)

be hardened after completion, it must be started, i.e. heated for several hours to temperatures of up to 350 °C.

Gold–copper alloys up to a fineness of Au508 ($AuCu_3$) is slightly tarnished. With Au333 even stress corrosion can occur. Corrosion products are quickly formed, due to mechanically induced lattice disturbances and tensions (forming by hammering, bending, rolling, stretching, etc.). These are evoked by skin exudates (NaCl, ammoniac NH_3, butanoic acid C_3H_7COOH) or by various acids in the soil environment (hydrochloric acid (HCl), sulphuric acid (H_2SO_4). They lead to the formation of dark red fine velvet coatings on the gold objects.

7.3.2.3 Platinum Group Minerals in Gold

Inclusions of platinum group minerals (PGM), also called inclusions of platinum group elements (PGE) in ancient gold artefacts is a supra-regional phenomenon (Ogden 1977; Meeks and Tite 1980; Junk and Pernicka 2003; Mongiatti et al. 2010; Lemasson et al. 2015). PGM inclusions are visible with the naked eye. They appear as silvery specks embedded in gold (Figs. 3.30 and 3.31). They were found in gold artefacts from various time periods, i.e. they are not age-specific, but they are specific phenomena for certain geological environments. PGM in gold were discussed before in Sect. 3.4.8.

PGM inclusions in gold objects got world famous first through countries of northern South-America (Colombia, Ecuador, Peru). They formed dense clusters in many gold objects where they changed the colours of gold to silvery shining lustres (Scott and Bray 1980; Scott 2012b). In every case, due to their extremely high melting points PGM were detected as unmelted inclusions embedded in a liquefied matrix of gold. However, PGM inclusions in gold objects were also found all over the Old World. The most famous discovery of gold coins with PGM comes from the Celtic oppidum of Manching (Bavaria) (3rd to 1st century BC). The hoard excavated comprised more than 450 Celtic gold coins (Gebhard 1995; Gebhard et al. 1995). It is the largest find of Celtic gold of the last century. A number of these Celtic gold coins contain PGM inclusion. This means that these gold coins were made from placer gold. On the other hand, a raw material source of primary gold cannot be excluded, as well as the use of recycled gold. The Celtic gold coins were cast in multiple coin moulds. Bendall et al. (2009) conducted trace elements and lead isotope analyses on a selection of Celtic coins to test the possibilities of gold provenance analyses. Junk and Pernicka (2003) pursued the same goal by analysing the osmium isotope pattern of PGM inclusions in Celtic gold coins. They very critically commented on the usefulness of osmium isotope analysis as a tool for provenancing gold.

More successful have been the investigations of Jansen et al. (2016) using osmium isotope analysis for provenancing gold from Ur. Artefacts from this excavation contained many inclusions of PGM (Figs. 3.29–3.31). Using analytical methods combined with archaeological observations the authors successfully suggested an origin from the area of the upper Amur Daria, the ancient Oxus river in northern Afghanistan/Tajikistan. From there, gold scattered with PGM along with the blue semiprecious stone lapis lazuli was traded over thousands of kilometres to Mesopotamia in the 3rd millennium BC. Gold objects with PGM inclusions were described also from the so-called Oxus Treasure of the 5th–4th century BC in the British Museum by Curtis (2012).

Microinclusions of PGM in (pre-) historic gold artefacts might be used as an important source of geoarchaeological information to provenance the source of gold. They originate from dunite or serpentinite rocks in (ultra-) basic rock units, in which they are sparsely and irregularly distributed. Generally, such rock units are part of ophiolithic complexes that often occur in geological formations of various ages. They are usually understood as an indication for an origin from placer deposits (Cabri et al. 1996). During erosion of these rocks, the PGM end up in alluvial environments where they could be mixed with particles of placer gold. PGM are geologically of different origin than gold deposits and are

Table 7.4 Classification of silver alloys according to the Wiener Convention (cf. Wolters 1986)

• Silver 999. This is the name for chemically pure silver, i.e. fine silver. It is described with 1000/1000 thousandths. Before 1871, it was called (or stamped as such) 16-soldered silver. The fine silver obtained after smelting and segregation is used as the starting metal for the coinage and jewellery industry.
• Silver 925. Sterling silver; it is an alloy of 92.5 wt.% silver and 7.5 wt.% other metals, usually copper. The fineness is 925/1000.
• Silver 830. A relatively high copper content gives this type of silver a slight yellow shade. It is hardly used anymore. The fineness is 830/1000.
• Silver 800 has only price advantages, but is much harder than the other alloys. High copper content may lead to the formation of green corrosion products. Copper content of the alloy can be removed superficially by repeated annealing (cf. depletion gilding).

associated only in a secondary environment. Their presence in artefacts confirms that the origin of the gold was from placer sources rather than from a primary source in solid rock (Ogden 1977; Meeks and Tite 1980).

This provenance criterion is valid for the origin of gold artefacts with PGM inclusions from many localities. Ur's gold comes from gold soaps close to Paleozoic ophiolite complexes in the area of Samti (Takhar) in northern Afghanistan. The PGM inclusions in gold artefacts from the Urals also come from the vicinity of extensive ophiolites in this mountain range (Zaykov et al. 2016). The frequent silver-coloured gold objects from northern South America are casually attributed to the gold soaps in the regionally extensive ophiolite belts on the continental margin of the western continent of America (Bourgois 2013).

7.4 Silver Alloys

Processing techniques for silver are basically similar to those of other metals. The metal is easy to cast but it does not oxidise, and due to its low hardness and exceptionally good ductility, it is easy to process into thin layers (e.g. thin silver bowls). Silver plating of objects by mechanically applied coatings was not uncommon in ancient times (La Niece 1993). Chemically initiated enrichment of silver on the surface of corresponding alloys (depletion silvering) has been used as well. In contrast to the variety of different gold alloys with silver and copper, according to colour, hardness and fineness, however, the number of silver alloys used for decorative purposes is very limited.

Because silver is a soft metal, it may have been alloyed with other metals to increase its hardness. It is generally assumed that it could have been a common practice in the Near and Middle East to add copper to silver in quantities of only a few percent to improve this effect. In addition, in the Early Bronze Age of the Middle East, silver–copper alloys, still difficult to explain, have been found to contain a few percent by weight of gold, as, e.g. in Egypt (Stos-Gale et al. 1995) and in the southern Levant (Rehren et al. 1996).

Silver–copper alloys are of importance. Silver has been used as a metal in coinage for about 2500 years, where it has been purposefully alloyed, particularly with copper (Bachmann and Hammer 2003).

In modern coinage and the jewellery industry, the fineness of silver is judged according to the Vienna Convention (Table 7.4).

7.4.1 Aurian Silver

It has already been mentioned in the Sects. 6.3 and 7.3 that gold–silver alloys were exclusively those with silver contents of maximal 50 wt.%. In most cases, however, silver contents are significantly higher. The problem of aurian silver, i.e. silver alloys with a gold content of a significantly lower level, is best discussed on the basis of artefacts from Egypt (Gale and Stos-Gale 1981b). Such alloys must be carefully distinguished from gold-containing silver, in which gold can only be detected as a trace element.

In the investigation of ancient silver in Egypt, the authors defined aurian silver as silver-containing gold in excess of about 5 wt.%. However, they are aware of the fact that in nature, gold–silver alloys predominantly occur with 40–50 wt.% silver (cf. Sect. 6.3). This is not directly compatible with the assessments of Wedepohl (1968–1978), that silver and gold in nature would form a continuous series of alloys, from silver through aurian silver to argentian gold. This is supported by experimental work of Hansen and Anderko (1958) who stated that gold and silver indeed form a complete solid solution. Alloys of all compositions between the two metals would be possible (cf. Sect. 4.2.1.1). This would mean that in nature, gold-bearing silver would occur with, e.g. >10 wt.% Au. Spiridonov and Yanakieva (2009) addressed this problem and believed that such alloys actually would occur in volcanogenic hydrothermal deposits.

By contrast, the specialist of gold deposits Boyle (1979) states that calculations of aurian silver with gold contents of <1–25 wt.% are not based on natural occurrences of homogeneous aurian silver, but on average calculations of various natural gold- and silver-ores associated in deposits as a mixture of native gold, electrum, argentite, ruby silvers (pyrargyrite and proustite), and of fahlore varieties etc. But if such a mixture is melted, it is very likely to yield gold-bearing silver, which may also contain copper. Kindermann (2005) refers in this sense to the occurrence of silver-bearing jarosites with gold contents in the upper regions of the copper deposits in the Troodos ophiolite complex in Cyprus, the so-called "Devil's Mud".This precious metal-bearing formation also occurs in the massive sulphide deposits of the Iberian Pyrite Belt (IPB) in the south-western Iberian Peninsula (Aljustrel, La Zarza, Tharsis, Rio Tinto, Almeria). It is conceivable that a pyrometallurgical use of such ores could also produce aurian silver.

Be that as it may, in Egypt there are silver objects low in gold with c. 1 wt.% from predynastic times and onwards. Gale and Stos-Gale (1981b) analysed numerous silver objects from the Old and Middle Kingdom (3000–2700 and 2100–1800 BC), and from the Late New Kingdom (1300–1100 BC) with often much more than 10 wt.% Au. On the other hand, many of the silver objects from the "el-Tod treasure" of the Middle Kingdom in Egypt contain only <0.2 wt.% of gold. In addition, lead well under 1 wt.%, and minute traces of copper were analysed. Some objects have up to 4 wt.% of copper. If such a case would represent intentional alloying, as the authors suggest, remains open. Here, the observations of Boyle should be kept in mind.

Since a separation of silver from natural gold–silver alloys at this early time apparently was not yet possible, Ogden (2000) concluded that such alloys must either have been obtained directly from silver-rich ores with varying proportions of gold, or they were extracted by cupellation from lead ores. These considerations were apparently supported by low levels of lead in the artefacts. As a rule, even then silver was extracted (Sect. 6.4) from lead-rich ores (galena, PbS, cerussite $PbCO_3$) by cupellation. This metallurgical process was already known in the Near East in the 4th millennium BC (Hess et al. 1998; Pernicka et al. 1998). Theoretically, silver-rich ores may have been easily available in ancient ore deposits. Examples of this are argentite (Ag_2S), pyrargyrite (Ag_3SbS_3) or proustite (Ag_3AsS_3), which can also always be geochemically connected with copper-bearing ores (fahlores).

7.4.2 Copper-Containing Silver

It is therefore rather questionable whether silver objects low in copper contents are intentional alloys. It may be a fact that, as already mentioned, the addition of only a few percent of copper increases the hardness of silver. Moorey's (1994) opinion that this could have increased the amount of rare and expensive noble metal, is hardly convincing.

Copper-containing silver artefacts can also be found at Arslantepe (eastern Anatolia), where a ring from level VI A was analysed to be composed of silver with several percent of copper and lead (2.7–9.3 wt.% Cu, up to nearly 3 wt.% Pb) (Caneva and Palmieri 1983). At Mahmatlar, a site close to Alaca Höyük, six silver ingots with up to 3 wt.% copper were part of an Early Bronze Age hoard (Sayre et al. 2001); from Middle Bronze Age Acemhöyük south of Ankara with a silver fragment with 1.66 wt.% copper (Sayre et al. 2001), and from Middle Bronze Age II Ebla where a copper-containing silver ingot with 2 wt.% copper and a bracelet with 4.4 wt.% copper were found (unpublished data Rome). This raises the question, whether the Cu–Ag alloy is the product of deliberate mixing of copper and silver, or whether it came about by smelting of suitable ores as it was suggested by Boyle (1979). It is a matter of fact that in archaeometallurgy geological and mineralogical questions of raw sources are sometimes neglected.

Silver artefacts from the Royal Tombs of Ur (c. 2500 BC) have copper concentrations between 1.5 and 6.7 wt.% (Salzmann 2019). These contents are explained with the use of copper-containing silver ores, possibly of fahlores, which have apparently also been gold bearing. In any case, according to measurements by mass spectrometry (SC-ICP-MS), the artefacts contain up to 0.57 wt.% gold. It was mentioned before (Sect. 7.3) that Jansen (2019) gives another explanation for impurities such as copper in noble metal artefacts, in his case for impurities in gold artefacts exemplified by the artefacts from the Royal Tombs of Ur, Mesopotamia. He points out that a comparison of gold artefacts with natural gold is not without problems, and the pattern of minor and trace elements could be changed significantly by metallurgical processing. It would be hardly possible to estimate it quantitatively. The same would be true for silver. Less is known on the organisation about the organisation of metal workshops: it cannot be excluded that places, houses, tools, vessels of a copper metal worker would be quite different from those of a gold or silversmith.

7.4.3 Silver–Copper Alloys

Clearly, among the most specific metal objects known in ancient times are those silver objects high in copper but with a silvery shining surface. Generally, silver alloys that high in copper (or vice versa) seem to be extremely rare, not only in the Old World but also in the New World and in the Far East.

In Mesopotamia it is one object, an arrowhead composed of 26 wt.% silver and 69.5 wt.% copper found in the "Riemchengebäude" at Uruk-Warka (Pernicka 1993) dating to the Late Uruk period (c. 3500–3300 BC). Two hair rings with 48 wt.% silver and copper each, were discovered in a Middle Bronze Age context at Arich, a settlement located at a tributary of the Araxes river (Transcaucasia, Armenia; Gevorkjan 1980). Further to the north, at Modinache, Satchere (Georgia), six rings of the same type, composition and the same age were found (unpublished data Bochum; see also Gambaschidze et al. 2001). In addition to these finds, there are four silver-coloured shaft-hole axes of the Kozarac type, part of a funerary gift dating to the first half of the 3rd millennium BC, probably from the Carpathian basin, which are also made from a silver–copper alloy. Two of the axes consist of 32 and 36 wt.% silver and 67 and 64 wt.% copper, respectively, two of 53 and 56 wt.% silver and 47 and 43 wt.% copper, respectively (Born and Hansen 2001).

Famous finds of such alloys were excavated at the "royal" tomb at Arslantepe, level VI B1 (Early Bronze Age, 3000–2900 BC) in eastern Anatolia (Frangipane et al. 2000; Hauptmann et al. 2002a). These were 28 artefacts (ornamental objects like bracelets, rings, diadems and a dagger) made of silver–copper alloys with silver contents ranging between 23 and 65 wt.% (Fig. 7.17). It appears plausible that this silver–copper alloy should be made intentionally with silver gained from the cupellation of lead.

The phase diagram silver–copper is discussed in Sect. 4.2 and displayed in Fig. 4.3. It shows that the eutectic composition of 28 wt.% copper and 72 wt.% silver marks the lowest melting point of the binary alloy at 779 °C. Even if the alloys presented here are different from this composition, they will start melting, or solidifying, approximately at this temperature. Compared with arsenical copper with, say, 5–10 wt.% arsenic and a melting point around 1000 °C, this is

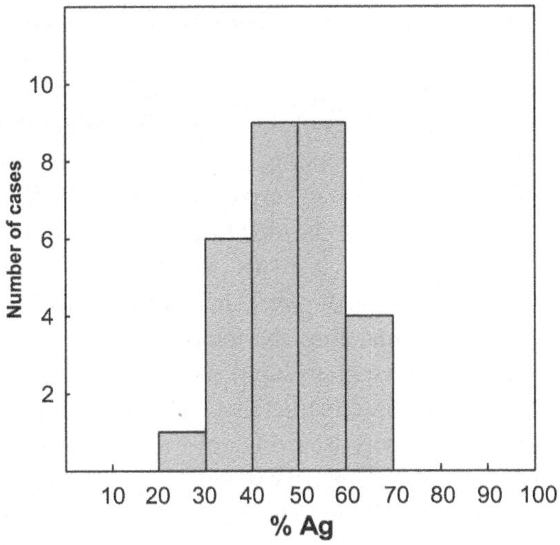

Fig. 7.17 Silver contents of copper–silver alloys from the "royal" tomb at Arslantepe, Eastern Anatolia. From Hauptmann et al. (2002)

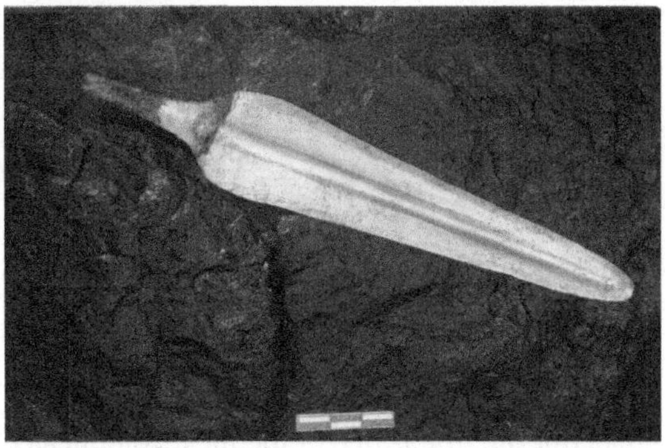

Fig. 7.18 Arslantepe (Anatolia), "Royal" Tomb, Early Bronze Age I (level VIB, 3000–2900 BC). Silvery shining blade of dagger (Inv.-no. 3680) made by depletion silvering. On the handle, the green coloured copper–silver alloy is visible which was not processed to deplete the copper out of the alloy. Length of the dagger 18.3 cm. Photo: Roberto Ceccacci. From Di Nocera et al. (2004)

easier to handle for melting and casting. When cooling under equilibrium conditions, the colour of such an alloy would be a mottled red copper, due to the precipitation of a copper-rich β-solid-solution phase embedded in a fine-grained matrix of silver and copper. Upon rapid cooling, as will presumably have occurred in antiquity, such copper–silver alloys, upon solidification, are of a bright silvery colour due to the formation of a homogeneous solid solution which is metastable.

Depletion Silvering

Heating the metal to below the melting point results in a pronounced increase in lustre and hardness, and in case an object is treated after casting by (repeated) hammering and annealing (as it is necessary to give it the final design or to harden it), the copper-rich part is selectively oxidised. After removing such copper oxide scale, either mechanically or with slight acids, silver is enriched at the surface. Thus, these metal alloys, perhaps of a reddish tint when cast, are bright silvery in colour after having been hammered into shape. Such techniques to produce silvery shining objects from silver–copper alloys are very similar to the enrichment of gold from a gold–silver–copper alloy. This technique is called depletion gilding, or in this case depletion silvering. It has been reported from later periods, from the Americas by Lechtman (1980, 1988) and from the Czech Republic by Kamarzin et al. (1996).

Indeed, after removing the corrosion layer and restoring the original surface, all objects from the "royal" tomb that consist of such a silver–copper alloy show a brilliant silvery colour. This is especially true for the silvery shining dagger that was made by depletion silvering (Fig. 7.18).

Certainly, these advantages were well recognised by the ancient metal workers, which may well have been the reason to prefer such alloys over pure silver. But if so why, then, should the composition of the alloys be so variable? Why produce alloys with 65 wt.% silver if one with half as much silver serves the same purpose to the extent that it is visually indistinguishable from the presumably twice as valuable silver-rich variety? And why should the distribution in time and space of such alloys be so limited? Perhaps their apparent absence is only a pseudo-problem due to the fact that many of the alleged silver objects, reported from all over ancient Europe and the ancient Near East, actually have not been completely analysed as of yet (Prag 1978; Born and Hansen 2001).

7.5 Iron Alloys

7.5.1 The Late Coming of Iron and Steel

Isolated use of iron or steel goes back far into human history before the beginning of the age named after this metal, the Iron Age. This was reported in Sect. 6.6. However, compared to other metals and alloys, such as those of gold, silver, copper or tin, iron is one of the most complicated materials, which challenged the technical skills of humanity. The reasons why iron or steel later prevailed over bronze as a mass metal par excellence, and thus marked the transition from the Bronze Age to the Iron Age, are likely multi-layered. Before this is explained, it should again be noted that the various, distinctively coloured iron raw materials that were known to mankind since the Palaeolithic as dyes, have obviously been deliberately not used for a long time.

The hypothesis that an increased need for charcoal in copper smelting, in contrast to iron production (Horne 1982) and related shortages could have caused the change from tin bronze to iron, is too brief. Nor was there any shortage of copper ores, as stated occasionally, so that people had to resort to a new material. There is sufficient evidence that even in the Iron Age copper was still produced not only from many smaller, but also from numerous supra-regionally significant sources of raw materials (Weisgerber 2007; Hauptmann 2007; Craddock 2013a; Kassianidou 2018; Reitmaier-Naef 2019 and others). This implied further significant fuel needs of charcoal. But it is almost impossible that the copper produced in the Iron Age has found wide use as metal alone. Rather, it was almost regularly alloyed with tin or other metals to bronzes, so that a collapse of long-distance trade relations and possibly resulting restrictions on resource trading, especially of tin, is not convincing.

The reason for the relatively late use of the metal iron and its alloys on a broader basis could probably have been the difficulty of a systematically controlled heat treatment of this metal in all stages of production, which has prevailed comparatively late. It was now possible to give this metal, which was "smelted" in the solid state (and not in the liquid state!) and was not too hard at first mechanical properties such as hardness and malleability. This far surpassed any craftsmanship skills that were needed to work gold, copper or bronzes. It was actually the discovery of a complex system of (re-) heating, quenching, long-lasting annealing and tempering which was connected with carburisation and the formation of characteristic microstructures. These were perhaps reasons why iron or steel was only able to take hold as one of the most widely used metals or alloys after a millennium of smelting experience.

Iron itself has an extremely high melting point at 1540 °C, which was hardly accomplishable with the technical possibilities in ancient times. Burning of charcoal in smelting furnaces could only exceptionally reach continuously c. 1400 °C. It was generally only possible to reduce largely, but not completely carbon-free iron in solid state. Depending on the heat treatment and annealing, that this metal was subjected to, i.e. slow cooling just in air or quenching in water, oil or other liquids, forging temperatures in the smithing hearth, the physical properties such as hardness, brittleness or malleability may have varied considerably from the surface of an artefact to its core. Next to hardening by forging, carburisation and quenching are the major advantages of iron in comparison to base metals and their alloys.

In all these processes, the colour of the individual alloys changed not very much, they are just showing different stages of light and darker grey. This is in contrast to other alloys, such as, e.g. those of gold, silver, copper, bronze or brass, in which the colour nuances in ancient times played a very important role very early. The colour of iron–carbon alloys was used by the technique of pattern welding for decoration purposes. This is a hammer welding technique in which sorts of iron with different compositions are physically distorted, joined together, forged, polished and etched to display often beautiful patterns on the surface (Buchwald 2005). This technique (in German *Damaszieren*) was observed on some Celtic swords, more frequently in the middle of the first millennium AD.

7.5.2 Iron, Steel and Cast Iron

The most important alloying component of iron is carbon in the range of up to 4 wt.%, and there are several sorts of alloys with different concentrations of carbon. Other elements suitable for alloying with iron are Mn, Cr, P, Ti, V, Ni, Co, Mo, W, V. Nb, Ta, Zr, Se and Ce.

In the bloomery process, which was common worldwide since the 1st millennium BC, low-carbon iron was produced predominantly. This sort of iron is called ferritic wrought iron. The early wrought iron was produced in the solid state as a bloom. However, it was a heterogeneous mass of iron with changing amounts of mechanical inclusions of slag, charcoal or even badly reduced ore. Such blooms may also occasionally had parts richer in carbon, which prove that by this process also steel, and sometimes cast iron could have been produced. This was impressively demonstrated experimentally by Straube (1996), when he explored the production method of the countless high-quality finds of *Ferrum Noricum* from the Celtic/Roman period of the Magdalensberg (Austria). These finds consisted mainly of carburised iron, i.e. steel.

He was able to prove that the ordinary bloomery process is capable of producing not only ferritic wrought iron, but also steel and sections of cast iron.

Wrought iron was tough, malleable, ductile and easily welded. Before the development of industrial methods of steelmaking in the 18th century AD, wrought iron was the most common form of malleable iron. It was given the name *wrought* because it was hammered, rolled or otherwise worked when it was hot enough. It had to be reheated and forged at temperatures 1000–1200 °C to squeeze out molten slag inclusions and to shut weld cavities. Internal gradients of carbon contents are homogenised by diffusion (Tylecote et al. 1971). Upon reheating, pure solid iron can only dissolve carbon at temperatures of 727 °C, in case of impurities this temperature may increase to about 740 °C.

The modern functional equivalent of wrought iron is a homogeneous mild or low carbon steel. Its carbon concentrations are usually c. <2 wt.%. As pointed out by Tylecote (1992a), steel can be hardened by quenching, mild steel does not contain enough carbon to be hardenable by heating and quenching.

Carburisation

The proper treatment of iron was to carburise a bloom or wrought iron in a controlled way and to create the right, suitable FeC phases by heat treatment. Several aspects are discussed.

If iron ore is smelted with a high excess of charcoal and powerful bellows and tuyères, absorption of the resulting iron with carbon, as mentioned, may occur. Compositions of iron with carbon <2 wt.% are called steel. At 2 wt.% carbon the iron starts to be liquefied at temperatures of about 1146 °C (Fig. 7.19). In the liquid state, iron absorbs carbon, or carbon monoxide, respectively, more rapidly. Iron with carbon contents of c. 2–4 wt.% is called cast iron. There are significant physical differences between steel and cast iron. Depending on its carbon concentration, steel is usually malleable while cast iron is brittle. It cannot be ruled out that cast iron was produced, and possibly also used, in smaller parts by accident during the Iron Age in the Old World (see below). In China, cast iron was regularly melted since the middle of the 1st millennium BC (overview in Jianli and Rubin 2013).

Generally, molten cast iron may be included in small quantities in a porous bloom, sometimes, if larger quantities are formed, they may collect at the bottom of a bloomery furnace. In such a case the concentration of carbon in iron may be up to 4 wt.%. Except for carbon, the formation of liquid iron is enhanced by concentrations of phosphorous (P_2O_5) introduced by the ore. Such an example of iron with 2.5 wt.% phosphorous and almost 3 wt.% carbon was found at a smelting site at Metzingen, Swabian Alb, from the 11th/12th century AD (Yalçın and Hauptmann 1995).

Rehder (1989) discusses the topic of carburisation of iron. He emphasises that, if the ratio of carbon in the fuel to iron in the ore is

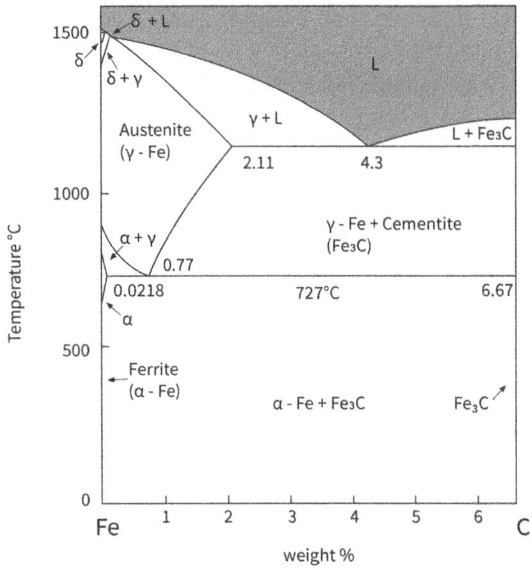

Fig. 7.19 Section of the iron–carbon phase diagram, i.e. the section Fe–Fe_3C (iron–cementite). Values are given in weight-%. Note the small area where ferritic iron (α-iron) is stable. Main area of compositions is covered by the more C-rich austenite (γ-iron) that decomposes at 723 °C either to ferrite + pearlite or, alternatively, at slightly higher C-concentrations to pearlite and cementite. For explanation see text

above a certain value, the reduced iron dissolves carbon, increasing in amounts as the ratio increases. Thus, iron made in a furnace may contain from nearly zero to well over four percent carbon. The agent which transfers carbon into the solid iron is usually carbon monoxide (CO) gas. Above about 740 °C, its reaction with iron is given by the equation:

$$Fe + 2\,CO \rightarrow Fe_3C + CO_2$$

According to this formula, carbon is successively absorbed by the iron in the solid state and is absorbed selectively or in layers, depending on the configuration of a bloom or an artefact.

Rehder thus addresses one of the fundamental issues of iron metallurgy, namely how the different levels of carbon come about: The problem, whether steel, that is the iron–carbon compounds, was already produced during smelting, or if a low-carbon alloy was not carburised before the bloom was treated in a smithing hearth, has already been pointed out.

It has already been mentioned that a bloom was generally inhomogeneous, with significant proportions of slag and charcoal inclusions, and was made up of ferritic iron, steel and cast iron. A simple but probably lengthy process for compacting and homogenising such a bloom was to crush it and subject it to a long annealing treatment, either in reheating hearths or in crucible shaped vessels. Such a practice is well known from ethnographic findings from iron smelters in African countries (Sect. 9.3; Celis 1991). Presumably, such a reheating of a bloom or its fragments has a long history. Also, Moesta (1983) reports on a long-lasting annealing of low-carbon iron pieces in a closed crucible or box of fired clay, which was filled with excess charcoal and could lead to a carburisation.

Many reheating hearths described in archaeology, which evidently date back to the Latène period, are technologically quite acceptable. In Pleiner (2000) the topic is discussed in great detail. He presented a number of these archaeological findings, but also emphasises very clearly that they can by no means always be interpreted irreproachably.

Decarburisation

If we have a look at the Far East, to China, we should also consider a third possibility of steel making in ancient times, namely a decarburisation of cast iron to steel. As a matter of fact, this really has to be considered. Rehder (1989) argues that in the opposite process (that is, the carburisation of a piece of ferritic iron, e.g. a bloom, which is placed in a charcoal bed of a smithing hearth or heated and forged repeatedly) iron could hardly be carburised to steel.

Rostoker et al. (1989b) also write that such an interaction of liquid cast iron and solid low-carbon iron was probably used in early metallurgy. By examining metal samples, they found that this could have happened with the well-known Japanese Tatara process. The Tatara process is a special form of the bloomery process that has been carried out in Japan in box-shaped smelting furnaces since the 7th century AD (Gowland 1899; Kubota 1970). It became known as a process in which wrought iron, steel and cast iron could be produced in a one-step process. The peculiarity of this process was that it was conducted for up to three days and reached very high temperatures. Rostoker et al. (1989b), however, are of the opinion that the formation of steel happened in two steps in this process. In a first step, a sintered, spongy low carbon wrought iron bloom was formed next to portions of liquid high carbon cast iron. In a second step, this liquid iron fills the capillary spaces formed by the voids of the solid bloom. Solid-state diffusions between the two metals lead to the formation of steel.

7.5.3 The Iron–Carbon Phase Diagram

The temperatures of the bloomery process did not reach the melting point of iron. In pure iron, it is at 1540 °C, at the eutectic of Fe–C at carbon contents of about 4.3 wt.% in the system Fe–Fe_3C, it is lowered to about 1147 °C. We are dealing with a metastable equilibrium diagram, which has a high cooling rate. Under these conditions solid iron and carbon as Fe_3C are formed. We have to deal with parts of the diagram that falls under the conditions low carburising + working temperatures not much exceeding 1200 °C.

The Fe–C phase diagram is discussed in detail in many publications. There is a very reasonable explanation in the German language edited by Horstmann (1985), for special issues in archaeometallurgy the explanations of Buchwald (2005) and of Giumlia-Mair and Maddin (2004) are recommended, which are also equipped with useful metallographic pictures. These are therefore of importance, because different smelting and forging processes produce different Fe- and Fe–C-phases and microstructures. Here, the understanding of the fundamentals of the Fe–C phase diagram is useful (Fig. 7.19).

Iron has a cubic crystal system, the atoms of iron are located at the corner points of a cube. The amount of carbon that can dissolve in the iron depends on the modification of the iron solidifies. In iron or steel metallurgy, the following solid solutions with different carbon concentrations are distinguished:

1. Ferrite α-solid solution (α-Fe). In this solid solution of iron, the carbon atoms form a body-centered cube (bcc). This modification forms up to a temperature of 911 °C. At 723 °C ferrite can only absorb up to 0.02 wt. % C. Therefore, the stability range of ferrite in the Fe–C system is very limited.
2. Austenite γ-solid solutions (γ-Fe). In this solid solution, the embedded carbon atoms are arranged in the centre of each cube surface. They form a face-centered cube (fcc). These crystals are stable in a range between 1392 and 911 °C. Austenite can absorb significantly more carbon than ferrite, up to 2.11 wt.%. The stability field of austenite is much larger than that of ferrite. Together with the cementite, austenite plays an essential role in the hardening of iron.
3. The (δ-Fe) solid solution (δ-ferrite), which has a face-centered lattice (fcc), plays a less important role in archaeometallurgy. This solid solution is only of importance in high-levelled modern sorts of steel. These crystals form in the temperature range of 1493 °C, this means shortly below the liquidus point of iron at 1536 °C.

All these phases are relatively soft but harder than pure iron due to the strain hardening of the solved carbon.

The iron–carbon compound cementite (Fe_3C) is crystallising also as a separate phase. However, this is not to be confused with the mentioned iron solid solutions, because this is a metastable phase, which decomposes under equilibrium conditions to Fe and C (graphite). It contains 6.67 wt.% C. It is extremely hard and brittle and contained in all steel grades. Variations in the grain size and shape of the crystals of cementite allow for manipulation of the hardening and other physical properties of the steel.

Depending on the carbon contents of the iron and depending on the cooling rates, typical Fe–C phases are formed, which are present in highly variable but very characteristic structures.

- Ferrite (Fig. 7.20a). Large parts of ancient blooms consist of (soft) ferritic iron. It contains up to 0.006 wt.% C.
- Pearlite (Fig. 7.20b) is a densely laminated structure consisting of α-iron (ferrite) and cementite (Fe_3C). It is a constituent of steel. If no cementite is formed then the structure indicates 0.02–0.8 wt.% C. This is a composition of a steel.
- Martensite (Fig. 7.20c) is a needle-like metasta ble structure of pearlite and cementite in variable amounts. This structure is caused by a quenching of austenite-rich iron. It leads to a considerable hardness. It contains 0.8–2.06 wt. % C. This is a steel.
- Ledeburite is a eutectic constituent of austenite and iron carbide (Fe_3C). Normally, a mixture of ledeburite (II) (formed < 723 °C) + Fe_3C + pearlite. It has carbon concentrations between 2 and 6.67 wt.%. This is called cast iron.
- Bainite consists similar to pearlite of the two phases ferrite and cementite. It is not easy to distinguish from martensite, but it has a slightly different history of cooling. It has in the German language the special name *Zwischenstufengefüge* (bainitic structure).

These structures are identifiable in light microscopy and provide crucial information about their formation. Therefore, when investigating iron artefacts, care should be taken to carry out appropriate extensive sampling (several mm^3 or more) in order to examine as many special parts of an artefact as possible. If this is not possible, it is likely that the result will be little more than a designation as an "iron artefact". The corrosion of old iron objects is very strong.

If, for example a piece of iron of a bloom with a total composition of 1 wt.% carbon slowly cools from about 1200 °C, nothing happens initially until the temperature has dropped down to about

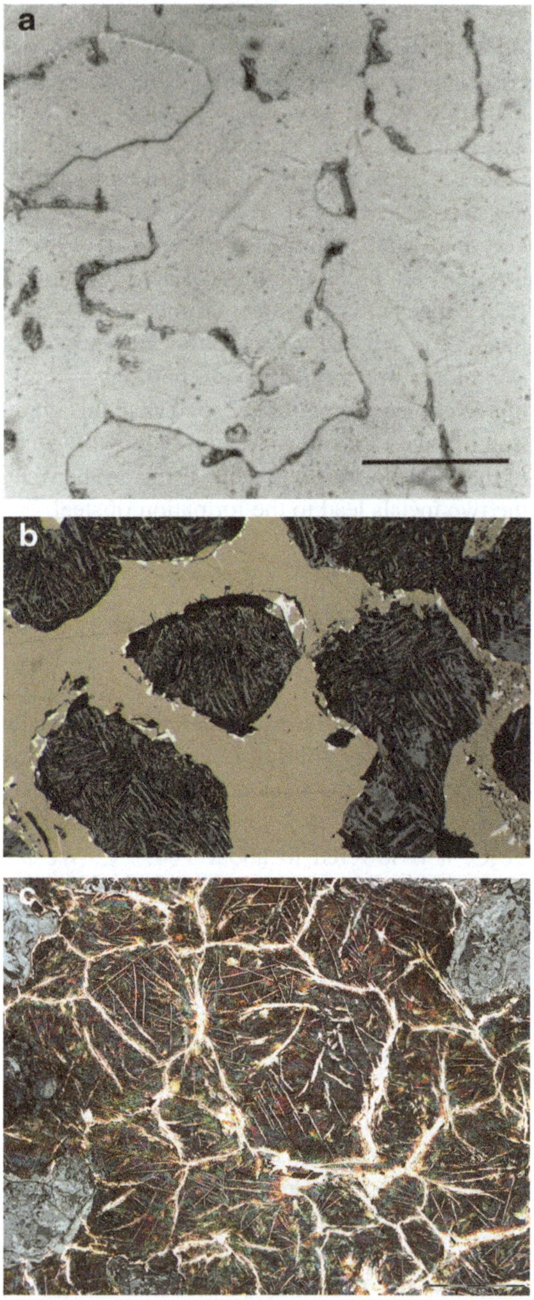

Fig. 7.20 (**a**) Roman settlement of Saalburg at the Upper German Limes near Frankfurt. Shaft hole of an axe, sample ZM 1270b Ferrite grains with small amounts of pearlite (dark) in between. The iron is only very slightly carburised. Scale bar 0.04 mm. From Maddin et al. (1991). (**b**) Metzingen-Kurleshau (Swabian Alb, Germany, 12th century AD), sample D-22/9. Structure of a hypoeutectic Fe–C alloy (wrought iron). The globules of pearlite are built up by light lamellae of cementite and dark

900 °C. Here the γ-iron can no longer keep the carbon content in solution. The very hard cementite (and some graphite) precipitate. At 723 °C, the remaining γ-iron is converted into α-iron (ferrite) and a eutectic mixture of cementite is formed, which is intensely intergrown with the soft α-iron. Pearlite is formed. Below 723 °C all mixed crystals are pearlite and ferrite. Because of its ferrite content, such slowly cooled iron is relatively soft and malleable so long as the hard and brittle cementite content does not become too high. The limit of the forgeability of iron is about 1.6 wt.% carbon, which corresponds to a content of c. 25% cementite.

By rapid quenching, when the iron is immersed in cold water, this phase transformation of γ-iron (austenite) into α-iron + cementite will not occur. The high carbon content of austenite leads to stresses in the crystal lattice during cooling, which results in a hardening of the steel. Such a steel can no longer be forged, it is brittle. This quenched version of the steel has a special needle-like microstructure called martensite (Fig. 7.20c). As pointed out by Smith (1967), in comparison to all other hardening mechanisms, e.g. as alloying of copper with tin, quenching of iron to martensite has the most significant effect on strength, as shown in Fig. 7.21.

In order to give a steel blade both properties, i.e. to forge it as optimally as possible, to give it a strong, tough core and at the same time the highest possible hardness, complicated crafting techniques are needed. This can include a number of very different, special tricks of cooling and

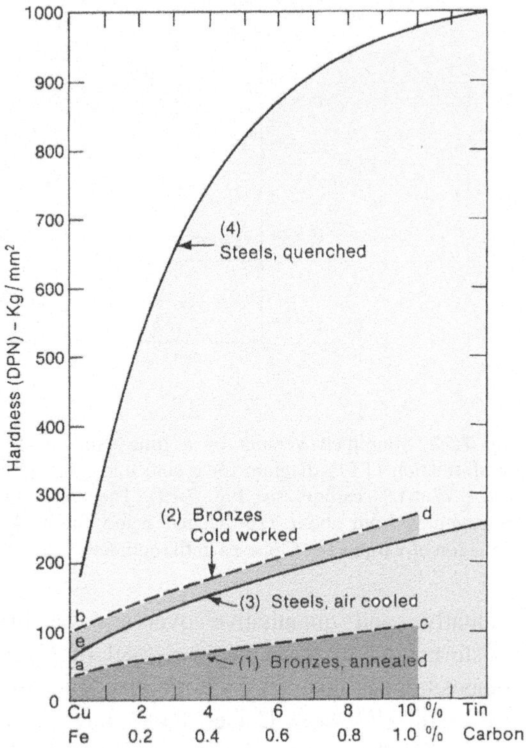

Fig. 7.21 Diagram showing the enormous effect of hardening of steel by quenching. For comparison, the hardness of air-cooled steel, of pure copper and of cold worked and annealed tin bronzes are shown. Shown is the hardness (DPN, measured with the Vickers diamond indenter, units kg/mm^2), and on the abscissa the tin contents of bronzes, and the carbon content in iron. While copper low in tin, tin bronzes and air-cooled steel hardly surpass 250 DPN, the hardness of quenched steel raises up to 1000 DPN! From Smith (1967)

reheating, which are difficult to grasp in detail. To identify these and to reconstruct such a They require careful metallographic investigations. These techniques are thus much more complicated than the shaping and production of non-ferrous metal artefacts and their alloys and require a considerable amount of experience with this special material.

So the structure of steels does not just depend on the carbon content of the iron. It also very much depends on the speed of cooling and the formation of specific structures. To illustrate this, the time–temperature–transformation (TTT) diagram was constructed (Fig. 7.22). It is a

Fig. 7.20 (continued) (corroded) ferrite. The light matrix consists of secondary cementite with some remains of white iron-phosphide (Fe_3P). SEM picture, backscattered mode, width of picture c.0.5 mm. Photo: Ü. Yalçın, Deutsches Bergbau-Museum Bochum. (**c**) Late Roman settlement of Widdersdorf, close to the modern city of Cologne Sample D-135/5. Fragment of a dense piece of iron, probably a fragment of a bloom. Etching of the sample shows a characteristic structure of martensite with white needles and intergranular cementite. In between extremely fine-grained, dark densely streaky pearlite. This is an example of high-grade carbon steel (<0.8 wt. %) which was quickly quenched, perhaps in water. Etched with HNO_3. Scale: 500 μm. From Hauptmann (2002)

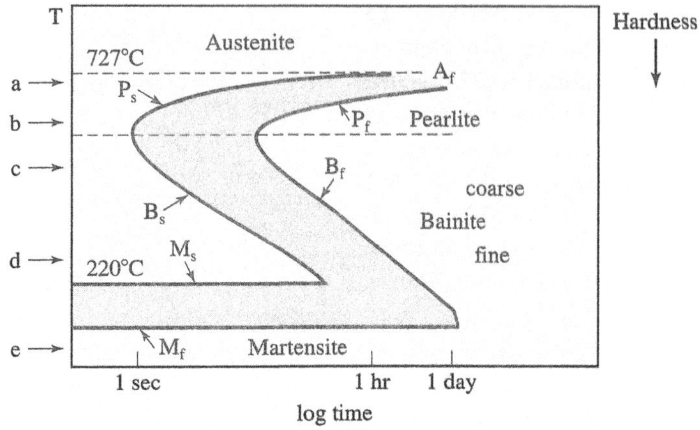

Fig. 7.22 Simplified version of a time–temperature–transformation (TTT) diagram for a eutectoid steel (iron with 0.77 wt.% carbon, see Fig. 7.19). The annealing temperatures from about 720 °C and below show the formation of various Fe–C phases until room temperatures are reached. This depends on the velocity of cooling, indicated at the *x*-coordinate (abscissae) from 1 second to 1 day. The scale on the right side points to increasing hardness of the alloys. Modified after Hummel (1998)

qualitative and quantitative overview on the transformation of austenite in a steel of given composition. In this case by a piece of eutectoid steel with 0.77 wt.% C (see Fe–C diagram in Fig. 7.19). The *y*-coordinate (ordinate) shows the temperature of transformation, and the *x*-coordinate (abscissa) shows the time of the transformation in the diagram in terms of seconds to days. The opposite *y*-coordinate shows increasing hardness with decreasing temperatures. This diagram indicates the consequences of phase transformation, which, depending on annealing, quenching and tempering in a unique multiple ways would alter and improve the mechanical properties of steel. Hence, it definitely depends on the large varieties of cooling rates.

Hummel (1998) described a few special cases. By quenching a eutectoid steel from above 727 °C, i.e. from the austenite region a little bit down to the level of "a", only a little bit of ferrite and cementite (= pearlite) are formed. The beginning of pearlite is indicated by the pearlite start time (P_s). The abbreviation P_f gives the pearlite finishing time at the same temperature when all the material is transformed into pearlite. Because the temperature is still high, and the diffusion may be long, coarse-grained pearlite will be formed. At "b" the situation is somewhat different because the austenite is quenched. A very fine-grained pearlite is formed which is harder. In case of level "d" fine bainite is formed which is harder again. If austenitic, eutectoid steel is very rapidly quenched to room temperature (between M_s and M_f). The transformation of austenite to α-iron and cementite is prohibited, a very hard and brittle body-centered needle-like tetragonal structure, called martensite is formed.

Nevertheless, at least on superficial layers of iron artefacts a slight carburisation may occur in a smithing hearth. The welding of layers of differently carburised steel has been found a wide application, e.g. in the production of Damascus blades. In this way, pattern welding allowed for the creation of certain patterns with beautiful aesthetic qualities. Such blades were also considered quality features.

7.5.4 Phosphorous Iron

Not rarely Fe-phosphides have been found in ancient iron where they were mainly gained from phosphorus-rich iron ores of sedimentary origin, such as bog ores (Yalçın and Hauptmann 1995; Ganzelewski 2000). Later, in the medieval period, mixed iron ores (magnetite with apatite

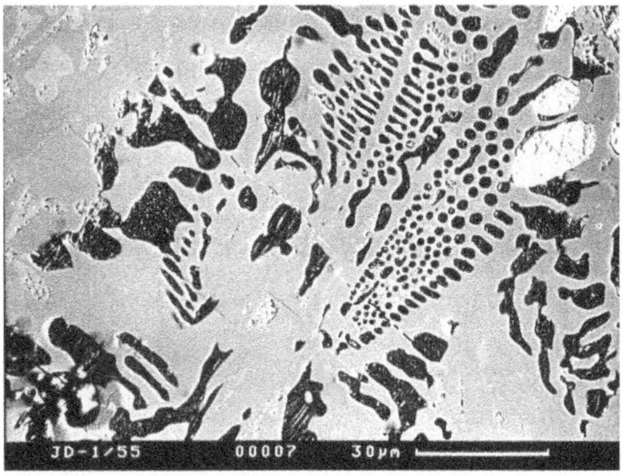

Fig. 7.23 Sample JD-1/55, Faynan 1 (Jordan). Inclusion of tabular Fe3P crystals with a Fe3P–Fe–phosphate eutectic intergrowth in copper. It was found at the Roman copper smelting site at Faynan. The microstructure is very similar to cast iron inclusions in copper from Timna (Craddock and Meeks 1987) or from Sa Sedda in Sardinia (Tylecote 1983) and had still been interpreted in that way by Hauptmann et al. (1993). Scanning electron micrograph, backscattered electron mode. Photo from Hauptmann (2007)

($Ca_5(Cl|(PO_4)_3)$) such as from Grängesberg (Sweden) were smelted (Buchwald 2005). The medieval smelting of P-bearing lead ores in the Black Forest (Germany) also led to the formation of iron-phosphides (Goldenberg 1996). Phosphorus as an indicator for provenance studies has not as yet been investigated in detail. However, this is possible in principle, but it makes only little sense because sedimentary iron ore deposits are widely distributed worldwide in a number of different geological formations (Schneiderhöhn 1962).

Theoretically, the element is easily oxidised and slagged, but Fe-phosphides have nonetheless been found in ingots, which had been remelted at least once. Examples are iron-containing copper ingots from En Yahav, Wadi Arabah (Israel), and from Deir Alla in Jordan, described by Roman (1990). This proves that phosphor can definitely be found in final objects. Fe-phosphides in copper have also been found in Faynan (Jordan) and in Timna (Hauptmann 2007; Roman 1990). The existence of Fe-phosphides in ancient copper, which had been produced in the surroundings of sedimentary ore deposits such as Faynan or Timna, is not surprising because the sandstones and other sedimentary host rocks in these regions are so high that they are commercially mined in the so-called phosphorous belt between Amman and Maan in enormous quantities (Abu-Ajamieh et al. 1988). In addition, due to the higher affinity of phosphorous to iron than to copper, inclusions of Fe-phosphides are not rarely to be found.

If the iron ore contains phosphorus it is not suitable for steel making. It produces, at best, a mixture of phosphorferritic and pearlitic structures (Buchwald 2005).

Figure 7.23 shows a eutectic intergrowth of Fe_3P + Fe-phosphate in Fe_3P, which displays an astonishing similarity to ledeburite (cementite + pearlite). The same structure has been described for inclusions of cast iron in ancient copper from Sa Sedda in Sardinia (Tylecote 1983) and from Timna (Craddock and Meeks 1987). This corresponds with the stability relations in the system Fe–C–P (Schürmann 1958), following which, carburisation of P-bearing iron phases is generally suppressed.

8 Macro- and Microstructure of Metals

Old crafting techniques used in the production of metal objects can be understood by structure analyses from both the macroscopic to microscopic scale. This mainly applies to copper with its alloy components tin, lead and zinc, furthermore to iron and/or steel or cast iron and gold with the alloy components silver and copper.

However, prior to performing metallographic investigations by their conventional invasive incisions into the material, important information on artefact fabrication steps may be obtained through a series of non-destructive procedures. For reasons of conservation, they have become increasingly important in archaeology, as here non-destructive analytical methods are generally preferred.

For imaging the macro- and microstructure of metal artefacts there are presently three major techniques employed:

1. Radiography and computed tomography using X-rays
2. Metallography: Reflected light optical microscopy (RL-OM) and scanning electron microscopical analysis (SEM/EDS)
3. Crystallographic texture analysis performed on neutron diffraction

8.1 Macrostructure

A first informative non-destructive macroscopic analysis of metal artefacts is possible through the use of X-rays. It is a conventional X-ray radiography and high-energy 3D-computed tomography.

These investigation methods mainly concern artefacts of larger dimensions, such as swords made of iron or steel or larger metal vessels. They can be easily examined in two dimensions, but also in (moving) three-dimensional space for their homogeneous or heterogeneous composition.

8.1.1 Conventional Radiography

Basic studies on conventional radiography are now possible in many museums and research institutions that either have their own X-ray facilities or have access to them. But also equipment with tomography capabilities, which allows the three-dimensional analysis of objects, have improved too.

Investigations of medieval pattern welded blades from England have been performed by Lang and Ager (1989). Later, Lang (2015) showed in a very well-arranged publication on iron artefacts from a fifth century AD Anglo-Saxon cemetery in England, how to reconstruct the exact structure of artefacts via a combined radiography, optical metallographic and scanning electron microscopical analysis (SEM/EDS). However, for this purpose, she had to cut through some artefacts from top to bottom. In this way, she was able to show that the medieval

A. Hauptmann, *Archaeometallurgy – Materials Science Aspects*, Natural Science in Archaeology,
https://doi.org/10.1007/978-3-030-50367-3_8

blacksmiths had made their objects from several types of iron: wrought iron, low carbon iron, phosphoric iron, steel and As–Ni-containing iron (Fig. 8.1). So they were able to distinguish between these varieties.

8.1.2 Computed Tomography (CT)

Lehmann (2016) characterised some early medieval double-edged swords from Westphalia (Germany), the so-called spatha. They dated to the sixth to eighth century AD. In the case of these iron blades, the question was whether they were pattern welded or not. This was solved using conventional X-ray radiography (tomography). The author was able to find out non-destructively that the blades of the swords, which are about 90 cm long, were artfully welded together from numerous individual parts. He was able to display the highly complex damascene patterns of the swords very well. In contrast to the investigations of Lang (2015), he was not permitted to take samples to distinguish the composition of the individual types of iron or steel.

The method can also be easily applied to other metals, such as tin bronzes. As an example, the ritual vessels of the Shang and Zhou dynasties in China (c. sixteenth century to eleventh century BC and twelfth century to eighth century BC) should be noted here, of which numerous examples from collections in the United States have been analysed. These up to metre scaled vessels are evidence for a high artistic and technological casting technology. At this time, the Chinese developed a multistage process for

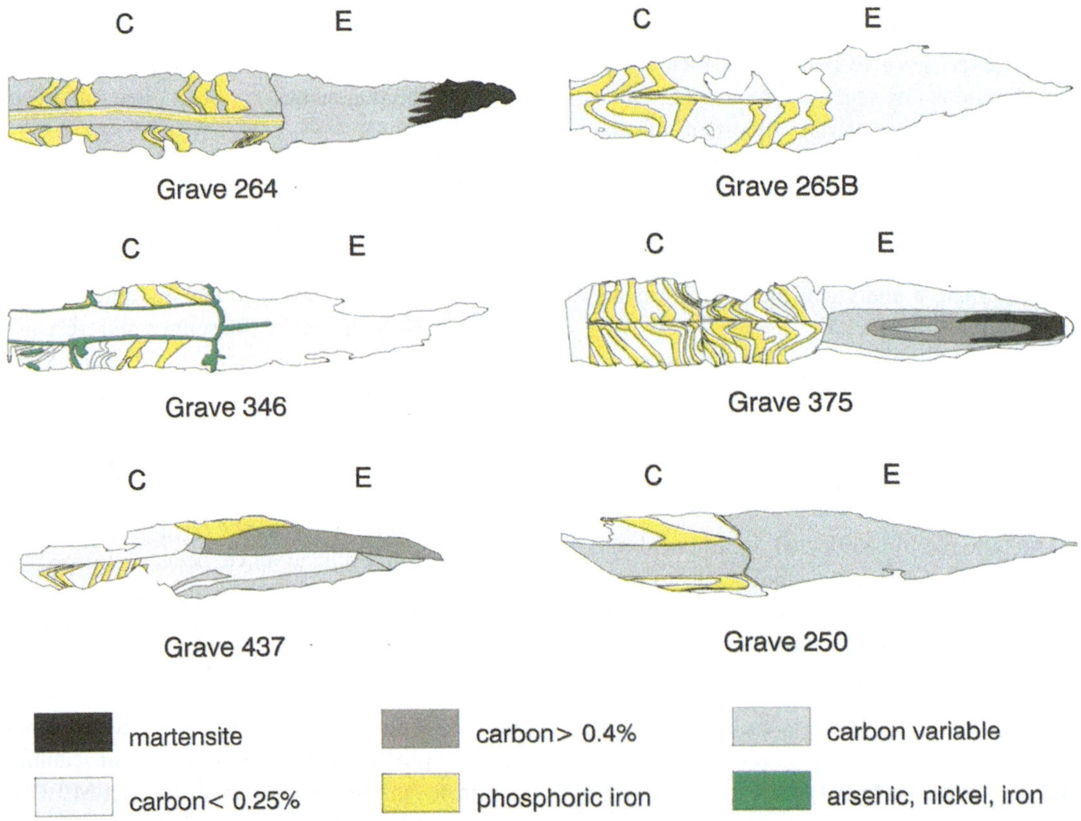

Fig. 8.1 Anglo-Saxon cemetery, Dover Buckland, Kent (UK, fifth century AD). Half-blade cross sections of the swords and weaving batten (not to scale) showing the distribution of carbon, phosphorus and the martensitic phase. Arsenic and nickel contents are mapped in the sword section from grave 346. From Lang (2015)

casting bronze in ceramic section moulds (piece-mould casting, overview in Meyers (1988). The vessels are made of ternary alloys of copper, tin and lead (between 6 and 16 wt.% tin, up to c. 20 wt.% lead).

Bonadies (1994) conducted radiographic studies on ritual vessels from the Cincinnati Museum using high-energy industrial computed tomography to shed light on basic casting techniques. Part of this were questions such as which parts of the vessels were made of solid metal, how was the variation of wall thickness, was there air entrapped in parts of the vessel, what was the extent of porosity, was there any trace of repairing by soldering or repairs by reassembling of legs and others. One important result of the investigations of Chinese ritual vessels generally is the observation that they were cast, but there is no evidence of shaping by hammering.

Since the energy of conventional X-rays is often not sufficient for tomographic analysis of artefacts made of very heavy metal, high-energy X-ray computed tomography has to be used to explore non-invasive textures of the inner parts of archaeological objects, especially when they are covered by corrosion incrustation.

Mille et al. (2004, 2005) were able to perform high-energy X-ray computed tomography (XCT) to investigate the heavy "Leopards weight" of Shahi-Tump (Baluchistan, Pakistan, late fourth/early third millennium BC) and thus to determine its internal structure. The egg-shaped "Leopards weight" measured about 13.5 × 6 cm and had a weight of >15 kg. It was examined at the Commisariat à l' Ènergie (CEA) with a high-power X-ray beam (of about 3 MeV, max 8 MeV). The making of this heavy and complex composed artefact is an example of the oldest lost wax casting known so far in this region. Its chemical composition was analysed using plasma spectrometry (ICP-AES). It had a core of pure lead and a thin jacket of copper.

Artioli et al. (2015) investigated by XCT and subsequent scanning electron microscopy (SEM-EDS) apparently worthless "slag"-samples from the Bronze Age excavations of a settlement at Santa Caterina Tredossi, Cremona (Italy) and could identify these samples a copper ingots.

8.2 Microstructure

8.2.1 Metallography

Metallography is the investigation as well as qualitative and quantitative description of the microstructure of metallic materials by reflected light optical microscopy (RL-OM). This means that preparations can be examined by an optical microscope by magnifications above 25 × up to c. 500 ×. In addition, scanning electron microscopy can be used to enlarge the magnification considerably, up to a few thousand times and to analyse chemical compositions at the same time by an energy dispersive system (SEM/EDS).

Hence, metallographic studies can provide important information about the chemical composition of the metal. Deforming and thermal treatments that have been used to turn a blank into an artefact can be analysed, no matter what metal or alloy it is.

Due to their very low melting points, no lead and tin, and of course not mercury are suitable metals for metallographic investigations.

Metallography is a technique based on the extraction of material from artefacts; it is, therefore, an invasive method. Sampling can be minimised by targeted sawing or drilling, but it usually still is of the order of several cubic millimetres. The method to be used depends on the size, hardness and material of the artefact itself. Due to conservational aspects in a museum environment, sampling for metallographic investigations can often cause problems, as artefacts are definitely damaged. The extraction of samples and the subsequent making of mounted samples for microscopy, however, offer outstanding opportunities for subsequent high-resolution chemical analysis of minor and trace elements and even the analysis of isotopes. For metallographic investigations, too much heating or deformation of the samples must be avoided in order not to blur any traces of thermal treatments or steps of shaping performed in ancient times.

Methodologically, the use of various etching methods, as described in the classic textbooks by

Scott (1991) or Oettel and Schumann (2011) is advisable in order to detect grain sizes and structures, twinning, formation of dendrites and strain lines. Very good overviews of metallographic investigations on many (pre-)historical metal objects with corresponding etching solutions can also be found in Scott (2012a, b). Metallography in archaeology and art are compiled in a book by Scott and Schwab (2019).

In addition to structural analysis in the traditional metallographic way, this can also be done by means of electron backscatter diffraction (EBSD). Only minute fragments of an object are needed for such an investigation, and, hence, it may be considered as a microinvasive method. Electron backscatter diffraction is a crystallographic technique that can be applied to analyse the texture of crystals (Schwartz et al. 2013). Typically, this technique is used to explore microstructures, revealing texture, defects, grain morphology and deformation. The system is usually combined with scanning or transmission electron microscopes (SEM, TEM) for chemical phase or material discrimination.

8.2.1.1 Casting Metals

The "as-cast" structure means that little or no subsequent treatment other than perhaps some abrasive surface cleaning, to remove casting burrs, etc., was undertaken after the casting of an artefact (Fleming et al. 2011).

The structure of the cold metal depends not only on the composition of the alloy or the purity of the metal but also on the temperature the melt achieved, the cooling rate and the nature of the mould, in which a melt is poured. These are factors that can lead to a large number of different crystallisations. Pure metals are less prone to dendritic growth. Instead, so-called cellular structures are formed.

Dendritic growth is particularly pronounced in the crystallisation of solid solutions (Oettel and Schumann (2011)). That is why most of the ancient cast metal artefacts have a dendritic structure. In metallo- and crystallography, dendrites (from the Greek déndron "tree") refer to fern- or treelike, shrubby, sharp or rounded skeletal crystal structures. They mostly arise from two-phase alloys (e.g. CuAs, CuSn, CuZn) or sometimes from three-phase alloys and can have different shapes. The cooling rate affects decisively the shape of dendrites. The faster the cooling, the smaller the dendrites.

Dendrites usually show a pronounced coring. In general terms, the centre of the dendrites is rich in that element which melts at higher temperatures (in our case it is copper), whereas the spaces between the dendrites contain less of this element. This mechanism is referred to as segregation or coring. Coring can be made visible by appropriate etching of a sample.

The remaining components of the alloy will deposit in the interstices in variable proportions. Depending on the cooling rates, these residual melts either crystallise in eutectic two-component intergrowths or solidify in metastable homogeneous crystallites (Fig. 8.2a,b). For ancient times it is almost certain that the cooling of liquid metals during casting has taken place very fast. Therefore, the solidification of metals or alloys did not occur in equilibrium conditions which means that the compositions of the single phases visible in mounted samples do not necessarily match the precise value that can be determined from a phase diagram. This fact has already been mentioned in the context of the solidification of siliceous slags (cf. Sect. 4.3).

When examining artefacts using the scanning electron microscope with energy dispersive equipment (SEM/EDS) or electron microprobe (EMS), care should be taken to measure a chemical composition. Bulk analyses and single-point analyses must be carefully distinguished, because significant differences may appear. During their analyses of Salezer axes, Kienlin et al. (2003) showed how the results of 9.4 wt.% antimony and 3.6 wt.% arsenic from bulk analyses could increase to about 40 wt.% and about 6 wt.% in single-point analyses of intermetallic phases.

8.2.1.2 Foreign Inclusions

In addition to the structural analysis of the metal components, an identification and spatial distributional analysis of oxidic, sulphidic and siliceous inclusions should be carried out. Such phases are not only present in ingots and iron

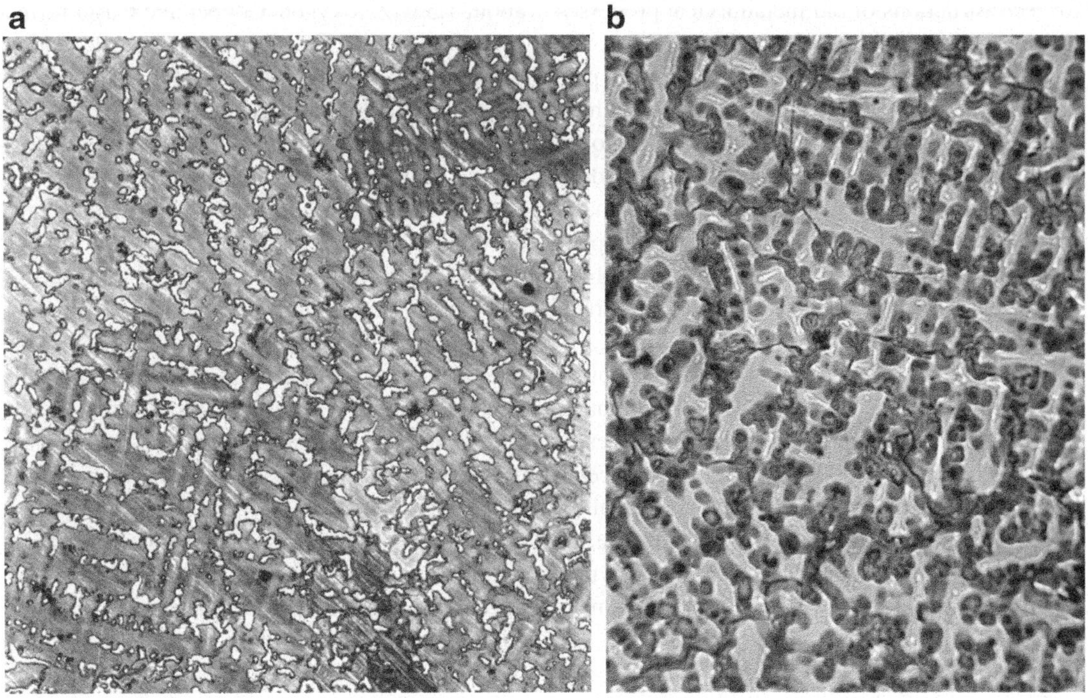

Fig. 8.2 (**a**) War Kabud, Western Iran, Iron Age III cemetery. Micrograph of a macehead made of a tin bronze (WK.B181-4; copper with 8.5 wt.% Sn; 0.62 wt.% As) from a v-shaped section cut from the rim of the longer ferrule that is decorated with three circlets. This microstructure shows fully formed cored sharp long-shaped dendrites and a high density of an interdendritic phase. Under equilibrium conditions, this phase would have crystallised as a two-phase α-δ eutectoid. This is not the case. The microstructure indicates that this artefact cooled fast during casting. The size of dendrites varies, though on the whole, they are fairly coarse. Magnification 200×. Etched by $K_2Cr_2O_7$ + alcoholic ferric chloride. From Fleming et al. (2006). (**b**) War Kabud, Western Iran, Iron Age III cemetery. Micrograph of a large bronze button (WK.A150-5; copper with 2.6 wt.% Sn; 0.41 wt.% As) from a v-shaped section cut from the edge. The microstructure has the array of cored dendrites that are characteristic of bronze casting. In addition, rounded interdendritic inclusions are present, but no visible α-δ eutectoid. The artefact must have gone through rapid solidification, perhaps because it is relatively thin, but not through a subsequent homogenising heat treatment that would have dissolved the non-equilibrium α-δ eutectoid. Magnification 100 x. Etched by NH_4OH · H_2O_2 + alcoholic ferric chloride. From Fleming et al. (2006)

blooms, but also in finished ancient artefacts of base metals or in iron/steel. In contrast to modern materials, they occur almost regularly in ancient ones and find their way into the metals as harmful inclusions during the smelting, casting and working processes. The identification of these phases must be done before etching the metal.

Oettel and Schumann (2011) differentiate between exogenic and endogenic inclusions. Exogenic inclusions are siliceous materials or particles coming from sand moulds or are remains from slag smelting or forging processes. Endogenic inclusions originate from metallurgical high-temperature reactions in a (liquid) metal or alloy melt.

Exogenic inclusions are a result of an incomplete separation of metal and slag during or even after smelting or even refining processes, i.e. in metallurgical operations in which metal was processed in the solid or liquid state. Inclusions of sulphides or speiss, on the other hand, give valuable hints on the nature of ores used, that is, they indicate whether and to what extent pure copper sulphides or polymetallic ores (chalcopyrites, galena, fahlores or other as-rich ores) were smelted. They are information carriers for reconstructions of old metallurgical processes and for provenance studies.

Siliceous exogenic inclusions were found particularly in the Late Bronze Age copper oxhide ingots from the Uluburun shipwreck (Anatolia) (Hauptmann et al. 2002a; Hauptmann and Maddin 2005) (Fig. 8.3). They indicate that not much attention was paid to a high quality of copper while preparing the metal for casting oxhide ingots, which entered the supra-regional trade from Cyprus.

Exogenous slag inclusions are also often found in iron or steel objects (Rehren and Hauptmann 1995; Yalçın and Hauptmann 1995; Schwab et al. 2006; Degryse et al. 2007). This is more acceptable since the extraction of this metal or this alloy has to predominantly occur in the solid state. In Fig. 8.4 such Fe-rich slag inclusions are shown in a hundred-weight block of iron from the Roman fortification of Saalburg, which was used for architectural purposes.

Endogenic inclusions are formed during cooling from a homogeneous high-temperature melt by exsolution. These are, e.g. inclusions of metallic lead in copper. These metals are only soluble in each other above 1000 °C and segregate in two components when cooled. Lead ores are often associated paragenetically with copper ores.

Inclusions of cuprite (Cu_2O) are often found in tools, weapons and jewellery made of copper, which show that a copper melt came into contact with oxygen in varying proportions. This can easily have happened during the casting of copper (Kienlin et al. 2003). In detail, Kienlin et al. (2008), who investigated the metallography of Early Bronze Age flat axes from the Northern Alpine region, however, have doubts as to whether such inclusions are indicative of a casting of the metal into open moulds as opposed to casting into a closed (bivalve) form. Rather, the presence of cuprite in copper may reflect a limited control over gas absorption during different steps, from (s)melting the metal to the casting itself, for instance by covering the crucible with a sufficient layer of charcoal or by cooling rates in different mould materials. In the liquid state at temperatures of c. 1200 °C, copper can absorb about 1.5 wt.% oxygen. Copper and Cu_2O form a eutectic system. The eutectic is at 1065 °C and an oxygen content of 0.39 wt.%. From the microstructure of the Cu_2O inclusions, it is easy to estimate the oxygen content of the copper. Figure 8.5 shows the microstructure of a hypoeutectic alloy altogether with only c. 0.09 wt.% oxygen: here, primary copper crystals were formed from the melt, followed by eutectic crystallisation of the Cu + Cu_2O—remaining melt. At 0.67 wt.% O_2, the microstructure of a hypereutectic copper–oxygen alloy contains more oxygen. From the melt, Cu_2O-dendrites crystallise primarily. Secondary, the eutectic solidification of the residual melt took place.

Cast copper often displays characteristic holes or porosity caused by micropores (in the few micrometre scale) and macropores (up to centimetre scale), the last ones, especially in bars and ingots. Porosity is caused by the effervescence of gases such as oxygen, carbon monoxide and dioxide, by water from burning charcoal as well as sulphur dioxide from oxidation of sulphide inclusions in the liquid state (Scott 1991; Laschimke and Burger 2011). These gases are segregated during cooling and, in the case of copper, also lead to the so-called scattering (in German *Spratzen*) or the formation of blister copper, which is characterised by a high proportion by volume of gas bubbles (Fig. 8.5). Examples of blister-copper are all the Late Bronze Age copper oxhide ingots in the Mediterranean (Hauptmann et al. 2015) and much smaller copper ingots found in the German low mountain ranges (Jansen and Löffler 2016).

8.2.1.3 Working of Metal

Often mechanical deformation of an artefact is necessary directly after casting the melt into an open or a closed (bivalve) mould. For example, an axe is hammered at room temperatures to smooth out casting flaws and casting defects, and to cut remains of casting protrusions. The artefact is then finally shaped by hammering. Because such "cold-working" always creates a certain amount of brittleness in the metal, it was then heated (annealed) again to release

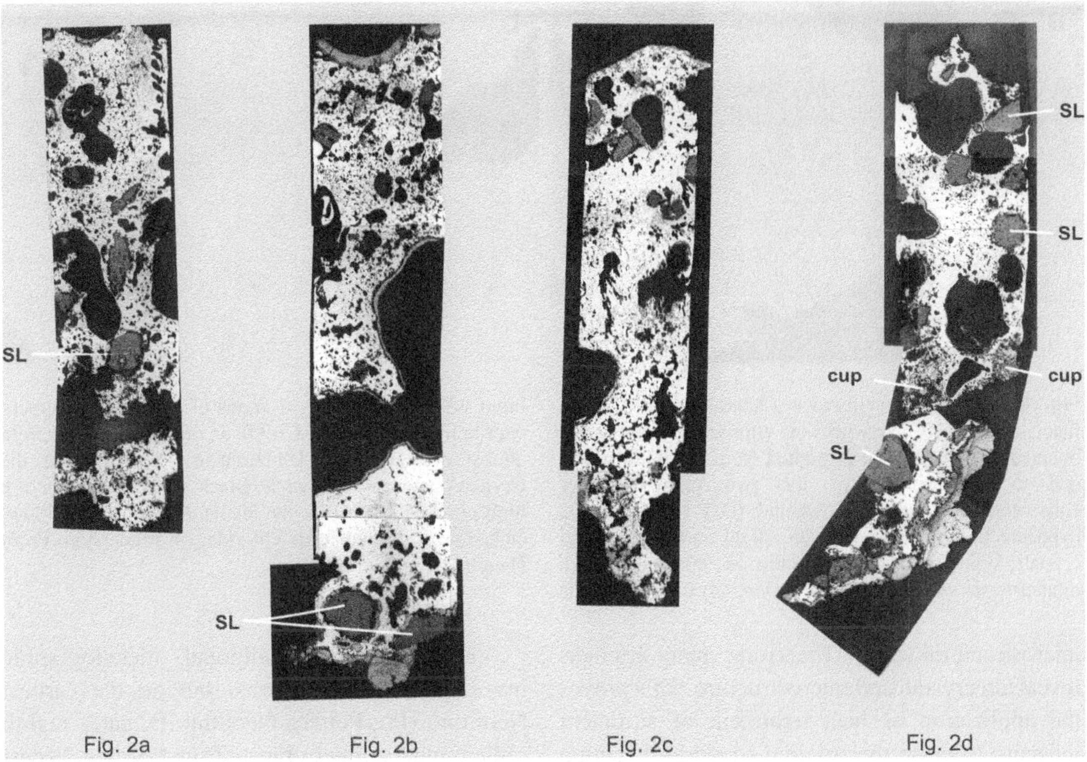

Fig. 8.3 Uluburun shipwreck (Anatolia), copper oxhide ingot KW 1548. Four cores (**a–d**) taken near the edges of the ingot. They have an extremely high porosity with large voids, often encrustated by cuprite (cup1) in common. Formation of primary cuprite at core KW 1548/d, middle (cup2). The cores show compositional heterogeneities: KW 1548/2d is extremely rich in angular slag inclusions (SL), while KW 1548/2c is made up of relatively dense metal, and in KW 1548/2b dense and porous metal are next to each other. Length of the samples: 2.4–3.4 cm. From Hauptmann et al. (2002)

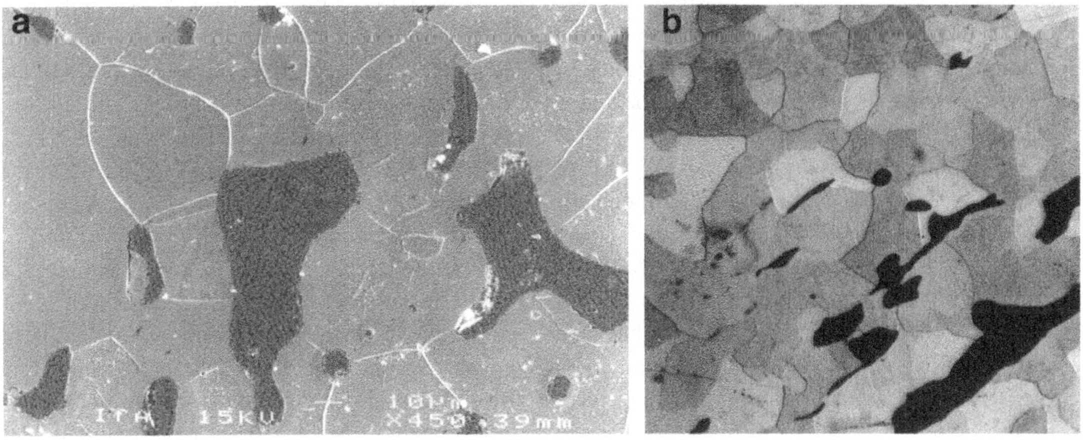

Fig. 8.4 (**a**) Roman castle of Saalburg, part of the Roman limes (Germany). Iron block S 1113. Fayalitic slag inclusions (see dark crystals). The iron is of ferritic composition. Cementite (Fe_3C) is limited to grain boundaries and knuckles. The slag inclusion does not show any deformation. SEM picture, secondary electron mode. Length of scale 0.01 mm. From Rehren and Hauptmann (1994). (**b**) Rio Tinto (Spain), probably Roman age hammer pick, sample RT10. Ferritic iron with elongated slag inclusions (dark) caused by welding. Magnification 100x. From Maddin et al. (1996)

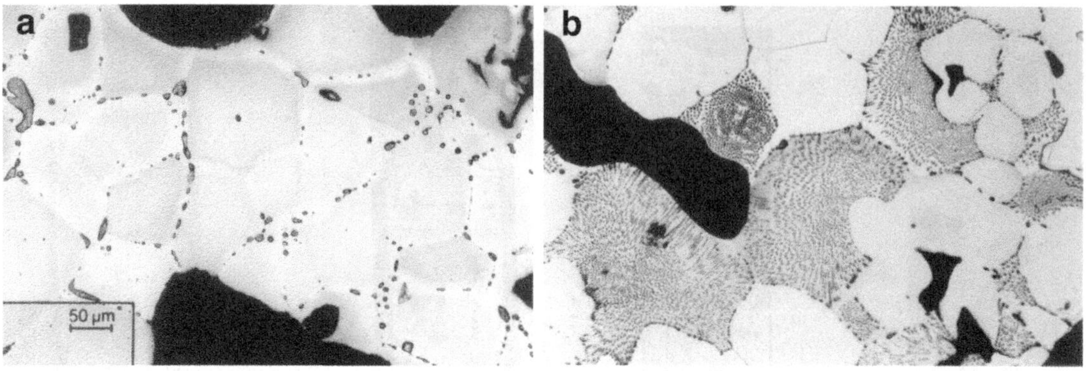

Fig. 8.5 Uluburun shipwreck (Anatolia). (**a**) Copper ingot KW 2730. Network of cuprite between grain boundaries of copper, precipitated from the liquid metal due to immiscibility of the two phases. Minor concentrations of oxygen (around 0.05 wt.% oxygen, hypoeutectic composition). Note slight coring of copper crystals. Voids are black. Magnification 200x, etched with diluted ferric chloride, oil immersion. (**b**) Copper oxhide ingot KW 1396. Equiaxed grains of copper and copper–cuprite intergrowth ($Cu–Cu_2O$), an omnipresent texture in almost all ingots from Uluburun investigated. Here, the oxygen content is 0.39 wt.%. Black voids are details of a high porosity typical for the ingots. Magnification 200×, etched with diluted ferric chloride; oil immersion. From Hauptmann et al. (2002)

microstructural stress. Therefore, many artefacts reveal a recrystallised microstructure. This proves the application of heat treatment of sufficient intensity to erase the original (dendritic) casting structure and causes the formation of new so-called "equiaxed" grains by recrystallisation in the solid state. Annealing twins indicate deformations in the hot state or prior to the heat treatment (hot working vs. cold working followed by annealing, see Scott 1991).

The change and interdependence of microstructures from a dendritic casting structure to cold and hot working and annealing with the formation of equiaxed grains, the formation of strains and twinning grains are shown for alloys such as tin bronzes in Fig. 8.6.

The formation of twinnings as a result of annealing at c. 300 °C and above, after mechanical deformation, is the characteristic microstructure for metals like copper and also for gold (Nocete et al. 2014) or for base metal alloys. It has to be pointed out, however, that the microstructures shown in Fig. 8.6 are ideal depictions. The transition of microstructures caused by cold working to those caused by an annealing process and vice versa can show all transitions, which can make interpretation difficult.

Very insightful traditional metallographic investigations were carried out on the earliest Neolithic (Pre-Pottery Neolithic B, early eighth millennium) copper objects from Çayönü Tepesi and Aşıklı Höyük (Fig. 8.7) (Maddin et al. 1999; Yalçın and Pernicka 1999). In combination with chemical analyses, it could be proven that they were produced from native copper. The cm-sized copper beads of Aşıklı Höyük were hammered alternately in cold and warm condition and annealed in between, to avoid embrittlement of the copper by recrystallisation. This implies the knowledge of work hardening and the reverse effect of increased temperatures on mechanical properties to restore deformability. The earliest copper artefacts were not melted.

Despite the early evidence of annealing in the Near East, this knowledge cannot be taken for granted everywhere. The early metalworking of the Iberian Peninsula seems to be an example of a tradition operating largely on the basis of cold-working as-cast objects without knowledge of, or interest in, the beneficial effects of annealing (Rovira and Gómez Ramos 2003). In his studies on Chalcolithic to Early Bronze Age flat axes form the north Alpine region; however, Kienlin et al. (2004, 2006) observed that final cold working was often stronger than the deformation

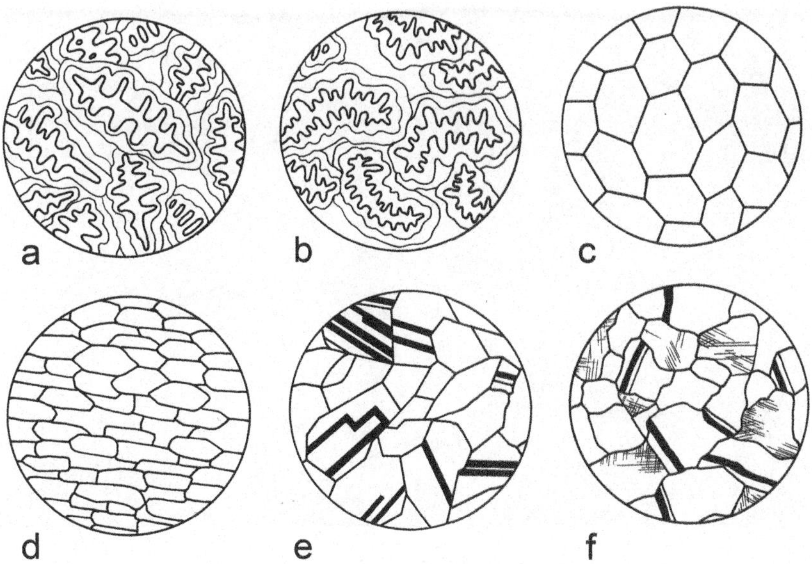

Fig. 8.6 Basic examples of microstructural features in solid solution alloys. (**a**) Cored grains with dendritic structures. This structure is exemplified by Fig. 8.2b. (**b**) Cold-worked cored grains. (**c**) Fully annealed homogeneous hexagonal equi-axed grains. (**d**) Cold-worked annealed metal with flattened grains. This structure is exemplified in Figs. 8.4b and 8.7. (**e**) Annealed after cold-working showing twinned grains. This structure is also exemplified in Fig. 8.7. (**f**) Cold-worked after annealing showing distorted twin lines and strain lines in the grains. After Scott (1991)

achieved in the previous step. This certainly implies that the metal workers were fully aware of a final shaping and actively interested in increasing the hardness of their axes.

The general increasing of hardness of a casting that results from alloying, e.g. copper with tin or antimony, and from cold working of cast metal, is quantified by measuring the Vickers Hardness Number (HV).

8.2.1.4 Iron and Steel

The metallography of iron and steel objects, which in ancient times consist exclusively of FeC (or FeP) alloys, has been the subject of considerable research, and is of decisive importance in modern times as well (cf. Figs. 7.20, Sect. 7.5). Here, the heat treatment of the material can be understood in great detail. Apart from cast iron, which had reached its liquid state, but was very rare in ancient times, the shaping of prehistoric iron or steel artefacts happened almost entirely in the solid state.

The extraordinarily complex range of different microstructures of Fe–C alloys has been extensively described in numerous papers (overview see Horstmann 1985). The microstructures of ancient Fe–C alloys, aside from cooling and working conditions, are predominantly affected by varying carbon concentrations, which reach up to 4 wt.% C. This means, that compositions of almost pure iron with ferrite (α-Fe with 0.02 wt.% C) to slightly carburised steel with austenite (γ-solid solution with max. 2.06 wt.% C) are existing, as well as phases in changing amounts intergrown with cementite (Fe_3C with 6.67 wt.% C). Frequently, pearlite can be observed (microstructure of 88 wt.% ferrite +12 wt.% cementite).

Metallographic analyses allowed not only to prove the existence of steel, e.g. in Iron Age and Roman blooms, and with it the primary production of steel in the bloomery process (Straube 1996; Gassmann et al. 2005), but it also allowed for meticulous investigations of the exceedingly difficult heat treatments of steel (Buchwald 2005).

In addition to the figures mentioned above, a scanning electron image of different FeC

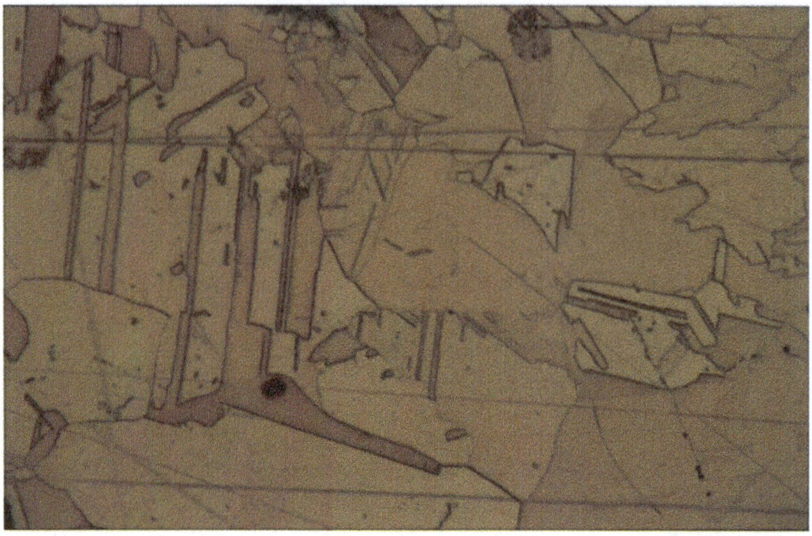

Fig. 8.7 Aşıklı Höyük, Anatolia, Pre-Pottery Neolithic B (first half eighth millennium). Bead made of native copper. The microstructure shows a number of recrystallisations. Every generation of recrystallisations has pronounced twinnings. This means that the bead was repeatedly heated and shaped by hammering. Width of the picture: 0.2 mm. Etched by $NH_4OH{\cdot}H_2O_2$. From Yalçın and Pernicka (1999)

compositions is inserted at this point (Fig. 8.8), showing the diffusion of Fe and C atoms in the solid state. This is the very finely formed lamellar structure of α-iron and cementite (Fe_3C), which is called pearlite. Cementite forms during the cooling of γ-Fe, which is stable at high temperatures and may contain larger amounts of carbon. Upon cooling, cementite is formed. It contains 6.67 wt.% C and is extremely hard and brittle and occurs in all grades of steel. Quantity, size and shape of cementite formation gives information about the degree of hardness of the steel.

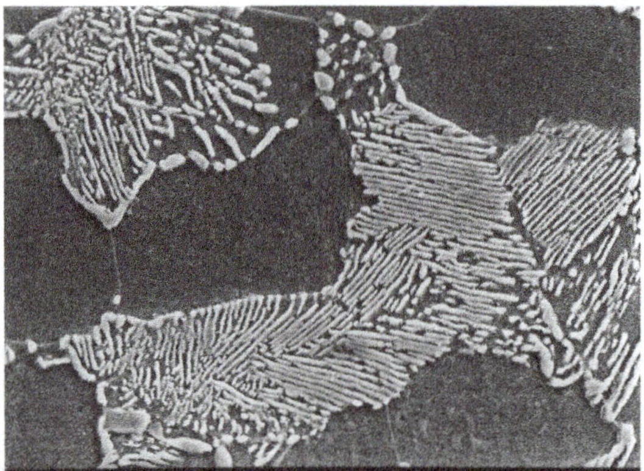

Fig. 8.8 Rio Tinto, slag heap, probably a pick of Roman age, sample RT4. The micrograph shows a starting formation of lamellae of cementite (Fe_3C) out of ferritic α-iron. Cementite forms light lamellae, ferrite shows dark lamellae and grains. From Maddin et al. (1996)

8.2.2 Neutron Diffractography

It was pointed out that for physical reasons conventional X-rays cannot penetrate all materials equally well. They essentially react with the electron shell of the atoms by absorption or scattering, but neutrons react with the atomic nuclei. One of the main advantages of neutron radiation is their high material penetration, in the order of centimetres in common metals such as copper and iron. Neutrons penetrate through coatings and corrosion layers.

This property also makes them ideal for non-destructive analysis not only of surface compositions, but also the interior of large museum objects or archaeological finds for which sampling is not acceptable (Kockelmann et al. 2006). In order to check the crystallographic phase and herewith, the bulk metal composition and its homogeneity in a non-destructive mode, we suggest using neutron diffraction (ND) analysis. This method of measurement is based upon the elastic scattering of thermal (~25 meV) neutrons from a crystalline object.

Recently, ND measurements were conducted on several Middle Bronze Age II duckbill axes in Israel (Caspi et al. 2009; Shalev et al. 2014) at the Nuclear Research Center—Soreq, Israel. The results demonstrated the ability of the method to distinguish between phases and their concentration in different parts of an as-cast partially corroded tin–bronze axe. In these researches, low-resolution studies of a few different types of axes (i.e. thick solid cast) and ribbed daggers (at least partially homogenised after casting) have been conducted to map the differences in composition and phases in different parts of the objects. The authors interpret the results as proof that the axes are the products of a highly controlled alloying of tin–bronzes and arsenical coppers in more or less optimal amounts of tin or arsenic. For a criticism on the possibility of such optimised alloying of arsenic and tin with copper, see Sects. 7.2.1 and 7.2.2. Furthermore, it was found that the tin contents are distributed homogeneously over the lengths of the axes. The axes had undergone special heat treatment for a homogenisation, which is necessary for subsequent hardening of the edge by hammering. Secondary intermetallic phases (e.g. the δ-phase in Cu-Sn; and Cu_8As or Cu_3As in Cu-As) were neither found in arsenical copper nor in tin bronzes.

Using ND, Haquet et al. (unpubl. data) were able to shed light on the production method of a Chalcolithic gold pearl of Mehrgarh (Baluchistan). The pearl (83.5 wt.% Au, 15.5 wt.% Ag, 0.6 wt.% Cu), due to being an artefact made of a noble metal, of course, had to be analysed in a non-destructive way. Neutron diffraction was carried out at the four-circle diffractometer of the Orphee reactor at Saclay, France. The results showed a very smooth wire texture which, despite the low level of anisotropy, was shown by previous experiments to be characteristic for gold hammering. Thus, texture analysis by ND analysis proved in this case to be the only effective non-invasive technique able to reveal the manufacturing process, namely casting and hammering.

9 Ethnographic Evidence and Artisanal Metal Production

9.1 General Remarks

Ethnoarchaeology is a branch of archaeology, in which social science studies are carried out to solve archaeological problems. Ethnoarchaeology includes the study of today social structures and processes, material objects and institutions and their possible transference to archaeological findings. It is, therefore, an indirect approach to understanding earlier historical societies (Renfrew and Bahn 2007). In archaeology, not all aspects of ancient life are comprehensible. Reconstructions rely to a considerable extent on material legacies, and this approach is therefore generally limited in its content.

This also applies to archaeometallurgy. As far as archaeometallurgy is concerned, the strength of the archaeological legacies lies in a comparatively large series of finds, for example ores, slags, remnants of melting vessels, fire pits and many other things, as explained more detailed in the previous chapters. However, their decoding and interpretation options just by analytical methods are not unlimited, so that knowledge and juxtapositions with ethnographic findings are useful to understand these findings, and thus to grasp the working and living conditions of ancient miners and metallurgical craftsmen.

One of the major sources of ethnographic observations in archaeometallurgy is artisanal and small-scale mining and metallurgy, primarily from non-European societies and societies prior to the Industrial Revolution. For this purpose, the book of Pieth (2019) is highly valuable. It is focused on the frameworks of modern artisanal gold exploitation in countries of South America and in western Africa. The compendium of Priester et al. (1993) provides material reflection. In their book "Tools for Mining. Techniques and Processes for Small Scale Mining" the authors note that small scale mining (and metallurgical) operations—from Africa to the Far East and to the Andes—are largely homogeneous. However, tools used for exploitation and processing of ores, while similar in function, may be shaped and decorated differently depending on the cultural background. What concerns the extension of exploitation the authors define "small" mines as those, where less than ten or so people are employed. Such small-scale operations depend, of course, primarily on the morphology and geometry of corresponding mineralisations and their ore composition.

Often, however, the extraction of metal is operated with significantly more massive personnel expenses. This is true, as will be shown, in the case of gold. In many countries industrial, governmental controlled and non-industrial, illegal mining and metallurgical processes still exist side by side today. Unfortunately, there are too many examples that these illegal, small-scale activities were performed by thousands of people under horrifying and scandalous social conditions (Salgado 2019; Pieth 2019).

A. Hauptmann, *Archaeometallurgy – Materials Science Aspects*, Natural Science in Archaeology,
https://doi.org/10.1007/978-3-030-50367-3_9

Werthmann (2000, 2009) describes that parallel to the establishment of industrial gold mines by Ghanaian and European entrepreneurs since the end of the nineteenth century, there continues to be the centuries-old artisanal and traditionally seasonal gold mining.

In many countries of the African continent, direct ethnographic observations on artisanal iron production and processing have also been made in the recent past, which were published in many documentaries. Here, unique opportunities to study the social role of miners and smelters within (village) communities arose. For example, whether these specialists were carried by the village as such, or were they specialists who were only engaged if there was a demand for iron objects? In addition, further research focused on metallurgical processing techniques, which provided insight into the formation of iron and steel and the processing of "molten" blooms to ingots and artefacts. The interest of archaeometallurgy is focused on the latter.

In the case of gold, however, the focus of ethnographic research has been on the effects that the most coveted metal has always had on humans. Unlike with iron, the focus here is on the effect of gold rushes, where thousands of people were and are involved all over the world. Werthmann (2009) conducted fundamental studies in Africa and investigated the principles by which populations can be mobilised at a very rapid pace. The metallurgical extraction of gold is still rather simple since it already occurs in nature in native form. On the other hand, considerable effort is required to mine and to enrich the metal by beneficiation, as it occurs only in very low concentrations in its deposits.

The ethnography of the archaeometallurgy also includes an evaluation of textual and pictorial representations that are available. These go back to the end of the Middle Ages. Theophilus Presbyter writes about the art of craftwork in the twelfth century AD (Theobald 1984), Lazarus Ercker covers mines, assaying and coins (Winkelmann 1968), and in 1540 AD Vannocio Biringuccio, in his work Pirotechnia, reports in ten books on the extraction and processing of ores and metals. Finally, in 1556 AD Georg Agricola publishes his famous work De re metallica XII, a uniquely illustrated compendium on this topic. But there are also pictorial representations from the late eighteenth century AD. Examples are the very detailed pictorial representations of the extraction of gold, copper, silver and lead in Japan and China, from the hanging scrolls about gold extraction on the island Sado around 1720 (Winkelmann 1964), in the woodcuts of the work *Kodō-zuroku* by Masuda Tsuna from the year 1801 or in the Chinese work Tian gong kai wu (= Exploitation of the work of nature: Chinese agriculture and technology in the XVII century) by Song (1980).

9.2 Gold Exploitation in Africa

Ethnoarchaeological observations from recent times on mining and the production of metal with simple, manual methods can be observed particularly impressive for gold.

Pieth (2019) reports on "Land Grabbing" in Ghana. To the large gold producers of Africa belongs this small country located on the western coast of the continent. The author is citing Reisenberger and Seifert (2011) who published that there would be c. 500,000 artisanal diggings, out of these are only 20,000 would be formalised.

According to Borg's (2014) experiences, in many countries, especially in Africa, many different political difficulties led to largely illegal, artisanal gold mining. When gold was discovered in remote areas of African countries, but also in Papua New Guinea or countries of South America (Peru, Columbia, Brazil) many thousands of people from all directions arrived in virtually no time to have a part in the anticipated immeasurable wealth (Fig. 9.1). Quickly, claims were marked and simple settlements emerged out of nowhere. In the gold-rich country of Burkina Faso, Werthmann (2000, 2009) closely observed the social backgrounds and structures in the formation of gold rushes after the first discovery of a nugget to the subsequent gold mining and production. In the south of Burkina Faso, which was known as a gold-bearing zone long before the colonial period, in May of 1998, a gold-bearing

Fig. 9.1 Bouloun-Djounga, Niger. In April 1992 a group of stampeders found gold. The picture shows the situation a few days after: hundreds of people arrived to join the gold rush. They marked claims, started to deepen shafts and to wash gold from the soil. Foto: Georg Jöbkes, Bonn, 1993. Archive Deutsches Bergbau-Museum Bochum

vein was found by a gold digger. Within a few weeks, some 5000 men (not only miners and gold diggers, but also traders, butchers, barkeepers, and prostitutes) from different adjacent countries invaded a small village and opened a (non-industrial) gold mine. She describes how, in the immediate vicinity of a village after the discovery of gold, gold miners invaded "like a swarm of locusts". Initially, the situation developed in a similar way to the classic conceptions of the "Wild West": A handful of outlaws established a reign of terror, robbing young gold diggers and raping girls and women. Brawls and stabbings were the order of the day. After about 2 months, the situation escalated to the point where the military intervened and shot four of the "desperados". The police presence was strengthened, the entrances to the settlement were controlled by guards etc. The often improper excavation of such a mine, which could reach depths of 100 m (Fig. 9.2), generally harboured great dangers and catastrophic landslides with numerous deaths occurred often.

Throughout history, gold rushes occurred time and time again, whether in 1848 in California or in 1874 on the Klondike River or in the Yukon Territory in Canada. Many gold rushes are also recorded from the late nineteenth century and earlier in countries like Australia, New Zealand, Africa and South America. One of the most impressive and shocking photographic documentations of such a situation is that of world-famous photographer Sebastiao Salgado on the work of the thousands of Garimpeiros in the open-pit mine of the Serra Pelada mine, which began in 1978 (Salgado 1993, 2019). We do not know anything about earlier gold rushes, but it would be surprising if they did not exist in (pre-) historical times.

Borg (2014), who worked in Tanzania as a geoscientist, documented the technological sequence of artisanal and small-scale mining and smelting of gold with numerous photos. These

Fig. 9.2 Tialkam, Niger. View into the large gold mine of Tialkam. Note the simply constructed and dangerously inclined haulage of the opencast mine. In the background, simple timbering is visible which was constructed as a protection against a collapse of the mine. Hundreds of men are towing gold-bearing lateritic sands and clays which are treated and washed at the surface to achieve a gold concentrate. Foto: Georg Jöbkes, Bonn, 1993. Archive Deutsches Bergbau-Museum Bochum

representations also illustrate the simple and dangerous methods a gold mine was deepened. It also sheds some light on the extraordinary effort required to enrich the gold, which usually only occurs in minimal concentrations in the hostrock ("invisible gold", cf. Sect. 3.5.2). For this, the gold-bearing rock must be ground to powder. This can be done dry in (wooden) mortars. The gold is then enriched from the powdered rock in sluices with water. Grinding and milling of gold-bearing rocks can also be done using other tools. Until recently, huge stone mortars were used in Peru to grind gold-bearing rocks, so-called "see-saw-crushers" (*quimbaletes*) (Fig. 9.3; Stöllner 2009b).

Like it is the case in many parts of the world, ground and powdered gold is washed out and then mixed with mercury in a washing pan to form an amalgam. The separation of the toxic mercury from the gold then takes place without any protection just over an open fire! In most cases, all activities in the gold mining chain, from mining to amalgamation, are carried out by men. Only the grinding of gold ore by manually operated saddle querns is performed by women. Saddle querns of similar constructions are well known from many archaeological sites related to metal production, but they are basically indistinguishable from querns used as flour mills.

9.3 Iron in Africa

In 1989, Harvard University's Department of Anthropology began a project to collect all the publications on African iron making, keeping abreast of the current research into the ethnography of this unique craft which was still practiced by local people in recent decades. At about the same time, the mining archaeologist Gerd Weisgerber carried out the same project at the Deutsches Bergbau-Museum Bochum. Today not only are appropriate publications available in this library, but also film documents, exhibitions of metallurgical materials collected in countries of Africa and other research material.

Due to the industrialisation, old artisanal crafting techniques of iron extraction, which, as shown in Sects. 6.6.3 and 7.5, could be complex and tricky, were lost in many countries, not just in Europe. Therefore, we know very little details from the early periods. In Africa, however, techniques have survived until recently, possibly identical or at least similar to those used in ancient Europe. Africa has offered a large role in reconstructing bloomery iron furnaces (and copper smelting furnaces as well) as there are still many regions and locations, where iron ore was smelted to iron and steel and blooms, were further treated in smithing hearths and these were further processed to artefacts. But there is also the

Fig. 9.3 Saramarca, Palpa (Peru). Beneficiation of gold-bearing rock using a *quimbalete* ("rocker crusher" or "see-saw crusher" after Priester et al. 1993) needs great physical effort which can only be slightly disburdened by a well-formed grinding surface. This beneficiation can be done both using water or without water. The method was and still is in use in Latin America. Photo: T. Stöllner, Deutsches Bergbau-Museum Bochum

possibility of observing simple mining and metallurgical processes, which are still used today in some Third World countries. In contrast to the cultural development in Europe, in many regions of Africa south of the Sahel the Iron Age did not develop as a logical consequence of the Bronze Age. Outside of the Nile valley and the Red Sea, a Bronze Age is virtually unknown in Africa.

Van der Merwe (1980) had already pointed out that all the ethnographic observations provide a record of iron smelting processes for the African continent which are without equal anywhere else. This is in part due to fact that, up until the recent past, the extraction and the forging of iron, especially in countries of western and southern Africa, was intimately connected with myths and religious, even magical rituals as well as sociological factors. De Barros (2000) reports in detail on this topic and illustrates how the individual steps of iron metallurgy, starting with the mining of the ore, the processing thereof, the selection of the most suitable clay for the building of a furnace and the needed nozzles, the smelting process itself and the work in the smithy could only be carried out by highly specialised men, which were indeed carried by the village community. The contribution of women to the actual metallurgical activities of iron production was limited, and as a rule, they were excluded. The process of smelting often had sexual aspects. The smelting furnace was seen as a female body, the formation of iron from the ore, and the formation of the blooms was often compared to a foetus and finally to a birth. In Burkina Faso, a bloomery furnace was inaugurated with an animal sacrifice. Only then

Fig. 9.4 Tyantyamana, Burkina Faso. Inauguration of a bloomery furnace with an animal sacrifice. Only afterwards can the smelting of the ores begin. From Celis (1991)

could the smelting of the ore commence (Fig. 9.4).

Most ethnologists were more attracted to these religious-magical rituals and to the incredible variety of metallurgical facilities than to technical details. That may perhaps be the reason why the underlying technology is often misunderstood. African iron smelting cannot be described in terms of either the so-called direct or the indirect processes, because its primary products often seemed to be neither pure wrought iron nor cast iron. Instead, blooms of high-carbon steel were produced directly from the smelting furnace and were subsequently decarburised in the forge (Schmidt and Avery 1978; Van der Merwe 1980). This statement is in striking harmony with the experiments made by Straube (1986, 1996) on the Celtic/Roman *Ferrum Noricum* of the Magdalensberg in Austria. He has produced both theoretical and experimental mixtures of high-carburised steel and wrought iron. This production of high-carbon steel does not necessarily have to be postulated as an invention from Africa, but has been known earlier and may have been more widespread in Europe than previously thought.

One of the prominent scientists involved in traditional techniques is Georges Celis (Celis 1991). He spent decades studying the iron smelting and processing technology in regions of West Africa (Côte d'Ivoire, Mali, Burkina Faso), Central Africa (Zaïre) and East Africa (Rwuanda, Burundi) near Lake Victoria. Like many other authors, he points out that the extraction of iron (and steel) in sub-Saharan Africa is "very old" in archaeological and historical terms (ca. fifth to third centuries BC; cf. Van der Merwe 1980) and achieved an enormous cultural and economic importance.

For the Senufo (Côte d'Ivoire, Burkina Faso), the construction of a bloomery furnace, as observed by Eckert (1986), takes about 2 months. Such a furnace has a height of 1.50 m and is constructed from a clay and straw mixture. The life span of such a furnace is 4–6 years. In the Koni village visited by Eckert in the spring of 1974, there were 13 furnaces, of which 12 were still in operation. The ore consisted of iron oxides and hydroxides (limonite, hematite) typical for widespread laterites and ferricrusts in Africa. It is mostly mined in simple pits and has to be extensively washed out of the clayey matrix. It is shaped into roughly palm-sized balls and roasted under straw.

Eckert precisely describes how a smelting furnace is preheated and how six nozzle pipes of about 50 cm in length are inserted into the lower part of the furnace shaft in a star-shaped pattern. The furnace is then successively filled with charcoal by 2–3 people and, depending on the size of the oven, with 100–200 ore balls. These men are specialists who inherited their craft from their fathers. The Senufo did not use bellows for air supply. Instead, their furnaces were natural draft furnaces. The actual smelting process took 20–22 h. Afterwards, the furnace shaft was opened at the bottom and the bloom was removed, which weighed 15–35 kg. It usually consisted of a mixture of slag, charcoal and metal.

Celis (1991) uses the example of an iron smelting in the western part of the Sahel, which is not too far removed from the Senufo described by Eckert, and which therefore can be used as a comparison to describe how the natives tackled the complex procedure of cleaning a bloom. Since the bloom consists of a mixture of pieces of metal, slag, charcoal and possibly ore pieces, it is broken down with a heavy hammer stone and pieces of metal are sorted out. This process can apparently be carried out systematically at appropriate places and leave interesting arrangements of small cup-shaped depressions or stamp bowls in the solid anvil rock (Fig. 9.5). This phenomenon is reminiscent of the shape of irregularly arranged egg boxes. They are also known from numerous other archaeological sites. They are formed when the ores, slags or other intermediate products of metallurgy are processed systematically with a hammer stone or mallet.

In the case of Sohono, the chunks of the crushed bloom are reheated in roughly cup-sized crucibles to remove slag residues and to weld the individual metal fragments into larger pieces (Fig. 9.6).

Fig. 9.5 Sohono bei Korhogo, Côte d'Ivoire. The egg boxes-like cup depressions or stamp bowls were made by hammering and cleaning blooms on an exposure of a suitable hard rock. This phenomenon is known also in archaeological context from other (pre-)historic smelting sites such as Faynan 5 and Ras en-Naqb in the Faynan area, where copper prills were extracted from slags (Hauptmann 2007), and from Ari in the Laurion district (Lohmann in press), where lead ores or litharge have been processed. From Celis (1991)

9.4 Copper Smelting in Himalaya

There are two publications about small-scale artisanal copper metallurgy that are of utmost importance for the understanding and interpretation of (pre-)historic archaeological field evidence. They discuss observations made in the Himalaya area of Nepal–Sikkim. Percy (1861) already reported

Fig. 9.6 Sohono bei Korhogo, Côte d'Ivoire. The bloom from a bloomery furnace still contains numerous inclusions of slag, charcoal and ore pieces. These have to be removed. For this reason, the bloom is crushed and the small pieces are re-melted in small crucibles. From Celis (1991)

on his participation in a copper production in Mahanudi Valley, Sikkim. The metallurgists came, he writes, from Nepal, where a field of expertise in copper technology existed. Anfinset (2011) reports on mining and smelting of copper by indigenous people in Okharbot, Western Nepal, which he experienced first-hand on multiple field trips in the 1990s. Copper was mined and smelted in several villages in this area as recently as 50 to 100 years ago. Nowadays, the knowledge of this technology has almost been lost, but in exceptional cases, it is still practiced today. In 1960, 100–150 miners and metallurgists worked in one village and produced 25 kg of copper per day, which they gave to the Indian Bidala Company. Already in the Middle Ages and during the following centuries, the copper certainly was a coveted material for craftsmen in the Kathmandu valley, where the Newar had especially outstanding skills in the crafting of Buddha statues (Furger 2017).

The reports of Percy and Anfinset, both in terms of sociological organisations, their ritual and mythical backgrounds, and especially in regards to the technological aspects of copper mining, have probably been identical or at least comparable at both localities. In Nepal, Anfinset and a camera team from Norwegian Television arranged an impressive documentation of the local copper extraction, which can be found online at https://vimeo.com/121204255.

In recent years, these reports and pictures have been used for the comparison with field work on Bronze Age and Iron Age smelting sites in the Alpine region in order to better classify the function of metallurgical facilities and slag finds and to reconstruct possible processing techniques (Reitmaier-Naef 2020; Silvestri et al. 2019). Goldenberg et al. (2011) conducted a series of smelting experiments as part of the Collaborative Research Centre on History of Mining Activities in Tyrol and adjacent areas (SFB HIMAT) at the University of Innsbruck, which were based on the reports and experiences of the two authors Percy and Anfinset and the construction of the copper smelting furnace from Nepal/Sikkim. These experiments are discussed in Chap. 10. In Grisons (Switzerland), Reitmaier-Naef (2020) reconciled field work and analytical work on slags and ores from this area with observations from the Himalayas and presented a convincing solution for a copper production process based on the descriptions of Percy and Anfinset.

The copper production in the Himalayas was carried out as follows:

After fire setting, ore was mined with simple tools. It consisted of chalcopyrite with pyrite with quartz gangue and embedded in quartz-hornblende-slates. The mines reached depths of a hundred metres. The subsequent dry processing of the ore was first tackled with stone mallets, after which the crushed ore was carefully washed in sluices or in simple troughs. This work was done by women. The ore concentrate was smelted in a shallow, box-shaped smelting furnace using angled nozzles (Fig. 9.7a). As fuel and reducing agent, charcoal was used, which was produced not far from the roofed smelting furnace with very simple means. During the first smelting process, a large part of the gangue and hostrock, which was

Fig. 9.7 (**a**, **b**) Traditional smelting furnaces for the production of copper seemed to be of comparable construction in the Himalaya. These smelting furnaces were rectangular, rather flat constructions. Air supply was organised from above by two angled tuyères. (**a**) Mahanudi Valley Sikkim, nineteenth century AD (Percy 1861). (**b**) Okharbot, Nepal, 1995 (Anfinset 2011)

still intergrown with ore concentrate, was slagged. The liquid slag was superficially cooled with water and pulled out of the furnace in disks. Underneath a matte was formed, which accumulated at the bottom of the furnace to larger units. Technologically, this process was a matte smelting, as described in Sect. 6.2. The matte was crushed, mixed with cow dung, formed into small balls, which—after the smelting!—were roasted. Contrary to the popular assumption in archaeometallurgy, it should be noted that the sulphidic ore indeed was not roasted first and then smelted again in a second step. In this way, copper was enriched until it precipitated from the matte in metallic form (Fig. 9.8). For explanation cf. Fig. 5.52c in Sect. 5.7.

9.5 Pictures of Metal Making in Japan and China

During and after his travels to Japan at the turn of the twentieth century, Gowland published a series of illustrated works (Gowland 1899, 1901, 1906, 1914, 1917/1918) in which he reported on the traditional melting of copper, tin, lead and silver in Japan. It was his intention and conviction to make important contributions to the processing, organisation and, above all, to the understanding of metallurgical processes in this country, which are also very useful for archaeological research.

In contrast to the copper smelting in Nepal, there is a very thorough pictorial documentation of various mining and metallurgical techniques from eighteenth-century Japan. These were used by Zschocke and Preuschen (1932) in their mining-archaeological investigations of early copper production at the Mitterberg in Austria. These are woodcuts of the writings of *Kodō-zuroku* of Masuda Tsuna from the year 1801. Smith (1983) and Lewin and Hauptmann (1984) commented on and interpreted them and thus provided important insights into traditional technologies of the small scale production of silver-bearing copper, the liquation or *Seigerhüttenprozess* by adding lead to copper, and the final separation of silver from lead by cupellation (cf. Sect. 6.4), which developed in Japan since the sixteenth century AD. In this process, the ore was first roasted in roasting pits at about 400–700 °C for about 30 days.

The Mabuki process, an old Japanese Convertor process (Okada 1911), which was also introduced in the sixteenth century AD, worked differently. In contrast to *Kodō-zuroku*, in the Mabuki process the sulphidic copper ore was not roasted before smelting, but it was first smelted to a copper (iron-) sulphide under very strong air supply to the upper part of the furnace, according to the principle of the oxidising roast reaction process, and then successively smelted to metallic copper. This process is very similar to Nepalese copper smelting.

As far as the technology of Japanese gold mining is concerned, Winkelmann (1964) published a hanging scroll created around 1720 AD on the island of Sado (Japan). Sado became the largest gold-producing mine in this part of the world as early as the beginning of the seventeenth century. The key technology of massive gold production was gold–silver parting by the cementation process with salt (Izawa and Nakanish 2014).

On 20 foldable panels, the various mining and metallurgical techniques and facilities are illustrated in great detail. The depictions how these processes were carried out by hand and, in most cases, how they were guided and supervised by multiple officials. There is no indication of any resemblance to a gold rush. All steps of gold production are obviously well organised and controlled. These representations are similar to the images and descriptions of Georg Agricola (1556), and they were based on experiences of past centuries. The Japanese scrolls show the extraction of gold both from placers and primary ore deposits, as well as the refining of natural gold–silver alloys by cementation with salt (cf. Sect. 6.3) and the extraction of silver from lead by cupellation (cf. Sect. 6.4).

Although the island of Sado in Japan was portrayed as a fabled gold country for sailors of the sixteenth and seventeenth century, in reality, it obviously did not have the scale of gold production known from European countries (Quiring 1948). Izawa (2009) believes, that the technology for gold–silver parting as well as the liquation process to extract silver from copper ores were transferred to Japan from the Portuguese or Spanish people at the end of the sixteenth century AD.

Two pictorial representations are selected in Figs. 9.9 and 9.10 to illustrate part of the processing and a portion of the more complex smelting for gold by salt cementation.

9.6 Bronze Casting in India

It is widely known that the art of bronze casting and the *cire-perdue* or lost wax casting process can be traced back to prehistoric times in India. In the northwest of the subcontinent, it dates back to the Mohenjo-Daro and Indus Valley Civilisation of the third millennium BC.

From there, this technology spread into the countries around Nepal, Tibet and China. Especially in the Kathmandu valley, a highly regarded minority of the Newar developed as specialists in the art of metal crafts, but also in agriculture and commerce, which Furger (2017) studied in great detail. He points out that at least since the Middle Ages, not only in India but also in the more northern countries of the southern Himalayas, a large variation of alloys for making cult figures in arts and crafts can be found, in which copper,

Fig. 9.8 Okharbot, Nepal. Cake of high-grade matte produced in a smelting furnace around 1995 in Nepal, as shown in the right picture in Fig. 9.7b. Its composition is adequate to bornite, Cu_5FeS_4, and digenite ($\sim Cu_2S$) with dissolutions of fine veins of metallic copper. Photo: courtesy of N. Anfinset, University of Bergen, Norway

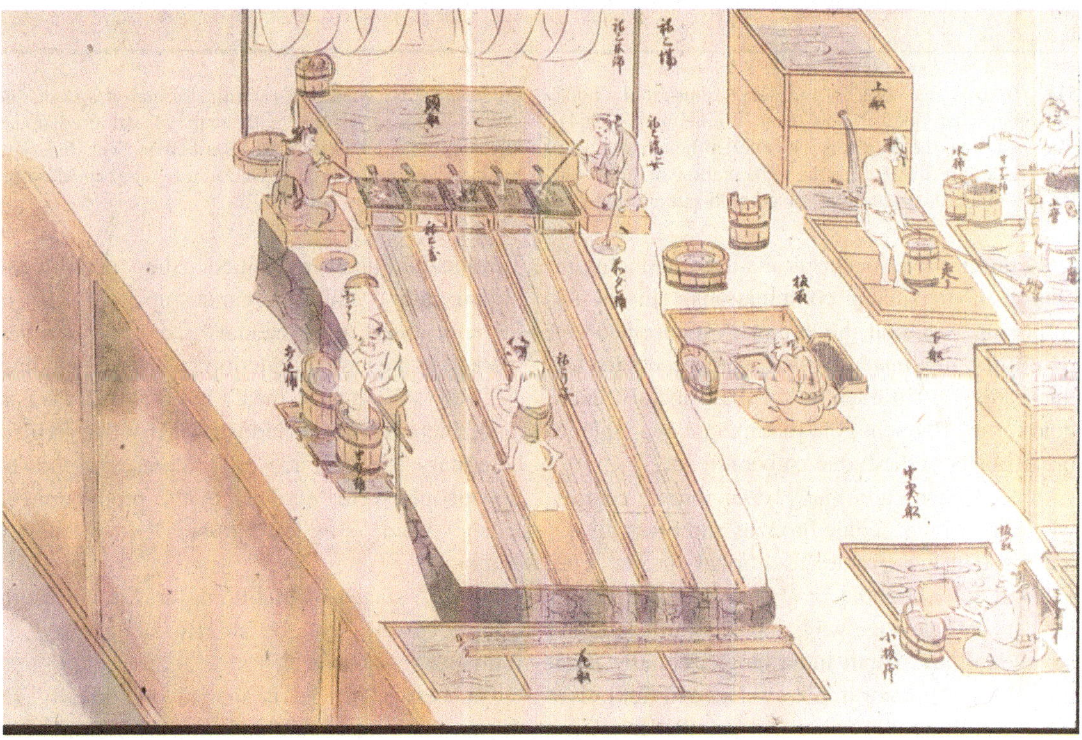

Fig. 9.9 Table XI-1 from the illustrated book "Altjapanischer Goldbergbau" from Winkelmann (1964). The picture shows some scenes from the smeltery "Tatsumi Guchi", where placer gold and crushed gold ore is washed out in various installations, mainly in three inclined sluice boxes. The water supply is located above the sluices. These are covered with tarpaulins on which heavy minerals including gold are concentrated. There is a remarkable similarity to installations of the ancient silver production at Laurion (cf. Sect. 3.8). The book is property of the Deutsches Bergbau-Museum Bochum

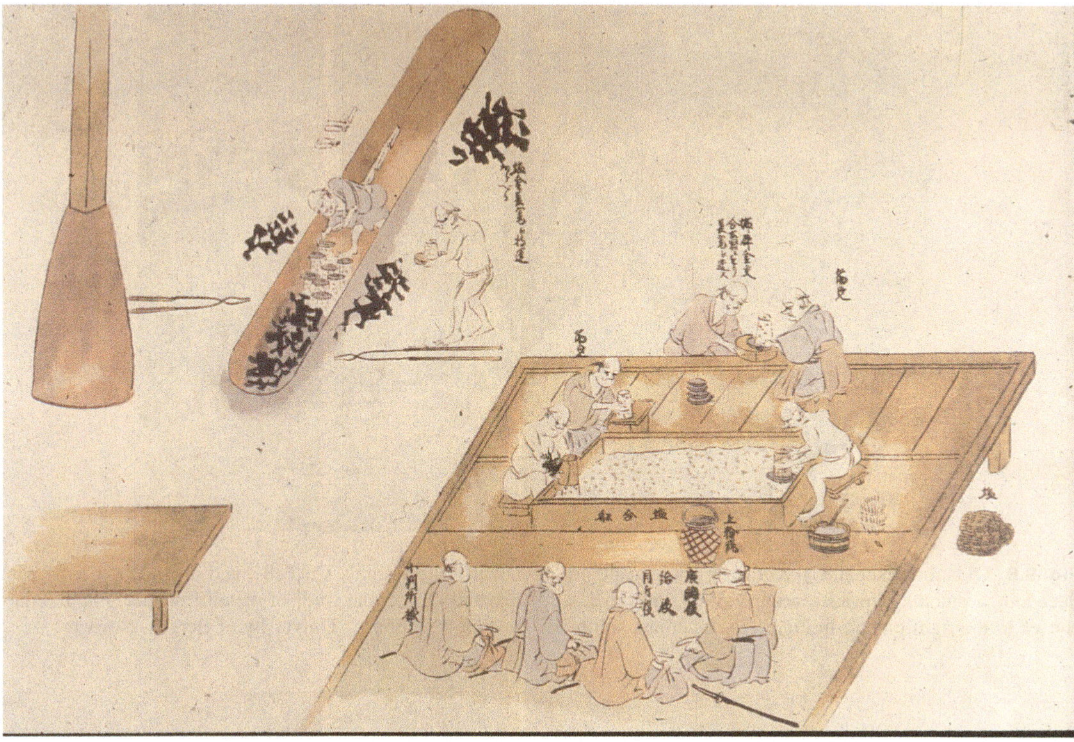

Fig. 9.10 Table XV from the illustrated book "Altjapanischer Goldbergbau" from Winkelmann (1964). The head of the smeltery, a mint official and two supervisors are overseeing the cementation of gold. On the podium, gold is carefully mixed with salt and filled in porcelain pots. These are carefully placed in an oblong-shaped furnace, covered with charcoal and fired under high heat. The procedure is described in Sect. 6.3. The book is property of the Deutsches Bergbau-Museum Bochum

zinc, tin and lead occur as main components in almost all imaginable combinations. This is not surprising, since all the countries mentioned are rich in the raw materials that were necessary for this trade, so that even today these metals, even gold for the frequent gold plating, are available in ore deposits of the Indian subcontinent.

Indeed there are many commercial centres where all sorts of iconic bronzes and brass alloys are made using simple, traditional means. The demand for cult objects always stimulated the industry in a positive way, and it continues to do so, even today. South India has an equally long prehistory as the north, but evidence of definitive metal working came to light later in the Iron Age, c. 1000–300 BC.

Levy et al. (2008) visited such a hereditary bronze casting workshop in the small town of Swamilamai in the southeast of India in Tamilnadu and examined the details of organisation and craftsmanship. In 1961, 26 specialised professionals from 17 families worked in the icon production business, including 17 "very skilled craftsmen". In 2007, there was a large increase in the industry. Fifty-two skilled hereditary craftsmen were active, and up to 300 people were involved in the production of icons, from simple "jobbers" to "master craftsmen".

All the stages of traditional craftsmanship in bronze castings can be shortly summarised as follows.

A beeswax model is shaped individually. In the case of mass production, clay moulds are sometimes used to cast such wax models. These wax models are covered with several layers of clay, the first of which must be made of extremely fine-grained clay to capture all the fine details of

Fig. 9.11 Swamimalai, Tamilnadu (South-west India). Recycled metal remains and swarfs won by the mud-cleaners from the soil in the back of the manufactory courtyard in the hereditary bronze casting workshop is cast into large open moulds on the ground to form ingots. When cool, the ingots are stored with other metal ingots in a special storeroom. From Levy et al. (2008)

the model. This step sounds banal, but is based on a lot of skill and experience and requires very specific raw material sources of clay. Depending on the size of the intended figure, several or even numerous runnels are incorporated into the clay casing to guarantee that later on during pouring the liquid metal fills all spots and corners of the figurine. Then the clay shell is built up into a more massive casting mould and the wax is burned out and collected. The casting mould is set into the ground. In the meantime, the bronze (or brass) was liquefied in several graphite crucibles in a deep-set furnace. The red-hot crucibles are transported to the casting mould and the molten metal is poured into the cavity through the runnels. In the workshop in Swamimalai, crucibles with a capacity of 1–60 kg of metal were used. After cooling, the casting mould is smashed with a hammer, so it is wholly destroyed, the fragments are thrown away. This is also the reason why generally no such clay moulds can be found in archaeological findings. Now the runnels need to be removed and the icon needs to be sanded and polished and engravings are added. This finishing work may only be carried out by skilled craftsmen. The workers are only lightly clothed and barefoot in all their activities like the handling of hot, liquid metal. They do not wear any safety clothing.

Depending on the thoroughness of these individual activities, varying amounts of scrap metal are produced. These were collected for re-melting and recycling.

Levy et al. (2008) describe two interesting aspects from the bronze casting of Swamimalai, which are largely unknown in archaeology. These concern the activities of the mud-cleaner, who conducts an on-site recycling of leftover waste in the workshop. This re-cycling not only involves the collection of scrap metal from casting, but also the scraping off of metal incrustations from the crucibles. The mud cleaner also cleans the foundry floor by washing the sediments to a depth of approximately 1 m in a prepared depression of the ground and thus recovering also amounts of mm-sized metal particles. The discovery of this individual task in the hereditary manufactory provides a set of important analogues for archaeometallurgical research. Possibilities of such recycling has hardly been considered in excavations so far. An example is large amounts of fine-grained silver-bearing litharge that can be found in the soils of the processing plants of the classic Laurion (Greece). They obviously originated from recycling processes (Papadimitriou 2016).

The work in the foundries of Swamimalai includes yet another interesting step, which is generally underestimated. This step involves the casting of metal or alloys. Levy et al. (2008) show in their pictures how the workers pour the melt from a 60 kg crucible, not into prepared ceramic moulds, but into several approximately 80 cm long depressions into the sand (Fig. 9.11).

10 Experimental Archaeometallurgy

10.1 General Remarks

Experimental archaeometallurgy is part of experimental archaeology. It includes experiments on mining, the processing of ores, the methodology of roasting and smelting to the extraction and processing of metals by casting, alloying and patinating. It includes the replication of possible crafting techniques from the entire *chaîne d'opératoire* to better understand ancient metallurgical steps and processes. Because in many cases attempts to reconstruct these they have been influenced, adopted and restricted by models of modern industrial metallurgy. Therefore, experimental work, carried out as closely as possible to archaeological evidence, is necessary. It often leads to new ideas and knowledge about early metallurgical procedures.

In Chaps. 5 and 6, a few selected examples from the wide range of individually shaped smelting furnaces and other metallurgical installations found in archaeological excavations or recent contexts have been presented. For this topic, more detailed descriptions have been compiled by Heeb and Ottaway (2014) or Pleiner (2000). They discussed the variety of shapes and designs of furnaces, the number, size and constellation of tuyères for air supply and the designs of facilities for roasting processes. One of the few Bronze Age roasting installations was found in Austria by Goldenberg (2004).

Reconstructing metallurgical installations from the archaeological field evidence is often difficult. They are usually more or less destroyed after the smelting process, to recover copper, matte, slag remains or iron blooms precipitated in the solid state. The peculiar Iron Age copper smelting furnaces from Timna 30, stratum I and II/III (Israel) or from Faynan 5 (Jordan) are well known and heavily discussed examples. During archaeological excavations at both locations at Timna 30 by Rothenberg (1990) and later at Faynan 5 by Gerd Weisgerber (Hauptmann 2007) similar, difficult-to-interpret constructions of flat, round furnaces with a height of only c. 0.6 m and a diameter of c. 1.5 m were excavated. For Timna, Bachmann proposed a pear-shaped reconstruction of these furnaces with a diameter of only 0.6 m but much higher with a variable number of tuyères (Bachmann and Rothenberg 1980). This reconstruction was certainly affected by the generally known model of vertically oriented shaft furnaces known from other ancient sites supplied with several tuyères for air supply.

Based upon this type of shaft furnaces Merkel (1990) carried out interesting smelting experiments and was able to produce similarly heavy cakes of tap slags as were found in the Iron Age Timna 30, Beer Ora and Faynan. Löffler (2017), on the other hand, was able to find an ethnographically similar case of flat furnace as described above, namely a traditional furnace construction for iron smelting in Nigeria and

A. Hauptmann, *Archaeometallurgy – Materials Science Aspects*, Natural Science in Archaeology,
https://doi.org/10.1007/978-3-030-50367-3_10

was, therefore, able to work out a plausible solution for the Wadi Arabah style.

Heeb and Ottaway (2014) also draw attention to another difficulty, when, for example, archaeological excavations reveal clear indications of metallurgical processes, e.g. by finds of slags or metal prills, but no corresponding metallurgical facilities are to be found. This was the case at excavations in Belovode (Serbia), dated to the beginning stages of metallurgy (c. 7000 years ago), where some pieces of semi-liquified slags were found but not any smelting installations (Radivojević et al. 2010). Faced with such difficult archaeological findings, smelting experiments may be useful. A possible explanation for such a difficult evidence is presented by Simon Timberlake (Timberlake 2007). He succeeded in smelting oxidic as well as sulphidic copper ores in very simple, clay-lined depressions in the soil and produced metallic copper. The archaeological evidence for such a simple processing in such simple constructions is likely to be very difficult to grasp after a few thousand years.

Empirically, smelting experiments have very good publicity and are therefore gladly carried out in public. Unfortunately, despite some excellent ideas, they are not always carried out seriously and consistently, and they are on a tightrope walk to serious science. Charles (1980) thinks "...it is interesting, if not amusing, to learn of the problems encountered by present-day experimenters in reproducing primitive smelting operations where the production of a sizable bun ingot from the hearth, rather than porous lumps from the furnace charge, presents a major difficulty..." (Fig. 10.1).

Many experimental works have been carried out mainly concerning the metallurgy of copper and iron. Experiments with copper were carried out by Tylecote and Boydell (1978), Tylecote and Merkel (1985), Merkel (1990), Herdits (2003), Kölschbach et al. (2000), Bourgarit et al. (2002), Rovira and Gutierrez (2003). Smelting tests for the reconstruction of iron or steel have mainly been carried out by Tylecote et al. (1971) and Straube (1986) based on Roman finds in the Southern Alps (Sect. 7.5). Pleiner (2000) discussed other experiments in summary. But above all, there are a number of recent ethnographic materials that have been documented in African countries (Sect. 9.3); in their importance, they are comparable to smelting experiments.

Fig. 10.1 A product of copper smelting in a wind-powered furnace reconstructed after Early Bronze Age furnaces at Faynan (Jordan, see Fig. 6.5) in 1998 (Kölschbach et al. 2000). One of the products was a piece of slag with a poorly agglomerated lump of copper with a weight of only 200 g. Pieces like that had to be re-melted to larger units if any artefacts should be shaped. Collection C. Hauptmann. Photo H.W. Voß, Deutsches Bergbau-Museum Bochum

Here are just a few examples have been selected that have contributed to the understanding of the earliest processing techniques or deciphered new, previously not deciphered technologies. This concerns the earliest extraction of the widespread copper, as well as that of tin and lead. Very little is known about the extraction technology of tin.

10.2 Early Copper Smelting

10.2.1 (Re-)Melting Native Copper

The finds of the earliest copper artefacts from the Pre-pottery Neolithic settlements of Çayönü Tepesi and Aşikli Höyük in south-eastern Anatolia have led to a series of studies investigating whether these artefacts could have been produced from native copper without a melting process, or whether they were made from smelted copper ores. The question was (and still is) in which regions, even outside of Anatolia, and up to which epochs was native copper used for the production of artefacts, and when was copper utilised which had been smelted from ores. These are central questions of archaeometallurgy, which affect the formulation of metallurgical development models. This touches on the problem of differentiating between these sorts of copper. This problem was already discussed in Sect. 6.2.

It was concluded in the 1990s that the numerous jewellery beads, hooks and needles of Çayönü Tepesi and Aşikli Höyük were made of native copper, which was forged without a previous melting process. This idea was supported by the fact that neither crucibles nor any remnants of blowpipes were known from that time. For the beads of Aşikli Höyük it was assumed that small pieces of native copper were hammered to sheets and then rolled up into small hollow cylinders. They usually have diameters of <1 cm.

There are apparently two ways of distinguishing native and (s)melted copper from one another, but these cannot always be used as criteria without a doubt. This is on a geochemical basis, as discussed in Sect. 6.2, and then according to metallographic criteria (Maddin et al. 1980; Rapp 1988; Pernicka 1990; Pernicka et al. 1997; Yalçın 1998; Yalçın and Pernicka 1999).

10.2.2 Microstructural Investigations

To work out possible differences between native and smelted copper, Wayman and Duke (1999) conducted a series of smelting experiments. They studied 12 samples of native copper from White Pine mine, Ontonagon County, Michigan, and observed microstructures that varied from coarse- to fine-grained sizes. In addition, twinnings were observed in these natural samples, which were caused by frequent metamorphic pressures that occur in this region, so they were based on geological facts. Except for occasional discrete silver granules and copper sulphides, they were free of any inclusions. According to the observations of the authors, cuprite was a weathering product formed only along grain boundaries.

The authors melted these samples completely using small crucibles, to work out criteria that would allow to distinguish native and (s)melted copper. Upon cooling, the metal solidified either in dendritically crystals or in equiaxed grains, as is typical for cast metals. At the same time, a slight interdendritic/intergranular porosity formed due to shrinkage and/or gas evolution. A Cu-Cu_2O eutectic texture was characteristic. This texture was never observed in native copper. It can, therefore, be considered a key criterion for (s)melted copper.

Such a Cu-Cu_2O eutectic texture was observed by Yalçın (2000a, b) in the study of copper artefacts from the Chalcolithic Mersin in southern Anatolia (c. 5000 BC, Fig. 6.9). Apart from the fact that the smelting of copper ores was well-known during this period, not only in Anatolia, the highlighting of this specific texture as a unique feature is justified in view of the fact that native copper frequently occurs in nature.

Yalçın (1998) further concludes that Can Hasan's macehead from southern Anatolia (c. 6000 BC) was not cast, as has long been postulated before. It was hammered from native copper in a just warm state. The metal contained discrete inclusions of silver that, in the liquid state, would have dissolved into the copper. Thus, he was able to exclude a liquification of the copper. He also identified twinning in the metal. However, he attributed this to artificial mechanical action (hammering).

The idea of the Aşikli Höyük beads being manufactured from native copper was not experimentally tested back then. Laschimke and Burger (2015) made up leeway in this regard, which lead them to propose a different version. They doubt that the most irregularly shaped pieces of native copper, which in nature may consist of mixtures of metal, copper oxide and malachite, as described in Sect. 3.5.1, could have been divided into materials suitable for forging and forming of beads etc. using Neolithic stone tools. Instead, they assume that pieces of native copper were first re-melted into dense, small grains that were easily forgeable. These were forged red hot to make a round circular blank. This was rolled to make a bead like that from the site. Alternatively, a copper bead was forged red hot to a sheet and wrapped around a stone bolt.

10.2.3 Crucible Smelting

Reconstructing early copper smelting techniques, such as from the late Neolithic or Chalcolithic periods is difficult due to the poor archaeological evidence. From many sites, there are only fragments of slagged ceramic crucibles, no tuyères and no remains of any blowpipes. There are only a few pieces of heated ores or partly liquefied slags, even actual metal artefacts are more common. The interpretation of these few finds is laborious, as is the reconstruction of a complete *chaîne d'opèratoire* and thus of any organisation patterns as well. Fireplaces, hearths and shallow pits are often overestimated to be related to metallurgical activities while they are just remains of everyday life, e.g. just for food supply. Craddock et al. (2007b) formulate the situation as follow: "The sparcity of smelting remains from the British Isles in the Bronze Age means that it is often necessary to argue from negative evidence, indeed making metal while leaving no durable evidence was one of the principal objectives of the experiments."

Based on the results of the excavations of the Chalcolithic settlement Zambujal in southern Portugal, Hanning et al. (2010) recognised the necessity of experimental studies and carried out 17 smelting experiments with copper ores in crucibles. Since 1930, excavations of the fortified settlement have brought to light some 900 copper objects with a total weight of 3–4 kg, yet just a few fragments of technical pottery used for metallurgical processes were scattered around.

The team performed smelting replications using mixed oxidic–sulphidic copper ores from sources located partly more than a hundred kilometres' distance from Zambujal. These ores were carefully selected and were high in copper, and low in gangue and host rock. Very simple means of smelting were used: flat opened clay crucibles heated with wood and charcoal, in simple stone-lined hearths. Six persons using clay-tipped blowpipes made of bamboo supplied the forced draft. The physical and mineralogical makeup of the small amount of smelting slags, poorly smelted and full of oxidic phases such as magnetite, cuprite and delafossite, and of copper reflected not only the poorly reducing atmosphere within the crucible. It also indicates that a considerable amount of copper in the ore was never liberated as metal. Metal precipitated in the liquid state from the ore was shaped as cm-sized irregularly or round globules and accumulations, largely mixed with slag and matte (Fig. 10.2).

The mechanism of copper-precipitation from matte is demonstrated by the formation of μm-sized whiskers, which, depending on the firing conditions, successively compact to droplets and agglomerations. Such whiskers have already been shown in Fig. 5.57 in Sect. 5.7. For this topic, Hanning and Pils (2011) has also conducted experiments for the smelting of sulphidic ores according to the Bronze Age model of the Mitterberg (Eastern Alp, Austria), and Craddock et al. (2007a, b) reached the same conclusions in their smelting experiments under primitive conditions.

The results of the experimental smelting experiments at Zambujal coincide well with those archaeological remains, from which due to their open design of the vessels and hearths created difficulties in providing a reduction environment and maintaining sufficient temperature over a longer period of time.

In addition, the total absence of blowpipes in the archaeological context of Zambujal was not surprising. During the experimental work, the wooden blowpipes easily started to burn and the (unburned) clay tipping cracked after one or two experiments. This could be a reasonable explanation for the almost total absence of blowpipes in the archaeological record of the Chalcolithic period. It is also interesting to observe that the smelting experiments led to various kinds of damage to the ceramic crucible wall. There were definitely deeply vitrified and slagged sections in the crucibles. But there were also parts of the crucibles where there were no discernible traces

Fig. 10.2 Sectioned crucible from smelting experiments at chalcolithic Zambujal, southern Portugal (Iberian Peninsula) to explore copper smelting technology 5000 years ago. Three distinct layers are visible with the naked eye. (**A**) A dark, porous layer of slag mixed with the melted ceramic body and small, finely interspersed copper prills. (**B**) A grey-silvery shining layer of matte (copper sulphides); (**C**) Globules and accumulations of metallic copper, partly mixed with slag. The crucible was heated from above by blowpipes. Width of crucible 11 cm. From Hanning et al. (2010)

of high-temperature treatment or "soiling". The authors believe that this needs to be kept in mind when archaeological material is interpreted.

With the production of small prills and agglomerations of copper in the order of a few grams, Hanning et al. (2010) finished their experimental work. They did not touch on the question of how and in what vessels a melting together of these smaller pieces was done in order to produce larger artefacts like ingots, double axes and flat axes, which were all part of the standard spectrum of metal artefacts in Europe.

Axes from the Late Neolithic canton of Thurgau (Pfyn Culture, fourth millennium BC, Switzerland) weighed in the order of several hundred grams up to one kilogram. For example, they were found at numerous sites in the Alpine foothills (Matuschik 1998). Leuzinger (1997) published ceramic vessels of various shapes, that he interpreted as crucibles. Fig. 6.1 shows three of those. These are one round vessel with a wide handle and a spout and two trough-shaped ones. Since these trough-shaped vessels do not have a handle or a spout, they could hardly have been crucibles for casting.

The caster specialists Laschimke and Burger (2017, 2018) suggest that such trough-shaped crucibles would not be suitable for casting liquid metal. According to their experiments, they suspect that these were used as melting vessels to re-melt copper prills in order to produce larger units. And indeed, at this early time, there is no evidence for a production of sufficient amounts of metal in a first melting step in order to cast an axe or an equivalent artefact. The authors believe that in the earliest copper smelting, metal was only produced in small amounts, similar to the smelting experiments in Zambujal. In order to produce copper in larger amounts suitable for tool making, many of these prills had to be melted together.

During their melting experiments, the authors preheated such a crucible in a charcoal bed to red heat and filled it with successively with copper prills (Fig. 10.3). They pointed out that the surface of the melt was contaminated by ashes, oxides and charcoal residues. Such impurities were not allowed to enter the liquid. They had to be removed. Even today, it is common practice in foundry technology to skim off a copper melt prior to casting by drawing the contaminants with a tool from the liquid copper surface and sliding it over the edge of the crucible. After cooling, a dense ingot suitable for forging a flat blade was removed.

The re-melting of prills merges seamlessly into a re-melting of flawed artefacts. Examples of this are finds that, however, date to later epochs. Finds of such copper pieces are known both from early Bronze Age Khirbat en-Nahas (Faynan area, Jordan; Levy et al. 2002), as well as in Shahdad (Iran; Meier 2015). They are shown in Fig. 5.12.

Geochemically, a re-melting leads to an average composition of several single (s)melting

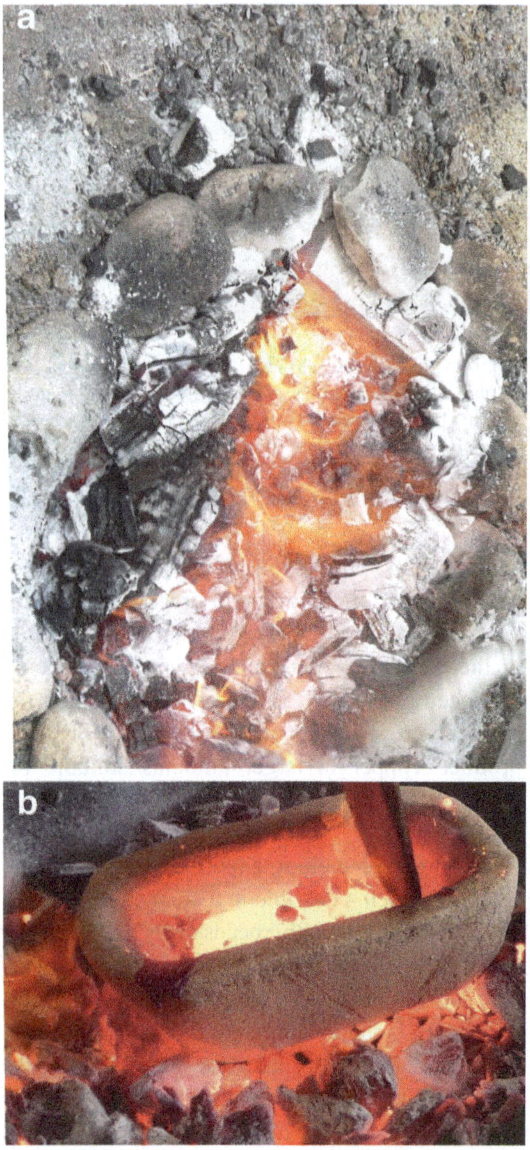

Fig. 10.3 (a) Heating a trough-shaped crucible on the model of the handleless crucible of Gachnang in Switzerland (Pfyn Culture, fourth millennium BC; see Fig. 6.1) in a simple stone setting filled with a glowing charcoal bed. (b) Skimming of impurities in such a crucible which was successively filled with copper prills to larger metal units. Photos from Laschimke and Burger (2018)

processes, and, hence, to an average composition of larger parts of ores from an ore deposit. It will be shown in Chap. 11 that this may offer considerable benefits by making provenance studies. However, the average compositions can also lead to difficulties if the number of ores and/or intermediates from an ore deposit analysed does not include a sufficient number of samples.

10.3 Casting Copper Oxhide Ingots

The manufacture of finished products such as weapons, jewellery or metal tools involves a number of different casting techniques, for which moulds of various shapes, materials and constructions have been used. A differentiation is made between simple open casting with one-piece moulds made of pottery or of stone. Casting with bivalve moulds means those consisting of two parts of a mould that fit together and can be sealed, which can simply be taken apart after the casting and can be reused. Then there is the lost wax casting (cire perdue), in which the desired shapes are formed with wax and covered with clay. Then the wax was burned out of the clay mould and the cavity filled with metal. An extended form is overcasting, in which, for example, a sword blade is provided with a handle in a mould. In Mesopotamia, lost wax casting has been known since the fourth millennium BC (Uruk period) and was then widely used in the early dynastic period (third millennium) for the production of complex shapes (Müller-Karpe 1990b). In Mesopotamia, on the other hand, there are no moulds of any kind for the numerous shaft-hole axes found there, so that one must assume that a different production technique was used, which left no traces or traces that are difficult to grasp archaeologically. One such possibility is another casting process, namely in a lost sand mould, as could be observed until modern times in Baghdad for the production of keys. The author had the production of shaft-hole axes experimentally reproduced in a metal foundry. He used a wooden axe model, the form of which was pressed into a moistened mixture of sand, clay and ground fire brick and connected to an appropriate casting channel for the introduction of the liquid metal. Casting in the lost sand mould was simple and quick, and a sharply set-off casting blank could be produced which, apart from the separation of the casting spigot, required little reworking.

With this process we approach the casting of ingots, which is first of all the basis for the explained production of finished products.

A difficult archaeological problem is posed by the question of the production of Late Bronze Age copper oxhide ingots from Cyprus. Although there are many hundreds of such ingots, especially in the Mediterranean region (Fig. 6.11), with the sole exception of a so-called Ras Ibn Hani mould carved in stone in Syria (Lagarce et al. 1983), no others are known. But how were these metal ingots cast? And what was the purpose of their characteristic rectangular shapes with differently shaped ears? Were these ingots produced by tapping metal from a smelting furnace or from a crucible?

Laschimke and Burger (2011) suggested that the shape of the ingots was a compulsory result of the technology of casting applied and that it was not chosen for practical reasons to transport them. As copper oxide ingots usually weigh around 25 kg, sometimes even up to 30 kg, it is very unlikely that an ingot could be cast from a single crucible, as no (ceramic) crucible with a stability to liquify such an amount of copper and to handle it for casting could have existed in the Late Bronze Age. In any case, the famous 17-1 reaction vessel from Late Bronze Age Enkomi (Kassianidou 2016) could hardly have been suitable.

The authors, therefore, decided to conduct experiments with several crucibles. The required amount of copper was distributed on batches in four crucibles and poured in liquid form simultaneously very carefully from four corners into a sand mould (Fig. 10.4). To achieve the desired composition of the moulding sand, one part of clay and two parts of powdered charcoal were mixed according to recipes by Georg Agricola (1556). This mixture was dried for a long time and then moistened for moulding. In this way, a proper gas expelling mixture could be prepared which is necessary for casting.

It was found that the water content of the moulding sand in contact with liquid metal abruptly steams from its upper layer and would

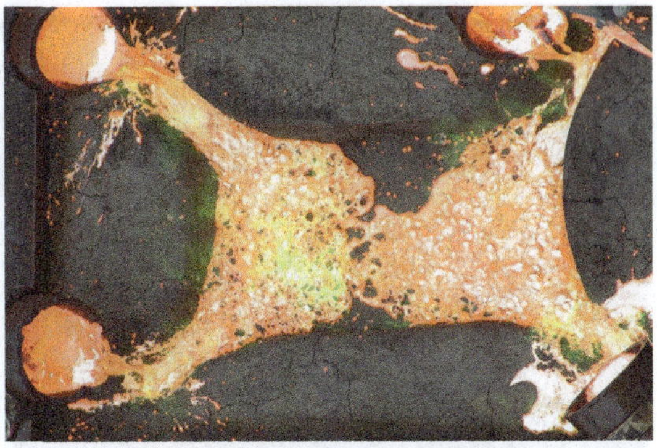

Fig. 10.4 Horizontally casting of liquid copper from four crucibles into a mould made of clay and crushed charcoal. The liquids will not be evenly mixed. Gases produced by reactions at the moulding face are inflamed in greenish-yellowish colour at the surface of the liquid. Photo: Laschimke and Burger, Sigmaringendorf

cause a porosity due to gas bubbles and a suitable permeability. At the freezing point, a cooling rim is formed at the surface of the liquid copper. Rising gas bubbles are now prevented from escaping and are bound at the surface (Fig. 10.5a). In addition, also at the bottom of the ingot, a cooling rim was formed and a large number of vertical tubular pinholes were formed (Fig. 10.5b).

Hauptmann et al. (2002) compared metallographic results obtained from sampling a number of copper oxide ingots from Uluburun and found that all these cooling effects observed at the experimental work of Laschimke and Burger (2011) are identical with archaeological copper ingots.

10.4 Early Tin Assaying and Production

The knowledge of prehistoric tin metallurgy is limited. There are hardly any slag finds or those of tin objects. This is astounding given the incredible mass production of tin bronze worldwide. Concerning this problem, Craddock et al. (2007a, b) point towards the ideas of Charles (1975) who speculated that perhaps tin metal was not normally produced at all, but instead the

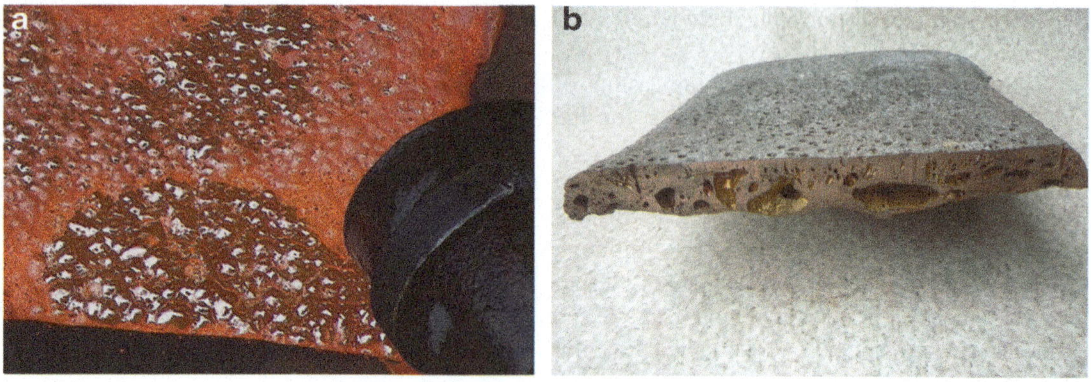

Fig. 10.5 (**a**) The surface of the experimentally cast ingot after the liquid metal was cast into the sand mould. A solidified surface forms an impermeable skin and prevents gases to escape. This causes large gas bubbles to swell under the surface and to arch upwards at many points. This structure is identical with many of the archaeological copper oxhide ingots. (**b**) Section of an experimentally produced copper oxhide ingot (upside down), showing the dense pattern of pinholes at the bottom (in the picture upside). At the surface of the ingot (in the picture below), flat tubular macropores are formed which caused a humpy skin. Photos: Laschimke and Burger, Sigmaringendorf

beneficiated tin ore could have been co-smelted with the copper ore or added straight to the molten copper and an alloy was produced directly.

10.4.1 Kestel and Göltepe

In the 1980s and 1990s, Aslihan Yener had excavated the prehistoric mine of Kestel and the nearby settlement of Göltepe in south-eastern Turkey. Her idea that this might be an Early Bronze Age tin mine and a corresponding metallurgical workshop where the ore was smelted to metal, was met with considerable criticism from the scientific community. Due to the fact that only very low concentrations of cassiterite were found in the mine (<0.1 wt.% Sn; Yener et al. 1989a; Yener et al. 2003), a mining of this ore and a metal production was considered to have been impossible at this site (Muhly et al. 1991). Hall and Steadman (1991) were convinced that the extraction of these exceedingly low tin concentrations in the deposit of Kestel was beyond the capabilities of Bronze Age miners and smelters. On the basis of chemo-physical considerations, they argued that a tin mineralisation, intergrown with various iron ores, was impossible. They definitely ruled out natural accumulations of tin ores in the mine, even though these do occur in the mineralogy of ore deposits (cf. Sect. 3.2), and are also possible in the mineralisation type of Kestel. Muhly et al. (1991) suggested instead that the mine was operated for the production of gold, although it was unlikely that this metal would occur in greater concentrations.

In all these heated discussions, however, some crucial points had not been taken into account. For one, the low concentrations of tin indicated were merely bulk analyses of the entire deposit, as modern geologists use to measure. Aspects of this topic have been explained in Sect. 3.7. Moreover, it is a common problem in the survey of ancient mines that the ores mined in the old days are usually no longer present in their original quality and quantity since they have simply been extracted. Nor had it been considered that these low metal contents could have been enriched by any kind of beneficiation, e.g. by simple grinding and vanning, maybe even occasionally with gold. It is widely known in mining archaeology and in archaeometallurgy that beneficiation is logically one of the earliest stages of metal extraction, although its archaeological evidence is sometimes difficult to grasp. Based on excavations in the Mitterberg area (Austria) Stöllner (2019b) convincingly demonstrated that ample and intensive beneficiation of ores was one of the most important steps of metal production everywhere.

A clear evidence of exceptionally intensive and extensive beneficiation close to the mine of Kestel and at Göltepe are already the masses of finely crushed ore and the incredible masses of 5000 hammer- and anvil stone tools excavated and 50,000 (!) estimated laying on the surface (Yener et al. 2003). Furthermore, in the Early Bronze Age settlement of Göltepe, thousands of fragments of crucibles have been excavated. Their insides were covered by thin slag skins, in which characteristic inclusions of cassiterite were identified (Yener and Vandiver 1993; Yener et al. 2003).

Due to the fierce discussion and in order to verify her theses, the excavator of Kestel and Göltepe has made the decision to reproduce the stages of tin production in experiments.

With a series of simple replication experiments, the mining engineer Bryan Earl showed how in the tin province of Cornwall, at least since the Middle Ages to the recent past, tin ores, both from placers and from primary deposits, could be tested for its tin contents by assaying. Also at Cornwall, he had to deal with ores with only low concentrations of tin. Earl (1985, 1986) reports that for centuries, tin ore was classified as very rich if it had 5 wt.% Sn, rich if it had about 3 wt.% Sn, and just good if it contained 1 wt.%. In a first step, Earl did some grinding of cassiterite tin ore consisting of hematite, limonite with minor amounts of tourmaline and chlorite to produce a fine-grained powder. In a second step, the vanning technique of assay, particularly suited to assessing the values in an ore, followed. Generally, vanning is a very old process of beneficiation. In this case, it was a tradition in Cornwall to do the vanning with a shovel. The third step was the fire assay to determine how a black tin concentrate would respond to melting. This activity was carried out in small crucibles and was in principle comparable to the mediaeval assaying ("micro-smelting"), which could also be performed with very small amounts using the blowpipe technique on a piece of charcoal. It is unknown how old this technique is.

The same steps of assaying tin ores were retraced in the 1990s in both Turkey and in the Oriental Institute of the University of Chicago, using ores from the Kestel mine (Earl and Yener 1995; Yener et al. 2003). In his experiments, Earl produced exactly the same results as they were excavated at the archaeological site. It could thus be convincingly shown that it is quite possible to extract the metal from ores with low tin contents. The same technology that was practiced in the laboratory on a small scale was possibly practiced for larger productions in the Early Bronze Age in south-eastern Turkey. Even if Kestel/Göltepe and surrounding tin deposits are no revival for the large tin deposits in Cornwall or in the countries of the Middle East (Tadjikistan, Uzbekistan, Kasachstan), where obviously the most part of ancient tin was produced it shows at least that also small tin deposits have been exploited very early.

10.4.2 Hampshire, Cornwall

An important aspect of prehistoric tin metallurgy is the extremely small amount of slag. It has already been discussed (Sect. 5.5.4) that it required at least good cassiterite concentrations. Craddock et al. (2007a, b) have made smelting experiments with processed cassiterite concentrates in Hampshire (Cornwall) (with 10–20 wt.% "iron" and 5–10 wt.% arsenic, caused by associated minerals). He was able to easily smelt tin, while also only producing very small amounts of slag. The tin still contained 0.9 wt.% arsenic and c. 0.2 wt.% iron. These elements were thus incorporated into the later produced bronze via the tin ore itself. Especially such an input of arsenic (which evaporated only in subordinated amounts during metallurgical operations) into tin and tin bronzes could simplify interpretations of possible origins of such artefacts.

In some of the experiments, however, there was a considerable volatilisation of the tin, evidenced by the white deposition of tin oxide on the sides of the smelting unit and on the tuyère.

10.5 Lead Smelting

Little is known about the (pre-)historical smelting of lead ores, and not many smelting experiments have been performed so far.

In the 1960s, several extensive "metallurgical expeditions" were organised by Theodor Wertime and Cyril Stanley Smith from Washington (Smith et al. 1967). Radomir Pleiner participated as a member of the team of scientists. During his visit to Tal-i-Iblis (Iran) Pleiner (1967) conducted two smelting experiments with lead ores. The first experiment was performed in a low shaft furnace (height 0.7 m), the second one in a small open hearth. In both cases, goatskin bellows were used for burning and heating. The purpose was to learn whether such simple constructions could develop efficient temperatures for suitable slag formations of the components in the ore and to reduce lead itself. The ore consisted of galena, cerussite, limonite and quartz, in another attempt the galena was additionally intergrown with sphalerite (ZnS). Obviously, the zinc content of the ore caused problems. Charcoal was used as a heating and reducing agent.

It was not a problem to reach temperatures near 1200 °C in the shaft furnace. After several hours of preheating, a charging of ore in several portions was done within 2.5 h keeping the level of this high temperature. The yield was poor. Just a very small amount of lead appeared in the depression below the hearth. Apparently, the zinc content of the ores was too high, and the remaining slag and lead solidified in the furnace when it was left to cool. At the level of the tuyere, there was a conglomerate of slag and lead. The base of the furnace was covered by ash layers.

The smelt in the simple hearth required only 1 h, and also here it was not a problem to reach temperatures of 1160 °C. The smelt was not successful. Neither slag nor metal was observed, only sintered ore.

Pleiner (1967) analysed several dozens of slags and ores used for the experiments for their chemical composition. In view of the low yield of lead, he points out that much more experience would be needed to smelt successfully in such a shaft furnace and on such a small hearth. In no way it would be comparable with the processes applied in ancient time, and with the amounts of metal produced. But he had at least the chance to test the furnace, the bellows, and the nature of the various components in operation, i.e. the temperature and the reducing conditions.

Hetherington (1980) performed a series of smelting experiments to smelt lead ores under very simple conditions. He constructed a deep bowl of approximately 0.23 m in diameter and 0.3 m in depth. Modern materials were used. A mullite tube was inserted through the furnace wall above the tap hole. Charcoal was used for burning. The charge consisted of a mixture of galena with something limestone and fluorite (CaF_2). Iron-containing fluxes and silica-rich sand was added for fluxing. It is interesting that the sulphur-containing galenite was added directly to the smelting furnace. He did not perform a roasting before the smelting process because it is relatively easy to reduce lead from a sulphide ore.

Following the basic experience of ethnoarchaeology, as well as from today's experience, the furnace was heated for 2 h before the experiments and then loaded with charcoal. Air supply was provided. After a short time, a mixture of powdered charcoal and ore was added. After about an hour, liquid lead and slags formed, which were sometimes tapped via the tap hole, and sometimes they solidified in the furnace. During the experiments up to 1300 °C were reached. In each individual experiment, Hetherington precisely documented the kinetics and thermodynamics between air supply, CO/CO_2 ratios (redox conditions) and temperature ranges. These are fundamentally controlled by the well known three reactions of burning carbon:

$$2\,C + O_2 \rightarrow 2\,CO$$

$$C + O_2 \rightarrow CO_2$$

$$2\,CO + O_2 \rightarrow 2\,CO_2$$

These reactions are all strongly exothermic. Reducing the air supply means that all reactions of the batch are slowed down and the temperatures drop.

Hetherington points out that these different redox conditions have a significant impact on the amount of metallic lead produced, as well as on the contents of Pb^{2+} in slags in the form of Pb silicates or Pb-rich glass. In addition, changing redox conditions have a decisive influence on the Fe^0, Fe^{2+} and Fe^{3+} contents of the melt and thus on changing phase contents of the Fe silicates and oxides. When the CO/CO_2 ratio in the furnace increases and the environment is more reducing, the lead content in the slag decreases as more PbO is reduced to Pb^0 which separates from the slag. On the other hand, there is also a considerable risk that the Fe^{2+} content is reduced and metallic iron precipitates as hardheads.

Provenance Studies 11

11.1 Introduction

The provenance of metal artefacts and the search for ore resources used in prehistoric and historic times for smelting processes is one of the basic questions of archaeology. The purpose of this research is the reconstruction of the obviously very complex trade relations, which, quite early on, reached near global geographic distances of more than a thousand kilometres. Supra-regionally, metal artefacts may occur in varying distributions. On the one hand, they are found in areas exceptionally rich in mineral deposits, such as the incredible masses of Bronze Age axes and other artefacts in the area of present-day Central Europe during the late third and second millennium BC (Tumulus culture, Urnfield culture; Parzinger 2016). On the other hand, they also occur in areas that do not have any ore deposits, as, e.g. in the case of the rich finds of gold, silver, copper and bronze in the Early Dynastic and Akkadic periods in Mesopotamia, also in the third millennium BC. In Mesopotamia, between the Euphrates and Tigris, there are only quaternary sediments, no ore deposits. Many of the metal objects found there were imported, sometimes from distant lands, e.g. from Makan (Oman), from northern Afghanistan or from Anatolia (Salzmann 2019; Jansen et al. 2016; Jansen 2019).

It has already been mentioned that the origin of Bronze Age tin bronzes is still one of the most fundamental questions of archaeology. To answer this, several research projects have been carried out, with focus on mining–archaeological field work (Garner 2014), as well as on analytical investigations (Berger et al. 2019a, b). Interestingly enough, the focus is again on countries of Central Asia (Afghanistan and countries such as Uzbekistan, Tajikistan, Kazakhstan; Stöllner et al. 2013a, b).

Similarly, the arsenic and antimony-containing artefacts from the treasure of Nahal Mishmar (Israel, late fourth millennium BC) were smelted from a still unknown and apparently quite distant ore deposit.

Written sources on this topic only date back to the middle of the third millennium BC. But even with the help of those sources, an identification of the mentioned geographic locations, as well as the material and metallurgical terms, succeeded only rarely (Waetzold 1985; Reiter 1997; Begemann and Schmitt-Strecker 2009; Paoletti 2016; Hauptmann et al. 2018).

It is a common knowledge that the origin of archaeometallurgical artefacts cannot in all cases be traced just by typological descriptions and with the help of the archaeological context, and thus shed light on early trade routes and relations. It has also been proven useful to apply indirect, scientific analyses, but these should always be performed in close context with archaeology.

The concerns of natural scientists working on archaeological issues and trying to find the provenance of metal artefacts date back to the late

A. Hauptmann, *Archaeometallurgy – Materials Science Aspects*, Natural Science in Archaeology,
https://doi.org/10.1007/978-3-030-50367-3_11

eighteenth century. In Germany, it was the mineralogist Martin Heinrich Klaproth (1795–1815), who dealt with the question of the composition of archaeological materials. He was the first to analyse larger series of archaeological, art historical and ethnological objects in order to answer cultural–historical questions. Above all, Klaproth analysed metal objects such as weapons, devices, coins and sculptures. Pernicka (2014b, c) writes on the concern of the chemist Karl Christian Göbel, who in 1842 presented a publication "Ueber den Einfluß der Chemie auf die Ermittelung der Vorzeit" (On the influence of chemistry on the determination of the past) and at the same time gave a short history of provenance analysis of archaeological objects.

For several decades now, proven methods for provenance studies have been established, which initially aimed to analyse "geochemical fingerprints" of a limited number of minor and trace elements in metal artefacts and later of ores. This focus has been preserved to this day, with the knowledge that the pattern of minor and trace elements undergoing significant changes in metallurgical processes on the way from ore to metal. In addition, to date, both the number of trace elements and their detection limit has been significantly increased by much more sensitive measurement methods, especially in mass spectrometry.

In the last several decades, the measurement of the isotopic composition of lead, which is contained in varying amounts in all these objects, has proven to be highly useful for the purpose of provenance studies of archaeological objects (lead isotope analysis will be abbreviated henceforth as LIA). The main principle of this method is the measurement of an isotopic fingerprint of the lead in metal artefacts (bars and ingots, slag and other intermediate products) and to then compare it to those of ore deposits, i.e. to search the deposits for a match between the isotopic fingerprint of the artefact and of the deposit. LIA is not limited to lead ores or lead artefacts. Lead can be found in nearly all oxide and sulphide ores, which play a role in archaeometallurgy. It exists, as a natural impurity, in, e.g. copper, silver and even in iron ores.

The key advantage of this method is that unlike the minor and trace elements, the lead isotopic fingerprint does not change much on its way from ore to metal: all chemical and physical processes along this long way do modify the isotopic composition of the lead only to a non-significant extent. Identification of possible ore sources from which metal artefacts have been produced requires reference databases of ores and metals, which in the meantime comprise several thousand analyses. The basics of LIA, analysis methods and types of mass spectrometers are described, e.g. by Gale and Stos-Gale (2000) or by Klein (2007).

The utilisation of alternative isotope systems of elements has become increasingly important in the last 20 years due to the development of high-resolution multiple collector inductively coupled plasma mass spectrometry (MC-ICP-MS). These are systems of tin, copper or osmium, iron, helium and strontium. They now can be investigated with high precision. The last three are non-traditional isotope systems which are of limited use in archaeometallurgy.

11.2 Trace Element Analysis

The analysis of trace element contents in archaeological metal artefacts is a central component for determining their so-called geochemical fingerprint. Trace elements differ from the main and minor elements due to their lower concentration (c. <0.1 wt.% or 1000 ppm).

The distribution of trace elements and of main and minor elements on the way from ore to metal is controlled by technological–metallurgical processes according to specific geochemical criteria. Pernicka (1999) was the first to correspondingly classify them for copper and its most important alloying components. Schmiderer (2008) expanded them for the extraction of gold.

11.2.1 Grouping of Elements

Pernicka (2014c) compiled the current version in three groups (Table 11.1).

Table 11.1 Grouping of selected chemical elements with regard to their importance in defining metallurgical techniques and provenance studies

Technology (lithophile elements)	Technology and/or provenance (smelting sensitive elements)	Provenance (robust elements)
Al, B, Ba, Be, Ca, Cr, Cs, Fe, Ga, Ge, Hf, K, Mg, Mo, Na, Nb, REE, Si, Sr, Ti, U, W	As, Bi, Cd, Co, Cr, Fe, Hg, In, Mn, P, Sb, Se, Te, Tl, Zn	Ag, Au, Bi, Co, Ir, Ni, Os, Pt, Pd, Rh, Ru, Cd, Hg, Sb
Sn > c. 1% Zn > c. 2% Pb > c. 5%	Sn < c. 1% Zn < c. 2% Pb < ca. 5%	

The table is mainly based on different reducibility of the elements from their oxides. Values for Sn, Pb and Zn are the approximate limit for deliberately produced alloys. Slightly modified following data from Pernicka (2014c)

The first group are elements that are separated from the originating metal during metallurgical processes. These are elements that are not reduced from their oxidic state under the redox conditions obtained in ancient times (cf. Fig. 4.15). They are called lithophile elements. They mainly offer technological information. Most of the elements analysed today in archaeometallurgical studies are of this group. Technological information about redox conditions can be obtained, e.g. from the degree of oxidation of iron (Fe^0, Fe^{2+} and Fe^{3+}), which can be present as a metal, as Fe^{2+}-siliceous phases (fayalite) or as Fe-oxides (magnetite). As main or minor elements they are components in oxidic and siliceous slags.

Note the conditions for the contents of tin and lead as alloying materials since the Bronze Age and of zinc since the late Hellenistic period, as mentioned in Table 11.1. In the left column (Technology) their values should respectively be >1, >2 and >5 wt.%, to be recognised as alloys. Accordingly, in the central column, their values must be lower to rule out an intentionally created alloy and thus a falsification of the element pattern of the original ore.

In the second group, there are elements listed that can be used for provenance studies or for technological reconstructions. They are called smelting sensitive (Schmiderer 2008). Their concentration decreases in the extraction and processing of metals, as they either were slagged or volatilise in dependence of redox conditions during (s)melting. Smelting-sensitive elements may also form already in ores characteristic geochemical fingerprints which are to recognise in the metal produced. It may provide information of ore types (oxidic, sulphidic).

The third group compile robust elements. These are suitable for provenance studies. These are not only main and minor elements such as copper, silver and gold but also a couple of trace elements as listed in Table 11.1. These are elements which are not oxidised and their concentrations do not change significantly from ore to metal. These offer significant information for provenance studies of metals.

Altogether, most of the elements shown in Table 11.1 can be used to gain information on technological aspects and only a few elements are useful for provenance studies.

11.2.2 Developments

A global overview of previous research activities on trace element analysis of archaeological metal artefacts has been compiled by Pernicka (2014a, b, c). Local focal points were the cities Halle and Stuttgart (Germany), Vienna (Austria), Moscow (Russia) and Baku (Azerbaijan).

In the 1930s, the scholars H. Otto and W. Witter started to analyse c. 1300 Neolithic to Bronze Age metal artefacts in Halle (Otto and Witter 1952). Herewith a significant stage of Archaeometallurgy was beginning. The aim of their research was to find out whether metal was produced using local raw material sources in prehistoric times in Germany. This was initiated by the fact that it was now possible to do chemical analyses on comparatively small amounts of metal, due to the newly established atomic emission spectroscopy. The authors analysed ten elements (Fe, Co, Ni, Cu, As, Sn, Ag, Sb, Pb, Bi), and in addition sulphur. They classified six groups, arranged in chronological order as follows: (1) very pure copper; (2) unalloyed copper with minor impurities; (3) arsenical copper; (4) fahlore metal (*Fahlerzkupfer*) high in arsenic, antimony, nickel and silver; (5) copper high in arsenic and nickel; (6) copper–tin alloys. They classified these as *Leitlegierungen*. Unfortunately, they had included only a few ore samples in their investigations, although they were well aware of the importance of such samples and the meaning and importance of mining–archaeological investigations. They furthermore assumed that ore deposits were homogeneous and that the trace element pattern would not be affected by smelting processes. They reached the conclusion, possibly influenced by political factors, that the majority of the analysed artefacts had their origin in deposits of Saxony. It should be emphasised that the work of Otto and Witter was a pioneering achievement both from the methodological approach and from the archaeological objective.

At around the same time, R. Pittioni and E. Preuschen did their research in Vienna (Preuschen and Pittioni 1937). They pursued the same goals, but in addition carried out mining–archaeological field work on old ore deposits in Austria. They analysed c. 2000 ore samples and c. 6000 metal artefacts from Bronze Age Europe. They were well aware of the difficulties of inhomogeneities of ore deposits and metallurgical processes and were therefore convinced that semi-quantitative analyses would be sufficiently informative. It would be important to find a characteristic pattern of elements. They classified the data into five groups and tried to coordinate it with ore deposits in the eastern Alpine region and in Slovakia. There were heavy debates about the proper methodological approach between Otto and Witter and the Vienna group which was deepened later by the following huge research project of S. Junghans, E. Sangmeister and M. Schröder in Stuttgart (Junghans et al. 1960, 1968, 1974).

In their opus *Studien zu den Anfängen der Metallurgie* ("Research on the beginnings of metallurgy", widely called the SAM project) they pursued a different concept. They analysed metal artefacts not to search for the origin of raw metals, but simply to use the chemical analyses of metal objects as an independent criterion for classification in addition to conventional typological descriptions. They did not try to compare metal artefacts with mineral deposits. They performed 22,000 analyses on Bronze Age objects distributed over Europe by atomic emission spectroscopy, as done previously by Otto and Witter (1952), and suggested a classification of 29 groups of metals to find chemically similar prehistoric copper projects. These groups were called the *Stuttgart Stammbaum*. It turned out that of all analysed elements, only five had any meaning and importance, namely Ag, Ni, As, Sb and Bi. One of the metal types widely used in Central Europe is the fahlore metal (*Fahlerzkupfer*), which was the source not only for the so-called ingot torques ("Ösenhalsringe") but also for the world-famous Early Bronze Age metal objects from Nebra (sky disc) (Pernicka 2010).

It must be emphasised that this project was done at a time when there were no computers yet. The enormous amount of data was only arranged in hand-drawn diagrams. Meanwhile, this dataset is considerably enlarged and comprises some 35,000 chemical analyses mainly of alloy artefacts from the Chalcolithic and Early Bronze Age of European countries and the Near East (Krause 2003; Junghans et al. 1960, 1968, 1974).

Independently, similarly large research projects were started by Selimkhanov (1967) in Baku (Azerbaijan), and by Chernykh (1966) in Moscow. Altogether, around 50,000 metal objects from all over Europe and Eurasia were analysed for their chemical composition.

Despite all criticism, these projects were able to provide important foundations for the development of prehistoric metallurgy. Accordingly, metallurgy in Europe and northern Asia began using native copper before the fifth millennium. Even afterwards, the extracted copper was very pure, presumably through the use of very pure ores. Arsenical copper followed in the fourth millennium. There is still some controversy as to whether this variety of copper was deliberately manufactured or automatically produced due to the supra-regionally enriched levels of arsenic at the surface of copper deposits and in soils (cf. - Chaps. 3.5.1 and 7.2.2).

Pernicka (1990) stated that changes of ores in chemical composition by metallurgical processes were actually of no major concern as long as metals from workshops nearby the original ore deposit were used for analyses where metals collected from smelting sites were processed and re-melted to larger units. This would provide a much better average composition than the original ore.

He argued on the basis of 25,000 analyses he obtained later and evaluated by cluster analysis. He then concluded that the groups of the *Stuttgart Stammbaum* would be valid. The workshop recipe sounds very reasonable and is confirmed by archaeological excavations and analytical measurements, e.g. of finds from the Early Bronze Age copper workshop at Khirbet Hamra Ifdan in the Faynan area (Levy et al. 2002; Hauptmann et al. 2015). At this large workshop copper ores exploited from the c. 150 mines in the Faynan area were smelted and re-melted and the herewith homogenised "Faynan copper" was cast to bar ingots.

11.2.3 Distributions of Trace Elements

Despite the modern technological advances for chemical analyses, in many cases, the compositions of artefacts are hardly enough to securely match specific ore sources. Ottaway (1994) identified key problems when doing compositional analyses to identify metal sources of objects. These were further discussed and deepened by Pernicka (2014a, b, c). These arguments are based on aspects of the geochemistry of ore deposits and on metallurgical effects. Mainly focused on copper, these are:

1. Ores from one and the same deposit may have differing minor and trace element compositions because as a rule, ore deposits are not homogeneously composed (see Chap. 3.3). Ore formation processes usually result in primary and secondary zoning. Hydrothermal veins of lead, silver, copper and zinc show a primary zoning. This is a primary genetic effect that is based on the precipitation of specific ores at different temperatures. This means that the mineral contents (and thus also the minor and trace element patterns) can vary depending on the depth, even within one vein. Due to the influence of climatics at the outcrops of ore deposits, primary (sulphidic) ores are forming oxidation zones with (leached) gossans or iron huts. This is secondary zoning. In this part of an ore deposit, various elements are dissolved and precipitated again in different proportions in the secondary enrichment zone at the groundwater level (see Figs. 3.46 and 3.58). Examples are copper, cobalt, nickel, silver and gold, while zinc and lead are stable in the surface area. The intensity of climatic agencies on outcrops of ore deposits is dependent on climatic regions, which vary globally and over the course of geological time scales.
2. Ores with characteristic trace element proportions may originate from locally different deposits. However, the origin of ore deposits, despite their enormous range of variation, follows global geological–geochemical processes. Lead–zinc–copper–silver and also gold deposits or copper ores high in arsenic and nickel may be comparable across regions, whether these are located in Middle Europe, Middle East, Central Asia or the Americas. The last group are mineralisations associated with ultrabasic rock complexes (ophiolites), such as those of Cyprus or Magan/Oman. Here there are typical ore deposits high in arsenic and nickel. They are located in the Tethyan Eurasian metallogenic belt (TEMB, Chap. 3.4.2) and may have similar age and genesis. Their age differs from those deposits that formed in the Precambrian cratons and Phanerozoic belts of the Arabian shield, northwest India or in China.
3. Not enough research about the behaviour of different trace elements during roasting, smelting, melting and re-melting and casting processes has been carried out. This is because

there are hardly any archaeological findings in which a complete *chaîne d'opératoire* can be reconstructed, even if that is often suggested. Of course, ores are smelted by different metallurgical processes, depending on whether they are oxidic or sulphur-containing ores. As shown in detail in Chap. 6 (6.2 especially for copper), the temperature and redox conditions of individual metallurgical processes can vary considerably. This has direct consequences on the distribution of minor and trace elements between the formed slag and metal. The Ellingham diagram discussed in Sect. 4.3.2.1 shows these relationships for a series of metals and semimetals.

4. The composition of copper objects with minor elements such as silver, arsenic, tin or lead is often not homogeneous. During casting, segregation effects show depletion or enrichment of these elements leading to gilding, silvering or tinning processes.
5. Metal is a recyclable material and any mixing and re-melting of scrap metal or objects originally from different deposits can bias the results.
6. The addition of another alloying metal changes the original composition of trace elements in objects. Deliberately produced alloys cannot always be clearly defined. This might be true for tin bronzes. Also, arsenic, lead or zinc concentrations in copper are difficult to understand whether added deliberately or not (see Chap. 7).

The discussion above is mostly related to copper. Trace elements of some other metals will be discussed below.

11.2.4 Silver and Lead

Native silver is rather rare in nature. Even though the noble meal appears as rich grave goods in Mesopotamia (Royal Cemetery of Ur, Zettler and Horne 1998) or on the Iberian Peninsula (Hunt Ortiz 2003), it was generally much less common in European and Middle Eastern prehistory.

Since the fourth millennium BC, silver has been extracted from silver-bearing lead ores. The most important silver-bearers were galena, cerussite, various fahlores or, in the case of Rio Tinto, jarosite was discussed as a source for silver. When galena was used, silver was extracted in a two-step process (see Chap. 6.4): Lead ore was first smelted using charcoal to produce silver-containing lead. This was liquefied under oxidising conditions and a lead oxide (litharge) was produced. The noble silver was separated from the lead oxide (cupellation) in its metallic state. Due to the processes conducted alternatingly under reducing and oxidising conditions, however, the trace element pattern changed drastically from the lead ore to the silver.

In case silver ore was low in lead, e.g. in the case of jarosite, foreign lead had to be added to separate the silver as described. The silver might still have the trace element signature of the ores smelted, but it will have the isotopic signature of the foreign lead added, and not the one of the original silver ore. It is suggested that large amounts of silver were extracted from jarosite in the first millennium BC at Rio Tinto by adding foreign lead from Cartagena (Craddock et al. 1985; Salkield 1987; Rothenhöfer et al. 2016; see Chap. 6.4).

When silver-bearing copper ore was mined, as was the case in the late Middle Ages in Central Europe, e.g. by the Fugger company in the Inn valley or in Slovakia, then a sort of copper was produced which contained silver. This was then liquefied again with the addition of lead. The lead acted as precious metal collector and separated from the copper. The principle of this procedure is called liquation (Chap. 6.2). Both in the Inn Valley and in Slovakia, lead from other reservoir districts was added. The de-silverised copper now no longer shows the trace element pattern of the copper deposits, but that of the lead. In the case of the liquation of the silver-bearing copper used in Slovakia, Hauptmann et al. (2016) were able to prove the origin of the used lead to be the ore district of Cracow.

The analysis of trace elements is therefore hardly sufficient for provenance studies on silver and lead objects. The analysis of lead isotope ratios is better suited for this purpose. Pioneering research on this topic was done within the scope of extensive mining–archaeological and archaeometallurgical fieldwork in Anatolia by scientists of the Max Planck institutes in Heidelberg and Mainz (Germany). The team of Gale and Stos-Gale in Oxford set up a databank of lead isotope analyses from Cyprus and many other countries in the Mediterranean. Their work on the systematics of lead isotope analysis are detailed in Sect. 11.3.1.

11.2.5 Gold

Natural gold is often relatively pure except for its almost regular silver content, which can reach several 10 wt.%. In addition, in many deposits gold is associated with copper ores, such as in the (Trans-)Caucasus (Hauptmann et al. 2010; Jansen et al. 2018b) or in Egypt and Nubia (Klemm and Klemm 2013), so that copper is hardly separated fully from the gold, even if it has its origin in placers. Whether copper is a specific (trace) element of ore deposits, or whether it was added intentionally by alloying, as is common practice, has to be examined carefully. This is by no means always clear. Such a differentiation also applies to lead, which can be introduced into the gold with copper.

Hartmann (1970, 1982) successfully used atomic absorption spectroscopy to analyse several thousand archaeological gold artefacts and to classify them into different groups based on their chemical composition. Their distribution in time and space was the basis of cultural–historical conclusions. He systematically analysed Ag, Cu, Sn and Ni. In some cases, he was able to detect Pb, Bi, Sb, Pt, Hg, Zn and Te. However, the analyses were critically accepted due to a lack of comparative analyses of natural gold, and thus due to a lack of knowledge about paragenetically associated elements with gold.

When it was discovered in the 1970s that prehistoric gold often contains inclusions of platinum metals (PGM with Pt, Ir, Os, Ru), new possibilities for provenancing of gold seemed to be emerging. Thus, it seemed advisable to determine the origin of gold artefacts by corresponding trace element analyses. Guerra and Calligaro (2003) followed this suggestion and recommended the chemical elements Pt, Pd, Ru, Sn, Zn, Te and Sb as tracers for provenance analyses. The authors performed elemental analyses via micro-beam particle-induced X-ray emission (PIXE) and particle-induced γ-ray emission (PIGE). Guerra and Calligaro (2007) analysed gold artefacts from the Visigothic treasure of Guarrazar and Visigothic coins from the seventh century AD in Spain and natural gold by PIXE and PIGE. They were able to find the provenance of the metal used for making the coins and objects from raw sources in the Iberian Peninsula. Unfortunately, neither chemical analyses nor the isotope analyses of osmium from PGM inclusions yielded in fact usable results. The scattering within the geological resources was too large so that no differentiation of deposits seemed possible. However, it will be shown in Sect. 11.3.4 that PGM can be dated geologically by means of Os-isotopy, thus determining the age of the associated ophiolite complexes. This implies possible regional–geographic constrictions.

Lead isotope analysis has been used as a tool to identify sources of copper, silver and lead. Unfortunately, it has hardly been used as a tool for studies on the origin of gold. There have been concerns that this method is hardly suitable because gold usually contains very little lead (Pernicka 2014a, b, c). However, in the meantime, the highly sensitive possibilities of laser ablation MC-ICP-MS have significantly increased the chances of sampling gold artefacts and gold nuggets in an almost non-destructive way. This was shown by Bendall (2003) and Bendall et al. (2009) in their investigation of Celtic gold coins (rainbow cups). Reconnaissance elemental analyses performed by Kamenov et al. (2013) revealed that vein gold and placer gold nuggets from a site in Arizona contain up to thousands of ppm lead and therefore would be suitable for lead isotope analysis.

In order to check possibilities and limitations of trace element analyses and of lead isotope analyses Jansen (2019) investigated several dozens of gold artefacts and natural gold both from the Royal Tombs of Ur, Mesopotamia, and from gold sources in the (Trans-)Caucasus area. He could use high-resolution mass spectrometry to analyse major elements (Ag, Cu, Ag) and 13 minor and trace elements (Co, Ni, Zn, As, Se, Pd, Cd, Sn, Sb, Te, Pt, Pb, Bi). The concentrations of these elements are exemplarily illustrated in Fig. 11.1. The diagram shows that gold artefacts from Early Dynastic IIIa Ur are especially high in tin (Sn), platinum (Pt) and lead (Pb).

Jansen came up with the conclusion that the composition of gold artefacts not only would be affected by the manifold heavy minerals especially of the Hinterland of placer gold deposits (cf. Fig. 3.62). It would be more affected and considerably modified when all the impurities in the placer gold would be melted together by metallurgical operations in workshops. With the exception of silver, all other impurities ended up in the gold artefacts not before the metallurgical processing. It might be therefore difficult to directly compare natural gold grains with gold of artefacts. It would make sense to address this problem with "metallurgical fingerprint" rather than "geochemical fingerprint".

Unfortunately, neither the chemical analyses nor the isotope analyses of osmium from PGM inclusions yielded usable results. The scattering within the geological resources was too large, so that no differentiation of deposits seemed possible. However, it will be shown in Sect. 11.3.4 that the PGM can be dated geologically by means of Os-isotopy, thus determining the age of the associated ophiolite complexes. This implies possible regional–geographic assignments.

With the advancement of analytical techniques by means of laser ablation and plasma mass spectrometry, it is now possible to analyse small gold objects almost non-destructively or to carry out meaningful multi-element analyses on very small samples. Since a non-destructive analysis of gold objects has become a mandatory requirement on the part of almost all museums due to their high cultural and historical value and for conservational reasons, this may open up new perspectives for gold provenance studies.

11.2.6 Iron

In comparison to copper and other base and noble metals, iron is the most widely distributed metal. Corresponding mineral deposits are densely located worldwide and theoretically may have been available everywhere as local sources of raw material for iron and steel.

Of course, there are outstanding examples of iron ore deposits that played important roles in research on early metallurgy in Europe, such as the island of Elba and opposite on the mainland Populonia (Benvenuti et al. 2013), the Noricum in Austria (Straube 1996), the numerous iron ore mineralisations in the northern European lowlands from Denmark over Germany to Poland, in Central and Southern Germany, the extensive ore districts in southern and central Sweden and others (see overview in Pleiner 2000). Many of these deposits are geochemically very similar since they are usually sedimentary bog iron ores, ferricrust and bean ores, which are genetically similar. But they can be different in their lead isotopic composition. This aspect is still largely unknown in archaeometallurgy.

Geochemical–archaeometallurgical characterisations of iron objects are evaluated differently. For instance, Beukes et al. (2003) defined hematite (and other Fe-oxide minerals) as "a mineral with . . . simple chemical composition that reveals little about its origin. . .". Even though an increasing number of articles in recent years have been devoted to the study of trace element geochemistry of the main iron ore minerals, including hematite, magnetite, Fe hydroxides (goethite) and siderite (e.g. Schwab et al. 2006; Desaulty et al. 2009; Benvenuti et al. 2013), other authors are of quite a different opinion and follow the statements of Beukes and coworkers. They continue to maintain that the trace element geochemistry is of little importance for provenance studies. Also, Blakelock et al. (2009) stated in a summary

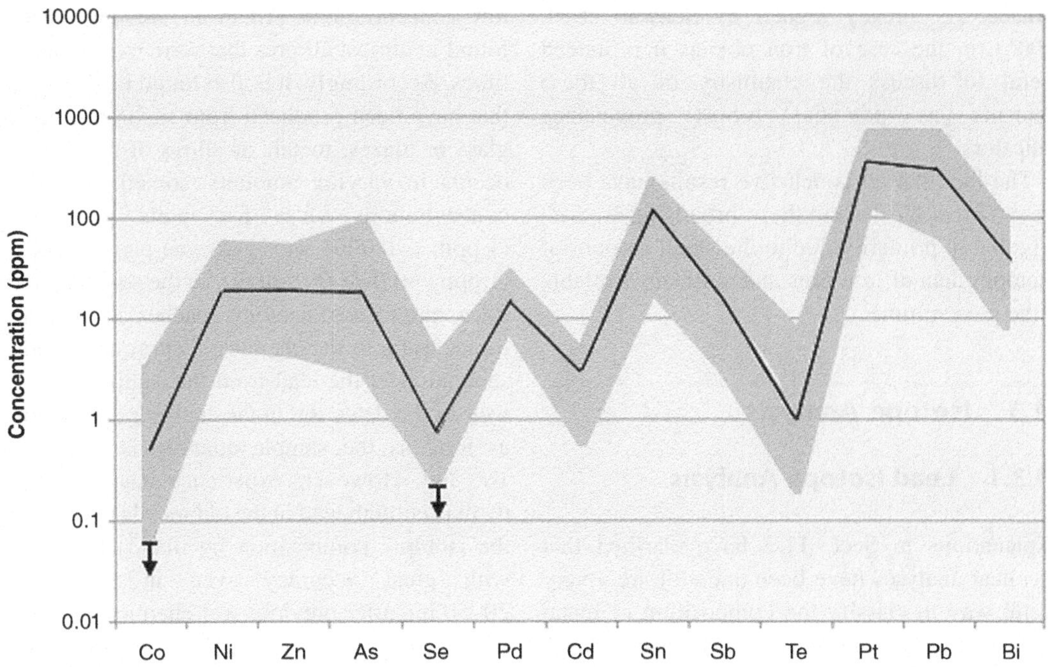

Fig. 11.1 Trace element pattern of gold artefacts ($n = 43$) from Early Dynastic IIIa Ur, Mesopotamia. Averages (continuous line) together with maximum and minimum values (dotted lines). Note the high values for tin (Sn), platinum (Pt) and lead (Pb). For Co and Se the minimum is below the detection limit (arrows). Analyses by LA-ICP-MS. From Jansen (2019)

that even the study of slag inclusion on a chemical basis have mostly proven inconclusive.

In contrast, Schwab et al. (2006) previously already pursued a multi-method approach in order to trace back the origin of iron used to make Celtic objects found at Manching in southern Germany. Like several previous scholars, they examined the chemical composition of siliceous slag inclusions in iron. For in the extraction of iron and steel, these are deposited predominantly in the solid state, so that, unlike with other metals, there is hardly any complete separation of metal and slag. It is almost the rule that blooms and iron artefacts show inclusions of slags in the micro-range, which can be used as indicators not only for associated ore minerals and gangues but also for host rocks. However, they used electron microprobe analysis not only on the main components of Fe-Si-Ca-Al-Mn-oxide inclusions but also on minor and trace elements such as Cr, Sr, Ce, Zr, Ti. Slag inclusions are believed to provide data with the potential for provenancing iron artefacts. Schwab et al. (2006), in addition, analysed lead isotope ratios of iron artefacts from Manching and of bog iron ores also from nearby sources. They found that local bog iron ores could easily have been the raw source for these iron artefacts. Degryse et al. (2007) suggested to combine lead isotope data of iron ores with strontium isotopes.

Another example is studies carried out by Benvenuti et al. (2013) on iron ore deposits of the Mediterranean island of Elba and the smelting sites of Populonia, located just opposite to the island on the Italian continent. The authors point out that the crystal chemistry of the Fe-oxides from this ore deposit would dictate an uptake of significant lithophile elements and minerals into the iron of quartz, remains of calcite, dolomite, feldspars, garnet, pyroxene, pyrite, pyrrhotite (Ca, Mg, Cr, Sr, W, Ti, Mo, U, V, REE, Th; Rb) etc. (see Table 11.1) as well as chalcophile elements of galena or sphalerite etc. (Cu, Zn, Pb, Sb etc.) of the ores charged into a smelting furnace. As already argued by Schwab et al. (2006), in the case of iron objects it is indeed useful to discuss the chemistry of all these elements as potential robust provenance indicators.

The fact that no conclusive results have been presented so far despite these promising studies, is probably primarily due to the small amount of (isotope) data of iron ores and artefacts available at the present time.

11.3 Isotope Analysis

11.3.1 Lead Isotope Analysis

Explanations in Sect. 11.2 have clarified that chemical analyses have been and still are a very useful way to classify the composition of metal artefacts. But there are problems in using chemical analyses alone to assign metal objects to particular raw sources, i.e. to perform provenance studies. These are basically the geochemical variability of ore deposits and the fractioning of trace elements during metallurgical processes from ore to metal.

The isotopic composition of lead is one geochemical device which does not have these disadvantages. Lead isotope analysis (LIA) is the most powerful tool for provenance studies of metal artefacts in archaeology. The principle of the method is to compare an isotopic fingerprint of lead in artefacts with that of lead from different deposits and to search among those deposits for those whose isotopic fingerprint coincides with that of the lead in the artefact. Naturally, such ore deposits, in which mining activity is attested at the same time as the production of the metal object and which are in closer cultural contexts, should be focused.

Lead is a ubiquitous element in ores and rocks. It accumulates either as an admixture of lead minerals or by incorporation into crystal lattices of rock-forming minerals. Lead is also a trace element in all kind of rocks. It is a dominant constituent of lead–silver–zinc deposits, but also occurs in arsenic, antimony, tin and mercury ores, whereas in iron and gold ores it occurs only in low concentrations. All in all, however, lead is found in almost all ores that were used in ancient times. Accordingly, it is also found in all products that have been produced from such ores, be they glass or glazes, metals or alloys of metals. Lead occurs in varying amounts, sometimes only in concentrations of a few parts per million (1 ppm = 0.0001 wt.%) or even parts per billion (1 ppb = 0.0000001 wt.%). In the so-called ultra-trace range, lead isotopy reaches its analytical limits, as far as the chemical ("chromatographic") separation of the lead from the sample matrix as well as the detection in the analyser is concerned, as long as the sample quantity is not several 100 mg. However, experience has shown that there is enough lead in the old metals to determine the isotopic composition by mass spectrometry with great accuracy even in samples of 20–50 mg after previous wet-chemical extraction of the lead.

Most copper, bronze, or silver artefacts contain lead at a level of 10–1.000 ppm (=µg/g) (Fig. 11.2). LIA is destructive, but due to the relatively small amount of material necessary for analysis it can be used for provenance studies of many valuable objects, even of coins, and in most cases still leaves enough material for a chemical analysis.

LIA has been used before in geosciences to understand the geological history of ore formation (plumbotectonic model of Doe and Zartman (1979). LIA was rapidly taken up by archaeology, initially to assign lead and silver artefacts to particular mining regions or ore deposits. The first archaeological objects, namely Roman lead pipes and ingots, were analysed by Grögler et al. (1966). A history of the application of LIA in archaeology has been compiled by Pernicka (1990, 2014c), Gale and Stos-Gale 2000) and Stos-Gale and Gale (2009). At the beginning, the scholarly acceptance of the LIA in archaeology was similar to that given to Libby's radiocarbon dating, which had initial problems but became very commonly used in general during the 1950s. Since at least the 1970s, LIA was greeted as a new, "secure" method for provenance studies of metal artefacts (Muhly 1995). However, its advantages have been soon contested.

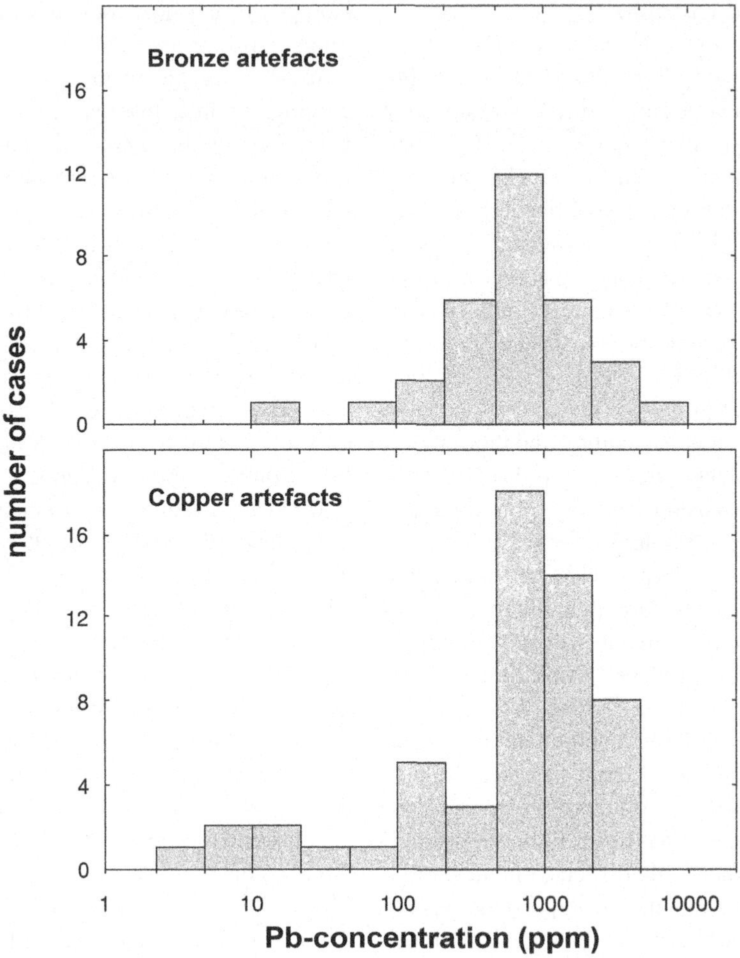

Fig. 11.2 The histogram shows lead concentrations in Chalcolithic and Bronze Age copper and bronze artefacts, and proves that lead isotope analysis is very appropriate for provenance investigations of these objects. Modified after Begemann and Schmitt-Strecker (2009)

In the last decades, it has been mainly the Heidelberg-Mainz group (Max-Planck-Institutes, F. Begemann, E. Pernicka, S. Schmitt-Strecker, G.A. Wagner) and the Isotrace Laboratory of the University of Oxford (Z.A. Stos-Gale, N. Gale) that carried out lead isotopes analyses of metal artefacts and ores from many sites and ore deposits in archaeological contexts not only in the Old World, though also, e. g. the Conservation Analytical Laboratory, Smithsonian Institution, in Washington (R.B. Sayre), the Geosciences Department, Università degli Studi di Padova (G. Artioli), and the FIERCE (Frankfurt Isotope and Element Research Center) at the Goethe University, Frankfurt am Main (S. Klein) have used this method.

The classification with only three or four isotope ratios is simple when compared to the concentrations of ca. 30 different chemical elements sometimes quantified in provenance studies, e.g. of pottery. But the small number, on the other hand, raises the probability that two different ore deposits might have matching isotope ratios and thus are frequently overlapping and make provenance studies problematic.

11.3.1.1 Lead Isotopes

Lead isotope analysis is based on four stable isotopes, namely ^{208}Pb, ^{207}Pb, ^{206}Pb and ^{204}Pb. The first three isotopes are formed by radioactive decay of uranium and thorium (^{238}U, ^{235}U, ^{232}Th) (Table 11.2). Minerals in the host rock containing uranium and thorium are the source of these three lead isotopes. ^{204}Pb is not radiogenic. It is primordial lead, and its abundance has been constant since the earth's formation some 4.6 billion years ago. The percentage of the four lead isotopes are approximately 52.4% (^{208}Pb), 22.1% (^{207}Pb), 24.1% (^{206}Pb) and 1.4% (^{204}Pb).

The initial content of uranium and thorium has changed due to their decay. This has led to a permanently increasing quantity of radioactive isotopes ^{206}Pb, ^{207}Pb and ^{208}Pb in the earth crust. If the lead resulting from the radioactive decay of thorium and uranium dominates in a lead-bearing sample, this is called "radiogenic lead". This is the case for $^{208}Pb/^{206}Pb < 2.04$, $^{207}Pb/^{206}Pb < 0.82$ and $^{204}Pb/^{206}Pb < 0.052$. However, a clearly defined delimitation does not exist (Begemann and Schmitt-Strecker 2008b). The relative frequency of these isotopes is not constant throughout the world, because uranium and thorium are unevenly dispersed in the earth's crust. The isotope abundances depend also on the length of time during which radiogenic isotopes were able to form in the host rocks. Generally, lead is geochemically separated from uranium and thorium during the formation of lead ore deposits.

As explained, e.g. by Gulson (1986) and Evans (1992), the solubility of lead in fluids exceeds the one of the parent isotopes uranium and thorium. Due to this fact, there is often already a separation between parent and daughter isotopes in the host rock. The solubility of uranium, thorium and lead also depends on which minerals they are incorporated in. If fluids are cooling off in a magma, the different chemical behaviour of the dissolved elements leads to a separation of the different cations and anions. During the crystallisation of galena, neither uranium nor thorium will be taken up in its crystal lattice. At this moment, lead isotope ratios are frozen and galena has a fixed ratio.

Table 11.2 The formation of the three radiogenic lead isotopes by radioactive decay of uranium and thorium, and their half-life

Decay	Half-life ($\times 10^9$ a)
$^{238}U \rightarrow {}^{206}Pb$	4.468
$^{235}U \rightarrow {}^{207}Pb$	0.704
$^{232}Th \rightarrow {}^{208}Pb$	14.01

LIA is a sensitive analysis of masses, and isotope ratios are measured with a mass spectrometer. Their high sensitivity also resulted in the utilisation of mass spectrometry in microanalyses. For routine measurements, 250 ppb lead (0.000025%) would be sufficient. The lead isotope standards NBS SRM 981 and 982 of the National Bureau of Standards are used for the LIA in all laboratories. In geoscience mass spectrometers and software are usually calibrated for ^{204}Pb, because this isotope contains no radiogenic particles. But the isotopes ^{206}Pb, ^{207}Pb and ^{208}Pb can be measured more accurately due to their higher concentration. It turned out that in archaeometallurgy, reasonable results for provenance studies were achieved when the following three isotope ratios were taken in comparison: $^{208}Pb/^{206}Pb$, $^{207}Pb/^{206}Pb$ and $^{204}Pb/^{206}Pb$. Today the relative standard deviation of measurements is less than 0.1%, depending on the abundances of the isotopes. In principle, all geological measurements normalised to ^{204}Pb can be compared with data published from archaeological studies if the appropriate precision and accuracy has been achieved. The development and introduction of Multicollector-Inductively Coupled Plasma Mass spectrometers (MC-ICP-MS) has dramatically improved the efficiency and thus the analytical quality of the data (Niederschlag et al. 2003; Barker et al. 2007; Albarède et al. 2012; Artioli et al. 2016, 2017).

The quality of a provenance study essentially depends on the completeness of the reference data and on the most precise possible characterisation of the lead isotope fields. The smaller and more homogeneous the lead isotope fields of deposits, the greater will be the precision of the assignment.

Intensive research has been conducted in both directions over the past decades.

11.3.1.2 Model Ages

The common lead dating method for model ages of rocks and ores is based on the evolution of lead since the time (T = age of the earth, 4.6 billion years) when the earth started to act as a closed system for the elements U, Th and Pb to the time when lead was isolated from these radioactive parent elements. The different half-lives of ^{235}U, ^{238}U and of ^{232}Th lead to the formation of the three lead isotopes ^{206}Pb, ^{207}Pb and ^{208}Pb (Table 11.2). These are suitable for determining the earth's age and geological developments, which have led, over millions of years, to the formation of ore deposits. There is a direct relationship between the isotope ratios and the age of an ore deposit.

Therefore, age determinations by lead isotopes allow an estimate of the approximate age of an ore mineralisation from which an artefact was made. And vice versa, the information from the isotopes is the basis on which to reject other deposits as possible sources for a particular object. One precondition for the viability of these results lies in the necessity to exclude with certainty that any other metal, including lead, had been alloyed during the production of this artefact (see below). Otherwise one would only describe the isotope characteristics of this added metal or those of the mixture between the lead of the initial and the added material.

The characterisation of a deposit by lead isotopes depends on the geological age, the percentage of ^{204}Pb in the ores and the concentration of the mother isotopes U and Th, but also on a possible remobilisation of the ore. Such a remobilisation appears particularly in epigenetic ores. It is therefore usual that the expected variations of the isotopic composition of the deposit are given as between 0.4% and 0.5% (Schmitt-Strecker et al. 1992; Stos-Gale 1993). This is clearly above possible experimental errors.

The age determination mainly of geologically very old raw sources can be calculated by the "Holmes–Houtermans" model (Houtermans 1960). The precondition for this model was based on the assumption that lead, which has to be dated, has a "one-stage-history", meaning a radioactive equilibrium and that the Pb/U ratio has not changed through time. But this model has often resulted in abnormally ancient dates, even in negative model ages, which have been caused by a mixing with radiogenic lead from uranium decay. This could be explained by, e.g. a closely located uranium deposit.

The single-stage model of Houtermans (1960), as well as the two-stage Pb evolution model of Stacey and Kramers (1975), depend on two parameters. The one is a measure of how U and Pb are separated through time in each geological province, $\mu = {}^{238}U/{}^{204}Pb$. μ-zoning is strong and robust, no matter how complex a metamorphic and hydrothermal history may be. The second parameter is $\kappa = {}^{232}Th/{}^{238}U$. They can distinguish geological provinces with high μ-values from those with low μ-values, and likewise tell apart geological provinces with high and low κ-ratios, μ and κ being highly characteristic geochemical parameters of metal sources (Villa 2009).

Stacey and Kramers (1975) therefore came to the conclusion that a development which happens in two chronological stages would be far more reliable in describing the percentage of the isotopes (Fig. 11.3). This is exemplified by the copper ore deposit in the Wadi Arabah Rift Valley and the regions on the Sinai Peninsula and the Eastern Desert in Egypt (Fig. 11.4). Ore samples of Faynan (Jordan) and Timna (Israel) are of only slightly different age. A cluster is between 500 and 600 million years, other compositions show up to c. 100 million years. Mineralisations of the Eastern Desert, Sinai, and of are showing even negative ages. The variety of compositions is due to a (repeated) remobilisation of copper in a uranium-containing sandstone environment.

For example, the geologically young circum-Mediterranean Alpine Belt (which is part of the global Tethyan Eurasian Metallogenic Belt (TEMB; Jovanović 1988) is mostly <90 Ma, but has incorporated geological old terranes with >> 100 Ma. This TEMB is in contrast to the old Hercynian basement (c. 300 Ma old) of the European Atlantic façade (Albarède et al. 2012).

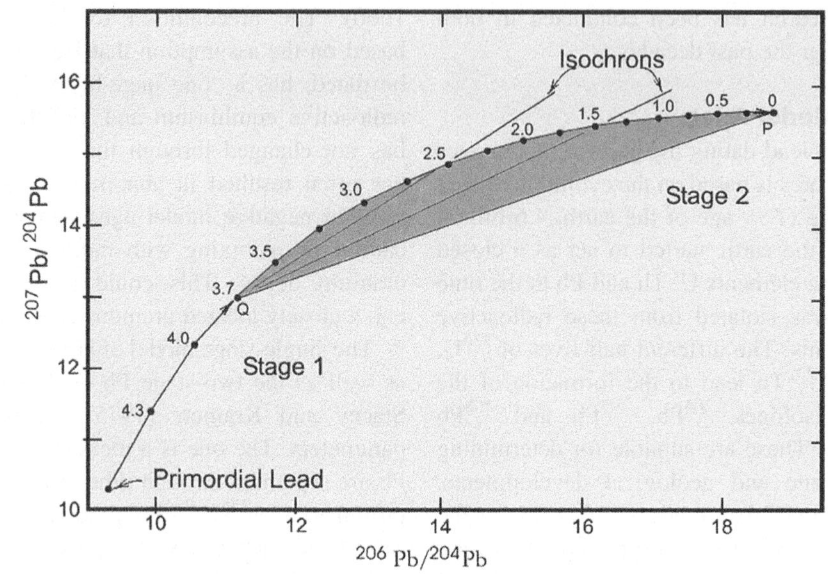

Fig. 11.3 Development of lead isotope abundance ratios from the decay of uranium dependent on the duration of time; depicted in the curved lines as a two-stage lead development model (after Stacey and Kramers 1975). Lead develops from a composition of the primordial lead of 4.57–3.70 Ba in a reservoir with $^{238}U/^{204}Pb$ (μ) = 7.19 (stage 1). At point Q (3.7 Ba) μ changes to 9.74 via geochemical differentiation. After that, lead develops undisturbed towards point P (present). The straight lines from point Q to varying points along the development line are isochrones. From Faure (1986)

The lead–silver ores of the Cyclade islands in the Aegean (Greece) were formed in recent geological times (<about 20 Ma ago), while the zinc–lead ores of Iglesiente (south-west Sardinia) were formed >500 Ma ago, during a much more distant geological past (Fig. 11.5). Accordingly, the lead isotope ratios of Greek ores and Sardinian ores are very different. Metal artefacts deriving from Greek mining, therefore, have a very distinct "lead isotopic fingerprint" differing from that of Phoenician mining in Iglesiente. Reconstructing trade routes is thus made possible by identifying the provenance of metal ores. LIA in the upper right corner of the diagram in Fig. 11.5 represent geologically old ore deposits, and the data which plots into the lower-left corner represents compositions from geologically young mineralisations.

Hence, LIA, in particular the model ages derived from LIA are useful for fingerprinting the tectonic event ages of geological formations of ore deposits.

11.3.1.3 Geoscientific Aspects of Lead Isotopy

In archaeometallurgy, the lead isotope composition of a sample is usually expressed through the isotope ratios $^{208}Pb/^{206}Pb$, $^{207}Pb/^{206}Pb$ and $^{204}Pb/^{206}Pb$, standardised for ^{206}Pb. In geosciences, however, the standardised ratios $^{208}Pb/^{204}Pb$, $^{207}Pb/^{204}Pb$ and $^{206}Pb/^{204}Pb$, standardised for ^{204}Pb, are commonly used. Here, lead isotopes prove in particular the model ages derived from the lead isotope ratios, they are at least useful for fingerprinting the ages of the geological ore formation (Albarède et al. 2012).

Among others, Baron et al. (2014) called for an increased consideration for the geological situation of the ore deposit in provenance studies. The authors suggest to consider the following points:

1. The genesis of an ore deposit is a process that can take geological ages to complete. This means that the precipitation of individual ores can take longer periods of time and can occur in individual waves over millions of years. A

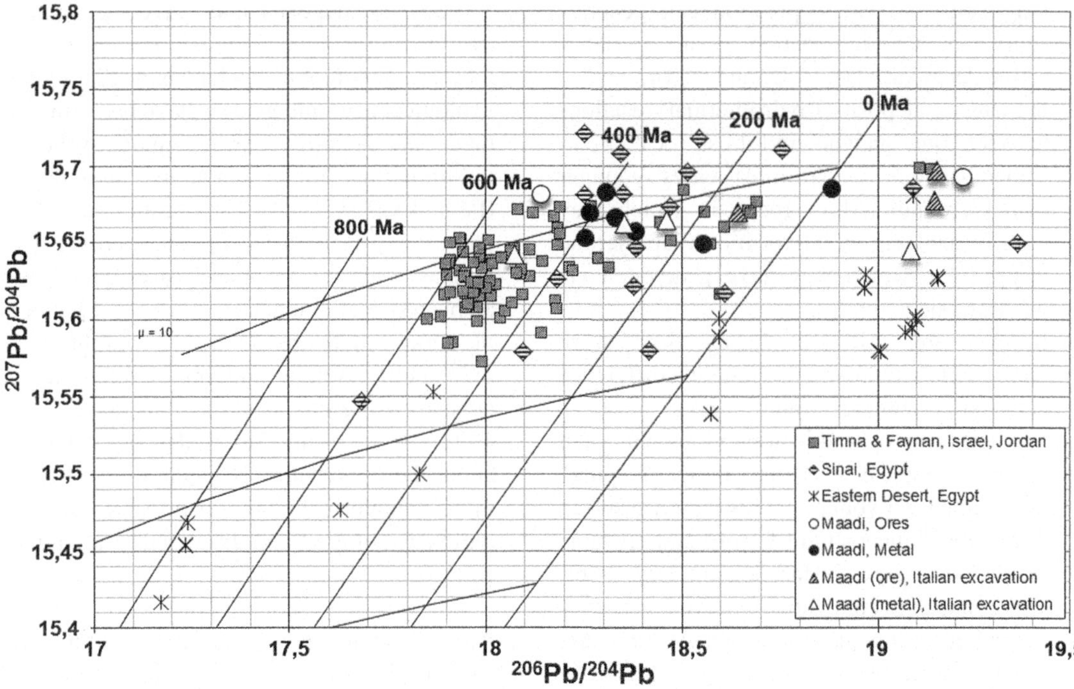

Fig. 11.4 Model ages of copper ore samples taken from the ancient mining areas at Timna and Faynan in the Wadi Arabah Rift Valley (Israel/Jordan). Plots after the two-stage model of Stacey and Kramers (1975). There is a cluster at the age of c. 500 million years which corresponds with a geological age of Cambrium. Varying ages between c. 600 and 100 million years indicate a (repeated) remobilisation of copper until the recent past in a slightly enriched environment of uranium-containing sandstone. For comparison isotope ratios of ores and metal objects from the Chalcolithic site of Maadi (Egypt) and of ores from the Sinai Peninsula and the Eastern Desert of Egypt. If lead is associated with uranium, the $^{206}Pb/^{204}Pb$ and $^{207}Pb/^{204}Pb$ ratios increase with time, from left to right along the curved μ-"development lines". This is especially true for ores and slags from the Eastern Desert of Egypt. Data from Gale et al. (1990), Hauptmann (2000b, 2007), Segal (2015), Asael (2010), Asael et al. (2012), Abdel-Motelib et al. (2012) and Ketelaer and Hauptmann (2016). From Hauptmann (2017)

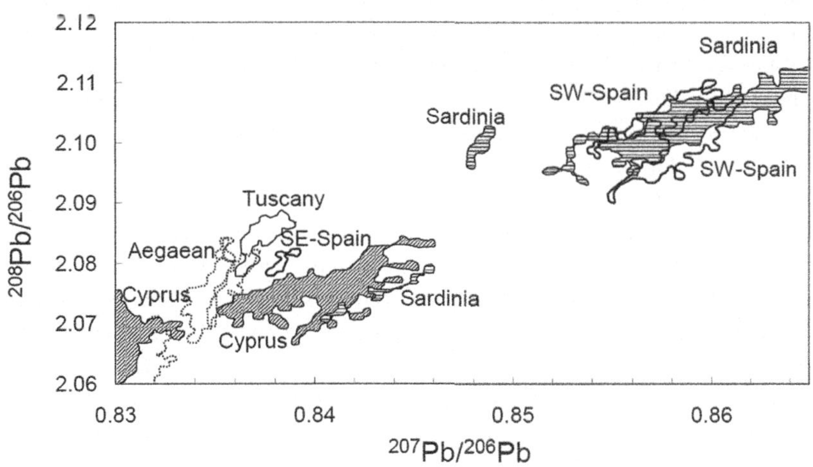

Fig. 11.5 Lead isotope abundance ratios of copper and lead ore deposits from various mining areas exploited in Roman times in the Mediterranean. The ages of the ore deposits range from Cambrian (high $^{207}Pb/^{206}Pb$ values) to Tertiary (low $^{207}Pb/^{206}Pb$ values). The apparent gap of lead isotope ratios between 0.847 and 0.856 corresponds to an age gap between the Variscan and the precursory of the Alpine orogeny in Europe. From Klein et al. (2009)

long period of ore deposition can generate significant isotopic variations related to radioactive decay of ^{238}U, ^{235}U, and ^{232}Th.

2. In many ore deposits, (hydrothermal) reactions between ore, gangue and host rock occur. Mineralisations are in most cases younger than surrounding host rocks in which they have formed, e.g. in the form of hydrothermal veins, or in which they have penetrated via metasomatic activities.
3. If ores are collected during field work intergrown with host rock (note the difference between gangue and host rock!), measurements of lead isotope ratios will yield higher geological ages than those of "pure" ores, which are younger. This would result in different isotopic patterns for one and the same ore deposit (see, e.g. Marcoux and Bril 1986). This can lead to confusion in provenance studies and raises the question of whether it would not be better anyway to carry out isotope measurements on metal inclusions in slags from nearby heaps for provenance studies. The argument of Pernicka (1990) is probably very useful, that in provenance studies it may be more useful to look for workshops than for ore deposits since it is much more likely to find a homogenisation of metal compositions in those places.
4. The formation of sulphide ore deposits leads to a separation of lead from its parent nuclides uranium and thorium. In contrast to lead, these cannot be included in sulphides. As a result, the lead isotopes are "frozen" at the time an ore deposit is formed (Villa 2009). Nevertheless, depending on the accuracy of the resolution of the mineral phases to be analysed, these deposits show either large variations in their isotopy or even multiple separate lead isotope fields. Another problem is deposits with "abnormal" lead isotopic data. They result from a subsequent input of radiogenic lead, e.g. via the addition of uranium and thorium by weathering processes or due to events that stop a complete separation of uranium and thorium (Hunt Ortiz 2003).
5. Various ore deposits may have identical or overlapping isotope compositions, even when they are geographically far apart. Since their geological ages and genesis have been similar, their lead isotopy does not show sufficient dissimilarities for a distinctive assignment (Pernicka et al. 1984). The question, which of those deposits served as the ore source, has to remain doubtful in those cases where no archaeological evidence could help. Members of American teams studied the problem of overlapping isotope patterns in different ore deposits using statistical methods (Sayre et al. 1992). It can be summarised that any statement about the provenance of metal artefacts is far more reliable if the area of possible origin is regionally limited and discussed in a cultural context.

Albarède et al. (2012) are pointing out that the set of lead isotope ratios mainly used in archaeometallurgy ($^{207}Pb/^{206}Pb$, $^{208}Pb/^{206}Pb$, $^{204}Pb/^{206}Pb$) would not be the ideal one. In the past, the choice of these ratios was motivated not by a particular interpretative property, but by the difficulty that has long prevailed in accurately measuring the less abundant isotope ^{204}Pb. It is well known in geochemical literature that in these plots modern magmatic events and ore formations tend to form very narrow alignments, which, in addition, tend to overlap strongly with each other and have limited resolutions. It would be useful to add the evaluation of $^{208}Pb/^{204}Pb$, $^{207}Pb/^{204}Pb$ and $^{206}Pb/^{204}Pb$ as it is practised by Artioli et al. (2017) exemplified in Fig. 11.6. This would allow a more robust definition of the major isotopic groups of and, hence, a better discrimination of metal extracted from these ore deposits.

Albarède et al. (2012) are afraid that the relationship between trace element concentrations remains largely descriptive and offer little correspondence with the particular geological background of the ores in question. This statement is not generally valid. It has been shown in Chap. 3 that there are several ore deposits that can already be classified well into specific geological units

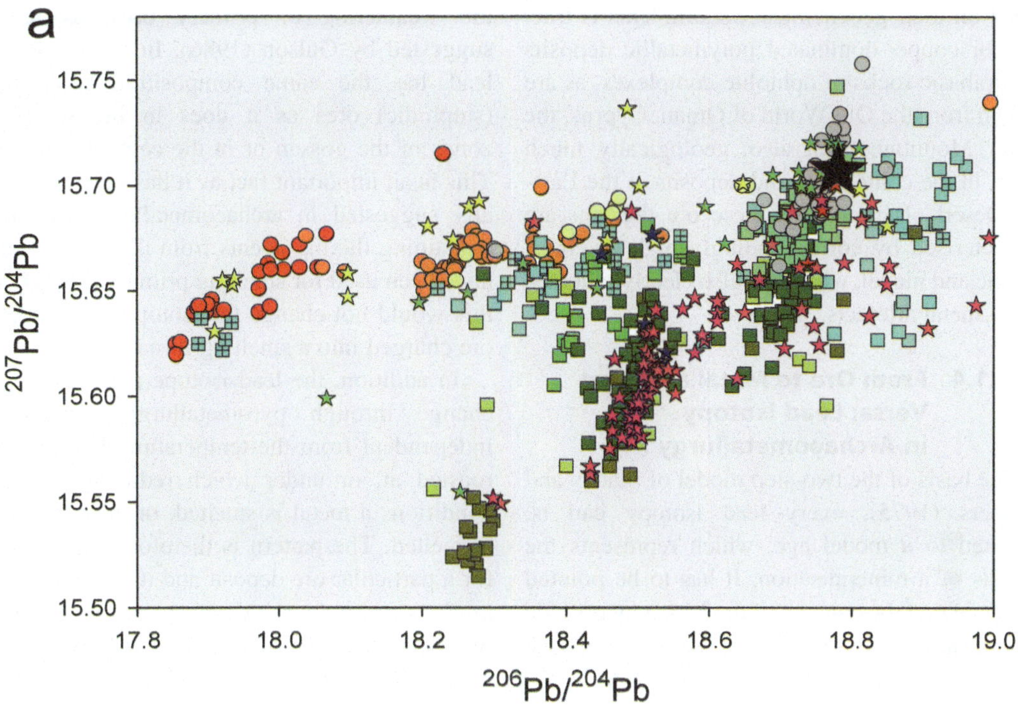

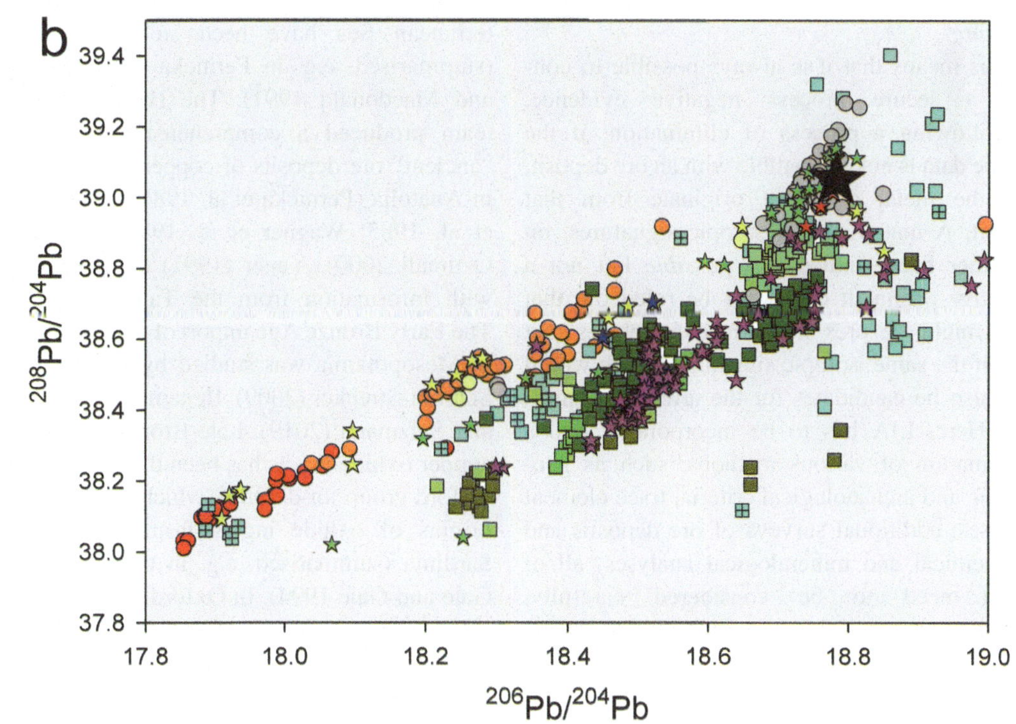

Fig. 11.6 Two-dimensional projections of lead isotope data of Neolithic and Copper Age artefacts from Italy, Tyrol and Serbia (late 5th millennium to the early 3rd millennium BC). Plots are showing the ratios (**a**) $^{206}Pb/^{204}Pb$ vs. $^{207}Pb/^{204}Pb$; (**b**) $^{206}Pb/^{204}Pb$ vs. $^{208}Pb/^{204}Pb$ plot. Note the plot of the copper axe of the iceman mummy from Ötztal (black star) which is compatible with the isotope ratios of copper deposits in Tuscany. Modified after Artioli et al. (2017)

based on their geochemical pattern. This is true, e.g. for copper-dominated polymetallic deposits in ultrabasic rocks of ophiolite complexes, as are known from the Old World of Oman, Cyprus, the Pamir Mountains, but also, geologically much older, in the context of gold deposits in the Eastern Desert of Egypt. All these ore deposits are characterised by correspondingly high levels of arsenic and nickel, which are also clearly reflected in the metal artefacts.

11.3.1.4 From Ore to Metal and Vice Versa: Lead Isotopy in Archaeometallurgy

On the basis of the two-step model of Stacey and Kramers (1975), every lead isotopy can be assigned to a model age, which represents the genesis of a mineralisation. It has to be pointed out again that the concept of the LIA may provide geological ages, but these ages are in no case the age of an archaeological artefact (Villa 2009). Due to this fact, deposits from different geological contexts and ages carry different lead isotope signatures.

This means that it is always possible to conduct a secure process negative evidence, i.e. following a process of elimination: if the isotope data is not compatible with an ore deposit, then the metal does not originate from that deposit. A match of the isotopic signatures, on the other hand, indicates a *possible* but not a definitive origin. It can never be ruled out that there might be ores from different places that exhibit the same isotopic signature, which would then also be candidates for the artefact in question. Here, LIA has to be incorporated into a combination of various methods, such as geographic and archaeological criteria, trace element analyses, additional surveys of ore deposits and geochemical and mineralogical analyses, all of which need to be considered carefully. Archaeometallurgical field work has more importance than ever in the study of evidence for mining and smelting processes.

As mentioned above, a defining advantage of LIA in comparison with provenance studies on the basis of trace elements is the fact that the isotope pattern of lead does not change during the weathering of primary ores, as it was suggested by Gulson (1986). In an ore deposit, lead has the same composition in primary (sulphidic) ores as it does in the weathered zones of the gossan or in the cementation zone. This is an important fact as it has been occasionally suggested in archaeometallurgy that iron-containing fluxing agents from the gossan might have been used for smelting primary ores. Hence, this would not change the isotopic pattern of an ore charged into a smelting furnace.

In addition, the lead isotope pattern does not change through pyrometallurgical processes, independent from the temperature that an ore is roasted at, or under which reducing–oxidising conditions a metal is smelted, or how often it is re-melted. The pattern is therefore characteristic for a particular ore deposit and its products, such as slags or artefacts. All have the same isotopic fingerprints. In the ideal case, it allows a secure assignment of the finished product to the initial raw material.

The metallogenic regions of the Eastern Mediterranean Sea have been studied intensively (summarised, e.g. in Pernicka 1987; Stos-Gale and Macdonald 1991). The Heidelberg–Mainz team produced a comprehensive overview of "ancient" ore deposits of copper, lead and silver in Anatolia (Pernicka et al. 1984, 2003; Seeliger et al. 1985; Wagner et al. 1986; Wagner and Öztunali 2000). Yener (1991) complemented it with information from the Taurus Mountains. The Early Bronze Age import of copper and silver in Mesopotamia was studied by Begemann and Schmitt-Strecker (2009), Begemann et al. (2010), and Salzmann (2019). Late Bronze Age trade of copper oxhide ingots has been the research of the Oxford group for decades, which also studied the origins of oxhide ingots from Cyprus and/or Sardinia (summarised, e.g. in Gale 1991; Stos-Gale and Gale 1994). In Oxford, more than 2000 metal objects and c. 1500 ore samples from deposits in numerous European countries have been analysed (Stos-Gale 1993). Isotope ratios of ores in the southern Levant (Wadi Arabah) and the Sinai Peninsula were published by Hauptmann et al. (1992), Hauptmann (2007), Segal et al. (2015) and Abdel Motelib et al.

(2012) and from Sardinia by Gale (1991); Sayre et al. (1992). One of the major studies in archaeometallurgy at FIERCE concerned copper on the Iberian Peninsula, coins in the Roman empire and the complex of problems of isotope geochemistry (Klein et al. 2012; Jansen et al. 2017).

Thousands of data sets have been collected, and databanks are accessible via the internet database OXALID. In addition, databanks are set up in the Curt Engelhorn Zentrum für Archäometrie, Mannheim, at the Deutsches Bergbau-Museum Bochum (both Germany), and at the University of Padova.

A database of copper and silver ores and from archaeometallurgical finds (slags, metal artefacts) from the western part of the Alpine mountain range and from northern Italy (Apennines) was built up as a fundamental reference frame for the diffusion and trade of metal from the Final Neolithic to the Bronze Age. Cattin et al. (2009), Guénette-Beck (2009); Giunti et al. (2009) and Artioli et al. (2016, 2017) from universities in Switzerland and from the University of Padova conducted intensive work on this topic. The team of Artioli also focused on the origin of the copper axe of the 5300-year-old Iceman, the Similaun man from the Ötztal Alps, which is the oldest prehistoric metal blade in the world that was found complete with ropes and wooden handle. This axe was most likely made from copper from the rich ore deposits of Tuscany (Italy) (Kaufmann 2014).

In their research, they targeted geological units in their region, which on the one hand are extremely rich in ore sources. On the other hand, this region forms a neat cultural barrier separating distinct metal circuits in the north and in the south. Not used in the final data analysis was lead isotope data from the Balkans, the Aegean Sea, Anatolia, the Near East, or the Iberian Peninsula because of the poor match with the measured isotope ratios in the Alps and because these regions are considered very unlikely sources for the metal objects found in the Alps during the Copper Age. For their measurements, high-sensitive Multicollector-Inductively Coupled Plasma Mass spectrometers (MC-ICP-MS) was used to determine precise lead isotope ratios ($^{206}Pb/^{204}Pb$, $^{207}Pb/^{204}Pb$, $^{208}Pb/^{204}Pb$) at the University of Bern (Switzerland). Furthermore, the feasibility of the $^{65}Cu/^{63}Cu$ isotope ratio and its eventual use as a possible ore tracer was tested.

In connection with the isotope analysis of lead and copper, extensive trace element analyses have been carried out too. For her evaluation, Cattin et al. (2009) used the data and the classification system of Junghans et al. (1960, 1968–1972) and was able to outline a complex trade system with copper in the mentioned time frames. Artioli et al. (2017) also analysed minor and trace elements. Advanced strategies based on multivariate analysis were then used to discriminate the ore mineral provenances.

Lead isotope abundance ratios are presented conventionally in two-dimensional fields (Figs. 11.5–11.7). In the beginning, ellipses were used, particularly by the team from Oxford, to form the borders of single regions. This form of illustration is a compromise. In reality, every analysis should be plotted in a three-dimensional space, so the ellipses can only be seen as sections through an ellipsoid.

Although omitting the third dimension by dropping ^{204}Pb does, admittedly, create some limitations, it does not necessarily impede the abilities to discriminate between lead sources and/or to make provenience determinations. Klein et al. (2004) were able to differentiate Roman lead and copper sources, and for Hauptmann et al. (2010) it was possible to characterise gold artefacts and ores from Georgia without using ^{204}Pb data. In fact, when Sangster et al. (2000) compared Pb isotope data from 151 lead deposits around the world, they found that statistical analyses suggested, that, while ^{204}Pb is critical for identifying a small proportion of environmental Pb sources, about 86% of the source discrimination power is due to the ^{206}Pb, ^{207}Pb and ^{208}Pb isotopes. Thus, the requisite analytical precision, rather than a lack of ^{204}Pb data, is the most critical issue with respect to unequivocal identification of Pb sources in most cases.

Shortly after its inception, the results of LIA already led to many new discoveries in archaeology. This led to remarkable results on the

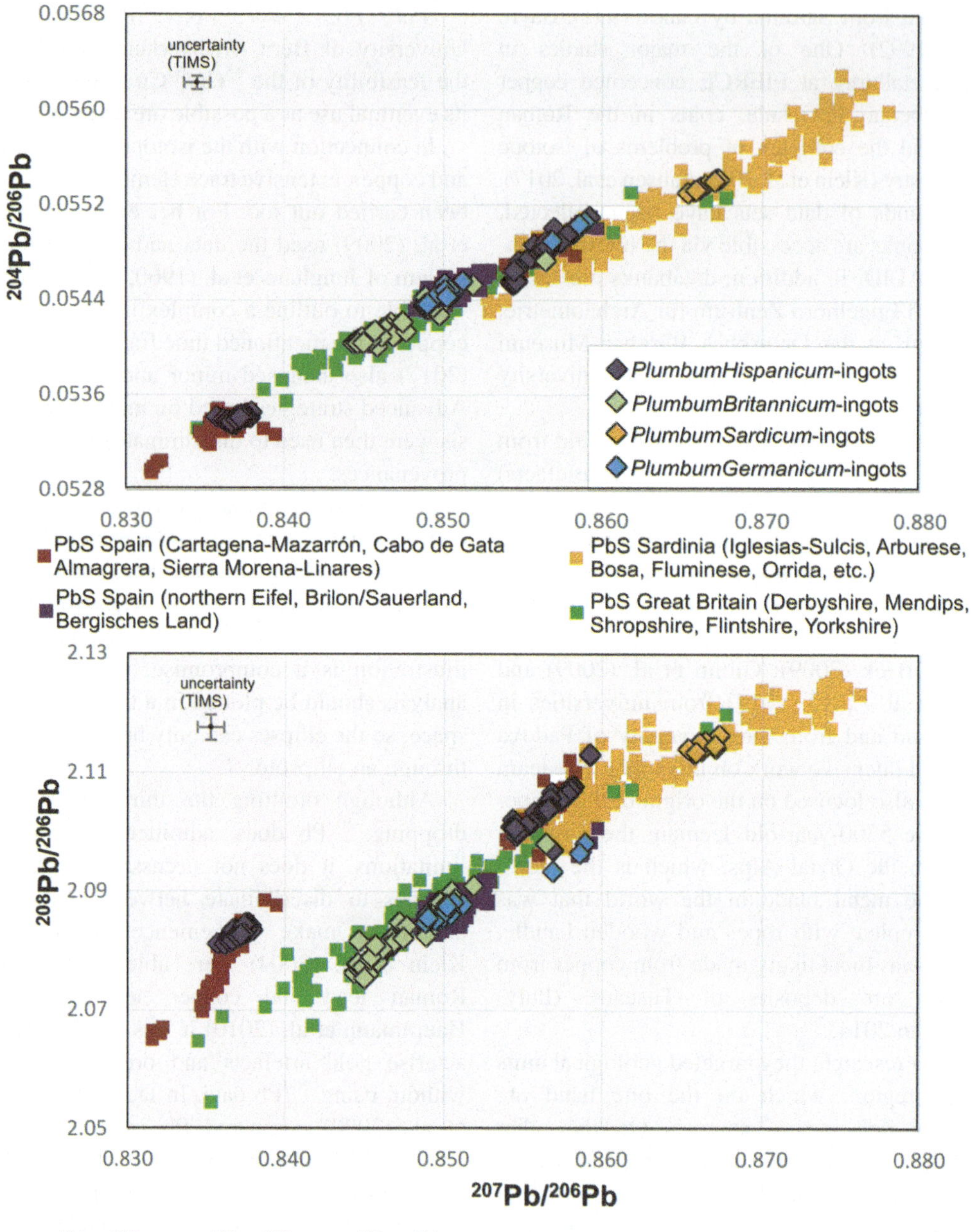

Fig. 11.7 $^{204}Pb/^{206}Pb$ and $^{208}Pb/^{206}Pb$ vs. $^{207}Pb/^{206}Pb$-diagram with lead isotope ratios of Roman lead ingots (differently coloured diamonds) in comparison with the lead ore deposits (differently coloured squares; cf. Bode et al. 2009) which are eligible candidates for lead ingot. Irrespective of the time position, all available isotope analyses of Roman lead ingots and their presumed regions of origin are presented. Modified after Bode (2013)

beginnings of metallurgy and the reconstruction of ancient trade ways especially in the Old World.

In all these endeavours, next to the geological aspects mentioned above, a number of archaeometallurgical aspects had to be considered. These are:

1. The assignment of metal artefacts to a raw material source could be hindered if ores from different sources were smelted together. This problem seems to be particularly relevant for the period before the full development of

the Bronze Age. During this early period, ores were transported to settlements not only from nearby located raw sources but also from quite distant deposits so that a mixing of different ores was easily possible. Numerous ancient settlements imported ores from elsewhere, e.g. in the southern Levant (Shiqmim: Shalev and Northover 1987; Abu Matar: Hauptmann 1989), in southern Anatolia (Arslantepe: Palmieri et al. 1993; Hauptmann et al. 2003) and in the upper Euphrates area (Hassek Höyük: Schmitt-Strecker et al. 1992).

2. Recycling of scrap metal: Investigations into the archaeometallurgical craftsmanship of Early Bronze Age III/IV Khirbat Hamra Ifdan (Levy et al. 2002; cf. Chap. 5) proved recycling and re-melting of copper objects. This resulted in a considerable homogenisation of metal. The issue becomes more problematic with the growing "industrialisation" of metal production in the later Bronze Age and the Iron Age. The discussion of the trade routes of the Cypriot oxhide ingots in the Mediterranean revolves around this problem (Muhly 1991; Gale 1991; Stos-Gale and Gale 1994). The question of whether these ingots might be made of metal from several different ore sources in Cyprus or even melted together from recycled scrap metal has developed into a sore point in the discussion (Budd et al. 1995a; Gale and Stos-Gale 1995; Hauptmann et al. 2002b).
3. LIA of alloys: Identification of originally used (copper) ores becomes impossible at the moment when an alloying constituent is added whose lead is overlapping the copper ore pattern. This is the case for Cu–Pb alloys (see Chap. 7.2). If lead dominates, provenance studies determine the origin of the lead and not that of the copper. The distinction between intentional lead alloys and naturally occurring lead in copper ores is not always clear. If copper deposits are lead-bearing, then the lead is representative of the deposit itself. Examples are known from the copper deposit of Faynan, where Pb contents of 6 wt.% were measured in the ore (Hauptmann 2007). In Early Bronze Age IV, copper ingots with around 1 wt.% Pb were produced regularly, some of which were also exported to the Levant (Hauptmann et al. 2015). At Sardinia, slightly leaded tin bronze artefacts from Nuragic times were found which were not only imported but were made locally too (Begemann et al. 2001). Their lead isotopy agrees with copper and lead deposits from Sardinia. The authors suggest that the lead could have been added from local sources in the Iglesiente-Sulcis district during the production of the artefacts.

11.3.1.5 Corpus Massarum Plumbearum Romanarum (CMPR)

Since 2009, several scholars of archaeology and of natural science in a joint venture worked on the distribution of lead ingots in the Roman Empire (Bode 2013; Rothenhöfer et al. 2013; Rothenhöfer and Hanel 2013). P. Rothenhöfer, formerly German Archaeological Institute (Munich); N. Hanel, Archaeological Institute, University of Cologne; and M. Bode, Deutsches Bergbau-Museum (Bochum) participated in the project which was funded by the Deutsche Forschungsgemeinschaft (DFG). The motive for this project was the enormous use of lead in the Roman Empire, which shows that huge amounts of lead were needed in architecture alone. As a selected example, the authors mention the need for around 10,000–15,000 t of lead just for the construction of aqueduct systems near Lyon. The demand for these quantities of lead was met by lead ingots, which were produced near the raw material sources and then were traded in this shape. The total number of known Roman lead ingots is approximately 2900 specimens from the time between 200 BC and 500 AD. Roman lead ingots are a special sort of finds. Their average weight is between 32 and 33 kg.

The majority of them had inscriptions with the names of mining owners, companies and mining provinces, merchants or emperors and thus can be considered as an important historical source for the organisation of mining, processing and trading of this metal. Where were the raw sources located of these huge amounts of lead?

Aside from this epigraphic information, the analysis of lead isotope abundance ratios of c. 500 lead ingots produced valuable information on the provenance of the ingots and thus enabled the localisation of the raw material source of the lead (Fig. 11.7). Importantly, as direct products of exploitation, they reflect the lead isotopic composition of lead ores of specific deposits, and not of scrap metal that has been poured together (Bode 2013). The most important provinces were located in the western part of the Roman Empire on the Iberian Peninsula near Cartagena-Mazarron and in the Sierra Morena, where silver was extracted in addition to lead. These metals were also extracted in Britain and on the Balkan Peninsula, in the province of Moesia Superior. Further occurrences were located in the south-western part of Sardinia and in the Germanic provinces near Mechernich, Wiesloch near Heidelberg and Brilon. The ingots date from the turn of the second/first century BC to the fifth century AD.

So far, no other metal—be it iron, copper or tin—has provided similarly important possibilities of epigraphic and at the same time scientific evaluation.

11.3.1.6 The Mitterberg Mining District: Source for Copper in the Bronze Age?

An extensive mining–archaeological–archaeometallurgical research project of the Deutsches Bergbau-Museum and the Curt-Engelhorn Centre for Archaeometry (Mannheim, Germany) dealt with the Bronze Age production of copper at Mitterberg (Austria) and its distribution in central Europe. In contrast to the principle of the CMPR project, in which the corresponding raw material sources were sought on the basis of analytical examinations of ingots, the question posed in this project is somewhat different: known is a major copper-rich mining district as well as an extraordinary amount of Bronze Age metal artefacts. Are these artefacts from the Mitterberg mining district?

It was part of the special research area SFB HIMAT of the University of Innsbruck. In the history and development of this project, the complexity of provenance studies is clearly evident when it comes to the exploration of larger mining areas, whose prehistoric features are considerably covered by modern activities, and also when it is a matter of the analytical processing of considerable quantities of Bronze Age metal artefacts, which could possibly originate from this mining area. Current compilations of the project were presented by Lutz (2016) and Pernicka et al. (2016).

This research project developed from several thematic complexes.

The first was the processing of prehistoric mining on the Mitterberg, exposed and cut by modern shaft sinking, drifting and haulage, which was already proven in the first half of the twentieth century to a depth of 120 m (summarised in Stöllner 2015, see also Chaps. 3. 5.1 and 6.2). By underground excavations, the strategy of prehistoric mining and also smelting activities could be mapped (Fig. 3.10). Several dozens of ^{14}C-data could prove these mining activities in the time of the eighteenth/seventeenth centuries BC with a clear emphasis on the fifteenth/thirteenth century BC, i.e. the time periods from which a large part of the metal artefacts originate. In addition, about 100 smelting sites from the Bronze Age were mapped in the Mitterberg district among all the find sites.

To enable a comparison of raw material—final artefact, extensive series of analyses have been carried out on Mitterberg ores in recent years. It has already been shown that there are six large ore veins with partly different genesis (Bernhard 1965), which were genetically connected with volcanism. The major component of the Mitterberg veins was chalcopyrite with concentrations of Ni, As, Sb, Ag and Co in the trace element level. Based on these trace elements, the Mitterberg ores can be distinguished from other deposits in the region, e.g. those at Schwaz and Brixlegg (Inn Valley), which have significantly higher Sb and Ag contents (Fig. 11.8). In contrast, it is difficult to define other resources with the main component

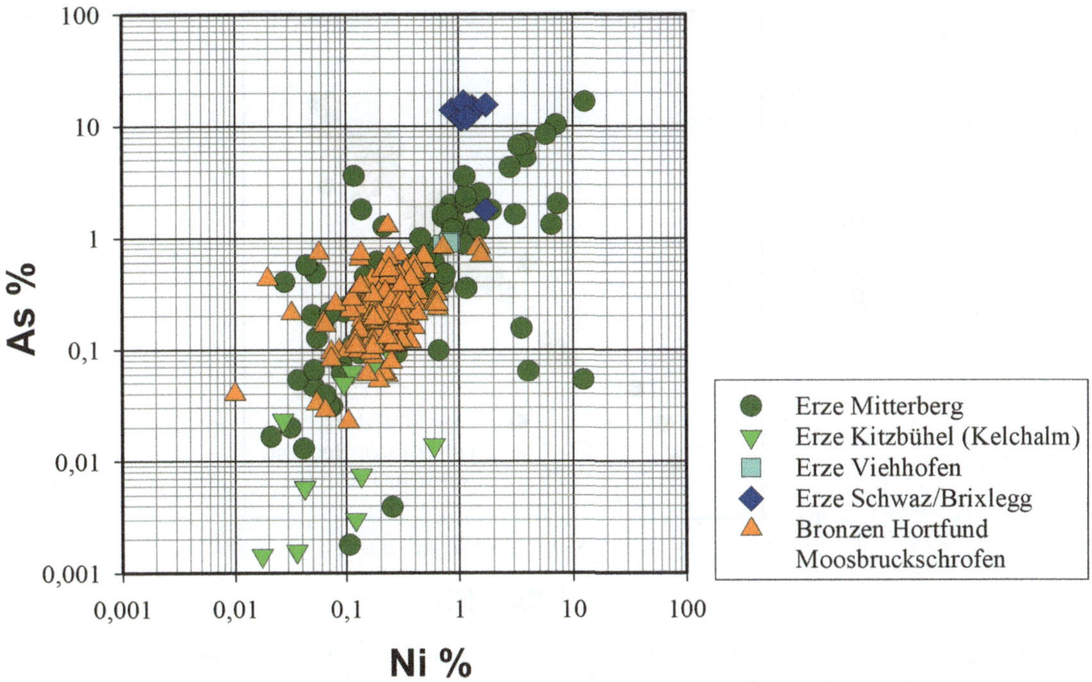

Fig. 11.8 Trace element pattern of ores from veins of the Mitterberg in comparison with artefacts of the Moosbruckschofen in Tyrol, exemplified by Ni–As ratios, are in good agreement and show the making of the Bronze Age artefacts from this ore deposit. Courtesy J. Lutz, Curt-Engelhorn-Zentrum Archäometrie, Mannheim

chalcopyrite. This applies, for example, to the copper deposits in Slovakia. In provenance studies in which these two deposits are discussed, geographical factors play a significant role.

During the last years, chemo-analytical measurements and characterisation of ores from the veins of Mitterberg were considerably intensified by the discovery of the Early Bronze Age (c. 1600 BC) Nebra Sky Disc and the hoard of bronze artefacts associated with it (Pernicka 2010). The trace elements of the Mitterberg ores were in excellent agreement with these finds, which could be further confirmed by lead isotope analyses.

In addition to the mining findings, numerous discoveries of plano-convex copper ingots ("cast cakes") and other copper artefacts were made very early. The ingots appear to be blister copper, as produced in the melting furnaces. In the plano-convex ingots, the chemical and isotopic signatures are still largely unaltered, since they have not yet been alloyed with tin. They are therefore most likely to be classified as ore deposits. In Pittioni (1957), among others, this led to the first considerations as to how the relationship between ore deposits and metal artefacts could be solved analytically. This question was pursued very intensively in the following decades, particularly by the research group in Mannheim. It should be emphasised that today much more precise measurements of trace elements as well as in isotope analysis are possible than in the 1950s and 1960s.

Pernicka et al. (2016) explain that in these discussions the composition of the lead isotope ratios should be measured in addition to chemical analysis. The wide variation in lead isotope values (Fig. 11.9) is due to a paragenesis of copper ores with uranium, so that in one ore batch "common lead" will be mixed with radiogenic lead in varying amounts. Other ore deposits in the Old World show similar large variations as well, e.g. Rudna Glava (Serbia), Faynan and Timna (Jordan, Israel), the Erzgebirge, Sardinia, and other (see above).

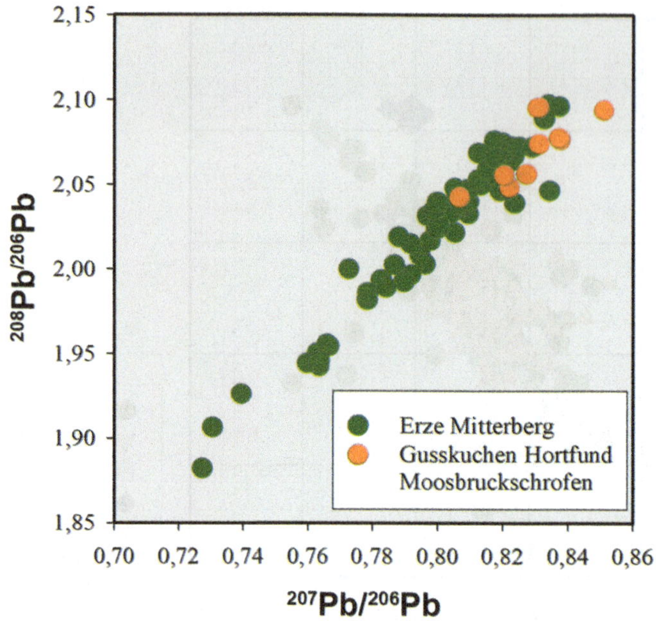

Fig. 11.9 Lead isotope ratios of ores from various vein lodes of the Mitterberg copper district in comparison with cast cakes and artefacts from the Moosbruckschrofen. Many of the ores from epigenetic main lodes are more radiogenic than the syngenetic ores from other veins and show a large range of compositions. The composition of the cast cakes is limited probably due to the use and smelting of a special sort of ores. Courtesy J. Lutz, Curt-Engelhorn-Zentrum Archäometrie, Mannheim

In 2001, about 360 Middle Bronze Age metal objects and numerous fragments of cast cakes of a hoard found at Moosbruckschofen in the Tyrolean Oberland were analysed chemically and in part for their lead isotopic composition. This is the largest hoard in Central Europe. A few artefacts date around 1600 BC. Lutz (2016) found that the pattern of trace elements Sb, Ni, As and Ag of ores from the Mitterberg correspond very well with the metal artefacts from Moosbruckschofen. This is especially true for nickel and arsenic (Fig. 11.8). Also, the comparison of the lead isotope values points towards a provenance from the Mitterberg district (Fig. 11.9). Herewith, the author presented an excellent example of a sequence from ore over cast cake as an intermediate product to the final product by a combined trace element and lead isotope analysis.

More problematic is the long-discussed question of the origin of Ösenringe, which is a form of copper widely used in large quantities since the Early Bronze Age. They are often referred to as ingots, i.e. semi-finished products similar to plano-convex copper ingots. However, this interpretation seems very questionable because of the much larger quantity of finished products of the same composition (Lutz 2016). This specific metal with the high contents of As, Sb, Bi and Ag was in the past called an Ösenring-copper. The chemical composition and the lead isotope analyses indicate a potential origin from the fahlore deposits of the Inn valley and exclude an origin from the Mitterberg district. However, to date, there is no field evidence of Early Bronze Age mining in the Inn Valley. This would have to be very extensive due to the large number of Ösenringe. Whether this early mining was destroyed by modern activities must remain open.

Currently, a data set of about 1800 analyses of ores and metal artefacts provides a very good statistical basis and allows the reconstruction of the development of copper metallurgy and the changing use of ore deposits in the Alpine region, at least in broad outline. Overall, however, the approach shows the necessity of joint mining–archaeological fieldwork and analytical laboratory work.

11.3.2 Tin Isotopy

It is one of the central questions of archaeologists, historians and scientists: Where did the tin of the Bronze Age come from? Even though this question has been discussed for over a century, and an origin in northern Persia and the Hindukush is suspected (von Baer 1792–1876; last in Alimov et al. 1998), possible answers never stepped out of the area of speculation. Conclusive evidence has not yet been presented by any party. Enormous masses of this metal were needed to enable the production of bronze artefacts from this epoch in Central Europe alone.

Tin bronzes are in regular use around 2500 BC in Mesopotamia, where the most prominent examples come from the Royal Tombs of Ur and Susa, as well as from Troy in Turkey. Mesopotamia itself is an area without tin deposits. For Central Europe, the use of bronze has been proven since about 2200 BC (Pernicka 1998). While there are numerous copper deposits in this region as well as evidence of prehistoric copper mining, such as in the Alps, until a short while ago the origin of early tin was still completely in the dark. The role of the two well known tin districts in Cornwall and the Ore Mountains as possible sources of Bronze Age tin origin is not yet absolutely clear (Berger et al. 2019a) even if there is growing evidence by recent research for such an early exploitation especially for the Ore Mountains (Tolksdorf et al. 2019). The deposits of Kestel or Göltepe and Hisarcik (Anatolia) as well as of Deh Hosein (Iran) and those of the Balkan peninsula (Mason et al. 2016) are rather limited. Extensive deposits of tin, however, are located in the countries of Central Asia (Afghanistan, Tajikistan, Kazakhstan, Uzbekistan; see Chap. 3.4.11).

11.3.2.1 Analytical Approach

The reason why until recently only relatively few insights into the origin of tin in the ancient world have been found is obvious. There were no scientific methods that allowed the tin to be traced back to the source. It was estimated that the lead isotope analysis, established since the mid-1960s of the last century for determining the origin of ancient metal, was not used because cassiterite, and thus the tin derived from it usually contains very little lead.

With tin bronzes, there is a much lower possibility that the alloying partner tin overwrites the original lead isotope pattern of the copper. Analyses of tin ores from 25 different deposits have shown that their lead content is <30 ppm (Begemann and Schmitt-Strecker 2008b). This means that an addition of about 10 wt.% of tin to copper, i.e. in the production of a “classic” tin bronze, contributes at most 3 ppm of lead per gram to the total lead of the bronze. This is negligible and would only insignificantly affect the isotope pattern of the original copper (Figs. 11.2 and 11.10). Comparably low lead contents were also measured in Late Bronze Age tin ingots, which were discovered in the shipwreck of Uluburun at the Turkish southern coast (Hauptmann et al. 2002a), as well as at Ha Hotrin, Hishuley Carmel, Kefar Shamir at the Israeli coast (see compilation in Begemann et al. 1999).

Therefore, the lead isotopic ratios in old bronzes are essentially determined by copper and its lead contamination rather than by tin. This means, however, that it may be possible to determine the origin of the copper, but not that of the tin.

11.3.2.2 Tin Isotopes

To find a solution to this tin problem, Gale (1997) was the first to propose a use of tin isotope ratios in archaeological questions. Then, in the Curt Engelhardt Center for Archaeometry, Mannheim, a method was developed in which the isotope ratios of the tin itself can be used to determine

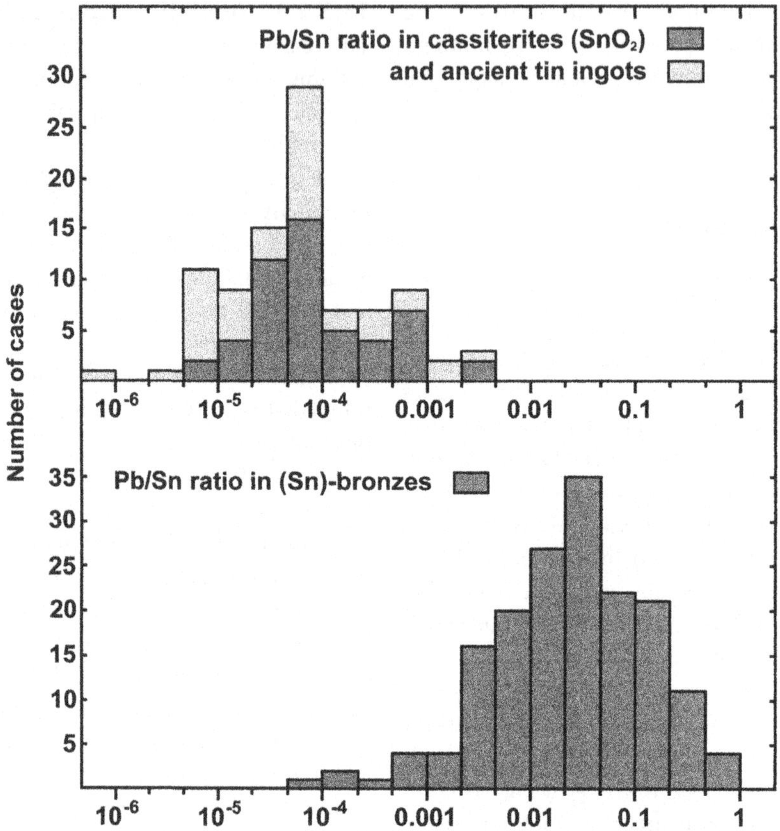

Fig. 11.10 The lead contents of cassiterite and ancient tin ingots, expressed as Pb/Sn ratio (upper panel), are much too low to account for the lead in ancient bronzes (lower panel). Obviously, most of the lead of the bronzes was not introduced together with the tin which precludes tracing the tin via the isotopic composition of the lead that might have been associated with the tin. From Begemann et al. (1999)

its origin. This first happened as part of a research project that included the sensational findings of Nebra, with particular attention to the "Nebra Sky Disk" (Haustein et al. 2010; Haustein 2013). In 2013–2019 in Mannheim, within the framework of a project funded by the European Research Council (ERC), the question of the origin of tin by means of tin isotopes was investigated. The goal of this project was to investigate the usefulness of isotope analysis for provenance studies in archaeology. It would lead to new insights into the origin and diffusion of bronze technology in the third and second millennium in the Middle East and Europe.

Tin has 10 stable isotopes ($^{112-124}Sn$). ^{120}Sn, with a proportion of 32.4% in natural tin, is the most abundant isotope, followed by ^{118}Sn at 24.1%, ^{116}Sn at 14.1% ^{124}Sn at 5.8%, ^{122}Sn at 4.7%, ^{117}Sn at 7.6% and ^{119}Sn at 8.6% (Haustein 2013). Tin isotope data are given in the δ-notation.

For a long time, a clear idea of the scale of the isotope variation of tin in nature could not be gained, and the question whether tin isotope ratios can be used as a tool for provenancing tin sources seemed to be difficult (see discussion in Haustein et al. 2010). The authors claimed that there were two reasons for this drawback. Too few samples were analysed, and the analytical preconditions were not suitable to trace the isotope ratios of tin isotopes.

For this purpose, a measurement methodology for tin isotopes based on multicollector mass spectrometry (MC-ICP-MS) was developed (Tin Isotope analysis abbreviated as TIA). The present method involves both the preparation and transfer of the cassiterite in solution as well as a possibly necessary cleaning of the measuring solution by means of ion chromatography.

In the work of Haustein (2013), 80 samples from the Ore Mountain area (Erzgebirge), the Vogtland and Cornwall were initially measured for the composition of the tin isotope ratios. The measurements of these ores should on the one hand show to what extent individual occurrences can be differentiated isotopically from each other, on the other hand, correspondences with archaeological finds, in particular, the "Nebra Sky Disc" should be worked out. For the comparison of these ores with the Sky Disc, the author chose the isotope ratios $^{117}Sn/^{119}Sn$ vs. $^{122}Sn/^{116}Sn$ (Fig. 11.11). The investigations show that it is quite possible to distinguish between different tin deposits based on their tin isotope ratios. In addition, it is likely that the origin of the tin contained in the famous "Nebra Sky Disc" does not lie, as possibly expected due to geographical reasons, in the nearby Vogtland or the Ore Mountains. Rather, an isotopic similarity between the "Sky Disc" and cassiterite samples from Cornwall could be noted.

Also, other Sn ratios of tin ores were measured which showed variations depending on their geological genesis. Marahrens et al. (2016) found that the isotopic ratios in cassiterite are highly variable, and the $\delta^{124}Sn/^{120}Sn$ ranges generally from −1.29 to +0.85‰. This variation is observed in samples from tin deposits in Central Asia which have an average $\delta^{124}Sn/^{120}Sn$ of

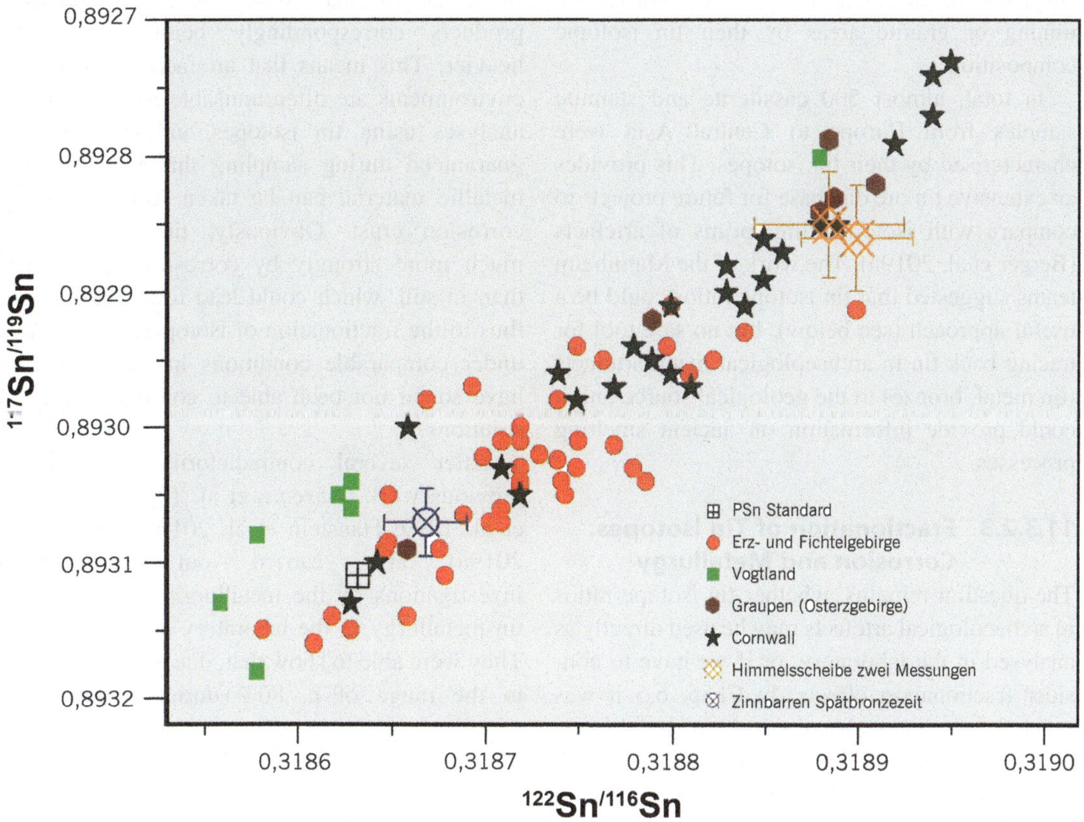

Fig. 11.11 Picture of tin isotope ratios ($^{117}Sn/^{119}Sn$ vs. $^{122}Sn/^{116}Sn$) of cassiterite deposits from the Ore Mountains, Vogtland (Germany) and Cornwall (England) in comparison with the established isotope signature of the tin contained in the "Nebra Sky Disc". The disc is isotopically similar to the ores of Cornwall as well as to the tin ores (*Graupen*) from the eastern Ore Mountains. From Haustein (2013)

−023 ± 0.46‰. The cassiterite from the tin provinces in Europe have on average a heavier isotopic composition of 0.09 ± 0.14‰. The two tin provinces of the Variscan fold belt (Cornwall and the Saxonian-Bohemian Ore Mountain) show the largest variation but the isotope ratios overlap to a large extent. Despite the large overlap, cassiterite from Spain (Spain $\delta^{124}Sn/^{120}Sn = -0.05‰$) and France ($\delta^{124}Sn/^{120}Sn = -0.04‰$) on average tends to have lighter isotopic compositions than that of Cornwall ($\delta^{124}Sn/^{120}Sn = -0.10‰$), Portugal ($\delta^{124}Sn/^{120}Sn = -0.07‰$) or the Saxonian-Bohemian province ($\delta^{124}Sn/^{120}Sn = -0.13‰$). In summary, because of the large overlap and the highly variable isotopic composition of cassiterites from all tin provinces, it is difficult to distinguish between them using the tin isotope ratios, even between samples from the Variscan fold belt. But on a more detailed scale within a single tin province, it is possible to discern several mining or granite areas by their tin isotopic composition.

In total, almost 500 cassiterite and stannite samples from Europe to Central Asia were characterised by their tin isotopes. This provides an extensive tin ore database for future projects to compare with isotopic fingerprints of artefacts (Berger et al. 2019a). The work of the Mannheim teams suggested that tin isotope ratios could be a useful approach (see below), but no safe tool for tracing back tin in archaeological metal artefacts (tin metal, bronze) to the geological source and it could provide information on ancient smelting processes.

11.3.2.3 Fractionation of Tin Isotopes: Corrosion and Metallurgy

The question remains, whether tin isotope ratios in archaeological artefacts may be used directly as analysed in the laboratory, or if we have to consider fractionation effects. In Chap. 6.5 it was pointed out that archaeological tin and bronze artefacts sometimes exhibit strong tendencies for corrosion and that many of them are covered with profound corrosion crusts made of (hydr-)oxidic minerals of both of these metals. As a result, their original metallic consistency has been changed, which raises the question of a possible isotopic change of the original metal (Berger et al. 2018). A prominent example of this question is the authenticity of the "Nebra Sky Disk".

Berger et al. (2019a) carried out experiments with bronze and tin in various corrosion solutions that simulated both weathering in terrestrial and in marine environment, because many archaeological metal finds, as Uluburun's example shows, have been found from shipwrecks. These experiments showed that there is no obvious change in the isotopic composition in the case of corrosion in the soil. This means that, to the present state of knowledge, artefacts recovered from soil excavations are suitable for provenance analysis.

The situation is different with tin objects corroded in the sea. There, a much more significant corrosion of tin artefacts can be observed. The corrosion layers differ up to $\delta^{124}Sn/^{120}Sn = 0.4‰$ from the original metal, with the corrosion products correspondingly being isotopically heavier. This means that artefacts from marine environments are often unusable for provenance analyses using tin isotopes, unless it can be guaranteed during sampling that actually pure metallic material can be taken from under the corrosion crust. Obviously, tin is mobilised much more strongly by corrosion in seawater than in soil, which could lead to a removal and thus to the fractionation of isotopes. Simulations under comparable conditions in the laboratory have so far not been able to contribute fully to solutions.

After several contradictorily experimental previous work (Kazenas et al. 1996; Begemann et al. 1999; Haustein et al. 2010; Berger et al. 2019a) also carried out experimental investigations on the metallurgical processes of tin metallurgy in the laboratory and in the field. They were able to show that, due to the loss of tin in the range of c. 80% during smelting by vaporisation or slagging, a fractionation of $\delta^{124}Sn/^{120}Sn = 0.07–0.15‰$ occurs from the starting ore to the tin metal. The metal is enriched with heavy isotopes, while the vapour is isotopically lighter. Basically, also other metallurgical operations, like casting and multiple (re-)melting of tin and bronzes may alter the tin isotope ratios.

This could have happened irregularly, as these steps could have happened under various redox conditions in variously constructed open or closed, wooden or ceramic casting moulds. In spite of this fractionation, TIA for provenance analyses are in principle still possible, because the isotope shifts in smelting are small compared to the natural variation in tin-mines within a single tin province (Berger et al. 2018, 2019a). These results lead to the conclusion that even in prehistory, there is no reason to expect large changes in the tin isotope pattern, for example in the case of re-melting of ingots, flawed casts or scrap metal.

11.3.2.4 Tin and Lead Isotopy

The investigations by Marahrens et al. (2016) have once again shown that, despite modern mass spectrometers (NeptunePlus), precise assignment of tin objects, e.g. to Cornwall or to Ore Mountains is not possible. Both deposits cannot be geologically differentiated. Therefore, another concern of the mentioned project in Mannheim was also to discuss analytical methods that could be supplementary to TIA.

One of these possible methods was via lead contents in tin objects. Due to their geochemical position, tin ingots only contain small amounts of lead, but Hauptmann et al. (2002a) were able to show for 32 tin ingots from the shipwreck of Uluburun that these at least contained lead concentrations in the range of 0.006–0.007 wt. %, sometimes even a little more. These lead contents have already been used by Stos-Gale et al. (1998a), who prepared lead isotope analyses of 57 tin bars from Uluburun. They could demonstrate that—next to a few lower $^{208}Pb/^{206}Pb$ ratios—a pronounced cluster exist between c. 2.05 and 2.07 and would have a model age from the Palaeozoic. The authors concluded that such an isotope pattern would exclude a provenance from the tin mineralisations in the Ore Mountains, but it was not possible to offer a possible provenance.

Berger et al. (2019a) are convinced that the combination of trace elements and both tin and lead isotopes will, for the very first time, allow a delineation of the origin of the Bronze Age tin. Analysing 38 tin ingots from Uluburun and other localities off the coast of the Mediterranean (Crete, Israel) and Salcombe (England), the authors were able to show that these, with a few exceptions, exhibited a linear correlation in $^{206}Pb/^{204}$ Pb and $^{207}Pb/^{204}Pb$, which allowed a calculation of the geological model ages of the tin ores used for these ingots. One of their results was that the tin ingots from the Israeli coast had a model age of c. 291 $\pm$ 17 Ma, i.e. they also were of Palaeozoic age. The deposits eligible for these bars are not located in nearby Kestel/Göltepe or Deh Hosein, they can be ruled out as their geological genesis is attributed to later time periods. On the other hand, a delivery area from Europe is possible, as comparable model ages could be measured from the tin ores in Cornwall (and Ore Mountains). The trace element patterns of the Israeli and British tin ingots lead to the same conclusion. Relevant elements for provenance are indium (In), antimony (Sb), lead (Pb) and bismuth (Bi). It has to be pointed out, however, that so far, no tin measurements on artefacts or tin ores from Central Asia could be carried out.

In a similar way, further provenance studies were conducted, with the authors emphasising that it was precisely through the combination of trace elements, lead and tin isotopes that it became possible to delineate the origin of tin. Berger et al. (2019b) and oral presentation (Miskolc, Hungary) are even of the opinion that the most important proxy for provenancing tin ingots is the lead isotope analysis.

11.3.3 Copper Isotopy

11.3.3.1 Copper Isotopes and Archaeometallurgy

For archaeometallurgy, the stable copper isotopes provide a promising approach for the identification of ore types used in ancient times to smelt metal, since copper and copper-based artefacts from primary (hypogene) sulphidic, secondary (supergene) sulphidic, or oxidised ores can be differentiated through their copper isotope signature (Klein et al. 2010). The method of copper isotope analysis can in a sense be called a refined method of lead isotope analysis: but it is not a method that allows a geographic localisation of an ore deposit, it is not suitable for provenance studies.

The shift of mining from colourful oxidised ores near the surface to sulphidic ores at greater depths of the same deposit would have required substantial technological innovations not only in mining exploitation, e.g. hauling, drainage and timbering to support deeper adits. In addition, the extraction of copper from sulphidic ores such as chalcopyrite requires advancements in metallurgical processing. Roasting of ores or intermediate products of smelting processes can be postulated, smelting of oxidic and sulphidic ores to matte and the transition of matte to metal. Thus, the identification of ore types used in prehistoric times is of substantial interest (Jansen et al. 2017). Under certain circumstances, (relative) chronological estimates can be made by analysis of copper isotopes. Conclusions can be made about the depth of the mining activities in ancient times. This is especially important when all other traces of ancient mining and metallurgy were destroyed by later activities. In this context, it is important to note what Woodhead et al. (1999) stated, namely that no fractionation of copper isotopes would occur from ore to metal by any of the metallurgical operations known in ancient times.

Analyses of copper isotopes in geosciences were first published by Walker et al. (1958). But only after the application of multicollector mass spectrometry with inductively coupled plasma (MC-ICP-MS) did a sustained research into the behaviour of copper isotopes, i.e. of a fractionation of isotopes in various contexts, begin.

Natural copper isotope variations have already been found useful as a geochemical tracer in ore geology (Larson et al. 2003) and were suggested as a possible tracer in archaeology by Gale et al. (1999) and Woodhead et al. (1999).

Jansen et al. (2018a) presented a very convincing paper about combined analyses of Cu- and Pb-isotopes of 39 Late Bronze Age copper oxhide ingots from the Aegean in which they discussed their ore sources and provenances on the island of Cyprus.

11.3.3.2 Copper Isotopes

Copper has different isotopes, from ^{53}Cu to ^{82}Cu. Of these, ^{63}Cu and ^{65}Cu are the natural isotope mixture, which is also used in archaeometallurgy. These two isotopes have a frequency of occurrence of 30.83% (^{63}Cu) and 69.17% (^{65}Cu) (Albarède 2004). Next to these isotopes, there are several short-life radioactive isotopes. The most stable one of these is ^{67}Cu which unfortunately has a half-life of less than 3 days. Therefore, the radioactive isotopes are not useful for archaeometric research.

Usually, Cu isotope data is given in the δ-notation. This is a value for the ratio of stable isotopes. $\delta^{65}Cu$ indicates the ratio of $^{65}Cu/^{63}Cu$. It indicates the deviation of a measured value from a standard in parts per thousand, the ratio of the sample being compared to that of a certified NIST standard (SRM-976). A positive value means an enrichment of the heavier isotope in the sample compared to the standard; a negative value means an enrichment of the lighter isotope. The advantage of this notation lies in the comparability of all values normalised to this standard as well as a comfortable calculation with the values.

11.3.3.3 Fractionation of Copper Isotopes in Copper Deposits

Fractionation processes of copper isotopes in deposits is one of the major issues in the research of ore deposits. Physical and especially chemical processes in copper deposits are leading to a fractionation of the isotopes, i.e. to a separation by mass and enrichment of isotopes in different phases. It was possible to show that primary (sulphidic) copper ores exhibit an isotope composition between $\delta^{65}Cu = -1‰$ and $+1‰$ (Kimball et al. 2009), while all ore minerals (oxidic and sulphidic) together showed a range from $\delta^{65}Cu = -16.49‰$ to $+9.98‰$. For copper isotope fractionation, the oxidation state (Cu^{+}, Cu^{2+}) is of primary relevance. In addition to these factors, a temperature dependence is active in the lower ranges. The higher the temperature, the lower the energy difference between the isotopes and the more similar both isotopes can behave (Schauble 2004).

Markl et al. (2006) and Li et al. (2010) noted that the variation of δ^{65}Cu inside one deposit can be larger than the variation between several deposits, which prevents a reconstruction of the origin of the raw material.

Due to redox sensitivity, a pronounced zoning in the near-surface parts of deposits happens during weathering of sulphidic copper ores. Sulphuric acid is formed. ^{65}Cu is dissolved and washed deeper into lower parts of ore deposits (Mathur 2009). At groundwater level, oxidised solutions enriched in ^{65}Cu encounter reducing conditions and precipitate. The longer this process is going on, the more ^{65}Cu is collected in the enrichment zone. In the oxidation zone, ^{63}Cu remains, making it isotopically lighter. By copper isotope analyses it is possible to reconstruct approximately from which part of a deposit an ore would come. Thus, the copper isotopes will become important in detailed studies on the use of primary versus secondary ores for the manufacturing of artefacts.

Jansen et al. (2017, 2018a) presented a sketch of the Cu isotope distribution for a weathered ore body at its outcrop (Fig. 11.12) based on the observation that secondary oxidised copper ores show a heavier copper isotope composition than primary ores and supergene sulphidic ores are tendentially lighter (Klein et al. 2010). The authors also observed significant differences in the copper isotope signature of different copper ore types from Cypriot deposits supporting the distribution presented in this sketch.

Asael et al. (2007, 2009, 2012) investigated oxidic and sulphidic copper ores from the stratabound sedimentary deposit of Timna (Israel) which was extensively exploited in ancient times. They found that the copper isotope signatures of oxidised copper minerals are regularly shifted to more positive values in relation to the sulphide minerals.

In the study on copper ores from the Iberian Pyrite Belt (IPB; southern Portugal-Spain) Klein et al. (2010) by MC ICP-MS in Frankfurt a.M. they stated that analysed primary sulphides mainly had a composition between $\delta^{65}\text{Cu} = -0.4$ and +0.3‰, oxidised ores (hydrocarbonates) usually had a heavier composition of $\delta^{65}\text{Cu} > +0.3$‰ (enriched in the heavier isotope ^{65}Cu) and supergene sulphides were lighter in their composition with $\delta^{65}\text{Cu} < -0.4$‰ (enriched in the lighter isotope ^{63}Cu). Archaeological artefacts can tangentially be assigned by this classification.

Prior to the study of Klein et al. (2010), Durali-Müller (2005) had analysed copper isotopes of primary and secondary copper ores from various localities. She observed that all secondary copper

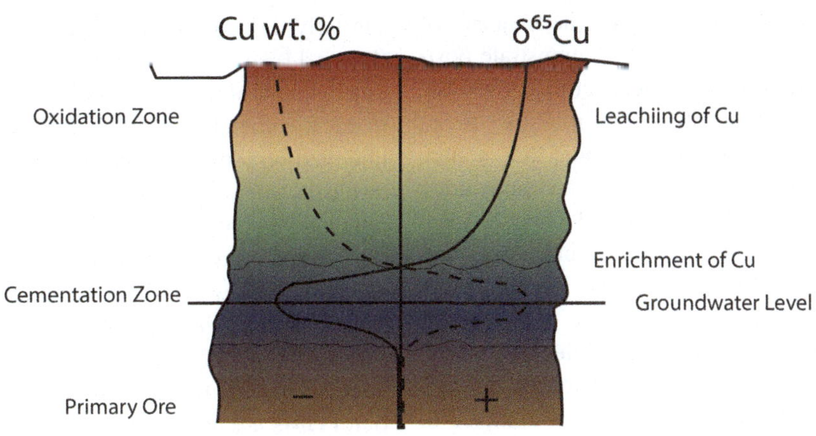

Fig. 11.12 Sketch showing a profile of the distribution of stable Cu isotopes of copper ores in the near-surface part of a hydrothermal sulphidic vein deposit. The sketch refers to copper ores only: in the oxidation zone, secondary copper ores like malachite show a heavy isotopic composition (positive δ^{65}Cu values) which changes in the cementation zone with supergene sulphidic ores like chalcocite and covellite to negative δ^{65}Cu values. In the lower parts of the vein, the non-altered primary ore with sulphidic minerals like chalcopyrite, formed at high temperature, shows a narrow homogenous isotopic composition. From Jansen et al. (2017, 2018a)

minerals had positive δ^{65}Cu values, whereas δ^{65}Cu values in the corresponding sulphides had negative δ^{65}Cu. The lead isotope ratios were identical in the sulphides and their corresponding secondary minerals. The secondary copper ores from the Sierra Morena (southwest Iberian Peninsula) are in agreement with these observations: They have positive δ^{65}Cu values from 0 to +2.7 but encompass a wide range of lead isotope ratios corresponding to their primary mineralisation.

Jansen et al. (2018a) combined copper and lead isotope analyses of Late Bronze Age copper ingots found at various sites and islands in the Eastern Mediterranean. The goal of his investigations was to not only determine the origin of the ingots as precisely as possible but also to determine the ore types that were used to make the ingots. The authors relied on the results of Stos-Gale and Gale (2009) who concluded after decades of research that all oxhide ingots found in the Mediterranean and dating after 1400 BC can be assigned to Cypriot deposits. In fact, most of the copper ingots could be assigned to ore deposits in the Solea graben region in the northern Troodos ophiolite complex through consistent lead isotope ratios.

The copper isotope characterisation of these ingots indicates the use of different ore mineral sources, possibly reflecting the exploitation of various zones within the same or multiple deposits. Besides a few outliers, the majority of ingots dating prior to 1300 BC demonstrate positive δ^{65}Cu values while ingots later than 1300 BC show negative δ^{65}Cu values. This might be due to a shift from oxidised to sulphidic ore source around 1300 BC. Tendencies in the chronology of the clusters could be clarified by additional measurements of ingots in the future. A general trend with a shift from oxidised ores before 1300 BC and sulfidic ores after 1300 BC is already visible. This data could reflect the accession of copper from individual mines through time, including a general trend from oxidised to sulfidic ore mineral sources.

The authors have shown that the copper isotopes can be successfully used to determine the origin of artefacts from either sulphide or hydrocarbonate ores. This is useful in following the mining activities in individual areas. In combination with lead isotopes, individual mines or mining areas may be deciphered together with the mining progress in these areas. The combination of copper and lead isotopes may be particularly useful in the future for Copper and Bronze Age objects and for the question of the beginning of the use of copper. The information gained from copper isotopes is a high potential innovation in archaeometallurgy. In the future, it may function as a crucial tool for better determination, e.g. of the different metallurgical stages of development as defined by Strahm and Hauptmann (2009).

11.3.4 Osmium Isotopy

11.3.4.1 Osmium Isotopes and Archaeometallurgy

With the increasing scientific interest in gold in archaeology, and simultaneously, with the available analytical methods of virtually non-destructive sampling for those highly prized and valuable artefacts, provenance studies for this metal too will become more and more central. The isotope analysis of osmium (Os) plays a special role, as this element is a major constituent of inclusions of platinum group placer minerals (PGM) in gold. The isotope analysis of osmium, similarly to lead, tin and copper isotope analysis (Sects. 11.3.1, 11.3.2 and 11.3.3), was first used as a dating tool in geosciences and later found use in archaeometallurgy. Unlike lead, which was liquefied on the way from ore to metal in metallurgical smelting processes, PGM inclusions remained as unaltered inclusions in the metal, in this case in gold, because their melting points are so high that they could not be liquefied in ancient times. Platinum has a melting point of 1772 °C, that of Os-containing compounds can lie well over 2000 °C. In archaeometallurgy, osmium-rich PGMs are not known in the context of other metals.

In nature, osmium bearing minerals and alloys of osmiridium and compounds containing other platinum group elements such as platinum and ruthenium (see Chap. 3.5.3) occur in ultramafic igneous rocks such as dunites or peridotites which derived from sources in the earth's mantle. They can be used as tracers of geological processes and developments, especially in (ultra-)mafic rocks or in meteorites. Hirt et al. (1963a) originally demonstrated that samples of osmiridium from several deposits of different ages have different $^{187}Os/^{186}Os$ ratios. The age determination of meteorites through the rhenium–osmium (Re/Os) dating method can potentially also be applied in archaeology to differ between iron artefacts made of terrestrial or of meteoritic iron. The age determination of gold artefacts, more precisely, means the common method of dating platinum group mineral inclusions in gold by $^{187}Os/^{186}Os$.

11.3.4.2 Osmium Isotopes

There are seven stable osmium isotopes between ^{184}Os and ^{192}Os. All other known and characterised osmium nuclides are unstable. In archaeometallurgy, the isotope ratios $^{187}Os/^{188}Os$ and $^{187}Os/^{186}Os$ are measured. The relative frequencies of these isotopes in the mixture of the stable osmium isotopes are $^{187}Os = 1.96\%$, $^{188}Os = 13.24\%$ and $^{186}Os = 1.59\%$.

^{187}Os is produced via the radioactive decay of ^{187}Re. The Re/Os ratio is a geological dating tool (Hirt et al. 1963b).

Osmium and Rhenium (Re) have differing geochemical properties: Re is exceedingly rare in terrestrial rocks, its estimated abundance in igneous rocks in the earth crust is c. 0.5 ppb (Wedepohl 1968–1978). It only appears in very few minerals, e.g. in molybdenite (MoS_2). This mineral is one of the few that is suitable for Re/Os dating. Os, on the other hand, is enriched in mantle rocks. In geological periods, fractionation of Re and Os occurs and different levels of ^{187}Os are produced. As the pattern of the Os isotopy depends on the ore source, it can be used for provenance studies. However, due to the limited amount of comparable data, this can potentially be difficult.

11.3.4.3 Platinum Group Placer Minerals (PGM) in Ancient Gold Artefacts

The osmium isotope analysis of gold artefacts is based on the following principles: In the ideal case, inclusions of minerals of the platinum group elements (PGM, see Chap. 3.6.4) are measured which, due to their high melting points, occur as discreet inclusions in gold. They usually contain higher osmium contents. Secondary gold deposits (placer gold deposits) exploited in ancient times are frequently related to (weathered) ultramafic rocks where these inclusions occur mixed with gold particles. In fact, it was already suggested by Hartmann (1970) that the presence of PGM inclusions in gold objects can be used as an indication for an origin from placer deposits. Such inclusions consist of alloys of iridium, osmium and ruthenium with low platinum contents in grain sizes of about 10–100 µm (Fig. 3.29).

The interpretation of Meeks and Tite (1980) that PGM inclusions in gold could be indicators for a placer deposit origin of the gold is therefore likely to be correct, just like their hint back then that using such inclusions for provenancing gold would be difficult. The following should be noted: as accessories together with chromite ($(Fe,Mg)Cr_2O_4$), titanomagnetite ($(Fe,Ti)_3O_4$) and ilmenite ($FeTiO_3$), PGM occur only in ultramafic igneous rocks, especially in dunites. They are thus regionally limited to corresponding ore deposits or rock complexes of ophiolites: Thus, identifying PGM inclusions in gold artefacts suggests that the origin of such artefacts may be in one (or more) regional environments of ophiolites. A good overview of platinum group mineral (PGM) placer deposits can be found at Cabri et al. (1996). Elevatorski (1996) described gold deposits of Europe and the Middle East. Gold placers with PGMs are known, e. g. from western Turkey, one of the major gold sources of the Persian Empire. PGMs also occur in ultramafic rocks in old Precambrian shields in the "greenstone belts" of Egypt (Klemm and Klemm 2013). In geologically young deposits, placer gold may be mixed with weathered ultramafic and basic volcanic rocks.

Thus, if PGMs are comparatively frequently mixed with gold in the sedimentary domain and then appear as inclusions in gold artefacts, it remains to be determined to what extent even isotope measurements in the osmium system of PGM inclusions can even be used as a tracer. There could potentially be a problem: Even though in geology the measurement of $^{187}Os/^{188}Os$ is used as an indicator of geological processes and developments especially of ultramafic rocks, Bowles et al. (2000) point out that possibly alluvial platinum group mineral grains primarily formed within ultramafic intrusions, followed by weathering into laterites could possibly (but not necessarily) suffer from a further alteration prior to alluvial concentration. This note is important to consider, because in the environment of ophiolites under suitable climatic conditions, nickel laterites and bauxites can be formed. Their use in ancient times cannot be ruled out.

The provenance of gold artefacts was tried by a chemical–analytical characterisation of PGM inclusions. This was done on several archaeological gold objects of different provenance. Young (1972) analysed a series of PGM inclusions in gold artefacts from the Royal tombs of Ur, Mesopotamia, and noted at the same time that Lydian gold coins from the Pactolus region in western Anatolia have PGM of comparable chemical composition. He concluded that the gold artefacts in Mesopotamia would originate from the Pactolus river and its tributaries. Using energy dispersive X-Ray fluorescence spectrometry in a scanning electron microscope, Meeks and Tite (1980) analysed over 100 PGM inclusions of gold artefacts from the collection of the British Museum, London, and mostly found alloys of iridium, osmium and ruthenium, while platinum only occurred in a subordinate fashion. They found, however, that the PGM inclusions would vary so much within a single object that they would not be suitable tracers for the identification of gold deposits. No further considerations about the origin of the gold artefacts were therefore made in the publication, in part also because no further comparable data was available.

Junk (2001) began using Os isotopes to analyse PGM inclusions in Celtic gold coins, based on a new analytical basis. He began to publish detailed data about virtually non-destructive spot analysis by laser connected with a LA-MC-ICP-MS.

Consequently, Junk and Pernicka (2003) investigated nine Celtic gold coins of the second century BC which circulated in central Europe, the so-called rainbow cup coins (German *Regenbogenschüsselchen*). With these coins, platinum group mineral (PGM) inclusions in the Celtic gold appeared, and they continued to analyse the isotope abundance ratios of $^{187}Os/^{188}Os$ vs. $^{187}Os/^{186}Os$. The authors performed 100 spot analyses on PGM inclusions. The results showed that the osmium abundance ratios varied considerably and indicated different types of PGM inclusions and, herewith, possibly various PGM deposits associated with different gold placers. The authors excluded the possibility of any bulk analyses for PGE in the matrix of gold artefacts. It was obviously difficult to define a special raw source for this special sort of artefacts which, due to their wide distribution all over Europe, certainly suffered from problems such as too many ophiolitic rocks in the Alpine regions, on the Balkans and in Anatolia, and, in addition, problems of recycling and re-melting in the late first century BC.

These aspects might have been more limited at the Early Dynastic III city of Ur, Mesopotamia. A large number of gold artefacts was excavated from the Royal tombs of Ur (Zettler and Horne 1998), and the typical feature of these gold artefacts was their close association with decorations of the beautiful unique semi-precious blue lapis lazuli. Jansen et al. (2016, 2018b) thoroughly studied the exceedingly abundant PGM inclusions of these artefacts for their chemical and their osmium isotope composition (Fig. 3.29). Using measurements of 79 isotope ratios of $^{187}Os/^{188}Os$, the authors were not only able to assign a geological characterisation to the PGM and noticed that the PGM minerals are connected to ophiolite complexes. They furthermore were also able to date these PGM: they were formed

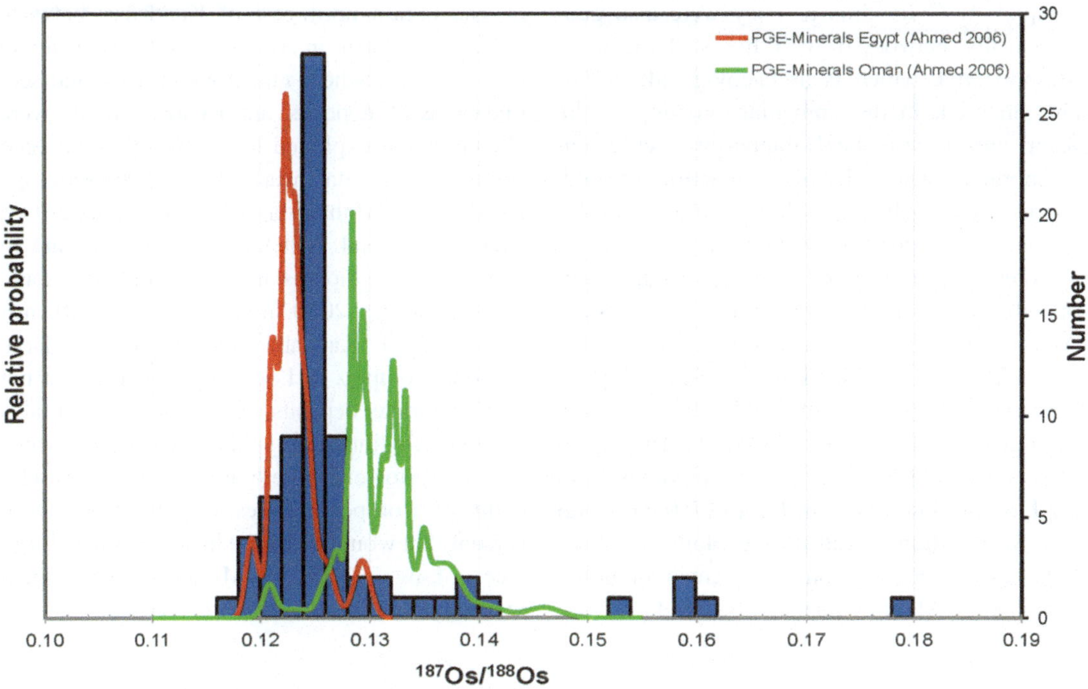

Fig. 11.13 Osmium isotopic composition of PGM inclusions in gold artefacts from the Royal tombs of Ur (blue histogram) and the composition of PGM from ultramafic rocks of the Eastern Desert of Egypt and the Samail ophiolite complex from Oman as relative probability (data by Ahmed et al. 2006). PGM inclusions from Ur show a frequency distribution that differs from that of PGM of Oman and the Eastern Desert. The data of the inclusions of Ur was not presented as Relative Probability due to the very small measurement error, which results in frequency spectra characterised by many narrow single peaks. From Jansen et al. (2016)

290–690 Ma ago (Fig. 11.13). They also indicate a potential origin of the gold in a geological context with (ultra-)basic rocks (ophiolites) with chromite ores. But also nickel laterites and bauxites, which frequently carry gold and PGM together, are eligible.

Their results together with archaeological observations allow for a very good regional delineation of the origin. Exceedingly rich gold placers in the geological vicinity of ophiolite complexes especially of this age, in striking vicinity of the world-famous enormous mines for lapis lazuli at Sar-i-Sang in north-eastern Afghanistan point to the die gold placers close to the Paleozoic ophiolites like Samti of Samti (Takhar) in northern Afghanistan (Bactria, in the region of the river Amur Darya) near the border to Tadzhikistan (Chap. 3.6.3 and Fig. 3.44b). It is commonly known that the deposit of lapis lazuli is unique in the ancient world, which means that it is only rational to search for a gold source in its vicinity. The gold deposits at Samti are currently worked by artisanal miners to a depth of c. 60 m (Chmyriov et al. 1973; Dronov et al. 1972). These placers are 8 km long, 900–1700 m wide and have a thickness of 28 m. The reserves are estimated to be around 20–25 t of gold. The PGM-leading gold of the Achaemenian "Oxus-treasure" probably also comes from Samti (Mongiatti et al. 2010) and possibly also that of the Hellenistic gold treasure of Tillya Tepe (Schiltz 2010). Also, the late third/early second millennium BC dated gold artefacts from Tepe Fullol (Tosi and Wardak 1972; Jarrige 2010) could possibly originate from this source. These settlements, as well as the ophiolitic gold deposits, are all located along old trade routes between Mesopotamia and the Indus Valley (Kohl and Lyonnet 2008), whose early phases are called Bactria-Margiana archaeological complex or Oxus culture. They were also geographically networked with the settlements in Mesopotamia, which is archaeologically verifiable after all. In the immediate vicinity of the Samti area is also the Harappa-time settlement of Shortugai, whose discovery spectrum of gold, copper and carnelian also shows a direct connection with the Indus Valley (Francfort 1989).

Alternatively, from the pure geological dating, ophiolitic rocks from Zarshouran (Iran, western Azerbaijan) should also be mentioned as possible sources, but the combination with archaeological facts makes this source rather unlikely. Due to the geological dating, other PGM bearing gold placers, e.g. old Neoproterozoic ophiolites from the Eastern Desert type in Egypt (750–900 Ma) and Saudi Arabia, and the geological young Mesozoic ophiolites from the Samail ophiolite complex (96 Ma) in Oman can be ruled out.

The region north-east of Kandahar, between Zabul and Ghazni in Afghanistan, also appears to be of interest, where numerous primary and secondary gold deposits are located, which were already being exploited in the twelfth–fourteenth century AD. In total, nearly 100 gold deposits were listed by the Afghanistan Geological Survey (see also Wolfart and Wittekindt 1980). They are in context with ophiolites and also with laterites and bauxites (Peters 2000; Benham et al. 2010). Just north-east of Kandahar the 2000 km^2 large ophiolite complex of Logar is situated, where not only chromite deposits can be found, but also economically interesting platinum group elements.

So far, however, there are very few trace element or isotope analyses of ores from these regions that would be of use in archaeometallurgy (see Law 2008). Mining–archaeological investigations could also not be carried out.

Glossary

Adit Drifting an adit from the surface was originally performed to follow a mineralisation exposed at the surface. In a second stage, an adit could be constructed for prospecting purposes and transportation of ore.

Age-Hardening The process of hardening spontaneously overtime at ambient conditions. Some steels age-harden as do other alloys. The process was first studied in aluminium-copper where coherent precipitates of different structure form as the first stage of the process.

Alloy A metallic mixture of two or more elements. An alloy can be formed by two or more metals, e.g. copper and tin to form bronze, or by a mixture of a metallic and non-metallic element, e.g. iron and carbon to form steel or cast iron. An alloy tends to be low in ductility and to be hard and brittle. It can be mixed deliberately, but it can be a natural alloy smelted by polymetallic ores. In archaeometallurgy, it is sometimes difficult to prove deliberate mixing.

Alluvial Deposit The local concentration of heavy minerals (gold, tin, > PGM) in old river deposits. These form when minerals weathered and transported from original solid geology deposits are locally concentrated in the bed of streams. These may be buried by quaternary gravels.

Alpha-Iron See Ferrite

Aludel Distillation apparatus. A pear-shaped ceramic vessel open at both ends so that one could fit over another, used for sublimation and distillation of (semi-) metals (mercury, antimony) with low melting and boiling points via the vapour phase.

Amalgam A compound or mixture of mercury with other metals. Mercury may form an amalgam with gold, silver, tin, zinc, lead, copper and other metals.

Amalgamation Gilding The process used for gilding of many copper alloys in ancient and historic times. Gold becomes pasty when mixed with mercury and may be applied as a paste over a surface. This can be extracted by heating to drive off most of the mercury.

Annealing Heat-treatment is carried out on metal or alloy, usually to soften the material after cold working to allow further deformation (e.g. hammering). The lowest temperature at which metal will soften varies with the degree of cold-working, greater amounts of work tending to reduce it.

Annealing Twin In face-centred cubic (fcc) metals, a process of recrystallisation metals worked and annealed in which a mirror plane in the crystal growth results in two parallel straight lines appearing across the grain when the metal is etched.

As-Cast Structure The metallurgical structure (distribution of phases and grain texture) formed during the solidification of the metal or alloys after casting. Such structures are often composed of heavily cored dendrites such as in bronzes. Grain sizes and shapes vary with distance from the mould surface. The chemical composition also varies as a result of elemental segregation during freezing. As a result of gravity segregation, inverse segregation, heavily cored structures can be produced. To make the metal suitable for use anneal and mechanical work is necessary to

A. Hauptmann, *Archaeometallurgy – Materials Science Aspects*, Natural Science in Archaeology,
https://doi.org/10.1007/978-3-030-50367-3

homogenise the metal and reduce the grain size and allow diffusion to even out the local chemical inhomogeneity.

Assaying To test the quality of ore or metal. In ancient times by making use of simple chemical reactions in small scales to the extracted metal in visible form. Multiple assaying can be serial production.

Austenite Phase transition of body-centred (bcc) to face-centred cubic configuration (fcc) of iron between 910 and 1390 °C. It will dissolve up to 2.1 wt.% of carbon, the presence of which expands the stability range to 723–1493 °C. Other elements (Ni and Cr) will expand the austenite stability range to room temperature.

Bainite A term used to describe the metastable microstructures consisting of a fine dispersion of carbide particles in ferrite. It occurs in rapidly quenched hardenable steels, which have been quenched at a rate intermediate between that necessary to produce fine pearlite and martensite.

Bellows Means of blow air into a furnace. They are not frequently preserved in the archaeological record. The main evidence of various bellows that could have been used in ancient times comes from historical or ethnographic sources. Types of bellows that could have been used: pot bellows, bag bellows, box bellows as well as the 'normal' smiths bellows. Initially, bellows were hand or foot powered. With the introduction of water-powered bellows furnace sizes and temperatures could increase if suitable refractory materials were available.

Beneficiation Methods of improving the concentration of ores from gangue and host rock before smelting. In ancient times this was done by crushing with (stone) tools, hand sorting or various washing techniques.

Black Copper Crude copper high in oxygen content. Black copper would require refining to remove the excess oxygen.

Blast Furnace Shaft furnace in which the charge and fuel descend from the top, with air is blown in near the base. The combustion of the fuel at the base provides both temperature and redox-conditions required for the reactions, and to melt both the desired product and slag.

Blister Copper Crude metallic copper produced by the decomposition of partially oxidised matte. It typically contained 80–90 wt.% metal with remains of sulphur, iron, oxygen.

Bloom Porous or spongy mass of ferritic iron with parts of steel mixed with liquid silicate slag and charcoal inclusions. It is an intermediate product that has to be refined and densed in smaller vessels. The carbon content is usually low. However, it was possible to produce high-carbon bloomeries with steel. Crude bloom must be extensively worked at welding heat to consolidate the iron and remove inclusions of slag and charcoal. It is usually hammered to larger blocks.

Bloomery Process Process in which a spongy mass of iron is produced in a (semi-) solid state directly as a result of the reduction (e.g. smelting) of iron ore in a bloomery (shaft) furnace. The slag produced with the bloom is liquid and usually tapped. In German language Rennfeuer-verfahren. Pure iron melts at 1535 °C, but bloomery iron has usually never been heated above c.1250 °C. The bloomery process was the main method by which metallic iron was obtained before the introduction of the indirect or blast furnace process.

Body-Centred Cubic (bcc) A unit cell in which atoms are arranged at each corner of a cube with one atom in the centre of the cube. Each atom at the corners is shared by each neighbouring unit cell. BCC metals (barium, chromium, iron, molybdenum, tantalum, vanadium, tungsten) are strong and tough and can accommodate small atoms within the lattice.

Boudouard Equilibrium Relationships between the reaction gases carbon monoxide (CO) and carbon dioxide (CO_2) generated during the combustion of carbon combustion during smelting. Boudouard equilibrium defines a temperature-dependent equilibrium position in the combustion of carbon.

Brass Alloy of copper and zinc (up to 40 wt.%). In ancient times with copper as a major alloying component and only 10–30 wt.%

zinc. Colour of brass changes with increasing zinc from copper red to pale yellow to white.

Bronze Copper-based alloy with metals except for zinc and nickel. The most important bronze in ancient times was an alloy of copper with tin, very often around 10 wt.%.

Carat Carat is a scale unit used in several ways. The metric carat is used for the weight of diamonds and other gems. 1 carat is 0.2 g. Then it is a scale unit for the purity or fineness of gold. Fineness of alloyed gold can be expressed in the number of parts of gold by weight that are contained in 24 parts of the alloy, e.g. 18-carat gold contains 18/24 parts of gold and is 75 wt.% gold or 750 fine. Pure gold is 24 carat or 999 gold, i.e. 99.9 wt.% gold (theoretically1000 fine but this is technically impossible).

Carburisation The process of increasing the carbon content of wrought or bloomery iron by heating the metal below its melting point with carbonaceous matter (charcoal). The process is slow as it is controlled by solid-state diffusion. Carburisation of iron may happen in the bloomer process where mostly ferritic iron is produced. If this is treated in a smithing hearth it will result in an uneven distribution of carbon with higher concentrations of carbon at the surface than the centre of the piece. Thus, to produce a steel with a more uniform carbon distribution it was necessary to form welded composite by forging the bar and welding it either back on itself or together with other bars.

Casting The operation of pouring liquid metal into a mould of the desired shape and allowing it to solidify. The simplest way was casting of raw metal into sand moulds as exemplified by copper oxide ingots. Clay shaped or stone open moulds are uncovered at the time of casting. They were used for simple early Bronze Age axes. Piece moulds are made of two or more fitting pieces in stone, bronze or sand-clay mixture. See also > Lost wax casting

Cementation—Gold/Silver Several meanings in archaeometallurgy. Cementation of gold alloys with salt in a crucible may remove silver leaving pure gold behind. Brass has been made by the interdiffusion of zinc vapour to metallic copper, called the cementation process. Cement copper is formed by the precipitation of fine copper on scrap iron from copper-bearing solutions.

Cementation Zone Metal-rich supergene enrichment zone or cementation zone occurs at the base of the surface near oxidised part of an ore deposit at the groundwater level. This area is characterised by oxygen deficiency, i.e. by a reducing environment. Copper sulphides are enriched there, gold, silver and nickel.

Chaîne Opératoire Ordered a chain of actions, materials and processes in a production sequence in metallurgy which led to the transformation of a given material towards the finished product. The concept is significant in allowing the archaeologist to infer back from the finished artefact to the procedures, the intentionality in the production sequence, and ultimately to the conceptual template of the maker.

Chamber-Pillar Mining This way of exploitation is applied in planar mining activity to exploit a horizontally mineralisation. To avoid collapsing of the hanging wall pillars have been left or are constructed in certain distances inside a mine.

Charcoal Wood heated by limited air supply, so that part of the wood is converted to pure carbon, driving off the water and other volatile components. It is possible to achieve higher furnace temperatures using charcoal than wood. Black coal in ancient metallurgy was not known.

Cire Perdue Term meaning > Lost wax casting.

Clarke Values (Clarkes) Basic values for geochemical comparison of rocks. They describe proportions of chemical elements in the lithosphere.

Component Every chemical material is used for making a chemical system. The gold-silver-copper-system consists of the components gold, silver and copper.

Converter Process Metallurgical process in steel making only used for oxidising refining, i.e. decarburisation of cast iron, and additionally for removal of phosphorus.

Craton A craton is an old and stable part of the continental lithosphere. Cratons are generally found in the interiors of tectonic plates. They are composed of geologically ancient crystalline basement rocks. They have a thick crust and deep lithospheric roots that extend as much as several hundred kilometres into the Earth's mantle. Cratons can be described as shields, in which the basement rock crops out at the surface, and platforms, in which the basement is overlaid by sediments and sedimentary rocks.

Crucible Small vessel for smelting metals from ores in the early stages of metallurgy. Made of 'normal' pottery. Also container for melting metal. Later made from refractory clay, often graphitized.

Crucible Steel Crucible steel is a generic term to describe all types of steel formed or melted in a crucible. There were three basic ways of making crucible steel. 1. Simply melting steel of suitable composition. 2. Melting a mixture of low carbon iron and cast iron in a crucible together with a flux to seal the melt to prevent decarburisation. 3. To seal a mixture of low carbon iron, a material high in carbon, and fluxing material in a crucible to form a seal once the mixture had started to melt. Then to heat the crucible so that the iron was carburised to the point that the melting point was reduced to the operating temperature of the furnace.

Crystal Homogeneous solid material each individual atomic particles are aligned in regular three-dimensional arrays. In most pure metals crystals consist of cubic face centred, cubic body centred or hexagonal repeat units. Intermetallic compounds, minerals and phases can much more complex atomic arrangements that define the crystal structure. Those materials which do not have a regular repeating arrangement of atoms are amorphous or glassy.

Cupel A porous ceramic, often made from bone ash or other refractory components. The cupel is used to extract or assay precious metals that have been dissolved in the metallic lead by the process of cupellation.

Cupellation The process used for extracting silver and gold from lead. The principle involves first the dissolution of the material to be tested in molten lead, then the selective oxidation of lead to litharge (PbO) in a shallow, dish-shaped crucible usually made of bone ash (> cupel), leaving the precious metals behind as a molten globule. A temperature of about 1000 °C is needed. The litharge is skimmed off or is combined with the bone ash in the cupel.

Depletion Gilding Surface enrichment of gold by removal of one or more base components. Well known from ancient South America for the gilding of > tumbaga, but also used in Early Bronze Age Mesopotamia.

Depletion Silvering Silver-copper alloys usually develop a scale of copper oxide when worked and annealed. Removing this oxide scale enriches the surface in silver creating a depleted copper zone and making the alloy silver in colour.

Diffusion The movement of one type of metal or alloy through another without reaction. Usually, heat is required for this process to occur. Many of the important processes in metallurgy are controlled by diffusion. Carburisation of iron is a diffusion. During the solidification of a liquid, the rate of diffusion can be controlled by rapid cooling (quenching).

Distillation This is the process of separating volatile components or substances from a heavy volatile material by boiling and subsequent condensation in closed vessels. In archaeometallurgy, highly volatile metals are arsenic, mercury and zinc.

Ellingham Diagram The diagram shows the stability of compounds as a function of temperature. In metallurgy, the Ellingham diagram is used to predict the equilibrium temperature between a metal, its oxide, and the reactions of metal with sulfur, nitrogen, and other non-metals, i.e. to predict the conditions under which ore will be reduced to its metal.

Eluvial (Residual) Ore Deposits In residual ore deposits, the interesting material is concentrated in situ, while weathering removes diluting parts of the rock. Hardpan and bog iron ores are eluvial deposits. Eluvial placers are bauxite, lateritic gold, platinum, iron (Ni, Co) and nickel ores, residual enrichment of subeconomic protore iron and manganese.

Enthalpy of Reaction (ΔH) The magnitude of ΔH is directly related to the amount of reactant used up in a process, but it is opposite in sign. Is it negative, energy will be free, the reaction is exotherm. Is it positive, energy has to be added, the reaction is endotherm.

Equiaxed Grain Crystal of equal dimensions or properties in all directions. Equiaxed grains have ideally hexagonal structures. The term is true for metal and any minerals.

Equilibrium Phase Diagram It is a diagram in two or three dimensions giving a plot of phases present at a given composition and temperature. Two-dimensional phase diagrams plot the composition of two elements (horizontal x-coordinate) against temperature (vertical y-coordinate), when three or more elemental are involved it becomes more complex to illustrate the phase diagram. Usually binary or triangular phase diagrams are used with the temperatures of phase boundaries or with the solidus or liquidus surfaces plotted as contour maps marked. Note, that the equilibrium phase diagram describes the (laboratory) situation at equilibrium. This often does not agree with (ancient) metallurgical realities and non-equilibrium phases (metastable phases such as martensite) are formed. See also > Micro-equilibrium.

Eutectic The eutectic is a fixed composition of two alloys or materials with the lowest melting point. During cooling of a liquid, the eutectic reaction converts one liquid phase directly into two distinct solid phases and the microstructure consists of an intergrowth of two or several solid phases ('view from above'). By firing charged ores (and fluxes) liquid will be formed at the fixed composition of the eutectic or a eutectic through ('view from below'). If the composition of the charge meets the eutectic in a system, the charge will theoretically completely liquefied.

Fire-Setting Removing rock or ore from a rock face by firing against it. It causes loosening and relaxation by dehydration effects and decomposition of minerals.

Gallery A horizontal or nearly horizontal underground passage in a mine. Drifting a gallery can be done within an ore deposit or in the adjacent host rock to open a mine.

Gangue The unwanted part of the ore that has to be removed by beneficiation. During smelting remains of gangue combines with any fluxing material embedded in ore or that may have been added to form slag.

Gibbs Energy (ΔG) Gibbs energy is a thermodynamically potential with the variables pressure and temperature and material.

Gossan The upper part and exposure of a metalliferous ore deposit from which metal has been leached down into the zone of secondary enrichment. This part of the vein is rich in iron, hence the German term Eiserner Hut (Iron Hat). Humid to semi-arid environments will forward the formation of gossans, in regions of glaciation the original gossan may have been removed.

Gunmetal Ternary alloy of copper, zinc and tin. Proportions vary but 88 wt.% copper, 8–10 wt.% tin, and 2–4 wt.% zinc is an approximation.

Hardhead This is an unwanted product very high in iron formed by strong reducing conditions during smelting of iron-containing ores of tin, copper, zinc and lead. It is formed in the solid state and may contain high concentrations of the base metal ($FeSn_2$).

Hydraulic Mining Mining by using water power to exploit fine-grained ores from sedimentary rocks. Mostly applied in ancient times for gold and tin ores.

Hypereutectic An alloy containing more of the alloying element than that required to form the eutectic composition. In steels, this would require more than 0.8 wt.% carbon, the amount needed to create a completely pearlitic structure, but less than about 2 wt.% C as above 2% the composition enters the range for cast irons.

Hypoeutectic An alloy containing less of an alloying elements than that required to form the eutectic composition. The microstructure may show some eutectic intergrowth and primary dendrites of the alloying material (solvent).

Isotope Analysis Isotopes are atoms of the same element that have an equal number of protons and unequal number of neutrons, giving them slightly different weights. Isotope analysis is the identification of the abundance of certain stable isotopes and chemical elements within organic and inorganic compounds, the isotopic signature. Stable isotope ratios are measured using mass spectrometry, which separates the different isotopes of an element on the basis of their mass-to-charge ratio. In archaeometallurgy, the ratios of lead isotopes (^{204}Pb, ^{206}Pb, ^{207}Pb, ^{208}Pb) are the most widespread and successful tools for provenance studies of any materials related to the production and distribution of metal. Next to lead isotopes those of tin (selection of $^{112\text{-}124}Sn$), copper (^{63}Cu, ^{65}Cu), and osmium ($^{187}Os/^{188}Os$ and $^{187}Os/^{186}Os$) are investigated.

Leached Capping Part of an oxidation zone of ore deposits. Enrichment of silica, often porous, and secondary metal minerals.

Ledeburite Ledeburite is the name applied to the cementite-austenite eutectic at 4.3% carbon which solidifies at 1130 °C. During cooling, the austenite in the eutectic may transform into a mixture of cementite and ferrite (which may be pearlitic).

Liquation The Liquation Process is the recovery of silver from other metals, notably from silver-containing raw copper. It is a considerable metallurgical innovation of the sixteenth century AD in Middle Europe. The basic principle of this process relies on the silver's strong affinity for lead and on copper-lead liquid–liquid separation at high temperature. Variable amounts of lead or lead oxide were added to liquid copper. The next step was the separation of the argentiferous lead from the solid copper-lead-silver cakes by melting at moderate temperatures. Lead promoted quick liquefaction, and the liquid now contained the silver originally solved in the copper. The liquated argentiferous lead ran off. It was further processed in the cupellation hearth. The lead was oxidised to PbO (litharge). It was skimmed off the cupellation hearth. When all lead was oxidised, the shine of the silver or *Silberblick* appeared.

Liquidus Line The boundary on a phase diagram that shows the temperature at which solidification begins during cooling from the melt. It separates the complete liquid state from the state with beginning crystallisations. In a ternary diagram, the liquidus is a surface, not a line. See also > Solidus line.

Litharge The lead oxide a PbO, typically yellow to reddish in colour. Formed as a result of cupellation to recover silver from argentiferous lead.

Lost-Wax Casting Casting from a wax model. An object to be made is shaped in wax and is covered in a clay mould. The mould is heated to melt and burn out the wax. The resulting space is then filled with molten metal.

Matte Product present at some stage in the production of copper from sulphide minerals. It consists of a liquid or solid mixture of sulphides, usually of FeS_x and Cu_2S, but other sulphides may be dissolved in the mixture too. Increasing copper concentrations in CuFe-matte is changing from yellow to violet to blue. Matte is rather quickly liquefied and separated from siliceous charged ore and from slag in its state of formation.

Microequilibrium The concept of micro- or local equilibria applies to slag formation systems under generally non-equilibrium, i.e. under conditions in small crucibles or smelting vessels. Local equilibria are established whenever individual mineral grains such as quartz, limonite and clay are in close contact with each other. Because firing in crucibles is a short-term event, the time during which a charge was subjected to maximum firing temperature is rarely sufficient to attain an equilibrium situation. Even if partially melted phases existed in a slag formation in situ the considerable viscosity of these silica-rich melts largely prevented the crystallisation of equilibrium phases. The charge, therefore, is left in a state of distinct non-equilibrium. The useful concept of microequilibrium explains why thermodynamically incompatible mineral phases such as quartz, gehlenite and wuestite can occur side-by-side.

Mohs Hardness In geoscience, the hardness of minerals is measured by a 10-graded test defined by a selection of minerals. The method used mainly in mineralogy, it is not suitable for metals and alloys because the differentiation is too low. See Vickers hardness.

Mother Alloy A mother alloy (or parent alloy) is an intermediate product that contains a high concentration of the alloying partner in the parent metal. It is melted with the base metal to produce the actual alloy.

Neumann Lines or Bands Twins in iron are caused by sudden shock impacts at low temperatures. Frequently observed in ancient iron artefacts caused by cold-hammering. In meteorites probably caused by shock on entering the earth's atmosphere. They disappear when the metal is heated above about 600 °C.

Niello Mixture of sulphides, usually copper and silver, are used as a black decorative inlay on silver and some other metals.

Open Mould A form of mould for flat or long object of relatively simple shape in which the top is left open to the air. Examples of the type of object that were cast in open moulds are Bronze Age flat axes, ingot moulds of all periods, cast-iron pigs and fire-backs and probably Roman mirrors.

Oxidation Chemical reaction in which a metal is converted to its oxide, or from one oxidisation state to a higher state (that is the equivalent of an increase in the number of oxygen atoms for each metal atom). Example: $Fe^0 \rightarrow Fe^{2+}O \rightarrow$, $Fe^{2+,3+}{}_3O_4 \rightarrow Fe^{3+}{}_2O_3$, i.e. iron is progressively oxidised from metal to its highest oxide, hematite. The metal oxide equilibria are compiled in the > Ellingham diagram.

Oxidation Zone The upper part of an ore deposit reaching sometimes down to the groundwater level. Deep weathering leads to the formation of oxidic ores such as malachite, cuprite, calamine, cerussite, limonite and others from primary sulphidic ores. The most upper part or outcrop is the > gossan.

Panning Mineral washing process in which the lighter unwanted mineral is removed from the wanted mineral in a shallow vessel (or pan). Panning was used as a prospecting method, and for small-scale ore processing of gold deposits.

Parent Alloy See Mother alloy

Parting Technology to extract silver from natural or artificial gold-silver alloys. The earliest way was done using salt (natrium chloride). The impure gold was heated in a matrix of sand and salt to convert the silver to silver chloride which diffused out of the metal. Earliest archaeological evidence to produce the gold for the Lydian coinage produced at Sardis in the sixth century BC. Other parting methods used to extract silver from gold included acid, and the antimony or sulphide parting.

Pearlite Eutectic phase mixture of ferrite (α-Fe) and cementite (Fe_3C) in alternate parallel lamellae. Formed by eutectoid decomposition of austenite by slow to moderate cooling.

Pewter Ancient pewter is an alloy of tin and lead, much used in Roman times, and from the Medieval to the eighteenth century. Ancient pewter may consist of 60–80% tin and 20–40% lead, modern pewter may consist of 15–30% copper, 5–10% antimony and 87–94% tin. The poisonous nature of lead has resulted in the replacement of lead with antimony, although antimony is also inadvisable in high amounts for cooking utensils.

PGE See > Platin group elements

PGM See > Platin group minerals

Phase Several Meanings:

1. Artificially produced components even of identical compositions are phases, contrary to natural minerals. Silver in solid solution with gold is also a single phase no matter what the composition. The same is true for Mg- and Fe-

rich olivines (fayalite). There are complete solid solution between the two materials.

2. Single chemical component having different physical and crystallographic states: water, ice and steam have the same chemical formula but are different phases. Due to decreasing temperatures various modifications of alloys due to decreasing. Notation with Greek letters (α, β, γ, δ, ε).

3. Not to change with the archaeological meaning of the chronological phase.

Phase Diagram (Equilibrium) A diagram with axes of temperature and composition describing the different phases that will be present in a chemical system or alloy at equilibrium at a given composition and temperature. A binary phase diagram consists of two elements. A ternary system consists of three elements; these are often plotted as a triangular diagram with temperature as contour lines. In modern material science discussions on temperatures of solidification. In contrast, discussions of the lowest liquation of charged ores along eutectic throughs ('view from below') interesting for ancient slag formation or the liquation process of desilvering copper.

Placer Deposits Placer deposits are formed by the release of heavy minerals (or metals in case of gold) by weathering of a primary ore body, and concentration by water. The most important placer deposits are of gold and tin minerals (cassiterite), although magnetite and platinum group minerals may be concentrated also. Placer deposits are mostly associated with geologically older rivers. Alluvial 'stream' tin deposits of Devon and Cornwall were buried several metres below present river levels. As well as alluvial deposits in river and stream valleys, placer deposits include eluvial deposits weathered out from primary ores.

Platinum Group Elements (PGE) The group of elements whose characteristics are similar to platinum, i.e. high densities and melting points, and unreactive. These elements are iridium, osmium, palladium, rhodium and ruthenium

Platinum Group Minerals (PGM) Platinum group minerals are alloys of osmium-iridium-ruthenium. They occur as small silver-grey particles in alluvial gold. As they were harder and more heat-resistant than the gold they will form specific inclusions in the precious metal after melting. Sometimes found in archaeological gold artefacts, useful indications for provenance studies.

Protor Primary (sulphidic) zone of an ore deposit covered by the metal-rich cementation zone and oxidation zone. > Zoning primary mineralisation.

Reduction Chemical reaction in which a metal is transformed from its oxidic or sulphidic state to the metallic form, or from higher to lower oxidic state ($Fe^{3+} > Fe^{2+}$).

Refining Refining is the separation of a metal from its impurities. As such it is applied to a wide range of different processes for the different metals. Copper after smelting from its ores was in general impure (black copper). It had several percent of iron, copper-sulphides and -oxides, either solved in the metal or as mechanical inclusions. It would have to be reduced by a refining process if the copper was to be useable for most purposes. This was done by melting prills and ingots under rather oxidising conditions to convert the iron to iron oxide (usually magnetite), and then under reducing conditions (poling) to remove oxygen. Examples for trading black copper are oxhide ingots from Cyprus. It is open when a deliberate refining process is developed in the history of metallurgy.

Refractory The term either describes the heat-resisting properties of a material, or the heat resistant objects such as tuyères, crucibles and furnace and hearth linings. A refractory material must have a high melting point, chemical inert with respect to the charge, and stability at high temperatures.

Roasting Heating of ores in open heaps or in roasting stalls. Ores are roasted under oxidising or very limited reducing conditions. Roasting was applied to drive off the water (constitutional water and dampness), to decompose carbonates, and to break up ore to make it more permeable to gases. Roasting may be used to partially convert sulphur contents to oxides prior to forming a matte.

Salamander See Hardhead

Secondary Enrichment Zone See > Cementation zone

Segregation It describes compositional inhomogeneity developing during the solidification of an alloy or impurities in a metal from a melt. This results in the liquid composition becoming enriched in the lower melting point component. A 'normal' segregation occurs in the casting when the metal freezes from the outside to the centre. The local composition of the metal on the outside of the casting will have a higher melting point than the centre. The centre of the casting will collect all the low melting point components. Inverse segregation means that the lower melting point constituents, such as the phases rich in tin, arsenic or lead in bronzes are concentrated toward the surface of the cast due to contraction of the casting forcing the remaining low melting point liquid from the centre to the edge of the casting.

Shaft Miners are sinking a vertical or diagonal shaft to follow a mineralisation (vein) exposed to the surface and to open a mine. Further, the construction of shafts served as vertical adits to climb down to an underground mine, for hauling and for ventilation.

Shaft Shaft mining or shaft sinking is carving a vertical or near-vertical tunnel from the top down. Shafts were sunk for several purposes: prospection to reach mineralisations to be exploited, hauling, vertical riding and climbing, ventilation of a mine. They often have the function of a central installation of a mine.

Sluice Box An elongated or large box with riffles transverse at the bottom to the inclined box. This is used to trap heavy gold particles while lighter minerals are washed away.

Solid Solution Single solid crystalline phase containing two or more chemical materials. Concentrations are limited by miscibility gaps of phase equilibrium.

Solidus Line The line in a phase diagram describes the temperature at which a material of given composition is completely solid. It is also a boundary between those regions that are completely solid and those with a mixture of solid and liquid. At specific temperatures and compositions, the liquidus and solidus lines may meet (the eutectic and peritectic points, and at the melting point of a pure phase).

Speiss The word speiss describes the compounds of the heavy metals iron, copper, nickel and cobalt with major amounts of arsenic and/or antimony. They are called arsenides or antimonides. In speiss gold, silver and platinum can be enriched. Nevertheless, speiss is a rather unwanted product in metallurgy formed in copper- and lead-silver-production. A recovery of precious metal in ancient times is doubtful. Speiss and sulphidic matte are only slightly soluble in each other.

Steel Steel is an alloy of iron and carbon in which the carbon content does not exceed about 2.1 wt.%. Low-carbon steel contains 0.09–0.2 wt.% carbon and is soft—the same composition range as bloomery iron. Medium carbon steels contain 0.2–0.4 wt.% carbon, high carbon steels > 0.4 wt.%. Soft steel may be hardened by heat treatment.

Structure Structure is the spatial arrangement of various components of comparable materials in the macro scale. Rock structure shows identical mineral groups, fragments of rocks, joint system, folding, slags arrangement, shape and size of bubbles, inclusions. Structure in the German language is synonym with > Texture.

TEMB see > Tethyan Eurasian metallogenic belt

Tethyan Eurasian metallogenic belt The Tethyan Eurasian metallogenic belt (TEMB) was formed by rifting in the area of the former Tethyan ocean on the southern margin of Eurasia with the Afro-Arabian and Indian plates. The TEMB is of global size. It extends from the western Mediterranean via the Alps to south-eastern Europe, Anatolia, Caucasus, Hindukush, Tibet, Burma, south-west

Indonesia. The geological geotectonic evolution was connected with the formation of an enormous number of ore deposits (Fe, Cu, Cu-Ni-Co, Pb/Zn, Au, Ag and others). The extraordinary wealth of ore deposits led to intensive mining activities in ancient times in these regions.

Texture The spatial relationships of the various single components of a rock, ore, slag, metal (crystals, parts of crystals, multigranular aggregates, glassy materials). The most common parameter to quantify texture is the crystal size distribution. In slags cooling rate and oxygen fugacity affects texture of crystals. Texture is observed in the hand specimen or microscopically. Texture in the German language is synonym with > structure.

Time Temperature Transformation Diagram (TTT) Heat treatments such as annealing, quenching, tempering alters and improves mechanical properties. Correlations of these treatments are presented in a time-temperature-transformation (TTT) diagram. It is especially used for describing the transformations of austenite, pearlite, bainite, martensite in steel metallurgy.

Tinning The application of a thin surface layer of tin for decoration (copper alloys), or corrosion protection (iron alloys).

Trace Elements Trace elements in ores or metals are a powerful tool for provenance studies. Useful are especially those elements that are stable from ore to metal through metallurgical processes (Au, Ag, Bi, Co, Ni, Pt and others). Trace elements differ from main and minor elements due to their lower concentrations (< c. 0.1 wt.%).

TTT Diagram See > Time Temperature Transformation diagram.

Tumbaga Name given in South America to the alloys of copper and gold. Large range of compositions and colours.

Vickers Hardness Microhardness test for all materials that uses a pyramid-shaped diamond. It is one of the most useful hardness testing scales. Hardness numbers internationally normalised.

Volcanic Massive Sulphide (VMS) Type of metal sulphide ore deposits, mainly copper-zinc, which are associated with and created by volcanic-associated hydrothermal events in submarine environments (> black smokers). Contain lead, gold, silver and other trace elements.

Whisker Needle-like, hair-like, fibrous crystals with diameters of a few micrometre and a length of hundred micrometre up to a millimetre. Whiskers are observed in (modern) copper smelting and experimental work.

White Metal Nickname for pure copper sulphide (Cu_2S) at matte smelting.

Widmanstätten Structure Texture formed by solid-state decomposition of a homogeneous phase into one or more other phases. This texture occurs in meteoritic iron, but also in low-carbon steels which cooled from high temperatures (~1000 °C) at a moderate rate. The texture in meteoritic iron is composed of an mesh-like octahedral pattern of lamellation of low-nickel kamacite, nickel-rich taenite and fine-grained intergrowths of the two, the plessite. Note that it is the proper name of the German language.

Zoning Primary Mineralisation Sequence in ore deposits including oxidation zone (above), cementation zone and protore (below, primary mineralisation).

Bibliography

Abdel-Motelib A, Bode M, Hartmann R, Hartung U, Hauptmann A, Pfeiffer K (2012) Archaeometallurgical expeditions to the Sinai Peninsula and the Eastern Desert of Egypt (2006, 2008). Meta 19(1/2):3–59

Abels BU (1972) Die Randleistenbeile in Baden-Württemberg, dem Elsaß, der Franche Comté und der Schweiz. C.H. Beck, München

Abu-Ajamieh MM, Bender FK, Eicher RN (1988) Natural resources in Jordan. Inventory-Evaluation-Development Program. Natural Resources Authority Amman

Addis A (2013) Late bronze age metallurgy in the Italian Eastern Alps: copper smelting slags and mine exploitation. PhD-diss, Dipt Geoscienze, Univers degli Studi di Padua

Addis A (2015) Late bronze age copper smelting slags from Luserna (Trentino, Italy): interpretation of the metallurgical process. Archaeometry 58(1):96–114

Addis A, Abgelini I, Nimis P, Artioli G (2016) Late bronze age copper smelting slags from Luserna (Trentino, Italy): interpretation of the metallurgical processes. Archaeometry 58:96–114

Addis A, Angelini I, Artioli G (2017) Late bronze age copper smelting in the southeastern Alps: how standardized was the smelting process? Evidence from Transacqua and Segonzano, Trentino, Italy. J Archaeol Anthropol Sci 9(5):985–999, https://doi.org/10.1007/s12520-016-0462-5

Ager DA (1980) The geology of Europe. McGraw Hill, London

Agricola Georgius (1556) De Re Metallica libri XII. Ed Fraustadt G, Prescher H. VEB Deutscher Verlag Wiss Berlin 1974

Ahmed AH (2007) Diversity of platinum group minerals in podiform chromitites of the the late Proterozoic ophiolite, Eastern Desert, Egypt. Ore Geol Rev 32:1–19

Ahmed AH, Hanghoj K, Kelemen PB, Hart SR, Araj S (2006) Osmium isotope systematics oft he Proterozoic and Phanerozoic ophiolitic chromites: in situ ion probe analysis of primary Os-rich PGM. Earth Planet Sci Lett 245(3–4): 777–791

Alamri Y, Hauptmann A (2013) The iron ore mine of Mugharet a-Wardeh, Jordan: excavation and new dating. In: Humphris J, Rehren T (eds) The world of Iron. Archetype Publ, pp 214–222

Albaréde F (2004) The stable isotope geochemistry of copper and zinc. Rev Mineral Geochem 55(1):409–427

Albarède F, Desaulty AM, Blichert-Toft J (2012) A geological perspective on the use of Pb isotopes in archaeometry. Archaeometry 54(5):853–867

Albertus Magnus (c 1269 AD) De re mineralibus et rebus metalliscis. Ed by Strunz H 1951/52, Acta Albertina Ratisbonensia (prev Ber Naturwissenschaftl Ver Regensburg) 20:19–39

Al-Ganad I, Lagny P, Lescuyer JL, Ramboz C, Touray JC (1994) Jabali, a Zn-Pb-(Ag) carbonate-hosted deposit associated with late Jurassic rifting in Yemen. Mineral Deposits 29:44–56

Algaze G (2005) The Sumerian takeoff. Struct Dyn eJ Anthropol Relat Sci 1(1):1–41

Algaze G (2008) Ancient mesopotamia at the Dawn of civilization. University of Chicago Press, Chicago, IL

Al-Hamdānī K (1968) Die beiden Edelmetalle Gold und Silber. Ed. and translated by C. Toll. Acta Univ. Uppsaliensis 1. Uppsala

Alimov K et al (1998) Zinnbergbau in Mittelasien. Eurasia Antiqua 4:137–199

Allan JA (1979) Persian metal technology, 700–1300 AD. Ithaca Press, London

Alpers CN, Brimhall GH (1989) Paleohydrological evolution and geochemical dynamics of cumulative metal supergene enrichment at La Escondida, Atacama Desert, northern Chile. Econ Geol 84:229–255

Alpers CN, Hunerlach MP, May JT, Hothem R (2005) Mercury contamination from historical gold mining in California. USGS Fact sheet 1314

Al-Saad Z (2000) Technology and provenance of a collection of Islamic copper-based objects as found by chemical and lead isotope analysis. Archaeometry 42 (2):385–397

Al-Shorman A (2007) Refractory ceramic through the ages: an archaeometric study on finds from Fenan, Jordan and other sites. PhD-diss. Ruhr University Bochum

Ambert P (1995) Les mines préhistoriques de Cabrières (Herault): quince ans de recherches. État de la question. Bull Société préhist Française 92(4):499–508

A. Hauptmann, *Archaeometallurgy – Materials Science Aspects*, Natural Science in Archaeology,
https://doi.org/10.1007/978-3-030-50367-3

Ambert P (1999) Les minerais de cuivre et les objets metalliques en cuivre à antimoine-argent du sud de la France. Preuves d'une exploitation minière et métallurgie du début du III° millenaire av. J.C. In: Hauptmann A, Pernicka E, Rehren T, Yalçın Ü (eds) The beginnings of metallurgy, Der Anschnitt Beih, vol 9, pp 193–210

Ambert P, Guendon J, Laroche M, Mischka D (2002) Donneés nouvelles concernant le plus vieil établissement de métallurgistes de France. La Capitelle de Broum (district minier de Cabri``eres Peret, Herault) Péret-Herault. Archéologie en Langedoque 26:45–53

Ambert P, Bouquet L, Guendon J, Mischka D (2005) La Capitelle de Broum (district minier de Cabriéres Peret, Herault): établissement industriel de l'aurore de la métallurgie francaise, 3100-2400 BC. In: Ambert P, Vauer J (eds) La Première Metallurgie en France et dans le pays Limitrophes, Mémoire 37 de la Soc Préhist France, Carcasonne, pp 83–96

Amorós J, Lunar R, Tavira R (1981) Jarosite: a silver bearing mineral of the gossan of Río Tinto (Huelva) and La Unión (Cartagena, Spain). Mineral Deposits 33:111–136

Amstutz GC (1960) A geometric classification of basic intergrowth patterns of minerals. Amer Geol Instit Date Sheet 21

Anderson CG (2012) The metallurgy of antimony. Chem Erde 72:3–8

Anfinset N (1996) Social and technological aspects of mining, smelting and casting copper. An ethnoarchaeological study from Nepal. MA-thesis, Dept. Archaeol, Univers Bergen, Norway

Anfinset N (2011) Social and technological aspects of mining, smelting and casting copper. An ethnoarchaeological study from Nepal. Publ Deutsches Bergbau-Museum Bochum 181

Anguilano L (2012) Roman lead silver smelting at Rio Tinto: the case study of Corta Lago. PhD-diss, University College London

Anguilano L, Timberlake S, Rehren T (2010a) An early lead smelting bole from Banc Tynddol, Cwmyrtwyth, Ceredigion. J Hist Metall Soc 44(2):85–103

Anguilano L, Rehren T, Müller W, Rothenberg B (2010b) The importance of lead in the silver production at Riotinto (Spain). Archéoscience 34:269–276

Anheuser K (1999) Im Feuer vergoldet. Geschichte und Technik der Feuervergoldung und der Amalgamversilberung. AdR Schriftenreihe Restaur u Grabungstechnik 4. Theiss, Stuttgart

Appel MH, Garges JA (1991) New evidence of the meteoritic origin of the Tettenwang Iron Ore. Zeitschr Deutschen Geol Gesellsch 142:29–35

Aramburu-Zabala J (2001) Prospección Arqueológica en el Archipiélago de Cabrera (Islas Baleares), 4

Ardaillon É (1867) Les mines du Laurion dans l'antiquité. Paris

Armbruster BR (2000) Goldschmiedekunst und Bronzetechnik. Studien zum Metallhandwerk der Atlantischen Bronzezeit auf der Iberischen Halbinsel Monogr Instrument 15

Armbruster BR (2001) Metallguß. In: Hoops J (ed) Reallexikon der Germanischen Altertumskunde. Gruyter, Berlin, pp 623–642

Armbruster BR (2012a) Wikingerzeitliches Goldschmiedehandwerk in Haithabu. In: Pesch A, Blankenfeldt R (eds) Goldsmith mysteries. Archaeological, pictorial and documentary evidence from the 1st millennium AD in northern Europe. Schrift. Archäolog, Landesmus, vol 8. Wachholtz, Neumünster, pp 195–213

Armbruster BR (2012b) Feinschmiedewerkzeuge vom Beginn der Metallurgie bis in die Römische Kaiserzeit. In: Pesch A, Blankenfeldt R (eds) Goldsmith mysteries. Archaeological, pictorial and documentary evidence from the 1st millennium AD in northern Europe. Schrift. Archäolog, Landesmuseum, vol 8. Wachholtz, Neumünster, pp 59–85

Armbruster BR (2016) Technological aspects of selected gold objects from Ur – preliminary results and perspectives. Meta 22(1):113–135

Artioli G, Angelini I (2011) Mineralogy and archaeometry: fatal attraction. Eur J Mineral 6:849–855

Artioli G, Dugnani M, Hansen T, Lutterotti L, Pedrotti A, Sperl G (2003a) Crystallographic texture analysis of the iceman and coeval copper axes by non-invasive neutron powder diffraction. In: Fleckinger A (ed) Die Gletschermumie aus der Kupferzeit, Bd. 2. Schrift, Südtiroler Archäologiemuseums, vol 3, Bozen, pp 9–22

Artioli G et al (2003b) Early copper alpine metallurgy. Archaeometallurgy in Europe. Proc AIM Milano 2:19–27

Artioli G, Angelini I, Nimis P, Addis A, Villa I (2013) Prehistoric copper metallurgy in the Italian Eastern Alps: recent results. J Hist Metall 47(1):51–59

Artioli G, Parisatto M, Angelini I (2015) High energy X-ray tomography of bronze age copper ingots. In: Hauptmann A, Modarressi-Tehrani D (eds) Archaeometallurgy in Europe III. Proc Internat Conf. Deutsches Bergbau-Museum Bochum June 2011, Der Anschnitt Beih, vol 26, pp 387–400

Artioli G, Angelini I, Nimis P, Villa IM (2016) A lead-isotope database of copper from the Southeastern Alps: a tool fort he investigation of prehistoric copper metallurgy. J Archaeol Sci 75:27–39

Artioli G, Angelini I, Kaufmann G, Dal Sasso C, Villa IM (2017) Long-distance connections in the copper age: new evidence from the Alpine Iceman's copper axe. PloS One 12(7)

Artzy M (1983) Arethusa of the tin ingot. BASOR 50:51–55

Aruz J (2003) Art of the first cities. The Third Millennium B.C. from the Mediterranean to the Indus. The Metropolitan Museum of Art, New York

Asael D (2010) Copper stable isotope fractionation in low-temperature geological systems. Unpubl. PhD-thesis, Hebrew University, Jerusalem

Asael D, Matthew A, Bar-Matthew M, Halicz L (2007) Copper isotope fractionation in sedimentary copper mineralization (Timna Valley). Chem Geol 243:238–254. https://doi.org/10.1016/j.chemgeo.2007.06-007

Asael D, Matthew A, Oszepalski S, Bar-Matthew M, Halicz L (2009) Fluid speciation controls of öow temperature copper isotope fractionation applied to the Kupferschieder and Timna ore deposit. Chem Geol 262:147–158. https://doi.org/10.1016/j.chemgeo.2009.01.015

Asael D, Matthew A, Bar-Matthew M, Harlavan Y, Segal I (2012) Tracking redox controls and sources of sedimentary mineralization using copper and lead isotopes. Chem Geol 310–311:23–35. https://doi.org/10.1016/j.chemgeo.2012.03.021

Ashkenazi D, Cvikel D, Stern A, Klein S, Kahanov Y (2014) Metallurgical characterization of brass objects from the Akko 1 shipwreck, Israel. Sci Dir Mater Charact 92:49–63

Askeland DR (1996) Materialwissenschaften. Grundlagen Übungen Lösungen. Spektrum Akad Verlag

ASM Handbook Committee (1973) Metals handbook, vol 8, 8th edn. American Society for Metals, Metals Park, OH

Asmus B (2012) Medieval copper smelting in the Harz Mountains, Germany. Montanregion Harz 10. Deutsches Bergbau-Museum Bochum

Asrih L, Ingener N, Meyer T (2019) Bergbau als technonaturales system – Ein Beitrag zur modernen Bergbaugeschichte. Anschnitt 1:2–18

Assereto R, Brigo L, Brusca C, Omenetto P, Zuffardi P (1976) Italian ore/mineral deposits related to emersion surfaces – a summary. Mineral Deposits 11:170–179

Aversa G, Balasonne G, Boni M, Amalfitano C (2002) The mineralogy of the "calamine" ores in SW Sardinia (Italy): preliminary results. Periodico di Mineralogia 71(3):201–218

Avery D (1982) The iron bloomery. In: Wertime TA, Wertime SF (eds) Early pyrotechnology. The evolution of the first fire-using industries. Smithsonian Inst Press, Washington DC, pp 205–213

Avner U, Carmi I, Segal D (1994) Neolithic to bronze age settlement of the Negev and Sinai in light of radiocarbon dating: a view from the Southern Negev. In: Bar-Yosef O, Kra RS (eds) Late quaternary chronology and paleoclimates of the Eastern Mediterranean. RadioCarbon, Arizona, pp 265–300

Baadsgaard A, Monge J, Cox S, Zettler R (2011) Human sacrifice and intentional corpse preservation in the Royal Cemetary of Ur. Antiquity 85:27–42

Baatz D (1990) Schleudergeschosse aus Blei. Eine waffentechnische Untersuchung. Saalburg-Jahrb 45:59–67

Bachmann HG (1977) Bleiglätte-fund aus der Nordeifel. Bonner J 177:617–622

Bachmann HG (1978) Schlacken: Indikatoren archäometallurgischer Prozesse. In: Hennicke HW (ed) Mineralische Rohstoffe als kulturhistorische Informationsquelle. Hagen, Verlag Deutscher Emailfachleute, pp 66–103

Bachmann HG (1982a) Copper smelting slags from Cyprus: review and classification of analytical data. In: Muhly JD, Maddin R, Karageorghis V (eds) Early metallurgy in Cyprus, 4000-500 BC. Nicosia, pp 143–151

Bachmann HG (1982b) The identification of slags from archaeological sites. Inst Archaeol, Univers London, Occ Pap 6

Bachmann HG (1982c) Archäometallurgische Untersuchungen zur antiken Silbergewinnung in Laurion. Erzmetall 35(5):246–251

Bachmann HG (1991) Archäometallurgie des Silbers. Die Geowissenschaften 1:12–17

Bachmann HG (1993) The archaeometry of silver. In: Francovich R (ed) Archaeologia delle Attività Estrattive e Metallurgiche. Ed all'Insegna del Giglio. Firenze, pp 487–495

Bachmann HG (1997) Zementation: Ein mehrdeutiger Begriff in der Metallurgie. Meta 4(2):45–48

Bachmann HG (1999a) Gold for coinage: history and metallurgy. In: Schmidtbaur H (ed) Gold: progress in chemistry, biochemistry and technology. Wiley, pp 3–37

Bachmann HG (1999b) On the early metallurgy of gold. Some answers and more questions. In: Hauptmann A, Pernicka E, Rehren T, Yalçın Ü (eds) The beginnings of Metallurg. Proc intern Conf "The beginnings of metallurgy", Bochum 1995, vol 9. Der Anschnitt Beih, pp 267–275

Bachmann HG (2001a) Goldscheide- und –probierkunst von den Anfängen bis zur Neuzeit. In: Ammerer G, Weiß AS (eds) Das Tauerngold im europäischen Vergleich, Archäologische und historische Beiträge des Internationalen Kongresses in Rauris, Oktober 2000, pp 81–88

Bachmann HG (2001b) Geschichte(n) einer mittelalterlichen Buntmetallschlacke aus dem Oberharz. Zeitschr Gesch Berg- und Hüttenwesen 7(2):27–36

Bachmann HG (2003) Bunt- und Edelmetalle aus mitteleuropäischen Komplexerz-Lagerstätten: Fahlerzverhüttung von der Bronzezeit bis zur Renaissance. In: Stöllner T, Körlin G, Steffens G, Cierny J (eds) Man and Mining – Mensch und Bergbau. Studies in honour of Gerd Weisgerber on occasion of his 65th birthday, vol 16. Der Anschnitt Beih, pp 25–35

Bachmann HG (2006) Mythos gold. 6000 Jahre Kulturgeschichte. Hirmer Verlag München

Bachmann HG, Hammer P (2003) Vergleichende metallanalytische Untersuchungen an römischen Denaren aus der 2. Hälfte des 2. Jahrhunderts n. Chr. In: Stribrny K (ed) Funktionsanalyse barbarisierter, barbarischer Denare mittels numismatischer und metallurgischer Methoden. Zur Erforschung der sarmatisch-germanischen Kontakte im 3. Jahrhundert n. Chr. Stud, vol 18. Fundmünzen der Antike (SFMA), Mainz, pp 107–133

Bachmann HG, Hauptmann A (1984) Zur alten Kupfergewinnung in Fenan und Hirbet en-Nahas im Wadi Arabah in Südjordanien. Ein Vorbericht. Anschnitt 35(4):110–123

Bachmann HG, Jockenhövel A (1974) Zu den Stabbarren aus dem Rhein bei Mainz. Archäol Korr-Blatt 4:139–144

Bachmann HG, Rothenberg B (1980) Die Verhüttungsverfahren von Site 30. In: Conrad HG, Rothenberg B (eds) Antikes Kupfer im Timna-Tal. Der Anschnitt Beiheft, vol 1, pp 215–236

Bachmann HG, Tsinov Z (2003) Placer gold in Bulgaria: past and present. Gold Bull 36(4):138–143

Bachmann HG, Everts H, Hope CA (1980) Cobalt-blue pigment on 18th dynasty Egyptian pottery. Mitt Deutsch Archäol Inst Kairo 36:33–37

Bachmann HG, Jockenhövel A, Spichal U, Wolf G (2002/2003) Zur bronzezeitlichen Metallversorgung im mittleren Westdeutschland: Von der Lagerstätte zum Endprodukt. Berichte der Komm Archäol Landesforsch Hessen 7:67–120

Baker T, Pollard P, Mustard R, Mark G (2005) A comparison of granite-related tin, tungsten, and gold-bismuth deposits: implications for exploration. Soc Econ Geol Newsl 61:5–17

Balmuth MS, Thompson CM (2000) Hacksilver: recent approaches to the study of hoards of uncoined silver. Laboratory analyses and geographical distribution. In: Kluge B, Weisser B (eds) XII. Numismatischer Kongress, Berlin 1997: Proc contribut, pp 159–169

Bamberger M, Wincierz P (1990) Ancient smelting of oxide copper Ore. In: Rothenberg B (ed) Researches in the Arabah 1959-1984, vol II: the ancient metallurgy of copper. Inst Archaeo-Metall Stud, London, pp 123–157

Bamberger M, Wincierz P, Bachmann HG, Rothenberg B (1986) Ancient smelting of oxide copper ore: archaeological evidence at Timna and experimental approach. Metallwissenschaft + Technik 40:1166–1174

Bar-Adon P (1980) The cave of the treasure: the finds from the caves in Naḥal Mishmar. Israel Exploration Soc, Jerusalem

Barker G, Gilbertson D, Mattingly D (2007) Archaeology and desertification. The Wadi Faynan landscape survey, southern Jordan, Wadi Faynan series 2. Levant Suppl. Series 6. Oxbow Books, Oxford

Barnes JW (1970) Ancient clay bars from Iran. J Hist Metall Soc 7:8–17

Barnes JW (1974) Some problems of bronze age geology in the Middle East. Intern. Manuscript of Oral presentation, University of Wales Collegiate Colloquium, Mineral Exploit Dept, Univ of Cardiff

Barnes JW (1988) Ores and minerals: introducing economic geology. Wiley, Open University Press

Barnes R, Stein EN, Diebold B (2015) Gold in early Southeast Asia. Selected Papers from the Symposium Gold in Southeast Asia. Yale University Art Gallery 2011

Baron S, Tămaş CG, Cauuet B, Munoz M (2011) Lead isotope analyses of gold-silver ores from Roşia Montană (Romania): a first step of a metal provenance study of Roman mining activity in Alburnus Maior (Roman Dacia). J Archaeol Sci 38(5):1090–1100

Baron S, Tămaş CG, Le Carlier C (2014) How mineralogy and geochemistry can improve the significance of Pb isotopes in metal provenance studies. Archaeometry 56:665–680

Barriga FJAS (1990) Metallogenesis in the Iberian Pyrite Belt. In: Dallmeyer RD, Martínez García E (eds) Pre-Mesozoic geology of Iberia. Springer, Berlin, pp 369–379

Bartelheim M (2007) Die Rolle der Metallurgie in vorgeschichtlichen Gesellschaften. Sozioökonomische und kulturhistorische Aspekte der Ressourcennutzung. Ein Vergleich zwischen Andalusien, Zypern und dem Nordalpenraum. Forsch. Archäometrie u. Altertumswiss 2. Rahden/Westfalen

Bartelheim M (2012) Detecting social structures in the bronze age of southeastern Spain. In: Kienlin TL, Zimmermann U (eds) Beyond elites, alternative to hierarchical systems in modelling social formations. R Habelt, Bonn, pp 339–354

Bartelheim M, Niederschlag E, Rehren T (1998) Research into prehistoric metallurgy in the Bohemian/Saxon Erzgebirge. In: Hänsel B (ed) Mensch und Umwelt in der Bronzezeit Europas. Kiel, pp 225–230

Bartelheim M, Eckstein K, Huijsmans M, Krauss R, Pernicka E (2002) Kupferzeitliche Gewinnung in Brixlegg, Österreich. In: Bartelheim M, Pernicka E, Krause R (eds) Die Anfänge der Metallurgie in der Alten Welt. Forsch. Archäometrie u. Altertumswiss. 1, Westfalen, pp 33–82

Bartelheim M, Contreras F, Montero L, Moreno A, Murillo M, Pernicka E (2012) Das Silber der südspanischen El Argar-Kultur: Produktion und Distribution. Trab Prehist 69:293–309

Bartels C (1992) Vom frühneuzeitlichen Montangewerbe zur Bergbauindustrie. Erzbergbau im Oberharz. Veröff. Deutsches Bergbau-Museum Bochum 54

Bartels C (1994) Erzbergbau in Westfalen – ein Überblick. In: Bartels C, Feldmann R, Oekentorp K (eds) Geologie und Bergbau im rheinisch-westfälischen Raum. Bücher aus der hist. Bibliothek des Landesoberbergamtes Nordrhein-Westfalen, Münster, pp 35–53

Bartels C (1996) Abbau, Weiterverarbeitung und wirtschaftliche Nutzung des Endorfer Bleis im 15. und 16. Jahrhundert. In: Westfälisches Schieferbergbaumuseum Schmallenberg-Holthausen (ed) Bergbau im Sauerland. Chronik des vorindustriellen Erzbergbaus und der Metallgewinnung im Raum Sundern, pp 141–154

Bartels C (1997) Strukturwandel in Montanbetrieben des Mittelalters und der frühen Neuzeit in Abhängigkeit von Lagerstättenstrukturen und Technologie. In: Gerhard HJ (ed) Struktur und Dimension. Festschr. H Kaufhold 65. Geburtstag Bd. 1: Mittelalter und Frühe Neuzeit, pp 27–70

Bartels C (2000) Sauerländisches Blei für die Mansfelder Hütten. Blei zur Entsilberung des Mansfelder Kupfers. Knappe 2000:105–120

Bartels C (2003) Die Entwicklung des Bergbaus im nordwestlichen Harz bis zum Beginn der Industrialisierung: Bergbau – eine "Großveranstaltung?". In: Denecke D, Egli HR, Gringmuth-Dallmer E, Irsigler F, Schenk W (eds) Siedlungsforschung. Archäologie – Geschichte – Geographie, vol 21, pp 97–111

Bartels C (2008) Conservation and presentation of historical European mining landscapes: the Rammelsberg and Goslar UNESCO world heritage site, and the North-Western Harz Mountains in Germany. In: Marchaetti N, Thuesen I (eds) ARCHAIA. Case studies on research planning, characterisation, conservation and management of archaeological sites. Brit. Archaeolog, Records international series 1877, Oxford, pp 241–246

Bartels C (2014) Bleiglanz als hauptsächliches Silbererz des Mittelalters und der frühen Neuzeit? – Zur Entstehung und Geschichte eines grundlegenden Irrtums. Anschnitt 6:190–213

Bartels C, Bingener A (2006) Der Bergbau bei Schwaz in Tirol im mittleren 16. Jahrhundert "1556 Perkwerch etc.", vol 3. Das Schwazer Bergbuch, Bochum

Bartels C, Klappauf L (2012) Der Aufschwung des Bergbaus unter den karolingischen und ottonischen Herrschern, die mittelalterliche Blüte und der Abschwung bis zur Mitte des 14. Jahrhunderts. In: Bartels C, Slotta R (eds) Der alteuropäische Bergbau. Von den Anfängen bis zur Mitte des, vol 18. Jahrhunderts. Aschendorff Verl, pp 111–248

Bartels C, Fessner M, Klappauf L, Linke FA (2007) Kupfer, Blei und Silber aus dem Goslarer Rammelsberg von den Anfängen bis 1620. Publ Deutsches Bergbau-Museum Bochum 151

Bartos PJ (2000) The pallacos of Cerro Rico de Potosi: a new deposit type. Econ Geol 95:645–654

Bar-Yosef O, Alon D (1988) Nahal Hemar Cave. The excavations. Altiqot 18:1–30

Bass G (1966) Troy and Ur: links between two ancient capitals. Expedition 8(4):26–29

Bass G (1967) Cape Gelidonya: a bronze age shipwreck. Am Philos Soc 57(8):1–177

Bass G (1970) A hoard of Trojan and Sumerian jewelry. Am J Archaeol 74(4):335–341

Bass G (1986) A bronze age shipwreck at Lu Burun (Kaş): 1984 campaign. Am J Archaeol 90:269–296

Bass G (1987) Oldest known shipwreck reveals bronze age splendors. Natl Geogr 172(6):691–733

Bassiakos Y, Catapotis M (2006) Reconstruction of the copper smelting process at Chrysokamino based on the analysis of ore and slag samples. In: Betancourt PP (ed) The Chrysokamino metallurgical workshop and its territory, Hesperia suppl, vol 36, pp 329–353

Baur E, Glaessner A (1903) Über die Einwirkung von Kohlenstoff, Kohlenoxid und Kohlensäure auf das Eisen und seine oxide. Zeitschr. Physikalische Chemie 43:354–368

Bayburtoğlu B, Yıldırım S (2008) Gold and silver in Anatolia. In: Yalçın Ü (ed) Anatolian metal IV, Der Anschnitt Beih, vol 21, pp 43–53

Bayley J (1984) Roman brass-making in Britain. J Hist Metall 18:42–43

Bayley J (1990) The production of Brass in antiquity with particular reference to Roman Britain. In: Craddock PT (ed) 2000 years of zinc and Brass, Brit mu. Oc. Pap, vol 50, pp 7–27

Bayley J (1991) Anglo-Saxon non-ferrous metal working: a survey. World Archaeol 23(1):115–130

Bayley J (1992) Metalworking ceramics. Medieval Ceramics 16:3–10

Bayley J, Butcher S (2004) Roman brooches in Britain: a technological and typological study based on the Richborough collection. Soc. Antiquaries of London

Bayley J, Eckstein K (1997) Silver refining: production, recycling and assaying. In: Sinclair A et al (eds) Archaeological sciences 1995, Oxbow monographs, vol 64, pp 107–111

Bayley J, Rehren T (2007) Towards a functional and typological classification of crucibles. In: La Niece S, Hook D, Craddock P (eds) Metals and mines. Studies in Archaeometallurgy. Archetype Publ ass Brit Mus London, pp 46–55

Bear LM (1963) The mineral resources and mining industry of Cyprus. Bull 1. Nicosia, Geol Surv Dept, Ministry Sciand Industry, Cyprus

Beck L (1884) Die geschichte des Eisens. Erste Abteilung. Von der ältesten Zeit bis um das Jahr 1500 n. Chr. Braunschweig

Beck R, Berg G (1922) Abriss der Lehre von den Erzlagerstätten. Borntraeger, Berlin

Beckmann S (2007) Kolluvien und Auensedimente als Geoarchive im Umfeld der historischen Hammerwerke Leidersdorf und Wolfsbach (Vils/Opf.). Regensburger Beitr Bodenkunde, Landschaftsökol u Quartärforsch 12

Begemann F, Schmitt-Strecker S (1994) Das Blei von Schiff und Ladung: seine Isotopie und mögliche Herkunft. In: Hellenkemper-Salies G (ed) das wrack: Der antike Schiffsfund von Mahdia, Bonn, pp 1073–1076

Begemann F, Schmitt-Strecker S (2008a) Zur Herkunft des Silbers aus tall Munbaqa (Nordsyrien). Prähist Zeitschr 79(1):111–117

Begemann F, Schmitt-Strecker S (2008b) Bleiisotopie und die Provenienz von Metallen. In: Yalçın Ü (ed) Anatolian metal IV, Der Anschnitt Beih, vol 21, pp 125–134

Begemann F, Schmitt-Strecker S (2009) Über das frühe Kupfer Mesopotamiens. Extrait d'Iranica Antiqua XLIV:1–45

Begemann F, Schmitt-Strecker S, Pernicka E (1992) The metal Finds from Thermi III-V: a chemical and Lead-isotope study. Studia Troica 2:219–239

Begemann F, Pernicka E, Schmitt-Strecker S (1994) Metal finds from Ilipinar and the advent of arsenical copper. Anatolica XX:203–219

Begemann F, Kallas K, Schmitt-Strecker S, Pernicka E (1999) Tracing ancient tin via isotope analyses. In: Hauptmann A, Pernicka E, Rehren T, Yalçın Ü (eds) The beginnings of metallurgy, Der Anschnitt Beih, vol 9, pp 277–284

Begemann F, Schmitt-Strecker S, Pernicka E, Loschiavo F (2001) Chemical composition and Lead Isotopy of copper and bronze from Nuragic Sardinia. Eur J Archaeol 4(1):43–85

Begemann F, Schmitt-Strecker S, Pernicka E (2003) On the composition and provenance of metal finds from Beşiktepe (Troia). In: Wagner G, Pernicka E, Uerpmann HP (eds) Troia and the Troad. Springer, pp 173–201

Begemann F, Haerinck E, Overlaet B, Schmitt-Strecker S, Tallon F (2008) An archaeometallurgical study of the early and middle bronze age in Luristan. Iran Extrait d'Iranica Antiqua XLIII:1–66

Begemann F, Hauptmann A, Schmitt-Strecker S, Weisgerber G (2010) Lead isotope and chemical signature of copper from Oman and its occurrence in Mesopotamia and sites on the Arabian Gulf Coast. Arab Archaeol Epigr 21:145–179

Beierlein PR, Lange A (eds) (1960) Lazarus Ercker, Beschreibung der allervornehmsten Mineralischen Erze und Bergwerksarten vom Jahre 1580, Freiberger Forschungsh. D34. Akademie-Verlag, Berlin

Belgiorno MR (2000) Project "Pyrame" 1998–1999: archaeological, metallurgical and historical evidence at Pyrgos (Limassol). Rep. Dept. Antiqities, Cyprus, pp 1–17

Bendall C (2003) The application of trace element and isotopic analyses to Celtic gold coins and their metal sources. PhD-diss, Universität Frankfurt a.M, FB Geowissenschaften

Bendall C, Wigg-Wolf DF, Lahaye Y, von Kaenel M, Brey GP (2009) Detecting changes of Celtic gold sources through the application of trace element and Pb isotope laser ablation analysis of Celtic gold coins. Archaeometry 51(4):598–625

Benham AJ et al (2010) Chromite and PGE in the Logar Ophiolite complex, Afghanistan. IMM Transact B 118 (2):45–58

Bennett A (2017) The ancient history of U thong city of gold. Sirivatana Interprint Public Comp. Ltd., Bangkok

Benvenuti M, Mascaro I, Costaliola P, Tanelli G, Romualdi A (2000) Iron, copper and tin at Baratti, (Populonia): smelting processes and metal provenances. J Hist Metall 34:67–76

Benvenuti M, Dini A, D'Orazio M, Chiarantini L, Corretti A, Costagliola P (2013) The tungsten and tin signature of Iron ores from Elba Island (Italy): a tool for provenance studies of Iron production in the Mediterranean region. Archaeometry 55(3):479–506

Benvenuti M et al (2016) Experimental smelting of iron ores from Elba (Tuscany, Italy): results and implications for the reconstruction of ancient metallurgical processes and iron provenance. J Archaeol Sci 70:1–14

Ben-Yosef E (2012) A unique casting mould from the new excavations at Timna 30 (Israel): evidence from western influence? In: Kassianidou V, Papasavvas G (eds) Eastern mediterranean metallurgy and metalwork in the second millennium BC. A conference in honour of James D. Muhly, Nicosia, October 2009, Oxbow Books, Oxford, pp 188–196

Ben-Yosef E (2014) The material culture of iron age copper production in Faynan. In: Levy TE, Najjar M, Ben-Yosef E (eds) New insights into the iron age archaeology of Edom, Southern Jordan II. Cotsen Inst. Archaeology Press, Univ. California, Los Angeles, pp 887–961

Ben-Yosef E et al (2008a) A new approach for geomagnetic archaeointensity research: insights on ancient metallurgy in the southern Levant. J Archaeol Sci 35 (11):2863–2879

Ben-Yosef E et al (2008b) Application of copper slag in geomagnetic archaeointensity research. J Geophys Res 113:B08101. https://doi.org/10.1029/2007JB005235

Ben-Yosef E et al (2010a) The beginning of Iron age copper production in the southern Levant: new evidence from Khirbat al-Jariya, Faynan, Jordan. Antiquity 84(325):724–746

Ben-Yosef E, Tauxe L, Levy TE (2010b) Archaeomagnetic dating of copper smelting site F2 in the Timna Valeey (Israel) and its implications for the modelling of ancient technological developments. Archaeometry 52(6):1110–1121

Ben-Yosef E, Najjar M, Levy T (2014) New iron age excavations at copper production sites, mines and fortresses in Faynan. In: Levy T, Najjar M, Ben-Yosef E (eds) New insights into the iron age archaeology of Edom, Southern Jordan II. Cotsen Inst. Archaeol. Press, Univ. California, Los Angeles, pp 767–885

Benzel K (2013) Puabi's adornment for the afterlife: materials and technologies of jewelry at Ur in mesopotamia. PhD-dissertation, Graduate School of Arts and Sciences, Columbia Univers

Benzel K (2015) What does Puabi want (today)? The status of Puabi as image. In: Chi JY, Azara P (eds) From ancient to modern. Archaeology and aesthetics. Exhibition catalogue, New York University, 2015. Instit. Study Ancient World at New York & Princeton Univers. Press, pp 132–193

Berger D, Schwab R, Wunderlich CH (2010) Technologische Untersuchungen zu bronzezeitlichen Metallziertechniken nördlich der Alpen vor dem Hintergrund des Hortfundes von Nebra. In: Meller H, Bertemes F (eds) Der Griff nach den Sternen. Wie Europas Eliten zu Macht und Reichtum Kamen. Tagungen des Landesmuseums für Vorgeschichte Halle (Saale) 5/II, pp 751–777

Berger D, Brügmann G, Pernicka E (2018) Fraktionierung von Zinnisotopen bei pyrometallurgischen Prozessen und bei Korrosion von Metallartefakten. In: Glaser L (ed) Archäometrie und Denkmalpflege. Publ. Verlag Deutsches Elektronen-Synchrotron, pp 60–63

Berger D, Brügmann G, Pernicka E (2019a) On smelting cassiterite in geological and archaeological samples: preparation and implications for provenance studies on metal artefacts with tin isotopes. J Archaeol Anthropol Sci 11(1):293–319

Berger D, Brügmann G, Pernicka E (2019b) Zum Stand der Zinnforschung: Neue Erkenntnisse zu Zinnherkunft und –handel anhand von Isotopen- und Spurenelement-Analysen Spätbronzezeitlicher

Zinnbarren. In: Herm C, Merkel S, Schreiner M, Wiesinger R (eds) Archäometrie und Denkmalpflege 2019, Metalla Sonderheft, vol 9, pp 192–195
Bernhard J (1965) Die Mitterberger Kupferkieslagerstätte Erzführung und Tektonik. Jahrb Geol BA 109. Wien
Berni GV et al (2014) The Serra Pelada Au-Pd-Pt deposit, Carajás, Brazil: geochemistry, mineralogy, and zoning of hydrothermal alteration. Econ Geol 109 (7):1883–1899
Bersch W (1898) Mit Schlägel und Eisen. Eine Schilderung des Bergbaus und seiner technischen Hilfsmittel. Reprint by W Kroker 1985. VDI Verlag Klassiker der Technik
Bertrand G, Guillou-Frottier L, Loiselet C (2014) Distribution of porphyry copper deposits along the western Tethyan and Andean subduction zones: insights from a paleotectonic approach. Ore Geol Rev 60:174–190
Betancourt PP (2006a) The Chrysokamino metallurgical workshop and its territory. Hesperia suppl 36
Betancourt PP (2006b) Joining techniques of early bronze age Trojan jewelry. Studia Troica 16:89–95
Betechtin AG (1968) Lehrbuch der speziellen Mineralogie. In: Rösler HJ (ed) VEB Deutsch Verl Grundstoffind Leipzig
Beukes NJ, Gutzmer J, Mukhopadhyay J (2003) The geology and genesis of high-grade hematite iron ore deposits, applied earth science. Trans Inst Mining Metall B 112:18–25
Bevins RE (2010) Mineralization in England and Wales: an introduction. In: Bevins RE, Young B, Mason JS, DAC M, Symes RF (eds) Mineralization in England and Wales, Geol Conservation Rev Ser, vol 36, pp 3–28
Beyschlag F, Krusch P, Vogt JHL (1910) Die Lagerstätten der nutzbaren Mineralien und Gesteine nach Form Inhalt und Entstehung 1. Ferdinand Wenke Verlag
Bielenin K (1976) Eingetiefte Rennöfen der frügeschichtlichen Eisenverhüttung in Europa. Archäol Austriaca Beih 14:13–27
Bielenin K (1983) Der Rennofen mit eingetieftem Herd und seine Formen in Polen. Dent Off 40:47–61
Bielenin K, Woyda S (1978) Zwei Eisenverhüttungszentren des Altertums im Weichselbogen (1.Jh. v.u.Z. - 4 Jh.u.Z.), Eisen + Archäologie. Eisenerzbergbau und -verhüttung vor 2000 Jahren in der VR Polen. In: Weisgerber G, Fober L, Hauptmann A (eds) , vol 14. Publ Deutsches Bergbau-Museum Bochum, pp 25–55
Bimson M (1980) Pigments from the "Royal Cemetery" at Ur. Iraq 42(1):75–77
Bingener A, Bartels C, Fessner M (2012) Die Grosse Zeit des Silbers. In: Bartels C, Slotta R (eds) Geschichte des deutschen Bergbaus 1, Der alteuropäische Bergbau. Von den Anfängen bis zur Mitte des 18. Jahrhunderts. Aschendorff Publ Münster, pp 317–452
Biringuccio V (1540) Pirotechnia. Translated by Cyril Stanley Smith and Martha Teach Gnudi 1943. American Institute of Mining, Metallurgical, and Petroleum Engineers, New York
Bisson MS (2000) Precolonial copper metallurgy: sociopolitical context. In: Vogel J (ed) Ancient African metallurgy. Altamira-Press, Walnut-Creek, pp 83–145
Biswas AK, Davenport WG (1980) Extractive metallurgy of copper. Elsevier
Björkenstam N, Fornander S (1985) Metallurgy and technology at Lapphyttan, Jernkontorets Forskning 34, 1985 (medieval Iron in society). pp 184–225
Blakelock E, Martinon-Torres M, Veldhuijzen HA, Young T (2009) Slag inclusions in iron objects and the quest for provenance. J Archaeol Sci 36(8):1745–1757
Blanchard I (1992) Technical implications of the transition from silver to lead smelting in twelfth century England. In: Willies L, Cranstone D (eds) Boles and Smeltmills. Hist Metall Soc, London, pp 9–12
Blömeke C (1885) Über die Erzlagerstätten des Harzes und die Geschichte des auf demselben geführten Bergbaus. Berg- und Hüttenmänn Jb XXXIII, 47ff
Bode M (2006) Neues aus dem alten Rom mit Hilfe der Archäometallurgie. In: Köhne R, Reininghaus W, Stöllner T (eds) Bergbau im Sauerland, Schrift Histor Kommiss Westfalen, vol 20, pp 32–44
Bode M (2008) Archäometallurgische Untersuchungen zur frühen Blei/Silbergewinnung im Germanien der frühen Kaiserzeit. PhD-diss, University of Muenster, Germany
Bode M (2013) Bleiisotope als Schlüssel zur Herkunftsbestimmung von Metallen – Die römischen Plumbum Germanicum Barren. Atuatuca 4:50–57
Bode M, Baumann A, Hauptmann A, Mezger K, Prange M (2003) Geochemical and lead isotope investigations for lead production in Roman times in the Rhineland and in Westphalia (Germany). In: Archaeometallurgy in Europe. Proc. Internat. Conf., Milano, pp 201–209
Bode M, Hauptmann A, Mezger K (2009) Tracing Roman lead sources using lead isotope analyses in conjunction with archaeological and epigraphic evidence – a case study from Augustean/Tiberian *Germania*. J Archaeolog Anthropol Sci 1(3):177–194
Bonadies SD (1994) Tomography of ancient bronzes. In: Scott DA, Podany J, Considine BB (eds) Ancient and historic metals. Conservation and scientific research. Proc. Symp. Paul Getty Museum & Inst. Nov. 1991, The Getty Conservation Inst., pp 63–84
Boni M (1979) Zur Paleogeographie, Mineralogie und Lagerstättenkunde der Paläokarsterze in Südwest-Sardinien (Iglesiente-Sulcis). PhD-diss, Univers Heidelberg
Boni M, Large D (2003) Nonsulfide zinc mineralization in Europe: an overview. Econ Geol 98:715–729
Boni M, Di Maio G, Frei R, Villa M (2000) Lead isotopic evidence for a mixed provenance for water Roman pipes from Pompeij. Archaeometry 42(1):201–208
Boni M, Gilg HA, Aversa G, Balassone G (2003) The "calamine" of Southwest Sardinia: geology, mineralogy, and stable isotope geochemistry of supergene Zn mineralization. Econ Geol 98:731–748
Borbein A, Hölscher T, Zanker P (2000) Klassische Archäologie. Reimer Verlag, Berlin

Borg G (1996) The reopened PGM-Au-placer deposits of the Middle Urals and their source rocks. Erzmetall 49 (11):667–675

Borg G (1997) Distribution of Osmium, Iridium and Ruthenium in platinum ores from Nishni Tagil, Middle Ural. In: GDMB (ed) Spurenelemente in Lagerstätten. Schriftenreihe Gesellschaft Bergbau, Metallurgie, Rohstoff- und Umwelttechnik, vol 80, pp 93–104

Borg G (2010) Warum in die Ferne schweifen? Geochemische Fakten und geologische Forschungsansätze zu Europas Goldvorkommen und zur Herkunft des Nebra-Goldes. In: Meller H, Bertemes F (eds) Der Griff nach den Sternen. Wie Europas Eliten zu Macht und Reichtum kamen. Tag. Landesmus. Vorgesch, Halle (Saale) 5/II, pp 735–749

Borg G (2014) "Gold is where you find it" – Zeitgenössischer artisanaler Goldbergbau in Afrika als Analogie (prä-)historischer Goldgewinnung. In: Meller H, Risch R, Pernicka E (eds) Metalle der Macht – Frühes Gold und Silber. Mitteldeutscher Archäologentag, Oct 2013, Halle (Saale), pp 53–69

Borg G, Pernicka E (2017) Goldene Zeiten? – Europäische Goldvorkommen und ihr Bezug zur Himmelsscheibe von Nebra. In: Meller H (ed) Jahresschr mitteldeutsche Vorgesch, vol 96, pp 111–138

Borg G, Piestrzyński A, Bachmann GH, Puttmann W (2012) An overview of the European Kupferschiefer deposits. In: Hedenquist JW, Harris M, Camus F (eds) Geology and genesis of major copper deposits and districts of the world. Soc Econ Geol. Spec Publ 16: Tribute to RH Sillitoe, Littleton, CO, pp 455–486

Born H, Hansen S (2001) Helme und Waffen Alteuropas. Sammlung Axel Guttmann. Philip von Zabern, Mainz

Born H (1984) Meisterwerke kaukasischer Bronzeschmiede. Staatl Mus Preuß Kulturbesitz. Berlin

Born H et al (2009) Granuliertes gold aus Troia in Berlin. Restaur u Archäol 2:19–30

Bornhardt W (1910) Über die Gangverhältnisse des Siegerlandes und seiner Umgebung (Tl. 1). Archiv Lagerstättenforschung 2, XII

Boroffka N et al (2003) Bronze age tin from Central Asia: preliminary notes. In: Boyle K, Renfrew C, Levine M (eds) Ancient interactions east and west in Eurasia, McDonald institute monographs, vol 9, pp 135–159

Bourgarit D (2007) Chalcolithic copper smelting. In: La Niece S, Hook D, Craddock P (eds) Metals and mines. Stud in archaeometall. Brit Mus & Archetype Publ, London, pp 3–14

Bourgarit D, Thomas N (2015) Ancient brasses: misconceptions and new insights. In: Hauptmann A, Modarressi-Tehrani D (eds) Archaeometallurgy in Europe III, vol 26. Proc 3rd Internat Conf June/July 2011, Deutsches Bergbau-Museum, Der Anschnitt, pp 255–261

Bourgarit D, Mille B, Burens A, Carozza L (2002) Smelting of chalcopyrite during chalcolithic times: some have done it in ceramic pots as vase-furnaces. In: Kars H, Burke E (eds) 33rd Internat Symp Archaeometry, Amsterdam, April 2002. Vrije Universiteit, Amsterdam, pp 297–302

Bourgarit D, Mille B, Prange M, Ambert P, Hauptmann A (2003) Chalcolithic Fahlore smelting at cabrières: reconstruction of smelting processes by archaeometallurgical finds. In: Archaeometallurgy in Europe Proc 1, Milano, pp 431–440

Bourgarit D, Rostan P, Burger E, Carozza L, Mille B, Artioli G (2008) The beginning of copper mass production in the western Alps: the Caint-Véran mining area reconsidered. J Hist Metall Soc 42(1):1–11

Bourgois J (2013) A review on tectonic record of strain buildup and stress release across the Andean Forearc along the Gulf of Guayaquil-Tumbes Basin (GGTB) near Ecuador-Peru border. Int J Geosci 4:618–635

Bovill EW (1968) The golden trade of the moors. Oxford University Press, Oxford

Bowell RJ (1994) Sorption of arsenic by iron oxides and oxyhydroxides in soils. Appl Geochem 9:279–286

Bowen NL, Schairer JF (1932) The system FeO-SiO2. Am J Sci 24:177–213

Bowen NL, Schairer JF (1935) The system MgO-FeO-SiO2. Am J Sci 229:151–217

Bowles JFW, Lyon IC, Saxton JM, Vaughan DJ (2000) The origin of platinum group minerals from the Freetown intrusion, Sierra Leone, inferred from osmium isotope systematics. Econ Geol 95:539–548

Boyle RW (1979) The geochemistry of gold and ist deposits. Geol Survey Canada Bull 280

Brenner IB, Mart J (1970) Geochemical prospecting for copper in the Wadi Rimthy area, Southern Sinai. Geol. Surv. Israel, Jerusalem

Brepohl E (2016) Theorie und praxis des Goldschmieds. Carl Hanser Verlag, München

Bronson B (1987) Terrestrial and meteoritic nickel in the Indonesian kris. J Hist Metall 21:8–15

Brothwell DR, Pollard AM (2001) Handbook of archaeological sciences. Wiley, Chichester

Brown MA, Blint-Stoyle AE (1959) A sample analysis of British middle and late bronze age materials using optical spectrometry. Proc Prehist Soc 25:188–208

Bruhns B, Fischer RX (2000) Crystallization of cristobalite and tridymite in the presence of vanadium. Eur J Mineral 12:615–624

Buchwald VF (1975) Handbook of iron meteorites. Their history, distribution, composition and structure. University of California Press

Buchwald VF (1992) On the use of iron by the Eskimos in Greenland. Mater Charact 29:139–176

Buchwald VF (2005) Iron and steel in ancient times. The Royal Danish Academy of Sciences and Letters. Historisk-filosofiske Skrifter 29

Buchwald VF, Leisner P (1990) A metallurgical study of 12 prehistoric bronze objects from Denmark. J Danish Archaeol 9:64–102

Budd P, Ottaway BS (1991) The properties of arsenical copper alloys: implications for the development of eneolithic metallurgy. In: Budd P, Chapman B, Jackson C, Janaway R, Ottaway BS (eds)

Archaeological sciences 1989, Oxbow monograph, vol 9, pp 132–142
Budd P, Pollard M, Scaife B, Thomas R (1995a) Oxhide ingots, recycling and the mediterranean metals trade. J Mediterr Archaeol 8(1):1–32
Budd P, Pollard AM, Scaife B (1995b) The possible fractionation of lead isotopes in ancient metallurgical processes. Archaeometry 37(1):143–150
Burnett A, Craddock PT, Meeks N (1986) Early Italian currency bars. In: Swaddling J (ed) Italian iron age artefacts in the British museum. Brit Mus, London, pp 127–130
Buschendorf F et al. (1971) Die Blei-Zink-Erzgänge des Oberharzes 1. Geologie des Erzgang-Reviers, Mineralogie des Ganginhalts und Geschichte des Bergbaus im Oberharz. Beih Geolog Jb 118
Butt CRM, Hough RM (2009) Why gold is valuable. Elements 5:277–280
Butt CRM, Cluzel D (2013) Nickel laterite ore deposits: weathered serpentinites. Elements 9:123–128
Cabri L (2002) The platinum-group minerals. Can Inst Min Metall Petrol Spec 54:13–129
Cabri L, Harris DC, Weiser TW (1996) Mineralogy and distribution of platinum-group mineral (PGM) placer deposits of the world. Explor Min Geol 5(2):73–167
Çağatay A (1978) Güneydoğu Anadolu bakır yatak ve zuhurlarının jeolojik-mineralojik etüdü sonunda elde edilen jenetik bulgular. Maden Tetkik ve Arama Dergisi 89:46–69
Cagatay A, Pehlivan N (1988) Mineralogy of the celaller (Nigde-Camardi) tin mineralization [in Turkish]. Jeoloji Muhendisligi 32–33:21–31
Caldwell JR (1967) Tal-i Iblis. Iran V: 144–146
Caley ER (1964) Orichalcum and related ancient alloys. Amer. Numismatic Soc., Notes and monographs 151
Çambel H, Braidwood R (1970) An early farming village in Turkey. Sci Am 3:50–57
Cambi L (1959) Problemi della metallurgia etruschi. Studi etruschi 27:63–80
Camizuli E (2013) Impact des anciens sites miniers et métallurgiques sur desècosystèmes terrestre et aquatique actuels. Etude comparative de deux moyennes montagnes: le Morvan et les Cévennes. PhD-diss, Univers Bourgogne
Camm GS (1995) Gold in the counties of Cornwall and Devon. Cornish Hillside Publications (St. Austell)
Camm GS, Croot DG (1994) Quaternary placer cassiterite deposits in Cornwall: the role of periglacial processes in their development. Note of poster display at the Annual Conference of the Ussher Soc
Camm S, Hosking KVG (1984) The Stanniferous placers of Cornwall, Southwest England. Geol Soc Malays Bull 17:323–356
Caneva C, Palmieri AM (1983) Metalwork at Arslantepe in late chalcolithic and early bronze age I: the evidence from metal analyses. Origini XII 2:637–654
Caneva C, Frangipane M, Palmieri A (1985) I metalli di Arslantepe nel quadro die più antichi sviluppi della metallurgia vicino-orientale. Quaderni 112:115–137
Caneva I, Frangipane M, Palmieri A (1987) Pre-dynastic Egypt: new data from Maadi. Afr Archaeol Rev 5 (1):105–114
Caponetti E, Martino DFC, Armetta F, Saladino ML (2017) First discovery of orichalcum ingots from the remains of a 6th century BC shipwreck near Gela (Sicily) seabed. Mediter Archaeol Archaeom 17 (2):11–18
Carbó-Nóver J, Williamson J (1967) The crystallisation and decomposition of SnO-SiO2 glasses. Phys Chem Glasses 8:164–168
Caspi EN, Ettedgui H, Rivin O, Peilstöcker M, Breitman B, Herskhol SS, Shalev S (2009) Preliminary neutron diffraction study of two fenestrated axes from the 'Enot Shuni' bronze age cemetery (Israel). J Archaeol Sci 36(12):2835–2840
Catt J (1992) Angewandte Quartärgeologie. Enke Publ Stuttgart
Cattin F, Villa I, Besse M (2009) Copper supply during the final Neolithic at the Saint Blaise/Bains des dames site (Neuchâtel, Switzerland). J Archaeol Anthropol Sci 1:161–176
Cauuet B, Ancel B, Ricoh C, Tămaş C (2003) Ancient mining networks. The French archaeological missions 1999-2001. In: Damian P (ed) , Alburnus major monographic series I, Bucarest, pp 467–526
Celauro A, Loepp D, Ferro D (2017) Ancient procedures of gold cementation and gold scorification: considerations on their reliability through experimental arcchaeology, interpretation of chemical reactions and thermodynamics. Acta rerum naturalium 21:177–200
Celis G (1991) Eisenhütten in Afrika. Les fonderies africaines du fer. Museum für Völkerkunde Frankfurt am Main. Henrich, Frankfurt a.M.
Cerny I (1989) Die karbonatgebundenen Blei-Zink-Lagerstätten des alpinen und außeralpinen Mesozoikums. Die Bedeutung ihrer Geologie, Stratigraphie und Faziesgebundenheit für Prospektion und Bewertung. Arch Lagerst forsch Geol B.-A. 11, Wien, pp 5–125
Černych EN (2003) Die vorgeschischtlichen Montanreviere an der Grenze von Europa und Asien: das Produkzionszentrum Kargaly. In: Stöllner T, Körlin G, Steffens G, Cierny J (eds) Man and mining – mensch und Bergbau. Stud honour of G Weisgerber for his 65th birthday, vol 16. Der Anschnitt Beih, pp 79–92
Chachutajšvili DA (2009) The manufacture of iron in ancient colchis. BAR Int Ser, Oxford Archaeopress
Chakrabarti DJ, Laughlin DE (1984) The Cu-Pb (copper-lead) system. Bull Alloy Phase Diagr 5(5):503–510
Chalkias S, Vavelidis M, Schmitt-Strecker S, Begemann F (1988) Geologische Interpretation der Blei-Isotopen-Verhältnisse von Erzen der Insel Thasos, der Ägäis und Nordgriechenlands. In: Wagner GA, Weisgerber

G (eds) Antike Edel- und Buntmetalgewinnung auf Thasos, Der Anschnitt Beih, vol 6, pp 59–74

Chang YA, Lee YE, Neumann JP (1976) Phase relationships and thermodynamics of the ternary copper-iron-sulfur system. In: Yannopoulos JC, Agrawal JC (eds) Extractive metallurgy of copper. Metall Soc & Americ Inst Min Metall & Petroleum Eng, New York, pp 21–48

Charles JA (1967) Early arsenical bronzes – a metallurgical view. Am J Archaeol 71:21–26

Charles JA (1975) Where is the *Tin*? Antiquity 49:19–24

Charles JA (1980) The coming of copper and copper-based alloys and iron: a metallurgical sequence. In: Wertime TA, Muhly JD (eds) The coming of the age of iron. Yale University Press, New Haven and London, pp 151–182

Charles JA (1985) Determinative mineralogy and the origins of metallurgy. In: Craddock PT, Hughes MJ (eds) Furnaces and smelting technology in antiquity, Brit. Mus. Press Occ. Papers, vol 48, pp 21–28

Chase TW, Notis M, Pelton AD (2007) New Eh-pH (Pourbaix) diagrams of the copper tin system. Metal 07(3):15–21

Chegini N, Helwig B, Parzinger H, Vatandoust A (2004) Eine prähistorische Industriesiedlung auf dem iranischen Plateau – Forschungen in Arisman. In: Stöllner T, Slotta R, Vatandoust A (eds) Persiens Antike Pracht, Vol. I, vol 128. Publ. Deutsches Bergbau-Museum, Bochum, pp 210–217

Chemikerausschuss Gesellsch Deutscher Metallhütten- und Bergleute (1964) Edelmetallanalyse. Probierkunde und naßanalytische Verfahren. Springer, Berlin Göttingen Heidelberg

Chernykh EN (1966) Istoriya drevneishei metallurgii Vostochnoi Evropy. Moscow

Chernykh EN (1992) Ancient metallurgy in the USSR. The early metal age. Cambridge University Press

Chernykh EN (2003) Die vorgeschischtlichen Montanreviere an der Grenze von Europa und Asien: das Produkzionszentrum *Kargaly*. In: Stöllner T, Körlin G, Steffens G, Cierny J (eds) Man and mining - mensch und Bergbau, vol 16. Der Anschnitt Beih, pp 79–92

Chernykh EN (2011) Eurasian steppe belt: radiocarbon chronology and metallurgical provinces. In: Yalçın Ü (ed) Anatolian metal V, Der Anschnitt Beih, vol 24, pp 151–171

Childe VG (1928) The most ancient east: the oriental prelude to European prehistory. Kegan Paul

Childs TS (2000) Traditional iron working: a narrated ethnoarchaeological example. In: Bisson MS, Childs ST, de Barros P, Holl AFC (eds) Ancient African metallurgy. Altamira Press, Roman Littlefield Publ, Walnut Creek, pp 199–253

Childs TS, Herbert E (2005) Metallurgy and its consequences. In: Stahl AB (ed) African archaeology. A critical introduction, Blackwell studies in global archaeology. Blackwell Publ., Oxford, pp 276–300

Chirikure S (2015) Metals in past societies. A global perspective on indigeneous African metallurgy. Springer, Heidelberg

Chirikure S, Sinamai A (2015) World history from the seabed: rescuing a Portuguese shipwreck off the coast of Namibia. In: Tripati S (ed) Shipwreck around the world. Revelations of the past. Delta Book World, New Delhi, pp 114–129

Chirikure S, Hall S, Miller D (2007) One hundred years on: what do we know about tin and bronze production in southern Africa? In: La Niece S, Hook D, Craddock P (eds) Metals and mines, Studies in archaeometallurgy. Archetype Publ Brit Mus, pp 112–119

Chirikure S, Heimann R, Killick D (2010) The technology of tin smelting in the Rooiberg Valley, Limpopo Province, South Africa, ca. 1650-1850 CE. J Archaeol Sci 37:1656–1669

Chmyriov VM et al. (1973) Mineral resources of Afghanistan (an explanatory note to the map of mineral deposits and occurrences of Afghanistan, scale 1:1 000 000). Kabul, DGMS

Cierny J (1997) Rame, stagno e bronzo. In: Endrizzi L, Marzatico F (eds) . Ori delle Alpi. 6 Quaderni della Sezione Archaeologica, Trento, pp 75–81

Cierny J (2008) Prähistorische Kupferproduktion in den südlichen Alpen, Region Trentino Orientale. Der Anschnitt Beih 22

Cierny J, Weisgerber G (2003) Bronze age tin mines in Central Asia. In: Giumlia-Mair A, LoSchiavo F (eds) The problem of early tin, BAR Internat. Ser, vol 1199, pp 23–31

Cierny J et al (1995) Endbronzezeitliche Kupferproduktion im Trentino. Anschnitt 47(3):82–91

Cierny J, Stöllner T, Weisgerber G (2005) Zinn in und aus Mittelasien. In: Yalçın Ü, Pulak C, Slotta R (eds) Das Schiff von Uluburun. Welthandel vor 3000 Jahren. Katalog Ausstellung Deutsches Bergbau-Museums, Bochum, pp 431–448

Ciliberto E, Spoto G (2000) Modern analytical methods in art and archaeology, Chemical analysis 155. Wiley, New York

Cincotti A, Massidda L, Sanna U (2003) Chemical and lead isotope characterization of lead finds from Santa Barbara nuraghe (Bauladu, Sardinia). J Cult Herit 4:263–268

Clark AH, Sillitoe RH (1971) Cuprian galena solid solutions, Zapallar mining district, Atacama, Chile. Am Mineral 56:2142–2145

Cleuziou S, Berthou T (1982) Early tin in the near east: a reassessment in the light of new evidence from Afghanistan. Expedition 24(3):14–19

Cleuziou S, Tosi M (2007) In the shadow of the ancestors. The prehistoric foundations of the early Arabian civilization in Oman. Ministry of Heritage and Culture, Muscat

Cleuziou S, Tosi M (2012) Reasoning about Magan … A geographical and archaeological perspective. In: Giraud & Gernez (ed) Aux marges de l'archéologie.

Hommage à Serge Cleuziou, TAP. De Boccard, pp 305–323
Clissold ME (2007) Aspects of the supergene geochemistry of copper, nickel and bismuth. PhD-diss, Univers Western Sydney
Coghlan HH (1977) Notes on prehistoric and early iron in the Old World. Pitt Rivers Mus Univers Oxford, Occ Pap Technol 8
Cohen CR, Van Buren M, Mills B, Rehren T (2010) Current silver smelting in the Bolivian Andes: a review of the technology employed. J Hist Metall 44 (2):153–162
Coleman RG (1981) Tectonic setting for ophiolite obduction in Oman. J Geophys Res 86(B4):2497–2508
Colls D, Domergue C, Laubenheimer F, Liou B (1975) Les lingots d'étain de l'épave port-Vendres II. Gallia 33 (1):61–94
Comşa E (1991) L'utilisation du cuivre en Roumanie pendant le Néolithic moyen. In: Éluère S, Mohen JP (eds) La Découverte du métal. Picard (Millénaire 2), Paris, pp 77–84
Conophagos C (1980) Le Laurium antique et la technique grecque de la production de l'argent. Ekdotike Hellados, Athen
Conrad HG, Fober L, Hauptmann A, Lieder W, Ordentlich I, Weisgerber G (1980) Untersuchung zur Bergbautechnik und ihre Interpretation. In: Conrad HG, Rothenberg B (eds) Antikes Kupfer im Timna-Tal, Der Anschnitt Beih, vol 1, pp 69–94
Constantinou G (1982) Geological features and ancient exploitation of the cupriferous sulphide orebodies of cyprus. In: Muhly JD, Maddin R, Karageorghis V (eds) Early metallurgy in cyprus, 4000–500 BC, Acta Intern Archaeol Symp, Larnaca 1981, Nicosia, pp 13–24
Contreras-Cortés F, Moreno-Onorato A, Bartelheim M (2014) New data on the origin of silver in the Argaric culture: the site of Peñalosa. In: Meller H, Risch R, Pernicka E (eds), Metalle der Macht – Frühes Gold und Silber. Tag. Landesmus. Vor- und Frühgesch. Halle 11 (1): 269 281
Cooke SRB, Aschenbrenner S (1975) The occurrence of metallic iron in ancient copper. J Field Archaeol 2:251–266
Cooke SRB, Nielsen BV (1978) Slags and other metallurgical products. In: Rapp G, Aschenbrenner SE (eds) Excavations at Nichoria in Southwest Greece I. Univ Minnesota Press, Minneapolis, pp 182–326
Cordella A (1869–1871) Le Laurium. Marseille Ref. Berg-Hüttenmänn. Zeitschr
Cornwall HR (1956) A summary of ideas on the origin of native copper deposits. Econ Geol 51:615–631
Corretti A, Benvenuti M (2001) The beginning of iron metallurgy in Tuscany with special reference to Etruria Mineraria. Proc. 1st Internat. Coll. Archaeology Africa and Mediterranean Basin, Geneva (1999). Mediterran Archaeol 14:127–145
Courcier A (2014) Ancient metallurgy in the caucasus from the sixth tot he third millennium BC. In: Roberts BW, Thornton CP (eds) Archaeometallurgy in global perspective, pp 579–664
Cox DP (1986) Descriptive model of W skarn deposits. In: Cox DP, Singer DA (eds) Mineral deposit models, US Geolog Surv Bull 1693, p 55
Craddock PT (1976) The composition of the copper alloys used by the Greek, Etruscan and Roman Civilisations. 1. The Greeks before the archaic period. J Archaeol Sci 3:93–113
Craddock PT (1977) Dto., 2. The archaic, classical and Hellenistic Greeks. J Archaeol Sci 4:103–123
Craddock PT (1978) Dto., 3. The origins and early use of Brass. J Archaeol Sci 3:93–113
Craddock PT (1984) Tin and tin solder in Sumer: preliminary comments. MASCA J 5:1–16
Craddock PT (1990) Zinc in classical antiquity. In: Craddock PT (ed) 2000 years of zinc and Brass, Brit Mus Press Occ Paper, vol 50, pp 1–6
Craddock PT (1993) A short review of the evidence for bronze age mining in the British Isles. In: Steuer H, Zimmermann U (eds) Montanarchäologie in Europa. Freiburger Forsch 1, Jahrtausend in Südwestdeutschland, vol 4, pp 37–56
Craddock PT (1995) Early metal mining and production. Edinburgh University Press 1995
Craddock PT (2000a) Assaying in antiquity. In: Ramage A, Craddock PT (eds) King Croesus' gold. Excavations at Sardis and the History of Gold Refining. Brit Mus Press, London, pp 245–250
Craddock PT (2000b) From hearth to furnace: evidences for the earliest metal smelting technologies in the Eastern Mediterranean. Paléorient 26(2):151–165
Craddock PT (2000c) Historical survey of gold refining. 1 Durface treatments and refining worldwide, and in Europe prior to AD 1500. In: Ramage A, Craddock PT (eds) King Croesus' gold. Excavations at Sardis and the History of Gold Refining. Brit. Mus. Press, London, pp 27–53
Craddock PT (2000d) Historical survey of gold refining. 2 post medieval period. In: Ramage A, Craddock PT (eds) King Croesus' gold. Excavations at Sardis and the History of Gold Refining. Brit. Mus. Press, London, pp 54–71
Craddock PT (2000e) The platinum group element inclusions. In: Ramage A, Craddock PT (eds) King Croesus' gold. Excavations at Sardis and the History of Gold Refining. Brit Mus Press, London, pp 238–244
Craddock PT (2009) Scientific investigation of copies. Elsevier, Fakes and Forgeries
Craddock PT (2013a) Tarteso y la Explotatión Minera. Local traditions and foreign contacts: innovation in Tartessian metallurgy. In: Alvar J, Campos Carrasco JM (eds) Acta Congreso Tartesso. Editores Tarteso. El emporio del metal. Almuzara, Huelva, pp 231–268
Craddock PT (2013b) Archaeometallurgy 1962-2013: the establishment of a discipline. J Hist Metall Soc 47 (1):1–12
Craddock PT (2013c) Two millennia of the sea-Bourne metalstrade with India. Ind J Hist Sci 48(1):1–37

Craddock PT (2017a) Survey and excavation of the surface sites at Zawar, Dariba and Agucha. In: Craddock PT (ed) Early Indian metallurgy. The production of lead, silver and zinc in North West India. Archetype Publ, London, pp 85–133

Craddock PT (2017b) The production of brass and zinc. In: Craddock PT (ed) Early Indian metallurgy. The production of lead, silver and zinc in North West India. Archetype Publ, London, pp 211–230

Craddock PT, Freestone I (1988) Debris from metallurgical activities at site 200. In: Rothenberg B (ed) The mining temple at Timna. Res in the Arabah 1959-1964 I, Inst Archaeo-Met Studies. UCL, London, pp 192–203

Craddock PT, Meeks N (1987) Iron in ancient copper. Archaeometry 29:187–204

Craddock PT, Picton J (1986) Medieval copper alloy production and West Africa bronze analyses – part II. Archaeometry 28(1):3–32

Craddock PT, Hook DR (1995) Copper to Africa: evidence for the international trade in metal with Africa. In: Hook DR, Gaimster DRM (eds) Trade and discovery: the scientific study of artefacts from post-medieval Europe and beyond, Brit Mus Press Occ Paper, vol 109, pp 181–193

Craddock PT, Eckstein K (2003) Production of Brass in antiquity by direct production. In: Craddock PT, Lang J (eds) Mining and metal production through the ages. Brit Mus Press, London, pp 216–230

Craddock PT, Freestone IC, Gale NH, Meeks ND, Rothenberg B, Tite MS (1985) The investigation of a small heap of silver smelting debris from Rio Tinto, Huelva, Spain. In: Craddock PT, Hughes MJ (eds) Furnaces and smelting technology in antiquity, Brit Mus Press Occ Papers 48, London, pp 199–217

Craddock PT, Freestone IC, Hunt Oriz M (1987) Recovery of silver from speiss at Rio Tinto (Spain). Inst Archaeometall Stud Newsl 10/11:8–11

Craddock PT, Freestone I, Gurjar LK, Middleton A, Willies L (1989) The production of lead, silver and zinc in ancient India. In: Hauptmann A, Pernicka E, Wagner GA (eds) Archäometallurgie der Alten Welt/Old World Archaeometallurgy, Der Anschnitt Beih, vol 7, pp 51–69

Craddock PT, Freestone I, Gurjar LK, Middleton A, Willies L (1990) Zinc in India. In: Craddock P (ed) 2000 years of zinc and Brass, Brit Mus Press Occ Paper, vol 50, pp 29–72

Craddock PT, Freestone I, Dawe C (1996) Casting metals in limestone moulds. J Hist Metall Soc 31(1):1–7

Craddock PT, Cowell M, Guerra MF (2005) Controlling the composition of gold and the invention of gold refining in Lydian Anatolia. In: Yalçιn Ü (ed) Anatolian metal III, Der Anschnitt Beih, vol 18, pp 67–77

Craddock PT, Freestone I, Middleton A, Van Grunderbeek MC (2007a) Early iron-smelting debris from Rwanda and Burundi, East Africa. J Hist Metall Soc 41(1):1–14

Craddock PT, Meeks N, Timberlake S (2007b) On the edge of success: the scientific examination of the products of the early mines research group smelting experiments. In: La Niece S, Hooks D, Craddock PT (eds) Metals and mines. Stud Archaeometall. Archetype Publ, The British Museum, London, pp 37–45

Craddock PT, Cartwright C, Eckstein K, Freestone I, Gurjar L, Hook D, Middleton A, Willies L (2013) Simple sophistication: mauryan silver production in Northwest India. Tech Res Bull Brit Mus 7:79–93

Craig JR (2001) Ore-mineral textures and the tales they tell. Can Mineralog 39(4):937–956

Craig JR, Kullerud G (1967) The Cu-Fe-Pb-S system. Carnegie Inst Wash Year Book 65:344–352

Craig JR, Scott SD (1974) Sulfide phase equilibria. In: Ribbe PH (ed) Sulfide mineralogy. Short course notes 1. Mineral Soc Amer, Southern Printing Co, Blacksburg, VA, pp 1–109

Cretu C, van der Lingen E (1999) Colours of gold alloys. Gold Bull 32(4):115–126

Crew P et al. (2011) Cast iron from a bloomery furnace. In: Hošek J, Cleere H, Mihok L (eds), The archaeometallurgy of iron. Recent developments in archaeological and scientific research. Dedicated to Professor R Pleiner. Inst Archaeol, Prague: 239–262

Çukur A, Kunç S (1989) Analyses of Tepecik and Tülintepe metal artifacts. Anatol Stud XXXIX:113–120

Curtis J (2012) The Oxus treasure. The British Museum Objects in Focus, London

Czedik-Eysenberg F (1958) Beiträge zur Metallurgie des Kupfers in der Urzeit. Archaeol Austriaca Beih 3:1–18

Dahm C, Lobbedey U, Weisgerber G (1998) Der Altenberg: Eine Bergbausiedlung des 13. Jahrhunderts im Siegerland 1, 2. Denkmalpflege u Forsch in Westfalen 34. Bonn, Habelt

Dalton OM (1964) The treasure of the Oxus with other examples of early oriental metal-work, 3rd edn, London

Dart RA, Beaumont P (1969) Evidence of iron ore mining in Southern Africa in the middle stone age. Curr Anthropol 10(1):127–128

Davey CJ, Edwards WI (2007) Crucibles from the bronze age of Egypt and Mesopotamia. Proc R Soc Vistoria 120(1):146–154

David S (1969) The unbound prometheus: technological change and industrial development in Western Europe from 1750 to the present. Press Syndicate University of Cambridge, Cambridge, NY, p 91

David N, Heimann R, Killick D, Wayman M (1989) Between bloomery and blast furnace: Mafa iron-smelting technology in North Cameroon. Afr Archaeol Rev 7:183–209

Davies O (1935) Roman Mines in Europe. Oxford, Clarendon Press. (1935) Antimony bronze in Central Europe. Man 91: 86–89

Day PM, Doonan RCP (2007) Metallurgy in the early bronze age Aegean. Sheffield studies in Aegean archaeology 7. Oxbow Books, Oxford, UK

Dayton J (1978) Minerals metals glazing & man. Harrap, London

De Barros P (2000) Iron metallurgy. In: Bisson MS, Childs TS, De Barros P, Holl AFC, Vogel JO (eds) Ancient African metallurgy, The sociocultural context II. Altamira Press, Lanham MD, pp 83–145

De Vos W et al (2005/2006) Metallogenic minerals provinces and world class ore deposits in Europe. In: Salminen R (ed) Geochemical atlas of Europe. Foregs (electronic version)

Deer WA, Howie RA, Zussman J (1967) Rockforming minerals 5, non-silicates. Longmans, London

Deer WA, Howie RA, Zussman J (1978) Rock forming minerals 2A, single-chain silicates. Longmans, London

Degryse P, Schneider J (2008) Pliny the elder and Sr-Nd isotopes: tracing the provenance of raw materials for Roman glass production. J Archaeol Sci 35:1993–2000

Degryse P, Schneider J, Kellens N, Waelkens M, Muchetz P (2007) Tracing the resources of iron working at ancient sagalassos (South-West Turkey): a combined lead and strontium isotope study on iron artefacts and ores. Archaeometry 49(1):75–86

Degryse P, Schneider J, Muchez P (2009) Combined Pb-Sr-isotopic analysis in Provenancing late Roman iron raw materials in the territory of Sagalassos/SW Turkey. J Archaeol Anthropol Sci 1:155–159

Degryse P, Vanhaecke F, Walton M, Shortland A (2015) Isotopic investigation into the raw materials of late bronze age glass making. J Archaeol Sci 62:153–160

Delaloye M, Desmet A, Desmons J, Gagny C, Rocci G (1980) Geochronological interpretation of the Troodos sheeted dike complex. Ofioliti 5:27–34

Delibes G et al (1991) Almizaraque, Almeria, Spain. In: Mohen JP, Éluère C (eds) Découverte du Métal, Millénnaire, vol 2, pp 303–315

Delouga P, Hill H, Lloyd S (1967) Private houses and graves in the Diyala Region, vol 88. Oriental Inst Chicago Publ

Derner K, Hrubý P, Schubert M (2016) Mittelalterliche Silberproduktion im wettinischen und přemysilidischen Regierungsraum: Neue archäologische Untersuchungen. Anschnitt 68 (6):216–241

Desaulty AM, Dillmann P, L'Héritiera M, Mariet C, Gratuze B, Joron JL, Fluzin P (2009) Does it come from the Pays de Bray? Examination of an origin hypothesis for the ferrous reinforcements used in French medieval churches using major and trace elements. J Archaeol Sci 36:2445–2462

Dever WG (1995) Social structure in the early bronze IV period in Palestine. In: Levy TE (ed) The archaeology of society in the Holy Land. Facts on File, New York, pp 282–296

Devos V, Senn-Luder M, Moor C, Salter C (2000) Laser ablation inductively coupled mass spectrometry (LA-ICP-MS) for spatially resolved trace analysis of early-medieval archaeological iron finds. Fresenius J Anal Chem:873–880

Di Nocera GM (2013) Organization of production and social role of metallurgy in the prehistoric sequence of arslantepe (Turkey). Origini 25:113–142

Di Nocera GM, Hauptmann A, Palmieri A (2004) I metallic della Tomba Reale e la metallurgia agli albori del III millennio. In: Frangipane M (ed) Alle origini del potere. Arslantepe, la collina dei Leoni. Electa, Roma, pp 123–143

Dies K (1967) Kupfer und Kupferlegierungen in der Technik. Springer, Berlin

Dietrich V (1972) Die sulfidischen Vererzungen in den Oberhalbsteiner Serpentiniten. Beitr. Geol. Schweiz. Geotechn. Ser., Lieferung 49

Dilabio RNW, Newsome JW, McIvor DF, Lowenstein PL (1988) The spherical form of gold: man-made or secondary? Econ Geol 83:153–162

Dilek Y, Moores EM, Delaloye M, Karson JA (1991) Amagmatic extension and tectonic denudation in the Kizildag Ophiolite, southern Turkey: implications for the evolution of Neotethyan oceanic crust. In: Peters T (ed) Ophiolite genesis and evolution of the oceanic lithosphere, Proc. conference, Muscat 1990. Kluwer, pp 485–500

Dill H (1993) Die Antimonvorkommen der mitteleuropäischen Alpiden und Varisziden. Zeitschr Deutsche Geol Gesellsch 144:434–450

Dill H (1995) Die Verwendung des Metalls Antimon in Vergangenheit und Gegenwart. Anschnitt 47:168–169

Dill H (2009) Pyrometallurgical relics of Pb-Cu-Fe-deposits in south-eastern Germany: an exploration tool and record of mining history. J Geochem Explor 100:37–50

Dill H (2010) The "chessboard" classification scheme of mineral deposits: mineralogy and geology from aluminum to zirconium. Earth Sci Rev 100:1–420

Dill H, Melcher A, Weber B, Bäumler W (2010) Post-Miocene and bronze age supergene Cu-Pb-arsenate-humate-axalate-carbonate mineralization at mega Livadi, Serifos, Greece. Can Mineral 48:163–181

Dillis S, Van Ham-Meert A, Degryse P (2019) Provenancing antimony in metal and vitreous materials

Diodor (1st cent BC) Siculi Bibliotheca historica

Dobler L (2000) Schwermetalltiefengradienten in Auensedimenten der Selke als Ausdruck der historischen Montanwirtschaft im Ostharz. In: Wippermann T (ed) Bergbau und Umwelt. Langfristige geochemische Einflüße, Berlin., (=GUG Schriftenreihe Geowissenschaften und Umwelt), pp 67–85

Doe BR, Zartmann RE (1979) Plumbotectonics, the phanerozoic. In: Barnes HL (ed) Geochemistry of hydrothermal ore deposits. Wiley, New York, pp 22–70

Doelter C (1912–1932) Handbuch der Mineralchemie. Steinkopff, Dresden

Domergue C (1970) Les mines antiques. Antiqua-Picard, Paris

Domergue C (1990) Les mines de la péninsule Ibérique dans l'Antiquité romain. (CollEFR 127)

Domergue C (2008) Les mines antiques. La production des métaux aux époques grecque et romaine. Picard, Paris

Donaldson CH (1976) An experimental investigation of olivine morphology. Contrib Mineral Petrol 57:187–213

Doonan RCP (1999) Copper production in the eastern Alps during the bronze age: technological change and the unintended consequences of social reorganization. In: Young S, Pollard AM, Budd P, Ixer RA (eds) Metals in antiquity, BAR Internat Ser, vol 792, pp 72–77

Doonan RCP, Day PM, Dimopoulou-Rethemiotaki N (2007) Lame excuses for emerging complexity in early bronze age crete: the metallurgical finds from Poros Katsambas and their context. In: Day PM, Doonen RCP (eds) Metallurgy in the early bronze age aegean, pp 98–122

Dougherty RC, Caldwell JR (1966) Evidence of early pyrometallurgy in the Kerman range. Iran Sci 153:27–40

Dronov VI, Kalimulin SM, Sborschchivok IM, Svezhentsov VP, Chistyakov AN, Zelansky ED, Cherepov PV (1972) The geology and minerals of the North Afghanistan. DGSM, Kabul

Drost E, Haußelt J (1992) Uses of gold in Jewellery. Interdiscip Sci Rev 17(3):271–280

Druitt TH, Francaviglia V (1992) Caldera formation on Santorini and the physiography of the island in the late bronze age. Bull Volcanol 54:484–493

Druitt TH et al (1999) Santorini volcano. Chapter 3: development of the Santorini volcanic field in time and space. Geolog Soc Memoir 19:13–57

Dschaparidze O (2001) Zur frühen Metallurgie Georgiens vom 3. bis zum 1. Jahrtausend v. Chr. In: Gambaschidze I, Hauptmann A, Slotta R, Yalçın Ü (eds) Georgien – Schätze aus dem Land des Goldenen Vlies. Ausstellungskatalog, Publ Deutsches Bergbau-Museum Bochum, vol 100, pp 92–119

Duberow E, Pernicka E, Krenn-Leeb A (2009) Eastern Alps or Western Carpathians: early bronze age metal within the Wieselburg culture. In: Kienlin TL, Roberts BW (eds) Metals and societies, studies in honour of Barbara S. Ottaway, Universitätsforsch Prähist Archäol, vol 169. Habelt, Bonn, pp 336–349

Ducret AC, Rankin WJ (2002) Liquidus temperatures and viscosities of FeO-Fe2O3-SiO2-CaO-MgO slags at compositions relevant to nickel matte smelting. Scandinav J Metall 31:59–67

Dumas F (1972) Trente Siècles sous la Mer, Paris

Dungworth D, Wilkes R (2009) Understanding hammerscale: the use of high-speed film and electron microscopy. J Hist Metall Soc 43(1):33–46

Dunlop DM (1957) Sources of gold and silver in Islam according to al-Hamdānī (10th century A. D.). Stud Islam 8:29–49

Dunning FW, Mykura W, Slater D (1989) Mineral deposits of Europe 2: Southeast Europe. The Mineralog. Soc. & Inst Mining Metall, London

Durali-Müller S (2005) Roman lead and copper mining in Germany—their origin and development through time, deduced from lead and copper isotope provenance studies. Available http://www.mineralogie.uni-frankfurt.de/petrologie-geochemie/forschung/dissertationen/diss0106/index.html

Dušan B (2009) Absolute dating of metallurgical innovations in the Vinča culture of the Balkans. In: Kienlin T, Roberts BW (eds) Metals and societies. Studies in honour of Barbara S. Ottaway, Universitätsforschungen zur prähistorischen Archäologie, vol 169. Habelt, Bonn, pp 191–245

Dušanić S (1977) Aspects of roman mining in Noricum, Pannonia, Dalmatia and Moesia superior. Aufstieg und Niedergang der Römischen Welt 2:52–94

Dutrizac JE (1976) Reactions in cubanite and chalcopyrite. Can Mineral 14:172–181

Dykstra FR (1967) Analysis of tin sample. In: Gelidony C et al. (eds), A bronze age shipwreck. Amer Philosoph Soc New Ser 57(8): 171–172

Dyson R (1964) Sciences meet at Hasanlu. Nat Hist 73 (8):16ff

Earl B (1985) "Melting tin" in the West of England: a study of an old art. J Hist Metall 19(2):153–161

Earl B (1986) Melting tin in the West of England: part 2. J Hist Metall 20(1):17–32

Earl B, Yener A (1995) Tin smelting at the oriental institute. Oriental Inst News Notes 146:2–5

Earl B, Özbal H (1996) Early bronze age tin processing at Kestel/Göltepe, Anatolia. Archaeometry 32 (2):289–304

Eaton ER, McKerell H (1976) Near Eastern alloying and some textual evidence for the early use of arsenical copper. World Archaeol 8(2):169–191

Echt R, Thiele WR, Ivanov IS (1991) Varna - Untersuchungen zur kupferzeitlichen Goldverarbeitung. In: Lichardus J (ed) Die Kupferzeit als historische Epoche, Saarbr. Beitr Altertumskunde, vol 2, pp 633–692

Eckert HE (1986) Bergbau, Verhüttung und Eisenschmieden bei den Senufo in Westafrika. Beobachtungen im Jahr 1985. Anschnitt 2:70–80

Eckstein K, Hauptmann A, Rehren T, Richter U, Schwabenicky W (1994) Hochmittelalterliches Montanwesen im sächsischen Erzgebirge und seinem Vorland. Anschnitt 46(4–5):114–132

Eckstein K, Craddock PT, Freestone IC (1996) Silber, Blei oder Zink? Frühe Bergbau- und Verhüttungsaktivitäten in Zawar, Dariba und Agucha (Nordwestindien). Meta 3(1):11–26

Economou-Eliopoulos M (2010) Platinum-group elements (PGE) in various geotectonic settings: opportunities and risks. Hellenic J Geosc 45:65–82

Efe T, Fidan E (2006) Middle bronze age metal objects from inland Western Anatolia: a typological and chronological evaluation. Anatolia Antiqua XIV:15–43

Eggert M (2001) Prähistorische Archäologie: Konzepte und Methoden. UTB, Francke Publ., Bern München

Ehrt D, Leister M, Matthai A (2001) Polyvalent elements Iron, tin and titanium in silicate, phosphate and fluoride glasses and melts. Eur J Glass Sci Technol B 42 (3):231–239
Ehser A, Borg G, Pernicka E (2011) Provenance of the gold of the early bronze age nebra sky disk, central Germany: geochemical characterization of natural gold from Cornwall. Eur J Miner 23(6):895–910
Eibner C (1972) Mitterberg-Grabung 1971. Anschnitt 24 (2):3–15
Eibner C (1974) Mitterberg-Grabung 1972. Anschnitt 26 (2):14–22
Eibner C (1979) Zum Stammbaum einer urzeitlichen Kupfererzaufbereitung. Berg-u Hüttenmänn Monatshefte 124:157–161
Eibner C (1982) Kupfererzbergbau in Österreichs Alpen. In: Hänsel B (ed) Südosteuropa zwischen 1600 und 1000 v. Chr., Prähist Archäol Südosteuropa, vol 1, pp 399–408
Eibner-Persy A, Eibner C (1970) Erste Großgrabung auf dem bronzezeitlichen Bergbaugelände von Mitterberg. Anschnitt 22(5):12–19
Eichmann R, Khalil L, Schmidt K (2009) Excavation at Tall Ḥujayrat al-Ghuzlan (Aqaba/Jordan). Excavations 1998-2005 and stratigraphy. In: Khalil L, Schmidt K (eds) Prehistoric Aqaba I, Orient-Archäologie, vol 23, pp 17–77
Einecke G, Köhler W (1910) Die Eisenerzvorräte des Deutschen Reiches. Königlich Preussische Geologische Landesanstalt, Berlin
Eiwanger J (1996) Barrenmessing – ein mittelalterliches Handelsgut zwischen Europa und Afrika? Beitr zur Allg und Vergleich Archäol 16:215–226
El Gayyar ES, Jones MP (1989) A possible source of copper ore fragments found at the old kingdom town of Buhen. J Egypt Archaeol 75:31–40
El Sharkawi MA, El Aref MM, Abdel Motelib A (1990) Syngenetic and paleokarstic copper mineralization in the Paleozoic platform sediments of West Central Sinai, Egypt. Spec Publ Int Sediment 11:159–172
El Shazley EM, Webb JS, Williams D (1957) Trace elements in sphalerite, galena and associated minerals from the British Isles. Trans Inst Min Metall 66:241–271
Elevatorski EA (1996) Gold resources of Europe and the Middle East. MINOBRAS Mining Services
Eliopoulos DG, Economou-Eliopolous M (2000) Geochemical and mineralogical characteristics of Fe-Ni and bauxitic laterite deposits of Greece. Ore Geol Rev 16:41–58
Eliopoulos DG, Economou-Eliopoulos M, Apostolikas A, Golightly JP (2012) Geochemical features of nickel-laterite deposits from the Balkan Peninsula and Gordes, Turkey: the genetic and environmental significance of arsenic. Ore Geol Rev 48:413–427
Ellingham HTJ (1944) The physical chemistry of process metallurgy. J Soc Chem Ind 63:125
Éluère C (1986) A prehistoric touchstone from France. Gold Bull 19(2):58–61
Éluère C, Raub CJ (1991) Investigation on the gold coating technology of the great dish from Varna. In: Mohen JP, Éluère C (eds) Découverte du Métal. Picard, Paris, pp 13–30
Emsley J (1991) The elements, 2nd edn. Oxford University Press, Oxford
Erb-Satullo NL, Gilmour BJJ, Khakhutaishvili N (2014) Late bronze age and iron age copper smelting technologies in the South Caucasus: the view from the ancient Colchis c. 1500-600 BC. J Archaeol Sci 49:147–159
Ercker L (1556) Das Kleine Probierbuch. In: Beierlein PR (ed) Vereinigung für Kunst und Kultur im Bergbau, Bochum, pp 31–144
Ercker L (1580) Beschreibung der allervornehmsten mineralischen Erze und Bergwerksarten. Translation and edition by P.R. Beierlein & A. Lange. Freiberger Forschungshefte D34. Akademie Verlag, Berlin 1960
Erdrich M (1995) Zur Herstellung von Hemmoorer Eimern. In: STA M (ed) Acta of the 12th international congress on ancient bronzes Nijmegen 1992. Nederland. Archeol Rapp 18. Rijksdienst voor het Oudheidkundig Bodemonderzoek, Amersfort, pp 33–38
Ericsson T, Filippidis A (1986) Cation ordering in the limited solid solution Fe_2SiO_4–Zn_2SiO_4. Am Mineral 71:1502–1509
Esin U (1993) Copper beads of Aşıklı. In: Mellink M, Porada E, Özgüç T (eds) Aspects of art and iconography: anatolia and its neighbours, Studies in honor of Nimet Özgüç, pp 173–183
Esin U (1996) Asikli, ten thousand years ago: a habitation model from central Anatolia. In: Housing and settlement in Anatolia: a historical perspective. Habitat, Istanbul, pp 31–42
Espelund A (1991) A retrospective view of direction production. In: Espelund A (ed) Bloomery ironmaking during 2000 years I, Trondheim, pp 71–99
Espelund A (1992) The Mellager site in Trondheim. In: Spelund A (ed) Bloomery ironmaking during 2000 years 2, Trondheim, pp 93–114
Espelund A (1996) Archäo-Metallurgie, von Norwegen aus betrachtet. FERRUM 6:43–53
Espelund A (1999) Luppenstudien in Norwegen. Archaeol Austriaca 82–83:528–536
Ettler V, Legendre O, Bodénan F, Touray JC (2001) Primary phases and natural weathering of old lead-zinc pyrometallurgical slag from Příbram, Czech Republic. Can Miner 39:873–888
Ettler V, Johan Z, Hradil D (2003) Natural alteration products of sulphidic mattes from primary lead smelting. C.R. Geoscience 335:1013–1020
Ettler V, Johan Z, Bezdička P, Drábek M, Šebek O (2009a) Crystallization sequences in matte and speiss from primary lead smelting. Eur J Miner 21:837–854
Ettler V, Červinka R, Johan Z (2009b) Mineralogy of medieval slags from lead and silver smelting (Bohutín, Pribram District, Czech Republic): towards estimation

of historical smelting conditions. Archaeometry 51 (6):987–1007

Ettler V, Johan Z, Selmi Wallisova M, Mihaljevič M, Šebek O (2015) Slag remains from the Na Slupi site (Prague, Czech Republic): evidence for early medieval non-ferrous metal smelting. J Archaeol Sci 53:72–83

Eugster HP, Wones DR (1962) Stability relations of the ferruginous Biotite, Annite. J Petrol 3:82–125

Evangelou VP (1998) Environmental soil and water chemistry, Principles and applications. Wiley, New York

Evans AM (1992) Erzlagerstättenkunde. Enke Verlag, Stuttgart

Exel R (1986) Sardinien. Geologie, Mineralogie, Lagerstätten, Bergbau, Samml Geolog Führer 80. Borntraeger, Berlin

Faber W (1936) Die Mineralien der Bleischlacken. Chem Erde 10:67–115

Falbe J, Regitz M (1996) Römpp-Chemielexikon, 10th edn. Georg Thieme Verlag, New York

Falconer D, Craw D (2009) Supergene gold mobility: a textural and geochemical study from gold placers in southern New Zealand. Econ Geol Spec Publ 14:77–93

Farnsworth M, Smith CS, Rodda JL (1949) Metallographic examination of a sample of metallic zinc from ancient Athens. Hesperia Suppl 8:126–129

Fasnacht W (1989) Excavations at Ayia Varvara-Almyras. Rep. Dept. Antiquities, Cyprus, pp 59–76

Fasnacht W (1996) Excavation at Agia Varvara-Almyras: fifth preliminary report. Rep. Dept. Antiquities, Cyprus, pp 95–125

Fasnacht W (1997) The Sia-Mathiatis-Agia Varvara survey project (SMASP): second preliminary report. Rep. Dept. Antiquities, Cyprus, pp 1–6

Fasnacht W (1999) Experimentelle Rekonstruktion des Gebrauchs von frühbronzezeitlichen Blasdüsen aus der Schweiz: Kupferverhüttung und Bronzeguss. In: Hauptmann A, Pernicka E, Rehren T, Yalçın (eds) The beginnings of metallurgy, Der Anschnitt Beih, vol 9, pp 291–294

Fasnacht W (2000) Excavation at Agia Varvara-Almyras: a review of twelve years of research. Rep. Dept. Antiquities, Cyprus, pp 179–184

Fasnacht W (2006) Die Tiegel von Niederwil. In: Hasenfratz A, Raemaekers DCM (eds) Niederwil – Eine Siedlung der Pfyner Kultur, Vol V: Anorganische Funde, Palynologie und Synthese. Archäologie im Thurgau, vol. 13, pp 137–148

Fasnacht W, Peege C, Hedley I (2000) Agia Varvara-Almyras – final excavation report. Rep. Dept. Antiquities, Cyprus, pp 101–115

Faure G (1986) Principles of isotope geology. Wiley, Chichester

Fellmann R (1991) Die Zinktafel von Bern-Thormebodenwald und ihre Inschrift. Archäologie der Schweiz 14:270–273

Ferguson RB (1950) Red gold from the San Antonio gold mine, Bissett, Manitoba. Am Mineral 35 (5–6):459–460

Fernandez-Miranda M (1991) Tartessos: Indigenas, Fenicios y Griegos en Huelva. Attis del II Congreso Internazionale di Studi Fenici e Punici. Consigl Nazional Ric, Rome, pp 87–96

Ferrand J et al (2014) On the origin of the green colour of archaeological bone artefacts of the Gallo-Roman period. Archaeometry 56(6):1024–1040

Fessner M, Bartels C (2012) Von der Krise am Ende des 16. Jahrhunderts zum deutschen Bergbau im Zeitalter des Merkantilismus. In: Bartels C, Slotta R (eds) Geschichte des deutschen Bergbaus 1: Der alteuropäische Bergbau. Aschendorff, Münster, pp 453–590

Filippova IB, Bush VA, Didenko A (2002) Middle Paleozoic subduction belts: the leading factor in the formation of the Central Asian fold-and-thrust belt. Russ J Earth Sci 3(6):405–426

Fiorini E (1991) L'impegio del piombo Romano nelle ricerche di eventi rari. Il Nuovo Saggiatore 7(5):29–39

Fischer T (2001) Die Römischen Provinzen. Eine Einführung in ihre Archäologie. Wissenschaftliche Buchgesellschaft Darmstadt

Fleischer M, Pabst A, Mandarino JA, Chao GY (1977) New mineral names. Am Miner 62:1057–1061

Fleming SJ, Pigott VC, Swann CP (2006) The archaeometallurgy of *War Kabud*, Western Iran. Iran Antiq 41:31–57

Fleming SJ, Nash S, Swann C (2011) The archaeometallurgy of period IVB bronzes at Hasanlu. In: De Schauensee M (ed) Peoples and crafts in period IVB at Hasanlu, Iran. Univers Pennsylvania Press, Philadelphia, pp 103–134

Flörke OW (1959) Über Kieselsäurekristalle in Gläsern. Glastechn Ber 32(1):1–9

Foord EE, Shawe DR, Conklin NM (1988) Coexisting galena, PbS_{ss} and sulfosalts: evidence for multiple episodes of mineralization in the round mountain and manharan gold districts, Nevada. Can Miner 26:355–376

Forbes RJ (1950) Metallurgy in antiquity. Brill, Leiden

Forbes RJ (1971) Metallurgy in antiquity 1. Early metallurgy, the smith and his tools, gold, silver and lead, zinc and brass. Brill, Leiden

Forbes RJ (1972) Metallurgy in antiquity 2. Copper and bronze, tin, arsenic, antimony and iron. Brill, Leiden

Fordyce G, Alchorne S (1779) An examination of various ores in the museum of Dr. William Hunter. Phil Trans:527–536

Förster T (1987) Glänzend wie gold. Gelbguss bei den Senufo, Elfenbeinküste. Veröffentl Mus Völkerkunde Berlin, Abt Afrika, IX

Frame LD (2012) Reconstructing ancient technologies: chalcolithic crucible smelting at Tal-i-Iblis, Iran. In: Jett P, McCathy B, Douglas JD (eds) Fifth Forbes symposium: studies of ancient Asian Asian metallurgy. Smiths, Washington DC, pp 181–202

Francaviglia VM (1999) Dating the ancient dam of Marib (Yemen). J Archaeol Sci 27:645–653

Francfort HP (1989) Fouilles de Shortugai. Recherches sur l'Asie Centrale Protohistorique. Memoires de la Mission Archeologique Francaise en Asie Centrale, Paris

Frangipane M (2004) Alle origine del potere. Arslantepe, la colina die leoni. Milano

Frangipane M, Di Nocera GM, Hauptmann A, Morbidelli P, Palmieri A, Sadori L, Schultz M, Schmidt-Schultz T (2001) New symbols of a new power in a "royal" tomb from 3000 BC Arslantepe, Malatya (Turkey). Paléorient 27(2):105–139

Frantz JH, Schorsch D (1990) Egyptian red gold. Archaeomaterials 4:133–152

French D (1962) Excavations at Can Hasan. First preliminary report. Anatol Stud 12:27–40

French M, Eugster HP (1965) Experimental control of oxygen fugacities by graphite-gas equilibriums. J Geophys Res 70(6):1529–1539

Frenzel G, Otiemann J, Kurtze W (1973) Über Cu-haltigen Bleiglanz und Pb-haltigen Covellin von Boarezzo (Varese) und ihre Sulfidparagenese. Schweiz Mineral Petrogr Mitt 53:217–229

Frölich K (1953) Goslarer Bergrechtsquellen des früheren Mittelalters, insbesondere das Bergrecht des Rammelsberges aus der Mitte des 14, Jahrhunderts. Gießen, Wilhelm Schmitz Verlag, pp 39–94

Frost R, Mavrogeneses JA, Tomkins AG (2002) Partial melting of sulfide ore deposits during medium- and high-grade metamorphism. Can Miner 40:1–18

Frotzscher M (2009) Geochemische Charakterisierung von mitteleuropäischen Kupfervorkommen zur Herkunftsbestimmung der Himmelsscheibe von Nebra. PhD-Diss. Halle. http://digital.bibliothek.uni-halle.de/hs/content/titleinfo/432396

Furger AR (2017) Der vergoldete Buddha. Traditionelles Kunsthandwerk der Newar-Giesser in Nepal. Librum Publ., Basel, Frankfurt a.M.

Furger AR (2018) Antike Schmelztiegel. Archäologie und Archäometrie der Funde aus Augusta Raurica, Beitrr Technikgesch 1. Librum Publ., Basel, Frankfurt a.M.

Furger AR, Riederer J (1995) Aes und aurichalcum: empirische Beurteilungskriterien für Kupferlegierungen und metallanalytische Untersuchungen an Halbfabrikaten und Abfällen aus metallverarbeitenden Werkstätten in Augusta Raurica. Jb Augst und Kaiseraugst 16:115–180

Gadalla AMM, Ford WF, White J (1962/1963) Equilibrium relationships in the system CuO-Cu_2O-SiO_2. Trans British Ceramic Soc 62:45–66

Gale NH (1991) Copper oxhide ingots: their origin and their place in the bronze age metals trade in the mediterranean. In: Gale NH (ed) Bronze age trade in the mediterranean, Jonsered, pp 197–239

Gale NH (1997) The isotopic composition of tin in some ancient metals and the recycling problem in metal provenancing. Archaeometry 39:71–82

Gale NH (1999) Lead isotope characterization of the ore deposits of Cyprus and Sardinia and its application to the discovery of the sources of copper for late bronze age oxhide ingots. In: Young SMM, Pollard AM, Budd P, Ixer RA (eds) Metals in antiquity, BAR Int series, vol 792, pp 110–121

Gale NH (2001) Archaeology, science-based archaeology and the mediterranean bronze age metals trade: a contribution to the debate. Eur J Archaeol 4(1):113–130

Gale NH (2006) Lead isotope studies – Sardinia and the mediterranean. Provenance studies of artefacts fund in Sardinia. Instrumentum 23:4–9

Gale NH, Stos-Gale ZA (1981a) Cycladic lead and silver metallurgy. Ann Brit School Athens 76:169–224

Gale NH, Stos-Gale ZA (1981b) Ancient Egyptian silver. J Egypt Archaeol 67:103–115

Gale NH, Stos-Gale ZA (1995) Comments on "Oxhide ingots, recycling, and the mediterranen metals trade". J Mediterr Archaeol 8:33–41

Gale NH, Stos-Gale ZA (2000) Lead isotope analyses applied to provenance studies. In: Ciliberto E, Spoto G (eds) Modern analytical methods in art and archaeology. Wiley, New York, pp 503–584

Gale NH, Stos-Gale ZA (2005) Zur Herkunft der Kupferbarren aus dem Schiffswrack von Uluburun und der spätbronzezeitliche Metallhandel im Mittelmeerraum. In: Yalçın Ü, Pulak C, Slotta R (eds) Das Schiff von Uluburun. Welthandelvor 3000 Jahren. Katalog der Ausstellung des Deutschen Bergbau-Museums Bochum, pp 117–131

Gale NH, Papastamataki A, Stos-Gale ZA, Leonis K (1985) Copper sources and copper metallurgy in the Aegean bronze age. In: Craddock PT, Hughes MJ (eds) Furnaces and smelting technology in antiquity, Brit Mus Occ Papers, vol 48, pp 81–102

Gale NH, Bachmann HG, Rothenberg B, Stos-Gale ZA, Tylecote RF (1990) The adventitious production of Iron in the smelting of copper. In: Rothenberg B (ed) Researches in the Arabah 1959–1984, II: the ancient metallurgy of copper. Inst Archaeo-Metall Stud, London, pp 182–191

Gale NH, Stos-Gale ZA, Lilov P, Dimitrov M, Todorov T (1991) Recent studies of eneolithic copper ores and artefacts in Bulgaria. In: Mohen JP, Éluère C (eds) La Découverte du Metal, Millenaire 2. Paris, pp 49–75

Gale NH, Stos-Gale ZA, Maliotis G, Annetts N (1997) Lead isotope data from the isotrace laboratory, Oxford: archaeometry data base 4, Ores from Cyprus. Archaeometry 39:237–246

Gale NH, Woodhead AP, Stos-Gale ZA, Walder A, Bowen I (1999) Natural variations detected in the isotopic composition of copper: possible applications to archaeology and geochemistry. Int J Mass Spectrom 184:1–9

Gale NH, Kayafa M, Stos-Gale ZA (2008) Early helladic metallurgy at Raphina, Attica, and the role of Lavrion. In: Tzachili I (ed) Aegean metallurgy in the bronze age. Ta Pragmata, pp 87–104

Gale NH, Kayafa M, Stos-Gale ZA (2009) Further evidence for bronze age production of copper from Ores in the Laurion Ore District, Attica, Greece. In: Proc. 2nd International Conference Archaeometallurgy in Europe 2007. Associazione Italiana di Metallurgia, Milano, pp 158–176

Galili E, Gale NH, Rosen B (2013) A late bronze age shipwreck with a metal cargo from Hishuley Carmel, Israel. Int J Naut Archaeol 42(1):2–23

Galili E, Bauvais S, Rosen B, Dillmann P (2015) Cargoes of iron semi-products recovered from shipwrecks off the carmel coast, Israel. Archaeometry 57:505–535
Gambaschidze O, Gambaschidze I (1995) Die Hauptergebnisse der Meskhet-Dschawareti Expedition von 1987. In: Polewije Archaeologitscheski Issledowanija w 1987 godu, Tbilissi. (Russian language)
Gambashidze I, Stöllner T (2016) The gold of Sakdrisi. Man's first gold mining enterprise. Publ Deutsches Bergbau-Museum Bochum 211
Gambaschidze I, Hauptmann A, Slotta R, Yalçın Ü (eds) (2001) Georgien. Schätze aus dem Land des Goldenen Vlies. Ausstellungskatalog des Deutschen Bergbau-Museums Bochum
Ganzelewski M (2000) Archäometallurgische Untersuchungen zur frühen Verhüttung von Raseneisenerzen am Kammberg bei Joldelund, Kreis Nordfriesland. In: Haffner A, Jöns H, Reichstein J (eds) Frühe Eisengewinnung in Joldelund, Kreis Nordfriesland. Ein Beitrag zur Siedlungs- und Technikgeschichte Schleswig-Holsteins, vol 59. Universitätsforsch Prähist Archäol, p 3
Garenne-Marot L (1984) Le Cuivre en Égypte Pharaonique: sources et Métallurgie. Paléorient 10 (1):97–126
Garfinkel Y, Klimscha F, Shalev S, Rosenberg D (2014) The beginning of metallurgy in the Southern levant: a late 6 millennium CalBC copper Awl from Tel Tsaf. PLoS One 9(3):e92591
Garner J (2010) Der latènezeitliche Verhüttungsplatz in Siegen-Niederschelden "Wartestraße". Meta 17
Garner J (2014) Das Zinn der Bronzezeit in Mittelasien II: Die montanarchäologischen Forschungen an den Zinnlagerstätten. Archäolog Iran Turan, Von Zabern, Darmstadt
Garrels RM (1954) Mineral species as functions of pH and oxidation-reduction potentials, with special reference to the zone of oxidation and secondary enrichment of sulphide ore deposits. Geochim Cosmochim Acta 5:153–168
Garrels R, Christ CL (1965) Solutions, minerals and equilibria. Freeman, Cooper & Company, San Francisco
Gassmann G (1995) Untersuchungen von Verhüttungsanlagen in Neuenbürg, "Schnaizteich", Enzkreis. Arch Ausgr Baden-Württemberg 1995:155–158
Gassmann G, Yalçın (2009) Die archäometallurgischen Arbeiten. In: Stöllner T, Garner J, Gassmann G, Kalje AJ, Röttger K, Stobbe A, Tegtmeier U, Yalçın Ü Latètenezeitliche Eisenwirtschaft im Siegerland. Interdisziplinäre Forschungen zur Wirtschaftsarchäologie Metalla 16(2):147–177
Gassmann G, Yalçın Ü, Hauptmann A (2005) Die archäometallurgischen Materialuntersuchungen zur keltischen Eisneverhüttung in Baden-Württemberg. In: Forsch. Ber. zur Vor- und Frühgesch (ed) In Baden-Württemberg (=Forschungen zur keltischen Eisneverhüttung in Südwestdeutschland), vol 92, pp 84–114
Gassmann G, Klein S, Körlin G (2015) The Roman mines near Ulpiana, Kosovo. In: Hauptmann A, Modarressi-Tehrani D (eds) Archaeometallurgy in Europe III. Proc 3rd Conf Bochum 2011, pp 33–43
Gatzweiler R, Akin H (1993) Sampling of ore deposits and evaluation of results. In: Kraft G (ed) Dampling in the non-ferrous metals industry. Concentrates and recycled commodities, Series on bulk materials handling 6. Trans Tech Publ, pp 147–157
Gauß R (2015) Früher Bergbau und Metallurgie auf der Iberischen Halbinsel. Faszikel 1: Zambujal und die Anfänge der Metallurgie in der Estremadura (Portugal). Technologie der Kupfergewinnung, Herkunft des Metalls und soziokulturelle Bedeutung der Innovation. Iberia Archaeologica 15.1. Wasmuth, Tübingen
Gauthier J, Fluck P, Disser A, Chateau C (2015) The Alsatian Altenberg: a seven-hundred year laboratory for silver metallurgy. In: Hauptmann A, Modarressi-Techrani D (eds) Archaeometallurgy in Europe III, Der Anschnitt Beih, vol 26, pp 271–278
Gebhard R (1995) Industry in Celtic Oppida – aspects of high temperature processes. In: Morteani G, Northover JP (eds) Prehistoric gold in Europe. Mines, metallurgy and manufacture, NATO ASI series E: applied sciences, vol 280. Kluwer Academic Publ, Dordrecht, pp 261–272
Gebhard R, Krause R (2016) Bernstorf. Archäologisch-naturwissenschaftliche Analysen der gold- und Bernsteinfunde vom Bernstorfer Berg bei Kranzberg, Oberbayern. Bernstorf-Forschungen 1/Frankfurter Archäol. Schriften 31
Gebhard R et al (1995) Coin moulds and other ceramic material: a key to celtic precious metal working. In: Morteani G, Northover JP (eds) Prehistoric gold in Europe. Mines, metallurgy and manufacture, NATO ASI series E: applied sciences, vol 280. Kluwer Academic Publ., Dordrecht, pp 273–301
Gebhard R, Krause R, Röpke A, Bähr V (2014) Das gold von Bernstorf – Authentizität und Kontext in der mittleren Bronzezeit Europas. In: Meller H, Risch R, Pernicka E (eds) Metalle der Macht: Frühes gold und Silber. Metals of power: early gold and silver, Tag. Landesmus. Vorgesch. Halle II, pp 761–776
Geçkinly A, Özbal H, Craddock PT, Meeks ND (2000) Examination of the Sardis gold and the replication experiments. In: Ramage A, Craddock PT (eds) King Croesus' gold. Excavations at Sardis and the history of gold refining. Brit Mus Press, London, pp 184–199
Geerlings W (1983) Ins Bergwerk verurteilt - die "damnatio ad metalla". Der Anschnitt 35 (4–5):130–136
Geerlings W (1985) Zum biblischen und historischen Hintergrund der Bergwerke von Fenan in Jordanien. Der Anschnitt 37(5–6):158–162
Geffroy J (1980) De l'antimoine à l'or: La Lucette (Mayenne). Monde Mineraux 39(11):20–26
Geller RF, Creamer AS, Bunting EN (1934) The system PbO-SiO2. National Bureau of Standards 13, Res. Paper RP705: 237–244
George L, Cook NJ, Cristiana L, Wade BP (2015) Trace and minor elements in galena: a reconnaissance LA-ICP-MS study. Am Mineral 100:548–569

Germann K, Schwarz T, Wipki M (1999) Verwitterungsbedingte Lagerstättenbildung in Sedimentbecken Nordost-Afrikas. In: Klitzsch E, Thorweihe U (eds) Nordost-Afrika: Strukturen und Ressourcen. Ergebnisse aus dem Sonderforschungsbereich "Geowissenschaftliche Probleme in ariden und semiariden Gebieten". Wiley, pp 250–298

Gevorkjan AZ (1980) Aus der Geschichte der alten Metallurgie des armenischen Berglandes, Jerewan

Ghirshman R (1938) Fouilles de Sialk 1. Paul Geuthner, Paris

Ghorbani M (2013) The economic geology of Iran, Mineral deposits and natural resources. Springer, Dordrecht

Gialli S (2013) The controversial polymetallic madneuli deposit, Bolnisi District, Georgia: Hydrothermal Alteration and Ore Mineralogy. MSc-thesis, Univers de Genève, Switzerland

Gialoglu G, Vavelidis M, Wagner GA (1988) Die antiken Silber-Belibergwerke auf Thasos. In: Wagner GA, Weisgerber G (eds) Antike Edel- und Buntmetallgewinnung auf Thasos, Der Anschnitt Beih, vol 6, pp 75–87

Giardino C (2000) The beginning of metallurgy in Tyrrhenian south-central Italy: the Eneolithic facies of Gaudo. In: Ridgway D, Serra Ridgway FR, Pearce M, Herring E, Whitehouse RD, Wilkins JB (eds) Ancient Italy in its mediterranean setting. Studies in honour of E Macnamara. Accordia Specialist

Gilchrist JD (1989) Extraction metallurgy, 3rd edn. Pergamon Press, Oxford, UK

Gilles JW (1957a) 25 Jahre Siegerländer Vorgeschichtsforschung durch Grabungen auf alten Eisenhüttenplätzen. Archiv Eisenhüttenwesen 28:179–185

Gilles JW (1957b) Alter der Eisengewinnung an der unteren Sieg. Stahl und Eisen 77(26):1883–1884

Gilles KJ (2002) Ein Depotfund römischer Rohkupferbarren aus Trier. Funde und Ausgrabungen im Bezirk Trier 34:73–76

Gilmour B, Khakhutaishvili N, Erb-Satullo N (2014) Ancient metal production sites from southwest Georgia in light of new archeological evidence. Phasis 17:137–158

Ginzburg IV, Lisitsina GA, Sadikova AT, Sidorenko GA (1962) Fayalite granitic rocks and their alteration products. (Kuramin Ridge, Soviet Central Asia). Trans MinMus Acad Sc USSR 13. In: Deer WA, Howie RA, Zussman J (eds) Rock-forming minerals 1A, Orthosilicates, London, pp 16–42

Giumlia-Mair A (2009) The history of mercury production in the mine of Idrija, Slovenia. In: Archaeometallurgy in Europe. Proc 2nd Int Conf Archaeometallurgy in Europe 2007, Aquileia, pp 68–78

Giumlia-Mair A, Craddock PT (1993) Corinthium aes. Das schwarze gold der Alchimisten. Antike Welt 24:1–62

Giumlia-Mair A, Maddin R (2004) The origins of iron. In: Nicodemi W (ed) The civilisation of iron. From prehistory to the third millennium. Olivares, Milano, pp 35–61

Giunti I, Artioli G, Giussani B, Marelli M (2009) Chemical and isotopic provenance tracers in ancient copper and bronze artifacts: a geochemical database of copper mines. Anal Chim Acta 185:1–17

Given M, Knapp B (2003) The Sydney cyprus survey project. Social Appoaches to regional archaeological survey, Monumenta Archaeologica 21. University of Los Angeles

Glocker EF (1829) Elemente der Mineralogie, Nürnberg

Glumac PD, Todd JA (1991a) Earl metallurgy in Southeast Europe: the evidence for production. MASCA Res Pap 8(1):9–19

Glumac PD, Todd JA (1991b) Eneolithic copper smelting slags from the middle Danubian Basin. In: Pernicka E, Wagner GA (eds) Archaeometry 1990. Birkhäuser, Basel, pp 155–164

Godrabelidze SA (1933) Mineralische Resssourcen der sozialistischen Republik Georgien. Tbilissi (Russian language)

Goettler GW, Firith N, Huston CE (1976) A preliminary discussion of ancient mining in the Sultanate of Oman. J Oman Stud 2:43–56

Golden J (2014) Who Dunnit? New clues concerning the development of chalcolithic metal technology in the Southern Levant. In: Roberts BW, Thornton CP (eds) Archaeometallurgy in global perspective, methods and syntheses. Springer, pp 559–578

Goldenberg G (1990) Die Schlacken und ihre Analysen - Relikte der Metallgewinnung und Metallverarbeitung. In: Erze, Schlacken und Metalle - Früher Bergbau im Südschwarzwald. Freiburger Universitätsblätter 109:147–172

Goldenberg G (1994) Die Gewinnung von "Antimon Crudum". Archäometallurgische Befunde seit dem späten Mittelalter bei Sulzburg im Südschwarzwald. Anschnitt 46(4–5):144–149

Goldenberg G (1996a) Archäometallurgische Untersuchungen zur Entwicklung des Metallhüttenwesens im Südschwarzwald (Blei, Silber, Kupfer - Frühgeschichte bis 19. Jahrhundert). Archäol u Gesch 8, Sigmaringen

Goldenberg G (1996b) Archäometallurgische Untersuchungen zur Entwicklung des Metallhüttenwesens im Schwarzwald. Blei- Silber- und Kupfergewinnung von der Frühgeschichte bis zum 19. Jahrhundert. In: Goldenberg G, Steuer H (eds) Archäometallurgische Untersuchungen zum Metallhüttenwesen im Schwarzwald. Jan Thorbecke Verlag, Sigmaringen, pp 9–274

Goldenberg G (2004) Ein Verhüttungsplatz der mittleren Bronzezeit bei Jochberg (Nordtirol). In: Weisgerber G, Goldenberg G (eds) Alpenkupfer – Rame delle Alpi, Der Anschnitt Beih, vol 17, pp 165–176

Goldenberg G (2013) Prähistorischer Fahlerzbergbau im Unterinntal – Montanarchäologische Befunde. In: Oeggl K, Schaffer V, Montanwerke Brixlegg AG:

Cuprum Tyrolense. 5550 Jahre Bergbau und Kupferverhüttung in Tirol. Brixlegg: 55–70
Goldenberg G (2014a) Kupfererzverhüttung in der ausgehenden Mittelbronzezeit bei Jochberg (Nordtirol) - ein montanarchäologischer Befund. Universität Innsbruck, Habilschrift
Goldenberg G (2014b) Spätbronzezeitliche Kupferproduktion aus Fahlerz im Unterinntal (Nordtirol) - Rekonstruktion der "chaîne opératoire". Universität Innsbruck, Habilschrift
Goldenberg G (2015) Prähistorische Kupfergewinnung aus Fahlerzen der Lagerstätte Schwaz-Brixlegg im Unterinntal, Nordtirol. In: Stöllner T, Oeggl K (eds) Bergauf Bergab 10.000 Jahre Bergbau in den Ostalpen. VML Verlag Marie Leidorf, Rahden/Westfalen, pp 151–163
Goldenberg G, Rieser B (2004) Die Fahlerzlagerstätten von Schwaz/Brixlegg (Nordtirol). In: Weisgerber G, Goldenberg G (eds) Alpenkupfer - Rame delle Alpi, Der Anschnitt Beih, vol 17, pp 37–52
Goldenberg G, Steuer H (2004) Mittelalterlicher Silberbergbau im Südschwarzwald. In: Markl G, Lorenz S (eds) Silber Kupfer Kobalt. Schriftenreihe des Mineralienmuseums Oberwolfach (= Veröffentl. Alemann Inst 72), pp 45–80
Goldenberg G et al (2011) Das Nepal-Experiment - experimentelle Archäologie mit ethnoarchäologischem Ansatz. In: Oeggl K, Goldenberg G, Stöllner T, Prast M (eds) Die Geschichte des Bergbaus in Tirol und seinen angrenzenden Gebieten. Proc 5th Milstone-Meeting SFB-HiMAT, October 2010. Innsbruck University Press, Mühlbach, pp 83–90
Gontscharov A (2019) Metall der bronzezeitlichen Kulturen aus Zentral- und Ostkasachstan. PhD-diss Ruhr Univers Bochum
Gopher A, Tsuk T, Shalev S, Gophna R (1990) Earliest gold artefacts in the Levant. Curr Anthropol 31:436–443
Gornostayev SS, Dodatko AD, Laajoki KVO, Mochalov AG (2000) Origin of platinum-bearing placers in the Aluchin Horst, Russian Far East. Econ Geol 95:540–558
Gowland W (1894) A Japanese pseudo-speise (shirome) and its relation to the purity of Japanese copper. J Soc Chem Ind 13(5):1–26
Gowland W (1899) The early metallurgy of copper, tin and iron in Europe, as illustrated by ancient remains and the primitive processes surviving in Japan. Archaeologia 56(2):267–322
Gowland W (1900) Remains of a Roman silver refinery at Silchester. Archaeologia 57(1):113–124
Gowland W (1901) The early metallurgy of silver and lead, part 1: lead. Archaeologia 57(2):359–422
Gowland W (1906) Presidential address: copper and ist alloys in prehistoric times. J Anthropol Inst Great Britain Ireland 36:11–38
Gowland W (1914) The metallurgy of the non-ferrous metals. London
Gowland W (1917/1918) Silver in Roman and earlier times. Archaeologia 69:121–160
Graham-Campbell J (1982) Viking silver hoards: an introduction. In: Farell R (ed) The Vikings, London, pp 32–41
Grant MR (1994) Iron in ancient tin from Rooiberg, South Africa. J Archaeol Sci 21:455–460
Greeves T (2016) Called home. The Dartmoor Tin Miners 1860-1940. Photographs & memory. Twelveheads Press, Truro
Grießl E (2014) Entwicklung einer Methodik zur Bewertung von Handlungsalternativen im internationalen Kleinbergbau. PhD-diss Fakultät Geowissensch, Geotechnik und Bergbau, Techn Univers Bergakademie Freiberg
Grögler N, Geiss J, Grünenfelder M, Houtermans FC (1966) Isotopenuntersuchungen zur Bestimmung der Herkunft von römischer Bleirohre und Bleibarren. Z Naturforsch 21:1167–1172
Grothe H (1971) Zur Geschichte des Zinks. Erzmetall 24:324–329
Groves D, Goldfarb RJ, Robert F, Craig JRH (2003) Gold deposits in metamorphic belts: overview of current understanding, outstanding problems, future research, and exploration significance. Econ Geol 98(1):1–29
Guénette-Beck B, Meisser N, Curdy P (2009) New insights into the ancient silver production of the Wallis area, Switzerland. J Archaeol Anthropol Sci 1:215–229
Guerra MF, Calligaro T (2003) Gold cultural heritage objects: a review of studies of provenance and manufacturing. Meas Sci Technol 14:1527–1537
Guerra MF, Calligaro T (2007) The treasure of Guarrazar: tracing the gold supplies in the Visigothic Iberian Peninsula. Archaeometry 49(1):53–74
Guerra MF, Rehren T (2009) AURUM: Archaeometry and authenticitation of gold, Archaeosciences. revue d'archéométrie 33. Presses Universitaires de Rennes, pp 13–18
Gugushvili VI, Akhvlediani R, Natsvlishvili M, Hart I (1999) Two stages of gold mineralization within Sakdrisi deposit (Bolnisi Mining District, Georgia). Geologica Carpathica spec. Issue, Vol. 53
Guidon N, Delibrias G (1986) Carbon-14 dates point to man in the Americas 32.000 years ago. Nature 321 (6072):769–771
Guiollard PC (1993) Les mines d'Or et d'Argent de Rouez. Exploitation du chapeau de Fer. 130 p. SAI Biarritz
Guirado MP, Téreygol F, Peyrat F (2010) Initial experiments on silver refining: how did a cupellation furnace work in the 16th century? J Hist Met 44(2):126–135
Gulson BL (1986) Lead isotopes in mineral exploration, Developments in Econ Geol Ser 23. Elsevier, Amsterdam
Habashi F (1978) Chalcopyrite - its chemistry and metallurgy. McGrawhill, New York
Haffner A, Jöns H, Reichstein J (2000) Frühe Eisengewinnung in Joldelund, Kr. Nordfriesland 1, 2. Universitätsforsch Prähist Archäol 59
Hahn HP (2014) Die Sprache des Glanzes: wert und Werte als Kontext von gold. In: Meller H, Risch R, Pernicka E (eds) Metalle der Macht - Frühes gold und Silber.

Metals of power – Early gold and silver. 6th Archaeol Conf of Central Germany, October 2013, vol 22. Tagungen des Landesmus Vorgesch Halle, Halle (Saale), pp 21–31

Hall M, Steadman SR (1991) Tin and Anatolia: another look. J Mediterran Archaeol 4(2):217–234

Hammer P (1995) Messing. In: Hoops J (ed) Reallex German Altertumskunde. Gruyter, Berlin, pp 609–615

Hammer P (1998) Verfahrenstechnische Untersuchung. Bericht zu Römische und germanische Bunt- und Edelmetallfunde im Vergleich. Archäometallurgische Untersuchungen ausgehend von elbgermanischen Körpergräbern. Ber Röm-German Kommisssion 79:179–199

Hammer P, Voß HU (1998) Gemeinsamkeiten und Unterschiede bei der Anwendung bevorzugter Feinschmiedetechnik. In: Voß HU et al (eds) Römische und germanische Bunt- und Edelmetallfunde im Vergleich, Archäometalurgische Untersuchungen ausgehend von elbgermanischen Körpergräbern. Berichte Röm German Kommission, vol 79, pp 292–306

Hammer P, Voß HU (2000) Zur Gruppierung von Kupferlegierungen – Der Terminus "Aes" bei Plinius. Meta 7(1):23–32

Hammer P, Voß HU, Lutz J (1998) Das Material – Die Verwendung von Bunt- und Edelmetallen bei römischen und germanischen Handwerkern. In: Voß HU et al (eds) Römische und germanische Bunt- und Edelmetallfunde im Vergleich, Archäometalurgische Untersuchungen ausgehend von elbgermanischen Körpergräbern. Ber Röm German Komm, vol 79, pp 276–288

Hanel N, Morstadt B (2014) Der Iglesiente – Eine unbekannte antike Montanlandschaft im Südwesten Sardiniens. Anschnitt 6:214–229

Hanel N, Bode M (2016) Messingbarren aus einem römischen Schiffswrack bei Aléria (Korsika). Der Anschnitt Beih 29:167–181

Hanel N, Rothenhöfer P, Bode M, Hauptmann A (2013a) Nach der Schlacht von Lugdunum (197 n. Chr.). Britannisches Blei auf dem Weg nach rom. Chiron 43:297–325

Hanel N, Rothenhöfer P, Bode M, Hauptmann A (2013b) Britannisches Blei auf dem Weg nach rom. Die Metallversorgung der Reichsmetropole am Beginn der Herrschaft des L Septimius Severus Skyllis 13:38–42

Hanilçi N, Öztürk H, Altuncu S (2017) Was Göl Tepe (Çamardı-Niğde, Turkey) Tin or Gold Processing Site: A Soil Geochemistry Study. Internat Symp Geoarchaeol, Ancient Mining and Metallurgy, November 2017, Istanbul. Abstract: 15

Hanning E (in prep) Smelting of sulfidic ore during the bronze age in the eastern Alpin region: a mining and experimental approach. PhD-diss Inst. Archaeol Sci, Ruhr Univers Bochum

Hanning E, Pils R (2011) Experimentelle Untersuchungen zur bronzezeitlichen Kupferverhüttung im ostalpinen Gebiet – Erste Ergebnisse. In: Oeggl K, Goldenberg G, Stöllner T, Prast T (eds) Die geschichte des Bergbaus in Tirol und seinen angrenzenden Gebieten. Proc 5th milestone meeting SFB-HIMAT, 2010, Mühlbach, pp 129–134

Hanning E (2012) Reconstructing bronze age copper smelting in the Alps: an ongoing process. Experimentelle Archäologie in Europa, Bilanz 2012:75–86

Hanning E (2019) Slag heap quantification: re-evaluating the Mitterberg smelting sites (Salzburg, Austria). In: Turck R, Stöllner T, Goldenberg G (eds) Alpine copper II - Rame delle Alpi II – Cuivre des Alpes II, New results and perspectives on prehistoric copper production. Der Anschnitt Beih, vol 42, pp 311–321

Hanning E, Gauß R, Goldenberg G (2010) Metal for Zambujal: experimentally reconstructing a 5000-year-old technology. In: Proc Symp Arqueometalurgia: Aspectos tecnológicos, económicos y sociales en la orehistoria reciente Europea. Trabajos de Prehistoria, vol 67, pp 287–304

Hanning E, Herdits H, Silvestri E (2015) Alpines Kupferschmelzen – technologische Aspekte. In: Stöllner T, Oeggl K (eds) Bergauf Bergab. 10.000 Jahre Bergbau in den Ostalpen. Veröffentl, Deutsches Bergbau-museum Bochum 207. Verlag Marie Leidorf Rahden, pp 225–231

Hannington MD, Galley AG, Herzig P, Peterson S (1998) Comparison of the TAG mound and Stockwork complex with Cyprus-type massive sulfide deposits. In: Herzig PM, Humphries SE, Miller DJ, Zierenberg RA (eds) Proceedings of the ocean drilling program, scientific results, vol 158, pp 389–415

Hansen S (2009) Kupfer, gold und Silber im Schwarzmeerraum während des 5. Und 4. Jahrtausends v. Chr. In: Apakidze J, Govedarica B, Hänsel B (eds) Der Schwarzmeerraum vom Äneolithikum bis in die Frühbronzezeit (5000-500 v. Chr.), Prähistorische Archäologie in Südosteuropa, vol 25, pp 11–50

Hansen S (2011) Technische und soziale Innovationen in der zweiten Hälfte des 4. Jahrtausends v. Chr. In: Hansen S, Müller J (eds) Sozialarchäologische Perspektiven: Gesellschaftlicher Wandel 5000-1500 v. Chr. Zwischen Atlantik und Kaukasus. Mainz, Philipp von Zabern, Archäologie in Eurasien, vol 24, pp 153–191

Hansen S (2016) Prähistorische Innovationsforschung. Das Altertum 61:81–132

Hansen-Streily A (2000) Early pottery kilns in the Middle East. Paléorient 26(2):69–81

Hansen M, Anderko K (1958) Constitution of binary alloys. McGraw-Hill, New York

Happ J et al (1996) Bilan des recherches en métallurgie expérimentale relatives à la production du cuivre au Chalcolithique et à l'Age du bronze. In: Mordant C, Pernot M, Rychner V (eds) L'atelier du bronzier en Europe, Neuchâtel/Dijon, vol 2, pp 83–92

Harris DC (1990) The mineralogy of gold and ist relevance to gold recoveries. Mineral Deposits 25:3–7

Harris DC, Cabri LJ (1991) Nomenclature of platinum-group-element alloys; review and revision. Can Miner 29(2):231–237

Harrison FA (1931) Ancient mining activities in Portugal. Min Mag 14:137–145

Hartley AJ, Rice CM (2005) Controls of supergene enrichment of porphyry copper deposits in the Central Andes: a review and discussion. Mineral Deposita 40 (5):515–525

Hartmann A (1970) Prähistorische Goldfunde aus Europa – Spektralanalytische Untersuchungen und deren Auswertung. Studien zu den Anfängen der Metallurgie 3. Berlin

Hartmann A (1982) Prähistorische Goldfunde aus Europa II – Spektralanalytische Untersuchungen und deren Auswertung. Studien zu den Anfängen der Metallurgie 5. Berlin

Hauptmann A (1985) 5000 Jahre Kupfer in Oman 1: die Entwicklung der Kupfermetallurgie vom 3. Jahrtausend bis zur Neuzeit Der Anschnitt Beih 4

Hauptmann A (1989) Chemical analyses of prehistoric metal Artefacts from the Indian subcontinent. Jb Röm German Zentralmus 36(1):261–267

Hauptmann A (2000a) Reconstructing ancient smelting processes: Applied mineralogy in archaeometallurgy. In: Rammelmair D, Mederer J, Oberthür Th, Heimann RB, Pentinghaus H (eds) Applied mineralogy in research, economy, technology, ecology and culture. Balkema, Rotterdam, pp 29–32

Hauptmann A (coord) (2000b) Early Pyrotechnology: The Evolution of the first fire-using industries. Paléorient XXVI(2): 204 p

Hauptmann A (2007) The early metallurgy of copper. Evidence fom Faynan, Jordan. In: Wagner GA, Herrmann B (eds) Natural science in archaeology. Springer, Heidelberg

Hauptmann A (2011) Slags from the late bronze age metal workshops at Kition and Enkomi, Cyprus. In: Ferrence S, Betancourt P (eds) Metallurgy: understanding how, learning why, Studies in honor of James D. Muhly. INSTAP Academic Press, Philadelphia, pp 189–202

Hauptmann A (2014) The investigation of archaeometallurgical slag. In: Roberts BW, Thornton CP (eds) Archaeometallurgy in global perspective, Methods and syntheses. Springer, pp 91–105

Hauptmann A (2017) Copper and ores from elsewhere and the question of metallurgy at Maadi. In: Bajeot J (ed) Predynastic Maadi in context. The research of the Italian expedition revisited (1977-1986), Studi di Preistoria Orientale, vol 4, pp 145–152

Hauptmann A, Gambaschidze I (2001) Antimon – eine metallurgische Besonderheit aus dem Kaukasus. In: Gambaschidze I, Hauptmann A, Slotta R, Yalcin Ü (eds) Georgien – Schätze aus dem Land des Goldenen Vlies. Deutsches Bergbau-Museum Bochum, Bochum, pp 150–155

Hauptmann A, Horowitz A (1980) Zur Geomorphologie und Paläomorphologie des Modellgebietes. In: Conrad HG, Rothenberg B (eds) Antikes Kupfer im Timna-Tal, Der Anschnitt Beih, vol 1, pp 57–68

Hauptmann A, Klein S (2016) Golden artifacts from the Royal Tombs of Ur, Mesopotamia. Meta 22 (1):84–97

Hauptmann A, Lewin B (1984) Kodo-zuroku "Illustrierte Abhandlung über die Verhüttung des Kupfers" 1801. Deutsches Bergbau-Museum Bochum

Hauptmann A, Löffler I (2013) Technological innovations and organisational structures of prehistoric mining and metal production – examples from Faynan, Jordan. In: Burmeister S, Hansen S, Kunst M, Müller-Scheeßel N (eds) Metal matters. Innovative technologies and social change in prehistory and antiquity, vol 12. Menschen – Kulturen – Traditionen. Forschungsclust Deutsches Archäolog Inst, VML Rahden/Westf, pp 65–89

Hauptmann A, Maddin R (2005) Die Kupferbarren von Uluburunl I: Qualitätsmetall für den Weltmarkt? In: Yalcin Ü, Pulak C, Slotta R (eds) Das Schiff von Uluburun – Welthandel vor 3000 Jahren. Catalogue, exhibition, Deutsches Bergbau-Museum Bochum 2006. Cuno, Calbe Publ, pp 141–147

Hauptmann A, Slotta R (1979) Zu den Denkmälern des Quecksilberbergbaus von Almaden. Anschnitt 31:81–100

Hauptmann A, Wagner I (2007) Prehistoric copper production at Timna: TL-dating and evidence from the east. In: LaNiece S, Hook D, Craddock P (eds) Metals and mines. Studies in Archaeometallurgy. Brit Mus & Archetype Publ, London, pp 67–75

Hauptmann A, Weisgerber G (1996) The early production of metal in the near east. In: Bagolini B, LoSchiavo F (eds) The copper age in the near east and Europe. Proc Internat Congress Prehist and Protohist Sc, Forli, pp 95–101

Hauptmann A, Weisgerber G (2001) Luppe. In: Beck H, Geuenich D, Steuer H (eds) Reallexikon der Germanischen Altertumskunde 19. de Gruyter, Berlin, pp 48–51

Hauptmann A, Yalçın Ü (2000) Lime plaster, cement and the first Puzzolanic reaction. Paléorient 26(2):61–68

Hauptmann A, Öffler I (2013) Technological innovations and organisational structures of prehistoric mining and metal production – examples from Faynan, Jordan. In: Burmeister S, Hansen S, Kunst M, Müller-Scheeßel N (eds) Metal matters. Innovative technologies and social change in prehistory and antiquity. Menschen – Kulturen – Traditionen, vol 12. Studien aus den Forschungsclustern des Deutschen Archäologischen Instituts, Rahden/Westf, pp 65–89

Hauptmann A, Pernicka E, Wagner GA (1988) Untersuchungen zur Prozeßtechnik und zum Alter der frühen Blei-Silbergewinnung auf Thasos. In: Wagner GA, Weisgerber G (eds) Antike Blei-Silbergewinnung auf Thasos, Der Anschnitt Beih, vol 6, pp 88–112

Hauptmann A, Maddin R, Weisgerber G (1991) Metallographic investigation of two Miner's tools from Wadi Khalid, Feinan. Ann Dept Antiquities Jordan XXXV:195–201

Hauptmann A, Begemann F, Heitkemper E, Pernicka E, Schmitt-Strecker S (1992) Early copper produced at Feinan, Wadi Araba, Jordan: the composition of ores and copper. Archaeomaterials 6(1):1–33

Hauptmann A, Pernicka E, Lutz J, Yalçın Ü (1993) Zur Technologie der frühesten Verhüttung von Kupfererzen im östlichen Mittelmeerraum. In: Frangipane M, Hauptmann H, Liverani M, Matthiae P, Mellink M (eds) Between the Rivers and over the mountains: Archaeologica Anatolica et Mesopotamica Alba Palmieri Dedicata. Univers di Roma "La Sapienza", pp 541–572

Hauptmann A, Rehren T, Pernicka E (1995) The composition of gold from the ancient mining district of Verespatak/Roşia Montană, Romania. In: Morteani G, Northover P (eds) Prehistoric gold in Europe. Cambridge, pp 369–381

Hauptmann A, Bachmann HG, Maddin R (1996) Chalcolithic copper smelting: new evidence from excavations at Feinan/Jordan. In: Archaeometry '94, Proc In. Symp Archaeometry, Ankara '94, pp 3–10

Hauptmann A, Maddin R, Prange M (2002a) On the texture and composition of copper and tin ingots excavated from the shipwreck of Uluburun. Bull Am Sch Orient Res 328:1–30

Hauptmann A, Schmitt-Strecker S, Begemann F, Palmieri A (2002b) Chemical composition and lead isotopy of metal objects from the "royal" tomb and other related finds from Arslantepe, eastern Anatolia. Paléorient 28 (2):43–70

Hauptmann A, Rehren T, Schmitt-Strecker S (2003) Early bronze age copper metallurgy at Shahr-i Sokhta, reconsidered. In: Stöllner T, Körlin G, Steffens G, Cierny J (eds) Man and mining. Studies in honour of G Weisgerber on occasion of his 65th birthday, Der Anschnitt Beih, vol 16, pp 197–213

Hauptmann A, Schmitt-Strecker S, Begemann F (2009) Evidence of late chalcolithic/early bronze age I copper production from Timna Ores at Tall al-Magass, Aqaba. In: Khalil L, Schmidt K (eds) Prehistoric Aqaba I, Orient-Archäol, vol 23, pp 295–304

Hauptmann A et al (2010) Gold in Georgien. Analytische Untersuchungen an Goldartefakten und an Naturgold aus dem Kaukasus und dem Transkaukasus. In: Hansen S, Hauptmann A, Motzenbäcker I, Pernicka E (eds) Von Majkop nach Trialeti – Gewinnung und Verbreitung von Metallen und Obsidian in Kaukasien im 4.-2, Jahrtausend v. Chr. Beitr Internat Symp Berlin June 2006, Habelt, Bonn. Kolloquien zur Vor- und Frühgeschichte, vol 13, pp 139–160

Hauptmann A, von Bohlen A (2011) Aurian silver and silver beads from tombs at El-Mahasna, Egypt. In: Friedman RF, Fiske PN (eds) Egypt at its origins 3. Proc 3rd Int Conf origin of the state, Orientalia Lovanensia Analecta, vol 205, pp 430–435

Hauptmann A, Schmitt-Strecker S, Levy TE, Begemann F (2015) On early bronze age copper Bar ingots from the southern Levant. Bull Am Soc Orient Res 373:1–24

Hauptmann A, Schneider G, Bartels C (2016) The shipwreck of Bom Jesus, AD 1533: Fugger copper in Namibia. J Afr Archaeol 14(2):181–207

Hauptmann A, Klein S, Paoletti P, Zettler RL, Jansen M (2018) Types of gold, types of silver: the composition of precious metal artifacts found in the Royal Tombs of Ur. Mesopotamia Zeitschr Assyriologie 108 (1):100–131

Hauser L, Nothdurfter H (1988) Bronzezeitliche Kupferschmelzöfen aus Fennhals. Denkmalpflege in Südtirol 1986:177–190

Haustein M (2004) Die Thermoluminszenzdatierung einer alten Metallhüttenschlacke aus dem Paltental. Res montanarum 33:31–33

Haustein M (2013) Isotopengeochemische Untersuchungen zu möglichen Zinnquellen der Bronzezeit Mitteleuropas. Forschungsber Landesmus Vorgesch Halle 3. Halle, Saale

Haustein M, Krbetschek MR (2002) The red thermoluminescence of quartz and its application in dating archaeometallurgical slags. Radiat Prot Dosim 101:375–378

Haustein M, Röwer G, Krbetschek MR, Pernicka E (2003) Dating archaeometallurgical slag using thermoluminescence. Archaeometry 45:519–530

Haustein M, Gillis C, Pernicka E (2010) Tin Isotopy – a new method for solving old questions. Archaeometry 52(5):816–832

Hedenquist JW, Harris M, Camus F (2012) Geology and genesis of major copper deposits and districts of the world: a tribute to Richard H. Sillitoe. Soc Econ Geol Spec Publ 16

Hedges ES (1964) Tin in social and economic history. Arnold, London

Hedges REM, Gowlett JAJ (1986) Radiokohlenstoff-Datierung mit Beschleuniger-Massenspektrometrie. Spektrum der Wissenschaft 3:110–121

Heeb J (2014) Copper shaft-hole axes and early metallurgy in South-Eastern Europe. Archaeopress Archaeology, Oxford

Heeb J, Ottaway B (2014) Experimental archaeometallurgy. In: Roberts BW, Thornton CP (eds) Archaeometallurgy in global perspective, Methods and syntheses. Springer, pp 161–192

Heide F (1988) Kleine Meteoritenkunde. Verständliche Wissenschaft. Springer, Berlin

Heimann R (1977) High-temperature and high-pressure modifications of silica. Miner Sci Eng 9:57–61

Heimann R (1989) Assessing the technology of ancient pottery: the use of phase diagrams. Archaeomaterials 3:123–148

Heimann R, Magetti M (2014) Ancient and historical ceramics. Materials, technology, art and culinary traditions. Schweizerbart Sc Publ, Stuttgart

Heimann R et al (1998) Archaeometallurgical investigations into the iron production technology in upper Lusatia, Saxony, from the early Iron age (Billendorf period) to the 12th century AD. Eur J Miner 10:1015–1035

Heimann R, Chirikure S, Killick D (2010) Mineralogical study of prehistoric (1650-1850 CE) tin smelting slags from Rooiberg, Limpopo Province, South Africa. Eur J Miner 22(5):751–761

Heimbruch G (1990) Archäometrie an Verhüttungsrelikten der Harzregion. PhD-diss, TU Clausthal

Heimbruch G, Koerfer S, Brockner W (1992) Possibilities for determination of the process temperature of archaeometallurgical finds. In: Antonacci-Sanpaola E (ed) Archeometallurgica ricerche e prospettive. CLUEB, Bologna, pp 565–578

Heimpel W (1987) Das Untere Meer. Zeitschr Assyriologie u vorderasiat Archäol 77:22–91

Hein A, Kilikoglu V, Kassianidou V (2007) Chemical and mineralogical examination of metallurgical ceramics from a late bronze age copper smelting site in Cyprus. J Archaeol Sci 34(1):141–154

Helke A (1964) Die Kupfererzlagerstätte Ergani Maden in der Türkei. N Jb Miner Abh 101:233–270

Helwing B (2009) Rethinking the Tin Mountain: patterns of usage and circulation of tin in greater Iran from the 4th to the 1st millennium BC. Tüb-Ar 12:209–221

Helwing B (2011) The small finds from Arisman. In: Vatandoust A, Parzinger H, Helwing B (eds) Early mining and metallurgy on the western central Iranian plateau. Report on the first five years of research of the joint Iranian-German research project. Arch Iran u Turan, pp 254–327

Helwing B (2013) Frühes Silber und seine Verwendung in den frühstaatlichen Gesellschaften Westasiens. In: Abstracts, Metalle der Macht. Frühes Gold und Silber. 6. Mitteldeutscher Archäologentag. Internat Tagung, Halle, p 47

Helwing B (2014) Silver in the early state societies of greater Mesopotamia. In: Meller H, Risch R, Pernicka E (eds) Metalle der Macht. Frühes gold und Silber. 6. Mitteldeutscher Archäologentag, vol 11/II. Internat Tagung, Halle,. October 2013. Tagungen des Landesmus Vorgesch Halle, pp 411–421

Hensen A (2001) Der Wieslocher Vicus – eine römische Landstadt am Leimbach, Wiesloch – Beitr zur Gesch 2, Ubstadt-Weiher, p 11

Hérail G, Pérez-Garcia LC (1989) Intéret archaeologique d'une étude geomorpho-gitologique: les gisements d'or alluvial du nord-ouest de L'Espagne. In: Domergue C (ed) Mineria y Metallurgia en las antiguas civilizationes Mediterraneas y Europeas, Coll. Madrid, pp 21–34

Herbenar AW, Siebert CA, Duffendack OS (1950) Zinc in solid alpha brasses. J Metals 188:323–326

Herbert E (1973) Aspects of the use of copper in pre-colonial West Africa. J Afr Hist 14(2):179–194

Herbert E (1998) Mining as microcosm in precolonial sub-Saharan Africa: an overview. In: Knapp B, Pigott VC, Herbert E (eds) Social approaches to an industrial past. The archaeology and anthropology of mining. Routledge, London, pp 138–154

Herdits H (1993) Zum Beginn experimentalarchäologischer Untersuchungen einer bronzezeitlichen Kupferverhüttungsanlage in Mühlbach, Salzburg. Archaeologia Austriaca 77:31–38

Herdits H (2003) Bronze age smelting site in the Mitterberg mining area in Austria. In: Craddock PT, Lang J (eds) Mining and metal production through the ages. The British Museum Press, pp 69–75

Herdits H, Löcker K (2004) Eine bronzezeitliche Kupferhütte im Mitterberger Kupferkies-Revier (Salzburg). In: Weisgerber G, Goldenberg G (eds) Alpenkupfer – Rame delle Alpi, Der Anschnitt Beih, vol 17, pp 177–188

Hermanns MH (2015) Ibiza, Spanien: Zum antiken Bleiglanzabbau am Puig de s'Argentera. https://publications.dainst.org

Herodotus (440 BC) Histories (VI, 46–47)

Hezarkhani Z, Keesmann I (1996) Archäometallurgische Untersuchungen an Kupferschlacken aus dem Zentraliran. Metalla 3(2):101–125

Herzig P et al (1991) Gold-rich sea floor gossans in the Troodos OPhiolite and on the mid Atlantic ridge. Econ Geol 86:1747–1755

Heskel D, Lamberg-Karlovski CC (1980) An alternative sequence for the development of metallurgy: Tepe Yahya, Iran. In: Wertime TA, Muhly JD (eds) The coming of the age of iron. Yale University Press, Mew Haven, pp 239–265

Hess K (1998) Zur frühen Metallurgie am oberen Euphrat: Untersuchungen an archäometallurgischen Funden vom Arslantepe aus dem 4. und 3. Jahrtausend v. hr. PhD-diss Goethe Univ., Frankfurt a.M.

Hess K, Hauptmann A, Wright H, Whallon R (1998) Evidence of fourth millennium BC silver production at Fatmali-Kaleçik. In: Rehren T, Hauptmann A, Muhly JD (eds) Metallurgica Antiqua. In honour of HG Bachmann and R Maddin, Der Anschnitt Beih, vol 8, pp 57–67

Hetherington R (1980) Investigations into primitive lead smelting and its products. In: Oddy WA (ed) Aspects of early metallurgy, Brit Mus Occ papers, vol 17, pp 27–40

Hezarkhani Z, Keesmann I (1996) Archäometallurgische Untersuchungen an Kupferschlacken aus dem Zentraliran. Meta 3(2):101–125

Higham T, Chapman J, Slavchev V, Gaydarska B (2007) New perspectives on the Varna cemetery (Bulgaria) – AMS dates and social implications. Antiquity 81 (313):640–654

Higham CFW, Douka K, Higham TFG (2015) A new chronology for the bronze age of northeastern Thailand and its implications for southeast Asian prehistory. PLoS One 10(9). https://doi.org/10.1371/J.pone.0137542

Hildebrandt LH (1989) Der mittelalterliche Blei-Zink-Silber-Bergbau im nordwestlichen Kraichgau südlich Heidelberg. In: Hauptmann A, Pernicka E, Wagner GA (eds) Archäometallurgie der Alten welt. Beiträge zum Internationalen symposium "Old World Archaeometallurgy", vol 7. Der Anschnitt Beih, Heidelberg, pp 241–246

Hildebrandt LH (1998) Schwermetallbelastungen durch historischen Bergbau im Raum Wiesloch. PhD-diss Univers Heidelberg

Hingst H (1983) Das Eisenverhüttungsrevier auf dem Kammberg-Gelände in Joldelund, Kreis Nordfriesland. Festschrift für H Hingst Offa 40:153–156

Hirt AM (2010) Imperial mines and quarries in the Roman world. Organizational aspects 27 BC - AD 235, Oxford classical monographs. Oxford University Press

Hirt B, Herr W, Hoffmeister W (1963a) Age determinations by the rhenium-osmium method. Radioactive Dating. Internat. Atom. Energy Agency, Vienna, pp 35–44

Hirt B, Tilton GRW, Herr W, Hoffmeister W (1963b) The half-life of 187Re. In: Geiss J, Goldberg ED (eds) Earth science and meteorites. North-Holand, Amsterdam, pp 273–280

Hochuli-Gysel A, Picon M (1999) Les creusets en graphite dècouverts à Avenches/Aventicum. Bull de l'Assoc pro Avertico 41:209–214

Holliday JS (1999) Rush for riches. Gold fever and the making of California. University of California Press, Berkeley

Hölscher T (2006) Klassische Archäologie. Grundwissen. Wissenschaftliche Buchgemeinschaft Darmstadt

Holzer HF, Ghasemipour R (1973) Geology oft he Nakhlak lead mine area (Anarak district, Central Iran). Contributions to the geology of mineral resources of Iran. Geol Surv Iran Rep 21:27–36

Höppner B et al (2005) Prehistoric copper production in the Inn Valley (Austria), and the earliest copper in Central Europe. Archaeometry 47(2):293–315

Horejs B, Mehofer B, Pernicka E (2010) Metallhandwerker im frühen 3. Jt v Chr – Neue Ergebnisse vom Çukuriçi Höyük Istanbul Mitt 60:7–37

Hornbogen E (1991) Werkstoffe. Aufbau und Eigenschaften von Keramik, Metallen, Polymeren und Verbundwerkstoffen. Springer

Hornbogen E, Warlimont H (1991) Metallkunde. Aufbau und Eigenschaften von Metallen und Legierungen. Springer, Berlin

Horne L (1982) Fuel for the metalworker: the role of charcoal and charcoal production in ancient metallurgy. Expedition 25:6–13

Horstmann D (1985) Das Zustandsschaubild Eisen-Kohlenstoff. Stahl & Eisen, Düsseldorf

Hosking KFG (1951) Primary ore deposition in Cornwall. Roy Geol Soc Cornwall Trans 18:309–356

Hošek J (2003) Metallography in the service of archaeology. Archaeologický Ústav Av ČR Praha, Technicka Universzitá Liberci

Hosking KFG (1979) Tin distribution patterns. Geol Soc Malays Bull 11:1–70

Hosking KFG (1988) The World's major types of tin deposits. In: Hutchison CS (ed) Geology of tin deposits in Asia and the Pacific. Springer, pp 3–49

Hosler D (1994) The sounds and colors of power. The sacred metallurgical Technology of Ancient West Mexico. MIT Press, London

Housley RA, Manning SW, Jones RE, Hedges RE (1999) Radiocarbon, calibration, and the chronology of the late Minoan IB phase. J Archaeol Sci 26:159–171

Houterman FG (1960) Die Bleimethoden der geologischen Altersbestimmung. Geol Rundsch 69:168–196

Huet MA (1878) Mémoires Sur le Laurium. Deuxième Mémoires la Société des Ingénieurs Civils de France: 530–566

Hummel RE (1998) Understanding materials science – history properties applications. Springer, New York

Humphries J (2010) An Archaeometallurgical investigation of Iron smelting traditions in southern Rwanda. PhD-disser UCL London, Inst. Of archaeology

Humphries J, Rehren T (2013) The world of iron. Archetype Publications, London

Humphris J, Carey C (2016) New methods for investigating slag heaps: integrating geoprospection, excavation and quantitative methods at Meroe, Sudan. J Archaeol Sci 70:131–144

Hunt Ortiz MA (2003) Prehistoric mining and metallurgy in south West Iberian Peninsula. BAR Internat Ser 1188

Huska Mason A et al (2014) Placer tin ores from Mt. Cer, West Serbia, and their potential exploitation during the bronze age. Geoarchaeology 29(6):477–493

Hutchison CS (ed) (1988) Geology of tin deposits in Asia and the Pacific, Mineral concentrations and hydrocarbon accumulations in the ESCAP region 3. Springer, Berlin

Huttenlocher H, Ramdohr P (1965) Mineral- und Erzlagerstättenkunde I, II. Sammlung Göschen vol. 1014/1014a. Walter de Gruyter, Berlin

Ikana E, Nakanish T (2014) Gold refining by cementation with salt at Sado in early seventeenth century Japan. ISIJ Int 54:1098–1105

Imris I, Komorova L (1993) Removal of arsenic, antimony and mercury from Tetraedrite concentrates by low temperature volatilizing roasting, Proc. of mater. Process. Technol. Environ. Prot. Min. Metall., pp 23–30

Imris I, Komorova L, Holmström A (1985) Behaviour of antimony during the roasting of tetraedrite concentrates. Proc. Symp. Extraction Metallurgy. The Inst. Min. Metallurgy, London, pp 1015–1033

Inanishvili G (2007) About the history of Iron production in Georgia. Meta 14(1):1–62

Indolev LN, Nevoisa GG (1974) Silver-lead deposits of Yakutia. Izd. Nauka, Novosibirsk

Ingo GM, de Caro T, Bultrini G (2004a) Microchemical investigation of archaeological copper based Artefacts disclosing an ancient witness of the transition from the value of the substance to the value of the appearance. Microchim Acta 144:87–95

Ingo GM, Angelini E, de Crao T, Bultrini G, Mezzi A (2004b) Microchemical investigation of archaeological copper-based artefacts used for currency in ancient Italy before the coinage. Surf Interface Anal 36:866–870

Inverno C et al. (2015) Introduction and Geological Setting of the Iberian Pyrite Belt. In: Weihed P (ed),

3D, 4D and predictive modelling of major mineral belts in Europe, Mineral Resource Reviews, Modelling of the Cala area (Ossa-Morena Zone). Springer Cham: 191–208

Ivanov IS (1991) Les objets métalliques de la nécropole chalcolithique de Varna. In: Mohen JP (ed) Découverte du Metall, vol 2. Millenaires, pp 9–12

Ixer RA (1999) Role of ore geology and ores in the archaeological provenancing of metal. In: Young SMM, Pollard M, Budd P, Ixer RA (eds) Metals in antiquity, BAR Internat Ser, vol 792, pp 43–52

Ixer RA (2004) The petrography of prehistoric pottery from Ross Island mine. Appendix 7. In: Ross Island, mining, metal and Society in Early Ireland, W.O'Brien. Dept Archaeol Galway, pp 643–650

Ixer RA, Pattrick RAD (2003) Copper-arsenic ores and bronze age mining and metallurgy with special reference to the British Isles. In: Craddock PT, Lang J (eds) . Brit Mus Press, Mining and metal production through the ages, pp 9–21

Izawa E (2009) The liquation process utilized in silver production from copper ore: the transfer to and development in Japan. In: Mei J, Rehren T (eds) Metallurgy and civilisation. Eurasia and beyond. Archetype Publ, London, pp 163–169

Izawa E, Nakanish T (2014) Gold refining by cementation with salt at Sado in early seventeenth century Japan. ISIJ Int 54(5):1098–1105

Jaez R et al (2003) The extractive metallurgy of copper from Cabezo Juré, Huelva, Spain: chemical and mineralogical study of slags dated to the third millennium B.C. Can Miner 4:627–638

Jak E, Degterov S, Hayes PC, Pelton AD (1989) Thermodynamic modelling of the system Al2O3-SiO2-CaO-Fe2O3 to predict the flux requirements for coal ash slags. Fuel 77:77–84

Jameson R (1821) Manual of mineralogy. Edinburgh, Archibald Constable & Co

Janković S (1997) The Carpatho-Balkanides and adjacent area: a sector of the Tethyan Eurasian metallogenic belt. Mineral Deposits 32:426–433

Jansen M (2019) Geochemie und Archäometallurgie des Goldes der Bronzezeit in Vorderasien. PhD-diss, Fac Geoscience, Ruhr Univers Bochum

Jansen M, Löffler I (2016) Das bronzezeitliche Gusskuchendepot von Kleinbardorf (Rhön-Grabfeldt). In: Körlin G, Prange M, Stöllner T, Yalçın Ü (eds) From bright ores to shiny metal, Festschrift for a Hauptmann. Der Anschnitt Beih, vol 29, pp 125–131

Jansen M et al (2016) Platinum group placer minerals in ancient gold artifacts – geochemistry and osmium Isotopy of inclusions in early bronze age gold from Ur/Mesopotamia. J Archaeol Sci 68:12–23

Jansen M, Hauptmann A, Klein S, Seitz HM (2017) The potential of stable cu isotopes for the identification of bronze age ore mineral sources from Cyprus and Faynan: results from Uluburun and Khirbat Hamra Ifdan. J Archaeol Sci 10(6):1485–1502. https://doi.org/10.1007/s12520-017-0465-x

Jansen M, Hauptmann A, Klein S (2018a) Copper and lead isotope characterization of late bronze age copper ingots in the eastern Mediterranean: results from Gelidonya, Gournia, Enkomi and Mathiati. In: LoSchiavo F, Giumlia-Mair A (eds) Festschrift for R Maddin and V Karageorghis, Monogr Instrumentum, vol 56, pp 552–577

Jansen M, Klein S, Hauptmann A (2018b) Zur Herkunft des Goldes aus den Königsgräbern von Ur. In: Yalçın Ü (ed) Anatolian metal VIII, Der Anschnitt Beih, vol 39, pp 123–137

Jarrige JF (2010) Der "Schatz" von Tepe Fullol. In: Gerettete Schätze Afghanistan Die Sammlung des Nationalmuseums in Kabul Ausstellung Kunst- u Ausstellungsh Bonn Oct 2010: 36–44

Jianli C, Rubin H (2013) Manufacturing techniques and dates of iron objects found recently at Chinese archaeological sites. In: Humphries J, Rehren T (eds) The world of Iron. Archetype Publications, London, pp 345–354

Jockenhövel A, Willms C (2005) Das Dietzhölzetal-Projekt. Archäometallurgische Untersuchungen zur Geschichte und Struktur der mittelalterlichen Eisengewinnung im LahnDill-Gebiet (Hessen). Münster Beitr Ur- u Frühgesch Archäol 1

Johnson D, Tyldesley J, Lowe T, Withers P, Grady M (2013) Analysis of a prehistoric Egyptian iron bead with implications for the use and perception of meteorite iron in ancient Egypt. Meteorit Planet Sci 48 (6):997–1006

Jöns H (1992/1993) Zur Eisenverhüttung in vor- und frühgeschichtlicher Zeit in Schleswig Holstein. Dent Off 49

Jöns H (1997) Frühe Eisengewinnung in Joldelund, Kr. Nordfriesland 1. Universitätsforsch Prähist Archäol 40. R Habelt, Bonn

Jones IW, Ben-Yosef E, Lorentzen B, Najjar M, Levy TE (2017) Khirbat al-Mana'iyya: an early Islamic-period copper-smelting site in south-eastern Wadi 'Araba, Jordan'. Arab Archaeol Epigraph 28:297–314. https://www.academia.edu/35239933/

Jovanović B (1988) Early metallurgy in Yugoslavia. In: Maddin R (ed) The beginning of the use of metals and alloys. MIT, Cambridge, pp 69–79

Junghans S, Sangmeister E, Schröder M (1960) Metallanalysen kupferzeitlicher und frühbronzezeitlicher Bodenfunde aus Europa. Studien zu den Anfängen der Metallurgie 1. Mann, Berlin

Junghans S, Sangmeister E, Schröder M (1968) Kupfer und bronze in der frühen Metallzeit Europas 1–3. Mann, Berlin

Junghans S, Sangmeister E, Schröder M (1974) Kupfer und bronze in der frühen Metallzeit Europas 4. Mann, Berlin

Junk SA (2001) Ancient artefacts and modern analytical techniques—usefulness of laser ablation ICP-MS demonstrated with ancient gold coins. Nucl Inst Methods Phys Res B 181(1–4):723–727

Junk M (2003) Material properties of copper alloys containing arsenic, antimony, and bismuth. The material of Early Bronze Age ingot torques. PhD-diss, TU Bergakademie Freiberg

Junk SA, Pernicka E (2003) An assessment of osmium isotope ratios as a new tool to determine the provenance of gold with platinum-group metal inclusions. Archaeometry 45:313–331

Kacharava D, Kvirkvelia G (2008) Wine, worship and sacrifice: the Golden graves of ancient Vani. Arthur Sackler Gallery & Inst. For the Study of the Ancient World at New York University. Inst for the Study of the Ancient World & Princeton University Press

Kakavoyannis E (2001) The silver ore processing workshops of the Lavrion region. Annual of the British School at Athens 96:365–380

Kalcyk H (1982) Untersuchungen zum attischen Silberbergbau. Gebietsstruktur, geschichte und Technik. Europäische Hochschulschriften, Reihe III. Geschichte und ihre Hilswissensch 160. Peter Lang, Frankfurt a M

Kamarzin L, Brezina J, Jelinkova D (1996) Metallography and Electron probe microanalysis of the material of silver Temple rings found in the Slavic cemetery at Musov, Czech Republic. Archaeol Aust 80:277–290

Kamenov GD, Melchiore EB, Ricker FN, DeWitt E (2013) Insights from Pb isotopes for native gold formation during Hypogene and supergene processes at Rich Hill, Arizona. Econ Geol 108:1577–1589

Kaniuth K (2006) Metallobjekte der Bronzezeit aus Nordbaktrien. Archäologie in Iran und Turan 6. Mainz

Kaniuth K (2007) The metallurgy of the late bronze age Sapalli culture (southern Uzbekistan) and its implication for the "tin question". Iran Antiq 42:23–40

Karageorghis V, Kassianidou V (1998) Metalworking and recycling in late bronze age Cyprus – the evidence from Kition. Oxf J Archaeol 18(2):171–188

Karim MM, Holland D (1995) Physical properties of glasses in the system SnO-SiO2. Phys Chem Glasses 36:206–210

Kassianidou V (1992) Monte Romero (Huelva), a silver producing workshop of the Tartessian period in SW Spain. Unpubl. PhD-diss Inst of Archaeol, UCL London

Kassianidou V (1994) Could Iron have been produced in Cyprus (?). Rep. Dept. Antiquities, Cyprus, pp 73–81

Kassianidou V (1998) Was silver actually recovered from speiss in antiquity? Reconsidering the evidence from Rio Tinto. In: Rehren T, Hauptmann A, Muhly JD (eds) Metallurgica Antiqua, In honour of HG Bachmann & R Maddin. Der Anschnitt Beih, vol 8, pp 69–76

Kassianidou V (1999) Small scale mining and smelting in ancient Cyprus. In: Knapp B, Pigott VC, Herbert E (eds) Social approaches to an industrial past. The archaeology and anthropology of mining. Routledge, London, pp 226–241

Kassianidou V (2003a) Archaeometallurgy: data, analyses and discussion. In: Given M, Knapp B (eds) The Sydney Cyprus survey project, Social approaches to regional archaeological survey. Monumenta Archaeologica, vol 21, pp 214–221

Kassianidou V (2003b) Early extraction of silver from complex Polymetallic ores. In: Craddock PT, Lang J (eds) Mining and metal production through the ages. Brit Mus Press, London, pp 198–206

Kassianidou V (2008) The formative years of the Cypriot copper industry. In: Tzachilli I (ed) Aegean Metallurgy in the Bronze Age. Proc. International Symposium University of Crete Rethymnon, Greece, November 2004, Athens. Ta Pragmata Publications, pp 249–267

Kassianidou V (2012) Metallurgy and metal work in Enkomi: the early phases. In: Kassianidou V, Papasavvas G (eds) Eastern Mediterranean metallurgy and metalwork in the second millennium BC, A conference in honour of JD Muhly, Nicosia October 2011. Oxbow Books, Oxford, pp 94–106

Kassianidou V (2016) Metallurgy and metalwork in Enkomi – Revisiting Porphyrios Dikaios' excavations. In: Bourogiannis G, Muhlenbock C (eds) Ancient Cyprus Today: Museum Collections and New Research, SIMA Pocket-book 184, Uppsala, pp 79–90

Kassianidou V (2018) Apliki Karamallos on Cyprus: the 13th century BCE miners' settlement in context. In: Ben Yosef E (ed) Mining for Ancient Copper. Essays in memory of Beno Rothenberg, Tel Aviv University – Sonia and Marco Nadler Instit. Archaeol, Monogr. Ser. 37. Tel Aviv: Emery and Claire Yass Publ. In Archaeol. And Winona Lake, Eisenbrauns, pp 345–356

Kassianidou V, Knapp B (2005) Archaeometallurgy in the Mediterranean: the social context of mining, technology and trade. In: Blake E, Knapp B (eds) The archaeology of mediterranean prehistory, Blackwell studies in global archaeology. Blackwell, Oxford, pp 215–251

Kassianidou V, Koukouli-Chrysanthaki C, Tylecote RF, Gialoglu G (1989) Precious metals extraction in Palaia Kavala. An Archaeometallurgical attempt to locate Skapte Hyle. In: Hauptmann A, Pernicka E, Wagner GA (eds) Old world archaeometallurgy, Der Anschnitt Beih, vol 7, pp 179–190

Kassianidou V, Rothenberg B, Andrews P (1995) Silver production in the Tartessian period. The evidence from Monte Romero. ARX 1(1):17–34

Kaufmann B (2013) Copper alloys from the 'Enot Shuni cemetery and the origins of bronze metallurgy in the EB IV – MB II Levant. Archaeometry 55(4):663–690

Kaufmann G (2014) L'ascia dell'Uomo venuto dal ghiaccio. Rivista di Scienze Preistoriche 64:57–81

Kavtaradze GL (1999) The importance of metallurgical data for the formation of a central Transcaucasian chronology. In: Hauptmann A, Pernicka E, Rehren T, Yalçın Ü (eds) The beginnings of metallurgy, Der Anschnitt Beih, vol 9, pp 67–101

Kayafa M, Stos-Gale S, Gale N (2000) The circulation of copper in the early bronze age in mainland Greece: the lead isotope evidence from Lerna, Lithares and Tsoungiza. In: Pare CFE (ed) Metals make the world go round. Proc Conf Univers Birmingham, June 1997, pp 39–55

Kazenas EK, Bol'shikh MA, Petrow AA (1996) Thermodynamics of processes of vaporization, dissociation, and gas-phase reaction in vapors over tin-oxygen system. Izv Ross Akad Nauk Met 3:23–29

Keesmann I (1985) Verfahrensbedingungen aus Phasenbestand und chemischer Analyse von alten Kupferschlacken (Afunfun, Agadez/Niger). Analyt Chemie 320:660

Keesmann I (1989) Chemische und mineralogische Detailuntersuchungen zur interpretation eisenreicher Schlacken. In: Pleiner R (ed) Archaeometallurgy of Iron. Proc Symp Comité pour la siderurgie ancienne de l'UISPP. Liblice 1987, Prague, pp 17–34

Keesmann I (1991) Rio Tinto: die Technik der Silbergewinnung zu Beginn des Mittelalters. In: Argent, plomb et cuivre dans l'histoire, Lyon, pp 1–13

Keesmann I (1993) Naturwissenschaftliche Untersuchungen zur antiken Kupfer- und Silberverhüttung in Südwestspanien. In: Steuer H, Zimmermann U (eds) Montanarchäologie in Europa. Archäologie und geschichte. Freiburger Forsch 1, Jt in Südwestdeutschland, vol 4. Thorbecke, Sigmaringen, pp 105–122

Keesmann I (1999) Eisen in antiken Schlacken des südwestiberischen Sulfiderzgürtels. Pallas 50:339–360

Keesmann I, Heege A (1990) Archäometallurgische Untersuchungen an material der Ausgrabung 1984 am "Steinbühl" bei Nörten-Hardenberg, Lkr. Northeim. Nachr Niedersachsens Urgesch 59:87–109

Keesmann I, Hellermann B (1989) Mineralogische und chemische Untersuchungen an Schlacken vom Morro de Mezquitilla. Madrider Mitt 30:92–117

Keesmann I, Hilgart T (1992) Chemische und mineralogische Untersuchung der Schlacken von Manching. In: Manching 15, Ergebnisse der Ausgrabungen 1984–1987 in Manching. Steiner, Wiesbaden, pp 391–413

Keesmann I, Moreno Onorato A (1999) Naturwissenschaftliche Untersuchungen zur frühen Technologie von Kupfer und Kupfer-Arsen-bronze. In: Hauptmann A, Pernika E, Rehren T, Yalçın Ü (eds) The beginnings of metallurgy, Der Anschnitt Beih, vol 9, pp 317–332

Keesmann I, Preuss J, Endres J (1983) Eisengewinnung aus latertitischen Erzen, Ruki-region, Provinz Wquateur, Zaïre. Offa 40 (Festschr H Hingst 75. Geburtstag): 183–190

Keim MF, Vaudrin R, Markl G (2016) Redistribution of silver during supergene oxidation of argentiferous galena: a case study from the Schwarzwald, SW Germany. N Jb Miner Abh 193(3):295–309

Kekelia SA, Kekelia MA, Kuloshvili SI, Sadradze NG, Gagnidze NE, Yaroshevich VZ, Asatiani GG, Doebrich JL, Goldfarb RJ, Marsh EE (2008) Gold deposits and occurrences of the greater Caucasus, Georgia Republic: their genesis and prospecting criteria. Ore Geol Rev 34:369–386

Kellogg HH, Basu SK (1960) Oxidation of galena to Lead Sulphate. Trans Metal Soc AIME 218:70

Kenoyer JM, Miller HML (1999) Metal technologies of the Indus Valley tradition in Pakistan and Western India. In: Pigott V (ed) The archaeometallurgy of the Asian Old World, Univers. Mus. Monograph 89. Univers. Pennsylvania, pp 107–151

Kense FJ (1983) Traditional African Iron working. African Occ pap 1. University of Calgary

Kepper J (2004) A hindered-settling model applied to the flat-washing platforms at Laurium, Greece. Hist Metall 38(2):75–83

Kepper J (2005) Third contact ore mineralogy at Laurium, Greece. Hist Metall 30(1):1–11

Ketelaer A, Hauptmann A (2016) In the shadow of Timna: the mining region of Wadi Amram. New Analytical and Archaeological Aspects Metalla 22(2):169–183

Khakhutaisvili DA (1987) Iron production in old Kolkhida. Tbilisi

Khalil L (2009) The excavations at tall al-Magaṣṣ. Stratigraphy and architecture. In: Khalil L, Eichmann R (eds) Prehistoric 'Aqaba I, Orient Archäol, vol 23, pp 5–15

Khalil L, Bachmann HG (1981) Evidence of copper smelting in bronze age Jericho. J Hist Metall Soc 15:103–106

Khalil KI, El-Shazly AK (2012) Petrological and geochemical characteristics of Egyptian banded iron formations: review and new data from Wadi Kareim. Geochem Explor Environ Anal 12:105–126

Khalil L, Riederer J (1998) Examination of copper metallurgical remains from a chalcolithic site at el-Magass, Jordan. Damaszener Mitt 10:1–9

Kiderlen M, Bode M, Hauptmann A, Bassiakos Y (2016) Tripod cauldrons produced at Olympia give evidence for trade with copper from Faynan (Jordan) to South West Greece, c. 950–750 BCE. J Archaeol Sci 8:303–313

Kienlin TL (2007) Tradition and innovation in copper age metallurgy: results of a metallographic examination of flat axes from eastern Central Europe and the Carpathian basin. In: LaNiece S, Hook D, Craddock PT (eds) Metals and mines. Studies in archaeometallurgy. Brit Mus & Archetype Publ, London, pp 1–21

Kienlin TL, Roberts BW (2009) Metals and societies. Studies in honour of B Ottaway. Universitätsforsch prähist Archäol 169. Habelt, Bonn

Kienlin TL, Bischoff E, Opielka H (2003) Zur Metallographie urgeschichtlicher Artefakte: Ergebnisse einer Untersuchung an Kupfer- und Bronzebeilen des nordalpinen Raumes. 37. Metallographie-Tagung der Deutschen Gesellsch. Materialkunde, September 2003, Seiten

Kienlin T, Bischoff E, Opielka, H (2004) Zur Metallographie urgeschichtlicher Artefakte: Ergebnisse einer Untersuchung an Kupfer- und Bronzebeilen des nordalpinen Raumes. In: Portella P (ed) Fortschr Metallographie. 37. Metallographie-

Tagung, September 2003, Berlin. Sonderbände Praktische Metallographie, vol. 35, pp 3–10

Kienlin TL, Bischoff E, Opielka H (2006) Copper and bronze during the Eneolithic and early bronze age: a metallographic examination of axes from the Northalpine region. Archaeometry 48(3):453–468

Kienlin TL, Bischoff E, Opielka H (2008) Tradition and innovation in copper age metallurgy: results of a metallographic examination of flat axes from eastern Central Europe and the Carpathian Basin. Proc Prehist Soc 74:79–107

Killick D (2014) Cairo to Cape: the spread of metallurgy through Eastern and Southern Africa. In: Roberts BW, Thornton CP (eds) Archaeometallurgy in global perspective methods and syntheses Springer: 507-527

Killick D (2015) Invention and innovation in African Iron-smelting technologies. Camb Archaeol J 25:307–319

Killick D, Fenn T (2012) Archaeometallurgy: the study of preindustrial mining and metallurgy. Annu Rev Anthropol 41:559–575

Killick D, Hayashida F (2012) Lung-powered copper smelting of the the middle Sicán period (ca. 900-1000 CE), north coastal Peru. 39th Internat Symp Archaeometry, Leuven, Belgium

Killick D, van der Merwe NJ, Gordon RB, Grébénart D (1988) Reassessment of the evidence for early metallurgy in Niger, West Africa. J Archaeol Sci 15:367–394

Killick D, Miller D, Thondhlana TP (2016) Copper mining and smelting technology in the northern Lowveld, South Africa, ca. 1000 CE to ca. 1880 CE. J Archaeol Sci 75:10–26

Kim HG, Sohn HY (1997) Minor-element behaviour and iron-partitioning during the cleaning of copper converter slag under reducing conditions. Can Metall Q 36:31–37

Kim HG, Sohn HY (1998) Effect of CaO, Al2O3 and MgO additions on the copper solubility, ferric/ferrous-ratio and minor-element behavior of iron-silicate slags. Metall Mater Trans 29B:583–590

Kimball BE et al (2009) Copper isotope fractionation in acid mine drainage. Geochim Cosmochim Acta 73:1247–1263

Kind HD (1965) Antike Kupfergewinnung zwischen Rotem und totem Meer. Zeitschr Deutsch Palästina-Verein 81:56–73

Kindermann A (2005) Mineralogisch-geochemische Charakterisierung hydrothermaler Goldvererzungen des Troodos-Ophiolith-Komplexes, Zypern Freiberg Forschungshefte C508

Kingery D (1991) Attic pottery technology. Archaeomaterials 5:47–54

Kingery WD, Vandiver P, Prickett M (1988) The beginnings of Pyrotechnology, part II: production and use of lime and gypsum plaster in the pre-pottery Neolithic near east. J Field Archaeol 15:219–244

Kirchheimer F (1977) Bericht über Spuren römerzeitlichen Bergbaus in Baden-Württemberg. Archäol Nachr Baden 19:16–24

Kirnbauer T (ed) (1998) Geologie und hydrothermale Mineralisationen im rechtsrheinischen Schiefergebirge, Nassau Verein Naturkunde spec vol 1, Wiesbaden

Kirnbauer T, Hucko S (2011) Hydrothermale mineralisation und Vererzung im Siegerland. Der Aufschluss 62(10):257–296

Kishida T et al (1998) Characteristic changes of several Iron and steel material during forging process of the Japanese Swordmaking. In: Proc 4th Int Conf beginning of the use of metals and alloys (BUMA IV), Shimane, Japan, pp 41–46

Kisnawi A, De Jesus P, Rihani B (1983) Preliminary report on the mining survey, northwest Hidjaz, 1982. Atlal 7:76–83

Klappauf L (2000) Spuren deuten – Frühe Montanwirtschaft im Harz. In: Arbeitshefte zur Denkmalpflege in Niedersachsen 21 (Auf den Spuren einer frühen Industrielandschaft. Naturraum-Mensch-Umwelt im Harz), pp 19–27

Klaproth MH (1795–1815) Beiträge zur chemischen Kenntnis der Mineralkörper 1–6. Posen/Berlin

Klaproth MH (1802) Chemische Untersuchungen des Gediegen-Spießglanzes von Andreasberg. Beiträge zur chemischen Kenntnis der Mineralkörper 3:169–172

Klaunzer M (2017) Frühe Prestigeobjekte aus Metall in Anatolien. Selbstdarstellung, Kommunikation und Metallkonsum im späten 4. Und 3. Jahrtausend v. Chr. Der Anschnitt Beih 35

Klein S (2007) Dem euro der Römer auf der spur – Bleiisotopenanalysen zur Bestimmung der Metallherkunft römischer Münzen. In: Wagner GA (ed) Einführung in die Archäometrie. Springer, pp 140–152

Klein S, Hauptmann A (2016) Ur, Mesopotamia: the Lead metal from pit X. Meta 22(1):136–140

Klein S, Hauptmann A (2020) Lead metal in Ur, Mesopotamia – The role of intergrown minerals. In press

Klein S, Von Kaenel HM (2000) The early Roman imperial Aes coinage: metal analysis and numismatic studies, part I. Schweizer. Numismat Rundsch 79:53–106

Klein S, Lahaye Y, Brey GP, Von Kaenel HM (2004) The early Roman Imperial Aes coinage II: tracing copper sources by Lead- and copper-isotope analysis – copper coins of Augustus and Tiberius. Archaeometry 46 (3):469–480

Klein S, Rico C, Lahaye Y, von Kaenel HM, Domergue C, Brey GP (2007) Copper ingots from the western Mediterranean Sea: chemical characterisation and provenance studies through lead- and copper isotope analyses. J Roman Stud 20(1):202–221

Klein S, Domergue C, Lahaye Y, Brey G, Von Kaenel HM (2009) The lead and copper isotopic composition of copper ores from the Sierra Morena (Spain). J Iber Geol 35(1):59–68

Klein S, Brey GP, Durali-Müller S, Lahaye Y (2010) Characterisation of the raw metal sources used for the production of copper and copper-based objects with copper isotopes. J Archaeol Anthropol Sci 2(1):45–56

Klein S, Von Kaenel HM, Lahaye Y, Brey GP (2012) The early Roman Imperial Aes coinage III: chemical and isotopic characterisation of Augustan copper coins from the mint of Lyons/Lugdunum. Schweiz Numismat Rundsch 91:63–110

Klein S, Gorecki J, Bollingberg HJ, Brey GP, Pearson G (2017) Metallkundliche und analytische Untersuchungen an den im Deposito Archeologico der Soprintendenza Archeologica di Pompei aufbewahrten Metallgefäßen. Germania 2017:161–336

Kleinheisterkamp H (1944) Über die Grundlagen *der* Speisebildung im Schachtofen. PhD-diss. TH Berlin

Kleinheisterkamp H (1948) Einige grundlegende Gleichgewichte bei der Bildung metallurgischer Speisen. Erzmetall 1:65–72

Klemm S (2015) Bronzezeitliche Kupfergewinnung in den Eisenerzer Alpen, Steiermark. In: Stöllner T, Oeggl K (eds) Bergauf Bergab. 10.000 Jahre Bergbau in den Ostalpen, vol 207. Publ Deutsches Bergbau-Mus, Bochum, pp 195–200

Klemm R, Klemm DD (2013) Gold and gold Mining in Ancient Egypt and Nubia. Geoarchaeology of the Ancient Gold Mining Sites in the Egyptian and Sudanese Deserts. Springer

Klemm DD, Klemm R, Murr A (2001) Gold of the Pharaos – 6000 years of gold mining in Egypt and Nubia. J African Earth Sci 33:643–659

Klemm DD, Klemm R, Murr A (2007) Die Herkunft der Pyramidenbausteine. In: Wagner GA (ed) Einführung in die Archäometrie. Springer, pp 154–178

Klemm DD, Murr A, Grimm A (2009) Analysenergebnisse altägyptischer Goldartefakte des Staatlichen museums Ägyptischer Kunst München. In: Hauptmann A, Stege H (eds) Archäometrie und Denkmalpflege. Kurzber 2009. Metalla, Sonderheft, vol 2, pp 180–181

Klimscha F (2013) Another great transformation: technical and economic change from chalcolithic to the early bronze age in the southern Levant. Zeitschr Orient-Archäol 6:82–112

Klockmann F (1908) Die eluvialen Brauneisenerze der nördlichen Fränkischen Alb. Stahl und Eisen 2:1913–1919

Klose O (1916) Die prähistorischen Funde vom Mitterberge bei Bischofshofen im Städt. Museum Carolinum-Augusteum zu Salzburg und zwei prähistorische Schmelzöfen auf dem Mitterberge. In: Kyrle G (ed) Urgeschichte des Kronlandes Salzburg, Österreich Kunst-Topographie XVII. Beitrag II

Knabe W, Noli D (2012) Die versunkenen Schätze der Bom Jesus. Sensationsfund eines Indienseglers aus der Frühzeit des Welthandels. Nicolai Verlag, Berlin

Knapp BA (1999) The archaeology of mining: fieldwork perspectives from the Sydney Cyprus survey project. In: Young SMM, Pollard AM, Budd P, Ixer R (eds) Metals in antiquity, BAR Internat Ser, vol 792, pp 98–109

Knau HL (1952) Neue Forschungen zur Frühgeschichte des Eisens. Der Märker 41:103–111

Knolle FE, Wilfried HO (2016) Montanwesen, Umweltbelastungen und Schwermetallflora im Harz. Zur UN-Dekade der Biodiversität (44), Braunlager Zeitung März: 10

Kockelmann W et al (2006) Neutrons in cultural heritage research. J Neutron Res 14(1):37–42

Kohl PL, Lyonnet B (2008) By land and by sea: the circulation of materials and Peolples, ca. 3500-1800 B.C. In: Olijdam E, Spoor RH (eds) Intercultural relations between south and Southwest Asia. Stud commemoration of E.C.L. during Caspers, Soc Arab stud Monogr 7, BAR Internat Ser 1826, pp 29–42

Kohlmeyer K (1994) Zur frühen geschichte von Blei und Silber. In: Wartke RB (ed) Handwerk und Technologie im Alten orient, Mainz, pp 59–66

Kolitsch U, Brandstätter F, Schreiber F, Fink R, Auer C (2013) Die Mineralogie der weltweit einzigartigen Schlacken von Waitschach, Kärnten. Ann Naturhist Mus Wien Ser A 115:19–87

Kölschbach S, Woelk G, Hauptmann A (2000) Experimente zur simulation prähistorischer Kupfergewinnung: Zur Verfahrenstechnik von Windöfen. Meta 7(1):5–22

Kongoli F, Yazawa A (2001) Liquidus surface of FeO-Fe2O3-SiO2-CaO slag containing Al2O3, MgO, and Cu2O at intermediate oxygen partial pressure. Metall Mater Trans 32B:583–592

Körlin G (2000) Römische Bleiverhüttung auf dem Lüderich im Bergischen land. Archäologie im Rheinland 2000, Stuttgart 2001: 88–90

Körlin G (2002) Die Römer auf dem Lüderich Grabungen 2000-2002. In: Bergbau im Bergischen land. Schriftenreihe Geschichtsverein Rösrath 32. Rösrath 2002:89–107

Körlin G, Gassmann G (2016) In: Körlin G, Prange M, Stöllner T, Yalçın Y (eds) From bright ores to shiny metals. Festschrift a Hauptmann occasion 40 years research in Archaeometallurgy and Archaeometry, Der Anschnitt Beih, vol 29, pp 187–204

Kortan O (1970) Zur Bildung der Schwefelkies-Kupferkies-Vorkommen Cyperns unter besonderer Berücksichtigung der Lagerstätte Skouriotissa. Dissertation, Technische Universität Clausthal

Körlin G, Gechter M (2003) Römischer Bergbau auf dem Lüderich – Vorbericht über die Grabungen 2000-2002. In: StöllnerT KG, Steffens G, Cierny J (eds) Man and mining – mensch und Bergbau. Studies in honour of G Weisgerber on occasion of his 65th birthday, Der Anschnitt Beih, vol 16, pp 237–248

Kosay H, Gültekin H (1949) Karaburun (Izmir), Cakmak Civa Madeninde Östarik Buluntulari, Türk Tarik. Arkeologya ve Ethnografya Dergisi 5:16–23

Köster T (2008) Die Metallressourcen in den Randgebieten Mesopotamiens und ihre mögliche Nutzung im Altertum. MA-thesis, University of Tübingen

Kossack G (1986) Verhältnis der Archäologie zu benachbarten Fächern. In: Kossack G, Behre KE, Schmidt P (eds) Archäologische und naturwissenschaftliche Untersuchungen an ländlichen und frühstädtischen Siedlungen im deutschen Küstengebiet vom 5, Jahrhundert v. Chr. Bis 11. Jahrhundert n. Chr. 1. Ländliche Siedlungen, Weinheim, pp 44–50

Koukouli-Chrysanthaki C, Weisgerber G (1999) Prehistoric ochre mines on Thasos. In: Koukouli-Chrysanthaki C, Müller A, Papadopoulos S (eds) Thasos: Matières Premières et Technologie de la Préhistoire à nos jours. De Boccard, Paris, pp 129–144

Kovacs S et al (2009) Characterization of calibration materials for trace element analysis and fingerprint studies of gold using LA-ICP-MS. J Anal Atomic Spectr 24:476–483

Krasa O (1955) Neue Forschungen zur vor- und frühgeschichtlichen Eisenindustrie im Siegerland. Westfäl Forsch 8:194–197

Krause R (1996) Zur Chronologie der frühen und mittleren BronzezeitSüddeutschlands, der Schweiz und Österreichs. Acta Archaeologia Kopenhagen 67:73–86

Krause R (2003) Studien zur kupfer- und frühbronzezeitlichen Metallurgie zwischen Karpatenbecken und Ostsee, Vorgeschichtl Forsch 24. Publ. M Leidorf, Rahden

Krause R, Pernicka E (1998) Frühbronzezeitliche Kupfersorten im Alpenvorland und ihr archäologischer Kontext. In: Mordant C, Pernot M, Rychner V (eds) L'Atelier du Bronzier en Europe du XXe au VIIIe Siècle avant notre ère. Actes du colloque international "bronze 96", Neuchâtel et Dijon, 1996, pp 191–202

Krawczyk E (1986) Petrologische und geochemische Untersuchungen an Kupferverhüttungsschlacken aus dem südlichen Teil des Wadi Arabah (Israel). Diploma-thesis, Fachb Geowissensch, Univers Mainz

Krawczyk E (1991) Die gallo-römische Eisentechnologie in der Montagen noir und im Massiv von Mouthoumet (Südfrankreich). PhD-diss, Fachb. Geowissensch. Univers Mainz

Krismer M, Goldenberg G, Tropper P (2011a) Die chemische Zusammensetzung von spätbronzezeitlichen Schlacken vom Schmelzplatz Mauken/ Brixlegg. In: Oeggl K, Goldenberg G, Stöllner T, Prast M (eds) Die Geschichte des Bergbaus in Tirol und seinen angrenzenden Gebieten. Proc. 5. Milestone Meeting des SFB HIMAT, Oktober 2010, Mühlbach, pp 185–192

Krismer M, Köhl B, Tropper P (2011b) Mineralogische Charakterisierung der co- und Ni-reichen Kupfererze des Geyer-Reviers bei Brixlegg (Tirol). Mitt Österr Mineralog Gesellsch 157:81–88

Krismer M, Vavtar F, Tropper P, Kaindl R, Sartory B (2011c) The chemical composition of tetrahedrite–tennantite ores from the prehistoric and historic Schwaz and Brixlegg mining areas (North Tyrol, Austria). Eur J Miner 23:925–936

Krismer M, Töchterle U, Goldenber G, Tropper P, Vavtar F (2013) Mineralogical and Petrological investigations of early bronze age copper smelting remains from the Kiechlberg (Tyrol, Austria). Archaeometry 55 (5):923–945

Krivsky WA, Schuhmann R (1957) Thermodynamics of the cu-Fe-S system at matte smelting temperature. Trans AIME 209:981–988

Królik H, Schild R (1999) Rydno, a final palaeolithic ochre mine and surrounding aggregation grounds. Tanged points cultures in Europe. Internat Archaeol Symp, Lublin, September 1993

Kronz A (1997) Phasenbeziehungen und Kristallisationsmechanismen in fayalitischen Schmelzsystemen – Untersuchungen an Eisen- und Buntmetallschlacken. PhD-diss Fakultät Geowiss, Mainz

Kronz A (2000) Self-mixing melt compositions in fayalitic slags – a key in understanding early metal production? In: Rammelmair et al (eds) Applied mineralogy in research, economy, technology, ecology and culture. Proc 6th internat congr applied mineralogy II, Rotterdam, pp 1005–1008

Kronz A (2003) Keltische und römische Eisengewinnung in der Eifel. In: Reger K (ed) 6. Internat Bergbauworkshop Rescheid/Eifel, Oktober 2003, Heimatverein Rescheid, pp 60–65

Kronz A (2005a) Erzbergbau, Buntmetallmineralisationen und Silber-Metallurgie im Bereich der mittleren Mosel. Zeitschr Gesch Berg- und Hüttenwesens – Fischbacher Hefte 1:4–33

Kronz A (2005b) Zur Geologie und Lagerstättenkunde. In: Jockenhövel A, Wilms C (eds) das Dietzhölze-Projekt. Archäometallurgische Untersuchungen zur geschichte und Struktur der mittelalterlichen Eisengewinnung im Lahn-Dill-Gebiet (Hessen), Münster Beitr Ur- und Frühgeschichtl Archäol, vol 1, pp 12–15

Kronz A, Keesmann I (2005) Fayalitische Schmelzsysteme – Ein Beitrag zur vorneuzeitlichen Eisen- und Buntmetalltechnologie im Dietzhölztal (Lahn -Dill – Gebiet, Hessen). In: Jockenhövel A, Willms C (eds) das Dietzhölzetal-Projekt. Archäometallurgische Untersuchungen zur geschichte und Struktur der mittelalterlichen Eisengewinnung im Lahn-Dill-Gebiet (Hessen), Münster Beitr ur- und frühgeschichtl Archäol, pp 403–499

Krupp RE, Weiser T (1992) On the stability of gold-silver alloys in the weathering environment. Mineralal Deposits 27:268–275

Krusch P (1907) Die Untersuchung und Bewertung von Erzlagerstätten. Ferdinand Enke Verlag, Stuttgart

Kruse R, Dörfler W, Jöns H (1997) Angewandte Prospektionsmethoden. In: Jöns H (ed) Frühe Eisengewinnung in Joldelund, Kr. Nordfriesland, 1. Universitätsforsch prähist Archäol 40:12–34

Kruse R, Dörfler W, Jöns H (2000) Angewandte Prospektionsmethoden. In: Jöns H (ed) Frühe Eisengewinnung in Joldelund, Kr. Nordfriesland 1. Universitätsforschungen zur Prähistorischen Archäologie 40 Kiel. R Habelt GMBH, Bonn, pp 12–34

Krysko WW (1979) Blei in geschichte und Kunst, Stuttgart

Kubaschewski O, Alcock CB (1979) Metallurgical thermochemistry, Intern Ser Materials Technology. Pergamon Press, Oxford
Kubota K (1970) Japan's original steelmaking and its development under the influence of foreign technique. Pont à Mousson, Int. co-op. Hist. Tech. Committee 6
Kuhn CG (1826) Trans. And ed: Claudii Galeni opera Omnia, Leipzig
Kühn H (1970) Verdigris and copper Resinate. Stud Conserv 15(1):12–36
Kuhnen HP, Bunk W (2003) Roman time copper ingots found at Trier: original and experiment. In: Associazione Italiana di Metallurgia (ed) Archaeometallurgy in Europe Vol. 2, Milano, pp 393–401
Kullerud G, Yund RA (1962) The Ni-S system and related minerals. J Petrol 3:126–175
Kuniholm PI et al (1995) Anatolian tree rings and the absolute chronology of the eastern Mediterranean, 2220-718 BC. Nature 381:780–783
Kunstmann L (2004) Ergebnisse der archäometallurgischen Untersuchungen zum Röstprozess in der spätbronzezeitlichen Kupfermetallurgie der Ostalpen. In: Res montanarum, pp 33–36
Kuntz J (1904) Kupfererzvorkommen in Südwestafrika. Z Prakt Geol 12:402–405
Kuxmann U, Fischer P (1974) Lead monoxide-aluminum oxide, lead monoxide-calcium oxide, and lead monoxide-silicon dioxide phase diagrams. Erzmetall 27(11):533–537
Kylander ME et al (2005) Refining the pre-industrial atmospheric Pb isotope evolution curve in Europe using an 8000 year old peat core from NW Spain. Earth Planet Sci Lett 240:467–485
La Niece S (1995) Depletion gilding from third millennium B.C. Ur. Iraq 57:1–7
Lacerda LD, Salomons W (1998) Mercury from gold and Silver Mining: a chemical time bomb? Springer, Berlin
Lagarce J (1986) The production of copper ingots at Ras Ibn Hani (Syria). Acta Praehist et Archaeol 18:85–90
Lagarce J, Bounni E, Saliby N (1983) Les fouilles à Ras Ibn Hani en Syrie (campagne de 1980, 1981 et 1982). In: Contribution Rendus de l'Académie des Inscriptions, pp 249–290
Laihunite Research Group, Guiyang Institute of Geochemistry, Academia Sinica and 101 Geological Team of Liaoning, Metallurgical and Geological Prospecting Company (1976) Laihunite - a new iron silicate mineral. Geochimica 1976(2):95–103
Lamb W (1936) Excavations at thermi in lesbos. University Press, Cambridge
Lane HC (1973) Field surveys and excavation of a Roman-British settlement at Scarcliffe Park, east Derbyshire. Research report 1, Derwent archaeological society
Lane HC (1986) The romans in Derbyshire, vol 2. Veritas, Bolsover
Lang J (2015) The Anglo-Saxon cemetery at Dover Buckland, Kent, UK and the technology of some of the iron artefacts. In: Hauptmann A, Modarressi-Tehrani D (eds) Archaeometallurgy in Europe III, Der Anschnitt Beih, vol 26, pp 185–192
Lang J, Ager B (1989) Swords of the Anglo-Saxon and Viking periods in the British museum: a radiographic study. In: Chadwick-Hawkes (ed) Weapons and warfare in Anglo-Saxon England, Oxford Univers. Comm Archaeol Monogr, vol 21, pp 85–122
La Niece S (1993) Silvering. In: S La Niece and P Craddock (eds) Metal Plating and Patination. Butterworth & Heinemann: 201–210
Lang JR, Baker T (2001) Intrusion-related gold systems: the present level of understanding. Mineral Deposits 36:477–489
Large D (2001) The geology of non-Sulphide zinc deposits – an overview. Erzmetall 54(5):264–276
Larson PB et al (2003) Copper isotope ratios in magmatic and hydrothermal ore forming environments. Chem Geol 201:337–350
Laschimke R (2013) Steinbeile im zentralen Bergland von Irian Jaya. Experimentelle Archäologie in Europa – Bilanz 2013:192–200
Laschimke R, Burger M (2011) Archäometallurgische Experimente zum Giessen von bronzezeitlichen Ochsenhautbarren aus Kupfer. Metall 3:86–92
Laschimke R, Burger M (2015) Die Anfänge der Kupfermetallurgie - archäometallurgische Versuche zur Verhüttung von Malachit am offenen Feuer. In: Weller U, Lessig T (eds) Experimentelle Archäologie in Europa, pp 10–20
Laschimke R, Burger M (2017) Neolithische Flachbeile aus Kupfer, die ältesten Metallwerkzeuge des menschen – archäometallurgische Experimente zu ihrer Herstellung. Metall 3:85–88
Laschimke R, Burger M (2018) Die Herstellung von kupferzeitlichen Äxten mit Schaftloch – archäometallurgische Experimente. Metall 6:239–243
Lauffer S (1979) Die Bergwerkssklaven von Laurion. Franz Steiner Verlag, Stuttgart
Lavender DS (1962) The story of Cyprus mines company. Hunting Library, California
Lavric JV, Spangenberg JE (2003) Stable isotope (C, O, S) systematics of the mercury mineralization at Idrija, Slovenia: constraints on fluid source and alteration processes. Mineral Deposits 38:886–889
Law R (2008) Inter-regional interaction and urbanism in the ancient Indus valley: a geologic provenience study of Harappa's rock and mineral assemblage. PhD-diss Univers Wisconsin – Madison
Law, R, Burton JH (2008) Nondestructive Pb-isotope sampling and Analyis of archaeological silver using EDTA and ICP-MS. American Laboratory, Application Note
Lawrance L (2009) The supergene behavior of gold: a natural enrichment and refinement Phenomnen. Econ Geol Spec Publ 14:147
Lawrence LM, Griffin BJ (1994) Crystal features of supergene gold at Hannan south, Western Australia. Mineral Deposits 29:391–398

Leblanc M (1991) Platinum-group elements and gold in Ophiolithic complexes: distribution and fractionation from mantle to ocean floor. In: Peters TJ, Nicolas A, Coleman RG (eds) Ophiolite genesis and evolution of the ocean lithosphere. Kluwer, Dordrecht, pp 231–260

Lazarovici G, Lazarovivici CM, Constantinescu B (2012) Despre analizele pieselor de Aur din atelierul de Bijuterii la Cheile Turzii – Pestera Caprelor/Pestera Ungureasca

Leach DL, Taylor RD (2009) Mississippi Valley-type Lead-zinc deposit model. U.S. Geological Survey, open-file report 2009–1213. Reston, Virginia

Leake RC, Chapman RJ, Bland DJ, Stone P, Cameron DG, Styles MT (1998) The origin of alluvial gold in the Leadhills area of Scotland: evidence from interpretation of internal chemical characteristics. J Geochem Explor 63:7–36

Lechtman H (1973) The gilding of metals in pre-Columbian Peru. In: Young W (ed) Application of science in examination of works of art. Boston Museum of Fine Arts, Boston, pp 38–52

Lechtman H (1979) A pre-Colombian technique for electrochemical plating of gold and silver on copper objects. J Metals 31:154–160

Lechtman H (1980) The Central Andes: metallurgy without Iron. In: Wertime T, Muhly JD (eds) The coming of the age of Iron. Yale University Press, New Haven, pp 267–334

Lechtman H (1988) Tradition and styles in central Andean metalworking. In: Maddin R (ed) The beginnings of the use of metals and alloys, Cambridge, MA, pp 344–378

Lechtman H (1996) Arsenic bronze: dirty copper or chosen alloy? A view from the Americas. J Field Archaeol 23(4):477–514

Lechtman H, Klein S (1999) The production of copper-arsenic alloys (arsenic bronze) by Cosmelting: modern experiment, ancient Practise. J Archaeol Sci 26:497–526

Leemans WF (1971) Gold. In: Ebeling F, Meissner B, Weinder E, von Soden W (eds) Reallexikon der Assyriologie, 3. Band. De Gruyter, pp 504–515

Leese M, Craddock P, Freestone IC, Rothenberg B (1985/1986) The composition of ores and metal objects from Timna, Israel. Wiener Ber Naturwiss Kunst 2 (3):90–120

Lehmann B (1990) Metallogeny of tin. Lecture Notes Earth Science 32. Springer, Berlin

Lehmann U (2016) Wurmbunte Klingen. Studien zu Konstruktion, Herstellung und Wertigkeit der frühmittelalterlichen Spatha in Westfalen. Veröffentl. Altertumskomm. Westfalen Landschaftsverb. Westfalen-Lippe 21

Lehrberger G (1995) The gold deposits of Europe. An overview of the possible metal sources for prehistoric gold objects. In: Morteani G, Northover J (eds) Prehistoric gold in Europe, Mines, metallurgy and manufacture. Kluwer Academic Publ, pp 115–144

Leistel JM et al (1998) The volcanic-hosted massive sulphide deposits of the Iberian Pyrite Belt. Mineral Deposits 33:2–30

Lemasson Q, Moignard B, Pacheco C, Pichon L, Guerra MF (2015) Fast mapping of gold jewellery from ancient Egypt with PIXE: searching for hard-solders and PGE inclusions. Talanta 143:279–286

Leonhard G (1860) Grundzüge der Mineralogie. Wintersche Publ, Leipzig & Heidelberg

Leroi-Gourhan (1943) L'Homme et la Matìere: evolution et Techniques (archive). Peris, Albin coll

Lesniak CP (1991) Thermodynamik und Metallographie von Arsen- und Antimon-Mangellegierungen am Beispiel archäologischer Artefakte. PhD-diss Univers Saarbrücken

Leusch V, Pernicka E, Armbruster B (2014) Chalcolithic gold from Varna – provenance, circulation and function. In: Meller H, Risch R, Pernicka E (eds) Metalle der Macht – Frühes gold und Silber, 6. Mitteldeutscher Archäologentag, Oct 2013, Halle (Saale), pp 165–182

Leuzinger U (1997) Die jungsteinzeitlichen Kupferfunde aus dem Kanton Thurgau. Arch Switz 20:51–53

Levy TE (2007) Journey to the copper age, Archaeology in the Holy Land. San Diego Museum of Man

Levy TE, Adams RB, Hauptmaqnn A, Prange M, Schmitt-Strecker S, Najjar M (2002) Early bronze age metallurgy: a newly discovered copper manufactory in southern Jordan. Antiquity 76:425–437

Levy TE, Levy AM, Sthapathy DR, Sthapathy DS (2008) Masters of fire. Hereditary bronze casters of South India. Publ Deutsches Bergbau-museum Bochum 162. WAZ-Druck Duisburg

Levy TE, Najjar M, Ben-Yosef E (2014) New insights into the Iron age archaeology of Edom, southern Jordan. Surveys, excavations, and research from the University of California, San Diego – Department of Antiquities of Jordan, Edom lowlands regional archaeology project (ELRAP). The Cotsen Inst Archaeol Press, Los Angeles. Vol. 1, 2

Levy TE, Bettilyon M, Burton MM (2016) The Iron age copper industrial complex: a preliminary study of the role of ground stone tools at Khirbat en-Nahas, Jordan. J Lithic Stud 3(3):1–23

Lewin B, Hauptmann A (1984) Kodo-Zuroku. Illustrierte Abhandlung über die Verhüttung des Kupfers, 1801. Zur geschichte der Kupfergewinnung in Japan. (= Veröffentl Deutsches Bergbau-museum. 29): 1–119

L'héritier M, Téreygol F (2010) From copper to silver: understanding the saigerprozess through experimental liquation and drying. J Hist Metall Soc 44(2):136–150

Li W, Pearson NJ, Jackson SE, Alard O (2010) The cu isotopic signature of granites from the Lachlan Fold Belt, SE Australia. Chem Geol 258(1):38–49

Lichardus J (1988) Der Westpontische Raum und die Anfänge der kupferzeitlichen Zivilisation. In: Fol A, Lichardus J (eds) Macht, Herrschaft und gold. Das Gräberfeld von Varna (Bulgarien) und die Anfänge einer neuen europäischen Zivilisation, Catalog Saarland-Mus Saarbrücken, pp 79–129

Lieder W (1980) Vermessung und Kartographie. In: Conrad HG, Rothenberg B (eds) Antikes Kupfer im Timna-Tal. 4000 Jahre Bergbau und Verhüttung in der Arabah, Der Anschnitt Beih, vol 1, pp 33–40

Linduff K, Mei J (2014) Metallurgy in ancient eastern Asia: retrospect and prospects. In: Roberts BW, Thornton CP (eds) Archaeometallurgy in global perspective, Methods and syntheses. Springer, pp 785–803

Lins PA, Oddy WA (1975) The origins of mercury gilding. J Archaeol Sci 2:365–373

Liu S, Rehren T, Pernicka E, Hausleiter A (2015) Copper processing in the oases of Northwest Arabia: technology, alloys and provenance. J Archaeol Sci 53:492–503

Lleras Perez R, Botero CI, Velez SL, Sanchez Cabra (2007), The art of gold. The legacy of pre-hispanic Colombia. Collection of the Gold Museum in Bogota. Banco de la Republica

Löffler I (2017) New thoughts about Iron age metallurgy in Faynan: a discussion. In: Eisenach P, Stöllner T, Windler A (eds) The RITaK conferences 2013–2014, Der Anschnitt Beih, vol 34, pp 151–162

Löffler I (in prep) das Montanrevier von Faynan, Jordanien. Studien zu einer prähistorischen Wirtschaftslandschaft. Der Anschnitt Beih

Lohmann H (1996) Die Santorinkatastrophe – ein archäologischer mythos? In: Olshausen E, Sonnabend H (eds) Naturkatastrophen in der antiken welt, Stuttgarter Koll hist Geographie Altertums, vol 6, pp 337–363

Lohmann H (2005) Prähistorischer und antiker Blei-Silberbergbau in Laurion. In: Yalçın Ü (ed) Anatolian metal III, Der Anschnitt Beih 18, pp 105–136

Lohmann H (2015) Ari – a classical Mining District at Anavysso (Attica). In: Sheedy K (ed) Mines, metals money in Attica and the ancient world. Proc Internat Conf epigraphic and numismatic museum, Athens

Löhneyß GE von (1690) Gründlicher und außführlicher Bericht von Bergwercken. Stockholm, Hamburg 1690

Lohrscheider F, Maass A, Steiniger D (2003) Frühe Kupferproduktion – archäologischer Befund und experiment: Versuche zur Fahlerzverhüttung in einem einzigen Ofengang. In: Stöllner T, Körlin G, Steffens G, Cierny J (eds) Man and mining – mensch und Bergbau. Studies in honour of G Weisgerber on ocasion of his 65th birthday, Der Anschnitt Beih, vol 16, pp 301–307

Lordkipanidze O (1991) Archäologie in Georgien. Von der Altsteinzeit bis zum Mittelalter. Quellen und Forschungen zur prähistorischen und provinzialrömischen Archäologie 5. VCH Verlagsgesellschaft Weinheim

Louis H (1911) Metallurgy of tin. McGraw-Hill Book, New York

Lovering JF, Nichiporuk W, Chodos A, Brown H (1957) The distribution of gallium, germanium, cobalt, chromium, and copper in iron and stony-iron meteorites in relation to nickel content and structure, Geochim. Cosmochim Acta 11:263–278

Lucas A (1928) Silver in ancient times. J Egypt Archaeol 14:313–319

Lucas A (2006) Wind, water, work, Ancient and Medieval Milling Technology. Brill, Leiden

Lucas A, Harris J (1967) Ancient Egyptian materials and industries. E Arnold Publ, London

Lupu A, Rothenberg B (1970) The extractive metallurgy of the early Iron age copper industry in the Arabah, Israel. Archaeol Aust 47:91–129

Lutz J (1990) Geochemische und mineralogische Aspekte der frühen Kupferverhüttung in Murgul/Nordost-Türkei. PhD-diss, Univers Heidelberg

Lutz J (2004) Röntgenfluoreszenzanalysen. In: Hauptmann H, Pernicka E (eds) Die Metallindustrie Mesopotamiens von den Anfängen bis zum 2. Jahrtausend v. Chr. Leidorf, Rahden, pp 109–149

Lutz J (2016) Alpenkupfer – die Ostalpen als Rohstoffquelle in vorgeschichtlicher Zeit. In: Bartelheim M, Horejs B, Kraus R (eds) Von Baden bis Troia. Ressourcennutzung, Metallurgie und Wissenstransfer, Oriental and European archaeology 3. Leidorf, Rahden/Westfalen, pp 333–358

Lutz J, Pernicka E, Wagner GA (1994) Chalkolithische Kupferverhüttung in Murgul, Ostanatolien. In: Wartke RB (ed) Handwerk und Technologie im Alten orient. Ein Beitrag zur Geschichte der Technik im Altertum. Zabern, Mainz, pp 59–66

Lutz J, Helwing B, Pernicka E, Hauptmann H (1995) Die Zusammensetzung einiger Metallfunde aus Uruk-Warka. Baghdad Mitt 26:1–23

Luyken W, Heller L (1938) Arsen in Eisenerzen und die Möglichkeit seiner Austreibung vor der Verhüttung. Archiv f d Eisenhüttenwesen 11(10):475–481

Lyonnet B (2001) Another possible interpretation of the Bactro-Margiana culture (BMAC) of Central Asia: the tin trade. In: Jarrige C, Lefèvre V (eds) South Asian archaeology I, Prehistory. Éditions Rech. Sur Civilisations, Pari, pp 191–200

Machatschki F (1953) Spezielle Mineralogie auf geochemischer Grundlage. Springer, Wien

Mackay E (1933) Decorated carnelian beads. Man 33:143–146

Maclean WH (1969) Liquidus phase relations in the FeS-FeO-Fe_3O_4-SiO_2 system and their application in geology. Econ Geol 64:865–884

Maclean PI, Scaife B, de Schauensee M (1992) Hasanlu lion pins: observations on the role of antimony in bronze-making in Iron age Iran. Text for Poster Session, Berkeley, CA

Maczek M, Preuschen E, Pittioni R (1952) Beiträge zum problem des Ursprunges der Kupfererzverwertung in der Alten welt. Archaeol Aust 10:61–70

Maddin R (1988) The beginning of the use of metals and alloys II. MIT Press, Cambridge/MA

Maddin R (2006) Archaeometallurgy in Sardinia: a general metallurgical evaluation. Instrumentum 23:2–3

Maddin R, Muhly JD (1974) Some notes on the copper trade in the ancient mid-east. J Metals 5:1–7

Maddin R, Muhly JD, Stech-Wheeler T (1980) Distinguishing artifacts made of native copper. J Archaeol Sci 7:221–225

Maddin R, Hauptmann A, Baatz D (1991) A metallographic examination of some iron tools from the Saalburgmuseum. Saalburg-Jahrb 46:5–23

Maddin R, Hauptmann A, Weisgerber G (1996) Metallographische Untersuchungen an römischem Gezähe aus Rio Tinto/Spanien. Metalla 3(1):27–44

Maddin R, Muhly JD, Stech T (1999) Early metalworking at Çayönü. In: Hauptmann A, Pernicka E, Rehren T, Yalçın Ü (eds) The beginnings of metallurgy, Der Anschnitt, Beih, vol 9, pp 37–44

Maggetti M, Baumgartner D, Galetti G (1991) Mineralogical and chemical studies on Swiss Neolithic M. crucibles. In: Pernicka E, Wagner G (eds) Archaeometry 1990, pp 95–104

Mahe-Le Carler C, Ploquin A, Fluck P (2010) Apport de la géochimie et de la pétrologie à la connaissance de la métallurgie primaire du plomb argentifère au Moyen age: les exemples du Mont Lozère (Cévennes) et de le Pfaffenloch (Vosges). Archéoscience, revue d'archéometrie 34:150–176

Mahfoud RF, Beck JN (1997) Copper Mineralizations in the Ophiolite of Oman: the genesis and emplacement relationship with the Orogenic movements of Serpentinized Peridotite. Int Geol Rev 39:252–286

Makkay J (1995) The rise and fall of gold metallurgy in the copper age of the Carpathian Basin: the background of the change. In: Morteani G, Northover JP (eds) Prehistoric gold in Europe, NATO ASI Ser E, Appl Sc, vol 280. Springer, Dordrecht, pp 65–76

Malchenko EG, Ermolov PV (2006) Metallogenetic summary of Central Kazachstan and adjacent areas. In: Shalov V, Seltmann R, Kremenetsky A, Lehmann B, Popov V, Ermolov P (eds) Granite-related ore deposits of Central Kasachstan. Int Ass Genesis Ore Deposits (IAGOD), CERCAMS, London, pp 67–81

Maltzan HF (1869) Reise auf der Insel Sardinien. Dyck'sche Buchhandlung, Leipzig

Manasse A, Mellini M (2002) Archaeometallurgical slags from Kutná Hora. N Jb Mineralogie Mh 8:369–384

Manasse A, Mellini M, Viti C (2001) The copper slags of the Capattoli Valley, Campiglia Marittima, Italy. Eur J Miner 13:949–960

Mangin M, Keesmann I, Birke W, Ploquin A (1992) Mines et métallurgie chez les Éduens – Le district sidérurgique antique et médieval du Morvan-Auxois. In: Ann Littéraire de l' Univers de Besancon, Paris, p 456

Marahrens J (2018) Tin isotopic composition of tin deposits in Europe and Central Asia. In: Pernicka E et al (eds) Bronze age tin. Geological sources, production, and distribution of tin in bronze age Eurasia, Internat. Workshop Mannheim, March 2017, Dent Abstr, p 16

Marahrens J, Berger D, Brügmann G, Pernicka E (2016) Vergleich der stabilen Zinnisotopenzusammensetzungen von Kassiteriten aus europäischen Zinn-Lagerstätten. In: Greiff S, Kronz A, Schlütter F, Prange M (eds) Archäometrie und Denkmalpflege 2016, Metalla Sonderheft, vol 8, pp 190–193

Marcoux E (1998) Lead isotope systematics of the giant massive sulphide deposits in the Iberian Pyrite Belt. Mineral Deposits 33:45–58

Marcoux E, Bril H (1986) Héritage et sources des métaux d'après la géochimie isotopique du plomb. Exemple des minéralisations filonniennes (Sb, Pb, Ba, F). du haut-Allier (massif central, France). Mineral Deposits 21:35–43

Maréchal JF (1985) Methods of ore roasting and the furnaces used. In: Craddock PT, Hughes MJ (eds) Furnaces and smelting Technology in Antiquity, Brit Mus Occ paper, vol 48, pp 29–41

Maréchal JF (1998) A propos de nouveaux lingots de fer d'epoque romaine découverts Sur la côte de Narbonnaise. Bull Soc National Antiquaires de France 1998:239–254

Marignac C et al (2003) Remobilisation of base metals and gold by Variscan metamorphic fluids in the south Iberian pyrite belt: evidence from from the Tharsis VMC deposit. Chem Geol 194:143–165

Marinos G (1982) Greece. In: Dunning FW, Mykura W, Slater D (eds) Mineral deposits of Europe. Mineralogical society. Inst Mining Metall, London, pp 233–253

Marinos GP, Petrascheck WE (1956) Laurion/Laurium. Publ Athēnai Institout Geōlogias

Markl G (2015a) Minerale und Gesteine. Mineralogie – Petrologie – Geochemie, 3rd edn. Springer Spektrum

Markl G (2015b) Schwarzwald - Lagerstätten und Mineralien aus vier Jahrhunderten 1: Nordschwarzwald und Grube Clara. Bode, Salzhemmendorf

Markl G (2016a) Schwarzwald - Lagerstätten und Mineralien aus vier Jahrhunderten. Vol. 2: Mittlerer Schwarzwald, Teil 1. Bode, Salzhemmendorf

Markl G (2016b) Schwarzwald - Lagerstätten und Mineralien aus vier Jahrhunderten. Vol. 3: Mittlerer Schwarzwald 2. Bode, Salzhemmendorf

Markl G (2017) Schwarzwald - Lagerstätten und Mineralien aus vier Jahrhunderten 4: Südschwarzwald. Bode, Salzhemmendorf

Markl G, Lahaye Y, Schwimm G (2006) Copper isotopes as monitors of redox processes in hydrothermal mineralization. Geochim Cosmochim Acta 70(16):4215–4228

Marqués Sierra AL (2018) Metallurgical treatment of calamine before scientific discovery of zinc metal. De Re Metallica 31(2):15–24

Marsh EE, Anderson EP, Gray F (2010) Nickel-cobalt laterites – a deposit model. USGS scientific investigation report 2010-5070-H

Marshall T (2015) die Macht der Geographie. Wie sich Weltpolitik anhand von 10 Karten erklären lässt. Dtv München

Martín-Crespo T, Gómez-Ortiz D, Martín-Velázquez S (2012) Geoenvironmental characterization of riverbeds affected by mine tailings in the Mazarrón district (Spain). J Geochem Exp 199–120:6–16

Martín-Crespo T, Gómez-Ortiz D, Martín-Velázquez S (2019) Geoenvironmental characterization of sulfide mine tailings. IntecOpen

Martinek KP (1996) Archäometallurgische Untersuchungen zur frühbronzezeitlichen Kupferproduktion und –verarbeitung auf dem Buchberg bei Wiesing, Tirol. Fundber Österreich 34:575–584

Martinón-Torres M, Rehren T (2002) Agricola and Zwickau: theory and practice of renaissance brass production in SE Germany. J Hist Metall 36:95–111

Martinón-Torres M, Rehren T, Freestone I (2006) Mullite and the mystery of hessian wares. Nature 444 (23):437–438

Mason AH, Powell WG, Bankoff HA, Mathur R, Bulatovic A, Filipovic V, Ruiz J (2016) Tin isotope characterization of bronze artifacts of the Central Balkans. J Archaeol Sci 69:110–117

Mata MP, Peacor DR, Gallart-Mart MD (2002) Transmission Electron microscopy (TEM) applied to ancient pottery. Archaeometry 44(2):155–176

Mathur R (2009) Exploration potential of cu isotope fractionation in porphyry copper deposits. J Geochem Exp 102:1–6

Matuschik I (1998) Kupferfunde und Metallurgiebelege, zugleich ein Beitrag zur Gechichte der kupferzeitlichen Dolche Mittel-, Ost- und Südeuropas. In: Mainberger M (ed) das Moordorf von Reute. Archäologische Untersuchungen in der jungeneolithischen Siedlung Reute-Schorrenried. Teraqua CAP, Staufen i. Brsg., pp 207–261

Maxwell-Hyslop KR (1971) Western Asiatic Jewellery, c. 3000-612 BC. In: Methuan' handbooks of archaeology. Methuen, London

Maxwell-Hyslop KR (1977) Sources of Sumerian gold. Iraq 39:83–86

McClenaghan MB, Cabri L (2011) Review of gold and platinum group element (PGE) indicator minerals methods for surficial sediment sampling. Geochem Explor Environ Anal 11:251–263

McDonald AS, Sistare GH (1978) The metallurgy of some carat gold Jewellery alloys. Gold Bull 11(3):66–73

McKerell HH, Tylecote RF (1972) The working of copper–arsenic alloys in the early bronze age and the effect on the determination of provenance. Proc Soc Prehist 38:209–218

Meeks ND (1986) Tin-rich surfaces on bronze – some experimental and archaeological considerations. Archaeometry 28(2):133–162

Meeks ND (1993) Surface characterization of tin bronze, tinned iron and arsenical bronze. In: La Niece S, Craddock P (eds) Metal plating and Patination. Cultural, technical and historical developments. Butterworth Heinemann, pp 247–275

Meeks ND (1995) Technical study of Roman bronze mirrors. In: Mols STAM et al (eds) Acta of the 12th international Congres on ancient bronzes, Nederlandse arch. Rapporten, vol 18, pp 179–183

Meeks ND (2000) In: Ramage A, Craddock PT (eds) King Croesus' gold. Excavations at Sardis and the history of gold refining. Brit Mus Press, London, pp 99–156

Meeks ND, Tite MS (1980) The analysis of platinum group element inclusions in gold. J Archaeol Sci 7:267–275

Meeks N, Laniece S, Estévez P (2002) The technology of early platinum plating: a gold mask of the Tolita culture, Ecuador. Archaeometry 44(2):273–284

Mehrabi B, Yardley BWD, Cann JR (1999) Sediment-hosted disseminated gold mineralization at Zarshuran, NW Iran. Mineral Deposits 34:673–696

Mehrabi B, Sianib MG, Azizi CH (2014) The genesis of the epithermal gold mineralization at north Glojeh veins, NW Iran. Int J Sci 5(1):479–497

Mei J (2000) Copper and bronze metallurgy in late prehistoric Xinjiang. BAR Internal Ser 865

Mei J (2009) Early metallurgy in China: some challenging issues in current studies. In: Mei J, Rehren T (eds) Metallurgy and civilisation: Eurasia and beyond. Archetype Publ, London, pp 9–16

Meier SW (1991) Der Blei-Fernhandel in republikanischer Zeit. In: Hellenkemper-Salies G (ed) das wrack: Der antike Schiffsfunde von Mahdia. Katalog Rhein. Landesmus, Bonn, pp 767–787

Meier SW (1995) Blei in der Antike. PhD-diss. Univers Zürich

Meier SW (2013) Römischer Goldbergbau auf Dolaucothi. Anschnitt 65(4):130–144

Meier DMP (2015) Traces of protohistorical metallurgical activities during the 3rd millennium BCE in East Iran with a special focus on the case of Shahdad in the Dasht-eh Lut (Kerman province). PhD-diss FU Berlin

Meignen L, Goldberg P, Bar-Yosef O (2007) The hearths at Kebara Cave and their role in site formation processes. In: Bar-Yosef O, Meignen L (eds) Kebara Cave, Mt. Carmel, Israel: the Middle and Upper Paleolithic archaeology. Part I. American School of Prehistoric Research, Peabody Museum, Harvard University, Cambridge, MA, pp 91–122

Meixner H, Paar W (1982) New observations on ore formation and weathering of the Kamariza deposit, Laurion, SE Attica (Greece). In: Amstutz GC, Frenzel G, Kluth C, Moh G, Wauschkuhn A, Zimmermann RA, ElGoresy A (eds) Ore Genesis. The State of the Art.Special Publ Soc Geol Appl Mineral Deposits. Springer, Berlin, pp 760–767

Meliksetian K, Schwab R, Kraus S, Pernicka E, Brauns M (2011) Chemical, lead isotope and metallographic analysis of extraordinary arsenic-rich alloys used for jewellery in bronze age Armenia. In: Hauptmann A, Modarressi-Tehrani D, Prange M (eds) Archaeometallurgy in Europe III. Abstracts Internat Conf, Deutsches Bergbau-Museum Bochum, June 2011, pp 211–212

Mellaart J (1966) Excavations at Çatal Höyük, 1965. Fourth preliminary report. Anatol Stud 16:165–191

Mellart J (1967) Çatal Hüyük, a Neolithic town in Anatolia. Thames & Hudson, London

Menic S (2016) Die latènezeitliche Eisenproduktion im Siegerland. Chaı�ne opératoire und Ökonometrie der Prozessschritte Der Anschnitt Beih 32

Merideth C (1998a) An archaeological survey for ancient tin mines and smelting sites in Spain and Portugal. Mid-central western Iberian geographical region 1990-1995, Brit Archaeol rep Internat Ser 714, Oxford

Merideth C (1998b) La Mina el Cerro de San Cristobal: a bronze age tin mine (Extremadura, Spain). Pap Inst Archaeol 9:57–69

Merkel J (1986) Ancient smelting and casting of "oxhide ingots". In: Balmuth MS (ed) Studies in Sardinian archaeology II, Ann Arbor, pp 251–271

Merkel J (1990) Experimental reconstruction of bronze age copper smelting based on archaeological evidence from Timna. In: Rothenberg B (ed) The ancient metallurgy of copper, Inst. Archaeo-Metall. Stud. University College, London, pp 78–122

Merkel J (2007) Imperial Roman production of Lead and silver in the northern part of upper Mopesia (Mt. Kosmaj area). J Serbian Archaeol Soc 23:39–78

Merkel S (2018) Archaeometallurgical investigations of a Viking brass ingot hoard from the Hedeby Harbor in northern Germany. J Archaeol Sci 20:293–302

Merkel J, Beno R (1997) The earliest steps to copper metallurgy in the western Arabah. In: Hauptmann A, Pernicka E, Rehren T, Yalçın Ü (eds) The beginnings of metallurtgy, Der Anschnitt, Beih, vol 9, pp 149–165

Merkel J, Bräutigam B, Klein S, Hauptmann A (2013) The analysis of slag from the Panjhīr mining region, Afghanistan: an investigation of (medieval) silver production technology. Archäol Mitt Iran u Turan 45:231–250

Merkel S, Hauptmann A, Kirnbauer T, Téreygol F (2016) Silver production at al-Radrad: Archaeometallurgical analysis of ore, slag, and technical ceramics from early medieval Yemen. Arab Archaeol Epigraph 27:107–126

Merwe N (1980) The advent of iron in Africa. In: Wertime TA, Muhly JD (eds) The coming of the age of iron. Yale University Press, pp 463–506

Metten B (2003) Beitrag zur spätbronzezeitlichen Kupfermetallurgie im Trentino (Südalpen) im Vergleich mit anderen prähistorischen Kupferschlacken aus dem Alpenrau. Meta 10 (1/2):1–122

Meyer M (2005) Das comeback der Rohstoffe: Neues vom gold. Geowissenschaftl Mitt 20:8–18

Meyers P (1988) Characteristics of casting revealed by the study of ancient Chinese bronzes. In: Maddin R (ed) The beginning of the use of metals and alloys. Proc 2nd Int Conf BUMA, Zhengzhou, China, Oct 1986. MIT Press, Cambridge, MA, pp 283–295

Meyers P (2003) Production of silver in antiquity: ore types identified based upon elemental compositions of ancient silver artifacts. In: Lambertus ZV (ed) Patterns and process. Festschr Honor of EV Sayre, Smiths Center for Materials Res and Educ, Suitland, MD, pp 271–288

Middleton AP (2017) The scientific examination of industrial materials from Zawar, Dariba and Agucha. In: Craddock PT (ed) Early Indian metallurgy. The production of Lead, silver and zinc in north West India. Archetype Publ, London, pp 135–164

Middleton AP, Hook DR, Humphrey MS (2000) Scientific examination of some ceramic materials and samples of litharge. In: Ramage A, Craddock PT (eds) King Croesus' gold. Excavations at sardis and the history of gold refining. Brit Mus Press, pp 157–168

Middleton AP, Cartwrigh CR, Craddock PT, Eckstein K, Freestone IC, Hook DR, Meeks ND (2017) The scientific examination of industrial materials from Zawar, Dariba and Agucha. In: Craddock PT (ed) Early Indian metallurgy. The production of lead, silver and zinc in North West India. Archetype Publ, London, pp 135–164

Mille B, Besenval R, Bourgarit D (2004) Frühes Gießen in verlorener form in Baluchistan (Pakistan): das "Leopardengewicht" aus Shahi-Tump. In: Stöllner T, Slotta R, Vatandoust A (eds) Persiens Antike Pracht. Bergbau- Handwerk – Archäologie. Ausstellungskatalog Deutsches Bergbau-museum Bochum Nov 2004/May 2005, Cuno Calbe, pp 274–280

Mille B, Bourgarit D, Haquet JF, Besenval R (2005) From the 7th to the 2nd millennium BC in Balochistan (Pakistan). The development of copper metallurgy before and during the Indus civilisation. In: Jarrige C, Levevre V (eds) South Asian archaeology 2001. Editions Recherche, Paris, pp 237–244

Miller D, Hall S (2008) Rooiberg revisited – the analysis of tin and copper smelting debris. J Hist Metal Soc 42 (1):23–38

Milton C, Dwornik E, Finkelman RB, Toulmin P III (1976) Slag from an ancient copper smelter at Timna, Israel. J Hist Metall Soc 10:24–33

Mirnejad H, Simonetti A, Molaslehi F (2011) Pb isotopic compositions of some Zn-Pb-deposits and occurrences from Urumieh-Dokhtar and Sanandaj-Sirjan zones in Iran. Ore Geol Rev 39:181–187

Mishara J, Meyers P (1974) Ancient Egyptian silver: a review. In: Bishay A (ed) Science and technology of materials, III, New York, pp 29–45

Miske WV (1908) Die prähistorische Ansiedlung von Velem St. Vid. Wien

Misra K (2000) Understanding ore deposits. Springer, Berlin

Modarressi-Tehrani D (2004) Ein ensemble frühlaènezeitlicher Metallverarbeitung aus der Siedlung von Eberdingen-Hochdorf (Lkr. Ludwigsburg). Meta 11 (1):1–148

Modin H, Modin S, Serning I (1985) Iron from Lapphyttan. A metallographic investigation and guide with archaeological comments. Meritförlaget, Johanneshov

Mödlinger M, Sabatini B (2016) A re-evaluation of inverse segregation in prehistoric as-cu-objects. J Archaeol Sci 74:60–74

Moesta H (1982) Bronzezeitliche Hüttenprozesse in den Ostalpen. Naturwissenschaften 70(3):142–143

Moesta H (1983) Erze und Metalle. Ihre Kulturgeschichte im Experiment. Springer, Berlin

Moesta H (1986) Erze und Metalle. Ihre Kulturgeschichte im Experiment. Springer, Berlin

Moesta H (1993) Daten zur Rekonstruktion der Arbeitsweise alter Kupferhütten. In: Francovich R (ed) Archeologia delle Attività Estrattive e Metallurgiche, Firenze, pp 515–531

Moesta H (2004) Bemerkungen zu bronzezeitlichen Metallen mit hohen Gehalten an Arsen und/oder Antimon, den sog. Fahlerzen. In: Weisgerber G, Goldenberg G (eds) Alpenkupfer- Rame delle Alpi, Der Anschnitt Beih, vol 17, pp 269–272

Moesta H, Franke PR (1995) Antike Metallurgie und Münzprägung. Birkhäuser, Basel

Moesta H, Rüffler R, Schnau-Roth G (1989) Zur Verfahrenstechnik der bronzezeitlichen Kupferhütten am Mitterberg. In: Hauptmann A, Pernicka E, Wagner GA (eds) Archäometallurgie der Alten welt, Der Anschnitt Beih, vol 7, pp 141–153

Moghadam HS, Stern RJ (2011) Late cretaceous forearc ophiolites of Iran. Island Arc 20:1–4

Mohen JP (1990) Metalurgie prehistorique: introduction a la Paleometalurgie. Masson, Paris

Molist M et al (2010) New metallurgical findings from the pre-pottery Neolithic: tell Halula (Euphrates valley, Syria). Paléorient 35(2):33–48

Möller B (1971) Thermodynamische Untersuchungen an Zinnschlacken. Neue Hütte 16(7):400–404

Möller P (1995a) Accumulation of gold by electrochemical processes. In: Morteani G, Northover J (eds) Prehistoric gold in Europe, NATO ASI Ser E, Appl Sc 280. Kluwer, Dordrecht, pp 145–157

Möller P (1995b) Electrochemical corrosion of natural gold alloys. In: Morteani G, Northover J (eds) Prehistoric gold in Europe, NATO ASI Ser E, Appl Sc 280. Kluwer, Dordrecht, pp 357–367

Momenzadeh M (2004) Metallische Bodenschätze in Iran in antiker Zeit. Ein kurzer Überblick. In: Stöllner T, Slotta R, Vatandoust A (eds) Persiens antike Pracht. Bergbau-Handwerk-Archäologie. Ausstellungskatalog Deutsches Bergbau-museum, Bochum, pp 8–21

Mondillo N, Boni M, Balassoni G, Grist B (2011) In search of the lost zinc: a lesson from the Jabali (Yemen) nonsulfidic zinc deposit. J Geochem Explor 108:209–219

Mongiatti A, Meeks N, Simpson SJ (2010) A gold four-horse model chariot from the Oxus treasure: a fine illustration of Achaemenid gold work. Brit Mus Technical Res Bull 4:27–38

Montero-Ruiz I, Perea A (2007) Brasses in the early metallurgy of the Iberian Peninsula. In: La Niece S, Hook D, Craddock PT (eds) Metals and Mines. Studies in Archaeometallurgy. Archetype Publ. Brit. Mus, London, pp 136–139

Montero Ruiz I, Ruíz Taboada A (1996) Enterramiento colectivo y metalurgia en el yacimiento Neolítico de Cerro Virtud (Cuevas de Almanzora, Almería). Trabajos de Prehist 53(2):55–75

Moon CJ (2001) Comparison of mineral deposits between Georgian and Turkish sectors of the Tethyan Metallogenic Belt. In: Piestrzynski A (ed) Mineral deposits at the beginning of the 21st century. Swets and Zeitlinger, pp 309–312

Moorey PR (1994) Ancient Mesopotamian materials and industries. The archaeological evidence. Clarendon Press, Oxford

Moorey PRS, Schweizer F (1972) Copper and copper alloys in ancient Iraq, Syria and Palestine: some new analyses. Archaeometry 14:177–198

Morimoto N et al (1988) Nomenclature of pyroxenes. Amer Miner 73(9,10):1123–1133

Morimoto N et al (1989) Nomenclature of pyroxenes. Can Miner 27:143–156

Moroni M, Girardi VAV, Ferrario A (2001) The Serra Pelada au-PGE deposit, Serra dos Carajás (Para state Brazil): geological and geochemical indications for a composite mineralising process. Mineral Deposits 36:768–785

Morteani G, Fuganti A (1988) The classification of gold deposits. In: Fuganti A, Morteani G (eds) Workshop on rare and precious metals, extended abstarcts, Castel Ivano, pp 21–37

Morteani G, Northover JP (1995) Prehistoric gold in Europe. Mines, metallurgy and manufacture, NATO ASI series, applied sciences, vol 280. Kluwer, Dordrecht

Morton V (2019) Brass from the Past. Brass made, used and traded from prehistoric times to 1800. Archaeopress Publ, Summertown

Morton GR, Wingrove J (1969a) Slag, cinder, and bear. J Hist Metall Soc 3(2):55–61

Morton H, Wingrove J (1969b) Constitution of Bloomery slags. Part I. Roman J Iron Steel Inst 207:1556–1564

Morton H, Wingrove J (1972) Constitution of Bloomery slags. Part II. Medieval J Iron Steel Inst 210:478–488

Muhly JD (1973) Copper and tin: the distribution of mineral resources and the nature of the metals trade in the bronze age. The Connecticut Academy of Arts and Sciences, New Haven

Muhly JD (1980) Bronze figurines and near eastern metalwork. IEJ 30:148–161

Muhly JD (1988) The beginnings of metallurgy in the old world. In: Maddin R(ed) The beginnings of the use and metals and alloys. Cambridge, MIT press, pp 2–20

Muhly JD (1991) The development of copper metallurgy in late bronze age Cyprus. In: Gale NH (ed) Bronze age trade in the Mediterranean, J Mediterran Archaeol, vol 90, pp 180–196

Muhly JD (1995) Lead isotope analysis and the archaeologist. J Mediterran Archaeol 8(1):54–58

Muhly JD (1996) The significance of metals in the late bronze age economy of Cyprus. In: Karageorghis V, Michaelides D (eds) The development of the Cypriot economy from the prehistoric period to the present day, Nicosia

Muhly JD (2006) Chrysokamino in the history of early metallurgy. In: Betancourt P (ed) The Chrysokamino metallurgical workshop and its territory, Hesperia suppl, vol 36, pp 155–177

Muhly JD (2011) Archaeometry and shipwrecks. A review article. Expedition 53(1):36–44

Muhly JD, Pernicka E (1992) Early Trojan metallurgy and metals trade. In: Herrmann J (ed) Heinrich Schliemann: Grundlagen und Ergebnisse moderner Archäologie 100 Jahre nach Schliemanns Tod. Akademie, Berlin, pp 309–318

Muhly JD, Maddin R, Stech T, Özgen E (1985) Iron in Anatolia and the nature of the Hittite Iron industry. Anatol Stud 35:67–84

Muhly JD, Begemann F, Öztunali Ö, Pernicka E, Schmitt-Strecker S, Wagner GA (1991) The bronze metallurgy of Anatolia and the question of local tin sources. In: Pernicka E, Wagner GA (eds) Archaeometry 1990. Birkhäuser, Basel, pp 209–220

Mukhopadhyay DK, Lindsley DH (1983) Reconnaissance study on the extent of solid solution in Ca-mg-Fe olivines. Geol Soc Am Abstr Progr 14:572

Muan A (1955) Phase Equilibria in the system FeO-Fe_2O_3-SiO_2. J Miner Metals Mater Soc 7(9):965–976

Muan A (1958) Phase Equilibria at high temperatures in oxide systems involving changes in oxidation states. Am J Sci 256:171–207

Müller R, Rehren T, Rovira S (2004) Almizaraque and the early copper metallurgy of southeastern Spain: new data. Madrider Mitt 45:33–56

Müller R, Goldenberg G, Bartelheim M, Kunst M, Pernicka E (2007) Zambujal and the beginnings of metallurgy in southern Portugal. In: La Niece S, Hook D, Craddock PT (eds) Metals and mines. Archetype Publications in ass. with Brit Mus, pp 15–26

Müller R, Brey GP, Seitz HM, Klein S (2014) Lead isotope analyses on late republican sling bullets. Archaeol Anthropol Sci 7:473–485

Müller-Karpe M (1989) Neue Forschungen zur frühen Metallverarbeitung in Mesopotamien. Jahrb RGZM 36(1):179–192

Müller-Karpe A (1994) Altanatolisches Metallhandwerk. Offa-Bücher 75. Wachholtz Verlag, Neumünster

Müller-Karpe M (1990a) Metallgefäße des dritten Jahrtausends in Mesopotamien. Archaeol Korr Blatt 20:161–176

Müller-Karpe M (1990b) Der Guss in der verlorenen Sandform in Mesopotamien. Mitt Deutsche Orient-Gesellsch 122:173–192

Müller-Karpe M (1993) Metallgefäße im Iraq I. Prähistorische Bronzefunde II/14. Steiner-Verlag, Stuttgart

Murillo-Barroso M, Montero-Ruiz I, Bartelheim M (2014) Native silver resources in Iberia. In: Meller H, Risch R, Pernicka E (eds), Metalle der Macht – Frühes gold und Silber. Tagungen des Landesmus Vorgesch Halle 11/1: 257–267

Murillo-Barroso M, Montero-Ruiz I, Rafel N, Hunt Ortiz MA, Armada XL (2016) The macro-regional scale of silver production in Iberia during the first millennium BC in the context of Mediterranean contacts. Oxf J Archaeol 35 (1):75–100. https://doi.org/10.1111/ojoa.12079

Mussche Herman F (2006) More about the silver-rich Lead of ancient Laurion. L'antiquité classique 75:225–230

Naldrett AJ (1969) A portion oft he system Fe-S-O between 900 and 1080°C and ist application to sulfide ore magmas. J Petrol 10(2):171–201

Naldrett AJ (1989) Sulfide melts – crystallization temperatures, Solubilities in silicate melts and Fe, Ni and cu partitioning. Ore deposition associated with magmas. Rev Econ Geol 4:5–31

Negueruela Martínez I (2014) The Phoenician ships of Mazarrón. In: Aruz J, Rakic Y, Graff S (eds) Assyria to Iberia: at the dawn of the classical age. New York, pp 243–246

Negueruela IR et al (2004) Mazzarón-2: el barco fenicio del siglo VII a.C. Campañade noviembre-1999/marzo 2000. Estud Orient 5-6:453–484

Nesel B, Brügmann G, Pernicka E (2015) Tin isotopes and the sources of tin in the early bronze age Únětice culture. In: Mata-Perelló JM, Hunt Oriz M, Orche García E (eds) Patrimonio Geológico y Minero: De la Investigacion a la Difusón. Actas del XV Congreso Internacional sobre Patrimonio Geológico y Minero, Logosán, pp 1–19

Neukirchen F, Ries G (2016) Die welt der Rohstoffe. Lagerstätten, Förderung und wirtschaftliche Aspekte. Springer, Berlin

Neuninger H, Pittioni R, Siegel W (1964) Frühkeramikzeitliche Kupfergewinnung in Anatolien. Arch Austriaca XXXV:98–110

Neuninger H, Preuschen E, Pittioni R (1970) Versuch einer Fahlerzverhüttung nach dem alten Arbeitsgang auf der Insel Luzon. Archaeol Aust 47:87–89

Newbury BD, Notis M, Newbury DE (2005) Revisiting the zinc comosition limit of cementation brass. J Hist Metall Soc 29(2):75–81

Nezafati N (2006) au-Sn-W-cu-mineralization in the Astaneh-Sarband area, west Central Iran. PhD-dissertation, Geowissenschaftl. Fakultät, Universität Tübingen

Nezafati N, Pernicka E (2012) Early silver production in Iran. Iranian arch 3, 37: 46

Nezafati N, Hessari (2017) Tappeh Shoghali; a significant early silver production site in north Central Iran. Periodico Mineralogia 86:67–73

Nezafati N, Pernicka E (2006) The smelters of Sialk. Outcomes oft he first stage of archaeometallurgical research at Tappeh Sialk. In: MalekShahrmirzadi S (ed) The fisherman of Sialk. Sialk reconsideration Projec, report 4, Arch. Report Monogr. Ser., Tehran, pp 79–102

Nezafati N, Pernicka E, Momenzadeh M (2006) Ancient tin: old question and a new answer. Antiquity 80:308–311

Nezafati N, Pernicka E, Momenzadeh M (2008) Iranian ore deposits and their role in the development of the ancient cultures. In: Yalçın Ü (ed) Anatolian metal IV, Der Anschnitt Beih, vol 21, pp 77–90

Nezafati N, Pernicka E, Momenzadeh M (2009) Introduction of the Deh Hosein ancient tin-copper mine, Western Iran: evidence from geology, archaeology, geochemistry and Lead isotope data. Tüb-ar 12:223–236

Nicholson PT, Peltenburg E (2000) Egyptian faience. In: Nicholson PT, Shaw I (eds) Ancient Egyptian materials and technology. Cambridge University Press, pp 177–194

Nicholson PT, Shaw I (2000) Ancient Egyptian materials and technology. Cambridge University Press

Nickel D, Haustein M, Lampke T, Pernicka E (2012) Identification of forgeries by measuring tin isotopes in corroded bronze objects. Archaeometry 54:167–174

Nicodemi W (ed) (2004) The civilisation of Iron. From prehistory to the third Millennium. Olivares, Milano

Niederschlag E, Pernicka E, Seifert T, Bartelheim M (2003) The determination of Lead isotope ratios by multiple collector ICP-MS: a case study of early bronze AgeArtefacts and their possible relation with ore deposits of the Erzgebirge. Archaeometry 45 (1):61–100

Nielen HD (2006) Zink oder messing? Ein Beitrag zu den metallurgischen Tätigkeiten im Legionslager Neuss. Meta 13:1

Nieweglowski A (1995) Produktion und Herkunft römischer Gefäße aus bronze und messing vor dem Hintergrund metallkundlicher Analysen. In: Mols STA (ed) Acta of the 12th international congress on ancient bronzes in Nijmegen 1992, Nederlandse Archeologische Rapporten, vol 18, pp 203–206

Nixon S, Rehren T, Guerra MF (2011) New light on the early Islamic west African gold trade: coin moulds from Tadmekka, Mali. Antiquity 85:1353–1368

Nocete F et al (2008) The smelting quarter of Valencia de la Concepción (Seville, Spain): the specialised copper industry in a political Centre of the Guadalquivir Valley during the third millennium BC (2750-2500 BC). J Archaeol Sci 35:717–732

Nocete F et al (2011) Direct chronometry (14C AMS) of the earliest copper metallurgy in the Guadalquivir Basin (Spain) during the third millennium BC: first regional database. J Archaeol Sci 38:3278–3295

Nocete F, Sáez R, Bayond MR, Nieto JM, Peramo A, López P, Gil-Ibarguchi JI (2014) Gold in the southwest of the Iberian Peninsula during the 3rd millennium BC. J Archaeol Sci 41:691–704

Noll W, Hangst K (1975) Zur Kenntnis der altägyptischen Farbpigmente. N Jb Miner Mh H5:209–214

Noll W, Heimann RB (2016) Ancient Old World pottery. Materials, technology, and decoration. Schweizerbart Sc Publ

Nomicos S (2020) Laurion. Montan- und siedlungsarchäologische Studien von der geometrischen bis in frühbyzantinische Zeit Der Anschnitt Beih 43

Northover P (1989) Properties and use of arsenic-copper alloys. In: Hauptmann A, Pernicka E, Wagner GA (eds) Old World Archaeometallurgy. Proc Internat Sympos, Heidelberg 1987, Der Anschnitt Beih, vol 7, pp 111–118

Notton JH (1974) Ancient Egyptian gold refining. Gold Bull 7(2):50–56

Nriagu JO (1983) Lead and lead poisoning in the antiquity, New York

Obenauer K (1954) Mikroskopische Untersuchungstechnik der Eisenhüttenschlacken. In: Freund H (ed) Handbuch der Mikroskopie in der Technik 2: Mikroskopie der Bodenschätze, Umschau Frankfurt/Main, pp 459–518

Oberthür R, Saager T (1986) Silver and mercury in gold. Particles from the Proterozoic Witwatersrand placer deposits of South Africa: Metallogenetic and geochemical implications. Econ Geol 81:20–31

O'Brien W (1999) Arsenical copper in Irish metallurgy. In: Young SMM, Pollard AM, Budd P, Ixer R (eds) Metals in antiquity, BAR Internat Ser, vol 792, pp 33–42

O'Brien W (2013) Bronze age copper mining in Europe. In: Fokkens H, Harding A (eds) Oxford handbook of the European bronze age

Oddy A (1981) Gilding through the ages. Gold Bull 14 (2):75–79

Oddy A (1983) Assaying in antiquity. Gold Bull 16 (2):52–59

Oeggl K, Prast M (2009) Die geschichte des Bergbaus in Tirol und seinen angrenzenden Gebieten. Proc. Zum 3. Milestone meeting des SFB-HIMAT vom 23–26.10.2008 in Silbertal, Innsbruck University Press

Oelsen W, Schürmann E (1954) Untersuchungsergebnisse alter Rennfeuerschlacken. Archiv Eisenhüttenwesen 25:507–514

Oettel H, Schumann H (2011) Metallografie. Wiley, Weinheim

Ogden JM (1976) The so-called 'platinum' inclusions in Egyptian goldworks. J Egypt Archaeol 62:138–144

Ogden JM (1977) Platinum group metal inclusions in ancient gold artifacts. J Hist Met Soc 2:53–72

Ogden JM (1993) Aesthetic and technical considerations regarding the colour and texture of ancient goldwork. In: La Niece S, Craddock P (eds) Metal plating and patination. Butterworth/Heinemann, London, pp 39–49

Ogden JM (2000) Metals. In: Nicholson PT, Shaw I (eds) . Cambridge Univ Press, Ancient Egyptian materials and technology, pp 148–176

Okada Y (1911) Der Mabuki-Prozeß: die japanische Gewinnungsmethode des metallisvchen Kupfers aus Kupferstein. PhD-diss. Bergakadem Freiberg

Okamoto H, Massalski TB (1985) The au–Pt (gold-platinum) system. Bull Alloy Phase Diagrams 6 (1):46–56

Okamoto H, Chakrabarti DJ, Laughlin DE, Massalski TB (1987) The au-cu (gold-copper) system. Bull Alloy Phase Diagrams 8(5):454–474

Okrusch M, Matthes S (2005) Mineralogie. Eine Einführung in die spezielle Mineralogie. Springer, Petrologie und Lagerstättenkunde

O'Neill HSC, Wall HJ (1987) The olivine-Orthopyroxene-spinel oxygen Geobarometer, the Nickel precipitation curve, and the oxygen fugacity of the Earth's upper mantle. J Petrol 28(6):1169–1191

Oren E (1989) Early Bronze Age settlements in north Sinai: a model for Egypto-Canaanite interconnections. In: de Miroschedji P (ed) L'urbanisation de la Palestine á l'age du Bronze ancien Bilan at perspectives des recherches actuelles, Acte du Colloque d'Emmaus 1986. BAR Internat. Ser 527

Orzechowski S (1994) The settlement background and development conditions of the holy Cross Mountains iron production. Archaeol Rozhledy 46:328–329

Osann B (1959) Eisenhüttenmännische Aussagen der Rennofen-, Rennstahl- und Rennschlackenfunde von Salzgitter-Lobmachtersen. Stahl u Eisen 78:1201–1211

Osann B (1971) Rennverfahren und Anfänge der Roheisenerzeugung: zur Metallurgie und Wärmetechnik der alten Eisengewinnung. Fachausschussber. Düsseldorf, Verein Deutsch Eisenhüttenl

Osborn EF, Muan A (1964) In: phase equilibrium diagrams of oxide systems. Amer ceramic Soc and Orton E, jr., ceramic foundation, Columbus, Ohio; cf. also Levin/Robbins/McMurdie, phase diagrams for ceramists. Amer. Soc., Columbus, Ohio (1964), fig. 696 and 82

Osborn EF, De Vries RC, Gee KH, Kraner HM (1954) Optimum composition of blast furnace slag as deduced from Liquidus data for the quaternary system CaO-MgO-Al2O3-SiO2. Trans AIME v 200:33–45

Ostwald D, Lieber W (1957) Blei- und Zinkminerale von Nußloch nahe Wiesloch. Der Aufschluss 8:156–157

Ottaway B (1989) Interaction of some of the earliest copper using cultures in Europe. In: Hauptmann A, Pernicka E, Wagner GA (eds) Old world archaeometallurgy, Der Anschnitt Beih, vol 7, pp 19–24

Ottaway BS (1994) Prähistorische Archäometallurgie. M Leidorf, Espelkamp

Ottaway BS (1999a) The settlement as an early smelting place for copper. In: 4th intern. Conf. On the beginnings of the use of metals and alloys (BUMA-IV). Shimane, Japan, pp 165–172

Ottaway BS (1999b) Innovation, production and specialization in early prehistoric copper metallurgy. Eur J Archaeol 4(1):87–112

Ottaway BS (2003) Experimental archaeometallurgy. In: Stöllner T, Körlin G, Steffens G, Cierny J (eds) Man and mining. Studies in honour of G Weisgerber on occasion of his 65th birthday, Der Anschnitt Beih, vol 16, pp 341–348

Ottaway BS, Quanyu W (2004) Casting experiments and microstructure of archaeologically relevant bronzes, BAR Internat Ser 1331, Oxford

Otto H, Witter W (1952) Handbuch der ältesten vorgeschichtlichen Metallurgie in Mitteleuropa. Barth, Leipzig

Özbal H, Adriaens A, Earl B (1999) Hacınebi metal production and exchange. Paléorient 25(1):57–65

Özdoğan M, Özdoğan A (1999) Archaeological evidence on the early metallurgy at Çayönü Tepesi. In: Hauptmann A, Pernicka E, Rehren T, Yalçın Ü (eds) The beginning of metallurgy. Proc. Int. Conf. Bochum 1995, Der Anschnitt, Beih, vol 9, pp 13–22

Öztunali Ö (1989) Lagerstättenkundliche Probleme der Archäometallurgie. Der Anschnitt, Beih, vol 7, pp 293–297

Paar WH (1993) Geologische Einführung und Mineralogie der Kupfererzlagerstätten im Raum Mühlbach am Hochkönig-Bischofshofen. In: Günther W, Eibner C, Lippert A, Paar W (eds) 5000 Jahre Kupferbergbau Mühlbach am Hochkönig - Bischofshofen. Gemeinde Mühlbach am Hochkönig

Palmieri A, Hauptmann A (1995) Archaeometallurgy at Arslantepe (Malatya, Turkey). A Case study for Provenance Studies and Technology of Early Metal Production. In: Proc 1st Int Congr Science and Technology for the Safeguard of Cultural Heritage in the Mediterranean Basin 1, Catania, pp 591–595

Palmieri A, Hauptmann A (2000) Metals from Ebla: chemical analyes of metal artefacts from the bronze and iron ages. In: Matthiae P, Peyronel L, Pinnock F (eds) First Internat Congr Archaeology of the Ancient Near East. Proc Rom, Rome, pp 1259–1272

Palmieri A, Sertok K, Chernykh E (1993) From Arslantepe metalwork to arsenical copper Technology in Eastern Anatolia. In: Frangipane M, Hauptmann H, Liverani M, Matthiae P, Mellink M (eds) Between the Rivers and over the Mountains. Archaeologica Anatolica et Mesopotamica Alba Palmieri Dedicata, Roma, pp 573–599

Palmieri A, Hauptmann A, Sertok K, Hess K (1996a) Minerali e Scorie Rinvenuti nei Livelli Chalcoliti e del Bronzo Antico di Arslantepe (Turchia), Actes Colloque Périgeaux 1995, Suppl. à la rev. d'Archéometrie, pp 141–145

Palmieri A, Hauptmann A, Sertok K, Hess K (1996b) Archaeometallurgical survey in 1994 at Malatya-Arslantepe and its surroundings. XI. Arkeometri Sonuçlari Toplantasi, Ankara, pp 105–115

Palmieri A, Hauptmann A, Hess K (1998) Les Objets en Métal du Tombeau Mounumetal d'Arslantepe de 3000 AV. J.-C. (Malatya, Turquie). Revue d'Archéometrie 22:39–43

Palmieri A, Frangipane M, Hauptmann A, Hess K (1999) Early metallurgy at Arslantepe during the late chalcolithic and the early bronze age IA – IB periods. In: Hauptmann A, Pernicka E, Rehren T, Yalçın Ü (eds) The beginnings of metallurgy, Der Anschnitt Beih, vol 9, pp 141–148

Palumbo G (2008) The early bronze age IV. In: Adams RB (ed) Jordan. An archaeological reader. Equinox, London, pp 227–262

Pampaloni A (2017) The tin of Campigliese: 40 centuries of usage. Rel: 1–25

Panayiotou A (1980) Cu-Ni-Co-Fe Sulphide mineralization, Limassol forest, Cyprus. In: Panayiotou A (ed) Ophiolites. Proc Intern Ophiolite Symp Cyprus 1979. Nicosia, pp 102–116

Paoletti P (2016) "Raffiniertes" gold? In: Gebhard R, Krause R (eds) Bernstorf. Archäologische und naturwissenschaftliche Analysen der gold- und Bernsteinfunde vom Bernstorfer Berg bei Kranzberg, Oberbayern. Abh. u. Bestandskataloge Archäol. Staatssamml, vol 33, pp 177–185

Papadimitriou GD (2008) The technological evolution of copper alloys in the aegean during the prehistoric period. In: Tzachili I (ed), Aegean Metallurgy in the Bronze Age Proc. Internat. Symp. Univers of Crete, 2004. Ta Pragmata Publ. Athens: 271–287

Papadimitriou GD (2012) Litharge: waste or useful raw material? An intriguing material revisited. In: Zacharias N, Georgakopoulou M, Polykreti N, Fakorellis G, Vakoulis T (eds) Proc 5th Symp Hellenic Soc Archaeometry, Athens 2008, Univers of Peloponnesus, human ser. Papazissis, Athens, pp 799–819

Papadimitriou GD (2016) The so-called helicoidal ore washeries of Laurion: their actual function as circular mills in the process of beneficiation of silver and lead contained in old litharge stocks. In: Proc. 6th Symp Hellenic Soc Archaeometry, Athens 2013, BAR Internat Ser 2780, pp 113–118

Papadimitriou GD (2017) Ore washeries and water cisterns in the mines of Laurion-Attica. In: Proc. 16th Internat. Conf. Hist. Water management and hydraulic engineering Mediterr. Region, Athens 2015. Schrift Deutschen Wasserhist Gesellsch 27(2): 395–416

Paparoni G, Webster JD, Walker D (2010) Experimental techniques for determining tin solubility in silicate melts using silica capsules in 1 atm furnaces and rhenium capsules in the piston cylinder. Am Mineral 95 (5–6):776–783

Paracelsus (Theophrast von Hohenheim) (1922–1933). Sämtliche Werke I. In: Sudhoff K (ed) Medizinische, naturwissenschaftliche und philosophische Schriften. Druck & Verlag R Oldenbourg, München-Berlin

Paradis S, Hannigan P, Dewing K (2007) Mississippi Valley-type lead zinc deposits. In: Goodfellow WD (ed) Mineral deposits of Canada. A synthesis of major deposit-types, district Metallogeny, the evolution of geological provinces and exploration methods, Geol Ass Canada, Mineral Deposits Div, Special Publ 5, pp 185–203

Pare C (2000) Bronze and the bronze age. In: Pare CFE (ed) Metals make the world go round. Oxbow Books, Oxford, pp 1–38

Park JS, Shinde V (2013) Bronze technology of the ancient megalithic communities in the Vidarbha. Region of India. J Archaeol Sci 40:3811–3821

Parker AJ (1992) Ancient shipwrecks of the Mediterranean & the Roman provinces, Brit Archaeolog rep Internat Ser 580, Oxford

Parzinger H (2016) Abenteuer Archäologie. Eine Reise durch die Menschheitsgeschichte. Beck Publ

Parzinger H, Boroffka N (2003) Das Zinn der Bronzezeit in Mittelasien I. die siedlungsarchäologischen Forschungen im Umfeld der Zinnlagerstätten. Archäol Iran u Turan 5. Philipp v. Zabern, Mainz

Patterson CC (1971) Native copper, silver, and gold accessible to early metallurgists. Am Antiq 36:286–321

Patterson CC (1972) Silver stocks and losses in ancient and medieval times. Econ Hist Rev 25:205–233

Pearce M (2015) The spread of early copper mining and metallurgy in Europe: an assessment of the diffusionist model. In: Hauptmann A, Modarressi-Tehrani D (eds) Archaeometallurgy in Europe III, pp 45–54

Pearce M (2016) Archaeology and archaeometallurgy: some unresolved areas in the interpretation of analytical data, STAR: Science & Technology of Archaeol Res, pp 46–53

Pearce M (2018) The curse of the pXRF: the negative Conseuqences of the popularity of handheld XRF analysis of copper-based metal Artefacts. Meta 24 (2):81–85

Peissel M (1984) The ants gold. The Discovery of the Greek El Dorado in the Himalayas. Collins, ISBN 978-0-00-272514-9

Penhallurick RD (1986) Tin in antiquity. The Institute of Metals, London

Percy J (1861) Metallurgy: fuel, fire clays, copper, zinc, Brass. J Murray, London

Percy J (1864) Metallurgy: iron; steel. J Murray, London

Percy J (1870) Metallurgy: lead, including silver from lead. J Murray, London

Perea A, Armbruster B (2011) Tomb 100 at Cabezo Lucero: new light on goldworking in fourth-century BC Iberia. Antiquity 85:158–171

Perea A, Garcia Vuelta O, Freire F (2010) El Proyecto au. Estudio arqueométrico de la producción de oro en la peninsula Ibérica, Bibl Prehist Hispana 27, Madrid

Pereira F (2015) Effects of long-term aging in arsenical copper alloys. Microsc Microanal:1–7. https://doi.org/10.1017/S1431927615015263

Pérez M, Romero A, Hernández A, Almaguer I, Benavides R (2012) Distribution of lead and silver under lead blast furnace conditions. Rev Metal 48(3):213–222

Pérez-Garcia LC, Sanchez-Palencia FJ, Torres-Ruiz J (2000) Tertiary and quaternary alluvial gold deposits of Northwest Spain and Roman mining (NW of Duero and Bierzo basins). J Geochem Explor 71:225–240

Pernicka E (1981) Archäometallurgische Untersuchungen zur antiken Silbergewinnung in Laurion. I. Chemische Analysen griechischer Blei-Silbererze Erzmetall 34 (7/8):396–400

Pernicka E (1987) Erzlagerstätten in der Ägäis und ihre Ausbeutung im Altertum. Jb Röm-German Zentralmus 34(2):607–714

Pernicka E (1990) Gewinnung und Verbreitung der Metalle in prähistorischer Zeit. Jb Röm-German Zentralmus 37(1):21–129

Pernicka E (1993) Analytisch-chemische Untersuchungen an Metallfunden von Uruk-Warka und Kis. In: Müller-Karpe M (ed) Metallgefäße im Iraq, Prähistorische Bronzefunde II, vol 14, pp 312–316

Pernicka E (1998) Die Ausbreitung der Zinnbronze im 3. Jahrtausend. In: Hänsel B (ed) Mensch und Umwelt in der Bronzezeit Europas. Oetker-Voges Verlag, Kiel, pp 135–147

Pernicka E (1999) Trace element finger printing of ancient copper: a guide to technology or provenance? In: Young SMM, Pollard MA, Budd P, Ixer RA (eds) Metals in antiquity, BAR Internat Ser, vol 792, pp 163–171

Pernicka E (2004) Kupfer und Silber in Arisman und Tappeh Sialk und die frühe Metallurgie in Iran. In: Stöllner T, Slotta R, Vatandoust A (eds) Persiens Antike Pracht. Ausstellungskatalog Deutsches Bergbau-Museum Bochum, pp 232–239

Pernicka E (2010) Archäometallurgische Untersuchungen am und zum Hortfund von Nebra. In: Meller H, Bertemes F (eds) Der Griff nach den Sternen. Wie Europas Eliten zu Macht und Reichtum Kamen. Tagungen des Landesmuseums für Vorgeschichte Halle (Saale) 5/II, pp 719–734

Pernicka E (2013) Die Ausbreitung der Metallurgie in der Alten welt. In: Bronzezeit (ed) Europa ohne Grenzen. 4.-1. Jahrtausend v. Chr. Ausstellungskatalog St. Petersburg, Moskau, Berlin. St. Petersburg publishing house, pp 66–78

Pernicka E (2014a) Zur Frage der Echtheit der Bernstorfer Goldfunde. In: Meller H, Risch R, Pernicka E (eds) Metalle der Macht: Frühes gold und Silber. Metals of power: early gold and silver, Tagungen des Landesmus Vorgesch Halle I, pp 247–256

Pernicka E (2014b) Possibilities and limitations of provenance studies of ancient silver and gold. In: Meller H, Risch R, Pernicka E (eds) Metalle der Macht: Frühes gold und Silber. Metals of power: early gold and silver, Tagungen Landesmus Vorgesch Halle I, pp 153–164

Pernicka E (2014c) Provenance determination of archaeological metal objects. In: Roberts BW, Thornton CP (eds) . Springer, Archaeometallurgy in global perspective, Methods and syntheses, pp 239–268

Pernicka E (2016) Echt oder falsch? Ein Zwischenstand zum Disput um die Funde von Bernstorf. Archäologie in Deutschland 3:62–65

Pernicka E (2018) Science versus archaeology? The case of the Bernstorf fakes. Meta 24(2):73–80

Pernicka E, Bachmann HG (1983) Archäometallurgische Untersuchungen zur antiken Silbergewinnung in Laurion III. Das Verhalten einiger Spurenelemente beim Abtreiben des Bleis Erzmetall 36:592–597

Pernicka E, Hauptmann A (1989) Chemische und mineralogische Analyse einiger Erz- und Kupferfunde von Maadi. In: Rizkana J, Seeher J (eds) Maadi III. The non-lithic smallfinds and the structural remains of the predynastic settlement. Mainz, pp 137–140

Pernicka E, Wagner GA (1985) Alte Goldgruben auf Sifnos. In: Silber, Blei und gold auf Sifnos. Prähistorische und antike Metallproduktion, Der Anschnitt Beih, vol 3, pp 174–185

Pernicka E, Wunderlich CH (2002) Naturwissenschaftliche Untersuchungen an den Funden von Nebra. Archäologie in Sachsen-Anhalt 1:24–30

Pernicka E, Seeliger T, Wagner GA, Begemann F, Schmitt-Strecker S, Eibner C, Öztunali Ö, Baranyi I (1984) Archäometallurgische Untersuchungen in Nordwestanatolien. Jb Röm-German Zentralmus 31 (2):533–599

Pernicka E, Lutz C, Bachmann HG, Wagner GA, Elizsch C, Klein E et al (1985) Alte Blei-Silber-Verhüttung auf Sifnos. In: Wagner GA, Weisgerber G (eds) Silber, Blei und gold aus Sifnos. Prähistorische und antike Metallproduktion, Der Anschnitt Beih, vol 3, pp 185–199

Pernicka E, Begemann F, Schmitt-Strecker S, Grimanis AP (1990) On the composition and provenance of metal artefacts from Poliochni on Lemnos. Oxford J of Arch 9(3): 263–297

Pernicka E, Begemann F, Schmitt-Strecker S, Wagner GA (1993) Eneolithic and early bronze age copper artefacts from the Balkans and their relation to Serbian copper ores. Prähist Zeitschr 68:1–54

Pernicka E, Begemann F, Schmitt-Strecker S, Todorova H, Kuleff I (1997) Prehistoric copper in Bulgaria. Eurasia Antiqua 3:41–180

Pernicka E, Rehren T, Schmitt-Strecker S (1998) Late Uruk silver production by cupellation at Habuba Kabira. In: Rehren T, Hauptmann A, Muhly JD (eds) Metallurgica Antiqua, In honour of HG Bachmann and R Maddin. Der Anschnitt Beih, vol 8, pp 123–134

Pernicka E, Schmidt K, Schmitt-Strecker S (2002) Metallhandwerk. In: Schmidt K (ed) Norşuntepe. Kleinfunde II, Archaeologica Euphratica, vol 2. Zabern, Mainz, pp 115–137

Pernicka E, Eibner C, Öztunali Ö, Wagner GA (2003) Early bronze age metallurgy in the north-East Aegean. In: Wagner GA, Pernicka E, Uermann HP (eds) Troia and the Troad. Scientific approaches. Springer, Berlin, pp 143–172

Pernicka E et al (2011) Archaeometallurgical researches at Arisman in Central Iran. In: Vatandoust A, Parzinger H, Helwing B (eds) Early mining and metallurgy on the central Iranian plateau, Report on the first five years of research of the joint Iranian-German research project. Archäol Iran Turan, vol 9, pp 633–687

Pernicka E, Lutz J, Stöllner T (2016) Bronze age copper produced at Mitterberg, Austria, and its distribution. Archaeol Aust 100:19–55

Perrot J (1989) Réflexion Sur l'état des recherches concernant la Préhistoire récente du Proche et du Moyen-orient. Paléorient 26(1):5–28

Perry MA, Coleman DS, Dettman DL, Al-Shiyab AH (2009) An isotopic perspective on the transport of byzantine mining camp labors into southwestern Jordan. Am J Phys Anthropol 140:429–441

Perry MA, Coleman DS, Dettman DL, Grattan JP, Al-Shiyab AH (2011) Condemned to metallum? The origin and role of 4th – 6th century a.D. Phaeno mining campresidents using multiple chemical techniques. J Archaeol Sci 38:558–569

Peters T (2000) Formation and evolution of the Western Indian Ocean as evidenced by the Masirah Ophiolite: a review. Geol Soc Am Spec Pap 349:525–536

Petrascheck W, Pohl W (1982) Lagerstättenlehre. Schweizerbarth, Stuttgart

Petrikovits HV (1958) Bergbau und Hüttenwesen in der Rheinzone. Zeitschr Erzbergbau Metallhüttenwesen 11:594–600

Petrikovits Hv (1981) Die Spezialisierung des römischen Handwerks II. Habelt Publ

Petrović VP (2007) Pre-Roman and Roman dardania: historical and geographical considerations. Balcanica XXXVII:7–24

Pfaff K, Hildebrandt LH, Leach DL, Jacob DE, Markl G (2010) Formation of the Mississippi-Valley-type Zn-Pb-Ag deposit in the extensional setting of the upper Rhinegraben in the Wiesloch area, SW Germany. Mineral Deposits 45(7):647–666. https://doi.org/10.1007/s00126-010-0296-5

Pfaffenberger B (1992) Social anthropology and technology. Ann Rev Anthropol 21:491–516

Pfeiffer K (2009) The technical ceramic for metallurgical activities in tall Hujayrat al-Ghuzlan and comparable sites in the southern Levant. In: Khalil L, Schmidt K (eds) Prehistoric Aqaba. Orient Archäologie, vol 23, pp 305–338

Pfeiffer K (2013) Neue Untersuchungen zur Archäometallurgie des Sinai. Die Entwicklungsgeschichte der innovation "Kupfermetallurgie". Menschen-Kulturen-Traditionen. ForschungsCluster 2. Leiden, Rahden

Pfister U (2010) Silber. In: Enzyklopädie der Neuzeit 12: Silber – Subsidien, Stuttgart, pp 1–8

Photos E (1989) The question of meteoritic iron versus smelted nickel-rich iron: archaeological evidence and experimental results. World Archaeol:402–421

Photos E, Koukouli-Chrysanthaki C, Tylecote RF, Gialoglu G (1989) Precious metal extraction in Palaia Kavala, N.E. Greece. An Archaeometallurgical attempt to locate Skapte Hyle. In: Hauptmann A, Pernicka E, Wagner GA (eds) Old world archaeometallurgy, Der Anschnitt, Beih, vol 7, pp 179–190

Piaskowski J (1965) Correlation between the phosphorus content in Iron ore or slag and that in Bloomery. Arch Polonia 8:83–103

Piaskowski J (1982) A study of the origin of the ancient high-Nickel Iron generally regarded as meteoritic. In: Wertime TA, Wertime SF (eds) Smithsonian Instit, Washington DC, pp 237–243

Piaskowski J (1992) Technical studies on high Nickel irons, with special reference to the Indonesian Kris. Archeomaterials 6:35–52

Piccottini G (1994) Gold und Kristall am Magdalensberg. Germania 72(2):467–477

Picon M, Nezet-Celestine L, Desbat A (1995) Un type particulitiare de grands récipients en terre réfractaire utilisés pour la fabrication du laiton par cémentation. In: Rivet L (ed) Productions et importations dans le Nord-Ouest de la Gaule et relations avec la Bretagne romaine. Actualité des recherches céramiques, Soc Française d'Étude Céramique Antique en Gaule, Actes Congrès Rouen May 1995, Marseille, pp 207–215

Pickin J (2010) Early lead smelting sites in southern Scotland. J Hist Metall 44(2):81–84

Piel M, Hauptmann A, Schröder B (1992) Naturwissenschaftliche Untersuchungen an bronzezeitlichen Kupferverhüttungsschlacken von Acqua Fredda/ Trentino. Universitätsforsch prähist Archäol 8:463–472

Pieth M (2019) Goldwäsche. Elster & Salis Zürich

Pigott VC (1996) Near eastern archaeometallurgy: modern research and future directions. In: Cooper JS, Schwartz GM (eds) . The study of the ancient near east in the twenty-first century, Eisenbrauns, pp 139–176

Pigott VC (1999) The development of metal production on the Iranian plateau – an Archaeometallurgical perspective. In: Pigott VC (ed) The archaeometallurgy of the Asian old world, Univers Pennsylv Mus, pp 73–106

Pigott VC (2003) Reconstructing the copper production process as Practised among prehistoric mining /metallurgical communities in the Khao Wong Prachan Valley of Central Thailand. In: Young S, Pollard M, Budd P, Ixer RA (eds) Metals in antiquity, BAR international series, vol 792, pp 10–21

Pigott VC, Lechtman H (2003) Chalcolithic copper-based metallurgy on the Iranian plateau: a new look at old evidence from Tal-i Iblis. In: Potts T, Roaf M, Stein D (eds) Culture through objects: ancient near eastern studies in honour of P.R.S. Moorey. Griffith Institute, Oxford, pp 291–312

Pigott VC, Natapintu S (1988) Archaeological investigations into prehistoric copper production: the Tailand archaeometallurgy project 1984–1986. In: Maddin R (ed) The beginning of the use of alloys. MIT, Cambridge, MA, pp 156–162

Pigott VC, Rogers HC, Nash SK (2003a) Archaeometallurgical investigations at Malyan: Banesh period finds from ABC and TUV. In: Sumner WM (ed) Malyan excavation reports III: early urban life in the land of Anshan: excavations at Tal-e Malyan in the highlands of Iran, Univers Mus rep 117. Univers Pennsylvania Mus Archaeol Anthropol, Philadelphia, pp 94–102

Pigott VC, Rogers HC, Nash SK (2003b) Archaeometallurgical investigations at Malyan: the evidence for tin-bronze in the Kaftari phase. In: Miller NF, Abdi K (eds) Yeki bud, yeki nabud: essays on the archaeology of Iran in honor of WM Sumner, Monogr 48. Cotsen Instit Archaeol. Univers Californi, Los Angeles, pp 161–175

Pike A (2002) Appendix: analysis of Caucasian metalwork – the use of Antimonal, arsenic and tin bronzes in the late bronze age. In: Curtis J, Kruszyskinna M (eds) Ancient Caucasian and related material in the British museum, Brit Mus Occ pap, vol 121, pp 87–92

Pike AWG, Cowell MR, Curtis JE (1996) The use of antimony bronze in the Koban culture. J Hist Metall Soc 30(1):11–16

Pilz R (1917) Beitrag zur Kenntnis der Kupfererzlagerstatten in der Gegend von Arghana Maden. Zeitschr Prakt Geol 25:191–198

Pittioni R (1957) Urzeitlicher Bergbau auf Kupfererz und Spurenanalyse. Beiträge zum Problem der Relation Lagerstätte Fertigobjekt Archäol Austriaca Beih 1

Platteeuw JC, Meyer G (1956) The system tin + oxygen. Trans Faraday Soc 67:1066–1073
Pleiner R (1967) Preliminary evaluation of the 1966 metallurgical investigations in Iran. In: Caldwell JR (ed) Investigations at Tal-Iblis, Illinois: state museum preliminary report, vol 9, pp 340–405
Pleiner R (2000) Iron in archaeology: the European Bloomery smelters. Helvetica & Tempora, Praha
Pleiner R (2003) European iron blooms. In: Nørbach LC (ed) Prehistoric and medieval direct iron smelting in Scandinavia and Europe, aspects of technology and society. Aarhus University Press, pp 183–189
Pleiner R (2006) Iron in archaeology: early European blacksmiths. Helvetica & Tempora, Praha
Pleiner R, Bjorkman JK (1974) The Assyrian iron age: the history of Iron in Assyrian civilization. Proc Am Phils Soc 118(4):283–313
Plenderleith HJ (1934) The preservation of antiquities. The Museums Association, London
Plinius Secundus, Naturalis Historia
Pliny P (1984) Plinius der Ältere über Kupfer und Kupferlegierungen. Schriftenreihe der Georg Agricola Gesellschaft, Düsseldorf
Pliny P (1989) Plinius der Ältere über Blei und Zinn. Werkheft, Attempto Verlag, Tübingen
Pliny P (1993) Gold und Vergoldung bei Plinius dem Älteren, Werkheft 10. Attempto Verlag, Tübingen
Plotinskaya OY, Kovalenker VA, Seltmann R, Stanley CJ (2006) Te and se mineralogy of the high-sulfidation Kochbulak and Kairagach epithermal gold telluride deposits (Kurama ridge, middle Tien Shan, Uzbekistan). Mineral Petrol 87:187–207
Pohl WL (2005) Mineralische und Energie-Rohstoffe. Eine Einführung zur Entstehung und nachhaltigen Nutzung von Lagerstätten. Schweizerbarth, Stuttgart
Pohl W, Belocky R (1994) Alpidic metamorphic fluids and Metallogenesis in the eastern Alps. Mitt Österr Geol Gesellsch 86:141–152
Pollard AM, Thomas RG, Williams PA (1990) Experimental smelting of arsenical copper ores: implications for early bronze age copper production. In: Crew P, Crew S (eds) Earliy mining in the British Isles, Snowdonia, pp 72–74
Ponting MJ (2002) Roman military copper-alloy artefacts from Israel: questions of organization and ethnicity. Archaeometry 44:555–571
Ponting MJ (2010) Scientific analysis and interpretation of the copper-alloy metalwork. In: Geva H (ed) Jewish quarter excavation in the Old City of Jerusalem 4: the burnt house of area B and other studies, Israel Explor Soc and Inst Archaeol. Hebrew Univers Jerusalem, pp 261–275
Popov H, Jockenhövel A (2010) At the northern Borders of the Mycenean world: Thracian gold mining from the late bronze and the early Iron age at Ada Tepe in the eastern Rhodopes. Anodos Stud Ancient World 10:265–281
Potts TF (1993) Patterns of trade in the third millennium BC Mesopotamia and Iran. In: Oates J (ed), Ancient trade: new perspectives world archaeology 24 (3): 379–402
Potts T (1994) Mesopotamia and the east. An archaeological and historical study of foreign relations ca. 3400-2000. Oxford University Committee for Archaeology Monograph 37
Potts DT (1997) Mesopotamian civilization. The material foundations. Cornell University Press
Pouit G, Castel G (1999) Les exploitations pharaoniques, romaines et arabes de cuivre, fer et or. L_exemple des ouadi Dara (désert oriental d'Egypte). In: L'or dans l'Antique, pp 113–144
Pourbaix M (1977) Electrochemical corrosion and reduction. In: Brown BF, Burnett HC, Chase WT, Goodway M, Kruger J, Pourbaix M (eds) Corrosion and metal artefacts. A dialogue between conservators and archaeologists and corrosion scientists, vol 479. NBS spec Publ, pp 1–16
Prag K (1978) Silver in the Levant in the fourth millennium BC. In: Moorey J, Parr P (eds) Archaeology in the Levant: essays for Kathleen Kenyon, Warminster, pp 36–45
Prange M (2001) 5000 Jahre Kupfer in Oman, band II: Vergleichende Untersuchungen zur Charakterisierung des omanischen Kupfers mittels chemischer und isotopischer Analysenmethoden. Meta 8(1/2):1–126
Prange M, Ambert P (2005) Charactérisation géochimique et isotopique des minerais et des métaux base cuivre de Cabrières. Mémoire XXXVII des la Societé préhistorique française
Prange M, Götze HJ, Hauptmann A, Weisgerber G (1999) Is Oman the ancient Magan? Analytical studies of copper from Oman. In: Young SMM, Pollard AM, Budd P, Ixer RA (eds) Metals in antiquity, Brit Archaeol rec Internat Ser, vol 792, pp 187–192
Preßlinger H (2002) Der Bau metallurgischer anlagen in der Spätbronzezeit. Res montanarum 28:5–10
Preßlinger H, Eibner C, Sperl G (1980) Ergebnis der Erforschung urnenfelderzeitlicher Kupfermetallurgie im Paltental. Berg- Hüttenmänn Monatsh 125 (3):131–142
Preßlinger H, Eibner C (1993) Prähistorischer Kupferbergbau und die Verhüttung der Erze. In: Preßlinger H, Köstler HJ (eds) Bergbau und Hüttenwesen im Bezirk Liezen (Steiermark), Kleine Schriften der Abt. Schloß Trautenfels am Steiermärkischen Landesmuseum Joanneum, vol 24, pp 25–36
Prechtl JJ, Karmarsch K (1838) Technologische Encyklopädie oder alphabetisches Handbuch der Technologie, der technischen Chemie und des Maschinenwesens 9: Kupfer – Metallgießerei. Cotta, Stuttgart
Press F, Siever R (2003) Allgemeine Geologie. Spektrum Akadem Verlag, Heidelberg
Preuschen E, Pittioni R (1937) Untersuchungen im Bergbaugebiete Kelchalpe bei Kitzbühel, Tirol. Erster Bericht über die Arbeiten 1931-1936 zuir Urgeschichte des Kupferbergwesens in Tirol. Mitt Prähist Komm 3

Priesner C, Figala K (1998) Alchemie. Lexikon einer hermetischen Wissenschaft (in German). C.H. Beck, München. ISBN 978-3-406-44106-6

Priester M, Hentschel T, Benthin B (1993) Tools for mining. Techniques and processes for small scale mining. Vieweg, Lengerich

Primas M (1995) Gold and silver during the 3rd millennium Cal. BC. In: Morteani G, Northover JP (eds) Prehistoric gold in Europe, NATO ASI series E: applied sciences 280. Kluwer, Dordrecht, pp 77–93

Primas M (2007) Innovationstransfer vor 5000 Jahren: Knotenpunkte an land- und Wasserwegen zwischen Vorderasien und Mitteleuropa. Eurasia Antiqua 13:1–19

Primas M, Pernicka E (2002) Der Depotfund von Oberwilfingen. Germania 76(1):25–65

Prince A (1995) Silver-gold-copper. In: Effenberg G, Aldinger F, Prince A (eds) Ternary alloys, vol 12. VCH, Weinheim, pp 1–21

Projektgruppe Pliny (1998) Silberbergau und –verhüttung in der Antike nach Texten von Plinius, Diodor und Dioskorides. Schriftenreihe der Georg Agricola Gesellschaft 22, Bochum

Pryce W (1778) Mineralogia Cornubiensis: a treatise on minerals, mines and mining. B. Barton, Truro 1972

Pryce TO, Natapintu S (2010) Smelting iron from laterite: technical possibility or ethnographic aberration? Asian Perspect 48(2):249–264

Pryce TO, Bassiakos Y, Catapotis M, Doonan RC (2007) "De Caerimoniae" technological choices in copper smelting furnace design at early bronze age Chrysokamino, Crete. Archaeometry 49(3):543–557

Pryce TO et al (2014) More questions than answers: the southeast Asian Lead isotope project 2009–2012. J Archaeol Sci 42:273–294

Pryce TO et al (2017) High-tin bronze bowls and copper drums: non-ferrous archaeometallurgical evidence for Khao Sek's involvement and role in regional exchange systems. Archaeol Res Asia 13:50–58

Pryor RN, Rhoden HN, Villalón M (1972) Sampling of Cerro Colorado, Rio Tinto, Spain. Trans Mining Metall A 81:143–159

Pulak C (1998) The Uluburun shipwreck: an overview. Int J Naut Archaeol 27(3):188–224

Pulak C (2005) Das Schiffswrack von Uluburun. In: Yalçın Ü, Pulak C, Slotta R (eds) Das Schiff von Uluburun. Ausstellungskatalog Deutsches Bergbau-Museum Bochum, pp 55–103

Punz W (2001) Von den Erzpflanzen zu den Metallophyten. Symp Gesch Erdwissensch in Österreich 2001, Ber Geol Bundesanst 56, Wien/Hallstatt, pp 87–88

Quiring H (1945) Die metallischen Rohstoffe, ihre Lagerungsverhältnisse und ihre wirtschaftliche Bedeutung 7: Antimon. Enke, Stuttgart

Quiring H (1946) Die Metallischen Rohstoffe, ihre Lagerungsverhältnisse und ihre wirtschaftliche Bedeutung 8: Arsen. Enke, Stuttgart

Quiring H (1948) Die geschichte des Goldes. Die goldenen Zeitalter in ihrer kulturellen und wirtschaftlichen Bedeutung. Enke, Stuttgart

Rademakers FW, Rehren T, Pusch EB (2018) Bronze production in Pi-Ramesse: alloying technology and materials use. In: Ben-Yosef E (ed) Mining for ancient copper. Essays in memory of Beno Rothenberg. Tel Aviv Univ Monogr Series, vol 37. Eisenbrauns Publ, pp 503–525

Radivojević M (2013) Archaeometallurgy of the Vinča culture: a case study of the site of Belovode in eastern Serbia. J Hist Metall 47(1):13–32

Radivojević M, Rehren T, Pernicka E, Šljivar D, Brauns M (2010) On the origins of extractive metallurgy: new evidence from Europe. J Archaeol Sci 37:2775–2787

Radivojević M, Rehren T, Kuzmanović-Cvetković J, Marija J, Peter N (2013) Tainted ores and the rise of tin bronzes in Eurasia, c. 6500 years ago. Antiquity 87:1030–1045

Radivojević M et al (2017) Repealing the Çatalhöyük extractive metallurgy: the green, the fire and the 'slag'. J Archaeol Sci 86:101–122. https://doi.org/10.1016/j.jas.2017.07.001

Radivojević M et al (2018) The provenance, use, and circulation of metals in the European bronze age: the state of debate. J Archaeol Res 27(2):131–185. https://doi.org/10.1007/s10814-018-9123-9

Raistick A (1927) Notes on lead mining and smelting in West Yorkshire. Trans Newcomen Soc 7:81–96

Ramage A (2000) The excavations and finds. In: Ramage A, Craddock PT (eds) King Croesus' gold. Excavations at sardis and the history of gold refining. Mus. Press, London, pp 72–98

Ramage A, Craddock PT (2000) King Croesus' gold. Excavations at Sardis and the history of gold refining. Brit Mus Press, London

Ramdohr P (1975) Die Erzmineralien und ihre Verwachsungen. VEB-Verlag, Leipzig

Ramdohr P, Strunz H (1978) Klockmanns Lehrbuch der Mineralogie. Enke Verlag, Stuttgart

Ramis D, Hauptmann A, Coll J (2005) Réduction de cuivre dans la Prehistoire de Majorque. In: Ambert P, Vaquer J (eds) La preière métallurgie en France et dans les pays limitrphes, Actes du Colloque Internat Carcassonne, September 2002, pp 217–224

Rapp G Jr (1977) Trace elements as a guide to the geographical source of tin ore; smelting experiments. In: Franklin AD, Olin JS, Wertime TA (eds) The search for ancient tin. Smiths Inst, Washington DC, pp 59–63

Rapp G (1988) On the origins of copper and bronze alloying. In: Maddin R (ed) The beginnings of the use of metals and alloys, Cambridge, MA, pp 21–27

Rapp GR (2009) Archaeomineralogy. Natural science in archaeology. Springer

Rapp G, Rothe R Jr, Jing Z (1999) Using neutron activation analysis to source ancient tin (Cassiterite). In: Young SMM, Pollard AM, Budd P, Ixer RM (eds) Metals in antiquity, BAR international series 792. Archaeopress, Oxford, pp 153–162

Rapson WS (1990) The metallurgy of the coloured carat gold alloys. Gold Bull 23(4):125–133

Rath vom G (1879) Ueber eine sternförmige Zwillingstafel von gediegen Silber. Ztschr Kristallogr Mineral:385–444

Raub CJ (1986) Was kann der Archäologe von der Metallkunde erwaten? Fundber. Baden-Württemberg 10:343–366

Raub CJ (1993) How to coat objects with gold – Pliny, Leyden papyrus X, Mappae Claviculae and Theophilus seen with a modern chemist's eye. In: Eluère C (ed) Outils et ateliers d'orfèvres des temps ancients, Antique nationales mémoire, vol 2, pp 101–110

Raub C (1995) The metallurgy of gold and silver in prehistoric times. In: Morteani G, Northover JP (eds) Prehistoric Gold in Europe. NATO ASI Ser, vol 280, pp 243–259

Ravich IG, Ryndina NV (1995) Early copper-arsenic alloys and the problem s of their use in the bronze age of the North Caucasus. Bull Metals Mus 23:1–18

Ravitz SF, Fisher KE (1936) Equilibrium in lead smelting. AIME Trans 121:119–126

Read J (2008) The Indus-Mesopotamia relationship reconsidered. In: Olijdam E, Spoor RH (eds) Intercultural relations between south and Southwest Asia, Studies in commemoration of E.C.L. during Caspers. Soc. Arabian stud. Monographs 7, BAR Internat. Series 1826, pp 12–18

Recknagel R (1908) On some mineral deposits in the Rooiberg District. Trans Geol Soc South Africa 11:83–106

Redman CL (1999) Human impact on ancient environments. The University of Arizona Press, Tucson

Rehder JE (1989) Ancient carburization of Iron to steel. Archaeomaterials 3:27–37

Rehder JE (2000) The mastery and uses of fire in antiquity. McGill-Queens University Press, Montreal

Rehren T (1995) Kupfer vom Tor zur welt. In: Kissipenny und Manilla, Ausstellung Duisburg, 19, Duisburger Akzente. Afrika – Wurzeln und Visionen, pp 59–64

Rehren T (1996) A Roman zinc tabula from Bern, Switzerland – reconstructing the manufacture. In: Demirci Ş, Özer AM, Summers GD (eds) Archaeometry 1994. Proc 29th Internat Symp Archaeometry, Ankara 1994, Tübitak, Ankara, pp 35–45

Rehren T (1998) Medieval lead-silver smelting in the Siegerland (West Germany). J Hist Metall 33:73–84

Rehren T (1999a) Antike Fayenceproduktion im Kontext früher Pyrotechnologien. In: Busz R, Gehrke P (eds) Türkis und Azur. Minerva, Kassel, pp 122–125

Rehren T (1999b) Small size, large scale – Roman brass production in Germania inferior. J Archaeol Sci 26:1083–1087

Rehren T (2000) Rationales in Old World Base glass compositions. J Archaeol Sci 27:1225–1234

Rehren T (2001) Ores, crucibles and Cupels: past and present possibilities of scientific analysis. In: Weisshaar HJ, Roth H, Wijeyapala W (eds) Ancient Ruhuna – Sri Lanka, Cahier d'archaeologie du CELAT, vol 10, pp 65–71

Rehren T (2003) Crucibles as reaction vessels in ancient metallurgy. In: Craddock PT, Lang J (eds) Mining and metal production through the ages: 207–215. Brit Mus Press, London

Rehren T (2004) Die Kupfersulfid-Krusten der Pfyner Schmelztiegel (Schweiz). In: Weisgerber G, Goldenberg G (eds) Alpenkupfer – Rame delle Alpi, Der Anschnitt Beih, vol 17, pp 273–280

Rehren T (2009) From mine to microprobe – the Neolithic copper melting crucibles from Switzerland. In: Shortland A, Freestone I, Rehren T (eds) From mine to microscope. Oxobowbooks, Oxford, pp 155–162

Rehren T (2011) Zur frühen Gewinnung von Silber in Südamerika. In: Slotta R, Schnepel I (eds) Schätze der Anden - Chiles Kupfer für die welt. Deutsches Bergbau-Museum, Bochum, pp 257–264

Rehren T, Hauptmann A (1995) Silberraffinations-Schlacken aus der CUT (Xanten), insula 39: Mineralogische Untersuchung und archäometallurgische interpretation. Xantener Ber 6:119–139

Rehren T, Klappauf L (1995) "...ut oleum aquis". Vom Schwimmen des Silbers auf Bleiglätte. Meta 2:19–28

Rehren T, Kraus K (1999) Cupel and crucible: the refining of debased silver in the Colonia Ulpia Traiana, Xanten. J Roman Archaeol 12:263–272

Rehren T, Kraus K (2000) Cupel and crucible: the refining of debased silver in the Colonia Ulpia Traiana, Xanten. J Roman Archaeol 12:263–272

Rehren T, Pusch E (1997) New kingdom glass-melting crucibles from Qantir-Piramesses. J Egypt Archaeol 83:127–141

Rehren T, Hess K, Philip G (1996) Auriferous silver in Western Asia: ore or alloy? J Hist Metall 30(1):1–10

Rehren T, Nixon S (2014) Refining gold with glass – an early Islamic technology at Tadmekka, Mali. J Archaeol Sci 49:33–41

Rehren T, Pernicka E (2014) First data on the nature and origin of the metalwork from tell el-Farkha. In: Maczyńska A (ed) The Nile Delta as a Centre of Cultural Interactions between upper Egypt and the southern Levant in the 4th millennium BC, Poznań, pp 237–252

Rehren T, Pusch EB, Herold A (1998) Glass coloring works within a copper-centered industrial complex in late bronze age Egypt. In: McCray P, Kingery WD (eds) The prehistory and history of glassmaking technology. Ceramics and cilization III, pp 227–250

Rehren T, Schneider J, Bartels C (1999a) Medieval lead-silver smelting in the Siegerland, West Germany. J Hist Metall 33(2):73–84

Rehren T, Vanhove D, Mussche HF, Oikonomakou M (1999b) Litharge from Laurion. A medical and metallurgical commodity from South Attika. L' antiquité classique 68:299–308

Rehren T, Pusch E, Herold A (2001) Qantir-Piramesses and the organisation of the Egyptian glass industry. In: Shortland A (ed) The social context of technological change. Egypt and the near east, 1650–1550 BC. Proc. St. Edmund Hall, Oxford, pp 223–238

Rehren T, Vanhove D, Mussche H (2002) Ores from the ore washeries in the Lavriotiki. Meta 9(1):27–46

Rehren T, Boscher L, Pernicka E (2012) Large scales melting of speiss and arsenical copper at early bronze age Arisman. Iran J Archaeol Sci 39(6):1717–1727

Rehren T et al (2013) 5000 years old Egyptian iron beads made of hammered meteoritic iron. J Archaeol Sci 40:4785–4792

Reich M, Vasconcelos PM (2015) Geologic and economic significance of supergene metal deposits. Elements 11:305–310

Reich M et al (2008) Atacamite formation by deep saline waters in copper deposits from the Atacama Desert, Chile: evidence from fluid inclusions, groundwater geochemistry, TEM, and ^{36}Cl data. Mineral Deposits 46:663–675

Reiff W, Böhm M (1995) Die Eisenerze und ihre Gewinnung im Bereich der östlichen und im Vorland der Schwäbischen Alb. In: Beiträge zur Eisenverhüttung auf der Schwäbischen Alb Forsch und Ber Vor- u Frühgesch Baden-Württemberg, vol 55, pp 15–37

Reisenberger B, Seifert T (2011) Schwarzbuch gold. Deuticke Verlag, Wien

Reiter K (1997) Die Metalle im Alten orient unter besonderer Berücksichtigung altbabylonischer Quellen. Ugarit-Verlag, Münster

Reith F, Lengke MF, Falconer D, Craw D, Southam G (2007) The geomicrobiology of gold. ISME J 1:567–584

Reitmaier-Naef L (2019) Copper smelting slag from the Oberhalbstein (Canton of Grisons, Switzerand). Methodological considerations on typology and morphology. In: Turck R, Stöllner T, Goldenberg G (eds) Alpine copper II – Alpenkupfer II – Rame delle Alpi II – Cuivre des Alpes II, new results and perspectives. Prehistoric copper production. Der Anschnitt Beih, vol 42, pp 229–244

Reitmaier-Naef L (2020) Vom Erz zum Metall. Die chaîne opératoire der prähistorischen Kupfergewinnung im Oberhalbstein Graubünden. Der Anschnitt Beih, in press

Relvas JMRS et al (2006) Hydrothermal alteration and mineralization in the Neves-Corvo volcanic-hosted massive sulfide deposit, Portugal. I. Geology, mineralogy, and geochemistry. Econ Geol 101:753–790

Renfrew C (1969) The autonomy of the south-east European copper age. Proc Prehist Soc 36:12–47

Renfrew C (1992) Kings, tree rings and the Old World. Nature 381:733–734

Renfrew C, Bahn P (2000) Archaeology: theories, methods and practice. Thames and Hudson, London

Renfrew C, Bahn P (2007) Basiswissen Archäologie. Theorien, Methoden, Praxis. Philipp von Zabern

Renock D, Becker U (2011) A first principles study of coupled substitution in galena. Ore Geol Rev 42 (1):71–83

Renzi M, Montero-Ruiz I, Bode M (2009) Non-ferrous metallurgy from the Phoenician site of La Fonteta (Alicante, Spain): a study of provenance. J Archaeol Sci 36(11):2584–2596

Riederer J (1976) Kupfergeräte aus Habuba Kabira und Mumbaqat. Untersuchungsbericht. Mitt Deutsche Orientgesellsch Berlin 108:23–24

Riederer J (1980) Bibliographie zu material und Technologie kulturgeschiuchtlicher Silberobjekte. Berliner Beitr Archäometrie 5:229–239

Riederer J (1991) The scientific examination of forgeries of mesopotamian bronze heads. In: Mohen JP (ed) Découverte du Métal, Millénaires, vol 2, pp 439–446

Riederer J (1995) Die Verwendung zinkhaltiger Kupferlegierungen in römischer Zeit. In: STA M (ed) Acta of the 12th International Congress on Ancient Bronzes Nijmegen 1992, Nederlandse Archeologische Rapporten 18. Rijksdienst voor het Oudheidkundig Bodemonderzoek, Amersfort, pp 207–213

Riederer J (1999) Die Aussagekraft der Spurenelementkonzentrationen in mittelalterlichen Kupferlegierungen. Berliner Beitr Archäometrie 16:21–41

Riederer J (2001) Die Berliner Datenbank von Metallanalysen römischer Objekte. Berliner Beitr Archäometrie 18:139–259

Riederer J (2002) Die Berliner Datenbank von Metallanalysen kulturgeschichtlicher Objekte. Antike Objekte Mitteleuropas, des Mittelmeerraumes, des Nahen und Mittleren Ostens. Berliner Beitr Archäometrie 19:72–225

Riederer J (2008) Die Metallanalyse von Funden aus Silber und Kupferlegierungen. In: Künzel E (ed) die Alamannenbeute aus dem Rhein bei Neupotz. Plünderungsgut aus dem römischen Gallien 1, Monogr. Röm German Zentralmus 34 (1.2), pp 407–446

Robb L (2005) Introduction to ore-forming processes. Blackwell Publishing, Oxford

Roberts N (1998) The holocene. An environmental history. Blackwell Publ, Oxford

Roberts B, Ottaway B (2003) The use and significance of socketed axes during the late bronze age. Eur J Archaeol 6(2):119–140

Roberts BW, Thornton CP (2014) Archaeometallurgy in global perspective, Methods and syntheses. Springer, Heidelberg

Roberts S, Sofaer J, Kiss V (2008) Characterization and textural analysis of middle bronze age Transnubian inlaid wares of the Encrustated pottery culture, Hungary: a preliminary study. J Archaeol Sci 35:322–330

Roberts BW, Thornton CP, Pigott VC (2009) Development of metallurgy in Eurasia. Antiquity 83:1012–1022

Robins NA, Hagan NA (2012) Mercury production and use in colonial Andean silver production: emissions

and health implications. Environ Health Perspect 120 (5):627–631
Roden C (1985) Montanarchäologische Quellen des ur- und frühgeschichtlichen Zinnbergbaus in Europa – ein Überblick. Anschnitt 37(2–3):50–80
Roden C (1988) Blasrohrdüsen. Ein archäologischer Exkurs zur Pyrotechnologie des Chalkolithikums und der Bronzezeit. Anschnitt 40(3):62–82
Rohl BM (1996) Lead isotope data from the Isotrace laboratory, Oxford: archaeometry data base 2, galena from Britain and Ireland. Archaeometry 38:165–179
Roja JM, Baonza VG, San Andres M (2007) Application of Raman microscopy to the characterization of different verdigris variants obtained using recipes from older treatises. Spectrochim Acta A 68:1120–1125
Roman I (1990) Copper ingots. In: Rothenberg B (ed) Researches in the Arabah 1959–1984, Vol II: the ancient metallurgy of copper. Inst Archaeo-Metall Stud, London, pp 176–181
Rose D, Endlicher G, Mucke A (1990) The occurrence of "iscorite" in medieval iron slags. J Hist Metall Soc 24 (1):27–32
Rosenqvist T (1974) Principles of extractive metallurgy. McGraw-Hill edition
Rosman KJR, Chisholm W, Hong S, Candelone JP, Boutron CF (1997) Lead from Carthaginian and Roman Spanish mines identified in Greenland ice dated from 600 B.C. to 300 a.D. Environ Sci Technol 31(12):3413–3416
Rossignol-Strick M (1993) Late quaternary climate in the eastern Mediterranean region. Paléorient 19 (1):135–152
Rostoker W, Dvorak JR (1991) Some experiments with co-smelting to copper alloys. Archaeomaterials 5:5–20
Rostoker W, Sadowski M (1980) The carbon reduction of fully oxidized chalcopyrite (copper) ores. J Hist Metall Soc 14(1):38–42
Rostoker W, McNallan M, Gebhard ER (1983) Melting/smelting bronze at Isthmia. J Hist Metall Soc 17 (1):23–27
Rostoker W, Pigott VC, Dvorak JR (1989a) Direct reduction to copper metal by oxide-sulfide mineral interaction. Archaeomaterials 3(1):69–87
Rostoker W, Bronson B, Dvorak JR (1989b) Smelting to steel by the Japanese Tatara process. Archaeomaterials 3(1):11–25
Rosumek P (1982) Technischer Fortschritt und Rationalisierung im antiken Bergbau. Habelts, Bonn
Rothenberg B (1972a) Timna. Thames and Hudson, London
Rothenberg B (1972b) Survey im Sinai. Hadashot Arkheologiot 43:29–31
Rothenberg B (1973) Timna. Das Tal der biblischen Kupferminen. Lübbe, Bergisch Gladbach
Rothenberg B (1988) Researches in the Arabah 1959-1984 vol. 1: the Egyptian mining Temple. Inst Archaeo-Metall stud, Univers London
Rothenberg B (1990a) Researches in the Arabah 1959–1984, vol 2: The Ancient Metallurgy of Copper. Inst Archaeo-metall Stud, Univers London
Rothenberg B (1990b) Copper smelting furnaces, Tuyeres, Slags, Ingot-Moulds and Ingots in the Arabah: the archaeological data. In: Rothenberg B (ed) The ancient metallurgy of copper. Archaeology- Experiment-Theory. Inst Archaeo-Metall Studies. Thames & Hudson, London, pp 1–77
Rothenberg B, Blanco-Freijeiro A (1981) Ancient mining and metallurgy in south-West Spain. Inst of Archaeometall Stud (IAMS), London
Rothenhöfer P (2003) Geschäfte in Germanien: Zur Ausbeutung von Erzlagerstätten unter Augustus in Germanien. Zeitschr. Papyrologie und Epigraphik 143:277–286
Rothenhöfer P, Bode M (2012) Römische Bleigewinnung und Bleihandel im Licht neuer epigraphischer und naturwissenschaftlicher Forschungen. In: Olshausen E, Sauer V (eds) Die Schätze der Erde – Natürliche Ressourcen in der antiken welt, Stuttgarter Kolloquium zur Historischen Geographie des Altertums, vol 10, pp 345–361
Rothenhöfer P, Hanel N (2013) The romans and their Lead – tracing innovations in the production, distribution, and secondary processing of an ancient metal. In: Burmeister S, Hansen S, Kunst M, Müller-Scheeßel N (eds) Metal matters. Innovative technologies and social change in prehistory and antiquity, Menschen – Kulturen – Traditionen; Forschungscluster 2 (12), Rahden/Westf, pp 273–282
Rothenhöfer P, Hanel N, Bode M (2013) Auf den Spuren des Bleis der Römer: das Forschungsprojekt Corpus der Römischen Bleibarren. Atvatvca 4:68–73
Rothenhöfer P, Hanel N, Bode M (2016) Old finds - new insights: remarks on two roman lead ingots from minas de Rio Tinto (Huelva, España). Onuba 4:127–134
Rovira S (1999) Early slags and smelting by-products of copper smelting in Spain. In: Bartelheim M, Pernicka E, Krause R (eds) Die Anfänge der Metallurgie in der Alten welt, Forsch Archäometrie u Altertumswiss, vol 1, pp 83–98
Rovira S (2004) Technología metalúrgica y cambio cultural en la Prehistoria deula Ibérica. Norba Revista de Historia 17:9–40
Rovira S, Ambert P (2002) Les céramique à réduire le minerai de cuivre: Une technique métallurgique utilisée en Ibérie, son extension en France méridionale. Bulletin de la Société Prehistorique Francaise 99 (1):105–126
Rovira S, Gómez Ramos P (2003) Las Primeras etapas metalúrgicas en la Península Ibérica 3. Estudios metalográficos, Madrid
Rovira S, Gutierrez A (2003) Toro 2001. Cronica de un proceso de fundicion experimental de minerales de cobre'. Mineros y Fundidores en el inicio de la Edad de los Metales. El midi francés y el norte de la Península Ibérica, pp 70-81
Rovira S, Montero I (2003) Natural tin-bronze alloy in Iberian Peninsula metallurgy: potentiality and reality. In: Giumlia-Mair A, LoSchiavo F (eds) The problem of early tin, Brit Archaeol rec Int Ser 1199, pp 15–22
Rovira S, Renzi M, Montero Ruiz I (2009) Experimental co-smelting to copper tin alloys. In: Kienlin T, Roberts

BW (eds) Metals and societies. Studies in Honour of Bs Ottaway, vol 169. Universitätsforschungen zur Prähistorischen Archäologie, pp 407–414

Ruvalcaba Sil JL, Torres L, Franco F, Ortiz Díaz E (2004) Artifact rich gold surfaces: depletion gilding or natural surface corrosion. In: Perea A, Montero I, García-Vuelta O (eds) Ancient gold technology: America and Europe. CSIC, Madrid, pp 41–47

Ruvalcaba Sil JL, Peñuelas Guerrero G, Contreras Vargas J, Ortiz Díaz E, Hernández Vázques E (2009) Technological and material features of the gold work of Mesoamerica. ArchéoSciences 33:289–298

Ryndina N (2009) The potential of metallography in investigations of early objects made of copper and copper-based alloys. J Hist Metall Soc 43 (1):1–18

Sáez R, Almodovar GR, Pascual E (1996) Geological constraints on massif sulphide genesis in the Iberian Pyrite Belt. Ore Geol Rev 11:429–451

Saez R, Nocete F, Nieto J, Capitán À, Rovira S (2003) The extractive metallurgy of copper from Cabezo Jurè, Huelva, Spain: chemical and mineralogical study of slags dated to the third millennium BC. Can Mineral 41:627–638

Sainsbury CL (1969) Tin resources of the world. US geological survey bull. 1301. Washington

Sajó IE et al (2015) Core-shell processing of natural pigment: upper palaeolithic red ochre from Lovas, Hungary. PLoS One 10(7):e0131762. https://doi.org/10.1371/J.pone.0131762

Salgado S (1993) Workers. An archaeology of the industrial age. In: An aperture book. Asia One, Hongkong

Salgado S (2019) Gold. Taschen Publ Cologne, Serra Pelada Gold Mine

Salkield LU (1987) A technical history of the Rio Tinto mines: some notes on exploitation from pre-Phoenician times to the 1950s. Inst Mining Metall, London

Salmang H, Scholze H, Telle R (2007) Keramik. Springer, Berlin

Salvatori S, Vidale M, Guida G, Masioli E (2009) Ilgynly-depe (Turkmenistan) and the 4th millennium BC metallurgy of Central Asia. Paléorient 35(1):47–67

Salzmann E (2013) Provenienzstudien an frühem Eisen im Siegerland. Unpubl. MA, Ruhr University Bochum

Salzmann E (2019) Silver, copper, and bronze in early dynastic Ur, Mesopotamia: a high-resolution analysis approach Der Anschnitt Beih 41

Samida S, Eggert MKH (2013) Archäologie als Naturwissenschaft? Eine Streitschrift. Reihe Pamphletliteratur 5. Vergangenheitsverlag Berlin

Samusikov VP (2002) Relationships between Ag and au in au-Ag natural systems. Russ Geol Geophys 43 (8):746–753

Sanadze VV, Labartkava NI (1961) Phase transitions in the gold — platinum system. J Struct Chem 2 (6):647–654

Sánchez Cabra E, Botero CI (2007) The gold museum. In: Botero CI, Lleras Pérez R, Véléz SL, Sánchez Cabra E (eds) The art of gold. The Legacy of Pre-Hispanic Colombia, Collection of the Gold Museum in Bogotá. Skira ed, Milano, pp 260–265

Sánchez-Palencia Ramos FJ (1979) Römischer Goldergbau in NordWest Spaniens Ein Forschungsbericht. Anschnitt 2–3:38–61

Sánchez-Palencia FJ (2000) Las Médulas (León). Un paisaje cultural en la Asturia Augustana. Instituto Leonés de Cultura. León

Sánchez-Palencia FJ, Pérez LC, Orejas A (2000) Geomorphology and archaeology in the Las Médulas archaeological zone (ZAM) (Leon, Spain): evaluation of wastes and gold production. In: Vermeulen F, de Dapper M (eds) Geoarchaeology of the landscapes of classical antiquity, Int Coll Ghent, Oct 1998, pp 167–177

Sánchez-Palencia FJ, Currás Refojos B (2012) Minería romana en zonas interfronterizas de Castilla y León y Portugal. In: Sánchez-Palencia F-J (ed) Mineria romana en zonas interfronterizas de Castillay León y Portugal (Asturia y NE de Lusitania), Documentos Pahis, a & M Graficas S.L, Spain, pp 13–16

Sánchez-Palencia FJ, Sastre Prats I, Orejas Saco del Valle A, Ruiz del Árbol M (2017) Augusto y el control administrativo y territorial de las zonas mineras del Noroeste hispano. Consejo Superior de Investigaciones Científicas 35:863–874

Sangster DF (1988) Breccia-hosted Lead-ZincDeposits in carbonate rocks. In: James NP, Choquette PW (eds) Paleokarst. Springer, pp 102–116

Sangster DF, Outridge PM, Davis WJ (2000) Stable lead isotopes characteristics of lead ore deposits of environmental significance. Environ Rev 8:115–147

Santosh M, Philip R, Jacob MK, Omana PK (1992) Highly gold placer formationin the Nilambur Valley, Wynad Gold field, southern India. Miner Depos 27:336–339

Sarver JF, Hummel FA (1962) Solid solubility and eutectic temperature in the system Zn_2SiO_4–Mg_2SiO_4. J Am Ceram Soc 45(6):304

Sato M (1992) Persistency-field eh-pH diagrams for sulfides and their application to supergene oxidation and enrichment of sulfide ore bodies. Geochim Cosmochim Acta 56:3133–3156

Sayre EV, Yener A, Joel EC (1992) Statistical evaluation of the presently accumulated Lead isotope data from Anatolia and surrounding regions. Archaeometry 34 (1):73–105

Sayre EV, Joel EC, Blackman MJ, Yener KA, Özbal H (2001) Stable lead isotope studies of Black Sea Anatolian ore sources and related bronze age and Phrygianartefacts from nearby archaeological sites. Appendix: new central Taurus ore data. Archaeometry 43(1):77–115

Schade M (2001) Gold in Thüringen. Thüringer Wald, Schiefergebirge, Frankenwald. Herkunft, Entstehung, Fundorte. Thüringer Landesanst Geol Weimar

Schäfer MW (1983) Measurements of iron (III)-rich fayalites. Nature 303:325–327

Schäfer A (2013) Zur Eisenverarbeitung im Oppidum von Manching. Schlacken und Herdfragmente der Ausgrabung Altenfeld 1996–1999. In: Sievers S,

Leicht M, Ziegaus B (eds) Ergebnisse der Ausgrabungen in Manching-Altenfeld 1996–1999. Reichert Wiesbaden, pp 295–333

Schairer JF, Yagi K (1952) The system FeO-Al2O3-SiO2. Am J Sci:471–512

Schalk E (1998) Die Entwicklung der prähistorischen Metallurgie im nördlichen Karpatenbecken. Eine typologische und metallanalytische Untersuchung. Internationale Archäologie. Naturwissenschaft und Technologie 1, M Leidorf, Rahden Westf

Schattner T (2017) Between state economy and private entrepreneurship: framework for innovations in the Roman mining industry of southwestern Hispania. Vortrag Clustertreffen DAI: Innovationen Metall - technisch, sozial. Lugo 4.10.2017

Schauble EA (2004) Applying stable isotope fractionation. Theory to new systems. Rev Mineral Geochem 55:65–111

Scheffer C (2017) The Lavrion Pb-Zn-Fe-cu-Ag detachment-related district (Attica, Greece): structural control on hydrothermal flow and element transfer-deposition. Tectonophysics 717:607–627

Schiegl S (1997) Feuernutzung durch den Frühmenschen. In: Wagner GA, Beinhauer KW (eds) Homo heidelbergensis von Mauer. Das Auftreten des Menschen in Europa. Winter, Heidelberg, pp 298–303

Schiltz V (2010) Tillya Tepe, der Goldhügel. Eine Nekropole der Nomaden In: Gerettete Schätze Afghanistan Sammlung des Nationalmuseums in Kabul Kunst- u Ausstellung Bonn Oct 2010: 57–63

Schlegel H, Schüller A (1952) Die Schmelz- und Kristallisationsgleichgewichte im system Kupfer-Eisen-Schwefel und ihre Bedeutung für die Kupfergewinnung. Freiberger Forschungsh B 2:3–32

Schlichtherle H (2012) Die Jungsteinzeit. In: Der nördliche Bodenseeraum, Theiss, pp 65–75

Schlosser W (2002) Zur astronomischen Deutung der Himmelsscheibe von Nebra. Archäol Sachsen-Anhalt 1:21–23

Schmiderer A (2008) Geochemische Charakterisierung von Goldvorkommen in Europa. PhD-diss Naturwiss. Fak. Univers Halle-Wittenberg

Schmidt P (1996) The culture and technology of African iron production. Symbolism, science, and archaeology. University Press of Florida

Schmidt C (2005) Überregionale Austauschsysteme und Fernhandelswaren in der Ur III-Zeit. Baghdader Mitt 36:7–156

Schmidt P, Avery HD (1978) Complex iron smelting and prehistoric culture in Tanzania. Science 201 (4361):1085–1089

Schmitt-Strecker S, Begemann F (2005) Kupfer- und bronzezeitliche Artefakte vom Westbalkan: Zur Frage nach den Quellen ihres Kupfers. Prähist Zeitschr 80 (1):49–64

Schmitt-Strecker S, Begemann F, Pernicka E (1992) Chemische Zusammensetzung und Bleiisotopenverhältnisse der Metallfunde vom Hassek Höyük. In: Behm-Blancke MR (ed) Hassek Höyük – Naturwissenschaftliche Untersuchungen und lithische Industrie, Istanbuler Forsch 38, pp 108–123

Schmitt-Strecker S, Begemann F, Pernicka E (1994) Untersuchungen zur Metallurgie der Späten Uruk- und Frühen Bronzezeit am oberen Euphrat. Resumé. In: Wartke B (ed) Handwerk und Technologie im Alten orient. Zabern, Mainz, pp 97–98

Schnabel C (1890) Lehrbuch der Allgemeinen Hüttenkunde. Springer, Berlin

Schnabel C (1901) Handbuch der Metallhüttenkunde: Kupfer-Blei-Silber-gold. Springer, Berlin

Schnabel C (1905) Handbook of metallurgy: copper-Lead-silver-gold. Macmillan, London

Schneider G (2008) The roadside geology of Namibia, Samml Geolog Führer 97. Borntraeger, Berlin

Schneiderhöhn H (1962) Erzlagerstätten. Fischer, Stuttgart

Schoop UD (1995) Die Geburt des Hephaistos. Technologie und Kulturgeschichte neolithischer Metallverwendung im Vorderen orient, Intern Archäol 24, Leidorf, Rahden, Westfalen

Schreiner M (2002) Mineralogical and geochemical investigations into prehistoric smelting slags from Tepe Sialk/Central Iran, TU Bergakademie Freiberg. Institut für Mineralogie

Schubert E (1981) Zur Frage der Arsenlegierungen in der Kupfer- und Frühbronzezeit Südosteuropas. In: Lorenz H(ed) Studien zur Bronzezeit. Festschrift für Wilhelm Albert v Brunn. Philipp von Zabern, Mainz, pp 447–459

Schubert M, Pernicka E (2013) Die frühbronzezeitliche Kupferverarbeitung auf dem Buchberg im Unterinntal, Tirol. In: Oeggl K, Schaffer V (eds) Cuprum Tyrolense. 5550 Jahre Bergbau und Kupferverhüttung in Tirol. Montanwerke Brixlegg, pp 123–152

Schulz O, Schroll E (1997) Pb-Zn-(Fahlerz-) Erzbezirk Anis Nordtiroler Kalkalpen. In: Weber L (ed) Handbuch der Lagerstätten der Erze, Industriemineralien und Energierohstoffe Österreichs 19, Geol Bundesanstalt Wien, pp 355–358

Schürmann E (1958) Die Reduktion des Eisens im Rennfeuer. Stahl und Eisen 19:1297–1308

Schwab R, Heger D, Höppner B, Pernicka E (2006) The provenance of iron artefacts from manching: a multi-technique approach. Archaeometry 48(3):433–452

Schwartz AJ, Kumar M, Adams BL, Field DP (2013) Electron backscatter diffraction in materials science. Springer, Cham, p 17

Schweizer F (1994) Bronze objects from Lake sites: from Patina to "biography". In: Scott F, Podany J, Considine B (eds) Ancient and historic metals. Conservation and scientific research, Proc Symp Getty conservation Inst 1991. Getty Conservation Inst, Los Angeles, pp 33–50

Schwinden L (2001) Die Porta Nigra. In: Kuhnen HP (ed) Das römische Trier. Theiss, Stuttgart, pp 143–157

Scott D (1983) Depletion gilding and surface treatment of gold alloys from the Nariño area of ancient Columbia. J Hist Met Soc 17(2):99–115

Scott D (1991) Metallography and microstructure of ancient and historical metals. Getty Conservation Institute/Archetype Books, Los Angeles

Scott D (2002) Copper and bronze in art. Corrosion, colorants, conservation. Getty conservation Inst, Los Angeles

Scott D (2012a) Ancient metals: microstructure and metallurgy I. Copper and copper alloys, Createspace

Scott D (2012b) Ancient metals: microstructure and metallurgy II. Gold and platinum metallurgy of ancient colombia and ecuador, Createspace

Scott D, Bray W (1980) Ancient platinum technology in South America. Its use by the Indians in pre-hispanic times. Platinum Metals Rev 24(4):147–157

Scott DA, Doehne E (1990) Soldering with gold alloys in ancient South America: examination of two small gold studs from Ecuador. Archaeometry 32(2):183–190

Scott D, Schwab R (2019) Metallography in archaeology and art. Springer, Cham

Scott D, Seeley NJ (1983) The examination of a pre-Hispanic gold chisel from Colombia. J Archaeol Sci 10:153–163

Scott D, Podany J, Considine BB (1994) Ancient & historical metals. Conservation and Scientific Research. Proc. Symp. J. Paul Getty Museum and the Getty Conservation Institute, November 1991. Getty Conserv. Inst., Los Angeles

Seeliger T, Pernicka E, Wagner GA, Begemann F, Schmitt-Strecker S, Eibner C, Öztunali Ö (1985) Archäometallurgische Untersuchungen in Nord- und Ostanatolien. Jb Röm-German Zentralmus 32:597–659

Segal I, Halicz L (2014) An advanced method of laser ablation MC-ICP-MS for provenance studies in Archaeometallurgy: chalcolithic metal objects from Israel as a case study. Atiqot 79:1–10

Segal I, Bar-Matthews M, Harlavan Y, Asael D (2015) Provenance of ancient metallurgical artifacts: implications of new Pb isotope data from Timna ores. In: Hauptmann A, Modarressi-Tehrani D (eds) Archaeometallurgy in Europe III. Proc. 3rd Internat Conf, Bochum, vol 26. Der Anschnitt, Beih, pp 221–227

Segebade C (1996) Remarks on European damascene Steelware. Proc. on the forum for the fourth Int. Conf. Of the use of metals and alloys (BUMA IV), Shimane, Japan, pp 133–149

Segerstrom K, Stotelmeyer RB (1909) Mineral resources of the White Mountain wilderness and adjacent areas. New Mexico Geol Surv Bull:1453

Segev A (1987) The age of the latest Precambrian volcanism in southern Israel, northeastern Sinai and southwestern Jordan - a re-evaluation. Precambrian Res 36:277–285

Segev A, Sass E (1989) Copper-enriched Syngenetic Dolostones as a source for epigenetic copper mineralization in sandstones and Shales (Timna, Israel). Geol Assoc Can Spec Pap 36:647–658

Segev A, Beyth M, Bar-Matthews M (1992) The geology of the Timna Valley with emphasis on copper and manganese mineralization - updating and correlation with the eastern margins of the Dead Sea rift. Geol Survey Israel Rep No GSI 14:1–31

Selimkhanov IR (1967) Spektralanalytische Untersuchungen von Metallstücken des 3. Und 2. Jahrtausends aus dem östlichen Transkaukasien. Archaeol Aust 28:71–78

Selimkhanov IR (1975) Sur l'etude du fragment de vase de Tello… et le problème del'utilisation de l'antimone dans l'antiquité. Annales du Laboratoire de Recherche des Musées de France: 45

Selimkhanov IR (1977) Zur Frage einer Kupfer-Arsenzeit. Germania 55:1–6

Sella Q (1999) Sulle Condizioni Dell'Industria Mineraria Nell'Isola di Sardegna. Nuoro Illissio

Seltmann R, Porter TM (2005) The porphyry cu-au/Mo deposits of Central Eurasia 1: tectonic, geologic and Metallogenic setting and significant deposits. In: Porter TM (ed) Super porphyry copper & gold deposits: a global perspective, vol 2. PGC Publ, Adelaide, pp 467–512

Senn M, Gfeller U, Guènette-Beck B, Lienemann P, Ulrich A (2010) Tools to qualify experiments with bloomery furnaces. Archaeometry 52(1):131–145

Serneels V (1993) Archéométrie des scories de fer. Recherches Sur la siérurgie ancienne en Suisse occidentale. Cahiiers d' Archaeologie Romande 61. Lausanne

Serneels V, Perret S (2003) Quantification of Smithing activities based on the investigation of slag and other material remains. In: ass. Italiana Metall. (org.) (ed) Archaeometallurgy in Europe I. ass. Italiana Metall., Milano, pp 469–478

Serning I (1979) Prehistoric iron production. In: Clarke H (ed) Iron and man in prehistoric Sweden. Jernkontoret, Stockholm, pp 50–99

Serning I, Hagfeldt H, Kresten P (1982) Vinarhyttan. A late Medieval iron production site at Lake Haggen Norrbärke parish, Dalarna, county, Sweden. Jernkontor Forsk H 21. Med Bilaga Av Birgitta Hulthè

Servelle C, Servelle G (1991) Premier mètal, ultime pierredans le Sud-Ouest de la France. In: Mohen JP, Eluère C (eds) Découverte du Metal. Paris, pp 229–250

Shalev S (1991) Two different copper industries in the chalcolithic culture of Israel. In: Mohen JP, Éluére C (eds) Découverte du Métal. Picard, Paris, pp 413–419

Shalev S (1993) The earliest gold artifacts in the southern Levant: reconstruction of the manufacturing process. In: Éluère C (ed) Outils et ateliers d'orfèvres des temps ancients. antiquités nationales mémoires, vol 2, pp 9–12

Shalev S, Braun E (1997) The metal objects from Yiftahel II. In: Braun E (ed) Yiftah'el. Israel Antiquit. Author. Rep. 2, pp 92–96

Shalev S, Northover PJ (1987) Chalcolithic metal and metalworking from Shiqmim. In: Levy TE (ed) Shiqmim I, BAR intern Ser 356, pp 357–371

Shalev S, Caspi EN, Shilstein S, Paradowska AM, Kockelmann W, Kan-Cipor Meron T, Levyy (2014) Middle bronze age II Battle axes from Rishon Lezion, Israel: archaeology and metallurgy. Archaeometry 56 (2):279–295

Shanks RI (1936) Shism al Tasa 26/37D. DGMR Rep No. 204/6023, Jeddah

Shareq A et al. (1977) Mineral resources of Afghanistan, 2nd ed. afghan geological and mines survey, United Nations development support project, AFG/74/012. Kabul

Shcheka G, VrzHosek A, Lehmann B, Tolstykh N (2004) Association of Platinum-group minerals from the Zolotaya Gold Placer, Primorye, Russia Far East. Can Mineral 42:583–599

Shemakhanskaya MS (1991) New aspects of the origin of ancient bronze. Bull Metals Museum 16:3–9

Sherratt S (2000) The captive Spirit: catalogue of Cycladic antiquities in the Ashmolean museum. Oxford:76–87

Shimada I, Craig AK (2013) The style, technology and Organization of Sicán Mining and Metallurgy, northern Peru: insight from holistic study. Chungara Revista de Antropología Chilena 45(1):3–31

Shirakawa Z, Dibner B, Smith CS (1983) Kodo Zuroku. Illustrated book on the smelting of copper by Masuda Tsuna. Burndy Libr, Norwalk, Conn

Shortland AJ (2002) An antimony bead from Jerablus Tahtani. J Hist Metall Soc 36(1):1–5

Shugar A (2000) Archaeometallurgical investigation of the chalcolithic site of Abu Matar, Israel: A re-assessment of technology and its implications for the Ghassulian Culture. PhD-diss Univers London

Shugar A, Rehren T (2002) Formation and composition of glass as a function of firing temperature. Proc XIX Int Congr glass, Edinburgh, July 2001. Glass Technol 43C:145–150

Siebenschock M, Goldenberg G, Wagner H (1996) Archäometallurgische Untersuchungen zur Verhüttung von Antimonerzen im Schwarzwald. Gewinnung von Antimon Crudum im Spätmittelalter und in der Neuzeit bei Sulzburg. In: Goldenberg G, Otto J, Steuer H (eds) Archäometallurgische Untersuchungen zum Metallhüttenwesen im Schwarzwald. Thorbecke, Sigmaringen, pp 275–336

Sillitoe RH (1972) A Plate tectonic model for the origin of porphyry copper deposits. Econ Geol 67:184–197

Sillitoe RH (1979) Metallogenic consequences of late precambrian suturing in Arabia, Egypt, Sudan and Iran. In: Al-Shanti AMS (ed) Proc symp evolution and mineralization of the Arabian-Nubian Shield, Jeddah 1978, 1. Pergamon Press, Oxford, pp 109–120

Sillitoe RH (2009) Superfgene silver enrichment reassessed. In: Titley SR (ed) Supergene environments, processes, and products, Soc econ Geol spec Publ 14, pp 15–32

Sillitoe RH (2012) Copper provinces. In: Hedenquist JW, Harris M, Camus F (eds) Geology and genesis of major copper deposits and districts of the world: a tribute to Richard H. Sillitoe, vol 16. Soc Econ Socy, Spec Publ, pp 1–18

Silvestri E, Hauptmann A, Bellintani P, Mottes E, Nicolis F (2015) Bronzezeitliche Kupferverhüttung in Trentino. In: Stöllner T, Oeggl K (eds) Bergauf Bergab. 10.000 Jahre Bergbau in den Ostalpen. Verlag M Leidorf, Bochum, pp 201–208

Silvestri E, Bellintani P, Hauptmann A (2019) Bronze age copper ore mining and smelting in Trentino (Italy). In: Turck R, Stöllner T, Goldenberg G (eds) Alpine copper II – Rame delle Alpi II – Cuivre des Alpes II. New results and perspectives on prehistoric copper production, Der Anschnitt Beih, vol 42, pp 261–277

Simon C (1865) Über das Herdschmelzen der arsenikalischen Kupfererze von Mancayan auf der Insel Luzon (Manila). Berg- Hüttenmänn Zt, Freiberg, ref in: Zschocke K, Preuschen E, das urzeitliche Bergbaugebiet von Mühlbach-Bischofshofen. Mater Urgesch Österreichs 6:257–262

Sitschick H, Ludwig F, Wetzel E, Luckert J, Höding TH (2005) Raseneisenerz – auch in Brandenburg ein mineralischer Rohstoff mit bedeutender wirtschaftlicher Vergangenheit. Bog iron ore – also in Brandenburg a raw material with important economical history. Brandenburg Geowiss Beitr 12 (1–2):119–128

Slag Atlas des Vereins Deutscher Hüttenleute (1995)

Slotta R (1982) Einführung in die Industriearchäologie. Wissenschaftliche Buchgesellschaft Darmstadt

Slotta R (2012) Der (Silber-) Bergbau als Kunst-Katalysator. In: Tenfelde K, Berger S, Seidel HC (eds) Geschichte des deutschen Bergbaus 1, C. In: Bartels C, R. Slotta (eds), Der alteuropäische Bergbau. Aschendorff Verlag, Münster, pp 591–618

Slotta R, Wollmann V, Dordea I (2001) Silber und Salz in Siebenbürgen 6, Einblicke in eine unbekannte Welt-Fotografien von Arthur Oskar Bach, Albert Schotsch und Bazil Roman. Ausstellung im Deutschen Bergbau-Museum Bochum 2002-2003. Publ Deutsches Bergbau-Museum 101

Skarpelis N (2007) The Lavrion deposit: geology, mineralogy, and minor elements chemistry. N Jb Miner Abh 183(3):227–249

Skarpelis N, Argyraki A (2009) Geology and origin of supergene ore at the Lavrion Pb-Ag-Zn deposit, Attica. Resour Geol 59:1–14

Smettan HW (1995) Archäoökologische Untersuchungen auf dem Albuch. In: Beiträge zur Eisenverhüttung auf der Schwäbischen Alb Forsch u Ber Vor- und Frühgesch Baden-Württemberg, vol 55, pp 37–146

Smirnov SS (1954) Die Oxidationszone sulfidischer Lagerstätten. Akademie-Verlag, Berlin

Smith FG (1961) Metallic Sulphide melts as igneous differentiates. Can Mineral 5:663–669

Smith CS (1967) The interpretation of microstructures of metallic artifacts. In: Young WJ (ed) Application of science in examination of works of arts. Res Lab Mus Fine Arts, Boston, pp 20–52

Smith CS (1973) An examination of the arsenic-rich coating on a bronze age bull from Horoztepe. In: Young WJ (ed) Applicat Sc Examin arts, pp 96–102

Smith CS (1974) Historical notes on the Colouring of metals. In: Bishay E (ed) Recent advances in the science and technology of materials 3. Plenum Press, New York, pp 157–167

Smith CS (1981) A search for structure. Selected essays on science, art and history. MIT Press, Cambridge, MA

Smith CS (1983) Introduction. In: Kodō Zuroku by Masuda Tsuna 1801. Burndy Libr Norwalk, pp 10–27

Smith R (1996) An analysis of the process for smelting tin. Bull Peak Distr Mines Hist Soc 13(2):91–99

Smith CS, Wertime TA, Pleiner R (1967) Preliminary reports of the metallurgical project. In: Caldwell JR (ed) Investigations at Tal-i-Iblis, Illinois state museum preliminary reports 9. Springfield, Ill, pp 318–326

Smuts J, Steyn JGD, Boeyens JCA (1969) The crystal structure of an iron silicate, iscorite. Acta Cryst B25:1251–1255

Snyder DA, Carmichael SE (1992) Olivine-liquid-equilibria and chemical activities of FeO, NiO, Fe_2O_3 and MgO in natural basaltic melts. Geochim Cosmochim Acta 56:303–318

Soetbeer A (1879) Edelmetall-Produktion und Werthverhältnis zwischen gold und Silber seit der Entdeckung Amerikas bis zur Gegenwart. Gotha 1879

Sokol E, Sharygin V, Kalugin V, Volkova N, Nigmatulina E (2002) Fayalite and kirschsteinite solid solutions in melts from burned spoil heaps, south Urals. Russ Eur J Mineral 14:795–807

Song YS (1980) Tian gong Kai wu (= exploitation of the work of nature: Chinese agriculture and technology in the XVII century). Cinese culture series 2-3. Imprint Taipei, Taiwan, China

Sönnecken M (1977) Forschungen zur spätmittelalterlich-frühneuzeitlichen Eisenherstellung in Kierspe, Märkischer Kreis. Fachausschussbericht 9006 VDEh Düsseldorf

Sperl G (1979) Zur ehemaligen Kupfergewinnung in der Radmer und im Johnsbachtal. In: Tagungsband geschichte des Erzberggebietes. Montanhist Verein Österr, Leoben, pp 135–145

Sperl G (1980) Über die Typologie urzeitlicher, frühgeschichtlicher und mittelalterlichen Eisenhüttens-chlacken. Studien zur Industrie-Archäologie VII, Wien

Sperl G (1990) Urgeschichte des Bleis. Z Met 81:799–801

Spindler K (1971) Zur Herstellung der Zinnbronze in der frühen Metallurgie Europas. Acta Praehist Archaeol 2:199–253

Spiridonov S (1993) Silvering. In: La Niece S, Craddock PT (eds) Metal plating and Patination. Butterworth & Heinemann, Oxford, pp 201–210

Spiridonov E, Yanakieva D (2009) Modern mineralogy of gold. In: Guerra MF, Rehren T (eds) Authentication and analysis of Goldwork, Archaeo-Sc, vol 33, pp 67–73

Srejović D (1975) Lepenski Vir. Eine vorgeschichtliche Geburtsstätte europäischer Kultur. Gustav Lübbe Publ

Srinivasan S (1999) Lead isotope and trace element analysis in the study of over a hundred south Indian metal icons. Archaeometry 41:91–116

Stacey JS, Kramers JD (1975) Approximation of terrestrial Lead isotope evolution by a two-stage model. Earth Planet Sci Lett 26:207–221

Stech T (1999) Aspects of early metallurgy in Mesopotamia and Anatolia. In: Pigott VC (ed) The archaeometallurgy of the Asian old world. The university museum, University of Pennsylvania, Philadelphia, pp 59–71

Stech Maddin R (1988) Reflections on early metallurgy in Southeast Asia. In: Maddin R (ed) The beginning of the use of alloys. MIT, Cambridge MA, pp 163–174

Stech T, Pigott V (1985) The metals trade in Southwest Asia in the third millennium B.C. , Iraq, pp 39–64

Stech-Wheeler T, Maddin R (1976) The techniques of the early Thai Metalsmith. Expedition 18(4):38–47

Stech-Wheeler T, Maddin R (1980) Metallurgy and ancient man. In: Wertime TA, Muhly JD (eds) The coming of the age of iron. Yale University Press, New Haven, pp 99–126

Stedingk K (2004) Die Minerallagerstätten des Harzes und seines Umfeldes. In: Tagungsband 7. Internationaler Bergbauworkshop, Clausthal-Zellerfeld: Sankt Andreasberger Verein Gesch u Altertumskunde, pp 13–34

Steuer H (1987) Der Handel der Wikingerzeit zwischen Nord- und Westeuropa aufgrund archäologischer Zeugnisse. In: Untersuchungen zu Handel und Verkehr der vor- und frühgeschichtlichen Zeit in Mittel- und Nordeuropa IV: Der Handel der Karolinger- und Wikingerzeit. Abh Akad Wiss Göttingen, vol 156, pp 13–197

Steuer H (1990) Zum Umfang der Silbergewinnung im mittelalterlichen Europa nach der schriftlichen Überlieferung. Freiburger Universitätsblätter 109:79–83

Steuer H (1998) The ore deposits in middle Asia and Viking age silver in Scandinavia, Hammaburg, Neue Folge 12. Habelt, Bonn, pp 111–123

Steuer H, Stern WB, Goldenberg G (2002) Haithabu und die frühe Stadtentwicklung im nördlichen Europa. Schriften Archäol Landesmus 8:133–165

Števko M, Sejkora J (2012) Supergene arsenates of copper from the Piesky deposits, Špania Dolina, Central Slovakia, Acta Mineral-Petrogr, Abstr Ser, Szeged, vol 7, p 130

Stoffers P, Hannington M, Wright I, Herzig P, de Ronde C (1999) Elemental mercury at submarine hydrothermal vents in the Bay of Plenty, Taupo volcanic zone, New Zealand. Geology 27(10):931–934

Stöllner, T (2003a) Bemerkungen zur bronzezeitlichen Gewinnung von sog. Fahl- und Kupferkieserzen in Salzburg und Nordtirol. Beitrag zu Bachmann, H.-G., bunt- und Edelmetalle aus mitteleuropäischen Komplexerz-Lagerstätten: Fahlerzverhüttung von der Bronzezeit bis zur renaissance. In: Stöllner T, Körlin G, Steffens G (eds), Man and mining. Der Anschnitt Beih 16: 30–35

Stöllner T (2003b) Mining and economy – a discussion of spatial Organisations and structures of early raw material exploitation. In: Stöllner T, Körlin G, Steffens G (eds) Mensch und Bergbau, Der Anschnitt Beih, vol 16, pp 415–446

Stöllner T (2004) Prähistorischer und antiker Erzbergbau in Iran. In: Stöllner T, Slotta R, Vatandoust A (eds) Persiens Antike Pracht. Bergbau, Handwerk, Archäologie. Ausstellungskatalog Deutsches Bergbau-Museum Bochum, pp 44–63

Stöllner T (2008a) Mining landscapes in early societies – imprinting processes in pre- and protohistoric economies? In: Bartels C, Küppers-Eichas C (eds) Cultural Heritage and Landscapes in Europe. Proc Intern Conf Bochum 2007. Veröff. Deutsches Bergbau-Museum Bochum, vol 161, pp 65–92

Stöllner T (2008b) Montan-archaeology and research on old mining: just a contribution to economic history? In: Yalçın Ü (ed) Anatolian metal IV, Der Anschnitt Beih, vol 21, pp 149–178

Stöllner T (2009a) Prähistorische Montanreviere der Ost- und Südalpen – Anmerkungen zu einem Forschungsstand. In: Oeggl K, Prast M (eds) Die Geschichte des Bergbaus in Tirol und seinen angrenzenden Gebieten . Proc 3rd Milestone-Meeting, SFB HIMAT 10/2008, pp 37–60

Stöllner T (2009b) Latènezeitliche Eisenwirtschaft im Siegerland. Interdisziplinäre Forschungen zur Wirtschaftsarchäologie. Metalla 16:2

Stöllner T (2009c) Gold in Southern Peru? Perspectives of research into mining archaeology. In: Reindel A, Wagner GA (eds) New technologies for archaeology. Multidisciplinary investigations in Palpa and Nasca, Peru. Springer, Heidelberg, pp 393–407

Stöllner T (2009d) Der bronzezeitliche Bergbau im Südrevier des Mitterberggebietes. Bericht zu den Forschungen der Jahre 2002 bis 2006. Archaeol Austriaca 90:87–137

Stöllner T (2011a) Präkolumbischer Bergbau in den Anden. In: Slotta R, Schnepel I (eds) Schätze der Anden. Chiles Kupfer für die Welt. Katalog der Ausstellung des Deutschen Bergbau-Museums Bochum, Mai 2011 to February 2012. GCC graph. Centrum Calbe, pp 181–212

Stöllner T (2011b) Tin from Kazakhstan: steppe tin for the west? In: Yalçın Ü (ed) Anatolian metal V, Der Anschnitt Beih, vol 24, pp 231–251

Stöllner T (2013) Der Mitterberg als Großproduzent für Kupfer in der Bronzezeit. In: Stöllner T, Oeggl K (eds) Bergauf Bergab. 10.000 Jahre Bergbau in den Ostalpen. Wissenschaftl. Band Ausstellung Deutsches Bergbau-Museum Bochum, October 2015 to April 2016. Leidorf, Rahden/Westf, pp 175–185

Stöllner T (2019a) What is an ore deposit? Approaches from geoscience and archaeology in understanding the usage of deposits. Metalla 24(2):87–110

Stöllner T (2019b) Between mining and smelting in the bronze age – beneficiation processes in an alpine copper Production District. Results of 2008 to 2017 excavations at the "Sulzbach-moos"-bog at the Mitterberg (Salzburg, Austria). Der Anschnitt Beih 42:165–190

Stöllner T, Gambaschidze I (2011) Gold in Georgia II: the oldest gold mine in the world. In: Yalçın Ü (ed) Anatolian metal V, Der Anschnitt Beih, vol 24, pp 187–199

Stöllner T, Samašev Z (2013) Unbekanntes Kasachstan. Archäologie im Herzen Asiens. Exhibition Catalogue Deutsches Bergbau-Museum Bochum

Stöllner T, Weisgerber G, Momenzadeh M, Pernicka E, Shirazi AS (2004a) Die Blei/Silbergruben von Nakhlak und ihre Bedeutung im Altertum. Anschnitt 56 (2-3):76–97

Stöllner T, Slotta R, Vatandoust A (2004b) Persiens Antike Pracht. Bergbau Handwerk Archäologie. Katalog Ausstellung Deutsches Bergbau-Museum Bochum 2005

Stöllner T, Gambaschidze I, Hauptmann A, Mindiaschvili G, Gogočuri G, Steffens G (2010) Goldbergbau in Südostgeorgien – Neue Forschungen zum frühbronzezeitlichen Bergbau in Georgien. In: Hansen S, Hauptmann A, Motzenbäcker I, Pernicka E, Majkop bis Trialeti V (eds) Gewinnung und Verbreitung von Metallen und Obsidian in Kaukasien im 4.-2. Jt.v.Chr. Proc Internat Symp Berlin Juni 2006R. Habelt, Bonn, pp 103–138

Stöllner T, Thomas P, Maass A, Röttger K (2009) Großproduktion für Kupfer im Raum Mitterberg in der Bronzezeit – Forschungsbericht für die Jahre 2007 und 2008. In: Oeggl K, Prast M (eds) Die Geschichte des Bergbaus in Tirol und seinen angrenzenden Gebieten. Proc 3rd Milestone Meeting SFB-HIMAT Oct 2008, pp 231–241

Stöllner T et al (2013a) Zinn und Kupfer aus dem Osten Kasachstans. Ergebnisse eines deutsch-kasachischen Projekts 2003-2008. In: Stöllner T, Samašev Z (eds) Unbekanntes Kasachstan. Archäologie im Herzen Asiens. Ausstellungskatalog Deutsches Bergbaumuseum Bochum, pp 357–382

Stöllner T et al (2013b) Metall und Metallgewinnung der bronze- und Früheisenzeit in Zentral- und Ostkasachstan. In: Stöllner T, Samašev Z (eds) Unbekanntes Kasachstan. Archäologie im Herzen Asiens. Ausstellungskatalog Deutsches Bergbau-Museum Bochum, pp 383–398

Stöllner T et al (2014) Gold in the Caucasus: new research on gold extraction in the Kura-Araxes culture of the 4th millennium BC and early 3rd millennium BC. With an appendix of Jansen M, Stöllner T, Courcier A. In: Meller H, Pernicka E, Risch R (eds) Metalle der Macht, Tagung Landesmus Vorgesch Halle, vol 11, pp 71–110

Stöllner T, von Rüden C, Hanning E, Lutz J, Kluwe S (2016) The Enmeshmaent of eastern alpine mining communities in the bronze age. In: Körlin G, Prange M, Stöllner T, Yalçın Ü (eds), From bright ores to shiny metals. Festschr a Hauptmann on occasion of 40 years res in Archaeometallurgy and Archaeometry. Der Anschnitt Beih 29: 75–102

Storetvedt KM (1990) The Tethys Sea and the Alpine-Himalayan orogenic belt; mega-elements in a new global tectonic system. Phys Earth Planet Inter 62(1–2):141–184

Stos-Fertner Z, Gale NH (1978) Chemical and Lead isotope analysis of ancient Egyptian gold, silver and Lead. In: Proc 18th Internat Symp Archaeometry Archaeol prospection, Bonn, pp 299–314

Stos-Gale ZA (1993) Isotopic analyses of ores, slags and Artefacts: the contribution to Archaeometallurgy. In: Francovich R (ed) Archaeologia delle attività estrattive e metallurgiche. Firenze, all'Insegna del Giglio, pp 593–627

Stos-Gale ZA (2001a) Minoan foreign relations and copper metallurgy in Protopalatial and Neopalatial Crete. In: Shortland A (ed) The social context of technological change. Egypt and the near east, 1650-1550 BC. Proc St Edmund Hall, Oxford 2000, pp 195–210

Stos-Gale ZA (2001b) The impact of the natural sciences on studies of Hacksilber and early silver coinage. Am Numism Soc 24:53–75

Stos-Gale ZA (2011) "Biscuits with ears:" a search for the origin of the earliest Oxihide ingots. In: Betancourt PP, Ferrence SC (eds) Metallurgy: understanding how, Learning why. Studies in Honor of JD Muhly. Instap Academic Press, Philadelphia, pp 221–229

Stos-Gale ZA, Gale N (1979) Chemical and lead isotope analysis of ancient Egyptian gold, silver and lead. Archaeophysica 10:299–315

Stos-Gale ZA, Gale N (1980) Sources of galena, lead and silver in predynastic Egypt. Revue d'Archeometrie III:285–295

Stos-Gale ZA, Gale N (1992) New light on the provenience of the copper ingots found on Sardinia. In: Tykot RH, Andrews TK (eds) Sardinia in the Mediterranean; A Footprint in the Sea. Studies in Sardinian Archaeology presented to Miriam Balmuth. Sheffield Academic Press, Sheffield, pp 317–346

Stos-Gale ZA, Gale NH (1994) Metals. In: Knapp B, Cherry JF (eds) Provenience studies and bronze age Cyprus, Monographs in world archaeology 21. Prehistory Press, Madison/Wisc, pp 92–121

Stos-Gale ZA, Gale N (2009) Metal provenancing using isotopes and the Oxford archaeological lead isotope database (OXALID). J Archaeol Anthropol Sci 1:195–213

Stos-Gale ZA, MacDonald CF (1991) Sources of metals and trade in the bronze age Aegean. In: Gale NH (ed) Bronze age trade in the Mediterranean, Stud Mediterran Archaeol XC, pp 249–284

Stos-Gale Z, Gale N, Houghton J (1995) The origin of Egyptian copper. Lead isotope analysis of metals from El-Amarna. In: Davies WV, Schofield L (eds) Egypt, the Aegean and the Levant. Interconnections in the second millennium BC. Brit. Mus. Press, pp 127–135

Stos-Gale ZA, Maliotis G, Gale NH, Annetts M (1997) Lead isotope characterization of the Cypriot copper ore deposits applied to provenance studies of copper oxhide ingots. Archaeometry 31(1):83–123

Stos-Gale ZA, Gale NH, Bass G, Pulak C, Galili E, Sharvit J (1998a) The copper and the tin ingots of the late bronze age Mediterranean: new scientific evidence. In: Proc IVth Internat Conf beginning of the use of metals and alloys. Matsue, Shimane, pp 115–126

Stos-Gale S, Maliotis, G, Gale N (1998b) A preliminary survey of the Cypriot slag heaps and their contribution to the reconstruction of copper production on Cyprus. In: Rehren T, Hauptmann A, Muhly JD, Metallurgica Antiqua. In honour of HGt Bachmann and R Maddin. Der Anschnitt Beih 8: 235–262

Strahm C (1994) Die Anfänge der Metallurgie in Mitteleuropa. Helvetia Archaeol 25:2–39

Strahm C (2007) L'introduction de la métallurgie en Europe. In: Guilaine J (ed) Le Chalcolithique et la Construction des Inegalités, Collection des Hesperides, pp 47–71

Strahm C, Hauptmann (2009) The metallurgical developmental phases in the old world. In: Kienlin TL, Roberts BW (eds) Metals and societies studies in honour of Barbara S Ottaway, Univers Forsch Prähist Archäol, vol 169. Habelt, Bonn, pp 116–128

Straube H (1986) Kritische Gegenüberstellung der Theorien über die Metallurgie des Rennfeuers. Ferrum 57:20–28

Straube H (1996) Ferrum Noricum und die Stadt auf dem Magdalensberg. Springer, Wien

Strauss GK, Gray KG (1986) Base metal deposits in the Iberian Pyrite Belt. In: Friedrich GH (ed) Geology and metallogeny of copper deposits. Springer, p 304

Ströbele F, Wenzel T, Kronz A, Hildebrandt LH, Markl G (2010) Mineralogical and geochemical characterization of high medieval lead-silver smelting slags from Wiesloch near Heidelberg (Germany) – as an approach to process reconstruction. J Archaeol Anthropol Sci 2:191–215

Ströbele F, Hildebrandt LH, Baumann A, Pernicka E, Markl G (2015) Pb isotope data of Roman and medieval objects from Wiesloch near Heidelberg, Germany. J Archaeol Anthropol Sci 7:465–472

Stromer W (1995) Die Saigerhüttenindustrie des Spätmittelalters. Entwicklung der Kupfer-Silber-Scheidekünste zur "ars conflatoria separantia argentum a cupro cum plumbo". Dent Tech 62:187–219

Stumpfl EF, Clarke AM (1964–1965) Electron probe microanalysis of gold-Platinoid concentrates from Southe East Borneo. Trans Inst Min Metall 74:933–946

Subramanian PR, Laughlin DE (1998) The as-cu (arsenic-copper) system. Bull Alloy Phase Diagr 9(5):605–617

Subramanian PR, Perepezko JH (1993) The Ag-cu (silver-copper) system. J Phase Diagr 14:62–63

Sugaki A, Shima H, Kitakaze A, Harada H (1975) Isothermal phase relations in the system cu-Fe-S under hydrothermal conditions at 350 °C and 300 °C. Econ Geol 70:806–823

Sugaki A et al (1983) Ggeological study on the polymetallic ore deposits in the Potosi district, Bolivia. Sci Rep Tohoku Univ 15:409–460

Suhling L (1976) Der Seigerhüttenprozeß. Riederer, Stuttgart

Sutherland DG (1985) Geomorphological controls on the districbution of placer deposits. J Geo Soc Lond 142:727–737

Sverchkov LM (2009) A history of research on ancient mining in Uzbekistan. Archäol Mitt Iran und Turan 41:141–164

Swart C (2006) Edles gold und schnelles Silber – was Sind Amalgame? In: Born H, Völling E (eds), Gold im Alten orient. Technik – Naturwissenschaft – Altorientalistik. Nachr Martin von Wagner Mus Univers Würzburg, Reihe : 71-72

Szakáll S, Papp G, Sajó I, Kovács A (2000) Antimony oxide minerals from Hungary, Acta Mineralogica-Petrographica, Szeged, XLI, pp 31–62

Tadmor M (2002) The Kfar Monash hoard again – a view from Egypt and Nubia. In: van den Brinck E, Levy TE (eds) Egypt and the Levant – interrelations from the 4th through the early 3rd millenium B.C.E. Continuum, London, pp 239–251

Tadmor M, Kedem D, Begemann F, Hauptmann A, Pernicka E, Schmitt-Strecker S (1995) The Nahal Mishmar hoard from the Judean Desert: technology, composition, and provenance. Atiqot XXVII:95–148

Tafel V (1929) Lehrbuch der Metallhüttenkunde 2 (Wismut, Blei, Zinn, Antimon, Zink, Quecksilber, Nickel, Aluminium). Hirzel, Leipzig

Tafel V, Wagenmann K (1951) Lehrbuch der Metallhüttenkunde 1 (gold, Silber, Platinmetalle, Kupfer, Quecksilber, Wismut). Hirzel, Leipzig

Taggart AF (1945) Handbook of mineral dressing. Ores and industrial minerals. Wiley, New York

Tanelli G, Lattanzi P (1985) The cassiterite-polymetallic sulfide deposits of Dachang (Guangxi, People's Republic of China). Mineral Deposits 20(2):102–106

Taylor RG (1958) Geology of tin deposits, Developments in economic geology 11. Elsevier, Amsterdam

Taylor RG (1979) Geology of tin deposits 11. Elsevier, Amsterdam

Taylor SR (1964) Abundance of chemical elements in the continental crust; a new table. Geochimica et Cosmochim Acta 28(8):1273–1285

Taylor R (2011) Gossans and leached cappings. Field assessment. Springer, Berlin

Taylor NW, Williams FJ (1935) Reactions between solids in the system CaO-MgO-SiO2 in the temperature range 600-1200°C. Bull Geolog Soc 46:1121–1135

Tchernia A (1969) Recherches archéologiques sous-marines. Gallia 27(2):465–499

Telle R, Thönnissen M (2006) Prähistorische feuerfeste Werkstoffe und ihre Weiterentwicklung in keltischer und römischer Zeit. Prakt Metallograph 43(2):55–87

Téreygol F (2002) Frühmittelalterlicher Bergbau und Silber-Produktion von Melle in Frankreich. Anschnitt 54:253–266

Thalhammer O, Stumpfl EF, Panayiotou A (1986) Postmagmatic, hydrothermal origin of sulfide and arsenide Mineralizations at Limassol Forest, Cyprus. Mineral Deposits 21:95–105

Thalheim K (1990) Das Silber von 1477 in der Sammlung des museums für Mineralogie und Geologie Dresden. Mitt Senckenberg Naturhist Samml Dresden:1–16

Theobald W (1984) Technik des Kunsthandwerks im zwölften Jahrhundert des Theophilus presbyter Diversarum Artium Schedula. VDI Verlag, Düsseldorf

Theophilus P (c1125 AD/1984) Diversarum Artium Schedula. Translated and annotated by W. Theobald. VDI Verlag, Düsseldorf

Tholander E (1986) Metallurgy and technology at Lapphyttan. Disc: 66–69, 105–108, 127–128. In: Jernkontorets Forskning H39, Stockholm

Thomas R, Pott DT (1996) Atacamite pigment at tell Abraq in the early Iron age. Arab Archaeol Epigr 7:13–16

Thompson CM (2009) Three 20th dynasty silver horads from the Egyptian garrison. In: Panitz-Cohen N, Mazar A (eds) Excavations at Tel Beth-Shean 1989–1996. Israel Explor Soc, Jerusalem, pp 597–611

Thomsen R (1975) Et meget m⍰rkelight metal. Varde Staalv⍰rk A/S

Thorndycraft VR, Pirrie D, Brown AG (2004) Alluvial Records of Medieval and Prehistoric tin Mining on Dartmoor, Southwest England. Geoarchaeology 19 (3):219–2004

Thornton C (2007) Of brass and bronze in prehistoric Southwest Asia. In: La Niece S, Hook D, Craddock PT (eds) Metals and mines. Studies in archaeometallurgy, pp 123–135

Thornton C (2009a) The emergence of complex metallurgy on the Iranian plateau: escaping the Levantine paradigm. J World Prehist 22:301–327

Thornton C (2009b) The chalcolithic and early bronze age metallurgy of Tepe Hissar, Northeast Iran: a challenge to the "Levantine paradigm". PhD-diss Univers of Philadelphia

Thornton C (2010) The rise of arsenical copper in southeastern Iran. Iranica Antiqua XLV:31–50

Thornton CP (2014) The emergence of complex Metallurgy on the Iranian Plateau. In: Roberts BW, Thornton CP (eds) Archaeometallurgy in global perspective. Methods and syntheses. Springer, New York, pp 665–696

Thornton C, Ehlers CB (2003) Early brass in the ancient near east. Inst Archeometall Stud Newslett 23:3–8

Thornton C, Rehren T (2009) A truly refractory crucible from fourth millennium Tepe Hissar, Northeast Iran. J Archaeol Sci 36:2700–2712

Thornton C, Rehren T, Pigott V (2009) The production of speiss (iron arsenide) during the early bronze age in Iran. J Archaeol Sci 36:308–316

Timberlake S (2007) The use of experimental archaeology/archaeometallurgy for the understanding and reconstruction of early bronze age mining and smelting technologies. In: La Niece S, Hook D, Craddock PT (eds) Metals and mines. Brit Mus, London, pp 27–36

Timberlake S (2018) Experimental archaeology in bronze age mining and smelting- hard rock, hot metal, new ideas. In: Eisenach P, Stöllner T, Windler A (eds) The

RITAK Conf 2013–2014, Der Anschnitt Beih, vol 34, pp 125–139

Tite MS, Freestone IC, Bimson M (1983) Egyptian Fayence: an investigation of their methods of production. Archaeometry 25(1):17–27

Tizzoni M (1999) Etruscan tin: a legend? Pallas 50:89–100

Todd M (1996) Pretia victoriae? Roman lead and silver mining in the Mendipp Hills, Somerset, England. Münster Beitr antike Handelsgeschichte 15(1):1–18

Tolksdorf JF, Hemker C, Schubert M (2019) Bronzezeitlicher Zinnseifenbergbau bei Schellerhau im östlichen Erzgebirge, Sachsen. Anschnitt 71:223–233

Tosi M (1983) Prehistoric Sistan 1. Istituto Italiano il Medio ed Estremo Oriente e Istituto Universitario Orientale di Napoli. ISMEO, Rome

Tosi M, Vidale MT (1990) 4th millennium BC lapis lazuli working at Mehrgarh, Pakistan. Paléorient 16 (2):89–99

Tosi M, Wardak R (1972) The Fullol hoard: a new find from bronze age Afghanistan. East West 22:9–17

Treister MJ (1996) Die trojanischen Schätze. In: Der Schatz aus Troja. Schliemann und der mythos des Priamus-Goldes. Katalog Ausstellung Moskau 1996/1997. Belser Verlag, Stuttgart Zürich, pp 197–236

Tretner A (2002) Sorptions- und Redoxprozesse von Arsen an oxidischen Oberflächen. Experimentelle Untersuchungen. PhD-diss, Univers Heidelberg

Trevor Hodge A (1983) Siphons in Roman Aquaeducts. Pap Br Sch Rom 51:174–221

Trojer F (1951) Delafossit in einer Kupferverblaseschlacke. Mikroskopie 7(8):232–236

Trojer F (1963) Die oxydischen Kristallphasen der anorganischen Industrieprodukte. Schweizerbart, Stuttgart

Tsaimou C, Tsakiridis PE, Oustadakis P (2015) Analytical and technological evaluation of ancient Lead slags from Lavrion, Attika, Greece. Mediterran Archaeol Archaeom 15(2):113–127

Tsuna M (1801) Kodo Zuroku. Illustrated book on the smelting of copper. Edited 1983 from Zenryu Shirakawa. Burndy Libr. Norwalk, Connecticut and 1984 from B. Lewin and A. Hauptmann, Deutsches Bergbau-museum Bochum

Tubb KW (1985) Preliminary report on the Ain Ghazal statues. Mitt Deutsche Orientges 117:117–134

Turck R, Stöllner T, Goldenberg G (eds) (2019) Alpine Copper II – Rame delle Alpi II – Cuivre des Alpes II. New Results and Perspectives on Prehistoric Copper Production. Der Anschnitt Beih 42

Twaltschrelidze A (2001) Erzlagerstätten in Georgien. In: Gambaschidze I, Hauptmann A, Slotta R, Yalçın Ü (eds) Georgien – Schätze aus dem Land des Goldenen Vlies. Exhibition Catalogue Deutsches Bergbau-Mus, Bochum, pp 78–89

Tylecote RF (1962) The archaeology of metal fabrication in the British Isles. Brit Weld J 9:512–522

Tylecote RF (1976) A history of metallurgy. The Metal Soc, London

Tylecote RF (1980) Summary of results of experimental work on early copper smelting. In: Oddy WA (ed) Aspects of early metallurgy, Brit Mus Occ pap, vol 17, pp 5–12

Tylecote RF (1982a) Early copper slags and Copper-Base metal from the Agadez region of Niger. J Hist Metall Soc 16(2):58–64

Tylecote RF (1982b) Metallurgical crucibles and crucible slags. In: Olin JS, Franklin AD (eds) Archaeological ceramics. Smithsonian Inst, Washington DC, pp 231–243

Tylecote RF (1983) Scottish antimony. Proc Soc Ant Scot 113:645–646

Tylecote RF (1986) The prehistory of metallurgy in the British Isles. The Institute of Metals, London

Tylecote RF (1987) The early history of metallurgy in Europe. Longman Archaeology Series, London

Tylecote RF (1991) Recent highlights in archaeometallurgy. In: Budd P, Chapman B, Jackson C, Janaway RC, Ottaway B (eds) Archaeological science 1989. Oxbow, Oxford, pp 194–201

Tylecote RF (1992a) A history of metallurgy. Inst Materials, London

Tylecote RF (1992b) Extraction metallurgy; historical development and evolution of the processes. In: Antonnacci Sanpaolo E (ed) Archeometallurgia Ricerche e Prospettive, Bologna, pp 25–51

Tylecote RF, Boydell PJ (1978) Experiments on copper smelting based upon early furnaces found at Timna. In: Rothenberg B (ed) Archaeometallurgy: chalcolithic copper smelting, Inst Archaeo-Metallurg stud Monogr 1, London, pp 27–49

Tylecote RF, Merkel JF (1985) Experimental smelting techniques: achievements and future. In: Craddock PT, Hughes MJ (eds) Furnaces and smelting technology in antiquity, Brit Mus Occ papers, vol 48, pp 13–20

Tylecote RF, Austin JN, Wraith AE (1971) The mechanism of the Bloomery process. J Iron Steel Inst 5:342–363

Tylecote RF, Rothenberg B, Lupu A (1974) The examination of metallurgical material from Abu Matar, Israel. J Hist Metall Soc 8(1):32–34

Tylecote RF, Ghaznavi HA, Boydell PJ (1977) Partitioning of trace elements between the ores, fluxes, slags and metal during the smelting of copper. J Archaeol Sci 4(4):305–333

Tylecote RF, Balmuth MS, Massoli-Novelli R (1983a) Copper and bronze metallurgy in Sardinia. J Hist Metall Soc 17:63–78

Tylecote RF, Balmuth MS, Massoli-Novelli R (1983b) Copper and bronze metallurgy in Sardinia. In: Balmuth MS, Rowland RJ Jr (eds) Studies in Sardinian archaeology. University of Michigan, pp 115–162

Tylecote RF, Photos E, Earl B (1989) The composition of tin slags from the south-west of England. World Archaeol 20(3):434–445

Valera PG, Valera RG (2005) Outline of geology and mineral deposits of Sardinia. In: Lo Schiavo F,

Giumlia-Mair A, Sanna U, Valera R (eds) Archaeometallurgy in Sardinia from the origins to the early Iron age, Monographies instrumentum, vol 30, pp 35–42
Valera RG, Valera PG, Lo Schiavo F (2002) Lead in Nuragic Sardinia: ores, isotopy, and archaeology. In: Bartelheim M, Pernicka E, Krause R (eds) Die Anfänge der Metallurgie in der Alten welt. The beginnings of metallurgy in the Old World, Forsch Archäometr u Alterumswiss, vol 1. M Leidorf, Rahden/Westf, pp 359–377
Valera RG, Valera PG, Rivoldini A (2005) Sardinian ore deposits and metals in the bronze age. In: Lo Schiavo F, Giumlia-Mair A, Sanna U, Valera R (eds) Archaeometallurgy in Sardinia from the origins to the early Iron age, Monographies instrumentum, vol 30, pp 43–87
Vallespìn Gómez O (1986) The copper wreck (Pecio del cobre). Internat J Nautical Archaeol Underwater Explor 15:305–322
Van Buren M, Cohen C (2010) Technological changes in silver production after the Spanish conquest in Porco, Bolivia. Boletin del Museo Chileno de Arte Precolombino 15(2):29–46
Van der Merwe NJ, Avery DH (1987) Science and magic in African technology: traditional Iron smelting in Malawi. J Internat Afr Inst 57(2):143–172
Van Lerberghe K (1988) Copper and bronze in Ebla and in Mesopotamia. In: Waetzold H, Hauptmann H (eds) Wirtschaft und Gesellschaft von Ebla, Akten der Int tag Heidelberg Nov 1986. Heidelber Stud Alter Orient, vol 2, pp 253–255
Vatandoust A (2004) Alter Bergbau und Metallurgie in Iran – Vergangenheit und Zukunft einer Forschungsperspektive. In: Stöllner T, Slotta R, Vatandoust A (eds) Persiens Antike Pracht 1. Exhibition-catalogue Deutsches Bergbau-museum Bochum, pp 2–21
Vatandoust A, Parzinger H, Helwing B (2011) Early mining and metallurgy on the western central Iranian plateau. Report on the first five years research on the Joint Iranian-German Research Project Archäologie in Iran und Turan 9
Vavelidis M, Bassiakos I, Begemann F, Patriarcheas K, Pernicka E, Schmitt-Strecker S, Wagner GA et al (1985) Geologie und Erzvorkommen. In: Wagner GA, Weisgerber G (eds) Silber, Blei und gold auf Siphnos. Prähistorische und antike Metallproduktion, Der Anschnitt Beih, vol 3, pp 59–80
Vavelidis M, Pernicka E, Wagner GA (1988) Die Goldvorkommen von Thasos. In: Wagner GA, Weisgerber G (eds) Antike Edel- und Buntmetallgewinnung auf Thasos, Der Anschnitt Beih, vol 6, pp 113–124
Velasco Roldan F, Herrero JM, Suarez S, Yusta I (2013) Supergene features and evolution of gossans capping massive sulphide deposits in the Iberian Pyrite Belt. Ore Geol Rev 53:181–203
Venerandi-Pirri I, Zuffardi P (1982) The tin deposit of Monte Valerio (Tuscany, Italy): Pneumatolytic-hydrothermal or sedimentary-remobilization processes. In: Amstutz GC et al (eds) Ore genesis. Springer, Berlin, pp 58–65
Verein Deutscher Eisenhüttenleute (1995) Slag atlas. Verlag Stahleisen, Düsseldorf
Vickers M, Kakhidze A (2001) The British-Georgian excavation at Pichvnari 1998: the "Greek" and "Colchian" cemeteries. Anatol Stud 51:65–90
Villa IM (2009) Lead isotopic measurements in archaeological objects. J Archaeol Anthropol Sci 1:149–153
Vitruvius MVP (1st cent BC) Ten Books on Architecture
Vlachović J (1977) Die Kupfererzerzeugung und der Kupferhandel in der Slowakei vom Ende des 15. Bis zur Mitte des 17. Jahrhunderts. In: Kellenbenz H (ed) Schwerpunkte der Kupferproduktion und des Kupferhandels in Europa 1500–1650. Böhlau Verlag, Köln, pp 148–171
Voß HU, Hammer P, Lutz J (1998) Römische und germanische bunt- und edelmetallfunde im Vergleich. Archäometallurgische Untersuchungen ausgehend von elbgermanischen Körpergräbern. Ber Röm-German Komm 79:107–382
Vogel H (2015) Wer wurde in der Grabkammer der Royal Tomb 1054 im Royal Cemetery in Ur bestattet? Zeitschr Orient-Archäol 8:38–60
Von Baer KE (1792–1876) Findbuch zum Nachlass von Karl Ernst von Baer. Ed. by Tammiskaar E 1999. Gießen
Von Naumann CF (1852) Elemente der Mineralogie. Publ W Engelmann, Leipzig
Voss O, Smekalova T (2003) Using magnetic prospecting for investigation of Roman age iron smelting sites with slag blocks in Denmark. In: Associazione Italiana di Metallurgia (ed) Archaeometallurgy in Europe I. Conf Proc Milano, Italy. Ass. Italiana di Metallurgia, Milano, pp 53–62
Waetzold H (1985) Rotes gold? Oriens Antiquus XXIV (1–2):1–16
Waetzold H, Bachmann HG (1984) Zinn- und Arsenbronzen in den Texten aus Ebla und aus dem Mesopotamien des 3. Jahrtausends. Oriens Antiquus XXIII(1–2):1–18
Wagenbreth O, Wächtler E (1986) Der Freiberger Bergbau. Technische Denkmale und Geschichte. Deutscher Verlag Grundstoffindustrie, Leipzig
Wagenbreth O, Wächtler E, Becke A, Douffet H, Jobst W (1990) Bergbau im Erzgebirge. Technische Denkmale und Geschichte, Deutsch Verlag Grundstoffind, Leipzig
Wagner G, Öztunali Ö (2000) Prähistorische Kupferquellen in der Türkei. In: Yalçın Ü (ed) Anatolian metal I, Der Anschnitt Beih, vol 13, pp 31–67
Wagner T, Schneider J (2002) Lead isotope systematics of vein-type antimony mineralization, Rheinisches Schiefergebirge, Germany: a case history of complex reaction and remobilisation processes. Mineral Deposits 37:185–197
Wagner GA, Pernicka E, Seeliger T, Lorenz I, Begemann F, Schmitt-Strecker S, Eibner C, Öztunali

Ö (1986) Geochemische und isotopische Charakteristika früher Rohstoffquellen für Kupfer, Blei, Silber und gold in der Türkei. Jb Röm-German Zentralmus 33:723–752
Wagner GA et al (1989) Archäometallurgische Untersuchungen an Rohstoffquellen des frühen Kupfers Ostanatoliens. Jb Röm-German Zentralmus 36:637–686
Wagner GA, Wagner I, Öztunali Ö, Schmitt-Strecker S, Begemann F (2003) Archäometallurgischer Bericht über Feldforschung in Anatolien und bleiisotopische Studien an Erzen und Schlacken. In: Stöllner T, Körlin G, Steffens G (eds) Mensch und Bergbau, Der Anschnitt, Beih, vol 16, pp 475–494
Wahl J (1988) Três Minas. Vorbericht über die archäologischen Untersuchungen im Bereich des römischen Goldbergwerks 1986/87. Madrider Mitt 29:221–244
Wahl J (1993) Treŝ Minas. Vorbericht über die archäologischen Ausgrabungen im Bereich des römischen Goldbergwerks 1986/87. In: Steuer H, Zimmermann U (eds) Montanarchäologie in Europa. Thorbecke, Sigmaringen, pp 123–152
Wahl-Clerici R, Wiechowski A, Helfert M, Ramminger B (2012) Die Golderzaufbereitung im römischen Bergwerksbezirk von Três Minas und campo de Jales in Nordportugal. Anschnitt 64(2–3):109–118
Walach G (2004) Die bronzezeitliche Kupfergewinnung im Palten- und im Liesingtal, Prospektionsbefunde und vergleichende Bewertung. Res montanarum 33:15–22
Walach G (2008) Archäometrische Prospektion. In: die Produktion von Ferrum Noricum am Hüttenberger Erzberg. Die Ergebnisse der interdisziplinären Forschungen auf der Fundstelle Semlach/Eisner in den Jahren 2003-2005. Austria Antiqua 2:15–27
Walach G (2017) Geophysical prospection at the Hüttenberger Erzberg. In: Chech B (ed) The production of Ferrum Noricum at the Hüttenberger Erzberg. Austria Antqua, vol 6, pp 27–45
Waldbaum JC (1980) The first archaeological appearance of Iron and the transition to the Iron age. In: Wertime TA, Muhly JD (eds) The coming of the age of iron. Yale University Press, New Haven, pp 69–98
Waldbaum JC (1999) The coming of the age of iron in the eastern Mediterranean. In: Pigott VC (ed) The archaeometallurgy of the Asian old world, University museum monograph 89. The university museum, Univers Pennsylvania, pp 27–57
Walker EC, Cuttitta F, Senftle FE (1958) Some natural variations in the relative abundance of copper isotopes. Geochim Cosmochim Acta 15:183–194
Waniczek K (1986) Ein Beitrag zur Zinnmetallurgie der Bronzezeit. Alt-Thüringen 21:112–135
Watson H (1786) A delineation of the strata of Derbyshire, Sheffield 1811. Reprint little Haywood, Staffordshire: moorland reprints 1973 (ed TD Ford)
Wayman ML (1987) On the early use of iron in the Arctic. Part I: meteoritic iron. CIM Bull 80(899):147–150
Wayman ML, Duke MJ (1999) The effects of melting on native copper. In: Hauptmann A, Pernicka E, Rehren T, Yalçın Ü (eds) The beginnings of metallurgy. Proc Internat Conf, Bochum 1995, Der Anschnitt Beih, vol 9, pp 55–63
Wayman ML, Smith RR, Hickey CG, Duke MJM (1985) The analysis of copper artifacts of the copper Inuit. J Archaeol Sci 12:367–375
Weber L (1997) Handbuch der Lagerstätten der Erze, Industrieminerale und Energierohstoffe Österreichs. Archiv für Lagerstättenforschung, band 19. Wien
Weber L, Pausweg F, Medwenitsch W (1972) Zur Mitterberger Kupfervererzung (Mühlbach/Hochkönig). Mitt Geol Gesellsch Wien 65:137–158
Wedepohl KH (1968–1978) Handbook of geochemistry. Springer, Berlin
Weeks L (1997) Prehistoric metallurgy at tell Abraq, U.A.E. Arab Archaeol Epigr 8:11–85
Weeks L (1999) Lead isotope analyses from tell Abraq, United Arab Emirates: new data regarding the "tin problem" in Western Asia. Antiquity 73:49–64
Weeks L (2004) Early metallurgy of the Persian Gulf. Technology, trade, and the bronze age world, American school Prehist res Monogr Ser (Lamberg-Karlovsky CC, Pilbeam D, Bar-Yosef O (eds)). Brill academic Publ, Boston
Weiler E, Nover L (2008) Allgemeine und molekulare Botanik, Stuttgart
Weinhold G (2002) Die Zinnerzlagerstätte Altenberg / Osterzgebirge. Bergbau in Sachsen 9. Union Druckerei Dresden
Weiser T (2004) Platinum-group minerals (PGM) from placer deposits in the mineral collection of the Museum of Natural History, Vienna, Austria. Ann Naturhist Mus Wien 105A:1–28
Weiser T, Bachmann HG (1999) Platinum-group minerals from the Aikora river area, Papua New Guinea. Can Mineral 37:1131–1145
Weisgerber G (1978) A new Kind of copper slag from Tawi 'Arja, Oman. J Hist Metall Soc 12(1):40–43
Weisgerber G (1979) Das römische Wasserheberad aus Rio Tinto in Spanien im British museum London. Der Anschnitt 2–3:62–79
Weisgerber G (1980) Patterns of early Islamic metallurgy in Oman. Proc Seminar Arab Stud X:115–126
Weisgerber G (1981) Mehr als Kupfer in Oman – Ergebnisse der Expedition 1981. Der Anschnitt 33(5–6):174–263
Weisgerber G (1985) Die Blei-Silbergruben von Agios Sostis. In: Wagner GA, Weisgerber G (eds) Silber, Blei und gold aus Sifnos. Prähistorische und antike Metallproduktion, Der Anschnitt Beih, vol 3, pp 113–158
Weisgerber G (1987a) Montanarchäologie – ein Weg zum Verständnis früher Rohstoffversorgung. In: Pörtner R, Niemeyer HG (eds) Die großen Abenteuer der Archäologie 19. Andreas Verlag, Salzburg, pp 3503–3540

Weisgerber G (1987b) Archaeological evidence of copper exploitation at Arja. J Oman Stud 9:145–237
Weisgerber G (1989) Montanarchäologie. Grundzüge einer systematischen Bergbaukunde für Vor- und Frühgeschichte und Antike. Anschnitt 6:190–204
Weisgerber G (1991) Die Suche nach dem altsumerischen Kupferland Oman. Das Altertum 37(2):76–90
Weisgerber G (1993) Römischer Erzbergbau in Deutschland. In: Steuer H, Zimmermann U (eds) Alter Bergbau in Deutschland. Sonderheft Archäologie in Deutschland. Theiss, pp 55–62
Weisgerber G (1996) Montanarchäologie – mehr als Technikgeschichte: das Beispiel Fenan (Jordanien). In: Kroker W (ed) Montantechnologie an historischen Schnittstellen, Schriftenreihe Georg-Agricola-Gesellschaft, vol 20, pp 19–34
Weisgerber G (1995) Aufgaben der Montanarchäologie. Archäologie Österreichs 6(2):23–29
Weisgerber G (2003) Spacial organisation of mining and smelting at Feinan, Jordan: mining archaeology beyond the history of technology. In: Craddock P, Lang J(eds) Mining and metal production through the ages. The British Museum Press, pp76–89
Weisgerber G (2004a) Prähistorischer und historischer Bergbau in Afghanistan 1. Anschnitt 4:126–151
Weisgerber G (2004b) Prähistorischer und historischer Bergbau in Afghanistan 2. Der Anschnitt 5–6:190–211
Weisgerber G (2004c) Schmelzanlagen früher Kupfergewinnung: ein Blick über die Alpen. In: Weisgerber G, Goldenberg G (eds) Alpenkupfer, Der Anschnitt Beih, vol 17, pp 15–37
Weisgerber G (2006) The mineral wealth of ancient Arabia and its use I: copper mining and smelting at Feinan and Timna – comparison and evaluation of techniques, production, and strategies. Arab Archaeol Epigr 17:1–30
Weisgerber G (2007) Roman brass and lead ingots from the western Mediterranean. In: La Niece S, Hook D, Craddock PT (eds) Metals and mines. Studies in archaeometallurgy. Brit Mus and Archetype Books, London, pp 148–158
Weisgerber G, Al-Shanfari AAB (2013) Archaeology in the Arabian Sea. Masirah and Al Hallaniyyat Islands. Sultanate of Oman, Muscat
Weisgerber G, Cierny J (2008) Zu paläolithischer Gewinnung roter Farbmineralien auf der Insel Thasos. In: Yalçın Ü (ed) Anatolian metal IV, Der Anschnitt Beih, vol 21, pp 179–190
Weisgerber G, Goldenberg G (eds) (2004) Alpenkupfer – Rame delle Alpi. Der Anschnitt Beih 17
Weisgerber G, Heinrich G (1983) Laurion und kein Ende? Kritische bemerkungen zum Forschungsstand über eines der bedeutendsten antiken Bergreviere. Anschnitt 6:190–200
Weisgerber G, Pernicka E (1995) Ore mining in prehistoric Europe. In: Morteani G, Northover JP (eds) Prehistoric gold in Europe. Kluwer Academ Publ, pp 159–182
Weisgerber G, Yule P (2003) Al-Aqir near Bahla – an early bronze age dam site with planoconvex "copper" ingots. Arab Archaeol Epigr 14:24–33
Wenk HR, Bulakh A (2004) Minerals. Their constitution and origin. Cambridge University Press
Werner O (1981) Analysen mittelalterlicher Bronzen und Messinge II und III. In: Ankner D, Schauer P, Waurick G (eds) Archäologie und Naturwissenschaften 2. RGZM, Mainz-Bonn, pp 106–170
Werthmann K (2000) Gold rush in West Africa. The appropriation of "natural" resources: non-industrial gold Mining in South-Western Burkina Faso. Sociologus 50(1):90–104
Werthmann K (2003a) The president of the gold diggers: sources of power in a gold mine in Burkina Faso. Ethnos 68(1):95–111
Werthmann K (2003b) Cowries, gold and "bitter money". Gold Mining and notions of ill-gotten wealth in Burkina Faso Paideuma 49:105–124
Werthmann K (2006) Gold diggers, earth priests, and district heads: land rights and gold Mining in Southwestern Burkina Faso. In: Kuba R, Lentz C (eds) Land and the politics of belongings in West Africa. Brill, Leiden, pp 119–136
Werthmann K (2009) Bitteres gold. Bergbau, Land und Geld in Westafrika. Rüdiger Köppe Verlag Köln
Wertime TA (1967) A metallurgical expedition through the Persian Desert. In: Caldwell R (ed) Excavation at Tal-I Iblis, Illinois state museum, preliminary reports, vol 9, pp 327–339
Wertime TA (1968) A metallurgical expedition through the Persian Desert. Science 159:927–935
Wertime TA (1973) The beginnings of metallurgy: a new look. Science 182:875–887
Wertime TA, Muhly JD (1980) The coming of the age of Iron. Yale University Press, New Haven/London
Westermann E (1986) Zur Silber- und Kupferproduktion Mitteleuropas vom 15. Bis zum frühen 17. Jahrhundert. Über Bedeutung und Rangfolge der Reviere von Schwaz, Mansfeld und Neusohl. Anschnitt 38:187–211
Westner K, Klein S, Gassmann G (2016) Roman to medieval precious and base metal smelting near Ulpiana (Kosovo): evidence for complex multistage extraction of silver. In: Körlin G, Prange M, Stöllner T, Yalçın Ü (eds) From bright ores to shiny metals, Festschrift a Hauptmann. Der Anschnitt Beih, vol 29, pp 205–218
Westphalen P (1989) Eisenschlacken aus Haithabu. Ber Ausgr Haithabu 26
Widdowson M (2007) Laterite and Ferricrete. In: Nash DJ, SJ ML (eds) Geochemical sediments and landscapes. Wiley, Oxford, UK, pp 46–94
Wiegels RR (2014) Zwei gestempelte Silberbarren und die spätantiken Goldbarren aus Siebenbürgen. Akston Verlag, München
Wilke A (1952) Die Erzgänge von St. Andreasberg im Rahmen des Mittelharz-Ganggebietes, Beih Geol Jb 7, Hannover, pp 1–228

Wilke A (1975) Verfahren zur Probenahme voon Erzen (Konzentraten und ähnlichen Rohstoffen). In: Analyse der Metalle 3: Probenahme. Springer, Berlin, pp 82–187

Wilkinson TC (2014) Tying the threads of Eurasia. Sidestone Press, Leiden

Willers H (1901) Die römischen Bronzeeimer von Hemmoor, Hannover/Leipzig

Williams D (1934) The geology of the Rio Tinto mines, Spain. Bull Inst Min Metall 43:593–678

Williams D (1950) Gossanized breccia-ores, Jarosites and jaspers at Rio Tinto, Spain. Bull Inst Min Metall 526:1–12

Williams PA (1990) Oxide zone geochemistry. Ellis Horwood, Chichester

Willies L (1982) Lead and lead mining. Shire Album 85, Aylesbury

Willies L (1992) Report on the 1991 archaeological survey of Kestel mine, Turkey. Bull Peak District Mines Hist Soc 11(5):241–247

Wilmsen EN, Campbell AC, Brook GA, Robbins LH, Murphy M (2013) Mining and moving secular haematite in Botswana, ca. 200-1300 SD. In: Humphries J, Rehren T (eds) The world of iron. Archetype Publications, London, pp 35–45

Wilsdorf H (1952) Bergleute und Hüttenmänner im Alterum bis zum Ausgang der römischen Republik. Freiberger Forschungshefte D, 1

Wilson L, Pollard AM (2001) The provenance hypothesis. In: Brothwell DR, Pollard AM (eds) Handbook of archaeological sciences. Wiley, Chichester, pp 507–517

Wimmenauer W (1985) Petrographie der magmatischen und metamorphen Gesteine. Enke Publ, Stuttgart

Winkelmann H (1964) Altjapanischer Goldbergabu. Eisenhütte Westfalia, Lünen

Winkelmann H (1968) Lazarus Ercker. Drei Schriften: Das Kleine Probierbuch von 1556. Vom Rammelsberge, und dessen Bergwerk, ein kurzer Bericht von 1565. Das Münzbuch von 1563. Vereinigung Freunde Kunst Kultur, Bochum

Wolf D et al (2011) Geoarchäologische Untersuchungen der Goldvorkommen von Sotk und Fioletovo, Armenien. In: Meller H, Avetisyan P (eds) Archäologie in Armenien, Veröff. Landesamt Denkmalpflege und Archäologie Sachsen-Anhalt – Landesmus Vorgeschi, vol 64, pp 51–68

Wolfart R, Wittekindt H (1980) Geologie von Afghanistan, Beitr Regionale Geol Erde 14. Borntraeger, Berlin

Wollmann V (1976) Römischer Goldbergbau in Alburnus Major (Roşia Montană) in Dakien. Anschnitt 28 (6):182–191

Wolters J (1986) Der gold- und Silberschmied. Rühle-Diebener-Verlag, Stuttgart

Woodhead AP, Gale NH, Stos-Gale AZ (1999) An investigation into the fractionation of copper isotopes and its possible application to Archaeometallurgy. In: Young SMM, Pollard AM, Budd P, Ixer RA (eds) Metals in antiquity, BAR Internat Ser, vol 792, pp 134–139

Woolley CL (1934) The royal cemetery, Ur excavations, vol 2. Trustees of the British Museum and of the Museum of the University of Pennsylvania, London

Woolley CL (1938) The excavations at Al Mina, Sueidia II. J Hellenic Soc 58:133–170

Woolley CL (1955) Ur excavations, the early periods IV. Trustees Briti Mus and Mus Universi Pennsylvania, London

Wotruba H, Riedel F (2006) Sensor-based sorting of Metallferous ores - an overview. XXIII International Mineral Processing Congress, Istanbul, Turkey, pp 21–30

Wright PA (1982) Extractive metallurgy of tin, 2nd edn. Elsevier, Amsterdam

Wunderlich CH, Lockhoff N, Pernicka E (2014) De Cementatione oder: Von der Kunst, das Gold nach Art der Alten zu reinigen. In: Meller H, Risch R (eds) Metalle der Macht – Frühes Gold und Silber. 6, Mitteldeutscher Archäologentag, October 2013, Halle (Saale), pp 353–366

Wynne EJ, Tylecote RF (1958) An experimental investigation into primitive iron smelting technique. J Iron Steel Inst:339–348

Yahalom-Mack N, Segal I (2018) The origin of the copper used in Canaan during the late bronze age/Iron age transition. In: Ben-Yosef (ed) Mining for ancient copper. Essays in memory of B Rothenberg. Eisenbrauns, Indiana & Emery and Claire Yass Publ, Tel Aviv, pp 313–331

Yahalom-Mack N et al (2014) New insights into Levantine copper trade: analysis of ingots from the bronze and Iron ages in Israel. J Archaeol Sci 45:159–177

Yahalom-Mack N, Langgut D, Dvir O, Tirosh O, Eliyahu-Behar A, Erel Y (2015) The earliest Lead object in the Levant. PLoS One 10(12):e0142948. https://doi.org/10.1371/J.pone.0142948

Yakubchuk A, Seltmann R, Shatov V (2001) The Altaids: tectonic evolution and metallogeny. Soc Econ Geol Newslett 46:7–14

Yakubchuk A, Cole A, Seltmann R, Shatov V (2002) Tectonic setting, characteristics, and regional exploration criteria for gold mineralizetion in the Altai Orogenic collage: the Tien Shan Province as a key example. Soc Econ Geol Spec Publ 9:177–201

Yalçın Ü (1998) Der Keulenkopf von can Hasan (TR). Naturwissenschaftliche Untersuchung und neue interpretation. In: Rehren T, Hauptmann A, Muhly JD (eds) Metallurgica Antiqua, In honour of Hans-Gert Bachmann and Robert Maddin. Der Anschnitt Beih, vol 8, pp 279–289

Yalçın Ü (1999) Early iron production in Anatolia. Anatol Stud 49:177–187

Yalçın Ü (2000a) Anfänge der Metallverwendung in Anatolien. In: Yalçın Ü (ed) Anatolian metal I, Der Anschnitt Beih, vol 13, pp 17–30

Yalçın Ü (2000b) Frühchalkolithische Metallfunde von Mersin-Yumuktepe: Beginn der extraktiven Metallurgie? Tüb-Ar III:109–128

Yalçın Ü (2000c) Zur Technologie der frühen Eisenverhüttung. Arbeits- und Forschungsberichte zur Sächsischen Bodendenkmalpflege 42:307–316

Yalçın Ü (2008) Ancient metallurgy in Anatolia. In: Yalçın Ü, Özbal H, Günhan Paşamehmetoğlu AG (eds) Ancient mining in Turkey and the eastern Mediterranean. Atilim University, Turkey, pp 15–40

Yalçın Ü (2016a) Zinn für die Königin. Ein Barrenfragment aus Alacahöyük und seine Deutung. In: Körlin G, Prange M, Stöllner T, Yalçın Ü (eds) From bright Orest o shiny metals. Festschrift a Hauptmann on the occasion 40 years research in archaeometallutgy and archaeometry, Der Anschnitt Beih, vol 29, pp 69–74

Yalçın Ü (2017) The beginning of metal use in West Asia. In: Tsuneki A, Yamada S, Hisada K (eds) Ancient West Asian Civilisation. Geoenvironment and Society in the Pre-Islamic Middle East, Springer, pp 115–130

Yalçın Ü (2020) Prähistorische Kupfergewinnung in Derekutuğun 2: Die Bergbausiedlung in Derekutuğun. Der Anschnitt Beih, in prep.

Yalçın Ü, Hauptmann A (1995) Zur Archäometallurgie des Eisens auf der Schwäbischen Alb. In: Beiträge zur Eisenverhüttung auf der Schwäbischen Alb. Forsch Vor- und Frühgesch Baden-Württemberg, vol 55, pp 269–309

Yalçın Ü, Hauptmann A (2000) Zur Technologie der frühen Eisenverhüttung. Arbeits- und Forschungsber Sächs Bodendenkmalpfl 42:307–316

Yalçın Ü, Hauptmann A (2003) Archäometallurgie der früh- und hochmittelalterlichen Eisenverhüttung im Vorland der Schwäbischen Alb. In: Abbau und Verhüttung von Eisenerzen im Vorland der Schwäbischen Alb. Forsch. Vor- und Frühgesch Baden-Württemberg, vol 86, pp 127–158

Yalçın Ü, İpek Ö (2016) Prähistorische Kupfergewinnung in Derekutuğun, Anatolien 1: Montanarchäologische Forschungen in den Jahren 2009-2011. Ein Vorbericht Der Anschnitt Beih 30

Yalçın Ü, Maass A (2013) Prähistorische Kupfergewinnung in Derekutuğun, Anatolien. In: Yalçın Ü (ed) Anatolian metal VI, Der Anschnitt Beih, vol 25, pp 153–194

Yalçın Ü, Özbal H (2009) Ein neues Zinnvorkommen in Kayseri-Hisarcık, Zentralanatolien: Ein Vorbereicht. Tüb-Ar 12:117–122

Yalçın Ü, Pernicka E (1999) Frühneolithische Metallurgie von Aşıklı Höyük. In: Hauptmann A, Pernicka E, Rehren T, Yalçın Ü (eds) The beginnings of metallurgy. Proc Internat Conf "the beginnings of metallurgy", Bochum 1995, Der Anschnitt Beih, vol 9, pp 45–54

Yalçın Ü, Yalçın G (2008) Der Hortfund von Tülintepe, Ostanatolien. In: Yalçın Ü (ed) Anatolian metal IV, Der Anschnitt Beih, vol 21, pp 101–123

Yalçın Ü, Yalçın G (2013) Reassessing anthropomorphic metal figurines of Alacahöyük, Anatolia. Near Eastern Archaeol 76(1):38–49

Yalçın Ü, Pulak C, Slotta R (2005) Das Schiff von Uluburun. Welthandel vor 3000 Jahren. Ausstellungskatalog Deutsches Bergbau-museum Bochum

Yamaguchi T, Shirashi T (1971) Kinetic studies of eutectoid Decompostion of $CuFe_5O_8$. J Am Ceram Soc 54 (11):556–558

Yazawa A (1956) Fundamental studies on copper smelting V: mutual dissolution between matte and slag produced in the system Cu2S – FeS – SiO2. J Min Inst Jpn 72:305–311

Yener A (2000) The domestication of metals – the rise of complex metal industries in Anatolia. Leiden, Brill

Yener A (2008) Revisiting Kestel mine and Göltepe: the dynamics of local provisioning of tin during the early bronze age. In: Yalçın Ü, Özbal H, Pasamehmetoglu AG (eds) Ancient mining in Turkey and The Eastern Mediterranean, Ankara, pp 57–64

Yener A (2009) Strategic industries and tin in the ancient near east: Anatolia updated. Tüb-Ar 12:143–154

Yener A, Vandiver P (1993) Tin processing at Göltepe, an early bronze age site in Anatolia. Am J Archaeol 97:207–238

Yener A, Özbal H, Kaptan E, Pehlivan AN, Goodway M (1989a) Kestel: an early bronze age source of tin ore in the Taurus Mountains, Turkey. Science 244:200–203

Yener A et al (1989b) Stable isotope studies of central Taurus ore sources and related artifacts from eastern Mediterranean chalcolithic and bronze age sites. J Archaeol Sci 18:541–577

Yener A, Sayre EV, Joel EC, Özbal H, Barnes IL, Brill RH (1991) Stable Lead isotope studies of central Taurus ore sources and related artifacts from eastern Mediterranean chalcolithic and early bronze age sites. J Archaeol Sci 18:541–577

Yener A, Adriaens A, Earl B, Özbal H (2003) Analyses of Metalliferous residues, crucible fragments, experimental smelts, and ores from Kestel tin mine and the tin processing site of Göltepe, Turkey. In: Craddock PT, Lang J (eds) Mining and metal production through the ages. Brit Mus Press, London, pp 181–197

Yener A et al (2015) New tin mines and production sites near Kültepe in Turkey: a third millennium BC highland production model. Antiquity 89(345):596–611

Yigit O (2006) Gold in Turkey – a missing link in Tethyan metallogeny. Ore Geol Rev 28:147–179

Young WI (1972) The fabulous gold of the Pactalos valley. Bull Mus Fine Arts 70(359):5–13

Young T (2011) Some preliminary observations on hammerscale and its implications for understanding welding. J Hist Metall Soc 45(1):26–41

Yule P (1989) The copper hoards of the Indian subcontinent. Preliminaries for an Interpretation With Appendices by Hauptmann A and Hughes MJ Jahrb RGZM 36:193–274

Yule P, Weisgerber G (2001) The metal hoard from 'Ibrī/ Selme, Sultanate of Oman. Prähist Bronzefunde XX, 7, Stuttgart

Yund RA, Kullerud G (1966) Thermal stability of assemblages in the cu-Fe-S system. J Petrol 7 (1):454–488

Zaykov VV et al (2016) Platinoid microinclusions of a native osmium Group in Ancient Gold Artifacts from

Siberia and the Urals as a source of Geoarchaeological information. Archaeol Ethnol Anthropol Eurasia 44 (1):93–103

Zeiler M, Garner J, Golze R (2018) Der Altenberg bei Müsen, Kreis Siegen-Wittgenstein. Montanarchäol Westfalen 1LWL Münster

Zettler R (1984–1985) Inanna temple publication report. Oi.uchicago.edu: 47–50

Zettler R (1990) Metalworkers in the economy of Mesopotamian in the late third millennium B.C. MASCA J Res Pap Sci Archaeol 7:85–88

Zettler R, Horne L (1998) Treasures from the Royal Tombs of Ur. University of Pennsylvania Museum, Philadelphia

Zimmermann U, Goldenberg G (1991) Urgeschichtlicher Hämatitbergbau im Südschwarzwald. Anschnitt 43 (1):2–10

Zschocke K, Preuschen E (1932) Über das Herdschmelzen der arsenikalischen Kupfererze von Mancayan auf der Insel Luzon. In: das urzeitliche Bergbaugebiet von Mühlbach-Bischofshofen. Materialien zur Urgeschichte Österreichs. Anthropol Gesellsch Wien 6:257–262

Zuffardi P (1982) Giacimentologia e prospezione mineraria: Bologna. Editrice, Pitagora

Zwicker U (1980) Investigations on the extractive metallurgy of cu/Sb/as ore and excavated smelting products from Norsun Tepe (Keban) on the upper Euphrates (3500-2800 B.C.). in: Oddy WA (ed), Aspects of early metallurgy. Brit Mus Occ papers 17. London: 13-26

Zwicker U (1985) Investigation of samples from the metallurgical workshops at Kition. In: Karageorghis V, Demas M (eds) Excavations at Kition V. The Pre-Phoenician Levels, pp 403–417

Zwicker U (1991) Natural copper-arsenic alloys and smelted arsenic bronzes in early metal production. In: Mohen JP, Èluère C (eds) Découverte du métal. Picard, Paris, pp 331–340

Zwicker U, Viridis P, Ceruti ML (1980) Investigation on copper ore, prehistoric copper slag and copper ingots from Sardinia. In: Craddock PT (ed), Scientific studies in early mining and metallurgy. Brit. Mus. Occ. Paper 20, London: 135-163

Zwicker U, Greiner H, Hofmann KH, Reithinger M (1985) Smelting, refining and alloying of copper and copper alloys in crucible furnaces during prehistoric up to Roman times. In: Craddock PT, Hughes MJ (eds), Furnaces and smelting technology in antquity. Brit Mus Occ papers 48, London: 103-115

Subject Index

A
Acanthite, 88, 142, 143
Aes rude, 407–409
Aes signatum, 408
Africa-Eurasia convergent plate, 80
Agricola, G., 20, 79, 107, 155, 164, 167, 218, 252, 281, 289, 297, 300, 313, 320, 328, 340, 341, 344, 347, 352, 353, 370–373, 377, 446, 454, 465
Alite, 257, 267
Alluvial, 24, 56, 60, 61, 64–69, 71, 144–147, 149, 150, 152, 154, 166, 254, 320–322, 327, 351, 411, 419, 504
Alter Mann, 132, 171
Aludels, 371, 379
Amalgamation, 58–60, 80, 150, 323, 326, 327, 344, 378, 448
Amalgamation, 58–60, 80, 150, 323, 326–327, 331, 344, 378, 448
Andesites, 33, 105
Andremeyerite, 251, 257, 265, 266
Andronovo-Fedorovka culture, 228, 353
Anglesite, 75, 86, 88, 92, 94, 96, 139, 142, 278, 337
Anorthite, 254, 257, 265, 353
Antimony
  antimonites, 127, 128, 255, 370–372, 394, 395
  antimonium crudum, 371, 372
  crude, 371, 373
  needles, 255, 371
Antlerite, 40, 278
Anvil stones, 108, 162, 163, 200, 239, 467
Argentite, 75–77, 87, 95, 160, 337, 339, 421
Argentum vivum, 377
Arsenic, 7, 10, 21–23, 27, 41, 70, 83, 93, 104, 107, 116, 122–128, 130, 143, 159, 173, 179, 193, 203, 225, 249, 252, 271, 272, 280–285, 295, 296, 298, 299, 306, 310, 313, 316, 317, 324, 326, 339, 344, 354, 355, 359, 363, 364, 375, 378, 381, 383, 385–395, 398, 399, 401, 422, 467
Arsenical copper (CuAs), 17, 36, 42, 47, 125, 280, 283, 284, 306, 348, 381–393, 396–400, 422, 436, 474
Arsenides, 40, 69, 70, 125, 226, 256, 280, 281, 285, 286, 393
Arsenopyrites, 41, 46, 58, 62, 80, 100, 101, 105, 110, 111, 123–125, 128, 141, 172, 279, 282–285, 323, 390, 392, 401
Assaying, 166, 225, 227–228, 300, 342, 343, 353, 401, 413, 446, 465–467
Atacamite, 12, 13, 15, 40, 55, 57, 245, 309, 311, 391
Augite, 257, 264
Augustus, 149, 403
Aurian silver, 58–59, 75, 327, 381, 415, 420, 421
Aurichalcite, 88–90
Aurichalcum, 384, 385, 401–404
Auricupride, 22, 58, 59, 320, 416
Austenite, 426–430, 441
Azurite, 13, 15, 40, 83, 102, 138, 173, 298, 309, 311

B
Bacteria, 135
Bainite, 428, 430
Banded Iron Formation (BIF), 21, 26, 39, 60, 117
Bars and ingots
  bipyramidal bars, 360
Barite, 32, 75, 92, 96, 131, 140, 188, 224, 242, 245, 250, 251
Baur Glaessner diagram, 365
Bauxites, 24, 116, 504–506
The Beginning of the Use of Metals and Alloys (*BUMA*), 2, 9
Beneficiations, 5, 19, 29, 30, 32, 45, 58, 89, 97, 98, 109, 154, 161–169, 253, 255, 287, 313, 323, 329, 346, 351, 355, 366, 446, 449, 466, 467
Berggold, 60, 415
Black smokers, 25, 33, 47, 62
Blast furnaces, 121, 178, 233, 246–248, 264, 366
Blister copper, 18, 232, 317, 438, 493
Bloomery
  losses, 246, 366, 367
  processes, 120, 121, 196, 214, 216, 222, 233, 247, 315, 360, 363–367, 369, 425–427, 441, 449
  slags, 180, 196, 214, 216, 222, 233–235, 244, 246–248, 259, 315, 353, 363–368, 425, 452
  yield of iron, 366
Bloomery smelting, 248, 315
Blooms
  split, 360, 362
Blow pipes, 192, 200, 225, 226, 253, 299–301, 311, 343, 369, 461, 462, 467
Bodrogkeresztúr-culture, 301

A. Hauptmann, *Archaeometallurgy – Materials Science Aspects*, Natural Science in Archaeology,
https://doi.org/10.1007/978-3-030-50367-3

Bom Jesus, 341
Bornite, 29, 40, 41, 46, 48, 51, 231, 236, 272–276, 278, 309, 311, 312, 314, 455
Boudouard-equilibrium, 190–192, 365
Bouloun-Djounga, 447
Bowl furnaces, 231, 316
Brasses, 23, 27, 90, 92, 94, 95, 97, 143, 225, 296, 300, 374–376, 381, 382, 384, 385, 398, 401–406, 424
Braunite, 195
Brochantite, 40, 278, 391
Bronze Age
  Early Bronze Age, 16–18, 20, 27, 38, 44, 45, 51–57, 62, 80, 108, 109, 151, 152, 156, 159, 163, 172, 189, 197, 203, 206, 207, 209, 210, 219, 220, 225–228, 231, 232, 244, 248, 253, 259, 260, 263, 267–271, 283, 285, 300, 303, 306, 308, 315, 316, 318, 322, 328, 334, 338, 353, 356, 388, 389, 392, 393, 399–401, 407, 420, 422, 423, 438, 440, 460, 463, 466, 467, 474, 475, 488, 491, 493, 494
  Late Bronze Ages, 35, 44, 54, 57, 108, 128, 140, 173, 175, 204, 206, 217–219, 221, 227, 230, 231, 233, 235–240, 245, 249, 256, 258, 262, 270, 285, 303, 304, 308, 313, 314, 318, 349, 350, 360, 364, 370, 386, 392, 394, 395, 397, 399, 407, 438, 465, 488, 495, 500, 502
  Middle Bronze Ages, 47, 209, 308, 347, 395, 410, 422, 443, 494
Bronzes, 3, 6, 10, 11, 15, 17–20, 25, 27, 30, 33, 36, 39, 40, 43–46, 48, 52, 54, 55, 58, 64, 72, 74, 76, 83, 92–95, 98, 101–110, 123, 126, 128, 131, 138, 145, 159, 162, 167, 168, 170, 173, 181, 187, 188, 196, 199, 202–204, 208, 210, 217, 218, 225–227, 234, 235, 239, 241, 242, 249, 251, 253, 254, 258, 262, 264, 272, 276, 279, 280, 285, 286, 296, 299, 303, 304, 308, 313–318, 320–322, 330, 348–350, 354, 356–358, 381–383, 385, 386, 388–391, 393–401, 404, 406, 407, 410, 413, 416, 424, 429, 434, 440, 443, 456, 459, 462, 465–467, 471, 476, 495, 496, 498
Buddha, 452
Buffer equilibrium, 193, 197, 355
Bulk densities, 369
Buttermilcherz, 22, 76

C
Caesar, 323, 403
Calamine, 23, 27, 82, 83, 89–94, 143, 201, 375, 403–406
Calamine terre, 94
Carat, 412
Carbon
  carbon dioxide (CO_2), 190–192, 194, 345, 375
  carbon monoxide (CO), 191, 192, 194, 295, 311, 315, 345, 346, 363–365, 375, 425, 426, 438
Carnelian, 13, 73, 320
Cassiterites, 15, 19, 24, 30, 38, 43, 65, 69, 80, 97–111, 125, 143–146, 152, 166, 195, 227, 229, 253–255, 298, 350, 351, 353–356, 398, 401, 411, 466, 467, 495–498
Cassius'scher Goldpurpur, 64
Cast iron, 112, 247, 281, 296, 360, 363, 365, 381, 425–428, 431, 433, 441, 450
Casting moulds, 20, 200, 232, 301, 322, 457, 499
Castings, 3, 6, 14, 16, 17, 19, 225, 226, 228, 232, 281, 283, 296, 308, 309, 319, 320, 331, 356, 381, 384, 385, 387–389, 394, 395, 399, 401, 403, 404, 406, 408, 414, 417, 423, 434–438, 440, 441, 443, 454–459, 463–466, 475, 476, 498
Celsian, 251, 257, 266
Celtic, 72, 74, 238, 323, 360, 365, 369, 374, 413, 419, 425, 450, 477, 479, 504
Cementation zones, 7, 48, 51, 53, 74, 76, 84, 132, 134, 139–143, 157, 159, 315, 339, 488, 501
Cementite, 113, 114, 365, 426–431, 439, 441, 442
Cerussite, 75, 77, 78, 82, 83, 86–88, 92, 94–96, 142, 278, 286, 287, 293, 311, 332, 336–338, 344, 345, 421, 468, 476
Cervantite, 255, 256
Chaîne opératoire, 1, 19–20, 161
Chalcocite, 29, 40, 42, 48, 51, 77, 138, 139, 268, 272–277, 279, 299, 312, 501
Chalcolithic periods, 20, 217, 244, 284, 285, 289, 298, 310, 314, 386, 462
Chalcophile, 40, 69, 74, 86, 111, 123, 305, 479
Chalcopyrites, 15, 19, 22, 29, 30, 32, 37, 40, 41, 43, 44, 46–48, 51, 52, 61, 62, 71, 77, 80, 82, 87, 89, 93, 97, 98, 100, 101, 106, 110, 111, 116, 121, 122, 125, 128, 135, 137–139, 144, 170, 174, 208, 218, 231, 236, 253, 272–275, 278, 280, 299, 309, 311, 312, 314, 316, 318, 345, 358, 401, 437, 452, 492, 500, 501
Channel sample, 170
Charcoals, 178, 190–195, 200, 214, 219, 223, 225–227, 231–233, 237, 239, 268, 269, 295, 299, 300, 310, 311, 316, 317, 341, 345, 351, 356, 363–366, 368, 369, 371, 373, 375, 405, 407, 411, 424–426, 462–465, 467, 468
Chlorargyrite, 22, 75, 76, 88, 337
Chromites, 70, 71, 114, 117, 188, 224, 242, 503, 505, 506
Chrysocolla, 13, 40, 49, 51, 55, 57, 138, 311
Cicero, 403
Cinnabar, 26, 122, 130–132, 371, 377–379
Climates, 11, 16, 36, 76, 117, 145–146, 377
Clinopyroxenes, 196, 223, 224, 229, 231, 249, 251, 256, 257, 262–266, 269, 292
Coinage metals, 57, 333
Coin mould, 419
Colours of gold, 58, 412–413, 419
Conterfei, 377
Contrafeth, 377
Copper
  deposits, 13, 17, 18, 24, 26, 27, 32–36, 38–44, 46–52, 54–57, 59, 62, 63, 69, 83, 93, 94, 97, 98, 102, 104, 106, 124–126, 128, 129, 131, 132, 134, 136–138, 140, 141, 154, 156, 159–161, 167, 170, 171, 173, 194, 205, 207–209, 222, 231,

242, 245, 249, 259, 264, 278, 300–302, 306–309, 313, 316, 320, 322, 327, 349, 357, 359, 364, 368, 370, 382, 390, 394, 396, 400, 471, 474–476, 478, 487, 489, 491–495, 500–502, 505
isotopes, 38, 47, 51, 52, 55, 83, 122, 207, 209, 252, 297, 335, 369, 382, 393, 401, 407, 416, 472, 481, 483, 485, 487–489, 493–495, 499–502
melting, 10, 12, 16, 27, 59, 69, 181, 184, 189, 190, 216, 227, 234, 261, 271, 272, 274, 278, 280, 283, 300, 303, 304, 310, 311, 313, 318, 331, 332, 340, 341, 348, 356, 357, 364, 384, 387–390, 394, 399, 414, 416, 417, 422, 423, 438, 460, 461, 463, 493, 502
slags, 2, 7, 9, 15, 16, 18, 20, 30, 39, 47, 51, 52, 54, 55, 57, 63, 85, 124, 137, 162, 173–175, 181, 187, 189, 190, 194, 196, 197, 202, 204–210, 216, 217, 219–227, 229–245, 248–251, 256, 258–264, 267–278, 280, 281, 283, 285, 298–300, 303, 304, 308, 309, 311, 313–318, 324, 347, 364, 369–371, 390, 393, 398, 407, 409, 435, 437, 439, 451–453, 459, 460, 462, 463, 485, 489
smelting, 7, 14–18, 20, 30, 36, 39, 41, 44, 45, 47–51, 57, 124, 131, 170, 173–175, 187, 192, 194, 202, 205, 207–210, 216, 218–221, 223, 224, 226, 229–232, 234, 236, 238–242, 244, 245, 248, 250, 251, 256, 258, 259, 261–264, 268, 269, 271, 272, 277–279, 283–285, 297–301, 303–308, 310–319, 324, 339, 341, 343, 347, 357–359, 363, 366, 368, 371, 376, 386, 387, 390–393, 407, 408, 422, 424, 431, 437, 448, 451–454, 459–465, 471, 474, 494, 500
Copper minerals
atacamite, 51
azurite, 42
brochantite, 51
chalcocite, 40
chalcopyrite, 41–43, 51
covellite, 40
malachite, 41, 42, 51
native copper, 40
Copper ore deposits, 23, 27, 34, 36, 41, 47, 49, 125, 133, 138, 141, 205, 209, 278, 393, 483
Corpus Massarum Plumbearum Romanarum (CMPR), 335, 491, 492
Co-smelting, 7, 12, 96, 97, 296–299, 311, 315, 350, 390
Covellite, 29, 40–42, 48, 51, 101, 138, 139, 272, 273, 275, 314, 501
Cristobalite, 183, 184, 241, 243–245, 255, 257
Crucibles, 6, 15, 16, 18, 108, 109, 191, 193, 200, 219, 223–230, 232, 244–246, 253, 254, 256, 258, 271, 278, 283, 287, 295, 297, 299, 300, 308, 310, 311, 324, 328, 329, 342, 343, 353, 354, 356, 369–371, 373, 393, 398, 405–407, 416, 426, 438, 451, 452, 457, 458, 461–465, 467
Crude antimony, 371, 373
Cupellation, 168, 286–293, 299, 323–326, 333, 337–343, 345, 346, 421, 422, 454, 476
Cupels, 228, 286, 287, 289–292, 324, 333, 341–343, 347
Cuprite, 12, 15, 40, 48, 55, 138, 194, 195, 223, 224, 226–228, 230, 233, 257, 265, 267–270, 273, 280, 307, 310, 311, 314, 317, 357, 389, 438–440, 461, 462
Cuprospinel, 226, 268, 269
Cylindrite, 99
Cypro-classical time, 303

D
Dacites, 93, 105
Decarburisation, 361, 426–427, 450
Delafossite, 15, 16, 194, 195, 223, 224, 226, 227, 233, 257, 262, 268–270, 311, 314, 317, 462
Devil's mud, 49, 62, 79, 85, 142, 323, 421
Diopside, 257, 263
Dioscorides, 328, 336, 378
Distillation, 130, 295, 296, 299, 375, 376, 378, 379
Dolomite-Limestone-Shale Unit (DLS), 55, 56, 156, 157
Dolostones, 55, 90, 143, 156, 157
Dunites, 33, 70, 150, 419, 503
Dust tin, 97

E
Early Bronze Age, 18, 36, 37, 39, 51, 56, 83, 107–109, 156, 159, 173, 174, 189, 197, 206–210, 219, 220, 225–228, 231, 244, 253, 259, 263, 269, 270, 273, 283, 301, 308, 315, 349, 358, 370, 381, 386, 393, 422, 438, 440, 494
Egg boxes, 451
El Argar culture, 75, 338
Electrum, 58–59, 75, 141, 142, 327, 381, 389, 415, 421
Ellingham diagram, 193–195, 358, 375, 476
Enargite, 40–42, 124, 125, 139, 249, 309, 390, 391
English process, 211, 312, 314
Early Dynastic period, 464

F
Face centered lattice, 427
Fahlore
luzonite, 41
tennantites, 249
tetraedrite, 47, 249, 265
Famatinite, 41
Fayalites, 16, 177–181, 183–187, 195–197, 218, 220, 221, 223, 224, 226, 233–236, 238, 244, 247–249, 254, 256–263, 265, 266, 268, 269, 279, 315, 316, 346, 354, 355, 360, 366, 473
Ferricrusts, 118, 451, 478
Ferrifayalite, 257, 261, 262
Ferrite, 426–430, 441, 442
Ferritic iron, 19, 118, 247, 269, 360, 363, 364, 426, 428, 439
See also Iron
Ferropericlase, 269
Ferrosilite, 186, 195, 196, 263, 264
Ferrum Noricum, 122, 123, 202, 365, 425, 450
Fineness of gold, 106, 412
Fire gilding, 326, 331

Fischesserite, 58
Forsterite, 177, 179, 180, 254, 257, 259, 260, 268
Franckeite, 99
Furnaces, 1, 6, 18, 20, 152, 171, 178, 186, 191–196, 199, 200, 202, 204–206, 208, 209, 214, 215, 217, 218, 220, 222, 224, 226, 229, 232–234, 236–240, 245–248, 258, 261, 266, 267, 276, 280, 283, 284, 287, 295, 297–301, 303–305, 308, 313, 315, 318, 319, 324, 329, 340, 341, 344–347, 349, 351–354, 356, 358, 360, 362, 364–366, 369, 373–376, 379, 390, 398, 405, 407, 408, 410, 424–427, 448–457, 459, 460, 465, 468, 469, 480, 488, 493
Furnace slags, 199, 233, 236–238, 241, 246–248, 264

G
Galena
  argentiferous galena, 83, 88, 94, 338
Galenite, 22, 76, 77, 79, 80, 86, 92, 101, 172, 272, 274–276, 299, 311, 468
Gallium, 112, 113
Germanium, 112, 113
Gersdorffite, 43, 71, 124
Giants, 34, 35, 39, 49, 51, 52, 54, 61, 151, 201, 210
Gneisses, 61, 108
Goethite, 49, 77, 78, 111, 113–115, 117, 125, 137, 138, 140, 172, 242, 298, 339, 363, 478
Gold
  Amalgamation, 60, 150, 323, 326, 378
  authigenic gold, 63, 67–69, 72, 152
  electrum, 58, 59, 75, 141, 142, 381, 421
  mustard, 58, 62, 141
  native, 26, 30, 49, 57, 59, 62, 64, 75, 76, 129, 131, 139, 141, 296, 298, 318, 324, 338, 382, 411, 415, 421, 446
  only deposits, 26
  processing, 31, 58, 60, 64, 69, 73, 98, 109, 163, 165, 168, 180, 319, 322, 323, 331, 344, 411, 445, 454, 478
  silver contents, 58, 64, 66, 67, 209, 326, 328, 412, 416, 420, 477
Glacial, 11, 138, 141, 145
Goslarer Bergrecht, 199
Gossans, 3, 12, 16, 47–49, 51–53, 62, 75–77, 83–85, 87, 89, 90, 94, 125, 126, 132–142, 153, 159, 170, 172, 173, 190, 242, 252, 258, 302, 323, 339, 366, 390–392, 488
Grab samples, 169
Granites, 22, 99, 103, 105, 106, 108, 109, 128, 146, 152, 163, 498
Graphite, 406, 427, 429, 457
Greenstone belts, 60, 63, 503
Greisen, 99, 103, 105, 111, 351
Grinding, 161, 162, 164–166, 168, 239, 331, 385, 448, 449, 466, 467
Gunmetal, 381, 384, 402

H
Hammer stones, 85, 162, 163, 200, 239, 451
Hard-heads, 251, 253, 254
Hardystonite, 251, 264, 265
Hausmannite, 195, 197, 268
Hedenbergite, 196, 220, 251, 256, 257, 263, 264
Helicoidal washeries, 164
Hellenistic period, 250, 296, 321, 347, 473
Hematites, 21, 26, 30, 51, 63, 69, 77, 78, 83, 91, 95, 98, 108, 110, 111, 115–117, 120–122, 137, 138, 140, 144, 152, 159, 161, 173, 192, 195, 215, 227, 234–236, 242, 257, 265, 268, 300, 311–313, 339, 363, 364, 366, 367, 451, 467, 478
Hemimorphite, 83, 88–90, 92, 94–96, 405
Hemmoor buckets, 385, 403
Herzenbergite, 99
Hocartite, 99
Holzzinn, 99
Honeycomb texture, 66, 68
Huayrachina, 347, 349
Hyalophane, 251, 265, 266
Hydrargyrum, 377
Hydrothermal veins, 22–24, 27, 29, 30, 32, 37, 39, 43–45, 59, 67, 75, 79, 90, 91, 101, 103, 106, 116, 121, 123, 124, 129, 155–157, 172, 314, 336, 366, 475, 486
Hydrozincite, 83, 88–90, 92, 94, 405

I
Ilmenite, 69, 111, 144, 195, 254, 411, 503
Industrial phase, 12, 17–19
Ingots, *see* Bars
Inhibitors, 349
Inuit, 114
Interdisciplinary, 2–5, 45, 119, 121, 155, 202, 213, 214, 216, 335, 368, 369
Inverse segregation, 356, 388, 389, 391
Iridium, 69, 70, 72–74, 112, 113, 410, 411, 503, 504
Iron
  bloomery, 121, 191, 196, 214, 222, 233, 235, 244, 246–248, 315, 357, 360, 362, 364–366, 368, 369, 425–427, 441, 448–450
  casts, 374
  deposits, 3, 21, 24–27, 33, 55, 60, 70, 75, 91, 95, 111, 112, 115–122, 132, 135, 137, 153, 157, 172, 199, 214, 258, 311, 354, 358, 368, 408, 466, 472, 479, 480, 503
  ferritic, 365, 425, 439
  hut, 475
  isotopes, 114, 472, 479, 480
  pearlite, 426, 428, 429, 431, 441
  pig, 118, 248, 361, 363
  wrought, 233, 359–362, 364, 365, 425, 427, 428, 434, 450
Iscorite, 195, 234, 235, 257, 261–263, 267, 269
Isotopes
  analyses, 74, 481, 482, 499–501, 504
  copper, 297
  Leads, 86, 297, 481–484, 486, 496, 499, 500, 502
  osmium, 74, 502, 503
  tin, 495, 496, 499
Itabirite, 21, 26, 115, 117

J
Jarosites, 49, 63, 75, 77–80, 83–85, 91, 95, 137, 138, 140, 142, 241, 242, 299, 324, 329, 338, 339, 421, 476

K
Kalsilite, 228, 257
Kamacite, 110, 112, 113
Karst, 28, 80, 82, 83, 85, 87, 91, 94–96, 115, 118, 141, 143, 157, 250, 347
Karstic, 27, 65, 80, 83, 85, 87, 91, 95, 141, 143, 157, 250, 347
Karst cavities, 27, 28, 65, 94, 96, 115, 157
Kinnabari, 377
Kirschsteinite, 257, 259, 260, 268
Knebelite, 257, 259
Kodō-zuroku, 446, 454
Koller gang, 89, 164, 293
Koshthi, 375
Krennerite, 58, 60, 62

L
Laihunite, 195, 257, 261
Lapis lazuli, 73, 74, 102, 109, 110, 151, 320, 419, 504, 505
Larnite, 259, 260
Laterites, 24, 25, 64, 67, 114–117, 150, 359, 451, 504–506
Lauriotis, 374, 406
Lazarus Ercker, 20, 168, 169, 340, 341, 343, 370, 405, 406, 446
Leached cappings, 125, 132, 133, 135–137, 140
Lead
- carbonates, 77, 87, 89, 96, 102, 142, 143, 286, 298, 299, 332, 336, 338, 345, 347
- isotopy, 43, 480, 484–489, 491, 499
- ore deposits, 12, 24, 25, 30, 32, 43, 51, 52, 77, 83, 86, 92–94, 117, 125, 142, 143, 153, 171, 190, 201, 236, 239, 252, 298, 314, 336, 338, 339, 342, 345, 347, 401, 431, 454, 464, 472, 478, 480–486, 488, 490, 493, 500, 502
- half-life lead isotopes, 482

Lepidocrocite, 111, 114, 125, 137, 414
Leucite, 223, 233, 244, 249, 257, 266
Limonites, 12, 30, 41, 49, 58, 63, 65, 69, 75–78, 82–84, 87, 89, 90, 92, 94–96, 108, 114–117, 120–123, 125, 126, 135–138, 141, 142, 173, 186, 233, 242, 281, 298, 300, 311, 324, 329, 347, 357, 451, 467, 468
Liquation, 190, 218, 255, 337, 339, 340, 371, 454, 476
Listwaenites, 64
Litharges, 83, 85, 89, 96, 161, 162, 168, 200, 213, 286–293, 324, 329, 341–343, 346, 451, 458, 476
Lithargite, 86, 195, 286–288, 338, 342, 343, 345

M
Mabuki process, 454
Magistral, 344
Magnesiowuestite, 269
Magnetite, 15, 16, 24, 51, 69, 111, 113, 116, 122, 144, 148, 152, 153, 170, 188, 192, 195–197, 201, 220, 221, 223, 224, 226, 229, 231–236, 238, 242–244, 247, 248, 257, 259, 260, 262, 263, 265–270, 273, 274, 292, 298, 311–313, 316, 317, 354, 355, 363, 365–368, 430, 462, 473, 478
Maiolika, 254, 354
Malachite, 12, 13, 21, 30, 40, 41, 49, 51, 55, 57, 59, 62, 83, 90, 102, 133, 137, 138, 173, 270, 298, 309–311, 357, 390, 391, 398, 461, 501
Malayaite, 99
Mangellegierungen, 317, 393
Mantos, 81, 82, 94, 95
Martensite, 428–430
Massicote, 195
Massive Brown Sandstone (MBS), 55, 56
Matte smelting, 206, 228–231, 275, 276, 299, 312–315, 371, 453
Mattes, 29, 41, 162, 173, 199, 200, 206, 220, 228, 230, 231, 236, 239, 248, 256, 271–281, 284, 285, 298, 299, 304, 312–316, 324, 339, 346, 453, 455, 459, 462, 463, 500
Mawsonite, 99, 101
Medieval periods, 37, 75, 97, 107, 124, 164, 360, 378, 406, 430
Mercury, 21, 22, 26, 36, 52, 59, 60, 75, 80, 127, 129–132, 150, 295, 296, 299, 306, 326, 327, 331, 338, 344, 371, 372, 375–379, 435, 448, 480
Mercury ore deposits, 130, 131
Merwinite, 248, 257, 265
Mesozoic age, 33
Metallophytes, 154
Meteoric waters, 140, 153, 172
Meteorites, 112–114, 359
Meteoritic iron, 29, 112, 114, 116, 117, 298, 358, 359, 503
Micro-smelting, 467
Middle Bronze Ages, 208, 209, 312, 319, 333, 392, 395, 422, 443
Millerite, 43
Mississippi Valley-type deposit, 94, 95
Misy, 328, 329
Mithridates, 403
Mocksilver, 374
Mohenjo-Daro culture, 61, 454
Monazite, 98, 152
Monticellite, 257, 259, 260, 268
Moschellandsbergite, 326
Multidisciplinary, 5
Mushistonite, 42, 98, 99, 102, 143, 398, 401
Mustard gold, 58, 62, 141

N
Nagyagite, 62
Native metals, 12, 22, 40, 132, 172, 333
Neolithic periods, 11, 13–16, 45, 52, 153, 222, 300, 306, 386, 399, 405
Neutron diffractography, 442, 443

Nuggets, 30, 58, 63, 64, 66, 67, 70, 72, 145, 147, 148, 152, 155, 322, 351, 416, 446, 477

O

Obsidian, 13, 222
Oceanic crust, 33, 47–49
Olivenite, 42, 124, 391
Oolitic, 115
Ophiolites, 26, 33, 35, 47–50, 61, 63, 64, 69, 71, 73, 74, 92, 126, 151, 231, 359, 420, 475, 477, 478, 488, 503–506
Ore
  deposits, 1, 5–7, 11, 12, 16, 17, 19, 21–175, 201, 203, 204, 210, 213, 219, 249, 250, 268, 295, 310, 314, 325, 345, 377, 382, 391, 392, 395, 396, 398, 400, 401, 403, 405, 467, 471, 474, 475, 480, 481, 483, 484, 486, 488, 493, 497, 499, 503
  washeries, 83
Oreichalkos, 401
Osmiridium, 70, 502, 503
Osmium
  isotopes, 74, 151, 419, 477, 478, 503, 504
OXALID-databank, 1, 489
Oxibarometers, 195, 196
Oxidation zones, 23, 27, 34, 38, 40–43, 53, 58, 62, 63, 75–77, 83, 86, 88, 89, 91–94, 98, 101, 102, 122, 123, 125, 132, 133, 135, 137–143, 172, 278, 310, 311, 314, 315, 329, 336, 338, 344, 368, 390, 391, 398, 501
Oxhide ingots, 159, 160, 173, 308, 309, 350, 389, 399, 438–440, 464–466, 488, 491, 500, 502
Oxus treasure, 151, 419

P

Paleozoic age, 33, 38, 43, 47
Partridgite, 195
Patio process, 327, 344
Pegmatites, 23
Periclase, 224, 242, 257, 268
Peridotites, 25, 33, 70, 71, 117, 150
Pearlite, 426, 428–431, 441, 442
Petzite, 58, 60, 62
Pharaos, 60
Pharmacosiderite, 124, 173
Phase diagrams
  binary diagrams
    Au-Ag, 58, 62, 63, 65, 68, 382, 418
    Cu-As, 387
    Cu-Ag, 422
    Cu-Fe, 236
    Cu-Sn, 102
    Cu-O, 194, 197, 227, 303, 311, 313, 332
    Fe-C, 181, 427, 430, 441
    FeO- SiO_2, 183, 186–188, 247, 257
  ternary diagrams
    Au-Ag-Cu, 331, 410, 412–414, 417
    Cu-Fe-S, 250, 274–278, 284
    FeO-CaO-SiO_2, 15, 247
  quaternary diagrams
    CaO–FeO–Al_2O_3–SiO_2, 257
Phases, 12–17, 29, 43, 53, 65, 88, 96, 101, 113, 129, 131, 140, 141, 153, 296, 297, 300, 311, 313–315, 317, 318, 336, 354, 356, 375, 378, 379, 383, 386–390, 393, 394, 399–401, 407–409, 414, 418–420, 422, 423, 425–431, 473, 486, 500, 505
Philippines, 41, 71, 316
Phosphorous iron, 430, 431
Phyllite, 43
Pillow lavas, 48, 49
Placer deposits
  gold, 11, 26, 59, 60, 64, 66, 69, 71, 126, 144, 321, 419, 503
  platinum group minerals, 61, 144
  tin, 11, 107, 108, 144
Plano-convex bottom (PCB) slag, 219
Platinum
  platinum group elements (PGE), 69–71, 73, 122, 322, 411, 419, 502–504, 506
  platinum group minerals (PGM), 67, 69, 71–73, 150, 320, 419, 477, 502–504
Plessite, 113
Pliny, 20, 93, 108, 122, 148, 149, 165, 289, 292, 324, 326, 328, 336, 342, 344, 356, 357, 374, 376, 385, 401, 405, 406
Polymetallic ores, 26–27, 32, 37, 43, 46, 62, 89, 95, 103, 105, 107, 165, 186, 203, 213, 228, 241, 275, 280, 281, 382, 392, 394, 437
Porosity of gold, 320
Pozzolana, 13, 222
Precambrian, 37–39, 54, 57, 59–61, 66, 97, 475, 503
Pre-Pottery Neolithic, 9, 13, 51, 222, 306, 334, 440, , 442
Prospection, 49, 80, 153, 166, 171, 172, 201, 202, 416
Provenance studies, 2, 4–6, 25, 32, 34, 63, 72, 74, 86, 97, 101, 120, 122, 141, 143, 171, 218, 307, 369, 431, 438, 464, 471–506
Pseudargyros, 374
Psimythion, 336
Pyrite, 22, 29, 36, 37, 41, 43, 48, 51, 52, 58, 61–63, 69, 77, 80, 82, 87, 91, 93, 94, 96, 97, 101, 111, 115, 116, 122, 125, 128, 135, 137, 139, 141, 159, 172, 249, 250, 274, 279, 305, 311, 323, 329, 344, 364, 452, 479
Pyromorphite, 86, 88
Pyroxenes, 178, 221, 233, 248, 262–264, 313, 479
Pyrrhotite, 51, 61, 70, 71, 106, 111, 116, 170, 253, 272–276, 279, 280, 305, 312, 479

Q

Quartz, 12, 22, 30, 32, 43, 58, 62, 69, 75, 78, 79, 84, 100–103, 105, 106, 108, 109, 115–117, 122, 128, 133, 135, 137, 138, 144, 148, 149, 183, 184, 186–190, 194, 196, 197, 219, 220, 224, 226, 231, 232, 236–238, 240–245, 247, 249–251, 255–258, 313, 314, 316, 319, 320, 328, 355, 366, 452, 468, 479

Quartzite, 43, 131, 242
Quimbaletes, 448, 449

R
Radiocarbon-dating, 84, 219, 233, 480
Radiography, 359, 433, 434
Radiolites, 49, 51
Ramo secco bars, 407–409
Red ochre, 115, 116
Redox-conditions, 42, 63, 177, 178, 191, 195, 196, 199, 219, 398, 469, 473, 476
Reduction
  processes, 194, 298, 312, 315, 317, 345, 358, 360, 363, 407
Refractory gold, 58, 325
Regenbogenschüsselchen, 504
Rennfeuerschlacke, 366
Rennfeuerverfahren, see Slagging process; Bloomery smelting
Restites, 174, 188, 236, 238
Rhenium
  Re/Os-dating, 503
Ricchi in Argento, 94
Rhodonite, 197, 248, 263
Roasting
  dead, 312, 313, 315
  of ores, 18, 278, 300, 304, 312, 391, 398, 459, 500
  partial, 312, 313
Roman period, 35, 84, 94, 95, 119, 148, 156, 204, 214, 286, 288, 295, 315, 329, 339, 425
Round mills, 165
Ruthenium, 69, 70, 73, 74, 411, 502–504
Rutile, 98, 144, 152, 254, 411

S
Schreibersite, 113, 114
Secondary enrichment zones, 24, 30, 53, 77, 132, 139, 140, 143, 156, 160, 172, 392
SedEx-deposits, 90, 92
Sedimentary ore deposits, 39, 54, 141, 431
Seigerhüttenprozess, 339, 454
Se (Selenium), 69, 332, 425, 473, 478, 479
Senarmontite, 126, 127, 371, 372
Shaft furnaces, 192, 193, 214, 216, 231, 235, 253, 287, 299, 312, 313, 315, 339, 345–347, 360, 364, 459, 468
Shang and Zhou dynasties, 434
Silver
  aurian, 58, 59, 75, 327, 381, 415, 420, 421
  alloys, 17, 20, 57–59, 66–68, 72, 74, 75, 133, 179, 181, 225, 228, 253, 319–321, 326, 328, 329, 337, 340, 341, 357, 382, 387, 388, 393, 411–417, 420–423, 433, 454
  deposits, 21–23, 25, 30, 34–38, 40, 41, 45, 48, 53, 57–59, 62–64, 67, 74–77, 79, 80, 83, 85, 86, 90–96, 101, 105, 106, 131, 133, 139, 141–143, 159, 160, 209, 250, 252, 290, 292, 320, 323, 329, 333, 334, 336–340, 378, 415, 421, 477, 489, 492
  minerals, 22, 23, 40, 41, 58, 59, 62, 75–77, 79, 82–84, 87–93, 97, 99, 105, 129, 131, 133, 141–143, 160, 168, 252, 320, 324, 336, 338, 339, 372, 455, 475
  processing, 27, 32, 80, 89, 160, 164, 168, 213, 217, 310, 331, 340, 420, 422, 446, 453
Slag
  ancients, 51, 52, 54, 63, 137, 162, 163, 169, 173, 174, 178, 180, 184, 187, 190, 194, 200, 201, 209, 210, 213, 216, 218, 219, 222, 232, 235, 239, 242, 245, 251–253, 280, 297, 317, 339, 353, 407, 459
  archaeometallurgical, 15, 52, 121, 188, 216, 217, 219, 235, 256, 285, 478
  chemical bulk analyses, 222
  chemistry, 219, 228, 242, 251, 366
  compositions, 18, 51, 52, 177, 181, 184, 187, 190, 205, 216–219, 221–223, 225, 230, 231, 233, 236, 245, 247, 248, 252, 256–260, 263, 266, 282, 314, 353, 366, 367, 407, 439, 472, 478, 479
  copper content, 227
  Fe-rich, 247, 248, 251, 258, 259, 269, 407, 438
  furnace slag, 199, 233, 236–238, 241, 246–248, 264
  heaps, 2, 39, 51, 52, 54, 55, 63, 154, 160, 162, 163, 169, 171, 175, 199–210, 213–217, 219, 222, 231, 236–239, 249, 282, 299, 303, 304, 315, 339, 366, 369, 407, 442
  investigations, 16, 121, 201, 202, 215–219, 221, 222, 309, 313, 407
  phase contents, 178, 216, 228, 231, 248, 313, 366, 469
  phases, 15, 16, 177, 178, 180, 185, 187, 188, 190, 194, 216, 219, 221, 223, 228, 230, 231, 233, 241, 242, 247, 249, 251, 254, 256–261, 263, 266, 270, 271, 285, 353, 354, 462
  smithing, 219, 222, 244
  sulphide inclusions, 221, 231, 272–274, 276–280, 297, 307, 314, 438
  tap, 161, 206, 217, 218, 222, 233–237, 239, 305, 315, 366
  types, 162, 205, 206, 218, 222, 228, 229, 236, 238, 256, 267, 283, 299, 305, 314
Slagging process, 188, 203, 355, 406
Slagless metallurgy, 203, 226
Smelting
  crucibles, 199, 222–224, 226, 255, 269, 271, 297, 299, 354, 371, 390, 462, 463, 465
  reconstructions, 240, 249, 253, 297, 303–305, 313, 318, 347, 351, 369, 438, 459, 460, 462, 471, 473
Smithsonites, 83, 88–90, 92, 94, 95, 375, 405
Sphalerite, 22, 27, 32, 37, 41, 48, 51, 52, 61, 62, 77, 79, 82, 87–97, 101, 106, 139, 143, 172, 250, 253, 272, 274, 281, 344, 347, 375, 468, 479
Spinifex textures, 220, 237, 258
Spratzen, 438
Stamp mills, 154, 163, 164, 344
Stannite, 15, 19, 38, 42, 43, 80, 97–103, 105, 110, 111, 125, 143, 398, 498
Steels, 296, 300, 356–358, 360–365, 369, 374, 381, 382, 399, 424–431, 433, 434, 437, 438, 441, 442, 446, 448, 450
Steinarbeit, 19, 114, 159, 222, 225, 247, 261, 312, 313, 460, 478, 479

Stibiconite, 255, 256
Stibnite, 62, 127, 128, 394, 395
Strabo, 93, 122, 160, 214, 368, 374, 403
Supergene alteration, 61, 126
Sylvanite, 58, 60

T
Taenite, 111, 113, 114, 359
Tapslag, 263
Teallite, 99
Telescoping, 103, 106, 110, 121
Te (Tellurium), 69, 332, 473, 477–479
Tennantites, 40, 41, 46, 51, 75, 76, 124, 125, 249, 265, 309, 316, 382
Tenorite, 40, 195, 223, 233, 257, 270
Tension gashes, 91
Tephroite, 195, 197, 248, 257, 259
Terrestrial iron, 112–114
Tethyan Eurasian Metallogenic Belt (TEMB), 5, 11, 12, 17, 21, 26, 32, 34–39, 61, 71, 125, 145, 151, 359, 392, 399, 475, 483
Tetraedrite, 40, 41
Textures
  of ores, 29, 103
  of slags, 243, 258
  of metals, 104, 273, 383, 435
Theophilus Presbyter, 20, 218, 297, 315, 328, 329, 378, 403, 405, 446
Theophrast, 326
Thiobacillus ferrioxidans, 135
Tian gong kai wu, 446
Tile ores, 55, 138, 357
Time-temperature-transformation (TTT) diagram, 429, 430
Tin
  assaying, 166, 227, 300, 353, 401, 465–467
  belts, 25, 26, 97, 100, 103, 104, 109, 143, 396, 403, 498
  bronzes, 381, 382, 384–386, 388–390, 395–401, 403, 404, 411, 424, 429
  crackling, 349
  disease, 349
  dust tin, 97
  Greisen, 99, 104
  Holzzinn, 99
  isotope analyses, 107, 159, 350, 398, 401, 478, 495–497, 499, 502
  ore deposits, 2, 14, 15, 18, 19, 22, 24, 101, 105, 123, 132, 157, 159, 253, 399, 456, 466
  slags, 108, 109, 161, 203, 217, 222, 227, 228, 233, 253–255, 328, 351, 353–356, 398, 465, 467
  smelting, 10, 101, 106, 108, 152, 203, 217, 222, 225, 227, 253, 254, 300, 351–356, 381, 386, 390, 396, 398, 401, 407, 424, 467, 498, 502
  Stockscheider, 103
  tin and copper minerals, 43, 398, 400
  Visiergraupen, 99
  Zinngraupen, 99
  Zwitter, 99, 103, 351
Titanite, 69, 98, 411
Titanomagnetite, 111, 116, 122, 153, 503
Tomography, computed, 433–435
Trace elements, 25, 47, 52–54, 69, 113, 120, 128, 129, 131, 141, 147, 170, 171, 200, 218, 221, 222, 297, 298, 306, 316, 317, 335, 337, 346, 350, 355, 370, 386, 394, 395, 407, 410, 411, 419, 420, 422, 435, 472–480, 486, 488, 489, 492–494, 499, 506
Tracers, 63, 69, 71, 73, 74, 89, 122, 147, 477, 489, 500, 503, 504
Troy ounces, 412
Tumbaga, 331, 332, 381, 416–418
Tourmaline, 23, 98, 100, 103, 105–108, 152, 253, 255, 467
Tuyères, 18, 191, 193, 200, 204–206, 239, 240, 245, 247, 258, 303, 304, 316, 329, 425, 453, 459, 462

U
Ultrabasic rocks, 33, 70, 71, 73, 117, 475
Urnfield era, 300

V
Valentinite, 127, 255, 256, 371
Vanning, 351, 353, 466, 467
Varlamoffite, 98, 99, 102, 398
Volcanic massive sulfide deposit, 25, 62
Volcanic rocks, 35, 47–49, 61, 62, 79, 92, 105, 120, 131, 503

W
Weischanite, 58
Welsh process, 312, 314, 315
Willemite, 88, 90, 251, 257, 261, 265, 405
Wind powered furnaces, 18, 460
Wolframite, 98, 100, 103, 105, 107, 108, 110, 144, 152
Wollastonite, 257, 263, 264
Wrought iron, *see* Iron
Wuestite, 183, 184, 187, 192, 195–197, 201, 221, 234, 238, 244, 247, 248, 257–259, 262, 263, 266–269, 316, 365–367

Z
Zinc
  alloys, 19, 250, 295, 374, 381, 383–385, 401, 402, 404, 405, 473, 476
  koshthi, 375
  ore deposits, 23, 27, 52, 86, 88, 89, 92, 94, 97, 110, 154, 332, 374, 402, 405, 475
  ores, 7, 21, 23, 27, 52, 82, 83, 88–94, 97, 110, 129, 142, 143, 161, 250, 281, 295, 296, 332, 373–376, 381, 382, 402–405, 468, 473, 475
  lauriotis, 374, 406
  pseudargyros, 374
  vaporisation, 498
Zirkelite, 254
Zircon, 69, 254

Geographical Index

A

Aarja, 49, 273
Abu Matar, 20, 204, 223, 491
Acemhöyük, 358, 422
Acqua Fredda, 237, 238, 262, 314, 318
Ada Tepe, 64
Aegean, 18, 36, 52, 55, 78, 95, 157–159, 209, 210, 266, 292, 333, 346, 484, 489, 500
Afghanistan, 2, 11, 23, 30, 34, 66, 69, 71, 74, 78, 102, 103, 105, 110, 111, 132, 141, 151, 174, 252, 275, 276, 320, 334, 351, 382, 396, 397, 400, 401, 419, 420, 471, 495, 505, 506
Africa, 9, 14, 24, 26, 33, 36, 39, 59, 64, 66, 70, 71, 98, 104, 106, 115, 117, 118, 125, 132, 141, 143, 145, 155, 166, 201, 214, 217, 218, 226, 253, 254, 256, 261, 297, 322, 323, 334, 354, 355, 359, 360, 362, 363, 391, 396, 401, 412, 445–451
Agios Sostis, 95–96, 157, 158
Agpalilik (Greenland), 113
Agucha, 97, 343
Ai Bunar, 14, 306
Ain Ghazal, 13, 14
Alaca Höyük, 52, 61, 389, 422
Alaska, 71, 145, 155
Aléria, 95, 97, 402, 406
Alesia, 403
Algier, 25
Ali Kosh, 14
Al-Jabali, 92, 97
Aljustrel, 28, 37, 52, 421
Almaden, 26, 130, 131, 150, 327
Almeria, 10, 78, 87, 421
Al Maysar, 181, 308
Al Mina, 378
Alps, 5, 11, 26, 34, 43–45, 61, 91, 118, 122, 128, 141, 151, 188, 207, 221, 223, 235, 236, 238–240, 249, 299, 304, 309, 313, 314, 318, 460, 462, 489, 495
Altaids, 38–39, 61, 62
Altheim, 299
Alto Adige, 304
Amur Darya, 110, 151, 320, 505
Anarak, 36, 40, 128
Anatolia, 9, 10, 12–17, 19, 26, 30, 34, 35, 40, 47, 51–52, 63, 67, 71, 72, 78, 80, 91, 108–110, 114, 125, 126, 131, 138, 141, 151, 153, 161, 166, 188, 189, 203, 204, 209, 210, 222, 223, 226, 227, 232, 243, 252, 253, 259, 269, 277, 289, 292, 298, 306–309, 316, 319, 321, 324, 329, 332–334, 350, 351, 353, 356, 358, 360, 377, 386, 388, 389, 392, 400–402, 421–423, 438–440, 442, 460, 461, 471, 477, 488, 489, 491, 495, 504
Andes, 34, 41, 42, 278, 390, 391, 445
Andronovo-Fedorovka culture, 228, 353
Angola, 118
Annaberg, 252
Apliki, 160
Apuseni Mountains, 62
Arabian-Nubian shield, 60
Argentina, 59
Arisman, 18, 96, 204, 283, 289, 290, 393
Armenia, 59, 142, 386, 422
Arslantepe, 52, 61, 126, 188, 189, 231, 232, 243, 277, 283, 321, 332, 392, 421–423, 491
Aruchlo, 15
Ashanti mine, 173
Aşikli Höyük, 460, 461
Askaraly, 227–229, 253, 254, 351, 353, 354
Asqelon, 20
Assur, 370
Athens, 36, 80, 83, 227, 374
Atlantic Ocean, 32, 62
Australia, 14, 39, 59, 61, 66, 117, 132, 145, 148, 155, 391, 447
Aventicum, 406

B

Bactria, 110, 151, 505
Badakhshan, 66, 111
Baer Bassit, 35
Balkans, 5, 9, 10, 12, 14, 15, 17, 19, 41, 71, 91, 101, 108, 117, 141, 151, 153, 204, 221, 223, 306, 359, 400, 403, 489, 492, 495, 504
Baluchistan, 132, 435, 443
Ban Chiang, 397
Batán Grande, 239
Bayda, 49, 231
Belovode, 9, 15, 204, 223, 224, 226, 232, 460
Bir Nasib, 39, 57, 303

A. Hauptmann, *Archaeometallurgy – Materials Science Aspects*, Natural Science in Archaeology,
https://doi.org/10.1007/978-3-030-50367-3

Black Forest, 30, 32, 37, 79, 91, 115, 154, 157, 243, 370, 373, 431
Bleiberg (Kärnten), 91, 340
Bodrum, 309, 350
Bogotá, 418
Bolivia, 74, 80, 97, 99, 101, 105, 128, 129, 143, 347, 349
Bouloun-Djounga, 447
British Columbia, 148
British Museum, 74, 211, 320, 419, 504
Brittany, 37, 103, 128, 396, 398, 401
Brixlegg, 43–46, 208, 223, 249, 316, 339, 492
Buehl (Kassel, Germany), 112
Burkina Faso, 64, 145, 155, 165, 297, 323, 326, 446, 449, 450
Bushveld complex, 70, 253

C
Cabrières, 27, 43, 47, 91, 249, 316
California, 26, 34, 55, 72, 131, 145, 155, 211, 378, 447
Campiglia Marittima, 239
Canada, 39, 61, 70, 74, 127, 129, 447
Can Hasan, 14, 306, 307, 461
Cape Gelidonya, 350
Carinthia, 122, 322, 340
Carpathians, 11, 17, 36, 62, 128, 152, 166, 337, 422
Cartagena-Mazzarón, 93
Çatal Höyük, 15, 16, 204, 223, 334
Caucasus, 17, 62, 125, 127, 128, 167, 316, 320–322, 326, 367, 368, 370, 372, 382, 386, 392, 394, 395, 402, 477, 478
Çayönü Tepesi, 9, 13, 14, 51, 222, 231, 259, 306, 307, 400, 440, 460, 461
Central Asia, 2, 25, 37, 38, 98, 104, 105, 108, 109, 131, 132, 141, 153, 253, 351, 378, 382, 396, 398, 401, 402, 471, 475, 495, 497–499
Cernica, 14, 108
Cerro de San Cristobal, 108
Cerro Salomon (Rio Tinto), 53, 84
Cerro Virtud, 10
Chagai Hills, 132
Chah-i Ab, 151
Charterhouse on Mendip (England), 374
Chasséen, 14
Cheile Turzii (Rumania), 301
Chicago, 467
Chile, 12, 24, 34, 39, 59, 79, 125, 132, 141, 309, 327
China, 25, 26, 38, 39, 61, 74, 80, 97, 104, 109, 125, 127–129, 151, 286, 331, 378, 396, 425, 426, 434, 446, 453–455, 475
Chrysokamino, 393
Chuciquamata, 41
Circumpacific belt, 34, 41
Claydon Pike, 402
Cligga Head, 100, 103, 105
Colchis, 24, 62, 122, 153, 320, 321, 367, 368, 370, 395
Cologne, 287, 335, 406, 429, 491
Colombia, 69, 72, 416–419
Colonia Ulpia Traiana, 230, 343, 406
Coolgardie, 145
Cornwall, 22, 25, 37, 98–101, 103–107, 110, 111, 125, 146, 152, 166, 227, 253–255, 336, 351, 355, 382, 396, 398, 401, 467, 495, 497–499
Corsica, 92, 95, 97, 309, 402, 406
Corta Lago, 53, 79, 83, 84, 142, 204, 205, 210, 213, 214, 282
Côte d'Ivoire, 450
Cracow, 340, 376
Cracow-Silesia, 337
Crete, 114, 285, 393, 499
Curt Engelhorn-Zentrum für Archäometrie, 489
Cyprus, 5, 23, 25–28, 33, 35, 43, 47–49, 51, 52, 59, 62, 71, 79, 80, 85, 92, 116, 126, 134, 137, 138, 140, 142, 160, 171, 173, 174, 204–207, 214, 230, 231, 234, 258, 260, 270, 280, 303–305, 308, 323, 329, 366, 392, 421, 438, 465, 475, 477, 488, 491, 500
Czech Republic, 26, 37, 59, 127, 157, 168, 201, 221, 267, 281, 423

D
Dardistan, 151
Dariba, 97
Dartmore, 351
David Garedji, 15, 62
Dead Sea, 14, 316, 394
Delhi, 97
Derekutuğun, 40, 52
Devon, 254, 255
Disko (Greenland), 112

E
Ebla, 61, 318, 400, 422
Ecuador, 69, 72, 417, 419
Egypt, 12, 17, 25, 26, 33, 39, 56, 57, 59–61, 69, 72, 74, 114, 128, 132, 165, 187, 204, 304, 328–330, 333, 349, 359, 377, 392, 420, 421, 477, 483, 485, 488, 503, 505, 506
Eiblschrofen, 32, 46, 47
Eldorado, 418
El-Furn, 407, 409
Embiez, 402
En Yahav, 431
Ergani Maden, 26, 35, 48, 51, 52, 126
Essingen, 246
Euphrates river, 334
Eurasia, 5, 14, 15, 36, 38, 54, 60, 61, 107, 108, 153, 396, 474

F
Far East, 17, 59, 71, 331, 386, 422, 426, 445
Faynan, 16, 18, 20, 22, 26–28, 33, 39, 43, 47, 54–57, 156, 160, 161, 163, 172, 174, 175, 187, 188, 194, 197, 203, 206, 209–211, 219, 221, 225, 226, 239, 241, 243, 248, 250, 259, 264, 265, 267, 268, 270, 278, 298, 299, 301, 303, 306, 308, 309, 318, 357, 369, 382, 392, 407, 409, 431, 451, 459, 460, 463, 475, 483, 485, 491, 493
Fennhals, 304
Fergana, 132, 151

Fertile Crescent, 9, 11, 13, 14
Flintshire, 92
Flitzen (Austria), 313
Foraxi Nioi (Sardegna), 350
France, 11, 14, 27, 47, 79, 84, 85, 87, 91, 118, 126, 157, 249, 309, 316, 333, 350, 396, 402, 443, 498
Fuggerau, 340

G
Gallia Belgica, 93
Gänsekötigerz, 22, 76
Gela, 402
Gelidonya, 350
Georgia, 15, 24, 26–28, 30, 32, 51, 60, 62, 63, 66, 67, 69, 122, 128, 137, 138, 141, 159, 167, 171, 255, 256, 320, 321, 367, 368, 372, 395, 416, 422, 489
Ghana, 64, 173, 297, 323, 327, 446
Ghassul, 17
Ghazni, 506
Ghebi, 128
Golden Quadrangle, 36, 60, 62, 152, 337
Göltepe, 98, 99, 108, 109, 161, 227, 228, 253, 300, 353, 354, 382, 466, 467, 495, 499
Gorisziche, 128
Gornja Tuzla, 15
Goslar, 199, 212, 236, 377
Grängesberg, 431
Greece, 23, 27, 30, 34, 36, 59, 77, 80, 81, 88, 89, 91, 92, 95–96, 115, 117, 126, 137, 142, 143, 157, 158, 161, 164, 166, 207, 210, 213, 214, 235, 251, 256, 258, 265, 266, 282, 285, 288–290, 292, 293, 320, 321, 324, 334, 346, 391, 393, 407, 458, 484
Greenland, 112–114, 117
Grisons, 208, 237, 277, 314, 318, 452
Grootfontein (Namibia), 112, 113
Großkuchen, 259
Gujarat, 97

H
Ha Hotrin, 495
Haifa, 350
Haithabu, 243, 378
Haliköy ödemis, 131
Hallan çemi, 14
Harz Mountains, 22–24, 26, 28, 30, 37, 79, 143, 154, 157, 159, 160, 210, 236
Hassek Höyük, 51, 491
Hatay, 35
Hedeby, 403
Hellenide Domain, 80
Herat, 110, 111
Hermaringen, 246
Herrerias, 75
Himalayas, 5, 11, 32, 71, 141, 151, 359, 451–454
Hindukush, 5, 34, 71, 252, 495
Hisarcik, 495
Hishuley Carmel, 495
Hittites, 285, 358
Holy Cross Mountains, 214, 215, 235
Huancavelica, 80
Huelva, 53, 138, 242, 243
Huneberg, 212

I
Iberian Peninsula, 2, 10–12, 18, 19, 24, 26, 28, 30, 35, 36, 49, 52, 53, 66, 91, 93–94, 101, 103, 105, 107, 147, 149, 150, 153, 155, 165, 204, 205, 218, 221, 241, 269, 271, 281, 288, 299, 314, 322, 333, 338, 339, 344, 351, 391, 398, 401, 402, 411, 421, 440, 463, 476, 477, 489, 492, 502
Iberian Pyrite Belt, 37, 52–53, 83, 92, 126, 134, 138, 142, 154, 245, 411, 421, 501
Ibiza, 25, 27, 28, 96, 155, 157, 347
Idrija, 131
Ieli, 167
Iernut, 14
Iglesiente, 27, 36, 90, 94, 484
Ilipinar, 306, 386
Imperium Romanum, 90
India, 61, 64, 97, 103, 117, 341, 343, 363, 374–376, 454–458, 475
Indonesia, 39, 41
Inguri, 167
Inn Valley, 27, 41, 44, 46, 77, 79, 208, 249, 264, 316, 318, 476, 492, 494
Iolkos, 320
Iran, 2, 5, 10–13, 17, 23, 34, 35, 40, 71, 74, 78, 92, 96–97, 103, 108, 110, 111, 114, 125, 128, 132, 141, 151, 153, 154, 156, 162, 172, 187, 188, 201, 206, 220, 223, 225, 227, 229, 230, 235, 243, 259, 264, 267, 270, 277, 278, 280, 283, 285, 289, 290, 299, 306–308, 315, 332, 334, 338, 392–394, 400, 406, 407, 437, 463, 468, 495, 506
Iraq, 10, 12–14, 320, 334, 359, 360, 416
Ireland, 38, 46, 203
Iron Quadrangle (Brazil), 26
Israel, 14, 17, 22, 54, 56, 156, 175, 187, 201, 216, 245, 262, 278, 298, 306, 309, 316, 320, 329, 334, 350, 357, 382, 386, 387, 392, 394, 395, 398, 415, 431, 443, 459, 471, 483, 485, 493, 499, 501
Italy, 23, 36, 77, 80, 89, 91, 94–95, 108, 122, 131, 143, 162, 188, 208, 219, 236–241, 244, 245, 251, 262, 299, 304, 314, 318, 327, 333, 335, 350, 364, 377, 407–409, 435, 487, 489

J
Japan, 101, 105, 127, 129, 165, 166, 239, 297, 312, 328, 338, 339, 348, 351, 352, 363, 396, 427, 446, 453–455
Jericho, 13, 285
Jochberg (Austria), 208, 300, 302, 304, 318, 319, 347
Judean Desert, 17

K
Kambia mine, 48
Kandahar, 110, 151, 506
Kargaly, 38, 39, 43, 54, 55, 141, 153
Karnab, 98, 99, 109, 351

Kasbegi, 128
Kas, 350
Kathmandu, 452, 454
Kazakhstan, 33, 38, 109, 228, 253, 351, 353, 354, 396, 471, 495
Kazruleti, 60
Kefar Shamir, 495
Kestel, 30, 98, 108, 109, 111, 157, 159, 161, 166, 227, 253, 300, 351, 353, 354, 382, 466, 467, 495, 499
Kfar Samir, 350
Khaidarkan (Kirgistan), 132
Khirbat al-Hamra (Faynan), 20
Khirbat el-Jariye, 209
Khirbat en-Nahas (Faynan), 163, 209, 211, 249, 463
Khorsabad (Iraq), 359, 361
Kirmizitarla, 52
Kish, 61
Kitzbühel, 300, 302
Kizildağ, 35
Klondike, 145, 447
Kolar district Mysore, 61
Kongsberg, 59, 75, 159
Koni (Côte d'Ivoire, Burkina Faso), 451
Kosovo, 210, 211, 213, 221, 275, 280, 333, 347
Kutaisi, 128
Kwemo-Sakao, 128
Kyrgyzstan, 38, 61, 128, 132

L
La Capitelle du Broum, 47
Lahn-Dill area, 120, 121, 187, 214, 215, 217, 267, 368
Lake Superior, 40
La Lucette, 128
Landsberg, 130, 131
Langenau, 246
Languedoc, 27, 47
Lapphyttan, 248
La Roussignole, 47
Lasail, 49, 51, 136, 137, 209, 231
Las Médulas, 93, 147–150
Lauchheim, 269
Laurion, 23, 25, 27, 28, 30, 36, 77, 78, 80–83, 88, 89, 91, 92, 95, 142, 143, 154, 155, 157, 160–162, 164–166, 168, 169, 172, 210, 213, 235, 250–252, 256, 288, 290, 292, 293, 333, 336, 338, 342, 343, 346, 347, 374, 451, 455, 458
Lepenski Vir, 14
Les Magnons, 402
Levant, 13, 16, 17, 20, 26, 39, 54, 55, 141, 204, 223, 226, 259, 269, 285, 333, 382, 386, 392, 420, 488, 491
Licheta, 128
Limassol, 49
Linares, 32, 93, 346
Logar, 506
London, 74, 320, 504
Lüderich, 252, 286, 287
Luserna, 235–237
Luzon, 41, 316
Lyon, 336, 406, 491

M
Maadi, 20, 57, 485
Madneuli, 15, 26, 27, 62, 63, 142
Magdalensberg, 322, 365, 425, 450
Mahanudi Valley (Sikkim), 452, 453
Mahmatlar, 421
Mainz, 52, 403, 477, 488
Majdanpek, 14, 392
Malaysia, 25, 104, 109, 110
Mal di Ventre, 335
Mali, 64, 323, 450
Mallorca, 248
Mansfeld, 54, 76, 131, 340
Mari, 61, 413
MariaHilfsbergl, 45, 316
Masirah, 49
Mastau-Baj, 228, 353
Masua, 89
Mediterranean, 18, 25, 27, 33, 34, 36, 39, 47, 48, 57, 80, 85, 90, 91, 94, 96, 122, 131, 137, 143, 153, 155, 157, 159, 160, 163, 172, 173, 204, 207, 308, 322, 335, 338, 350, 355, 359, 363, 378, 398, 402, 438, 465, 477, 479, 485, 488, 491, 499, 502
Mega Livadi, 126, 172, 173, 391
Mehrgarh, 443
Mendip Hills, 92
Mersin, 14, 307, 461
Mesopotamia, 6, 12, 13, 21, 33, 34, 51, 55, 61, 67–69, 72–74, 101, 102, 109, 110, 114, 127, 128, 151, 159, 210, 222, 289, 308, 320, 321, 324, 329, 333, 360, 370, 377, 382, 389, 395, 401, 402, 410, 413, 415–417, 419, 422, 464, 471, 476, 478, 479, 488, 495, 504–506
Metzingen, 246, 425
Mexico, 74, 76, 79, 129, 159, 327, 332, 344, 384, 417, 418
Michigan, 461
Mississippi Valley, 25, 36, 90, 94, 95, 97, 143, 250
Mitterberg, 22, 24, 27, 28, 30, 32, 36, 43–45, 156, 161, 167, 168, 171, 208, 218, 277, 279, 304, 313, 318, 454, 462, 466, 492–494
Mondsee, 299
Monte Amiata, 36, 131, 377
Monteponi, 94
Monte Romero, 238, 240–242, 245, 266, 271, 281, 285, 324, 393
Moosbruckschrofen, 493, 494
Morvan, 214
Mullaq, 206, 231, 273
Murgul, 26, 35, 51, 52, 63, 188, 203, 223, 231
Muruntau, 38, 61
Mušiston, 101, 102, 109, 351
Mysore, 61

N
Nahal Hemar, 14
Nahal Mishmar, 17, 115, 283, 316, 382, 386, 387, 394, 395, 471
Nahal Qana, 320, 329, 382
Nakhlak, 77, 96, 156, 278, 299, 338

Near East
  geology, 2, 16, 108, 222, 386
  ore deposits, 14, 171, 218, 421
Nebra
  Nebra Sky Disc, 54, 493
Nepal, 235, 239, 297, 451–455
Neusohl, 337
Neuss, 406
Nevali çori, 188, 223, 231
Neves Corvo, 37, 38, 52, 106
New South Wales, 148
Nichoria, 227
Nimrud (Iraq), 360, 361
Nizhni Tagil (Ural), 70
Non Nok Tha (Thailand), 397
Noricum, 214, 478
Norsuntepe, 16, 51, 231, 283, 392
Nubia, 33, 60, 61, 477

O

Okharbot (Nepal), 452, 453, 455
Olympia, 227
Oman, 23, 25–28, 33, 43, 47, 49–52, 63, 71, 74, 92, 126, 134, 136–138, 140, 153, 163, 181, 187, 206, 209, 231, 234, 242–244, 248, 259, 260, 264, 273, 274, 276–278, 280, 285, 300, 302, 306, 308, 313, 401, 409, 416, 471, 475, 488, 505, 506
Ophiolites, 26, 33, 35, 47–50, 61, 63, 64, 69, 71, 73, 74, 92, 126, 151, 231, 359, 420, 475, 477, 478, 488, 503–506
Ophir, 148
Ore Mountains, 17, 22–24, 26, 27, 30, 37, 62, 76, 77, 79, 91, 99, 101, 103, 107, 110, 111, 125–129, 143, 154, 157, 159, 168, 266, 267, 336, 382, 396, 398, 401, 495, 497–499
Ötztal Alps, 489
Oxus river, 69, 419

P

Paotolus, 67, 69, 151, 329, 504
Pakistan, 11, 34, 71, 151, 166, 435
Paltental (Austria), 313
Panjihir, 252
Papua New-Guinea, 70, 71, 446
Peñarroya, 93
Pennine ore fields, 92
Peru, 34, 72, 79, 80, 97, 101, 105, 127, 141, 143, 147, 166, 239, 327, 332, 417, 419, 446, 448, 449
Peştera Caprelor (Rumania), 301
Pfyn, 226, 256, 299, 300, 463, 464
Philippines, 41, 71, 316
Pioch Farrus, 47
Platz von Mozze, 235, 236
Pločnik, 10, 15, 108, 401
Poland, 25, 26, 28, 39, 54, 116, 118, 119, 214–216, 235, 340, 478
Politiko Phorades, 206, 230, 231, 303
Pontids, 34, 209
Portugal, 9, 36, 52, 106, 108, 148, 154, 155, 163, 226, 311, 324, 325, 356, 411, 462, 463, 498
Potosi, 80, 99, 101, 105, 110, 128, 143, 347
Přibram, 127, 252

R

Racha, 128, 255, 256, 394, 395
Rajasthan, 97, 343, 375
Rakah, 49, 51
Rammelsberg, 32, 37, 89, 161, 212, 236, 376
Rhodopes, 64
Rioni, 128
Rio Tínto, 18, 93
Romania, 14, 60, 62, 127, 128, 164, 170, 333
Rooiberg, 98, 106, 253, 254, 354, 355, 401
Roşia Montană, 60
Ross Island, 27, 38, 43, 46–47, 316
Royal tombs, Ur, 74, 151, 222, 320, 322, 330, 377, 401, 410, 415–417, 422, 478, 495, 504, 505
Rudna Glava, 14, 171, 493
Russia, 39, 43, 59, 66, 70–72, 105, 117, 118, 131, 132, 153, 362, 473

S

Saalburg, 428, 438, 439
Sado, 165, 166, 328, 446, 454
Sagebi, 128
Sagebis Dsiris, 255, 256
Sahel zone, 449, 451
Sakdrisi, 30, 32, 60, 62, 63, 66, 67, 137, 138, 142, 159
Salcombe, 499
Samti, 66, 74, 151, 420, 505, 506
Saramarca (Palpa, Peru), 449
Sardinia, 23, 25, 27, 28, 36, 77, 80, 88–92, 94–95, 143, 213, 250–252, 335, 350, 408, 409, 431, 484, 488, 489, 491–493
Sardis, 19, 67, 72, 151, 286, 287, 289, 291, 323, 327, 329, 337, 342, 410
Sar-i Sang, 102, 110
Saxonian-Bohemian Ore Mountains, 498
Schwaz, 30, 32, 43–47, 208, 249, 265, 339, 492
Scotland, 60, 255
Segonzano, 240
Semdah, 209, 231
Senufo, 450, 451
Serbia, 9, 14, 15, 17, 36, 101, 108, 128, 131, 211, 223, 224, 232, 333, 392, 401, 460, 487, 493
Serifos, 126, 172, 391
Serra di Capivara, 21, 116
Serra Pelada, 145, 151, 327, 447
Shahdad, 225, 229, 308, 463
Shanidar, 14
Shar-i Sokhta, 227
Shiqmim, 20, 204, 491
Shortugai, 61, 506
Shropshire, 92
Siberia, 14, 38, 76, 101
Sícan, 239

Siegerland, 24, 27, 28, 37, 79, 80, 91, 121, 122, 169, 214, 215, 238, 242, 252, 267, 275, 367–369
Sierra Almagrera, 93
Sierra de Cartagena, 93
Sierra Morena, 91, 93, 344, 347, 492, 502
Sifnos, 80, 91, 95–96, 210, 252, 292, 324, 342
Sikkim, 451–453
Sinai, 2, 13, 18, 26, 39, 54, 56–57, 141, 187, 201, 209, 217, 248, 259, 303, 304, 392, 483, 485, 488
Sion, 14
Siphnos, 27, 28, 65, 157, 158, 346, 347
Skouriotissa, 140, 204, 205, 207, 209, 234, 329, 366
Skres, 265, 288, 407
Slovakian Ore Mountains, 62, 339
Słupia Stara, Opatów, 215
Snorup, Jutland, 214
Sohono (Korhogo, Côte d'Ivoire), 451, 452
Sopchito, 128
Spain, 9, 26–28, 30, 32, 36, 52, 64, 75, 78, 79, 84, 87, 91, 92, 108, 130, 131, 142, 148, 157, 161, 167, 187, 213, 240, 243, 248, 281, 285, 291, 309, 310, 324, 327, 333, 334, 338, 339, 342, 346, 377, 379, 411, 439, 477, 498
Špania Dolina, 126
Stahlberg, 131
Stolberg, 92
Sudan, 25, 26, 39, 118, 202
Sulzburg, 255, 370, 373
Sumer, 322, 370
Suplja Stena, 131
Susa, 333, 360, 361, 495
Svanetia, 62, 167
Swabian Alb, 25, 115, 118, 246–248, 259, 264, 265, 269, 365, 369, 425, 428
Swamilamai (India), 456
Swaziland, 115
Sweden, 86, 118, 120, 214, 247, 248, 360, 362, 368, 431, 478
Switzerland, 14, 59, 115, 118, 207, 226, 241, 256, 278, 299, 301, 374, 399, 452, 463, 464, 489

T
Tadzhikistan, 25, 101, 102, 109, 110, 505
Takhar, 66, 74, 151, 420, 505
Tal-I Iblis, 10, 16, 223, 299, 468
Tall Hujayrat al-Ghuzlan, 194, 231–232
Tall Magass, 226
Talmessi, 36, 40, 306
Tamilnadu (India), 456, 457
Tarim, 38
Tatsumi Guchi (Japan), 455
Tauern Mountains, 26
Taurids, 34, 35
Tawi Aarja, 231, 244, 302
Tawi Raki, 63
Tel Tsaf (Jordan valley), 15
Tepe Hissar, 61, 220, 235, 243, 283
Tepe Mochi, 229
Tepe Yahya (Kerman, Iran), 132
Terrasebis (Sardinia, Italy), 408
Tethyan Eurasian Metallogenic Belt (TEMB), 5, 11, 12, 17, 21, 26, 32, 34–39, 61, 71, 125, 145, 151, 359, 392, 399, 475, 483
Thailand, 25, 104, 109, 396, 397
Thasos, 23, 25, 28, 64, 80, 91, 115, 157, 210, 250, 265, 266, 275, 288, 292, 342, 346, 407
Thracia, 64
Tialkam, 448
Tibet, 454
Tien Shan, 38, 61, 132
Tigris river, 51, 471
Tillya Tepe, 151, 505
Timna, 2, 22, 26, 28, 33, 39, 43, 54–57, 156, 175, 187, 201, 216, 232, 240, 245, 248, 250, 259, 262, 264, 278, 298, 303, 306, 309, 357, 392, 407, 431, 459, 483, 485, 493, 501
Transcaucasia, 61, 392, 422
Trentino, 162, 188, 208, 219, 235–242, 244, 245, 261, 277, 304, 314, 318
Troiboden, 167, 168
Troodos
  Troodos Ophiolite, 35, 48, 49, 231, 421, 502
Tsumeb, 12, 132, 133, 138, 309
Tülintepe, 356, 400
Tunis, 25
Turan, 38
Tuscany, 26, 36, 105, 108, 111, 122, 128, 131, 377, 487, 489
Tyantyamana (Burkina Faso), 450

U
Ukraine, 131
Ulpiana, 213
Ulpia Traiana, 230, 343, 406
Uluburun, 19, 308, 309, 350, 355, 398, 399, 438–440, 465, 495, 498, 499
Umm Bogma, 26, 39, 54
Ur
  Royal tombs, 74, 151, 222, 320, 322, 330, 377, 401, 410, 415–417, 422, 478, 495, 504, 505
Urals, 26, 33, 38, 43, 54, 70, 71, 420
Uruk, 289, 324, 332–334, 413, 422, 464
Uzbekistan, 23, 25, 38, 61, 98, 99, 109, 110, 334, 351, 396, 467, 471, 495

V
Vallarade, 47
Varna, 17, 66, 298, 319–321, 330, 382, 415, 416
Vietnam, 129
Visiergraupen, 99
Vogtland, 497

W
Wadi Arabah, 17, 18, 39, 43, 54, 55, 141, 153, 187, 216, 217, 226, 303, 357, 407, 431, 460, 483, 485, 488
Wadi Fidan, 194, 209, 226, 248, 265, 270
Wales, 92, 164, 203
Wallis, 14

War Kabud, 437
Widdersdorf (Cologne), 429
Wiesloch, 92, 93, 201, 261, 267, 268, 282, 286, 288, , 492
Witwatersrand, 59, 66

X

Xanten, 228, 230, 343, 406

Y

Yarim Tepe, 12, 14
Yemen, 79, 92, 97
Yucatan, 418
Yukon, 101, 145, 447
Yumuktepe, 14, 307
Yunnan (China), 396

Z

Zabul, 151, 506
Zagros Mountains, 11, 16, 34, 111, 306
Zaïre, 450
Zarshuran (Iran), 392
Zawar district, 374
Zengővárkony, 15
Zhegovc, 213
Zinngraupen, 99

The manufacturer's authorised representative in the EU is Springer
Nature Customer Service Centre GmbH, Europaplatz 3, 69115 Heidelberg,
Germany. If you have any concerns regarding our products, please
contact ProductSafety@springernature.com

Printed and bound by CPI Group (UK) Ltd, Croydon, CR0 4YY
22/09/2026
02220956-0020